Biology Trending

Biological science is particularly relevant to the lives of all human beings, not just professional biologists or science majors. *Biology Trending* encourages students to think about biology in the context of current societal issues and its importance to their own lives and to the world in which we all live.

Visual simulations to demonstrate a variety of evolutionary processes and conditions such as Hardy-Weinberg equilibrium, natural selection, and genetic drift

Weblinks and bibliographic resources to expand knowledge and improve understanding

Longform and open response questions to test understanding and apply knowledge

Time-saving instructor resources

Go online to access these resources and more at:

biologytrending.routledge.com

BIOLOGY TRENDING

Biology Trending is a truly innovative introductory biology text. Designed to teach biological concepts in a social context using contemporary issues, *Biology Trending* encourages introductory biology students to think critically about the role that science plays in their world. This book features many current and relevant topics, including sea-level changes and ocean acidification; CRISPR/Cas9 gene editing; opioid abuse; Zika, Ebola, and COVID-19; threats to biodiversity; cancer immunotherapies; and more. The book is accompanied by online Instructor and Student Resources to support teaching and learning.

Key Features

- Adopts an "issues approach" to teaching introductory biology
- Up-to-date sections throughout, including climate change, CRISPR/Cas9 gene editing, new hominids, COVID-19, and new cancer therapies, among many others
- Suitable for both majors and nonmajors courses
- More succinct for ease in teaching and more affordable for students
- High-quality illustrations help to elucidate key concepts

This book is extended and enhanced through a range of digital resources that include:

- Long-form and open-response self-testing resources to test understanding and apply knowledge
- Visual simulations to demonstrate evolutionary processes
- Web links and bibliographic resources to expand knowledge
- Time-saving instructor resources such as PowerPoint slides, activity and assignment ideas, and comprehensive lesson plans

Eli C. Minkoff is Professor Emeritus of Biology at Bates College and Adjunct Professor at Worcester State University. He received his bachelor's degree, magna cum laude, from Columbia University and his MA and PhD from Harvard University. Dr. Minkoff has published several books as well as many articles in scholarly journals and encyclopedic reference works. His main field of research is the evolutionary biology of primates and other mammals.

Jennifer K. Hood-DeGrenier is Professor and Chair of Biology at Worcester State University. She received her bachelor's degree from Williams College, summa cum laude, and her PhD from Harvard University. She is a cell and molecular biologist with interests in genetics, biochemistry, cancer biology, and science writing. She has published in the areas of nuclear transport, cell cycle control, cell morphology, and biology education.

BIOLOGY TRENDING

A Contemporary Issues Approach

ELI C. MINKOFF
JENNIFER K. HOOD-DEGRENIER

CRC Press is an imprint of the
Taylor & Francis Group, an **informa** business

EXPLANATIONS OF COVER ILLUSTRATIONS (left to right):

Pitcher Plants: Illustrates an extreme plant adaptation for obtaining nitrogen (needed for protein synthesis) by attracting and digesting insects. Disappearing wetlands and insect population declines are challenges these unique plants face.

Spider monkey (*Ateles* sp.): Illustrates adaptations for grasping and climbing, including use of a prehensile tail and an opposable (thumblike) big toe. Slash-and-burn agriculture is eliminating habitat at an alarming rate.

Strawberry frog (*Oophaga* sp.): Bright colors deter predators by warning against poisonous toxins in the skin, but amphibian populations are sensitive indicators of deteriorating environments and many species are on the verge of extinction because of threats to its rainforest habitat. Climate change and rapidly spreading fungal diseases have caused dramatic declines and extinction of many frog species.

Cactus finch (*Geospiza scandens*): One of the many finch-like species of the Galapagos Islands whose unusual adaptations (such as its enlarged beak) impressed Charles Darwin during his voyage on H.M.S. *Beagle*. Populations are stable, but because this species lives on an island, climate change and introduced species can pose a potential threat to survival.

Dorcas gazelle (*Gazella dorcas*): Diverse adaptations in the shapes of defensive horns are a key feature of speciation and adaptive radiation among antelopes. Populations of this gazelle are threatening by drought, hunting, and habitat loss due to livestock overgrazing.

Coral reef in Red Sea: An extremely diverse marine habitat in which many species of corals and symbiotic algae provide shelter and habitat for numerous species of fishes, plankton, and invertebrates of all sizes, but global warming and ocean acidification threaten many such coral reef habitats. Changes in ocean temperatures and increased acidity cause coral bleaching.

First edition published 2024 (Please note, for the purposes of permissions, this is considered to be a Fourth Edition of *Biology Today*.)
by CRC Press

6000 Broken Sound Parkway NW, Suite 300, Boca Raton, FL 33487-2742

and by CRC Press
4 Park Square, Milton Park, Abingdon, Oxon, OX14 4RN

CRC Press is an imprint of Taylor & Francis Group, LLC

Library of Congress Cataloging-in-Publication Data
Names: Minkoff, Eli C., author. | Hood-DeGrenier, Jennifer K., author.
Title: Biology trending : a contemporary issues approach / Eli C. Minkoff, Jennifer K. Hood-deGrenier."
Description: First edition. | Boca Raton, FL : CRC Press, 2023. | Includes bibliographical references and index.
Identifiers: LCCN 2022058026 (print) | LCCN 2022058027 (ebook) |
ISBN 9781032488585 (hbk) | ISBN 9781032488042 (pbk) |
ISBN 9781003391159 (ebk)
Subjects: LCSH: Biology--Social aspects. | Bioethics.
Classification: LCC QH333 .M56 2023 (print) | LCC QH333 (ebook) |
DDC 570--dc23/eng/20230126
LC record available at https://lccn.loc.gov/2022058026
LC ebook record available at https://lccn.loc.gov/2022058027

ISBN: 978-1-032-48858-5 (hbk)
ISBN: 978-1-032-48804-2 (pbk)
ISBN: 978-1-003-39115-9 (ebk)

DOI: 10.1201/9781003391159

Typeset in Utopia
by KnowledgeWorks Global Ltd.

Access the Instructor & Student Resource by visiting this link: biologytrending.routledge.com

In memoriam

Neil B. Minkoff, MD
1971–2022

A brilliant man
A wonderful son
A wonderful husband
The best father in the world

CONTENTS

PREFACE

Biology Trending takes an issues-oriented approach to the teaching of biology, one that covers all of the major biological concepts. Our approach aims to educate citizens—biologists and nonbiologists alike—with an understanding that will enable them to evaluate scientific arguments and make informed decisions affecting their own lives and the well-being of society. We are devoted to the principle that science is not just a body of facts to be memorized, but a method of reasoning, accessible to all, by which all citizens can and do make evidence-based decisions. Biological science is particularly relevant to the lives of all human beings, not just professional biologists, and this book therefore emphasizes that many of the issues that citizens face are informed by the methods of science and basic knowledge of biology.

We are therefore committed to teaching science as a human activity that impinges upon other aspects of society and gives rise to social issues that require discussion. Individuals are increasingly called upon to deal with science-based issues throughout their lives. Each of us makes food choices daily and medical decisions nearly as often. DNA evidence is used more and more in solving crimes and in predicting susceptibility to disease. Stem cell technologies confront us with real possibilities only imagined a few years ago. Our waste disposal habits affect the environment in which we live, and our transportation and manufacturing choices affect the very air we breathe. Citizens, legislators, juries, and corporate managers need to make important decisions, affecting many lives, based in part on the findings of science. Everyone needs to be aware of science, the way that scientists work, and how science can be used and misused.

The issues themselves are not our focus, however; instead, they form a context in which to teach basic biology. This is very different from teaching the biology specific to a particular issue, which often leaves students thinking that the biological concept applies only to that case. Many other texts also incorporate current issues, but our approach is unique because in our text the issues are central to the pedagogy, not "add-ons" presented as case studies or in sidebars, boxes, or separate pamphlets. Consequently, the issues we have selected are those that are not only of current importance but also those that lend themselves as vehicles for teaching the major concepts of biology. For example, we use the chapter on the population explosion as a vehicle to teach the biology of reproduction, and osmosis and photosynthesis are covered in a chapter called Plants to Feed the World. Wherever possible, we have presented a concept in more than one part of the book, developing the concept in more than one context. Prokaryotic biology, for example, is introduced in the chapter on Classifying Nature's Diversity and is then further developed in the chapter on New Infectious Threats, while asexual reproduction is introduced in the chapter on Genes, Chromosomes, and DNA and is then further developed in both of these later chapters.

Like our earlier editions (published under the title *Biology Today*), *Biology Trending* engages beginning students to consider issues relevant to their own lives, even if they have no intention of becoming professional scientists. The principles of biological science are therefore taught in the context of issues that confront both scientists and nonscientists in the world today, and biological theories are taught as products of the evidence-based discoveries that have led scientists to adopt these theories.

Various groups of educators have developed lists of key concepts in biology, and each of these concepts is covered in our book. Biology itself evolves, so we have added new concepts as they have risen in importance. In this edition, we have brought the text up to date by expanding certain topics and adding new concepts. A new chapter on Aging and Alzheimer's Disease has been added (motivated by the projected continued increase in incidence of this devastating disease in the rapidly aging populations in many countries, including the United States), and a chapter on Sociobiology was dropped after many teachers using our book told us that they did not use this chapter in their teaching. Among new or expanded topics in this edition are extensive treatment of genetic testing and CRISPR/Cas9 gene editing in Chapter 3, genomics in Chapter 4, comparative genomics (and several of its findings) in Chapter 6, expanded discussion of cancer biology and modern cancer treatments in Chapter 10, COVID-19 and many other emerging diseases in Chapter 16, and newer and more plentiful evidence of global warming and other environmental threats in Chapter 19. Among the issues facing both individuals and the societies in which they live, we cover cloning and genetic engineering in Chapter 4, races and racism in Chapter 7, obesity and nutritional disorders in Chapter 8, cardiovascular problems in Chapter 8, reproductive health and also abortion in Chapter 9, mental illness in Chapter 11, care of the elderly in Chapter 12, HIV/AIDS in Chapter 15, public health issues in Chapter 16 (and elsewhere), agriculture in Chapter 17, habitat preservation in Chapter 18, and pollution and global warming in Chapter 19.

The issues have also been adjusted to stay current. Students (especially those not majoring in biology) are more likely to be interested in and develop an understanding of material if it is meaningfully related to issues of concern to them, and these change over time. COVID-19 and other infectious diseases are widely covered in the news, and these form the basis for Chapter 16. For issues of continuing interest, such as HIV/AIDS, cancer, drugs, and population control, we have incorporated the latest statistics as well as the latest biological advances. HIV/AIDS may not be at the forefront of public concern to the degree that it was a number of years ago, but it remains a significant threat and is a very useful topic for introducing a number of principles related to infectious diseases, virus biology, and public health; we have revised that chapter to include the many recent advances in drugs to control HIV transmission and prevent infection from progressing to AIDS, which brings this topic up to date. In the chapter called Drugs and Addiction, we have added significant content to reflect changing societal views towards marijuana, which have occurred in the United States and many other countries, as well as coverage of the opioid epidemic. We recognize that the average undergraduate student is now older than 22; therefore, we have picked issues with a wide age appeal. We have also attempted to use geographically diverse examples. We know our readers, regardless of where they attend college, are from around the world. Through examples cited in the text and in Thought Questions, we encourage students to think beyond themselves, both globally and locally.

In addition to its real-life appeal, the issues approach allows for a more comprehensive view of biology. As a discipline, biology has become fragmented to the extent that different perspectives on the same problem, for example, molecular perspectives and environmental perspectives, are often taught in separate courses with no reference to each other. In our book, the current understanding of each issue is covered from different perspectives, which often include cellular and molecular perspectives, organismal or individual perspectives, and global or population perspectives, combined as appropriate. Our approach accordingly helps students to experience the interdisciplinary nature of today's biology. Each chapter ends with a section called "Connections to Other Chapters" that further emphasizes this point.

Our approach also examines the intimate connections between biological and social issues. We have chosen to teach "facts" in a context that emphasizes how they are produced, organized, and used to solve problems. Other books often expose students to the *results* of biology without gaining understanding of biology as a *process of discovery*. Instead, we hope to instill in students an understanding and appreciation of this process. To help students, we have presented multiple interpretations or points of view as much as possible. Societal and ethical issues are mentioned wherever relevant, and part of the initial chapter is devoted to an examination of ethical principles. We encourage teachers to set aside time for class discussions to further stimulate student thought, or for students to set up such discussions among themselves informally. With *Biology Trending* we aim to stimulate critical thinking and questioning rather than memorization. Thought Questions, suitable for class discussions, are provided at the end of each section, and, in this new edition, new Thought Questions have been added.

Many new illustrations and tables have also been added in this edition. Some accompany new material, while others offer further support for concepts previously described only in the text. Where the vocabulary refers to key concepts, we have explained the new words more thoroughly in the text and have used most terms in the end-of-chapter summaries. All terms are defined in glossaries that now appear at the end of each chapter as well as in a comprehensive glossary online.

The online resources have been greatly expanded and are accessible through our web portal at http://biologytrending.routledge.com. Our website has an abundance of resources for students and teachers alike, including study guides, bibliographies, practice questions, test banks of exam questions, chapter outlines, animations, video links, lab resources, review quizzes, and much more.

In this newly revised edition, we have benefitted greatly from the advice of many people in producing a book that remains scientifically accurate and is optimally organized for student understanding. We would like to take this opportunity to thank the many people who reviewed portions of this new edition and provided us with helpful suggestions.

We would also like to thank the staff at CRC Press/Taylor & Francis, who were most helpful throughout the editing and production of this edition. They include Leah Burton, Chuck Crumly, Ana Lucia Eberhart, Kara Roberts, Marlena Sullivan, Jordan Wearing, with external help from Joan Whittemore and Suzanne Pfister. As in previous editions, Nigel Orme drew most of the illustrations. We could never have brought this project to fruition without their help.

Eli C. Minkoff
Jennifer K. Hood-DeGrenier

INSTRUCTOR AND STUDENT RESOURCES

Biology Trending is accompanied by an Instructor and Student Resource website, providing teaching, and learning resources that have been developed to improve learning outcomes. The website provides diverse and engaging content mapped to the textbook. All these resources are free to users of the book and have been developed by authors and subject specialists, providing accessible, high-quality content for both instructors and students for self-testing, revision, or independent study.

Access the resources at: biologytrending.routledge.com

Student Resources (for independent study, review, and self-testing)

- Chapter overviews and outlines to highlight key areas of focus.
- Links to selected videos, further readings, and other resources to expand knowledge.
- Text supplements for expanded treatment of selected subjects.
- Selected animations to enhance understanding of nerve impulses, membrane activities, and other topics.
- Computer simulations of natural selection, genetic drift, and related phenomena.
- Self-study and review questions to test understanding.
- Bibliography for each chapter.

Instructor Resources

- PowerPoint presentations for each chapter: a simpler version (series A) that includes all illustrations, and a more detailed version (series B) that also includes topic outlines and more.
- Selected ideas for classroom activities, class discussions, and outside assignments to save instructors time.
- A test bank of possible exam questions (password-protected), including both multiple choice questions and fill-ins.
- A series of lab activities, complete with ordering information for materials.

CHAPTER 1

Biology, Science, and Society

ISSUES

- How do we know what we know?
- How do scientists make discoveries and advance our knowledge?
- What constitutes a "discovery" in science?
- How is science creative?
- Does science contain absolute truths?
- How do ethics and morals fit into science?
- How do scientists make ethical decisions in a social context?
- How are decisions made on social issues, and to what extent can science help in these decisions?
- What rights do animals have? How do we safeguard those rights?
- How do we safeguard the rights of experimental subjects?

BIOLOGICAL CONCEPTS

- Properties of living organisms (organization, metabolism, selective response, homeostasis, growth and biosynthesis, genetic material, reproduction, populations)
- Hypotheses and theories
- Experimental science versus naturalistic science
- Normal science and paradigm shifts
- Science and society
- Biological ethics

CHAPTER OUTLINE

DOI: 10.1201/9781003391159-1

Biology is the scientific study of living systems. Our gardens, our pets, our trees, and our fellow humans are all examples of living systems. We can look at them, admire them, write poems about them, and enjoy their company. The Nuer, a pastoral community of Africa, care for their cattle and attach great emotional value to each of them. They write poetry about—and occasionally to—their cattle, they name themselves after their favorite cows or bulls, and they move from place to place according to the needs of their cattle for new pastures. They come to know individual cattle very well, almost as members of the family. The Nuer have also acquired a vast store of useful knowledge about many animal and plant species in their region. Many other people who live close to the land have a similar familiarity with their environment and the species living in it. Scientific understanding of the world around us grew out of this kind of familiarity with nature, supplemented by a tradition of systematic testing. In this chapter, we examine the methods of science in general and the application of those methods to the study of living systems.

Because living systems are complex and continually changing, an understanding of these systems often requires special methods of investigation or ways of formulating thoughts. This chapter describes the special methods that have come to be called **science**. Many people think that science is defined by its subject matter, but this is not correct. *Science is defined by its methods.*

The methods of science do not answer questions about values and, therefore, cannot address questions about whether certain types of research should be done or to what uses scientific results should be put. Such decisions often involve a branch of philosophy called ethics. Many issues confronting societies today have a scientific dimension. Policy decisions on such issues involve both science and ethics.

1.1 SCIENCE DEVELOPS THEORIES BY TESTING HYPOTHESES

1.1.1 Hypotheses

The essence of science is the formulation and testing of certain kinds of statements called **hypotheses**. At the moment of its inception, a hypothesis is a tentative explanation of events or of how something works. What makes science distinctive is that hypotheses are subjected to rigorous testing. Many hypotheses are **falsified** (rejected as false) by such testing. Eliminating one hypothesis often helps us to frame the next hypothesis. If a hypothesis is repeatedly tested and not falsified, it may be put together with related hypotheses that have also withstood repeated testing. Such a group of related hypotheses may become recognized as a **theory**.

Hypotheses must be statements about the observable universe, formulated in such a way that they can be tested. Observations of the material world that we make with our senses (aided in some cases by scientific instruments) are called **empirical** observations. To be a hypothesis, a statement must be capable of being tested by comparison with such empirical observations. Observations gathered for testing hypotheses are called **data**. Karl Popper, a philosopher of science, said that scientific hypotheses must always be tested in such a way that they might be rejected if they turn out to be inconsistent with observations—in his word, they must always be **falsifiable**.

Certain types of statements cannot be used as scientific hypotheses because they are not subject to testing and falsification. This includes esthetic judgments about what is valuable, beautiful, or likable. For example, I may like my kind of music, whatever you may say about it. Moral judgments and religious concepts are also not scientific hypotheses because observational data are not sufficient to test their truth. This means that a devout person's belief in God cannot be shaken by any demonstration of an empirical fact or observation. To a devout believer, no such demonstration is even possible. The same is true of strongly held beliefs about the goodness of human equality or the wrongfulness of inflicting death.

Hypotheses, therefore, must be (1) testable and (2) falsifiable. A third characteristic is that of simplicity—a problem should be stated in its most basic and simplest terms. When several hypotheses fit the facts of a problem, scientists usually choose the simplest hypothesis, a principle sometimes called **Occam's razor** or the **principle of parsimony**. An example is the appearance of crop circles: the simplest hypothesis—that human activity created the patterns—is more likely to be acceptable than the more complex hypothesis—alien activity—because in the latter case we would also need to explain who the aliens are, where they came from, how they came, and so on.

1.1.1.1 Specific versus general hypotheses

Hypotheses that are easy to verify generally tell us very little. For example, the hypothesis "this frog will jump if I touch it" can be tested by touching the frog and observing what happens. If the frog does jump, then our hypothesis is verified or confirmed; if the frog does not jump, then our hypothesis is falsified or disconfirmed. However, the confirmation of this hypothesis about a particular frog is far from an important scientific discovery. It is relatively unimportant because it is too specific, which is exactly what makes it verifiable.

Suppose, now, that we examine the much bolder hypothesis "all animals will react when stimulated." We can test this second hypothesis in the same way that we tested the first hypothesis, by touching or otherwise stimulating some animals, and we could also declare that the hypothesis would be falsified if one animal failed to respond. (Even then, we could never be sure, for it might respond in a way that we do not immediately notice, e.g., by remembering the event and responding at a later time.) But what if the animals tested do all respond? Does this verify that *all* animals will respond? Suppose we test 5 animals in a row or 5,000? No finite number of successes would be sufficient to verify the hypothesis for all animals and all circumstances. This is the kind of hypothesis that science usually examines: hypotheses that could potentially be falsified each time that we test them but cannot be absolutely verified for all possible occurrences.

Falsified hypotheses are rejected, and new hypotheses (which may in some cases be modifications of the original hypotheses) are suggested in their place. To extend the previous example, the hypothesis "all frogs will jump if touched" would be falsified if 2 out of 2,000 frogs did not jump. We could, however, modify the original hypothesis to one that is consistent with our data, e.g., "frogs will usually jump if touched." In practice, we might also want to be specific about the nature of the stimulus (how sharp the object, how firm the touch), the response (how frequent, how strong), and other particulars (the species or size of frog used, the temperature, and so on).

If testing a hypothesis does not reject it, we may want to generalize the hypothesis. For example, if a hypothesis tested using rats has not been falsified, we might want to apply the hypothesis to people as well or to all animals. However, we can never know how far we can extrapolate (generalize) results unless we continue to try to falsify our premise under different conditions. In this way, the testing of hypotheses allows us to draw conclusions about the observable world, but only to the extent that we have tested many possible circumstances and conditions.

1.1.1.2 Ways of devising hypotheses

There are myriad ways to devise hypotheses. One form of reasoning is called **deduction**, which is reasoning that guarantees the truth of a conclusion if we accept the truth of the premises. Deduction is frequently used to set up testable hypotheses: "*If* organisms of type X require oxygen to live, *then* this individual of type X will die if I put it in an atmosphere without oxygen." If the organism lives, then one of the premises must be rejected: either organisms of type X do not always require oxygen or else this individual was not of type X. Another type of reasoning, called **induction**, may be defined as reasoning that does not guarantee the truth of any conclusions drawn except in terms of probabilities. Science often uses induction to generalize from specific hypotheses, such as when we reason from five frogs to all frogs or from frogs and rats to all animals. Induction is also commonly used in everyday life: "I have liked pizza in each of the restaurants where I have ever eaten it; therefore, I will like pizza in any other restaurant." Induction allows us to reason beyond what we know with certainty, but in statistical terms only: I have always liked pizza, and I will probably like the next one that I try. However, since induction never guarantees the truth of any result, generalizations made by induction always need to be tested further. If I actually tried pizza in 450 restaurants that I had never tried before, I might discover that I only like the pizza in 442 of them and that there are 8 places that serve pizza that I don't like. The probability that I will like the pizza in a randomly chosen restaurant is thus 442 out of 450, or about 98.2%.

One common way of reasoning in science is to start with data already gathered, use induction to generalize and to formulate a hypothesis, use deduction to set up a test situation, and test the hypothesis through further observation. In our example, I can start with the five frogs that respond to touch, create the hypotheses that "all animals will respond to touch," reason by deduction that if all animals will respond to touch, then this rat and that starfish will respond, and finally set up an experiment to test the hypothesis and see if it works as expected. This process is often called the **scientific method**. In reality, few scientists adhere rigidly to this prescription.

Deduction and induction are only two of the many ways in which scientists go about the business of formulating hypotheses. Other ways include (1) intuition or imagination, (2) esthetic preferences, (3) religious and philosophical ideas, (4) comparison and analogy with other processes, and (5) serendipity, or the discovery of one thing while looking for something else. Moreover, these ways may be mixed or combined. For example, Albert Einstein declared that he arrived at his hypotheses about the physics of the universe by considering esthetic qualities such as beauty or simplicity and by asking, "If I were God, how would I have made the world?" Einstein also said that "imagination is more important than knowledge," a remark that is particularly true for the formulation of hypotheses (**Figure 1.1**). Nobel Prize–winning physicist Niels Bohr said that his initial hypothesis of atomic structure (the heavy nucleus in the center, with the electrons circling rapidly around it, "like a miniature solar system") first occurred to him by analogy with our solar system. Alexander Fleming found the first antibiotic as the result of a laboratory accident: on dishes of bacteria that should have been thrown away earlier, he observed clear areas where fungi had overgrown the bacteria. His hypothesis, that a product of the fungi had killed the bacteria, was validated by tests, and that fungal product is what we now know as penicillin. As these several examples show, *hypotheses are formed by all kinds of logical and extralogical processes*, which is one more reason why they must be subjected to rigorous testing afterward.

1.1.1.3 Biology: Hypothesis testing in living systems Animals, plants, and bacteria are complex and variable. So are other living systems, large and small, from ecosystems to individual cells. No individual animal or plant is exactly like any other animal or plant. At any moment, living systems that are otherwise similar may differ in external conditions, internal conditions, or in the way in which these conditions interact. Further, the same individual is not exactly the same from one day to the next. Because living systems vary, tests must be repeated. If the hypothesis is tested in one animal, or one cell, and a particular response occurs, the result is far less reliable as a means of prediction than if 10 animals, or 100 cells, all responded in the same way, which is why scientists always insist on large samples. What often happens, however, is that 9 out of 10 animals, or 94 of 100 cells, respond in one way and the remainder in another

Figure 1.1 **Imaginative hypotheses may originate from various logical or extralogical processes, especially from young scientists.** Does the idea shown here qualify as a scientific hypothesis? Why, or why not? Is it testable?

way. Interpretation of the results from tests on variable systems usually requires statistical treatment to ascertain whether the observed differences are "real" or can be explained by random variation. The differing responses may come from a source of variation that has not yet been identified, and scientists who study the anomalous cases sometimes discover new, previously overlooked phenomena.

1.1.1.4 A definition of science

Science may now be defined as a method of investigation based on the testing of hypotheses by organized comparisons with empirical evidence. Notice that the requirement of **evidence-based** testing makes scientific statements tentative, or provisional, and subject to possible falsification and rejection. Repeated exposure of our hypotheses to possible refutation increases our confidence in these hypotheses when test results agree with predictions but no amount of testing can guarantee truth or certainty.

Any hypothesis that is tested again and again, always successfully, is considered well supported and comes to be generally accepted. It may be used as the basis for formulating further hypotheses, so a cluster of related hypotheses soon arises, supported by the results of many tests, which is then called a **theory**. (Some people say that a widely accepted theory becomes a "law," but most scientists do not recognize a "law" as anything but a useful generalization.)

1.1.2 Hypothesis testing in science

Scientists test hypotheses by comparing them with the real world through empirical observations. Scientists differ from one another, however, in the ways in which hypotheses are tested. Some scientists do all their work in laboratories with

Figure 1.2 Scientists at work.

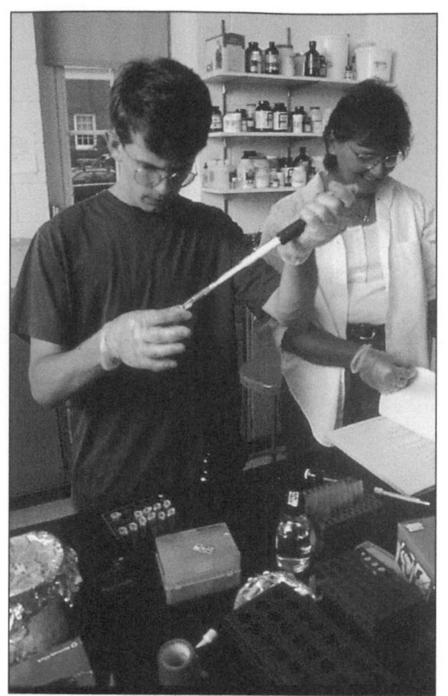

Transferring and examining solutions in a biochemistry laboratory.

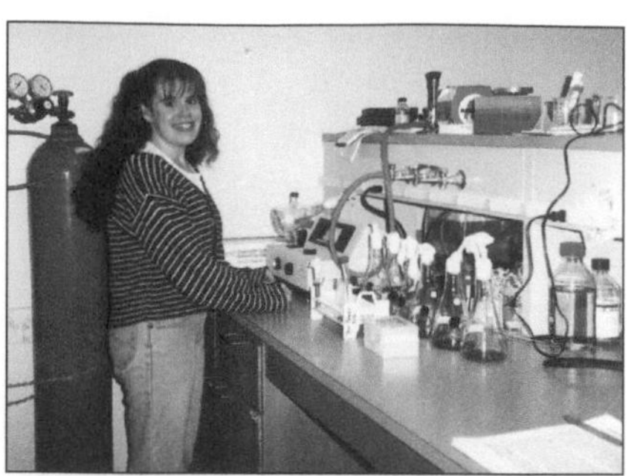

Preparing cultures in a bacteriology laboratory.

Examining cells with an electron microscope.

Surveying diversity in each square meter along a transect line perpendicular to a rocky shore.

specially designed equipment; other scientists gather data and specimens in the field for analysis and interpretation (Figure 1.2).

Some scientists test hypotheses by conducting **experiments**—artificially contrived situations set up for the express purpose of testing some hypothesis. Most **experimental sciences** aim to answer questions of the form "How does X work?" The scientist designs an experiment such that, if the hypothesis is true, a certain outcome is expected (or not expected). Then, the results of the experiment are determined "objectively," which means, in this context, *without bias either for or against the hypothesis being tested.*

In many experiments an experimental situation or group is compared with a control situation or with a **control group**. Ideally, the control group exactly matches the experimental group in all variables except the one being tested. For example, animals given a new drug are compared with a similar group of animals—the control group—that are not given the drug. The control group is given a substance similar to whatever is given to the experimental group, but lacking the one ingredient being tested. The two groups are selected and handled so as to be equivalent in every other way: similar animals, similar cages, similar temperatures, similar diets, and so on. Large samples are also required to allow for statistical tests.

As an example of the experimental approach, consider the following experiment in bacterial genetics that was conducted by Joshua and Esther Lederberg, part of the basis for a 1958 Nobel Prize. Most bacteria are killed by streptomycin, but the Lederbergs exposed the common intestinal bacterium *Escherichia coli* to this drug and were able to isolate a number of streptomycin-resistant bacteria. They allowed these bacteria to reproduce and were able to show that resistance to streptomycin was inherited by their offspring. In other words, the change to streptomycin resistance was a permanent genetic change; such changes are called mutations (see Section 3.1.4). The discovery of resistance gave the Lederbergs two hypotheses to test. The first hypothesis (the induced mutation hypothesis) was that the mutation had been induced, or caused, by exposure to the streptomycin. The second hypothesis (the prior mutation hypothesis) was that the bacteria had mutated before exposure to the streptomycin, in which case the mutation would be independent of the exposure. To distinguish between these hypotheses, the Lederbergs devised the experiment shown in Figure 1.3. In this experiment, a copy, or replica, of the original plate of bacteria was made. Only the replica, not the original, was exposed to streptomycin, and the position of each bacterial colony was noted. The induced mutation hypothesis predicted that bacteria exposed to streptomycin would mutate but that unexposed bacteria would not. In fact, most of the bacteria died, but a few survived and were thus identified as being streptomycin resistant. To test the prior mutation hypothesis, the Lederbergs went back and tested the colonies from the original plate. They discovered that the same colonies that were streptomycin resistant on the replica plate were also streptomycin resistant on the original plate. These findings support the prior mutation hypothesis for this particular sample of bacteria but falsify the hypothesis of induced mutations.

The prior mutation hypothesis for drug resistance had been tested and not falsified in the case of one mutation in one species of bacteria. How far could the finding be generalized? From this one experiment alone, one cannot tell. However, other investigators repeated the experiment for other mutations and other species of microorganisms. So far, the hypothesis of prior mutation has not been falsified. It is difficult to test the hypothesis in large or long-lived organisms, but most scientists are willing to assume the truth of the hypothesis for *all* organisms. There are many species (and thousands of mutations for each species) that have never been tested in this way, which leaves opportunities for the hypothesis to be falsified in the future.

1.1.2.1 Good experimental design (Part 1) The foregoing principles may be restated and elaborated as requirements for good experimental design:

1. The experiment should allow the possible observation of one outcome if the hypothesis were true and the possible observation of a different outcome if the hypothesis were false (or if a different hypothesis were true). Experimental conditions should not bias the results to favor one of these possibilities over the other. In the Lederbergs' experiment, different hypotheses predicted different outcomes.
2. Comparison should be made between material or subjects receiving the treatment being tested (the "experimental group") and those not receiving this treatment (the "control group"). In the Lederbergs' experiment, the experimental group were the colonies exposed to streptomycin and the control group were the colonies not exposed.
3. In order to isolate the one variable being tested, the experimental group and the control group should be subjected to the same set of controlled conditions to the maximum extent possible. For example, all the bacterial colonies grown by the Lederbergs were grown on the same nutrient culture under the same temperature and other environmental conditions.
4. To guard against unconscious bias, any financial or other interest of the experimenter should be disclosed. This was not an issue with the Lederbergs (they had no financial or other interest that favored either hypothesis), but a significant amount of research today is financially supported, e.g., by drug companies to test their products. Most scientific journals now insist that

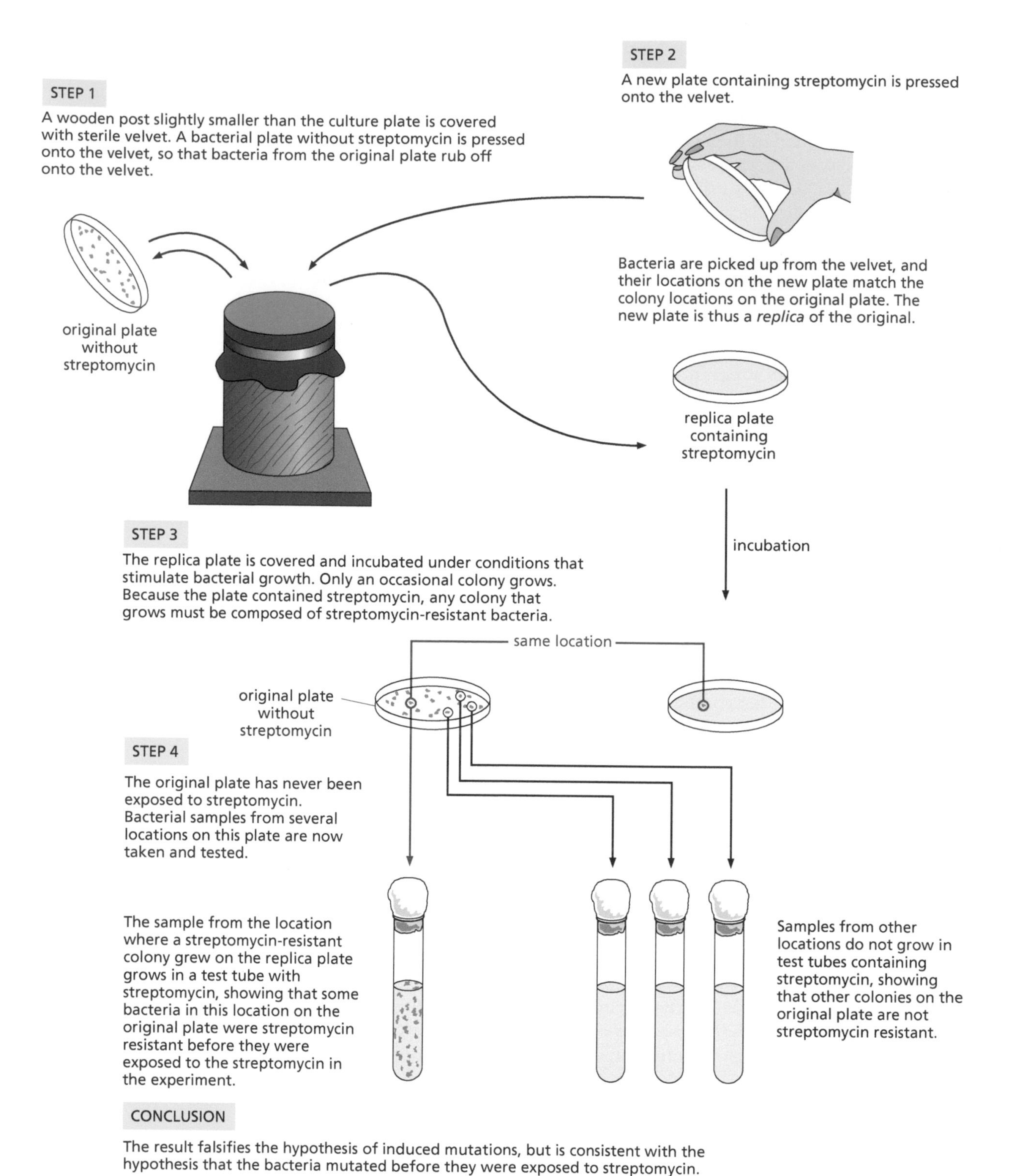

Figure 1.3 The replica-plating experiment of Lederberg and Lederberg.

authors disclose the funding sources for their research and any other financial or personal interest, such as owning stock in the company whose product they are testing.

5. Large sample sizes are required if statistical comparisons are made between the experimental and control groups in the percentage that showed one outcome over another or in the average value of some measured outcome.

1.1.2.2 An introduction to naturalistic science Another type of hypothesis testing is one in which direct experimental manipulation is either impossible or undesirable. For example, if an animal behaviorist wishes to study mating behavior *under natural conditions*, then any experimental manipulation that alters these natural conditions must be avoided. Thus, we see ornithologists hiding in blinds to study birds, while other naturalists photograph their subjects using telephoto lenses. The extinct species studied by paleontologists cannot be re-created in the laboratory to permit an experiment. We can refer to these sciences as **naturalistic (or "observational") sciences** because their method is based primarily on naturalistic observation rather than experiment. Naturalists do test hypotheses, but they do so by patient observation and record keeping. Naturalists often use control groups by comparing observations made when certain conditions are present with similar observations made when one of the conditions is not. The major difference between the experimental and the naturalistic sciences is that experimentalists set up and control their experiments, while naturalists can only observe and record those "experiments" that occur in nature. This often means that naturalists must search the world over for the right circumstances or must wait patiently for the right circumstances to occur.

The naturalistic sciences, moreover, are historical sciences. Any scientist seeking to understand why mammals differ from reptiles, or why the U.S. economy differs from the Japanese economy, will soon realize that the histories of animals, or of economies, form an important part of the explanation.

There are many types of questions in the naturalistic or historical sciences. The most characteristic type of question in these sciences is: "How did X get to be that way?" For example, a scientific team led by Rebecca Cann examined the DNA inside the mitochondria (the major energy-producing cell parts) of a large number of human populations. Mitochondrial DNA is always inherited from the mother, never from the father. Cann and her co-workers found that the chemical structure of the mitochondrial DNA in certain populations was very similar to its structure in other populations, allowing groups of related populations to be recognized. These scientists hypothesized that populations with similar mitochondrial DNA sequences share a close common descent through female lineages. This hypothesis explains the patterns of similarity among mitochondrial DNA sequences by a series of progressively "smaller" hypotheses about the past histories of a given set of populations: that the populations of the Americas all share a common descent, that the populations of New Guinea all share a common descent, and so on. Some of these smaller hypotheses are falsified by the data and must be replaced by modified hypotheses: New Guinea, for example, forms two clusters, and we can set up the hypothesis that it was colonized twice, with each line of descent forming a separate cluster. As modified in this manner, these smaller hypotheses of geographic dispersal are now interpreted as part of a common pattern of descent (Figure 7.7), with an area of origin in Africa. The data are consistent with a hypothesis that all human populations are descended from an ancestral African population, or from a single ancestral female, nicknamed "Eve" in the popular press. Like most other explanations in the natural sciences, the "Eve hypothesis" explains present conditions on the basis of their past history, an evolutionary or historical mode of explanation.

The requirements of "good experimental design," outlined earlier, apply as much to naturalistic observations as to experiments, with one major difference: scientists who cannot control "natural" conditions must seek far and wide for the proper comparison group (or "control"), and they must conduct extensive tests and measurements to ensure that the groups being compared are as similar as can be found for all variables except for the one being tested. Additionally, larger sample sizes and more extensive statistical tests are usually needed when there is more natural variation.

1.1.3 Theories

A **theory** is a coherent set of related and well-tested hypotheses that explain a broad set of observations and guide scientific research. Theories are usually developed by the testing of hypotheses, aided in many cases by mathematical and logical analysis of the resulting data. Theories then inspire future hypotheses and future tests. Most theories contain explanatory language that helps us understand some observed phenomena. One of the most important features of a good theory is that it may suggest new and different hypotheses. A theory of this kind stimulates further research and is sometimes called a productive theory. A theory may be productive for a while and then no longer stimulate new research. The theories that last are the ones that remain productive the longest, while the less productive theories may lie neglected without ever being disproved. In other cases, it is the falsification of one of its hypotheses (or the failure of a crucial test) that causes a theory to be rejected. Remember that the hypotheses that make up a theory are always subject to possible refutation. Even a long-cherished theory may be abandoned (or greatly modified) if it no longer holds predictive or explanatory power.

Public misconception of the term "theory" is widespread, as when people ridicule an idea (hypothesis) as "just a theory." Scientists use the term "theory" for well-established ideas that guide further research, like the atomic theory or the theory that germs can cause diseases. Scientists no longer doubt these theories—they are considered fundamental to ongoing research, along with the theory of relativity, the theory of evolution, and many others.

1.1.3.1 Theoretical language and models

A theory usually contains language that helps communicate its subject matter. Many theories also use a simplified mathematical or visual form, called a **model**. Such a model, while not a formal part of the theory, can nevertheless be an important teaching tool in helping communicate the theory to other people. For example, Bohr's conceptualization of the atom in terms of electrons circling around the nucleus "like a miniature solar system" was the model of atomic structure for generations of students. However, models are analogies. Like other analogies, models are comparable to the phenomena they describe only so far, and no further. Attempts to determine *how far* an analogy holds often suggest new hypotheses to test or new ways to test old hypotheses. The planetary model of atomic structure is a case in point. With the development of quantum physics, it became clear that the solar system model was inadequate to explain the behavior of subatomic particles. Scientific theories are tentative. Even the best-cherished theoretical models can be supplanted by other models—either because an important hypothesis is falsified or because a more satisfactory explanation or model is proposed. John A. Moore, an embryologist and science historian, stated that "great art is eternal, but great science tends to be replaced by even greater science."

Most theoretical models describe observable events in terms of underlying causes described in the special language of the theory. Atoms, for example, are described by a theory that explains much of the observable behavior of matter, and heredity was explained in terms of genes long before anyone really knew what genes were. Theoretical concepts are typically studied in terms of their observable effects: the properties of matter are described in terms of atoms, and heredity is described in terms of inherited genes. A scientist who describes heredity in terms of genes may never observe those particular genes, but the theory allows her to make further predictions regarding animals or plants and their inherited characteristics. In science, calling something "theoretical" simply means that, even if we cannot directly observe it, we can use the theory to make predictions and study the observable effects.

1.1.4 The properties of living systems

Animals, plants, and bacteria are examples of living systems that share many properties distinguishing them from nonliving systems. Each of these properties was initially a hypothesis about how living and nonliving systems differ. Each has been repeatedly tested and verified by observation across a wide variety of organisms, compared with various nonliving systems.

Table 1.1 Properties of living systems

Properties of living systems
Cellular organization
Metabolism
Selective response
Homeostasis
Growth and biosynthesis
Genetic material
Reproduction
Population structure

Properties shared by living systems, and the testable hypotheses about those properties, are summarized in Table 1.1 and listed as follows:

- *Cellular Organization*: The fundamental unit of life is called a cell. All living systems are composed of one or more cells.
- *Metabolism*: Living systems take energy-rich materials from their environment and release other materials that, on average, have a lower energy content. Some of the energy fuels life processes, but some accumulates and is released only upon death.
- *Selective response*: Living systems can respond selectively to certain external stimuli and not to others. Many organisms respond to offensive stimuli by withdrawing. Living organisms can distinguish needed nutrients from other chemicals and use only certain chemicals from among those available in their surroundings.
- *Homeostasis*: Living systems have at least some capacity to change potentially harmful or threatening conditions into conditions more favorable to their continuing existence, e.g., by converting certain toxic chemicals into less harmful ones.
- *Growth and biosynthesis*: Living systems go through phases during which they make more of their own material at the expense of some of the materials around them.
- *Genetic material*: Living systems contain hereditary information derived from previously living systems. This genetic material is a nucleic acid (either DNA or RNA) in all known cases.
- *Reproduction*: Living systems can produce new living systems similar to themselves by transmitting at least some of their genetic material.
- *Population structure*: Living organisms form populations. Populations can be defined retrospectively as groups of individual organisms related by common descent. Among organisms capable of sexual processes, a population is all those organisms that can interbreed with one another.

Implicit in this listing of properties is also the testable assertion that anything that is alive by one criterion usually meets the other criteria as well.

As new organisms were discovered, each of these hypotheses has been tested further and sometimes modified. For example, "breathing" was once considered an essential property of life, but this was modified several times to include other forms of gas exchange and metabolism. The invention of the microscope, around 1670, led to the discovery of bacteria, which caused us to expand our concepts of selective response and led to the cell theory in the 1830s. The discovery of viruses early in the twentieth century strained this theory even more: seven of the eight hypotheses apply, yet viruses are not cellular and cannot reproduce on their own—they must use the cellular machinery of other organisms to reproduce themselves and therefore fall into a gray area between the living and nonliving.

Together, these hypotheses form a working theory about the characteristics of living things that continues to be productive and to suggest new hypotheses to test. They are not rigid definitions that must be met. If something lacks one of the properties of life, we would not be forced to define it as nonliving; we could instead modify the hypotheses to define the limits of life more precisely. Viruses, for example, fulfill many of the properties of life but cannot reproduce without the help of another organism. We may therefore need to modify the criterion of reproduction to say that living things can bring about or can direct the production of new living things similar to themselves.

These hypotheses have been tested in a wide variety of living systems. We can therefore gain confidence that the properties summarized in Table 1.1 form a coherent theory about how living systems differ from the nonliving.

THOUGHT QUESTIONS

1. In a group, discuss the hypothesis shown in Figure 1.1. Is it testable? If you believe so, then explain what sorts of observations you would make to test it.
2. Scientists seek to provide evidence to "support" hypotheses. Why don't scientists say that their evidence "proves" hypotheses?
3. In the Lederbergs' experiment described earlier, what outcome would have been consistent with the induced mutation hypothesis?
4. Which of the following are experimental tests and which are naturalistic observations?
   a. Measurements made on the bones of an extinct species are compared with similar measurements made on the bones of a related living species.
   b. The activity of white blood cells in a blood sample taken from stressed rats is compared with the activity of white blood cells taken from unstressed rats.
   c. A group of animals is fed a certain chemical to see whether they will get cancer as a result.
   d. A list of the species found in a particular square meter near the coast is compared with another list of species found 20 meters farther inland.
5. Viruses strain any definition of living systems: they contain genetic material, yet they replicate only inside the cells of and with the help of some other organism. Should we think of viruses as alive and devise a theory (or definition) of life that includes them, or should we think of them as lifeless and devise a theory (or definition) that excludes them? Would one of these theories be right and the other wrong, or are we free to choose either option?

1.2 SCIENTISTS WORK IN PARADIGMS, WHICH CAN HELP DEFINE SCIENTIFIC REVOLUTIONS

In a book that was itself considered revolutionary when it was first published in 1962, Thomas Kuhn, a philosopher and science historian, proposed a new method of looking at the ways in which science accommodates to new discoveries. Kuhn's observations were based on his studies of historical revolutions in science.

1.2.1 Paradigms and scientific revolutions

According to Kuhn, everything that we have described thus far is part of **normal science**, science that proceeds by the piecemeal discovery and gradual accumulation of new but small findings. Normal science in Kuhn's theory is always channeled by what he calls a **paradigm** (pronounced "para-dime"). A paradigm is much more than a theory; it includes a strong belief in the truth of one or more theories and shared opinions as to what problems are important, what problems are unimportant or uninteresting, what techniques and research methods are useful, and so on. The research methodology and sometimes the instrumentation are important parts of the paradigm. Normal science proceeds cumulatively, in small steps, within the context of an existing paradigm. Paradigms, according to Kuhn, are best represented by science textbooks, which are written for the purpose of training new scientists within the paradigm. Students trained by these textbooks are taught not just facts, they are taught attitudes, approaches, values, and a vocabulary that teaches them to think in certain ways.

Once in a great while, says Kuhn, science proceeds in a very different way: a **scientific revolution** occurs and is marked by the emergence of a new paradigm, which often requires replacement of an older paradigm. Few scientists educated in the old paradigm support the new paradigm at first. Most support for the new paradigm comes from new scientists just beginning their careers, and the founder of the revolution is usually either young or a new entrant into that particular scientific field. Once a scientific revolution occurs, its new paradigm opens up a new field of investigation or rejuvenates an old one. Such an infrequent event is called a "**paradigm shift**." Scientists will not be attracted to a new paradigm unless they feel that it is somehow superior to the old paradigm, usually because it explains a wider variety of phenomena or because it explains certain new findings better than the old paradigm did. Often the vocabulary terms of the old paradigm are redefined or newer terms are adopted; terms that are no longer useful to the new paradigm may be abandoned. A paradigm shift is not just a triumph of logic or of experimental evidence. It is decided, at least in some measure, by a political-style process in which allegiances and influences shift. New paradigms succeed when scientists find them to be fruitful or productive of new approaches to research. Some examples of paradigm shifts are given in Table 1.2.

Paradigms are sometimes so powerful as to allow anomalies—observations that "do not fit"—to be ignored. To scientists working within a paradigm, anomalies are small problems that they agree to ignore, believing that the integrity and success of the paradigm are more important than trying to accommodate the unexplained anomaly. Scientists who become interested in the anomaly may become founders of their own new paradigms, and the increased attention that they draw to the anomaly may precipitate a scientific revolution. Paradigms become successful in large measure by the students that they attract. Paradigms that no longer attract students die out.

1.2.2 Molecular genetics as a paradigm in biology

As an example of a scientific paradigm in biology, we describe here the field of molecular genetics as it has existed since about 1950. Other examples of scientific paradigms are described in subsequent chapters, including Darwinian evolution in Chapter 5 and the connection between the mind and the body in Chapter 14.

The paradigm of molecular genetics (or molecular biology) emerged in the decades following the determination of the structure of DNA by James Watson and Francis Crick in 1953. The structure of DNA was itself simply a hypothesis that gained rapid acceptance because it explained many known facts and allowed many new predictions to be made. The "central dogma" of molecular biology (so named by the molecular biologists themselves) was that DNA was used to make RNA and RNA was used to make protein. (Further details of this process are described in Section 3.1.) Both DNA and RNA were said to contain information, and the making of one molecule from another was said to be a form of information transfer. As in other paradigms, the central dogma was more than just a theory because it also suggested a new vocabulary and drove a new research

Table 1.2 Some examples of paradigm shifts in biology

Old paradigm	New paradigm
▪ Natural theology, Lamarckism, and several other competing paradigms	Darwinism (since 1859)
▪ Blending inheritance and various folk ideas	Classical Mendelian genetics (since 1865 or 1900)
▪ *Various beliefs*: Bad humors, bad air or water, evil spirits, and many others	Germ theory of disease (Pasteur, Koch, since 1880)
▪ Competing paradigms, including Darwinism, mutationism, population genetics, and neo-Lamarckism	Modern evolutionary theory (since 1940)
▪ Classical Mendelian genetics	Molecular genetics (since 1950s)
▪ Various theories of territorial behavior, sexual behavior, etc., also psychological theories (gestalt, behaviorism, ethology)	Sociobiology (since 1975)
▪ Descartes' mechanistic theories and dualism	Mind–body connections (since 1980s)
▪ *Classic germ theory*: Pathogenicity as a characteristic of pathogens only	Pathogenicity as an interaction of pathogen and host (since 1990s)

program. The language used within a paradigm often reveals much about how the paradigm is understood by the scientists working within it. How did DNA make copies of itself? This was called *replication* long before any of its details became known. How was information from DNA transferred to RNA? This was called *transcription.* How was information from RNA transferred to protein? This was called *translation.* How replication, transcription, and translation occurred were among the major problems to be solved. The terminology in molecular biology, like that in many fields, was part of an elaborate analogy that drew its inspiration from a comparison with linguistics and included such new vocabulary words as *code* (the language itself) and *codons* (items in the code). Also, words such as *transcription* (rewriting within the same language) and *translation* (changing from one language to another) were deliberately chosen for literal meanings that matched the biological theory. Textbook descriptions were replete with verbs like *read, copy,* and *translate.* There were also a number of laboratory methods, inherited from the field of biochemistry, plus a few extra technical advances, such as the use of high-speed centrifuges. Together, this all formed an orderly paradigm that outlined not only what was known but also what remained to be discovered, what was thought to be important, and how the details were to be investigated and described. DNA was championed as the most important "master molecule," RNA was almost as important, and protein was important only until its synthesis was completed. Protein that was completely synthesized was no longer deemed interesting, except for a few enzymes that helped in the working of DNA or RNA. The paradigm thus defined the boundaries of the field.

The paradigm of molecular genetics guided research on DNA, RNA, and protein synthesis throughout the 1950s and 1960s; much of the work begun in those decades continues today. For its workers, the paradigm defined a set of shared beliefs (including the central dogma), a vocabulary, a set of research techniques, and, most of all, a set of problems to be solved. These problems included the mechanisms by which replication, transcription, and translation took place, as well as how to crack the genetic code. Once this last problem had been solved, the "coding dictionary" (i.e., the list of correspondences between RNA sequences and protein sequences, Table 3.1) was given a prominent place in every genetics book and most general biology texts.

As the molecular genetics paradigm matured, some of its early tenets were modified. Information flow, once thought to be unidirectional, is now understood to be more complex, sometimes flowing in both directions. Also, the idea of DNA as the "master"' molecule and RNA and protein as less important has been revised: although DNA does contain the instructions for making other molecules (RNAs and, indirectly, proteins), proteins comprise the class of molecules that perform the most diverse array of functions in cells, and some RNAs carry out roles beyond solely being intermediates between DNA and protein. Likewise, attention has shifted to new questions, such as how the environment of a cell influences that cell to transcribe certain portions of its DNA at certain times. Still more recently, technological advances have permitted the rapid determination of many DNA sequences, including the sequencing of whole genomes, which has allowed many additional questions to be raised. Thus, the molecular biology paradigm, like other paradigms before it, has gradually changed over time, although its core beliefs remain unshaken.

1.2.3 The scientific community

Is science something that only scientists can do? On the contrary, many people use scientific methods in their everyday lives. For example, if my car fails to start, I might formulate one hypothesis after another as to the possible cause. To test the hypothesis that the car is out of gas, I would examine the gas gauge. Additionally, I could add some gasoline to the tank and then try to start the car. If the car starts, I conclude that it was out of gas. The Swiss child psychologist Jean Piaget has written that children often behave as little scientists, formulating possibilities (hypotheses) in their minds and then testing them. "I can take the toy away from my little brother" can be tested by trying to take it away; the hypothesis would be falsified if the brother successfully resisted or if an adult intervened.

It is unusual for a single person to formulate a hypothesis, test it, and then critically evaluate the results. For this reason, it is important for scientists to communicate with one another so that all these steps can be performed. Early written examples of hypothesis testing are found in the writings of the Greek historian Herodotus (fifth century BCE). Early scientists in China, India, and elsewhere also wrote down their ideas by hand, but the spread of printing using movable type greatly speeded up the spread of scientific ideas after about 1500 CE. Early European scientists (Copernicus, Galileo, Harvey, Descartes, Newton, and others) wrote their ideas in the form of books, pamphlets, and private letters. A major advance, however, occurred in seventeenth-century England, with the founding of the Royal Society around 1660. This marked the first time in history that a permanent, *organized community* of scientists had communicated with one another and shared their results in a scientific journal *(Philosophical Transactions of the Royal Society)*. Now there was a written and permanent record of experiments performed and conclusions reached—a shared record that encouraged scientists to check one another's work in a systematic way. Because of this written record, scientists of the past continue to be part of the scientific community when their ideas are tested, even generations later. The scientists shown in **Figure 1.4**, whose accomplishments are each described elsewhere in this book, are all still part of this scientific community even though they published over a time span of about 150 years.

Many of the ways in which today's scientists behave toward one another may be viewed as efforts to maintain their ability to do the kind of systematic checking described earlier, including the ability to test hypotheses. Every test must be conducted in such a way as to make it possible for the hypothesis to be falsified, if indeed it is false, and the testing of hypotheses should be described as publicly as possible so as to permit the test to be repeated by other scientists. As David Hull (1988, p. 4) points out, "Scientists rarely refute their own pet hypotheses, especially after they have appeared in print, but that is all right. Their fellow scientists will be happy to expose these hypotheses to severe testing." Skeptics who doubt a particular result unless they have seen it themselves can best be won over by a tradition that allows them to hear about repetitions of the test or to repeat the test themselves and to make their own observations. For example, Galileo, the astronomer and early scientist, invited critics who doubted his observations to look for themselves through his telescope.

So, the process of science is conducted in the public forum as well as in the laboratory. The publishing and dissemination of results (both in print and increasingly on the Internet) and the repetition of observations and experiments by others are thus valued among scientists. This is why the human genome sequence and an increasing number of scientific journals are available through the Internet. Scientists are not expected to work in isolation, but rather to discuss their results with other interested scientists, allowing them to build upon the results of previous scientists. They can repeat experiments and confirm the results, but they do not need to start from scratch and repeat *all* earlier work in their field. Science is a cumulative process in which it pays for individual scientists to begin with some of the groundwork laid by others, rather than to start always from scratch. As Isaac Newton once said, "If I have seen further than others who have gone before me, it is because I have stood on the shoulders of giants."

THOUGHT QUESTIONS

1. Why is it so difficult for a scientist to work outside the prevailing paradigm? Give at least three reasons.
2. In what ways is the paradigm of molecular genetics more than just a scientific theory?
3. Many companies conduct what they call research and development, yet many of these companies zealously guard their results and do not publish them. Are they doing science?
4. Researchers have deciphered the complete genetic blueprints of over 800 viruses and many disease-causing bacteria. These sequences are available on the Internet. How might free access to the genetic makeup of disease-causing pathogens be used? How could it be abused?

(A) Charles Darwin (1809–1882), naturalist and evolutionary biologist. Darwin's theories are described in Chapter 5.

(B) Gregor Mendel (1822–1884), botanist and geneticist. Mendel's experiments in genetics are described in Chapter 2.

(C) Ernest Everett Just (1883–1941), marine biologist and embryologist. Just's contributions to understanding the process of sexual reproduction are described in Chapters 2 and 9.

(D) Barbara McClintock (1902–1992), agricultural geneticist and Nobel Prize winner. Some of her contributions to genetics are described in Chapter 2.

(E) Luis W. Alvarez (1911–1988), Nobel Prize-winning physicist, and Walter Alvarez (1940–), geologist. Walter's right hand rests on a layer of clay 65 million years old, at the boundary between the Age of Reptiles and the Age of Mammals. The Alvarezes' hypothesis to account for the extinction of dinosaurs and many other species across this boundary is described in Chapter 18.

(F) Jennifer Doudna (1964–), biochemist and molecular biologist. Together with Emmanuelle Charpentier, Doudna was awarded a 2020 Nobel Prize for developing the CRISPR-Cas9 method of gene editing explained in Chapter 4.

Figure 1.4 A few notable scientists.

1.3 SCIENTIFIC LITERACY AND SOCIETY

Science is, above all, a method of discovery. As outlined earlier, it is a method based on formulating hypotheses, testing them, and then presenting the evidence for public scrutiny. Anything less is not science. Science can only occur in a social context. An investigator who forever guards the secrecy of his or her observations is not contributing to the body of knowledge that the world shares and is therefore not conducting science. Scientists must at some point share their discoveries and also the evidence that guides them to their conclusions. The declaration of results without disclosing the evidence for these results is also not science.

Scientists should always be skeptical in evaluating evidence and should insist that hypotheses have been rigorously tested with good experimental designs. (This is especially true if life-and-death decisions will be based on the results.) Healthy skepticism, however, is vastly different than science denial. When scientific hypotheses are incorporated into well-tested theories based on abundant evidence from multiple lines of research, the claim that any other opinions are equally valid is tantamount to scientific malpractice and irresponsibility.

Scientific literacy is an awareness of science as a collective human endeavor based on the presentation of evidence. Scientific literacy includes the habit of evaluating evidence and making decisions based on this evidence. The National Science Education Standards describe scientific literacy this way:

> *Scientific literacy means that a person can ask, find, or determine answers to questions derived from curiosity about everyday experiences. It means that a person has the ability to describe, explain, and predict natural phenomena. Scientific literacy entails being able to read with understanding articles about science in the popular press and to engage in social conversation about the validity of the conclusions. Scientific literacy implies that a person can identify scientific issues underlying national and local decisions and express positions that are scientifically and technologically informed. A literate citizen should be able to evaluate the quality of scientific information on the basis of its source and the methods used to generate it. Scientific literacy also implies the capacity to pose and evaluate arguments based on evidence and to apply conclusions from such arguments appropriately.*
>
> [National Academy of Sciences, 1996,
> National Science Education Standards, p. 22]

Many of the decisions made by societies are informed by science. Each of us makes frequent decisions about what to eat and what medicines to take. Corporate leaders make additional decisions about the products they bring to market. Governments at all levels make decisions about what products or activities to restrict or inspect in the interests of public safety. Science plays a role in all these decisions, and all decision-makers must therefore be able to evaluate scientific evidence about the safety, health benefits, and other characteristics of the things we use and the ways in which we use them. Scientific literacy is therefore a responsibility of every citizen because we all make evidence-based decisions, and certain people (including doctors, food safety inspectors, legislators, and many others) make decisions that affect many other people.

1.4 SCIENTISTS OFTEN FACE SOCIAL ISSUES

Science itself can never tell us whether certain research should be done or how the results should be used by society; for those answers, we turn to the branch of philosophy called **ethics**. Many topics in this book have a social or ethical dimension. Are some applications of specific biological research morally right and other applications morally wrong? Should society place legal restrictions on scientific research? Should biologists concern themselves with the ethics, applications, and implications of their work? This section describes some of the ways in which individuals and societies make such decisions.

We each use beliefs concerning what is right or wrong, proper or improper, to guide our own behavior. It is right to come to class at the scheduled time and in general to keep appointments to which one has agreed. It is wrong to steal, to lie, to murder, or to park in the NO PARKING zone. It is proper to wait for the traffic light to turn green and to wait for one's turn in line. All these moral rules, or **morals**, are products of societies. Anthropologists who have compared societies from around the world tell us that moral rules differ from one society to the next; they also change over time as society changes.

Any personal decision about whether to follow a moral rule may be called a moral decision: for example, should I park in the NO PARKING zone? Moral decisions are often made with the knowledge that society will attempt to enforce the rules with penalties or sanctions. Formal sanctions include fines and jail time; informal sanctions, which operate more often, include being criticized or avoided by others and ending up with fewer friends.

1.4.1 Ethics

Ethics is a discipline dealing with the analysis of moral rules and the ways in which moral judgments are made and justified. Descriptive ethics, the study of how these judgments are actually made, is a social science that investigates human behavior using scientific methods. In contrast, normative ethics, a branch of philosophy, deals with the logical analysis of how ethical judgments *should* be made, an analysis for which observational data are insufficient—no data can either confirm or refute a moral law (such as "Thou shalt not kill"). In its simplest form, normative ethics is an attempt to reduce moral codes to a minimum set of basic rules (maxims). For example, I should come to class on time because my signing up for a course is like making an appointment. Appointments should be kept because they are promises or contracts. An ethic of keeping appointments is part of a larger ethic of keeping promises.

Some rules of conduct are simply inventions of a society for the convenience of its members, such as waiting for the green light, driving on the right side of the street (in North America), and the observing of NO PARKING zones (Figure 1.5). We cannot all drive through the intersection at the same time, and traffic lights are a

Figure 1.5 Would you park in this space? Give reasons to explain your decision.

convenient (if arbitrary) contractual way of arranging whose turn is next. The contractual nature of such agreements is obvious because there are usually publicly controlled processes (like city council meetings) to decide where to put NO PARKING zones. We promise to observe traffic laws when we apply for a driver's license, so following these laws may be viewed as another form of promise keeping.

Waiting one's turn in a line or waiting for a green traffic light are both ways of introducing order and fairness into a situation that would otherwise be chaotic and conducive to unnecessary disputes. A major difference, however, is that waiting one's turn in line is not enforced by law or traffic code. It is enforced informally by the tacit agreement of those who are present.

A simple moral code might therefore instruct us to keep our promises, not to interfere with the rights of others, and to observe the common social conventions. This could easily be expanded into a more general code of benevolence, cooperation, and mutual aid.

1.4.2 Resolving moral conflicts

There are occasions when conflicts arise within sets of moral rules. I know I should obey the traffic laws, but what if I am taking an injured person to the hospital and the person's life is in danger? Does the duty to save a life justify driving above the speed limit, driving through a red light, or parking in a loading zone? Can I justify disobeying traffic laws to keep an appointment? Does it matter how important I think the appointment is? Resolving conflicts of this kind is one of the major goals of ethics.

In most cases, the resolution of such moral conflicts is made by determining that one rule or goal is more important than another: saving a life is more important than obeying traffic rules, for example. Thus, there are *exceptions* to most moral rules: obey traffic rules and other useful conventions *except* when obeying them causes greater harm or violates a more important rule. Notice that this ranks certain rules as more important than other rules, allowing us to justify an exception to one rule by invoking a "higher" rule.

Although ethics is a branch of philosophy, ethical arguments arise in everyday life and also in science. For example, a scientific researcher may have financial interest in the success of a particular company, which might create a bias in scientific research regarding that company's products. This is one reason why most scientific journals now insist that each contributor must disclose such financial interests.

More and more, scientific endeavors are raising ethical issues that are of practical interest to people in all walks of life. Virtually all institutions that conduct research now have policies and procedures for managing conflicts of interest. The U.S. government has sponsored research, meetings, and publications in the field of biological ethics. Many government programs, notably the Human Genome Project (see Section 4.3), have set aside portions of their budgets for the examination of the ethical implications of science. The ethical approaches discussed in this chapter will support our examination of many issues with far-reaching ethical implications in the chapters that follow, including research on embryonic stem cells, altering genetic traits of individual people or of crop plants, protecting biodiversity, or mitigating climate change.

1.4.3 Deontological and utilitarian ethics

An ethical system is a set of rules for resolving ethical questions or for judging moral rules. Here we describe the two major types of ethical systems. Depending on whether one adheres to one or the other of these types of systems, one may be drawn to form different opinions about the various ethical issues that we will discuss in later chapters.

1.4.3.1 Deontological systems An ethical system is called **deontological** if each person has a duty to perform certain acts and to avoid others, but the rightness or wrongness of an act depends on the act itself and not on its consequences. To a deontologist, the wrongness of murder is in the act itself, not stemming from its results or its effects on society. Similarly, a deontologist who believes in keeping promises does so apart from any consequences.

The Bible, the Koran, and the sacred texts of other religions have been the source of many deontological moral codes. Philosopher Immanuel Kant (1724–1804) devised a deontological system without a religious basis. Kant based all ethical statements on a single precept: act only according to rules that you could want everyone to adopt as general legislation. Kant called this rule the **categorical imperative**. Kant's test of the morality of an act is whether the act can be universalized—that is, applied to all people at all times. Thus, killing is (always) wrong because I could not possibly want people always to kill one another—I would be willing my own death and the death of my loved ones. Keeping promises can be universalized, and promise keeping is therefore (always) moral. Respect for all human beings is an important part of Kant's system, and fundamental **rights** are based on respect for the dignity and autonomy of all persons. If you respect the dignity of all human beings, then you cannot ever will the death of any person, nor can you deny them their fundamental rights. If you respect their selfhood, then you cannot morally abridge their rights or their freedom.

To whom do these various rights apply? At various times in the past, certain groups of persons (including women, children, slaves, the lower classes of stratified societies, impoverished people, foreigners, members of various races, mental patients, and persons unable to speak for themselves) were denied the rights that were afforded to other members of society (**Figure 1.6**). Many people now invoke dignity and autonomy criteria in discussions of whether certain rights should also now be extended to unborn fetuses or to animals.

An argument against rights-based deontology is that there are many circumstances in which one right conflicts with another, resulting in a moral dilemma. Unless there is a clear way of deciding between conflicting rights, moral dilemmas are inevitable. An obvious way out is to declare one particular right (such as the right to life or the right to freedom of action) supreme over all others. Aside from the problem that different deontologists would choose different rights to take

Figure 1.6 **Do you have a deontological reason for agreeing or disagreeing with the premise that all persons share the same basic rights?** This woman was protesting the fact that U.S. President Woodrow Wilson supported the rights of poor Germans in World War I, while women in the United States were denied the right to vote.

precedence over the others, there is the more serious objection that insistence on a single right lead to the dangers of absolutism. Historically, many atrocities have been perpetrated by the followers of systems that put absolute adherence to a single principle above all others.

1.4.3.2 Utilitarian systems A **utilitarian** system of ethics is one in which acts are judged right or wrong according to their consequences: rightful acts are those whose consequences are beneficial, whereas wrongful acts are those with harmful consequences. To a utilitarian, murder is wrong because the death of the victim is an undesirable outcome under most circumstances. Also, on a larger scale, murder is additionally wrong because it produces a society in which people live in fear.

A challenge to all utilitarian systems is to find a way of measuring the goodness or badness of consequences. Utilitarian philosophers have proposed different criteria by which to judge consequences: the greatest happiness for the greatest number of individuals, the greatest excess of pleasure over pain, and so on. All utilitarian systems require that value judgments be made between outcomes that are difficult to measure and quantify (Figure 1.7).

Utilitarianism can be summarized by the rule "always act so as to maximize the amount of good in the universe." The first major utilitarian philosopher was Jeremy Bentham (1748–1832), who said that we should always strive to bring about "the greatest good for the greatest number." To decide which actions produce the greatest good, Bentham suggested a type of cost-benefit analysis that he called "a calculus of pleasures and pains." Other notable utilitarian philosophers include John Stuart Mill (1806–1873) and G.E. Moore (1873–1958).

One major criticism of utilitarianism is that its cost-benefit approach reduces the status and dignity of human beings and, in some cases, violates their rights. Can the killing of one person be justified if it results in saving the lives of other people? Deontologists argue that certain individual rights must be protected regardless of any benefit to society as a whole. According to these critics, even if a larger benefit can be demonstrated in a given instance, it is still unethical to violate an individual's fundamental right, because "the ends do not justify the means."

1.4.4 Ethical decision-making

Individuals who are faced with moral choices are "moral agents," meaning that they are held responsible for their decisions; they are praised for good decisions and criticized or punished for bad decisions, at least in a just world. Scientists are moral agents because they encounter moral decisions in their work. Science has also presented society at large with moral choices concerning the ways in which the findings of science may be used. Some moral decisions are left up to individual choice, while other decisions are made by institutions or by society as a whole. Many moral decisions involve conflicts between the rights and interests of individuals and the broader interests of large institutions or of society at large. Decisions that affect many people at once are usually made collectively, at least in

Figure 1.7 **What benefits could come to society from this nuclear power plant?** What costs or risks are present? How would a utilitarian argue in favor of this power plant? How might another utilitarian argue against it?

Table 1.3 Some moral imperatives

▪ Nonmaleficence ("above all, do no harm")
▪ Beneficence (doing good to others)
▪ Veracity (telling the truth)
▪ Fidelity (being faithful, honoring commitments, and keeping promises)
▪ Autonomy (respecting the rights and individual integrity of others)
▪ Justice (treating all people with equal respect)
▪ Preserving important social institutions
▪ Preserving life
▪ Setting a good example for others
This list is not exhaustive and should not be taken as a priority ordering.

Modified from Ruth Purtilo, *Ethical Dimensions in the Health Professions,* Philadelphia, W.B. Saunders Company, 3rd edition, 1999.

a democracy, and many people may seek to influence such collective decisions. Individuals making moral choices are often guided by principles of duty or moral imperatives, such as those listed in **Table 1.3**.

In facing a moral decision, a person using deontological reasoning asks, "What are my duties?" If several duties are in conflict, the question then becomes, "Which is the more important or higher duty?" In a utilitarian approach, the costs (or risks) and benefits of each possible choice must be compared, including the example that may be set for others. The question then becomes, "Which set of costs and benefits is preferable?" meaning "Which set achieves a maximum of benefits at a minimum of costs?"

In the case of a moral dilemma—where no option is entirely good—we often seek further guidance in reaching a decision. The following procedural steps can help us in reaching moral choices: (1) Gather all the facts and check them for correctness; (2) identify the ethical problem; (3) identify all the parties (stakeholders) that would be affected; (4) analyze the problem (e.g., with a deontological or utilitarian approach); (5) present the alternatives in priority order; (6) make and implement the decision; (7) if possible, reevaluate the decision as its consequences unfold. Several chapters in this book have links to ethical case studies for you to consider. You may wish to refer to these steps to help you complete the assignments or to help make moral choices in the real world.

1.4.4.1 Collective ethical decisions Moral choices are in many cases made by groups of people rather than by individuals. These groups can be legislative bodies, professional groups, or private corporations. Many of these choices can affect large numbers of people or society as a whole. Individuals play an important role in such group decisions; however, the decisions made by a particular group can sometimes conflict with one's own self-interest or one's own moral judgment. How, then, should ethical choices be made by collective groups?

In culturally homogeneous societies in which people share common values and the same religious beliefs, it may be possible to reach consensus about which acts are wrong and which are right. However, most societies today are *pluralistic* in the sense that their populations include people with differing cultural and religious backgrounds who will likely be diverse in their opinions, values, and ethical approaches. Reaching collective decisions on ethical issues is more difficult in pluralistic societies. In a pluralistic society, people come to the public forum with viewpoints aligned with varying ethical frameworks. One way of reconciling these different views is to have a public debate (so that all views are heard) and then vote. The voting procedure should be structured in a way that all parties recognize as fair. **Fairness** is the principle that all people should be treated impartially and equally (Rawls, 1972, 2001). Fairness is a way of ensuring that rights are not violated. In most cases, a system that works in this way results in the least displeasure with the decision. Of course, total agreement on a decision is hard to achieve in any large pluralistic society.

1.4.4.2 Policy decisions **Social policy** includes all those laws, rules, and customs that people follow in making individual decisions in a society. The making or changing of a new social policy is called a **policy decision**. Most social policies and policy decisions involve ethical considerations. An increasing number of policy decisions also involve some aspect of science, and it is often convenient to divide these into three phases: scientific issues, science policy issues, and policy issues.

- *Scientific issues*: What possible explanations (hypotheses) exist that might explain the available data? Can these hypotheses be tested? Do the tests support or falsify each hypothesis? Are alternative explanations available? What additional data are needed to evaluate these hypotheses? These are often characterized as "purely" scientific issues on which scientists of different political or ethical persuasions may be expected to agree if sufficient data are available.
- *Science policy issues*: What would be the consequences of this or that particular legislation or policy change? What would be lost or gained from each proposed plan of action? Can the probability of uncertain consequences be estimated? The probability of any outcome, especially one not desired, is called a **risk**. Can we calculate or estimate the risks? In a cost-benefit analysis, what are the costs and what are the benefits? How certain are we of the estimated values? These are still scientific issues in the sense that they are evaluated from data, but disagreements on the data or their significance are expected between experts with different political or ethical viewpoints. If the experts disagree, how shall we evaluate their respective positions?
- *Policy issues*: Once we have evaluated the possible consequences of various possible policies, which one should we choose? These are ethical decisions in which values have a prominent role (e.g., is it worth risking the disruption of the economy to stop global warming?). In general, is some predicted but uncertain benefit worth the calculated risks? If a proposed change can be made only by incurring a certain cost to society, is this social cost worth the intended benefit? In all these cases, the cost-benefit analysis is used not to make the final decision, but rather to provide the necessary data on which a policy decision can be intelligently based.

Who makes the decisions? In most societies, scientific issues are frequently decided by scientists with little input from interested citizens. Science policy issues are often decided in the court of public opinion by an interplay of scientists and other "experts" under the scrutiny of policy advocates for one side or the other. Public discussion and disclosure of evidence is useful in exposing and eliminating faulty information. Although evidence is used, it is more like courtroom evidence, obtained and evaluated by cross-examining witnesses, than like the evidence of science, obtained by the formulation and testing of alternative hypotheses.

As for the final policy decisions, who makes them: the scientists, the public, the media, or the government? In most democracies, the decisions are made either by the public or by government agencies acting (in theory at least) in the public interest. Decision-makers are often influenced by scientists, the media, particular interest groups, and public pressures of various forms—marketplace decisions (decisions by individuals about where to spend their money), letters and email, organized demonstrations (Figure 1.8), enthusiasm at public gatherings, public opinion surveys, and direct votes on referendum questions. All of these influences certainly have a role in what is essentially a political process.

Figure 1.8 **Why do people have a right to express their opinions through public demonstrations like this one?** What role do such public demonstrations have in shaping social policy?

THOUGHT QUESTIONS

1. Together with other students, make a list of five to ten laws or rules that are generally followed on your campus or in your community. For each, try to discover the following:
   a. Why is such a rule considered important (or why *was* it considered important when the rule was adopted)?
   b. Is there a more general moral concept of which this rule is just a special case?
   c. Is society better off with rules of this general kind than without them, and, if so, why?
2. Try to justify the wrongness of the following acts under the two ethical systems discussed in this section:
   - Murder
   - Rape
   - Bank robbery
   - Failure to repay a debt
   - Racial segregation
   - Driving over the speed limit
   - Parking in the handicapped space shown in Figure 1.5 if you are not handicapped

   Which ethical system makes it easy to explain the wrongness of these acts? Under which ethical system are these explanations difficult?
3. A scientist is paid by a drug company to test a new drug for safety and effectiveness. What are her duties to the drug company? What ethical conflicts might arise? What safeguards are needed? Does it make a difference if she owns stock in the drug company?
4. What scientific issues are raised by the nuclear power plant pictured in Figure 1.7? What science policy issues? What data would you seek on which to base a policy decision one way or the other?

1.5 EXPERIMENTAL SUBJECTS IN BIOLOGY

We now turn to the use of experimental subjects in biological research. This issue involves moral choices at all the levels we have discussed: individual, institutional, and societal. First, we consider animal rights and the uses of animals in scientific experiments. Second, we contrast animal experimentation with experiments on humans. Of the many ethical issues surrounding biology today, few are as divisive as those touched upon here.

1.5.1 Uses of animals

Human societies have kept animals at least since the origin of agriculture. There are few societies in which animals are not kept as food, as pets, or as workmates. Most societies that practice agriculture use animals for all three purposes. Love of animals and use of animals can go hand in hand.

By far the largest number of animals used by most societies are raised for consumption as food for humans. Animal products are used for clothing. Animals are also the targets of recreational hunting, fishing, and trapping, even in many industrial societies. Many people keep pets or "companion animals." Work animals pull and carry loads, help in police work, and help handicapped people. Finally, animals are often used in research, although the number is only a tiny fraction of the numbers consumed as food.

Nearly all new drugs, cosmetics, food additives, and new forms of therapy and surgery are tested on animals before they are tested on humans. Many people regard animal research as critical to continued progress in human health. Of the 216 Nobel Prizes in medicine and physiology awarded through 2019, 180 of them were for research that used experimental animals (https://fbresearch.org/medical-advances/nobel-prizes/). Organ transplants, open heart surgery, and various other surgical techniques were performed and perfected on animals before they were performed on humans. All vaccines were tested on animals before they were used on human patients. In most cases, animals are used in research as stand-ins for humans. If we did not use animals for these purposes, humans would be the experimental subjects tested. Most animal testing is limited to the initial development of a drug or surgical procedure, but the human benefit continues for many generations or longer.

Few people object to a use of animals that saves human lives. Most people also agree that animals should not suffer unnecessarily, whether they are pets, work animals, or research subjects. Scientists need to use healthy, well-treated animals in research, and the U.S. Guide for the Care and Use of Laboratory Animals reflects this concern. There are standards such as cage sizes that must be followed by scientists using animals. All research using live animals must, by law, be scrutinized and approved by supervisory committees, and the committees require the investigators to minimize both the number of animals used and the amount of pain that those animals experience and to substitute other types of tests where possible. According to statistics from the U.S. Department of Agriculture, 62% of animals used in research experienced no pain and another 32% were given anesthesia, painkillers, or both to alleviate pain. Only 6% of animals suffered pain without benefit of anesthesia. Federal law in the United States requires the use of anesthesia or painkillers in animal research wherever possible. Exceptions are allowed only when the use of anesthesia would compromise the experimental design and when no alternative method is available for conducting the test.

1.5.2 The animal rights movement

Those concerned with animal rights vary from traditional humane societies such as the Society for the Prevention of Cruelty to Animals (SPCA) and various national, state, and local humane societies, through groups such as People for the Ethical Treatment of Animals (PETA, founded in 1980), to groups such as the Animal Liberation Front (ALF, founded in 1972). Some animal rights advocates use a utilitarian ethic to advocate their position; others are deontologists.

Do animals have rights? Bernard Rollin, an American philosopher who supports animal rights, argues that there is no good reason for drawing an ethical distinction between mentally competent adult humans, other human beings (including children, comatose patients, or mentally ill or brain-damaged persons), and animals. Animals are therefore, in his view, worthy of any moral consideration that would normally be given to babies, comatose patients, and other people unable to speak for themselves or to articulate their own viewpoints.

Historically, animals have been treated legally as property. Animal owners have property rights such as the right to sue for damages if their animals are killed or injured, but the animals themselves have no legal rights. Many philosophers have attempted to justify this approach by asserting that animals cannot be held morally responsible for their actions and are therefore not moral agents. Only moral agents are considered to be capable of entering into contractual agreements or of having any rights.

The question of animal rights is actually part of a broader question: how far do we extend the scope of any rights that we recognize? Many societies have historically denied even the most basic of rights to certain classes of persons on the basis of economics, gender, race, ethnicity, or religious beliefs. The extension of certain basic rights to all humans, including children and convicted criminals, is now considered so fundamental to the ethical sensibilities of most people that we refer to these as *human* rights. International agreements, such as the Geneva Convention on the treatment of prisoners of war and the International Convention on Human Rights (the Helsinki Accord), attest to the importance given to these human rights in world affairs. But should we stop there? Animal rights advocates say that we should extend these same rights to all beings capable of sensing pleasure and pain. A few people go even further, asserting that even trees have such rights as the right to go on living or not to have their air and soil poisoned. To go still further, a few people assert that habitats themselves, including mountains and forests, have rights not to be despoiled.

1.5.3 Humans as experimental subjects

Many animal experiments are undertaken to determine the effects of some new drug or other therapy. The real question is usually about what the effects would be in humans, and the animals are merely used as stand-ins. Is it safe to extrapolate to humans results that were obtained from experimentation on nonhuman species? In most cases in which data are adequate to answer this question, human physiological reactions have turned out to be comparable to those of experimental animals. Even when differences between humans and other species are known, they are often known in sufficient detail that the different responses to testing can help us understand the human system better, which still makes the animal tests valuable.

A few instances are known in which humans and certain commonly used experimental animal species respond differently. Saccharin, for example, causes cancer in rats, but has never been shown to cause harm to humans.

Direct experimentation on humans avoids the question of comparability between species, meaning that the results can be used more directly than results obtained from other species. Also, results obtained from psychologists, epidemiologists, and others who study humans with naturalistic methods can be applied even more directly. For example, one could not ethically force-feed cholesterol to an experimental group of human subjects, but one could observe the diets that different people choose on their own and study how people with high-cholesterol diets differ from people with low-cholesterol diets. The diets in such a study are more directly comparable to the diets of other humans than are the diets of experimental animals fed with different amounts of cholesterol in their food. As explained in many later chapters, more than one type of method must be applied to answer many scientific questions. We should not view experimental animal studies versus naturalistic studies of humans as either/or choices; in most cases, both approaches are needed.

Among possible experimental subjects, humans have a special status. On the one hand, inflicting pain on human subjects or exposing them to experimental risks raises more ethical objections than does the similar treatment of nonhuman animals. On the other hand, human subjects can tell us how they feel or when and where they experience discomfort or pain. Certain types of drug side effects, like headaches or impairment of problem-solving ability, are difficult to assess without using human subjects.

1.5.3.1 Voluntary informed consent A purely ethical consideration is that human subjects can voluntarily consent to serve as experimental subjects, which is something that nonhuman animals cannot do. Humans are considered to be autonomous beings who have the right to consent to putting themselves at risk, whether in a space capsule, a bungee jump, or an experiment. An important consideration, however, is that the person serving as a subject must give consent voluntarily. This is a legal as well as a moral issue, because persons who did not consent voluntarily can sue for damages if any harm comes to them. Consent is usually obtained in writing on a form that informs the person of the possible benefits and risks of the experimental procedure and is therefore called **informed consent**. If an experiment may bring direct benefit to a subject (as when a disease or its symptoms are being treated), potential subjects may be more willing to undergo certain risks than they would otherwise.

Special questions arise in the case of persons who may not have the full capacity to understand all the possible risks and benefits, including mentally deficient persons, unconscious persons, or children. Most people would now consider it unethical to use such a person as an experimental subject unless there was obvious great promise of direct benefit and only minimal risk of harm. In most jurisdictions, parents are considered to have the legal right to make such decisions on behalf of their children. In past decades, prison populations were often used as sources for experimental subjects, but this practice is now frowned upon because the consent of a prisoner may not be truly voluntary if she or he thinks—rightly or wrongly—that cooperating might result in a sentence reduction.

The Tuskegee syphilis study, begun in Macon County, Alabama, in 1932, serves as an example of a study performed at a time before modern ethical guidelines were widely adopted. Study subjects included 600 African American males, of whom 399 had active syphilis. Subjects were told that they would be treated for "bad blood," but the syphilis itself was never treated. Subjects were enticed to participate with free medical care and burial expenses upon death, both of which presented very strong financial incentives to these subjects, most of whom were very poor. Subjects were not told of the true purpose or conditions of the study, nor was informed consent ever obtained. Even after penicillin became widely available as a cure for syphilis in 1947, this treatment was withheld from them, and they were not permitted to withdraw from the study in order to receive it. The study was terminated in 1972, amid much criticism, and a presidential apology was issued in 1999 and delivered to the surviving participants and to the families of the deceased. More information on this study can be found at https://www.cdc.gov/tuskegee/timeline.htm and https://www.tuskegee.edu/about-us/centers-of-excellence/bioethics-center/about-the-usphs-syphilis-study.

1.5.3.2 Guidelines for human experimentation As a safeguard against possible abuses, research on human subjects is now usually reviewed by institutional committees set up for that purpose. As is true in reviews of animal experimentation, the review process ensures that someone other than the researchers evaluate the ethics of the proposed experiment. Federally sponsored research in the United States and in many other countries requires that such committees authorize all experiments in which humans are used as subjects. In addition to ensuring that proper voluntary consent has been obtained, such committees also have the obligation to suggest ways in which risks can be reduced or benefits increased without impairing the validity of the experiment. Many scientists work within the ethical tradition in which exposing humans to experimental risks is more objectionable than exposing animals of other species to those same risks. Guidelines have been developed that specify testing to be carried out on animals first, then on small numbers of carefully chosen and carefully monitored human subjects, and only last on large and diverse human populations. In the United States, federally sponsored research and research on new drugs seeking federal approval are required to follow this procedure.

1.5.3.3 Avoiding gender bias Before the 1990s, many animal and clinical trials were done only on male subjects; one reason frequently given for this practice was to avoid the variable of hormonal fluctuations of the female reproductive cycles. Many test results based on male-only studies were extrapolated to women, and drug doses and other treatments were prescribed for women on this basis. In several cardiovascular conditions, including heart attacks and strokes, it now appears that men and women respond differently to certain drugs and that drug doses calculated for men may be inappropriate for many women. When Dr. Bernadette Healy was head of the National Institutes of Health, she criticized a number of studies that had been done only on men because she thought they would have been more appropriately done on women alone or on both sexes. One of the studies, for example, was based on the observation that pregnant women almost never have heart attacks. To test whether estrogen was the cause of this protective effect, the effects of estrogen therapy on heart attack rates was measured in *men*! Largely as the results of Dr. Healy's efforts, National Institutes of Health guidelines now require studies to be done on both sexes when appropriate. Although this is an issue of good experimental procedure, the issue was first raised as an ethical question of unfairness to women. Dr. Healy and others claimed that women were getting potentially substandard medical care if they were treated with drugs that had been tested on men only and with doses calibrated for male patients.

1.5.3.4 Good experimental design (Part 2) In addition to the guidelines explained earlier in this chapter (Section 1.1.2), several other features of good experimental design involve ethical principles when living subjects are being tested:

6. Humane practices should be followed as much as possible: pain should be minimized, anesthesia should be provided, and animal or human subjects should be kept as healthy as possible. This is good practice scientifically as much as morally, because results may be invalid or irrelevant if based on pathological conditions or on subjects experiencing pain.

7. If some new product or intervention might cause harm or suffering (or if nothing is known to the contrary), then safety testing should be done, wherever possible, on laboratory cultures before living animals, and on animals before human subjects. This is, in fact, required by federal law in the United States and in many other countries. Also, results from animal studies should be extrapolated with care and appropriate caution to any other species (including humans) that were not studied.
8. If human subjects are used, informed consent must be obtained first, and persons judged not competent to give informed consent must be excluded.
9. If something is being tested that may eventually be used by the general public, then all types of people, representative of different genetic backgrounds, age groups, and both sexes, should be included among the experimental subjects. Moreover, if children or pregnant women are not tested, then any marketed product must be clearly labeled with precautionary statements to that effect.
10. Research using live animals or human subjects should first be approved by an ethical review board to make sure that the appropriate guidelines are followed and to discuss necessary exceptions and special cases (e.g., the conditions under which children or pregnant women may be tested). Decisions on such matters often follow utilitarian principles such as minimal possibility of harm and maximum possibility of beneficial results.

THOUGHT QUESTIONS

1. For each of the following acts, try to give:
   a. A deontological argument against the act
   b. A deontological argument justifying the act
   c. A utilitarian argument against the act
   d. A utilitarian argument justifying the act
      - Beating your horse
      - Taking a canary into a coal mine so that if it dies from toxic gases, miners would be warned to evacuate
      - Raising broiler chickens or beef cattle for human consumption
      - Testing a drug on rats (or cats) before giving it to humans
      - Testing a drug on human prison convicts
2. How widely can experimental results be extrapolated? If a drug is tested on inbred male rats, is it certain that the results are applicable to humans? Is it likely? Is the drug likely to have similar effects on both sexes? What issues of methodology or of ethics are raised by experiments that used only inbred male rats?
3. Is it ethical to infect a few people with a deadly disease to study its effects in the hopes of saving many more lives in the future? How do you justify your answer? Give both deontological and utilitarian reasons.
4. What ethical principles were violated in the Tuskegee syphilis study described in the preceding section? What rules for good experimental design were not followed?
5. One hundred patients are to be enrolled in a study of a new drug. Half of the enrolled patients will be given the new drug, while the other half will be used as a control and will therefore not receive the new drug. How many of the patients need to give informed consent? Why?

CONCLUDING REMARKS

In science, we know what we know through a process often called the scientific method. Scientists formulate tentative ideas (hypotheses) about living systems and about the world in general, and they submit these ideas to extensive testing by comparing their ideas with observations made in the material world around them. Scientific knowledge is forever tentative and is never "proved" because it is always subject to change if new observations do not fit our existing theories. The language of science is often metaphorical. Scientists often use words with specialized meanings. Scientific paradigms give science its vocabulary, imagery, attitudes, and value judgments. Science is conducted in a social context that includes a community of scientists sharing their ideas and testing each other's hypotheses.

Ethical decisions can be made either by judging actions themselves (deontological ethics) or by judging actions on the basis of their consequences (utilitarian ethics). Science may sometimes confront individuals with moral choices. Many scientific issues have implications for large numbers of people or for society as a whole. All citizens, not just scientists, should help make these decisions collectively. However, scientists should bear some responsibility for educating others about the science issues and science policy issues that can inform these decisions. Scientists should educate themselves as to the ethical dimensions of their work, including both the treatment of experimental subjects and the possible uses and misuses of scientific findings.

CHAPTER SUMMARY

- **Science** is based on the testing of **hypotheses** under conditions in which it is always possible to observe results inconsistent with the hypotheses.
- Hypotheses originate by many types of reasoning and inspiration, including both **induction** and **deduction**. Hypotheses must always be **falsifiable**.
- A group of well-tested hypotheses forms a **theory**. A theory may be communicated by special vocabulary or by a descriptive analogy (a model).
- **Biology** is a science because biologists use hypothesis testing to study living systems.
- Living systems exhibit cellular organization, growth, metabolism, homeostasis, and selective response. They contain genetic material, and they belong to populations, most of whose members are capable of reproducing.
- Biologists test hypotheses either by studying natural conditions as they occur (**naturalistic science**) or by conducting **experiments** under conditions that the scientists help to create (**experimental science**). After data are gathered, the interpretation of results usually involves comparison with **control groups**, often with the help of statistical methods.
- Normal science proceeds by testing hypotheses one at a time, under the guidance of a **paradigm**. A new paradigm may replace an older one if it can explain things better than the old one did; such a shift in paradigms brings about a scientific revolution.
- **Scientific methods** are used in everyday life, but scientists use these methods more often and more systematically.
- Certain values are held by members of the scientific community that ensure them the continuing ability to test, falsify, and change each other's hypotheses.
- **Morals** are rules that guide our conduct. **Ethics** is the discipline that examines moral rules and attempts to explain or justify them.
- Two major types of ethical systems are **deontological** and **utilitarian**. Deontologists judge the rightness or wrongness of an act by characteristics of the act itself, apart from its consequences. Utilitarians judge the rightness or wrongness of an act on the basis of its consequences. Utilitarian analysis often includes a comparison of the undesirable effects (costs) of an act with its desirable effects (benefits).
- Science can confront individuals and societies with moral choices.
- Individual moral choices can be guided by principles such as nonmaleficence, beneficence, autonomy, and **fairness**.
- In facing collective ethical issues involving scientific questions, it is often useful to distinguish between scientific issues, science policy issues, and policy issues.
- Animals are used in our society for food, for labor, for companionship, and for laboratory experimentation. Biological experiments often use living organisms as subjects. In many cases, laboratory animals are used as stand-ins for humans, and their use is often justified on a utilitarian basis (the cost–benefit ratio is lower if animals are tested before humans) or on a deontological basis (humans have rights and animals do not, or human rights supersede animal rights).
- Before any experimentation on animals or humans can take place, the proposed experiments must pass an ethics review. If humans are used, their voluntary **informed consent** must be obtained.

BIBLIOGRAPHIC REFERENCES

Bibliographic references to material in this chapter can be found online at biologytrending.routledge.com/chapter1

GLOSSARY: KEY TERMS TO KNOW

These key terms are defined at the end of each chapter as an aid for student review.

Biology The scientific study of living systems.

Categorical imperative Kant's ethical criterion that an act is good or bad depending on whether you wish everyone to copy it.

Control group A group used for comparison in an experiment. For example, if experimental animals are exposed to a drug, then a control group might consist of similar animals not exposed to the drug but treated the same in every other way.

Data Information gathered so as to permit the testing of hypotheses.

Deduction Logically valid reasoning that guarantees a true conclusion whenever the premises are true.

Deontological A type of ethics in which the rightness or wrongness of an action is judged without reference to its consequences.

Empirical Based on observations of the material world that we make with our senses, aided in some cases by scientific instruments.

Ethics A discipline dealing with the analysis of moral rules and the ways in which moral judgments are made and justified.

Evidence-based Decision-making informed by the testing of hypotheses in comparison with empirical data.

Experiment An artificially contrived situation in which hypotheses are tested using empirical data in comparison with some known condition called the control condition.

Experimental sciences Sciences in which hypotheses are tested by conducting experiments and analyzing the results.

Fairness The ethical principle that all individuals in similar circumstances should receive similar treatment.

Falsifiable Capable of being proven false by experience.

Falsified Proven false by experience (as by comparison with empirical data).

Hypothesis A suggested explanation that is both testable and falsifiable.

Induction Reasoning from specific instances to general principles, which can sometimes be unreliable, as in "these five animals have hearts, so all animals must have hearts."

Informed consent A voluntary agreement to submit to certain risks by a person who knows and understands those risks.

Model A mathematical, pictorial, or physical representation of how something is presumed to work.

Morals Rules governing human conduct.

Naturalistic sciences Sciences in which hypotheses are tested by the observation of and comparison among naturally occurring events under conditions in which nature is manipulated as little as possible.

Normal science Science that proceeds step by step within a paradigm.

Occam's razor (principle of parsimony) The principle that the simplest explanation (or the one with the fewest assumptions) should always be preferred.

Paradigm A coherent set of theories, beliefs, values, and vocabulary terms used to organize scientific research.

Paradigm shift The replacement of one paradigm with another.

Policy decisions Decisions that must be made in terms of human preference, especially ethical preference, for one set of consequences over another.

Rights Any privilege to which individuals automatically have a just claim or to which they are entitled out of respect for their dignity and autonomy as individuals.

Risk The probability of occurrence of a specified event or outcome.

Science An endeavor in which falsifiable hypotheses are systematically tested.

Scientific literacy Familiarity with the methods, findings, and limitations of science.

Scientific method A method of investigation in which hypotheses are tested by comparison with empirical data.

Scientific revolution The establishment of a new scientific paradigm, including the replacement of earlier paradigms.

Social policy A formal or informal set of rules under which people make decisions in individual cases.

Theory A coherent set of well-tested hypotheses that guide scientific research.

Utilitarian A system of ethics in which the rightness or wrongness of an act is judged according to its consequences.

CONNECTIONS TO OTHER CHAPTERS

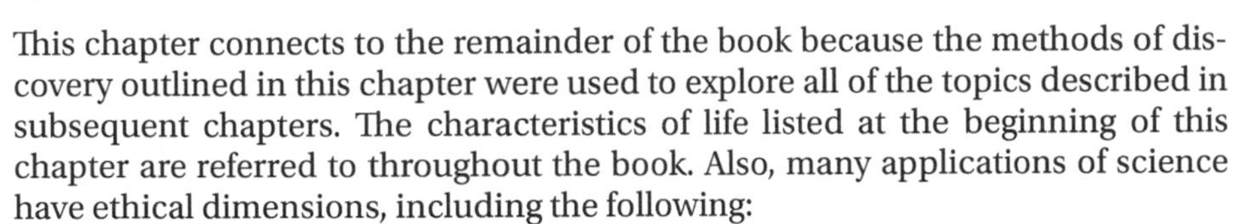

This chapter connects to the remainder of the book because the methods of discovery outlined in this chapter were used to explore all of the topics described in subsequent chapters. The characteristics of life listed at the beginning of this chapter are referred to throughout the book. Also, many applications of science have ethical dimensions, including the following:

Chapters 3–4: The Human Genome Project and human genetic testing raise ethical questions.
Chapter 7: Ethical objections have been raised against the ways in which biology has supported racism.
Chapter 8: Patterns of food consumption and distribution raise ethical issues.
Chapter 9: The need for population control conflicts with the ethic of allowing reproductive freedom.
Chapter 10: Cancer research often involves the use of animal experimentation.
Chapter 11: Brain research involves animal experimentation and also the use of fetal tissue.
Chapter 12: The care of dependent individuals, including those with diminished mental capacity, involves ethical decisions.
Chapter 13: Drugs are usually tested on animals before giving them to people.
Chapter 15: Many ethical issues surround the transmission of AIDS, testing for AIDS, and the prevention of AIDS.
Chapter 16: During pandemics and other public health crises, restrictions on individual freedoms raise ethical issues.
Chapter 17: Some people object to genetically engineered foods on ethical grounds.
Chapter 18: Many ethical issues are raised by habitat destruction and species extinction.
Chapter 19: Global patterns of pollution raise ethical issues.

PRACTICE QUESTIONS

1. Which property of life is exhibited by each of the following?
   a. The frog jumps around when I touch it.
   b. The bread rises because the yeast has given off carbon dioxide bubbles.
   c. Blood samples from healthy humans always have about the same pH and salt concentration.
   d. Wherever I find one mosquito, I usually find many.
   e. Only a few rabbits were brought to Australia, but now there are millions.
   f. Puppies usually resemble their parents.
   g. Baby animals get bigger and become adults.
   h. A bright light at night always attracts moths.

2. Which of the following are testable hypotheses? For each statement that you think is testable, explain what sort of test you might conduct and what possible outcome would falsify the hypothesis.
   a. Taylor Swift is a better singer than Lady Gaga.
   b. In a maze that they have never seen before, rats will turn right just about as often as they will turn left.
   c. If these two plants are crossed, approximately half of the offspring will resemble one parent and half will resemble the other.
   d. It is wrong to inflict pain on a cat.
   e. Restaurant A is better than restaurant B.
   f. The average science major at this school gets better grades than the average humanities major.
3. Which of the following examples of reasoning use induction? Which use deduction?
   a. If all adult female birds lay eggs, then this female chick will lay eggs if raised to maturity.
   b. If all known species of birds are egg-laying, then the next species to be discovered will be egg-laying too.
   c. If all known enzymes are made of protein and I discover a new enzyme, then it too will be made of protein.
   d. If the amounts of protein X are increased under stress, then I should be able to increase the amount of protein X in these frogs by subjecting them to stressful conditions.
   e. If this species mates in April, then I should be able to observe more mating on April 10 than on June 10.
4. Which of the following reflect the community nature of science?
   a. A scientist presenting a talk at a scientific meeting.
   b. Another scientist asking a question at that same meeting.
   c. A field naturalist tracking a rare species.
   d. The same field naturalist publishing her findings.
   e. A scientist feeding a new chemical to mice to study its effects.
   f. A bacteriologist using techniques developed by Louis Pasteur and Robert Koch for growing bacteria in laboratory cultures.
   g. A scientist displaying his experimental results over the Internet.
5. For each of the following, identify (a) whether the argument is based on utilitarian or deontological ethics and (b) whether any assumptions are made about whether animals have no rights, some rights, or rights equal to those of human beings.
   a. Hunting is wrong because the victim is part of nature, and it is wrong to interfere with nature.
   b. Hunting is justified because the death of one animal makes such a small difference to most hunted species.
   c. Hunting is wrong because it makes the hunter more prone to future violence.
   d. Hunting is justified if the animal is used as food but not as a trophy.
   e. Raising beef cattle for human consumption is justified because people need to eat.
   f. Raising beef cattle for human consumption is wrong because cows are sacred.
   g. Raising beef cattle for human consumption is wrong because people would be healthier if they ate more plant foods instead.
   h. Raising beef cattle for human consumption is wrong because it causes pain and suffering to the animals.
   i. Be kind to your pet because you will be rewarded with loving companionship.

CHAPTER 2

Genes, Chromosomes, and DNA

ISSUES

- How have our concepts of genes developed?
- What are the limitations of Mendelian genetics?
- Does Mendelian genetics explain inheritance in all species?
- How is hereditary material copied?

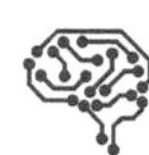

BIOLOGICAL CONCEPTS

- The gene
- Patterns of inheritance
- Mitosis and meiosis
- DNA (the genetic material, DNA structure, DNA replication)

CHAPTER OUTLINE

DOI: 10.1201/9781003391159-2

How do offspring come to resemble their parents physically? This is the major question posed by the field of biology called genetics, the study of inherited traits. Genetics begins with the unifying assumption that biological inheritance is transmitted by structures called genes. The discovery of what genes are and how they work has been the subject of many years of research. Among the earliest findings was the fact that the same basic patterns of inheritance apply to most organisms. The inheritance of many human traits can be explained by the same hypotheses first formulated from the study of pea plants. These basic principles are discussed in this chapter. The inheritance of other human traits, such as sex determination and susceptibility for many diseases, involves further complications that are addressed in Chapter 3.

We now know that genes are made of DNA, and we will discuss how we know this later in the chapter. We begin, however, with experimental evidence for the existence of genes and their role in inheritance.

2.1 MENDEL OBSERVED PHENOTYPES AND FORMED HYPOTHESES

No two individual organisms are exactly alike. Folk wisdom going back to ancient times taught people that a child or an animal resembles both its mother and its father by showing a mixture of traits derived from the two sides of the family. This suggested a concept that came to be called "blending inheritance," in which heredity was compared to a mixing of fluids, often identified as "blood." The research we are about to describe caused this explanation to be abandoned.

2.1.1 Traits of pea plants

During the nineteenth century, Gregor Mendel, a Czech scientist living under Austrian rule, worked out the principles of inheritance (or **genetics**) for simple traits that he described in "either/or" terms. Mendel (see Figure 1.4B) was a priest who grew pea plants (*Pisum sativum*, kingdom Plantae, phylum Anthophyta also called Angiospermae) in the garden of his monastery. Mendel was curious about differences that he observed among several different varieties of pea plants, so he decided to breed them and keep careful records.

Why were the peas a good species for Mendel's experiments? Pea plants have many distinctive traits (**Figure 2.1**) and other practical advantages: Mendel could easily grow them in the monastery garden, and many varieties were locally available, including some with yellow peas and others with green peas, and some with round peas and others with wrinkled peas. Mendel knew that peas undergo **sexual reproduction**—that is, a new individual forms by the union of an egg from a female with a sperm from a male, an event called **fertilization**. Unlike most animals, an individual plant may have both female and male reproductive organs. This is true of many plant species, including peas. Within the pea flower are male structures called anthers that produce pollen grains, each of which contains a sperm. The pea flower also has a female structure called a stigma that receives pollen grains (a step called "pollination") and permits the sperm to travel to the ovary to reach the eggs (**Figure 2.2**). Self-pollination occurs when pollen from a plant is deposited on a stigma of the same plant. In some of his experiments, Mendel sewed together the margins of one large petal to enclose the flower and ensure self-pollination. At other times, he cross-pollinated his peas by dusting pollen from a flower of one plant onto the stigma of a flower of another plant, after first removing the anthers from the recipient plant. Thus, he was able to study inheritance under very controlled conditions where he knew with certainty which plants were the parents of each set of offspring.

Mendel organized his work so as to answer specific questions, a procedure that we recognize today as good experimental design, including large sample sizes.

	Seed shape	Seed color	Flower color	Flower position	Pod shape	Pod color	Plant height
One form of trait (dominant)	round	yellow	violet-red	axial flowers	inflated	green	tall
A second form of trait (recessive)	wrinkled	green	white	terminal flowers	pinched	yellow	short

Figure 2.1 The seven traits studied by Mendel in peas.

Unlike most of his predecessors who failed to discover how plant offspring inherit their parents' traits, Mendel followed certain careful procedures:

1. First, for each of the traits he studied, Mendel used peas belonging to pure lines. A "pure line" is one that breeds true from generation to generation, always producing offspring that express the same form of the trait as the parents. For example, tall parents from one pure line always produce tall offspring, and short parents from another pure line always produce short offspring.
2. He chose which plants would mate by either cross-pollinating (crossing) the flowers or closing up the flower parts to ensure self-pollination (Figure 2.2).

pea flower with petals enclosing sexual parts inside

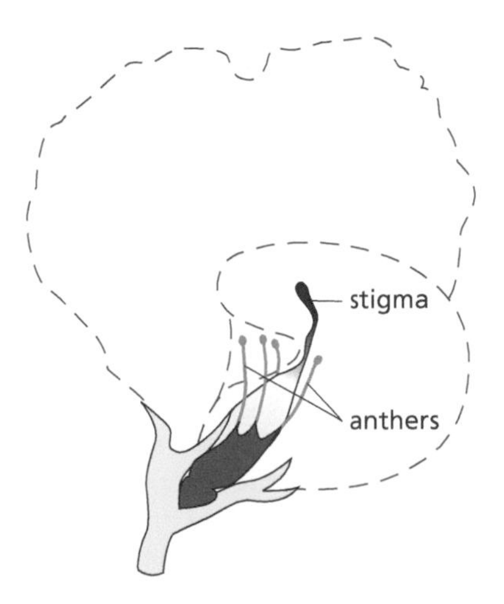

pea flower with petals removed, to show male and female flower parts

Figure 2.2 The structure of a pea flower. A more complete view of the sexual reproductive structures of flowering plants is shown in Figure 6.13.

3. He first studied only one trait (one variable) at a time until he understood its pattern of inheritance. Later, he studied two and three traits at a time. His predecessors, on the other hand, often began by examining several or many traits at once.
4. He counted large numbers of offspring from each cross and was thus able to recognize ratios between them. (Those few of his predecessors who looked at single traits never counted the offspring of each type and thus failed to find recurring ratios.)
5. He continued each experiment through several pea generations.

Mendel studied one trait at a time. For example, he crossed plants having white flowers with plants having violet-red flowers but similar in other traits. He found that all the first-generation offspring plants (symbolized as F_1) had violet-red flowers; none of the plants had white flowers. When he crossed tall and short plants, all the offspring were tall. Mendel introduced the term **dominant** for the form of the trait that appeared in the first-generation offspring of his initial cross; the trait that did not show up he called **recessive**. Thus, violet-red flowers are dominant to white, and tall is dominant to short. Mendel found for each of the seven pairs of either/or traits that one form of the trait was dominant and one was recessive. The forms of the traits shown in the upper row of Figure 2.1 are dominant; those in the lower row are recessive. The traits did not blend: Violet-red-flowered plants crossed with white-flowered plants produced violet-red-flowered offspring, not pink-flowered offspring, and tall plants crossed with short plants produced offspring of the height of the tall parents, not an intermediate height.

2.1.2 Genotype and phenotype

The tall F_1 plants in Mendel's experiments were just as tall as their tall parents. They looked the same as their tall parents; that is, their **phenotype** (appearance) for height was tall. But Mendel realized that the hereditary makeup of the F_1 plants was different from that of their parents: The tall parents had come from a pure line, so all their hereditary makeup had been tall, but each F_1 plant had both tall and short parents. Could this difference in hereditary makeup, or **genotype**, be made visible? Would it show up in future generations? Mendel answered this question by mating the plants of the F_1 generation with themselves (self-pollinating them) and raising a second generation, symbolized as F_2. He obtained both tall plants and short plants in the F_2, but no plants of intermediate height. When he counted them, he found that the tall plants were approximately three times as numerous as the short ones. Mendel conducted similar experiments with other traits, such as flower color. In each case, he discovered the same thing: In the F_2 generation produced by self-pollinating the F_1 individuals, the dominant phenotype outnumbered the recessive phenotype in the approximate ratio of 3:1; in other words, three-quarters of the F_2 plants had the dominant phenotype and one-quarter had the recessive one.

2.1.3 Mendel's explanation for inheritance of single traits

Mendel proposed a multipart hypothesis to explain his results. Using modern terminology and some modern understanding, we can list the points covered by his hypothesis:

1. The inheritance of traits is controlled by hereditary factors; today these factors are called **genes**.
2. Each individual has two copies of the gene for each trait (a "diploid" condition). If each gene is represented by a letter, then the genetic makeup (genotype) of an individual for a trait can be represented as two letters.
3. Each gene exists in different forms; these variant forms of the same gene are called **alleles**. For example, the gene for flower color in peas has an allele that produces white flowers and another allele that produces violet-red flowers. Alleles that are dominant produce only the dominant phenotype even when

the recessive allele is present. Mendel designated the dominant and recessive alleles controlling the same trait by a single letter of the alphabet, using a capital letter for the dominant allele and the lowercase of the same letter for the recessive allele. For example, the allele *V* for violet-red flowers is dominant to the allele *v* for white flowers. (Today, many genes are designated by two- and three-letter combinations and different alleles of the same gene by superscripts.)

4. An individual whose genotype contains two identical alleles, such as *VV* or *vv*, is said to be **homozygous** for the trait. An individual whose genotype combines dissimilar alleles, such as *Vv*, is said to be **heterozygous**.
5. Dominant alleles always show up in the phenotype, but recessive alleles are masked by their dominant counterparts. When a dominant and a recessive allele for the same trait are present in a heterozygous individual, the dominant allele produces the phenotype. Recessive alleles produce the phenotype only when they are homozygous—in other words, when the corresponding dominant allele is absent.
6. The genes behave as "particles" that remain separate instead of blending. Recessive genes are masked during the F_1 generation, which shows the dominant phenotype, but they are not functionally changed by being in a heterozygous F_1 individual.
7. When the F_1 individuals produce eggs and sperm, the dominant and recessive alleles separate from one another, or "segregate." The eggs and sperm are called **gametes**, and each gamete receives only one allele of each gene after they segregate. The principle of **segregation** of alleles is often stated as Mendel's first law. (The term "law" was used more frequently in the past; in this case, it simply means a concept devised to explain the data.)
8. During sexual reproduction, the gametes combine so that each F_2 individual has two alleles, one contributed by the egg and one by the sperm. This explains both why the recessive phenotype disappears in the F_1 and why it reappears in about one-quarter of the F_2 individuals.

Figure 2.3 shows Mendel's multipart hypothesis applied to one of the traits he studied. A pure-line plant with violet-red flowers and the genotype *VV*, which produces *V* gametes, is crossed with a pure-line, white-flowered *vv* plant, which produces *v* gametes. All F_1 plants, having received *V* from one parent and *v* from the other, have violet-red flowers and the heterozygous genotype *Vv*. Half of the eggs produced by the F_1 carry the dominant allele *V* and half carry the recessive allele *v*, and the same is true for the sperm, as indicated in the margins of the square in Figure 2.3. When the gametes from the self-pollinated F_1 combine at random to form the F_2 generation, one out of every four new individuals is homozygous for the dominant allele *VV* and has violet-red flowers, two out of four are heterozygous *Vv* with violet-red flowers, and one out of four is homozygous for the recessive allele *vv* and has white flowers. The ratio of violet-red to white phenotypes is thus 3:1. A square like this, showing how gametes combine to produce genotypes and phenotypes, is called a **Punnett square**.

2.1.3.1 Independent assortment After Mendel had investigated the inheritance of single traits, he proceeded to study the inheritance of two traits together. In one of his experiments, the parents differed in both seed shape and seed color. For these traits, yellow seed color is dominant over green seed color and round seed shape is dominant over wrinkled seed shape. The parents in one group were pure-breeding for yellow, round seeds (*YYRR*) and thus produced all *YR* gametes. The other parents were pure-breeding for green, wrinkled seeds (*yyrr*) and thus produced all *yr* gametes. The first-generation offspring (F_1) were all *YyRr*, heterozygous for both traits. All these F_1 plants produced seeds that were yellow (because yellow is dominant to green) and round (because round is dominant to wrinkled). No new principles were involved so far.

Mendel expected half of the gametes produced by the F_1 to contain the dominant allele *Y* and the remainder the recessive allele *y*. He also expected half of the gametes to contain allele *R* and the other half the allele *r*. But would the fact that a

Figure 2.3 One of Mendel's experiments with peas differing in a single trait.

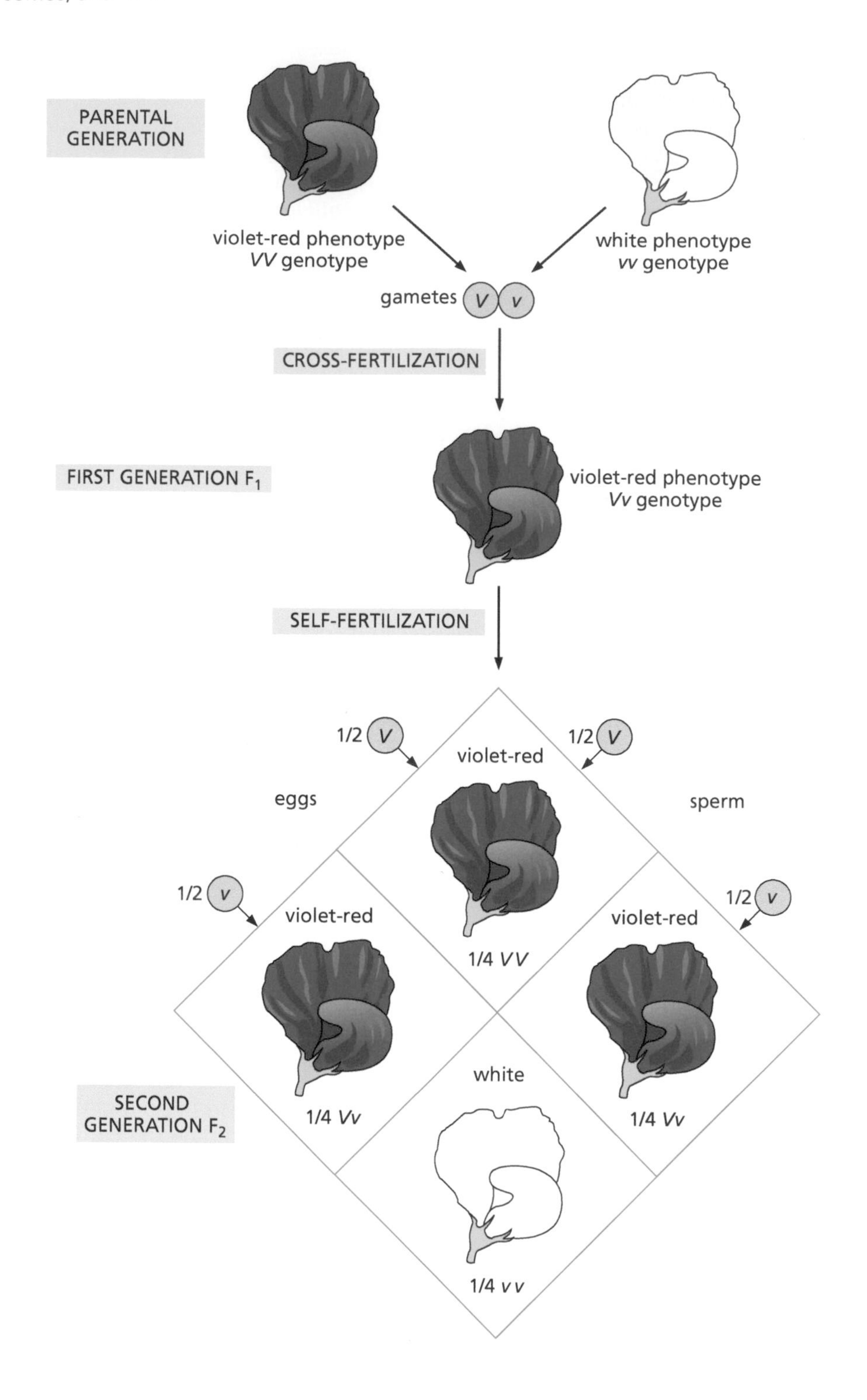

gamete received *Y* rather than *y* influence whether it received *R* or *r*? To find out, Mendel raised an F_2 generation by self-pollinating some F_1 plants. He obtained the 9:3:3:1 ratio of phenotypes shown in **Figure 2.4**: 9/16 of the F_2 individuals showed both dominant traits (both yellow and round); 3/16 were round but green, 3/16 were yellow but wrinkled; and only 1/16 showed both recessive traits (green and wrinkled). Mendel then reasoned that this ratio could be explained if he assumed that all four possible types of gametes (*YR*, *Yr*, *yR*, and *yr*) were produced in equal proportions (as shown in the margins of the square in Figure 2.4) and combined at random. From this, Mendel concluded that the inheritance of the trait of seed color has no influence on the inheritance of the trait of seed shape. This principle is called **independent assortment** and is known as Mendel's

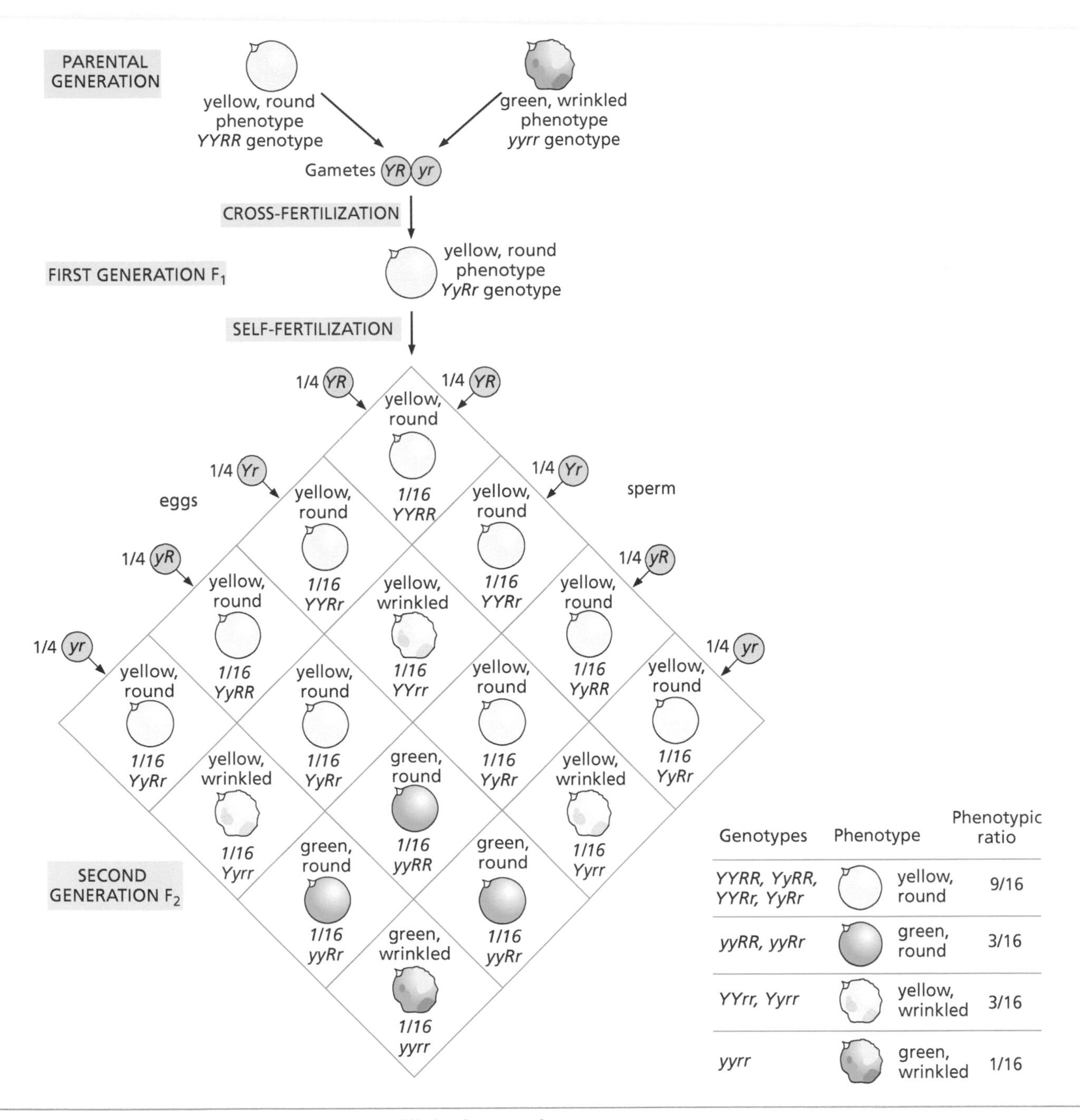

Genotypes	Phenotype	Phenotypic ratio
YYRR, YyRR, YYRr, YyRr	yellow, round	9/16
yyRR, yyRr	green, round	3/16
YYrr, Yyrr	yellow, wrinkled	3/16
yyrr	green, wrinkled	1/16

Figure 2.4 One of Mendel's experiments with peas differing in two traits.

second law. All of the traits that Mendel studied assorted independently when involved in two-trait controlled crosses; however, as we will soon see, there are some exceptions to Mendel's second law.

Mendel's results, published in 1865, were ignored by most scientists. The reasons for the lack of impact of his theories are many, but one contributing factor was that he presented his theories in the language of mathematics, to which most botanists of his generation were not accustomed. Another factor was that Mendel's findings were published in an obscure journal that was not widely read. Mendel's work was rediscovered in 1900, not once but three times: Three scientists in three different countries each conducted experiments similar to Mendel's, reached similar conclusions, and then subsequently discovered Mendel's earlier work. Simultaneous discovery of this kind often happens when the scientific community is ready for a new idea: Examples include Wallace's discovery of natural selection in 1858, a year before Darwin's *Origin of Species* (Section 5.1), and the independent confirmation of the chromosomal theory of inheritance in 1931 by researchers in Germany and the United States, described later in this chapter.

THOUGHT QUESTIONS

1. Mendel's experiments distinguished between two alternative hypotheses: Either traits blend, so that the offspring have traits intermediate between those of their parents, or traits are inherited as discrete particles that do not blend. Which hypothesis did Mendel's results support? Are Mendel's results a total proof of either hypothesis and a total disproof of the other? Explain.
2. An experimenter cross-pollinates flowers of tall pea plants with pollen from flowers from short pea plants. She harvests the seeds and plants them, and all of the plants that grow in the F_1 generation are as tall as the tall parent plants. The F_1 plants produce flowers, each containing many sperm and many eggs. Each sperm and each egg carry only one allele of each gene. What are the possible genotypes for plant height among the sperm from the F_1 plants? What are the genotypes among the eggs? What fractions of the gametes possess each genotype? Would the distribution of types of gametes in the F_1 plants be the same or different if they were produced by cross-pollinating flowers of short pea plants with pollen from tall pea plants?
3. To study two traits at a time, Mendel first crossed one line of plants that bred true for two traits with another line that bred true for different alleles of the same two genes. F_1 flowers were then self-pollinated and the F_2 peas produced as before. If you counted 16 of the F_2 peas, how many would you expect to be the doubly dominant phenotype, round and yellow? If you counted 1600 F_2 peas, how many round and yellow ones would you expect? Might your actual results deviate from these expectations? Comment on the difference between large and small sample sizes.
4. Why are certain traits studied in some species and not in others? Why were pea plants a good choice for Mendel's experiments?

2.2 THE CHROMOSOMAL BASIS OF INHERITANCE EXPLAINS MENDEL'S HYPOTHESES

Notice that some of Mendel's assumptions raise questions that Mendel himself did not answer:

1. Where are the genes located?
2. Why do the genes exist in pairs?
3. Why do different traits assort independently?
4. What are the genes made of?

Answers to the first three of these questions were suggested by a young American scientist, Walter Sutton, who read about the rediscovery of Mendel's work in 1900. By this time, it was already well known that the cells that make up the bodies of plants and animals contain a central portion called the **nucleus** and a surrounding portion called the **cytoplasm** (Figure 2.5A; for more on cell structure, see Chapters 6 and 10). Plants and animals are able to develop and grow because their cells divide. One cell divides to become two cells. Scientists who looked at dividing cells through their microscopes saw that the division of the

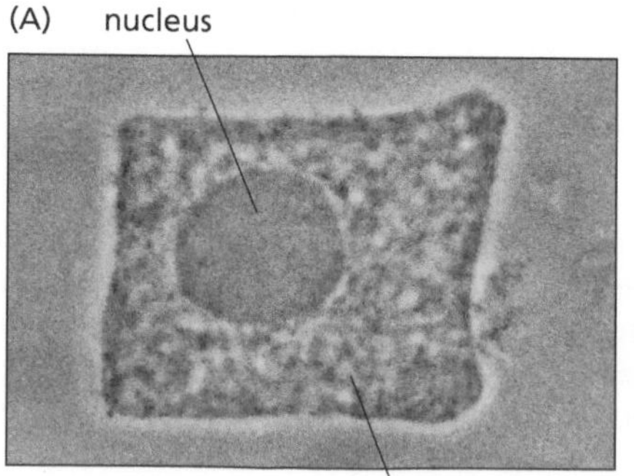

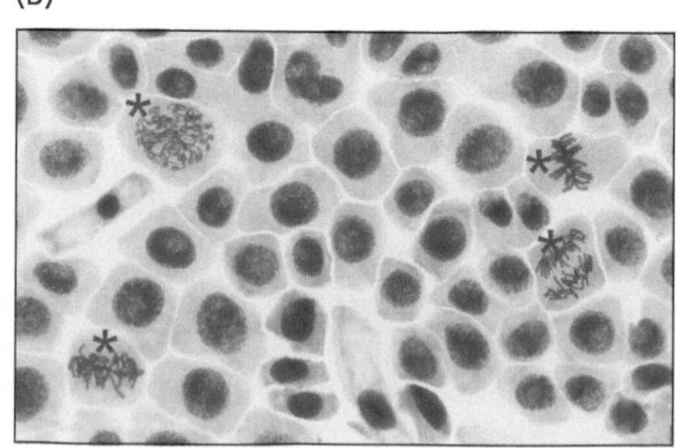

Figure 2.5 **(A) Structure of a nucleated cell. (B) Individual chromosomes can only be seen in cells undergoing division (marked with red asterisks).** The images shown are for onion root tip cells.

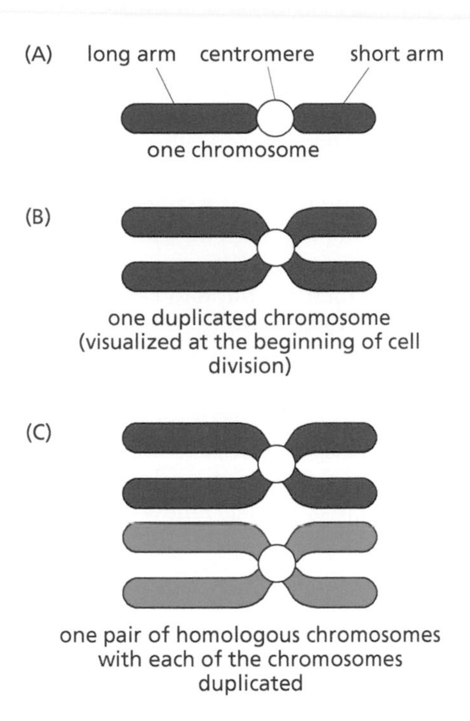

Figure 2.6 Comparison of (A) a single chromosome, (B) a duplicated chromosome, and (C) a homologous pair of duplicated chromosomes.

cytoplasm is a very simple affair but that the dividing nucleus undergoes a complex rearrangement of rod-shaped bodies inside it called **chromosomes**. Distinct chromosomes cannot be seen except in cells that are undergoing the process of division; at other times, the chromosomes are dispersed throughout the whole space of the nucleus (**Figure 2.5B**). When chromosomes are visible during cell division, differences in their structure can be seen. Each chromosome has a narrow constriction (the **centromere**) that divides it into two portions, called the "long arm" and the "short arm" (**Figure 2.6**). By measuring the lengths of these arms, we can distinguish different chromosomes from one another (Figure 3.8)

The number of chromosomes in a gamete is called the **haploid** number (N); this corresponds to one chromosome of each type for a given organism. All other body cells, called **somatic cells**, have twice as many chromosomes, called the **diploid** number ($2N$), with two chromosomes of each size and shape (**Figure 2.7**). In diploid cells, each chromosome has a partner that matches it in overall length,

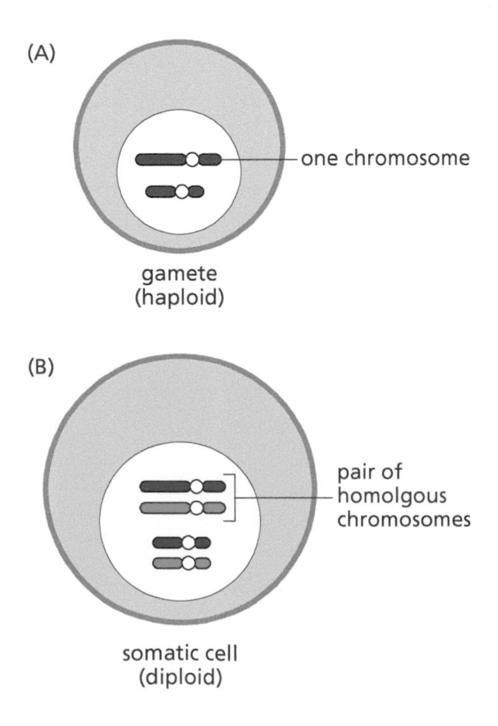

Figure 2.7 Comparison of chromosome numbers in (A) a gamete (haploid) (N = 2); (B) a somatic (diploid) cell (2N = 4) from a diploid species that has its DNA divided between two distinct chromosomes.

in the lengths of the long and short arms, and in other features, including the genes that are located on the chromosome and the order in which they occur. Each pair of chromosomes that looks alike is called a **homologous pair**, and the chromosomes of diploid cells characteristically occur in such pairs. The two chromosomes in a homologous pair are physically separate (not attached), but they do pair up with one another during meiosis, a special form of cell division that produces gametes, as described later. Though they look alike under a microscope and carry the same genes, they often carry different alleles, and thus different hereditary information. Gametes, which are haploid cells, have one chromosome from each homologous pair. Upon fertilization, a diploid cell is re-formed, with one chromosome of each homologous pair coming from the egg and one from the sperm.

Sutton noticed that eggs in most species are many times larger than sperm because they have a greater amount of cytoplasm (Figure 9.8). The nuclei of egg and sperm are approximately equal in size, and these nuclei fuse during fertilization. From these facts, Sutton reasoned as follows:

1. The genes are probably in the nucleus, not the cytoplasm, because the nucleus divides carefully and exactly, whereas the cytoplasm divides inexactly. Also, if genes were in the cytoplasm, the larger amount of cytoplasm in the egg would lead one to expect the egg's contribution always to be much greater than the sperm's, contrary to the observation that parental contributions to heredity are usually equal.
2. No other structures in cells, apart from chromosomes, are known to exist singly in gametes and in duplicate in somatic cells. The known arrangement and movement of chromosomes exactly parallel what Mendel had postulated for genes. Therefore, genes must be located on chromosomes.
3. The genes that Mendel studied assort independently because they are located on different chromosomes. However, there are only a limited number of chromosomes (4 pairs in fruit flies, 23 pairs in humans), but thousands of genes. Sutton predicted that Mendel's law of independent assortment would apply only to genes located on different chromosome pairs. In contrast, genes located on the same chromosome pair would be inherited together as a unit, a phenomenon now known as **linkage**.

Sutton's idea that genes are located on chromosomes came to be called the **chromosomal theory of inheritance**. To understand the chromosomal theory of inheritance and Sutton's prediction that some pairs of traits would not follow Mendel's law of independent assortment, we must understand more about chromosome movements in dividing cells. There are two types of cell division: **mitosis**, the process by which somatic cells divide to produce additional diploid cells, and **meiosis**, the division process that produces haploid cells (gametes) for sexual reproduction. Note that not all organisms reproduce sexually all the time; those that can undergo **asexual reproduction**, such as the unicellular budding yeast and the multicellular, water-dwelling hydra shown in Figure 2.8, use mitosis both

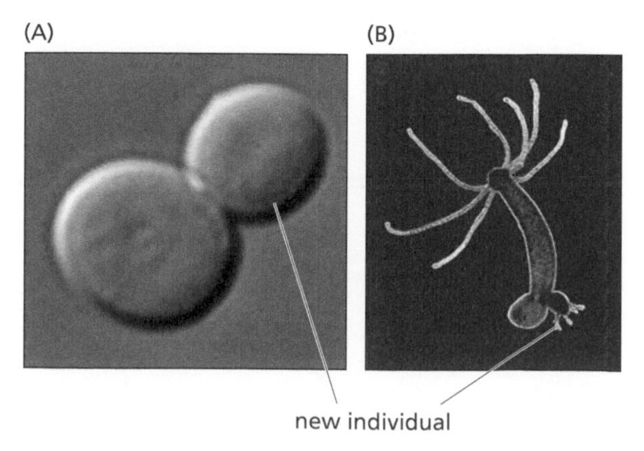

Figure 2.8 Asexual reproduction via mitosis. Many single-celled organisms, like the budding yeast *Saccharomyces cerevisiae*. **(A)** and even some multicellular organisms, like hydra **(B)** reproduce asexually via mitotic cell division to produce a new, genetically identical individual (clone) that separates from the parent.

for growth and reproduction. In asexual reproduction, one parent gives rise to offspring that are genetically identical to the parent, also referred to as **clones**. We now consider the two types of cell division, mitosis and meiosis, in turn.

2.2.1 Mitosis

In mitosis, a somatic cell divides into two cells in such a way that each new cell receives a diploid set of chromosomes—no more, and no less. Each offspring cell (usually called a "daughter cell") that results from mitosis is thus genetically identical to the original (or "mother") cell.

Prior to cell division, the cell is said to be in **interphase**. During interphase, a cell enlarges in size, increasing the volume of its cytoplasm by making more of the molecules that it contains. Also during interphase, the cell duplicates its chromosomes in a process called **replication**. (More on the process of replication is covered later in this chapter, after we discuss DNA, the main constituent of chromosomes.) Recall that individual chromosomes are not visible in the nucleus during interphase, but only once the cell has begun the process of division. Figure 2.7 shows a comparison between a single chromosome, a duplicated chromosome, and a homologous pair of chromosomes, representing the chromosomes in the condensed form in which they would be present during cell division.

Mitosis proceeds in four stages: prophase, metaphase, anaphase, and telophase. The example used in Figure 2.9 shows a cell with two pairs of chromosomes, for a total diploid number of 4.

- With the start of **prophase**, the chromosomes shorten and thicken and thus become visible under a microscope. Each chromosome was duplicated during interphase into two **sister chromatids**, which are still attached at the centromere. By the end of prophase, the membrane that surrounds the nucleus breaks down.
- During **metaphase**, each chromosome, consisting of two sister chromatids, lines up in the middle of the cell. Chromosome movement is organized by protein fibers that make up a structure called the **spindle**, part of the cell's cytoskeleton (Section 6.2.4). The homologous chromosomes of each type are not in contact with one another; rather, they are randomly spread along the midline of the cell. At the end of metaphase, all duplicated chromosomes are attached to the spindle in a "bipolar" manner, with one of the sister chromatids attached to the spindle fibers originating from one side of the cell and the other sister chromatid attached to the spindle fibers originating on the other side of the cell.
- In **anaphase**, the centromere of each chromosome splits and the sister chromatids are pulled apart by the spindle fibers to opposite ends of the cell.
- In **telophase**, all the sister chromatids have been pulled to opposite ends of the elongated cell. Therefore, each end of the cell now has the same (diploid) number of chromosomes as the original "mother" cell. During telophase, a nuclear membrane reappears around each diploid set of chromosomes, forming two nuclei—one at each end of the cell—and the chromosomes return to their loosened, interphase state, in which they can no longer be distinguished under a microscope. This completes the process of mitosis.

Notice that the steps of mitosis did not include the actual division of one cell into two. Mitosis specifically refers to division of the nucleus. In most cases, mitosis is linked to cytoplasm division (or **cytokinesis**), and cytokinesis occurs at the same time as telophase or shortly thereafter. At the completion of both mitosis and cytokinesis, both the nucleus and the cytoplasm of the original, diploid "mother" cell have been divided between two daughter cells, which are also diploid, having two copies of each chromosome, each now consisting of just one

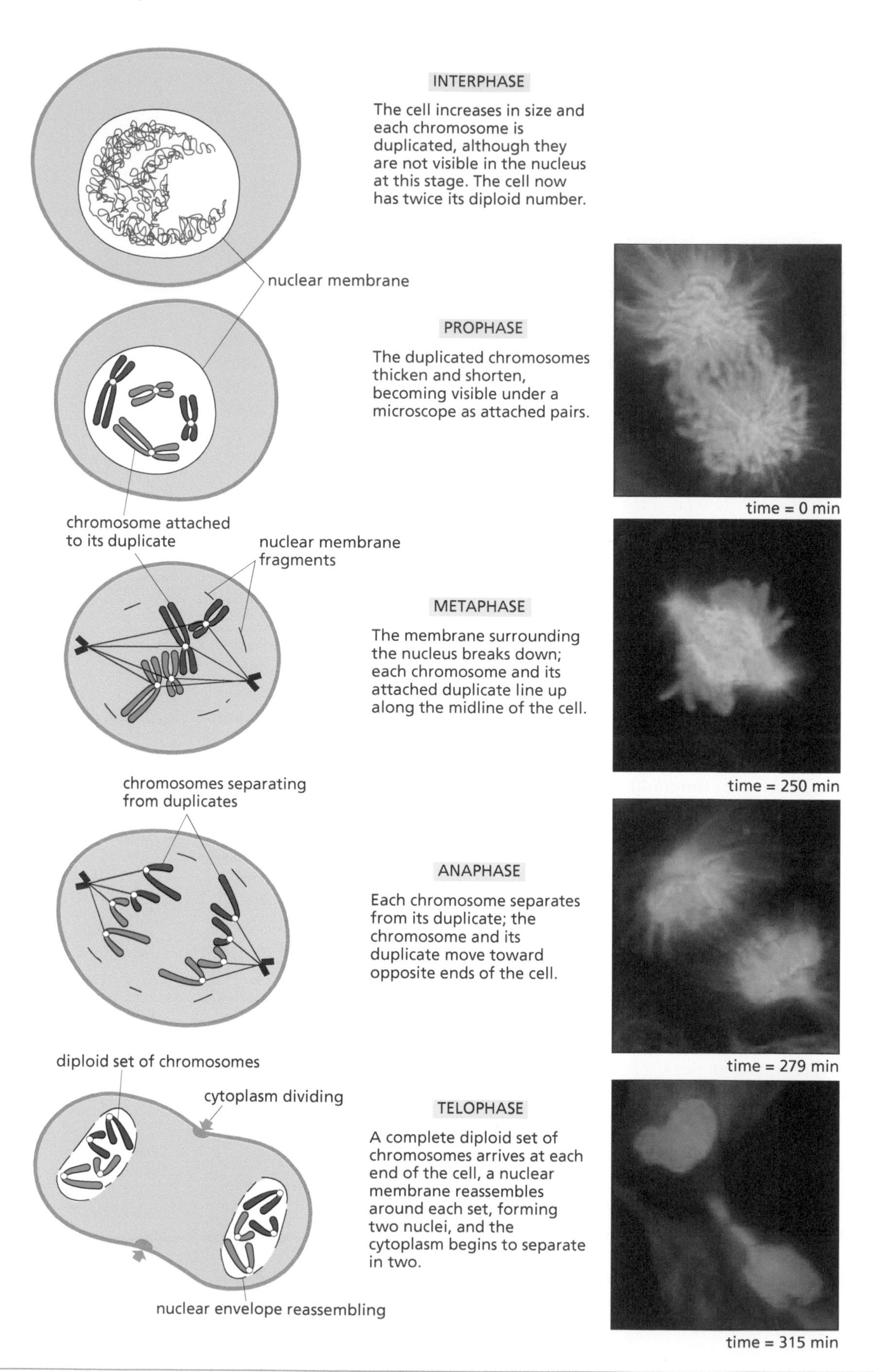

Figure 2.9 Mitosis. The cell divides in such a way that each of the two offspring cells contains the same number of paired chromosomes as the parent cell did. The drawings show a cell like that in Figure 2.6, with a diploid number of 2N = 4 chromosomes (two pairs). The photographs are fluorescent images that show both chromosomes and spindle fibers.

chromatid. (The chromosomes will not form a new pair of sister chromatids until the daughter cell again progresses through the DNA replication stage of interphase.)

2.2.2 Meiosis

Recall that pea plants reproduce sexually. In all species that reproduce sexually, two haploid gametes join to form a diploid cell called a **zygote** that can become a new individual organism. The gametes are produced in specialized cells in a process called **meiosis**.

In meiosis, as in mitosis, the chromosomes are first duplicated (in interphase) to form sister chromatids, and the cell then begins to divide. However, in meiosis, the original "mother" cell divides twice, with the chromosomes lining up differently in the two divisions. The end result of meiosis is thus four cells, each with a haploid number of chromosomes, rather than the two diploid products that result from mitosis.

The beginning of meiosis proceeds the same as mitosis: During prophase, each duplicated chromosome thickens and becomes visible, and the nuclear membrane breaks down. However, when the chromosomes line up in the middle of the cell at the first metaphase of meiosis, their pattern differs sharply from that of mitosis. Recall that in mitosis each chromosome, consisting of two sister chromatids, lines up on the midline separately from its homologous chromosome. In the first division of meiosis, however, the two homologous chromosomes, each with two sister chromatids, pair up next to each other at the cell midline, as shown in Figure 2.10. The chromosomes become attached to the fibers of the spindle, as in mitosis, but in this case, it is the two homologous chromosomes that are attached to the two ends of the spindle, not the sister chromatids. When anaphase occurs, one homologous chromosome (consisting of two sister chromatids) goes to each end of the cell, pulled by the forces of the spindle, but the sister chromatids do NOT separate from each other at this point. Telophase follows, during which a nuclear membrane reappears around each separated set of chromosomes, and the cell undergoes cytokinesis, separating into two offspring cells. This completes the first meiotic division. At this stage of meiosis, each of the two daughter cells contains one chromosome of each homologous pair and is therefore haploid, but each chromosome has its two sister chromatids still attached. The first meiotic division is now followed by a second meiotic division of each cell, in which the duplicated chromosomes behave as they do in mitosis (with the important difference that the total number of chromosomes is half what it would be in mitosis). The duplicated chromosomes align individually at the midline of the cell in metaphase of the second meiotic division, with the two sister chromatids attached to the two ends of the spindle, then the sister chromatids separate from each other in the second anaphase. The final result is four cells (gametes), each with a haploid number of unduplicated chromosomes.

To recap, three major differences distinguish meiosis from mitosis: (1) Meiosis has two cell divisions rather than the single division of mitosis; (2) the end result of meiosis is four haploid cells, each with half the diploid number of chromosomes; (3) during metaphase of the first meiotic division, the homologous pairs of duplicated chromosomes line up together as quadruplicates (or "tetrads"), rather than as individual, duplicated chromosomes. As we will soon see, at the point when the homologous chromosomes are lined up as quadruplicates, an event called **crossing over** can occur. During crossing over, one chromosome from each attached pair of sister chromatids can exchange part of itself with the corresponding part on the homologous chromosome, as indicated by the exchange of red and blue segments in Figure 2.10. Studies of this phenomenon allowed the discovery of gene linkage, which is discussed later.

2.2.3 Sexual life cycles

After the haploid gametes are produced by meiosis, they can be brought together by sexual reproduction, forming a diploid zygote (or fertilized egg),

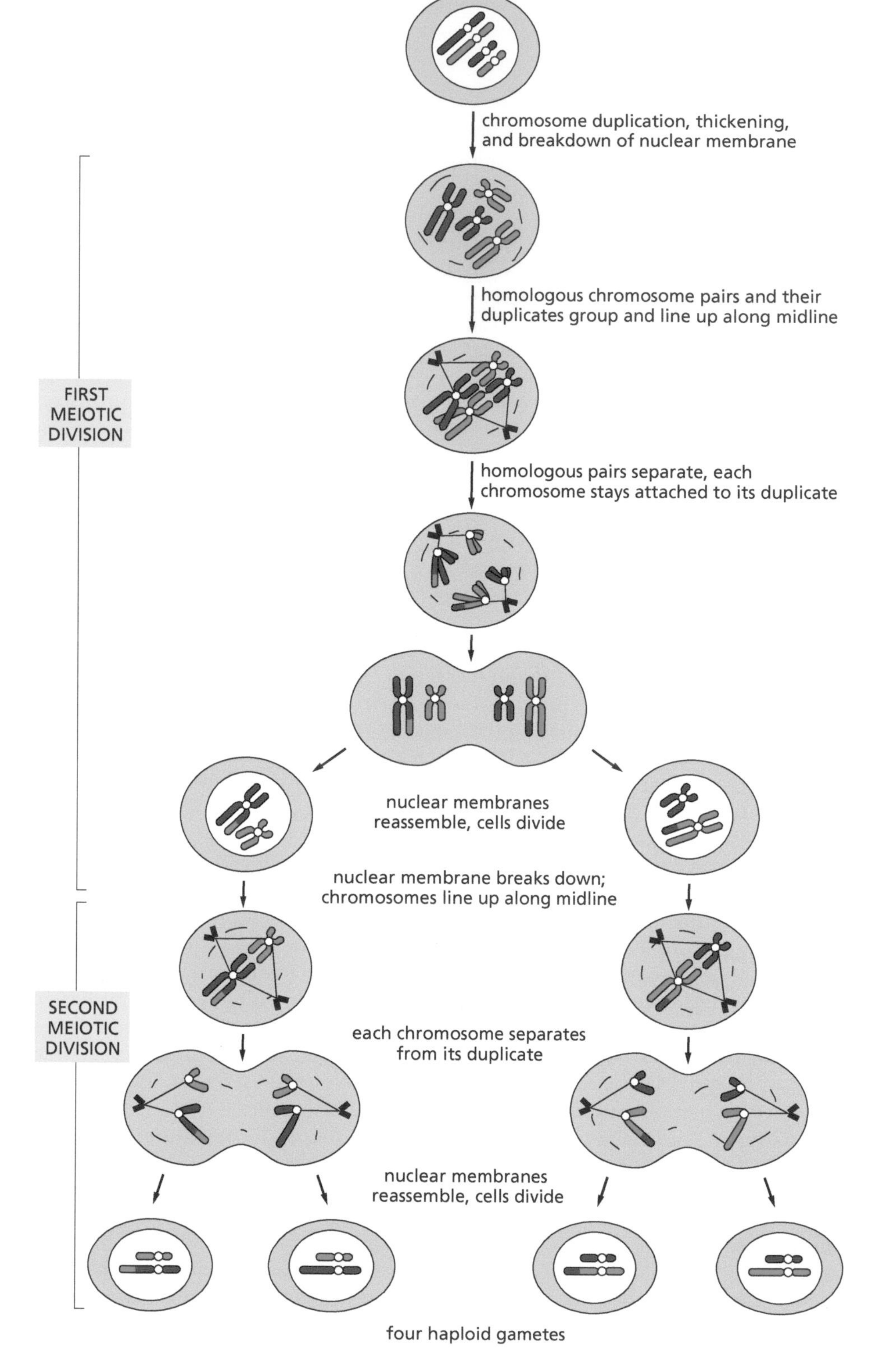

Figure 2.10 Meiosis. A cell with a diploid number of chromosomes (2N = 4 in this example) divides twice to produce four haploid gametes (here with two chromosomes each).

a process investigated by the Austrian zoologist Theodor Boveri and the American zoologist Ernest E. Just (Figure 1.4C). **Figure 2.11** shows a sexual life cycle. After the gametes fuse to form a zygote, the zygote undergoes repeated rounds of mitosis. In animals and plants, repeated mitosis produces a multicellular adult, but in some unicellular organisms like yeasts, the

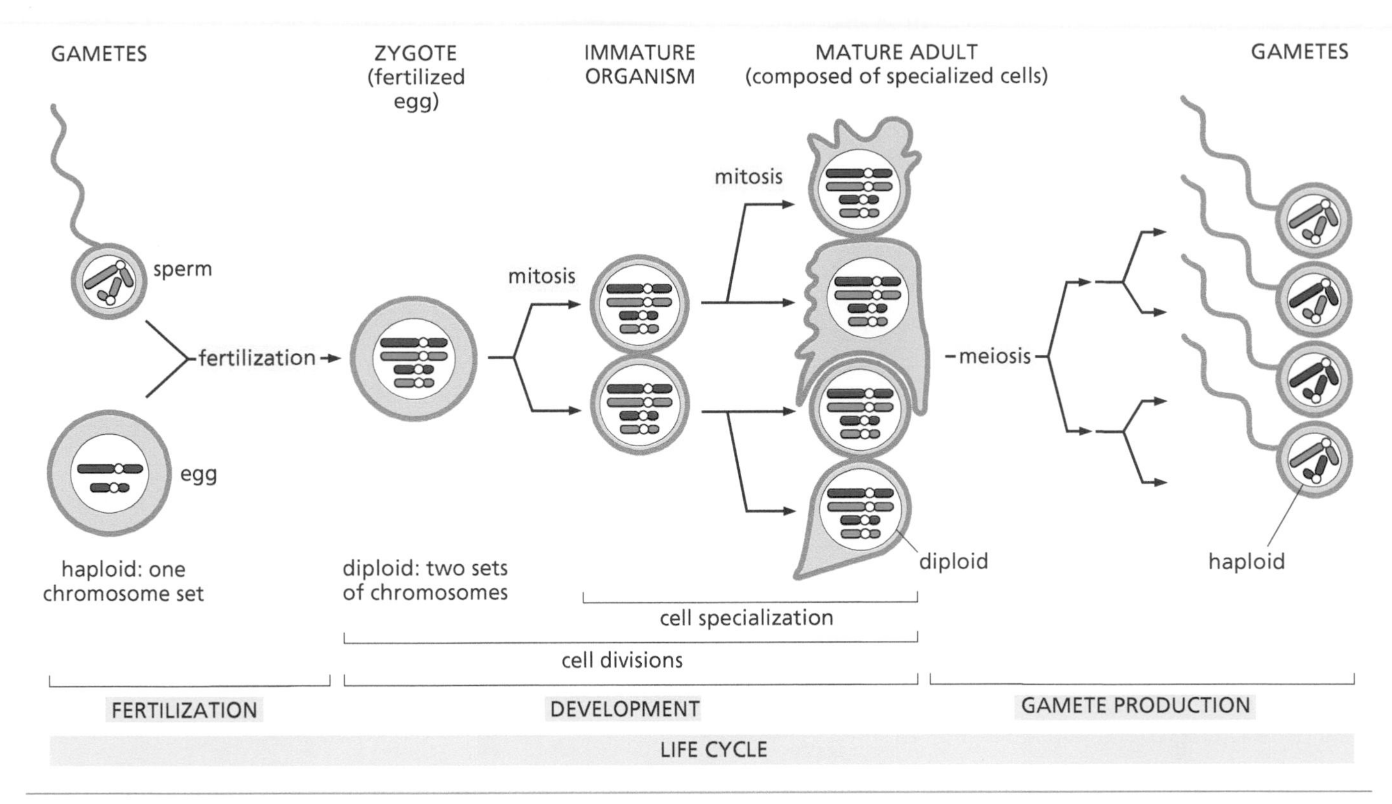

Figure 2.11 Sexual life cycles. In sexual reproduction, haploid gametes join by fertilization to form a new diploid individual with one of each pair of homologous chromosomes coming from each parent. In multicelled organisms the diploid zygote divides by mitosis to form the adult organism. Each of the somatic (body) cells contains a set of chromosomes the same as that in the zygote. In a male organism, meiosis produces sperm, as shown. In a female organism, meiosis produces eggs.

mitotic divisions that follow sexual reproduction yield new individuals rather than a multicellular organism; thus, yeasts can reproduce both sexually and asexually. Each somatic cell of a multicellular organism contains the full diploid set of chromosomes, and thus, all somatic cells are genetically identical. As the organism develops, different somatic cells specialize for different functions (Section 10.1), although they all retain the full diploid set of chromosomes. Some of these cells specialize to undergo meiosis, producing new haploid gametes and completing the sexual life cycle. Most, but not all, multicellular organisms have a sexual life cycle, alternating between haploid gametes and diploid somatic cells (the multicellular hydra shown in Figure 2.8 is one exception). The main advantage of sexual reproduction over asexual reproduction (via mitosis alone) is the generation of new combinations of gene alleles, some of which may give an individual an evolutionary advantage in the form of enhanced survival and reproduction. These new combinations of alleles result from three processes: (1) independent assortment of homologous chromosomes during anaphase 1 of meiosis (i.e., chromosomes that came from an individual's mother and father do not remain together as "sets"); (2) the process of crossing over (discussed further later); and (3) the bringing together of different gene alleles from the egg and sperm during fertilization.

2.2.4 Gene linkage

Sutton predicted that independent assortment would not apply to all pairs of genes and that, instead, genes that were located on the same chromosome would often be inherited together. Other investigators quickly confirmed his prediction, reporting evidence of linked genes in garden peas (the species Mendel had used) as well as in fruit flies (*Drosophila*), corn (*Zea mays*), and other species, showing that Mendel's laws and Sutton's theory applied to all

Figure 2.12 A cross between pure-line corn plants having different alleles for two linked genes located on the same chromosome.

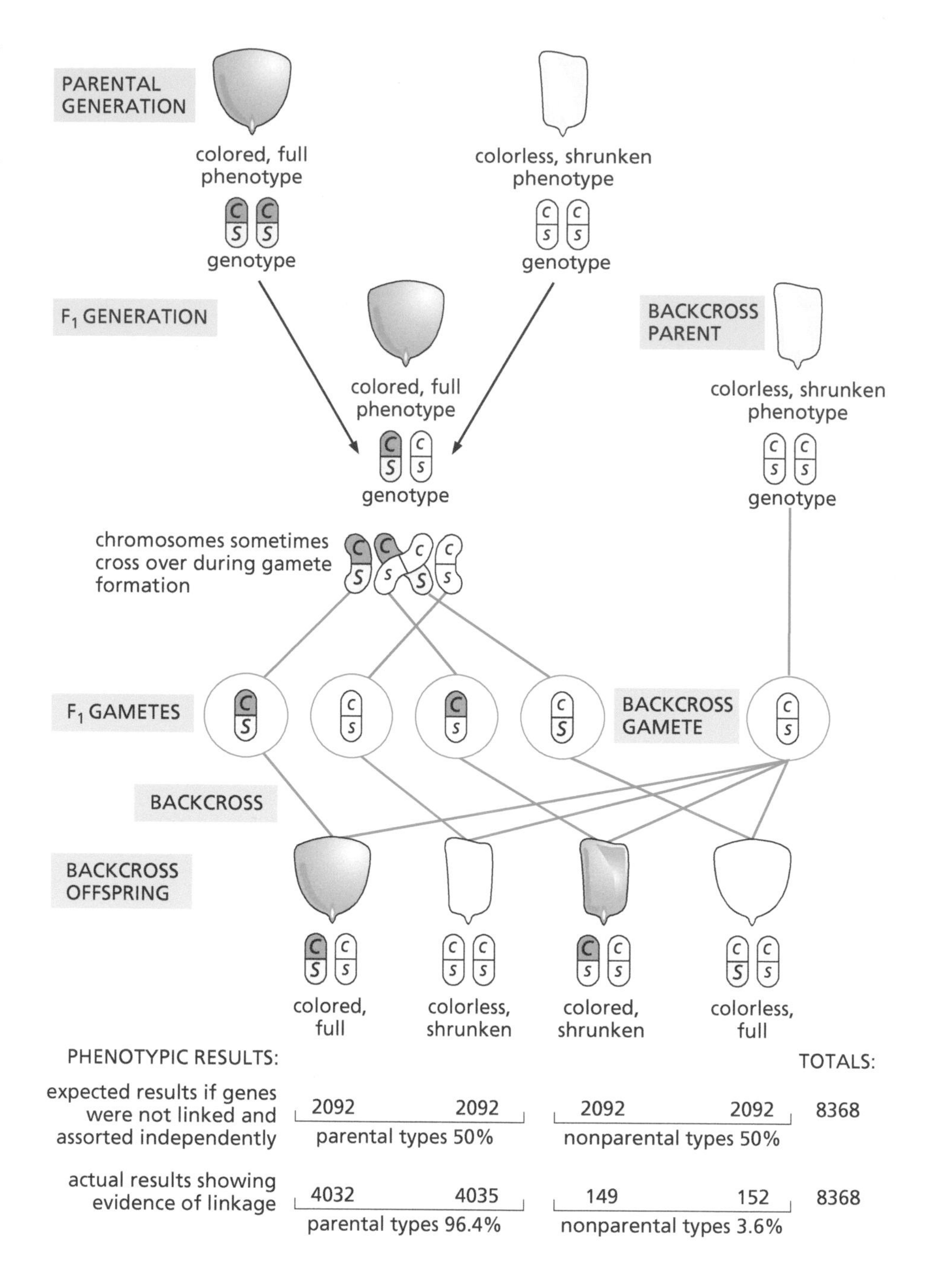

these species. Figure 2.12 shows a cross revealing that two genes are linked in corn. As expected from what we have learned previously, plants of the genotype *CCSS* (colored, full seeds) crossed with *ccss* plants (colorless, shrunken seeds) produced all seeds of the colored, full phenotype among the F_1, according to the principle of dominance. However, in this experiment, the F_1 plants (*CcSs* heterozygotes) were not self-fertilized as in Mendel's experiments (Figures 2.3 and 2.4); rather, the F_1 heterozygotes were fertilized by plants of the doubly recessive *ccss* genotype. This type of cross, in which an F_1 is crossed with one of the parental types, is called a "backcross." Applying Mendel's principle of independent assortment, we would expect to obtain approximately equal numbers of offspring with the four different combinations of phenotypes for the seed color and seed shape traits. In actuality, in the backcross offspring, most of the seeds that were colored were also full, and most of the seeds that were colorless also had the shrunken phenotype. Thus, the combinations of seed color and seed shape phenotypes that were present in the parents generally stayed together in the offspring, rather than recombining randomly as predicted by the principle of independent assortment.

In genetic terminology, the genes for these two traits are said to be **linked**. Their linkage could be explained by assuming that the genes for the two traits were on the same chromosome and were thus usually inherited together.

Most of the progeny in Figure 2.12 show the parental linked phenotypes. However, notice that small numbers of the backcross progeny were not of the colored and full or colorless and shrunken seed phenotypes, but instead had colored and shrunken or colorless and full seeds. These atypical plants had new (nonparental) combinations of the phenotypes for the traits; the underlying recombinant genotypes were hypothesized to have arisen from a process called **crossing over**, in which chromosomes break and recombine by exchanging equivalent pieces that contain the same genes but may contain different alleles of those genes. Further microscopy studies identified X-shaped arrangements of the chromosomes during meiosis (Figure 2.12) that were consistent with crossing over.

2.2.5 Confirmation of the chromosomal theory of inheritance

In 1931, Harriet Creighton and Barbara McClintock (Figure 1.4D) were able to demonstrate that, for linked genes (genes on the same chromosome), genetic **recombination** (the rearranging of genes) was always accompanied by **crossing over** (the rearranging of chromosomes), thus confirming the chromosomal theory in corn. Later that same year, Curt Stern observed the same thing in fruit flies.

The frequency of recombination between linked genes is roughly a measure of the distance between them along the chromosome: Recombination between closely linked genes (those physically near each other on a chromosome) is a rare event, while recombination between genes farther apart is more frequent. By making crosses between individuals having different alleles for pairs of linked genes, geneticists were able to determine the linear arrangement of many genes and the approximate distances between genes on the chromosomes of many species. Such studies produced maps of genes along chromosomes, such as the one shown in **Figure 2.13**, long before it was possible to determine the complete DNA sequence of a chromosome, as described in Chapter 4.

An interesting footnote to Mendel's work was provided in 1936 by the British geneticist R.A. Fisher, who noticed that garden peas have seven pairs of chromosomes. As it turned out, the seven traits Mendel had studied assorted independently because each is on a different chromosome! Because the probability of this occurring by chance is extremely remote, Fisher concluded that Mendel likely studied many more traits and only reported the results for the seven independently assorting traits (Figure 2.1) whose inheritance he could understand.

(?) THOUGHT QUESTIONS

1. In a cross between pea plants of genotype *YYRR* (yellow, round seeds) and *yyrr* (green, wrinkled seeds), the F_1 plants are all *YyRr*. Make a series of drawings showing the movements during mitosis of the chromosomes carrying these genes in cells of an F_1 heterozygous plant.
2. For the same F_1 heterozygous plants (*YyRr*) described in question 1, make a series of drawings showing the way in which the chromosomes separate in the first division of meiosis and in the second division of meiosis. Label each diagram with the symbols *Y*, *y*, *R*, and *r* to show how all four types of gametes originate.
3. Would you expect the cross *YYRR* × *yyrr* to give the same results as *YYrr* × *yyRR*? Why or why not?
4. Two test crosses are carried out, one between *AaBb* and *aabb* parents and the other between *CcDd* and *ccdd* parents. In the resulting progeny, the phenotypes of the parents (dominant for both traits or recessive for both traits) are overrepresented compared to the nonparental phenotypes (dominant for one trait and recessive for the other), but the *AaBb* × *aabb* cross produces more frequent nonparental type progeny than the *CcDd* × *ccdd* cross. Do these crosses involve linked genes? How do you know? Can you tell which pairs of genes (*A* and *B* or *C* and *D*) are located closer together along the chromosome?

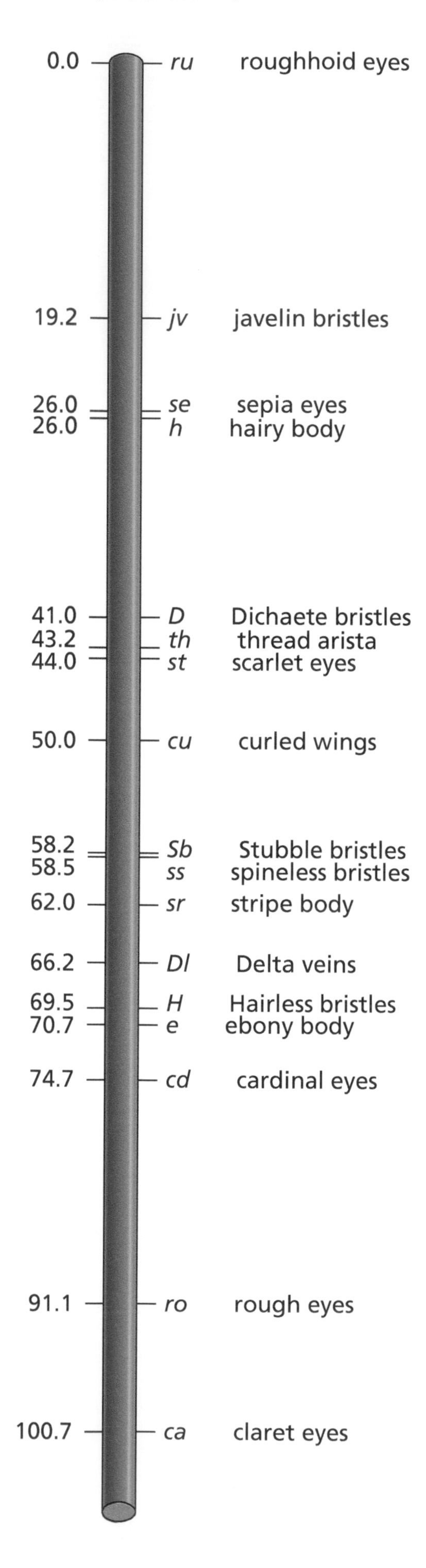

Figure 2.13. **Linkage map of chromosome III in the fruit fly *Drosophila melanogaster*, showing just a few of the many genes known.** Genes for recessive traits are in lowercase; genes for dominant traits are capitalized. Distances along the chromosome are in units called "centimorgans"; a distance of one centimorgan corresponds to a 1% probability of a crossover between closely linked genes.

2.3 OUR UNDERSTANDING OF DNA FURTHER EXPLAINS MENDEL'S HYPOTHESES

We now address the last of the four questions posed earlier: What are genes made of? In other words, what chemical substance transfers genotypes from parents to offspring?

The story that ultimately answered this question begins with a curious experiment carried out in 1928 by Frederick Griffith, a bacteriologist attempting to develop a vaccine against pneumonia. Griffith worked with two strains of bacteria that differed in their outer coats. Strain S ("smooth") had an outer coat that gave a smooth appearance when masses of bacteria (called colonies) were grown on agar in dishes. The smooth-colony bacteria were **virulent**, which means that the bacteria caused a disease (pneumonia in this case). Strain R ("rough") of the same bacterial species lacked the outer coat, which gave the colonies of this strain a rough appearance, and they were **nonvirulent**. When strain S was injected into mice, all the mice died of pneumonia. Strain R, when injected, did not kill mice, nor did bacteria of strain S that had been killed by heat. The surprising result, shown in Figure 2.14, was that a mixture of live bacteria of strain R and heat-killed bacteria of strain S *did* kill mice. Furthermore, living S bacteria were isolated from all mice that died this way. Griffith interpreted this experiment as indicating that something in the dead S bacteria had somehow transformed the living R bacteria into virulent S bacteria. This type of change came to be known as **transformation**. The bacteria had been altered genetically, not just phenotypically. A change that was only phenotypic would not be passed on to future generations, but Griffith demonstrated that descendants of the transformed bacteria were also of strain S and continued to kill mice.

What Griffith had done in his experiment was transfer a genetic trait from one bacterial strain to another. What was the chemical substance that had been transferred? Griffith's use of bacteria as an experimental species and the unequivocal evidence that genetic material had been transferred in his experiment are significant because they formed the background for work two decades later that demonstrated what the chemical substance is. We will now describe the research that revealed the chemical substance of genetic material, the composition of this substance, and the manner of its replication.

2.3.1 DNA and genetic transformation

From the beginning of the twentieth century and into the 1940s, most researchers thought that proteins were the most likely candidates to be the chemical substance of genes. They thought this because proteins were known to be complex and varied, while most other molecules were thought to be too simple to specify hereditary information. In an attempt to discover whether protein was indeed the chemical substance of heredity, in 1944 three bacteriologists, Oswald Avery, Colin MacLeod, and Maclyn McCarty, conducted a chemical study of the bacteria Griffith had injected into his experimental mice. First, they were able to show that a chemical extract of heat-killed S bacteria transformed strain R into strain S. They then separated the strain-S extract into different fractions, each containing different types of chemical molecules (i.e., proteins in one fraction and nucleic acids in another). They found that the fraction containing the **nucleic acids** transformed strain R into strain S, but the protein fraction did not. This showed that nucleic acids—NOT proteins—contained the hereditary information.

Finally, Avery, MacLeod, and McCarty distinguished between the two major types of nucleic acids (DNA and RNA) by using enzymes. **Enzymes** are biological molecules (nearly always proteins) capable of speeding up chemical reactions without themselves getting used up in those reactions (Section 8.2.3). Enzymes control many biological processes, and the action of most enzymes is very restricted in that specific enzymes act on specific types of molecules. The enzyme deoxyribonuclease (DNase) specifically cuts apart the long, linear structure of **deoxyribonucleic acid** (**DNA**) into its individual

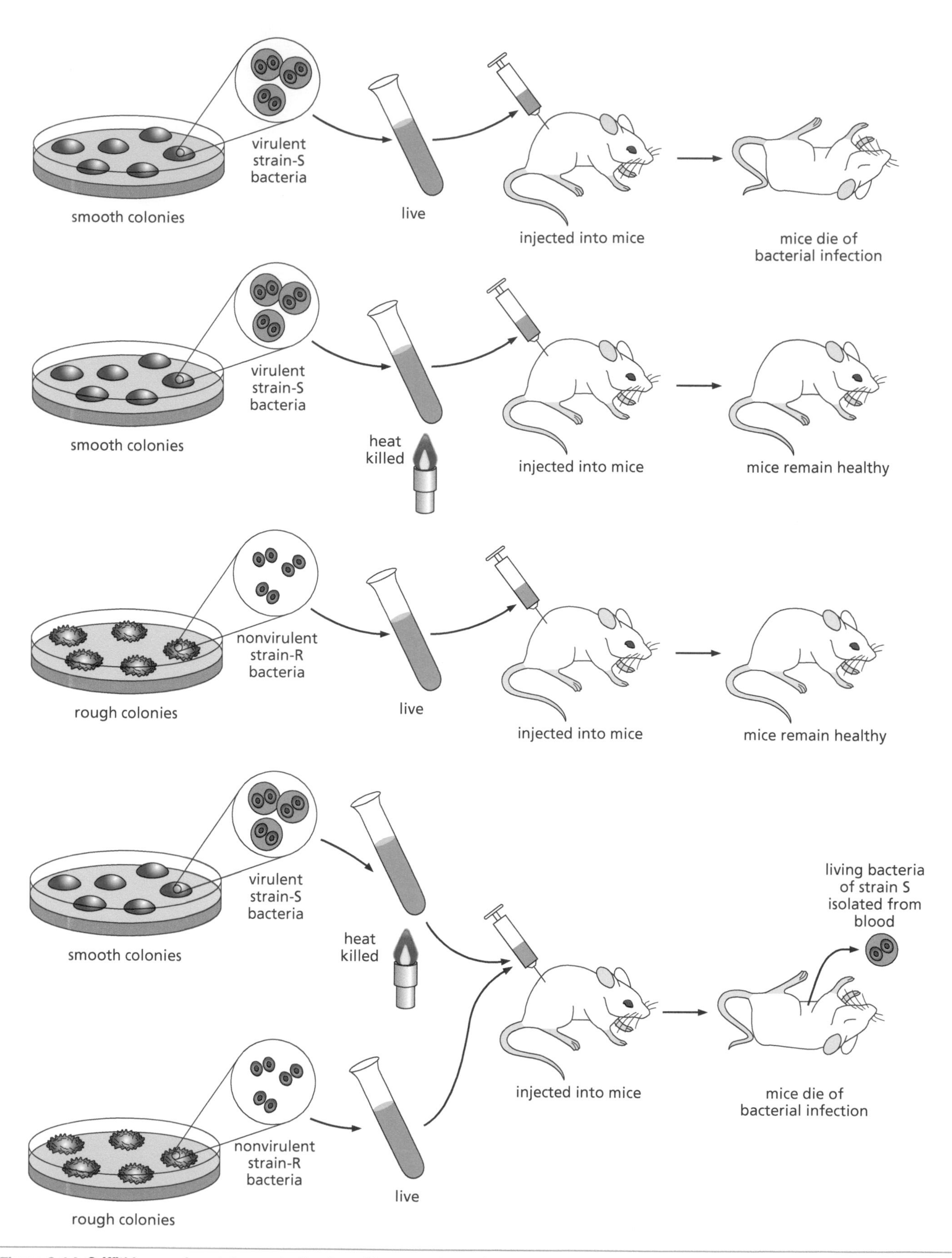

Figure 2.14 Griffith's experiment demonstrating hereditary transformation in bacteria.

building blocks, destroying the function of DNA. DNase treatment destroyed the ability of the strain-S extract to transform the strain-R bacteria. On the other hand, the enzyme ribonuclease, which specifically cuts apart **ribonucleic acid (RNA)**, had no effect on the ability to transform. Together, these results demonstrated that the chemical that carries the genetic material is, in fact, DNA and is not RNA.

The discovery made by Avery and his co-workers did not receive the attention it deserved at the time. Doubters still remained. As is frequently the case with large shifts in scientific understanding (so-called "paradigm shifts"—see Section 1.2.1), it took about a decade for many scientists to accept that DNA is the genetic material because they did not understand how the relatively simple structure of DNA could encode complex genetic information. Full acceptance of this fact did not occur until it was validated in a different experimental system.

In 1952, two American virologists, Alfred Hershey and Martha Chase, published the results of a landmark experiment that confirmed the finding that DNA, not protein, is the genetic material. For their experiment, Hershey and Chase used a virus that infects bacteria and reproduces within them. They infected bacteria with the viruses and studied the viral offspring. It was known that this virus consisted only of protein and DNA. Was it the protein or the DNA that carried the genetic material?

In preparation for their experiment, Hershey and Chase grew some viruses in a medium containing radioactive phosphorus (^{32}P) and others in a medium containing radioactive sulfur (^{35}S), atoms that the virus needs to make new viruses. Because DNA contains phosphorus but protein does not, the new viruses grown with ^{32}P had radioactive phosphorus in their DNA but no radioactivity in their protein. In contrast, the proteins, but not the DNA, of viruses grown with ^{35}S were radioactive because proteins contain sulfur but DNA does not. Because radioactivity is easily detected, material prepared in this way is said to be "radioactively labeled." Hershey and Chase applied the radioactive labels so they would be able to "see" what happened to the viral DNA and protein when the viruses infected bacteria.

Hershey and Chase exposed *Escherichia coli* bacteria to the radioactively labeled viruses for long enough to permit the viruses to infect the bacteria. During infection, part of the virus enters the cells, while another part remains outside the cells. The viral component that enters the cells directs the replication of more viruses, producing thousands of new virus particles and eventually killing the bacteria, breaking the cells open to release the new viruses (**Figure 2.15A**). Was the injected material that was carrying the viral genotype DNA or protein? Hershey and Chase devised a way to interrupt the viral cycle after the infection period by using a kitchen blender to knock the attached virus capsules off the bacterial surfaces. These detached capsules could easily be separated from the bacteria by spinning the mixture in a centrifuge. (The rapid spinning of the centrifuge causes the heavier bacteria to form a pellet at the bottom of a test tube, while the lighter viral particles remain suspended in the liquid above.) When ^{35}S-labeled viruses were used, the radioactive proteins remained outside the bacteria (**Figure 2.15B**); the new viruses eventually synthesized and released when the cells broke open were not radioactive, so they had not used radioactive protein to replicate themselves. However, when the ^{32}P -labeled viruses were used, the radioactive DNA entered the bacteria, making the bacteria radioactive. The viral offspring released when these bacteria broke open were also radioactive (**Figure 2.15C**), showing that they had used some of the radioactive DNA in their reproduction.

This experiment upheld the hypothesis that the genetic material of the virus was DNA. It falsified the hypothesis that the viral genetic material was made of protein. Finally, scientists at large accepted the fact that DNA was the universal genetic material in all organisms (although a few viruses use RNA instead). However, the structure of DNA remained a mystery, as did the means by which it could specify the complex information needed to make a whole organism.

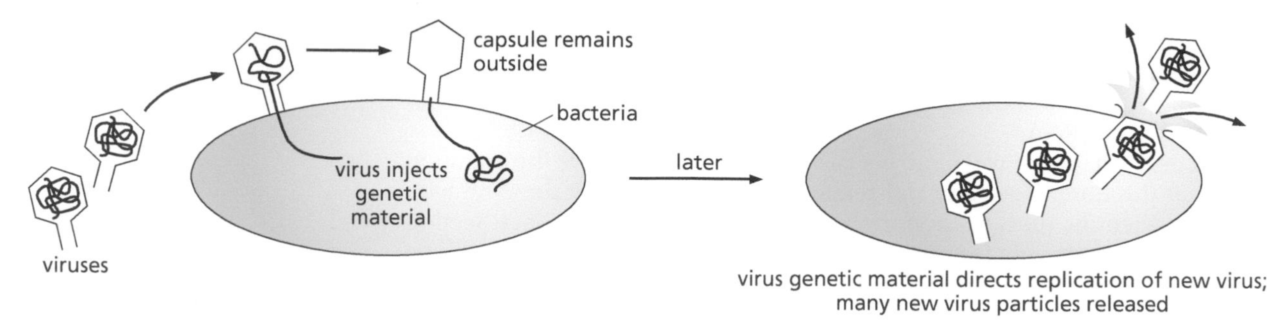

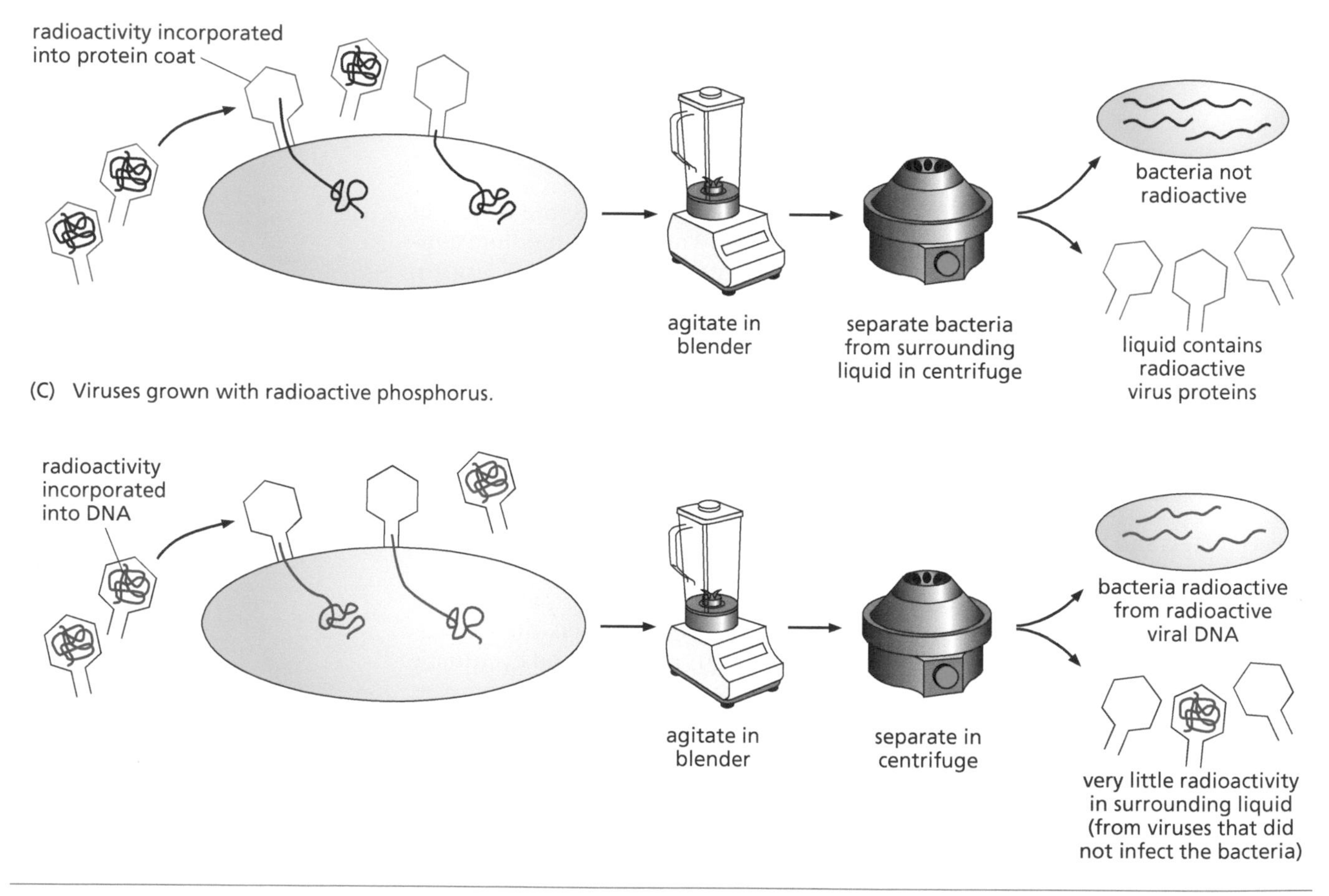

Figure 2.15 The Hershey–Chase experiment. This experiment confirmed DNA as the genetic material.

2.3.2 The composition and structure of DNA

Chemical breakdown of DNA into its parts showed that it was made of phosphate groups, deoxyribose (a sugar), and four nitrogen-containing bases called adenine (A), guanine (G), thymine (T), and cytosine (C) (Figure 2.16A). Biochemists soon realized that the deoxyribose could form a middle link between the phosphate groups and the nitrogenous bases, creating units called **nucleotides** (Figure 2.16B). Beyond this, the structure of DNA was unclear, and it was not at all obvious how the structure could give DNA the ability to carry genetic information.

2.3.2.1 Chargaff's rules Some researchers suggested that the order of bases in DNA repeated regularly: AGTCAGTCAGTC. . . . If this were true, then the amounts of the four nitrogenous bases should be equal: There should be 25% of each of the bases in a DNA molecule. To test this hypothesis, biochemist Erwin Chargaff of Columbia University took DNA from various sources, cut it into

Figure 2.16 The nucleotides of DNA.

(A) Two of DNA's nitrogenous bases, adenine (A) and guanine (G), are larger than the other two, thymine (T) and cytosine (C).

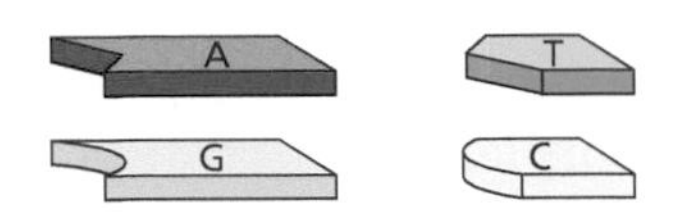

(B) With the addition of deoxyribose (a five-sided sugar molecule) and a phosphate group, each of these nitrogenous bases forms a nucleotide.

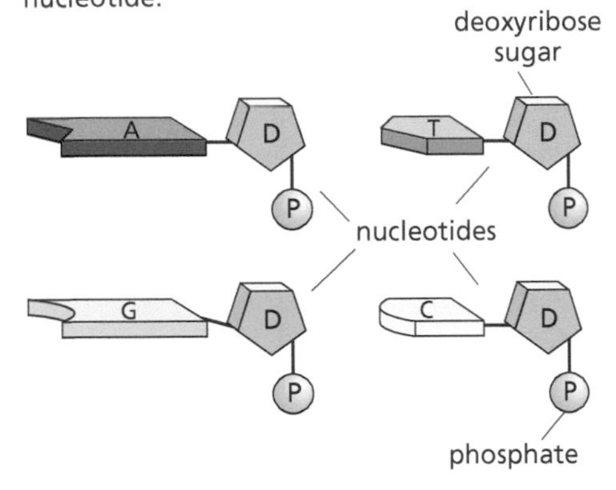

individual nucleotides using enzymes, and measured the relative amounts of the four nitrogenous bases. His findings were as follows.

1. The proportions of the four nitrogenous bases are constant for all cell types within a species. For example, all human cells contain about 31% adenine (A), 19% guanine (G), 31% thymine (T), and 19% cytosine (C), regardless of whether the DNA is from brain cells, liver cells, kidney cells, or skin cells.
2. Although the proportions are constant within a species, they differ from one species to another. All humans, for example, have the same proportions of the four bases. Proportions are different in rats and in bread molds, but all rats have the same proportions as one another and so do all bread molds.
3. The most unexpected finding, and the hardest to explain, was that the proportion of adenine was always the same as the proportion of thymine (within the limits of experimental error), and the levels of cytosine and guanine were also equal. These findings (symbolized as A=T and G=C) became known as **Chargaff's rules**.

2.3.2.2 The double helix structure of DNA In 1953, James Watson and Francis Crick, two geneticists working in Cambridge, England, proposed a structure for DNA that explains Chargaff's rules and also explains how DNA carries genetic information. They did this with the help of X-ray diffraction data obtained from the Cambridge laboratory of biochemist Maurice H.F. Wilkins, with whom they later shared a Nobel Prize. The technique of X-ray diffraction analyzes chemicals that have been gradually dried out such that they align in a very ordered way in crystals (similar to the way salt crystals form when seawater evaporates); these crystals are then placed in an X-ray beam, and the pattern generated when the X-rays bounce off the atoms in the crystal can be used to determine the arrangement of atoms in individual molecules within the crystal (Figure 4.17). The crucial X-ray diffraction data from Wilkins's laboratory were obtained and interpreted by Rosalind Franklin, a biochemist; tragically, Franklin's contribution never received the recognition it deserved, as she died from cancer at a young age, and Nobel Prizes are not awarded posthumously. Anne Sayre has written a biography of Rosalind Franklin that documents her role in the discovery of DNA structure.

The X-ray diffraction data suggested certain dimensions and distances for the repetition of structures within the DNA molecule. From this information,

Watson and Crick constructed the model of DNA structure summarized in the following list:

1. Each phosphate in a DNA molecule is attached to a deoxyribose sugar, which in turn is attached to a nitrogenous base. The three parts together constitute a **nucleotide** (Figure 2.16B).
2. The phosphate group of one nucleotide is connected to the deoxyribose sugar of the next nucleotide. The alternation of phosphates and sugars thus forms a "backbone" that holds the nucleotides together in a strand with the nitrogenous bases pointing inward (**Figure 2.17A**).
3. Each strand is a linear (nonbranching) sequence of nucleotides that is twisted in the shape of a corkscrew (or "helix") because of the angles of the chemical bonds between the atoms in the nucleotides. It was Rosalind Franklin who first realized that the DNA molecule had this helical shape.
4. DNA has two strands, wound around each other, forming a **double helix**, with the bases arranged in the interior, like steps in a spiral staircase (**Figure 2.17B**).
5. The strands run in opposite directions, with a phosphate at the end of one strand and a sugar at the equivalent position on the opposite strand. This is referred to as an **antiparallel** arrangement.
6. An adenine on one strand is always paired with a thymine on the other strand, and vice versa. Also, cytosine on one strand is always paired with guanine on the other strand, and vice versa (**Figure 2.17C**). These specific pairings of bases explain Chargaff's rules.
7. A and T fit together and C and G fit together because their shapes are complementary (Figure 2.17); thus, the two strands of DNA are said to exhibit **complementary base pairing**. Because of this complementary base pairing, each strand contains all the information necessary to determine the structure of the complementary DNA strand, a feature that is crucial for the copying of DNA when a cell divides (see the next section).

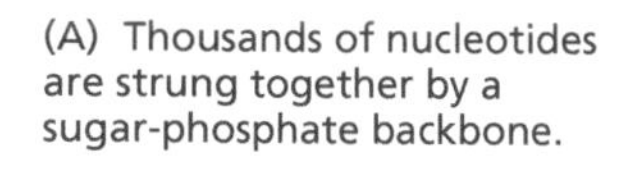

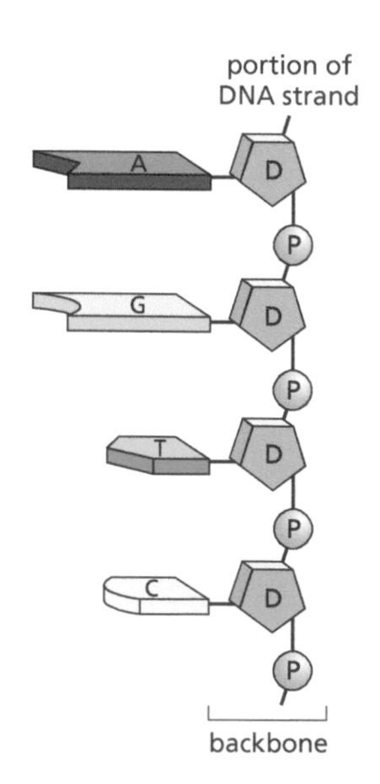

(B) Two strands of DNA twist around one another to form a double helix. A straightened portion of this double helix resembles a ladder with the paired (complementary) bases forming the rungs.

(C) Two nucleotide sequences running in opposite directions pair with one another, with each adenine (A) pairing with a thymine (T), and each guanine (G) pairing with a cytosine (C).

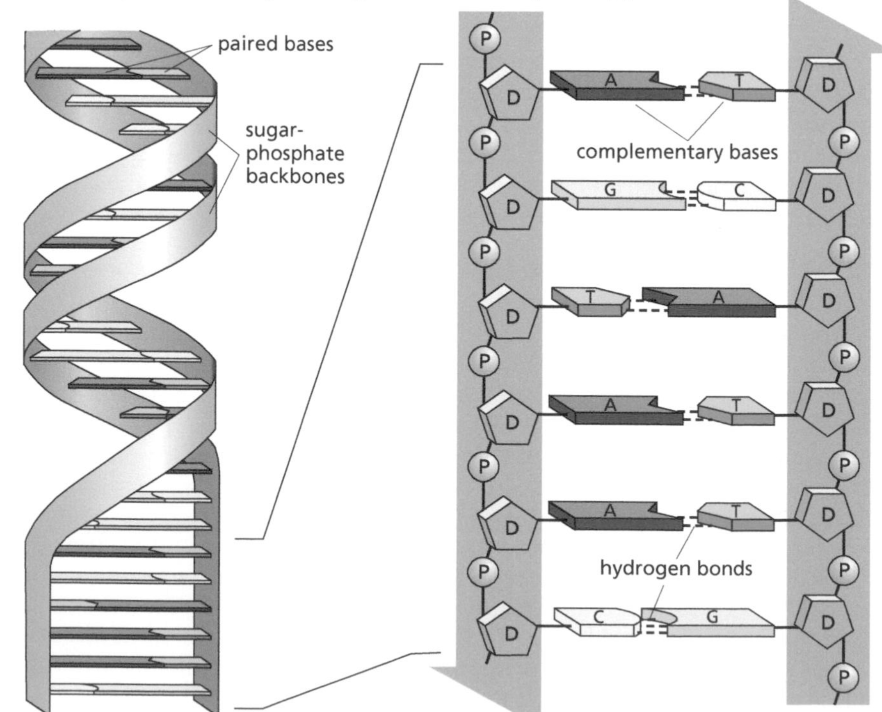

Figure 2.17 The three-dimensional structure of DNA as determined by the technique of X-ray crystallography.

2.3.3 DNA replication

Watson and Crick's model for the structure of DNA immediately suggested a hypothesis for how DNA might be copied, or replicated, and this hypothesis was quickly confirmed experimentally. Before **DNA replication**, the two strands of the double helix unwind and separate from each other, as shown in Figure 2.18A. This generates a "replication fork," like two roads diverging from a common point. After separating, the strands are bound by an enzyme (**DNA polymerase**), which actually begins the replication process. New strands of DNA are synthesized one nucleotide at a time, with one of the existing strands serving as a template (the pattern to be copied). If the next unmatched base on the template is adenine (A), then a thymine (T) nucleotide (adenine's complementary base) is added to the growing new strand opposite the adenine. In like manner, G is added opposite C, C opposite G, and A opposite T (Figure 2.18B). The other existing strand is simultaneously acting as a second template, and other complementary bases pair with their respective nucleotides. The backbone of the new strand is formed by joining (bonding) the phosphate group of the incoming nucleotide to the deoxyribose sugar of the previous nucleotide (Figure 2.18C).

Notice the orientation of the five-sided deoxyribose molecules in the orange-colored backbone, which shows that the two strands run in opposite directions (antiparallel; Figure 2.17C). DNA polymerase can only add a new nucleotide onto the sugar end of another nucleotide. Because of this limitation, copying of the two DNA template strands that are arranged antiparallel to one another must proceed in opposite directions relative to the replication fork. One new strand of DNA

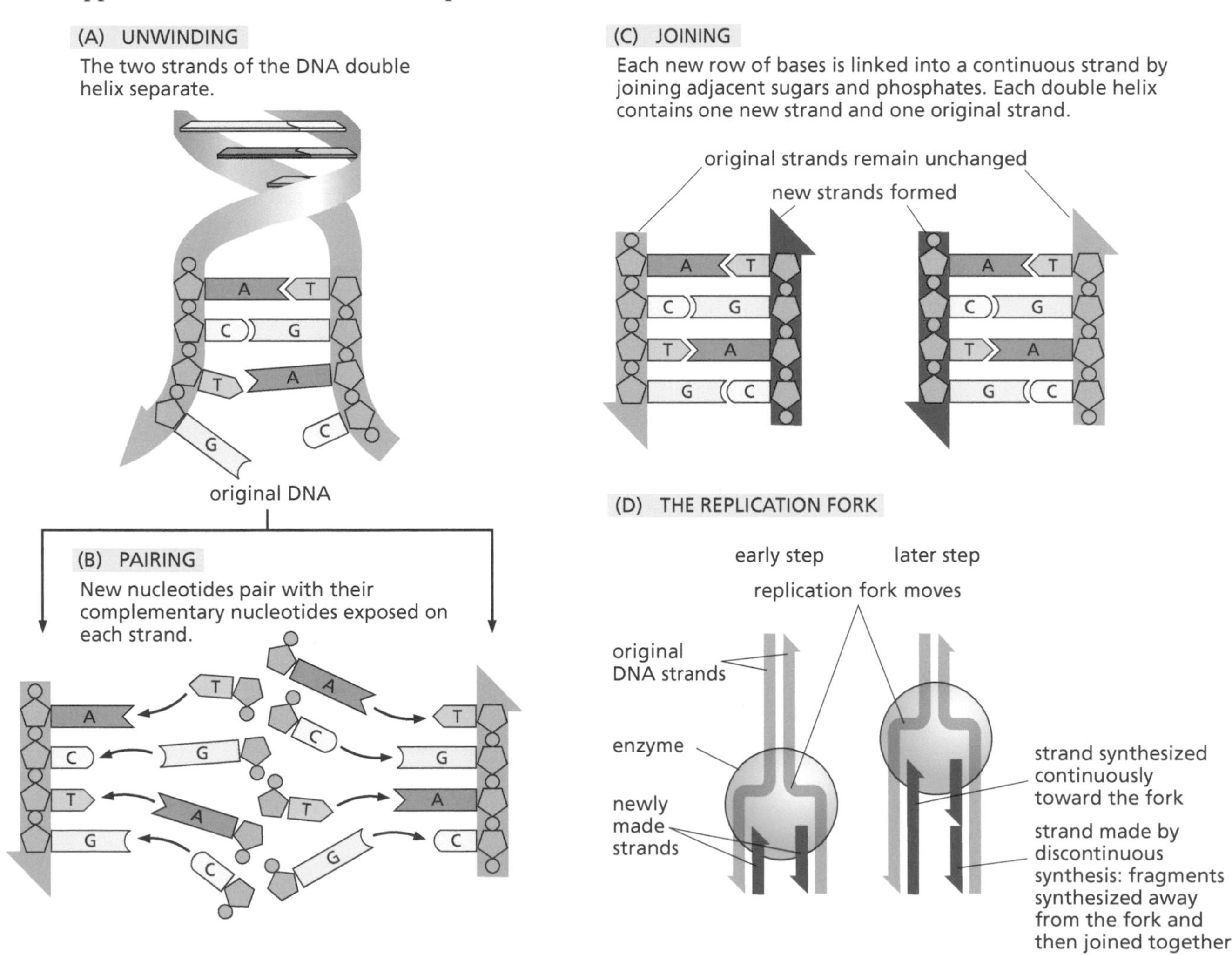

Figure 2.18 DNA replication. The two strands run in opposite directions, as denoted by the arrowheads on the orange (old) backbones. The direction of synthesis is indicated by the arrowheads on the red (new) backbones. Note that either strand contains all the information needed to synthesize the other (complementary) strand.

is synthesized continuously, with the direction of synthesis running towards the replication fork, the place where the two original strands are separating. The other strand, however, is synthesized in the opposite direction, away from the replication fork. This strand must be synthesized in short fragments that are later joined together (Figure 2.18D). Overall, the net progress of replication on both strands moves toward the replication fork. When an entire chromosome is replicated prior to cell division, the replication begins at many points along the chromosome, and all the different replication forks eventually meet, ultimately yielding two complete copies of the chromosome, each containing one old strand and one newly synthesized strand of DNA, each with the identical nucleotide sequence.

THOUGHT QUESTIONS

1. Before either mitosis or meiosis, the DNA in each chromosome is replicated, forming two identical chromosomes that are attached as a pair at the beginning of mitosis or meiosis. Does one of these contain the old DNA and the other the newly replicated DNA? Or does each contain some of the old DNA and some of the newly replicated DNA? If you need help to answer this question, you may wish to look up the 1958 experiment by Meselson and Stahl.
2. Is the DNA sequence in two sister chromatids identical? Is the DNA sequence in the two chromosomes of a homologous pair identical?
3. If Griffith had mixed together heat-killed nonvirulent R cells and living virulent S cells and injected them into a mouse, would you expect the mouse to live or die? Why? In the language of genetics, which phenotype is dominant: R or S?
4. A given chromosome contains 37% A nucleotides. What are the percentages of the other three nucleotides in this chromosome (T, C, and G)?

CONCLUDING REMARKS

In this chapter, we have seen how scientists, beginning with Mendel, used observation and experimentation to understand the patterns of inheritance of simple either/or traits. The same rules that work for pea plants work for other species, including humans. The units that assort and segregate in inheritance have come to be known as genes. Genes are located on chromosomes, a hypothesis that was first suggested because the numbers, locations, and movements of chromosomes could explain the observed patterns of inheritance. Genes were later found to be composed of DNA, a molecule that consists of long chains of nucleotides. The double-stranded structure of DNA accounts for its ability to be replicated accurately. In the next chapter, we will examine the process of **gene expression**, which is how the information encoded in the sequence of DNA is accessed to make the functional molecules that determine an organism's phenotypes. We will specifically focus on the genetics of human diseases, some of which exhibit the simple patterns of inheritance discovered by Mendel, while others are more complex, involving many different genes and also being influenced by environmental factors.

CHAPTER SUMMARY

- The study of the inheritance of biological traits is called **genetics**.
- Hereditary information is carried in the form of DNA segments known as **genes**, which are parts of **chromosomes**.
- An **allele** is a variant of a gene. **Dominant** alleles show up in the phenotype when either **homozygous** or **heterozygous**. **Recessive** alleles are only expressed when they are homozygous. An organism's hereditary makeup, or **genotype**, is the sum of its alleles.
- Plant and animal cells consist of two major compartments: a central compartment called the **nucleus**, which contains the chromosomes, and a surrounding compartment called the **cytoplasm**.
- Most cells in sexually reproducing species have the **diploid** number of chromosomes, and the chromosomes exist as **homologous pairs**. **Gametes** are an exception because they have the **haploid** number, including only one chromosome of each pair.

- In sexual reproduction, gametes fuse to form a **zygote** with the diploid number of chromosomes.
- Chromosomes are replicated and separated in **mitosis** during cell division, maintaining the diploid number and the full genotype.
- During gamete formation, **meiosis** halves the chromosome number to the haploid value and results in segregation of the alleles into different gametes.
- When two genes are located on different chromosomes, their alleles segregate independently during meiosis, undergoing **independent assortment**. When two genes are on the same chromosome, their alleles show **linkage**, staying together in meiosis unless **crossing over** occurs.
- **Deoxyribonucleic acid** (**DNA**) and **ribonucleic acid** (**RNA**) are two types of nucleic acids.
- Each chromosome contains two very long spiral molecules of DNA, wound around each other and forming a double helix. The two strands are composed of complementary, not identical, nucleic acids. Genes are segments within one of the two strands.
- Prior to mitosis, DNA undergoes **replication** to produce two identical double helices. Replication requires **enzymes**, biological molecules that speed up chemical reactions but are not themselves changed in the reaction. The replication enzymes move along each DNA strand separately, facilitating the addition of one new nucleotide at a time to produce a new strand complementary to one of the old strands.

BIBLIOGRAPHIC REFERENCES

Bibliographic references to material in this chapter can be found online at biologytrending.routledge.com/chapter2

GLOSSARY: KEY TERMS TO KNOW

These key terms are defined at the end of each chapter as an aid for student review.

Allele One of the alternative DNA sequences of a gene.

Anaphase The stage of cell division (mitosis or meiosis) in which sister chromatids or paired homologous chromosomes separate and move to opposite sides of the cell.

Antiparallel Oriented in opposite directions while remaining a fixed distance apart; in double-stranded DNA, one strand begins with a phosphate group where the opposing strand has a sugar.

Asexual reproduction Reproduction by one parent, producing a clone of genetically identical offspring.

Centromere A constriction point within a chromosome that divides the chromosome into two arms and serves as the place where the spindle becomes attached to the chromosomes during cell division.

Chargaff's rules Among the nucleotides in a DNA sample, the quantities containing adenine (A) and thymine (T) are always equal, and the quantity containing guanine (G) and cytosine (C) are always equal.

Chromosomal theory of inheritance The theory, now undisputed, that chromosomes contain an organism's hereditary information in the form of genes.

Chromosomes Long individual DNA molecules that together make up an organism's total genetic material. Prokaryotic cells have a single circular chromosome, while eukaryotic cells have multiple linear chromosomes that are located in the nucleus and contain protein as well as DNA.

Clone The genetically identical progeny derived from a single cell or individual by mitotic cell division or asexual reproduction.

Complementary base pairing Among nucleotides, a type of chemical bonding that always matches C and G with one another and A and T with one another (except that U substitutes for T in RNA).

Crossing over The rearrangement of linked genes when homologous chromosomes break and recombine; one form of genetic recombination.

Cytokinesis Division of the cytoplasm into two distinct cells following mitosis or following either of the divisions of meiosis.

Cytoplasm The portion of the cell outside the nucleus but within the plasma membrane.

Deoxyribonucleic acid (DNA) A nucleic acid containing deoxyribose sugar and usually occurring as two complementary strands arranged in a double helix.

DNA polymerase An enzyme that builds a new strand of DNA complementary to a preexisting strand used as a template.

DNA replication The copying of a DNA molecule to make two double-stranded molecules, each with one old and one new DNA strand.

Diploid Possessing chromosomes and genes in pairs, as in all somatic cells of most multicellular organisms.

Dominant A trait that is expressed in the phenotype of heterozygotes; an allele that expresses its phenotype even when only one copy of the allele is present.

Double helix The shape of a DNA molecule, with two paired, corkscrew-shaped strands running in opposite directions.

Enzyme A chemical substance (usually a protein, but sometimes an RNA) that speeds up a chemical reaction without getting used up in the reaction; a biological catalyst.

Fertilization The combining of an egg and a sperm to form a diploid zygote.

Gamete A reproductive cell (egg or sperm), containing one copy of each chromosome (haploid DNA content).

Gene A portion of DNA that contains the information for synthesis of a single protein chain (polypeptide) or sometimes a functional RNA that is not translated. In earlier use, a hereditary particle.

Gene expression Transcription and translation of a gene into its protein product (or, for genes that encode functional RNAs, transcription and possible processing of the RNA).

Genetics The study of heredity, including genes and how traits are passed from parents to offspring.

Genotype The hereditary makeup of an organism, as revealed by studying its offspring or, for a given gene, the combination of alleles present in an individual.

Haploid Containing only unpaired chromosomes, as in gametes or prokaryotic organisms.

Heterozygous Possessing two different alleles of the same gene in a genotype.

Homologous pair A set of two chromosomes that come together during the first division of meiosis, which contain the same sets of genes in the same order but possibly different alleles of those genes.

Homozygous Possessing two identical alleles of the same gene in a genotype.

Independent assortment (law of) Genes on different chromosomes are passed to offspring independently of one another; the separation of alleles for one trait has no influence on the separation of alleles for traits carried on other chromosomes. Also called "Mendel's second law."

Interphase The long interval between one mitosis and the next, during which cell metabolism is active but chromosomes are not visible under light microscopy.

Linkage An exception to the law of independent assortment in which genes located on the same pair of chromosomes tend to be inherited together, with the parental combinations of alleles predominating.

Linked genes Genes located on the same chromosome that tend to be inherited together.

Meiosis A form of cell division in which the chromosome number is reduced from the diploid to the haploid number; used to generate gametes for sexual reproduction. Compare to *mitosis*.

Metaphase The phase of cell division in which all chromosomes line up in the center of the cell attached to the fibers of the spindle before separating in the next phase (anaphase).

Mitosis The usual form of cell division, in which the number of chromosomes does not change; used for growth, development, tissue renewal, and asexual reproduction. Compare to *meiosis*.

Nonvirulent Not capable of producing an infectious disease.

Nucleic acids DNA and RNA; linear chains of nucleotide subunits that can be single-stranded (typical of RNA) or double-stranded (typical of DNA).

Nucleotide Subunit of a nucleic acid molecule consisting of a phosphate group linked to a five-carbon sugar and then to a nitrogen-containing base.

Nucleus (plural, nuclei) The central compartment of a eukaryotic cell (such as an animal or plant cell) that contains the chromosomes in a double-layered nuclear envelope, separate from the cytoplasm.

Phenotype The visible or biochemical characteristics or traits of an organism.

Prophase The first and longest stage of mitosis or meiosis, in which chromosomes condense and attach to the spindle before being fully lined up in the center of the cell at metaphase.

Punnett square A box diagram used to predict the proportions of offspring with various genotypes based on the genotypes of the parents.

Recessive A trait that is not expressed in heterozygotes; an allele that expresses its phenotype only when no dominant allele of that same gene is present.

Recombination Production of new combinations of alleles for different genes in the offspring, due either to independent assortment or to crossing over.

Replication See *DNA replication.*

Ribonucleic acid (RNA) A nucleic acid, usually existing in single-stranded form, containing nucleotides with ribose sugar.

Segregation (law of) When a heterozygous individual produce gametes, the individual alleles separate so that some gametes receive one allele and some receive the other, but no gamete receives both. Also called "Mendel's first law."

Sexual reproduction Production of offspring by a process that includes genetic recombination and the fusion of two haploid gametes.

Sister chromatids Paired duplicate strands within a chromosome, formed during late interphase and separated from one another during anaphase.

Somatic cell Any cell other than an egg or sperm; a diploid body cell.

Spindle A structure made from protein fibers of the cell's cytoskeleton that directs the movements of chromosomes during nuclear division by either mitosis or meiosis.

Telophase The last phase of mitosis or meiosis, during which the nuclear membrane re-forms and chromosomes begin to return to their interphase state; usually occurs when cytoplasmic division (cytokinesis) is beginning.

Transformation In bacteria or yeast, a hereditary change caused by incorporation of DNA fragments from outside the cell.

Virulent Capable of causing an infectious disease.

Zygote A fertilized egg (diploid), formed by the combining of an egg and sperm (both haploid).

CONNECTIONS TO OTHER CHAPTERS

Chapter 1: Mendel gave genetics its first paradigm; the structure of DNA is the basis for the current paradigm.

Chapter 3: Genes result in phenotypes by serving as blueprints for protein synthesis. The same laws of inheritance that apply to pea plants apply to humans. Many diseases have a genetic component.

Chapter 4: Genetic engineering requires a thorough understanding of genes and their structure.

Chapter 5: Mutations and other genetic changes supply the raw material for evolutionary change. Evolution takes place whenever the frequencies of different alleles change.

Chapter 6: The structure of genes and the processes of gene expression and of transmission across the generations are the same, or very nearly the same, in all known organisms.

Chapter 7: Human populations differ in the frequencies of many alleles.

Chapter 8: DNA contains the information needed to make proteins, which perform many different functions in cells, including as biological catalysts (enzymes).

Chapter 10: Cancer results from mutations that change the function of genes involved in many cellular regulatory processes, and these changes are passed from one cell generation to the next when a cell divides by mitosis. Some cancers have a hereditary component.

Chapter 12: Certain genes can influence the aging process and the risks for Alzheimer's disease.

Chapter 15: HIV is a retrovirus that contains an RNA genome.

Chapter 17: The genetic traits of plants can be used to improve agricultural yield.

Chapter 18: Conserving genetic diversity is an important aspect of protecting biodiversity.

PRACTICE QUESTIONS

1. A pea plant that is homozygous tall (*TT*) produces male gametes of what genotype? What is the genotype of the female gametes it produces?
2. If pollen from a homozygous tall pea plant fertilizes eggs from a homozygous short pea plant (*tt*), what are the expected genotype(s) of the F_1 offspring? What are their phenotype(s)? What are the expected F_1 genotype(s) and phenotype(s) if the pollen is from a homozygous short plant and the eggs are from a homozygous tall plant?
3. If two F_1 plants from the previous question are crossed, what genotypes are expected among the F_2? In what proportions will they be expected to occur?
4. What would be the offspring's genotype(s) and phenotype(s) if pollen from the F_1 in question 2 fertilizes eggs from a homozygous short pea plant?
5. What are the F_1 genotypes if eggs from a homozygous tall pea plant are fertilized by pollen from a heterozygous tall pea plant? What are the F_1 phenotypes?
6. What is the height of the F_1 plants when plants that are homozygous for both tall height and yellow peas are crossed with plants that are homozygous for both short height and green peas? What is their pea color?
7. If the F_1 plants from the previous question are self-fertilized, what are the genotypes, phenotypes, and phenotypic ratios of the F_2 plants?
8. If plants that are heterozygous for tall height and yellow peas are fertilized with pollen from plants homozygous for short height and green peas, what are the genotypes, phenotypes, and phenotypic ratios of the resulting offspring?
9. If *V* represents the dominant allele for violet-red flower color and *v* represents the recessive allele for white flower color, what genotypes of offspring would be produced in each of the following crosses and in what proportions?
   a. *Vv* × *vv*
   b. *Vv* × *Vv*
   c. *vv* × *vv*
   d. *VV* × *VV*
   e. *Vv* × *VV*
10. If *N* is the number of structurally different chromosomes in a mammalian species, how many chromosomes does a stomach cell or skin cell from this species have before it enters mitosis? How many chromosomes does one of the offspring cells have when mitosis is finished?
11. If *N* is the number of structurally different chromosomes in a reptilian species, how many chromosomes does a cell of this species have before it enters meiosis? How many chromosomes does one of the offspring cells have when meiosis is finished?
12. How many of the sister chromatids go to each end of the cell during anaphase of mitosis? How many of the two sister chromatids go to each end of the cell during the first cell division of meiosis?
13. Do homologous pairs of chromosomes line up together on the midline of the cell during metaphase of mitosis? Do homologous pairs of chromosomes line up together on the midline of the cell during the first cell division of meiosis?
14. How many times is DNA replicated in one round of cell division by mitosis? How many times is DNA replicated in a complete cycle of meiosis (two rounds of cell division)?
15. If a diploid cell containing 10 chromosomes divides by mitosis, how many centromeres are present:
    a. in the interphase cell?
    b. at prophase?
    c. at anaphase?
    d. in each daughter cell?
16. In Griffith's experiment, which of the results were most unexpected? Why?

CHAPTER 3

Human Genetics and Gene Expression

ISSUES

- How can we use the study of genetics to fight disease?
- Are there any inherent dangers of genetic research that we should be aware of?
- Is every genetic variant a defect?
- What are my genetic risks? What are my baby's risks? How can I find out?

BIOLOGICAL CONCEPTS

- Patterns of inheritance (trait, phenotype; sex-linked traits; sex determination)
- Molecular genetics (genetic markers and genes)
- Human health and disease (disease predisposition; genetic testing)

CHAPTER OUTLINE

DOI: 10.1201/9781003391159-3

A nervous couple sits in the waiting room, anxiously anticipating the results of a test. Their last child had lived her short life in almost constant pain and had died, blind, at age three, a victim of Tay-Sachs disease. The couple wants another child, but their previous experience was a heart-wrenching nightmare that they don't want to repeat. They are awaiting the results of an amniocentesis, a technique that you will read about later in this chapter. A doctor enters the room with good news: The enzyme that her technicians were testing for is present in the amniotic fluid. The mother-to-be is carrying a child who will not get Tay-Sachs disease. The couple can look forward to raising a healthy child in a happy home.

Scenes like the one just described are happening more often with each passing year. An increasing number of couples are undergoing medical procedures that did not exist when they themselves were born, seeking assurances that would have been unthinkable a mere 25 years ago. Tay-Sachs disease is one of a growing number of inherited conditions that can now be diagnosed before birth. This type of information can allow parents to make informed decisions about whether to consider terminating a pregnancy if the fetus has a severe condition that will not allow for a good quality of life or to prepare in advance for the care of a child with special needs.

Some traits follow the simple Mendelian patterns of inheritance that we studied in the previous chapter. Many more traits, however, follow much more complex patterns because they are governed by multiple genes, each contributing a small amount to the trait. How do scientists find which genes are associated with which trait? Molecular biology has greatly changed how scientists go about the search. It has also changed what we mean by the concepts of "trait" or "phenotype" and "mutation" and has led to many new uses for genetic information.

3.1 GENETIC INFORMATION INFLUENCES PHENOTYPES THROUGH THE PROCESS OF GENE EXPRESSION

3.1.1 Overview of gene expression

We learned in the previous chapter that DNA is the molecule that stores hereditary information. We also learned that an organism's full complement of DNA (its **genome**) is split between multiple chromosomes that undergo a carefully orchestrated process of segregation during cell division to ensure that the resulting daughter cells each receive a complete copy of the genome—one of each chromosome type in the case of haploid cells generated by meiosis, or two of each chromosome type in the case of diploid cells generated by mitosis. Here we will discuss further the nature of chromosomes and the process of gene expression, which is how the stored genetic information is accessed and how it influences specific functions of cells and phenotypic traits of organisms, usually by making proteins. The basic fact that genes make proteins was initially the clever insight of a British physician, Dr. Archibald Garrod, who was investigating human metabolic disorders. Garrod discovered that a number of these disorders were linked to defects in particular enzymes (proteins that catalyze chemical reactions), which in turn were caused by changes in the DNA sequences (i.e., mutations) of specific genes. We will explore this link later in this chapter, after we have described the steps in gene expression.

Each chromosome in a genome contains two very long strands of base-paired DNA. (In humans the length of DNA in each cell is about 1 meter or 3 feet.) In most organisms, chromosomal DNA is surrounded by DNA-binding proteins, which help to compress the long DNA strand so that it can fit into the cell's nucleus (which is less than one-millionth of a meter across) and also influence gene expression. A gene is a segment of the DNA strand (a subset of bases within the linear sequence) that results in a product, which is usually a protein but sometimes an RNA end-product. Each person or pea plant (or any diploid individual) has two chromosomes of each type, and thus has two genes for each hereditary trait, one on each chromosome of a homologous pair. Within a species, there may be several possible alleles for each gene, but each individual can only have two, one on each chromosome of a homologous pair.

The process of gene expression is summarized in what is often called the "Central Dogma of Molecular Biology," which can be stated as "DNA makes RNA, and RNA makes protein" (**Figure 3.1A**). The nucleic acids DNA and RNA can be compared to blueprints that contain instructions for building proteins. Genes (made of DNA) are the "master" copies of the blueprints that are stored in the "library" that is the nucleus. In the first stage of gene expression, **transcription**, the information contained in the DNA is copied into RNA, which can be thought of as a photocopy of the information that can be taken out of the "library" for use in the cytoplasm, where proteins are made. Most genes undergo a second stage of gene expression, called **translation**, in which the instructions contained in the RNA "photocopy" are used to make proteins. Proteins perform the vast majority of functions in cells, and the way they perform these functions results in observable phenotypes. Often, several proteins (and therefore several genes) interact to produce a phenotype. Although the end-products of gene expression are usually proteins, in some cases they can be RNA molecules instead. Some RNAs function as enzymes, catalyzing chemical reactions, some are critical components of the protein synthesis machinery, and others block the expression of other genes. Thus, gene expression always involves transcription of DNA into RNA and, for the majority of genes that encode proteins, it also involves translation of RNA into protein.

RNA (**Figure 3.2**) differs from DNA in several ways: (1) The RNA backbone contains the sugar ribose rather than deoxyribose, hence the name ribonucleic acid; (2) it is mostly single-stranded, although complementary base pairs can form within a single RNA strand; (3) RNA molecules are usually much shorter than DNA; and (4) the nitrogen-containing base uracil (U) replaces thymine (T), although the other three nitrogen bases, adenine, guanine, and cytosine, are the same. In RNA, C pairs with G and A pairs with U.

(A) The central dogma of molecular biology, as it was understood in the 1960s: information flows from DNA to RNA and then to protein.

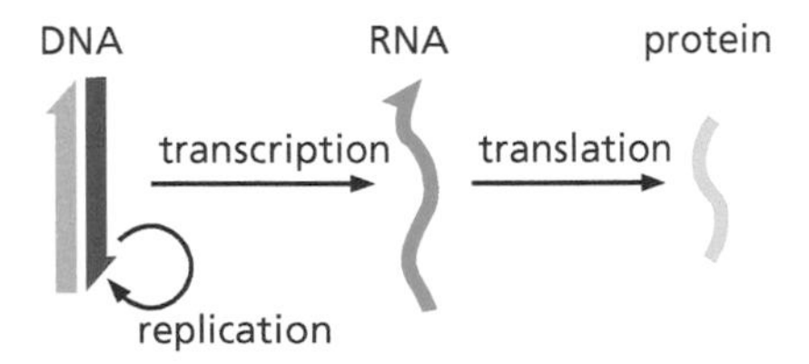

(B) Many exceptions are now known. For example, some viruses, including the one that causes AIDS, have a "reverse transcription" process in which RNA is used to make DNA. Also, many proteins influence the timing and amount of gene transcription and RNA translation.

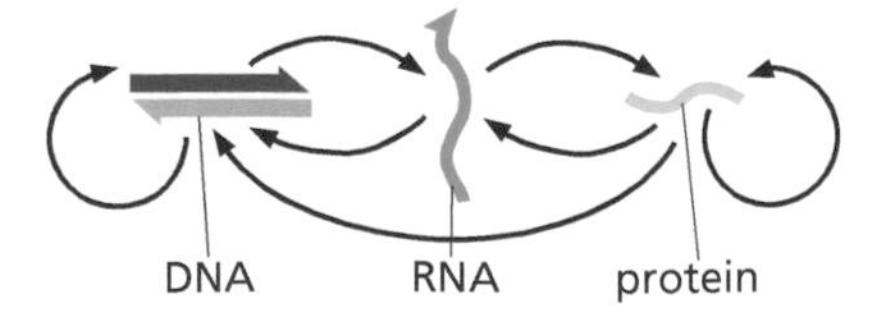

Figure 3.1 Changing concepts of the flow of genetic information.

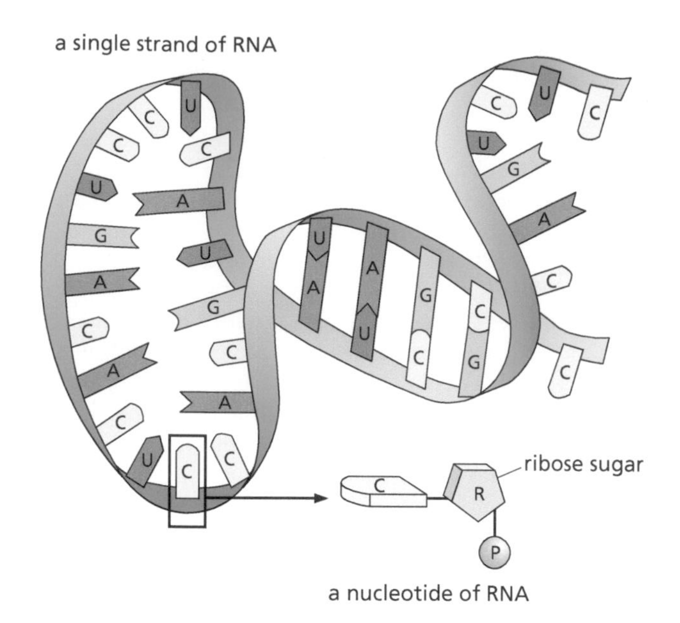

Figure 3.2 The structure of RNA. The molecule as a whole is usually single-stranded, but short portions of some RNA molecules can base-pair with other portions of the same molecule.

Before we consider the details of the two stages of gene expression, it is important to note that the Central Dogma depicted in Figure 3.1A summarizes the major processes, but it does not fully depict all forms of information flow in biological systems. Specifically, information does not flow only in one direction. Proteins can affect the transcription of DNA and the translation of RNA, and, in some viruses (like HIV, which causes AIDS—see Chapter 15), RNA can be a template for DNA synthesis. In addition, as we have already mentioned, RNAs are the functional end-products of some genes, and finally, some functional proteins consist of the products of more than one gene. Our current concepts of information flow are more accurately represented as a network (Figure 3.1B).

3.1.2 Transcription from DNA to RNA

During transcription, a segment of DNA is used as a template to make a single-stranded RNA (Figure 3.3). There are several differences between transcription and DNA replication (Figure 2.18). In DNA replication, the whole length of both DNA strands in a chromosome are copied to make two new and complete strands of DNA. In transcription, however, a small, discrete part of a single DNA strand is the template for the synthesis of RNA. The portions of a DNA strand that contain the necessary information to make different proteins (or RNA end-products) are the genes. A particular gene is transcribed from only one of the DNA strands, but, at a different place on the same chromosome, a different gene may be transcribed from the other DNA strand. Transcription begins when an enzyme (**RNA polymerase**) attaches to a short DNA sequence marking the beginning of a gene (Figure 3.3). (In Figure 3.3, RNA polymerase is represented as a single protein, but in eukaryotic cells like ours, it actually consists of a large number of proteins.) RNA polymerase causes the DNA of the gene to unwind, allowing RNA nucleotides to pair with the DNA nucleotides of the gene. Nothing happens on the other DNA strand. The RNA nucleotides become bonded together to form the backbone of the RNA strand. Eventually, RNA polymerase reaches a DNA sequence that signals the end of the gene; at this point, the RNA strand comes off of its DNA template, allowing the DNA to twist back into its double helical shape.

For most genes, the product of transcription is **messenger RNA (mRNA)**, which carries the information for protein synthesis. Two other forms of RNA function in protein synthesis: **transfer RNA (tRNA)** and **ribosomal RNA (rRNA)**; these are also transcribed from DNA but are not themselves translated into proteins. Many proteins are known that can either inhibit or enhance transcription of certain genes, providing a means by which the amount of RNA being made can be controlled (Figure 10.4). In addition, some RNAs, such as microRNAs, can also regulate gene expression, either by stimulating degradation of certain mRNAs or regulating the rate of translation.

STEP 1

RNA polymerase enzyme binds to DNA double helix at the beginning of a gene.

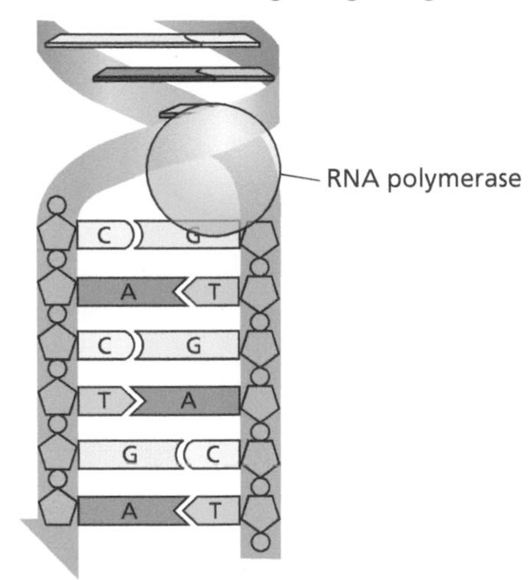

STEP 2

RNA polymerase (along with helper proteins in eukaryotes) sepatates the two DNA strands at the start of the gene. Free RNA nucleotides form base pairs with the DNA nucleotides on the DNA strand that serves as the template for RNA synthesis.

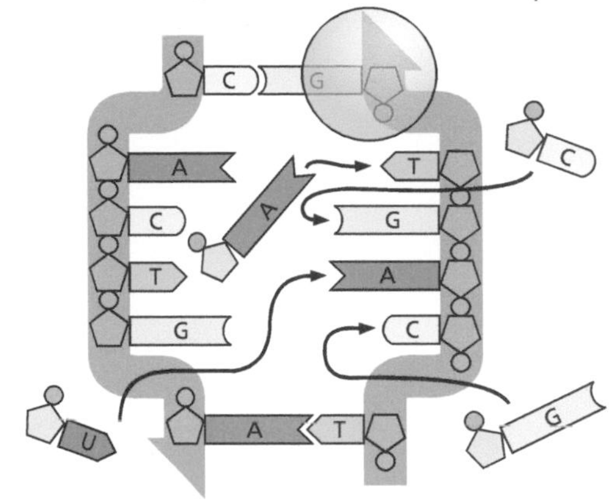

STEP 3

RNA polymerase joins the RNA nucleotides into a continuous strand, moving along the gene as transcription proceeds.

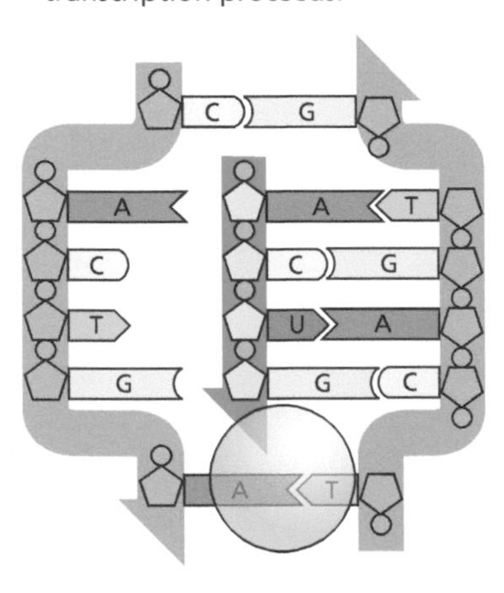

STEP 4

Once the RNA polymerase reaches the end of the gene, the enzyme and RNA strand detach from the DNA and the two DNA strands pair once more.

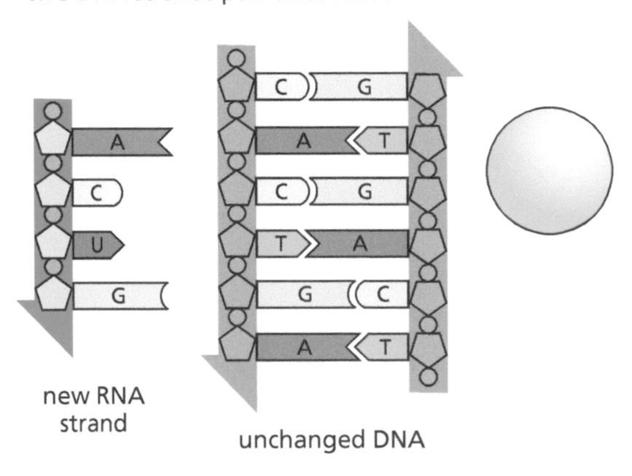

Figure 3.3 The transcription of DNA into RNA.

3.1.3 Translation from RNA to protein

Transcription of an mRNA is followed by translation, in which protein "factories" called **ribosomes** use the instructions contained in the mRNA to make a protein (Figure 3.4). Just as DNA is made up of a sequence of nucleotides, proteins are made up of a sequence of building blocks called **amino acids.** There are 20 different amino acids that are linked together in linear, unbranched chains to make proteins. During translation, the nitrogenous bases in the mRNA are interpreted in groups of three, called **codons.** As each codon is recognized, the amino acid it specifies is added to the growing protein chain. After the chains of amino acids are synthesized by translation, they fold into complex shapes that determine their function. How they fold depends on the sequence of amino acids. (For more on protein structure, see Figures 8.7 and 8.8).

In addition to the mRNA that contains the three-nucleotide codons to make a protein, two other types of RNA play crucial roles in protein synthesis: Transfer RNAs (tRNAs) serve as adapters to match each mRNA codon with its appropriate amino acid, while ribosomal RNAs (rRNAs) are structural components of the

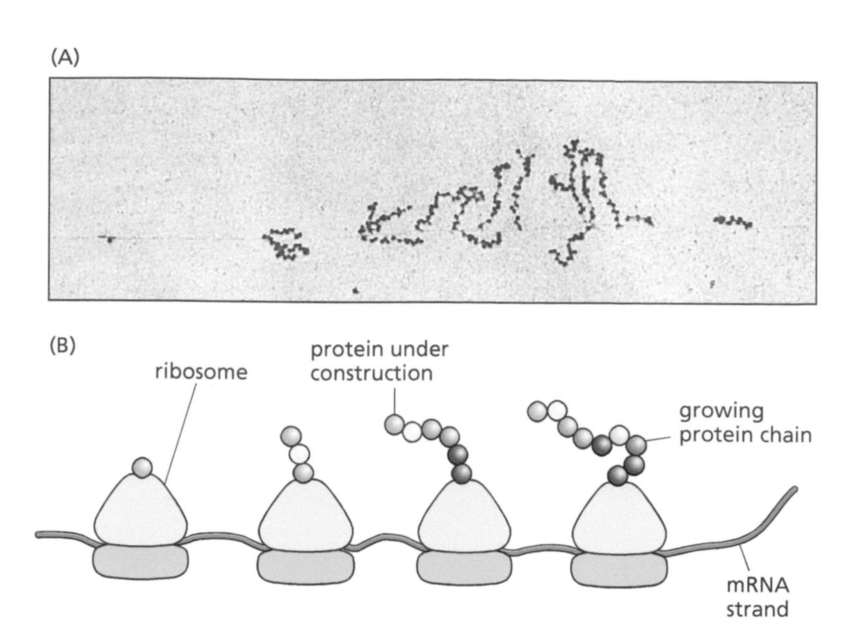

Figure 3.4 Translation of an mRNA into protein. The two subunits of a ribosome (shown in light and dark green) bind to the mRNA and move along it, interpreting the genetic code to link amino acids together into a protein chain. The protein chain elongates as the ribosome moves along, adding one amino acid at a time. **(A)** In the electron micrograph, notice that the mRNA strand appears as a very thin straight line. Both here and in the schematic diagram, **(B)** notice that multiple ribosomes can be engaged in different stages of translation of the same mRNA at one time, leading to synthesis of multiple protein molecules.

ribosomes that catalyze the chemical reactions that join amino acids together. Multiple segments of nucleotides within tRNA molecules form base-pair interactions to cause the tRNA to fold into a shape that resembles a three-leaved clover (Figure 3.5A). The middle "leaf" of this clover structure contains a three-nucleotide **anticodon** that forms complementary base pairs with mRNA codons that specify one amino acid; a specific enzyme recognizes the anticodon of each tRNA and joins the appropriate amino acid to the "stem" of the tRNA. The tRNA and the

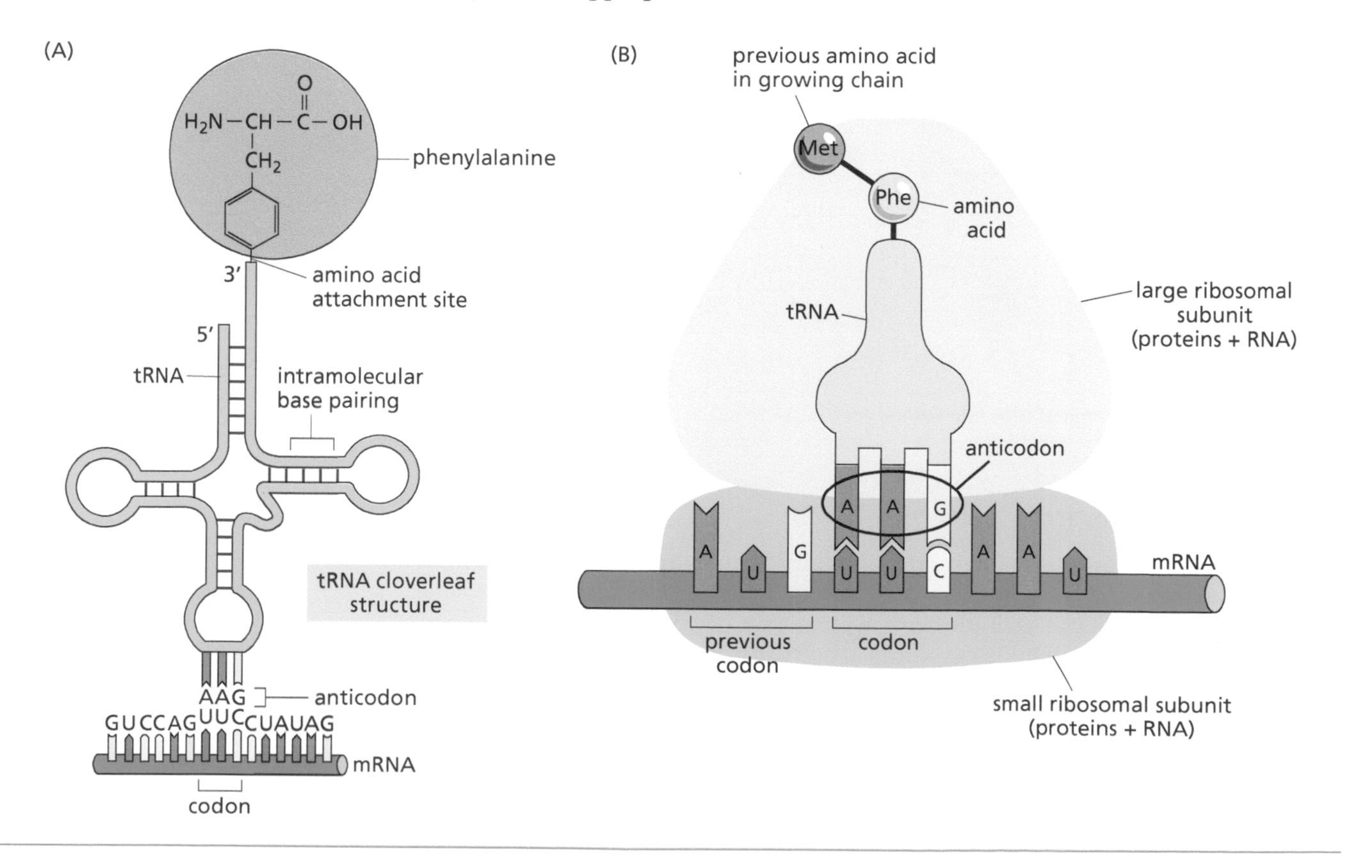

Figure 3.5 The structure of tRNA and its role in translation. (A) tRNA molecules make internal base pairs to form a cloverleaf structure with the three-letter anticodon at the tip of the middle "leaf" and a "stem" that becomes attached to the amino acid corresponding to the mRNA codon that the anticodon recognizes. In the example shown here, the mRNA codon UUC matches the tRNA anticodon AAG, and this tRNA is attached to the amino acid phenylalanine (abbreviated Phe) that is specified by the UUC codon. **(B)** The ribosome (containing ribosomal proteins and rRNA) holds the tRNA in place on the mRNA. Here, the tRNA cloverleaf is shown in a more schematic structure for simplicity.

Table 3.1 The genetic code (coding dictionary)

First nucleotide (letter)	Second nucleotide (letter)								Third nucleotide (letter)
	U		C		A		G		
U	UUU	Phenylalanine	UCU	Serine	UAU	Tyrosine	UGU	Cysteine	U
	UUC		UCC		UAC		UGC		C
	UUA	Leucine	UCA		UAA	STOP	UGA	STOP	A
	UUG		UCG		UAG	STOP	UGG	Tryptophan	G
C	CUU	Leucine	CCU	Proline	CAU	Histidine	CGU	Arginine	U
	CUC		CCC		CAC		CGC		C
	CUA		CCA		CAA	Glutamine	CGA		A
	CUG		CCG		CAG		CGG		G
A	AUU	Isoleucine	ACU	Threonine	AAU	Asparagine	AGU	Serine	U
	AUC		ACC		AAC		AGC		C
	AUA		ACA		AAA	Lysine	AGA	Arginine	A
	AUG	Methionine (START)	ACG		AAG		AGG		G
G	GUU	Valine	GCU	Alanine	GAU	Aspartate	GGU	Glycine	U
	GUC		GCC		GAC		GGC		C
	GUA		GCA		GAA	Glutamate	GGA		A
	GUG		GCG		GAG		GGG		G

mRNA codon that it "reads" are brought together on the surface of a ribosome. The ribosome is composed of two subunits (one small subunit shown in darker green in Figure 3.5B and one large subunit shown in lighter green). Each subunit contains at least one rRNA molecule and multiple proteins.

The correspondence between mRNA codons and amino acids is called the **genetic code**. Experiments performed in the early 1960s used cell extracts containing ribosomes and tRNAs loaded with amino acids to translate simple, repetitive RNAs of known sequences (e.g., poly-U, consisting only of the U nucleotide). Based on the proteins that were made in these extracts, scientists were able to match some codons with the amino acids they specified. For example, in the case of poly-U, the resulting protein contained only the amino acid phenylalanine; therefore, the scientists concluded that the mRNA codon UUU specified phenylalanine. Additional experiments like these and others that directly matched three-letter mRNA codons with the tRNAs that bound to them eventually led to the determination of the complete genetic code (Table 3.1). This "coding dictionary" is a nearly universal code that is the same for all organisms in which it has been studied (except for rare exceptions for single codons in a small number of cases).

Of the 64 possible three-letter codons formed from the four RNA nucleotides ($64 = 4^3$), 61 specify one of the 20 amino acids used in proteins. Most amino acids are specified by multiple codons (sometimes as many as six); for example, we already saw that the codon UUU specifies the amino acid phenylalanine, but so does UUC (Figure 3.5B). The amino acid methionine is coded for by just one codon, AUG. This codon also has a special function in signaling the start of the protein coding sequence within an mRNA and is therefore called the "start codon." Because the start codon specifies the amino acid methionine, all proteins are originally made with methionine as the first amino acid in the chain, although this is sometimes later removed in a processing step. The start codon can be thought of as the capital letter that signifies the beginning of a new sentence. Three codons that do not specify amino acids are called "stop codons" because they signal the end of a protein-coding sequence within an mRNA, similar to a period at the end of a sentence.

The synthesis of a protein begins when the anticodon of a tRNA loaded with methionine binds to the AUG start codon, held in place by the two ribosomal subunits (Figure 3.6, step 1). The assembled ribosome is positioned over three mRNA

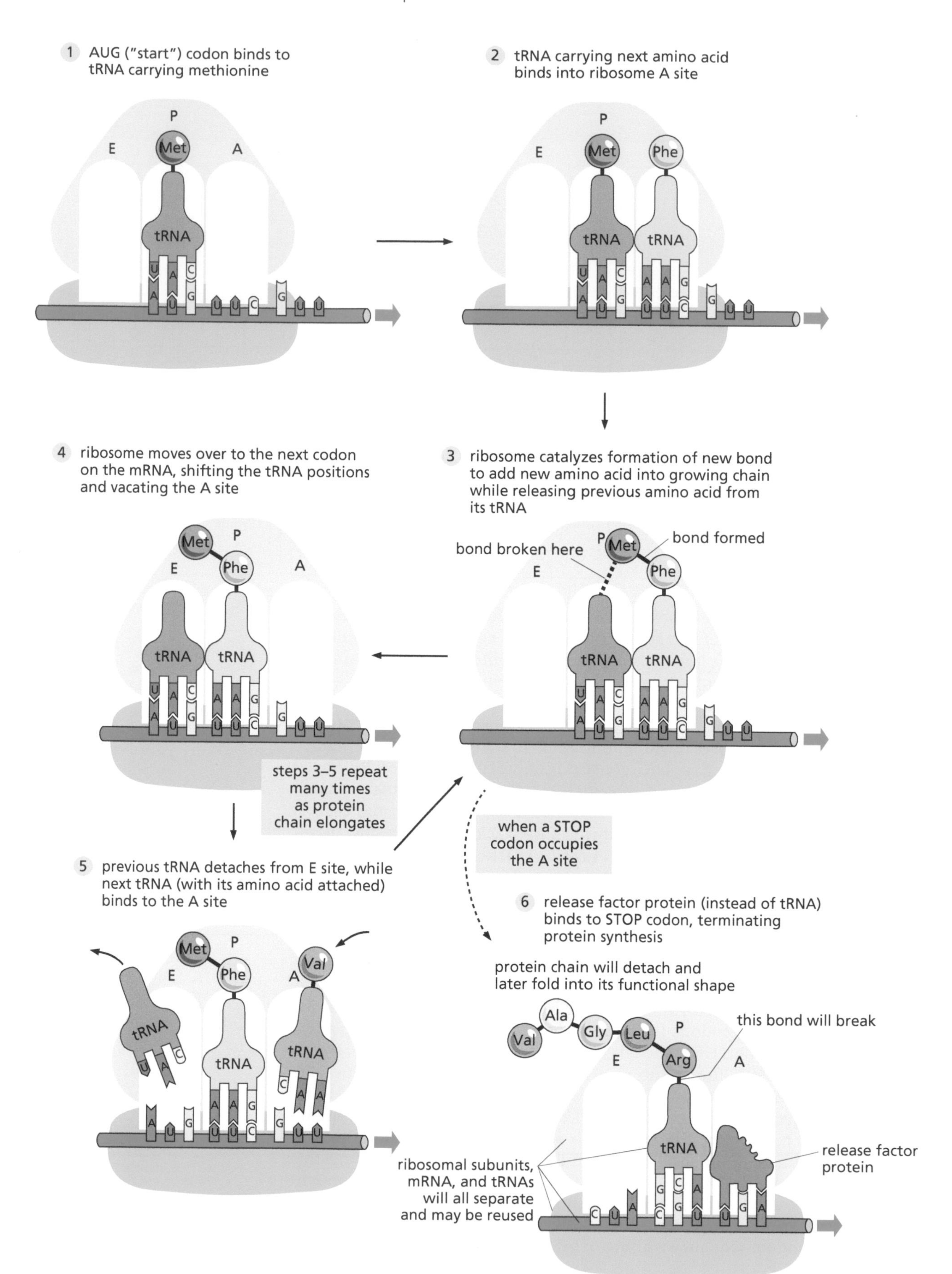

Figure 3.6 **Steps in translation.**

codons and has three positions that can hold tRNAs (labeled E, P, and A in Figure 3.6). When protein synthesis begins, the tRNA bearing the methionine amino acid is located in the middle (P, for "peptide" or "protein") position (Figure 3.6, step 1). In the next step, a tRNA with an anticodon matching the next mRNA codon binds in the A (A for "acceptor") site next to the methionine-bearing tRNA (Figure 3.6, step 2). The rRNAs of the ribosome catalyze a reaction that forms a bond between the methionine and the second amino acid of the protein, breaking the bond between the methionine and the tRNA in the P site and generating a two–amino acid chain attached to the tRNA in the A site. In order for protein synthesis to continue, the ribosome must then move three nucleotides along the mRNA (to the right in Figure 3.6, step 4) to access the next codon. The tRNAs do not move with the ribosome, so this shift in ribosome position (translocation) causes the tRNAs to shift to new binding sites (to the left in Figure 3.6). The tRNA bound to the growing amino acid chain is now in the P site, and the process of protein synthesis can continue, with the tRNA that recognizes the next codon now in the A site, bringing its amino acid to be added next in the chain (Figure 3.6, step 4). The previous tRNA now occupies the E site (E for "exit"; Figure 3.6, step 5), where it briefly pauses and then detaches from the ribosome. It can then be reloaded with another corresponding amino acid to participate again in translation. (The matching of tRNA anticodons with mRNA codons is driven by diffusion of tRNA molecules within the cell, but it happens very rapidly, with several amino acids added per second to a growing protein chain.) The process of protein elongation (steps 3–5) repeats until the A site of the ribosome is occupied by one of the three stop codons; when this occurs, a protein called a "release factor" (instead of a tRNA) binds to the stop codon and causes the ribosome to release both the mRNA and the completed protein chain (Figure 3.6, step 6). The protein can then fold into its functional, three-dimensional shape, and the mRNA and ribosomal subunits can go on to participate in additional rounds of translation.

3.1.4 Mutations

Changes in the nucleotide sequence of DNA are called **mutations**. These changes may occur for a number of different reasons. Some of them result from rare errors made by DNA polymerase during DNA replication (Section 2.3.3), while others result from DNA damage caused by chemicals or by radiation (Section 10.6.4). Most mistakes and damage are immediately fixed by various self-correction (repair) mechanisms and are therefore not passed on when a cell divides. A small number of mistakes are not repaired, however, and persist in the cells in which they occur, as well as in cells produced when the mutated cells divide. Mutations that persist in somatic (body) cells can cause problems (such as cancer) in the individual carrying the affected genetic material (Chapter 10), but these changes will not be passed on to the next generation. In contrast, mutations in cells that give rise to gametes may be passed on to future generations. Ultimately, mutations are what generate the different alleles of a gene.

The simplest kind of mutation is a single-base mutation (called a "point mutation"), such as the substitution of one nucleotide (A, G, C, T) for another. Substitutions of this kind that occur in a protein-coding region of DNA will change an mRNA codon; therefore, they may result in the wrong amino acid being inserted into a protein sequence (**Figure 3.7A**), which is termed a "missense" mutation. Alternatively, because multiple codons specify the same amino acid, a single nucleotide change (especially in the third position of a codon) may not alter the protein sequence; such a mutation is referred to as "silent." A more extreme case occurs when a single nucleotide change transforms a codon that encodes an amino acid to one of the stop codons; such a mutation is called a "nonsense" mutation and will result in the premature termination of translation, generating a shortened protein that is usually nonfunctional.

Larger changes occur when one or two extra nucleotides are inserted into a DNA sequence or when one or two are deleted from the sequence. Mutations that change the number of nucleotides can change all of the codons that follow the point of insertion or deletion. This is because the DNA or RNA code contains no

Figure 3.7 Examples of two types of mutations. The DNA sequences shown are the template strands from which the mRNA codons are copied.

(A) Single nucleotide substitution. When, for example, a C is substituted for a G in the DNA strand, the mRNA codon matches with the tRNA carrying the amino acid glycine rather than the tRNA carrying the amino acid alanine.

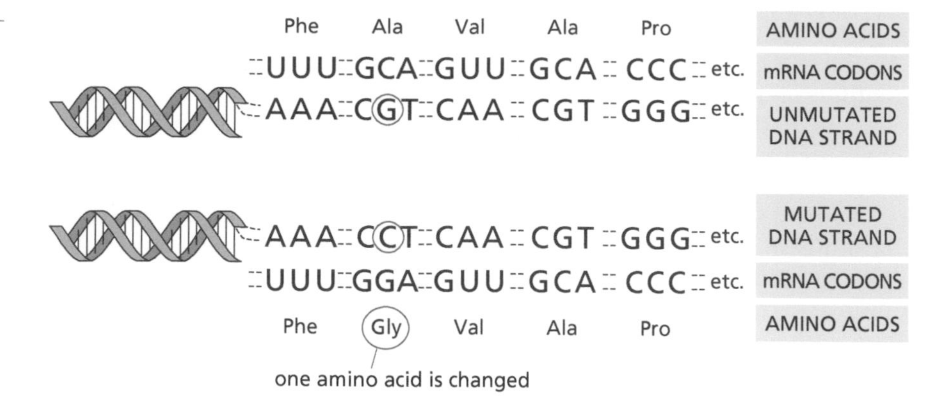

(B) Change in the number of nucleotides. When, for example, a G is deleted from the DNA strand, the codon that had the G and all subsequent codons are misread and different amino acids are placed in the chain. (The mRNA codons are not shown for the sake of simplicity.)

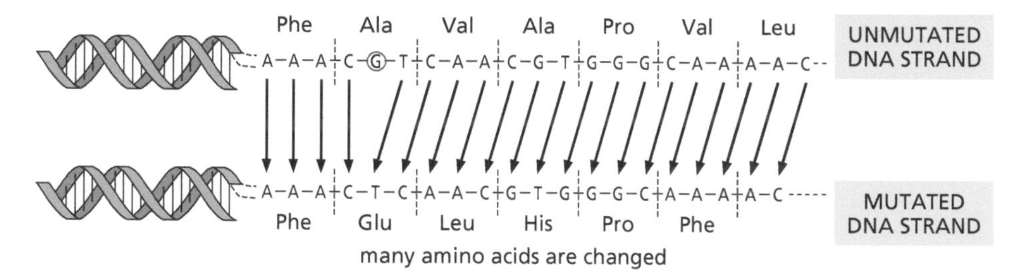

"commas" or other "punctuation" to signify where a new codon starts. Each new codon is simply the next three bases after the previous codon, so the first codon determines the starting point for the second, and so on. If an extra base is inserted or a base is deleted, that codon and all codons that follow the mutation are changed or shifted over. This type of mutation is often called a "frameshift mutation," and we say that the "reading frame" has shifted. In the example shown in **Figure 3.7B**, a deletion of a guanine nucleotide (G) causes a change in mRNA codons from that point on, leading to the wrong amino acids being added to a growing protein during translation. CGT in DNA is transcribed to the complementary GCU in mRNA, the codon for the amino acid alanine (Ala). By removing G, the DNA sequence becomes CTC, which is transcribed to the complementary mRNA codon GAG. GAG is the codon for the amino acid glutamine (Glu), not Ala. Note that the "C" that would have been part of the next codon is now part of the "Glu" codon. All succeeding codons are likewise shifted, leading to a protein with different codons and therefore a different amino acid sequence from that coded by the unmutated DNA strand. Because so many amino acids are affected, most mutations of this kind result in nonfunctional proteins. Note that an insertion or deletion of exactly three nucleotides (or any multiple of 3) does not cause a frameshift, but rather may cause deletion or insertion of one or more amino acids without affecting the rest of the amino acid sequence following the site of the mutation. This may or may not affect the function of the protein, depending on in which part of the protein the amino acids are inserted or deleted and also on the identities of the specific amino acids affected.

As mentioned earlier, protein function depends on the shape of the folded protein molecule. A substituted amino acid may alter the protein shape and therefore change or impair the protein's function. The more amino acids that are changed, the more likely it is that this change alters protein function. Phenotypic consequences of inserting a changed amino acid span the spectrum from those that are undetectable to those that are fatal, but changes that affect many amino acids at once are more often harmful or fatal.

THOUGHT QUESTIONS

1. DNA has been called the master molecule because it controls (or determines) RNA sequences, which in turn control protein sequences, which control most of the cell's other activities. Is the term "master molecule" an accurate description? Does the language of control (e.g., DNA controls the type of RNA produced) say more about the molecules or about the scientists? Does the use of a word such as "master" suggest a hierarchical approach in which information flows in one direction only?
2. DNA is a double helix, two complementary strands that wind around each other. A particular gene is on one of the two strands. A different gene at a different location may be on the same strand or on the other strand. In transcription, the strand with the gene acts as a template for mRNA synthesis. What happens on the other strand during transcription? What happens on the other strand during DNA replication?
3. Would it be worse for a mutation to occur in a body cell like a skin cell or in a gamete (egg or sperm)? Explain. Does your answer potentially depend on whether you are considering the possible outcome for the person whose cell sustains the mutation or that of their potential children?

3.2 GENES CARRIED ON SEX CHROMOSOMES DETERMINE SEX AND SEX-LINKED TRAITS

Not long after Sutton proposed that chromosomes carry genes (Section 2.2), chromosomes and their abnormalities were studied in both humans and insects, and much has been learned since from the study of chromosomes. In particular, some unusual conditions are associated with alterations of the chromosomes, including variations in chromosome number and large-scale chromosome rearrangements.

During mitosis and meiosis, the double-helical DNA molecule becomes further coiled around supporting proteins and then coiled around itself until the structure becomes thick enough to be visible under a light microscope. (It is to this thickened structure that the name "chromosome" was first applied.) When chromosomes are in this condensed state, they can be easily seen and photographed using a light microscope. Individual chromosomes can be identified based on their lengths and banding patterns. Geneticists can line up the photos of all the chromosomes, putting the homologous chromosome pairs together. Such an arrangement is called a **karyotype**.

The karyotypes of a female human and a male human are shown in **Figure 3.8**. The chromosomes of nearly every person can be arranged in a karyotype similar

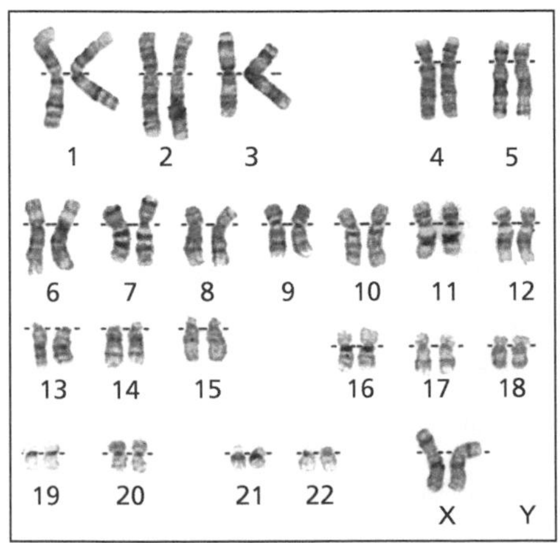

the XX karyotype (female)

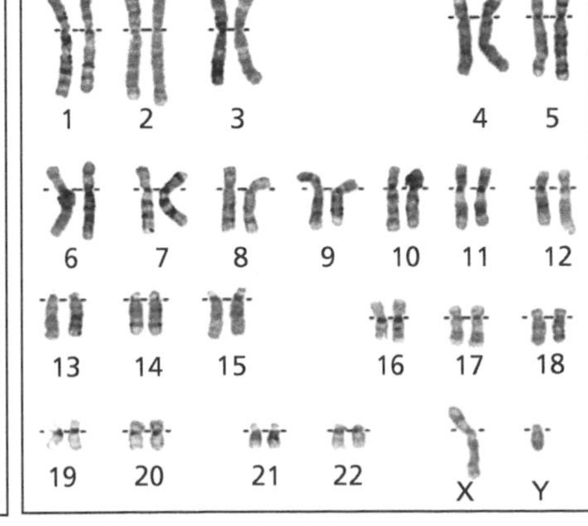

the XY karyotype (male)

Figure 3.8 **Human female and male karyotypes.**

to one of these, with 46 chromosomes arranged in 23 pairs. Of the 23 pairs of chromosomes, 22 pairs are the same in both sexes; these are called **autosomes**. The two chromosomes in each of these pairs are homologous, meaning that they carry the same set of genes (although possibly two different alleles of any given gene). The 23rd pair of chromosomes (labeled X and Y in Figure 3.8) are called the **sex chromosomes** because they differ in males and females and have a role in determining a person's sex. The X and Y chromosomes are only partly homologous: They do pair during meiosis, but they share only small regions of DNA sequence similarity.

3.2.1 Sex determination

Human females typically have two similar sex chromosomes, symbolized as XX. Human males typically have one X chromosome and one Y chromosome and thus are symbolized as XY (see Figure 3.8).

Not all human females are XX and not all human males are XY. There are unusual situations in which a cross-over during meiosis between an X and a Y chromosome is followed by an exchange of chromosome pieces. Approximately 1 in 20,000 normal males is chromosomally XX, but one of his X chromosomes contains a small piece of the Y chromosome. About the same frequency of normal females are chromosomally XY but are *missing* the same small piece of Y chromosome. One such XY female had 99.8% of the Y chromosomal DNA, indicating that a male-determining factor was located in the 0.2% portion of the Y chromosome that she did not have. Examination of this 0.2% portion of the Y chromosomal DNA led to the identification of a gene, now called *SRY*, that was hypothesized to induce development as a male.

The *SRY* gene helps determine sex by producing a protein, the SRY protein, which is also known as testis-determining factor (TDF). TDF is a protein that binds to DNA to cause the transcription of other genes, including a gene called *SOX9*; the Sox9 protein then turns on expression of multiple genes that lead to the development of the male sex organs, the testes (**Figure 3.9**). The developing testes produce hormones, including testosterone, that further trigger male development. (Hormones are small molecules that are used for cellular communication—see Chapters 9 and 10.) Embryos that do not express the SRY protein instead express factors that stimulate transcription of a different set of genes that stimulate development of female sex organs and female hormone profiles. Embryos with an *SRY* gene thus become males, while embryos without an *SRY* gene become females. However, there are also cases of individuals without *SRY* who are phenotypically male, as well as individuals with *SRY* who are phenotypically female, demonstrating that other genes in addition to *SRY* are involved in sex determination.

3.2.2 Sex-linked traits

The Y chromosome contains very few genes. In contrast, many more genes are located on the X chromosome. Genes that are located on either sex chromosome exhibit a **sex-linked** pattern of inheritance. Only males will exhibit phenotypes associated with the small number of genes on the Y chromosome, since females

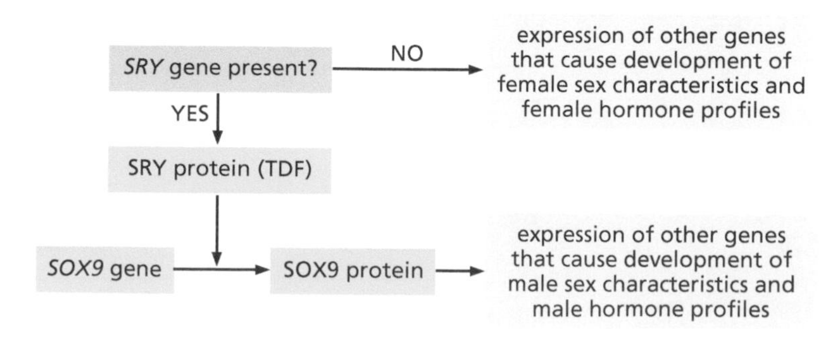

Figure 3.9 The relationship (simplified) between the *SRY* gene and sexual development in humans. The SRY protein is also known as testis-determining factor (TDF).

do not have this chromosome. Females have two X chromosomes, so they can be either homozygous or heterozygous for X-linked traits. If the allele for an X-linked trait is recessive, then a heterozygous woman has the dominant phenotype but is said to be a **carrier** of the trait. Because a male has only a single X chromosome, he has only a single allele for each X-linked trait, and this one allele therefore determines his phenotype for that trait. The inheritance of such a recessive, X-linked trait—red-green color blindness—is shown in Figure 3.10. If a woman is heterozygous for this trait, she shows the dominant phenotype and is not color blind. However, a male who carries the recessive allele for color blindness on his X chromosome will be color blind.

As already discussed, female mammals generally possess two X chromosomes; however, in a given cell, only one of them has active genes that make a product or express a phenotype. This is because, at an early stage in development, one X chromosome becomes inactivated in each cell in a female embryo. In some cells the X chromosome derived from the mother will be inactivated, while in others, the X chromosome derived from the father will be inactivated. Females thus express one phenotype (from their mother's X chromosome) in some cells and another phenotype (from their father's X chromosome) in other cells. This expression of two phenotypes at the cellular level is called **mosaicism**. All females are mosaics for those X-linked genes for which they are heterozygous. For example, in a female heterozygous for an X-linked allele for color blindness, patches of cells within the retina of the eye (Section 11.2.1) cannot respond to color. Other patches of cells, which express the normal allele on the other X chromosome, respond normally, however, and the overall result is normal vision.

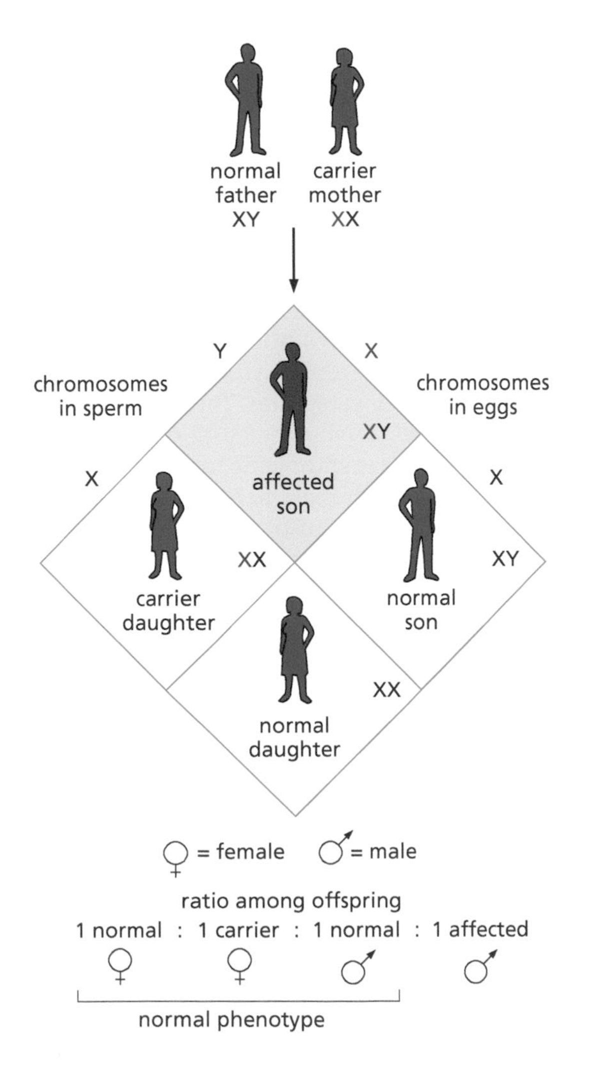

Figure 3.10 Inheritance of red-green color blindness, a sex-linked recessive trait.

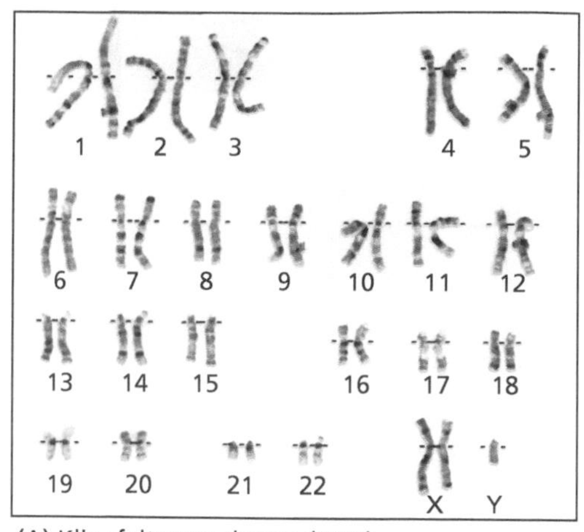

(A) Klinefelter syndrome (XXY)

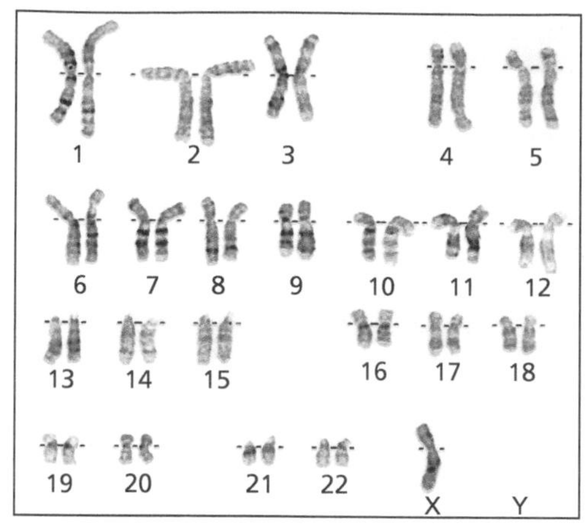

(B) Turner syndrome (XO)

Figure 3.11 Two variations in human X and Y karyotypes.

3.2.3 Chromosomal variation

As previously described, most humans have 46 chromosomes, consisting of 1 pair of sex chromosomes and 22 other pairs (autosomes). Other chromosomal patterns have multiple consequences, called syndromes, usually named after the physicians who identified them. Several such variations of the sex chromosome number are known. For example, the XXY chromosomal type (Figure 3.11A) results in **Klinefelter syndrome**; persons with this condition have male phenotypes but are sterile. Some of the symptoms of Klinefelter syndrome can be successfully treated with hormones. In contrast, **Turner syndrome** results from the XO chromosomal type, in which only one X chromosome is present, with the O representing its missing partner (Figure 3.11B). Persons with Turner syndrome develop as females; however, their ovaries do not produce female hormones. Puberty does not take place and gametes do not develop, resulting in infertility. Fertility treatments can sometimes allow women with Turner syndrome to conceive children, but there are some risks for individuals who have heart conditions that may also be associated with the syndrome. Hormone treatments can alleviate many of the other symptoms of Turner syndrome with much success. Clearly male and female are not either/or categories. There are persons who fall between, either because of chromosomal variation or variation in alleles at particular genes, such as *SRY* or the testosterone-receptor gene mentioned later in this chapter.

Turner and Klinefelter syndromes are believed to result from the same cause, an abnormal meiosis of the sex chromosomes. In this abnormal meiotic division, the two sex chromosomes fail to separate, resulting in some egg cells having two of the mother's X chromosomes and some having none. Abnormal separation of chromosomes during gamete production has actually been observed, partly confirming the hypothesized series of events shown in Figure 3.12. Also supporting the hypothesis is the very rare XXX chromosomal abnormality; most XXX females are sterile and developmentally disabled, with very low intelligence. The Y-only type of embryo, also predicted by the hypothesis, has never been observed, presumably because it dies at a very early stage of development. Recall that the X chromosome contains many more genes than the Y chromosome; because these are needed for both males and females to develop, embryos lacking an X chromosome are nonviable. Abnormal chromosome separation can also take place during the formation of sperm cells, resulting in XY sperm and O sperm. When these fertilize a normal X egg, again either Klinefelter syndrome (XXY) or Turner syndrome (XO) can result.

There are also variations in number for chromosomes other than the sex chromosomes. The most common of these causes **Down syndrome**, marked by facial characteristics, heart abnormalities, and a variable impact on intellectual development. Down syndrome usually results from an extra chromosome 21

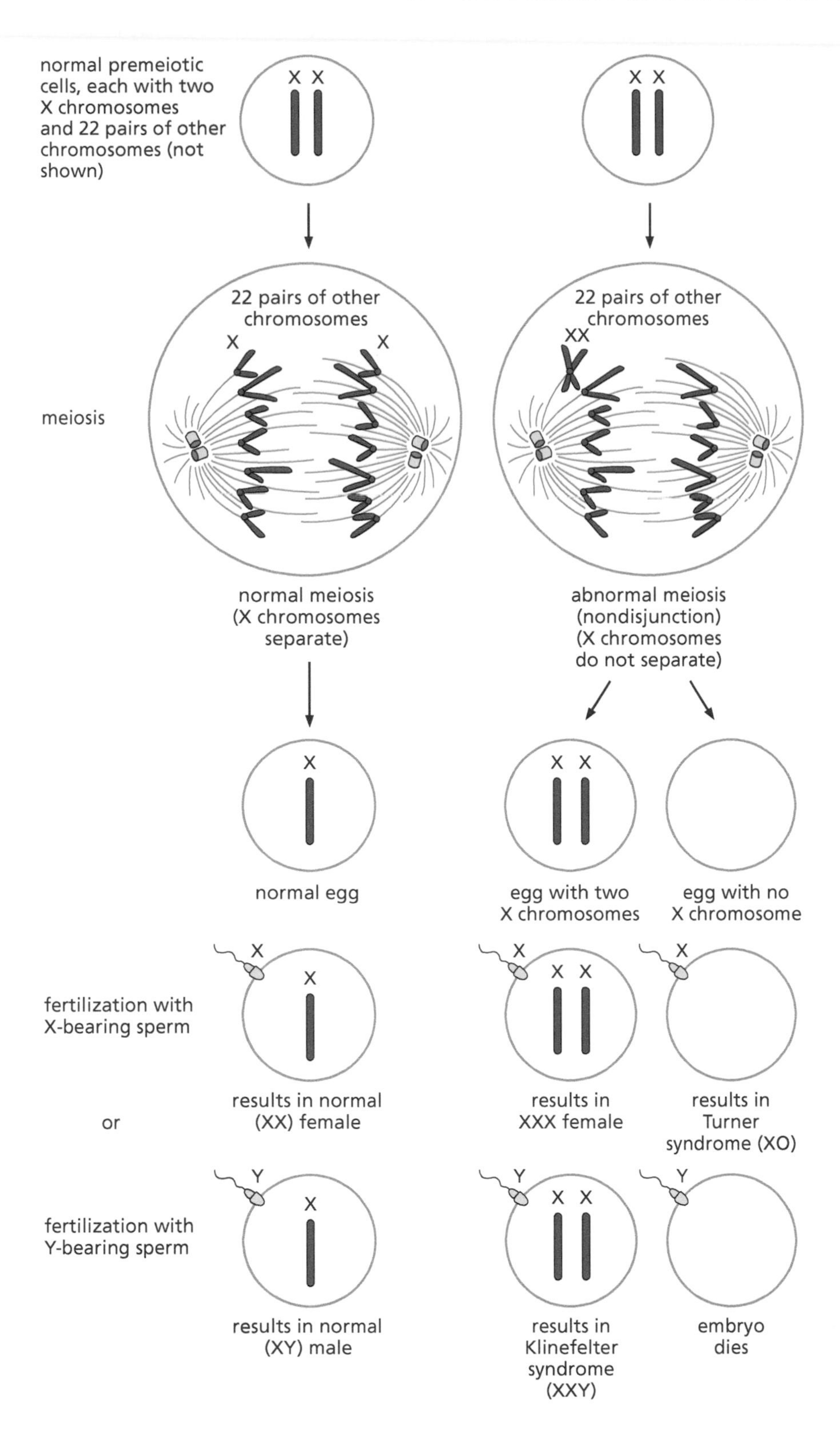

Figure 3.12 Abnormal meiosis during egg production, showing how certain chromosomal variations may arise. (Nondisjunction can occur during either the first or second division of meiosis; in either case, the resulting chromosomal configurations are as shown here.)

(Figure 3.13). Other chromosome abnormalities are less common. For example, Patau syndrome (three copies of chromosome 13) results in a small head with severely impaired brain development, extra fingers and toes, and usually death by one year of age. Instead of extra chromosomes, part or all of a chromosome can be missing. The *cri du chat* syndrome is caused by deletion of the short arm of chromosome 5 and results in a small head, a catlike cry, and impaired brain development. Like embryos lacking an X chromosome, embryos with many types of variations in autosomal chromosome number are nonviable; such chromosome abnormalities are frequently associated with miscarriage (spontaneous loss of pregnancy).

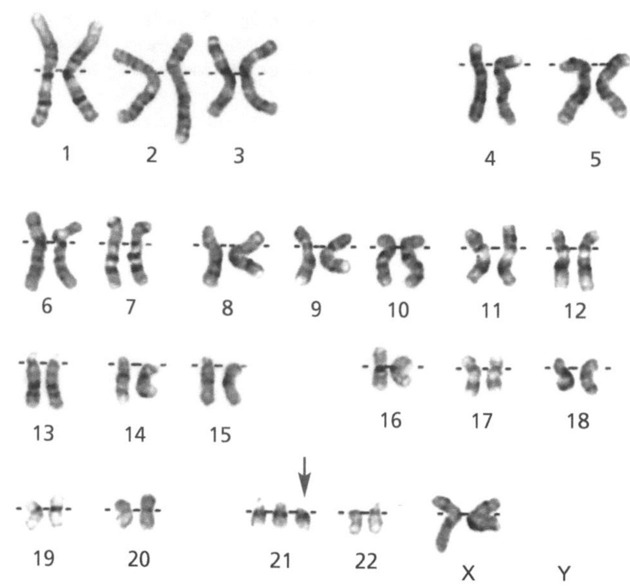

The karyotype with an extra chromosome 21 associated with most cases of Down syndrome

A child with Down syndrome

Figure 3.13 Down syndrome.

In addition to these changes in chromosome number, there are several kinds of large-scale changes involving chromosome fragments. Chromosome fragments may become duplicated (repeated); they may become attached at a new location, possibly on a different chromosome; or they may be lost entirely. A chromosome fragment may also be turned end-to-end and reinserted at its former location. Of these several types of chromosomal changes, end-to-end inversions are the most frequent and have the most limited effects, while the other types may result in nonviable phenotypes when the rearranged fragments are long. Although these chromosomal changes are often classified as mutations, many geneticists prefer to limit the use of the term "mutation" to changes in single genes and to refer to the larger changes as **chromosomal rearrangements** or "**chromosomal aberrations**." Both chromosomal aberrations and changes in chromosome number are frequently observed in cancer cells, which have typically lost the capacity to maintain genetic order (Chapter 10).

3.2.4 Social and ethical issues regarding sex determination

As we have seen, the determination of a person's sex is not always unambiguous. Although there is a strong societal expectation that each person should be categorized as either male or female, about 17 out of every 1000 people cannot be so easily categorized. Some are people with XY chromosomes but female anatomy or XX chromosomes but male anatomy. In rare instances, babies are born with partly male and partly female anatomy. In certain cases, such babies are often treated with plastic surgery and/or hormones to make them conform to one sex or the other. The ethics of beginning such treatments early in life are not straightforward, since the individuals themselves are not able to participate in making decisions that may have very important consequences for them later in life. Sometimes these procedures have also been performed without the informed consent of the parents, raising further ethical issues.

Realization that there are XX males and XY females forced the International Olympic Committee to reexamine the stipulation that only XX individuals could compete in female sporting events. If chromosome appearance is not sufficient to determine which individuals are male, should the presence of the *SRY* gene or the hormone testosterone be used as the test? Either turns out to be problematic. Females also have testosterone, although generally in smaller amounts than males. Also, there are rare XY individuals who have the functional *SRY* gene and produce male concentrations of testosterone but are nevertheless phenotypically female because they lack a functional allele of a different gene, the gene that encodes the protein receptor that recognizes and responds to testosterone (the androgen receptor). People with variations in the relationship between their sex chromosome karyotype and sexual phenotype are not defective individuals; they just do not fit a previously held view of what determines the sex of an individual. In 2018, the International Association of Athletic Federations (IAAF) recommended that female athletes with naturally high levels of testosterone be barred from competing in certain athletic events as women unless they take hormones sufficient to lower their testosterone levels to below 0.5 nmol/L. The International Olympic Committee currently uses the presence or absence of the functional *SRY* gene to decide the sex of Olympic athletes, but this is clearly not a perfect determinant. Other genes located on autosomes, like the androgen receptor gene, and possibly also some relatively uncharacterized genes on the Y chromosome, have a role in sexual development. Thus, even though we refer to X and Y as the sex chromosomes, many genes on many other chromosomes are also involved.

The relationship between sex and gender creates further complications. Sex refers to the biological phenotype of a person, which, as already mentioned, is usually male or female, but can be ambiguous in certain cases. Gender is a social concept that links certain behaviors, attitudes, and cultural roles to an individual's sex. Cultures around the world vary in their acceptance of gender-nonconforming individuals. In Samoa, at least 1% of the population is recognized as "fa'afafine," considered neither male nor female but accepted within Samoan society as a third gender. Several other Polynesian societies similarly recognize third-gender people, and many indigenous cultures in North America use the term "two-spirit" to refer to individuals who may not consider themselves male or female. Increasingly, individuals from many societies choose to identify themselves as nonbinary or nongendered based on a lack of identification with a single, sex-based gender.

THOUGHT QUESTIONS

1. Sports officials have repeatedly asked female athletes to submit to testing to confirm their femaleness, but comparable proof is seldom demanded of males. Why do you think this disparity exists? Can you think of other ways besides sex in which athletes might be classified? Could we use age as a criterion? Could we use fat-to-muscle ratios?
2. Individuals with Klinefelter syndrome or Turner syndrome are fully functional aside from their infertility, yet many are "treated" with hormones to give them a less indeterminate appearance of secondary sexual characteristics such as breasts. Why do humans tend to think of such differences as "abnormalities" needing "treatment" rather than as normal variation within a trait?

3.3 SOME DISEASES AND DISEASE PREDISPOSITIONS ARE INHERITED

Early in the twentieth century, pioneering geneticists discovered that Mendel's rules, formulated on the basis of experiments with pea plants, could also explain the inheritance of many human traits, such as the trait of albinism. Albinism is a deficiency of melanin pigment in the skin, eyes, hair, and internal organs, producing the characteristic white skin and hair of albino individuals. One of the normal functions of melanin is to block ultraviolet light; thus, albino individuals sunburn easily, and their eyes are very sensitive to bright lights. All geographical races of humans have albino individuals, as do many other species.

The inheritance of albinism is shown in Figure 3.14. A recessive allele is responsible for this condition; thus, it can be transmitted without detection through many successive generations of normally pigmented individuals (Figure 3.14A). However, matings between heterozygous individuals have a one-in-four chance of producing an albino child (Figure 3.14B). The likelihood that both parents are heterozygotes is increased if mates are chosen from among related persons such as cousins (see Figure 3.14C). This is because the same rare recessive alleles are more likely to be present in other family members than they are in the general public. Diagrams showing the pattern of mating and descent, as in Figure 3.14C, are called **pedigrees**.

(A) A family in which only one parent is heterozygous; none of the children are albinos.

PARENTS

AA normal pigment — Aa normal pigment

GAMETES all A — $\frac{1}{2}$ A $\frac{1}{2}$ a

AA — Aa

CHILDREN (male or female)

$\frac{1}{2}$ AA normal pigment — $\frac{1}{2}$ Aa normal pigment

(B) A family in which both parents are heterozygous; each child has a 1 in 4 chance of being an albino.

PARENTS

normal pigment Aa — normal pigment Aa

GAMETES $\frac{1}{2}$ A $\frac{1}{2}$ a — $\frac{1}{2}$ A $\frac{1}{2}$ a

AA Aa Aa aa

CHILDREN (male or female)

$\frac{1}{4}$AA $\frac{1}{2}$ Aa $\frac{1}{4}$ aa

$\frac{3}{4}$ of children normally pigmented — $\frac{1}{4}$ of children albino

(C) Human pedigree for a family with albinism; the short horizontal lines between a male and female represent matings that produced the children in the row below.

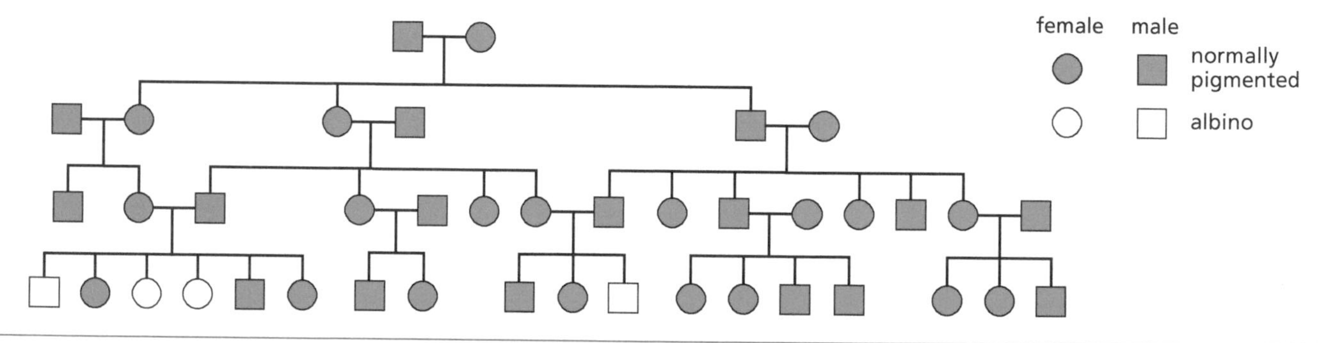

Figure 3.14 An example of simple Mendelian inheritance in humans. Albinism, a recessive trait, arises in most cases from matings between heterozygotes, although it could also arise from matings between a heterozygous person and someone who is homozygous recessive.

3.3.1 Inborn errors of metabolism

Shortly after the rediscovery of Mendel's laws, an English physician, Archibald Garrod, made an important discovery: Mendel's laws applied not only to visible characteristics such as eye colors and albinism but also to certain medical conditions. Garrod's identification of the genetic basis of a condition called alkaptonuria, described next, was an important milestone. In his studies of albinism, alkaptonuria, and several other conditions, Garrod was able to trace each disease to a defective (nonfunctional) enzyme made of protein. This was historically the first realization that gene products were usually proteins. Also important was the discovery of an important principle: Dominant alleles produce functional proteins; recessive alleles produce nonfunctional proteins instead.

3.3.1.1 Alkaptonuria **Alkaptonuria** is a rare condition in which a patient's face and ears may be discolored and in which their urine turns black upon exposure to air. Archibald Garrod tested the urine of these patients and discovered that the color is caused by an acid. We now know that this substance, homogentisic acid, is formed in the course of breaking down the amino acid tyrosine. In most individuals, the homogentisic acid can be broken down harmlessly with the help of an enzyme. However, in patients with alkaptonuria, the necessary enzyme is missing or defective, and Garrod realized that an error in an important biochemical (metabolic) process was responsible. Garrod (1908) called this type of condition an "inborn error of metabolism." He studied the families of individuals with alkaptonuria and three other such conditions, including albinism, and found a common pattern: Each of these inborn errors of metabolism was inherited as a simple Mendelian trait, and in each case the lack of a functional enzyme was recessive. Many other inborn errors of metabolism have since been discovered and their biochemical defects identified. Each of these inborn errors is caused by a recessive allele, the product of a DNA mutation that, when transcribed and translated to its protein product, results in a nonfunctional enzyme in place of a functional one. Alkaptonuria, albinism, and the condition called **phenylketonuria** (PKU), which is described next, all arise from errors in a series of closely related metabolic pathways (**Figure 3.15**).

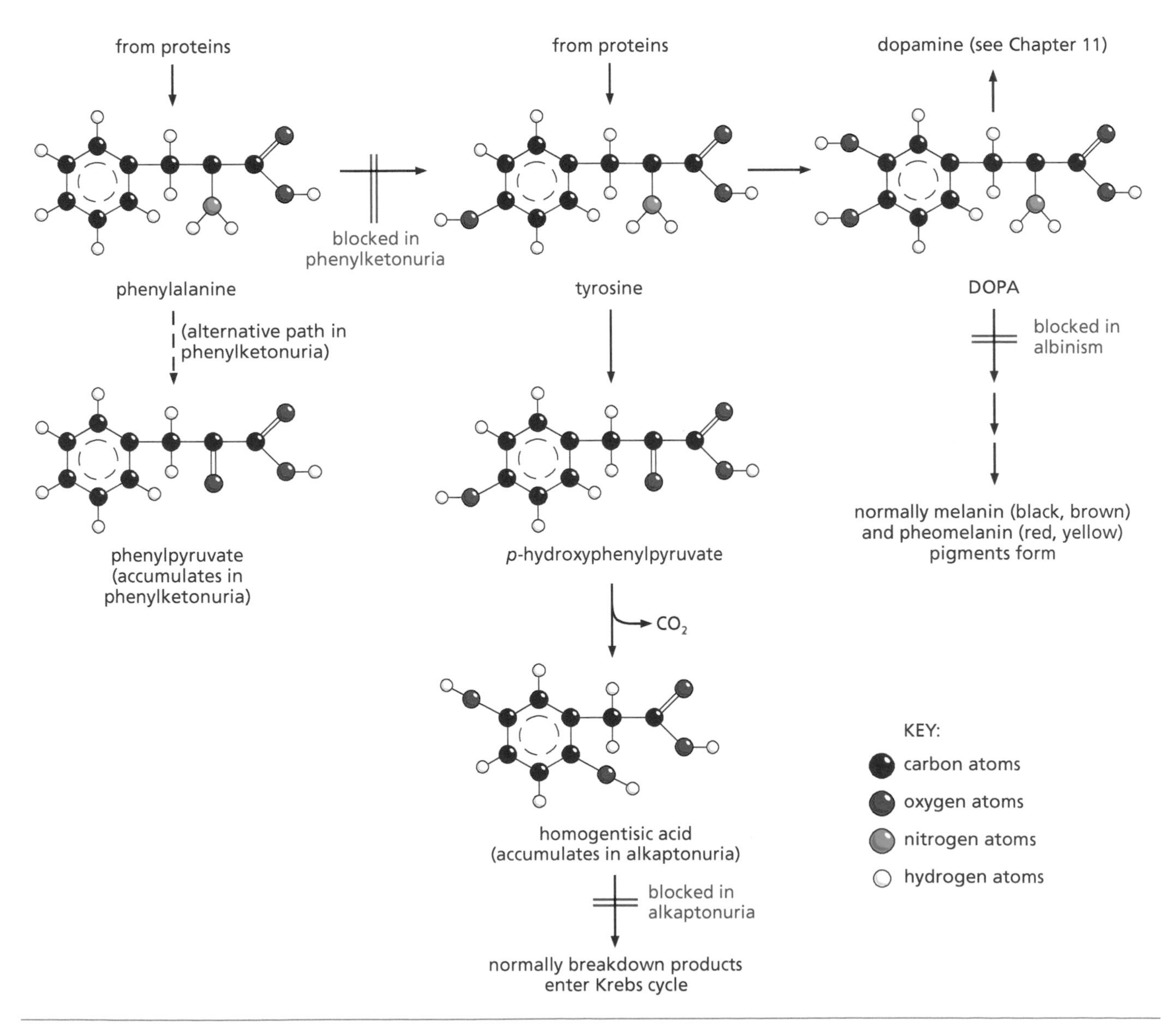

Figure 3.15 Biochemical pathways for three inborn errors of metabolism: phenylketonuria, alkaptonuria, and albinism.

3.3.1.2 Phenylketonuria Phenylketonuria (PKU) is a genetically controlled defect in amino acid metabolism. The amino acid phenylalanine, which is present in most proteins, is normally converted by an enzyme into another amino acid, tyrosine; the tyrosine is then broken down by the pathway shown in Figure 3.15. A defect in the enzyme that usually converts phenylalanine to tyrosine causes the phenylalanine to be processed by an alternative pathway. A product of this alternative pathway accumulates in the blood and in all cells, acting as a poison that causes most of the debilitating symptoms of the disease: insufficient development of the insulating layer (myelin) around nerve cells, uncoordinated and hyperactive muscle movements, retarded brain development, retarded bone growth, defective tooth enamel, and a life expectancy of 30 years or less. Thus, a change in one gene (and one enzyme) can have many phenotypic consequences throughout the body.

Fortunately for people carrying this genetic defect, a simple test for its presence exists, and this test is now routinely administered to infants within 24–72 hours after birth in most industrialized countries. If PKU is detected at birth or earlier, it is possible to avoid the symptoms of the disease by greatly limiting those foods that contain phenylalanine (nearly all proteins, including breast milk) and diet foods and soda containing the artificial sweetener aspartame, which breaks down to yield phenylalanine. Small amounts of phenylalanine are essential in protein synthesis, but the diet must be carefully monitored to guard against the larger amounts of phenylalanine whose breakdown products would be in toxic amounts that cause the disease symptoms.

Alkaptonuria and PKU are inherited via single recessive alleles that follow Mendel's laws. Yet each trait is a medical condition, rather than a visible trait like those studied by Mendel. Garrod's findings thus broadened the concept of a "trait." The concept has been broadened further by the discovery of a genetic basis for diseases that are not just present or absent but that show a range of severity in different people.

3.3.2 Identifying genetic causes for traits

Before searching for genes that cause a particular disease or trait, we first need to know whether there is any basis for thinking that the disease is inherited. As we will see, some diseases are the direct result of particular gene mutations. In other diseases, genetic effects are indirect and contribute to disease susceptibility—the likelihood that a person will get a disease. Susceptibility to many diseases seems to have a hereditary component. However, certain disease susceptibilities and many nondisease traits are the result of the interaction of multiple genes, not the result of mutations in single genes. Several kinds of studies that are used in answering the question of whether a disease or other trait is inherited are described next.

3.3.2.1 Pedigrees Geneticists have several ways of studying human hereditary traits. One of the most basic methods is to present the available data in pedigrees, as in Figure 3.14C. Pedigrees are most useful when they span many generations and hundreds of people or when separate pedigrees are available for hundreds of different families. The study of pedigrees can help to identify whether a condition is inherited and permit us to determine which traits are dominant, which are recessive, and which have a more complex genetic basis. If the genetic basis of a trait is complex or not fully known, then pedigrees can also help in an empirical determination of risks, including medical risks. For example, a child has a greatly increased risk of having type 1 diabetes (see p. 306) if one or both of the child's parents has the disease. The term **risk** has a precise statistical meaning: It is the probability that a particular condition will occur or that a particular condition will be inherited.

3.3.2.2 Studies of twins and of adopted children Studies of twins are sometimes useful in suggesting the extent to which the presence of a trait can be explained genetically, rather than environmentally. In such studies, numerous twin pairs are located in which at least one twin has the condition being investigated. For pairs of this kind, the frequency with which the other twin also has the condition is studied. This frequency is called the **rate of concordance**. For the vast majority of traits, studies on twins find that a mixture of both heredity and environment is involved. Traits under strong genetic control usually have higher rates of concordance for monozygotic twins (identical twins, derived from a single fertilized egg) than for dizygotic twins (fraternal twins, derived from two separate eggs). In contrast, traits with mainly environmental causes have similar rates for both types of twins, which would also be similar to the rates of concordance for nontwin siblings.

Adoption studies can also provide important clues about whether a particular trait is heritable: If adopted children show a higher rate of concordance with their birth parents than with their adoptive parents, then the hypothesis of a genetic cause is made stronger. Some researchers, however, have criticized this type of study because adoption agencies do not place children at random, but purposely try to match children with adopting parents whose backgrounds are similar to the backgrounds of the children's birth parents. This practice introduces a statistical bias that raises the concordance rates between adopted children and the parents who adopted them. Another complication is that many children are adopted by relatives, and such adoptions make it very difficult to sort out which similarities are environmental and which are genetic.

3.3.2.3 Linkage studies and DNA markers In early genetic studies, genes that did not assort independently were found to be located on the same chromosome; we refer to these as **linked** genes. The frequency of crossing over (recombination) during meiosis can be used as a measure of the distance (number of nucleotides) between linked genes on a given chromosome (Figure 2.13). If linkage to a visible chromosomal abnormality could be established, then a linked group of genes could be assigned to a particular chromosome. These classical genetic techniques were developed in other species, such as fruit flies, using hundreds or thousands of offspring in each generation to assess recombination frequencies. In general, these techniques were ineffective for mapping human genes because humans have such small family sizes and such a long generation time, and also because humans cannot be bred for experimental purposes. Before 1980, very few human genes had been mapped to their chromosomal locations.

Once evidence suggests that a trait has a genetic basis, **linkage studies** can help locate the relevant gene or genes. To do a linkage study, geneticists first need a set of **DNA markers**, pieces of chromosomes that are visibly different under the microscope or short sequences of DNA that can be revealed by the molecular techniques discussed later in this chapter. If they can locate a DNA marker whose pattern of inheritance is the same as the pattern of inheritance of the trait, then they can conclude that a gene associated with the trait is located near the marker. One problem is that researchers may have no idea where to look, so they need many markers, scattered across all the chromosomes. Also, large pedigrees are needed to carry out this type of analysis. After a linkage between DNA markers and a trait is found, other molecular techniques are used to close in on the actual gene. Finally, after the gene has been located, its full sequence can be analyzed in individuals that do and do not have the trait to identify specific genetic changes that may be responsible for the trait.

Several marker systems not dependent on gross chromosomal abnormalities were developed in the 1980s and thereafter for the purpose of studying human genetics. In addition to genes, DNA contains noncoding regions that vary from one individual to another in their length or their sequence. Many of the markers used in human genetics are in these noncoding regions of DNA. The first DNA marker system, developed in 1980, is based on "**restriction fragment length polymorphisms**" (abbreviated **RFLPs** and pronounced "riflips"). Restriction fragments are short pieces of DNA produced by enzymes called restriction enzymes, which cut the DNA at specific sequences; "polymorphism" simply means "different forms," or in this case differences in the lengths of the fragments created by the enzyme cutting process. (We will learn more about restriction enzymes in Section 4.1.1.) RFLPs served as useful markers because they are different lengths in different people; therefore, one person's DNA can be distinguished from another's. However, the occurrence of known RFLPs across the human genome was not frequent enough to map many human genes. More recently, other DNA markers have been discovered that occur with greater frequency throughout the genome than do RFLPs. Each of these is a short, unique DNA sequence with a known location in the genome. Each DNA marker sequence exists in different alleles, the sequence of which varies slightly from one person to another, again allowing one person's DNA to be distinguished from another's. For some DNA markers, called **microsatellite markers**, the same sequence is repeated a variable number of times. For other DNA markers, the identities of individual nucleotides within a short sequence differs from one allele to another. These include **expressed sequence tags (ESTs)**, small transcribed portions of genes that vary from one person to another, and **single-nucleotide polymorphisms (SNPs)**, sequences located either in genes or in noncoding regions that differ by a single nucleotide.

Each marker can be detected by a specific DNA **probe**, a piece of DNA with a sequence complementary to the marker sequence. When a radioactive DNA probe is added to a DNA sample, if it contains the marker sequences that can pair with the probe, the DNA picks up the radioactivity. DNA probes highlight only those fragments that have sequences complementary to the probe sequence. For markers such as microsatellites or RFLPs that vary in length, the alleles can be distinguished by their lengths after separation by a technique called **electrophoresis** (Figure 3.16). For markers that differ by sequence, a different DNA probe is required for each

(A) DNA CONTAINING A MICROSATELLITE

The microsatellite differs in length depending on the number of repeats that exist within it. In this example, the allele from the father is shorter because it has fewer repeats than the allele from the mother, which is longer because it has more repeats.

DNA from a pair of chromosomes
chromosome from father
...TCTGAGAGAGGC...
chromosome from mother
...TCTGAGAGAGAGAGGC...
repeat sequence

(B) AMPLIFICATION OF MICROSATELLITE BY PCR

(using techniques shown in Figure 3.19.)

(C) SEPARATION BY ELECTROPHORESIS

The mixture of microsatellite PCR products is placed on a gel and exposed to an electric field. Because DNA has a charge, the pieces move toward the electrode of opposite charge. In the time that the current is on, smaller pieces travel further through the gel than the larger ones do. None of these pieces is visible yet.

agrose gel electrophoresis
electric field
⊖
⊕

(D) DETECTION WITH A PROBE

None of the pieces can be seen; however, they can be detected with a variable-repeat probe (bands shown in color). The probe is a small piece of DNA with a sequence complementary to the sequence of that variable repeat, so the probe will bind to those pieces of DNA containing that variable repeat. The probe thus does two things: it identifies pieces with that specific repeat and it allows scientists to determine whether the sequence is repeated a few times (to give a short DNA piece that travels farther) or many times (to give a long piece that travels a shorter distance within the gel).

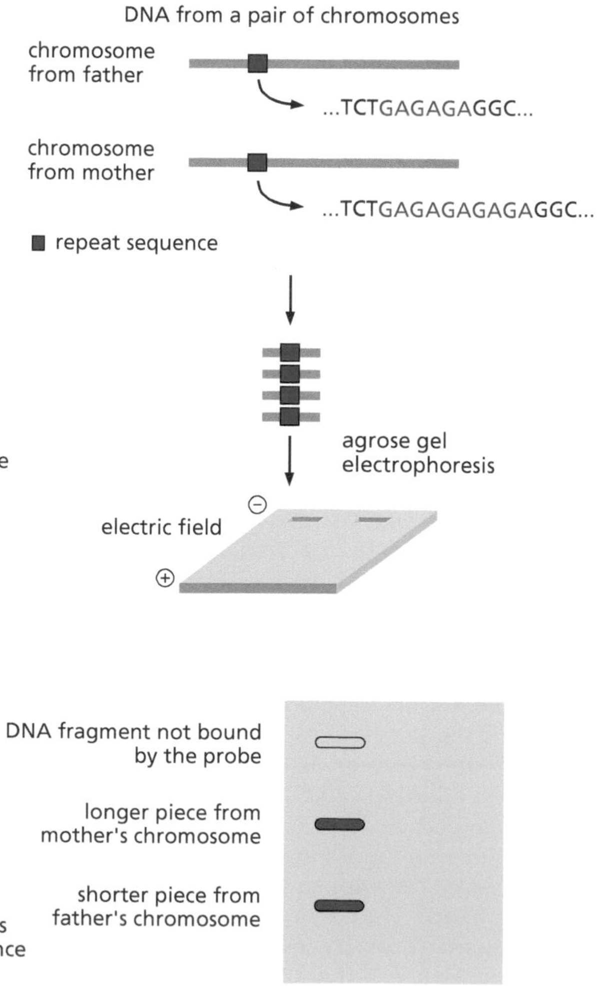

DNA fragment not bound by the probe
longer piece from mother's chromosome
shorter piece from father's chromosome

Figure 3.16 Microsatellite DNA marker alleles can be distinguished by the distance they travel during electrophoresis.

allele. These DNA probes may be clustered onto a solid surface called a DNA chip, or microarray. A fluorescently labeled sample of DNA is added. If the DNA sample contains the complement to the DNA probe, it binds, and its fluorescence is detected by a microscope attached to a computer. Techniques have been developed for placing 1 million DNA probes on a one-square-centimeter microarray, so that two alleles of each of a half-million markers can be tested at the same time.

The types of marker sequences just described have been identified throughout the human genome, and their chromosomal locations and genetic map positions have been established. Geneticists use all of these different types of markers to develop linkage maps. Because the chromosomal locations of the markers are known, geneticists can determine the positions of presumed genes located near the markers by finding a pattern of linkage between the trait and the marker. DNA samples are collected from members of a family with a pedigree in which the trait appears. Scientists then search the DNA samples for a marker, among the thousands of markers known, that is inherited within the family in the same pattern as the trait of interest. If all the individuals within a large pedigree who have a particular trait have a specific DNA marker and all those without the trait do not have the marker, the gene for the trait is presumed to be located near that marker (**Figure 3.17A**). For the other markers that are not in the vicinity of a gene linked to the trait, the mother's and father's markers can be found in the children, but the pattern of bands does not follow the pattern of inheritance of the trait (**Figure 3.17B**). In this way, many more presumed gene locations are being discovered at an accelerating rate.

(A) Marker 1 detected by probe 1, is linked to the trait

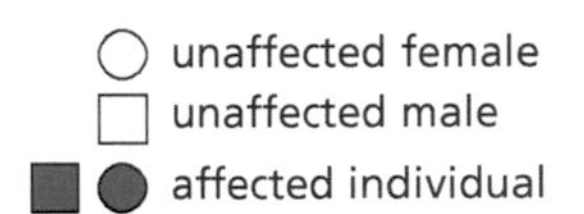

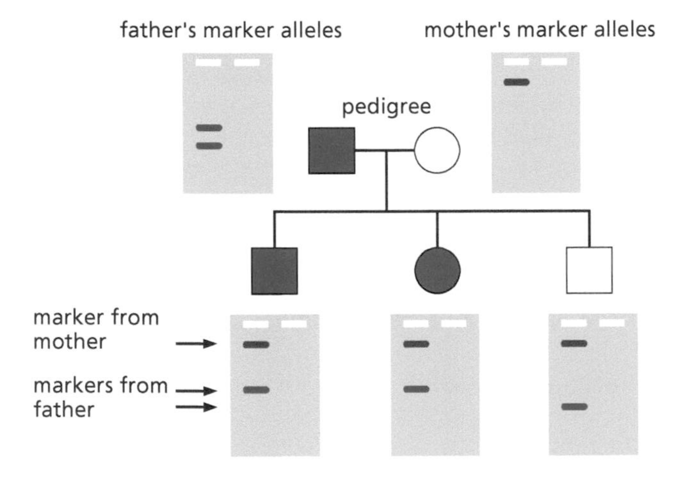

The father has the trait and is heterozygous at marker 1. The mother does not have the trait and is homozygous at marker 1; the mother's marker allele has a greater number of repeats than either of the marker alleles from the father's DNA.

The children who have the trait have one of their father's length of microsatellite marker. The child without the trait does not. The marker lengths detected with probe 1 thus follow the pattern of inheritance of the trait, indicating that an associated gene may be nearby.

(B) Probe 2 detects a different microsatellite marker, and shows that it is not linked to the trait

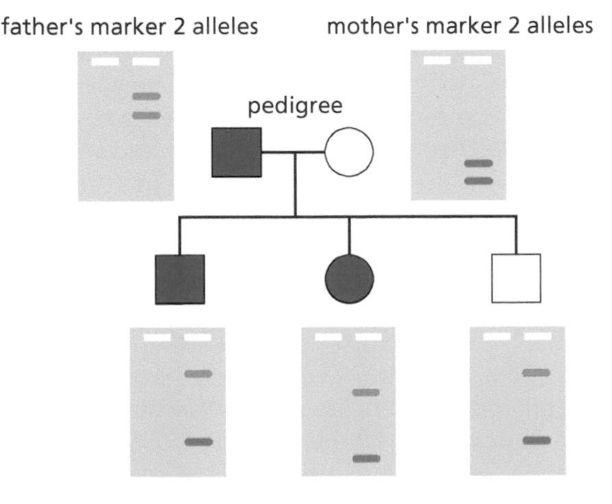

A child with the trait and a child without the trait have the same band pattern for the marker detected by probe 2. The marker detected by probe 2 is therefore not near any gene associated with the trait.

Figure 3.17 Using DNA markers to establish a linkage between a DNA region and an inherited phenotype.

3.3.2.4 Identifying a specific gene as the cause of a trait It is important to understand that DNA markers are not the genes themselves. In other words, the existence of linkage to a marker indicates the *approximate location* of a gene of interest. It does not tell us what the gene is, what its function is, or what alleles are associated with disease or nondisease states.

Often, when such a location has been found, news reports are published claiming that "the gene for X" has been discovered, but no individual gene has yet been identified, only a DNA region that maps with the trait. When a linkage area has been identified, sequence databases can be consulted to see what genes are known in this area of the chromosome. If genes are known, their protein products can sometimes be deduced from the nucleotide sequence, and those proteins can be further investigated for their connection with the trait or disease. Often, however, the genes within the linkage area have not yet been characterized, and neither their protein product nor their function is known.

The first gene to be located using DNA molecular markers was the gene for Duchenne muscular dystrophy, which is located on the X chromosome. Other diseases for which linkage areas were located and identified during the 1980s and 1990s include Huntington's disease (chromosome 4), cystic fibrosis (chromosome 7), Alzheimer's disease (chromosome 21), one form of hereditary colon cancer (chromosome 2), and two forms of manic depression (chromosome 11 and the X chromosome). Of these linkage areas, individual genes went on to be identified that cause Huntington's disease, cystic fibrosis, one form of early-onset Alzheimer's disease, and a high risk (but not a guarantee) of colon cancer. The genes associated with manic depression have remained elusive, as different studies have failed to replicate each other's results, and the disease is likely multifactorial. The ultimate goal of studying the genetics of human disease is obviously to develop therapies or cures. Of the diseases mentioned earlier, significant progress has been made in this direction only for cystic fibrosis. Multiple drugs are now available that treat the specific defects associated with several alleles of the gene responsible for cystic fibrosis, which is called *CFTR* and encodes a chloride ion transport protein. Different mutations are known for the *CFTR* gene; each responds to a different medicine, and some do not respond to any medicine currently known. As these examples indicate, the path from identification of a region of the genome associated with a human genetic disease to developing drugs or other treatments that can help people with that disease can be very long; however, technical advances in many aspects of molecular and cellular biology are currently accelerating this process.

Many examples of human diseases follow simple Mendelian genetics. Some were identified early in the twentieth century based on biochemical studies of enzymes whose lack of function cause the diseases. Several others were discovered more recently using DNA markers.

3.3.2.5 Duchenne muscular dystrophy Duchenne **muscular dystrophy** is an X-linked genetic disorder that causes muscles to become weak and nonfunctional. In most cases, the inability of the muscles of the diaphragm to keep the patient breathing leads to death during the teenage years or in the early twenties. After the gene responsible for this disease was located by linkage studies using DNA markers, its protein product, dystrophin, was found. So far, this discovery has not led to any new therapies that can slow or reverse the disease, but the research on dystrophin is greatly adding to our understanding of normal muscle contraction, and it is hoped that this knowledge will lead to effective treatments.

3.3.2.6 Cystic fibrosis **Cystic fibrosis** is the most common genetic disease among the Caucasian population in the United States and much of Western Europe. Cystic fibrosis is characterized by thickened fluids, especially the fluids that line the lungs. Normally, as these fluids are cleared from the lungs, they remove respiratory bacteria, preventing infections. The thicker-than-normal fluids of cystic fibrosis are not cleared from the lungs as they should be, so people with cystic fibrosis have breathing difficulties and frequent lung infections.

Cystic fibrosis was mapped by linkage studies to a region on chromosome 7. Later, the *CFTR* gene was identified as being responsible for the disease. As mentioned earlier, the product of the *CFTR* gene is a membrane transporter, a protein that carries chloride ions into or out of cells based on the balance of water inside vs outside the cell. Defects in this balance lead to the thickening of mucous secretions that is characteristic of the disease. The most common mutation in the *CFTR* gene, comprising 70% of all mutations seen in patients, causes the CFTR protein to fold slowly. A quality control pathway in the cell targets this slow-folding protein for degradation, so it is eliminated from the cell, even though it could function properly if it had time to fold and reach the cell surface. Several drugs have been developed that support folding of this mutant protein, with the most recent being a three-drug cocktail approved by the U.S. Food and Drug Administration in late 2019. Another drug that targets a less common mutation that interferes with opening of the CFTR ion channel was approved in 2012. The *CFTR* gene was discovered in 1989, so more than 20 years of research were required before the discovery yielded any concrete patient benefits. However, whereas the median age of survival for patients with cystic fibrosis was less than 20 years in 1980, it is now over 40 years, illustrating the clear value of genetic research.

3.3.2.7 Huntington's disease **Huntington's disease** (or Huntington's chorea) is a neurological disorder that begins between the ages of 40 and 50 with uncontrollable spasms or twitches of the hands or feet. As the disease progresses, the spasms become more pronounced, and the patient gradually loses conscious control of all motor functions and of mental processes. The disease progresses slowly but is invariably fatal. American song writer and balladeer Woody Guthrie died of Huntington's disease. Although it is always lethal, Huntington's disease does not appear until after its victims have lived through their prime reproductive years, during which they may have passed the mutation on to their children. Studies of family trees show that Huntington's disease is inherited as a dominant trait, meaning that only one mutant copy of the gene is required to inherit the disease. The gene responsible for Huntington's disease was mapped to chromosome 4 and fully identified in 1983, but the function of the normal gene has remained elusive. Recently, evidence has accumulated to implicate the Huntingtin protein that is encoded by the normal gene in a pathway that recycles and renews old cellular components.

Huntington's disease is associated with a different type of DNA mutation, called a "dynamic mutation." Most mutations are rare, stable, and inherited unchanged from one parent or the other. In dynamic mutations, a three-nucleotide sequence is repeated many times within a gene; during DNA replication or repair of DNA damage, the number of repeats can increase. There are several hypotheses as to how this occurs, but the precise mechanism is still being investigated. This increase in the number of three-nucleotide repeats can happen in cells that are undergoing either mitosis or meiosis. The former case would result in different cells within an individual having different repeat numbers within the affected gene, while the latter would result in a child inheriting more copies of the repeat than is present on either parent's chromosome.

The allele that causes Huntington's disease differs from the nondisease alleles of the gene by having many extra repetitions of the three-nucleotide sequence AGC. Persons with fewer than 30 repeats are unlikely to get the disease. Persons with more than 38 repeats are almost certain to get the disease, with the age of onset being younger when there are more than 50 repeats. A test for the allele responsible for Huntington's disease has been devised, but its interpretation can be somewhat ambiguous, since there is an overlap in the number of repeats associated with a particular outcome. The "normal" range in number of repeats is 9–37, but some people with as few as 30 repeats have become ill. Because the number of repeats can increase during meiosis, people with 30–37 repeats may pass the disease to their offspring. Increases in the number of repeats in cells dividing by mitosis may affect the timing of disease onset for a given individual.

Huntington's disease is one of several diseases, called trinucleotide repeat diseases, associated with dynamic mutations. Another is fragile X syndrome, the most common type of hereditary developmental brain disability, in which the repeat is CCG in a gene on the X chromosome. The gene responsible for fragile X syndrome has been identified, and the protein it encodes is implicated in regulating the expression of other genes in neurons. There is a similar situation for fragile X syndrome as for Huntington's disease, wherein normal individuals can exhibit a range of numbers of CCG repeats in the gene, and the severity of the disease increases with the number of repeats once a threshold value above the normal number is reached.

3.3.2.8 Genes increasing susceptibility to disease The transmission of human traits controlled by single genes follows the rules that Mendel developed for peas. The situation is more complex for phenotypic traits such as height or skin color that are controlled by many genes at once and that are also influenced by environmental variables such as nutrition.

Some genes have been identified that do not lead directly to a trait (such as a disease), but rather increase the probability that some trait or disease will develop. Having a particular allele of these genes does not mean that a person is certain to get a particular disease, only that the likelihood of their doing so is higher than that in people with other alleles. These genes, like most others, code for proteins. The proteins associated with susceptibility are often regulatory proteins that change the body's response to some environmental factor. The genes associated with predisposition to cancer, discussed in Section 10.6.1, are examples. As we mentioned earlier for other multigene traits, we must be cautious not to think of these "susceptibility genes" as simple either/or Mendelian alleles. For some cancer susceptibility genes, the likelihood of developing a particular form of cancer when you have a certain allele is close to 100%, but for most others, it is much less than 100%.

In many cases, all we know at present is a statistical association between some trait and some DNA markers; the genes themselves have not yet been identified. Traits such as bone density or obesity in mice and behavioral traits such as ethanol consumption by rats have been found to each be associated with several DNA markers, but no genes have yet been identified. Despite this, news stories have overplayed such statistical associations, referring, for example, to the "gene for obesity." Such terms are premature and surely oversimplified. Each of these traits is influenced by multiple genes, so we cannot think of them as we think of Mendelian either/or traits. These traits vary along a continuum and are not just present or absent like the traits studied by Mendel. In fact, the vast majority of human traits are influenced by more than one gene and often also by environmental factors, so the relationship between genotype and phenotype is often not easy to predict.

THOUGHT QUESTIONS

1. What is the difference between a genetic marker and a gene?
2. Is research on genetic diseases important for society even if it does not lead to new methods of treatment?
3. With our thinking influenced by genetic diseases like PKU or Huntington's disease, we have been accustomed to thinking of "mutations" as "defects." Not every mutation is a defect, however. With genetics research now turning to the identification of disease susceptibility, it is likely that we will all turn out to have a hereditary predisposition to something. Do new research directions necessitate reconceptualizing the term mutation as "variation," rather than "defect"?

3.4 GENETIC INFORMATION CAN BE USED OR MISUSED IN VARIOUS WAYS

After a genetic basis has been identified for a particular trait, what happens next depends in large part on the values that individuals and society place on that trait. We have considered in this chapter many traits that at least some people consider undesirable, although not all of them impair health or longevity. In many cases, but not others, we can identify a specific gene that fails to make a functional protein of some sort. These are often called genetic defects, a category that includes all inborn errors of metabolism such as those shown in Figure 3.15. The term "genetic defect" thus means that a specific allele and its product are defective; it does not mean that the person bearing the gene is defective. For this reason, many people prefer terms such as genetic disease or genetic condition, rather than genetic defect.

Humans can deal with hereditary conditions and hereditary risks in many ways. We can conveniently describe four broad categories of response: gathering and sharing information through genetic testing and counseling, changing individual genotypes, changing the gene pool at the population level, and changing the balance between genetic and environmental factors. Many of the methods in these four categories, which we consider next in turn, raise important ethical questions. The ethical questions can be summarized as follows:

- Who decides who should be tested?
- Who has access to the results of the test?
- What are the responsibilities of a person who carries a gene for a hereditary disease?
- Do we have a responsibility to maintain genetic diversity?
- Who determines what traits (if any) are called "defects"?

Think about these ethical issues as you read the rest of the chapter.

3.4.1 Genetic testing and counseling

Advances in medical genetics have led to better ways of detecting genetic diseases and to ways of detecting them earlier. Identification of chromosomal variations, mutated alleles of genes, or products of mutated alleles may allow the detection of a disease at the earliest possible stage.

3.4.1.1 Prenatal detection of genetic conditions Some conditions can be detected before birth, in utero (literally, "in the womb"). From conception until about the eighth week of pregnancy, a pregnant woman is carrying an **embryo**; from the eighth week until birth, the term **fetus** is used rather than embryo. In the technique of **amniocentesis** (Figure 3.18A), a small amount of fluid (amniotic fluid) is withdrawn from the sac in which the fetus is developing in the mother's uterus. The fluid itself is analyzed for the presence or absence of certain enzymes that might indicate a genetic defect in the fetus. Also, amniotic fluid usually contains cells that have been shed from the surface of the fetus; these can be grown in the laboratory, and their DNA can be isolated and analyzed if there is a concern that a genetic condition may have been inherited.

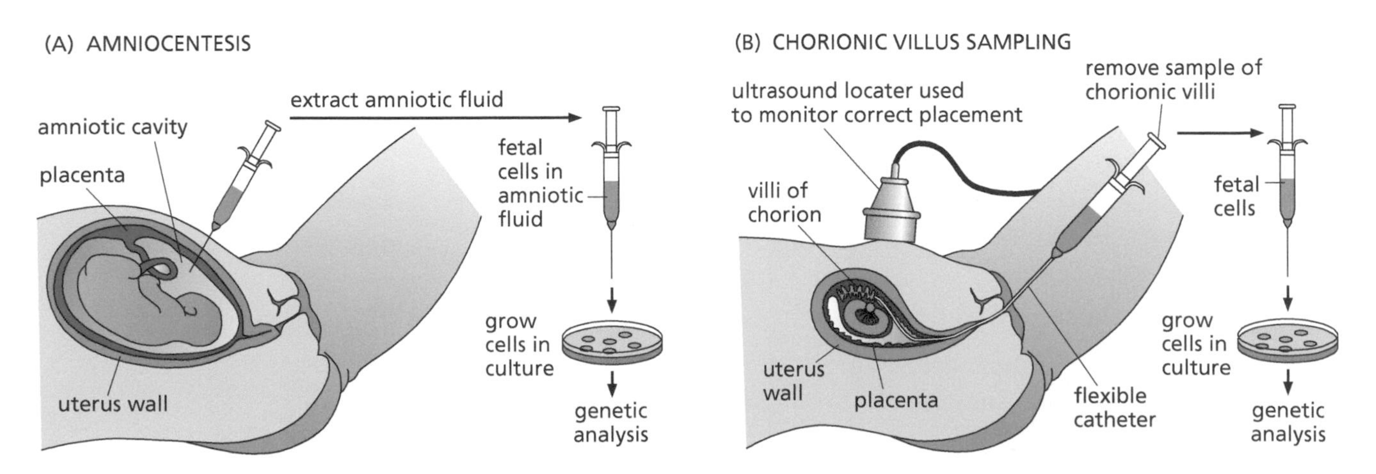

Figure 3.18 **Techniques for prenatal detection of genetic conditions.**

Instead of amniocentesis, **chorionic villus sampling** is sometimes used for prenatal detection. This technique is a type of biopsy (removal of living tissue for examination) of the placenta (Figure 3.18B). The placenta is the structure by which the fetus attaches to the wall of the uterus (Figure 13.10). Because the part of the placenta biopsied is tissue derived from the fetus, not from the mother, the cells sampled by this technique are fetal cells. Chorionic villus sampling can be performed earlier in pregnancy than amniocentesis can. Certain low, but nonzero, risks are associated with both amniocentesis and chorionic villus sampling, including a risk of mechanical injury to the growing fetus and a risk that the pregnancy will be prematurely terminated (miscarriage). Because of these risks and other reasons, these tests are not performed routinely on every expectant mother, but are offered to mothers who are at increased risk for genetic abnormalities due to family history or age.

After fetal cells have been obtained by either amniocentesis or chorionic villus sampling, chromosomes (karyotypes) from these cells are analyzed for evidence of Down, Turner, or Klinefelter syndromes. Also, if there is a reason to study a particular gene, its sequence can be determined by comparing it with a known sequence for that gene. This cannot be done on the minute amounts of DNA that exist in the cells unless these amounts are first increased, or amplified. Amplification of DNA is accomplished by the **polymerase chain reaction (PCR)** (Figure 3.19). Because of the nonzero risks associated with the invasive methods of amniocentesis and chorionic villus sampling, advances in DNA sequencing technology have recently been harnessed to analyze fetal DNA circulating in the mother's bloodstream. This type of test is called "cell-free fetal DNA screening" and is a form of lower-risk noninvasive prenatal testing (NIPT), in contrast to the more invasive, older techniques. NIPT can be performed as early as 10 weeks in a pregnancy and involves a simple blood draw from the mother. If NIPT reveals a cause for concern, a more definitive test (like amniocentesis) can be ordered as a follow-up.

The polymerase chain reaction is often used to detect genetic conditions by using DNA from eight-cell embryos prior to implantation. These embryos are derived from in vitro fertilization (literally "in glass"), meaning that the fertilization of the egg by the sperm took place in laboratory glassware rather than inside

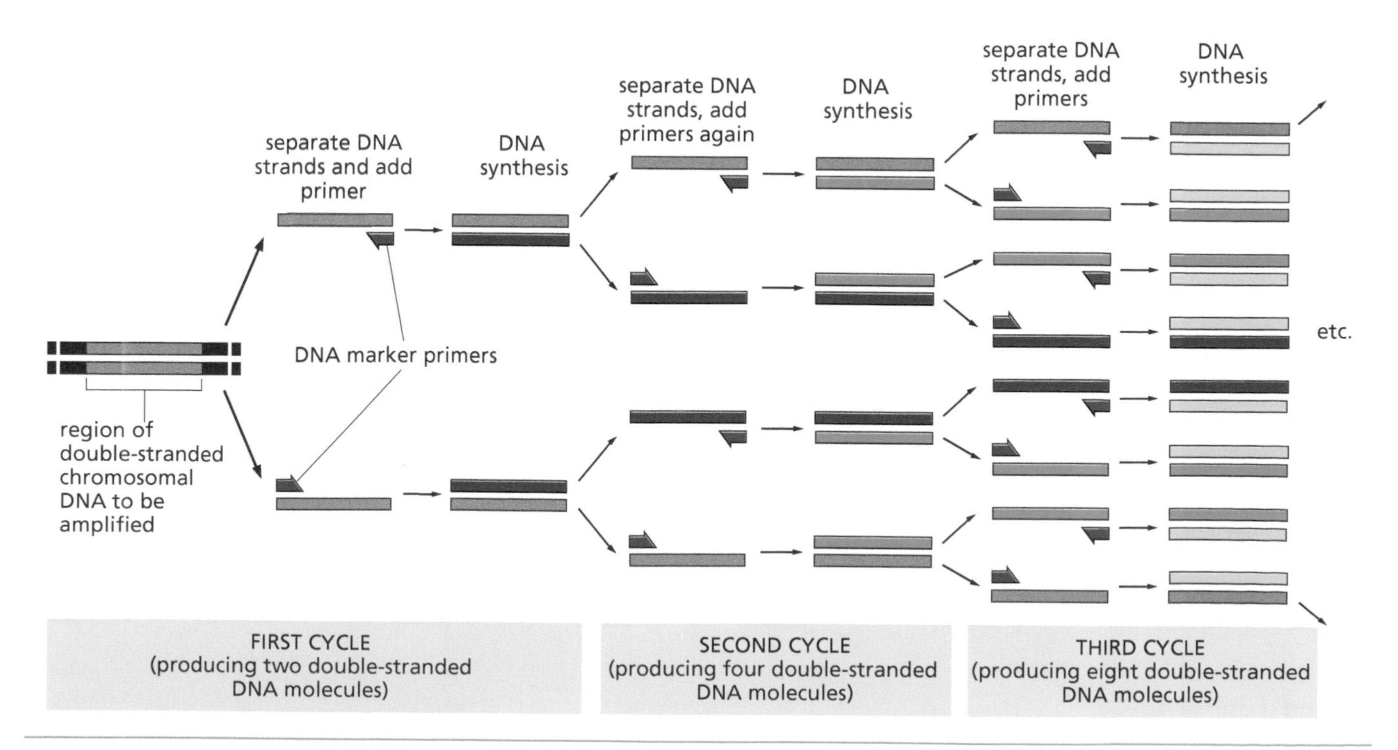

Figure 3.19 The polymerase chain reaction (PCR).

the body (in vivo). One of the eight cells can be removed for genetic testing, and the other seven can be implanted into a woman's uterus to grow to term. For couples undergoing in vitro fertilization due to fertility problems or a family history of genetic disease, preimplantation genetic testing can provide the best chance for a healthy baby. Several dozen genetic diseases are now detectable through prenatal tests, including Tay-Sachs disease, cystic fibrosis, and phenylketonuria.

3.4.1.2 Testing newborns or adults Other tests are done on newborns or on adults. Tests that are simple and inexpensive can be used for mass screening. For example, as mentioned earlier, all infants are now routinely screened at birth for phenylketonuria by a blood test to detect the amino acid phenylalanine (not by a DNA test). Such screening is considered ethical because it is done on all infants and it is a clear benefit to the infant for the information to be known. Tests that detect heterozygosity for defective alleles (e.g., testing for carriers of the allele causing sickle cell anemia, as described in Section 7.3.2) can be performed on adults before they become parents. Persons undergoing this type of screening must first give their **informed consent**. They must sign a form stating that they understand the nature of the test, the possible outcomes (including the conditions that the test can detect and the likelihood that the genotype will result in a disease phenotype), the possible risks of the procedure, and the possible benefits.

People in these situations often consult a genetic counselor to help them understand the test and the risks before giving their consent. For example, they can be advised that if they are identified as being heterozygous and they have children with a person also heterozygous for the same allele, each of their offspring has a 25% chance of being homozygous for the recessive condition. To understand why this is true, refer to **Figure 3.20A**. The extent to which homozygosity predicts disease severity differs with the disease.

3.4.1.3 Who should be tested? Genetic testing is expensive, and it would not be reasonable to test everyone for all genetic disorders. Therefore, an effort is made to identify persons at higher risk of having certain defective alleles. Figure 3.20 shows two ways in which risk is estimated. First, family history may prompt testing because having a family history of a genetic condition increases the probability that a family member carries the defective allele (see Figure 3.20A). However, because recessive traits show up phenotypically only when the genotype is homozygous or sex-linked, a recessive trait may not show up in a pedigree, and a person may not know their complete family history. A second way of estimating

(A) FAMILY PEDIGREE

Each child of two parents heterozygous for a recessive trait has a 25% probability of being homozygous recessive

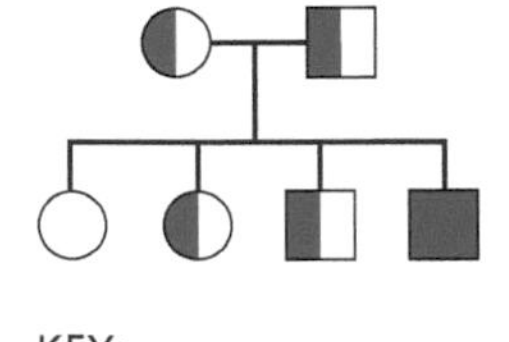

KEY:

males females

homozygous for dominant trait

homozygous for recessive trait

heterozygous

(B) POPULATION FREQUENCY

If 1% of a population expresses a trait known to be recessive (meaning that those who express the trait are assumed to be homozygous), 18% can be assumed to be heterozygous. Out of 100 individuals shown here, 1 is homozygous for the recessive trait and 18 are heterozygous.

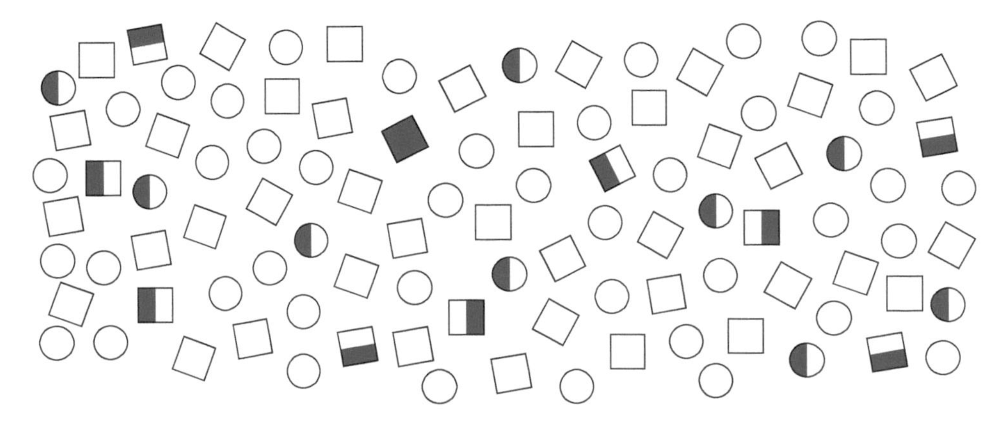

Figure 3.20 Two ways of assessing risk for a recessive trait with single-gene, simple Mendelian inheritance.

risk is by the population frequency of a trait (that is, how many people have the trait in a given population). The likelihood of carrying a recessive allele for a particular condition is higher for a person from a population in which the allele is more frequent (Figure 3.20B). Diagnostic testing for a particular genetic trait is sometimes recommended specifically to persons from those populations or ethnic groups known to have a greater prevalence of the trait. When the frequency of an allele is higher in a particular group, the probability of having an offspring homozygous for the trait is higher if both parents are from the same group than if one marries outside the group.

Examples of genetic testing for recessive alleles because of within-group risk include the following:

- Many African Americans now seek testing to see whether they carry a sickle cell allele because the frequency of sickle cell anemia is higher among African Americans (see Section 7.3.2).
- People of Mediterranean or Southeast Asian descent may seek testing to see whether they carry an allele for thalassemia (see Section 7.3.3) because the frequency of this disorder is higher in these groups.
- Ashkenazi Jews (those of Eastern European descent) commonly seek testing for the recessive allele that causes Tay–Sachs disease, a fatal disorder of brain chemistry, because the frequency of the disease is higher in their group.
- People of Western European (especially Irish) descent may seek testing for mutations in the *CFTR* gene responsible for cystic fibrosis because the frequency of the disease is higher in this group.

Each of these four diseases is a single-gene trait with recessive inheritance. There may be more than one defective allele for a disease and a range of disease severity depending on the exact mutation in the allele; cystic fibrosis is an example. Therefore, determining by genetic testing that a person is homozygous recessive does not necessarily tell you the severity of future disease, or even whether disease will actually develop.

People outside any higher-risk group can also inherit each of these diseases, but their probability of doing so is lower simply because the frequency of the recessive allele is lower in their groups. These traits are rare even in the higher-risk groups where they are "more frequent"—far rarer than the 1% shown in Figure 3.20B. Therefore, most people, even those in a higher-risk group, are not heterozygous carriers of these rare recessive traits. (The frequency of heterozygous carriers can be calculated from the frequency of the recessive trait using the formulas discussed in Box 7.2.)

Genetic testing of this sort should only be done on a voluntary, informed-consent basis and, in general, only when it is of potential benefit to those being tested or to their children. Community leaders of various ethnic groups, including many clergy, have helped to organize genetic testing programs and have encouraged people to participate.

3.4.1.4 Using information from genetic tests When a genotype for a disease is detected, the decision about what to do is left up to the person, or to his or her parents if the person is a child. A patient's decision should be based on a clear knowledge of the possible choices, their consequences, and the extent to which an outcome can or cannot be predicted by the test. Genetic counselors help people to understand these choices, but the code of ethical conduct of genetic counselors prohibits them from making a decision on behalf of their clients: Clients could rightfully resent any counselor who has pressured them into a decision.

Some decisions that must be made after genetic testing are difficult for the people making them. Couples who know they are at risk of bearing children with a genetic disease may decide to adopt children instead. In other cases, knowledge of a genetic condition permits medical intervention at the earliest possible stages, when chances of successful treatment may be better. For conditions that cannot be treated, some couples may choose to abort the fetus bearing the genetic defect. However, people committed to a pro-life position believe that the potential benefits to those being tested or to any future children can never justify what they view as the murder of a fetus.

Genetic testing has already led to some highly inventive mixtures of tradition and modern technology. The Hasidic Jews of Brooklyn, New York, who are mostly descended from the Ashkenazi Jews of Eastern Europe, have a relatively high population frequency of the allele for Tay-Sachs disease. Marriages are traditionally arranged within the Hasidic community, and marriages outside the community are rare; this pattern generally increases the rate at which recessive alleles come together and produce recessive phenotypes. Because they are ethically opposed to all abortions, the Hasidim do not permit genetic testing in utero. The availability of a test that detects Tay-Sachs heterozygotes has, however, allowed the Hasidic community to set up a computerized registry under their strict control. Testing of all persons within the community is encouraged, and the results are entered into the registry under a code number that guarantees confidentiality. The registry permits the traditional matchmakers to check potential couples before proposing a match; if both partners are carriers for Tay-Sachs disease, the matchmaker is warned of this fact and the match is never made. Before this registry was set up in 1984, the Kingsbrook Jewish Medical Center in Brooklyn, which serves the Hasidic community, had an average of 13 Tay-Sachs children under treatment at any one time; after just 5 years, the number of Tay-Sachs children under treatment in the hospital dropped to 2 or 3.

Another option for some couples at risk for having a child with a genetic disease is in vitro fertilization with **preimplantation genetic testing**. Some individuals are morally opposed to in vitro fertilization because it results in the production of more embryos than are actually implanted in the mother, and those embryos are eventually destroyed. However, for couples who do not have this moral objection, a single cell can be removed from an embryo produced by in vitro fertilization, and several different methods can be used to determine whether that embryo has the genetic defect for which the couple is at risk. Thus, only embryos who do not have the genetic defect will be implanted in the mother's uterus, and the risk of a child being born with the genetic defect the test detects is essentially eliminated. There is always the possibility that the embryo could have another, unanticipated genetic defect, however, since the testing is focused on the particular condition for which the couple is known to carry a risk.

3.4.1.5 The ethics of genetic testing Genetic testing is sometimes a mixed blessing. If a genetic defect can either be cured or phenotypically suppressed, or if heterozygote detection permits at-risk couples to decide against having children, then a genetic test can be justified on the grounds that it relieves future suffering. However, most genetic defects cannot be cured. What is the point of testing a person for a condition such as Huntington's disease that can neither be controlled nor cured? One reason is that it permits people who carry a genetic condition to decide whether or not to have children, given that they could pass on the condition to their children or might not live to see their children grow to adults. However, some people would rather live their lives not knowing because they feel the burden of knowing they are highly likely to develop a lethal disease would damage their quality of life. Other questions may also be raised by the results of a genetic test. Will a person who tests positive for such a genetic disease be denied insurance or employment on the basis of the test results? Will a woman choose to abort a fetus if a genetic disease is detected in utero? Box 3.1 examines some of the ethical questions that arise in connection with various forms of prenatal and at-birth testing.

Another ethical issue concerns the use of prenatal screening not for the purpose of detecting a disease-associated allele, but to find out whether the fetus is a boy or a girl, something easily determined from examining the chromosomes. Will couples use this technology to select the sex of their offspring? This already happens in India, where abortion is legal and determination of the sex of the fetus by ultrasound is widely available to those who can afford (or can borrow) a fee of

Box 3.1 Ethical issues in medical decision-making regarding genetic testing

Should society influence the private decisions of individuals? To what extent do (or should) financial considerations limit the choices available? Suppose a child is born with a birth defect or other congenital condition. Is it ever ethical to withhold treatment? (Similar ethical issues are raised by conditions resulting from injuries, infectious diseases, poor maternal nutrition, or other causes.) What if the same disease is diagnosed in a fetus in utero—is it ethical to abort the fetus? The decision to abort a fetus or to withhold treatment from a child with a genetic disease raises important ethical questions. Here are some questions to consider:

1. Tay-Sachs disease is a genetically controlled disease whose victims are in constant pain and never survive beyond about 4 years of age. Does it make sense to spend thousands of dollars on the medical care of a child who has no chance of living beyond age 4, or even of enjoying those few years free from pain? Would it make a difference if a few people with the disease were capable of surviving? What if we were dealing with a disease that people could survive, but only with some disability?
2. When genetic testing has been done, people often meet with a genetic counselor to have the results of the test explained to them. The code of ethics of genetic counselors includes the ethic that they give information about "health risks" in a nondirective and "value-neutral" way. Is this possible? In what ways does a person's concepts of "health" influence how they understand information about health risks?
3. The involvement of third-party insurance policies raises more issues. Should insurance policies pay for genetic testing? Should insurance policies pay medical expenses for genetic diseases that could have been avoided after screening? Some insurance policies will pay for medical treatment, but not for the testing that might have avoided the need for the treatment. Do you think insurance policies should cover genetic screening?
4. Should genetic screening be covered for certain ethnic groups but not others just because the risks differ? For example, thalassemia is more prevalent among Italians, Greeks, and certain Southeast Asians; should insurance cover testing for this condition in a person of Italian descent but not in a person of English or Danish descent? In the United States, ethnic descent is often mixed or unknown—should genetic testing be covered in those cases?
5. If a genetic disease is detected during pregnancy, should insurance policies cover termination of the pregnancy if desired by the parents? If parents elect not to terminate a pregnancy and a child is born with a genetic disease, should insurance policies cover any specialized medical care that might be necessary? Should insurance companies be allowed to deny coverage or raise premiums if a genetic disease is discovered?
6. Screening for some inherited diseases such as PKU is done on all newborns. Informed consent is not gathered, as the screening is universal and diet can prevent the harmful effects of the mutation. As tests are developed for other genetic diseases, should screening for these become routine? Do parents have the right to refuse to be told the results of such tests? What if there is no cure for the condition detected?
7. What role might science have in answering questions of medical ethics that are related to genetics? For example, can science help in assessing the benefits and risks of genetic screening? How reliable are the genetic tests? (For example, the test to detect a cystic fibrosis allele is currently less than 90% accurate.) To what extent does the detection of an allele predict a harmful phenotype? What constitutes sufficient evidence that a condition is genetically determined? What are the limits of the ability of science to contribute answers to these questions of medical ethics?
8. Direct-to-consumer genetic testing services offered by companies such as 23andMe provide customers with information about genetic variants they possess that may be associated with a wide range of health conditions and personal attributes, although many of the phenotypes they report on are influenced by a large number of genes as well as the environment. The results are provided without the guidance of a physician or genetic counselor, as would occur in a medical setting. What are the potential dangers of such tests? What are the potential benefits?

a few hundred dollars. A 1988 study of 8000 abortions in India's clinics showed that 7997 were female and only 3 were male. In the United States, clinics offering prenatal genetic testing have found that over one-fourth of the couples who come to them are motivated by the possibility of choosing their baby's sex.

As genetic testing becomes more common, it is inevitable that test results will occasionally be misused. In one case, school officials were told that a child needed to be kept on a special diet because he had phenylketonuria. Although he was functioning normally, the child was placed in a class for the learning disabled. The school officials apparently knew that PKU could interfere with brain development, but were unaware that this outcome could be averted by the special diet. As this case shows, the misuse of genetic information may result from ignorance.

Discrimination by employers or insurers represents another possible misuse of genetic information. In 2008, the Genetic Information Nondiscrimination Act (GINA) was passed by the U.S. Congress and signed into law by President George W. Bush. At the time, it was not known how many cases might have occurred in which employers refused to hire or promote, or insurance companies refused to insure, persons who were known to have or suspected of having a genetic condition. This legislation was intended to preempt such possible discrimination in light of the increasing number of available genetic tests. GINA prohibits discrimination in employment and health insurance based on a broad definition of genetic information, which includes the results of genetic tests as well as a person's family history. There are some limitations in its provisions: Employers with fewer than 15 employees are exempt, and the law does not address long-term care, life, or disability insurance. Some states have passed additional laws that broaden the protections against genetic discrimination in their states. As an example of protection provided by GINA, one mining company was found to be in violation of the law because they required prospective hires to undergo a medical exam (permissible under the law because of physical requirements of the job) and to answer questions about their family health history (unacceptable under the law).

So far, we have mostly been discussing genetic testing performed in medical contexts when there is some reason to believe a person might be at greater risk for a certain genetic condition. However, in the past decade, so-called direct-to-consumer genetic testing has come onto the scene and has rapidly gained popularity. Companies such as 23andMe allow individuals to submit their genetic material (through saliva samples) and, for a relatively low cost, to receive personalized reports that provide information on ancestry as well as disease risk and other dimensions of health. Certain testing kits now widely marketed in drugstores further promise personalized information about topics such as skin care, allergies, and weight control. The information provided by these direct-to-consumer genetic testing reports can certainly be interesting, but there is a danger of consumers making decisions about their health that are based on incomplete understanding of the information they receive from these companies. Because many disease and personal traits are influenced by a large number of genes as well as environmental factors, the fact that a person has a particular gene variant that has shown some association with a given trait may have little predictive value on its own. Therefore, people who decide to take these tests should be sure to consult their physicians before making any lifestyle changes based on the information they receive.

3.4.2 Altering individual genotypes

Recent developments in gene editing, such as CRISPR (discussed in Section 4.1.3), are rapidly bringing us to the point where it is sometimes possible to correct certain genetic defects by direct alteration of the individual genotype in the affected tissue. Using such gene editing techniques in embryos (which would affect all cells in the body, not just a specific tissue) is still controversial and potentially dangerous (Section 4.1.3), but, over time, this may also become accepted as a means to correct severe, life-threatening genetic conditions. We will cover the details of gene editing and other types of gene therapy in the next chapter.

3.4.3 Altering the gene pool of populations

Some people have proposed that, instead of treating people one at a time, we should alter the genetic makeup of populations (the entire gene pool) by changing the frequencies of certain genotypes. One difference between this approach and the approaches already described has to do with who is perceived to reap the benefits. Genetic testing, counseling, and the altering of individual genotypes are justified in terms of the pain and suffering that may be spared to individuals. In contrast, all attempts to alter the gene pool carry with them notions of harm or benefit to society rather than to the individual.

3.4.3.1 Positive eugenics The altering of the gene pool through selection is called **eugenics**, from Greek words meaning "good birth." This idea is not at all new: Plato's *Republic* (Book 5) suggests that the best and healthiest individuals of both sexes be selected to be the parents of the next generation, much as we breed our horses and cattle. Plato's type of eugenics is called "**positive eugenics**," meaning an attempt to alter the gene pool by selectively *increasing* the genetic contributions of certain chosen individuals or genotypes. Positive eugenics was also proposed in the twentieth century by the Nobel Prize–winning geneticist H.J. Muller, who advocated setting up sperm banks to which selected male donors would contribute. Muller thought that women would eagerly seek artificial insemination with these sperm in the hopes of producing genetically superior children. Several entrepreneurs have established sperm banks (and a smaller number of egg banks) offering to infertile couples (and others) the gametes of people thought to carry desirable traits. The system is not regulated, however, by any public agency, and many sperm banks and egg banks seem to be motivated more by profit than by any desire to change the gene pool. In 1999, a photographer in California began advertising the eggs of several fashion models, offering them at auction to the highest bidder over the Internet.

Ethical and other questions raised by positive eugenics usually center on the lack of an agreed-upon standard for human excellence. The traits most often discussed by those who favor eugenics are intelligence and athletic ability. However, these traits are genetically complex and are highly influenced by education, training, and other environmental variables. Studies attempting to demonstrate a genetic influence on these and other traits were, in many cases, poorly done, leading many scientists to doubt the existence of any reliable evidence concerning the genetic control of human intelligence and other complex traits.

The complexity of the human genotype raises other issues: What if Einstein had been heterozygous for some genetic disease? If a society wanted to use his germ cells to breed people of superior intelligence, they would also unwittingly be selecting whatever other traits he happened to possess, possibly including a genetic defect in the process. Suppose a person inherited the manic-depressive disorder of Robert Schumann or Vincent Van Gogh, instead of their creative talents? What liability or what responsibility would a sperm bank face if a descendant were born with a genetic defect? What constitutes "superiority" in an individual, and who should have the power to make such choices?

3.4.3.2 Negative eugenics Most discussions of eugenics have centered on "**negative eugenics**," the prevention of reproduction among people thought to be genetically defective or inferior. Founded by Francis Galton (1822–1911), a cousin of Charles Darwin, the modern eugenics movement has generally tended to emphasize negative measures. Galton and his supporters were very much interested in measuring intelligence, and they developed some of the early versions of what we now call IQ tests. Through the use of these and other tests, supporters of eugenics have long sought scientific respectability for their attempts to label certain people as genetically defective or inferior.

The Nazis instituted a program of negative eugenics in Germany, beginning with the forced sterilization of mental "defectives," deaf people, homosexuals, and others. The eugenics program soon grew into a program for the mass killing of all those millions who did not belong to Hitler's "master race." By 1945, the Nazis had killed millions in the name of racial purity and Aryan superiority. The Nazis also practiced positive eugenics by encouraging German women with certain traits to have more children.

In the United States, the eugenics movement started as a series of attempts to identify, segregate, and sterilize mental "defectives." The movement soon found allies among racists and especially among those who sought to curb the new waves of immigration during the period from about 1890 to 1920. During the 1890s, one Kansas doctor sterilized 44 boys and 14 girls at the Kansas State Home for the Feeble-Minded, while Connecticut passed a law prohibiting marriage or sexual relations between any two people "either of whom is epileptic, or imbecile, or feeble-minded." A 1907 Indiana law required the sterilization of "confessed criminals, idiots, imbeciles, and rapists in state institutions when recommended by a board of experts." Fifteen other states passed similar laws, as often for punitive as for eugenic reasons. From 1909 to 1929, 6,255 people were sterilized under such laws in California alone.

The writings of the American eugenicists became increasingly racist and anti-immigrationist in tone during this period. One eugenicist, for example, wrote in 1910 that "the same arguments which induce us to segregate criminals and feebleminded and thus prevent breeding apply to excluding from our borders individuals whose multiplying here is likely to lower the average [intelligence] of our people." In the 1960s, H.J. Muller wrote several articles warning against the practice of protecting and extending the lives of the "genetically unfit," those whom natural selection would tend to eliminate from the population. According to Muller, our medical intervention would only perpetuate genetic defects in our gene pool. Muller spoke pessimistically of a population divided into two groups, one so enfeebled from genetic defects that their very lives had to be sustained by extraordinary means, and the other group consisting of phenotypically normal people who had to devote their entire lives to the care and sustenance of the first group. Muller's views have not been substantiated by any evidence.

3.4.3.3 Biological objections to eugenics Biological arguments against negative eugenics are based on the realization that eugenic measures could be expected to produce only small changes at great cost. Most known genetic defects are both rare and recessive, and selection against rare, recessive traits can only proceed very slowly no matter what the circumstances. As the trait gets increasingly rare, selection against it becomes increasingly ineffective. For example, the gene for albinism has a frequency of about 1 in 2,000 in many human populations. If a eugenic dictator ordered all albinos to be killed or sterilized, theoretical calculations show that it would require about 2,000 generations (about 50,000 years) of constant vigilance just to reduce the frequency of this trait to half of its present value. The reason why the process works so slowly is that most individuals carrying the gene for a rare, recessive trait are heterozygous and their phenotype does not reveal the presence of the gene. In contrast, modern techniques that allow the detection of the gene in heterozygous form would greatly increase the effectiveness (and hence the dangers) of negative eugenic measures.

For characteristics such as height or IQ, which are controlled by many genes and influenced strongly by environmental factors, estimates are that eugenic selection would be so slow as to be barely perceptible. One geneticist calculated that it would take about 400 years of constant, unrelenting, and totally efficient selection to raise IQs by about 4 points; the same improvement could be achieved through education in as little as four years, and with far less cost.

Finally, eugenic measures can, at best, address only a small percentage of undesirable conditions, because most physical disabilities and medical conditions result from accidents, from infectious illnesses, or from exposure to toxic substances in the environment, not from inherited genetic makeup, and eugenic measures are powerless to alter these nongenetic causes. Genetic conditions may also result from new mutations, rather than from the inheritance of defective genes, and eugenic measures have no capacity to eliminate newly mutated genes or to depress the frequency of any gene below the mutation rate.

There is no biological basis for the claims of any eugenics movement that their methods could in any way improve humankind other than at great cost. The risks of negative eugenics are especially great and include the possibility of genocide—the attempted extermination of a race or ethnic group. In addition, there are no biological benefits. We are coming to know that population health and stability depend on genetic diversity; thus, narrowing a gene pool eugenically makes the population more vulnerable to infectious disease. Infectious disease is a far more widespread and common cause of sickness and death than is genetic disease.

3.4.4 Changing the balance between genetic and environmental factors

Although many traits are inherited, most are also influenced by the environment. Phenotypes result not just from genes but also from the interactions between genes and their environment. In addition, far more disability is caused entirely environmentally through accidents and illnesses than is caused genetically. Even for most conditions that are caused genetically, elimination of an allele from the population is hardly the only option. Most genetic conditions can be modified or accommodated in several different ways.

3.4.4.1 Euphenics **Euphenics** (literally, "good appearance") includes all those techniques that either modify genetic expression or alter the phenotype to produce a modified phenotype. Plastic surgery, such as to repair body parts, is a form of medical intervention that alters the phenotype. Other examples include the installation of pacemakers in defective hearts, the giving of insulin to people with diabetes, and the dietary control of phenylketonuria. Although the genes remain unchanged, euphenics modifies or compensates for their phenotypic expression in such a way that they no longer cause harm.

Many leading geneticists have argued, as an alternative to eugenics, that there is nothing wrong with altering the phenotype or the environment so that formerly disabling genotypes are no longer so harmful or debilitating. A leading advocate of this viewpoint was Theodosius Dobzhansky (1893–1975), who favored measures to permit people with hereditary "defects" to overcome their handicaps and become phenotypic copies (phenocopies) of normal, healthy human beings. Once phenotypes could be controlled culturally, said Dobzhansky, the presence of formerly defective genotypes would cease to be the subject of any great concern. Euphenic intervention is already common practice for a number of genetic conditions. As our ability to modify phenotypes increases (e.g., with advances in corrective surgery), this type of practice is likely to become more common.

3.4.4.2 Euthenics Another type of intervention is called **euthenics**. In this form of intervention, both genotype and phenotype remain unchanged, but the environment is modified or manipulated so that the phenotype is no longer as disabling as before. (In euphenics, by contrast, the phenotype is modified.) Examples of euthenic measures include canes, crutches, wheelchairs, and wheelchair ramps for those who cannot walk unaided; guide dogs and Braille for the sight-impaired; eyeglasses for the nearsighted; and so on. Conditions that are improved or assisted by euthenics may be either genetic or not.

Most people with disabilities support research that would prevent the recurrence of their condition in other people, especially if pain or paralysis is involved. However, medical research is expensive, its results are uncertain, and its benefits may take many years to become widely available. Euthenic measures are often less expensive and more quickly made available once they are developed. Many of the people who use euthenic devices feel that they would be better served by simple improvements in the devices (e.g., better wheelchairs) than they would be if medical research were our only emphasis.

3.4.4.3 Eupsychics Many people with uncommon traits or conditions (whether genetic or not) feel that they are best served by being accepted as they are and do not necessarily want to be "cured" (Box 3.2). Social and behavioral measures, or **eupsychics**, may lessen the impact of or compensate for disabling conditions. Included are the special education of handicapped individuals, mainstreaming (education of the handicapped in a regular public school setting), and the education and social conditioning of nonhandicapped members of society so that they will better understand and accommodate the needs of all citizens.

Genetic research seeks to understand the molecular mechanisms underlying normal physiology and health, but genetic testing and counseling emphasize diseases. The assumptions underlying genetic testing have at times included viewing variations as defects and desiring to apply to humans some arbitrary standard of perfection. Society may be better served by an emphasis on the abilities, rather than the disabilities, of each individual. All individuals should be encouraged to develop their talents and abilities to the fullest. Whenever a person is discouraged from trying to develop a certain skill, ability, or talent, both the individual and the society are the losers in the long run.

Box 3.2 Who decides what is considered a "defect"?

Very often, groups of people that society wants to "help" are given little or no voice in how society will treat them or "help" them. The deaf are a case in point. Some forms of deafness are inherited, yet many deaf people consider it abhorrent if genetic counseling is used to avoid having deaf children.

The following excerpt is from a speech by I. King Jordan delivered in 1990 to an international symposium on the genetics of hearing impairment. Research funded by the Human Genome Project is aimed at identifying genetic causes of deafness,

(*Continued*)

Box 3.2 continued

and there is an underlying assumption on the part of geneticists who are not deaf that people would then want to avoid this trait. A deaf man, Jordan was the first deaf president of Gallaudet University in Washington, D.C., established in the 1800s to educate deaf people. He served in that role from 1988 to 2006. He explains that he is deaf both medically and culturally. He speaks for most deaf people when he tries to explain that deafness is not a trait that he would want to eliminate.

> *For about 18 years, I have taught a course on the psychology of deafness. One of the first things we discuss in the class is the difference between viewing deafness as a pathology that should be cured or prevented and viewing it as a human condition to be understood. I call these two perspectives the medical and cultural points of view. Individuals from these two groups agree on audiological definitions, but disagree on the emphasis that should be given to social and rehabilitative services. I adhere to the social or cultural point of view.*
>
> *What I mean by this is that I, personally, and many of the people I know well, have accepted the fact that deafness is one aspect of my individuality. I do not spend any time or energy thinking about curing my deafness or restoring my hearing, but I do spend substantial time and energy trying to improve the quality of life for all people who are deaf.*
>
> *For some reason, people who hear have a very difficult time understanding this concept. If you will permit me to digress for a moment, I will give you an example. I was interviewed by Ms. Meredith Vieira for the television show "60 Minutes." During the interview, she asked me this question: "If there was a pill that you could take and you would wake up with normal hearing, would you take it?" I told her that her question upset me. I told her that it was something I spent virtually no time at all thinking about, and I asked her if she would ask me the same question about a "white" pill if I were a black man. Then I asked if, as a woman, she would take a "man" pill. Our conversation continued long after the videotaping was done, and we have had several subsequent conversations. But she never understood. She still does not. She still thinks only from her own frame of reference and imagines that not hearing would be a terrible thing. Deafness is not simply the opposite of hearing. It is much more than that, and those of us who live and work and play and lead full lives as deaf people try very hard to communicate this fact.*
>
> *As you can see, this is an emotional issue for me. It is a much more emotional issue for many other deaf people. Is that relevant here? Yes, I believe it is, because the genetic study of deafness and genetic counseling have a great deal of significance for the deaf community generally. Many deaf people, particularly those who consider themselves members of the deaf community, do not consider themselves to be defective, rather, they consider themselves to be different, normal but different. In particular, this difference has a cultural or sociological basis and is expressed most saliently in the use of sign language. If deaf people are not defective or dysfunctional then, at least in their own eyes, it follows that they would be suspicious of attempts to eradicate deafness. . . .*
>
> *Genetic counseling and screening with respect to potential deafness must differ, therefore, in a fundamental way from screening for "birth defects."*

[I. King Jordan, "Ethical Issues in the Genetic Study of Deafness," *Annals N. Y. Acad. Sci.* 630: 236–237.]

THOUGHT QUESTIONS

1. Are all "birth defects" genetic defects? Do the same ethical questions regarding diagnosis and counseling apply to both genetic and nongenetic traits?
2. In many hospitals, screening for phenylketonuria is often performed on all infants. Does this violate the principle of informed consent? Is this practice ethical or not? How is it commonly justified? Do you feel that the justification is adequate?
3. Some groups opposed to abortions have also begun to object to certain kinds of genetic testing. What good, they ask, can come from knowing that a fetus suffers from a particular genetic or chromosomal defect if the parents are opposed to abortion of the fetus on religious or similar grounds? For such situations, discuss the costs, benefits, and ethical status of genetic testing. Does it matter what kind of testing is performed? Does it matter what genetic or chromosomal defect is being tested for?
4. Do unborn children have a "right" to inherit an unmanipulated set of genes? Do they have a right to inherit "corrected" genes if such a possibility exists? What kind of informed consent can we expect on behalf of unborn generations? Can a person make decisions that affect the genotype of all of his or her progeny? Do we need safeguards to protect future generations against the selfish interests of the present generation?
5. In what sense is "positive" eugenics positive? In what sense is it negative? Should human populations be bred as we breed domesticated animals?
6. If public funds will be spent for the care and treatment of a person with a genetic condition, does that alter the ethical balance between the rights of the individual and the rights of society? If a euphenic measure is available, should public funds be used to "correct" the condition? Should the person have the right to refuse such treatment? Should public funds be withheld from a person who refuses such euphenic measures?
7. How much access to genetic information about a subscriber or employee *should* an insurance company or an employer have? How much access *do* they have now?

CONCLUDING REMARKS

Genes go through transcription to RNA and translation to protein, which was once thought to lead directly to phenotypes. Scientists now realize that no gene works independently of its cellular environment and that phenotypes for many traits are modifiable.

As we have seen in many contexts in this chapter, most of human genetics is more complex than the either/or traits that Mendel studied in pea plants. It is not that Mendel's laws do not apply to humans—they do—and there are some human conditions that do follow simple Mendelian inheritance with a mutation in a single gene leading to the trait. However, since Mendel's time, we have learned that the inheritance of most traits in all organisms, not just in humans, is more complex than Mendel and scientists early in the twentieth century envisioned. Many traits, including sex determination and disease susceptibility, are influenced by multiple genes. The expression of the genotype as a phenotype is also influenced by the environment. Most biologists today would agree that phenotypes are far more malleable than was assumed in the past, contradicting earlier ideas of biological determinism that assumed that genotypes solely determined phenotypes. In addition, phenotypes can be modified by technologies and other aspects of cultures.

We are now in an era in which people are able to find out a lot about their own genotype and the genotypes of their children. Identifying a genetic predisposition to a chronic disease may allow a person to make healthier choices about his or her diet and lifestyle. As we saw with the couple in the introduction to this chapter, in many cases, people now have the power to find out whether a fetus carries a genotype for a fatal illness. This new knowledge gives people choices that they did not have in the past, but oftentimes these are still not easy choices to make.

CHAPTER SUMMARY

- DNA, a nucleic acid, is the template for the synthesis of another nucleic acid, RNA, in a process called **transcription**.
- **mRNA** is the template for protein synthesis in a process called **translation**. Three mRNA bases form a **codon**, directing the addition of one **amino acid** to a protein.
- Changes in the DNA sequence (**mutations**) are reflected as changes in the mRNA sequence that usually result in changes in the amino acid sequence of proteins. Such changes may affect the folded shape and therefore the function of the proteins. Gene mutations create new alleles.
- Phenotypes result from the activity of proteins, often of several proteins acting together.
- A **karyotype** is the full set of chromosomes from a cell, photographed during mitosis and arranged in homologous pairs by size, shape, and banding pattern.
- In humans and most animal species, one pair of chromosomes differs between males and females; these chromosomes are called **sex chromosomes**.
- Human sex development is coded for by many genes, including a gene on the Y chromosome.
- Genes that are located on either the X or the Y chromosome are called **sex-linked genes**.
- Human genes are now being identified at an increasingly rapid pace through **pedigrees** and linkage with DNA markers.
- Many genetic diseases result from alleles that code for nonfunctional proteins. Some of these are inherited as simple Mendelian alleles; others show more complex patterns of inheritance.
- **Risk** is the probability of a condition's occurring; some genotypes are associated with increased risk.
- Genetic diseases can be diagnosed prenatally by **amniocentesis**, by **chorionic villus sampling**, and by other techniques, including those that use

the **polymerase chain reaction (PCR)** to amplify DNA sequences into many copies. Testing for genetic diseases can also be done on children or adults. In any case, testing requires prior **informed consent**.
- Altering the gene pool at the population level is called **eugenics**.
- For some genetic diseases, an enzyme product can be supplied artificially, or some substitute or mechanical compensation can be engineered.

BIBLIOGRAPHIC REFERENCES

Bibliographic references to material in this chapter can be found online at biologytrending.routledge.com/chapter3

GLOSSARY: KEY TERMS TO KNOW

These key terms are defined at the end of each chapter as an aid for student review.

Alkaptonuria A genetic condition (inborn error of metabolism) in which urine turns dark upon exposure to air because of the body's inability to break down a compound called homogentisic acid.

Amino acid A small molecule containing both an amino group ($-NH_2$) and a carboxylic acid group (-COOH) that is capable of being joined with other similar molecules to form long protein chains.

Amniocentesis A procedure that involves removing a small amount of amniotic fluid from a woman's uterus to test for possible genetic abnormalities in a developing fetus.

Anticodon A three-nucleotide sequence in a transfer RNA molecule that pairs with a messenger RNA codon.

Autosome A chromosome not involved in sex determination.

Carrier In genetics, an individual carrying a recessive allele in their genotype but not showing its phenotypic effect in their phenotype.

Chorionic villus sampling Removal of a small amount of tissue from the placenta to test for possible genetic abnormalities in a developing fetus.

Chromosomal aberration A change in a chromosome sequence large enough to contain many genes. Examples include chromosomal inversions, translocations, duplications, and deletions.

Chromosomal rearrangement Same as a chromosomal aberration.

Codon A coding unit of three successive nucleotides in a messenger RNA molecule that together determine an amino acid.

Concordance (rate of) In studies of twins or other matched individuals, the fraction of individuals with a certain trait whose twin (or matched individual) also has the trait.

Cystic fibrosis An inherited disease, caused by a defective CFTR chloride-ion transport protein, that results in the buildup of thickened fluids and frequent infections in the lungs, among other symptoms.

DNA marker Any part of DNA whose chromosomal location is known, permitting it to be used to help locate genes.

DNA probe See *Probe*.

Down syndrome A complex syndrome that includes varying degrees of altered brain development, plus eyes with epicanthic folds and, in some cases, heart malformations, arising usually from trisomy of chromosome number 21 or, less often, from other chromosomal abnormalities.

Electrophoresis A technique used to separate DNA or protein fragments by size according to their movement through a gel matrix, driven by an electric field.

Embryo The earliest stage of development. In humans, it lasts from fertilization to about eight weeks of gestation.

Eugenics An attempt to change allele frequencies through selection or changes in fitness. Raising the fitness of desired genotypes is called "positive" eugenics; lowering the fitness of undesired genotypes is called "negative" eugenics.

Euphenics Measures designed to alter phenotypes (producing phenocopies) without changing genotypes.

Eupsychics Social and educational measures that accommodate people with differences.

Euthenics Measures designed to assist people to overcome some of the consequences of their phenotypes. Wheelchairs and eyeglasses are examples.

Expressed sequence tag (EST) A short segment of an RNA (or its cDNA) that can be used as a genetic marker to indicate the level of expression of the corresponding gene, which may vary between individuals.

Fetus A developing embryo after all of its organs have formed; in humans, from about eight weeks postconception to birth.

Genome The total genetic makeup of an individual, including its entire DNA sequence.

Genetic code The correspondence between three-nucleotide codons and the amino acids (or translation stop signals) that they specify. See Table 3.1.

Huntington's disease An inherited condition, usually beginning at midlife, with uncontrolled twitches or spasms and resulting in a progressive and fatal neurological deterioration.

Informed consent A voluntary agreement to submit to certain risks by a person who knows and understands those risks.

Karyotype The chromosomal makeup of an individual.

Klinefelter syndrome A condition arising from the chromosomal arrangement XXY, resulting in a sterile male, often thin, with underdeveloped genitalia and varying degrees of lowered intelligence and breast development.

Linkage An exception to the law of independent assortment in which genes located on the same pair of chromosomes tend to be inherited together, with the parental combinations of alleles predominating.

Linkage study A strategy for identifying the gene(s) associated with a phenotype (such as a disease) by following genetic markers that tend to be inherited with the phenotype.

Linked genes Genes carried on the same pair of chromosomes. See *Linkage*.

Messenger RNA (mRNA) A strand of RNA that leaves the nucleus after transcription and passes into the cytoplasm, where it functions in protein synthesis.

Microsatellite marker A type of DNA marker in which a short nucleotide sequence is repeated a different number of times in different individuals.

Mosaicism The existence of cells or patches of cells that differ genetically from one another within an organism because of changes that took place during that individual's development.

Muscular dystrophy An X-linked genetic disorder in which muscles become weak and nonfunctional, usually resulting in a shortened lifespan.

Mutation Any sudden, heritable change in a DNA sequence or gene.

Negative eugenics Any attempt to change allele frequencies by lowering the fitness of those with undesired traits.

Pedigree A chart showing inheritance of a genetic trait within a family.

Phenylketonuria An inherited condition (inborn error of metabolism) in which the amino acid phenylalanine cannot be broken down and instead produces toxic products that impair health.

Polymerase chain reaction (PCR) An iterative method for amplifying (making many copies of) a specific DNA sequence using a template containing that sequence, plus a heat-stable DNA polymerase enzyme and primers that specify the ends of the sequence to be amplified.

Positive eugenics Any attempt to change allele frequencies by raising the fitness of those with desired traits.

Preimplantation genetic testing When blastocyst-stage embyros generated by in vitro fertilization are checked for chromosomal or other genetic abnormalities prior to transfer to the uterus; performed by removal of a small number of cells from the part of the embryo that will become the placenta.

Probe A piece of DNA with a sequence complementary to the marker sequence that one wishes to detect and capable of being revealed by a fluorescent or radioactive tag.

Rate of concordance See *Concordance*.

RFLPs (Restriction Fragment Length Polymorphisms) Differences in the lengths of the restriction fragments made by the same endonuclease among different individuals.

Ribosomal RNA (rRNA) An RNA sequence that combines with protein to make a ribosome.

Ribosome An intracellular particle containing RNA and protein and serving as the site of protein synthesis during translation.

Risk The probability of occurrence of a specified event or outcome.

RNA polymerase An enzyme that transcribes the DNA template strand of a gene into a complementary RNA strand.

Sex chromosomes One of the chromosomes that differ between the sexes, usually distinguished as X and Y.

Sex linked Carried on the X chromosome (more common) or on the Y chromosome (much less common).

Single-Nucleotide Polymorphism (SNP) A nucleotide position within a genome that shows variation across a population and can be used as a DNA marker.

Transcription A process in which DNA is used as a template to guide the synthesis of RNA; the first step in gene expression.

Transfer RNA (tRNA) A small RNA molecule that inserts an amino acid during translation based on matching its anticodon with a messenger RNA codon.

Translation A process in which amino acids are assembled into a polypeptide chain (part or all of a protein) in a sequence determined by codons in a messenger RNA molecule; the second step in gene expression.

Turner syndrome A condition arising from a single unpaired X chromosome (XO), resulting (if untreated) in a sterile female with immature genitals, widely spaced breasts that do not develop fully, webbing of skin at the neck, and varying degrees of lowered intellectual functioning.

CONNECTIONS TO OTHER CHAPTERS

Chapter 1: Molecular genetics is forcing a change in our concepts of phenotypes and traits.
Chapter 1: Manipulating human heredity has raised several ethical concerns.
Chapter 2: The same basic laws of genetics apply to humans as to other animals and to plants.
Chapter 4: The sequencing of the human genome has provided new information to assist in identifying human genes linked to particular diseases or phenotypes.
Chapter 4: Human genes are often very similar to those from other species, and the study of these relationships can tell us about a species' evolutionary history.
Chapter 4: CRISPR gene editing offers the potential to correct a range of genetic defects.
Chapter 5: Evolution takes place whenever allelic frequencies change.
Chapter 7: Human populations differ in the frequencies of many alleles.
Chapter 8: The functions of enzymes and other proteins are based on their three-dimensional structures, which can be altered by mutations. "Inborn errors of metabolism," which were among the first links made between genotype and phenotype, affect various metabolic enzymes.
Chapter 9: Population control seeks to manage the size of populations; eugenics, in contrast, seeks to alter the gene pool.
Chapter 10: Predispositions for many cancers are hereditary, and cancer cells accumulate large numbers of mutations as the disease develops.
Chapter 11: Huntington's disease is one of several brain disorders for which a genetic basis has been identified.
Chapter 12: Certain genes can influence the aging process and the risks for Alzheimer's disease.
Chapter 17: Identifying plant genes that influence particular phenotypes can allow genetic engineering to be used to improve the traits of commercially important plant species.
Chapter 18: Conserving genetic diversity is an important aspect of protecting biodiversity.

PRACTICE QUESTIONS

1. How many genes are involved in sex determination in humans? How many chromosomes are involved?
2. Do you and your mother have the same karyotype? Why or why not? What about you and your brother?
3. Do all people have the same number of chromosomes?
4. Specify what type of individual would be formed from a human zygote containing each of the following:
   a. 46 chromosomes, including 2 X
   b. 47 chromosomes, including 2 X and 1 Y
   c. 45 chromosomes, including 1 X and no Y
   d. 46 chromosomes, including 1 X and 1 Y
   e. 47 chromosomes, including 2 X and 3 copies of chromosome number 21
5. When a man and a woman who are both albino have children, what percentage of their children will be albino?
6. Under what circumstances would it be true that all of the children of a man with red-green color blindness would also be color blind?
7. Are DNA markers genes? Why or why not?
8. A particular disease is suspected of being a genetic disease. A person with the disease tests positive for 15 DNA markers, each detected with a specific DNA probe, and negative for 20 other markers. What would be the next step in trying to determine which of the markers might be located near a gene responsible for the disease?
9. What is the risk of a child's having a recessive genetic trait when both parents are from a population in which the frequency of the recessive allele is 1 in 1,000 (0.1%)? What is the risk when one parent is from a population with an allelic frequency of 0.1% and the other parent is from one in which the frequency is 1 in 100,000 (0.001%)?
10. How many copies of a DNA fragment are synthesized in 20 rounds of amplification by PCR (assuming that each step works correctly)? How many will be synthesized in 30 rounds?
11. How many different genes are mutated in cystic fibrosis? Do all people with cystic fibrosis have the same mutation?

CHAPTER 4

Genetic Engineering and Genomics

ISSUES

- Does genetic engineering fundamentally change the biology of an organism?
- When should gene therapy or CRISPR gene editing be used? When should they not be used?
- Do DNA tests positively identify individuals?
- Why did the U.S. government fund the Human Genome Project?
- What benefits have been derived from the Human Genome Project?
- How could the results of the Human Genome Project be misused? How can we guard against such misuse?
- How has next-generation DNA sequencing changed the way genetic information is used in daily life?

BIOLOGICAL CONCEPTS

- Biotechnology (The Human Genome Project; genetic engineering)
- Molecular biology (genomics; bioinformatics)
- Structure–function relationships (proteomics)

CHAPTER OUTLINE

DOI: 10.1201/9781003391159-4

A **genome** is all of the DNA sequences that comprise the hereditary information of a given organism. The **Human Genome Project** was a tour de force of international scientific collaboration involving governmental and private funding that spanned more than a decade, from its inception in 1990 to its declared completion in 2003. The project determined the nucleotide sequence of each of the human chromosomes and mapped the locations of all the predicted genes on those chromosomes. All of this information is stored in an enormous database that is publicly available for use by any scientist in the world. The scope of this accomplishment rivals that of the 1969 lunar landing in the amount of collaboration required among hundreds of scientists and other specialists. What is even more astonishing, however, is the pace of advances in DNA sequencing technology, genetic engineering, molecular medicine, and biotechnology that have occurred in the years since the completion of the Human Genome Project, jumpstarted by the wealth of data it produced and the enticement of putting this knowledge to practical use.

4.1 GENETIC ENGINEERING MAKES TARGETED CHANGES TO GENETIC INFORMATION

Direct alteration of individual genotypes is called **genetic engineering**, recombinant DNA technology, or gene splicing. These terms are used interchangeably. As we will soon describe, human genes can be inserted into human cells for therapeutic purposes (gene therapy). In addition, because nearly all species carry their genetic information in DNA and use the same genetic code, genes can be moved from one species to another. This chapter covers genetic engineering and its applications in human medicine; genetic engineering in plants is discussed separately in Chapter 17.

4.1.1 Methods of genetic engineering

Whether the "engineered" gene is one from the same species or a different species, the techniques are much the same; furthermore, they all depend on being able to cut, copy, and reassemble the genetic material in predictable ways. These are made possible by the use of special enzymes called **restriction endonucleases** and by the **polymerase chain reaction** (**PCR**), which makes many copies of specific DNA sequences.

4.1.1.1 Restriction endonucleases Restriction endonucleases, also called **restriction enzymes**, are enzymes that cut the DNA backbone at specific sites. A nuclease is any enzyme that cuts DNA, and *endo*nucleases cut in the middle of a DNA sequence rather than removing nucleotides from the end of a DNA molecule. There are about 3,000 restriction enzymes currently known that can cut DNA at over 230 different nucleotide sequences. Each type of restriction enzyme is a normal protein made by a particular bacterial species, and most are named for the bacteria from which they are derived. Thus, in Figure 4.1, *Hae*III is an enzyme from the bacteria *Haemophilus aegypticus*, and *Eco*RI is from *Escherichia coli* (*E. coli*). They are called restriction enzymes because their normal function within the bacteria is to restrict replication of bacterial viruses, also known as bacteriophage. Bacteria express DNA-modifying enzymes that make their own DNA resistant to degradation by their restriction enzymes; however, because the bacteriophage DNA is not modified, it can be cut, thus preventing the phage from replicating and killing the bacterial cells.

Several other nucleases exist that cut DNA randomly, but an enzyme that acts indiscriminately is of little use in genetic engineering. Restriction enzymes are useful because they cut with specificity. The target sites of restriction enzymes are generally about four to eight nucleotides long (Figure 4.1). Most, but not all,

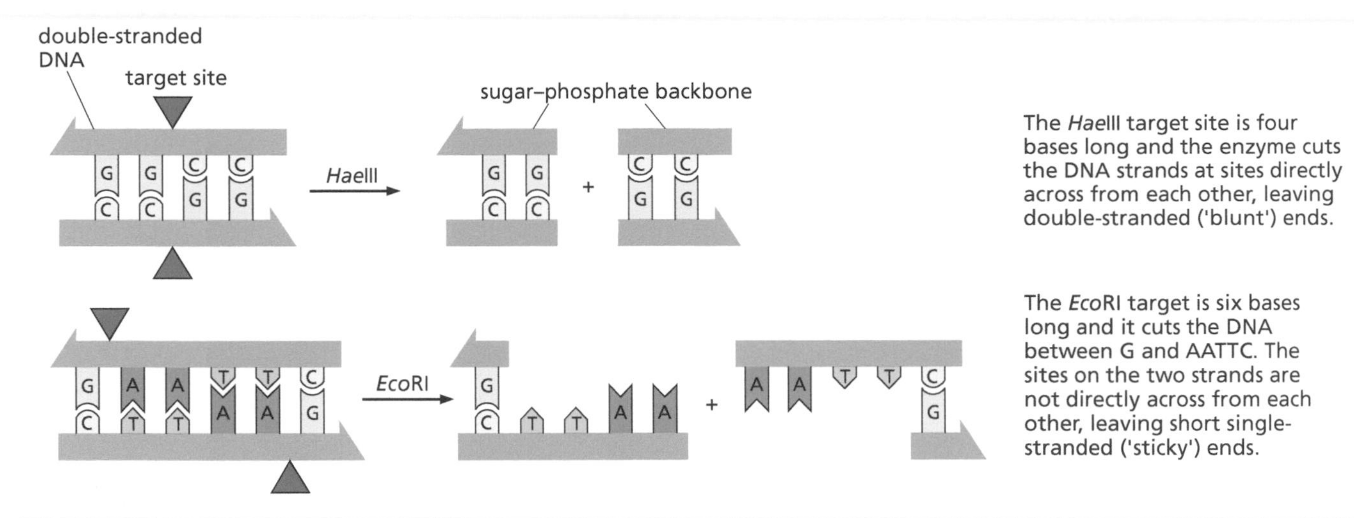

Figure 4.1 Restriction enzymes. The nucleotide sequences recognized and cut by the restriction enzymes *Hae*III and *Eco*RI are shown.

restriction enzyme recognition sites are DNA **palindromes**; this means that their nucleotide sequence reads the same in the direction of synthesis (5' to 3') on each of the two strands of the double-stranded DNA molecule or that the sequence is capable of base-pairing with itself when folded in half. The palindromes that are recognized by the *Eco*RI and *Hae*III enzymes are shown in Figure 4.1. A restriction enzyme will cut a sample of DNA at all sites that match its recognition sequence, producing a set of smaller pieces called **restriction fragments.** A given restriction enzyme mixed with the same sequence of DNA always produces the same number and sizes of fragments, so the pattern of restriction fragments produced can be thought of as a "fingerprint" that can be used to identify that DNA sequence. Before the discovery of restriction enzymes in 1970, breaking chromosomal DNA into pieces was done mechanically, producing different numbers and sizes of pieces every time, making the DNA fragmentation impossible to reproduce from one experiment to the next. Because restriction enzymes always cut at the same sites, they can be used in genetic engineering.

4.1.1.2 Restriction enzymes in genetic engineering

The first step in genetic engineering is to isolate the gene of interest. A specific gene can be isolated by using restriction enzymes to snip out a segment of DNA that contains the gene from a larger DNA molecule. The most useful restriction enzymes are those that cut the two DNA strands of palindromic sequences at locations that are not directly across from each other, producing short sequences of single-stranded DNA known as "sticky ends" (Figure 4.1B). For example, the commonly used restriction enzyme *Eco*RI always targets the sequence 5'-GAATTC-3', cutting it between G and AATTC, thus breaking the two-stranded sequence into fragments that have short stretches of overhanging single-stranded DNA. These overhangs are called "sticky ends" because they can stick together by base-pairing with another molecule containing the complementary overhang. In fragments cut with *Eco*RI, the single-stranded 5'-AATT-3' sequences can pair with one another in the antiparallel arrangement of double-stranded DNA (see Section 2.3.2) and then be joined permanently using an enzyme called **DNA ligase**, which seals the junctions in the DNA backbone. An enzyme such as *Hae*III that cuts at sites directly across from each other forms "blunt," rather than sticky, ends, as shown in Figure 4.1A.

Cutting an entire chromosome (or genome) with a restriction enzyme will produce many fragments, only one of which contains the target gene. The fragments of different lengths are separated by agarose gel electrophoresis (**Figure 4.2A**), a procedure explained in Chapter 3. Because DNA carries a negative charge, it moves toward the positive end of an electric field. When a DNA sample that has been cut into fragments is loaded onto a gel and electric

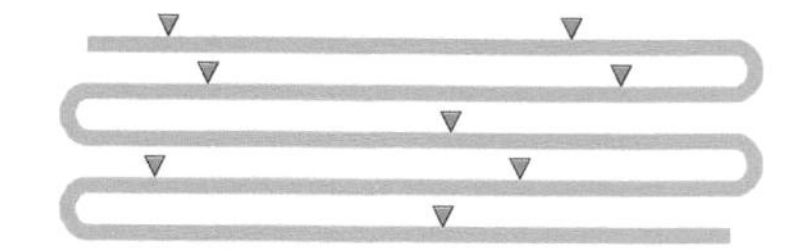
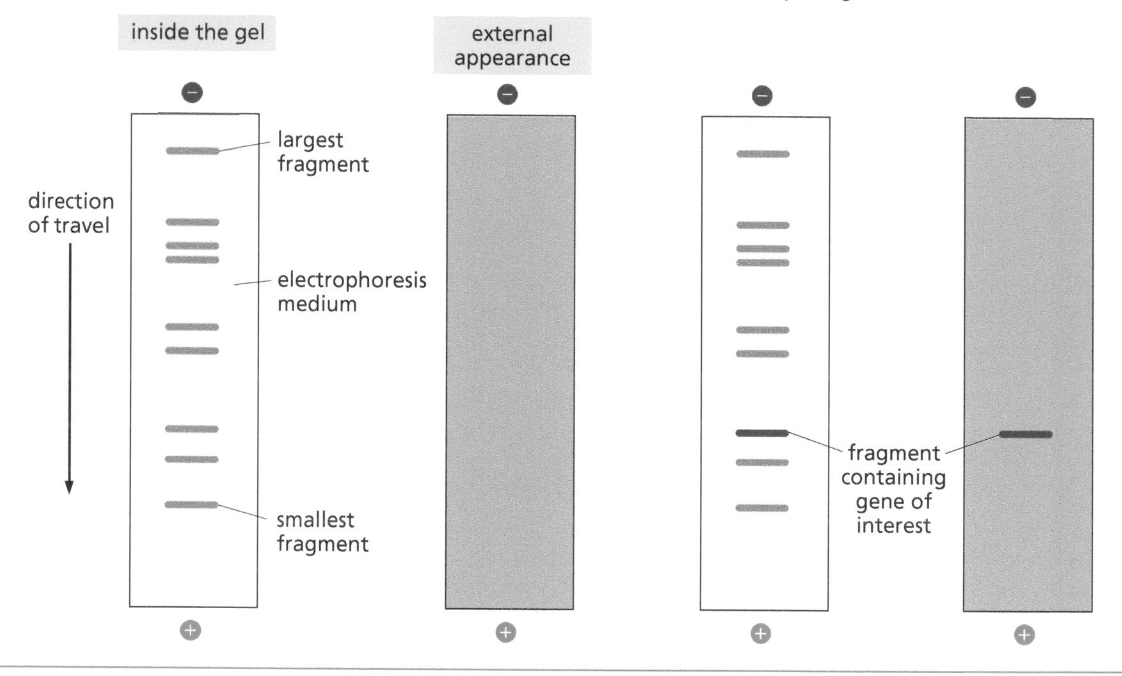

Figure 4.2 Cutting DNA with restriction enzymes and separating the fragments by gel electrophoresis.

current is applied, the fragments move. The gel material retards the movement of the fragments somewhat, and the larger the fragment, the more its movement is retarded by the gel. In the time that the electric current is on, smaller fragments will therefore move farther than large fragments. When an entire genome is subjected to restriction enzyme digestion, identifying the one fragment that contains a gene of interest is like finding a needle in a haystack. A **DNA probe** specific for the gene can be used to home in on the fragment containing the gene of interest, and that fragment can then be isolated from the rest (Figure 4.2A). As we described in Section 3.3.2, a probe is a complementary DNA strand that carries a radioactive or chemical tag. Before the probe is applied, the DNA fragments are usually transferred from the gel to a paper-like material called a membrane; this technique is called Southern blotting, named after the scientist who invented it, Edwin Southern. (This technique is further illustrated in Figure 4.10.)

Probes are still used for some applications; however, the polymerase chain reaction (PCR), first used in 1985, has rapidly become the preferred method for producing isolated DNA fragments for most recombinant DNA experiments (Figure 4.2B). As described in Section 3.4.1 and shown in Figure 3.19, PCR can be used to make many copies of a specific segment of DNA contained within a larger DNA sequence (such as a genome). Short, single-stranded segments of DNA called **primers** define the segment of DNA that is copied, or "amplified." Restriction enzyme recognition sites are often engineered into the ends of the

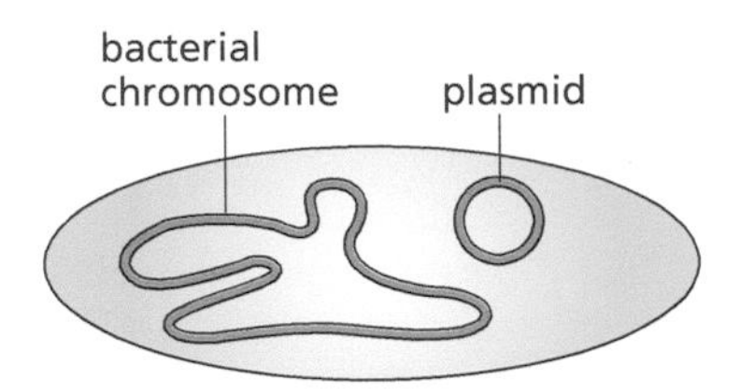

Figure 4.3 **A bacterial cell showing its single large, circular chromosome and one small, circular plasmid.**

primers to enable the insertion of the PCR product into another segment of DNA, as described later.

Once a gene of interest has been isolated, either by cutting it out of a larger DNA molecule or amplifying it by PCR, the next step in genetic engineering is to insert the gene into a circular piece of DNA called a "**vector**." The most common vectors are small, circular DNA molecules called "**plasmids**," which naturally occur in bacteria as additional genetic information that is separate from the larger, circular bacterial chromosome (Figure 4.3). The sticky ends of a gene segment cut out of a larger DNA molecule or copied by PCR and then digested with restriction enzymes can be joined with the end of a vector that has been digested with the same enzymes (Figure 4.4). Restriction enzymes that produce blunt ends can also be used to generate recombinant DNA molecules, but the ligation reactions that join blunt segments are not specific and are less efficient than those that join sticky ends, so enzymes that produce sticky ends are preferred.

Once a recombinant DNA molecule has been generated, it can be introduced into bacteria, yeast, or another "host" organism, including various types of animal cells. This is done using procedures called **transformation** (for bacteria and yeast; see Section 2.3.1) or **transfection** (for other types of cells). During a transformation or transfection procedure, the plasmid only enters a small percentage of host cells. These cells need to be retained, while the other untransformed or untransfected cells need to be eliminated. To achieve this, a DNA sequence called a **selective marker** is engineered into the plasmid (Figure 4.5A). For example, the plasmid might contain the gene for an enzyme that gives the host cells resistance to an antibiotic. If the cells subjected to a transformation or transfection procedure are grown in medium containing the antibiotic, only those that have received the plasmid will be able to grow, while the rest will be killed. In other words, the

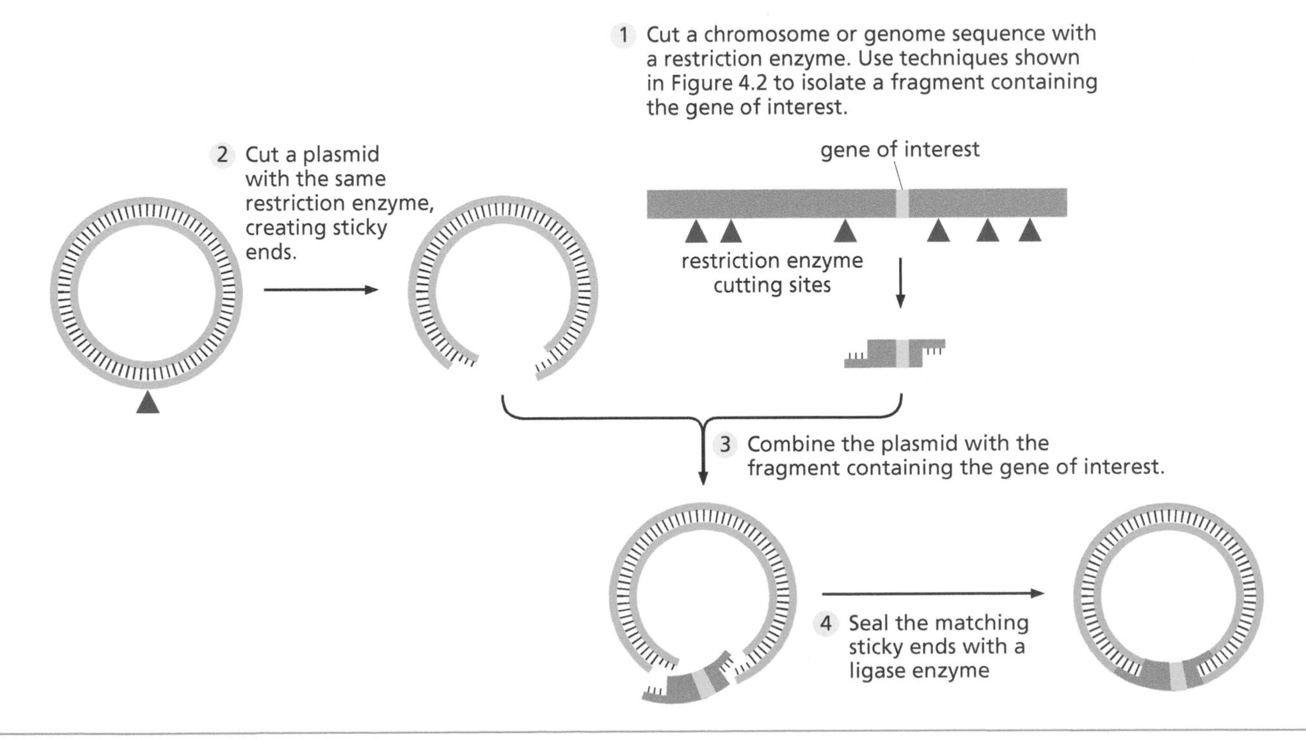

Figure 4.4 **Inserting a gene into a plasmid by joining complementary sticky ends.** The "insert" fragment may be generated by restriction enzyme digestion or by the polymerase chain reaction (PCR).

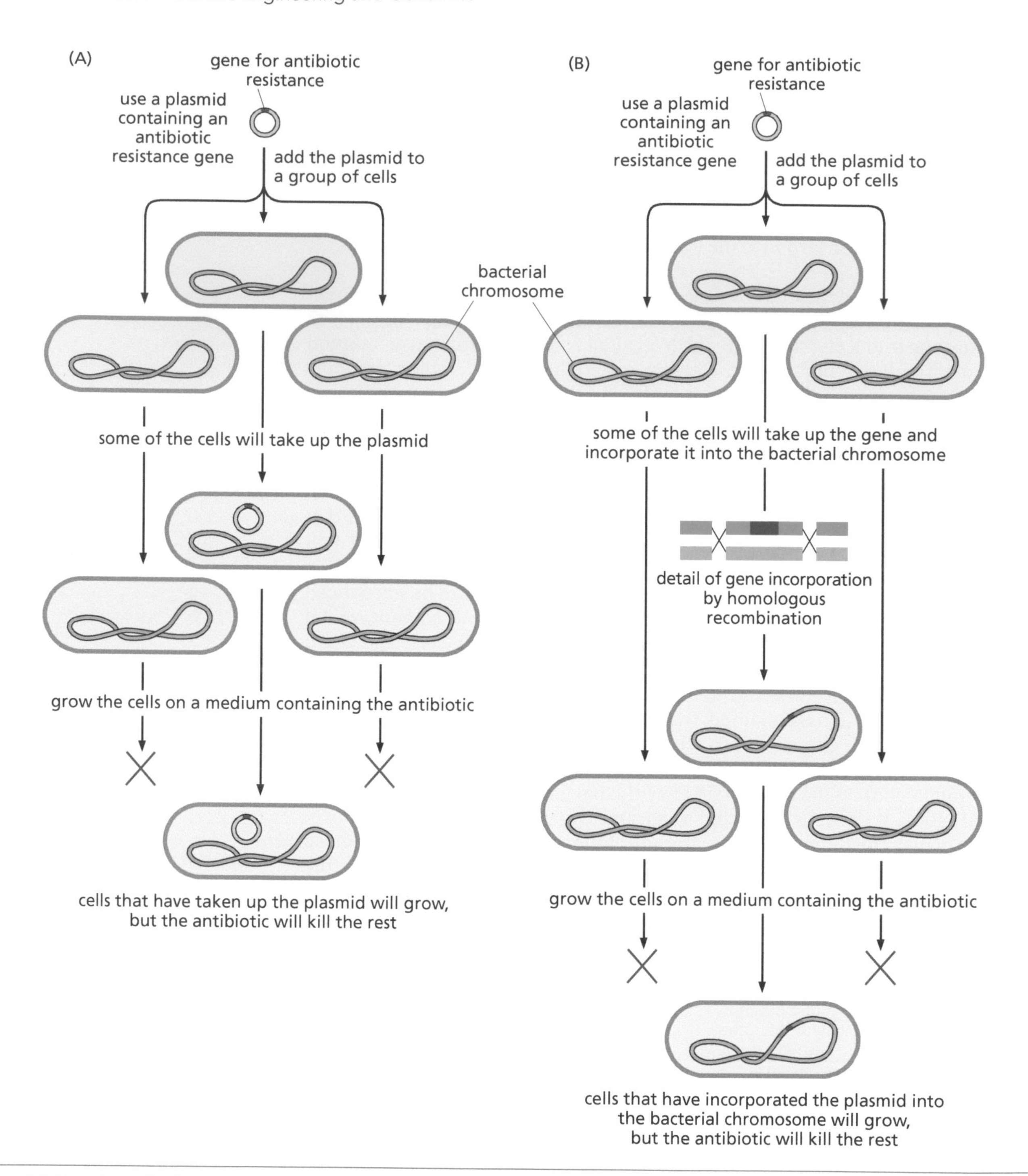

Figure 4.5 Two strategies for introducing and maintaining foreign DNA in host cells.

antibiotic-containing medium will "select" for the growth of the genetically modified cells. Circular plasmid vectors contain DNA sequences that allow them to be copied by the host cell's DNA replication machinery and passed on to the daughter cells when a cell divides (Figure 4.5A). However, the distribution of plasmids between daughter cells is less controlled than for chromosomes. Consequently, the antibiotic selection must always be maintained to ensure that all host cells still have the gene of interest. In another strategy, a plasmid can be linearized by cutting with a restriction enzyme prior to being introduced into a cell; in this case it can become inserted (integrated) into the genome of the host cell through the process of homologous recombination (**Figure 4.5B**). Cells that have successfully integrated the foreign DNA are again selected using an antibiotic resistance marker, but once resistant cells have been isolated, the selection does not need to be continued, since the recombinant gene is transmitted to daughter cells as part of the chromosome into which it has integrated.

Viruses can also be used as vectors for the introduction of linear DNA sequences that will integrate into the host genome; if a viral vector is used, the process of introducing the new DNA is called **transduction**.

4.1.2 Genetic engineering in medicine

Recombinant DNA technology has been used to produce many human gene products used as medicine. In many cases, this has involved inserting human genes into bacteria, but yeast, animal, or even plant cells may also be used. The reasons for using genetic engineering in medicine are largely practical: Human gene products that are therapeutically useful are more readily produced in large amounts inside genetically engineered bacteria than inside people. For example, the hormone somatostatin, also called growth hormone, is highly valued for the treatment of certain types of dwarfism; however, the hormone is difficult to obtain from human sources. The traditional way is to extract it from the pituitary glands of dozens of cadavers, which is very expensive. Insulin, the hormone needed by people with diabetes, is another example of a human gene product. Both of these hormones could be obtained from sheep or pigs or other animals, but the animal hormones are not as active in humans as the human hormones, and some patients are allergic to hormones obtained from other species. Genetic engineering provides a cost-effective way of manufacturing large amounts of these human hormones.

4.1.2.1 Genetically engineered insulin Human insulin was the first commercially produced genetically engineered product, approved by the U.S. Food and Drug Administration in 1982. Figure 4.6 illustrates the original method used for insulin production in recombinant bacteria. A recombinant plasmid containing the human insulin gene was transformed into *E. coli* bacteria, and a selective marker for antibiotic resistance was used to isolate the bacteria that received the plasmid. While maintaining antibiotic selection, the genetically altered bacteria were allowed to multiply asexually, producing vast numbers of genetically identical copies of themselves and the engineered plasmid. As these engineered bacteria grew and divided, they transcribed and translated the human gene just like one of their own genes, thus producing large amounts of the human insulin protein inside the bacterial cells. The bacterial cells were then broken open, and the insulin was separated from all the other molecules present in the bacteria. This recombinant human insulin was then available to be given to diabetic patients.

All of the steps for generating recombinant bacteria (or other host cells) that express a human gene of interest only need to be performed once. The resulting recombinant cells can be frozen at very low temperature and brought out of the freezer at any point to start a fresh culture.

In the years since insulin was first produced through recombinant DNA technology, scientists have made various revisions to the production process. For example, one biotechnology company developed recombinant yeast cells that transport, or **secrete**, human insulin into the growth medium. The insulin can then be purified from the growth medium without having to break open the cells, so the process is now more efficient. Insulin produced from both recombinant *E. coli* and yeast is still used to treat patients today.

4.1.2.2 Gene therapy Instead of using bacteria to manufacture human insulin, as shown in Figure 4.6, genetic engineering could theoretically be used to introduce the insulin gene into human cells that do not possess a functional copy, thus curing some types of diabetes. This type of genetic engineering is called **gene therapy**. To date, scientists have not successfully used gene therapy to replace a deficient insulin gene in the human pancreas, and progress on some alternative strategies for generating functional insulin-producing cells to treat diabetes is proceeding more rapidly. However, human gene therapy has been used successfully to treat **severe combined immune deficiency (SCID)**, a severe and usually fatal disease in which a child is born without a functional immune system.

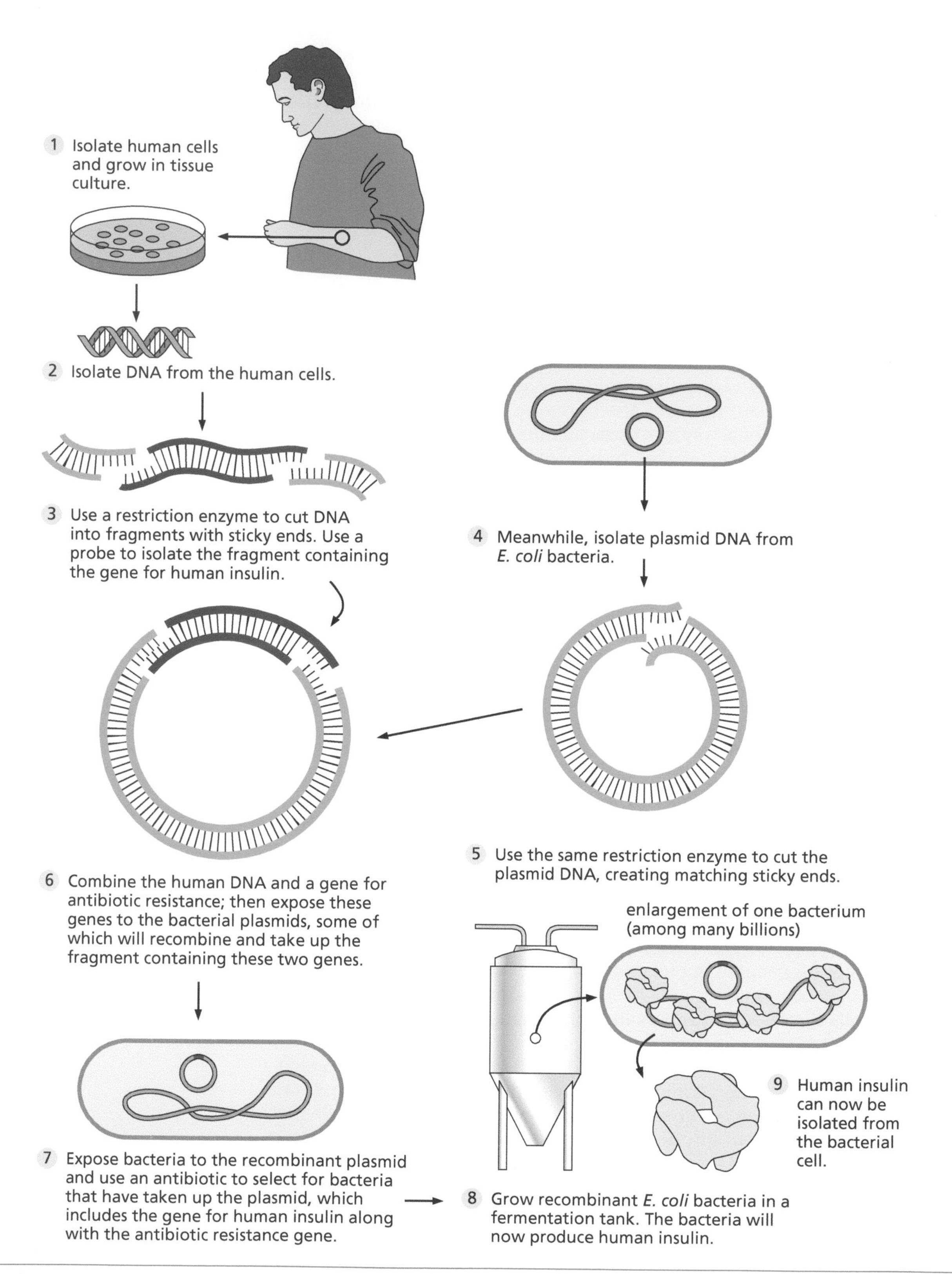

Figure 4.6 Production of human insulin by recombinant bacteria.

SCID is often referred to as the "bubble boy (or girl)" disease because, unable to fight infections, these children will die from the slightest minor childhood disease unless they are raised in total isolation (i.e., in a "bubble"). The enzyme that controls one form of SCID has been identified; it is called adenosine deaminase (ADA), and its gene is located on chromosome 20. A rare, homozygous recessive condition results in a deficiency of this enzyme, which in turn causes the disease. Gene therapy for this condition was first performed at the National Institutes of Health in Bethesda,

Maryland, in 1990, in a four-year-old girl (Bory et al., 1991). That first patient is still healthy, as are a number of others who were treated subsequently. The steps required to perform this gene therapy are described here and illustrated in Figure 4.7.

1. The functional gene for ADA is isolated from healthy human genomic DNA using the techniques described in this chapter.

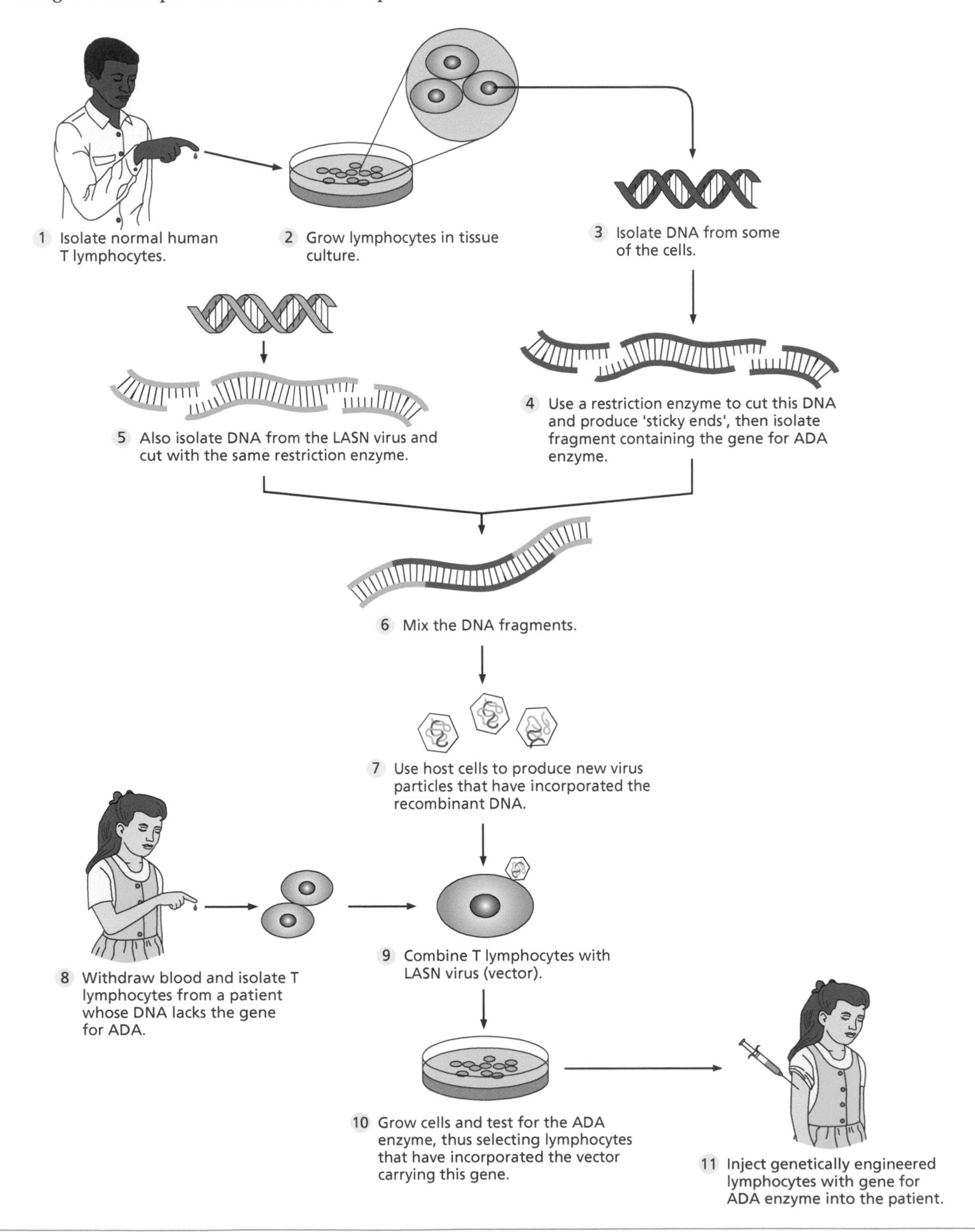

Figure 4.7 An example of gene therapy showing the transfer of the human gene responsible for adenosine deaminase (ADA).

2. The ADA gene is inserted (or "cloned") into a viral DNA vector, with DNA sticky ends joined by DNA ligase. The virus used for this therapy is called LASN. Most of its viral genes were removed in the engineering of the vector, but those needed by the virus to enter human cells and to assemble viral particles were retained.
3. The engineered viral vector is then transfected into cells that have been specially engineered to copy and assemble the virus.
4. The engineered virus is then allowed to infect T lymphocytes collected from the patient's blood. These lymphocytes, like all the other cells from this person, are ADA-deficient because they do not possess a functional ADA allele.
5. In the infected cells, the viral DNA enters the cell nucleus and inserts itself into the cell's genome. The position of this insertion is largely random; to be successful, the gene must incorporate into the cell's DNA in a location where it will be transcribed and where it does not break up some other necessary gene sequence.
6. The lymphocytes are tested to see which ones can produce a functional ADA enzyme, showing that they have successfully incorporated the functional ADA allele.
7. Genetically engineered lymphocytes with the functional ADA allele are injected into the patient, where they are expected to outgrow the genetically defective lymphocytes because the ADA-deficient cells do not divide as fast as cells with the ADA enzyme.

Technical difficulties in gene therapy are numerous. Many human genes are quite large and transferring large pieces of DNA into cells is difficult. Moreover, inserting a gene in a location in the DNA where its protein product will be transcribed and translated in a normal way is even more difficult. The gene therapy described earlier provides a functional gene that is transcribed and translated by the patient's lymphocytes, providing functional enzymes to these crucial immune cells. However, because lymphocytes are not the only cells that need the ADA enzyme, the patient must also receive injections of the ADA enzyme coupled to a molecule that permits it to enter cells. The enzyme controls the symptoms of the disease, but it is not a cure because the underlying disease is still present. In addition, because the genetically engineered cells are mature lymphocytes, which have only a limited lifetime, repeated injections of genetically engineered cells are needed.

To get around some of the problems with the initial ADA gene therapy, and in the hope of bringing about a more lasting cure, some Italian researchers tried using both genetically engineered lymphocytes (as described earlier) and genetically engineered bone marrow stem cells (Ferrari et al., 1992). Stem cells divide to form all the developed types of blood cells (see Section10.1), and they maintain this ability throughout life. Therefore, after repaired lymphocytes die off, stem cells with repaired DNA can divide to provide new, ADA-functional lymphocytes, possibly for the lifetime of the individual. This type of therapy was begun on a 5-year-old boy in 1992, and since then more than 20 other children have received this treatment.

4.1.2.3 Questions of safety and ethics There are legitimate safety concerns with human gene therapy. For example, any virus used as a vector must be capable of entering human cells. Might such a virus cause a disease of its own? To preclude this possibility, the viruses used in human gene therapy are from viral strains with genetic defects that render them incapable of reproducing and spreading to other cells. Might random insertion into the host DNA destroy some other gene? Methods are being developed for directing the insertion location, but it is still largely a random event. Gene therapy for another type of SCID, called X-linked SCID, which involves a different gene, was performed in London and Paris in the late 1990s. It was successful, but several of the patients later developed leukemia, which is a blood cell cancer. For some of these patients, the leukemia was clearly linked to the site of insertion of the gene therapy virus into their DNA, since it had inserted near genes known to be involved in cancer. These cases of gene therapy–linked leukemia occurred around the same time that an 18-year-old in the United States died from a severe immune reaction after receiving a viral vector for gene therapy for a metabolic disease. The boy's father testified at a U.S. Senate hearing that the boy and his family were not fully informed of the dangers of the experiment. This case led the U.S. Food and Drug Administration to issue new safety and informed consent guidelines for gene therapy trials. In the years since, progress on gene therapy has been slow but steady. In addition to the immune deficiency diseases that were the first targets of gene therapy, success has been achieved for a certain type of hereditary blindness, the blood diseases hemophilia and β-thalassemia, and Parkinson's disease.

In addition to safety concerns, gene therapy raises other ethical dilemmas. New recombinant DNA procedures are very expensive to develop. This raises ethical issues of fairness: Will the benefits of genetic engineering be available only to those who can afford them? Should government programs provide them through Medicare and Medicaid? Should insurance cover their use? How can society's health care resources best be distributed? If medical resources are limited, should an expensive procedure used on one person take up needed resources that could cover inexpensive treatments of other diseases for many people? These particular questions are not unique to genetic engineering; they apply to any expensive form of medical treatment.

Genetic engineering may someday become commonplace in human cells. In theory, gene therapy could be performed either on somatic cells or on gametes. When performed on somatic cells, the effects of the gene therapy last as much as a lifetime, but no longer. For example, insertion of the functional allele for insulin into the pancreatic cells of patients with diabetes might cure them of the disease, but they would still pass on the defective alleles to their children. Many medical ethicists agree that using gene therapy on somatic cells is justified if it is used for the purpose of treating a serious disease.

If successful gene therapy is performed on germ cells, then the genetic defect will be cured in the future generations derived from those germ cells. In addition to all the ethical questions raised earlier, gene therapy on germ cells raises many additional ethical questions, including the possibility of "designer babies" with a potential blurring of the line between what is medically necessary and what is desirable to a patient or to parents expecting a child. Most medical ethicists today advise caution and waiting in the case of germ-cell gene therapy on humans until we have more experience with gene therapy on somatic cells or in other species and until we devise clear guidelines as to what types of genetic manipulations are ethical and which should be forbidden.

4.1.3 New frontiers: CRISPR gene editing and direct base editing

In gene therapy, a normal (functional) gene is inserted somewhere in the genome to cover for the lack of function of a mutated (nonfunctional) gene. As described in the previous section, gene therapy trials have been conducted for over 25 years, but forward progress has been slow. Much more recently, new techniques for **gene editing** have made it possible to quickly and cheaply alter the DNA sequence of a gene located at its normal place in the genome. The primary method of gene editing is called **CRISPR-Cas9**, a technique that exploded onto the research and biotechnology scene in 2012 and has been rapidly adopted in laboratories around the world, with well over 1000 published papers using it by 2019.

CRISPR stands for Clustered Regularly Interspaced Short Palindromic Repeats, which are sequences discovered in bacteria in the 1990s that serve as a primitive form of adaptive immune system to protect bacteria against viruses. The normal function of CRISPR in bacteria is illustrated in Figure 4.8. When viral DNA enters a bacterial cell, segments of the viral DNA are incorporated into the CRISPR region of the bacterial genome in a process that is not well understood. These segments of viral DNA are inserted as "spacers" between repeated DNA segments in the CRISPR region, or "locus." The viral DNA segments "catalogued" in the CRISPR locus serve as a "reminder" of viruses to which the bacterium has been exposed,

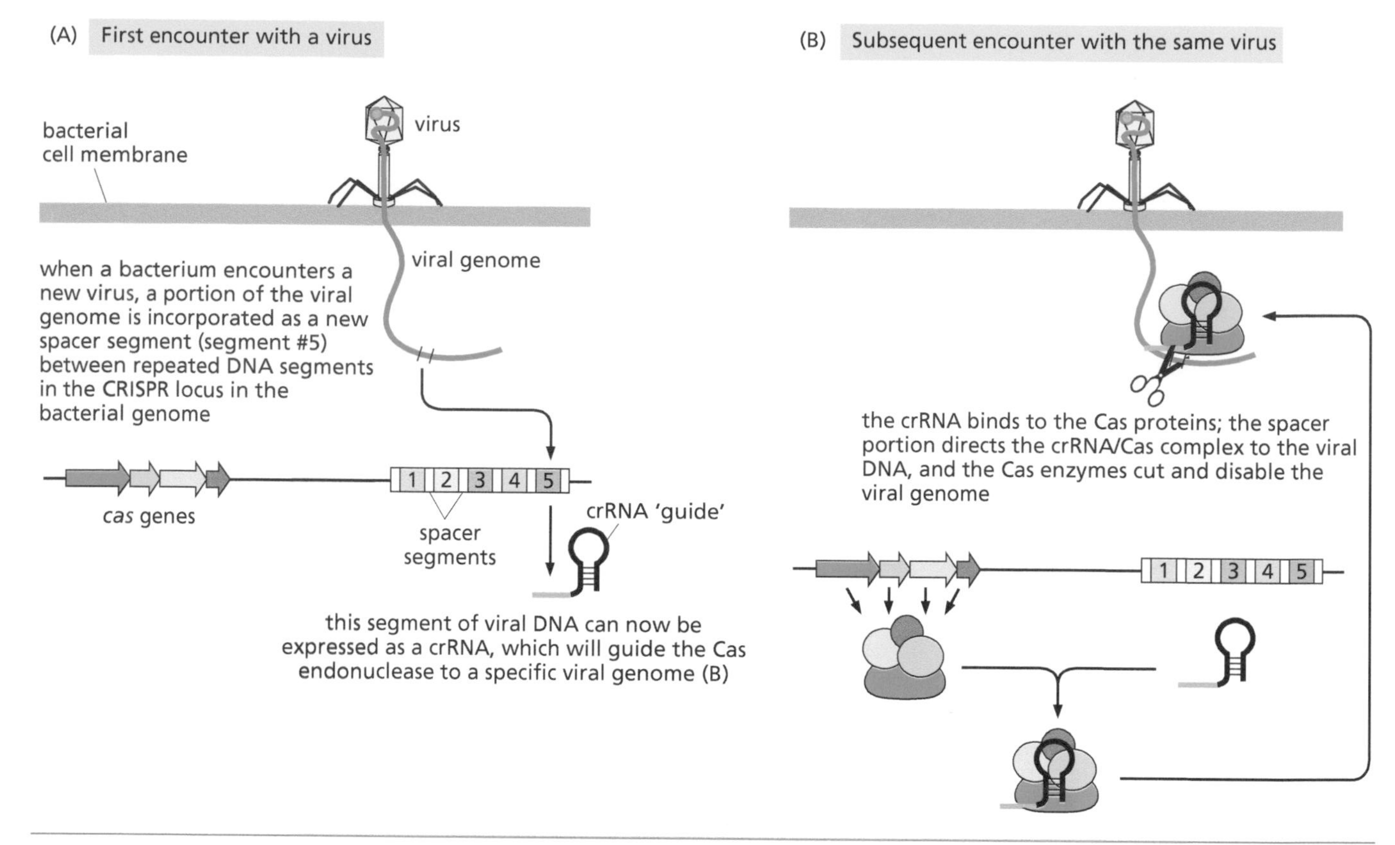

Figure 4.8 CRISPR RNAs function as a viral defense mechanism in bacteria.

serving a function like that of the memory B- and T-lymphocytes in mammals described in Chapter 14. The CRISPR locus is continually transcribed, resulting in the production of crRNAs, or "guide" RNAs, consisting of virus-specific spacers joined to CRISPR repeats, which form stem-loop structures (Figure 4.8A). The stem-loop portions allow the crRNAs to bind to Cas9, an endonuclease enzyme that can cut DNA. If there is invading viral DNA in the bacterial cell that contains a sequence complementary to the spacer portion of the crRNA, the crRNA-Cas9 complex will recognize that DNA, and the Cas9 enzyme will cut through the DNA backbone on both strands, protecting the bacterium by destroying the function of the viral DNA (Figure 4.8B).

A scientific breakthrough came in 2012, when a team that included Jennifer Doudna (Figure 1.4F), Emanuelle Charpentier, and others, realized that the basic mechanics of the bacterial CRISPR-Cas9 system could be applied to both bacterial and other genomes, targeting normal cellular genes in almost any organism (Doudna and Charpentier, 2014), and a Nobel Prize was awarded for this work in 2020. Instead of just disabling viral genes, CRISPR-Cas9 could now be used in genetic engineering as a directly targeted "molecular scissors" to cut a DNA sequence at almost any desired location. As illustrated in **Figure 4.9**, a recombinant DNA molecule is first made that contains a short sequence from the gene that one wants to "edit" in front of the CRISPR stem-loop sequence. A plasmid bearing the engineered crRNA coding sequence is introduced into the cells to be modified, either in a model organism or in human tissue culture, and the cell's gene expression machinery transcribes it. The bacterial Cas9 enzyme must also be expressed in these cells (from the same plasmid or a second plasmid). The Cas9 enzyme binds to the crRNA and then "finds" the genomic DNA sequence that is complementary to the crRNA "guide." Cas9 then cuts both strands of the genomic DNA in a very specific way.

What happens next depends on the DNA repair pathways of the cell. There are two possible outcomes (Figure 4.9B–D). First, the DNA double-strand break introduced by the Cas9 enzyme can be "fixed" by a so-called error-prone DNA repair mechanism called nonhomologous end joining (NHEJ) (Figure 4.9B).

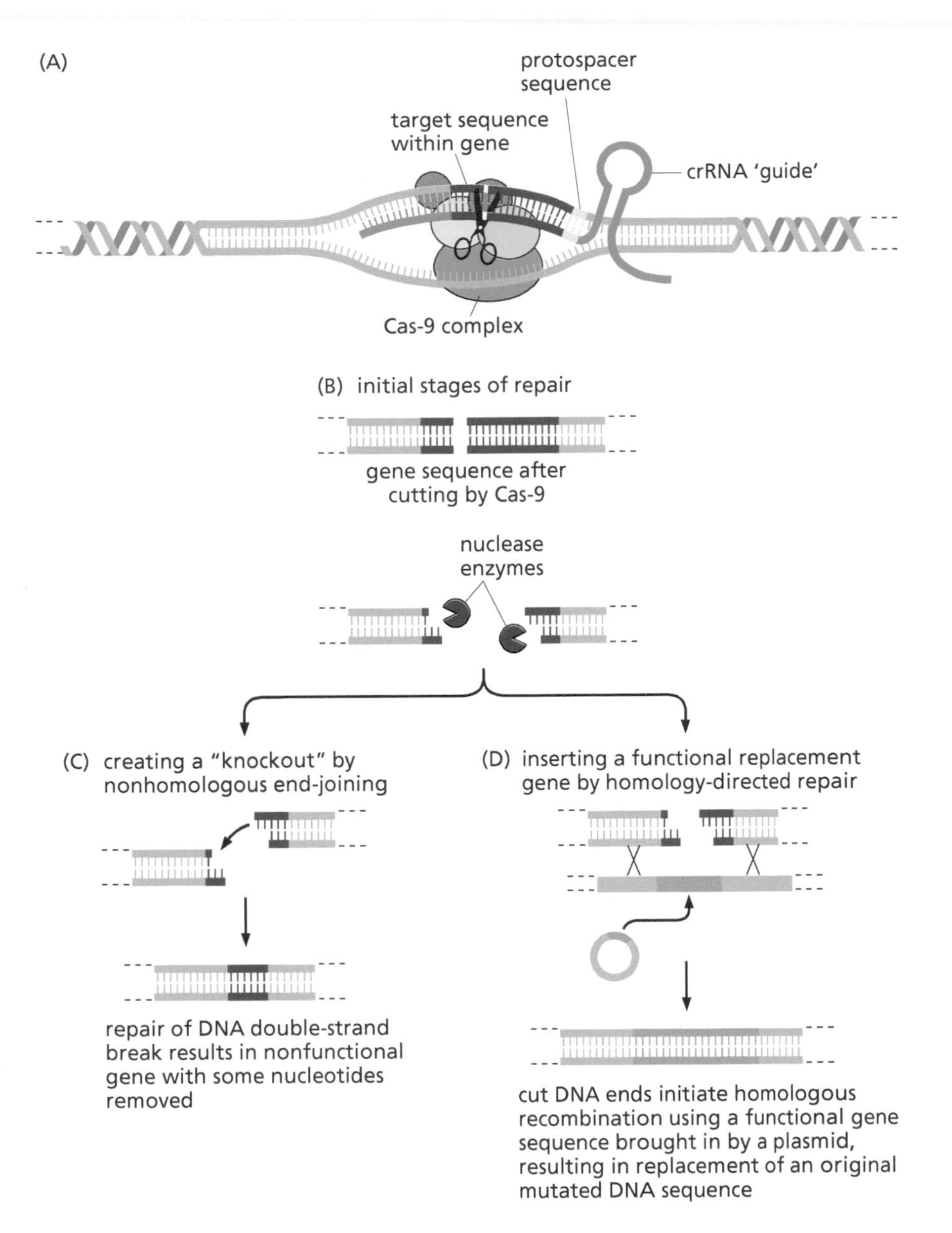

Figure 4.9 Application of the CRISPR-Cas system to genetic engineering.

Through this pathway, the broken ends of the DNA are "chewed back" by a nuclease, generating sticky ends similar to those created by restriction enzymes. When some small complementary region of DNA is uncovered by the nuclease, the two DNA strands will base-pair, and other enzymes will fill in the gaps and seal the breaks in the DNA backbone. This pathway is "error prone" because some nucleotides are lost in the process. Thus, the result of the "repair" is usually a nonfunctional gene, or what geneticists refer to as a **knockout** of gene function.

A CRISPR knockout can be very useful for studying the normal function of a targeted gene, but for applications in medicine, physicians usually want to replace a mutated, or disease-associated, copy of a gene with a functional, or "good" copy. This can happen if a second plasmid bearing a normal copy of the gene is introduced into the cells along with the CRISPR-Cas9 plasmid. In this case, the cells can use a different pathway to repair the double-strand break caused by Cas9. This pathway is called homology-directed repair (HDR) (Figure 4.9C). In HDR, a nuclease again removes some nucleotides from the ends of the broken DNA strands, as for the NHEJ pathway. The two strands of the broken DNA molecule then separate partially and base-pair with the complementary strands of the "good" copy of the gene on the plasmid. This functional gene is used as a template to extend the ends of the DNA strands of the broken chromosome. In some, but not all, cases, the product of repair will be a "fixed," functional gene within the cell's chromosome. A major advantage of this technique is that both copies of a gene on a cell's two homologous chromosomes can be "fixed" at the same time. If a mutant copy of a gene is provided on a plasmid, this same method can be used to introduce specific mutations (DNA sequence changes) into both copies of a given gene, which can be useful for creating models of human diseases or studying a gene's function. Several promising clinical trials are underway that aim to use CRISPR-Cas9 to modify a patient's cells outside of their body to repair a disease mutation and then to reintroduce the modified cells into the patient. Such trials include those to treat two blood diseases described in Section 7.3, sickle cell anemia and β-thalassemia (Coffey, 2019). Another milestone was reached in early 2020 when doctors introduced the CRISPR-Cas9 reagents directly into the eye of a patient to treat an inherited (and rare) form of blindness, Leber congenital amaurosis (Ledford, 2020). This approach was different than the previous trials because the CRISPR-Cas9 editing was performed in the patient's body rather than on cells removed from the body and reintroduced. Such an in vivo editing approach might be applied to many more types of body tissues, whereas the in vitro approach is best suited to blood cells, though it could potentially be used for other tissues as well.

Even as the CRISPR-Cas9 methods described so far have been widely applied in basic research and biotechnology, scientists have developed additional versions of the technique that allow more efficient and more precise gene editing and modulation of gene expression. For example, the CRISPR targeting system can be used to silence expression of a gene without altering its DNA sequence or to cause direct chemical modification of DNA bases within a gene to introduce specific sequence changes. The latter is called "direct base editing," and it may prove to be much more powerful than the original CRISPR technique.

As with gene therapy, there are many important ethical issues associated with our newfound ability to alter genetic sequences in living organisms. Scientists around the world were shocked and generally outraged in late 2018 when Chinese researcher He Juankui announced that he had used CRISPR-Cas9 to alter a gene called *CCR5* in twin girls born through in vitro fertilization. One mutation in this gene (referred to as *delta 32*) naturally occurs at a low frequency in Northern Europeans and, when present in two copies (the homozygous state), has been found to render individuals resistant to infection with the human immunodeficiency virus (HIV), the virus that causes acquired immunodeficiency syndrome (AIDS) (see Section 15.1). Dr. He reported that he edited the *CCR5* gene in a manner similar to (but *not* identical to) the *delta 32* mutation, with the assumption that the engineered edits would also prevent HIV infection. However, this was not shown directly in the published report of the study; moreover, although both copies of the *CCR5* gene were edited in one of the twins, only one was edited in the other, which would likely still leave her partially susceptible to HIV infection. Dr. He stated that he attempted the *CCR5* gene editing because the twins' father was HIV positive; however, there would have been other options to ensure that HIV was not passed on to the man's children (including sperm washing, which was performed prior to the in vitro fertilization). A majority of scientists and medical ethicists see Dr. He's actions as an irresponsible application of gene editing technology because the potential dangers from unanticipated consequences of gene editing have not been sufficiently explored. Several recent nonclinical studies of CRISPR-Cas9 gene editing in early-stage embryos reported that large genome rearrangements occurred at a relatively high frequency along with the intended CRISPR-specific editing. The result was described as "chaos" in the genome, which would likely result in many unintended consequences if the embryos were allowed to develop further. Thus, most scientists and medical ethicists support a ban on embryonic gene editing for clinical purposes, at least currently. Some experts have raised concern that the parents of the gene-edited twins may not have been fully informed of the risks associated with the embryonic gene editing and may have agreed to participate mainly to gain access to the in vitro fertilization procedure; in addition, there is controversy about who may have known or had suspicions about the Chinese experiment in advance of it being executed, including the Chinese government and several U.S. scientists working in the same general field. In the end, due to all these serious questions, Dr. He was fined nearly $1 million and sentenced to three years in prison (Regaldo, 2019a,b), with unlikely prospects of ever being able to work again in medical research.

Another application of CRISPR-Cas9 technology that has shown recent success (and is much less ethically fraught than the *CCR5* deletion strategy) aims to cure HIV-infected people permanently, rather than just preventing infection. As discussed in Chapter 15, a major problem with HIV/AIDS therapy is that, in order to stay healthy, patients typically need to remain on antiviral drugs for the rest of their lives once they are infected, which is both expensive and inconvenient. This is because the virus's genetic material becomes integrated into the infected person's genome and, even after all viral particles have been eliminated by antiviral drugs, the "latent" copies of the virus that remain in some cells of the person's body can eventually result in renewed virus production. A CRISPR-Cas9 approach has been tested in so-called "humanized" mice, meaning mice that have been genetically engineered to possess several human genes—in this case, genes that specify features of the human immune system that are relevant for HIV infection. In these mice, CRISPR-Cas9 is used to delete a segment of the HIV genome in cells that contain the virus integrated into their own DNA. When coupled with a special, slow-release version of antiviral drug therapy, this application of CRISPR-Cas9 has been shown to eliminate HIV from the infected animals, thus providing a true cure. Only two human patients have ever been documented to have been completely cured of HIV infection worldwide through standard drug treatments. Although it is still in the very early stages, this new gene editing approach has the potential to make a major impact on the global HIV crisis.

In addition to the many potential medical uses, scientists also hope to use CRISPR and other gene editing techniques to address environmental, public health, and agricultural issues. For example, researchers in Italy are working on a CRISPR-based approach to eradicate the species of mosquitoes that spreads malaria in Africa. The genetic changes they are introducing will, when homozygous, make female mosquitoes unable to bite or to lay eggs (Lovett, 2019). The particular technique they are testing is called a "gene drive." This refers to an engineered gene allele that will be preferentially passed on to the next generation (in comparison to the normal allele), thus allowing the genetic change to spread rapidly through a population. Although eradicating malaria would be very beneficial to humans, the decision to release mosquitoes bearing the CRISPR gene drive into the wild will have to be thoroughly researched and debated because of the possible significant unanticipated ecological consequences that we describe in Chapters 17 and 18.

THOUGHT QUESTIONS

1. The use of growth hormone for the treatment of shortness (not dwarfism) in otherwise healthy children is controversial, but its testing for this purpose was approved in 1993 by the Food and Drug Administration. When does a phenotypic condition unwanted by its bearer become a disease to be treated? Who decides? Should the use of human growth factor produced by engineered bacteria to increase someone's height be allowed? Is this simply another form of cosmetic surgery, similar to breast implants or facelifts?
2. If a person dissatisfied with his or her phenotype suffers from lack of self-esteem on that account, does the lack of self-esteem justify a procedure to correct the phenotype? (This same argument is raised to justify traditional forms of cosmetic surgery.) Do parents have the right to anticipate for a child what the future effects on self-esteem will be with and without corrective procedures? For a phenotype such as height that develops over a period of years, at what age is it appropriate (if ever) to evaluate the phenotype and decide upon corrective measures?
3. A procedure such as gene therapy is expensive. Who should pay for it? Is gene therapy a limited resource? Does giving gene therapy to one patient thereby deprive another of medical care?
4. Should CRISPR-Cas9 gene editing of embryonic cells for reproductive purposes be banned? Why or why not? Should research into CRISPR-Cas9 editing of embryos for research purposes (without the intention of producing genetically modified babies) continue? What relevant evidence would help inform these decisions?

4.2 MOLECULAR TECHNIQUES HAVE LED TO NEW USES FOR GENETIC INFORMATION

With the exception of identical twins, who share the same DNA, each person has a unique DNA sequence. Some of the same tools that form the basis of genetic engineering can also be used to distinguish differences between people. Identifying these differences can be useful in many contexts. For example, DNA analysis can be helpful as evidence in crime cases, for identifying individuals who are at risk for certain diseases, for establishing paternity, or for discovering features of one's ancestry.

Until recently, it was prohibitively expensive and highly impractical to sequence a person's whole genome as a means of identification. To that point, the complete sequencing of just one version of the human genome (discussed later in this chapter) took over 10 years and cost almost $3 billion. However, soon after the completion of the Human Genome Project, several new technologies for sequencing DNA came onto the market. Collectively referred to as "next-generation DNA sequencing," these have made sequencing entire genomes much faster and much cheaper. The current cost of sequencing a human genome using next-generation techniques is less than $1,000, and it can be completed in only one hour. Prior to these advances in DNA sequencing, various types of DNA "markers" were used to differentiate DNA samples from different people without sequencing their entire genomes. Many of these are still used today for applications where full DNA sequences are not needed. Here, we describe several of these marker systems before examining how DNA is sequenced, both using the traditional method and next-generation technologies.

4.2.1 DNA markers: Restriction-fragment length polymorphisms and beyond

The first markers that were used to differentiate DNA from different individuals were **restriction fragment length polymorphisms**, or **RFLPs**, as described in Section 3.3. Because restriction enzymes cut DNA at very specific sequences, and because these DNA sequences may differ in length from person to person, the technique can generate fragments of different sizes (Figure 3.16). These differences can be due to single nucleotide differences that eliminate or introduce a restriction site, insertions or deletions of nucleotides between two restriction enzyme cut sites, or chromosomal rearrangements that alter the positions of cut sites.

4.2.2 Using DNA markers to identify individuals

One application of DNA markers, including RFLPs, is for criminal investigations. To use RFLPs to differentiate possible suspects in a criminal investigation, various genomic DNA samples are cut with restriction enzymes, as shown earlier in Figure 4.2. The resulting fragments are then separated according to size by gel electrophoresis and then transferred to a membrane for Southern blotting. Radioactively labeled (or fluorescently labeled) probes complementary to known DNA sequences are then used to detect the fragments containing particular variable repeats. These fragments appear as bands, with their location indicating the fragment length. Several probes can be used at once so that many bands show up, not just one or two as in the example shown in Figure 4.2, in which just one probe was used. Bands at the same position indicate fragments of the same length in samples being compared. If the band patterns are not the same, then it can be stated with certainty that two samples did not come from the same person. In the example from a criminal investigation shown in Figure 4.10, person 1 can be eliminated as a suspect because the band pattern from the evidence is not the same as that from sample 1. The reverse is not true, however; band patterns that are the same are not an absolute guarantee that the samples came from the same individual. What are being visualized are chunks of DNA of variable lengths, not the DNA sequences of the chunks. A score is calculated that indicates how likely it is that a randomly chosen person, other than the one tested, could have the same band pattern.

The likelihood that another, randomly selected person could have the same banding pattern is made very small in two ways. First, the DNA probes selected are

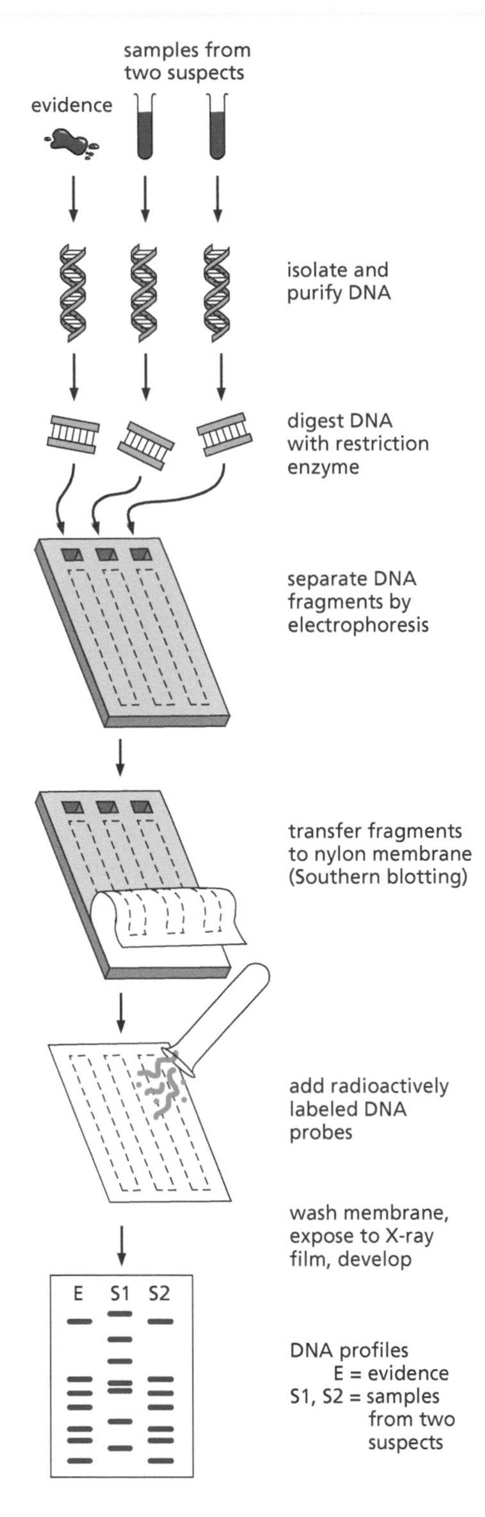

Figure 4.10 Forensic DNA analysis using RFLPs and DNA probes (Southern blotting). In this example, the evidence sample shows the same pattern of bands as DNA from suspect 2. There is therefore a high probability that the DNA in the evidence is from that suspect. The person from whom sample 1 was taken can be eliminated as a suspect.

those that pick up specific DNA markers that are rare in a given population. Also, several DNA probes are used, one after another, to produce a composite banding pattern. The probability that the bands produced with just one DNA probe are the same for two people is equal to the frequency of that DNA marker in the population. If more than one DNA probe is used, the probability of both band patterns matching is equal to the population frequency of the first DNA marker multiplied by the population frequency of the second, and so on for multiple DNA probes and markers.

There are many ways in which the banding pattern can yield flawed or ambiguous results if samples are not properly processed. In samples from crime scenes, there is often DNA from mixed sources, including DNA from several people and from bacteria or fungi. Protein material in the sample may slow the movement of a restriction fragment in the electrophoresis, making the DNA fragment appear as if it were larger than it is. Other chemicals in the samples, such as the dyes in cloth, can interfere with the restriction enzymes cutting the DNA. However, when the tests are done properly and with the proper controls, they can be very reliable. In addition to linking suspects to material taken from crime scenes, the methods can be used to settle questions of disputed parentage. The methods can also be used to identify the dead when an intact corpse is not available, as in the aftermath of the September 11, 2001, terrorist attacks in the United States.

4.2.3 Additional DNA markers

Since RFLPs were first used as genetic markers in 1980, a variety of other sequence markers have been identified, such as **expressed sequence tags (ESTs)**, **single-nucleotide polymorphisms (SNPs)**, and **microsatellite markers** (Figure 3.16). ESTs are short sequences within transcribed RNAs that can easily be assayed to indicate whether a given gene is expressed in a particular individual and at what level. Variations in the levels of ESTs expressed in people with different characteristics—such as people who have a particular disease vs those who don't—can be used to identify genes that may be responsible for those characteristics. SNPs are single nucleotide positions in the genome where there is known to be variation in which letter is present at that position in different individuals. These can occur either in coding or noncoding DNA regions, and sometimes they create a RFLP. In addition to genes, DNA contains noncoding regions that vary in length from one individual to another. As described later, these noncoding sequences make up a much larger portion of the human genome than the genes themselves. As mentioned in Section 3.3, some of these noncoding segments (short sequences of nucleotides, 3–30 bases long) are repeated over and over anywhere from 20 to 100 times. These are called microsatellites, microsatellite markers, or short tandem repeats (Figure 3.16); several thousand different such repeats are now known in humans, each with a unique sequence not found elsewhere in the genome. These can be detected as RFLPs or as differently sized products when PCR is used to amplify a given region of the genome.

4.2.4 Using DNA testing in historical controversies

An unusual use of this technique helped shed new light on a historical controversy involving Thomas Jefferson, the third president of the United States. DNA markers were used to investigate whether Thomas Jefferson could have been the father of children borne by one of his slaves, Sally Hemings. Two oral traditions exist: Descendants of Hemings's sons, Eston Hemings Jefferson and Thomas Woodson, believe that Jefferson was their ancestor, while descendants of Jefferson's sister believe that one of her children, Jefferson's nephew, fathered Sally Hemings's later children. Researchers compared Y chromosomal DNA from descendants of two of Sally Hemings's sons with DNA from descendants of one of Thomas Jefferson's uncles. No Y chromosomal DNA was available from Thomas Jefferson's direct descendants because he had no sons who survived to have children.

The DNA data showed that a set of 19 markers (collectively called the haplotype) was shared by all 5 of the descendants of Jefferson's uncle who were tested and by the descendants of Eston Hemings Jefferson. The haplotype was not shared by descendants of Hemings's other son, Thomas Woodson, or by the descendants of Jefferson's nephew, nor was it found in almost 1,900 unrelated men. Thus, Jefferson may definitively be ruled out as the father of Thomas Woodson. In the case of the positive match, however, the evidence supports, but does not prove, the idea that Thomas Jefferson could have been Eston Hemings Jefferson's father. As we explained earlier, positive matches indicate probabilities, not definite identity. The researchers state that because "the frequency of the Jefferson haplotype is less than 0.1%," their results are "at least 100 times more likely if the president was the father of Eston Hemings Jefferson than if someone unrelated was the father." They also state that they "cannot completely rule out other explanations of our findings," but that "in the absence of historical evidence to support such possibilities, we consider them to be unlikely." Interestingly, after the authors were very precise in the text of their article, the title, "Jefferson fathered slave's last child," overstated their results (Foster et al., *Nature* 396: 27, 1998).

THOUGHT QUESTIONS

1. Thomas Jefferson had daughters who survived to have children. Why was the DNA of their descendants not used in the study to determine the paternity of Eston Hemings Jefferson and Thomas Woodson?
2. The authors of the Jefferson study state that they "cannot completely rule out other explanations of our findings." What other explanations are biologically possible?
3. Think about the study done on DNA from descendants of Jefferson's family and Sally Hemings's sons. Why is the title of the study, "Jefferson fathered slave's last child," an overstatement of the results?
4. In the study on Jefferson's descendants, why did the researchers test DNA at 19 DNA marker sites, rather than just at 1 or 2 sites?

4.3 THE HUMAN GENOME PROJECT HAS CHANGED BIOLOGY

As described in the opening of this chapter, sequencing of the human genome was declared complete in 2003. It should be noted, however, that although we talk about "the human genome sequence," the DNA sequence of each person is unique. There is no one DNA sequence that is representative of every human, just as no one person could be said to represent all humans in any other method of describing people. It is estimated that one person differs from another in about 0.1% of the 3 billion base pairs in the human genome. People share many of the same genes, but the nucleotide sequences of those genes vary according to the specific alleles of each that a person inherited from their parents. There is even more sequence variation in the regions of the genome that are not genes. When people are grouped in different ways (by race, ethnicity, or sex, for example), there is always more genetic variation within a group than there are differences between groups. (The topic of the genetics of populations is covered in greater detail in Chapter 7.)

What was sequenced in the Human Genome Project was one copy of each segment of the genome from the DNA provided by several volunteer donors. The Human Genome Project mapped the positions of genes along the 23 human chromosomes and determined the nucleotide sequence of both the genes and the intergenic, or noncoding, regions for one complete genome "read," with parts of the total sequence coming from each of the different donor genomes.

4.3.1 Classical DNA sequencing methods were used to sequence the first human genome

The large task of determining the sequence of the first complete human genome was accomplished using what are now considered "classical" sequencing methods. Spurred by the recognition of the extraordinary value of genome sequencing, technologies have advanced massively since the human genome was completed. We will first describe the methods used in the Human Genome Project and then describe some of the benefits of the "next-generation" sequencing methods that have been developed since the project was completed.

Because the amount of DNA in even one chromosome is enormous (48 million nucleotides in the smallest human chromosome, chromosome 21), it is not practical to work with the whole length of a chromosome in determining sequences. The classical sequencing methods used for the Human Genome Project were able to generate continuous sequence stretches of only 500–700 bases at a time. Therefore, to prepare for sequencing, the human chromosomes were separated, and each was cut into overlapping pieces with restriction enzymes. Each piece was inserted ("cloned") into a plasmid vector, which was then transformed into bacterial cells. The bacteria were allowed to divide repeatedly, making large quantities of one small piece of the human genome at a time, as we saw earlier in Section 4.1.2 for bacterial production of human insulin.

The nucleotide sequence of each of the chromosome pieces was then determined using an established method (called the Sanger, or dideoxy method) based on DNA synthesis. In Sanger sequencing, the DNA is used as a template for the synthesis of new DNA strands in a test tube, as outlined in **Figure 4.11**. A short

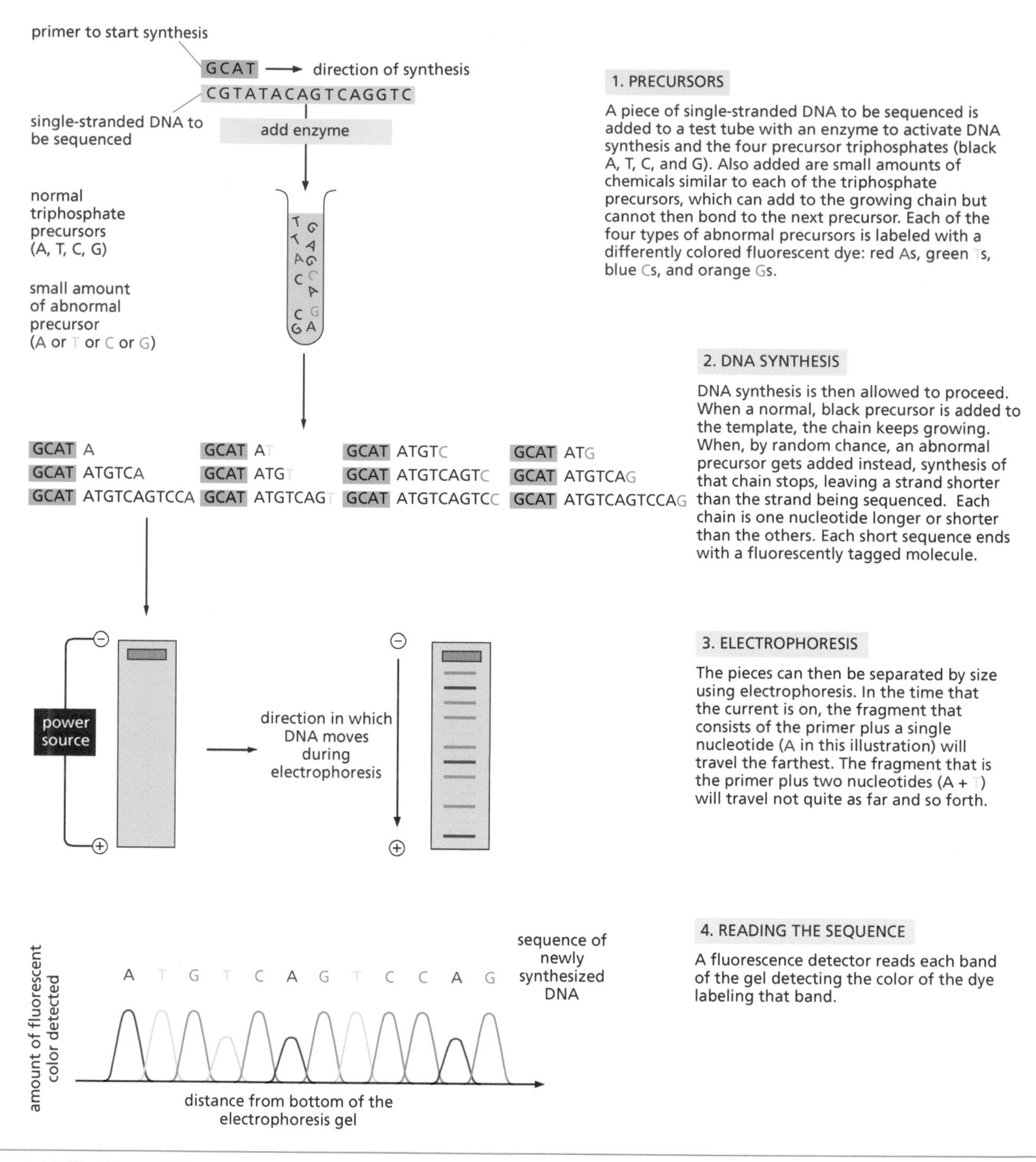

Figure 4.11 The Sanger (dideoxy) method for determining the sequence of DNA.

DNA primer that is complementary to one of the strands of the DNA to be sequenced binds to that strand through base-pairing. The end of this fragment is then extended by a DNA polymerase enzyme using a mixture of normal nucleotides and special dideoxy nucleotides that are labeled with fluorescent dyes, with the dye color corresponding to the identity of the nucleotide (G, A, T, or C). Whenever one of the fluorescent dideoxy nucleotides is added onto the end of a growing DNA strand, it terminates the strand, since "dideoxy" means that it is missing the oxygen atom that would normally be connected to the next nucleotide in the DNA strand. This is referred to as "chain termination." The color of the dideoxy nucleotide that is incorporated will correspond to the identity of the nucleotide at that position in the strand. The overall result is the production of a series of pieces of newly synthesized DNA of different sizes that all begin with the primer and have different numbers of additional nucleotides with one fluorescent dideoxy nucleotide at the end. These pieces are then separated by electrophoresis in a sequencing machine that has small capillaries containing a gel matrix (instead of a large, flat gel, as depicted in Figure 4.2). The shortest piece will travel most rapidly through the capillary. At one position in the machine, it will pass through a laser, which can excite the fluorescent molecules at the ends of the DNA strands. The color of fluorescence will be recorded by a detector. In this way, the sequence of the DNA immediately following the primer is read by the machine one nucleotide at a time as each strand passes through the detector in order by size.

After the sequence of each piece has been determined, the pieces must be arranged in their original order to get the overall sequence. Remember, this sequence analysis has been carried out on only one fragment of a chromosome at a time. The next challenge is to piece together the sequenced fragments, which is part of the mapping procedure discussed later.

4.3.2 The race to finish the human genome draft sequence

In February 2001, two groups simultaneously announced completion of a draft of the sequence of the human genome. One group, the International Human Genome Sequencing Consortium (IHGSC), made up of laboratories from the United States, Britain, Japan, France, Germany, and China, published their results in *Nature* (vol. 409: 860–921). The other group, a biotechnology company called Celera Genomics, published their results the same day in *Science* (vol. 291: 1304–1351). This simultaneous publication of the two draft sequences was the culmination of a race between public and private scientists, initiated in 1998 by Celera's founder, Dr. Craig Venter, who thought that the public effort was proceeding too slowly and at too high a cost. The two groups used the same basic sequencing technology (the Sanger method) but two different mapping strategies. The IHGSC first separated the genome into large chunks that they cloned into plasmids, ordered these relative to one another based on DNA markers, and then sequenced smaller, internal segments of each plasmid. Celera, in contrast, used a faster method that skipped the first step of separating the genome into large, ordered chunks, and instead went straight to sequencing smaller fragments generated by restriction enzyme digestion of the genomic DNA and relying on computer power to assemble them into a continuous sequence. This is referred to as the "whole-genome shotgun method," and it is modeled in Figure 4.12. In this figure, the largest piece of DNA contains 40 bases; actual DNA pieces for sequencing are around 500

(A) In this example, a DNA sequence of 150 bases is cut with two different restriction enzymes, producing the following fragments, each of which has been sequenced.

Fragments from the first restriction enzyme:

GGTCGGCTATGTAACGAGTTGCC
TCTTGTTCCTAGCTTGTCAACCGGGGATGAATGTTTACTG
CACGCGGACCGTCGGTTCAT
GTCGCAGAGCCTATTGCGAGAAGT
GCCCACCTT
TTATTGAGTTGATGCTCGACGTAGCCAGACTTAA

Fragments from the second restriction enzyme:

ACCGGGGATGAATGTTTACTGGTCGCAGAG
CCTATTGCGAGAAGTGGTCGGCTA
CTTGTCA
TGATGCTCGACGT
CGTCGGTTCAT
AGCCAGACTTAACACGCGGAC
TGTAACGAGTTGCCGCCCACCTTTTATTGAGT
TCTTGTTCCTAG

Try to piece these fragmentary sequences together and determine the entire sequence of 150 bases, before you turn the page.

Figure 4.12 Model of the whole-genome shotgun method of genome assembly. (A) Here are the fragments of a sequence cut with two different enzymes. Can you piece them together to reconstruct the complete sequence? Don't turn the page until you've tried it!

(B) When the two sets of fragments are lined up in this way, the order of the bases in the first row is the same as the order of the bases in the second row.

and the complete sequence is therefore as follows:

TCTTGTTCCTAGCTTGTCAACCGGGGATGAATGTTTACTGGTCGCAGAGC-
TCTTGTTCCTAGCTTGTCAACCGGGGATGAATGTTTACTGGTCGCAGAGC-
TCTTGTTCCTAGCTTGTCAACCGGGGATGAATGTTTACTGGTCGCAGAGC-

CTATTGCGAGAAGTGGTCGGCTATGTAACGAGTTGCCGCCCACCTTTTAT-
CTATTGCGAGAAGTGGTCGGCTATGTAACGAGTTGCCGCCCACCTTTTAT-
CTATTGCGAGAAGTGGTCGGCTATGTAACGAGTTGCCGCCCACCTTTTAT-

TGAGTTGATGCTCGACGTAGCCAGACTTAACACGCGGACCGTCGGTTCAT deduced sequence
TGAGTTGATGCTCGACGTAGCCAGACTTAACACGCGGACCGTCGGTTCAT fragments from first enzyme
TGAGTTGATGCTCGACGTAGCCAGACTTAACACGCGGACCGTCGGTTCAT fragments from second enzyme

Figure 4.12 (*continued*) **(B)** Here is the complete sequence.

bases in length. Because the pieces are so much longer and there are so many of them, computers are needed to line up the overlaps. The accuracy of the method increases with the length of the overlapping region. The longer the sequence of the overlap between two pieces, the higher the probability that the sequence will appear only once in the genome, allowing the unambiguous assignment of the position of the two pieces relative to each other. Those regions of the genome not included in either the IHGSC or Celera draft sequence were finally deciphered and published in 2021 (Zimmer, 2021).

The whole-genome shotgun method used by Celera is modeled in Figure 4.12A. The same long piece of DNA is cut with two different restriction enzymes, and each of the pieces is sequenced; computers are then used to determine how the two sets of pieces overlap. Figure 4.12A shows two sets of fragments of DNA produced by cutting a DNA sample with different restriction enzymes. The first restriction enzyme cut the DNA into six pieces only; the second resulted in eight pieces. The bases in the sequences of each of the eight pieces can be lined up to match the bases in the six pieces. Can you see how you would use this idea to determine the order the six pieces had originally been in? Now turn the page and look at **Figure 4.12B**.

Before beginning to sequence the human genome, Celera had successfully used the whole-genome shotgun method to sequence the genome of the bacterium *Haemophilus influenzae*, completed in 1995. This approach was also used successfully on the genomes of the 599 viruses, 31 bacteria, and 7 archaebacteria that were sequenced between 1995 and 2002. But scaling up the approach to mapping the human genome was still a challenge. One obstacle was size: The human genome is about 25 times larger than any that had been sequenced previously, although it is far from being the largest genome known. (One species of single-celled amoeba has a genome 200 times larger than humans!) Another obstacle to accurate reassembly is the fact that much of the noncoding DNA in the human genome is composed of repeated sequences of nucleotides (see later). This enormously complicated the job of putting pieces into unambiguous order. Species whose genomes had previously been sequenced do not contain these repeats, so it was much easier to determine which piece went where in these genomes.

4.3.3 Lessons from the Human Genome Project

Completion of the human genome draft sequence supported some previously established hypotheses, but also produced some surprises. Some key results are:

1. *Over 98% of the human genome represents noncoding DNA, a large proportion of which is composed of repetitive sequences.* Less than 2% of the human

genome is composed of genes that code for proteins. The rest of the genome consists of so-called "noncoding" DNA. However, just because it is noncoding does not mean it is unimportant. The noncoding regions of DNA contain various types of regulatory elements that control gene expression, structural elements of chromosomes such as centromeres (see Section 2.2) and telomeres (see Section 10.3.3), and sequences that are transcribed into RNA but never translated into protein, which may also affect the expression of other genes in several different ways. Still other parts of the genome—approximately 50% of the total—consists of repetitive DNA sequences of varying lengths. Because the lengths of these repeats can vary between individuals, they are often used as DNA markers. It has been known for a while that the complexity of an organism does not correlate with the size of its genome. Much of the excess size is due to noncoding repeat sequences. Detailed knowledge of these sequences is opening a new resource for studying evolution. These sequences can be likened to living fossils carried within each of us. They are already used in population genetic studies examining the migrations of human populations.

2. *The actual number of genes is smaller than previously estimated.* In humans, it is difficult to predict which sequences represent genes, for reasons that we discuss later. The estimate of the number of human protein-coding genes is currently between 19,000 and 22,000. (Previous estimates had been between 50,000 and 100,000 genes.) This is similar to the numbers of genes identified in the roundworm (*Caenorhabditis elegans*) genome, and only 1.5 times the number in the fruit fly (*Drosophila melanogaster*). Therefore, not only does organism complexity not necessarily correlate with genome size; it also does not track directly with the number of protein-coding genes (see point 3).
3. *The protein products of many human genes remain unknown and may vary from tissue to tissue.* It has been found that the RNA transcripts of many of the known human genes can be processed and translated in different ways to produce alternative protein variants from the same gene (see Figure 4.14). Thus, although we may not have more separate genes than roundworms, those genes may encode many more many different proteins. These different protein forms encoded by the same gene are usually expressed in different tissues, fine-tuning the function of certain cell types in those tissues (see Section 10.1).
4. *A very high percentage of our genes are not unique to humans, but are closely similar to comparable genes from other species.* In fact, only 1% of human genes have no sequence similarity to any other organism. Our genes are similar to 46% of the genes in yeast, among the simplest organisms whose cells have a nucleus. Changes within genes over time provide clues to rates and paths of evolution.
5. *More than 200 human genes and their protein products have been found to have significant similarity to those in bacteria.* These genes are not found in intermediate organisms such as fruit flies, and one school of thought suggests that these genes jumped from bacteria to humans or vice versa.
6. *Mutation rates differ in different parts of the genome.* They are also higher in males than in females, although the reason for such a difference is not known.
7. *Within each gene, there is an average of 15 sites at which different individuals carry a different nucleotide or at which the same individual may have a different nucleotide on each chromosome in a pair.* These variations, called single-nucleotide polymorphisms (SNPs), are greatly expanding how many alleles we think are possible for different genes. In addition, these small changes may affect the physiology of the organism possessing them. Some of these polymorphisms are associated with disease; most are not, but are instead associated with small changes in protein function or regulation. Knowledge of such small-scale variations continues to challenge our concepts of terms such as heterozygous, dominant and recessive, and allele. It also makes it clear that, as mentioned, there is no such thing as *the* human genome sequence. The genome sequence within each individual is unique.

4.3.4 Some ethical and legal issues

Having access to the complete human genetic sequence has greatly accelerated the incorporation of genetic information into medical diagnosis and treatment. For example, doctors can now order genetics tests that screen for variations in panels of genes that are implicated in complex diseases like cancer (see Chapter 10). Companies, such as 23andMe, have also seized on the opportunity to market genetic tests that provide information about ancestry or health risks directly to consumers, bypassing any medical professional. Many of the ethical and legal issues already covered in Chapter 3 regarding genetic testing will become more commonplace as molecular genetics continues to change medicine and enter into people's daily lives. How does an individual's right to privacy balance against a family member's desire to know the results of genetic tests or an insurance carrier's or employer's desire to cover or to hire only those individuals who will remain healthy? How does an individual's desire to control their own reproduction balance against possible eugenic aims of society or against further stigmatization of disabled people? How can genetic counseling be value-free while providing education about genetics and not just about the testing procedure itself? How will consumers react to receiving unexpected information about their health or ancestry through a genetic testing service?

When the Human Genome Project was funded, scientists saw the need for examination of the ethical, legal, and social issues (anticipated and unanticipated) that would be raised by the research. One percent of the funding was set aside for this effort. The issues just mentioned are among those being studied, but there are many others. Social workers, anthropologists, ecologists, ethicists, and others are working together to examine the issues raised by the study of genetic variation in human populations and by the integration of genetic information into health care as well as into nonclinical settings. Others are studying the ways in which socioeconomic factors, race, and ethnicity influence people's understanding, interpretation, and use of genetic information. Simultaneously, new genetic information continues to change our concepts of race and ethnicity (see Chapter 7), and new genetic knowledge and capabilities may raise challenges for different philosophical and theological traditions. Many of the working groups have composed reports with their answers to many of these questions and their guidelines for the use of genetic information. These reports are available at the Web site www.genome.gov.

In addition, the data derived from the Human Genome Project raise questions of ownership and patent rights. Who owns the human genome or the sequence of any particular gene? If a researcher localizes a gene to a particular chromosome, can that researcher patent the information? Can a gene sequence be copyrighted in the manner of a book? Can the genes themselves be patented? Rapid, public disclosure of all data was a central principle of the IHGSC that was responsible for the public human genome sequencing effort. This contrasted with the stated goal of Celera Genomics, which required others to pay for access to their genome databases. The IHGSC's decision to publish a draft sequence as fast as possible was driven, in their words, by "concerns about commercial plans to generate proprietary databases of human sequences that might be subject to undesirable restrictions on use" (*Nature* 409: 863).

In the decade following the completion of the Human Genome Project, many biotechnology companies sought patents for human genes from which they hoped to profit through the production of genetic tests, drugs for human diseases, and so forth. On June 13, 2013, the U.S. Supreme Court ruled (in the case of *the Association for Molecular Pathology v. Myriad Genetics, Inc.*) that human genes could not be patented because they were "products of nature" and therefore not intellectual property. This decision invalidated over 4,300 human gene patents that had previously been granted. The Supreme Court's decision did allow for the patenting of human DNA sequences that have been subject to genetic engineering, however. This leaves open, therefore, the possibility of patenting DNA constructs used for gene editing or to produce genetically engineered medicines, as described earlier in this chapter.

THOUGHT QUESTIONS

1. To what extent do you agree with James Watson's statement that sequencing the human genome will tell us what it means to be human? Suppose you knew the exact gene sequence of part or all your genome; what would you really know about yourself?
2. If only stretches of DNA 500–700 bases long can be sequenced at a time, how many of these small sections of DNA must be sequenced to determine the sequence of the entire human genome? (Think also about the overlaps required to piece the sequences together; assume an average of 10% overlap.)
3. Will the DNA sequence of the human genome tell us what traits are controlled by each part of the sequence? Will it tell us which sequences represent genes and which sequences represent spacers?
4. If you have a certain rare genetic condition and scientists use cell samples from your body to determine the gene's DNA sequence, what rights (if any) does this give you to the information? Do the scientists have the right to publish your gene sequence or any part of it? Is it an invasion of your privacy? Can the scientists sell the information? If they do, are you entitled to a share of the profits?

4.4 GENOMICS IS A NEW FIELD OF BIOLOGY THAT GAINED PROMINENCE WITH THE HUMAN GENOME PROJECT

The Human Genome Project also funded the sequencing of the genomes of many other species. This may seem odd at first, because the name of the project specifies the *human* genome, but there were several reasons for including these other species. The study of the genomes of species has become an entire new area of biology called **genomics**. This field has arisen in part to help unfold the mysteries of human genes now that the complete human genome sequence is known. One focus of genomics is the identification of individual human genes. The combination of molecular biology and computer science that has been necessary to navigate through the tremendous amounts of data produced by the various genome projects is called **bioinformatics**.

4.4.1 Bioinformatics

Just as the NASA space program led to many unexpected "spin-off" technologies in the 1960s and 1970s, the Human Genome Project has done so as well, with new computer technologies and genetic engineering having wide applications outside genetics. DNA sequencing and mapping would not have been practical before the advent of modern computers. Although the techniques for determining sequences of short pieces of DNA are rather simple (Figure 4.11), finding the overlaps that indicate how the small sequenced pieces were originally arranged (Figure 4.12) requires massive computer power. Further, once the longer sequences have been determined, storing the data has necessitated the development of larger and larger computer databases and new methods for searching them. Genomics requires the development of new types of computer software. The need for people who are trained in both molecular biology and computer science who can work with these data has made bioinformatics a fast-growing new field of employment.

One research objective within the field of bioinformatics has been the development of computer programs to locate genes within a genome. In the past, as we have seen, scientists started with an observable trait and worked backward to find the gene that controlled it. Now that the genome sequence is complete for many species, including humans, the method of gene discovery has changed. Researchers can now examine the DNA sequences and try to determine which parts may be genes, without any prior knowledge of a trait or a function for those genes. Genome sequencing projects in a wide range of species have led to the identification of many such "orphan genes," so-named because, at the time of their discovery, they could not be associated with any particular function or biological process. (The identification of the functions of orphan genes is part of another branch of genomics called functional genomics, which is described later.)

Researchers working in bioinformatics are programming computers to scan the sequence data to locate genes, meaning areas that code for RNAs and proteins.

To do so, programmers must discern "rules" of the genetic code: What characteristics of a sequence distinguish a coding region from a noncoding region? The computer search for genes within sequences is called "gene scanning."

4.4.1.1 Gene scanning in different organisms

Interestingly, most genes start with the codon ATG and end with one of three "stop codons": TAA, TAG, or TGA. If the nucleotides A, T, G, and C were distributed randomly, each of the stop codon triplets would be expected to occur, on average, every 4^3 or 64 bases. But nucleotides are not distributed randomly within genes; they are retained in a nonrandom pattern as a result of evolution because they code for a product conferring advantage to the organism. In bacteria, genes are typically 300–500 codons long, are contiguous, and do not overlap. In addition, bacteria have very little noncoding DNA (Figure 4.13A). These factors make gene scanning in bacteria relatively easy. A computer can scan the sequences that follow any ATG and find those areas where the next stop codon occurs a few hundred bases farther along.

Gene scanning is much more difficult in more complex organisms, namely the nucleated, eukaryotic organisms (see Section 6.2.4). In contrast to bacterial species, they have long, noncoding stretches of nucleotides, called **introns**, dispersed among much shorter regions that correspond to codons (Figure 4.13B). While the coding regions, called **exons**, are roughly the same lengths in different species, the size of the noncoding introns is much greater in humans than in other species (Figure 4.13C).

Only 1%–2% of the total human genome consists of protein-coding sequences. The remainder comprises introns, intergenic sequences, and sequences that are transcribed but not translated into proteins. This makes it very difficult to use raw sequence data to predict which nucleotide regions represent human genes. Gene-scanning programs are continually refined to include up-to-date knowledge about the characteristics that distinguish eukaryotic genes from noncoding regions based on the concept of "departures from randomness"—i.e., sequence features that would have a very low probability of occurring if nucleotides were ordered randomly, without selection for a biological function. One such departure from randomness is called "codon bias": Not all codons are used equally by a given species. For example, the amino acid alanine can be coded by four different codons in humans; furthermore, three of those are used much more frequently than the fourth. In a noncoding region of the genome, all of the codons have an equal probability of being represented, but in a coding region, the one codon is present less frequently.

The presence of noncoding regions within genes (introns) is clearly a complication for gene scanning. Of what benefit could it be to an organism for its genes to be interrupted by such noncoding regions? Over the course of evolution, it is these noncoding stretches that have allowed the shorter coding segments to recombine

(A) Gene structure and transcription in bacteria

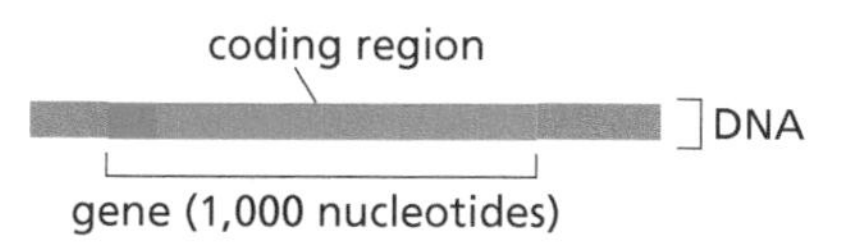

(B) Gene structure, transcription, and RNA splicing in eukaryotes

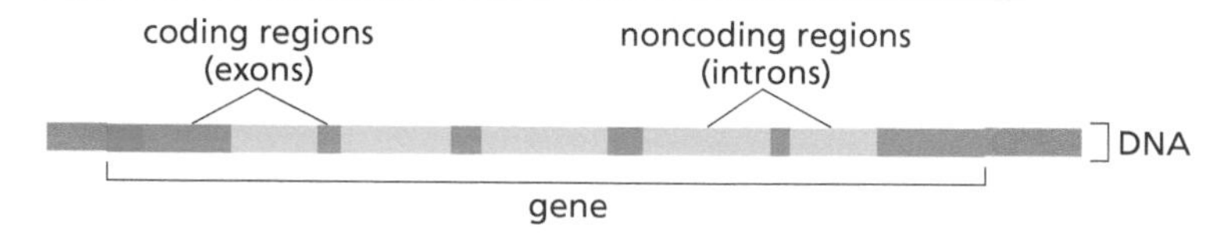

(C) Complex exon/intron structure of a human gene

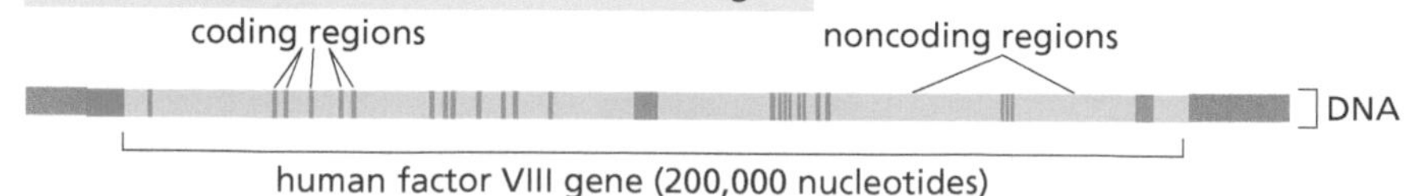

Figure 4.13 **Differences in gene structures in bacteria compared to eukaryotic cells. (A)** Bacterial protein-coding genes contain only coding regions, meaning that all of the RNA transcribed from the DNA is then translated to protein. **(B)** In eukaryotic cells, noncoding regions (introns) are interspersed between coding segments (exons) of genes. Both exons and introns are transcribed into RNA, but the introns are removed by a process called RNA splicing to generate the mature mRNA that is then translated to protein. **(C)** In humans (a eukaryotic species), the amount of noncoding DNA (introns) is much greater than the amount of coding DNA (exons), and some genes have a large number of exons and introns.

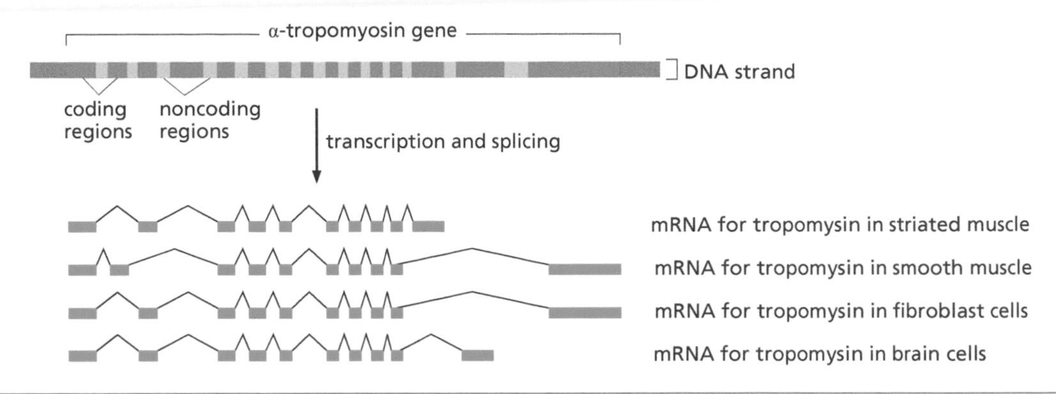

Figure 4.14 A single human gene can give rise to multiple proteins depending on which exons are included in the final mRNA that is translated. Different protein isoforms are expressed in different tissues and have specialized functions.

to form new genes. In other words, new genes are produced by the novel assembly of existing parts. This provides a mechanism for rapid genetic change (more rapid than by simple base-pair mutations). There is also another way in which the division of eukaryotic genes into many coding regions (exons) is adaptive: It provides a mechanism by which slightly different versions of a protein can be made in different tissues, adapted to the particular cellular environment and function of that tissue. An example is shown in Figure 4.14. The human gene for a protein called α-tropomyosin contains 15 coding regions (exons) scattered among noncoding regions (introns). This gene can give rise to as many as 20 distinct versions (or isoforms) of α-tropomyosin protein, depending on which cell type is being considered. This occurs in two ways: First, the gene may be transcribed from two different start points, yielding a precursor mRNA (pre-mRNA) that contains slightly different exon and intron sequences; second, the process of RNA splicing, which removes introns and joins exons to create a mature, protein-coding mRNA, can occur differently in different cell types. In this way, slightly different subsets of coding regions are selected from the total set of coding segments in each tissue. This results in different mRNAs being present in the different cells, and therefore in slightly different proteins being produced from those mRNAs after translation. Each protein is still α-tropomyosin, but with a slightly different amino acid sequence and, therefore, a slightly different functional capability.

Although scientists think these large noncoding regions within genes are adaptive for the organism, they do present a significant obstacle to identifying genes by gene scanning. Another complicating factor is the recent discovery of microproteins, proteins that consist of just a small number of amino acids that are encoded by genes that are therefore much smaller than the typical threshold length that was previously used to characterize protein coding regions. In fact, it does not appear that gene-scanning programs alone will be able to identify all the genes in a eukaryotic genome. Hence, the Human Genome Project also funded work on the genomes of other species, so that human genes could be located by comparison to the genes of other species, a field now known as **comparative genomics** (see later).

4.4.2 Next-generation DNA sequencing

The value of having a complete human genome sequence was immediately clear as the Human Genome Project neared completion. Moreover, the rapid development of bioinformatics tools has greatly facilitated genomic discoveries in recent years. Therefore, it is not surprising that the Human Genome Project stimulated a large amount of research and development into new technologies for sequencing DNA that would be faster and cheaper than the traditional Sanger method. Together, the new methods are referred to as **next-generation DNA sequencing** (also known as high-throughput or massively parallel sequencing). Several companies have developed their own proprietary next-generation sequencing platforms, with names such as 454, Illumina, and Ion Torrent, but all are distinct from Sanger sequencing in that they generate many sequence reads simultaneously from DNA samples that may be highly complex, as opposed to the single DNA template used for Sanger sequencing. For example, next-generation DNA sequencing can be used to estimate the number of different species of bacteria present in a soil sample from mixed genomic DNA extracted from the soil without isolating individual bacterial strains. This type of sequencing also allows for quantification of the relative levels of different template DNA molecules, so the relative abundance of different bacterial species in a soil sample can be estimated, with each unique DNA sequence for a given gene representing a different species.

The Illumina technology has become the most popular, so we will briefly describe it here. This method is also illustrated in Figure 4.15. The first step in an

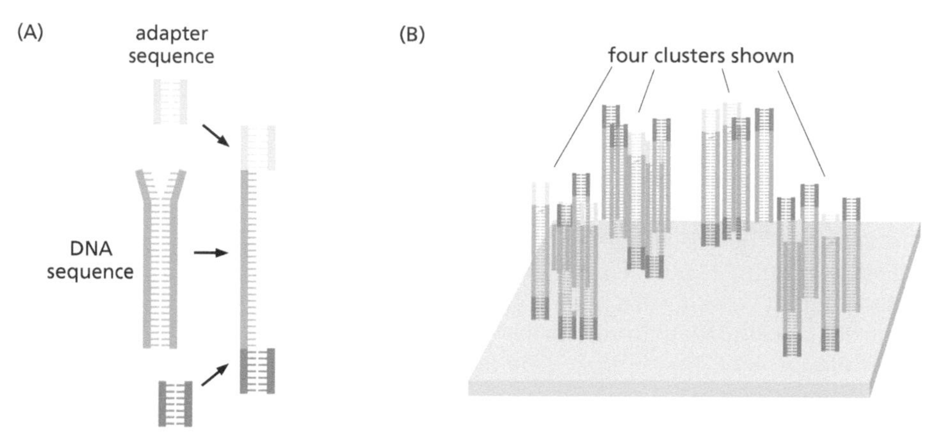

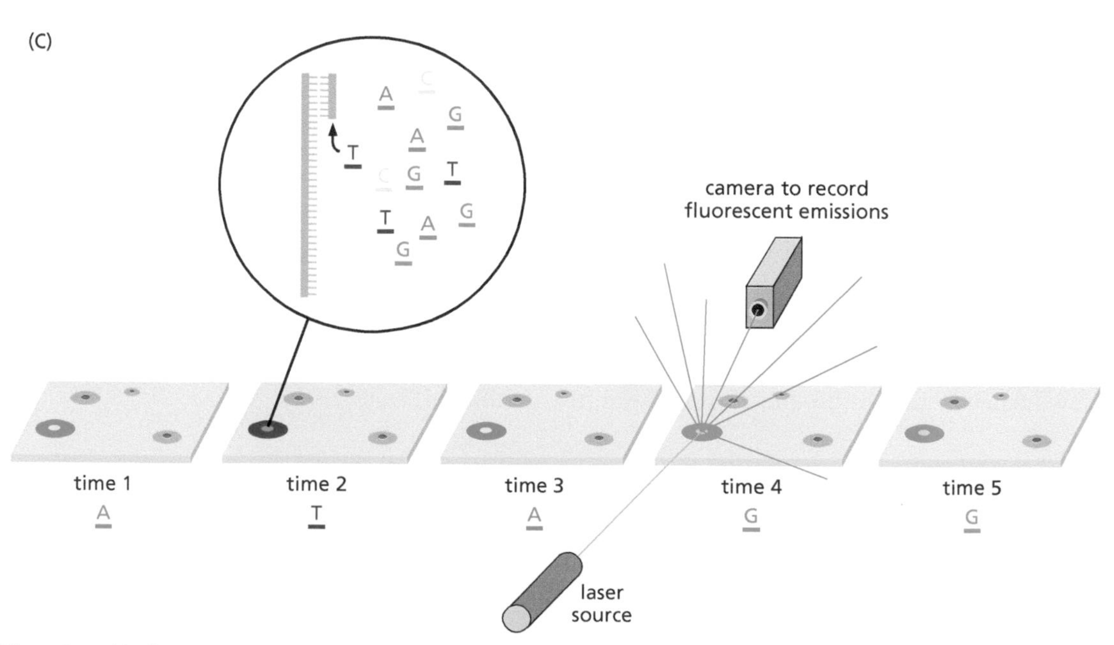

Figure 4.15 The Illumina method of next-generation DNA sequencing.

Illumina sequencing experiment is to prepare a library from the DNA sample you want to sequence. The DNA is processed into fragments (usually 200–500 base pairs long) using either physical or enzymatic means, and the two DNA strands are separated. Short adapter DNA molecules are then added to the ends of the DNA fragments using a ligase enzyme. The adapters are physically anchored to a surface (like a glass slide), and short, single-stranded DNA primers base-pair with the adapter sequences. These primers are used to make many copies of each DNA fragment that cluster together on the glass slide, thus amplifying the signal from each fragment. After the amplification step, special, fluorescently labeled nucleotides are used to extend the primers one nucleotide at a time. Like dideoxy nucleotides, these are chain terminators (ensuring that only one nucleotide is added at a time), but unlike dideoxy nucleotides, a chemical reaction can be used to "unblock" the chain termination. At each strand extension step, before the terminator is unblocked, a camera takes a picture of the slide, recording the color of each DNA fragment cluster. Then, after the unblocking step, the next fluorescent nucleotide can be added, thus extending the sequence "read" by another nucleotide. This method is referred to as "sequencing by synthesis."

Next-generation DNA sequencing methods have dramatically accelerated the sequencing of many different organisms' genomes; the National Center for Biotechnology Information of the U.S. National Institutes of Health (https://www.ncbi.nlm.nih.gov/) tallies over 12,000 eukaryotic genomes, about 250,000 prokaryotic genomes, and about 39,000 viral genomes that have been sequenced as of July 2020. Of equal importance, these next-generation techniques have made sequencing fast enough and cheap enough to allow comparisons between the genomes of many different individuals of the same species. This enables geneticists to study genetic variation among populations at the whole-genome scale. As an example of how this could be used, imagine two sets of people, one group that has a particular disease and one that is disease free. By sequencing the complete genomes of many representatives from each of the two groups and using statistical techniques to analyze the results, geneticists can identify particular versions (alleles) of certain genes that may correlate positively or negatively with occurrence of the disease. This type of study is called a **genome-wide association study**.

4.4.3 Comparative genomics

When scientists compare sequences of genes from one species to another, they are working in the field of comparative genomics. The size of the genome of many species has been determined (Table 4.1). As we saw earlier, genome size does not always correlate with the complexity of the organism. This is due to the very great differences in the amount of noncoding DNA in various genomes, as well as to the potential for one gene to encode more than one version (isoform) of a protein in multicellular organisms, thus allowing greater functional complexity than might be predicted simply by gene number.

One of the smallest genomes known belongs to *Mycoplasma genitalium*, a bacterium responsible for a sexually transmitted infection discussed in Section 16.2. The complete genome sequence, some 580,070 bases long, was first published in 1995. In 2008, a team led by Hamilton Smith at the J. Craig Venter Institute succeeded in synthesizing an entire artificial genome similar to that of *M. genitalium*, ushering in a new era of what is now called **synthetic biology**. These researchers wanted to determine the smallest amount of genetic information needed to sustain life; they also wanted to create a minimal "recipient" genome into which beneficial genes might be inserted to engineer microbes that might synthesize biofuels, clean up chemicals in the environment, or produce valuable drugs and other products. Synthetic biology of this kind represents a new branch of biotechnology.

As we have discussed earlier, sequences representing genes are much easier to identify in some species than in others. Once a gene has been identified and its sequence determined in one species, there is often enough sequence similarity for its counterpart gene (or genes) to be located in other species. This is the major

Table 4.1 Sizes of genomes of different species

Species	Genome size in millions of bases (Mb)
Bacterial species	
Mycoplasma genitalium	0.58
Bacillus megaterium	30
Escherichia coli	4.64
Nonbacterial species	
Saccharomyces cerevisiae (yeast)	12.1
Caenorhabditis elegans (roundworm)	100
Drosophila melanogaster (fruit fly)	140
Locusta migratoria (locust)	5,000
Fugu rubripes (pufferfish)	400
Mus musculus (mouse)	3,300
Homo sapiens (human)	3,000
Oryza sativa (rice)	565
Zea maize (corn)	5,000
Triticum aestivum (wheat)	17,000
Fritillaria assyrica (ornamental flower)	120,000

Adapted from Genomes, Terrence A. Brown, Wiley-Liss, N.Y., 1999

reason why other species' genomes were also examined as part of the Human Genome Project. Another reason was that sequencing the genomes of other species allowed scientists to develop the technology that was later used to analyze human genome sequences.

Many human genes were first located by their similarity to yeast genes. A yeast cell, like a human cell, has a nucleus, and many of its genes have remained very similar to the counterpart genes in humans. Animals are even more similar to humans, and one animal that has proved to be quite useful in comparative genomics is the pufferfish, *Fugu rubripes*. Its genome is only one-seventh the size of the human genome (see Table 4.1), yet it is estimated to have the same number of genes. Because of its small genome size, mapping gene locations is much easier in pufferfish, so its genome, completed in 2002, was useful as a reference for mapping human genes. The mouse genome, the sequencing of which was completed around the same time as the *Fugu* genome, was also useful in this respect. Many known genes in mice are located in the same order on their chromosomes as they are on human chromosomes, and this correspondence is extremely helpful in mapping genes. The mouse genome, however, is even larger than the human genome (see Table 4.1), so the problems of working with a large genome still pertain.

Aside from its usefulness in locating human genes, comparative genomics has produced new data for evolutionary biologists. Species that have a recent common ancestor have more genes and more nucleotide sequences in common than species that do not. Comparative genomics has thus made the reconstruction of family trees (phylogenetics) much easier, providing many new insights into the evolutionary history of large groups of organisms (Chapter 6). Unfortunately, the scientists working on a particular species have often independently devised the database for each species' genome. Consequently, another goal of bioinformatics is to devise ways of making the different databases compatible and interactive, thereby facilitating comparative genomics.

In addition to finding similarities between species, comparative genomics has led to the realization that, within a species, there are groups of genes that share large portions of their sequences. These "gene families" are presumed to have evolved from a common ancestral gene. Finding one gene in the family enables the location of the others, and most often the identification of the different protein products of the family members. For example, the human hormones oxytocin and vasopressin (both proteins) belong to the same gene family, and they have very similar amino acid sequences and genes that code for them. The same is true of the oxygen-carrying proteins hemoglobin and myoglobin.

4.4.4 Functional genomics

In Section 3.3, we described how Archibald Garrod and other scientists studied inborn errors of metabolism, disease conditions caused by changes in biochemical pathways. The study of similar changes in bacteria or yeast have often led to the discovery of entire chains of biochemical reactions. In the past, scientists looking for the molecules involved in such a biochemical pathway would start with a trait and work backwards to identify a protein and the gene that encodes it. Pedigrees, such as we saw in Chapter 3, would be linked with different forms of a protein. After purifying the protein and discovering its amino acid sequence, its gene sequence could be inferred. Gene sequencing turns this whole process around. Genes are found by linkage to DNA markers, and only later is the protein product found. However, finding a gene, mapping its location, sequencing it, and even deriving the amino acid sequence of its protein product will not tell us its function.

New sequences can be compared with those whose function is already known. This is the field of **functional genomics**. Species that can be easily manipulated experimentally have been most useful in discovering gene functions. The zebrafish (*Danio rerio*) is a vertebrate that reproduces rapidly, and many of its internal structures are visible in the living fish because overlying structures are transparent. For these reasons, zebrafish have become an experimental species of great interest to scientists working on the genetics of development. An even simpler species, baker's yeast (*Saccharomyces cerevisiae*), has been found to share many genes with humans. Gene functions that were discovered in yeast have proved to have parallels with disease-associated gene mutations in humans (Table 4.2). The functions of the yeast genes can be discovered by several methods: For example, one might examine which genes are transcribed to mRNA when the yeast undergoes a particular response or function; alternatively, one could inactivate (mutate) a gene and see what effect this has. Once the sequence of a gene is known, it is relatively easy to mutate it by manipulating the DNA causing a change in the protein product, which is now not functional. The opposite approach can also be used: Extra copies of the gene can be inserted, and observations made of the change in function under different environmental conditions. These approaches are not confined to yeast, but can also be used to discover gene functions in mice and other species.

Earlier in this chapter, we saw how CRISPR-Cas9 gene editing could be used to disrupt a gene ("knocking it out," or rendering it nonfunctional) or to introduce specific changes to a gene, such as correcting a mutation or introducing a specific mutation to determine its phenotypic outcome. Other methods of gene modification rely on the ability of a segment of double-stranded DNA to be inserted into a specific location in the genome based on homologous recombination (Figure 4.16). This works in organisms as diverse as yeast through humans, but it is less efficient than CRISPR-Cas9. Gene knockout experiments are often carried out in mice. If a gene is knocked out in embryonic stem cells at a very early stage of mouse development, an entire organism can develop that is missing the gene and its protein product in many or even all of its cells. A variation on this strategy, called the Cre-lox method, can be used to knock out a gene only in a specific tissue of an animal.

Table 4.2 Some of the human disease genes that have high similarity to yeast genes

Human disease	Yeast gene	Function in yeast
Amyotrophic lateral sclerosis	*SOD1*	Protection against superoxide (O_2^-), a toxic form of oxygen
Colon cancer	*MSH2*	DNA repair
Cystic fibrosis	*YCF1*	Metal resistance
Myotonic muscular dystrophy	*YPK1*	Codes for an enzyme (protein kinase)

Adapted from Genomes, Terrence A. Brown, Wiley-Liss, N.Y., 1999

Figure 4.16 Creation of a gene disruption ("knockout") by homologous recombination of a selectable marker.

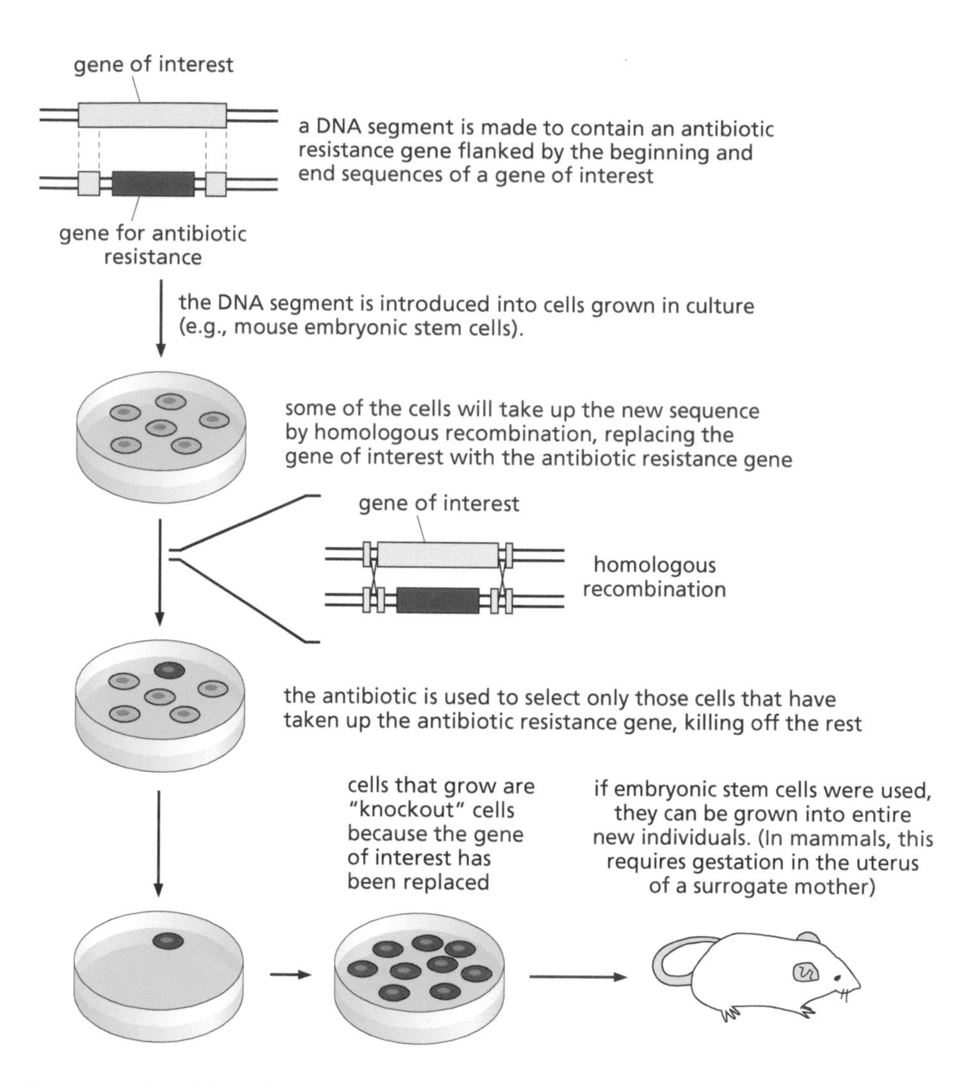

For example, although a certain neurotransmitter receptor might be expressed in many different cells in the brain, one might want to study the effect of its elimination in just the hippocampus. Knockout mice or mice with other types of modifications to specific genes have revealed important clues to the functions of human genes.

Within a given species' genome, families of genes have been found that express structurally related protein products that often have quite different functions. This has led scientists to realize that, during evolution, gene duplication often occurs first, and further mutation of the different gene copies eventually leads to new functions. Of course, a gene that is present in just a single copy cannot change to a new form (possibly with a new function) without giving up its original form and function. A duplicated gene, however, can undergo changes in one copy (possibly evolving new functions) while the other copy remains unchanged. Interestingly, the genome of baker's yeast is thought to have undergone a complete duplication at one point during evolution, so many of its genes exist in pairs that have similar but slightly specialized functions.

4.4.5 Proteomics

Just as the complete DNA sequence of an organism is its genome, the complete protein content of that organism is its proteome. **Proteomics** is the study of how the protein content changes over time in a cell and in an organism, how it differs in different tissues, and how it relates to the health and function of the organism. Proteins are synthesized as a result of **transcription** of genes and translation of mRNA, as we saw in Chapter 3. However, there are further modifications to a protein after it has been translated that affect both its activity and its concentration. No protein stays in a cell forever; all are eventually degraded. You will learn more about these aspects of protein function in cells in Chapter 10.

Rather than studying one protein at a time, proteomics also has another goal: to study all of a cell's proteins in the aggregate. Such a goal was unattainable in the past and is a big factor explaining why reductionism (reducing a problem to its simplest form or its smallest component parts) has been a widespread experimental approach in biology. New technologies that apply the tools of analytical chemistry to the study of proteins have greatly advanced the field of proteomics. The 2002 Nobel Prize in Chemistry was awarded to three scientists for their contributions to the analysis of proteins. Two of the winners developed techniques for identifying and quantifying proteins using mass spectrometry, and the third applied nuclear magnetic resonance (NMR) to determine the three-dimensional structures of proteins, a component of proteomics that we address next.

Knowing the DNA sequence of genes allows for the prediction of the amino acid sequences of proteins encoded by those genes. A protein's function is the direct result of the three-dimensional structure into which its amino acid chain folds, which is dictated by chemical bonding between different parts of the amino acid chain. Most proteins adopt a single lowest energy structure, or conformation. Computer programs can use what is known about the bonding behavior of amino acids to predict how portions of that protein will fold, but it is still very difficult to accurately predict the entire three-dimensional structure of a complex protein. In most cases, any computer-based predictions must be verified experimentally using techniques of structural biology, such as X-ray crystallography, nuclear magnetic resonance (NMR), or specialized electron microscopy methods (e.g., cryo-electron tomography). The task of predicting a protein's structure using a computer is simplified if the structure of a related protein (e.g., from another organism) has already been determined experimentally.

Knowing the detailed structures of proteins is very important because it can lead to new strategies for the design of medicines. In the past, natural products and synthetic compounds were randomly tested in functional assays to see which would work for a particular need. Now, small molecules can be designed to exactly fit a critical enzymatic site within a protein. A notable example is Gleevec®, a drug designed to target a mutant enzyme implicated in one type of leukemia (**Figure 4.17**). (See also Sections 10.5.2 and 10.7.2.) A few other examples of

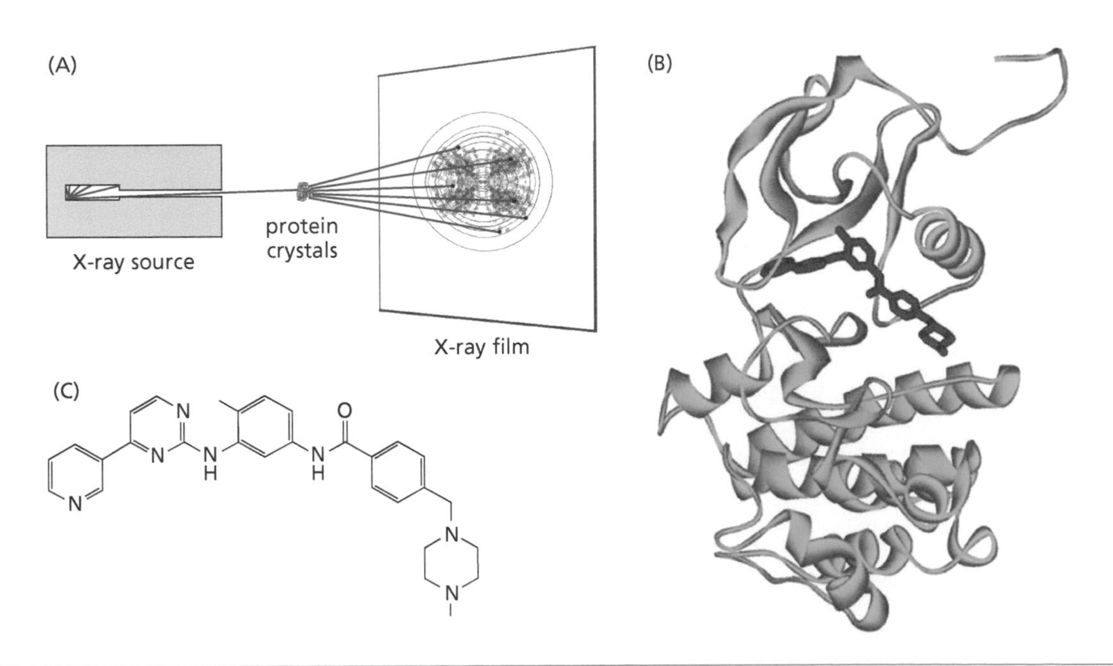

Figure 4.17 Protein structures determined by X-ray crystallography can be used to design drug molecules that inhibit proteins responsible for disease. (A) Protein crystals can be bombarded with X-rays, and the scattering patterns can be interpreted by computer programs to determine the three-dimensional protein structure. **(B)** An example showing the hybrid Bcr-Abl protein (in green) responsible for a type of leukemia (see Section 10.5.2), and a drug molecule (Gleevec®, in red) designed to fit into this protein and render it harmless. **(C)** The molecular structure of Gleevec®.

Table 4.3 Some examples of drugs developed by rational drug design

Drug name (generic name in parentheses)	Protein targeted	Disease treated	Year approved by U.S. FDA
Capoten® (captopril)	Angiotensin-converting enzyme (ACE)	Hypertension (high blood pressure)	1981
Pepcid® (famotidine)	Histamine H_2-receptor	Acid reflux	1986
Gleevec® (imatinib)	Bcr-Abl fusion protein	Chronic myelogenous leukemia	2001
Sabril® (vigabatrin)	Gaba transaminase	Epilepsy	2009
Vitrakvi® (larotrectinib)	Neurotrophic receptor tyrosine kinase fusion proteins	Multiple solid tumor types	2018

rationally designed drugs are listed in Table 4.3. Once a candidate drug molecule is designed by computer simulation, medicinal chemists then synthesize the molecule and biologists test to see if it has the desired outcome of blocking the protein's function. This process is usually iterative: If a candidate molecule is shown to have biological activity, chemists then synthesize a panel of similar molecules with slight variations, and biologists test each to try to optimize the effect on the target protein. This strategy is referred to as **rational drug design**. The action of drugs designed in this manner is far more specific than that of traditionally developed drugs, and these drugs will often have fewer side effects as a result, for reasons we will study in Chapter 14.

To synthesize a protein with even a slightly different structure from its component amino acids can be very difficult and costly. However, once the sequence for the gene coding for that protein is known, it becomes relatively easy to modify the protein by changing the sequence of its gene. Roughly the same technique that we saw in Figure 4.9 is used, but the inserted piece of DNA differs from the normal piece by just a few chosen nucleotides. Such modifications can, for example, lead to the development of proteins that are stable under a wider variety of conditions. These proteins find a variety of industrial applications. Stain removers in laundry detergents, altered enzymes for food processing, and enzymes that function in pollution remediation are just a few examples.

4.4.6 Transcriptomics

Beyond rapid sequencing of genomes, another common application of next-generation sequencing is called **RNA sequencing** (or RNA-seq). This name is a bit misleading, since it's still DNA that is sequenced, but the DNA in question is complementary DNA (**cDNA**) that has been generated by reverse transcribing RNA isolated from cells or tissues. **Reverse transcriptase** enzymes that can copy RNA into cDNA are encoded by the genomes of retroviruses such as HIV, which have an RNA genome that is copied into DNA once the virus infects a host cell (see Section 15.1). The RNA-seq method can be used to analyze the relative levels of expression (i.e., transcription) from all genes in an organism at once, an approach sometimes referred to as **transcriptomics**. RNA-seq is usually used to compare levels of RNA transcription under two different experimental conditions or in individuals who have different phenotypes. From the mid-1990s to mid-2000s, DNA microarrays (mentioned in Chapter 3) were used for this same purpose, but the data acquired from RNA-seq are much more accurate.

THOUGHT QUESTIONS

1. What are the arguments in favor of funding genomic studies in a wide range of organisms? What makes the human genome more difficult to study than genomes of some other organisms?
2. What technological advances have enabled the new "omics" fields (genomics, proteomics, transcriptomics) as well as the new fields of bioinformatics and synthetic biology?
3. How have genome sequencing and other advances stimulated by it allowed for new ways of identifying the functions of proteins?

CONCLUDING REMARKS

As genomics has discovered genes with useful properties within humans and other species, genetic engineering has given us the tools to transfer those genes into other species. The Human Genome Project has discovered that transfer of genes from one species to another also occurs in nature. Viral and bacterial genes are found in the human genome, for example. Because we have almost always studied the effect of one gene or one protein at a time, transferring a gene into a new genomic environment may lead to different results than we expect. As we further develop tools to alter genomes, proteomics may give us ways to study the effects of such changes throughout the cell. We also need to be mindful of effects at the level of the whole organism and effects of genetic engineering on ecosystems as well, which we will explore further in Chapter 17.

CHAPTER SUMMARY

- **Restriction enzymes** cut DNA into fragments with known sequences at their ends. Restriction enzymes that produce fragments with single-stranded sticky ends are used in genetic engineering to splice new genes into genomes.
- Variations in the lengths of fragments produced by restriction enzymes are called **restriction fragment length polymorphsims** or **RFLPs**. RFLPs were the first DNA markers used to map genes to locations within the genome.
- Along with RFLPs, genetic markers called **expressed sequence tags (ESTs)** and **single-nucleotide polymorphisms (SNPs)** can differentiate individuals.
- **Genetic engineering** consists of inserting functional genes into cells, thereby altering the cell's genotype. The recipient cells may be bacterial cells that may then acquire the ability to make certain human proteins, or they may be human cells that acquire a functional allele and are injected into a patient as **gene therapy**.
- Bacterial **plasmids** and viral **vectors** are used to carry genes into a new species.
- A **genome** is the total genetic information carried by a particular organism. The Human Genome Project was completed in 2003; many thousands of genomes have been sequenced in the years since, due to the advent of **next-generation DNA sequencing** methods that are fast and inexpensive.
- **Genomics** is the study of the genomes, including the comparison of genomes of different species and the discovery of gene functions.
- **Bioinformatics** combines computer science and molecular biology in the analysis of genomes and the identification of genes within a genome.
- **Proteomics** is the study of all of the proteins present within a cell.
- **Transcriptomics** is the study of levels of all mRNA transcripts present within a cell.

BIBLIOGRAPHIC REFERENCES

Bibliographic references to material in this chapter can be found online at biologytrending.routledge.com/chapter4

GLOSSARY: KEY TERMS TO KNOW

These key terms are defined at the end of each chapter as an aid for student review.

Bioinformatics The study of large collections of molecular sequences (such as nucleic acid or protein sequences) with the use of computer-based methodologies to catalog, search, and compare these sequences.

cDNA (complementary DNA) A strand of DNA produced by transcription from an RNA template molecule using a reverse transcriptase enzyme.

Cas9 An enzyme used in genetic engineering (or by bacteria) to cut DNA sequences at desired locations identified by CRISPR sequences.

Comparative genomics Comparison of genomes from different species and their use to answer questions about evolution and gene function.

CRISPR (clustered regularly interspaced short palindromic repeats) Repeated DNA sequences used by certain bacteria to recognize and defend against previously encountered viruses, or similar sequences used to target selected genes in genetic engineering.

CRISPR-Cas9 A method of precise gene editing that relies on CRISPR RNAs designed to target genes of interest and a CRISPR-directed DNA nuclease, Cas9. This method mimics a natural system used by certain bacteria as a protection against previously encountered viruses.

DNA probe A short DNA sequence, complementary to a sought-after "target" sequence, linked to a fluorescent or radioactive tag that can be made visible to show the location of the target sequence.

DNA ligase An enzyme that creates a continuous DNA strand by joining separate DNA fragments; used to create a recombinant DNA molecule from restriction enzyme fragments.

Expressed sequence tag (EST) A short segment of an RNA (or its cDNA) that can be used as a genetic marker to indicate the level of expression of the corresponding gene, which may vary between individuals.

Exon A segment of a protein-coding gene that is joined together with the gene's other exons to generate the complete mRNA that is then translated into protein.

Functional genomics The methods used to assign biological function to the nucleic acid sequences contained in a sequenced genome.

Gene editing Making targeted DNA sequence changes to a gene at its normal position in the genome, most often using the CRISPR-Cas9 method.

Gene therapy Introduction of a functional copy of a gene to overcome a genetic deficiency, especially when used to treat hereditary diseases.

Genetic engineering The intentional altering and manipulation of genotypes.

Genome The total genetic makeup of any individual, including its entire DNA sequence.

Genome-wide association study Statistical comparison between the genomes of individuals with and without a certain condition, used as an attempt to understand the genetic basis of the condition.

Genomics The scientific study of entire genomes.

Human Genome Project A large-scale project to map and sequence the human genome and, subsequently, the genomes of many other species.

Intron A segment of a protein-coding gene that is removed between exons during mRNA splicing and thus does not contribute sequence information for protein synthesis.

Knockout A genetically altered cell or organism in which a particular gene is deleted or disabled, usually in an attempt to study the gene's function.

Microsatellite marker A type of DNA marker in which a short nucleotide sequence is repeated a different number of times in different individuals.

Next-generation (next-gen) DNA sequencing High-throughput, massively parallel DNA sequencing methods developed after the completion of the Human Genome Project that have greatly accelerated genome sequencing due to their speed and their low costs compared to traditional Sanger sequencing.

Palindrome A sequence that reads the same forwards or backwards, as in the sentence "Madam, I'm Adam," or a DNA sequence that can base-pair with itself when folded in half, as in AACCCTTTGCAAAGGGTT.

Plasmid A small, circular DNA molecule, separate from the chromosomal DNA, that can be replicated and passed on through cell division and can sometimes integrate into chromosomal DNA. Found mostly in bacteria but also used in genetic engineering as a vector to transfer DNA sequences between different organisms.

Polymerase chain reaction (PCR) An iterative method for amplifying (making many copies of) a specific DNA sequence using a template containing that sequence, plus a heat-stable DNA polymerase enzyme, and primers that specify the ends of the sequence to be amplified.

Primer A short, single-stranded segment of DNA that can base-pair with the complementary DNA sequence that is used for PCR or DNA sequencing.

Probe See *DNA probe.*

Proteomics The study of the total set of proteins present in an organism.

Rational drug design Use of structural information about target molecules (such as proteins implicated in diseases) to guide creation of molecules that will interact with those targets for a medicinal purpose.

Restriction endonuclease (restriction enzyme) An enzyme that cuts a nucleic acid only at places with a specified sequence of bases.

Restriction enzyme See *Restriction endonuclease.*

Restriction fragment A linear DNA molecule released from a larger molecule by digestion (cutting) of that larger molecule by a particular restriction endonuclease.

Reverse transcriptase An enzyme that can make a complementary DNA copy of an RNA template; found naturally in retroviruses such as HIV.

RFLPs (restriction fragment length polymorphisms) Differences in the lengths of the restriction fragments made by the same endonuclease among different individuals.

RNA sequencing (RNA-seq) A method for studying the transcriptome that involves isolating RNA from cells, generating complementary DNA (cDNA) using a reverse transcriptase enzyme, then sequencing the cDNA using next-generation DNA sequencing methods.

SCID (severe combined immune deficiency) A rare genetic condition in which the entire immune system is nonfunctional.

Secrete To produce and release a cellular product to the environment outside of the cell.

Selective marker A gene that permits the survival of cells carrying the gene under conditions that will kill cells lacking the gene, such as a gene that confers resistance to an antibiotic.

Single nucleotide polymorphism (SNP) A nucleotide position within a genome that shows variation across a population and can be used as a DNA marker.

Synthetic biology A new field of biology that aims to extend traditional genetic engineering, creating modular DNA sequences that perform novel biological functions and even engineering entire genomes to create "designer" microbes that may perform useful functions.

Transcription A process in which DNA is used as a template to guide the synthesis of RNA; the first step in gene expression.

Transcriptomics The study of all RNA sequences expressed in an organism through transcription of DNA to RNA.

Transduction Introducing new DNA into a cell using a viral vector.

Transfection In animal or plant cells, a hereditary change caused by incorporation of DNA fragments from outside the cell.

Transformation In bacteria or yeast, a hereditary change caused by incorporation of DNA fragments from outside the cell. (Note: Transformation in cancer biology, discussed in Chapter 10, is different.)

Vector In molecular biology, a plasmid or other DNA sequence that can be used to introduce a gene into a target cell. (Note: A different meaning applies in epidemiology; see Chapter 16.)

CONNECTIONS TO OTHER CHAPTERS

Chapter 1: Genetic engineering and gene therapies raise many ethical issues.
Chapter 2: The genome is the blueprint for the proteome.
Chapter 3: We have learned a lot about human genetics from the Human Genome Project and comparative genomics.
Chapter 6: Comparative genomics is a new tool for discovering the evolutionary relationships among organisms.
Chapter 7: The amount of possible variation within each gene is much greater than was previously thought.
Chapter 10: Genetic engineering and bioinformatics are being used to develop novel cancer therapies.

Chapter 15: CRISPR-Cas9 gene editing techniques are being studied as a means of inactivating the HIV virus in infected individuals.
Chapter 16: The same DNA testing techniques that are used to identify individual humans can also identify the bacteria or viruses involved in new infectious outbreaks and can sometimes also identify their sources.
Chapter 17: Genetic engineering of crop species is increasing agricultural productivity and enhancing nutritional qualities of food plants.
Chapter 18: Comparative genomics is increasing our knowledge of biodiversity.

PRACTICE QUESTIONS

1. If one individual human differs from another in 0.1% of the genome, how many bases are different?
2. In the following stretch of DNA, how many fragments will result from digestion with the *Hae*III restriction enzyme shown in Figure 4.1? How many will result from digestion with *Eco*RI?

   strand 1
   ATCCGTAGGCCTAACCATCCTAGTGC
   TAGGCATCCGGATTGGTAGGATCACG
   strand 2

3. Why are restriction enzymes that produce fragments with "sticky ends" more useful in genetic engineering than restriction enzymes that produce fragments with "blunt ends"?
4. When a plasmid is being cut with a restriction enzyme in preparation for inserting a DNA fragment, the plasmid needs to be cut with the same restriction enzyme as was used to make the DNA fragment. Why?
5. Can DNA marker band patterns be used to identify maternity as well as paternity?
6. Can DNA marker testing be used to identify individual organisms in other species besides humans?
7. Sickle cell anemia is a blood disease cause by mutations in the β-globin gene. An individual must possess two mutant copies of the gene (i.e., no normal copies) in order to have the disease. One sickle cell mutation eliminates a cut site for the restriction enzyme *Dde*I within the β-globin gene sequence. The figure that follows shows the results of a Southern blotting experiment in which genomic DNA was cut with *Dde*I, the restriction fragments were separated by size using agarose gel electrophoresis, and a probe specific for the β-globin gene was used to identify bands containing sequences corresponding to that gene. The first two lanes from the left show the banding patterns for the normal and disease versions of the gene, respectively, while the remaining lanes show the results for four different individuals. Which of these individuals (1–4) would be expected to have sickle cell anemia? What are the genotypes of the others?

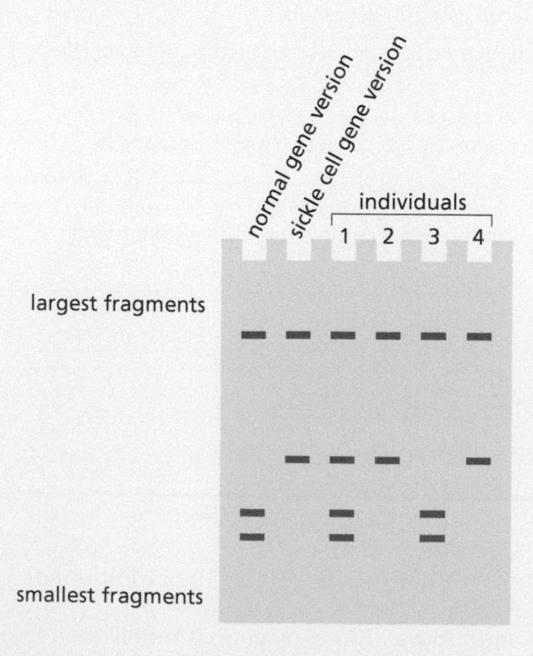

8. List ways that genetic engineering and genomics have impacted human medicine.
9. Although the Human Genome Project was a monumental accomplishment, it can be said that it was only the beginning step toward understanding our genome. Why?

WEB SITES FOR SEQUENCE DETAILS

The entire human DNA sequence, updated daily, is available on the World Wide Web. One site, maintained by the University of California at Santa Cruz, is available at http://genome.ucsc.edu/. Other sites are listed through the U.S. National Institutes of Health site at https://www.ncbi.nlm.nih.gov/genome/. This site also lists various browsers that have been developed to allow users to scroll along each chromosome and to look at the genome in different scales.

http://www.ncbi.nlm.nih.gov/Omim
This is the online version of a book called Mendelian Inheritance in Man, which contains information about human genes and disease.

http://www.genome.gov
A website devoted to information on the Human Genome Project, including the ethical, legal, and social issues (ELSIs) raised by human genome research.

ETHICAL CASE STUDY

It is a few decades in the future. Gene therapy of somatic cells has become commonplace and has begun to be carried out on germ cells as well. A family member of yours who has the disease cystic fibrosis is considering having gene therapy performed on some of his sperm. The sperm could then be used for in vitro fertilization of his wife's eggs without the fear of passing the genes for cystic fibrosis to his offspring. Assume that the success rate of such germline gene therapy is 15% and that it is very expensive. This close relative of yours has asked for your advice. What would you advise him to do? What types of ethical reasoning are you using?

CHAPTER 5

Evolution

ISSUES

- Did life evolve?
- Is science compatible with religion?
- How does evolution relate to genetics?
- How do new species originate?
- Is life still evolving?
- Will life continue to evolve?
- Is the science of evolution static or changing?

BIOLOGICAL CONCEPTS

- Evolution (descent with modification, natural selection)
- Fossils and geologic time
- Environmental influences on species
- Adaptation
- Form and function
- Species and speciation
- Origin of life

CHAPTER OUTLINE

DOI: 10.1201/9781003391159-5

Ask any biologist to name the most important unifying concepts in biology, and the theory of evolution is likely to be high on the list. As geneticist Theodosius Dobzhansky explained, "nothing in biology makes sense, except in the light of evolution." However, many people in the United States are unaware of the importance of evolution as a unifying concept: Public opinion surveys reveal that 25%–40% of Americans either do not believe in evolution or think that evidence for it is lacking. (The percentage varies depending on how the question is worded.) In this chapter we examine both the theory of evolution and the opposition to it.

As explained in Chapter 1, scientists use the word "theory" for a coherent cluster of hypotheses that has withstood many years of testing. In this sense, evolution is a thoroughly tested theory that has withstood over a century and a half of rigorous testing. Scientific evidence for evolution is as abundant as, and considerably more varied than, the evidence for nearly any other scientific idea. To refer to evolution as "just a theory" is thus a grave misunderstanding of both scientific theories in general and evolutionary theory in particular. When physicists speak of the atomic theory or the theory of relativity, or when medical professionals speak of the germ theory of disease, they are speaking of great unifying principles. These principles are now well established, but they have withstood repeated testing for somewhat fewer years than the theory of evolution. Educated people no longer doubt the existence of atoms or of germs, and nobody refers to any of these concepts as "just a theory." In the way that the atomic theory is a unifying principle for much of physics and chemistry, the theory of evolution is a unifying principle for all of the biological sciences.

5.1 THE DARWINIAN PARADIGM REORGANIZED BIOLOGICAL THOUGHT

Arguably the most influential biology book of all time was published in 1859. *On the Origin of Species by Means of Natural Selection*, written by the English naturalist Charles Darwin (1809–1882, Figure 1.4A), contains at least two major hypotheses and numerous smaller ones, along with an array of evidence that Darwin had already used to test these hypotheses. Both hypotheses deal with **evolution**, the process of lasting change among biological populations. Together, these hypotheses offer explanations for the origins and relationships of organisms, the great diversity of life on Earth, the similarities and differences among species, and the adaptations of organisms to their surroundings.

The first major hypothesis, **branching descent**, is that species alive today came from species that lived in earlier times and that the lines of descent form a branched pattern resembling a tree (Figure 5.1). Darwin used this hypothesis, which he called "descent with modification," to explain similarities among groups of related species as resulting from common inheritance. The second major hypothesis is that parents having genotypes that favor survival and reproduction leave more offspring, on average, than parents having less favorable genotypes for the same traits in a given environment. Darwin called this process **natural selection**, and he hypothesized that major changes within lines of descent had been brought about by this process. Both of these hypotheses are falsifiable, and they have been tested many thousands of times—without being falsified—since Darwin first proposed them in 1859. Darwin's two hypotheses made sense of several previously noticed but unexplained regularities in anatomy, classification, and geographic distribution. As both a unifying theory and a stimulus to further research, Darwin's *Origin of Species* fits the concept of a scientific paradigm expounded by Thomas Kuhn and explained in Section 1.2.1. Modern evolutionary thought is still largely based on Darwin's paradigm of branching descent and natural selection, expanded to include the findings of genetics.

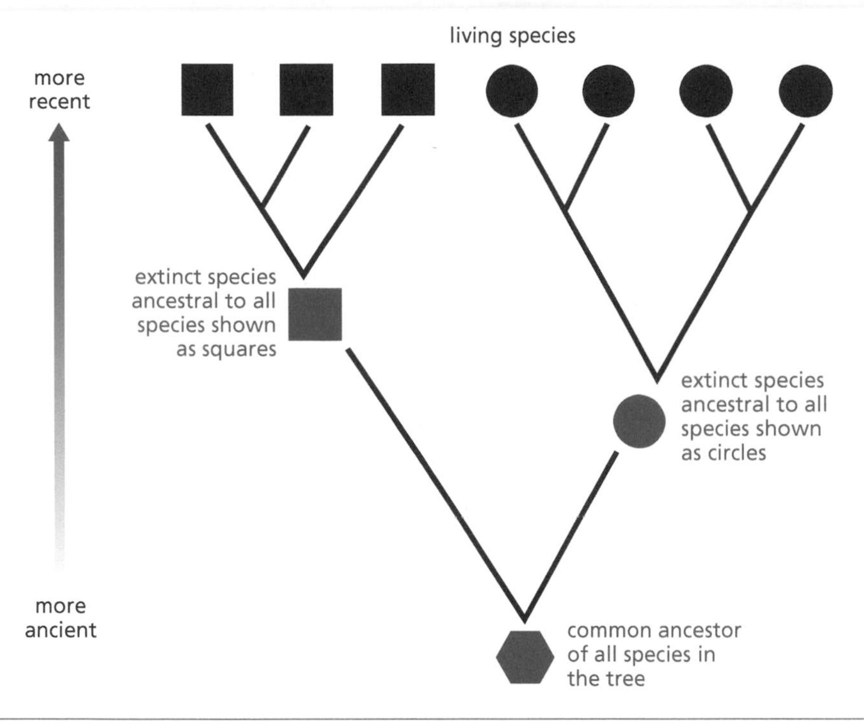

Figure 5.1 The pattern of branching descent. Living species in the top row are descended from the ancestors below them. The red circle represents the common ancestor to all other circles, and the red square is likewise ancestral to all the squares. The red hexagonal shape at the bottom is ancestral to all species shown in this family tree. In a classification, all the squares would be placed in one group and all the circles in another.

5.1.1 Pre-Darwinian thought

Darwin's evolutionary theory was not the first. An earlier theory had been proposed by the French zoologist Jean-Baptiste Lamarck in 1809, and this was based on a still earlier concept, the Scale of Being (Box 5.1). Lamarck believed in what he called *la Marche de la Nature* (the parade of nature), a single straight line of evolutionary progress. Lamarck also noticed that species were adapted to local environments. An **adaptation** is any feature that enables a species to survive in circumstances in which it could not survive as well without the adaptation. Adaptations had been observed since ancient times, but scientists of Lamarck's generation were among the first to propose explanatory hypotheses to explain adaptation. Along with several contemporaries, Lamarck was an environmental determinist, meaning that he believed in the almost limitless ability of adaptation to mold species to their environments and achieve a perfect match. Each environmental determinist favored a different explanation for adaptation. Lamarck's own explanation was based on the strengthening or increase in size of body parts through repeated use or their weakening or decrease in size through disuse. (A frequently cited example is that a giraffe would stretch its neck and thus lengthen it, especially during a time of scarcity when all the lower leaves had already been browsed from the trees.) Lamarck thought that such changes, acquired during the life of an individual, would be passed on to the next generation, but we now know that these **acquired characteristics** are not inherited and do not contribute to evolution. Other scientists, including Darwin, recognized adaptation to the local environment as an important phenomenon. However, Darwin differed from the determinists in seeing important limitations on the ability of adaptation to modify species.

British naturalists had quite different explanations for adaptation. The Natural Theology movement, led by the Reverend William Paley, sought to prove the existence of God by examining the natural world for evidence of perfection. By careful examination and description, British scientists found case after case of organisms with anatomical structures so well constructed, so harmoniously combined with

Box 5.1 The scale of being (*scala naturae*)

Lamarck's theory was a modification of an earlier nonevolutionary concept, that of a scale of being, also variously called chain of being, *échelle des êtres*, or *scala naturae*. Originally, this was a hierarchy of static, unchanging perfection, with animals below people and plants below animals. Theologians said that God had placed man above the animals and below the angels, and each species had its assigned place in this never-changing order. Racist, sexist, and class-based interpretations put the European population above other races, men above women, masters above slaves (as in the writings of Aristotle), and upper castes or classes above lower ones. People who were listed at lower levels were taught to accept their position as "natural" and unchangeable. The scale of being was thus a strong social concept that was used to justify many types of social inequality and oppression. As a theological concept, the scale of being was conceived as a continuum, from God to nothingness, celebrated in verse by Alexander Pope in his "Essay on Man":

> Vast chain of being, which from God began,
> Natures aetherial, human, angel, man,
> Beast, bird, fish, insect! what no eye can see,
> No glass [microscope] can reach! from Infinite to thee,
> From thee to Nothing!- On superior pow'rs
> Were we to press, inferior might on ours:
> Or in the full creation leave a void,
> Where, one step broken, the great scale's destroy'd:
> From Nature's chain, whatever link you strike,
> Tenth or ten thousandth, breaks the chain alike.

Because the continuum was created by God, any challenge to the unbroken chain was considered a challenge to established religion and to the social order. In biology, the scale of being long served as a unifying theory describing nature and predicting that each newly discovered species would somehow fit into its assigned place in the continuum, filling a "missing link" in the chain. In fact, the scale of being was the single most important unifying concept in biology from the time of Aristotle (fourth century BCE) into the early part of the nineteenth century. It was a great stimulus to exploration and research by those seeking to fill in the missing links, and many discoveries of the 1600s and 1700s did seem to confirm the continuum because newly discovered species could be made to fit into it somewhere. (The duck-billed platypus, for example, was hailed as a "missing link" between birds and mammals.) Gaps were explained by saying that further discoveries would fill them in. The terms "higher" and "lower," still occasionally heard in biological discussions, are really references to this pre-evolutionary concept.

Lamarck made the chain of being into a moving escalator, which he called nature's parade (*la marche de la nature*). He said that the lowest forms of life, such as bacteria, were formed by spontaneous generation from lifeless matter and that each species would slowly change (i.e., evolve) into the next higher species on the scale without ever leaving any gaps. Later scientists, especially Cuvier, challenged this concept, and Darwin's concept of branching descent put an end to it altogether.

one another, and so well suited in every detail to the functions that they served that one could only marvel at the degree of perfection achieved. Such harmony, design, and detail, they argued, could only have come from God. Paley offered well-planned adaptation as proof of God's existence: "The marks of *design* are too strong to be gotten over. Design must have a designer. That designer must have been a person. That person is God." (Paley, *Natural Theology*, end of Chapter 23; page numbers vary among many editions.) In a nation in which many clergymen were also amateur scientists, it became fashionable to dissect organisms down to the smallest detail—all the better to marvel at the wondrously detailed perfection of God's design. A large series of intricate and sometimes amazing adaptations were thus described, which Darwin would later use as examples to argue for an evolutionary explanation based on natural selection.

The publication of *On the Origin of Species* challenged many scientific ideas, including those of Lamarck and Paley, and it thus caused controversy among scientists. Some people felt that it also challenged social and religious views that had been taught for centuries. Darwin presented massive evidence to show that branching descent and natural selection explained the observed patterns of the natural world much better than any earlier theory, and thousands of scientists continue to find new forms of evidence supporting Darwin's explanations. Even so, there are still many antievolutionists today, and we discuss their ideas later in this chapter.

5.1.2 The development of Darwin's ideas

From 1831 to 1836, Charles Darwin traveled around the world aboard the ship H.M.S. *Beagle*. His observations in South America convinced him that the animals and plants of that continent are vastly different from those inhabiting comparable environments in Africa or Australia. For example, all South American rodents are relatives of the guinea pig and chinchilla, a group found on no other continent. South America also had llamas, anteaters, monkeys, parrots, and numerous other groups of animals, each with many species inhabiting different environments throughout the continent, but different from comparable species elsewhere (Figure 5.2). This was definitely not what Darwin had expected! Environmental determinist theories such as Lamarck's had led Darwin to expect that regions in South America and Africa that were similar in climate would have many of the same species. Instead, he found that most of the species inhabiting South America had close relatives living elsewhere on the continent under strikingly different climatic conditions. They had no relationship, however, to species living in parts of Africa or Australia with similar climates. The animals inhabiting islands near South America were related to species living on the South American continent. Fossilized remains showed that extinct South American animals were related to living

Figure 5.2 **An assortment of South American mammals.** These species are very different from the mammals found on other continents, even where climates are similar.

South American species. "We see in these facts some deep organic bond, prevailing throughout space and time, over the same areas of land and water, and independent of their physical conditions. The naturalist must feel little curiosity, who is not led to inquire what this bond is. This bond, on my theory, is simply inheritance, that cause which alone, as far as we positively know, produces organisms quite like, or. . . nearly like each other" (Darwin. *Origin of Species*, 1859, p. 350).

5.1.2.1 The Galapagos Islands The Galapagos Islands are a series of small volcanic islands in the Pacific Ocean west of Ecuador. Darwin's visit to these islands proved especially enlightening to him. In this archipelago, a very limited assortment of animals greeted him: No native mammals or amphibians were present; instead, there were several species of large tortoises and a species of crab-eating lizard. Most striking were the land birds, now often called "Darwin's finches" (phylum Chordata, class Sauropsida, order Passeriformes). These birds formed a cluster of more than a dozen closely related species (Figure 5.3), each living on only one or a few islands and differing in size, food preferences, beak shape, and habitat. The tortoises also differed from island to island, despite the clear similarities of climate throughout the archipelago. Darwin hypothesized that each species cluster had arisen through a series of modifications from a single species that had originally colonized the islands. The islands, Darwin noted, were similar to the equally volcanic and equally tropical Cape Verde Islands in the

Large ground finch (*Geospiza magnirostris*)

Common cactus finch (*Geospiza scandens*)

Isla Genovesa ground finch (*Geospiza acutirostris*)

Small ground finch (*Geospiza fulginosa*)

Woodpecker finch (*Camarhynchus pallidus*)

Figure 5.3 **Five species of Darwin's finches from the Galapagos Islands.** (All photos courtesy of Jonah Benningfield.)

Atlantic Ocean west of Senegal, which Darwin had also visited, but the inhabitants were altogether different. Darwin concluded that the Galapagos had received its animal colonists (including the finches) from South America, while the Cape Verde Islands had received theirs from Africa, so in each example the closest relatives were found on the nearest continent, not on geologically similar or climatically similar but distant islands. Geographic proximity, in other words, was often more important than climate or other environmental variables in influencing which species occurred in a particular place.

5.1.3 Descent with modification

As the result of his studies of species distribution on continents and on islands, Darwin concluded that each group of colonists had given rise to a cluster of related species through a process of branching descent. Darwin called this process "descent with modification," and he emphasized that each species in the cluster had been differently modified from a common starting point. Darwin was the first evolutionary theorist to emphasize that clusters of related species indicated a branching pattern of descent, a series of treelike branchings, which species correspond to the finest twigs, groups of species to the branches from which these twigs arise, larger groups to larger branches, and so forth (Figure 5.1). In this diagram, each branch point represents a time of species formation and genetic divergence, and the base of the tree represents the common ancestor of all the species that arose from it. Darwin used a very similar treelike diagram, the only illustration in his book.

Darwin found many large groups of related animal species inhabiting each continent. These groups were unrelated to the very different groups inhabiting similar climates on other land masses. Several large land areas with similar climates had flightless birds, but they differed strikingly from one continent or island to the next: rheas in South America, kiwis and extinct moas in New Zealand, emus and cassowaries in Australia, extinct elephant birds on Madagascar, and ostriches in Africa. Theories of environmental determinism (such as Lamarck's) could not explain these differences, nor could theories of divine creation explain why God had seen fit to create half a dozen distinct types of flightless birds instead of the best one everywhere; Darwin's examples thus falsified these earlier theories.

Before Darwin's time, biological classifications had already taken their modern hierarchical form, as described in the next chapter and as illustrated in Figure 5.1. Darwin explained this hierarchy as the natural result of branching descent with modification, a process that produces the similarities and differences that biologists have used in classifying organisms. Classifications, in other words, made sense because of evolution.

5.1.4 Natural selection

When Darwin returned to England, he began reading about the ways in which species could be modified. How, he wanted to know, could a single colonizing species produce a whole cluster of related species on a group of islands? To help find clues to answer this question, Darwin contemplated the results of plant and animal breeding in Britain. During the preceding hundred years, British breeders had produced many new varieties of plants, such as roses and apples, and animals, including dogs, sheep, and pigeons, by careful breeding practices. Through these same practices, the breeders had greatly improved wool yields in sheep and milk yields in cattle. By methodically selecting the individuals in each generation with the most desired traits and breeding these individuals with each other, the breeders had modified a number of domestic species through a process that Darwin called **artificial selection**. This process simply took advantage of the natural variation that was present in each species, yet it produced breeds that were strikingly different from their ancestors. Darwin remarked that some of the domestic varieties of pigeons or dogs differed from one another as much as did natural species, despite the fact that the domestic varieties had been produced within a short time from a known group of common ancestors. Could a similar process be at work in nature?

At about this same time, Darwin read Thomas Robert Malthus's *Essay on Population* (see Section 9.1). Malthus emphasized that, in the natural world, each species produces more offspring than are necessary to maintain its numbers. This overproduction is followed in each generation by the premature death of many individuals and the survival of only a few. When Darwin compared this process with the actions of the animal breeders, he concluded that nature was slowly bringing about change in each species. Individuals varied in every species, and those that died young in each generation differed from those that survived to maturity and mated to produce the next generation. In this "struggle for existence," Darwin hypothesized that

> *. . . individuals having any advantage, however slight, over others, would have the best chance of surviving and of procreating their kind. . . . On the other hand, we may feel sure that any variation in the least degree injurious would be rigidly destroyed. This preservation of favourable variations and the rejection of injurious variations, I call Natural Selection. . . . Natural selection . . . is a process incessantly ready for action, and is as immeasurably superior to man's feeble efforts, as the works of Nature are to those of Art.*
>
> [Darwin. Origin of Species (1859, p. 61, 81)]

All modern descriptions of natural selection are stated in terms of the concepts of genetics outlined in Chapter 2. New genotypes originate by mutation and by recombination, both of which act prior to any selection. Darwin, of course, knew nothing of mutations or of modern genetics, but he did realize that heritable variation had to come first and that "any variation which is not heritable is unimportant to us."

Natural selection may be defined as consistent differences in what Darwin called "success in leaving progeny," meaning the proportion of offspring that different genotypes leave to future generations. The relative number of viable individuals that each genotype contributes to the next generation is called its **fitness**, and *natural selection favors any trait that increases fitness*. Darwin's theory of natural selection is the basis for all modern explanations of adaptation. We have developed a computer program to demonstrate natural selection and several additional evolutionary phenomena; this program is currently available online at biologytrending.routledge.com/chapter5

Many agents of natural selection operate in nature. Often, the selecting agents are predators. Selection by predators may be convenient to study, but many other agents of selection are known. Any cause of death contributes to natural selection if it reduces the opportunity for reproduction and if some genotypes are more likely to die than others. Some genotypes may be more susceptible to particular diseases or parasites and die in greater numbers from these causes, while other genotypes might be more resistant and thus survive more readily. Starvation and weather-related extremes of cold, dryness, or precipitation may also be agents of selection if some genotypes can survive these conditions better than others. These and other causes of mortality are all agents of selection if there are differences in the death rates for different genotypes.

Not every agent of selection causes death, however. Natural selection also favors those genotypes that reproduce more and leave more offspring. A special type of selection, called **sexual selection**, operates on the basis of success (or lack of success) in attracting a mate and reproducing. For example, animals of many species attract their mates with mating calls (such as bird songs), visual displays (as in peacocks), or special odors (as in silkworm moths or many other invertebrate animals). Individuals that do not perform well enough to attract a mate may live long lives but leave no offspring.

5.2 A GREAT DEAL OF EVIDENCE SUPPORTS DARWIN'S IDEAS

As previously stated, Darwin proposed two major hypotheses that explained how evolutionary change takes place: branching descent ("descent with modification") acting over long periods of time and natural selection as a mechanism promoting change. In the years since Darwin proposed these two hypotheses, many scientists have used scientific methods to conduct thousands of tests of both hypotheses. The results of these tests have yielded much evidence that supports the hypotheses and none that falsifies them.

5.2.1 Mimicry as evidence for natural selection

One of the earliest tests of Darwin's hypothesis of natural selection involved the phenomenon of **mimicry**, in which one species of organisms deceptively resembles another. In one type of mimicry, a distasteful or dangerous prey species, called the model, gives a very unpleasant and memorable experience to any predator that attempts to eat it. Predators always avoid the model following such an unpleasant experience. A palatable prey species, the mimic, secures an advantage if it resembles the model enough to fool predators into avoiding it as well.

Selection by predators explains mimicry rather easily. Any slight resemblance that causes a predator to avoid the mimic as well as its model is favored by selection and passed on to future generations of the mimic species, while individuals not protected in this way are eaten in greater numbers and are less likely to pass on their genes. Predator species differ greatly in their abilities to distinguish among prey species, so a resemblance that fools one predator might not fool another. Any advantage that increases the number of predators fooled is favored by selection, causing closer and closer resemblance of the mimic to the model to evolve with the passage of time.

Sometimes several species that resemble each other are all distasteful to predators. Predators learn to avoid distasteful species, but a certain number of prey individuals are killed for each predator individual that learns its lesson. If each prey species looked different, then each would sustain this loss separately. Mimicry among prey species allows predators to learn the lesson with fewer individuals of each prey species dying in the process. The mimicry therefore benefits each prey species and is thus favored by natural selection.

Mimicry often varies geographically. Some wide-ranging tropical species mimic different model species in different geographic areas. The deceptive resemblance is always to a species living in the same area, never to a far-away species. Environmentalist theories such as Lamarck's had no way to account for the evolution of mimicry, and the patterns of geographic variation could not be explained by either environmentalist theories or by Paley's natural theology. Natural selection, however, explains all of these examples of variation as resulting from selection by predators.

In a well-known case of mimicry, the model is the monarch butterfly, a distasteful species that feeds on milkweed plants. An unrelated species, the viceroy, is similar in superficial appearance, and is thus avoided by many predators (Figure 5.4), even if they might have no actual taste aversion to the viceroy.

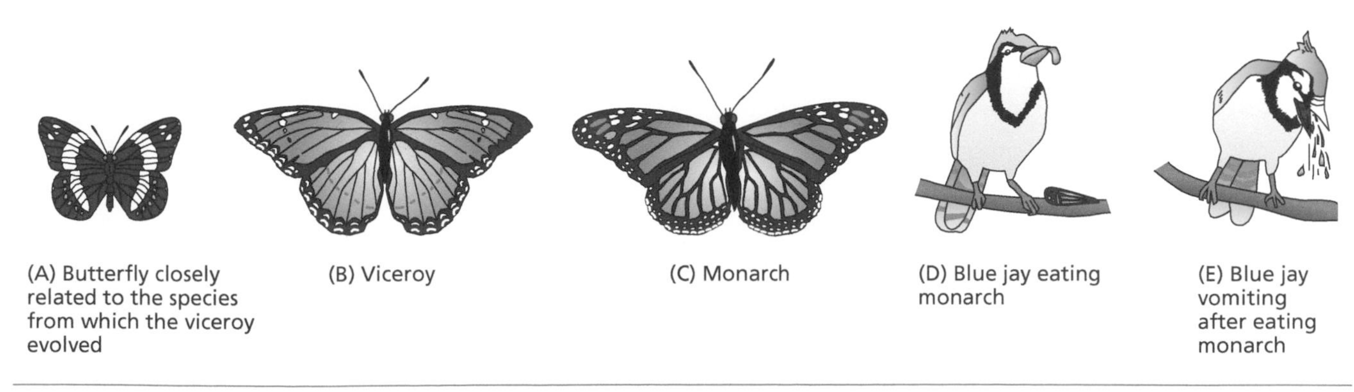

Figure 5.4 An example of mimicry. (A) *Limenitis arthemis*, a nonmimic relative of **(B)** *Limenitis archippus*, the viceroy. The viceroy resembles the unrelated monarch butterfly **(C)**, *Danaus plexippus*, the model. The monarch is avoided by predators after just a single unpleasant experience **(D, E)**. The warning color pattern of the monarch helps predators learn to avoid it; the viceroy is protected because its color pattern mimics that of the monarch.

5.2.2 Industrial melanism as further evidence for natural selection

The power of natural selection is also demonstrated by a phenomenon called **industrial melanism**, when darker colors have evolved in areas polluted by industrial soot in species that are usually light in color elsewhere. In the British Isles, a species known as the peppered moth, *Biston betularia* (phylum Arthropoda, class Insecta, order Lepidoptera), had long been recognized by an overall light gray coloration with a salt-and-pepper pattern of irregular spots. A black variety of this species was discovered in the 1890s. The black moths increased until they came to outnumber the original forms in some localities (Figure 5.5). The British naturalists E.B. Ford and H.B.D. Kettlewell studied these moths for several decades from about

Figure 5.5 Industrial melanism in peppered moths in the British Isles.

Geographic variation in the frequency of melanic moths in the 1950s, which reached as high as 100% in polluted localities downwind from major industrial centers.

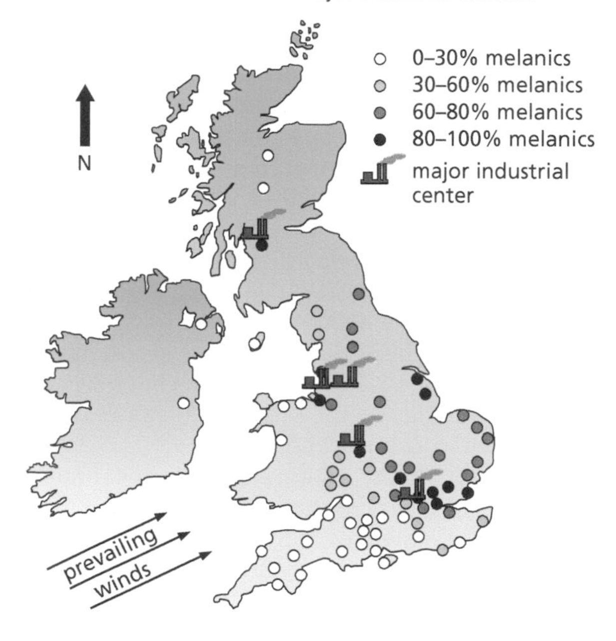

The melanic (black) variety and the original 'peppered' variety (below the right wing-tip of the melanic moth) on a light, lichen-covered tree trunk.

The same two varieties on a dark, soot-covered tree trunk.

1944 onward. Downwind from the major industrial areas, the woods had become polluted with black soot that killed the lichens growing on the tree trunks. Most of the moths living on the darkened tree trunks in these regions were black. However, where the woods were untouched by the industrial soot, the tree trunks were still covered with lichens and the moths had kept the light-colored pattern. Ford and Kettlewell hypothesized that the moths resembling their backgrounds would be camouflaged and thus harder for predators to see. To test this hypothesis, they pinned both light and dark moths on dark tree trunks in polluted woods, and they also pinned both types on lichen-covered tree trunks in unpolluted woods. They observed that birds ate more of the dark moths in the unpolluted woods (favoring the survival of the light-colored pattern), but birds in the polluted woods ate more of the light-colored moths, not the dark ones. These observations and the geographical patterns of variation (Figure 5.5) were easily explained in terms of natural selection by predators. In addition, since the experiments were first conducted, laws to control smokestack emissions and other forms of pollution were passed and enforced, and many of the woods affected by pollution have returned to their former state. In these woods, the lichens have returned to the tree trunks, and most of the moths in these places again have the original light-colored pattern.

Industrial melanism in insects demonstrates that the characteristics of a species can change in response to changes in the environment instead of each species being created with permanent, unalterable traits. Lamarckian mechanisms fail to explain industrial melanism because, in the moth species studied, the adult colors do not change once they are formed, and there is nothing that individuals can do that would alter their color. The experiments with birds as predators clearly show natural selection at work, resulting in a rapid change in the characteristics of a species in a particular, altered environment.

5.2.3 Evidence for branching descent

Darwin's contemporaries immediately recognized that his concept of "descent with modification" could be used to make sense out of a variety of observations not easily explained by other means. The branching pattern of descent explained the formation of groups or clusters of related species in particular geographic areas. Moreover, it explained the arrangement of these groups into a hierarchy of smaller groups within larger groups. Biologists before Darwin had been making classifications this way for about a century, but it was Darwin's theory of branching descent that explained why this type of classification made sense. Darwin predicted that biological classifications would increasingly become genealogies (that is, maps of descent similar to Figure 5.1) as more and more details about the evolution of each group of organisms became known. Darwin's prediction came true as scientists engaged in the comparative study of the anatomy, biochemistry, physiology, and embryology of different species and found more and more evidence showing that relationships among species arise in branching patterns of descent.

5.2.3.1 Homologies The construction of family trees is based in large measure on the study of shared structures or gene sequences. Under Darwin's paradigm, shared similarities are evidence that the organisms in question share a common ancestry. In a sense, a shared similarity is a falsifiable hypothesis that the several species sharing it are related to one another by descent. By itself, one such similarity reveals very little, but a large number of similarities that fit together into a consistent pattern strongly suggests shared ancestry. When the evidence for shared ancestry is compelling, the similarity is called **homology**.

Darwin noted the similarities among the forelimbs of mammals: "What can be more curious than that the hand of a man, formed for grasping, that of a mole for digging, the leg of the horse, the paddle of the porpoise, and the wing of the bat, should all be constructed on the same pattern, and should include the same bones, in the same relative positions?" (**Figure 5.6**). Darwin wondered why similar leg bone structures appeared in the wings and legs of a bat, used as they are for such totally different purposes. Why should a crustacean that has more mouthparts have correspondingly fewer legs, or why should those with more legs have fewer mouthparts?

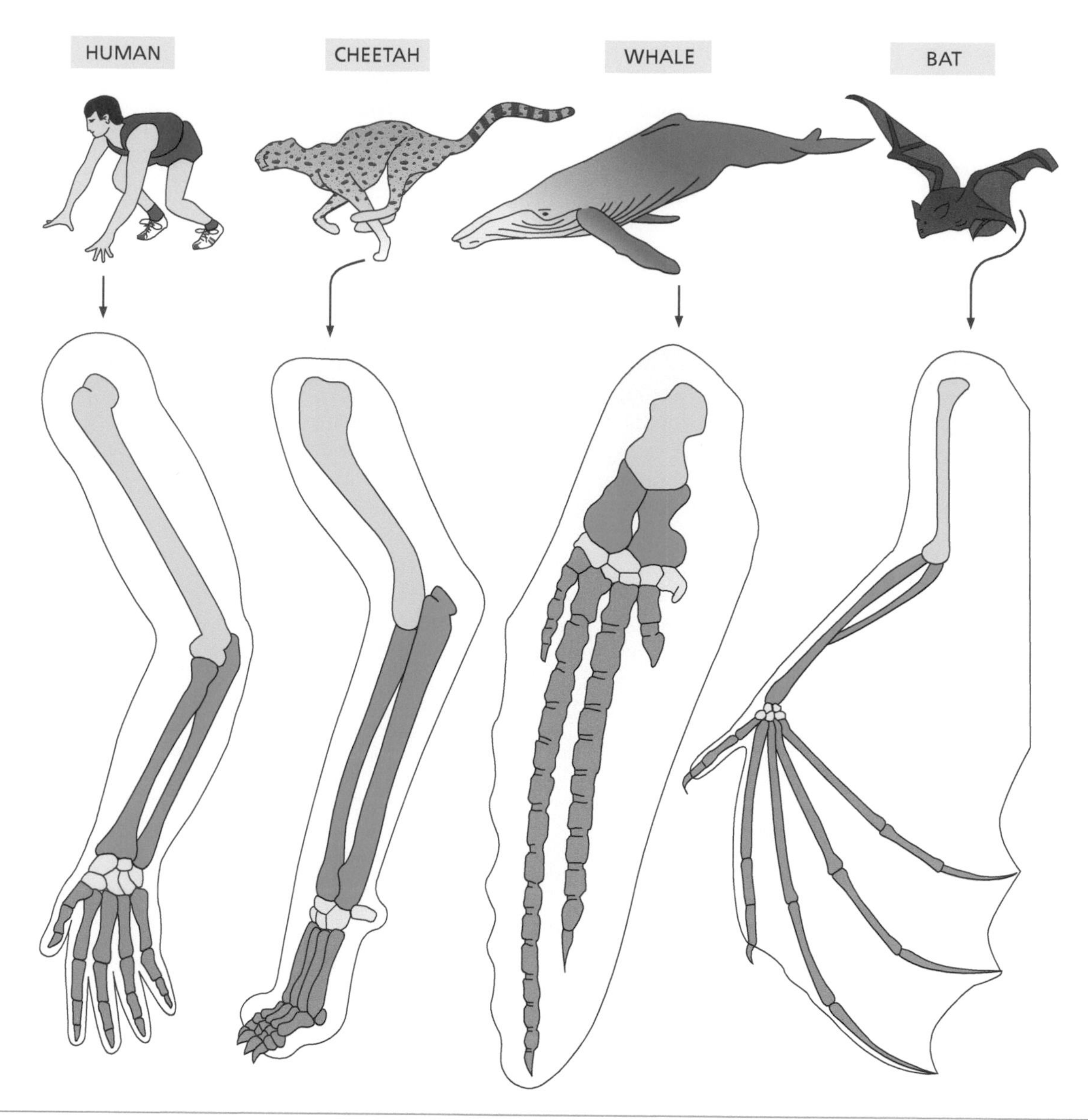

Figure 5.6 Homologies among mammalian forelimbs adapted to different functions.

Darwin's answer is that all these structures arose by modification of the same type of repeated part. Crustaceans, for example, have mouthparts, legs, and various other structures that are all derived from a common set of leglike appendages because natural selection favored different structures for different uses. That is, when the structure of these appendages varied (through mutation or through changes in developmental patterns), individuals with better-functioning mouthparts near the mouth, or with better-functioning legs near the center of gravity, left more offspring, and the proportions of the responsible genotypes increased in each population. As a consequence, when more appendages were used around the mouth and more mouthparts therefore evolved in these animals, there were fewer appendages remaining to be used as legs. An omnipotent God, however, would face no such limitation on the total number of appendages and could have added mouthparts without taking away legs, leaving Darwin to declare, "How inexplicable are these facts on the ordinary view of creation!" (Darwin. *Origin of Species*, 1859, p. 437.)

5.2.3.2 Vestigial structures Structures whose function has been lost in the course of evolution tend to diminish in size. Often, they persist as small, functionless remnants, called **vestigial structures**. A good human example is the coccyx, a set of two or three vestigial tail bones at the base of the spinal column, homologous to the tails of other mammals. The Darwinian paradigm of natural selection and branching descent explains these vestigial structures as the remnants of structures that were formerly functional. Neither Lamarck nor the creationists had any explanation for the presence of vestigial structures, and certainly not for the homologies between vestigial structures in many species and their functional counterparts in related species.

5.2.3.3 Convergence Similarities that result from common ancestry (that is, true homologies) should also be similar at a smaller level of detail, and even similar in embryological derivation, meaning that they should grow from the same source tissues. A hypothesis of homology can thus be falsified if two similar structures turn out to be dissimilar in detailed construction or in embryological derivation. There are also cases in which several hypotheses of homology are in conflict because they require different patterns of relationships for different characters. In such cases, evolutionists examine all the similarities more closely and repeatedly to see whether a reinterpretation is possible for one set of similarities.

Convergence is an evolutionary phenomenon in which similar adaptations evolve independently in lineages not closely related. Any resemblance resulting from convergent adaptation is called an **analogy**. If a hypothesis of homology is falsified by more careful scrutiny of similar structures, the similarities are often reinterpreted as convergent adaptations that evolved independently in unrelated lineages. Distinguishing homology from analogy in this way is an ongoing aim of evolutionary classification. For example, the wings of bats and insects are analogous, rather than homologous, structures. They are constructed in different ways and from different materials, and their common shapes (which they also share with airplane wings) reflect adaptation to the aerodynamic requirements of flying. Although bat wings are not homologous to insect wings, they are homologous to human arms, whale flippers, and the front legs of horses and elephants. These all have similar bones, muscles, and other parts in similar positions despite their very different shapes and uses, while insect wings have no bones and their muscles are very differently located.

5.2.3.4 Cephalopods as an example One frequent test of the hypothesis of branching descent is to identify a group of organisms that share some particular character, such as an anatomical or biochemical peculiarity. The general hypothesis of branching descent then gives rise to a more specific hypothesis, that these organisms all share a common descent from a common ancestor. An example of this type of reasoning can be illustrated by the Cephalopoda, a group of mollusks that includes the squids, octopuses, and their relatives. All cephalopods can be recognized by the presence of a well-developed head and a mantle cavity beneath (Figure 5.7). The mantle cavity contains the gills, the anus, and certain other anatomical structures. Other mollusks have mantle cavities, but only in the Cephalopoda is the mantle cavity located beneath the head and elongated into a nozzle-like opening known as the hyponome. Knowing this, we can formulate the specific hypothesis that all cephalopods share a common descent.

If our hypothesis is true, then we should be able to find, as evidence, some additional similarities among cephalopods not shared with other mollusks. Such similarities do exist: All cephalopods have beaklike jaws at the front of the mouth and a muscular part, called the "foot" in other mollusks, subdivided into a series of tentacles (Figure 5.7). Moreover, cephalopods have an ink gland that secretes a very dark, inky fluid. When a squid or octopus feels threatened by a predator, it releases this fluid into its mantle cavity and quickly squirts the contents of the mantle cavity forward through its nozzle-like hyponome. This action hides the animal and propels it backwards, in a direction not expected by its predator. The predator's attention is meanwhile held by the puff of black, inky fluid. By the time the color dissipates, the squid or octopus has vanished. Nearly all members of the Cephalopoda have this elaborate and unusual escape mechanism, including squids, cuttlefishes, and octopuses. The hypothesis of a common descent for all the Cephalopoda is thus consistent with the known data, meaning that the hypothesis has been tested and not falsified.

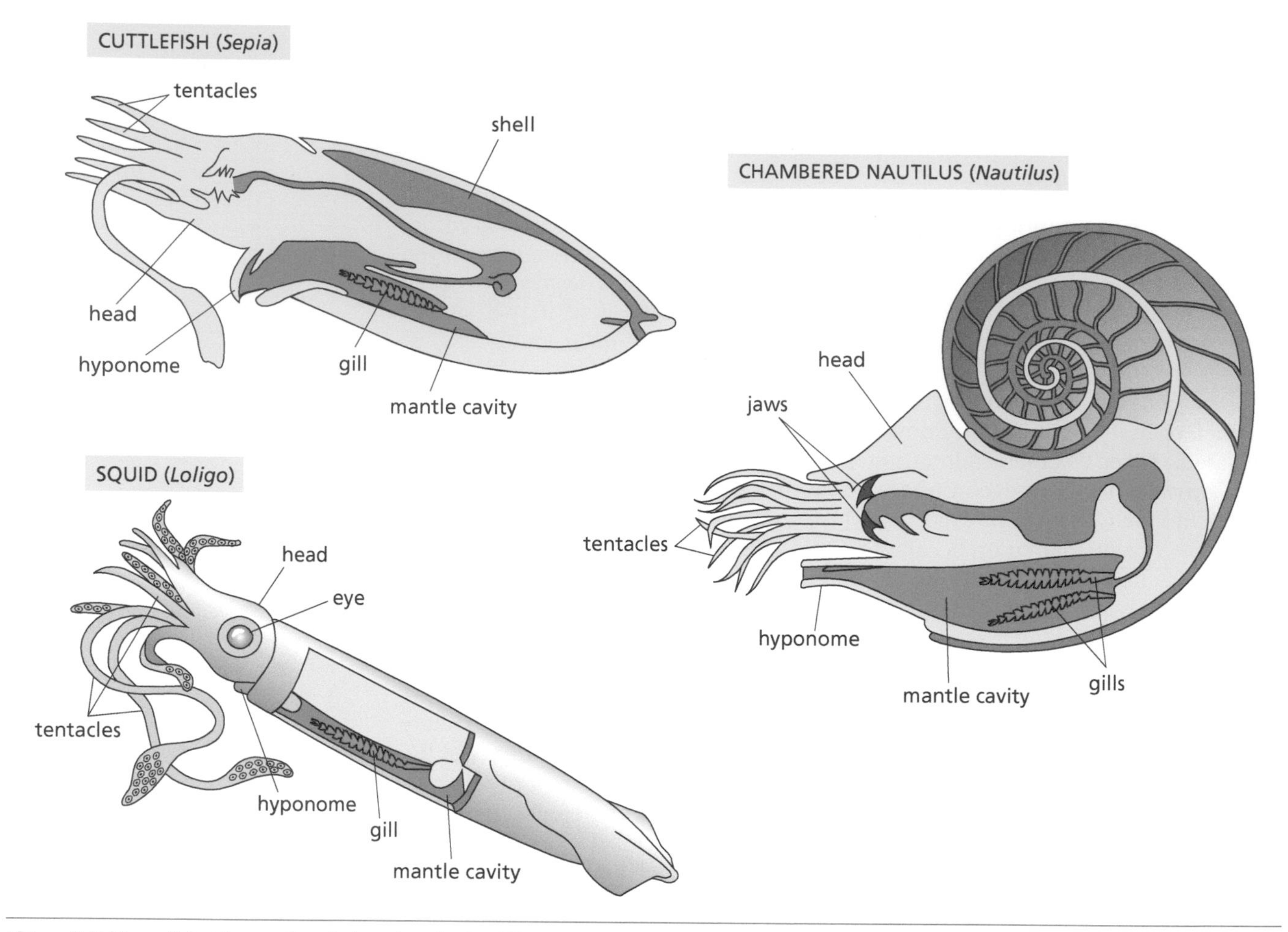

Figure 5.7 Three living types of cephalopod mollusks (Kingdom Animalia, phylum Mollusca, class Cephalopoda): the cuttlefish, the squid, and the chambered nautilus.

5.2.3.5 Further comparisons among species Since Darwin's time, many additional types of similarities have been discovered among organisms. The comparative study of embryonic development has resulted in the discovery of many new similarities among distantly related species, some of which are described in Chapter 6. Comparative genomics (Chapters 4 and 6) and comparative studies in biochemistry have revealed the detailed structure of protein chains, DNA and RNA sequences, and other large molecules of biological interest. As with anatomical similarities, similarities in biochemistry or in embryology group related species together, and small groups are contained within larger groups at many hierarchical levels. Each new type of similarity has brought new evidence of branching descent with modification: In most cases, the groups established by older methods are reaffirmed when newer methods result in the same groupings. In a few cases, new groupings are discovered, and sometimes these are later corroborated by further evidence such as new fossil discoveries.

On the basis of anatomical, embryological, and biochemical comparisons, hypotheses of common descent have been tested and confirmed for cephalopod mollusks and for many other groups of animals and plants. This increases our confidence in the larger hypothesis that all species of organisms have evolved from earlier species in patterns of branching descent. The many facts of comparative anatomy, comparative physiology, embryology, biogeography, and animal classification are all consistent with the hypothesis that modern species have evolved from ancestors that lived in the remote past, and they make little sense otherwise.

Since Charles Darwin published his evolutionary ideas in 1859, thousands of tests have been made of his twin hypotheses of branching descent and natural selection. Because these thousands of tests have failed to falsify either hypothesis, both now qualify as scientific theories that enjoy widespread support. The Darwinian paradigm continues to this day as a major guide to scientific research.

5.2.4 Further evidence from the fossil record

The history of life on Earth is measured on a time scale encompassing billions of years. This geological time scale (Figure 5.8) was first established by studying **fossils**, the remains and other evidence of past life forms. Most fossils are

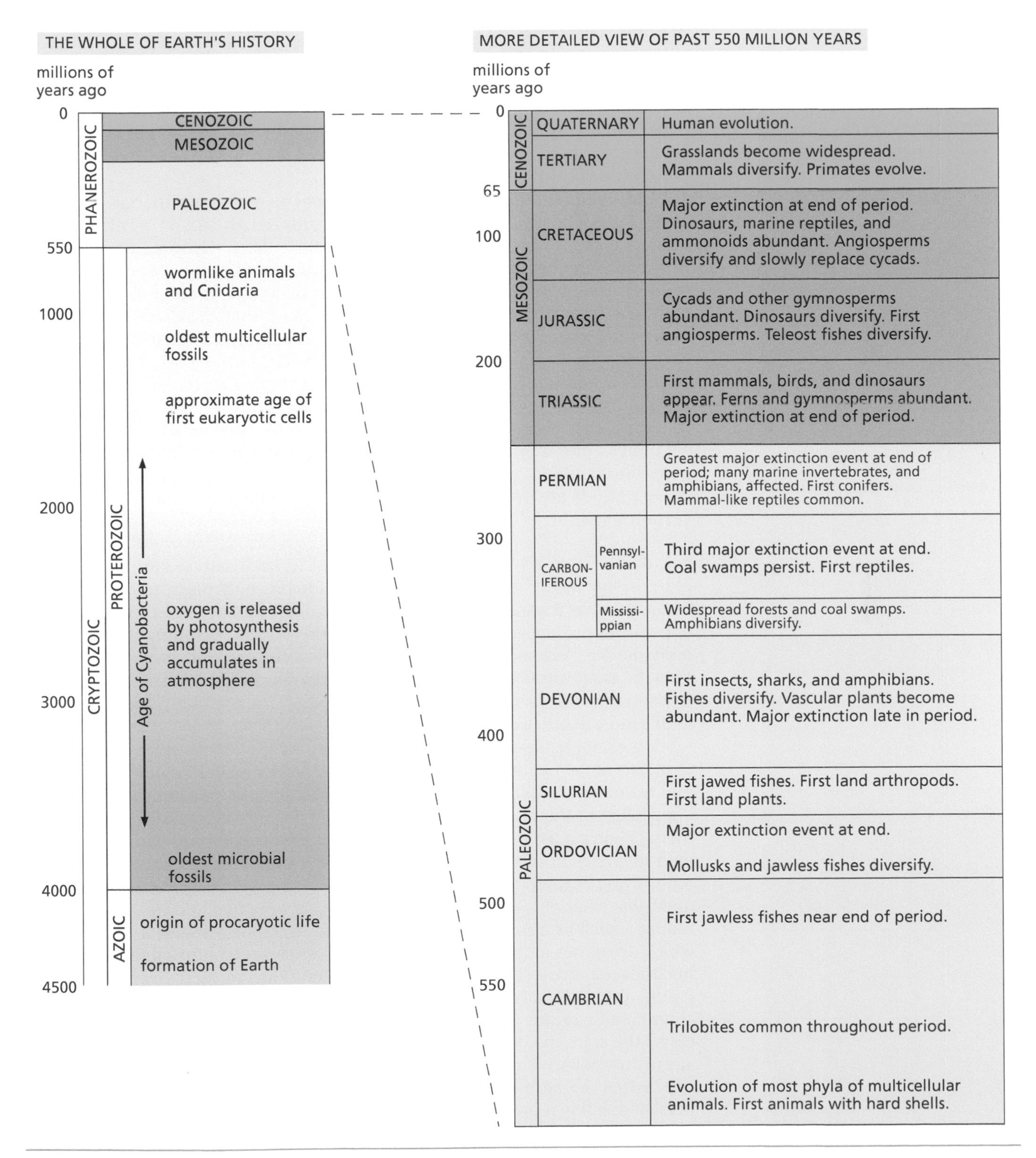

Figure 5.8 The geological time scale.

formed from the burial of plants or animals in sediment. The soft parts of these organisms are often consumed or decomposed, but they may leave imprints in soft sediments if buried rapidly. Scientists had recognized since 1555 that most fossils were the remnants of species no longer living, thus providing clear-cut evidence for extinction. Comparisons of fossils with living species provide evidence of change over time and thus play a role in supporting the theory of descent with modification and the concept of evolution more generally.

5.2.4.1 Stratigraphy The geological time scale was first established through the study of layered rocks, called **stratigraphy**. One of the first principles established in the study of these rocks was that when rock layers have not been drastically disturbed, the oldest layers are on the bottom and successively newer layers are on top of them. Using this principle, geologists can identify the rock formations in a particular place as part of a local sequence, arranged chronologically from bottom to top.

Local sequences from different places can be matched with one another in several ways, but the most reliable of these proved to be the study of their fossil contents. Two rock formations are judged to be from the same time period if they contain many of the same fossil species (the principle of **correlation by fossils**). The rocks do not need to be similar in composition or rock type—one can be a limestone and the other a shale—but if their fossil assemblages are similar, they are judged to be equally old. A single species of fossil is never sufficient; several fossil species are needed, and they must occur together with some consistency. Using this technique, paleontologists (scientists who study fossils) have been able to match up formations of the same age from many different localities, enabling them to assemble the world's various local sequences into a "standard" worldwide sequence, which is the basis for the complete sequence of time periods shown in Figure 5.8. The dates assigned to these periods are determined by measuring the rates of radioactive decay in certain rocks.

5.2.4.2 Family trees The age of a fossil, by itself, tells us very little about its place in any family tree. The relative ages of fossils only begin to have meaning when we study a group of organisms represented by many fossils. The family tree or genealogy of any group, called its **phylogeny**, fits into a pattern like that shown in Figure 5.1. Any such family tree is a hypothesis that biologists use to explain how the anatomical and other characteristics of each species are related, which leads to the classification of the group as a whole. In any family tree, the known fossils must fit into a consistent framework.

For example, there are many fossils of cephalopod mollusks, permitting further tests of the hypothesis of a common descent for all the Cephalopoda. Living and extinct cephalopods can be arranged into a family tree consistent with our knowledge of the characteristics of each species and the relationships among them (**Figure 5.9**). Differences among the living cephalopods can be explained with reference to this family tree. The chambered nautilus is very different from other living cephalopods because it is fully housed within a coiled shell and has four gills, while the squids and octopuses have only two gills and a very small, reduced shell or else none at all. One would therefore imagine a family tree in which octopuses and squids have a common ancestor that the chambered nautilus does not share. The fossil Cephalopoda conform to these expectations. The group of cephalopods with the oldest fossil record are the nautiloids, of which the chambered nautilus is the only living remnant. A second group of cephalopods, called the ammonoids (see Figure 18.6), flourished in Mesozoic times, during the age of dinosaurs. A small third group had an internal shell that became reduced in size. When the ammonoids became extinct, this third group, the Dibranchiata, persisted and is represented today by the squids and octopuses. Thus, the fossil record of the cephalopod mollusks, including both the anatomy and age relationships of fossil forms, confirms the relationships hypothesized on the basis of the anatomy of the living forms.

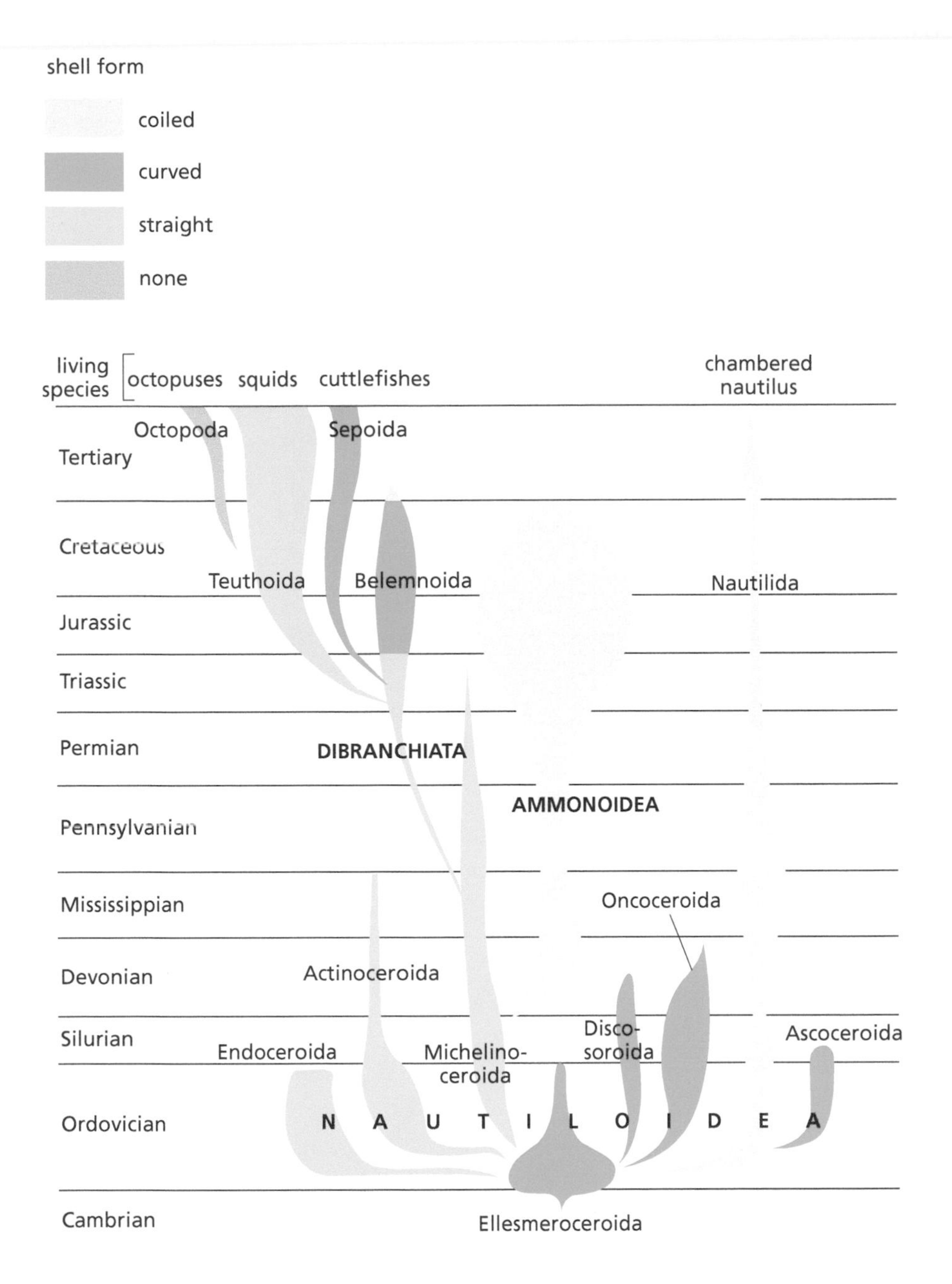

Figure 5.9 Family tree of the class Cephalopoda (phylum Mollusca) showing branching descent over time. Horizontal width represents number of species in each group; vertical distance represents time.

The fossil record has repeatedly confirmed hypotheses of descent for particular living species. For example, Thomas Henry Huxley, one of Darwin's early supporters, studied the anatomy of birds and declared them to be "glorified reptiles." The interpretation of birds as descendants of the reptiles was strengthened by the discovery of *Archaeopteryx*, a transitional fossil with many birdlike and also many reptilian features. Among the reptilian features of *Archaeopteryx* were a long tail, simple ribs, a simple breastbone, and a skull with a small brain and tooth-bearing jaws (**Figure 5.10**). Despite these reptilian features, *Archaeopteryx* had well-developed feathers and was probably capable of sustained flight. The discovery of transitional forms like *Archaeopteryx*, coupled with the fossils of other early birds and of feathered dinosaurs close to bird ancestry, strengthens our confidence in the hypothesis that birds evolved from reptiles, and particularly from dinosaurs.

Other transitional forms are known, such as those between older and more modern bony fishes, between fishes and amphibians, between reptiles and mammals, and between terrestrial mammals and whales. The transition between fishes and amphibians is now well documented by such transitional fossils as *Eusthenopteron*, *Tiktaalik*, and *Ichthyostega*. Instead of being exactly intermediate

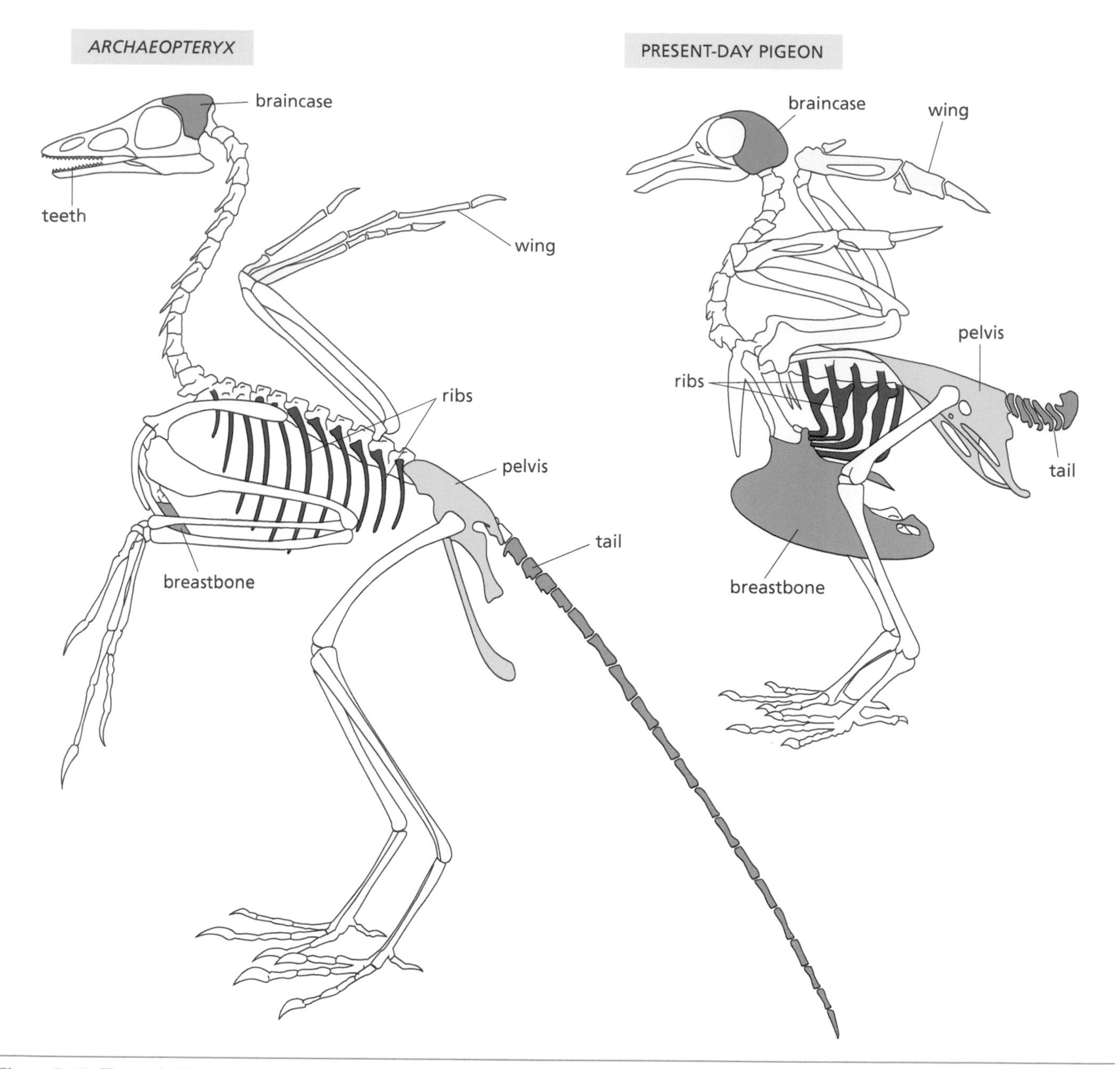

Figure 5.10 The early bird *Archaeopteryx* compared with a modern-day pigeon. Modern birds lack teeth, and evolution has enlarged the braincase and strengthened other parts (wing, ribs, breastbone, pelvis, tail) highlighted here.

in each trait, transitional fossils like *Archaeopteryx* or *Tiktaalik* usually exhibit a mix of some advanced (newer and more innovative) characteristics and some primitive (older) characteristics. *Tiktaalik*, for example, had fishlike scales and gills, a fishlike tail, and a streamlined body with a flat head. On the other hand, it had lungs and four powerful limbs. Although the limbs lacked individually movable fingers, they were certainly strong enough to allow *Tiktaalik* to creep or wiggle its way across mud flats or river banks. Descendants of *Tiktaalik* spent more time on land and evolved mobile fingers and stronger limbs, but modern amphibians still, to this day, lay their eggs in water and breathe with gills in their early life stages.

5.2.5 Post-Darwinian thought

One of the hallmarks of science is that hypotheses are subjected to rigorous and repeated testing. Darwin's hypotheses have been thoroughly and repeatedly tested for over a century and a half, and the general outlines of branching descent and natural selection have been repeatedly and consistently confirmed. A second hallmark of science is that theories are extended and modified as new data are discovered. Here again, Darwin's ideas have been extended and supplemented by newer findings. Many additional details are now known, none of which contradict the basic concepts of natural selection and branching descent. A third hallmark of a scientific theory is that it acts as a spur to further research; Darwin's two theories have stimulated more scientific research than almost any other theory in the history of biology, with the possible exception of Mendel's theories in genetics. Evolution guides our thinking in nearly every field of biology, which is why "nothing in biology makes sense, except in the light of evolution."

During the period 1860–1940, scientists who doubted the effectiveness of natural selection proposed many other hypotheses to explain evolutionary change. Our modern theory of mutations originated from one such hypothesis. In Czarist Russia, scientists of nearly every political stripe (from conservative monarchists to socialists and anarchists) found the British idea of competition very distasteful. They were therefore reluctant to embrace the concept of natural selection, which they felt was based on a model of competition. (Darwin had emphasized that he meant competition in a "large and metaphorical sense," but his Russian readers still found the similarity with capitalist economics distasteful.) As an alternative, the anarchist Petr Kropotkin and the novelist and pacifist Leo Tolstoy developed theories of "mutual aid" or "mutualism" as an important evolutionary mechanism. According to this view, organisms succeed (and leave more offspring) if they cooperate with one another instead of competing, as among social insects. Modern biologists now view mutualism as an adaptive interaction between species that may evolve as the result of natural selection. Natural selection favors any change that increases reproductive success, and this frequently includes changes that benefit other species directly. The theory of mutualism has thus been accepted into the mainstream of Darwinian evolutionary thought and is no longer viewed as something incompatible with natural selection. Darwin himself gave several examples of cooperative interactions between species.

Although Darwin's theories of natural selection and branching descent continue to guide biological research to this day, the early 1940s saw an expansion of the evolutionary paradigm called the **modern synthesis**. Darwin's ideas are retained in this expanded paradigm, but the findings of genetics are also incorporated and are used to explain the source of heritable variation. Natural selection is well documented as an important cause of evolutionary change, but it is by no means the only cause. Chance alone ("accidents of sampling," or random circumstances that determine which individuals die, which live, and which reproduce) can cause erratic changes in the allele frequencies of natural populations, especially small ones. This phenomenon, called **genetic drift**, is discussed further in Section 7.2. The changes produced by genetic drift are often nonadaptive, and they increase the chances that a small population will die out. Later in this chapter, we will also discuss the importance of geographic isolation, a nonselective phenomenon important in the evolution of new species.

Beginning in the 1970s, Niles Eldredge and Stephen Jay Gould advocated a theory that they viewed as an alternative to Darwinian thought. Darwin had frequently emphasized that evolutionary change was gradual, but Eldredge and Gould proposed instead that species remain static for long periods and then change abruptly. According to their theory, a new species begins as a small, isolated population on the geographic periphery of the original species. The small size of the isolated population allows it to undergo rapid change, producing a new species. Once the new species becomes successful, its numbers and geographic range may increase to the point where it invades the geographic range of the original species from which it evolved. If the new species successfully outcompetes the original one, the original species may become extinct. What we often see in the fossil record is the abrupt replacement of one species by another rather than a gradual change. Gould always claimed that this **punctuated equilibrium** theory is an alternative to Darwin's gradualism, but certain other evolutionists (such as Ernst Mayr) view the two theories as fully compatible.

THOUGHT QUESTIONS

1. For a family tree such as that shown in Figure 5.9, what kinds of fossil evidence (be specific) would falsify the descent pattern shown? What kinds of evidence would cause paleontologists to modify the family tree but continue to believe in a process of descent with modification? What kinds of evidence would falsify the hypothesis of descent with modification?
2. One of the Galapagos finches studied by Darwin has woodpecker-like habits and certain woodpecker-like features: It braces itself on vertical tree trunks with stiff tail feathers in the manner of true woodpeckers and drills holes for insects with a chisel-like bill. However, it lacks the long, barbed tongue with which true woodpeckers spear insects; it uses cactus thorns for this purpose instead. How would Lamarck have accounted for this set of adaptations? How would Paley? How would Darwin? Which of these explanations accounts for the absence of the barbed tongue in the woodpecker finch? How would each hypothesis account for the absence of true woodpeckers on the Galapagos Islands?
3. Is the study of evolution static or changing? Find some recent news articles dealing with new fossil discoveries or other new findings that deal with evolution.
4. Explain antibiotic resistance in bacteria using Darwin's concept of natural selection.

5.3 CREATIONISTS CHALLENGE EVOLUTIONARY THOUGHT

Opposition to the idea that life evolves has come from various quarters. Many opponents of evolution have been creationists, people who believe in the fully formed creation of species by God. In this section we discuss a variety of creationist ideas, along with the creationist opposition to evolution.

Creationists, by definition, believe that God created biological species. The majority of creationists believe that God created species much as we see them today and that they did not evolve. Creationists are usually devout believers, and most of them are Christian, but beyond these similarities, creationists do not always share all of the same beliefs. Some creationists have been practicing scientists who conduct research and follow scientific methodology, while others are strongly antiscience and may even seek the destruction of science and of scientific institutions. Creationist thought remains strong in the United States, where surveys reveal that up to 25% of adults reject Darwinian evolution.

Three major groups of creationists stand out:

1. Bible-based creationists, who insist on the biblical account of creation. These creationists work outside of science and reject any scientific theory that conflicts with scripture; some of them are openly hostile to science.
2. Intelligent design creationists, who try to work within the framework of science to find evidence of design in nature. They claim that biological systems are so complex and so well adapted to their functions that only an intelligent (and benevolent) designer could have made them, an argument reminiscent of the writings (circa 1800) of William Paley.
3. "Theistic evolutionists," who believe that God created the universe and all life but that species evolved subsequent to that time and that evolution is one of God's creative processes. Several practicing scientists and various clergy adhere to this view.

5.3.1 Bible-based creationism

In the United States, most creationists have based their beliefs on a literal interpretation of the Bible. Believing in the inerrant truth of their ideas, these creationists reject all science and all scientific evidence that contradicts their beliefs. These creationists therefore work outside of the boundaries of science and do not use scientific methodologies. Some of them are openly hostile to science. Some insist that the six days of creation were each 24 hours in length, while many others interpret the Hebrew word for "day" (*yom*) metaphorically, consistent with its use elsewhere in the Bible. Almost all of these creationists are Protestant Christians, but they represent a small minority within the Protestant tradition.

The large, established denominations accept the evidence for evolution as fully compatible with their religious beliefs.

5.3.1.1 Early fundamentalism In the early twentieth century, most opposition to evolution came from certain Protestants, mostly in the United States, who declared that evolution conflicted with the account of creation given in the Bible. These people founded a number of societies, including the Society for Christian Fundamentals (the origin of the term *fundamentalist*). The fundamentalists persuaded several state legislatures to pass laws restricting or forbidding the teaching of evolution in schools. Some of these state laws remained until the 1960s.

In 1925, a famous court case was brought in Tennessee by the fledgling American Civil Liberties Union. A teacher, John H. Scopes, was arrested for reading a passage about evolution to his high school class. The trial attracted worldwide attention. Scopes lost and was assessed a $100 fine. Upon appeal, the case was thrown out because of the way in which the fine had been assessed; the merits of the case were never really debated. The Scopes trial did, however, have a chilling effect on the textbook publishing industry: Books that mentioned evolution were revised to take the subject out, and most high school biology textbooks published in the United States between 1925 and 1960 made only the slightest reference, if any, to Charles Darwin or any of his theories.

5.3.1.2 Creationism in recent decades The Soviet launch of the Earth-orbiting satellite Sputnik in 1958 set off a wave of self-examination in American education. Groups of college and university scientists began examining high school curricula with renewed vigor, and several new high school science textbooks were written. Most of the new biology texts emphasized evolution, or at least gave it prominent mention.

Alarmed in part by the new textbooks, a new generation of creationists began a series of attacks on the teaching of evolution. These new creationists tried to portray themselves as scientists, calling their new approach "creation science" even though they never conducted experiments or tested hypotheses. Instead of making their studies falsifiable, the new creationists claimed that they held the absolute truth:

> *Biblical revelation is absolutely authoritative. . . . There is not the slightest possibility that the facts of science can contradict the Bible and, therefore, there is no need to fear that a truly scientific comparison . . . can ever yield a verdict in favor of evolution. . . . The processes of creation . . . are no longer in operation today, and are therefore not accessible for scientific measurement and study.*
>
> [H.M. Morris (Ed.). Scientific Creationism. San Diego: Creation-Life Publishers, 1974, pp. 15–16 and 104]

> *We do not know how the Creator created, what processes He used, for He used processes which are not now operating anywhere in the natural universe.. . . We cannot discover by scientific investigations anything about the creation process used by the Creator.*
>
> [D.T. Gish. Evolution: The Fossils Say No! San Diego: Creation-Life Publishers (1978, p. 40)]

In contrast, Darwin knew that his theories were—rightly—subject to empirical testing and possible falsification (see the quotation on p. 168).

Some creationist writings contain faulty explanations of scientific concepts. One such misinterpretation uses the second law of thermodynamics. According to this law, a closed system (one in which energy neither leaves nor enters) can only change in one direction, to that of less order and greater randomness. Thus, a building may crumble into a pile of stones, but a pile of stones cannot be made into a building without the expenditure of energy. Creationists have claimed that this law precludes the possibility of anything complex ever evolving from something simpler. The second law of thermodynamics does not, however, rule out the building up of complexity; rather, it states that making something complex out of something simple requires an input of energy. The second law of thermodynamics *does* apply to all biological processes. If the Earth were a thermodynamically closed system, life itself would soon cease. However, the Earth is *not* a thermodynamically closed system because energy is constantly being received from the sun, and this energy allows life to persist and evolve. Creationist claims on this point may have originated as an innocent error, but the point has been so well refuted that its continued use can only be a deliberate misrepresentation that lies outside the bounds of science or of honest debate.

In the 1960s, because many of the laws in the United States forbidding the teaching of evolution had been declared unconstitutional, one group of creationists, led by Henry Morris, Duane Gish, and John Slusher, decided on a new approach. Evolution could be taught in the schools, they argued, but only if "creation science" was taught along with it and given equal time. (The concept of "equal time" was originally a measure to ensure fairness in political campaigns.) A few state legislatures passed laws inspired by this new group of creationists. An Arkansas law known as the Balanced Treatment Act (Public Law 590) was finally declared unconstitutional in 1981, and a similar Louisiana law was declared unconstitutional a few years later. Interestingly, in the challenges to these laws, the scientific issues *were* raised in court, and prominent scientists were called upon to testify. Specifically, in the Arkansas and Louisiana cases, the U.S. Court of Appeals was asked to rule on what is scientific and what is not. The court finally ruled that evolution is a scientific theory and may be taught, whereas "creation science" is not science at all because it involves no testing of hypotheses and because its truths are considered to be absolute rather than provisional. Instead, "creation science" was found to be a religious idea and to include so many religious concepts (creation by God, Noah's flood, original sin, redemption, and so forth) that it could not be taught in a public school without violating the U.S. Constitution's historic separation of church and state.

In the 1990s, a new group of creationists emerged, preaching the view that modern science, particularly evolution, is the basis for the materialistic philosophy that they claim is responsible for all that is wrong in today's society. The avowed aim of this group, called the "Wedge group," is to destroy all of science, and evolution in particular, by driving in a thin "wedge" and then continuing to drive it in deeper and deeper until the body of science is split asunder. Small but well-financed, this group is the guiding force behind the Discovery Institute and the Center for the Renewal of Science and Culture (CRSC), both of which support intelligent design creationism.

Creationists continue to exert influence today. Despite state laws that have been declared unconstitutional, creationists continue to pressure local school boards and state education departments to support their approach. These efforts have sometimes been successful. In 1999, the Kansas State Board of Education approved, at the urging of creationists, a statewide science curriculum that included no mention of evolution. They also approved a statewide program for testing scientific knowledge and understanding that completely omitted evolution. Two years later, this same Board of Education reversed its earlier decision and restored evolution to the science curriculum in Kansas. Opposition of this kind to the teaching of evolution continues to this day and is largely an American phenomenon. Biologists in most countries other than the United States have not faced similar opposition.

5.3.2 Intelligent design

In the eighteenth and nineteenth centuries, many creationists were also scientists who proposed and tested hypotheses. For example, Reverend William Paley and his supporters proposed that biological adaptations were the work of a benevolent God. In 1996, American biochemist Michael Behe resurrected Paley's pre-evolutionary arguments and revised them in the new language of cell biology and biochemistry.

5.3.2.1 Paley's natural theology Paley sought to prove the existence of God by examining the natural world for evidence of perfection in design. The anatomical structures examined by Paley and his supporters were so well suited to the functions that they served and were, in his view, so perfectly designed that they could only have come from God. Paley's school of *Natural Theology* was very influential in Britain in the early nineteenth century, and the young Charles Darwin was educated in its lessons.

Paley and his supporters had paid much attention to complex organs such as the human eye. The eye, they pointed out, was composed of many parts, each exquisitely fashioned to match the characteristics of the other parts. What use would the lens be without the retina, or the retina without a transparent cornea? An eye, they argued, would be of no use until all its parts were present; thus, it could never have evolved in a series of small steps, but must have been created, all at once, by God.

Paley pointed to the structure of the heart in human fetuses as containing features that adaptation to the local environment could not account for. In adult mammals, including humans, the blood on the left side of the heart is kept separate from the blood on the right side of the heart (see Section 8.4). In fetal mammals, the blood runs across the heart from the right side to the left, bypassing the lungs, which are collapsed and nonfunctional before birth. As the blood enters the left side of the heart, it passes beneath a flap that is sticky on one side. When the baby is born, its lungs fill, and blood flows through them. The blood returning to the heart from the lungs now builds up sufficient pressure on the left side that the flap closes. Because it is sticky on one side, it seals shut. No amount of adaptation to the environment, said Paley, could endow a fetus with a valve that was sticky on one side so that it would seal shut at birth. Only a power with foresight could have realized that the fetus would need a heart whose pattern of blood flow would change at birth and thus designed the sticky valve. Paley attributed the foresight to God, and he insisted that no other hypothesis could explain such an adaptation to future conditions.

What is most interesting is that Paley and his many supporters understood the nature of science and used the methods of science to argue their case. Paley in particular sought scientific proof of God's existence and benevolence by arguing that no other hypothesis could explain the evidence as well. This example shows that good science is certainly compatible with a belief in God or a rejection of evolution. In fact, the best scientists of the period from 1700 to 1859 were, with few exceptions, devout men who rejected the pre-Darwinian ideas of evolution on scientific grounds.

5.3.2.2 Darwin's response Darwin was quite familiar with Paley's arguments, and he offered evolutionary explanations for many of the intricate and marvelous adaptations that Paley's supporters had described. In each case, Darwin argued that the hypothesis of natural selection could account for the adaptation at least as well as the hypothesis of God's design.

To counter Paley's argument about complex organs such as the eye, Darwin pointed out that the eyes of various invertebrates can be arranged into a series of gradations, ranging in complexity from "an optic nerve merely coated with pigment" to the elaborate visual structures of squids, approaching those of vertebrates in form and complexity. A large range of variation in the complexity of visual structures is found within a single group of organisms, the Arthropoda, which includes barnacles, shrimps, crabs, spiders, millipedes, and insects. All the visual structures, regardless of their degree of complexity, are fully functional adaptations, advantageous to their possessors. It would therefore be quite reasonable, argued Darwin, to imagine each more complex structure to have evolved from one of the simpler structures found in related animals. Eyes, in other words, could have evolved through a series of small gradations.

As an additional argument against Paley, Darwin also pointed out several adaptations that were *less* than perfect or that seemed to be "making do" with the materials at hand. The gills in barnacles are modified from a brooding pouch that once held the eggs. The milk glands of mammals are modified sweat glands.

The giant panda, evolved from an ancestor that had lost the true thumb, developed a new thumblike structure made from a little-used wrist bone. (This last example was not known in Darwin's time, but fits well into Darwin's argument.) These many adaptations seem more easily explained by natural selection than by God's design because the design is imperfect and God could presumably have "done better." Natural selection is limited to the use of the materials at hand, and then only if there is variation; an omnipotent God could have made barnacle gills from entirely new material without taking away the brood pouches and could have given pandas a true thumb instead of modifying a wrist bone. Darwin and his supporters used examples like these to show that the evolutionary explanation fitted the available evidence better than Paley's explanation of divine planning. For example, natural selection perpetuates only those hearts whose flaps seal properly at birth.

5.3.2.3 "Irreducible complexity" With today's knowledge of cell biology and biochemistry has come a return to Paley's argument from design at the molecular level. The major proponent of this argument is biochemist Michael Behe. Behe begins with the claim that every living cell contains many sophisticated molecular systems that he calls "irreducibly complex." An irreducibly complex system, according to Behe, is any system that is nonfunctional unless all of its parts are present and functional. Behe's argument, which echoes Paley's, is that no irreducibly complex system could evolve by small, piecemeal steps. According to this creationist argument, natural selection can only improve upon a functioning system, and so could never create a system that requires many parts in order to function at all. Thus, if a complex system cannot function without a minimum of five components, then natural selection could never bring about the evolution of a second component when only one existed, or a third component when only two existed, because none of these changes would improve anything if the system remained nonfunctional with fewer than all five of the required components. Paley had earlier made the same argument with regard to the several parts of the eye, as did the British zoologist St. George Mivart in Darwin's time. Darwin himself realized the power of this argument, for he wrote:

> *If it could be demonstrated that any complex organ existed, which could not possibly have been formed by numerous successive, slight modifications, my theory would absolutely break down.*
>
> [C. Darwin. Origin of Species. London: John Murray (1859, p. 186)]

There are at least two responses to counter the argument of irreducible complexity. One is to show that the system is not, in fact, irreducibly complex and that a partial system with only one or a few of its components does function in some capacity and represents an improvement over the same system with fewer components or none at all. If one component is an improvement over none and two are an improvement over one, then the entire system can evolve piecemeal, step by step, because each step is an improvement over the previous ones, and natural selection favors each small, successive change.

The other response to irreducible complexity is to recognize the role played by changes in function. A system may be incapable of its present function unless fully formed, and thus be described as irreducibly complex. However, the system, or some of its parts, may originally have served a different function, and thus could have evolved by a series of small steps as long as each step improved some ability to serve some function. This can be illustrated by the evolution of insect wings, which developed from external folds of the thoracic wall. By building models of insects with no folds, tiny folds, medium-size folds, and folds large enough to function as wings, scientists were able to show that an increase in the size of the fold from none to small or from small to medium would hardly have improved flying ability. Natural selection would probably not have been able to bring about the early increases in the size of the folds, based upon their usefulness in flying. On the other hand, the function of the folds in cooling the body was also considered. Muscular activity generates heat, and an animal would be in danger of cooking its own tissues if it exercised vigorously without somehow dissipating heat.

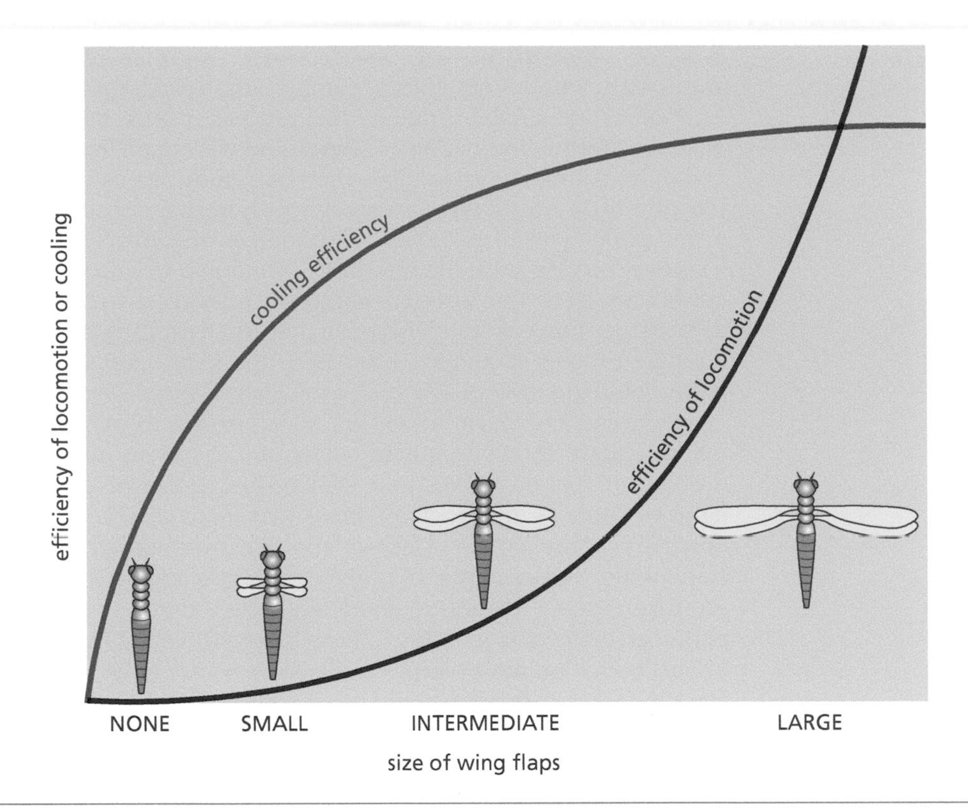

Figure 5.11 The evolution of insect wings. The efficiency of thoracic folds in primitive insect-like arthropods was measured according to two criteria: efficiency in cooling the body down by dissipating heat and efficiency in airborne locomotion by adding to downward air resistance and to lift. Up to a certain size, increments in the size of the folds improved cooling ability but had little effect on locomotion. Thus, early increases in fitness among small to moderate wing sizes depended on improved cooling; however, later increases in fitness depended more on improvements in flying ability.

The efficiency of the flaps in cooling the body also varies with their size, as shown in Figure 5.11. Most of the improvement comes in the smaller sizes, with medium flaps dissipating more heat than small ones, which in turn dissipate more heat than no flaps at all. Large flaps, on the other hand, are scarcely more efficient than medium-size flaps in their cooling ability. Thus, the early stages in the evolution of the wing flaps are thought to have been selectively favored because they improved the body's ability to exercise more without overheating. Only after the flaps had reached a certain medium size did their function as wings became more important than their function in dissipating heat. Thus, the early stages were selectively improved because they helped dissipate heat, while the later stages were selectively improved because they functioned as wings (Kingsolver and Koehl, 1985).

As this example shows, the early stages in the evolution of a structure may have been useful for a totally different reason than the function that they now serve. Half-built structures or systems with only a few components may have improved the ability of their possessors to pass on their genes even without fulfilling their present function. Many structures are now known to have changed their function in the course of evolution.

With these evolutionary counterarguments in mind, we can now examine Michael Behe's claims of irreducible complexity for several systems that function within cells. All of them could have evolved gradually, step by step, despite Behe's insistence to the contrary. For example, the clotting of blood is a multistep process that Behe argues is "irreducibly complex" because none of it would work unless all of it were present. In fact, blood can clot upon exposure to air, and the many chemicals that improve clotting ability could certainly have evolved one at a time, each representing a piecemeal improvement. Natural selection would favor the evolution of any protein or other compound that aided in the clotting process and reduced the chances of bleeding to death. Because blood chemistry does not fossilize, we have no proof of how blood clotting evolved, but a gradual evolution is certainly plausible.

5.3.2.4 Evolution of antibodies Another of Behe's claims is that antibody function is irreducibly complex and could not have evolved piecemeal. Antibodies are soluble proteins that are part of the immune system, which is described further in Chapter 14. The antibodies attach to disease-causing bacteria and other pathogens and reduce their harmfulness in a number of different ways (Section 14.2). Antibodies vary significantly in shape, and each is effective against a particular type of pathogen. Fortunately, the number of possible antibody shapes is very large (in the hundreds of billions) because the genes that code for antibodies have the unusual ability to rearrange themselves by cutting out different portions of their DNA each time, thus multiplying greatly the number of possible protein sequences (and thus antibody shapes) produced. How, asks Behe, could this system ever have evolved by natural selection?

Behe claimed that antibody function is irreducibly complex. However, immunologists have shown that many parts of this system function independently of the rest and therefore could have evolved in a piecemeal manner. In fact, the comparative study of immunology shows that many species survive quite well with only parts of the complete system and that antibody function did indeed evolve step by step. For example, DNA recombination and splicing give us the ability to produce many different kinds of antibodies that can combine with different pathogen molecules, and this splicing is accomplished by an enzyme that appears in backboned animals (vertebrates). Thus, a very large change in capability was acquired by the emergence of one enzyme. Animals other than vertebrates have proteins very similar to basic antibody molecules, and these proteins are useful for other reasons, for example, as molecules that hold cells together. Thus, if we allow for changes in function, molecules that hold on to other cells within the body could well have evolved into antibodies that hold on to pathogens and incapacitate them. Natural selection certainly favors such changes.

In addition to their ability to bind to many different bacterial molecules, antibodies come in different types that show clear evolutionary patterns. Cartilaginous fishes, including sharks, have only the immunoglobin M (IgM) type; reptiles and amphibians have IgM and another variant called IgG. Birds have IgM, IgG, and a third type called IgA. Finally, mammals have IgE in addition to the three earlier types. Each type is a variant on the molecular theme, not a completely new type of molecule. In other words, each was the result of a gene duplication and mutation in the DNA, and each remained in the population because each offered an additional immune capability not afforded by the others. Scientists use comparative genomics (Section 4.4.3) to trace the molecular evolution of antibody molecules in many species.

Any protein that could immobilize any invading pathogen would be an asset to the organism that made it and would be favored by natural selection, meaning that individuals having such a protein would more often live (and pass on to future generations the ability to make such a protein), while individuals not having such a protein would more often be killed and so would not pass on their genes. The system would not need to be perfect at first, or even very good—just a small benefit that worked against only a few types of pathogens would be favored by natural selection. Once such a system was in place, any improvement in its effectiveness would be favored by natural selection. First of all, natural selection would favor any increase in the number of different pathogenic substances recognized. Second, natural selection would also favor a more effective attack on, or immobilization of, the pathogens once they are recognized, and any gradual improvement in the system would be passed on to future generations. As is the case of evolved systems, the immune system is far from perfect—there are many bacteria and other pathogens that are not recognized (they do not provoke an immune response), and the system sometimes attacks the body's own cells (in so-called autoimmune diseases). A system designed by an omnipotent and intelligent designer could certainly work better than this, with more efficiency and fewer errors.

5.3.2.5 Conclusions It is important to note that Behe has conducted no research and provided no scientific evidence to support his claims of irreducible complexity or to test any of his other hypotheses related to intelligent design. Perhaps we should describe his claims as philosophical rather than scientific.

One important measure of a scientific theory is the amount of research that it stimulates. Behe's concept of irreducible complexity has stimulated no research that supports any of his claims, but many arguments have been offered to show that the systems that Behe discusses are not, in fact, irreducibly complex.

All of the systems that Behe claims to be irreducibly complex can be explained as the products of gradual, step-by-step evolution, especially if changes in function are considered. Various biologists have examined Behe's claims and none, to our knowledge, support them. Of the systems that Behe describes, none withstands scrutiny as an argument against evolution.

5.3.3 Reconciling science and religion

A majority of scientists are religious, and a majority of devout people of all religions also accept scientific findings. There are many ways of reconciling religious and scientific viewpoints, and a majority of theological seminaries of all faiths teach that science and religion are fully compatible. The following are examples of the ways in which some people have reconciled religious beliefs and science.

René Descartes was the originator of a dualistic philosophy that separates science and religion as operating in different spheres. In this view, science informs us about the physical world, including the human body, while religion informs us about the spiritual world, including both God and the human soul. Questions about the body can be answered by science, while questions about the soul or about God can be answered only theologically. A separation between science and religion, based on this dualism, has become the official view of the Roman Catholic Church.

The Protestant theologian Reinhold Niebuhr defines religion as the study of the "ultimately unknowable." In this view, advances in science have expanded the frontiers of knowledge—the study of what is knowable—but religion is the study of what remains, which is ultimately unknowable. Thus, religion and science operate in separate spheres, and there is no possibility of an incompatibility between them.

Various scientists have expressed the view that God should be excluded from scientific theories whenever it is possible to do so. One such scientist was the French astronomer and mathematician Pierre Simon LaPlace, one of the authors of the idea that galaxies and solar systems form from swirling masses as the result of natural gravitational forces. When he published his book on this "nebular hypothesis," he presented a copy to the emperor Napoleon, who asked him why he had not mentioned God in his book, LaPlace replied, "I have no need of that hypothesis." A similar attitude caused the British geologist Charles Lyell to exclude all miracles from his geological theories.

According to some twentieth-century versions of a theory called "operationalism," God's presence in certain scientific explanations may not be needed. Thus, the statement, "The Grand Canyon was formed by the action of running water over long periods of time" is indistinguishable from the statement, "God formed the Grand Canyon by the action of running water over long periods of time." Any evidence that could support either of these statements would also support the other, and any evidence against either statement would also be evidence against the other. The two statements are operationally equivalent (or indistinguishable) because no evidence could possibly distinguish between them. According to this view, God is not a necessary part of the explanation, in line with the view that LaPlace had expressed earlier.

A number of scientists and religious thinkers have reconciled science with their religious beliefs by accepting the findings of science as an explanation of the ways that God operates. God created the world along with the natural laws that govern it, and science attempts to discover these natural laws. Isaac Newton, William Paley, and Albert Einstein all expressed views along these lines. One version of this approach is that God set natural laws in place but then withdrew to allow the universe to unfold according to the workings of these natural laws. Another version is that God occasionally intervened to set things right by making exceptions to natural laws. Einstein, who favored the non-interventionist interpretation, ridiculed this second approach in his statement, "I can't believe that God plays dice with the world."

Theistic evolution represents an attempt along these lines to reconcile evolution with a creationist viewpoint, either with or without divine intervention. The Jesuit philosopher and paleontologist Pierre Teilhard de Chardin believed that evolution, including human evolution, was part of God's method of creation in accordance with natural law. Francis Collins, a physician and genomic scientist, led the portion of the Human Genome Project (see Section 4.3), centered at the U.S. National Institutes of Health (NIH) and was NIH director from 2009 to 2021. He describes himself as a devout Christian but forcefully rejects most creationism, including intelligent design. He has described scientific discoveries as "an opportunity for worship" and is a prominent modern example of a scientist who sees his religious faith as compatible with his scientific principles, a viewpoint that he described in his 2007 book, *The Language of God: A Scientist Presents Evidence for Belief.* Collins's beliefs align most closely with theistic evolution. Charles Lyell, a geologist who had inspired Darwin's early thinking, and Alfred Russell Wallace, a naturalist who discovered natural selection independently of Darwin, both came late in life to the belief that evolution was the consequence of natural laws but that divine intervention had been necessary to bring about the evolution of human beings. Most scientists, however, see no need for any involvement of God to explain human evolution.

THOUGHT QUESTIONS

1. In what ways did William Paley use scientific evidence? Did he use testable hypotheses? Which of today's creationists use falsifiable hypotheses to support their claims?
2. How much time should be devoted in science classes to alternative explanations or theories that have been tested and rejected? Should time be given to explanations that are not testable? Should all explanations be given equal time? How much (if any) of a science curriculum would you devote to divine creation as an alternative to evolution? To astrology as an alternative to astronomy? To the theory that disease is caused by demons or evil spirits?
3. Does the teaching of unpopular or rejected theories encourage students to think critically? Does it encourage attitudes of fairness? Does it increase students' understanding of what science is and how science works?
4. Do you think that the concept of intelligent design should be taught in high schools as an alternative to evolution by natural selection? Why or why not?

5.4 SPECIES ARE CENTRAL TO THE MODERN EVOLUTIONARY PARADIGM

The evolutionary paradigm known as the modern synthesis was based largely on the fusion of genetics with Darwinian thought. The cornerstone of the modern synthesis paradigm is a theory of speciation, the process by which one species branches into two species.

5.4.1 Populations and species

A biological **population** consists of those individuals within a species that can mate with one another in nature. If we look backward in time, we realize that any two individuals in a population share at least some of their genetic information because of common descent. If we look into the future, we see that any two opposite-sex individuals in a population are potential mates. Membership in a population is determined by descent and by the capacity to interbreed.

Biological populations within a species may exchange hereditary information (alleles) with one another. The combining of genetic information from different individuals or the exchange of genetic information between populations is called **interbreeding**. The existence of biological barriers to such exchange is called **reproductive isolation**. Interbreeding between populations of the same species takes place when members of different populations mate and produce offspring; reproductive isolation inhibits such matings to varying degrees.

Species are defined as *reproductively isolated groups of interbreeding natural populations*. There are several points to note in this definition. Physical characteristics (morphology) are not part of the definition of species; species are defined by breeding patterns instead. Populations belonging to the same species will

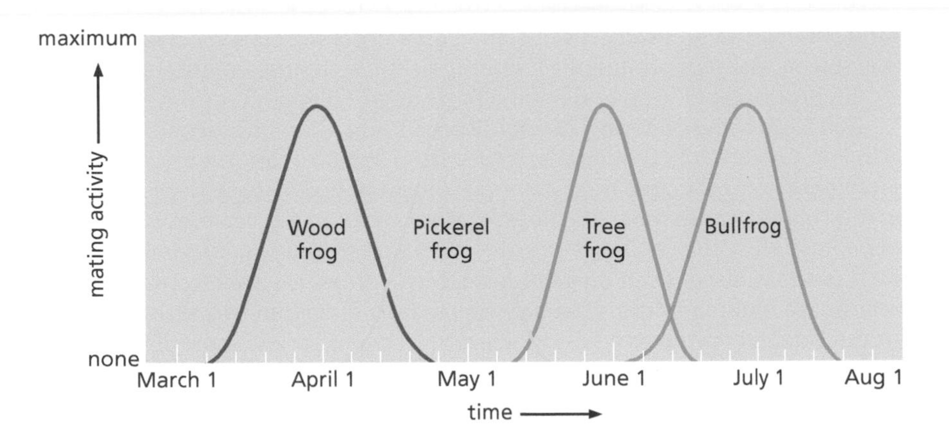

Figure 5.12 Reproductive isolation of several frog species by season of mating, an ecological means of preventing mating between species.

interbreed whenever conditions allow them to. Populations belonging to different species are reproductively isolated from one another and will thus not interbreed. Any biological mechanism that hinders the interbreeding of these populations is called a **reproductive isolating mechanism**, as explained later. Species are composed of natural populations, not of isolated individuals. Thus, the mating behavior of individuals in captivity can only serve as *indirect evidence* of whether *natural populations* would interbreed under natural conditions.

The many reproductive isolating mechanisms fall into two broad categories: those that prevent mating and those that interfere with development after mating has occurred. Mating is prevented when potential mates never encounter each other, which may occur for various reasons: They may live in different habitats, or they may be active at different times of the day or in different seasons, or they may not be physiologically capable of reproduction at the same time. Figure 5.12 shows that wood frogs are fully isolated ecologically from tree frogs and bullfrogs by breeding at different seasons; they are partly isolated from pickerel frogs because the breeding seasons overlap only slightly. Mating can also be prevented by differences in behavior, allowing potential mates with different courtship rituals to live together in the same place without mating. For example, different species of fireflies (phylum Arthropoda, class Insecta, order Coleoptera, family Lampyridae) use different flashing patterns and flight patterns (Figure 5.13) as

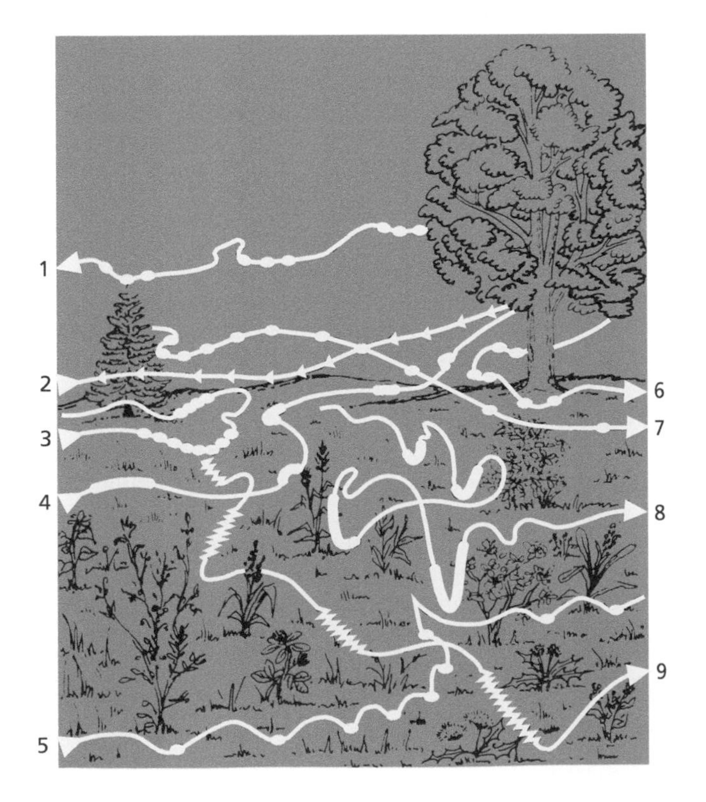

Figure 5.13 Flashing patterns used as mating signals by different species of fireflies. Species 1–9 are reproductively isolated from one another by the behavioral differences shown in these patterns. Details in this form of behavioral isolation include the duration of each flash, the number of repetitions, and the location of the insect when it flashes. A firefly will respond only to the flashing pattern of its own species.

mating signals. In addition, insects and some other animals have hardened and inflexible sexual parts (genitalia); mating of these animals requires a "lock and key" fit, and mating is prevented if the parts do not fit together properly.

There are other isolating mechanisms in which mating occurs but the offspring do not develop. In animals, sperm from a male of another species may die in a female's reproductive tract before fertilization takes place. In plants, the pollen may fail to germinate on the flowers of another species. If a mating takes place between species, the fertilized egg may die after fertilization. Incompatible chromosomes may disrupt cell divisions and developmental rearrangements, leaving the embryo or larva to die. Alternatively, hybrid individuals may live for a while but not reach reproductive age, or they may be sterile. For example, a mule is a sterile hybrid between a horse and a donkey. The sterility of mules keeps the gene pools of horses and donkeys separate, so they remain separate species.

5.4.2 How new species originate

To explain how a new biological species has come into existence, we need to explain how it has become reproductively isolated from closely related species. The origin of a species is thus the origin of one or more new reproductive isolating mechanisms.

In the vast majority of cases, new species have come into existence through a process of **speciation** that includes a period of **geographic isolation** in which populations are separated by some sort of barrier such as a mountain range or simply an uninhabited area that the organisms do not cross. The essence of the theory is that reproductive isolating mechanisms originate during times when such barriers separate populations geographically. Geographic isolation is not by itself considered to be a reproductive isolating mechanism; rather, it sets up the conditions under which the separated populations may evolve along different lines, resulting in reproductive isolation.

What happens depends in part on the length of time for which the populations are geographically isolated—more time allows more chances for reproductive isolating mechanisms to evolve. Another factor is that natural selection must favor different traits on the two sides of the geographic barrier. That is, conditions must be different enough for one set of traits to increase fitness on one side of the barrier and for a different set of traits to increase fitness on the other side. If the populations on the two sides of the barrier are subjected to different selective pressures for a long enough period of time, then one or more reproductive isolating mechanisms may evolve between the two sets of populations and separate them into different species (**Figure 5.14**). If the populations later come into geographical contact again, the reproductive isolating mechanisms that have evolved during their separation will keep them genetically separate as two species. For example, frog or cricket populations isolated on opposite sides of a mountain chain or a large body of water may develop different mating calls. Because the animals respond only to the mating calls of their own population, the two populations will be reproductively isolated and thus become separate species.

The geographic theory of speciation predicts that examples of incomplete speciation may be discovered. If two populations are separated for a very long time (or if selective forces on opposite sides of a barrier differ greatly), then the populations are likely to split into two species. If the separation is brief, then speciation is unlikely. These two situations lie at opposite ends of a continuum. Somewhere along this continuum lies the situation in which populations have been separated by a geographic barrier long enough for reproductive isolation to begin evolving, but not yet long enough for the reproductive isolation to be perfected. Partial or imperfect reproductive isolation between two populations would lessen the chances of interbreeding between them, but not prohibit it entirely. Such situations have indeed been found, for example, among the South American fruit flies known as *Drosophila paulistorum*. Crosses among divergent populations of *D. paulistorum* produce fertile hybrid females but sterile hybrid males. Dobzhansky and Spassky (1959), the geneticists studying these flies, referred to them as "a cluster of species *in statu nascendi*" ("in the process of being born").

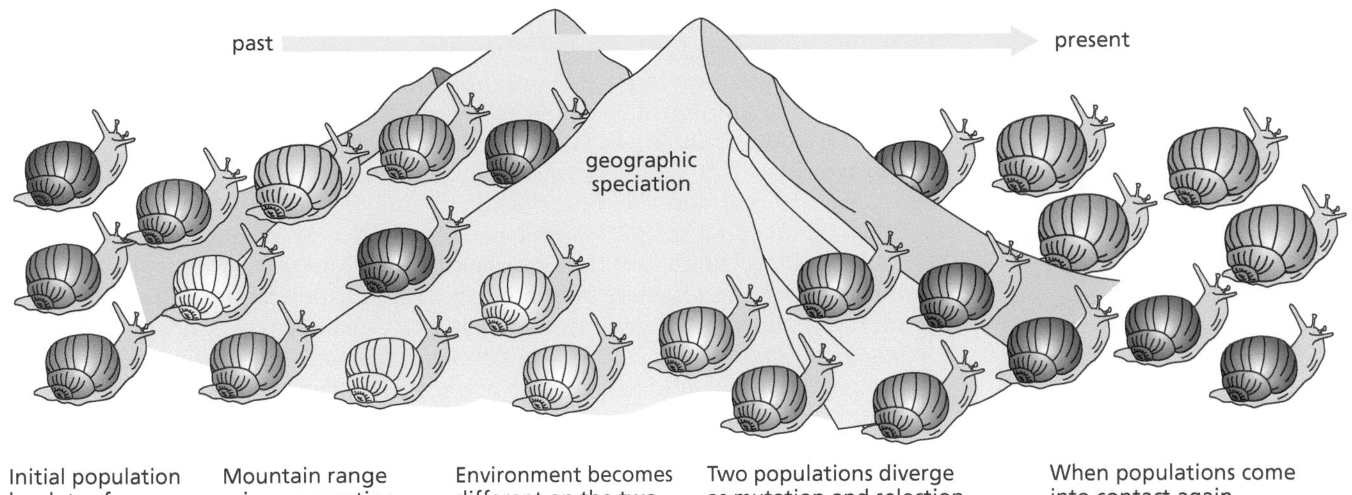

Figure 5.14 **Geographic speciation: the evolution of reproductive isolation during geographic isolation.** Genetically variable populations that spread geographically can develop locally different populations that are capable of interbreeding with one another initially. If the populations are separated for a long enough time by a barrier such as a mountain range or a deep canyon, they may develop differences that prevent interbreeding even after contact is resumed.

People intuitively group similar species together and give names to many collective groups: birds, snakes, insects, pines, orchids, and so forth. Biologists organize these collective groups into a classification that reflects the degree of evolutionary relatedness among species, using methods described in Chapter 6.

THOUGHT QUESTIONS

1. What kinds of reproductive isolating mechanisms might prevent related species of antelopes from interbreeding? Answer the same question for related species of birds, related species of trees, and related species of butterflies.
2. Several kinds of organisms reproduce asexually, producing offspring without combining gametes from two parents. Can asexually reproducing organisms belong to species? Can the definition of species be modified to apply to asexually as well as sexually reproducing organisms?

5.5 LIFE ORIGINATED ON EARTH BY NATURAL PROCESSES AND CONTINUES TO EVOLVE

In addition to explaining how species change and how new species arise, modern evolutionary theory also accounts for the origins of life on Earth. The origin of life and the early history of life on Earth are discussed next, and the effects of life on Earth's atmosphere are discussed in Section 19.1.1. Evolution also continues at present to change populations and create new species, and these processes can be studied as they occur.

5.5.1 Oparin, Miller, and the origin of life

5.5.1.1 Oparin's "origin of life" theory Modern thoughts on the origin of life on Earth began with a biological theory first proposed by the early twentieth-century Russian biochemist Aleksandr Oparin. Most scientists at the time accepted the French microbiologist Louis Pasteur's conclusions that living organisms always came from preexisting organisms. But where had the first living organisms come from? Oparin, a Marxist, rejected the possibility of a divine or other miraculous creation. Had life always existed? The evidence from astronomy at that time favored the view that the Earth originated under extremely hot conditions that could not have supported life. Oparin also knew that the conditions in interplanetary space, such as extreme cold (around -270°C), utter dryness, and constant bombardment by high levels of ultraviolet radiation, were incompatible with all forms of life. These facts convinced Oparin that life could not have come through space from anywhere else. Life must have originated on planet Earth.

Pasteur had said, "there is now no circumstance known in which it can be affirmed that microscopic beings came into the world without. . . parents similar to themselves." From this, Oparin reasoned that if present conditions do not permit organisms to originate from nonliving matter, then life must have originated on Earth under conditions very different from those that prevail today, at a time when the atmosphere had a very different composition from what it has now. From studying what was known about the chemical composition of the solar system, Oparin hypothesized an early atmosphere rich in hydrogen, the most abundant material in the Solar System and in the Universe as a whole. Chemists refer to such hydrogen-rich conditions as **reducing conditions**, in contrast to the **oxidizing** (high-oxygen) **conditions** of our present atmosphere. Under the early reducing atmosphere hypothesized by Oparin, no free oxygen was present. Instead, each element existed largely in its most reduced form, in combination with hydrogen. Thus, most carbon was combined with hydrogen into methane (CH_4), most nitrogen was combined with hydrogen into ammonia (NH_3), and most oxygen was combined with hydrogen to form water vapor (H_2O).

Working with the assumption of an early atmosphere consisting of the gases hydrogen (H_2), ammonia (NH_3), methane (CH_4), and water vapor (H_2O), Oparin deduced several chemical reactions that could have produced simple biological molecules such as sugars and organic acids. He theorized that these molecules would have built up in the primitive oceans once the Earth's temperature permitted water to become liquid. By the slow accumulation of these molecules, the world's oceans would have become like a "hot, dilute soup." Oparin published his detailed hypothesis in book form in 1935, and the book was soon translated into English under the title *The Origin of Life*.

5.5.1.2 The origin of life: Testing Oparin's ideas In 1952, an American biochemist named Stanley Miller decided to test Oparin's model experimentally. He built several types of apparatus, such as the one shown in Figure 5.15. He filled his apparatus with hydrogen, ammonia, methane, and water vapor, the four gases that had been postulated by Oparin. The gases then reacted in a chamber in which electric sparks simulated atmospheric lightning. Reaction products were cooled so that water became liquid, simulating rain. Compounds that formed in the reaction chamber dissolved in the water, which collected in the lower part of the apparatus, simulating the ponds and oceans of the primitive Earth. A heat supply vaporized some of the liquid and returned it to the reaction chamber. Miller circulated his reaction mixtures for several days and withdrew samples to analyze the results. He found that the compounds that had formed included amino acids (building blocks for proteins), simple sugars, and most of the building blocks for DNA and RNA. The constituents for all major biological molecules had thus been formed, without the aid of any organisms, under conditions simulating those of the primitive Earth.

Miller then proceeded to repeat his experiment under somewhat altered conditions, and several other scientists have also done so. They found that differences in the starting materials did not greatly affect the experimental results, as long as sources of hydrogen, oxygen, nitrogen, and carbon were all present. For example, carbon dioxide or another carbon compound could substitute for methane, and nitrogen gas or nitrogen oxides could substitute for ammonia. An energy source was also needed, but the particular kind of energy was not important. An ultraviolet light source could successfully substitute for the electric sparks, as could natural sunlight, even with no other heat source.

To show that the reaction products were produced by the experimental reactions and not by any biological contaminants, Miller also did a control experiment: He repeated all experimental conditions but omitted the electric sparks.

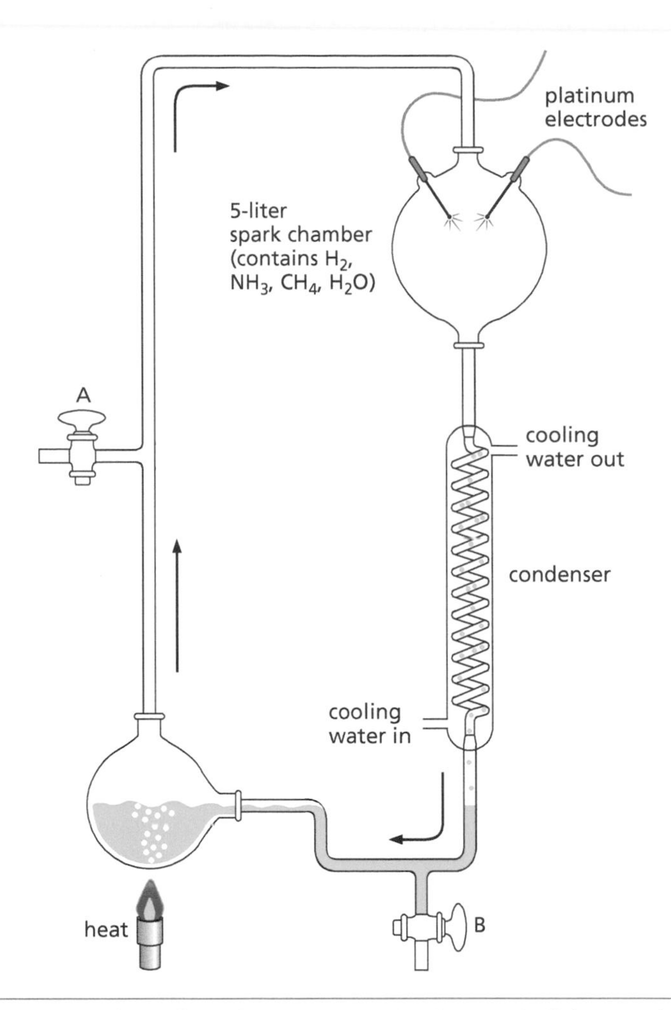

Figure 5.15 Stanley Miller's experiment in which amino acids and other molecules used by living organisms were produced. Heating the flask at the lower left boils the water and keeps the mixture circulating in the direction shown by the arrows. Reactions take place in the spark chamber, and reaction products are condensed and recirculated. Valve A is used to sterilize the apparatus and to introduce the starting materials; valve B is used to withdraw samples of the reaction products.

Without the energy source provided by these electric sparks, no reaction products were detected.

Experiments simulating primitive Earth conditions have succeeded in producing so many biologically important products that the following conclusion is now inescapable: All of the molecules important to life *could have been produced* in a lifeless environment on the primitive Earth, starting with nothing more than a few basic gases mixed together as a simulated reducing atmosphere. However, while these experiments show us what *could* have happened, they cannot show us what actually *did* happen.

5.5.1.2 Protein synthesis and the "RNA world" Proteins are made by stringing amino acids together in a process that chemists call polymerization. The necessary amino acids were formed in Miller's apparatus. Polymerization would not be expected in a "hot, dilute soup," but various mechanisms have been proposed that would concentrate the dissolved materials and thus favor polymerization. For example, tidal pools commonly form along the margins of the world's oceans, and evaporation within these tidal pools concentrates whatever is in them. Molecules with moderate to long hydrocarbon chains could surround water droplets and support conditions inside the droplet that could become very different (and more concentrated) than the conditions outside. Competition among such droplets could favor the growth of whatever droplet made more of itself, and the shaking apart of any large droplet would break it up into smaller droplets, a process akin to reproduction. In particular, any protein capable of functioning as an **enzyme** that made more of its droplet's contents would be favored by such "protoselection," and the number and variety of self-replicating droplets would slowly increase over time. If all this is true, then natural selection is even older than life itself!

The building blocks of RNA include nitrogenous bases and ribose sugar, both synthesized in Miller's apparatus. Phosphate groups, the only missing component, occur naturally in a variety of phosphate minerals. Spontaneous synthesis of RNA under abiotic conditions is therefore possible. Although in today's organisms most chemical reactions are catalyzed by protein-based enzymes, Nobel Prize-winning biochemists Thomas Cech (pronounced "keck"), Jack Szostak, and others have shown that some RNA molecules can also catalyze chemical reactions. In fact, some critical processes, such as protein synthesis in all cells and RNA splicing in eukaryotes, are still carried out by catalytic RNAs in collaboration with proteins. Thus, a small fraction of the RNA molecules that might have spontaneously formed under Earth's primordial conditions could have been capable of promoting the further synthesis of more RNA and of proteins. Moreover, natural selection among these molecules would have favored any that could make more of themselves, either directly or indirectly by making proteins that could serve as RNA-synthesizing enzymes. Cech and Szostak have used this reasoning to suggest the abiotic evolution of an "**RNA world**," long before the evolution of DNA. In this RNA world, proteins would be favored if they helped replicate their droplets, and RNA molecules would be favored if they helped synthesize desirable proteins.

The RNA world hypothesis has been tested in various experiments seeking to simulate "evolution in vitro" (Crisp and Carell, 2018). In particular, RNA sequences with synthetic abilities have been produced over time from spontaneous, random mixtures, and selection among RNA sequences has favored those with self-synthesizing capabilities!

5.5.2 Evidence of early life on Earth

Is there any evidence that the events re-created in Miller's flask or outlined in the preceding text actually did happen on the primitive Earth? There is some evidence, but it is incomplete and indirect. Meteorites falling to Earth from elsewhere in the solar system sometimes contain organic (carbon-containing) compounds indicative of a Miller-style synthesis in other parts of the universe, including compounds that organisms now on Earth produce only rarely or not at all. Ancient rocks on Earth over a billion (10^9) years old contain various kinds of "chemical fossils," compounds that give us clues to the conditions that prevailed at the time when these compounds were formed. Our present atmosphere differs greatly from the primordial atmosphere postulated by Oparin because free oxygen (O_2) is now abundant. The oldest rocks on Earth contain compounds that would not have persisted under oxygen-rich conditions and thus seem to have been deposited at a time when the atmosphere contained little or no oxygen.

One important finding is that chemicals formed from the breakdown of chlorophyll molecules are not present in the oldest rocks. Chlorophyll is a key molecule in photosynthesis, a reaction by which plants, algae, and blue-green bacteria (Cyanobacteria) produce O_2 (see Section 17.1.2). The breakdown products of chlorophyll first appeared in the geological record at about the time that the reducing atmosphere began to change slowly (over a period of about a billion years) to the oxidizing conditions that now prevail.

If the atmosphere now differs so drastically from the primordial atmosphere postulated by Oparin, how and when did the change occur? In particular, how did the present oxidizing conditions replace the earlier reducing conditions? Most scientists who have investigated this question have concluded that life itself is primarily responsible.

5.5.2.1 The first forms of life The first forms of life had to live under reducing conditions in which no oxygen was present, conditions that are called **anaerobic**. A variety of bacteria are capable of living under anaerobic conditions. The first bacteria or archaea would have been enclosed by membranes (which form spontaneously from their lipid components due to the behavior of lipids in water; see Section 8.2.2) and would have contained nucleic acids capable of passing on genetic information, but their mechanisms of obtaining and using energy would have been very different from the methods used by most organisms alive today. Most of the early bacteria would have been **heterotrophs**, meaning that they had to derive all their energy from the high-energy organic molecules that they found in their surroundings. Nowadays, such molecules are mostly produced by other organisms, but the first organisms would have needed to rely on the organic molecules that had formed without life (abiotically) under conditions like the ones simulated by experiments similar to Miller's. For many thousands or maybe millions of years, the supply of these molecules may have been adequate for heterotrophic life to expand and perhaps to flourish. (We can't tell for sure, because such organisms leave very few fossil traces of their existence.) Eventually, however, the organisms expanded to the point that the limited amount of energy-rich chemicals in the environment were just not enough. This was possibly the first global environmental crisis in the history of life on Earth.

We can imagine several possible responses to this crisis. Some organisms may have discovered a way to attack and devour other organisms, getting their nutrients from their prey. Organisms of this type would have continued to eat one another, but the total quantity of living organisms (the total biomass) which the planet could support would have remained limited. Those organisms more efficient at eating one another or at making do with what little food they could find would have done better than their less efficient competitors. Quite possibly, none of the organisms were able to adapt to this way of life, and all eventually perished. In that case, a second abiotic synthesis would have taken place all over again, and perhaps a third, until finally some group of organisms succeeded in evolving the means to feed on other organisms.

Even for organisms that ate other organisms, however, the possibilities would have been strictly limited. Other organisms would be encountered only so often, and only some of their constituents would be usable. Also, after a period of evolution had elapsed, more and more organisms would have evolved defenses against predators, and the amount of biomass would still have been limited. The problem may even have gotten worse because some of the waste products of metabolism included gases like CO_2, which simply escaped into the atmosphere, taking carbon out of reach of most organisms.

5.5.2.2 Evolution of photosynthesis and its atmospheric effects A more permanent solution to the limited supply of energy-rich chemicals occurred much later, when some organisms produced the first chlorophyll-like molecules, about 4 billion years ago. Such a molecule would have enabled life forms to use solar energy in the form of light, which would eliminate the problem of the limited number of other organisms to be consumed in heterotrophic metabolism. All forms of photosynthesis require the use of some chemical to supply hydrogen atoms and act as an electron acceptor (see Section 17.1.2). Most bacterial forms of photosynthesis use hydrogen sulfide (H_2S) for this purpose, but others use iron compounds or organic molecules such as nicotinamide adenine dinucleotide (NAD); importantly, NAD was already present in organisms, derived from the nucleic acids. The primitive forms of photosynthesis likely did result in some minor changes in the atmosphere, perhaps including the use of some atmospheric CO_2 as a raw material.

The greatest change to the Earth's atmosphere resulted from the evolution of a new and more efficient kind of photosynthesis, using a new and different hydrogen donor. The new source of hydrogen was water, the most abundant hydrogen source on Earth. The splitting of water in photosynthesis generated a new atmospheric gas: oxygen (O_2). The first organisms to evolve this kind of photosynthesis were blue-green bacteria (Cyanobacteria). During the next 2 billion years or so, these blue-green bacteria became the dominant form of life on Earth, reducing the abundance of atmospheric CO_2 to a small fraction of its former level and slowly generating more and more oxygen (Schopf, 1974). Calculations of the photosynthetic capabilities of these blue-green bacteria show that they were well capable of generating all the oxygen in the Earth's atmosphere within half a billion years or so.

As the Earth's atmosphere became more and more oxygen-rich, additional changes began to occur. Capture of certain wavelengths of ultraviolet (UV) light by oxygen breaks up the O_2 molecules into a pair of highly reactive oxygen atoms (O). These will then react with the nearest other molecule to produce new compounds. If oxygen atoms react with water, they produce hydrogen peroxide (H_2O_2), which is toxic to most bacteria. However, some bacteria are capable of breaking down small amounts of hydrogen peroxide with the aid of enzymes that are chemically similar to chlorophyll and other pigments. If oxygen atoms (O) react with oxygen molecules (O_2) in the presence of UV light, they produce ozone (O_3). Eventually, the production of ozone in this manner would slowly give rise to an ozone layer in the stratosphere that would screen out additional UV light from reaching the Earth's surface (see Section 19.3.5). Minimizing UV radiation is generally important for the evolution of life because of the damage it causes to nucleic acids and other macromolecules (see Section 10.6.4).

The simplest (prokaryotic) organisms flourished for a billion years or more as the only forms of life, but other, more complex life forms evolved later. The eukaryotic cells described in the next chapter allowed more complex organisms to evolve, and the evolution of sexual recombination (see Section 2.2) greatly contributed to their increase in diversity and more rapid evolution. In the last 550 million years or so, some eukaryotic organisms evolved hard parts that fossilized more easily, leaving a more complete record of their history than had previously been possible. We shall examine this history, and the resulting diversity, in Chapter 6.

5.5.3 Evolution as an ongoing process

Evolution is a process that continues to take place today. The evolutionary changes within species (at the population level) and the changes that create new species can be studied as they occur.

During the twentieth century, the peppered moths of some locations in England changed from predominantly light-colored to almost all dark and back again. In one species of Galapagos ground finches, *Geospiza fortis*, the average bill size changes back and forth. Small-beaked birds that eat soft seeds survive and produce the most offspring in years when rainfall is adequate, but birds with larger beaks are at an advantage in drought years because they can open large, tough old seeds. The average bill size of birds within the population thus increases following drought years and decreases after wet years. In fruit flies, different chromosomal variations (inversions) are favored in different seasons. We see that evolution responds adaptively to fluctuating environmental conditions. Different alleles are selected by different environmental conditions at different times because their phenotypes are more adaptive in those conditions. Such changes, reinforced by geographic isolation, may lead to rapid speciation: Different bill sizes and food preferences are among the features that distinguish species among the Galapagos finches.

Selection also continues to operate in human populations and in bacteria. For example, infant mortality is much higher among babies born under about 3 kg (7 lb.) in weight, even with all that modern medicine can offer. Natural selection thus favors higher birth weight, but only up to a point: Babies weighing more than 5.5 kg (12 lb.) at birth suffer higher mortality as well, unless delivered by cesarean section. Before the widespread availability of cesarean sections in countries like the United States, natural selection also favored women with wider hips and wider pelvic openings that would permit larger babies to pass through the birth canal. Selection also favors certain human genotypes in certain environments (Chapter 7) and during epidemics (Chapter 16). The use of antibiotics has favored the evolution of antibiotic resistance among bacteria, with new strains of bacteria appearing every few years that are resistant to newly developed antibiotics (Chapter 16). As all these examples show, evolution continues as an ongoing process. Biologists continue to study this ongoing process as well as the beginnings of life on Earth and the myriad results of branching evolution over the last several billion years.

THOUGHT QUESTIONS

1. Which step in the origin of life do you think was the most important? How much do we know about each step in the process of life's origin? What means of investigation do you think will bring us additional knowledge?
2. How do you think the properties of life listed in Chapter 1 originated?
3. Are there limitations on the types of organisms that could evolve on Earth? How would we investigate such limitations? What limitations, if any, would apply to organisms on other planets or in other solar systems? How similar to organisms on Earth would such extraterrestrial organisms necessarily be?
4. In the future course of human evolution, do you think we will evolve to be taller? What evidence would you seek to test this hypothesis? Does it matter that human height is also affected by nutrition?

CONCLUDING REMARKS

Considerable evidence now shows that evolution has taken place in the past and that organisms continue to evolve today, though often slowly. Modern evolutionary theory is the cornerstone of biology, and nothing in biology can be fully explained without explaining how it evolved. The origin of life itself can now be explained as the result of natural processes, and the search continues for evidence of those processes in Earth's past and also elsewhere in the universe. The ways in which species resemble one another and are related to one another reflect branching patterns of descent, of which we continue to learn further details. Evolutionary change is brought about by natural selection, a process that operates whenever some genotypes leave more offspring than others. Species are reproductively isolated from one another, and the splitting of species therefore requires the evolution of new reproductive isolating mechanisms. All species, including humans, arose by speciation and are products of evolution. Our attempts to classify the resulting diversity of species are explained in Chapter 6.

CHAPTER SUMMARY

- **Evolution** is the central, unifying concept of biology.
- Darwin's major contributions included his theories of **branching descent** ("descent with modification") and **natural selection.**
- Only inherited traits contribute to evolution and bring about **adaptation**; acquired characteristics do not.
- **Evolution** operates through **natural selection**: There is heritable variation in all **species**, and different genotypes differ in **fitness** by leaving different numbers of surviving offspring.
- Forces of **natural selection** include predators, disease, and **sexual selection.**
- **Mimicry** is easily explained by natural selection but not by any alternative hypothesis.
- Branching descent with modification accounts for **homology** between species.
- **Fossils** provide important evidence for evolution, as does the comparative study of anatomy, biochemistry, and embryological development.
- The **modern synthesis** combines Mendelian genetics with Darwinian evolution. It describes the evolution of genes and phenotypes in populations, and it includes a theory for the formation of species through **geographic isolation.**
- **Speciation** occurs through the buildup of genetic differences between **populations** arising primarily during times of geographic isolation. Over time, this results in **reproductive isolation**, which prevents **interbreeding** between species.
- Life on Earth probably originated under **reducing conditions** in an atmosphere containing ammonia, methane, water vapor, and an abundance of hydrogen. The first organisms were **heterotrophs** that probably used RNA to reproduce before DNA evolved.
- Evolution continues today in all species. In many cases, we can detect ongoing change from year to year.

BIBLIOGRAPHIC REFERENCES

Bibliographic references to material in this chapter can be found online at biologytrending.routledge.com/chapter5

GLOSSARY: KEY TERMS TO KNOW

Acquired characteristics Physiological or other changes developed during the lifetime of an individual but not inherited.

Adaptation Any trait that increases fitness or increases the ability of a population to persist in a particular environment.

Anaerobic Conditions in which free oxygen (O_2) is not present, or organisms that can live under such conditions.

Analogy Resemblance resulting from similar evolutionary adaptation but not indicative of a shared lineage, as in wings of similar shape made of different materials.

Artificial selection Consistent differences in the contribution of different genotypes to future generations, brought about by intentional human activity.

Branching descent (descent with modification) Relationships among species resulting from a common ancestor giving rise to multiple new species, with repetition creating a treelike pattern overall.

Convergence Independent evolution of similar adaptations in unrelated lineages, a process that gives rise to analogy.

Correlation by fossils Judging geological formations to be of the same age if they contain fossils of many of the same or similar species.

Descent with modification: Darwin's name for his theory of branching descent.

Enzyme A chemical substance (usually a protein, but sometimes an RNA) that speeds up a chemical reaction without getting used up in the reaction; a biological catalyst.

Evolution The process of permanent change in living systems, especially in genes or in the phenotypes that result from them.

Fitness The ability of a particular individual or genotype to contribute genes to future generations, as measured by the relative number of viable offspring of that genotype in the next generation.

Fossils The remains or other evidence of life forms of past geological ages as preserved in geological deposits.

Genetic drift Changes in allele frequencies in populations of small to moderate size as the result of random processes.

Geographic isolation Geographic separation of populations by an extrinsic barrier such as a mountain range or an uninhabitable region.

Heterotroph An organism not capable of manufacturing its own energy-rich organic compounds and therefore dependent on consuming such compounds as food.

Homology Shared similarity of structure resulting from common ancestry.

Industrial melanism The evolution of protective dark coloration in soot-polluted habitats.

Interbreeding The mating of unrelated individuals or the exchange of genetic information between populations.

Mimicry A situation in which one species of organisms derives benefit from its deceptive resemblance to another species.

Modern synthesis The modern evolutionary paradigm since about 1940 based upon Darwinian natural selection and the geographical theory of speciation.

Natural selection A naturally occurring process by which different genotypes consistently differ in fitness (i.e., in the number of copies of themselves that they pass on to future generations).

Oxidizing conditions Oxygen-rich conditions or other conditions in which electrons are readily removed from molecules.

Phylogeny A family tree or history of a group of organisms, forming a branching pattern of descent in most cases.

Population A group of organisms capable of interbreeding among themselves and usually sharing a common descent; a group of individuals within a species living at a particular time and place.

Punctuated equilibrium A theory that describes species as remaining the same over long periods of time and then changing suddenly and giving rise to new species.

Reducing conditions Hydrogen-rich conditions in which no free oxygen is present.

Reproductive isolating mechanism Any biological mechanism that hinders the interbreeding of populations belonging to different species.

Reproductive isolation The existence of biological barriers to interbreeding.

RNA world A theory that RNA-dependent protein synthesis prevailed for a long time before DNA evolved.

Sexual selection A process by which different genotypes leave unequal numbers of progeny to future generations on the basis of their success in attracting a mate and in reproducing.

Speciation The process by which a new species comes into being, especially by a single species splitting into two new species.

Species Reproductively isolated groups of interbreeding natural populations.

Stratigraphy The study of layered sedimentary rocks.

Vestigial structures Organs reduced in size and nonfunctional but often showing resemblance (by homology) to functional organs in related species.

CONNECTIONS TO OTHER CHAPTERS

Chapter 1: Darwinian evolution and modern evolutionary theory are both good examples of successful paradigms.

Chapter 1: Presenting creationist ideas in school classrooms raises several social policy issues.

Chapter 2: Gene mutations provide the raw material for evolution.

Chapter 3: Rare, recessive, disease-causing mutations persist in human populations despite being associated with decreased fitness.

Chapter 4: The CRISPR-Cas9 system used in gene editing evolved in bacteria as a defense against viruses.

Chapter 4: Genomic and proteomic data contribute to the determination of evolutionary relationships among species.

Chapter 6: Branching descent and other evolutionary processes have produced a great diversity of species that have been described and classified, and many others that await discovery and description.

Chapter 7: Differences have evolved and continue to evolve both within and among human populations.

Chapter 8: The building blocks of cells likely formed from abiotic processes in Earth's early atmosphere.

Chapter 9: Successful species may increase so rapidly in numbers that they outstrip the available resources.

Chapter 10: The development of cancer is an evolutionary process that occurs within an individual organism, spurred by mutations that provide a growth advantage to cancer cells over normal cells.

Chapter 11: Differences in brain anatomy in different species provide good evidence of evolution.

Chapter 13: The existence of brain receptors to certain drugs (opioids, cannabinoids) led scientists to hypothesize that such receptors would never have evolved unless there were a benefit in detecting naturally circulating compounds, and this hypothesis led to the discovery of internally produced substances such as endorphins and anandamide.

Chapter 14: The body's immune defenses are products of evolution by natural selection.

Chapter 15: Viruses may evolve disease-causing strains, as well as strains resistant to certain medicines. The rapid evolution of HIV has made it difficult to develop a vaccine.

Chapter 16: New infectious threats arise when new pathogenic strains of organisms or viruses evolve. Over time, pathogens evolve adaptations to the life cycles of their hosts. Bacteria often evolve antibiotic resistance through natural selection.

Chapter 17: Plant characteristics resulting from evolution include the presence of chloroplasts and vascular tissues.
Chapter 18: Speciation increases biodiversity, whereas extinction diminishes biodiversity.
Chapter 19: Earth's atmosphere changed over billions of years as the result of the evolution of new life forms, especially those carrying out photosynthesis.

PRACTICE QUESTIONS

1. Match the ideas in the left column with the people on the right. One name needs to be used twice.
   a. Evolution is a branching process
   b. Adaptations should be studied carefully as a way of understanding God's creation
   c. Evolution should never be taught
   d. New species originate by a process that includes geographic isolation
   e. Adaptation occurs by use and disuse
   f. Organisms with successful adaptations will be perpetuated, whereas those with unfavorable characteristics will die out
   g. Evolution and creation science should be given equal time in science classes

   i. Creationist supporters of the "Balanced Treatment Act"
   ii. Creationists of the period 1890–1940
   iii. Charles Darwin
   iv. Jean Baptiste Lamarck
   v. William Paley
   vi. Modern evolutionary biologists
2. The Bahamas are a group of islands in the Atlantic, made mostly of coral fragments. The closest mainland is North America, but political ties are to Great Britain. According to Darwin's reasoning, the birds and other species living on these islands should have their closest relatives in:
   a. Other islands of similar composition in the Pacific
   b. Islands such as the Canary Islands in the Atlantic at a similar latitude
   c. North America
   d. England
3. Which theory had no way of explaining the sticky flap in the fetal heart?
   a. Darwin's
   b. Paley's
   c. Lamarck's
4. In mimicry, the mimics and their models always:
   a. Live in similar climates, although they may be far away
   b. Live close together
   c. Taste the same to predators
   d. Are camouflaged to resemble their backgrounds
5. Which of these is NOT considered a reproductive isolating mechanism?
   a. Two geographically separated species
   b. Two species breeding in different seasons
   c. Two species that produce infertile hybrids when they mate
   d. Two species with different mating calls
   e. Two species whose external genitalia cannot fit together
6. Name the four gases used in Miller's initial experiment. Also name the four elements needed to produce biological molecules. Did the gases provide all of these elements?
7. Give a clear definition of the term *species*.
8. What is the basic argument used by supporters of intelligent design? What kinds of evidence can be used against this argument?

CHAPTER 6

Classifying Nature's Diversity

ISSUES

- Why is life so diverse?
- Why have some groups proliferated into so many species?
- Why is classification important?
- How do our classifications reflect evolution?
- How are classifications established?
- Did humans evolve? How and why? Where and when?

BIOLOGICAL CONCEPTS

- Evolution (adaptation, descent with modification)
- Form and function
- Species and speciation
- Products of evolutionary change (phylogenetic classification, biodiversity)
- Types of cells (prokaryotic, eukaryotic)
- Compartmentalization
- Specialization of function
- Hierarchy of organization
- Classification of organisms (three domains and many kingdoms)
- Sexual recombination
- Major groups of organisms
- Plant structure and adaptations (vascular tissues, seeds, flowers, etc.)
- Animal structure and adaptations (bilateral symmetry, sense organs, head, assembly-line digestion, body cavities, segmentation, etc.)

CHAPTER OUTLINE

DOI: 10.1201/9781003391159-6

Life on Earth is extremely diverse, with an estimated 10 million species or more. Why is life so diverse and so prolific? How do we describe this diversity? How can we begin to understand it? As Charles Darwin observed, every living species has so great a tendency to reproduce that, if left unchecked, there would soon be no standing room left on Earth for all its progeny. As we saw in Chapter 5, species can also undergo speciation: splitting and thus creating additional species. This process has been going on so long and so frequently as to have produced the many millions of species alive today, plus an even greater number of species that have become extinct in the past. In this chapter, we will describe certain innovations that greatly increased the ability of biological species to succeed in life and to speciate further.

6.1 WHY CLASSIFICATION IS IMPORTANT

In order to describe and understand life's diversity, we need a system of classification. People intuitively group all insects together and all birds together. All people have common names for collective groups of similar species: birds, snakes, insects, pines, orchids, and so forth. Some collective groups, such as beetles, are contained within larger groups, such as insects. Biologists organize these collective groups into a **classification**, an arrangement of larger groups that are subdivided into smaller groups, each reflecting their degree of evolutionary relatedness. Classifications help us to catalog and describe life's diversity, which is the first step on the road to understanding this great diversity.

Any collective group of similar organisms, such as insects or orchids, is called a **taxon**, and **taxonomy** is the study of how these taxa (plural of *taxon*) are recognized and how classifications are made. In a biological classification, species are grouped into successively more inclusive groups, or "higher taxa": related species are grouped into genera (singular, **genus**), related genera into **families**, related families into **orders**, related orders into **classes**, related classes into **phyla**, and related phyla into **kingdoms**, such as the animal or plant kingdoms. All these are arranged as groups within groups, with the less inclusive (smaller) groups sharing more characters and the more inclusive (larger) groups sharing fewer characters. Thus, species within a genus share more characters in common than do families within an order.

For example, human beings constitute the species *Homo sapiens*. **Figure 6.1** shows, beginning on the right, that *Homo sapiens* is grouped together with *Homo erectus* and certain other fossil species into the genus *Homo*. (A genus always has a one-word name that is capitalized; a species has a two-word name in which the first word is the name of the genus; after the two-word name has been introduced, the genus may subsequently be abbreviated, for example, *H. sapiens*.) The genus *Homo* is grouped together with the extinct genera *Australopithecus* and a few others into the family Hominidae. This family is included in the order Primates, which also includes apes (Pondigae), monkeys (Cebidae and Cercopithecidae), and lemurs (Lemuridae). The primates are grouped together with rodents (Rodentia), carnivores (Carnivora), bats (Chiroptera), whales (Cetacea), and over two dozen other orders into the class Mammalia, including all warm-blooded animals with hair or fur that feed milk to their young. Mammals are one of several classes in the phylum Chordata, a group that includes all vertebrates (animals with backbones) and a few aquatic relatives such as the sea squirts and amphioxus. The Chordata and several dozen other phyla are together placed in the animal kingdom (Animalia). Animals are one of the several kingdoms to be discussed later in this chapter.

6.1.1 "All those names"

Communication among scientists requires a common vocabulary. Even hunters, gatherers, farmers, and other nonscientists need names for all the types of

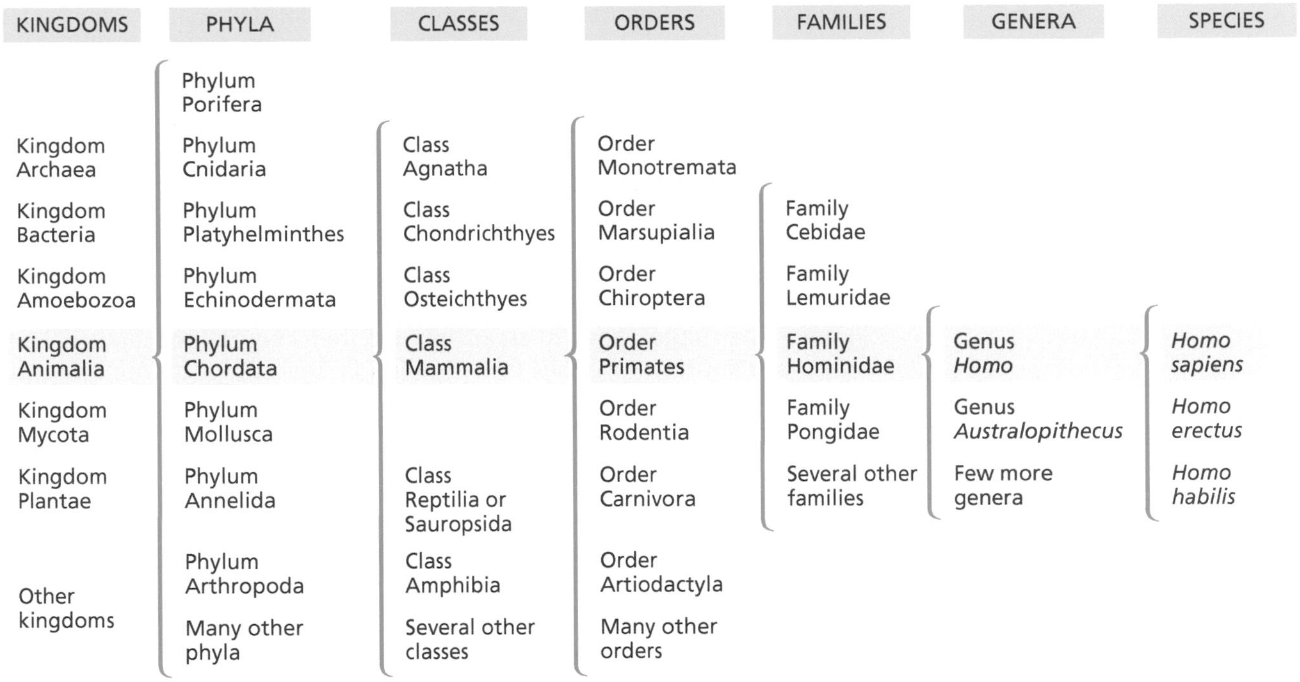

KINGDOMS	PHYLA	CLASSES	ORDERS	FAMILIES	GENERA	SPECIES
	Phylum Porifera					
Kingdom Archaea	Phylum Cnidaria	Class Agnatha	Order Monotremata			
Kingdom Bacteria	Phylum Platyhelminthes	Class Chondrichthyes	Order Marsupialia	Family Cebidae		
Kingdom Amoebozoa	Phylum Echinodermata	Class Osteichthyes	Order Chiroptera	Family Lemuridae		
Kingdom Animalia	Phylum Chordata	Class Mammalia	Order Primates	Family Hominidae	Genus *Homo*	*Homo sapiens*
Kingdom Mycota	Phylum Mollusca		Order Rodentia	Family Pongidae	Genus *Australopithecus*	*Homo erectus*
Kingdom Plantae	Phylum Annelida	Class Reptilia or Sauropsida	Order Carnivora	Several other families	Few more genera	*Homo habilis*
Other kingdoms	Phylum Arthropoda	Class Amphibia	Order Artiodactyla			
	Many other phyla	Several other classes	Many other orders			

Figure 6.1 The place of the species *Homo sapiens* in the classification of organisms. Reading from right to left shows the increasingly more inclusive taxa to which our species belongs. Reading from left to right focuses on taxa of increasingly narrow scope.

organisms familiar to them. However, in a linguistically diverse world, different people speak many different languages. Many years ago, the scientists of Europe discovered that they could best communicate with scientists in other countries by using Latin names. Thus, while the common names of animal and plant species differ from one language to another (e.g., dog, hund, cao, perro, chien), the scientific name, *Canis*, is the same for scientists the world over. Moreover, while modern languages change (modern English differs from Shakespeare's and even more from Chaucer's English), scientific names are based mostly on Latin and Greek roots that do not change over time.

The modern hierarchical form of classification, which we have just described, originated with the eighteenth-century Swedish naturalist Carl von Linné, who wrote under the name Linnaeus. Linnaeus also began a system in which each species has a two-word name, such as *Homo sapiens.* The use of two words for species names also allows us to recognize millions of species with far fewer names. Thus, *Canis familiaris* is the domestic dog, *Canis lupus* is the wolf, *Canis latrans* is the coyote, and *Canis aureus* is the jackal. Any person encountering any of these names for the first time immediately recognizes that they are all related species placed in the same genus. The name of each family is always based on the name of a well-known genus (e.g., Canidae for the dog family, comprising *Canis* and its relatives).

Larger collective groupings, such as orders and classes, also have names. Most of these names are descriptive of some important characteristic of the group. Thus, the class Mammalia (from the Latin word *mamma,* meaning a breast) includes all species whose young are nourished with their mother's milk, and the class Amphibia (from Greek roots *amphi* and *bio,* literally meaning "both lives") contains frogs, salamanders, and other animals that live in water as gill-breathing larvae before they transform into lung-breathing adults. The names, in other words, have a meaning that relates to the organisms and that makes the name easier to learn and remember, at least when you only learn a few at a time. Unfortunately for many beginners, there are many names to know, and students who see them all at once may be overwhelmed by the sheer number of unfamiliar terms. We suggest that you begin by simply learning to recognize the same name on repeated encounters—it is, after all, just a name. If you learn the meaning of the name (which often means learning its Latin or Greek roots), it may help you associate it with an important characteristic of the group. Remember that a classification is a tool designed to make communication easier.

6.1.2 Taxonomic theory

It is easy for students to look upon classifications as fixed and unchanging, but this is a false impression. Each classification is really just a hypothesis about how best to describe the variation among the organisms being discussed. If you read several accounts of the classification of the same group of organisms, you will probably find that not all authorities follow exactly the same classification. How, then, are classifications made? Sometimes, one person may become an expert on a particular group of organisms and then everyone accepts their classification as authoritative, but this situation is uncommon. More often there are multiple researchers, each following a slightly different classification, and each attempting to attract followers to their way of thinking. So how do scientists determine that one classification is preferable to another for the same group of organisms?

One important goal of classification is to summarize and communicate what we know about groups of organisms (taxa). Thus, one criterion of a good classification is that it should aid in describing variation among taxa, summarizing both their differences and their similarities. Early classification systems were based on common physical structures, and members of each group were therefore expected to share visible characteristics. Many familiar groups recognized by ordinary people were given formal names and recognized in the classification, for example, birds (Aves). Fishes, insects, clams, ferns, and orchids were also recognized in many early classifications; some of these groups are still recognized today. One reason that these taxa are useful is that there are many shared similarities that unite all their members and that distinguish these taxa from related taxa. However, other groupings, such as "worms," were shown to be heterogeneous when the included organisms were studied more closely; such unnatural groups are usually abandoned.

Another goal of classification is to describe evolution. In order to describe all the different descendants of some ancient species, we clearly need some collective name for the group. Such a group is called a **clade** (from a Greek word meaning "branch"), and the study of branching patterns in evolution, from common ancestors to their descendants, is therefore called **cladistics**. On a family tree such as the one shown in Box 6.1 or in Figure 5.1, a clade (also called

Box 6.1 Evolution and classification of the land vertebrates

The following example presents a simplified version of cladistics, the procedure commonly used to formulate a family tree and base a classification upon it. The organisms in this example are the land vertebrates, including amphibians (frogs, salamanders, etc.), turtles, lepidosaurs (lizards and snakes), crocodilians, birds, monotremes (egg-laying mammals such as the platypus), marsupials (pouched mammals), and placental mammals (those retaining the young inside the uterus throughout gestation). The groups (taxa) being classified are compared to an "outgroup" of organisms outside the assemblage being classified but related to it. Characters present in this "outgroup" are presumed to be primitive (or ancestral) for the assemblage in question. In this example, lungfish are the outgroup.

First, a character matrix is prepared. In this matrix, each taxon being classified is listed as either having or lacking each of the character traits used in the study. In our example, we are only using 14 characters; professional studies usually use many more than this. Our characters are as follows:

1. An egg containing certain internal membranes (amnion, etc.) capable of being laid on land
2. Hair or fur
3. Maintenance of a constant, warm body temperature by using heat generated through physiological activity (endothermy, commonly called "warm blooded")
4. A single centrale bone in the ankle
5. A skull with two window-like openings behind the eye
6. A shell enclosing most of the body
7. A window-like opening in the lower jaw
8. A transverse, slit-shaped anus
9. Feathers

(*Continued*)

10. An opening between the major arteries as they exit from the heart, permitting the blood to be diverted right or left as needed to balance pressure when the animal dives
11. Giving birth to live young
12. A jaw-opening muscle belonging to a superficial layer encircling the neck and used instead of a deeper muscle beneath the jaws
13. A placenta formed by two embryonic membranes (chorion and allantois) and attaching to the inner wall of the uterus during gestation
14. A pouch in which the young is nursed following its birth

Our character matrix is as follows:

CHARACTERS \ TAXON	Lungfish	Amphibians	Turtles	Lizards and snakes	Crocodilians	Birds	Monotremes	Marsupials	Placental mammals
1 egg with amnion	no	no	yes	yes	yes	yes	yes	yes	yes
2 hair	no	no	no	no	no	no	yes	yes	yes
3 warm blooded	no	no	no	no	no	yes	yes	yes	yes
4 single centrale bone	no	no	yes	yes	yes	yes	no	no	no
5 two-windowed skull	no	no	no	yes	yes	yes	no	no	no
6 shell	no	no	yes	no	no	no	no	no	no
7 opening in lower jaw	no	no	no	no	yes	yes	no	no	no
8 slit-shaped anus	no	no	no	yes	no	no	no	no	no
9 feathers	no	no	no	no	no	yes	no	no	no
10 major arteries connected	no	no	no	no	yes	no	no	no	no
11 live birth	no	no	no	seldom	no	no	no	yes	yes
12 superficial jaw-opening muscle	no	no	no	no	no	no	yes	no	no
13 placenta	no	no	no	seldom	no	no	no	no	yes
14 pouch	no	no	no	no	no	no	no	usually	no

For this matrix of characters, the family tree shown here is the simplest one possible, meaning that it requires the fewest evolutionary changes. Each labeled transition on the tree marks the evolution of a trait shared by all taxa on all of the branches to the right of that point. The family tree can be constructed from the character matrix by hand or by a computer program, using the following basic protocol:

a. Allow all the taxa sharing a derived trait (a "yes" in the matrix of characters) to form a single branch of the tree. Allow the various branches all to be nested in one another. For example, the land-based egg (character 1) is presumed to have evolved only once in a common ancestor of all the animals possessing this trait, that is, all of the animals to the right of the branch labeled character 1. If all of the characters evolved only once, then the tree is finished.
b. Where the first strategy does not result in a tree with each character showing up on only one branch of the proposed tree, we choose the simplest family tree as being the most likely. When a few characters do not fit in consistently, it may indicate that the character trait evolved but then was lost by some, but not all, species in a group, or that the character evolved more than once (by convergence). When this happens, select the tree that requires the fewest evolutionary changes, minimizing the number of character traits for which we must assume multiple origins or acquisition and subsequent loss. In the family tree shown here, warm body temperature (3) is shown as having evolved independently in birds and in mammals. A different tree would result in many more traits for which multiple origins must be assumed.

The tree formed by this protocol allows us to recognize the clades (branches of the family tree) that form the basis for the classification. Some of the taxonomic groupings supported by this particular family tree are shown as brackets on the right of the diagram. "Reptiles" are in quotation marks because they are a group that does not correspond to a clade in this arrangement.

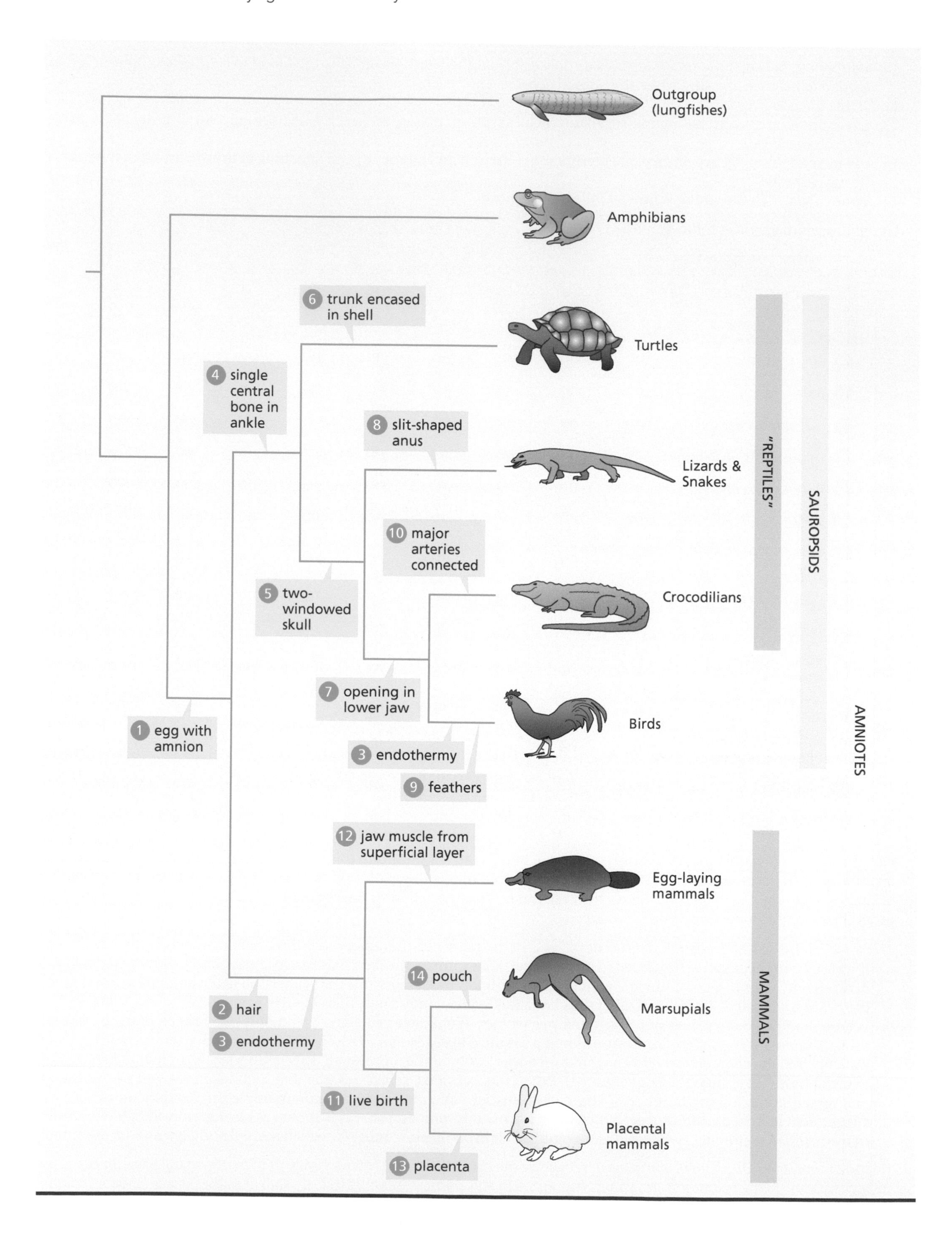
Outgroup (lungfishes)
Amphibians
6 trunk encased in shell
Turtles
4 single central bone in ankle
8 slit-shaped anus
Lizards & Snakes
10 major arteries connected
5 two-windowed skull
Crocodilians
7 opening in lower jaw
1 egg with amnion
Birds
3 endothermy
9 feathers
12 jaw muscle from superficial layer
Egg-laying mammals
14 pouch
Marsupials
2 hair
3 endothermy
11 live birth
Placental mammals
13 placenta
"REPTILES"
SAUROPSIDS
AMNIOTES
MAMMALS

a **monophyletic** group) consists of any ancestral species and all the lines branching out from it, representing its descendants. In traditional practice, the taxa and their groupings were decided upon first, and family trees were then offered as explanatory hypotheses that justified the classification. Cladistics reverses this order, deriving the family tree first and then basing the classification on the tree.

Biologists who follow cladistics develop family trees and classifications in which each larger taxon corresponds as closely as possible to a clade. After years of intensive study of a large and varied group of organisms, they make lists of important anatomical features and other characters that they think will help in classification. Such characters may include the occurrence of particular plant pigments, reproductive adaptations such as seeds or hard-shelled eggs, the presence of openings in the skull, the number and shape of teeth, the presence of certain glands and their secretions, or the anatomical structure of the limbs and other body parts. Often, characters present only in larvae or in embryonic life stages are used. After many such characters are listed for the many species being studied, the biologist will deduce (sometimes with the aid of a computer algorithm) the family tree that can explain the evolution of these characters with the smallest number of evolutionary changes. That is, if five species all possess a certain shared character, five changes would be required if each acquired the character independently, but only one change would be required if the character were acquired by an ancestor common to all five. From information of this kind, the five species are assigned to a clade, meaning a common branch of the family tree. This approach usually produces a consistent pattern, with smaller clades nested within larger clades. If, however, one character does not fit consistently with the others (that is, if it suggests clades that cannot be nested with the clades suggested by other characters), then that character tends to be discarded as misleading and the family tree is then based on the remaining characters. Once a family tree is established, a classification is then drawn up based on this family tree by giving a name (e.g., Insecta or Diptera) and a rank (e.g., class or order) to each clade. As an example, Box 6.1 shows how a family tree and classification of land vertebrates may be derived.

Molecular biology (including genomics, discussed in Section 4.4) provides further evidence that can be used to construct classifications. First, the DNA sequences or protein sequences of various species are compared. Older molecular approaches simply compared all pairs of species for the overall similarity of their molecular sequences and grouped species into a classification based on the percentage of overall similarity (or percent homology). The newer approaches use the methods of cladistics to construct family trees with the smallest number of evolutionary changes needed to explain the patterns or variation, as explained earlier.

We have said that one goal of classification is describing groups with many physical characters in common, and another goal is describing evolutionary history. Thankfully, these two goals usually result in very similar classifications. When they do not, more investigation is needed to determine the cause. If coherent groups, sharing many characters in common, do not correspond to clades, then an explanation is needed. Perhaps two unrelated groups evolved similar adaptations by convergence (see Section 5.2.3), developing many resemblances by analogy. Another possibility is parallel evolution, in which similar trends occurred independently several times among closely related groups. Perhaps some investigators were mistaken in describing or evaluating characters, for example, by misinterpreting one bone as a different bone or by failing to recognize that a structure that appeared to be the same in two species had different embryological origins and was therefore a different structure in each case. One thing that is certain is that taxonomic disagreements often spur further research when biologists attempt to discover the reasons behind the disagreement. After further research, classifications often change. For instance, the barnacles that grow on rocks (or on ships) were once thought to be related to clams or other filter-feeding animals with hard shells, but a study of their embryology revealed that they were built from the same parts as lobsters, shrimp, and other members of the class Crustacea, with which they are now classified.

THOUGHT QUESTIONS

1. The class Mammalia includes whales and bats along with the more numerous terrestrial species (rodents, monkeys, cats, bears, deer, elephants, humans, and many others). Why is this grouping a useful one? Think of as many reasons as you can.
2. Think of a large group of organisms that you know something about. How would a biologist decide which of several possible classifications is best for this group? What would she or he look for? What kind of evidence is relevant? (These are among the most basic questions of taxonomy.)

6.2 PROKARYOTIC CELLS DIFFER FUNDAMENTALLY FROM EUKARYOTIC CELLS

As we introduced in Chapter 1, all living organisms are composed of one or more compartments called **cells.** This tenet is part of the **cell theory** that was put forth in the early 1800s by the first cell biologists, who observed the compartmentalized structure of many different organisms under microscopes. The first organisms consisted of individual cells performing all life functions in one single compartment, with little or no spatial separation among their functions. Bacteria and certain other organisms living today have retained that simple cellular structure. Simple cells of this type are called **prokaryotic** cells (Box 6.2). Over time, some of the ancestral prokaryotic cells evolved to have greater complexity, dividing their functions among multiple membrane-bound **organelles** and establishing the **eukaryotic** branch of the tree of life (Box 6.2; see also Figure 6.3). Protists and some fungi are examples of single-celled eukaryotic organisms that, like all eukaryotes, have their DNA enclosed in a true nucleus and also possess many other organelles with specialized functions. Further evolution gave rise to larger organisms composed of many individual cells, including animals and plants. Every modern classification of organisms recognizes that a great gulf separates the major types of organisms on the basis of the structure of their cells and gives prominence to this fundamental distinction. The details of prokaryotic and eukaryotic cell structures are discussed later in this chapter; however, we will first address the advantages of **compartmentalization**—whether into the multiple organelles of a single-celled eukaryote or, further, into the many different types of cells that make up a multicellular organism.

6.2.1 Compartmentalization

Compartmentalization of various life processes into specialized organelles allows those processes to happen more efficiently, since the molecules involved are concentrated into a smaller space and can interact with each other more readily. Similarly, compartmentalization of most organisms into multiple cells permits those organisms to become much larger than they could be as single cells because the ratio of a cell's surface area to its volume imposes a physical restriction. The requirements for energy and the production of wastes both increase in proportion to the volume of an organism, while the organism's ability to absorb nutrients and to release wastes varies with its surface area. However, as an organism enlarges, its volume grows faster than its surface area (Box 6.3); therefore, to meet the energy and waste disposal requirements of each cell, larger organisms must have more cells rather than larger cells. Compartmentalization of the interior volume of an organism into cells keeps the volume the same but increases the effective surface area. This maintains an efficient ratio of surface area to volume so that nutrient intake stays balanced with metabolism.

Box 6.2 Prokaryotic and eukaryotic cells compared

The great difference between prokaryotic (bacterial) cells and eukaryotic cells (including both animal and plant cells) are shown in the accompanying drawings and also in chart form.

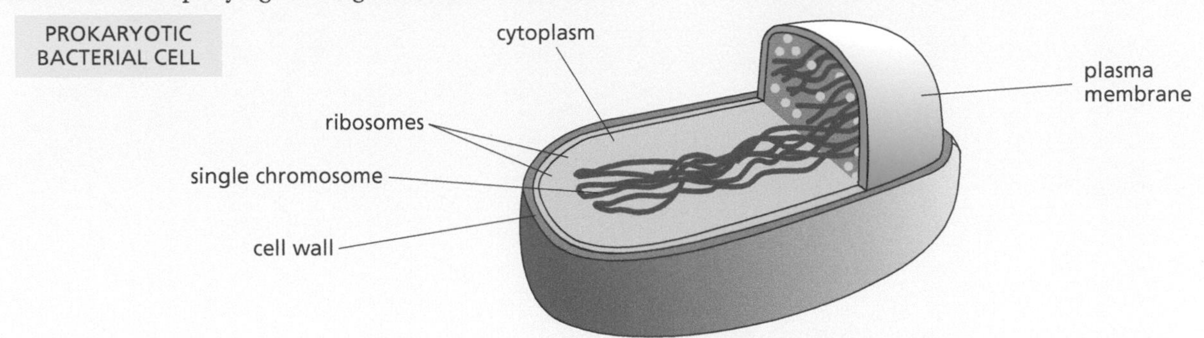

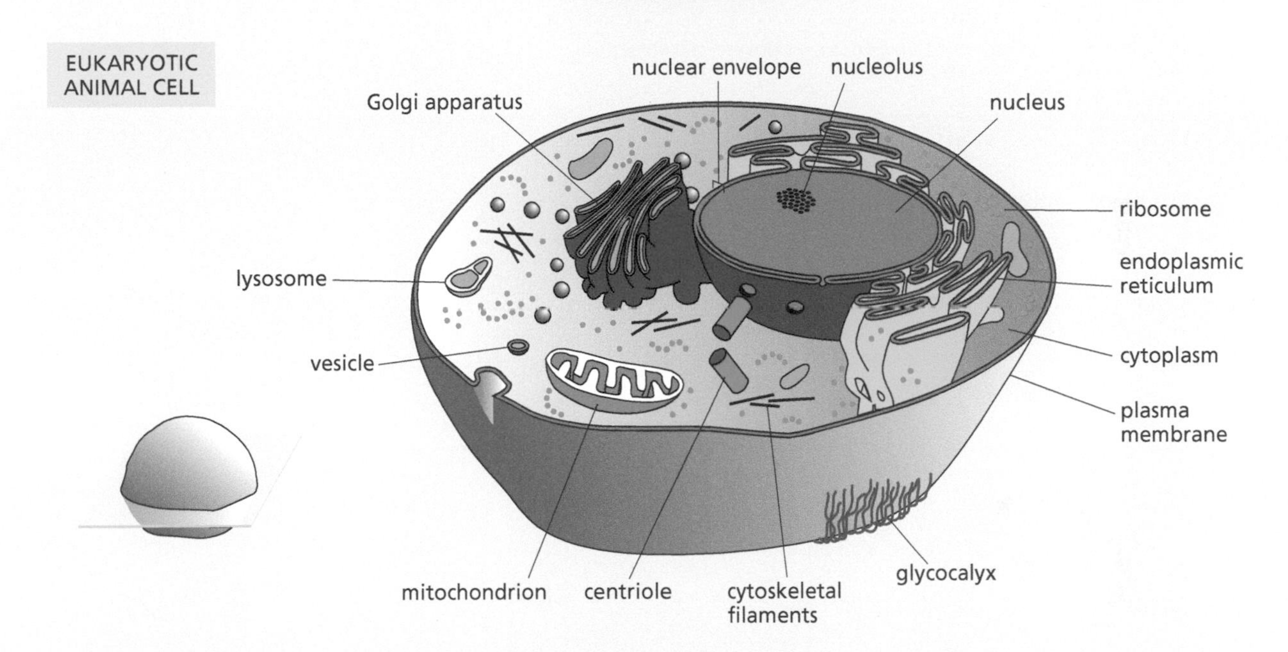

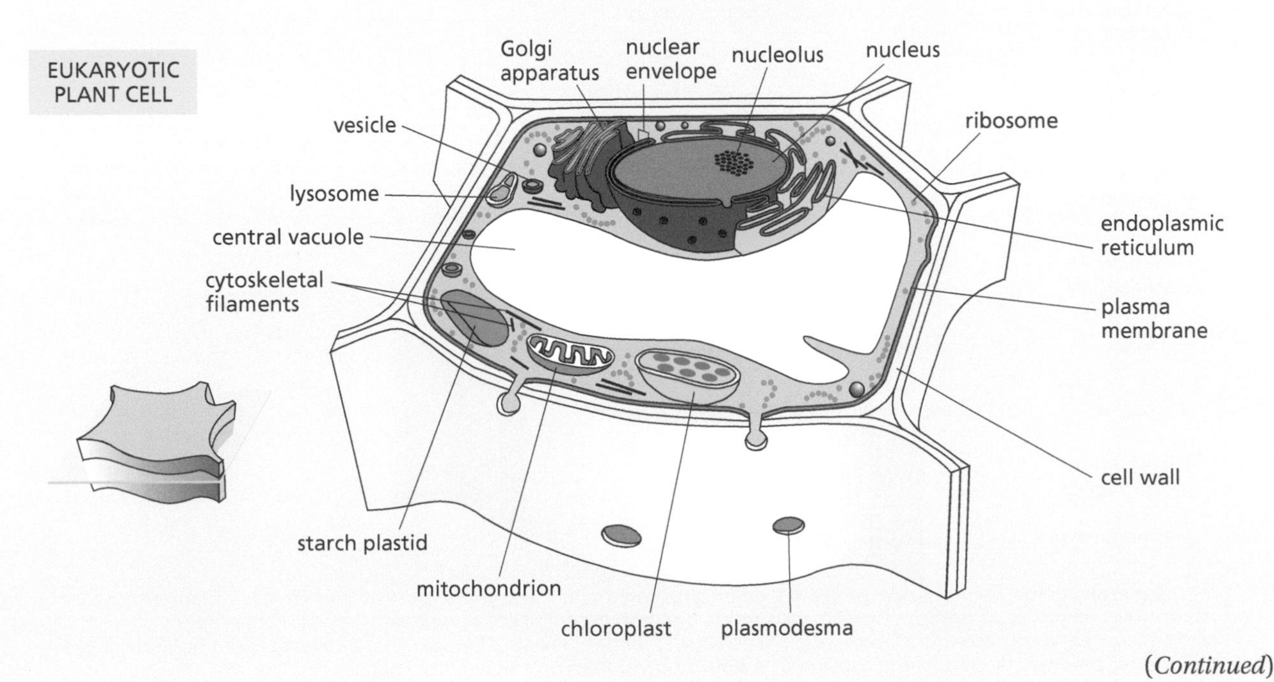

(*Continued*)

Structure	Function	present in prokaryotic cells	present in eukaryotic cells	
			plant cells	animal cells
plasma membrane	protection; communication; regulates passage of materials	✓	✓	✓
DNA	contains genetic information	✓	✓	✓
nuclear envelope	surrounds genetic material		✓	✓
several linear chromosomes	contain genes that govern cell structure and activity		✓	✓
cytoplasm	gel-like interior of cell	✓	✓	✓
cytoskeleton	aids in cell and organelle movement and in maintaining cell shape	*	✓	✓
endoplasmic reticulum	transport and processing of many proteins		✓	✓
Golgi apparatus	adds sugar group to proteins and packages them into vesicles		✓	✓
ribosomes	protein synthesis (translation) along mRNA	✓	✓	✓
lysosomes	contain enzymes; aid in cell digestion; have a role in programmed cell death		✓	✓
mitochondria	provide cellular energy		✓	✓
chloroplasts and other plastids	capture sunlight; produce energy for cell		✓	
central vacuole	maintains cell shape; stores materials and water	**	✓	
flagella (whiplike appendages)	cell movement	simple propeller type	complex undulating type	complex undulating type
cilia (hairlike appendages)	cell movement; present only in certain types of cells			✓
cell wall	protects cell; maintains cell shape	✓	✓	
glycocalyx	surrounds and protects cell			✓
pili (hairlike appendages)	mating; adherence	✓		
plasmodesmata	cell-to-cell communication		✓	

*Prokaryotes do not have exactly the same type of cytoskeleton as eukaryotic cells, but they do have filamentous structures composed of proteins homologous to the building blocks of eukaryotic cytoskeletal filaments.
**Some prokaryotes, the sulfur bacteria in particular, do contain a central vacuole that allows them to be much larger than most prokaryotic cells (see discussion of subdividing volumes on the next few pages).

Box 6.3 Cell Size is limited by the ratio of surface area to volume

Look at the cubes A and B shown here. What are their volumes? What are their surface areas? What is the ratio of surface area to volume for each? To calculate the volume of a, you must multiply three numbers: height times width times depth (a cubed function). To calculate surface area, you must multiply only two numbers—height times width (a squared function)—and then add up the products for each of the cube's faces. As you hopefully determined, the surface area/volume ratio for cube A is 6, but the surface area/volume ratio of cube B is only 2 because volume increases as a cubed function, while surface area increases as a squared function. Now, determine the volume/surface area ratio for cube C, which has the same overall size as cube B but is subdivided into 27 individual cubes, each with the same surface area as cube A. Subdivided cube C has the same volume as cube B, but the same surface area/volume ratio as cube A. Subdividing an organism into cells achieves the same effect and ensures that sufficient surface area is maintained to allow for efficient exchange of materials between the cell and its surroundings.

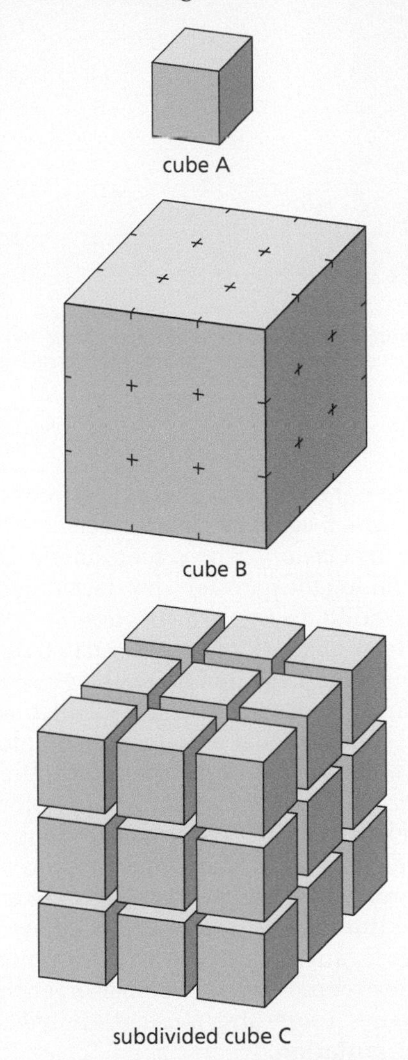

6.2.2 Specialization

Even unicellular organisms sometimes form multicellular "communities" or "aggregates" under certain environmental conditions (Figure 6.2). Examples include bacterial **biofilms** that can form on many different surfaces, including rocks in natural bodies of water, on medical devices (such as catheters), and on your teeth (as plaque). In these communities, individual cells develop different morphological features (shapes) and functions, and the different types of cells are often separated from each other in layered structures. The differences between cell types in such a community are caused by distinct patterns of gene expression (see Section 3.1), leading to different sets of proteins being present in each cell type. An advantage for both multicellular aggregates and multicellular organisms

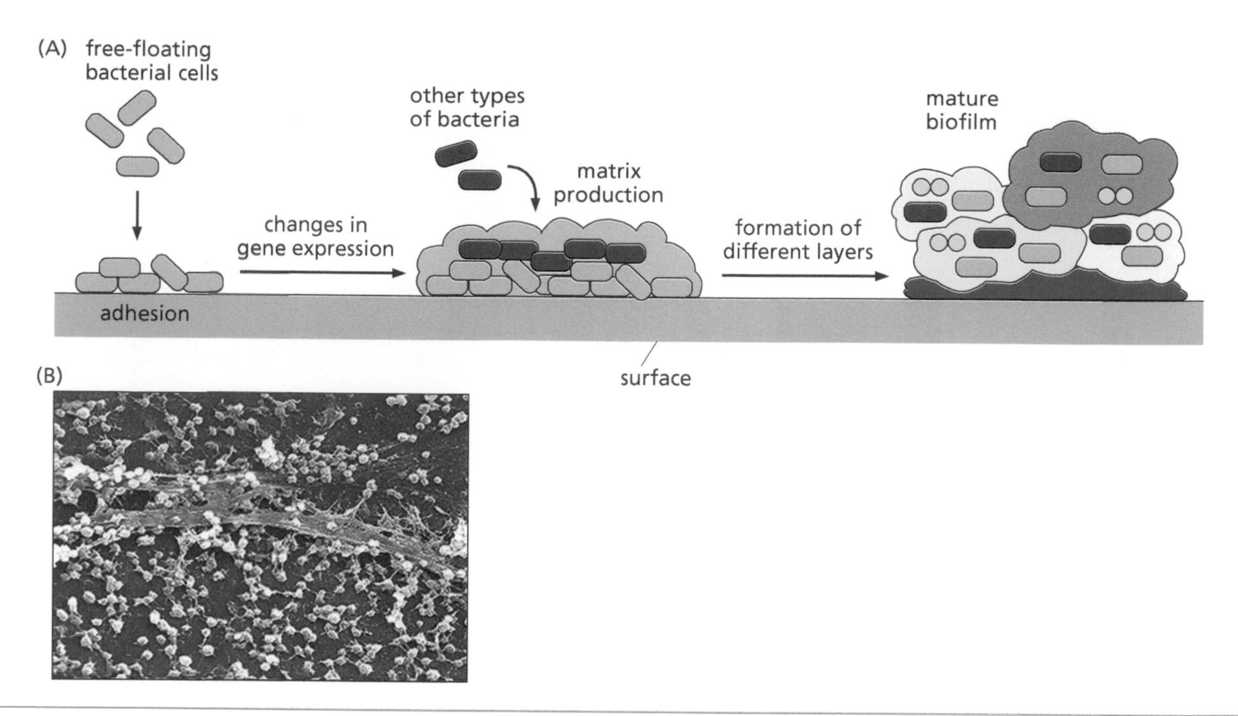

Figure 6.2 **Bacteria can sometimes form aggregates where different cells have specialized features and functions. (A)** Formation of a biofilm containing multiple bacterial species in distinct functional layers. **(B)** Scanning electron microscope image of a mature biofilm.

is that not every cell needs to perform every function. This allows for specialization, which again increases efficiency. This is similar to the concept of division of labor in a factory, where, for example, certain employees may be highly skilled in machining the parts for a product, others in assembling the product, and still others in adding decorative touches.

In sponges (kingdom Animalia, phylum Porifera), cells are specialized but are not organized into tissues (see also Figure 6.15). The cells of all other animals do form **tissues**, which are distinct structures consisting of similar cells and their products that function together to execute a particular biological process. Thus, tissues are both *structurally* and *functionally* integrated. The inner and outer cell layers of animals in the phylum Cnidaria, also called coelenterates, which include jellyfish, sea anemones, and corals (see also Figure 6.15), are separate tissues with different functions. Although the cells in the two layers are different, each cell is still changeable, so that the Cnidaria are able to regenerate an entire organism from a small piece. The cellular flexibility of Cnidaria (and other organisms capable of regeneration from parts) contrasts with the situation in more complex multicellular organisms. In these organisms, the fate of a cell becomes more and more restricted as it divides, through a process called **differentiation**.

6.2.3 Prokaryotic cells

The first organisms to evolve were simple cells with no internal compartments and thus no nucleus. A membrane called the **plasma membrane** formed an outer boundary of the cell and kept its contents inside. Prokaryotic cells lack most of the complex internal structures possessed by more advanced (eukaryotic) cells. Prokaryotic cells have a single chromosome, consisting of a DNA double helix joined end to end to form a circle, resembling a closed necklace. Prokaryotic chromosomes are not organized into chromatin, unlike the eukaryotic chromosomes to be described later; they are, however, associated with DNA-binding proteins that introduce and maintain coiled structures ("supercoiling") within the chromosome so that it is sufficiently compact to fit inside the cell. The region of the cell containing the prokaryotic chromosome is not surrounded by a membrane or set apart from the rest of the cell in any other way. Many prokaryotes also have smaller, circular DNA molecules, called **plasmids**, which can be replicated and passed on when a prokaryotic cell divides and can also sometimes become inserted into the chromosome (Section 4.1.1).

Bacteria are prokaryotic cells, and a majority of prokaryotes are bacteria. Two other groups of organisms that are prokaryotic cells are the blue-green photosynthetic organisms (Cyanobacteria) and the Archaea. Multiple lines of evidence suggest that many of the traits of the Archaea are the original ones from which the contrasting traits of other organisms likely evolved; in this sense, the Archaea are considered to be the most primitive of all organisms.

6.2.4 Eukaryotic cells

Plants, animals, fungi, and protists are examples of organisms composed of eukaryotic cells (Box 6.2). Some eukaryotic organisms are single-celled (unicellular), while others are multicellular; in contrast, all prokaryotic organisms are unicellular. Eukaryotic cells have various internal parts, called organelles, that are bounded by intracellular membranes separating the various functions of the cell into different compartments. A defining characteristic of eukaryotic cells is the presence of a **nucleus** surrounded by a double membrane called the **nuclear envelope** (the name *eukaryotic* means "true nucleus"). The nucleus contains linear chromosomes composed of DNA and proteins called histones that wrap the DNA into a regular, repeating structure called **chromatin**. The compaction of DNA into chromatin allows all of a eukaryotic cell's DNA to fit into the nucleus, just as the supercoiling of prokaryotic chromosomes allows them to fit inside a prokaryotic cell. Many types of cell organelles are shown and explained in Box 6.2.

As described earlier, the organelles of eukaryotic cells allow for the compartmentalization of specific cellular processes, which generates greater efficiency (division of labor) than if these processes were all carried out in the same space. Two eukaryotic organelles, mitochondria and chloroplasts, are devoted to energy metabolism; the evolution of these organelles is discussed later in this chapter, and their metabolic functions are described in Chapters 8 and 17. A number of the other organelles in eukaryotic cells are organized into a network called the "endomembrane system." These organelles include the endoplasmic reticulum (ER), the Golgi apparatus (also referred to as the Golgi complex), and lysosomes. Cellular materials are transported between these organelles in small, membrane-bound sacs called **vesicles**. Proteins that are secreted (expelled) from cells to the outside or proteins that are embedded in the outer membrane (plasma membrane) of cells are synthesized by ribosomes (protein factories; see Section 3.1.3) associated with the ER; these proteins move in vesicles from the ER to the Golgi apparatus and then to secretory vesicles that eventually fuse with the plasma membrane, delivering their contents to the cell exterior or cell surface. Along the way, these proteins are modified ("accessorized") by the addition of sugar molecules in chemical reactions catalyzed by enzymes that reside in the ER and Golgi compartments. Lysosomes are another type of organelle integrated into the endomembrane system. They contain enzymes that break down cellular molecules and even whole organelles to execute a cellular recycling program. The enzymes that function in lysosomes are made in the ER and trafficked through the Golgi apparatus to the lysosomes in vesicles; vesicles formed from the plasma membrane can also move materials from outside the cell or from the cell surface to lysosomes for recycling.

In addition to their membrane-bound organelles, eukaryotic cells also have an extensive, well-characterized internal network of protein filaments called a **cytoskeleton**. These filaments determine the shape of the cell and keep many organelles in their positions. Contraction of these filaments can help to move the whole cell. In addition, components of the cytoskeleton are essential for eukaryotic cell reproduction (see Section 2.2), forming the spindle structure that separates chromosomes and also executing cytokinesis, the division of one cell into two.

For many years, it was thought that prokaryotic cells lacked a cytoskeleton, but, in 1992, three laboratories simultaneously discovered that prokaryotes do have filamentous structures made of proteins homologous to those that form the eukaryotic cytoskeletal filaments. However, these prokaryotic filaments are still less well characterized than their eukaryotic counterparts.

Box 6.2 illustrates the generalized structures of bacterial, animal, and plant cells, emphasizing their similarities and differences. Keep in mind, however, that there are a vast number of variations on these cell types, both among species and among the various specialized cells within multicellular organisms; there is probably no actual cell that exactly matches the accompanying diagrams.

THOUGHT QUESTIONS

1. When the weight of some material in solution remains the same, how do changes in cell volume influence the concentration of the material?
2. How could multicellularity in animals or plants have evolved through natural selection?
3. The ratio of surface area to volume places a size restriction on all cells, yet most eukaryotic cells are significantly larger (approximately 10-fold) than most prokaryotic cells; how can eukaryotic cells still retain efficiency in their cellular processes?
4. Of what type of cells are the following organisms composed? Draw a typical cell for each organism and name its major distinguishing characteristics.
   a. *Bacillus anthracis*, the bacteria responsible for causing anthrax in humans
   b. Elephant
   c. Venus fly trap

6.3 THE TWO DOMAINS OF PROKARYOTIC ORGANISMS

The largest taxa of all have traditionally been called kingdoms. The animal and plant kingdoms have been recognized for centuries, and early classifications of organisms recognized only these two kingdoms. Plants were distinguished as being nonmotile organisms with rigid cell walls, capable of using sunlight as a source of energy. Animals, in contrast, were recognized for their ability to move, their lack of cell walls, and their inability to derive energy directly from sunlight. This two-kingdom classification continued in use despite the discovery of animals that do not move and other organisms, such as bacteria or fungi, that do not fit well into either the plant or the animal kingdom.

Advances in our knowledge made possible by electron microscopy, particularly by the discovery of the profound structural differences between prokaryotic and eukaryotic cells, led to major classification changes. Prokaryotic organisms include the Archaea, the Bacteria, and the Cyanobacteria; all other organisms are eukaryotic. A five-kingdom classification system reflecting this difference was first proposed in 1963 and became widely followed, sometimes with a modification to include a sixth kingdom for the Archaea.

Organisms did not change, but our arrangement of them changed because classifications are socially constructed, devised by humans, and agreed upon as a matter of social convention. To say that a classification scheme is socially constructed does not mean that the process of establishing a classification scheme is arbitrary or that all schemes are equally valid. Classifications are now usually understood as hypotheses about how organisms are related by patterns of descent (see Figure 5.1). As more and more knowledge is being accumulated about organisms, that knowledge is used to test the hypotheses and to replace rejected hypotheses with new ones. The division of living things into five or six kingdoms was simply a widely accepted theory, not an unchanging set of facts.

Studies of DNA sequences have shown that the Bacteria (including the Cyanobacteria) have DNA sequences that differ as much or more from the Archaea as they differ from various eukaryotes. Accordingly, many biologists now sort all

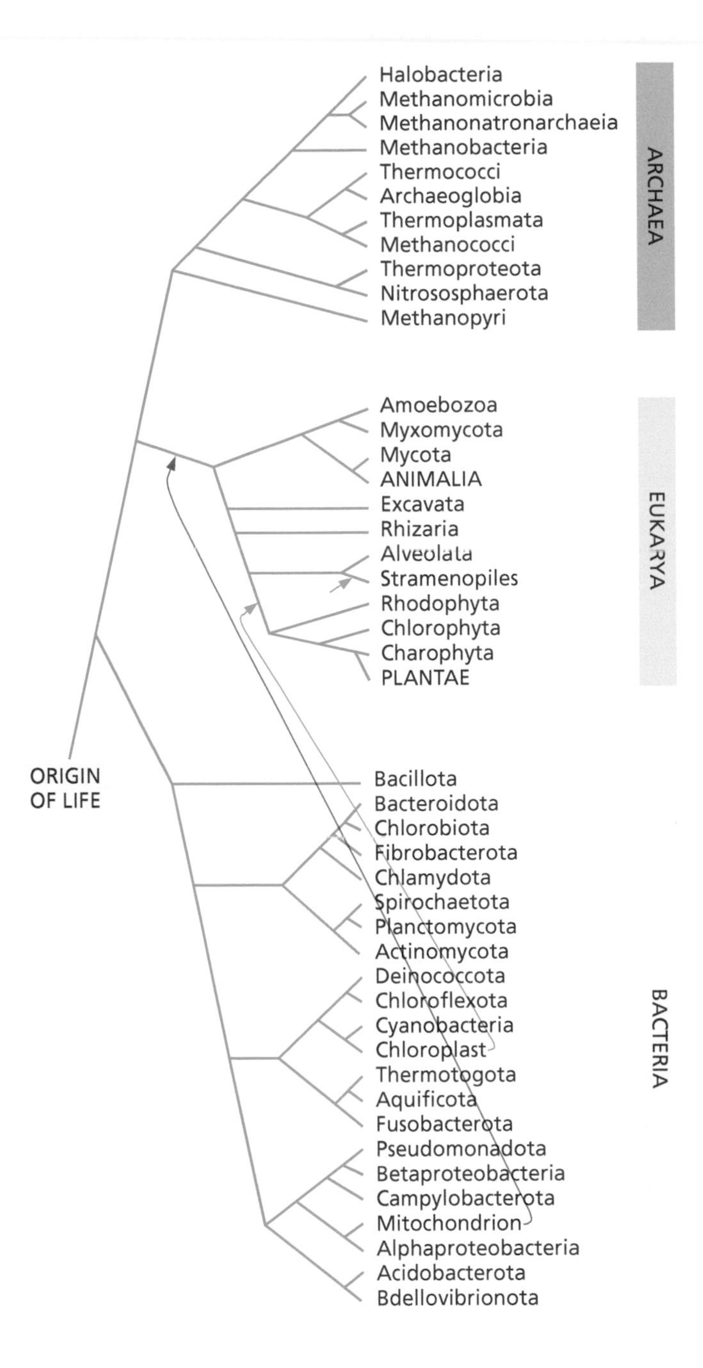

Figure 6.3 A family tree showing the arrangement of organisms into three domains, based on Ciccarelli et al. (2006), Yarza et al. (2008), and Parks et al. (2018), with names of bacterial phyla after Oren and Garrity (2021) wherever possible.

organisms into three "domains": one for the Archaea, one for the Bacteria, and a third for all organisms with eukaryotic cells—the Eukarya (Figure 6.3).

A classification of organisms as now recognized by biologists is given in Boxes 6.4 and 6.5, while the family trees of Figures 6.3 and 6.6 show current thinking about how these major groups are related. We now proceed to examine these several groups, as distinguished from one another by details of cell structure, development, nutrition, and overall morphology.

6.3.1 Domain and kingdom Archaea

The domain Archaea contains a single kingdom, called either Archaea or Archaebacteria. The Archaea are one of the two prokaryotic kingdoms and are characterized by the ability to live only in very special environments. Some archaea, the methane producers, live in oxygen-free environments, such as inside cows' guts, where, in the course of consuming energy, they perform chemical reactions that produce methane gas (CH_4) as a by-product. Others live in extreme environments

Box 6.4 A brief classification of organisms

A large group of scientists maintains a much more detailed classification, the "Tree Of Life," at tolweb.org

I. DOMAIN AND KINGDOM ARCHAEA

Phylum ARCHAEA or ARCHAEBACTERIA. A small group of simple, bacteria-like organisms with prokaryotic cells. Nucleic acid sequences set these organisms sharply apart from the other prokaryotes.

Methanogens: Methane-producing organisms.
Extreme Thermophiles: Organisms adapted to living in hot springs and similar environments, with high optimal growing temperatures.
Extreme Halophiles: Organisms adapted to extremely salty environments.

II. DOMAIN AND KINGDOM BACTERIA OR EUBACTERIA

This group includes the vast majority of prokaryotic organisms, with nucleic acid sequences showing homology (similarity by common descent) among the bacteria, but not to those prokaryotic organisms placed in the Archaea.

Phylum Proteobacteria: The vast majority of bacteria, lacking the characteristics of the next few groups. Subgroups are currently called alpha, beta, gamma, delta, and epsilon.
Phylum Chlamydia: Energy parasites that cannot generate their own ATP and can only survive inside animal cells.
Phylum Spirocheta: Motile, corkscrew-shaped bacteria.
Unnamed Phylum: *Gram-Positive Bacteria*: A possibly heterogeneous grouping of bacteria with a distinctive type of cell wall (a single membrane plus a very thick peptidoglycan layer).
Phylum Cyanobacteria: Prokaryotic organisms structurally similar to other bacteria, but possessing chlorophyll *a* and releasing oxygen during photosynthesis.

III. DOMAIN EUKARYA

Eukaryotic organisms with cells subdivided by membranes into multiple compartments, including true nuclei (surrounded by a double-membrane "nuclear envelope") and many different internal organelles that usually include mitochondria, endoplasmic reticulum, and contractile filaments.

Subdomain Unikonta: Eukaryotic organisms containing motile cells that move either by lobelike pseudopods or a posterior (trailing) flagellum.
- *Kingdom Amebozoa*: Single-celled eukaryotes that move by extending lobelike pseudopods.
- *Kingdom Myxomycota*: Slime molds, composed of motile, amoeboid cells that aggregate together into creeping colonies that practice absorptive, fungus-like nutrition and that reproduce using spores.
- *Kingdom Mycota (Fungi)*: Eukaryotic organisms possessing cell walls but no chloroplasts, usually reproducing by spores, and carrying out absorptive nutrition. Most fungi possess branched, threadlike absorptive filaments (hyphae).
- *Kingdom Choanoflagellata*: Single-celled organisms with a whiplike flagellum surrounded by a collar.
- *Kingdom Animalia*: The animals are shown separately in Box 6.5.

Subdomain Rhizaria: Single-celled, eukaryotic, planktonic predators that use needle-like pseudopods to trap and engulf their prey.
- *Radiolaria*: Rhizaria with beautiful spherical shells composed of silica.
- *Foraminifera*: Rhizaria that secrete spiral shells, usually made of calcium carbonate with numerous tiny pores.

Subdomain Excavata: Single-celled energy parasites with no mitochondria.

Subdomain Alveolata: Single-celled eukaryotes with fluid-filled bubbles beneath the cell membrane.
- *Ciliata*: Single-celled eukaryotes coated with numerous hairlike cilia that beat to create movement.
- *Apicomplexa*: Single-celled parasites that use a sharp-pointed "apical complex" to enter host cells.

Subdomain Stramenopiles: Photosynthetic eukaryotes containing cell walls composed of either silica or cellulose, chloroplasts with chlorophylls *a* and *c*, and two unequal flagella. This large group includes single-celled diatoms, single-celled dinoflagellates, and multicellular brown algae.

Subdomain Archaeplastida: A large group of photosynthetic eukaryotes containing plastids with chlorophyll *a* and either chlorophyll *b* or *d* (but not *c*) and cell walls that contain cellulose.
- *Kingdom Rhodophyta*: Red algae, with plastids containing chlorophylls *a* and *d*.
- *Kingdom Chlorophyta*: Green algae, with plastids containing chlorophylls *a* and *b*.
- *Kingdom Plantae*: Eukaryotic organisms possessing chloroplasts, with chlorophylls *a* and *b*, and cell walls containing cellulose. Fertilized eggs (zygotes) are surrounded by sterile, nonreproductive tissue.
  - *Bryophyta*: Nonvascular plants, including mosses and liverworts.
  - *Tracheophyta*: Vascular plants.

(*Continued*)

Divisions in which seeds are not present:
Psilophyta: Psilophytes.
Lepidophyta: Club mosses and other lepidophytes.
Arthophyta or Sphenopsida: Horsetails and their relatives.
Pterophyta: Ferns.
Divisions possessing seeds:
Pteridospermophyta: Seed ferns.
Cycadophyta: Cycads.
Ginkgophyta: Ginkgos.
Coniferophyta: Pines, spruces, and other cone-bearing plants (conifers).
Gnetophyta: *Gnetum*, *Ephedra*, and *Welwitschia*.
Angiospermae or Anthophyta: Flowering plants, the largest and most successful group.

once thought incompatible with life, such as areas with high salt concentrations, the edges of thermal vents in the ocean floor, or in hot springs. Scientists reasoned that adaptation for such extreme environments must require very different enzymes from those previously known, which proved to be so. One enzyme that can operate on DNA at very high temperatures is now used in the polymerase chain reaction, central to much work in biotechnology (see Chapters 3 and 4).

6.3.2 Domain and kingdom Bacteria

The domain Bacteria contains a great deal of diversity. There are a handful of cell shapes: spherical, rodlike, gently curved (banana-like), and even spiral (like a corkscrew) (**Figure 6.4**). There are also differences in the structure of the cell wall, as revealed by commonly used staining techniques such as the Gram stain. An even greater diversity exists at the biochemical level. Some bacteria can tolerate oxygen in their environment and others cannot; some bacteria can metabolize certain sugars and others cannot. The cyanobacteria can use cellular pigments to synthesize sugars with the aid of sunlight (a process called photosynthesis), while most bacteria cannot. Some bacteria can move with the use of a simple flagellum, while others cannot. Many bacteria are free-living (in soil, for example), but many other species live only in a narrow range of host species; for example, most bacteria that use dogs as hosts cannot live in humans.

Molecular genetics is now providing new information revealing that species diversity is immense within both the Bacteria and the Archaea. Although only

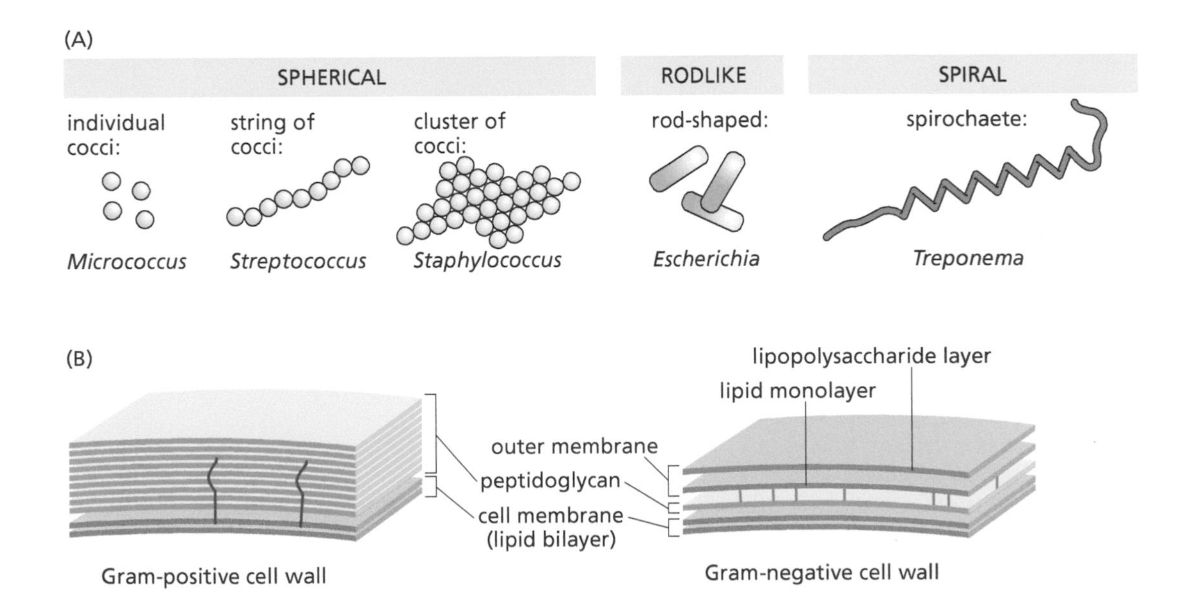

Figure 6.4 Bacterial diversity (A) in shape and (B) in structure of cell walls.

Box 6.5 A Brief classification of the animal kingdom

Kingdom Animalia: Multicellular eukaryotic organisms, usually motile, lacking chloroplasts and developing from a hollow ball of cells (blastula).

Phylum Porifera: Sponges. Animals possessing specialized cells but no organized tissues.
Phylum Mesozoa: A small group of species containing organisms with only a few cells each, without organized tissues.
Phylum Cnidaria: The coelenterates or cnidarians, a diverse group possessing stinging cells and including corals, anemones, hydroids, and jellyfish. Two tissue layers (endoderm and ectoderm) are present.
Phylum Ctenophora: "Comb jellies," animals with a "biradial" symmetry, like that of a two-armed pinwheel.
Phylum Platyhelminthes: Flatworms, including planarias, parasitic flukes, and parasitic tapeworms. Tissues formed by three germ layers (ectoderm, mesoderm, endoderm) but no body cavity present.
Phylum Rhynchocoela or Nemertea: Ribbon worms; similar to flatworms but with a protrusible head structure (proboscis).
Lophotrochozoa: Includes the next 10 phyla:
Phylum Rotifera: Microscopic aquatic organisms containing a ring of cilia that beat in a circular pattern resembling a wheel.
Phylum Acanthocephala: A group of parasitic worms with hook-studded heads.
Phylum Entoprocta: A small group of filter-feeders (animals that strain small particles from the water).
Phylum Phoronida: A small group of filter-feeding wormlike animals.
Phylum Bryozoa (*Ectoprocta*): Small filter-feeding "moss animals," usually living in colonies.
Phylum Brachiopoda: Filter-feeding animals with a shell composed of two unequal parts and a stalk that attaches the adults to a fixed location.
Phylum Mollusca: A large and diverse group of animals possessing a cavity lined with a layer of cells (called a mantle) that usually secretes some kind of an inflexible shell of calcium carbonate. Includes snails, clams, octopus, squid, and related species.
Class Monoplacophora: Primitive mollusks with dome-shaped shells.
Class Polyplacophora: Chitons.
Class Gastropoda: Snails and slugs.
Class Bivalvia: Clams.
Class Scaphopoda: Tusk shells.
Class Cephalopoda: Nautiloids, octopus, squids, etc.
Phylum Annelida: Segmented worms, including earthworms and sandworms.
Phylum Sipunculida: Peanut-shaped worms.
Phylum Echiurida: A small wormlike group.
Ecdysozoa: Includes the next eight phyla:
Phylum Nematoda: Roundworms, an abundant group including both free-living and parasitic species.
Phylum Kinorhyncha: Parasitic worms related to roundworms.
Phylum Gastrotricha: Small wormlike animals related to roundworms.
Phylum Gordiacea: Elongated horsehair worms.
Phylum Pentastomida: Small parasitic worms whose head and claws give the appearance of a five-branched head.
Phylum Tardigrada: Tiny aquatic "water bears," segmented animals with clawed appendages.
Phylum Onychophora: Segmented, wormlike organisms with clawed appendages.
Phylum Arthropoda: Animals with jointed legs, protected by an external skeleton that permits hinge-like movements between its more rigid parts. The largest phylum of all.
Class Trilobita: Extinct trilobites.
Class Crustacea: Lobsters, crabs, shrimp, barnacles, etc.
Class Merostomata: Horseshoe crabs.
Class Pycnogonida: Sea spiders.
Class Arachnida: Scorpions, spiders, mites, etc.
Class Chilopoda: Centipedes.
Class Diplopoda: Millipedes.
Class Insecta (*Hexapoda*): Insects.
Deuterostomia includes the next four phyla:
Phylum Chaetognatha: Arrow worms.
Phylum Echinodermata: Crinoids, starfishes, sea urchins, etc., possessing a water-vascular system, numerous tubelike feet, and in most cases a five-part radial symmetry in the adult stage.
Phylum Hemichordata: Acorn worms, pterobranchs, and graptolites, related to the Chordata but not sharing all chordate characteristics.

(*Continued*)

Phylum Chordata: Animals with a notochord (a stiff, flexible, rodlike structure), a dorsal hollow nerve cord, and gill slits, each at some stage of life.
- *Subphylum Urochordata*: Sea squirts (tunicates), salps, and their relatives, with actively swimming larval stages and generally with nonmotile filter-feeding adults.
- *Subphylum Cephalochordata*: Sea lancets such as amphioxus, with motile filter-feeding adult stages.
- *Subphylum Vertebrata*: Animals with a backbone. See also Box 6.1.
  - *Class Agnatha*: Jawless fishes.
  - *Class Placodermi*: Extinct, armored fishes.
  - *Class Chondrichthyes*: Cartilage fishes (sharks, etc.).
  - *Class Osteichthyes*: Bony fishes (the vast majority of fishes).
  - *Class Amphibia*: Frogs, salamanders, and other amphibians.
  - *Class Sauropsida*: Reptiles and birds.
  - *Class Mammalia*: Mammals.

4500 prokaryotic species have thus far been characterized and described, it is now estimated that millions exist.

We often think of bacteria as the "germs" that cause disease, as many of them do. However, a far greater number of bacterial species are beneficial to humans, to other organisms, or to entire ecosystems. For example, bacteria decompose dead material into chemical forms that other organisms can then use to sustain life. Without such decomposition by bacteria, all other forms of life would soon cease. Certain bacteria (and a few cyanobacteria) are also important in the reactions of the nitrogen cycle (see Section 17.3.2), which also sustains nearly all other species on Earth. Biotechnology uses bacterial plasmids and bacterial enzymes, and many industrial processes, including the making of cheese, yogurt, and sauerkraut, depend on chemical reactions performed by bacteria.

All Bacteria are single-celled organisms, but some grow in colonies or films and filaments of many individual cells attached to a substrate or to one another. Some of these colonies have characteristics that differ from characteristics of single cells of the same species (Figure 6.4); they thus exhibit what may have been the first step in the evolution of multicellularity.

6.4 EVOLUTION AND CLASSIFICATON OF THE DOMAIN EUKARYA

All the remaining organisms have eukaryotic cells and are therefore placed in the domain Eukarya. The earliest eukaryotes, sometimes called **protists**, remained small and in most cases unicellular. Multicellularity evolved independently several times among the Eukarya. One group of eukaryotes, the fungi (Mycota), developed an absorptive type of nutrition in which nutrients are directly absorbed into cells without a specialized structure. Although some fungi remained single-celled, most are now multicellular and carry out their absorptive nutrition with the aid of thin, absorptive filaments (hyphae). Several other groups of eukaryotic organisms developed photosynthetic abilities; the most successful among these became plants (kingdom Plantae). Multicellular embryos (protective structures for the egg cells) evolved among land plants, and further evolutionary advances, described later, included vascular tissues, seeds, and flowers. In yet another large group of eukaryotes, motility became increasingly developed, and multicellular animals (kingdom Animalia) evolved from this group. Many important innovations within the animal kingdom evolved later on, including bilateral symmetry, body cavities, segmented body plans, and, in our own phylum, backbones.

6.4.1 Endosymbiosis and the evolution of eukaryotes

In Section 5.5, we outlined some of the evidence for the origin of life under hydrogen-rich (reducing) conditions. Chemical fossils left by the earliest organisms provide evidence for the presence of chlorophyll (and thus photosynthesis) during the time when the atmosphere changed to its present oxygen-rich composition.

Simple structural fossils of the earliest organisms, between 4.0 and 4.2 billion ($4.0–4.2 \times 10^9$) years old, show that they were single cells about the size of modern prokaryotic cells, or roughly one-tenth the size of eukaryotic cells. Larger cells, comparable in size to modern eukaryotic cells, do not appear in the fossil record until about 1.5 billion years ago. But how did eukaryotic cells originate?

6.4.1.1 From prokaryotes to eukaryotes According to a theory first championed by the American cell biologist Lynn Margulis in the 1970s, eukaryotic cells arose from prokaryotic cells by a process called **endosymbiosis** (literally meaning, "living together inside"). According to this theory, large prokaryotic cells incapable of performing certain energy-producing chemical reactions (those of the Krebs cycle, see Section 8.3.3) engulfed smaller prokaryotic cells able to carry out these reactions. The larger (host) cells, which many biologists now believe were archaebacteria, could obtain energy by digesting the smaller cells, but they could obtain even more energy if they allowed the smaller cells to go on living inside them and then used the products of the energy-producing reactions. In this situation, host cells that allowed the smaller cells to persist were favored by natural selection over host cells that digested the smaller cells. Over time, the smaller cells became energy-producing cellular organelles called **mitochondria** (Figure 6.5).

Eukaryotic cells that are capable of photosynthesis have additional organelles called **chloroplasts**. The pigments that carry out photosynthesis are contained within these organelles. The chloroplasts are believed to have evolved by a process similar to the process just described for mitochondria. In this case, however, the smaller cells were cyanobacteria capable of photosynthesis. The larger cells achieved greater growth potential by harboring these smaller cells rather than digesting them. The large cells containing photosynthetic organelles did better and reproduced in greater numbers than similar cells that had none, and cells with these organelles persisted while many without chloroplasts died out. Many experts now think that the symbiotic capture of chloroplasts occurred independently several times.

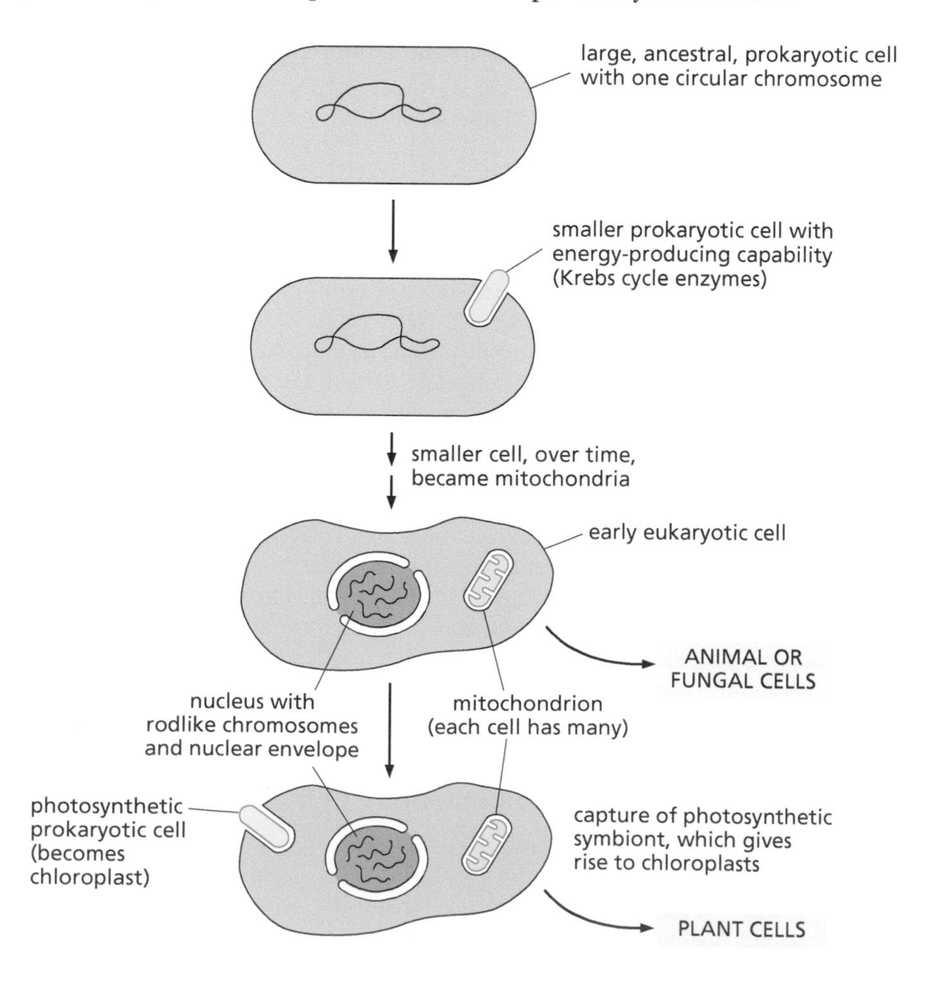

Figure 6.5 The origin of eukaryotic cells according to the widely accepted theory of endosymbiosis.

There is substantial evidence to support the theory of endosymbiosis. This evidence includes the fact that both chloroplasts and mitochondria have distinct types of membranes and their own DNA. The DNA of mitochondria and chloroplasts is separate and different from the DNA of the host cell that contains them, but similar to the DNA of prokaryotic organisms; the organelle membranes also share features with prokaryotic membranes that differ from eukaryotic membranes. In modern-day organisms, some of the proteins that function in mitochondria and chloroplasts are encoded by nuclear genes, but others are produced by transcription of the organelles' own DNA and translation by ribosomes located within the organelles; the details of transcription and translation in mitochondria and chloroplasts more closely resemble the prokaryotic versions of those processes than the eukaryotic ones. Finally, mitochondria and chloroplasts divide to produce more of themselves in a process called "binary fission," which is also the way prokaryotic cells reproduce.

6.4.1.2 Eukaryotic diversity As we have seen, evolutionary change occurs when natural selection acts on genetic diversity. Species evolve when new reproductive isolating mechanisms arise. Eukaryotic organisms such as plants and animals have speciated very often, so that most of the known species alive today are eukaryotic. In this section we examine the evolutionary advances that have taken place among eukaryotic organisms.

A number of important evolutionary advances were key to the success of eukaryotic organisms. Of these, **sexual reproduction** and multicellularity evolved early, and perhaps repeatedly, among organisms whose bodies were still small and simple. The vast majority of eukaryotic organisms reproduce sexually, meaning that new individuals are formed only after a process of genetic recombination (see Chapter 2). In most cases, this means that new individuals derive their genes from two parents. Sexual recombination may have initially evolved because it was a simple yet efficient way of generating new gene combinations. If an organism produces a thousand offspring by **asexual reproduction**, all of them will, in the absence of a mutation, be genetically identical to the parent. Even if a mutation occurs in one or a few genes, the offspring will still be similar to the parents in all other genes. Sexual recombination changes this. If an organism produces a thousand offspring through sexual recombination, they will all differ genetically from the parent and also among themselves. Given the uncertainties of future conditions, the chances of a few offspring having better (i.e., more highly adapted) gene combinations than the parent are greatly increased if reproduction is sexual. Also, among the many offspring, more than one favorable combination of genes may arise, and these may eventually result in different kinds of organisms (i.e., different species). Thus, in eukaryotic organisms, sexual recombination supplies a mechanism to generate increased diversity, and it is therefore not surprising that there are so many different species of sexually reproducing eukaryotic organisms in the world today.

Multicellularity is thought to have evolved more than once independently, giving rise to different lineages within the domain Eukarya. One route by which multicellularity likely evolved was by way of colonial organization. Single-celled organisms could in some cases maintain hospitable conditions better if they clumped together into small colonies. The aggregation into colonies would reduce the surface area of each cell exposed to the outside environment because much of its surface would be in contact with its neighbors. The reduced surface area would aid in maintaining homeostasis. At first, all the cells in such a colony would be the same, and each would carry out all life processes, as before. As colonies grew larger, however, some interior cells would begin to lose contact with the outside environment altogether, and this could initiate a simple division of labor between surface cells and interior cells. The interior cells would now be in a more protected position, but they could no longer meet their own needs without having nutrients supplied to them or wastes taken away by other cells. The surface cells would help supply nutrients to the interior cells and remove their wastes. In addition, certain functions (such as feeding or defense) could in most cases be carried out more efficiently by surface cells, while other functions (such as reproduction) could often be carried out better by the interior cells. Over time, the functions of cells in different parts of the organism would become increasingly different from one another, thus increasing the complexity of the organism in most cases. Bacterial biofilms, mentioned earlier in this chapter, mimic this process to some extent, but they are not true multicellular organisms because each cell can also live independently, which is not the case for cells from a multicellular organism.

6.4.2 Classification of eukaryotic organisms is now based on nucleic acid sequences

The earliest eukaryotes were simple, aquatic, single-celled organisms. Among the early eukaryotes, mechanisms evolved to ensure that, when cells divided, all their chromosomes would be present in the offspring cells. Mitosis and meiosis (see Section 2.2) first evolved among early eukaryotes, as did sexual recombination. By having haploid gametes that joined during fertilization to produce diploid fertilized eggs (**zygotes**), the eukaryotes became able to generate new genetic combinations, thus producing great genetic variation in every generation. Because variation is the raw material on which natural selection works, sexual recombination increased the rate of evolution among eukaryotes.

The early eukaryotes and their immediate descendants were formerly placed in a kingdom Protista, but most experts now view this assemblage as unnatural and heterogeneous instead of representing a monophyletic group or clade. Partly as a consequence of the Human Genome Project discussed in Section 4.3, nucleic acid sequences of many one-celled eukaryotes became known, beginning with the budding yeast *Saccharomyces cerevisiae* in 1996. Lengthy portions of these genetic sequences are shared among different eukaryotic species, and these shared resemblances are used to indicate close relationships. Species that share nucleic acid sequences are hypothesized to have derived such shared sequences from a common ancestor, with greater proportions of a shared sequence indicating a more recent common ancestor. Using these assumptions, scientists have used shared DNA and RNA sequences to construct a new family tree of the Eukarya (**Figure 6.6**) and to reclassify the Eukarya into the following major groups.

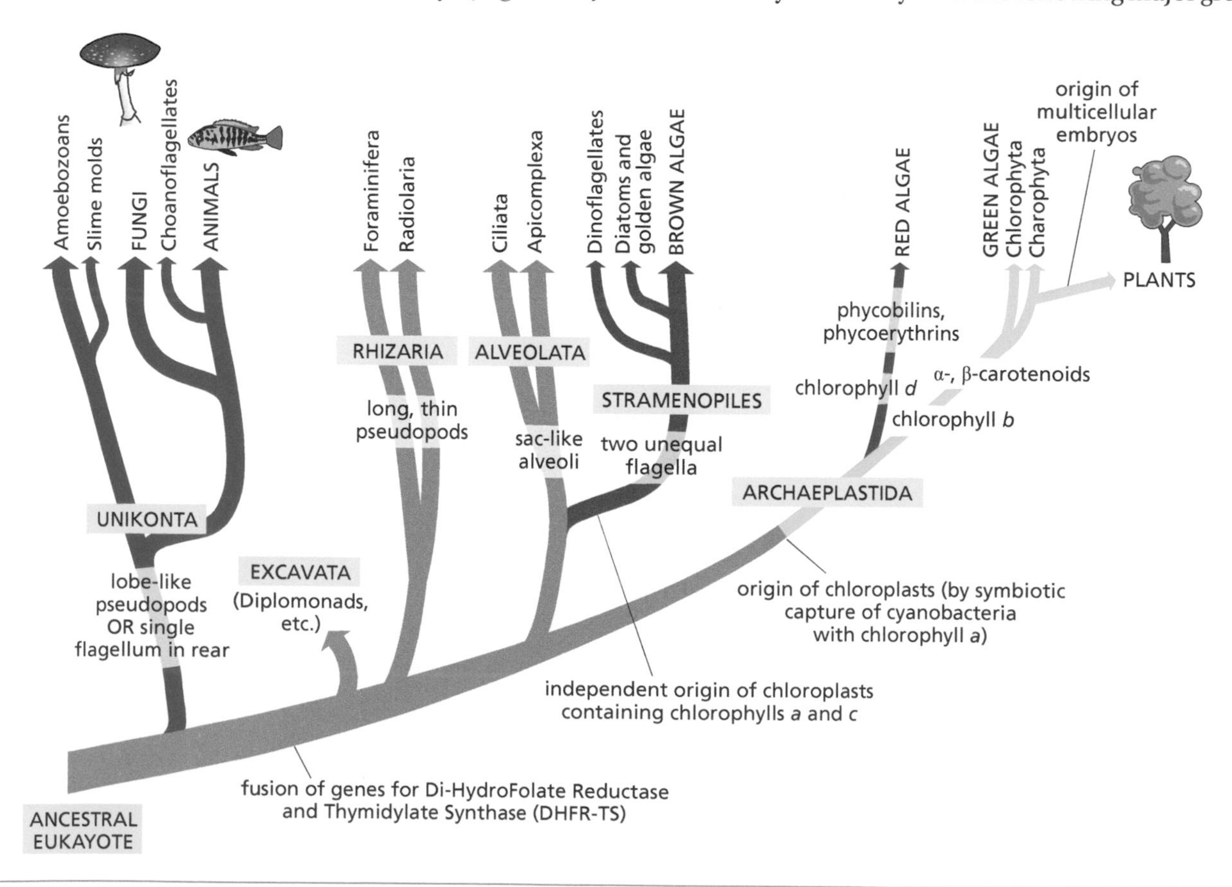

Figure 6.6 A family tree of Eukarya as currently conceived.

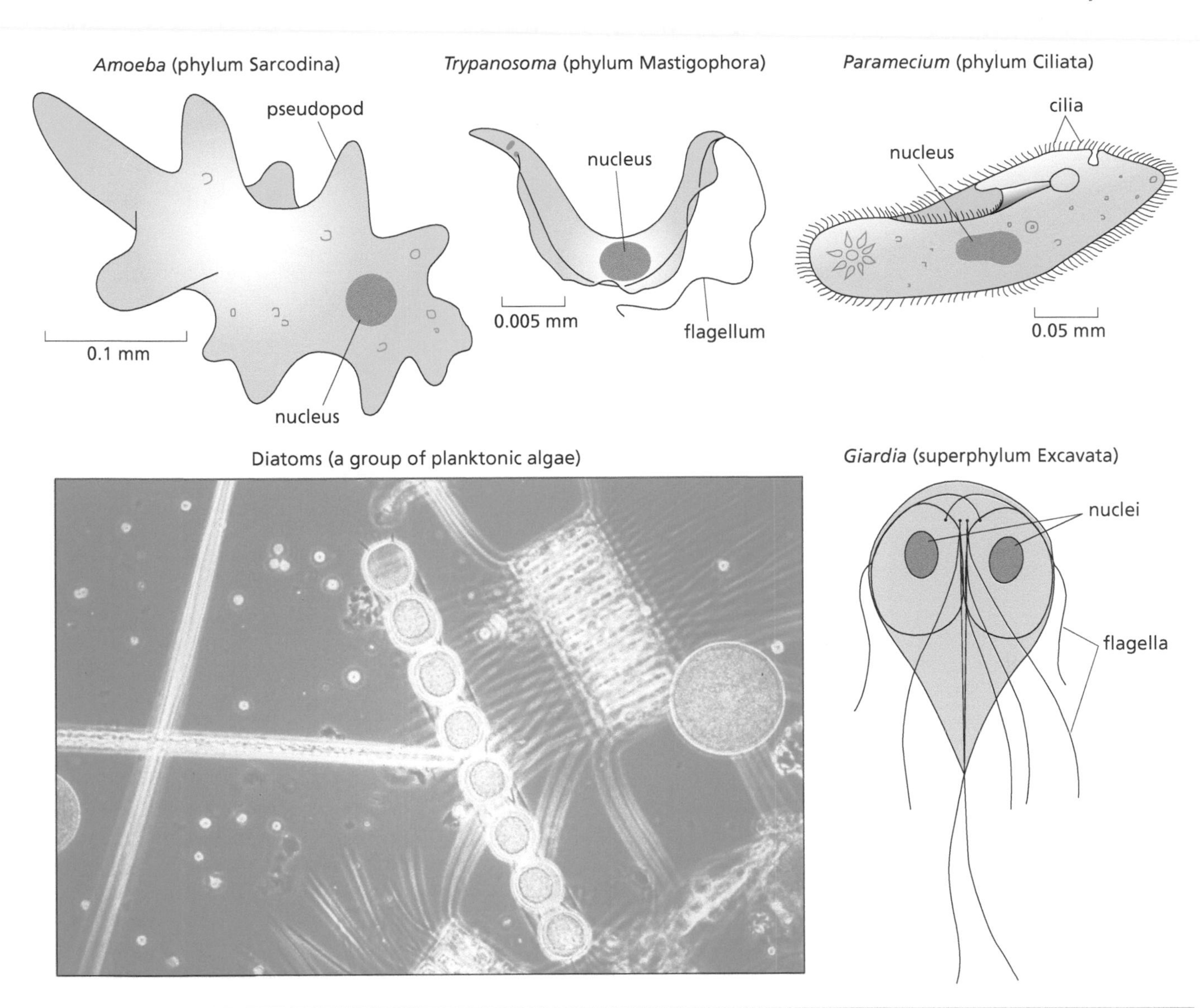

Figure 6.7 Representative single-celled eukaryotes.

6.4.2.1 Unikonta The early eukaryotes were single-celled (Figure 6.7). Among these organisms, there must have been a great selective advantage in being able to move from place to place to find food or to escape from unfavorable conditions, and several different mechanisms for motility (movement) evolved. The earliest protists had contractile protein filaments that allowed the cells to change shape and creep through their surroundings. One large group, the Amoebozoa (containing *Amoeba* and its relatives), can change their body shape to create movement. These protists move by sending out extensions called **pseudopods**; the flow of cytoplasm into the pseudopod determines the direction of movement. Other protists achieve motion through a whiplike structure called a **flagellum**. The **Unikonta** are a large group of eukaryotic organisms that emphasized motility during their early evolution, using either a single rear-facing (posterior) flagellum or lobelike pseudopods or both. Included in the Unikonta are the Amoebozoa, the fungi (Mycota), and the entire animal kingdom, along with a few lesser-known species. Common gene sequences provide evidence for the unity of this group.

Slime molds such as *Dictyostelium* (Figure 6.8) have multicellular reproductive stages that look like fungi and carry out absorptive nutrition, but at other times they live as motile, amoebalike individual cells. Thus, they are thought to resemble an early stage in the evolution of multicellularity. Slime molds were once thought to be related to more typical fungi, but most experts now consider them to be more closely related to amoebozoans and animals.

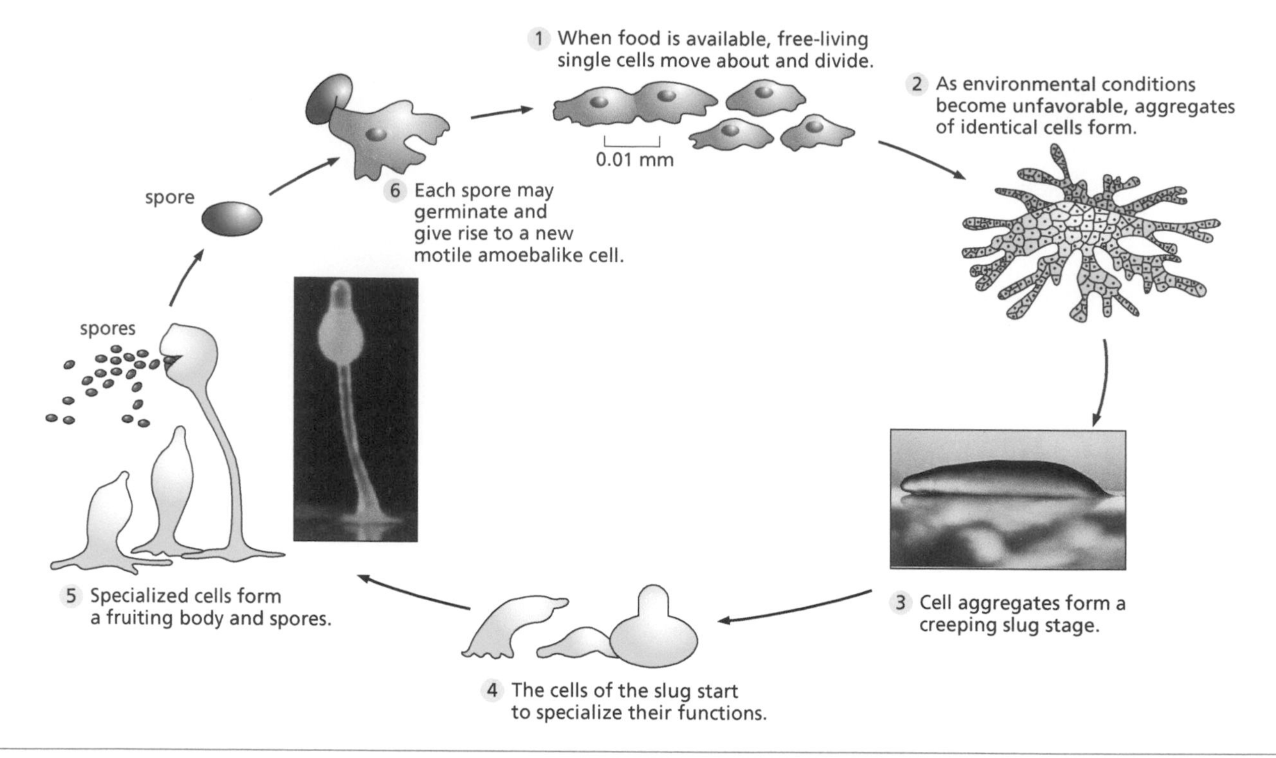

Figure 6.8 Life cycle of the slime mold *Dictyostelium*.

Organisms of the kingdom **Mycota** are commonly known as fungi (Figure 6.9). These organisms typically live on dead or decaying organic matter that they absorb through threadlike extensions called **hyphae**. Fungi have mitochondria, but not chloroplasts, and so they do not carry out photosynthesis. They have cell walls and are usually nonmotile, except for some primitive fungi that have flagellated gametes.

Among fungi, yeasts, morels, and certain molds belong to the phylum Ascomycota, which includes both single-celled and hyphal growth forms. Most ascomycetes can reproduce either sexually or asexually. Mushrooms belong to the phylum Basidiomycota. Their feeding structure consists of a branched network of fine hyphae; the familiar mushrooms (composed of tightly packed hyphae) are their reproductive structures. Most fungi reproduce with the aid of spores, minute haploid gametes that are not distinguishable as eggs or sperm. Fungi useful to humans include edible mushrooms, yeasts (used in brewing, baking, and biological research), and the mold *Penicillium*, the source of the antibiotic penicillin.

Except for the Unikonta, all other eukaryotes share the fusion of two genes (for the enzymes DiHydroFolate Reductase and Thymidylate Synthase, abbreviated as DHFR-TS) as a common derived trait. The groups that share this gene fusion are thus thought to form a natural (monophyletic) group or clade.

6.4.2.2 Rhizaria The **Rhizaria** include predators that use fine protoplasmic filaments to capture and engulf prey. Their cell bodies are usually encased in tiny shells, composed of silica in one group (the Radiolaria) and calcium carbonate in another (the Foraminifera). All Rhizaria are **planktonic**, meaning that they drift about at the mercy of the currents. Their shells frequently become fossilized, so the Rhizaria are very abundant in the fossil record, and their fossils are used to help determine the age of many rock formations, especially during the search for oil (petroleum) deposits.

6.4.2.3 Excavata The **Excavata** are obligate parasites, meaning they cannot live independently; they require a host to meet their energy needs. This is because they have no mitochondria; however, they do have proteins and other molecules that have been interpreted as the degenerate remnants of mitochondria that their unknown ancestors once possessed. They attach to their hosts by means of a feeding groove, excavated along one side, the trait for which they are named.

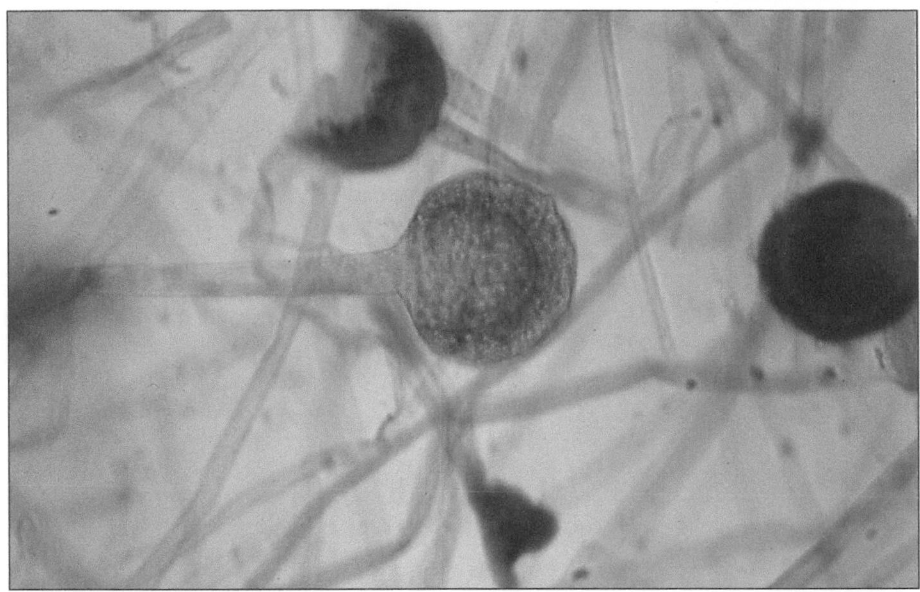

The black bread mold, *Rhizopus* (phylum Zygomycota). In this microscope slide, notice the many thin, absorptive filaments (hyphae)

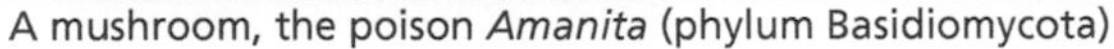

A mushroom, the poison *Amanita* (phylum Basidiomycota)

A morel, *Morchella* (phylum Ascomycota)

Figure 6.9 **Some types of fungi (kingdom Mycota).**

Giardia, discussed in Section 16.4.3, is an example; *Trichomonas*, which causes a sexually transmitted disease, is another.

6.4.2.4 Stramenopiles The **Stramenopiles** comprise a large group of species that possess cell walls containing cellulose, plastids (chloroplasts) containing chlorophylls *a* and *c* (presumably captured by endosymbiosis independently of other photosynthetic groups), and two unequal flagella (one simple and whiplike, the other with branches in a feather-like pattern). Included here are the brown algae, the diatoms, the dinoflagellates, and several other species. Brown algae are mostly multicellular; they include the kelps, ranging in size from a few inches across to the giant kelp, over 30 meters (100 feet) long. Diatoms (Figure 6.7), with shells composed of silica (SiO_2), are extremely abundant in all aquatic ecosystems and form the base of nearly all aquatic food chains. All diatoms are planktonic. Dinoflagellates have their two flagella oriented perpendicular to one another, each nestled in a groove. Some dinoflagellates produce a toxin that accumulates in the shellfish that eat them, and this toxin can cause illness—even death—in people who eat the contaminated shellfish. When water conditions trigger abundant growth of these dinoflagellates (and therefore unsafe shellfish harvesting), it is referred to as "red tide."

6.4.2.5 Alveolata The **Alveolata** are possibly related to the Stramenopiles but are nonphotosynthetic. They include the ciliates, predators like *Paramecium* that move by means of numerous hairlike structures called **cilia**. Also included are the Apicomplexa, protists with a set of sharp, piercing structures, called an **apical complex**, that allows them to penetrate the cells of other organisms and become parasitic. *Plasmodium*, the parasite that causes malaria (see Section 7.3.1), is an example of the latter. Both groups of alveolates feature a series of tiny fluid-filled vesicles, called alveoli, just under their plasma membrane.

6.4.2.6 Archaeplastida One of the great achievements of the early eukaryotic organisms was the acquisition of chloroplasts, which allowed these organisms to increase their energy production through photosynthesis. Photosynthetic organisms that do not develop from multicellular embryos are called **algae** (Figure 6.10). The large assemblage called **Archaeplastida** includes

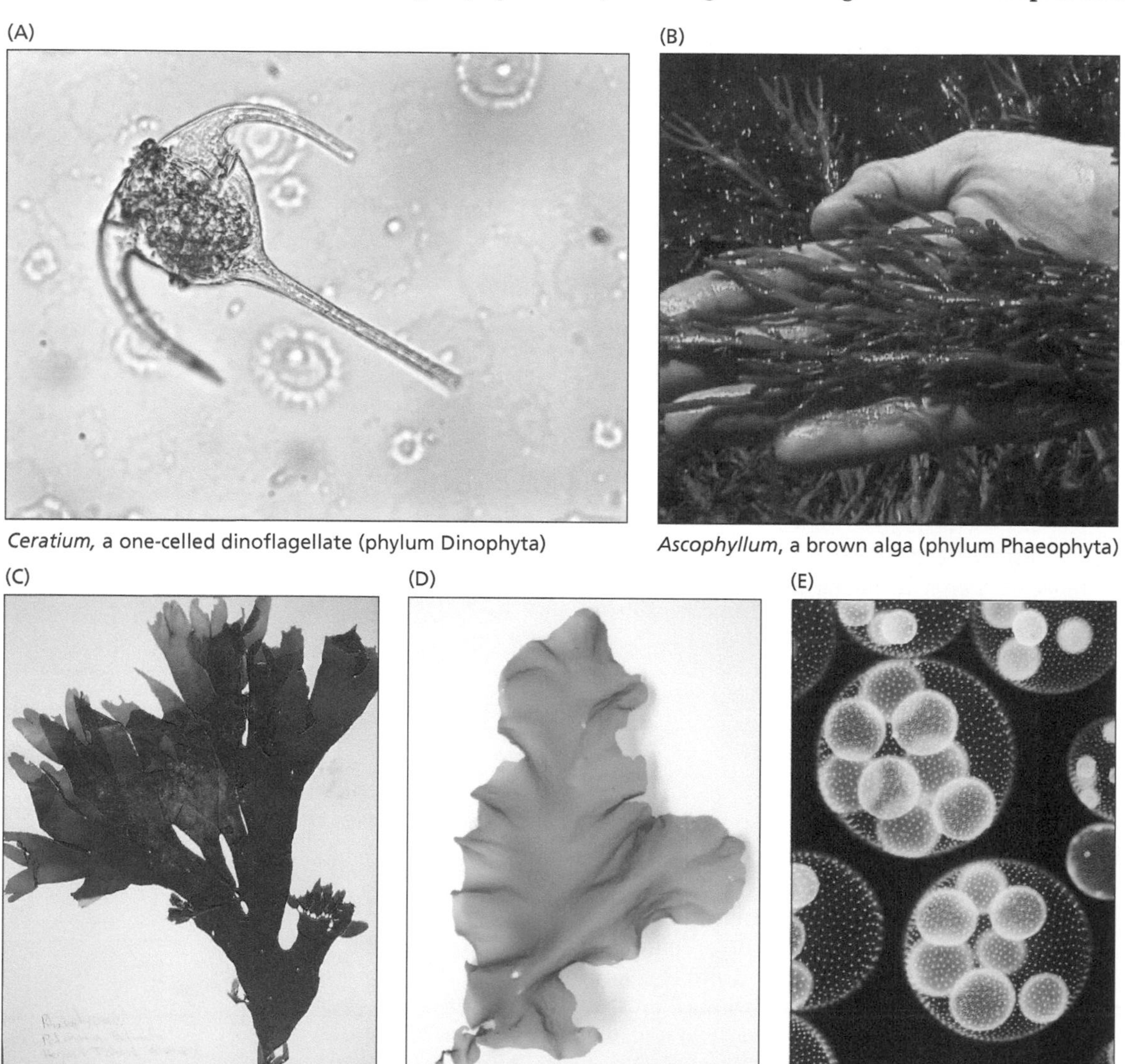

(A) *Ceratium*, a one-celled dinoflagellate (phylum Dinophyta)

(B) *Ascophyllum*, a brown alga (phylum Phaeophyta)

(C) Dulse (*Palmaria*), a red alga (phylum Rhodophyta)

(D) Sea lettuce (*Ulva lactuca*), a common green alga (phylum Chlorophyta)

(E) *Volvox aureus*, a green alga (phylum Chlorophyta) that forms ball-shaped colonies, within which daughter colonies may form and bud off

Figure 6.10 Various algae. Their different photosynthetic pigments are responsible for many different colors.

all photosynthetic eukaryotes with chloroplasts, except those with chlorophyll *c*. (DNA sequences of the Stramenopiles and the Archaeplastida are very different, suggesting that chloroplasts were acquired separately in the two groups.) Within the Archaeplastida, the red algae possess chlorophyll *d* and certain pigments (phycobilins and phycoerythrins) otherwise unknown among eukaryotes; they have also lost all flagellated cells, even among gametes. Other Archaeplastida, including both green algae and land plants, have chlorophylls *a* and *b* as well as α- and β-carotenoid pigments.

Within the Archaeplastida, but also independently in several other groups, multicellular aggregations evolved. At first, these aggregations were just colonies of similar cells, similar to those formed by the green alga *Volvox* (Figure 6.10). Then, colonies began to function as multicellular single organisms. Cells located in different places within evolving organisms began to develop differently. Evolved differences between surface cells and those in the interior, or between cells near the top and the bottom of a former colony, allowed the organisms to take advantage of the differences in environment between the various locations.

Please review the family tree of Eukarya shown in Figure 6.6. Remember that any family tree is simply a hypothesis that seeks to explain the known resemblances and differences. If further resemblances and differences among these groups are discovered, then the family tree (and the associated classification) will change.

6.4.2.7 The plant kingdom The land plants, which comprise the **Kingdom Plantae**, are distinguished from all other eukaryotes by developing from a multicellular **embryo** that includes a fertilized egg (zygote) surrounded by nonreproductive cells. Plants are now considered a subgroup of Archaeplastida. The simplest plants, the bryophytes (mosses, liverworts, and hornworts), have a layer of sterile, nonreproductive cells that surround and protect their egg cells; this adaptation permitted these plants to emerge from aquatic environments and colonize the land, although all bryophytes still live in relatively moist environments. Most botanists believe that bryophytes evolved from green algae because important photosynthetic pigments and other characteristics are shared by both groups. Three species of bryophytes are shown in Figure 6.11.

Bryophytes do not have deep underground parts because all parts of the plant carry out photosynthesis and therefore need to be in the light. They cannot grow very tall because they lack the **vascular tissues** (vessels or channels) that conduct fluids in larger plants, and they also lack the roots and stems needed to provide anchorage and support. They are, therefore, nonvascular plants.

Conocephalum, a thalloid liverwort

Marchantia, another thalloid liverwort

Polytrichum, a moss

Figure 6.11 Representative bryophytes (phylum Bryophyta). These are land plants with protective cells surrounding the egg, but their lack of vascular tissues keeps them small.

The earliest land plants were small bryophytes that were restricted to moist habitats, but some plants evolved vascular tissues, which allowed them to grow much taller, and these became the **vascular plants** (Tracheophyta). The most familiar and ecologically dominant plants are all vascular plants, and they include the largest and most conspicuous organisms in most terrestrial habitats. Vascular plants come in many shapes and sizes; a sample of this diversity is shown in Figure 6.12.

In algae and simple plants, each cell carries out its own photosynthesis, absorbs its own nutrients, and gets rid of its own waste products. The increasing specialization of cells in vascular plants allows different parts of each plant to perform different functions efficiently. Groups of similar cells are organized into tissues, and groups of tissues are organized into **organs** such as leaves and roots. For example, each leaf is an organ, while each cell layer within a leaf is a tissue. The simplest plants containing separate types of tissues are bryophytes, but the diversity and complexity of tissue types increase dramatically among vascular plants. Photosynthesis is carried out principally in the leaves, and other plant parts are also specialized for particular functions. Roots are specialized for water absorption and fruits for reproduction and dispersal. The division of labor among different parts of the plant would not be possible without the specialization of plant tissues, particularly the vascular (conducting) tissues that efficiently transport materials from one part of the plant to another. Underground roots, for example, are only possible because vascular tissues can bring photosynthetic products to underground parts where light never penetrates, something that is impossible among bryophytes (restricting them to moist habitats). We will examine vascular plants further in Chapter 17.

Figure 6.12 An assortment of vascular plants (Tracheophyta), belonging to several groups.

Nephrolepis, a fern (phylum Pterophyta), showing reproductive structures on the underside of the leaf

Equisetum, a horsetail (phylum Arthrophyta)

Pinus, the white pine (phylum Coniferophyta)

Flowering plants or angiosperms (phylum Anthophyta)

daisy (*Chrysanthemum*)

trillium (*Trillium*)

rose (*Rosa*)

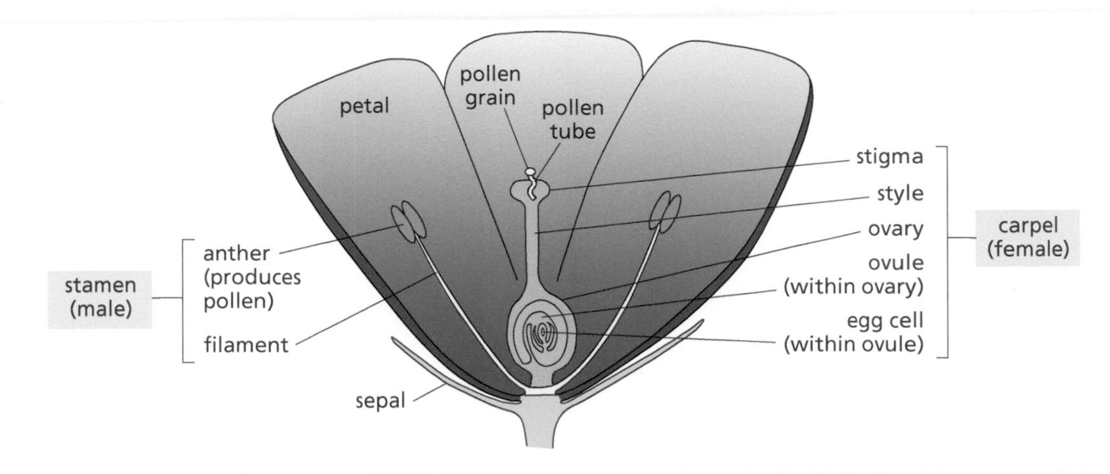

Figure 6.13 **Diagram of a complete flower, containing both male and female parts together.** After fertilization and ripening, the ovary becomes a fruit. Variations in the number and structure of the parts shown here are among the most useful characters in plant classification. Some plants have incomplete flowers in which there are separate male flowers with undeveloped female parts and female flowers with undeveloped male parts.

The most highly evolved vascular plants reproduce with the aid of seeds, which contain small diploid embryos capable of being dispersed to new locations away from the parent plant. These plants have branching roots and leaves with multiple veins. The largest and most diverse, as well as ecologically dominant, group of seed-producing plants are the flowering plants (Angiospermae or Anthophyta). The seeds of flowering plants develop within elaborate reproductive structures called flowers (Figure 6.13). Eggs, each with a haploid set of chromosomes, are produced by the female part of the flower within the ovary. The male part of the flower produces haploid gametes (sperm) within pollen grains in the anthers. Pollination is the introduction of the pollen onto the stigma, the female receptive surface of the flower (Figure 6.13). A pollen tube grows from the pollen grain to the ovary, where the sperm fertilizes the egg, resulting in a diploid zygote. Many flowers are pollinated by wind, but a much larger number are pollinated by insects. The relations between flowering plants and the insects that pollinate them are often quite elaborate and are key to much of the diversity and also the evolutionary success of both flowering plants and insects (see Chapter 18).

After fertilization, the zygote undergoes cell division to become an embryo, and the structures surrounding the zygote mature into a seed. The seed-bearing structures in a flower ripen into **fruits**, defined as ripened ovaries that contain seeds. In addition to the seeds themselves, fruits often contain tissues that attract various animals by means of conspicuous colors, special odors, carbohydrate nutrients, or a combination of these. Animals that eat the fruits may disperse the seeds in their feces, often far from the parent plant. Seeds are also dispersed in other ways (Figure 6.14). Among the fruits that humans eat are many that we commonly recognize as fruits (e.g., apples, peaches, melons, walnuts) and others that we do not always think of this way (e.g., grains, cucumbers, tomatoes, peppers, eggplants).

6.4.3 The animal kingdom

Animals (Kingdom Animalia) are multicellular organisms with eukaryotic cells and an embryonic life stage consisting of a hollow ball of cells called a **blastula**. Most animals have at least some motility during some stage of their life cycle.

As we have seen, there are many highly successful forms of life that are not animals. Even within the animal kingdom, most animals are very different from the group to which we belong. In fact, most animals lack a stiffening backbone and are called **invertebrates**. The animal kingdom also includes a great diversity of life strategies, and each is biologically successful within the habitat that it occupies.

The animal kingdom is divided into nearly 30 phyla (Box 6.5). Experts differ on the exact number of phyla because they are not in agreement about how to classify some animals. Some small phyla are not included in the survey that follows.

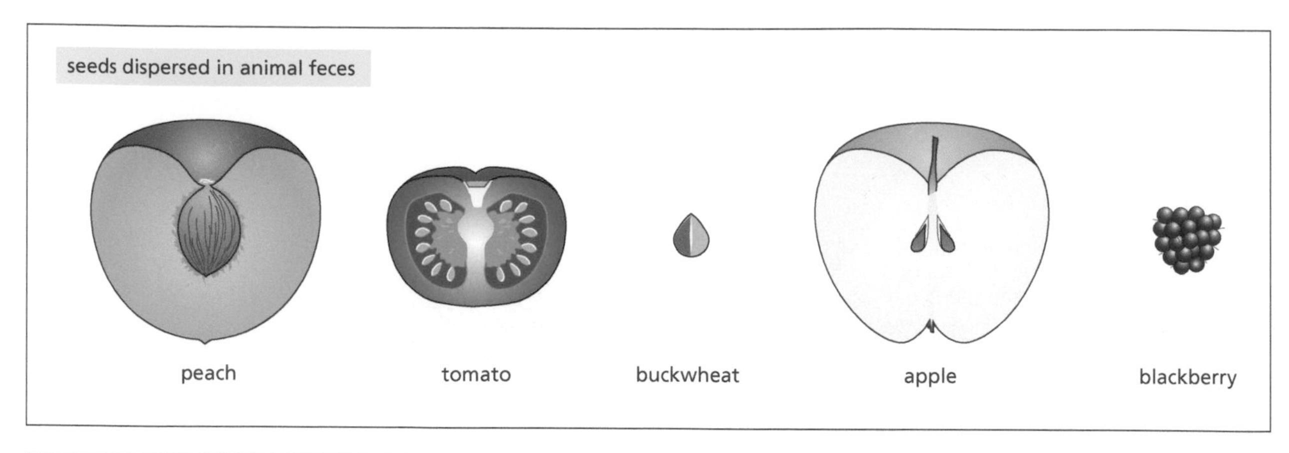

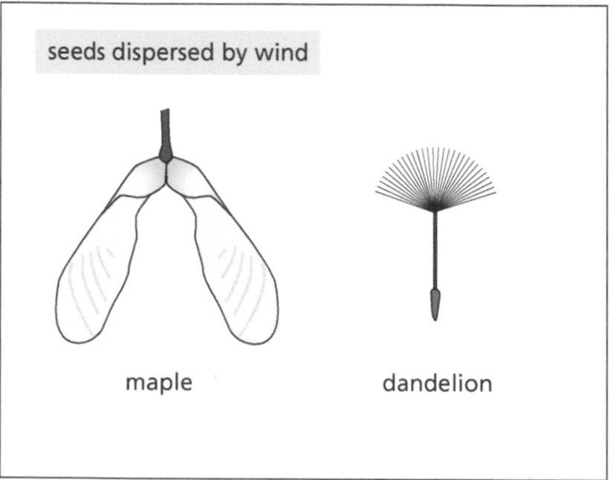

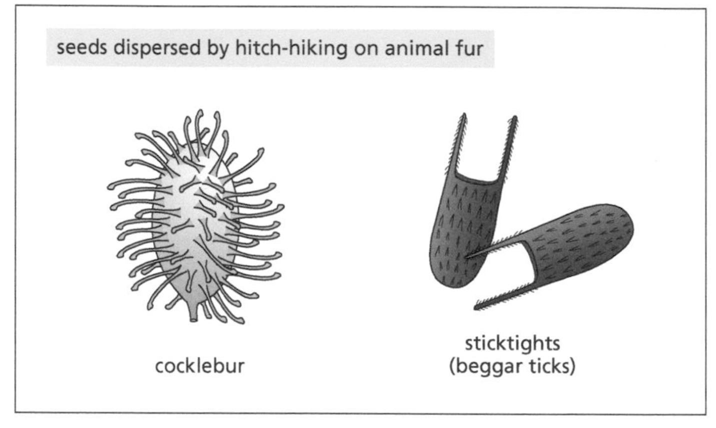

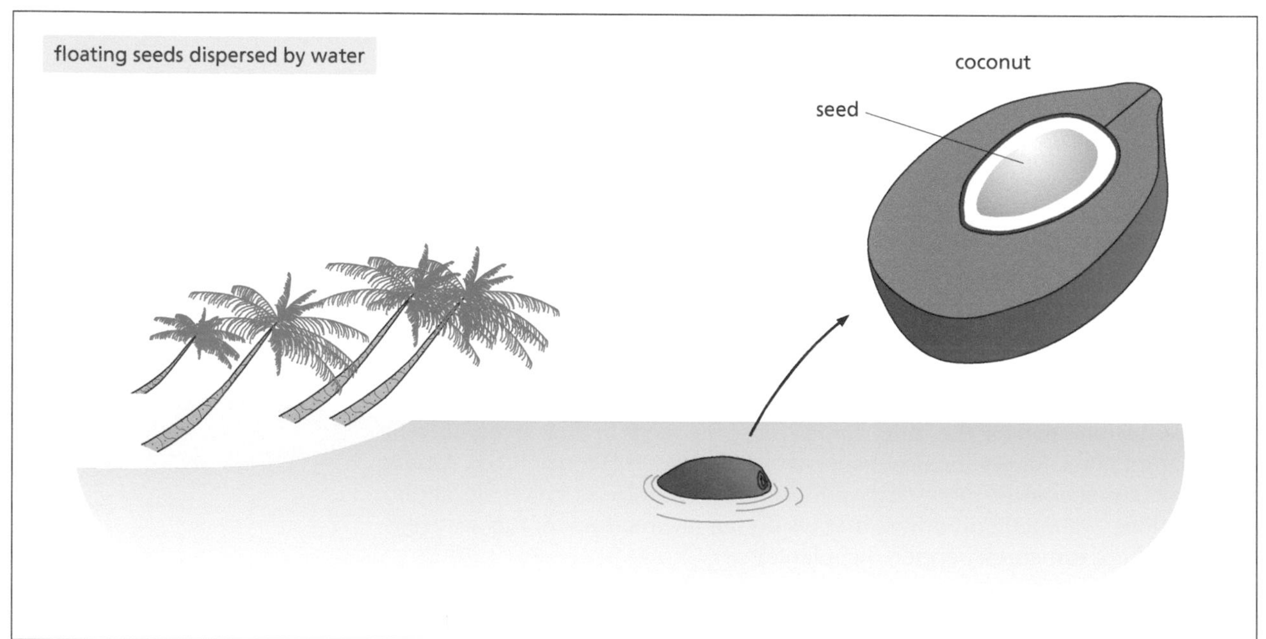

Figure 6.14 The dispersal of seeds in different kinds of fruits.

6.4.3.1 Minimal organization and the sponges Multicellular organization in animals takes several forms. The simplest animals are sponges (phylum Porifera) (**Figure 6.15**). Various cell types are present in these aquatic animals, including wandering amoeba-like cells, barrel-shaped cells with hollow interiors, and cells with a whiplike flagellum surrounded by a "collar." These cells, however, are not organized into different tissue layers, as

Phylum Porifera

Sponge

Phylum Cnidaria

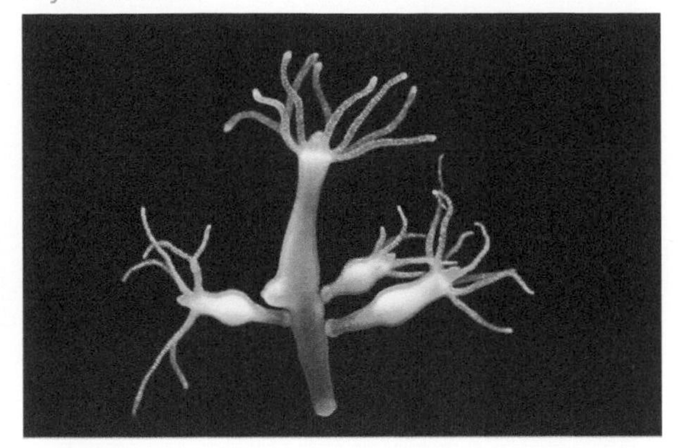

Hydra vulgaris

Phylum Cnidaria

jellyfish medusae

anemone polyp (*Actinia*)

Figure 6.15 Sponges (animals not organized into tissue layers) and cnidarians (animals with radial symmetry and with two tissue layers, an outer ectoderm and an inner endoderm).

they are in all other animal phyla. Sponges have a variety of adaptations that deter predators who would otherwise feed upon them: All sponges have sharp, needlelike structures (spicules), and many sponges also secrete poisonous chemicals. The sponges that lacked these defenses disappeared long ago.

6.4.3.2 Tissue layers and the phylum Cnidaria The simplest animals with cells organized into tissues are found in the phylum Cnidaria (Figure 6.15). Like the sponges, these are aquatic animals. In the Cnidaria and in the embryos of all other animals except sponges, one portion of the hollow blastula puckers inward and turns inside-out. The resultant cup-shaped structure, called a **gastrula**, contains two distinct layers of cells: the layer that has moved to the inside is called the **endoderm** and the layer that remains on the outside is called the **ectoderm**. These structures are shown in Figure 6.16.

The ectoderm and endoderm form two distinct tissue types, the beginnings of the differentiation of multicellular animals into a variety of such tissues. As in plants, tissues are groups of similar cells that form sheets or other integrated structures, each specialized to perform a different function. We discuss the processes that give rise to this specialization in Section 10.1. In addition to the ectoderm and endoderm, the gastrula contains an endoderm-lined central cavity, open to the outside. The fact that all animals (except for sponges) go through such a gastrula stage in their development is strong evidence that they all share a common ancestry.

The two tissue layers are arranged in two basic body plans among the Cnidaria. One plan (called a **polyp**) has the central cavity opening upward. Cnidaria with this body plan usually grow attached to the ocean bottom or to other animals; many live in large colonies that we recognize as corals. The other body plan (called a **medusa**) has the central cavity opening downward. Cnidaria with this body plan float freely in the water, and most can contract portions of their body to control

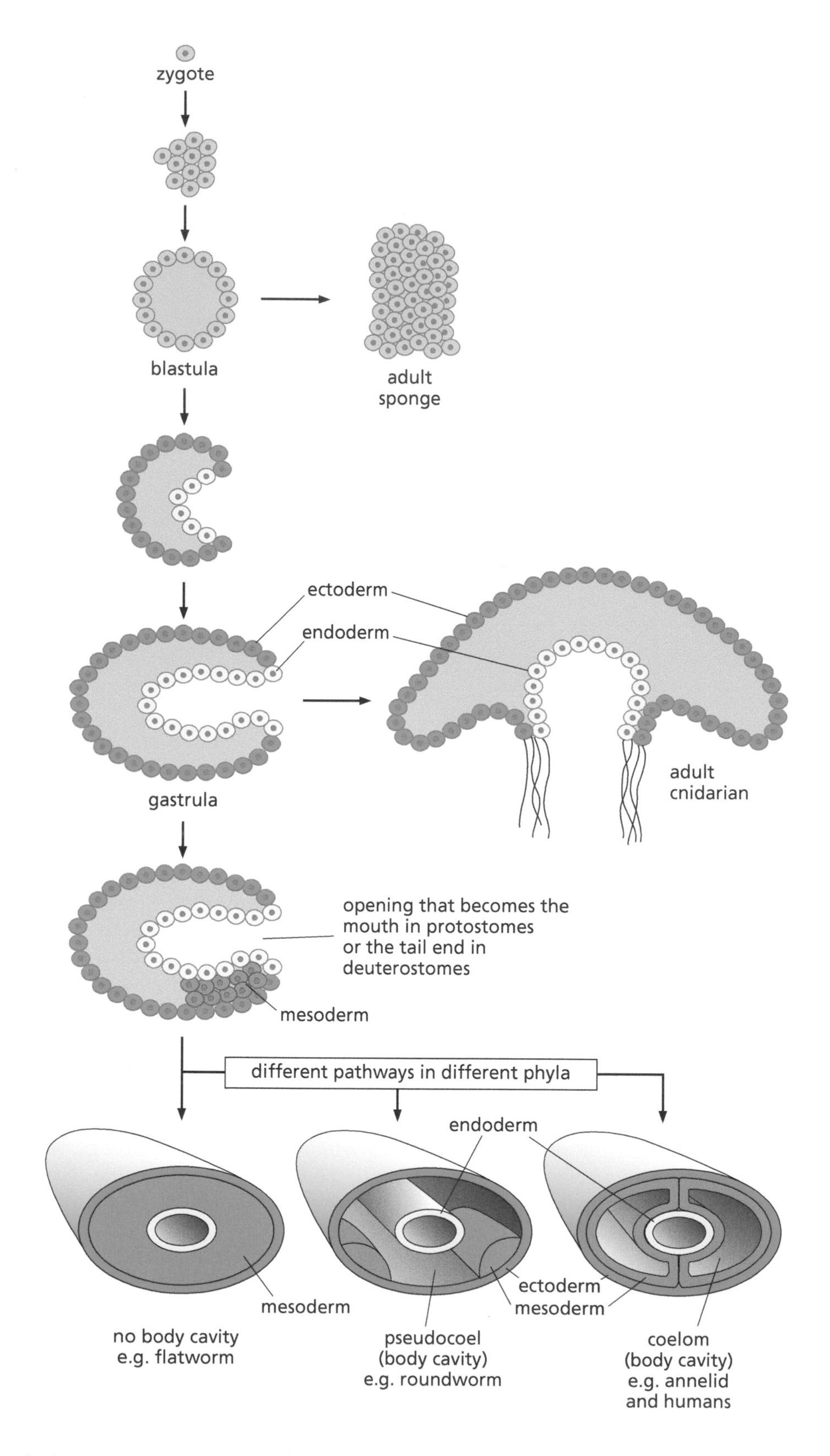

Figure 6.16 Early stages in the embryology of animals. Adult sponges develop directly from a modified blastula stage. All other animals go through both blastula and gastrula stages. Members of the phylum Cnidaria form adult stages that still resemble gastrulas, whereas most other animals develop a middle layer (mesoderm) that gives rise to additional internal organs.

their movement. Because of the large amount of jellylike material that lies between the outer and inner layer of cells, most of these Cnidaria are popularly known as "jellyfish." The major subgroups of Cnidaria are distinguished on the basis of whether their life cycle includes only one of these body plans or both of them. Both cnidarian body plans have a series of tentacles surrounding the opening of their central cavity. These tentacles contain specialized stinging cells, which are used to defend the animal against predators.

6.4.3.3 Bilateral symmetry and the flatworms Most sponges, and a few other types of animals that live attached to the ocean bottom, have irregular body shapes that show no symmetry. Other sponges and all members of the Cnidaria have a radially symmetrical body plan, that is, a body plan arranged in a circle. (If you look down on them from above, you will see the same anatomical details repeated over and over around the edge of this circle.) Regardless of body plan, a cnidarian has the same chance of finding something nutritious, or something dangerous, in any direction, and natural selection has therefore favored radial body plans among these animals.

In contrast, the vast majority of animals, belonging to over two dozen phyla (Figure 6.17), are characterized by **bilateral symmetry**, meaning that their body can be divided by a central plane such that structures on the left side of this plane are mirror images of corresponding structures on the right side. Bilateral symmetry is believed to have evolved in animals as an adaptation that came along with forward movement. Imagine an animal that creeps along the ocean bottom in such a way that one end of its body is in a forward position. Movement would be made easier by a streamlined or elongated body. New discoveries, whether of food or of danger, would be more likely to be made with the front end. Under these conditions, natural selection would favor

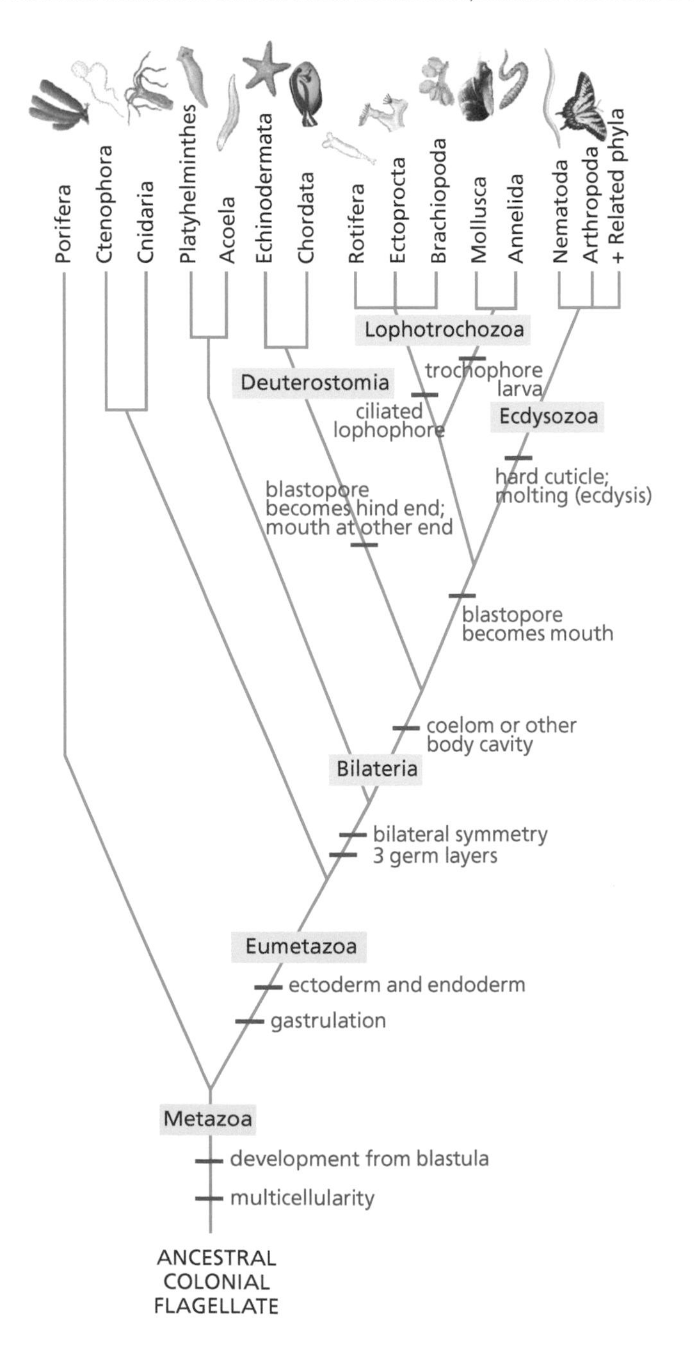

Figure 6.17 **A family tree of the major animal phyla.**

the development of a front end with sense organs (eyes, feelers, taste organs, sound and motion detectors), feeding organs, and possibly also aggressive weapons.

An animal that creeps along the ocean bottom may also be expected to react differently to the water above than to the sediment below, and so natural selection would tend to favor organisms having structures on the top (dorsal) surface that differ from those on the underside (ventral). However, any structure or ability that is adaptive on the right side is equally adaptive on the left, and organisms would have no selective advantage if their right side differed from their left. As a result of selection under these conditions, body plans that are bilaterally symmetrical are common in the animal kingdom and present in many different phyla.

The simplest animals with bilateral symmetry are the flatworms of the phylum Platyhelminthes (Figure 6.18). They have somewhat elongated bodies, with sense organs concentrated at one end, which is recognizable as a head. The body is flattened, with broad upper and lower surfaces that in many species differ from one another in coloration and in other ways. The body plan is bilaterally symmetrical, with right and left halves of the body being mirror images of one another. Flatworms have a middle layer of tissue (called the **mesoderm**) in addition to the ectoderm and endoderm (Figure 6.16). All the animals yet to be described in this chapter have tissues derived from these three basic layers.

6.4.3.4 Assembly-line digestion One way in which flatworms are similar to cnidarians is in their digestive system, which is just a sac with a single opening that serves as both entrance and exit. With this arrangement, much of what is discarded as waste is immediately taken in again as food, making the system very inefficient. A further inefficiency is that every region of the digestive tract, and every group of cells in the digestive lining, must be capable of performing the entire digestive process from beginning to end. With this arrangement, cells cannot specialize to carry out early or late steps of digestion.

A more efficient arrangement is an assembly-line digestive tract with an entrance, the mouth, at the front end and an exit, the anus, at the hind end. With this arrangement, selection can favor organisms in which the cells near the front end can perform the early stages of digestion more efficiently and those near the hind end are more efficient in completing the later stages. With this adaptation, certain parts of the digestive tract can specialize in the processing of different types of nutrients or of hard substances requiring mechanical break-up (see Chapter 8). Waste products are now discharged more efficiently because they are released from the hind end and left behind as the animal moves forward.

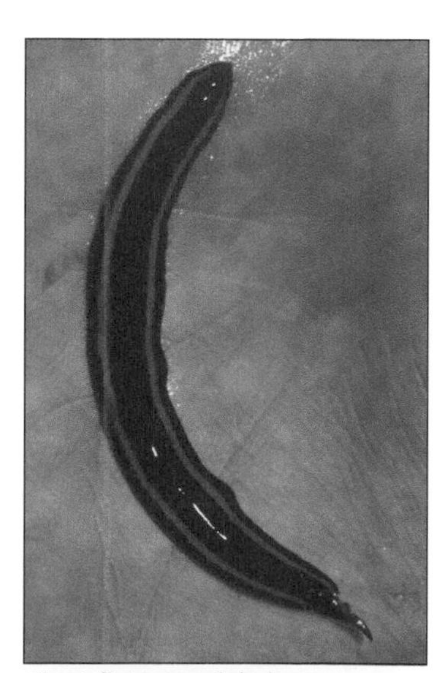

Giant flatworm (phylum Platyhelminthes)

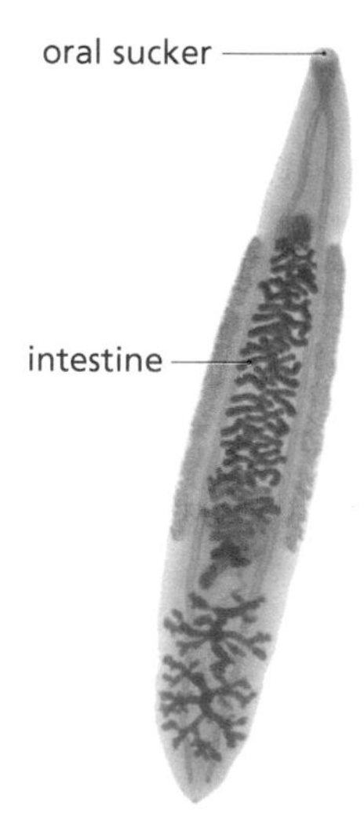

Asian liver fluke

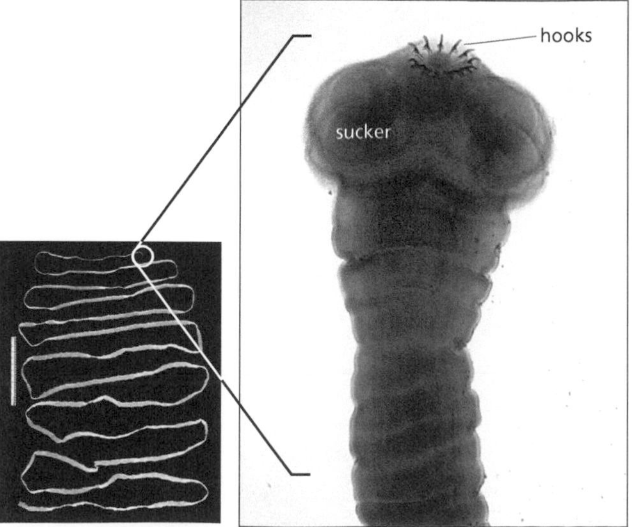

Tapeworm, which can sometimes grow to 3 m (10 ft) or more in total length

Figure 6.18 **Flatworms (phylum Platyhelminthes).**

6.4.3.5 The evolution of body cavities In the course of creeping forward along the ocean floor, some animals occasionally found reason to burrow into the bottom sediment. Burrowing is usually accomplished by a mechanical process of wedging the front of the body farther forward, then forcefully inflating part of the body to make it wider, then repeating the process. Forceful widening of the body thus alternates with forceful elongation, both in time and space. At any moment when the narrow portions of the body are pushing forward, the other parts of the body are widening to give the parts in front of them something against which to push.

This alternation of widening and elongating can be done much more efficiently if the body contains one or more fluid-filled cavities. Because fluids like water are not compressible, squeezing a fluid-filled bag (like a water balloon) in one place or in one direction causes it to bulge elsewhere. Thus, any fluid-filled cavity can be forcefully widened by contracting muscles running front to back, while the same cavity can be forcefully elongated by contracting muscles that encircle its girth.

In the course of evolution, various types of fluid-filled cavities of different constructions and different embryological derivations evolved in different phyla. These phyla are classified in part according to the nature of the body cavity and the cells lining its interior. Roundworms, horsehair worms, rotifers, and a handful of other animal phyla are characterized by body cavities lined with cells derived from several embryonic layers, including endoderm. The remaining phyla described next all have body cavities entirely lined with mesoderm; such a body cavity is technically called a **coelom** (Figure 6.16).

6.4.3.6 Protostome phyla Animal phyla with body cavities and assembly-line digestion are separated into two large groups, the protostomes and deuterostomes, which evolved in different directions. The **protostomes** include the mollusks, annelids, arthropods, roundworms (nematodes), and several smaller groups (Figures 6.19 and 6.20). All of them share certain embryological

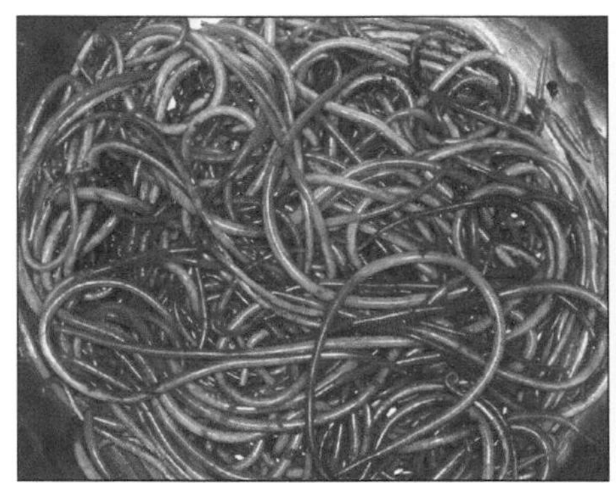
Ascaris (phylum Nematoda)

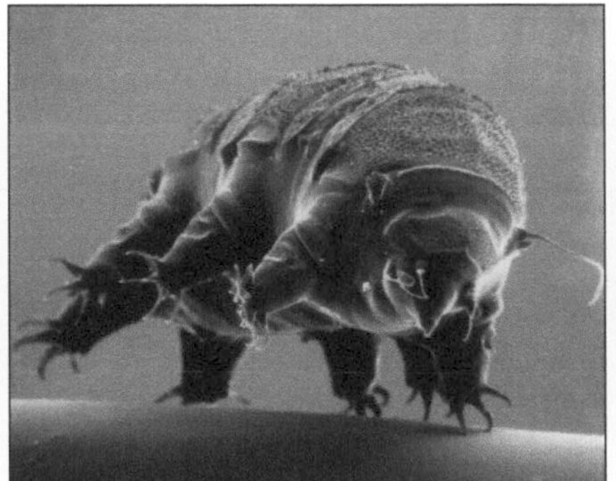
A water bear (phylum Tardigrada)

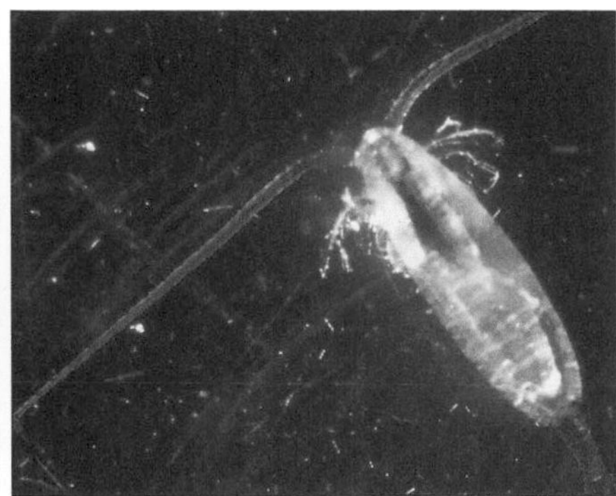
Copepod (phylum Arthropoda)

Crab (phylum Arthropoda)

Grasshopper (phylum Arthropoda)

Butterfly (phylum Arthropoda)

Figure 6.19 Representative invertebrate animals belonging to the Ecdysozoa.

Figure 6.20 Representative invertebrate animals belonging to the lophotrochozoan phyla.

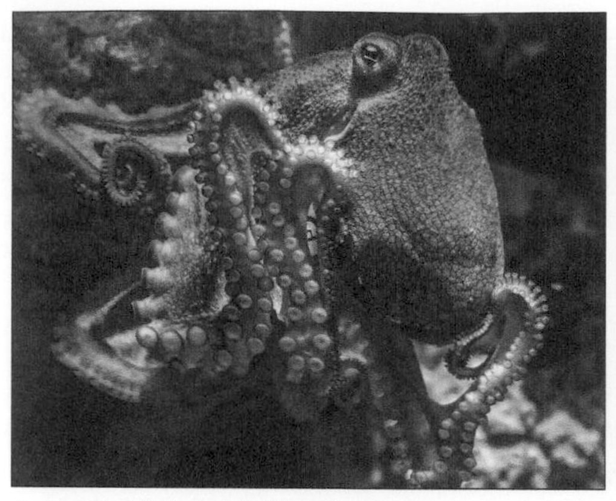

Octopus (phylum Mollusca)

Nautilus (phylum Mollusca)

Tree snail (phylum Mollusca)

Tropical earthworm (phylum Annelida)

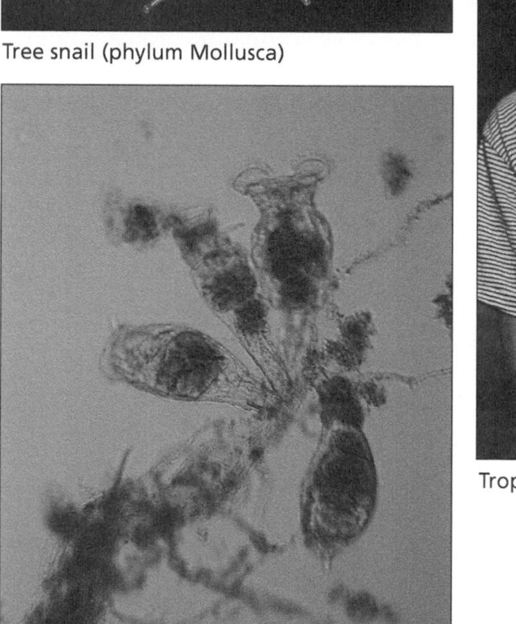

Rotifers (phylum Rotifera)

similarities, such as the way in which their body cavity develops and the derivation of the mouth.

At some point in protostome evolution, the mesoderm and its body cavity became subdivided into a series of individual blocks or pouches (somites). These blocks of tissue were arranged from front to rear, setting the stage for the evolution of segmentation of the body. Some of these animals, such as the annelid worms, are thoroughly segmented in both larval and adult stages, but others, including most mollusks, have lost most of their segmentation as adults.

Animals of the phylum Annelida, of which the earthworm is a familiar example, are anatomically arranged as a series of repeated segments; all the body segments are similar to one another in size and in anatomical structure. Segmentation permits parts of the body to work as self-contained units, allowing rhythmic swimming or crawling motions. The worms crawl through soil or sediment by using rhythmic waves of muscle contraction squeezing against the fluid-filled body cavity of each segment. A few body segments elongate while the next few widen. Waves of elongation alternate with waves of widening of the body segments, and these alternating waves pass down the length of the body from the front end to the rear. The annelid worms have no legs, but they do have tiny bristles (called setae) that stick out of their sides and anchor the widened, nonmoving segments of the body to the surroundings, giving the elongated segments something to push against as they move forward.

The phylum Arthropoda is the largest and most diverse phylum of the entire animal kingdom. The Arthropoda include lobsters, crabs, shrimp, barnacles, spiders, scorpions, centipedes, millipedes, and insects. Insects alone account for over two-thirds of the animal kingdom and over half of all living species on Earth. The body segments of arthropods are fewer than in annelids, and these segments are more specialized, differing from one another in both size and anatomy. Most notably, the arthropods have a series of leglike structures that differ in most cases from segment to segment. Arthropods have a strong, protective outer coating (called an **exoskeleton**), making each segment rigid. The rigid segments are separated by flexible hinge regions. The legs are also arranged as a series of rigid segments separated by flexible, hinged joints, giving the phylum its name (from the Greek, *arthro* meaning "hinged" or "jointed," plus *pod* meaning "leg" or "foot").

Sequencing of various animal genomes has revealed that the arthropods are closely related to the roundworms and to certain smaller phyla that all possess a hard, outer protective layer called a **cuticle**. In order for growth to take place, animals with a cuticle must break out of this protective layer, hide themselves while they are naked and vulnerable, grow quickly to a new, larger size, and develop a new, larger cuticle for protection. Coordination of all these complex changes, in the right order, requires the action of steroid hormones called **ecdysones**, and the process is called molting or ecdysis. All the animals that go through this process are now grouped together as Ecdysozoa (Figure 6.19).

Protostomes that do not undergo molting are quite varied in their morphology, but their genome sequences show that they are all related. In their early evolution, these phyla were characterized by a larval stage, called a trochophore, featuring bands of cilia, which some of their descendants still possess. In several phyla of this group, the bands of cilia form a feeding structure, called a lophophore, that strains tiny suspended food particles out of the water. Phyla featuring these bands of cilia, either in their larvae or their feeding structures, include the Mollusca, Annelida, Bryozoa, Brachiopoda, Rotifera, and many smaller groups (Figure 6.20).

Animals of the phylum Mollusca are in most cases protected by a hard, outer shell secreted by a special layer called the mantle. Part of the mantle is retracted at the rear of the animal to form a mantle cavity that contains both an anus and gills that allow respiration in water. Anyone who has admired seashells has some idea of the tremendous variety of species of mollusks. In addition to the familiar snails, clams, and oysters, mollusks also include cephalopods such as the squid and octopus, in which the shell is hidden inside or has been lost entirely (Figure 5.7). The creeping movements of snails and the digging movements of clams are both very similar to the waves of contraction used by annelids.

6.4.3.7 Evolution of the deuterostome phyla The remaining phyla of the animal kingdom are **deuterostomes**. All deuterostomes have body cavities completely lined with mesoderm, but they have evolved separately from the protostomes described earlier. One major difference is in the manner in which the body cavity usually develops; another difference lies in the embryologic formation of the mouth. In protostomes, the gastrula opens to the outside by an opening that becomes the future mouth. That same opening in deuterostomes ends up near the hind end of the animal, just above the anus, while the mouth develops as a secondary structure at the other end. (*Protostome* means "first mouth," while *deuterostome* means "secondary mouth.") Thus, in a very real sense, your head corresponds to an insect's hind end (they are homologous), and an insect's head corresponds to your rear.

Crinoid or sea lily

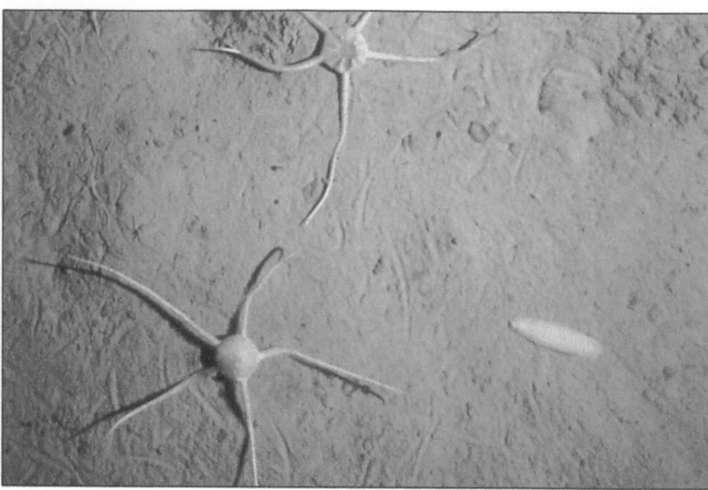
Brittle stars

Figure 6.21 Echinoderms (deuterostomes of the phylum Echinodermata).

The deuterostomes and protostomes evolved separately, but some convergent adaptations have appeared. For example, a form of segmentation of the muscles and certain other body systems evolved independently in both groups.

The deuterostomes include four phyla, all evolved from bilaterally symmetrical ancestors. All animals in these four phyla go through early (larval) stages of development that are bilaterally symmetrical, but a few become irregular (asymmetric) or radially symmetrical as adults. Two phyla, the echinoderms and chordates, are large groups.

The phylum Echinodermata (Figure 6.21) includes sea stars (starfish), brittle stars, sea urchins, sand dollars, crinoids (sea lilies), and sea cucumbers. The living species of echinoderms show a five-fold symmetry as adults, but their larvae are bilaterally symmetrical. Their other characteristics include a bumpy or spiny skin protected by calcium carbonate deposits and a water-vascular system through which sea water circulates in a series of tubes.

Humans and other animals with backbones belong to the phylum Chordata (Figure 6.22). Also included in this phylum are several small and less familiar sea creatures such as the sea squirts (also called tunicates) and the small sea lancet or amphioxus. The common ancestors of echinoderms and chordates probably lived attached to the ocean bottom, either directly or by means of a stalk. Many extinct groups of echinoderms grew this way (crinoids still do; see **Figure 6.21**), but most living echinoderms and chordates have a free-living, unattached way of life. The transition from attached to free-living is best shown by the tunicates in the phylum Chordata (**Figure 6.22**). These small animals generally spend their adult lives attached to a rocky bottom. Here they sit and pump water through a large basket-like structure (the pharynx) whose numerous slits strain the water through while suspended food particles collect on a sticky, ciliated surface coated with mucus. The attached filter-feeding existence of adult tunicates was the ancestral way of life for deuterostomes. Larval tunicates, by contrast, are actively free-swimming animals that resemble tadpoles. They swim by means of a long tail that sweeps back and forth like the tail of a fish, and the muscle blocks and nerves of this tail are segmentally organized. Free-swimming members of the phylum Chordata, including all fishes, have more in common with the tunicate "tadpole" than with the filter-feeding adult.

Although humans are obviously different from tunicates in many ways, as members of the Chordata they share major characteristics not found in any other group of organisms. Characteristics shared by all chordates include (1) a body axis containing a stiff, flexible rod (called a **notochord**); (2) a hollow nerve cord along the back; and (3) a series of openings called gill slits, located behind the mouth region. These three characteristics originate early in the embryo and are not always retained in the adult stages. For example, fish keep their gill slits

Drawing of a tunicate (class Urochordata)

Queen angelfish (class Osteichthyes)

Strawberry frog (class Amphibia)

Salamander (class Amphibia)

King snake (class Sauropsida)

Alligator (class Sauropsida)

Penguins (class Sauropsida)

Jack rabbit (class Mammalia)

Impala gazelle (class Mammalia)

Figure 6.22 **Representatives of the phylum Chordata, another deuterostome phylum.** All of these animals have embryonic notochords and gill slits.

throughout life and use them to breathe, but humans lose their gill slits long before birth, keeping only a few remnants here and there, such as the tube that connects the throat to the middle ear (Figure 11.12).

6.4.3.8 Animals with backbones (Vertebrata) Among the members of the phylum Chordata, the majority are backboned animals that make up the subphylum Vertebrata. The stiff, flexible rod found in other chordates is functionally replaced in adult vertebrates by a backbone made of a series of individual bones or cartilages. Included in the vertebrates are four classes of fishes: (1) the jawless fishes; (2) the extinct, armored Placodermi; (3) the fishes with cartilage skeletons (including the sharks and rays); and (4) the fishes with skeletons of true bone (the group to which most fishes belong). There are more living species of bony fishes than of all the other vertebrate classes combined. All fishes are aquatic vertebrates that use gills to breathe and that swim by waving their hind end from side to side.

The remaining vertebrate classes evolved, directly or indirectly, from the bony fishes, and thus all have bony skeletons. First of these are the amphibians (class Amphibia), which include the frogs, toads, and salamanders. These animals lay eggs in water, and the eggs develop into aquatic, gill-breathing larvae, commonly called tadpoles. After a while, the tadpoles undergo a rapid developmental change (metamorphosis) into adults that have legs and, in most cases, lungs. Fossil amphibians are known from the Devonian period to the present day.

Derived from the amphibians are the reptiles (class Reptilia or Sauropsida), which include turtles, snakes, lizards, crocodiles, and many extinct species, including dinosaurs. Unlike the amphibians, the reptiles have dry, scaly skin, and they lay their eggs on dry land (except for a few species that retain the egg inside the mother and give birth to live young). Fossil reptiles are known from the Pennsylvanian period to the present, but the Mesozoic era was populated by so many reptiles that it is often called the Age of Reptiles.

One group of reptiles, the Archosauria, included the dinosaurs and other dominant reptiles of the Mesozoic era. The birds (Aves) are derived from this group of reptiles; birds are distinguished by their possession of feathers. Most bird adaptations have to do with flying; these include adaptations that lighten the body (such as hollow bones and loss of one ovary) and those that allow for the high metabolism (and thus the high internal body temperature) that flying requires. Feathers do double duty as a flight surface and as insulation.

Another group of ancient reptiles had mammal-like features, and the class Mammalia, to which we belong, evolved from them. Mammals maintain a high and fairly constant body temperature, made possible by an insulating layer of hair or fur, supplemented in some cases by fat or blubber. A four-chambered mammalian heart prevents oxygen-rich blood from the lungs from mixing with oxygen-poor blood returning from other parts of the body. Also characteristic of mammals is the fact that they supply their young with milk, a secretion of the female's mammary glands. Mammals include kangaroos, shrews, monkeys, humans, bats, rats, squirrels, rabbits, whales, dogs, cats, bears, seals, elephants, horses, pigs, sheep, cattle, and many other species.

Humans are mammals because we share such mammalian characteristics as hair, a four-chambered heart, and the feeding of milk to our young. We are also chordates because we share in the embryonic gill slits and other characteristics that unite us with tunicates, amphibians, and other Chordata. We also share with all deuterostomes (chordates and echinoderms) the way our mouths develop embryologically. We share with many more phyla the anatomical structure and embryonic derivation of our body cavity, and with all animals the presence of motile cells and development from a blastula. We share the eukaryotic type of cell with plants, fungi, and many other species, putting us in the domain Eukarya. The evolution of species over billions of years accounts for these patterns of shared characteristics.

THOUGHT QUESTIONS

1. Why are algae sometimes considered protists? Why were they sometimes considered plants? How would you decide which is the better approach? If chloroplasts evolved only once, how would this affect your answer? What if chloroplasts evolved many times, independently?
2. Find four or more books on botany or general biology. List the phyla or divisions of the plant kingdom that each book recognizes. What similarities do you find? What differences do you find? How do you account for the differences?
3. Many bilaterally symmetrical animals have a long, thin "worm-like" body shape. What advantages do you think such a body shape can confer? What problems do you think can arise from such a body shape?

6.5 HUMANS ARE PRODUCTS OF EVOLUTION

As we saw in the last section, humans are one species among many. At what point in evolutionary history did our ancestors evolve into something we could call "human"? Answers to this question are reconstructed from fossils. The fossils help us to reconstruct our family tree, although there are frequent disagreements among scientists as to where a particular new fossil fits in.

6.5.1 Our primate heritage

Along with monkeys, apes, and lemurs, humans belong to the mammalian order Primates (Figure 6.23). We share many characteristics with other primates, but we did not evolve from any present-day species. Most adaptations shared by primates are related to the requirements of life in trees. Most primates live in trees today, and those that do not had ancestors that did. Nonprimate mammals whose ancestors never lived in trees do not share these adaptations. Primate characteristics directly related to the requirements of locomotion in trees are (1) the

Ring-tailed lemurs (*Lemur*)

Slow loris (*Loris*)

Spider monkey (*Ateles*)

Langur monkeys (*Presbytis*)

Young chimpanzee (*Pan*)

Mountain gorilla (*Gorilla*)

Figure 6.23 **Representative primates (phylum Chordata, class Mammalia, order Primates).**

independent and individual mobility of the fingers, (2) the ability of the thumb to oppose the action of the other fingers, and (3) the presence of friction ridges on the palm of the hand and the sole of the foot. Primates have also retained some primitive characteristics that many other mammals have lost in the course of their evolution, including the five-fingered hand, the collarbone (clavicle), and the ability to rotate the two bones of the forearm. Unlike those mammals that rely heavily on the sense of smell, primates rely heavily on vision. Primates have vision that merges images from both eyes to give three-dimensional information (binocular vision). This binocular vision is possible because the eyes moved forward to the front of the skull during early primate evolution, so that the visual fields from the two eyes overlap. The portions of the brain related to vision are expanded in primates, especially when compared to those mammals whose eyes are located on the sides of their heads and whose right and left visual fields are largely separate. Also, the outer surface of the brain (the cerebral cortex; see Chapter 11) is more complex in primates than in other mammals. The increased complexity of the primate brain is associated with an increased complexity of learned behavior. The reliance on learned behavior would be impossible without a lengthy period of very intensive parental care. Primates typically give birth to one offspring at a time. Primate nipples are restricted to a single pair in the chest region, while other mammals have many pairs. Other mammals have two uteri, but in primates these are fused into a single uterus. Primates include lemurs, lorises, galagos, tarsiers, monkeys, apes, and humans. Among these, humans are most closely related to apes, but differ from all nonhuman primates in habitually walking upright.

6.5.2 Early hominins

In 1925, a fossilized child's skull was discovered in a cave near Taung, South Africa, and was named *Australopithecus africanus.* Although the skull had both apelike and human features, most experts treated it as just another ape. Additional fossils of *A. africanus* were discovered subsequently (Figure 6.24). These fossils included skulls with a very low opening for the emergence of the spinal cord from the base of the brain, showing that the skull balanced on top of an erect vertebral column. Parts of the foot, the pelvis, and the lower part of the vertebral column were also present, and these structures confirmed that *Australopithecus* walked upright and was therefore more like humans than like apes. Direct evidence for upright walking comes from the discovery of a set of footprints at Laetoli in Tanzania, approximately 4.1 million years old. Primates that walk upright are placed in the family Hominidae, subfamily Homininae, and are referred to as **hominins.**

Scientists have since unearthed the remains of several other species of *Australopithecus* and related early hominins. The relationships among these hominins are shown in Figure 6.25. The oldest species is *Sahelanthropus tchadensis*, approximately 7 million years old. This earliest hominin had a very small brain

Figure 6.24 Fossils of the genus *Australopithecus.*

Adult *Australopithecus africanus* (Sterkfontein, South Africa)

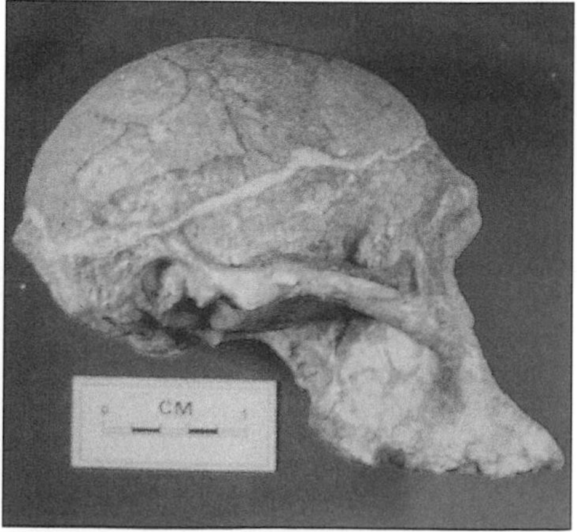

Side view of *Australopithecus africanus* (Sterkfontein, South Africa)

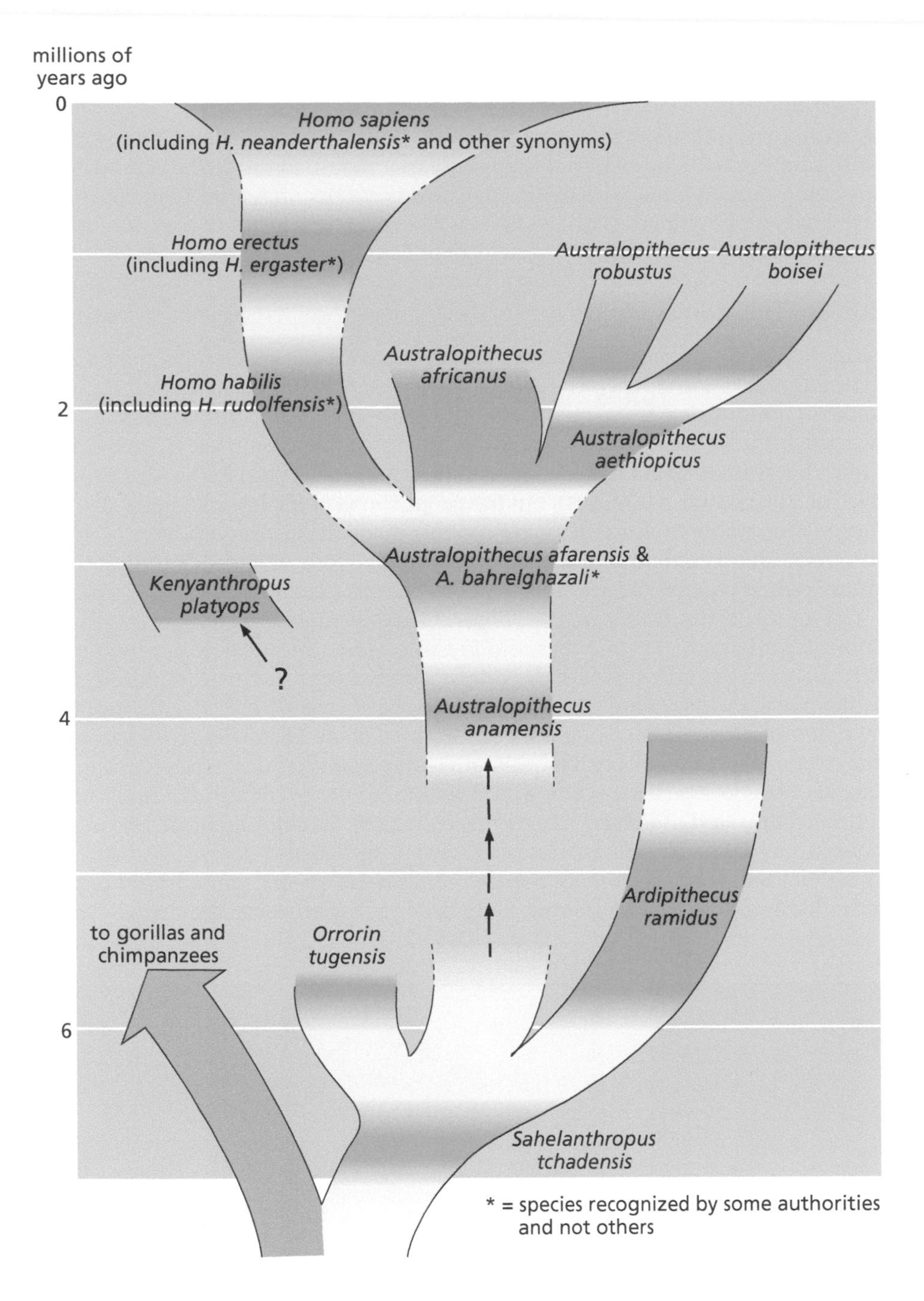

Figure 6.25 A family tree of the family Hominidae. Dark orange areas show the known time range of species represented by fossils. Some dates are approximate.

and certain apelike dental features. Another early hominin, *Australopithecus anamensis*, lived about 4 million years ago in Kenya. *A. anamensis* is thought to be the ancestor of all later species of *Australopithecus* and a close relative of the small hominin *Ardipithecus ramidus*. Another species, *Australopithecus afarensis*, about 3.5 million years old, is represented by the skeleton well known as Lucy, a female about 1.3 meters in height, slightly over 4 feet. Enough of Lucy's skull is preserved to permit us to estimate the size of her brain in proportion to her body size, and these proportions are consistent with the hypotheses that *A. afarensis* was the common ancestor of the genus *Homo* and of several later species of *Australopithecus*. Two of these later species, *A. robustus* and *A. boisei*, were considerably larger than the better-known *A. africanus*, which lived from approximately 3.0 to 2.0 million years ago.

Two other early hominins are *Orrorin tugensis*, an early species estimated to be about 6 million years old, and *Kenyanthropus platyops*, a flat-faced species about 3.3 million years old. In both cases, the dates are a bit uncertain. Few fossils of either species are known, and these fossils are fragmentary. For these reasons, the exact relationship of either of these two species to the better-known hominins is unclear.

Of the species we have described, *S. tchadensis*, *A. anamensis*, and *A. afarensis* were probably along the line leading to *Homo*, but the various other species were probably side branches of the family tree that died out without leaving any surviving descendants. The earliest *Australopithecus* appeared about 4 million years ago, well before the earliest known *Homo*. Later species of *Australopithecus* persisted side by side with *Homo*, at least in East Africa.

6.5.3 The genus *Homo*

Modern humans (*Homo sapiens*) and at least two extinct species are placed in the genus *Homo* (Figure 6.26). The oldest species of *Homo* was *Homo habilis*, which lived in East Africa from about 3.5 to 1.7 million years ago, coexisting with *Australopithecus boisei* and several other *Australopithecus* species. *H. habilis* had a brain that was small in absolute terms (about 400 cm^3, compared with 1200–1500 cm^3 for most modern humans), but the proportions of the brain to body size were more comparable to those of *Homo* than to those of *Australopithecus*. *H. habilis* has been found contemporaneously with certain types of tools, including simple stone tools. It is generally presumed that *H. habilis* was the maker of these tools.

A later species, *Homo erectus* (**Figure 6.26**), is now known from fossils from about 1.5 million to 300,000 years old in China, Java, Europe, and several parts of Africa. A cave site at Zhoukoudian, China (near Beijing), has heat-fractured stones indicative of the use of fire. There is also evidence of round or oval tents supported by poles and held down along the margins by a circle of stones. *H. erectus* varied geographically; the European populations are sometimes described as a separate species (*Homo ergaster*), while a population of very small individuals on the Indonesian island of Flores, sometimes called the "Hobbit man" or *Homo floresiensis*, may have survived to as late as 17,000 years ago and stood only about 3 feet (1 meter) tall. (The modern *H. sapiens* inhabitants of this same island are also remarkably small, leading some scientists to hypothesize that small stature

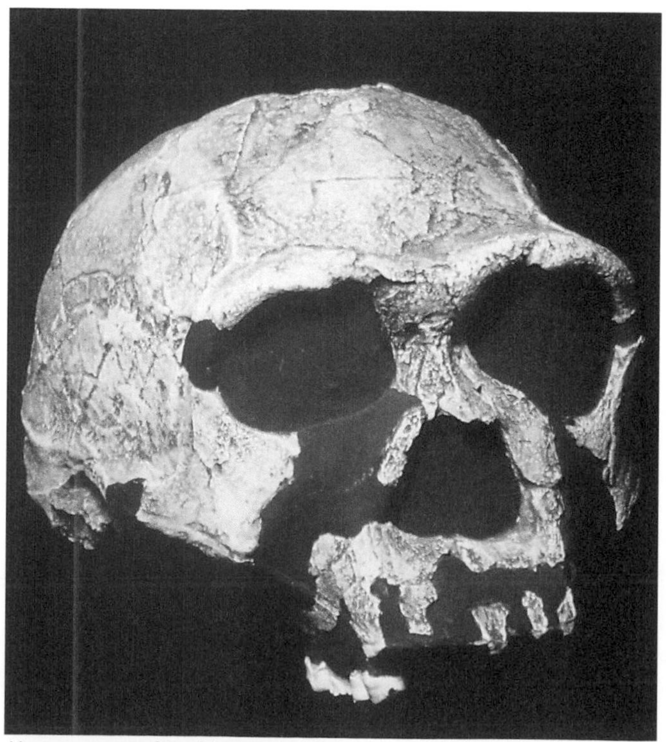

Homo erectus (Koobi Fora, Kenya)

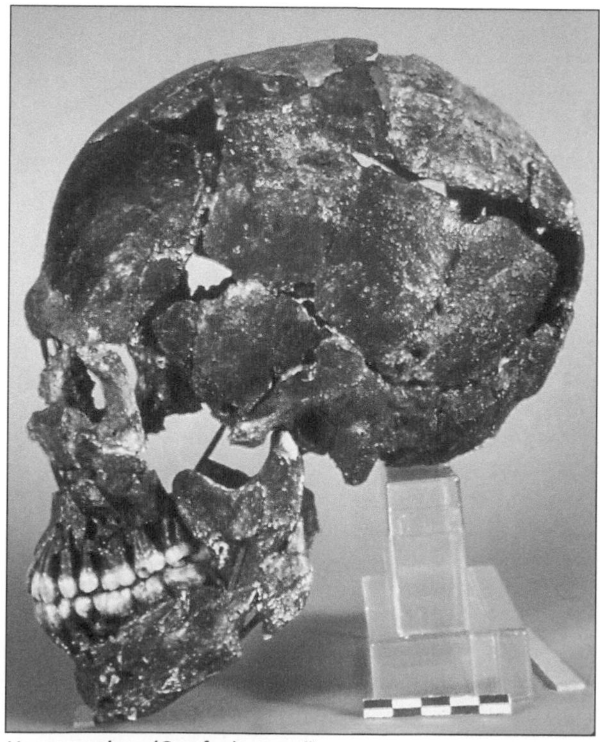

Homo sapiens (Quafzeh, Israel)

Figure 6.26 Fossils of the genus *Homo*.

evolved repeatedly as an adaptation to limited food resources available on the island.) In 2013, a cave in South Africa revealed many skeletons of a newly discovered species, *Homo naledi.* Now dated between 236,000 and 335,000 years old, these fossils show some peculiarly apelike traits in the shoulder region, along with an unexpectedly smaller brain. The distinctive traits show that multiple species of *Homo* may have coexisted over long periods of time.

H. erectus was the ancestor of *H. sapiens*, the species to which living humans belong. Current evidence suggests that *H. sapiens* originated in Africa around 320,000 years ago and spread outward from there, possibly more than once, beginning first around 210,000 years ago. As *H. sapiens* evolved, tools became more sophisticated and were in many cases mounted on wooden shafts. The *H. sapiens* who lived in Europe from about 150,000 to 50,000 years ago are called Neanderthals. Neanderthals hunted deer, horses, and even rhinoceroses and mammoths. Healed surgical wounds show that these skilled hunters took care of sick companions, set broken bones, and even performed simple brain surgery. They buried their dead and decorated the graves with flowers of preferred colors, mostly white or cream-colored. The decoration of graves is thought by several anthropologists to indicate a belief in an afterlife. Although opinions differ on whether Neanderthals should be classified as a separate species, genetic evidence now shows that they did interbreed with other *H. sapiens* populations and are therefore distinct only as a subspecies (see Section 5.4). Complicating this story further is the 2010 discovery in Denisova cave (in the Altai Mountains of southern Russia) of yet another human subspecies, called Denisovans, whose DNA shows evidence of some interbreeding with Neanderthals.

The more modern *H. sapiens* who replaced the Neanderthals were Upper Paleolithic people who lived from about 50,000 to 15,000 years ago. They had an even greater variety of tools, including fishhooks and harpoons. They hunted wooly mammoths and large herd animals. They also left records of their activities in the form of cave paintings, showing their interest in hunting and their understanding of both animal anatomy and physiology. By prominently drawing the heart and singling it out as a target, these hunters showed that they understood how vital this organ was. Their drawings of pregnant deer and of mating rituals show that they knew enough reproductive biology to understand the relationships between mating, birth, and subsequent herd sizes.

The discovery of agriculture ushered in a new phase of human history called the Neolithic. With the planting and harvesting of crops, humans began to settle down into villages, which later grew into towns and cities. Civilization has greatly changed the ways in which we live our lives. The rapid pace and power of cultural change leave many people wondering whether biological evolution of *H. sapiens* has become a thing of the past. If we need to travel faster, the argument goes, our species tames horses or builds automobiles instead of evolving longer legs for faster running. Evolution by natural selection is much slower than cultural innovation. In this view, the future development of our species resides more in our technology than in our bodies.

No one questions that cultural changes in human beings have far outstripped biological ones as the most rapid and far-reaching changes taking place today. Cultural innovation spreads rapidly, in part because there are no species barriers to prevent transmission from one human group to another. (Language barriers and geographic barriers can always be crossed, especially in the age of the Internet and jet travel.) The ease of travel and the global spread of people and their culture has brought us to an era in which there is no significant geographic isolation of human populations. Without geographic barriers, no reproductive isolating mechanisms will evolve, and all humans will remain one species. Cultural change is also more rapid than biological evolution because new inventions and other culturally acquired characteristics *are* inherited, although not genetically. Each generation inherits the stored knowledge of past generations (in libraries and museums, for example), along with tools (from tractors to telephones to satellites) and the technology needed to design and build new and better tools in the future.

Although natural selection continues to take place (see Chapter 5), the environment, and therefore the traits favored by selection, have changed because of our own culture. Many of the selection forces that shaped human evolution in the past, including famines, epidemics, and predators, have been greatly diminished in modern times. Many traits that were once disadvantageous have become much less so. For example, poor eyesight is no longer an important barrier to survival and reproduction in societies that supply eyeglasses.

Human evolution has not stopped, however. The evolution of lactose tolerance has permitted many European (and some African) populations to consume milk into adulthood, an adaptation that provides vitamin D to populations receiving little direct sunlight in winter months. A gene called *EPAS1* increases the production of hemoglobin, allowing physiological adaptation to high altitudes in Tibetan and other high-altitude populations. Also, natural selection continues to operate whenever the chances of survival or reproduction differ among people as a consequence of their genotypes. Many genetic conditions such as cystic fibrosis, Tay–Sachs disease, muscular dystrophy, and others (see Chapter 3) continue to cause numerous deaths before reproductive age, despite the best that medical technology has to offer. Other diseases that are generally survivable may reduce reproductive capacity, which decreases fitness. For example, chondrodystrophy is a rare disease, controlled by a dominant gene, in which the cartilage tissue turns bony at an early age, resulting in a form of dwarfism. Most chondrodystrophic dwarfs enjoy fairly normal health as adults, but have only about one-fifth as many children as their non-dwarf siblings. Lowered reproductive rates are also found in people with diabetes. As these examples show, natural selection continues to affect the human species. Biological evolution thus continues to operate and to interact with cultural evolution in all human populations.

THOUGHT QUESTIONS

1. The Neanderthals were similar to us in many respects, though their skulls had a somewhat more "rugged" appearance, with brow ridges and cheek bones protruding. How should we decide whether Neanderthals should be placed in their own species, separate from *Homo sapiens*?
2. In Europe, Upper Paleolithic culture replaced the culture of the earlier Neanderthal populations rather suddenly. Do you think that the replacement of one set of tools and traditions by another took place mostly by conquest, by intermarriage, or by some combination of the two? What evidence would you look for to test one hypothesis against the others?
3. Is the study of evolution static or changing? Find some recent news articles dealing with new fossil discoveries or other new findings that deal with evolution.

CONCLUDING REMARKS

The classification of organisms into species and higher taxa is an important way to summarize many of the results of evolution. Classifications cannot be static because our understanding of evolution keeps improving.

A major distinction recognized in all modern classifications is that between prokaryotic and eukaryotic cells. Organisms with prokaryotic cells are placed in the domains Archaea and Bacteria, while organisms with eukaryotic cells are placed in the domain Eukarya. On the basis of nucleic acid sequences, the Eukarya are currently divided into approximately six major groups or subdomains, with animals and fungi in the Unikonta, green algae and plants in the Archaeplastida, and four other major groups. Plants are classified by the presence or absence of vascular tissues, of seeds, and of flowers. Animals are classified by the number of embryonic tissue layers, the presence or absence of bilateral symmetry, and various details of embryological development, in addition to groupings by nucleic acid sequences.

Biological evolution continues to act today in humans as it does in other species, although it now interacts with cultural evolution and technological revolution. Natural selection continues to operate by both differential mortality and differential reproduction, and continued selection will result in biological changes within all species, including humans. One frequent result of evolution within species is geographic variation. In the next chapter, we examine some of the reasons for geographic variation within the human species.

CHAPTER SUMMARY

- Scientists make **classifications** that group species together into **taxa** on the basis of their similarities and their evolutionary history.
- The theory behind classifications is called **taxonomy**. One important school of taxonomy, called **cladistics**, bases classifications on the sequence of branching points in family trees and on making taxa correspond to **clades**.
- All organisms are built of **cells** that maintain an efficient ratio of surface area to volume for the organism.
- Early cells were **prokaryotic** and contained no internal membrane-bounded compartments or internal structural fibers. Prokaryotic cells have no nucleus and only a single, circular chromosome compacted through supercoiling.
- **Eukaryotic cells** contain a variety of internal organelles (including a membrane-bounded **nucleus**), internal structural fibers, and multiple chromosomes containing **chromatin**, a regular, repeated arrangement of proteins associated with DNA.
- Highlights of early eukaryotic evolution include the origin of **organelles**, the acquisition of chloroplasts and **mitochondria** by **endosymbiosis**, the evolution of multicellularity, and the origin of sexual recombination of gametes.
- Most prokaryotic and some eukaryotic organisms practice **asexual reproduction** in which no genetic recombination takes place.
- **Sexual reproduction**, which involves genetic recombination, has greatly increased the possibilities for generating new diversity among eukaryotic organisms.
- Evolutionary highlights within the plant kingdom include the origin of protective layers around egg cells and the origin of **tissues**, including vascular tissues. **Vascular plants** have tissues that conduct water and allow plants to have different **organs** devoted to different functions in different parts of the plant. Vascular tissues also allow plants to grow tall. Seeds and **fruits** have permitted flowering plants to adapt to many habitats not previously available to simpler plants.
- Highlights of animal evolution include the origin of **bilateral symmetry**, the evolution of body cavities, the development of segmented body plans, and the evolution of the backbone.
- Early human fossils are placed in the genus *Australopithecus* and several related genera. The genus *Homo* includes various fossils and all living people.

BIBLIOGRAPHIC REFERENCES

Bibliographic references to material in this chapter can be found online at biologytrending.routledge.com/chapter6

GLOSSARY: KEY TERMS TO KNOW

These key terms are defined at the end of each chapter as an aid for student review.

Algae An informal name for photosynthetic eukaryotes that are not plants.

Alveolata A major subgroup of Eukarya that contains such non-photosynthetic unicellular organisms as the ciliates and Apicomplexa, characterized by small, fluid-filled bubbles (alveoli) just inside the plasma membrane.

Animal A multicellular organism that develops from a hollow ball of cells (blastula) and usually has well-developed motility or motile parts.

Apical complex A sharp structure in the group Apicomplexa that allows them to penetrate other cells in order to gain entry as parasites.

Archaeplastida A major subgroup of Eukarya that contains plants, red algae, and green algae, all of which are photosynthetic.

Asexual reproduction Reproduction without any genetic recombination.

Bilateral symmetry Body organization in which the left half and right half are mirror images of one another.

Biofilm A surface made of many bacterial (or similar) cells and their secretion products.

Blastula An early embryonic stage in animal development consisting of a hollow ball of cells.

Cell The smallest unit of living organisms that shows the characteristics of life; can either be free-living or part of a multicellular organism.

Cell theory The theory that cells are the building blocks and functional units of all organisms and that all cells originate from other cells.

Chloroplasts Cytoplasmic organelles that contain chlorophyll and carry out photosynthesis.

Chromatin The repeating structure of proteins (histones) associated with DNA in eukaryotic chromosomes, resulting in their compaction.

Cilia (singular, cilium) A series of many small contractile hairlike structures that coat the surface of *Paramecium* and its relatives (phylum Ciliata) and occasionally other cell surfaces among eukaryotic cells.

Clade A branch of a family tree or dendrogram, forming a monophyletic group.

Cladistics The study of family trees (phylogenies) and the basing of classification on the branching patterns in these trees.

Class A major taxon containing several related orders; examples include insects (Insecta) and mammals (Mammalia).

Classification An arrangement of species into a series of taxa, or "groups within groups."

Coelom A body cavity, especially one whose lining is made entirely of mesoderm.

Cuticle A tough, chemically resistant outer covering, typical of most Ecdysozoa.

Cytoskeleton A set of proteins that give shape to eukaryotic cells and in some cases permit movement.

Deuterostomes Animals in which the blastopore becomes the hind end and the mouth develops at the opposite end.

Differentiation The process of becoming different; a restriction of the set of future possibilities of a cell's progeny.

Ecdysones Hormones that control the process of molting (ecdysis).

Ectoderm The outermost germ layer in a developing embryo.

Embryo The earliest stage of development. In plants, this consists of a reproductive cell surrounded and protected by nonreproductive tissue.

Endoderm The innermost germ layer in a developing embryo.

Endosymbiosis The engulfing of a smaller cell by a larger cell, followed by the persistence of the smaller cell as a specialized compartment or organelle within the larger cell; also, the theory that eukaryotic cells originated by such a process.

Eukaryotic A type of cell possessing a well-formed nucleus and many other internal compartments (organelles) surrounded by membranes.

Excavata A subgroup of Eukarya containing single-celled organisms with a feeding groove excavated on one side.

Exoskeleton A supporting skeleton on the outside, as in insects or crabs.

Family A group of related genera.

Flagellum A eukaryotic organelle whose tubular fibers create a whiplike motion. Also, an analogous structure that produces a propeller-like motion in some bacteria.

Fruit A ripened ovary, containing seeds.

Gastrula An early embryonic stage in most animals containing two to three distinct cell layers, formed from a blastula by the invagination (tucking in) of some cells to form an inner cell layer (endoderm) that is usually separated from an outer cell layer (ectoderm) by an intervening middle layer (mesoderm, absent in Cnidaria).

Genus (plural, genera) A group of closely related species.

Hominin A human being capable of walking upright.

Hypha (plural, hyphae) A threadlike absorptive filament in a fungus.

Invertebrate An animal not possessing a backbone.

Kingdom A major group of organisms, such as the animal kingdom or plant kingdom.

Medusa A free-swimming body type in many Cnidaria, with the mouth facing down; commonly called "jellyfish."

Mesoderm The middle layer of cells in three-layered embryos, from which muscles and various other organs (circulatory, excretory, reproductive, etc.) are derived.

Mitochondria (singular, mitochondrion) Energy-producing organelles in eukaryotic cells, containing all the enzymes of the Krebs cycle along with an outer membrane and a highly folded inner membrane.

Mycota A kingdom of organisms, commonly called "fungi," characterized by absorptive nutrition.

Monophyletic group A taxonomic group corresponding to a clade, containing an ancestral species and all of its descendants.

Notochord A stiff but flexible rod that defines the body axis in the phylum Chordata.

Nuclear envelope A two-layered membrane that surrounds the nucleus in all Eukarya.

Nucleus (plural, nuclei) The central compartment of a eukaryotic cell (such as an animal or plant cell) that contains the chromosomes in a double-layered nuclear envelope, separate from the cytoplasm.

Order A group of related families.

Organ A structure composed of multiple tissues that work together to carry out a function needed by the organism.

Organelles Structures within the cytoplasm of eukaryotic cells, usually enclosed in a membrane.

Phylum (plural, phyla) A major group of organisms, including several related classes.

Planktonic Having little or no locomotor ability and thus drifting at the mercy of water currents.

Plant kingdom A kingdom of photosynthetic organisms that develop from a multicellular embryo in which a fertilized egg (zygote) is surrounded by nonreproductive tissue.

Plasma membrane The outer membrane enclosing a cell.

Plasmid A small, circular DNA molecule, separate from the chromosomal DNA, that can be replicated and passed on through cell division and can sometimes integrate into chromosomal DNA. Found mostly in bacteria but also used in genetic engineering as a vector to transfer DNA sequences between different organisms.

Polyp A body type in some Cnidaria, with the mouth facing upward, usually growing attached to a surface.

Prokaryotic A simple type of cell, such as a bacterium, without a well-defined nucleus or other internal membranous compartments.

Protists An informal designation of those eukaryotic organisms, typically unicellular, that are neither animals nor plants nor fungi.

Protostomes Animals developing from an embryo in which the entrance into the archenteron (the principal cavity of the gastrula stage) develops into the mouth.

Pseudopods Outgrowths from the surface that are capable of changing shape in many eukaryotic cells to aid in locomotion or in food capture.

Rhizaria A major subgroup of Eukarya that includes unicellular organisms with long, needle-like pseudopods.

Sexual reproduction Production of offspring by a process that includes genetic recombination and the fusion of two haploid gametes.

Stramenopiles A major subgroup of Eukarya that includes brown algae and other photosynthetic organisms possessing chlorophyll *c*.

Taxon A group of species at any level of classification.

Taxonomy The study of classifications and how they are made.

Tissue A group of similar cells and their extracellular products, built together (structurally integrated) and working together (functionally integrated).

Unikonta A major subgroup of Eukarya, including the animal kingdom, the fungi (Mycota), and other groups possessing either broad pseudopods or a posterior (trailing) flagellum or both.

Vascular plants Plants possessing vascular tissue.

Vascular tissue Tissue (xylem and phloem) that conducts fluids between different parts of plants.

Vesicles Small, membrane-bounded sacs in which materials are transported between organelles of the endomembrane system within eukaryotic cells.

Zygote A fertilized egg (diploid), formed by the combining of an egg and sperm (both haploid).

CONNECTIONS TO OTHER CHAPTERS

Chapter 1: Well-established classifications are theories (and new classifications are hypotheses) that help explain the patterns of diversity found in nature.

Chapter 2: Differences in DNA structure help us understand the relationships shown in our classifications.

Chapter 4: Different species carry different genomes, and these differences can be used in classification.

Chapter 5: Classifications reflect evolutionary history because differences among taxa are the results of branching evolution.

Chapter 7: Geographic variation within the human species is now understood to be based on the same evolutionary processes that operate within all species.

Chapter 8: Differences in metabolic pathways can be used in classification, especially among bacteria.

Chapter 10: The developmental pathways that lead to tissue differentiation (specialization) are often disrupted or improperly activated in cancer.

Chapter 11: Differences in brain anatomy among different species provide good evidence of evolution that can be used in classifications.

Chapter 13: The presence or absence of drug molecules in plants can be useful characteristics in their classification.

Chapter 15: Species close to us phylogenetically are susceptible to viruses similar to HIV.

Chapter 16: Newly emerging infections can reflect the evolutionary origin of new strains of pathogens.

Chapter 17: Plant characteristics include the presence of chloroplasts responsible for photosynthesis. Plant classifications highlight the differences among plants, such as different photosynthetic pigments and the development of vascular tissues.

Chapter 18: Biodiversity can best be understood by developing a comprehensive classification of all living species.

PRACTICE QUESTIONS

1. What are the major aims of classifications?
2. Name the major groups of Eukarya as currently recognized.
3. Name five organelles that eukaryotic cells possess but prokaryotic cells do not.
4. Name three to four evolutionary advances among plants that you consider most important in their classification. Why was each important?
5. Why are mollusks and segmented worms grouped together? What do they have in common? What important differences characterize each of these phyla?
6. What important similarities do arthropods, nematodes, and tardigrades share? Why do we consider these similarities to be important?
7. What selection pressures fostered the development of each of the following characteristics?
   a. Radial symmetry
   b. Bilateral symmetry
   c. Complete "assembly-line" digestive tract
   d. Body cavities
   e. Segmentation
8. Why are vertebrates grouped together with echinoderms? What evidence exists for this arrangement?
9. What is the major anatomical difference between humans and apes? What is the major anatomical difference between the genus *Australopithecus* and the later members of the genus *Homo*?
10. Examine Box 6.1 and identify three taxa that correspond to clades.

CHAPTER 7
Human Variation

ISSUES

- How can we describe and compare variation within and variation between populations?
- How is the study of population genetics related to human variation?
- Why do human populations differ biologically?
- Do human races exist?
- Is there a biological basis for the idea of race? Is biology the most accurate descriptor of race?
- Will changing biological concepts of race diminish racism? Why or why not?

BIOLOGICAL CONCEPTS

- Populations and population ecology
- Population genetics (genetic variation, Hardy–Weinberg equilibrium, blood groups, genetic drift)
- Patterns of evolution (adaptation, physiology)
- Forces of evolutionary change (natural selection, environmental factors, communicable diseases, parasitism, human health)
- Gene action (molecular structure, genetic polymorphism)
- Scaling (body size and shape)

CHAPTER OUTLINE

DOI: 10.1201/9781003391159-7

The human species is highly variable in every biological trait. Humans vary in their physiology, body proportions, skin color, and body chemicals. Many of these features influence susceptibility to disease and other forces of natural selection. Continued selection over time has produced adaptations of local populations to the environments in which they live. Much of human biological variation is geographic; that is, there are differences between population groups from different geographical areas. For example, Northern European peoples differ in certain ways from those from Eastern Africa, and those from Japan differ in some ways from those from the mountains of Peru. Between these populations, however, lie many other populations that fill in all degrees of variation between the populations we have named, and there is also a lot of variation within each of these groups.

Central to the study of human variation is the concept of a biological population, as defined in Chapter 5, p. 172, and as explained again later in this chapter. Both physical features and genotypes vary from one person to another within populations, but there is also a good deal of variation between human populations from different geographic areas as the result of evolutionary processes. How do populations come to differ from one another? How do alleles spread through populations? How do environmental factors such as infectious diseases influence the spread? Why are certain features more common in Arctic populations and other features more common in tropical populations? Why have we traditionally thought of some of these variations as defining "races"? These are some of the questions that are explored in this chapter.

7.1 THERE IS BIOLOGICAL VARIATION BOTH WITHIN AND BETWEEN HUMAN POPULATIONS

All genetic traits in humans and other species vary considerably from one individual to another. Some of this variation consists of different alleles at each gene locus; other variation results from the interaction of genotypes with the environment. The simplest type of variation governs traits such as those discussed in Section 3.3, in which an enzyme may either be functional or nonfunctional. The inheritance of these traits follows the patterns described in Chapter 2, which you may want to review at this time. In particular, be sure that you understand the meaning of dominant and recessive alleles and of homozygous and heterozygous genotypes. Many other traits, as we saw in Chapter 3, have a more complex genetic basis. In this section we examine how biological variation is described.

7.1.1 Continuous and discontinuous variation within populations

Many human traits vary over a range of values, with all intermediate values being possible; such variation is called **continuous variation**. Continuously variable traits, such as height, can often be measured in an individual and expressed as a numerical value. Other traits that vary continuously, such as hair curliness or skin color, are seldom expressed numerically, although theoretically they could be.

Continuous variation usually results from the cumulative effects of multiple genes, each of which by itself contributes a small effect. Dozens of known genes, perhaps even hundreds, influence height in one direction or another. If we make the simplifying assumption that these effects are independent of one another and that they add up, we can predict that a population of individuals will show a variation in height similar to the bell-shaped curve (**normal distribution**) of Figure 7.1. When we measure heights in any large population, we do in fact get a curve that closely matches this predicted curve. Many other continuous traits vary in much the same way as height. For most of these traits, a strong environmental component also exists. Height, for example, is strongly influenced by childhood

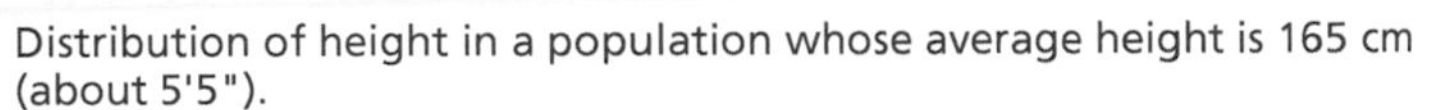

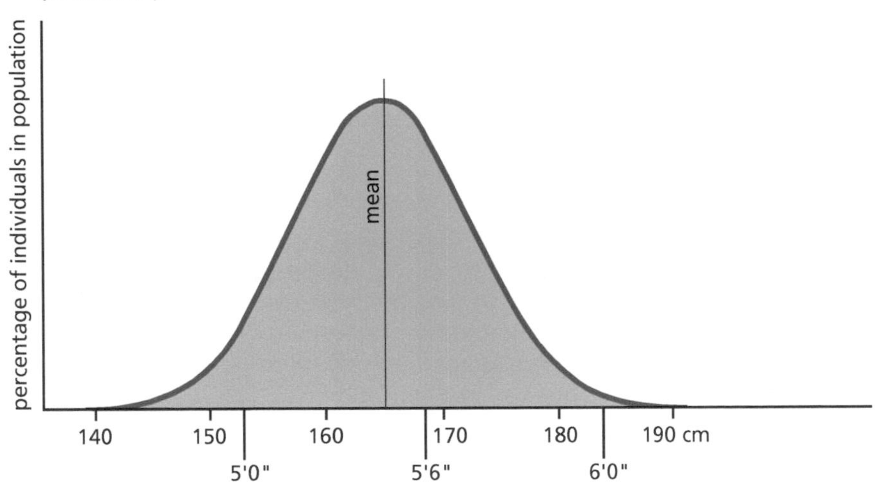

Figure 7.1 Continuous variation in a single population: All intermediate values are possible.

nutrition as well as by genes. Environmental components of traits also contribute to the formation of a bell-shaped curve.

A numerical description of continuous variation in a population requires the use of statistical concepts such as average (**mean**) values. The average values are characteristic of the population as a whole, not of any individual member within the group. For a particular group of people, we can calculate an average height, weight, or head breadth, but these averages are just statistical abstractions—there are perfectly normal individuals who differ from the average, perhaps even greatly, as can be seen in Figure 7.1. Thus, the group average for a continuously variable trait tells us little about any individual. Also, whereas height can actually be measured (and average height computed), concepts such as "tall" are relative: a height that is average in England may be considered tall in India or the Philippines.

Your individual traits result from both the genes that you inherited from your parents and the environmental factors to which you are exposed. What you inherit from your parents is a predisposition for a range of possible future variations in phenotype. For example, when a child is born, their exact height as an adult cannot be predicted, but if the mother and father are both significantly taller than average, then the child, if adequately nourished, will probably also be taller than average.

Discontinuous variation within a population is represented by traits that are either present or absent, with no intermediate values possible. Most of these traits have a simple genetic basis, so that someone's genotype may sometimes be deduced from their phenotype and the phenotype of their close relatives. Traits that vary discontinuously include blood groups and the presence or absence of conditions such as albinism or Tay-Sachs disease (see Chapter 3). A particular phenotype for such a trait is either present or not in a particular individual and is generally not altered by environmental influences.

To describe discontinuous variation in a population, we divide the number of people who have a particular phenotype by the total size of the population; the resulting fraction is the frequency of that phenotype. From these phenotypic frequencies, scientists can calculate the frequencies of the alleles responsible. These **allele frequencies** (originally called gene frequencies) are most easily studied for traits whose patterns of inheritance are known and simple. Like the average values of continuously variable traits, allele frequencies are characteristic of entire populations, not of individuals. All individuals have genotypes, but only populations can have allele frequencies.

7.1.2 Variation between populations

The study of genetic variation both within and between populations is called **population genetics**, and it includes the study of allele frequencies for discontinuous traits. The measuring of allele frequencies requires that the different genotypes, and the alleles responsible for them, can readily be distinguished from one another. It is for this reason that population geneticists often concentrate on those genes whose phenotypic effects are easy to tell apart from one another. Most of those genes control discontinuously variable traits that are either present or not. Differences in the average values for traits that vary continuously are also of interest to population geneticists, but the study of these traits is more difficult because the phenotypes of continuously variable traits are often altered by environmental influences such as nutrition.

One of the central tenets of modern biology is that evolution can occur only if populations are genetically varied. However, biologists did not always think in terms of evolving and variable populations. For over 2,000 years, biologists believed that species were constant, unvarying entities. Plato and Aristotle had declared that each species was designed according to an ideal form that they called an *eidos*, often translated as "type" or "archetype." Biologists following this view developed the **morphological species concept**. Each species was described as having certain fixed and invariant physical characteristics (morphology), and the "type" of that species was believed to be a cluster of "essential" (or "typical") characteristics.

Biologists now recognize that species are constantly evolving, largely as the result of natural selection working on the genetic variation that is present within populations (Chapter 5). The Human Genome Project (Section 4.3) has revealed that over 99.9% of the human genome is identical in all people. However, the remaining fraction of a percent varies geographically, meaning that populations from different locations differ from one another. We must have some clear way to describe this variation and to describe population groups.

To define a population or a larger group of populations, we could sort people by some physical trait, such as distinguishing between people who are tall, short, or average in height. For any trait that we could choose, much of the variation exists within each and every population. If we chose some other physical trait, such as eye color or hair curliness, we would find that each physical characteristic results in a different grouping of the same people. In addition, we find that groupings based exclusively and strictly on any single trait always group together people who are quite dissimilar in many other respects (especially on a worldwide basis). For these reasons primarily, biologists prefer not to base the definition of population groups on physical characteristics.

Instead of using physical characteristics to define populations, biologists use the term **population** to refer to all members of a species who live in a given area and therefore can interbreed with one another. Membership in a population is determined by geographical location and by mating behavior, not by physical characteristics. Populations that interbreed with one another under natural conditions belong to the same species (Section 5.4). All humans are placed in a single species, *Homo sapiens*, because all of them have the capacity to mate with one another and produce fertile offspring. However, people in different geographical locations belong to different populations. Genetic variation within any population is usually less than in the species as a whole. In past centuries, geographic isolation kept many human populations more distinct than they are now with worldwide transportation and migration. Population boundaries are not the same as national boundaries. Several different populations may live in the same geographic area, especially if cultural factors have maintained their separateness and inhibited matings between them. Sometimes, these populations are distinguishable by their derivation from geographically separate earlier populations.

Human populations in different places differ from one another in many physical traits. The average Canadian is taller than the average Southeast Asian, and the average African has darker skin than the average European. For natural selection, however, the characteristics that matter the most are those with the greatest impact on health and disease (or life and death). For example, cystic fibrosis and skin cancer are more frequent among people of European descent, but people of African descent have a higher risk of sickle cell anemia and are more susceptible

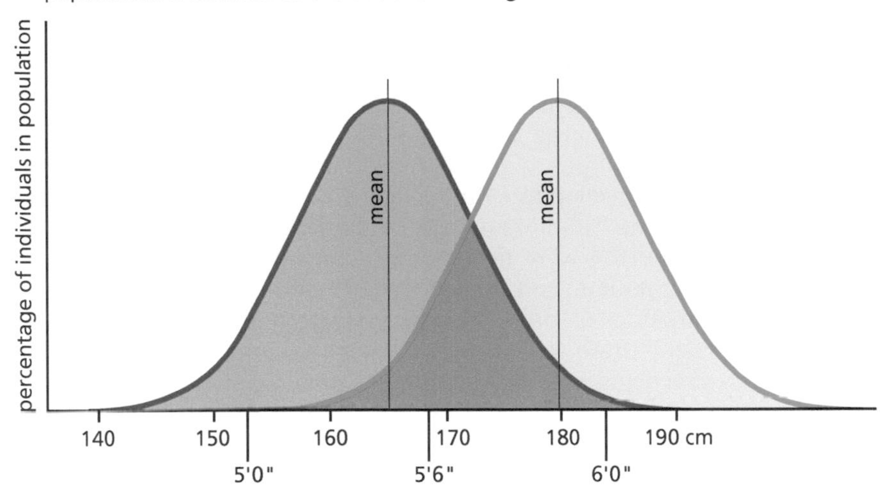

Figure 7.2 Continuous variation in two populations with different mean values.

to frostbite if exposed to very cold temperatures. Most discontinuously variable traits that are examined closely show differences in allele frequency from one human population to another. For continuously variable traits, the difference between the averages of two populations is much less than the variation within either population (**Figure 7.2**). For example, the average height in the United States is taller than in China, but many Americans are shorter than the Chinese average and many Chinese are taller than the American average.

Although it is easy to find human populations that differ from one another in both physical features (morphology) and genetic traits, it is usually very difficult to find sharp boundary lines dividing these populations from one another. If you were to walk from Asia to Europe and then to Africa, you would see populations differing only slightly, in most cases imperceptibly, from their neighbors, and you would meet representatives of the three largest population groups on Earth without finding any abrupt boundaries between them. Another way to say this is that variation between human populations is always continuous. This is true even when the trait in one individual is discontinuous. The population frequency of the allele responsible for the trait can vary continuously between zero (no one has the allele) and 100% (everyone has the allele in homozygous form). For discontinuous traits such as blood type, the allele frequencies of adjacent populations are generally close, just as is true for the average values of continuous traits, as seen in Figure 7.2.

7.1.3 Concepts of race

Humans have developed various ways of describing both themselves and the other human populations with which they have had contact. Biologists (who study all forms of life) and anthropologists (social scientists who study human populations and human cultures) have assisted in these descriptions by studying and measuring certain physical traits and allele frequencies. There are many means by which human variation can be described, and there are many ways in which these descriptions have been used. One of the most problematic has been the attempt to separate people into different **races**. As we will soon see, there are various different meanings to this term, all of them different from the term population. The term "population" always describes smaller and more cohesive units than the term "race." No physical features are used in defining populations, but some race concepts have been based on physical features. In this section we describe four different concepts of race in the order in which they originated. The older concepts have not entirely died out; they have, in many cases, persisted side by side with the concepts that came later.

7.1.3.1 Races based on cultural characteristics In the Bantu languages of Africa, the word for people is Bantu. Likewise, the Inuit word for people is Inuit. Every group of people has a name for itself and its members, and the name often means "people" or "human." Names that people apply to other groups of people may simply be descriptive, but value judgments are often implied as well. In some instances, the value judgment implicit in the choice of name has been used to justify widespread abuses against the negatively labeled population. Such was the case when land and labor shortages resulted from large-scale cereal agriculture, a problem that arose independently in many places. A commonly developed solution to these shortages was to conquer neighboring people (the "other") and confiscate their land. Slavery and several other systems of coercion were developed to secure the labor of conquered peoples. Slavery, oppression, and conquest all call upon the victorious people to practice certain atrocities on others that they would never tolerate within their own group. To justify these atrocities to themselves, and to protect their own members from practicing similar atrocities on one another, just about every conquering group has found it expedient to distinguish themselves from the "other" and, furthermore, to depict the conquered people as somehow inferior, subhuman, or deserving of their fate. Many of the groups that were culturally defined as races in the past are really language groups, cultural groups, or national groups that are hardly distinguishable on any biological basis from the group that traditionally oppressed them.

The imposition of social inequalities between "us" and "them" is now recognized as racism. **Racism** has many meanings, but all of them include the belief that some groups of people are better than others and that it is somehow justified or proper for the more powerful group to subdue and oppress the less powerful. In most cases, the motivation to conquer and oppress others came first; the racist ideology came later.

The "races" identified by the conquering group are socially constructed to serve the interests of the oppressors only. The distinctions and values of the oppressors are forcibly imposed on the oppressed, who are often taught to believe in their own inferiority. Most people now regard racism as unethical because it denies basic rights to many people and because it results in frequent crime, violence, and social conflict.

Separation based on race better serves the political and economic causes that have engendered it if the distinctions recognized are declared to be "natural" and unchangeable, as opposed to characteristics that can easily be changed by education or religious conversion. Scientists belonging to racist societies have therefore sometimes attempted to "prove"' that the traits characteristic of another race have an inherited basis that cannot easily be changed, an assertion called biological or genetic **determinism** (or hereditarianism). Behind such assertions is the view that a group identity (an "essence" or Platonic *eidos*) can be inherited, a view for which there is no basis in genetics. Anthropologist Eugenia Shanklin documents several instances in which scientists conducted "scientific" studies to help "prove" the values and prejudices of their own social group. In their genocidal campaigns of the 1940s, the Nazis exterminated many millions of Jews, gypsies, Slavs, and other groups, but not until they had declared each of them to be an inferior "race."

Racism and hereditarianism are not synonymous, but they often go together as attitudes shared by many of the same people. The supporters of eugenics (see Section 3.4.3) had many followers, including Nazis in Germany and anti-immigrationists in the United States. These followers sought ways to prove the inferiority, and especially the biologically unchangeable inferiority, of other people.

The other race concepts that we discuss next all differ from this earliest concept in their avoidance of language, customs, and other cultural traits in the delineation of races. However, racism is not confined to those societies that embrace the cultural concept of race. Many biologists and anthropologists have pointed out that racism is also built into the next concept: race delineated by body features.

7.1.3.2 The morphological or typological race concept Biologists who study plant and animal species often describe the geographical variation within a species by subdividing the larger species into smaller and more compact subgroups, each of which is less variable than the species as a whole. These subgroups are generally called **subspecies**, but within our own species they are called races. To bring the study of human variation more into conformity with that of other species, scientists began to restrict their attention to characters that could be studied biologically and to exclude personality traits, languages, religions, and customs more influenced by culture than by biology. Also, people who suffered oppression for multiple generations often learned the language of their oppressors and frequently converted to their oppressors' religion. Oppressors unwilling to grant these people full and equal citizenship began to insist that racial groups should be based on physical characteristics that could not so easily be changed.

Before the days of oceangoing vessels, most of the world's people had only a limited awareness of human variation on a worldwide scale. Each population, of course, knew about other populations nearby, but in most cases adjacent populations differ only slightly from one another. When trade extended over great distances, it usually did so in stages, so that none of the traders ever had to go more than a few hundred miles from home. The trade routes were also in most cases traditional, meaning that traders and migrants had generally come and gone over the same routes for centuries. This contributed to a gene flow or mixing of alleles that lessened the degree of difference between populations that would be noticed along the trade routes.

When explorers began to sail directly to other continents, they found people in other lands who differed more sharply from themselves in physical features. Many scientists subsequently became curious about the origin of these physical differences. Discussions of racial origins from about 1750 to 1940 tended to dwell on the origin of physical differences. A morphological definition of each race, based on physical features (morphology), was an outgrowth of the same thinking that had earlier resulted in a morphological species concept. At least initially, the major founders of this tradition were scientists who had no interest in oppressing the newly discovered peoples, so finding an excuse for racial oppression was less of a motive than was scientific curiosity. The emphasis was no longer on distinguishing only "us" from "them" but on distinguishing among many different racial groups.

By the 1700s, biologists were actively describing and categorizing the variation in all living species. The eighteenth-century naturalist Linnaeus (Carl von Linné) divided the biological world into kingdoms, classes, orders, genera, and species (see Section 6.1). He also divided humans into four subspecies: white Europeans, yellow Asians, black Africans, and red (native) Americans. The use of physical features such as skin color and hair texture to define subspecies was common among biologists using a **morphological race concept**. Other scientists in this same tradition recognized more races or fewer, but each race was always described on the basis of morphological characteristics such as skin color, hair color, curly or straight hair, and the occurrence of folds of skin over the eyes (epicanthic folds).

Under the morphological concept of race, each race was defined by listing its common physical features as though they were invariant. For example, when describing a feature such as color, only one color was given, as if this color were invariant throughout the group and throughout time. This approach, which classified races on the basis of "typical" or "ideal" characteristics, ignoring variation, is called **typology**. Morphological definitions of race were always typological. Africans, for example, were declared to have black skin and curly hair, overlooking the fact that both skin color and hair form vary considerably from place to place within Africa and even within many African populations. All of the morphological characteristics were assumed to be inherited as a whole; a person was assumed to inherit a Platonic *eidos* (a type or essence) for whiteness or redness, not just a white or red skin. Supporters of the typological concept of races were also supporters of a typological concept of species.

Years after morphological races had been defined, closer scrutiny revealed both variation within the morphological races and intergradation between them across their common boundaries. A few Europeans tried to save the morphological definitions by proposing that each race had originally been "pure" and invariant and that present-day variation within any population was the result of mixture with other races. One zoologist, Johann Blumenbach (1752–1840), divided up humans into American, Ethiopian, Caucasian, Mongolian, and Malayan races. He thought that each of these races was originally homogeneous (that is, "pure"), and he named each after the place that he identified as its ancestral homeland. For example, white-skinned people are called Caucasian because Blumenbach thought that this race originated in the Caucasus Mountains, east of the Black Sea.

There is no scientific support nowadays for the concept of originally pure races or for the concept of different ancestral centers of origin of different races; human populations have never been homogeneous and have always been quite variable. In some cases, however, Europeans and others who feared for the "purity" of their own group sought to pass laws limiting contacts, especially sexual contacts, between the races that they recognized. Most of these laws were brutal but still ineffective in stopping what were viewed as interracial matings. There is no scientific basis for the belief that such matings are in any way harmful. On the contrary, variation within any species confers a long-term evolutionary advantage because it provides the raw material that natural selection can use to adjust to changing environmental conditions.

However, hereditarian assumptions were even more strongly embedded in the morphological race concepts than they are in the culturally based race concepts. Lest one think that science has long since banished such attitudes in educated people, it is only necessary to point to the great storm of controversy that flourished over the subject of race and IQ in the 1970s. Arthur Jensen attempted to convince his readers that the mental abilities of African Americans were below those of other races and that these differences were fixed by heredity and unchangeable by educational means. A number of scientists, including Leon Kamin, Richard C. Lewontin, and Stephen Jay Gould, showed that his claims were unsupportable and based on fallacies and fabricated evidence. As recently as 1994, a book by Richard Herrnstein and Charles Murray once again brought up many of the same hereditarian arguments that had earlier been debunked (**Box 7.1**).

One of the strange ironies of a racist past is that many attempts at remediation, such as affirmative action, continue to require, at least for a time, the identification and naming of the same groups that were used previously for racially divisive purposes. Attempts to ensure fair and nondiscriminatory treatment for members of different socially recognized racial groups (in housing, employment, schooling, and so forth) require that we first identify and study the groups that we wish to compare. In this way, societies trying to overcome a history of racism find themselves using the very racial classifications of their racist past in order to redress the injustices of past generations.

7.1.3.3 Population genetics, clines, and race Modern studies of human variation are based in large measure on genetics. Genetic variation between populations is continuous, and all boundaries between groups of populations are arbitrary. Even for traits that vary discontinuously and for which an allele frequency can be calculated for each population, geographic variation in allele frequencies is continuous between populations.

A continuous increase or decrease in the average value, allele frequency, or phenotypic frequency of any one trait is called a **cline**, after a Greek word meaning "slope" (as in words like "incline" or "recline"). Clines are an accurate (but lengthy) way of describing the geographic variation in each trait, one trait at a time, and each cline could be shown on a map. For a dozen characteristics, a dozen different maps would be needed, because the patterns of variation would in general not coincide.

Box 7.1 Is intelligence heritable?

To address a question such as this, we must first define intelligence. Intelligence is not easily defined, but it includes the ability to reason and the ability to learn new ideas and new forms of behavior, the measurement of which is far from simple. The biological bases for these abilities are likely to be multifaceted, and genetic factors are likely to be the result of the interaction of many, many genes. Most discussions on the inheritance of human intelligence deal only with a single measure of this very complex trait, the IQ score, obtained from a test. IQ is not the same thing as intelligence and is, at best, an imperfect measure of mental abilities.

Also, to address this question, we must define the word "heritable." Heritability is defined in statistical terms as the proportion of the population's variation in some trait associated with genetic as opposed to environmental variation. Statistical association, or correlation, does not imply causation, and it certainly cannot be used to justify the claim that "there is a gene for" the trait in question. One way to determine heritability of a trait in a domesticated species is to compare the variability of that trait in the population at large with the variability of the trait among highly inbred, genetically uniform individuals. Another way to determine heritability is to compare the variability that a trait exhibits at large with the variability of that trait among individuals raised in a standardized, experimentally controlled environment. Neither of these methods can be applied to humans, and the measures that are used to study humans are all indirect, complicated, and subject to criticism on technical grounds. For these reasons, there is no agreement on the heritability of any important human ability, including intelligence. Moreover, because variation is a characteristic of populations and not of individuals, a term such as "60% heritable" would simply be a ratio of variation in one population to variation in another and would not tell us anything about the genotype or phenotype of any individual.

Numerous studies on IQ scores have shown the following:

- It is difficult to devise IQ tests that are free from cultural bias and from bias based on the language of the test, the gender and race of the test subjects, and the circumstances in which the test is administered.
- IQ scores seem to have both genetic and nongenetic components. Children's IQ scores correlate strongly with those of their parents. The IQ scores of adopted children usually agree more closely with their adoptive parents than with their birth parents, although studies on adopted children have been criticized for a variety of reasons (see Section 3.3.2).
- IQ scores can be greatly improved by environmental enrichment. They can also be adversely affected by poor nutrition, poor prenatal conditions, and a number of other environmental circumstances.
- Populations historically subject to discrimination, such as African Americans in the United States, Maoris in New Zealand, and Buraku-Min in Japan, have average IQ scores about 15 points below those of the surrounding majority populations. However, these lower average scores do not always persist in people who migrate elsewhere: Descendants of Buraku-Min living in the United States have, on average, IQ scores on a par with those of other people of Japanese descent.
- In the United States, IQ scores of whites and also of blacks (African-Americans) vary from state to state, in some cases more than the average 15-point difference between blacks and whites. Among African-Americans born in the South but now living in the North, IQ scores vary in proportion to the number of years spent in northern school systems.
- Transracial adoption studies show that African American children adopted at birth and raised by white families had IQ scores close to (in fact, slightly higher than) the white average.
- Careful studies of matched samples in schools in Philadelphia failed to show significant average differences in IQ scores between black and white schoolchildren if differences in background were controlled. "Matched samples" mean that children in the study were compared only with other children of comparable age, gender, family income level, parents' occupation, and similar variables.

Taken together, these data indicate that there is, at most, a small degree of heritability for IQ. They provide little support for the hereditarian claim that IQ is fixed and immutable or that observed differences in scores cannot be diminished. They provide no support whatever for predicting any individual's IQ score on the basis of their inclusion in any group.

The maps in Figure 7.3 show the clinal variation in the allele frequencies of three blood group alleles. Before such maps of allele frequencies can be drawn, local populations must first be identified and sampled. For example, blood groups must first be studied in many local populations, then the allele frequencies found in each geographic area can be drawn on maps such as those in Figure 7.3. From maps such as these, we learn that large continental areas usually show gradual clines. Thus, Figure 7.3A shows a gradual south-to-north increase in the frequency

(A) The distribution of allele *A*

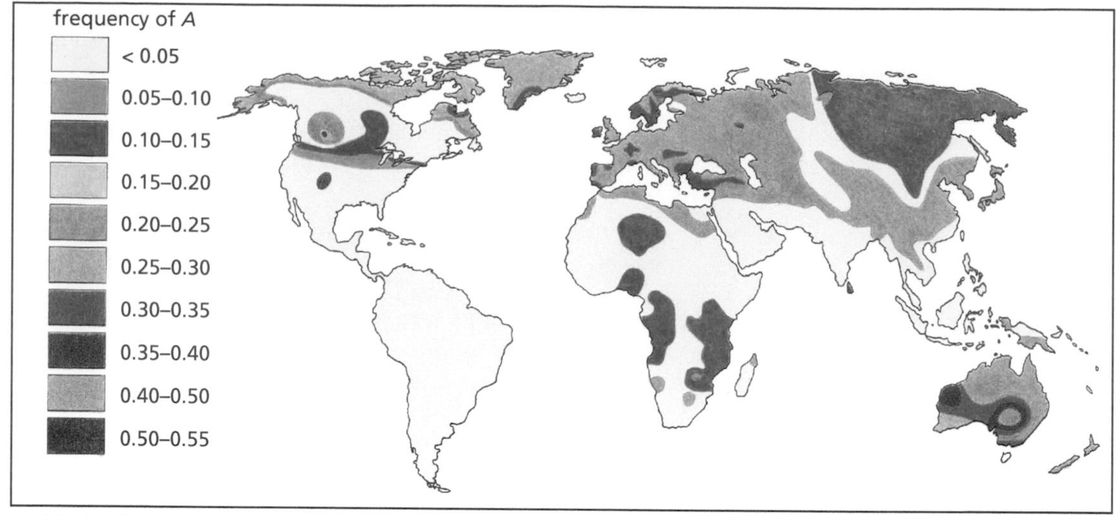

(B) The distribution of allele *B*

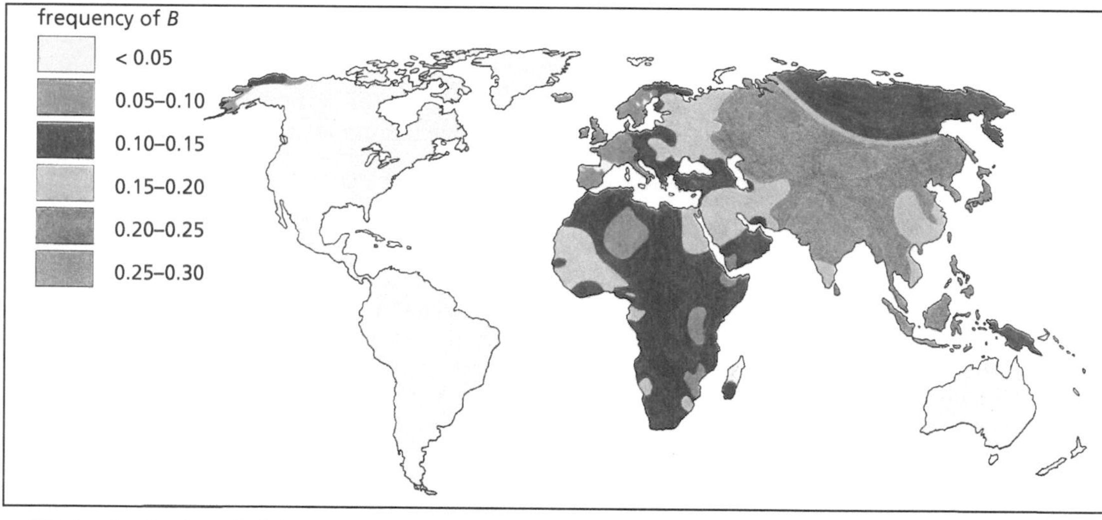

(C) The distribution of allele *o*

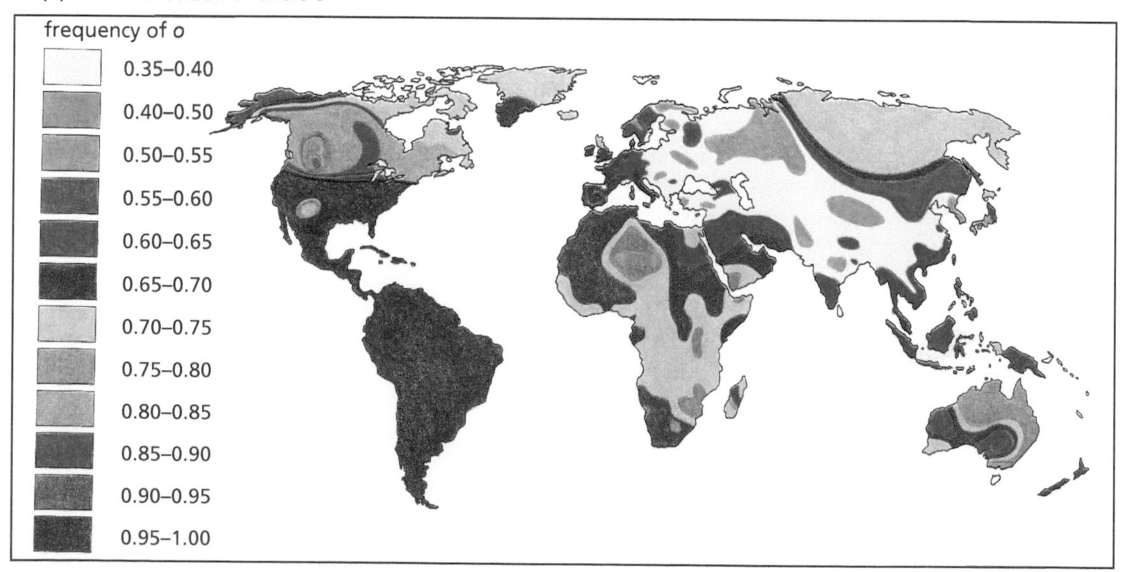

Figure 7.3 The clinal distribution of alleles for the ABO blood groups in indigenous populations of the world. Indigenous populations are those that have lived for hundreds or thousands of years in approximately the same region, to which they have had time to evolve adaptations. This includes Native Americans in the Western Hemisphere, Bantu and Xhoisan peoples in southern Africa, and Aborigines in Australia, but not the European colonists who came to these places after the year 1500.

of allele *A* across North America, and Figure 7.3B shows a gradual west-to-east increase in the frequency of allele *B* across most of Eurasia. In geographic variation, clines of this sort are gradual, and boundaries of population groups are therefore arbitrary. Abrupt changes are uncommon, and when they do occur, they generally coincide with geographic barriers that hinder both migration and gene flow. Examples of such barriers include the Sahara Desert, the Himalaya Mountains, and the Timor Sea north of Australia.

As can be seen in Figure 7.3, the frequencies of the blood group alleles *A*, *B*, and *o* vary greatly from one human population to another. The variations, however, do not necessarily coincide with other traits or with the groups recognized on the basis of morphology. Allele *B*, for example, reaches its highest frequency on mainland Asia, but is nearly absent from Native American populations or among Australian Aborigines. The frequency of allele *A* decreases from west to east across Asia and Europe. In Native American populations, allele *A* occurs mostly in Canada and is mostly absent from indigenous Central or South American populations. The allele for blood group O has a frequency of 50% or more in most human populations, but its frequency approaches 100% in Native American populations south of the United States. African populations generally have all three of the alleles for ABO blood groups at levels close to worldwide averages.

Since the clinal variation concept was introduced in 1939, it has become customary to describe human variation by drawing maps of one cline after another. In addition to cline maps of phenotypes and allele frequencies, the techniques of molecular genetics (such as the DNA marker techniques described in Section 4.2) are now being used to study clines at the molecular level. Clinal maps can also be drawn for continuous traits, in which case average values for the trait are calculated in each population. To describe the geographic variation in *Homo sapiens* or any other species, we could draw one map showing clinal variation in average body height, another showing variation in skin color or hair form, and so on. The population genetics approach encourages the scientific study and description of populations, including studies on the origins and former migrations of populations.

After the Holocaust (1933–1945), the fledgling United Nations felt the need to refute many Nazi claims about race. The result was the 1948 *Statement on Race and Racism*, written by a committee that included several prominent anthropologists and geneticists. The statement, which has been revised several times since 1948, correctly pointed out that nations, language groups, and religions have nothing to do with race and that no group of people can claim any sort of superiority over another. The statement went further, however, to proclaim a new definition of race that replaced older, morphological definitions based on the inheritance of Platonic "ideal types" with a new definition based on population genetics.

Under the population genetics definition, a race is a geographic subdivision of a species distinguished from others by the allele frequencies of a number of genes. A race could also be defined as a coherent group of populations possessing less genetic variation than the species as a whole. Either definition means that blood group frequencies are now considered more important than skin color in describing race and that races are groups of similar populations whose boundaries are poorly defined. It also means that one cannot assign an individual to a race without first knowing what interbreeding population that individual belongs to. "Race" is no longer a characteristic feature of any individual, because allele frequencies, like average phenotype values, characterize populations only, not individuals. Allele frequencies are consequences of population membership; they cannot be used to assign someone to a particular population or group. For this reason, racially discriminatory laws cannot and do not use population genetics; such laws rely invariably on the older morphological definitions or the still older social definitions of race.

Some writers maintain that racism is still contained in the population genetics race concept. They contend that studies that describe allele frequencies in geographic populations are merely "reinscribing" the racism of earlier concepts.

Although far fewer people see racism in population genetics than in the earlier race concepts, some wish to go even further than the UN statement goes.

7.1.3.4 The concept of "all one race" Some scientists went even further in rejecting the heritage of the racist past: led by the British American anthropologist M.F. Ashley Montagu (who had earlier contributed to the UN definition), they declared that they would not recognize different races at all. Among their arguments, one of the most compelling is that race concepts have always been misused by racists of the past and that the only way to rid the world of racism is to reject the entire concept of race. History is replete with examples of slavery, apartheid, discrimination, genocide, and warfare between racial groups. It is therefore easy to argue that the naming of races has in past generations done far more harm than good.

One stimulus to this new approach arises from the realization that there are no unique alleles or other genetic markers that could identify a person's race. Populations or population groups differ in the frequency of various alleles but do not have alleles that belong exclusively to that group. Even differences in allele frequencies have become less pronounced as a result of the great increase in international travel and migration that has occurred especially since World War II. To a certain extent, human populations have always mated with one another whenever there has been geographic contact between them; this is one reason why human population groups do not differ more than they do and why neighboring populations are so often similar. Since the advent of the jet age, frequent migrations have allowed more extensive contact and more opportunities for mating between people of different genetic backgrounds than ever existed before. Such matings even occur in societies that have tried to outlaw them. This type of mating will slowly but inevitably diminish the differences in the mix of alleles, or **gene pools**, of populations, making it progressively more difficult to identify any significant differences between populations.

7.1.4 The study of human variation

All studies of human variation carry the risk of being misused or misinterpreted by racists. Nevertheless, there are many good reasons for studying human variation, and this study serves as the basis for the entire field of "human factors engineering." To take a simple example, the design of a passenger compartment (for automobiles, aircraft, etc.) must accommodate a certain range in the size, sitting height, arm length, and other dimensions of its possible occupants. These and other accommodations must take into account the total range of human variation, including all races and both sexes. In airline cockpits and similar enclosures, controls should be both visible and reachable by persons of different sizes. Moreover, these features are often matters of safety as well as comfort. Vehicle seat belts and airbags, sports equipment, surgical equipment, wheelchairs and similar aids, boots, helmets, kitchen counters, telephone receivers, gas masks, toilets, and doorways all need to accommodate the range of dimensions of the human body. Variation in other human characteristics (breathing rates, sweating) must also be considered in the design of space suits, diving equipment, respiratory equipment for fighting fires, or protective clothing for other situations. Most of the variation relevant for human factors engineering is found within each population group, including variation by age and sex; variation between human populations is generally minor by comparison.

A further reason for studying genetic variation among human populations is that it can help us understand evolution. Population genetics has helped us recognize geographic patterns of disease resulting from natural selection acting on human populations. Studies of this kind can also help us to reconstruct the past history of particular human populations or of the human species as a whole. In the following sections of this chapter, we examine some of these studies.

THOUGHT QUESTIONS

1. Twentieth-century approaches to the description of human variation have in large measure been revolts against the earlier approaches. Against which of the earlier approaches was the "all one race" approach primarily directed? Against which earlier approach was the population genetics approach directed?
2. African Americans more often have high blood pressure and more often die from their first heart attacks than do white Americans. How would you decide whether this is the result of a difference in genes, in diets, in the availability of medical care, or in the lasting effects of discrimination in U.S. society? If people in rural Africa seldom have heart attacks or high blood pressure, what possible hypotheses are falsified?
3. To produce research results of the kind referred to in Thought Question 2, one must have a way of assigning an individual to a population group. How does one determine a person's membership in a biological population? Is it sufficient to know that they live in a particular place? Will asking people to name the racial or ethnic group in which they claim membership (self-identification) produce biologically meaningful results?

7.2 POPULATION GENETICS CAN HELP US TO UNDERSTAND HUMAN VARIATION

The geographic variation shown in Figure 7.3 deals with human blood groups. We know a lot about the genetic basis of blood groups, and a person's blood group is easily determined, making blood groups good candidates for study by population geneticists. We now look in more detail at human blood groups and what studying them has taught us about our own species.

7.2.1 Human blood groups and geography

In the days before reliable blood banks, blood transfusions were much riskier than they are today. Soldiers wounded in battle were generally treated in the field. If a transfusion was needed, it was done directly from the blood donor to a patient lying on an adjacent stretcher. Some transfusions were successful, but others resulted in death of the patient. Studies on the reasons for these different outcomes led to our knowledge of the existence of blood groups.

7.2.1.1 ABO blood groups During the Crimean War (1854–1856), a British army surgeon kept careful records of which transfusions succeeded and which did not. From his notes he was able to identify several types of soldiers, including two types that he called A and B. Transfusions from type A to type A were nearly always successful, as were transfusions from type B to type B, but transfusions from A to B or B to A were always fatal. Also discovered at this time was a third blood type, O, which was initially called "universal donor" because people with this blood type could give transfusions to anyone. These results were put to immediate practical use in treating battlefield injuries.

Karl Landsteiner, an Austrian pathologist who migrated to the United States, discovered the reason for these distinctions. Persons with blood type A make a carbohydrate of type A, which appears on the surfaces of their red blood cells. Persons with blood type B make a carbohydrate of type B; persons with type AB make both type A and B carbohydrates; and persons with blood type O make neither of these carbohydrates. The A and B carbohydrates are also called **antigens** because they are capable of being recognized by the immune system (see Chapter 14). The immune system of each individual makes antibodies against the blood group antigens that their own body does not make. In a person receiving a transfusion with incorrectly matched blood, these antibodies bind to the type A or B antigens, causing the blood cells to clump together within the blood vessels (Figure 7.4), often with fatal results. For explaining these immune reactions, Landsteiner received the Nobel Prize in 1930.

The A and B antigens allow all people to be classified into the four blood groups A, B, AB, and O. These blood groups are controlled by a gene that has three alleles: Allele *A* is dominant, and it contains information for producing antigen A (its phenotype); allele *B* is dominant, and it contains information for producing antigen B (its phenotype); allele *o* is recessive, and it functions as a "placeholder"

blood type	genotype	antigen	antibodies made	recipient ↓	donor			
					A	B	AB	O
A	*AA* or *Ao*	A	anti-B	A				
B	*BB* or *Bo*	B	anti-A	B				
AB	*AB*	A + B	neither anti-A nor anti-B	AB				
O	*oo*	neither	both anti-A and anti-B	O				

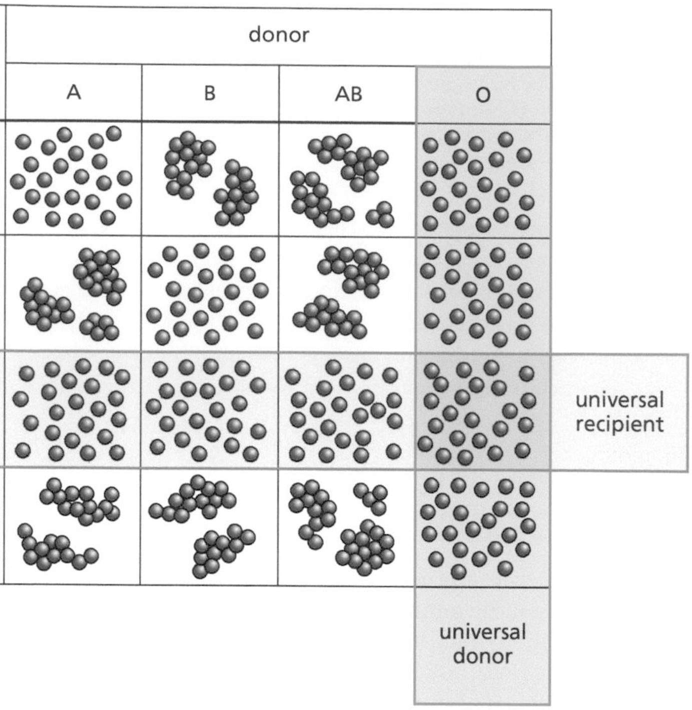

universal recipient

universal donor

Figure 7.4 The human ABO blood groups. If a person of blood type A, who makes antibodies against blood type B, receives a transfusion of type B or AB blood, those antibodies cause the donated blood cells to clump together. Transfusion with matched blood or with blood type O (no A or B antigens) does not cause clumping.

on the DNA but produces neither functional antigen. The *AA* and *Ao* genotypes both produce antigen A and are therefore assigned to blood group A. Likewise, both *BB* and *Bo* genotypes produce antigen B and result in the B blood type. Genotype *oo* produces neither A nor B antigens, which results in the O blood type (universal donor). Finally, genotype *AB* allows both alleles *A* and *B* to produce their respective antigens, resulting in the AB blood type. When they occur together, the *A* and *B* alleles are said to be **codominant** because the heterozygote shows both phenotypes.

For the purpose of matching blood donors and recipients, any person who shares your blood type is a good donor. It is therefore possible to collect blood in advance from many donors, sort the blood by blood type, and store it under refrigeration for use in an emergency. It is ironic that the doctor who developed this concept, an African American named Dr. Charles Drew (1904–1950), was denied its full benefits because many hospitals at the time kept separate blood banks for whites and nonwhite patients, a practice that has no biological foundation. Because the chemical composition of the allele products does not vary, type A antigen from an African American is identical to type A antigen from a Native American or from anyone else. A person with blood type A is therefore a good donor for almost any other person with blood type A.

7.2.1.2 Other human blood groups Karl Landsteiner also discovered several other blood group systems that are totally independent of ABO. One such system, called the Rh system, actually has three genes located very close together on the same chromosome: The first gene has alleles *C* and *c*, the second has alleles *D* and *d*, and the third has alleles *E* and *e*. Unlike the ABO system, in which alleles are codominant, *c*, *d*, and *e* are recessive to *C*, *D*, and *E*. In all, there are eight phenotypic possibilities, of which phenotype cde (genotype *ccddee*, homozygous recessive for all three genes) is sometimes called Rh-negative and the others Rh-positive. The CDe phenotype is the most frequent phenotype in most

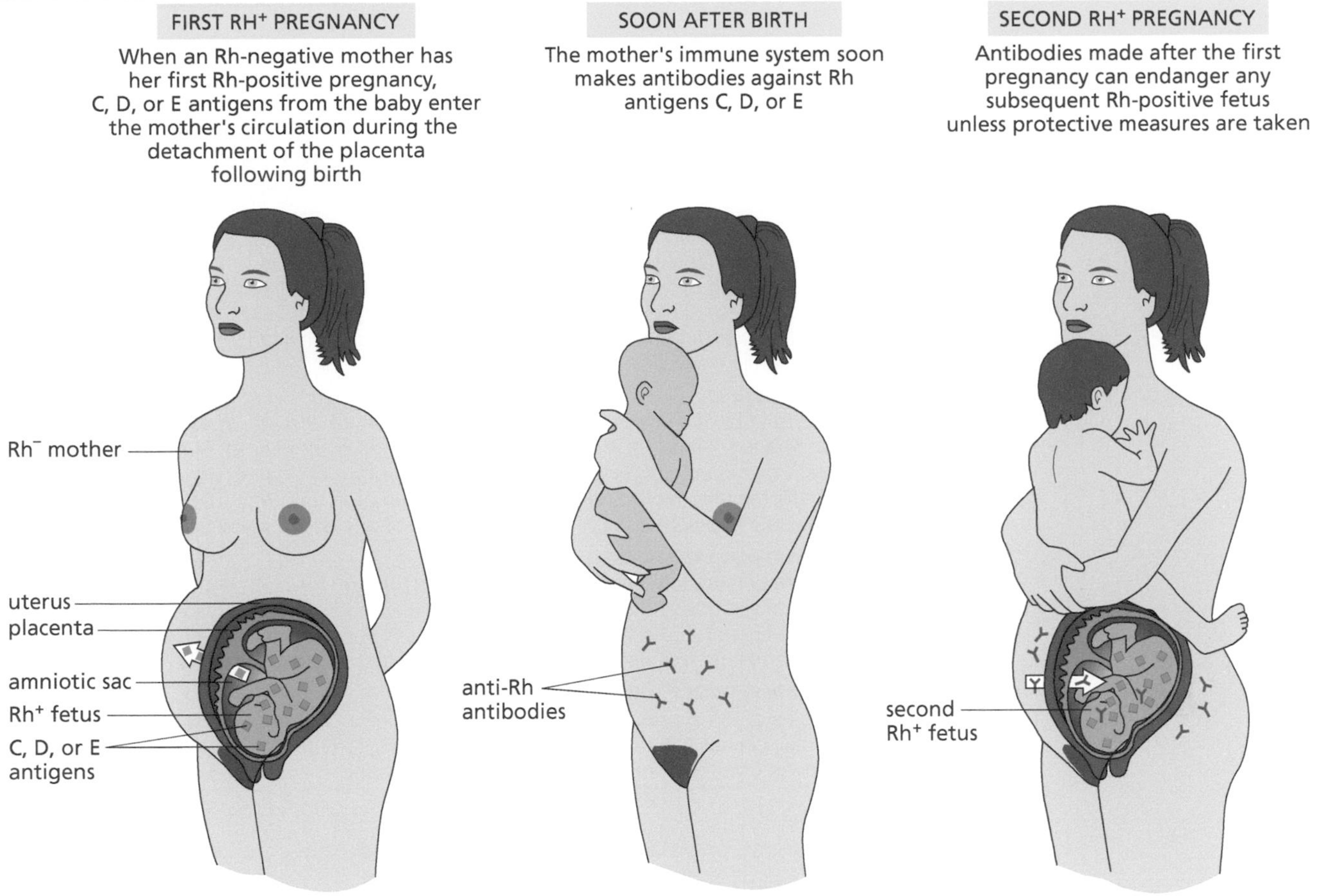

FIRST RH+ PREGNANCY
When an Rh-negative mother has her first Rh-positive pregnancy, C, D, or E antigens from the baby enter the mother's circulation during the detachment of the placenta following birth

SOON AFTER BIRTH
The mother's immune system soon makes antibodies against Rh antigens C, D, or E

SECOND RH+ PREGNANCY
Antibodies made after the first pregnancy can endanger any subsequent Rh-positive fetus unless protective measures are taken

Rh⁻ mother
uterus
placenta
amniotic sac
Rh+ fetus
C, D, or E antigens
anti-Rh antibodies
second Rh+ fetus

Figure 7.5 Rh incompatibility arising in an Rh-negative mother pregnant with an Rh-positive child.

populations, except in Africa south of the Sahara, where cDe predominates. The Rh-negative phenotype cde is the second most common Rh phenotype in Europe and Africa, but is rare elsewhere.

Problems arise when a mother with the cde Rh-negative phenotype is pregnant with a baby who has a dominant *C* or *D* or *E* allele and is therefore Rh-positive. In this case, the mother makes antibodies against the C, D, or E antigens on the baby's blood cells, especially in response to the tearing of blood vessels during the process of birth. Because these antibodies are made at the end of pregnancy, they usually don't affect the first Rh-positive fetus that the mother carries. However, once these antibodies are made, the mother's immune system attacks any Rh-positive fetus in a subsequent pregnancy, destroying many of the fetus's immature red blood cells, which can cause the death of the fetus (Figure 7.5). This problem can now be prevented by giving the Rh-negative mother gamma globulin (e.g., RhoGAM) at the time of the birth of any Rh-positive child; the globulin inhibits the formation of antibodies against Rh antigens, thereby protecting future pregnancies.

Separate from the ABO and Rh blood group systems are an MN system (with M most frequent among Native Americans and N among Australian Aborigines), a Duffy blood group system (with alleles *Fy*, Fy^a, and Fy^b), and many others.

7.2.1.3 Geographic variation in blood group frequencies

We saw earlier that the alleles for the ABO blood groups vary in frequency in different geographic locations (see Figure 7.3). Table 7.1 shows how the major geographic subgroups of *H. sapiens* differ in the frequencies of various blood groups and other genetic traits. It is important to remember that allele frequencies characterize populations only, not individuals. No blood group is

Table 7.1 Allele frequencies in the major geographic population groups

African populations Frequencies of blood group alleles *A*, *B*, and *o* in the ABO system and *M* and *N* in the MN system close to world averages; Rh blood group system with allelic combination *cDe* most frequent and *cde* second; allele *Fy* most common in the Duffy blood group system; allele P_1 more common than P_2 in the P blood group system; hemoglobin alleles Hb^S and Hb^C more frequent than in most other populations.
Caucasian (European and west Asian) populations Allele *A* in the ABO blood group system somewhat more frequent than in African or Asian populations; frequencies of *M* and *N* in the MN system close to world averages; Rh blood group system with allelic combination *CDe* most frequent and *cde* second; Duffy blood group system with allele Fy^a most frequent and Fy^b second; alleles P_1 and P_2 both common in P blood group system; alleles for G6PD deficiency and thalassemia more frequent than in most other population groups.
Asian populations High frequencies of allele *B* (and correspondingly less of allele *A*) in the ABO blood group system; *M* and *N* alleles at frequencies close to world averages; Rh blood group system with allelic combination *CDe* most frequent and *cde* rare or absent; Fy^a especially common in the Duffy blood group system; allele P_2 more common than P_1 in the P blood group system; some populations with high frequencies of alleles for thalassemia.
Native American (Amerindian) populations Very high frequencies of allele *o* and virtually no *B* in the ABO blood group system; very high frequencies of allele *M* in the MN system; Rh blood groups with allelic combination *CDe* most frequent and *cde* rare or absent; allele P_2 more common than P_1 in the P blood group system; high frequencies of Di^a in the Diego blood group system.
Australian and Pacific island populations Frequencies of blood group alleles *A*, *B*, and *o* in the ABO system close to world averages; high frequencies of allele *N* in the MN blood group system; Rh blood groups with allelic combination *CDe* most frequent and *cde* rare or absent.

unique to any population, so a person's blood type cannot identify them as a member of any population.

Frequencies of blood group alleles also vary on a smaller geographic scale. This is especially true among rural people who remain in their native villages or districts all their lives. The geneticist Luigi Cavalli-Sforza has documented variation in the ABO, MN, and Rh blood group frequencies from one locality to another across rural Italy. Similar results have been observed in rural populations in the valleys of Wales, in African Americans from city to city across the United States, and among the castes and tribes of a single province in India. These studies emphasize the hazards of assigning all people in a single country to a single population, especially when cultural barriers discourage random mating. However, populations that have become more mobile experience less of this microgeographic variation. As stated earlier, these are variations in allele frequencies and therefore cannot be used to establish clear-cut boundaries between populations.

7.2.2 Isolated populations and genetic drift

In large, randomly mating populations in which selection and migration are not operating, the frequencies of the genotypes in the population tend to remain the same. This principle, which operates in all sexually reproducing species, is called the **Hardy–Weinberg principle**, and the predicted equilibrium is called the **Hardy–Weinberg equilibrium** (Box 7.2).

One of the criteria for a Hardy–Weinberg equilibrium is that the population be large. In small populations, allele frequencies tend to vary erratically, in unpredictable directions, causing deviation from the expectations of the Hardy–Weinberg equilibrium. This phenomenon, called **genetic drift**, is defined as changes in allele frequencies in small to medium-sized populations due to chance alone.

The original model of genetic drift dealt with populations that remained small all the time, but other types of genetic drift were found to apply in particular situations. For example, if a large population became temporarily small and then large again, the random changes in allele frequencies that occurred when the population was small—the bottleneck—would be reflected in the allele

Box 7.2 The Hardy–Weinberg equilibrium

The Hardy-Weinberg principle can be stated as follows:

In a large, random-mating population characterized by no immigration, no emigration, no unbalanced mutation, and no differential survival or reproduction (that is, no selection), *the frequencies of the alleles (genotypes) tend to remain the same.*

Allele frequencies are fractions of the total number of alleles present. If a population of 500 individuals (or 1000 alleles at a single genetic locus) contains 400 alleles of type *A* and 600 alleles of type *a*, then we say that the frequencies of the two alleles are 0.40 and 0.60, respectively, or 40% and 60% of the total number of alleles in the gene pool. At a given locus, the allele frequencies always add up to 1, or 100% of the population's gene pool.

Under the conditions specified in the Hardy-Weinberg principle, as stated earlier, there is a simple equilibrium of unchanging allele frequencies. Let us consider the case of a gene locus that contains two alleles, *A* and *a*. If the frequency of allele *A* is called p and the frequency of allele *a* is called q (where $p + q = 1$), then the equilibrium frequencies of all three diploid genotypes is given by the Hardy-Weinberg formula:

Genotypes	*AA*		*Aa*		*aa*	
Frequencies	p^2	+	$2pq$	+	q^2	= 1

This formula predicts that the frequency of the homozygous dominant genotype *AA* will be p^2, the frequency of the heterozygous genotype *Aa* will be $2pq$, and the frequency of the homozygous recessive genotype *aa* will be q^2.

To show that these equilibrium frequencies remain stable over successive generations and do not tend to change in either direction, consider the production of gametes in a population already at equilibrium. All of the gametes produced by the dominant homozygotes *AA* carry allele *A*, so the frequency of *A* gametes from *AA* homozygotes is p^2. Half of the gametes produced by the heterozygotes *Aa* also carry allele *A*, so the frequency of *A* gametes from heterozygotes is half of $2pq$, which equals pq. The total proportion of *A* gametes is thus $p^2 + pq$. We can now use simple algebra, separating out the common factor and then applying the equation $p + q = 1$ to calculate the frequency of *A* gametes:

Frequency of *A* gametes:

$$\begin{aligned} p^2 + pq &= p(p+q) \\ &= p(1) \\ &= p \end{aligned}$$

In a similar fashion, the proportion of gametes carrying allele *a* is equal to pq (the other half of $2pq$) from the heterozygotes plus q^2 from the recessive homozygotes *aa*.

Frequency of *a* gametes:

$$\begin{aligned} pq + p^2 &= (p+q)q \\ &= (1)q \\ &= q \end{aligned}$$

So, the frequency of *A* and *a* gametes corresponds to the frequency of *A* and *a* alleles.

Combining the gametes in all possible combinations (to simulate random mating) produces the following results:

		Female gametes		
		A	*a*	←Gametes
		P	q	←Frequencies
Male gametes	A	*AA*	*Aa*	←Genotypes
	P	p^2	pq	←Frequencies
	a	*Aa*	*aa*	
	q	pq	q^2	

Taking the resulting genotypes from the chart (and adding the two heterozygous combinations together), we obtain:

AA		*Aa*		*aa*	
p^2	+	$2pq$	+	q^2	=1

(*Continued*)

This is the same equation that we started with, which shows that the frequencies have not changed. It can also be shown that a population that does not start out at equilibrium will establish an equilibrium in a single generation of random mating.

Notice all the assumptions of the model: The population must be closed to both emigration and immigration, and there must be no unbalanced mutation and no selection. The population must be large enough to permit accurate statistical predictions, and the population members must mate at random. In reality, most natural populations are subject to mutation, selection, and nonrandom mating (including inbreeding), and most usually experience emigration and immigration as well. The Hardy–Weinberg model, in other words, describes an idealized situation that is seldom realized in practice. The Hardy–Weinberg equilibrium is important to population genetics as an ideal situation with which real situations can be compared; if a population is *not* in Hardy–Weinberg equilibrium, one can ask why and then seek to measure the extent of the deviation from equilibrium. The same procedure is followed in other sciences as well. For example, "freely falling bodies without air resistance" are an ideal situation in physics, and air resistance can be measured as a deviation from this ideal.

The Hardy–Weinberg equation is useful in estimating allele frequencies for traits controlled by a single gene. For example, if a population of 1000 has 960 individuals showing the dominant phenotype (such as normal pigmentation) and 40 displaying the recessive phenotype (such as albinism), then q^2, the proportion of homozygous recessive individuals, is equal to 40/1000, or 0.04. From this, we can calculate $q = \sqrt{0.04} = 0.2$.

From the fact that $p + q = 1$, we can calculate $p = 1 - q$. Substituting the value of 0.2 that we found for q gives us $p = 1 - 0.2$ or $p = 0.8$. Then the proportion of homozygous dominants in the population is $p^2 = (0.8)^2 = 0.64$ and the proportion of heterozygous individuals is $2pq = 2(0.8)(0.2) = 0.32$.

frequencies of subsequent generations. This **bottleneck effect** is shown in Figure 7.6. Another type of genetic drift occurs if a small number of individuals become the founders of a new population. The allele frequencies in such a new population—whatever its subsequent size—will reflect the allele composition of this small group of founders, an influence known as the **founder effect.**

Several cases of genetic drift have been studied in isolated human populations. One well-studied example concerns the German Baptist Brethren, or Dunkers, a religious sect that originated in Germany during the Protestant Reformation. Forced to flee their native Germany, a few dozen Dunkers came to Pennsylvania in 1719 and started a colony that grew to several thousands and

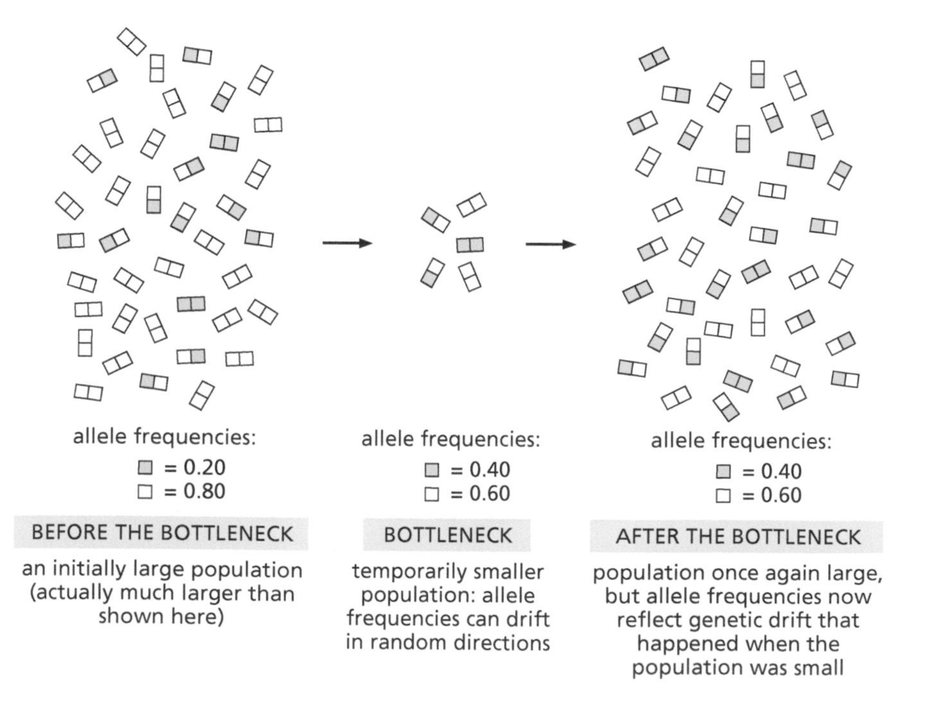

Figure 7.6 **The bottleneck effect, a form of genetic drift that originates when populations are temporarily small.**

spread to Ohio, Indiana, and elsewhere. Because their strict religious code forbids marriage outside the group, they have remained a genetically distinct population.

Allele frequencies among the Dunkers have been influenced by genetic drift, particularly by the founder effect. If the Dunkers were a representative sample of seventeenth-century German populations, we would expect similar allele frequencies to those of present-day German populations derived from the same source. If, however, natural selection had changed the Dunker populations as the result of adaptations to their new location, then we would expect their allele frequencies to come closer to those of neighboring populations of rural Pennsylvania. Neither of these predictions is correct. Allele frequencies among the Dunkers differ from populations of both western Germany and rural Pennsylvania in a number of traits that have been studied. Blood group B, for example, hardly occurs at all among the Dunkers, although the frequency of the *B* allele is around 6%–8% in most European-derived populations, including those of both Germany and Pennsylvania. Other genetically determined traits show similar patterns, including the nearly total absence of the Fy^a allele (from the Duffy blood group system) among Dunkers. The explanation that best agrees with the data is that the original founder population, known to have been made up of only a few dozen individuals, happened not to include anyone carrying Fy^a or the allele for blood group B. Additional alleles may have been lost by genetic drift while the population remained small. The result was a population that derived its allele frequencies from the assortment of alleles that happened to be present in the founders. We can test this assumption by looking for the rare Dunkers who do possess an allele such as Fy^a. In every case that has been investigated, the occurrence of such an allele among the Dunkers can be traced to a person who joined the group as a religious convert within the last few generations.

Because they are genetically isolated, except for occasional religious conversions, the Dunkers have kept a unique combination of unusual allele frequencies. In the absence of blood group B, they resemble Native American populations; in the absence of Fy^a, they resemble African populations. In most traits, however, their derivation from a European source population is evident. These findings show that population resemblances based on a single blood group or gene system may often be misleading and that distinctions among human populations, if used at all, should be based on a multiplicity of genetic traits.

The bottleneck effect has been used as a hypothesis to explain the near-total absence of blood group B among Native Americans and of cde (in the Rh blood groups) among Pacific Islanders. When the ancestors of these people first migrated from Asia, the random changes in allele frequency that occurred when the groups were small gave rise to distinct, isolated populations whose allele frequencies differed from those of the ancestral populations. Genetic drift of this kind would apply primarily to groups of people, like the Polynesians or Native Americans, whose founder populations were initially small. The effects of genetic drift are minimal in the larger and more widespread population groups of Africa, Europe, and mainland Asia.

7.2.3 Reconstructing the history of human populations

Allele frequencies and DNA sequences in modern populations can be used as clues to their evolutionary origins. For example, American molecular biologist Rebecca Cann and her co-workers studied mitochondrial DNA sequences in samples from over 100 human populations. Mitochondria are organelles in the cytoplasm of eukaryotic cells (see Section 6.4.1) that provide much of the cell's energy and that also contain small, circular DNA molecules independent of the DNA in the nucleus. Mitochondrial DNA is transmitted only maternally, from mother to both male and female offspring. Sperm from the father contain almost no cytoplasm and do not transmit mitochondrial DNA. Because the amount of mitochondrial DNA is much less than chromosomal DNA in the nucleus, it is easier to sequence and is thus ideal for tracing evolutionary patterns. On the basis of these DNA sequences, Cann and her colleagues proposed a family tree of human populations using a maximum-parsimony computer model: Of all possible family trees, the one shown in Figure 7.7 requires fewer mutational changes to have occurred than for any other tree. Another research team, headed by Luigi Cavalli-Sforza, used alleles of 120 genes to study the genetic similarities among 42 populations representing all the world's major population groups and many small ones as well. The findings of these two studies (and others) support the hypothesis of a divergence in the distant past between African and non-African populations, with the non-African populations later splitting into North Eurasian and Southeast Asian subgroups (see Figure 7.7). Australian Aborigines and Pacific Islanders are descended from the Southeast Asian subgroup, whereas Caucasians (Europeans, West Asians) and Native Americans (Amerind) are both descended from the

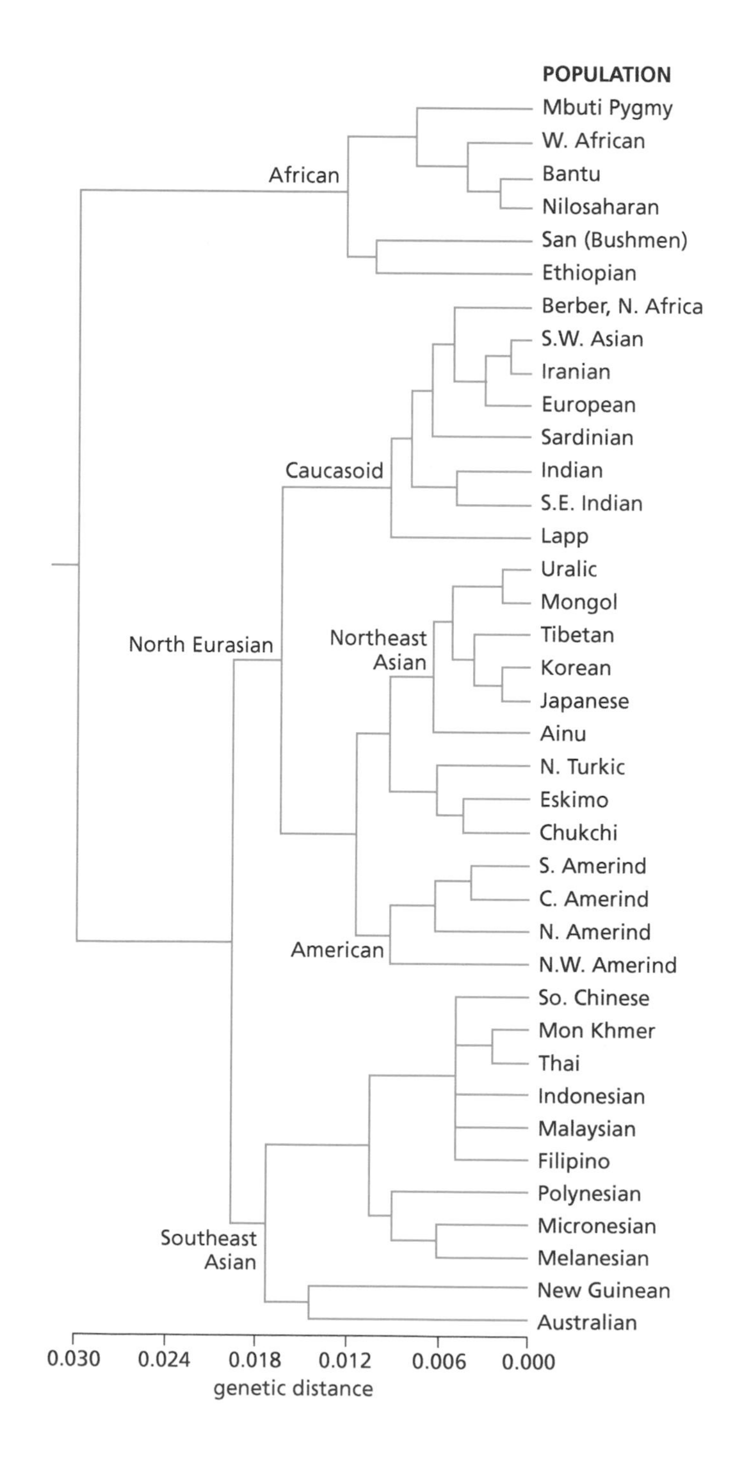

Figure 7.7 A family tree of human populations constructed on the basis of mitochondrial DNA sequences. "Genetic distance" refers to the fraction of mitochondrial DNA sequence not shared by two populations, so that a fork at a genetic distance of 0.006 means that the populations share 99.4% of their mitochondrial DNA sequences.

North Eurasian group, which also includes Arctic peoples. The groups suggested by this study are geographically coherent and confirm certain well-documented patterns of migration. Existing linguistic evidence also matches these groupings, except for a few cases of cultural borrowing, which can be documented historically. Cavalli-Sforza's group estimates, largely on the basis of archaeological evidence, that the split between African and non-African populations took place 92,000 or more years ago, but the spread of human genes outward from Africa may have occurred multiple times.

Studies such as those we have just described have sometimes been criticized for not being politically correct or for "reinscribing racism." A related criticism of the methodology is that geneticists with no training in anthropology are often tempted to lump together people who live close together even if there is good evidence that they have been historically and culturally separate. In other cases, people may maintain contact across considerable distances with other people who are culturally similar and speak the same language and may consider themselves as belonging to the same group, even if population geneticists list them as separate because of the geographical distance between them. Although a good deal of interbreeding between groups always takes place, people more often choose their mates from what they consider as their own group. In order to assess what population groups actually exist (or existed historically), population geneticists need to cooperate with anthropologists familiar with the people being studied.

Paleontological and anthropological studies show that *Homo sapiens* has always been geographically widespread, with early populations spread across three continents, from Indonesia to Zambia and Western Europe. The earlier species *Homo erectus* was also geographically widespread. Despite this geographic spread, however, neighboring populations have always maintained genetic contact. Adaptation to local environments has caused populations to evolve geographic differences from one another, while matings between populations have maintained enough gene flow to prevent populations from becoming even more different. These two opposing tendencies form the basis for what American anthropologist Milford Wolpoff has called the **multiregional model** of the human species, which asserts that human populations have always maintained genetic contact with one another despite the differences resulting from local adaptation. The genetic contact maintains all human populations as one species, while the local adaptations have prevented geographic uniformity.

The advent of next-generation DNA sequencing (see Section 4.4.2) has enabled more recent studies of human genetic diversity using whole-genome sequencing rather than just mitochondrial DNA; these have substantiated that all *H. sapiens* populations are derived from African populations rather than from geographically dispersed populations of *H. erectus*. The split (or the most recent such split) between African and non-African populations took place after *H. sapiens* had replaced *H. erectus*, at least in Africa. However, there is also good evidence for "multiregional" gene flow, back and forth and in multiple directions, between Neanderthal populations and the rest of *H. sapiens*, and even between Neanderthal and Denisovan populations: Nearly all non-African populations of the world have between 1% and 4% of Neanderthal DNA. Modern human populations interbreed wherever they come into contact (and contacts are frequent), making *H. sapiens* a single worldwide species, and lending strength to the "all one race" concept. Evidently this has been true all along; even when the advancing glaciers isolated Neanderthal populations from the rest of humankind, they still remained part of our species, according to the biological species definition discussed in Section 5.4.

The study of allele frequencies, and more recently of entire DNA sequences, has also been used to determine the origins of particular groups of people. One such study, for example, showed that Koreans are derived from a group that includes the Mongolians and Japanese but not the Chinese (Saha and Tay, 1992). Also, several studies have provided evidence for a Middle Eastern contribution (perhaps via Phoenician sailors) to the populations of both Sicily (Rickards et al., 1998) and Sardinia (Walter, Matsumoto, and DeStefano, 1991). Studies of Native Americans have shown that a minimum of three separate migrations were responsible for populating the Western Hemisphere, and more recent studies show that the situation may be far more complex than this.

Cost reductions in DNA sequencing now allow for the commercialization of individual DNA testing (see Section 3.4). Several companies will now test your DNA from a sample of your cells, usually cheek cells contained in a saliva sample. These tests are generally marketed as tests of your ancestry, allowing you to tell what part of Germany or Ireland your ancestors may have come from or what fraction of your DNA comes from elsewhere. Because these determinations depend on the testing company's database, they provide more detailed information for people whose ancestors came from certain European countries (like England, Ireland, Germany, and Italy) whose historic contribution to the U.S. population has been large and much less information about non-European ancestors. A more important use of these tests is as a convenient way of screening for various genes mentioned later in this chapter, such as those responsible for Tay–Sachs disease, sickle cell disorder, G6PD deficiency, and more.

How did the adaptations come about that led to the various population differences in allele frequencies? The next two sections attempt to provide some of the answers to this question.

THOUGHT QUESTIONS

1. Random mating in a sexual species means that any two opposite-sex individuals have the same chance of mating as any other two. If there are a million individuals of the opposite sex, then each should have an identical chance (one in a million) of being chosen as a mate. Do you think human populations mate at random? Have they ever mated at random? Why or why not?
2. Is there ever a real population (of any species) in which the conditions specified by the Hardy–Weinberg equilibrium exist? How close do particular populations come?
3. If language has nothing to do with race, why do you suppose that researchers attempting to reconstruct the past history of human populations use linguistic evidence?

7.3 MALARIA AND OTHER DISEASES ARE AGENTS OF NATURAL SELECTION

As any species evolves, biological differences among its populations arise largely through natural selection. Diseases are among the selective forces that can result in genetic differences among populations. In this section we consider some genetic traits that confer partial resistance to malaria. In malaria-ridden areas, natural selection acts to increase the frequency of alleles that confer partial resistance to malaria while decreasing the frequency of alleles that leave people susceptible to malaria. Many other selective forces have also operated over the course of human history, but resistance to malaria provides a series of well-studied examples.

New traits are produced by mutation (see Section 3.1.4) and are then subjected to natural selection, a process in which many traits die out in populations. The traits that survive natural selection are adaptive traits, or adaptations (Chapter 5)—traits that increase a population's ability to persist successfully in a particular environment. A good deal of human variation consists of adaptations that have resulted from natural selection operating over time, disease being a significant agent of that selective process.

7.3.1 Malaria

Over the last few centuries, **malaria** has caused more illness and deaths worldwide than most other diseases. Malaria also has a greater impact than most other diseases on the average human life expectancy because most of its victims are young, so many more years of life are lost for each death that occurs. Malaria is more prevalent in tropical and subtropical regions than in temperate climates. The threat of malaria has largely been eliminated in the industrially developed countries through mosquito eradication programs and the draining of swamps, but as late as the first half of the twentieth century, malaria claimed many thousands of victims in Southern Europe, Florida, Louisiana, Mississippi, and Virginia. In 2017, malaria caused over 600,000 deaths, mostly in Africa.

Historical and anthropological evidence confirms that malaria was rare (and therefore not a significant selective force) before the invention of agriculture. Even today, the disease is rare in undisturbed forests or in hunting-and-gathering societies. The clearing of forests for agricultural use opens up more swampy areas, and the building of irrigation canals or drainage ditches creates additional pools of stagnant water. The mosquitoes that carry malaria breed best in stagnant water open to direct sunlight. Agriculture therefore did much to change, in unintended directions, the agents of death (and thus the selective pressures) that act on human populations.

Malaria is caused by one-celled parasitic protists belonging to the genus *Plasmodium* (domain Eukarya, phylum Apicomplexa), which live in human blood and liver cells. The life cycle of *Plasmodium* and the mechanism of its transmission by mosquitoes are explained in Box 7.3.

7.3.2 Sickle cell anemia and resistance to malaria

One of the symptoms of malaria is anemia, an inability of the blood to efficiently transport oxygen around the body. There are many other types of anemia. A very serious type was first discovered in 1910 by a Chicago physician named Charles Herrick. This strange and usually fatal disease also produced abnormally shaped red blood cells that were described as resembling sickles. For this reason, Dr. Herrick called the disease **sickle cell anemia**.

A simple blood test was soon devised to test for the condition: A glass slide containing a bowl-shaped depression is used, and a drop of the patient's blood is placed inside the depression. A ring of petroleum jelly is placed around the margins of the depression and a cover glass is then applied, forming an airtight seal with the petroleum jelly. As the red blood cells use up the available oxygen in the

Box 7.3 Life cycle of the malaria parasite *Plasmodium*

Malaria is caused by one-celled parasitic protists of the genus *Plasmodium* (domain Eukarya, phylum Apicomplexa), which live in human blood and liver cells. Of the four species of *Plasmodium* that cause malaria, *Plasmodium falciparum* is the most virulent. All species of *Plasmodium* have a complex life cycle, spending different parts of their life cycle in two different host species, mosquitoes and humans. The *Plasmodium* sexual stages (male and female gametocytes) are intracellular parasites that inhabit human red blood cells. When a female mosquito of the genus *Anopheles* is ready to lay her eggs, she first takes a blood meal from a person during which she ingests large numbers of red blood cells. (Mosquitoes rarely bite otherwise.) If the red blood cells contain *Plasmodium*, the male and female gametocytes combine in the mosquito's gut to form zygotes (fertilized eggs). The zygotes develop asexually through several stages within the mosquito, culminating in the infective forms (sporozoites), which migrate into the mosquito's salivary glands.

The mosquito's thin mouthparts function like a tiny soda straw or hypodermic needle. Shortly before consuming a blood meal, the female mosquito injects her saliva into her victim. The saliva contains anticoagulants that prevent the human blood from clotting inside the mosquito's mouthparts. When the mosquito injects saliva into a new human host, any sporozoites present in her salivary glands are injected along with it. These sporozoites enter the human bloodstream and are taken up by the liver. Each parasite then develops into thousands more, which may remain in the liver for years. Some parasites periodically escape from the liver into the bloodstream and invade the red blood cells. The parasites reproduce asexually within the red blood cells, producing the disease symptoms. The parasites digest the cell's oxygen-carrying hemoglobin molecules, and one stage also ruptures the red blood cells. Any impairment of the ability

(*Continued*)

of the blood to carry oxygen to the body's tissues is called an **anemia**; all anemias leave their victims feeling run-down and weakened. In malaria, the anemia is caused by destruction of both the hemoglobin and the red blood cells. Cell rupture also brings on fevers, headache, muscular pains, and liver and kidney damage. Within a given host, the asexual cycle of *Plasmodium* continues again and again until the patient either recovers or dies. In the red cells, the parasites can also develop into the sexually reproducing gametocytes, which may be picked up by another mosquito in its next blood meal, spreading the disease.

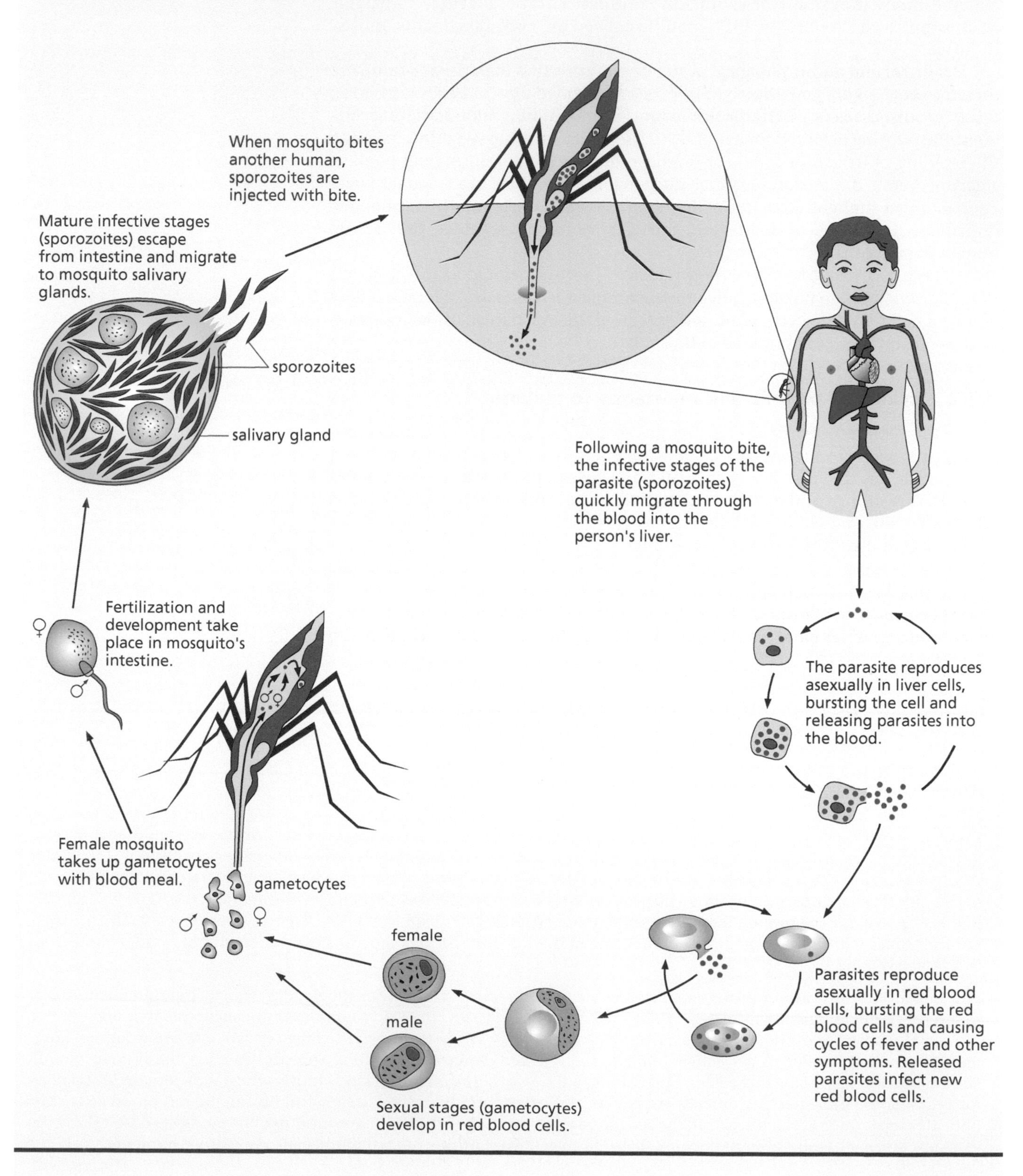

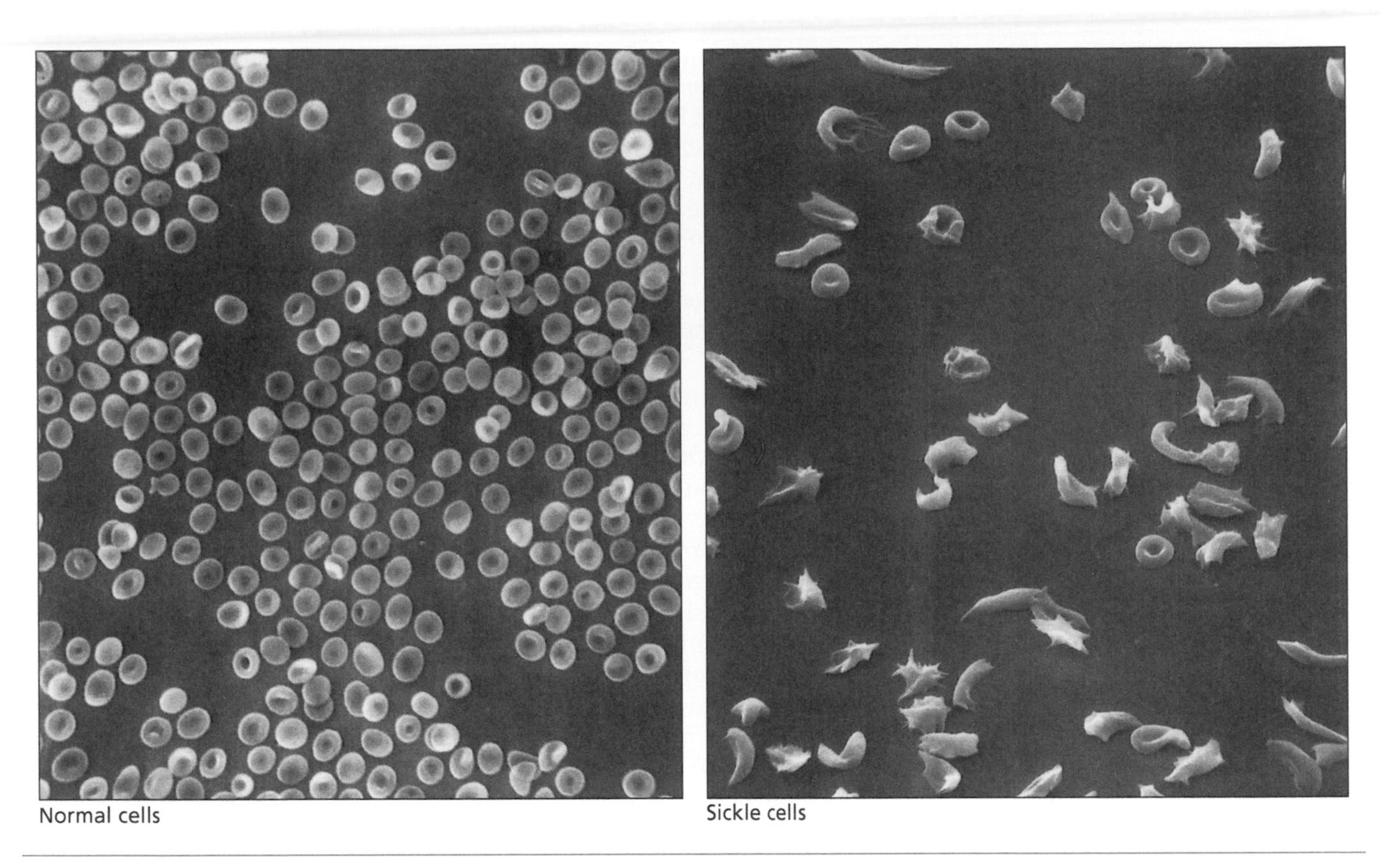

Figure 7.8 Normal red blood cells and red blood cells from a patient with sickle cell anemia.

depression, the oxygen level decreases. Under these conditions, the red blood cells of a person with sickle cell anemia assume their characteristic sickle-like shape, while normal red blood cells retain a circular biconcave shape (Figure 7.8). This blood test also allows the recognition of heterozygous carriers because their blood cells usually appear normal but take on a sickle shape when oxygen is very limited.

7.3.2.1 Normal and abnormal hemoglobin Sickle cell anemia is caused by an abnormality in the molecules (called **hemoglobin**) that carry oxygen within the red blood cells. The hemoglobin molecule consists of four protein chains (two each of two different proteins) surrounding a ringlike "heme" portion. Suspended in the middle of this ring is an iron atom that can bind one oxygen molecule (O_2), giving the hemoglobin its ability to transport oxygen and also its red color.

A change in a single amino acid, number 6 in one of the protein chains, is responsible for sickle cell anemia. Normal adult hemoglobin (hemoglobin A) has glutamic acid in this position in the chain, while sickle cell hemoglobin (hemoglobin S) has valine instead. This minute change makes the hemoglobin S molecules stickier; these molecules adhere to one another and also to the inside of the red cell membrane, deforming the cells into the characteristic sickled shapes. The sickled shape strains the ringlike heme part of the molecule so that hemoglobin S does not carry oxygen as well as hemoglobin A. The difference in the proteins is hereditary and is caused by an altered codon in the hemoglobin gene on the DNA.

7.3.2.2 The genetics of sickle cell hemoglobin Sickle cell anemia is inherited as a simple Mendelian trait. People who die from sickle cell anemia are always homozygous, and their parents are almost always heterozygous, as are a certain number of siblings and other relatives. The gene for hemoglobin is designated *Hb*, and the different alleles are designated by superscripts: Hb^A is the allele for normal hemoglobin and Hb^S is the allele for sickle cell hemoglobin.

In U.S. and Caribbean populations, the vast majority of people carrying the Hb^S allele for sickle cell hemoglobin are blacks of African ancestry. Tests of African populations also show high frequencies of the sickling allele, up to 25% in certain populations. In homozygous individuals (Hb^SHb^S), all the red blood cells are sickled at low oxygen concentrations, as commonly occurs during heavy exertion. Heterozygous individuals (Hb^AHb^S) have both types of hemoglobin. Because both alleles produce a phenotypic result in heterozygotes, they are codominant, as we described earlier in connection with the AB blood type.

7.3.2.3 Symptoms of sickle cell anemia Most of the debilitating symptoms of the disease are consequences of the deformed, sickle-shaped cells. The smallest blood vessels, capillaries, have a diameter only slightly larger than the diameter of blood cells. Because of their sickle shape and changed diameter, sickled cells cause flow resistance in the capillaries and thus impair microcirculation. In most of the body's organs, impaired microcirculation brings about reduced oxygen levels (hypoxia), which results immediately in a severely painful sickle cell crisis. These crises begin in infancy. Damaged cells collect in the capillaries of the joints and result in painful swelling. The sickled cells are also more easily disrupted and destroyed than the normal-shaped round ones, resulting in a decreased oxygen-carrying capacity (anemia). The anemia and impaired circulation result in tissue damage to many organs, eventually resulting in death (Figure 7.9). In African populations, the death of homozygous Hb^SHb^S individuals often occurs before adulthood, but in the United States and the

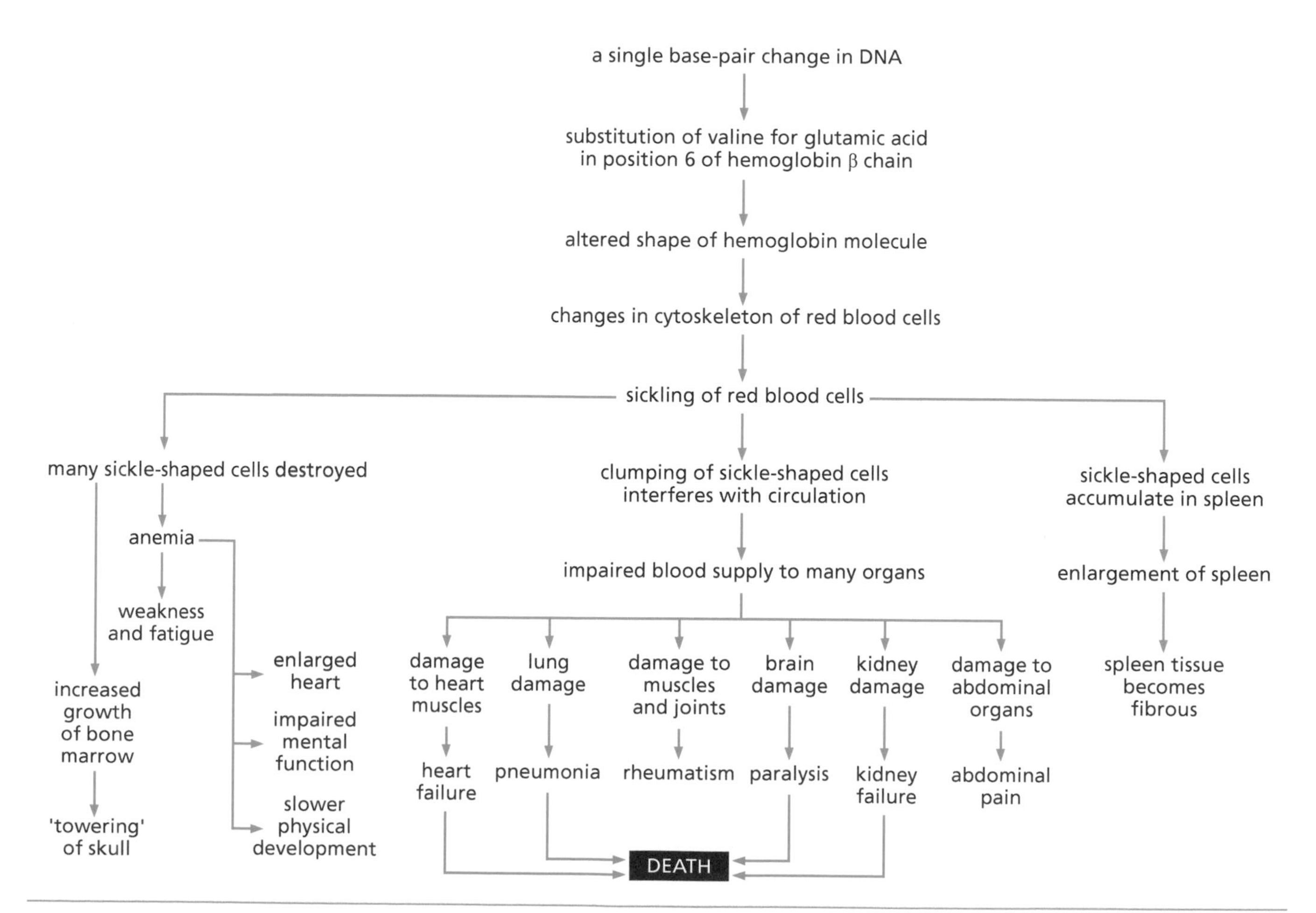

Figure 7.9 Development of the consequences of the Hb^S mutation in the hemoglobin gene. A small change in a gene can have many phenotypic consequences.

Caribbean, survival to reproductive age is now increasingly common. The reduction in red blood cell number and the sickle cell crises also occur among heterozygotes, but not as severely. The CRISPR gene editing technique (described in Section 4.1.3) has now been used, with apparent success thus far, to provide normal hemoglobin (and good health) to a few patients with sickle cell disease.

7.3.2.4 Population genetics of sickle cell anemia When geneticists realized that sickle cell anemia in the United States and Jamaica was largely confined to people of African descent, they began to investigate other populations. Using the blood test described earlier in this chapter, researchers investigated the frequency of the allele for hemoglobin S in many African and Eurasian populations. Over large parts of tropical Africa, researchers found remarkably high frequencies of the Hb^S allele, up to 25% or more. At first this appeared puzzling, because sickle cell anemia was nearly always fatal before reproductive age. An allele whose effects are fatal in homozygous form should long ago have been eliminated by natural selection because people having sickle cell children would have fewer children surviving to reproductive age.

Maps were made of the frequency of the sickle cell allele. From these maps and from other evidence, it was noticed that the areas where the sickle cell allele was frequent were also areas with a high incidence of malaria, particularly the variety caused by *Plasmodium falciparum* (**Figures 7.10A** and **B**).

Subsequent research confirmed the basic fact that the Hb^S allele, even in heterozygous form, confers important resistance to the most virulent form of malaria. Tests in which volunteers were exposed to *Anopheles* mosquitoes showed that the mosquitoes are far less likely to bite heterozygous Hb^A/Hb^S individuals than homozygous Hb^A/Hb^A individuals. Tests with the *P. falciparum* parasites showed that they thrive on the red blood cells of Hb^A/Hb^A individuals, who nearly always come down with a serious case of malaria after infection. However, when Hb^A/Hb^S heterozygotes or Hb^S/Hb^S individuals with sickle cell anemia are infected with *P. falciparum*, their malaria symptoms are mild and they recover quickly because the parasite cannot complete its asexual cycle in their sickled blood cells. The protection that the Hb^S allele affords against malaria is sufficient to explain its persistence in those populations in which the incidence of malaria is high.

Hemoglobin S thus decreases the fitness of homozygotes by causing sickle cell disease, but it increases the fitness of heterozygotes in areas where malaria occurs. In this way, malaria acts as an instrument of natural selection and has a dramatic influence on the allele frequencies of human populations.

In addition to hemoglobin A and hemoglobin S, several other genetic variants of hemoglobin have been discovered. Some of these, such as Hb^C, also occur principally in areas where malaria is present and are thought to confer some resistance to malaria.

7.3.3 Other genetic traits that protect against malaria

Sickle cell anemia is not the only heterozygous condition that protects against malaria. Others include thalassemia, G6PD deficiency, and ovalocytosis.

7.3.3.1 Thalassemia In many countries bordering the Mediterranean Sea (including Spain, Italy, Greece, North Africa, Turkey, Lebanon, Israel, and Cyprus), many people have suffered from a different debilitating type of anemia known as **thalassemia** (meaning "sea blood" in Greek). The disease also occurs farther east, especially in Southeast Asian countries such as Laos and Thailand (**Figure 7.10C**). Thalassemia is marked by a reduced amount of one or more of the protein chains in the hemoglobin molecule. The disease exists in a more serious, often fatal, homozygous form called thalassemia major and a less severe heterozygous form called thalassemia minor. Red blood cells containing nonfunctional hemoglobin are destroyed in the spleen, producing anemia.

The symptoms of thalassemia vary, but all forms result in some decrease in oxygen transport in the blood. The bone marrow compensates by overproducing

(A) Occurrence of *P. falciparum* malaria

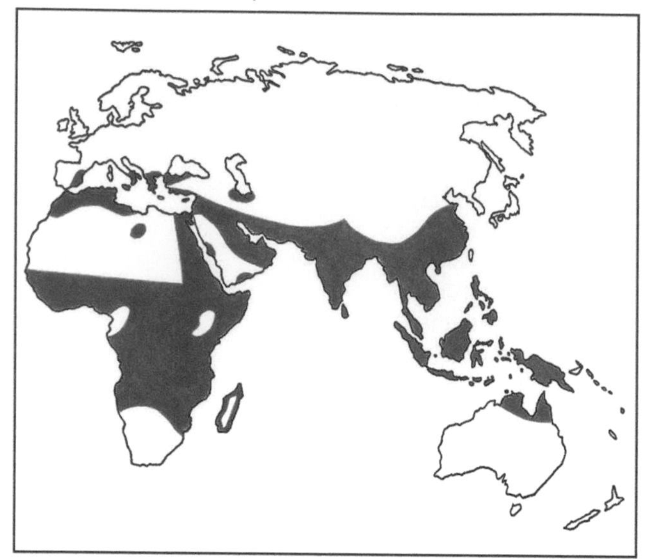

(B) Frequency of Hb^S allele

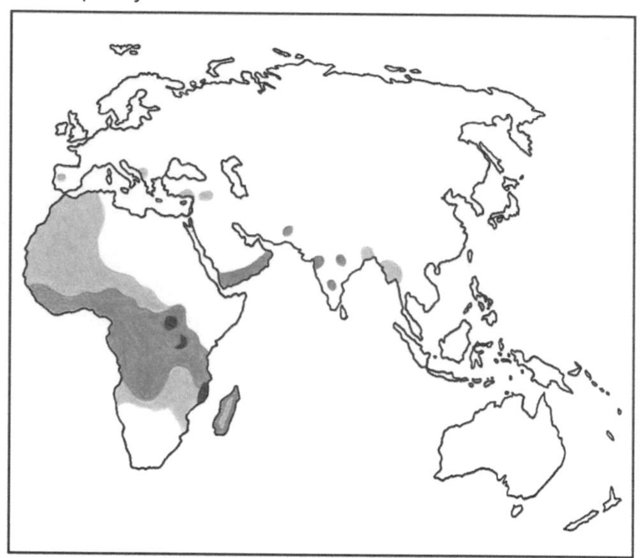

(C) Frequency of alleles for thalassemia (actually the sum for alleles that lead to several different forms of thalassemia)

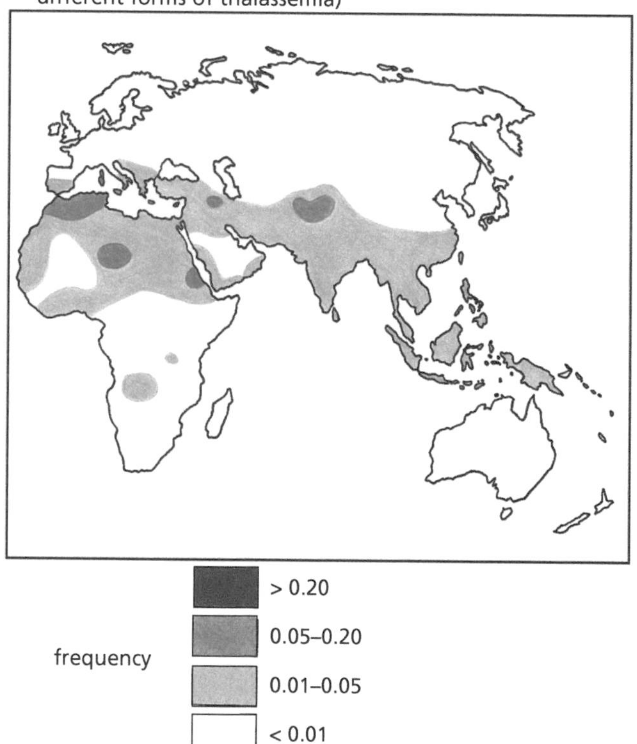

(D) Frequency of allele for G6PD deficiency

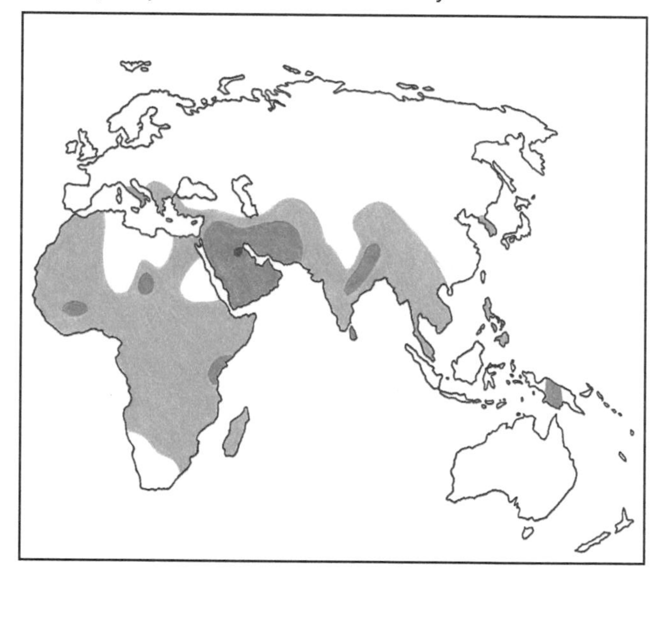

Figure 7.10 **Distributions in the Eastern Hemisphere of *Plasmodium falciparum* malaria and several genetic conditions that protect against it.**

red blood cells, and this overproduction robs the body of much-needed protein and results in stunted growth and smaller stature.

Populations in which thalassemia occurs can now be screened for the genotypes that cause the disease, and genetic counseling can be provided to those found to carry the trait. Screening programs and newer methods of treatment have greatly reduced the problems caused by this disease in Italy, Greece, and elsewhere in the Mediterranean.

The geographical distribution of thalassemia follows closely the distribution of malaria in countries where sickle cell anemia is infrequent or absent. For this reason, it has long been suspected that thalassemia confers a protective resistance to malaria, similar to that caused by sickle cell anemia. The evidence is indirect: If heterozygous individuals (those with thalassemia minor) did not have some selective advantage such as malaria resistance, then the deaths caused by thalassemia major would have caused the alleles for this trait to disappear long ago.

7.3.3.2 G6PD deficiency Blood sugar (glucose) is normally broken down within each cell in a series of reactions that begin with the formation of glucose-6-phosphate. Most of the glucose-6-phosphate is broken down into pyruvate (see Section 8.3.3) in a series of energy-producing reactions, but some is also used to make ribose (the sugar used in RNA) and to make reducing agents such as NADPH and glutathione. The removal of two hydrogen atoms from the glucose-6-phosphate molecule requires the enzyme glucose-6-phosphate dehydrogenase (G6PD). There are many people who have too little of this enzyme, a condition known as **G6PD deficiency** or favism. G6PD deficiency results from a mutation in the gene that encodes the G6PD enzyme.

Under many or most conditions, people with G6PD deficiency remain perfectly healthy, but they occasionally suffer from an anemia in which the red blood cells rupture, spilling their hemoglobin, which then becomes physiologically useless but easy to detect by simple lab tests. This type of anemia, which is potentially fatal, can occur in G6PD-deficient people as a reaction to certain drugs (aspirin, quinine, quinidine, chloroquine, hydroxychloroquine, chloramphenicol, sulfanilamide, tafenoquine, and others), in response to certain illnesses, or after eating fava beans (*Vicia faba*), a common legume of the Eastern Mediterranean and Middle East. The anemia may also exist chronically in a nonfatal form in people with G6PD deficiency.

G6PD deficiency has been shown to offer protection against *P. falciparum* malaria. It affects some 10 million people and is thus the most common disorder offering protection against malaria. Most importantly, heterozygous carriers of the deficiency are also malaria-resistant, but the exact mechanism of the resistance has yet to be explained.

G6PD deficiency occurs mostly in Mediterranean populations from Greece to Turkey and from Tunisia to the Middle East and among Sephardic Jews. It also occurs south of this area into Africa and eastward across Iran and Pakistan to Southeast Asia and southern China (Figure 7.10D). The Greek mathematician Pythagoras may have suffered from this disorder, for his aversion to beans (one of the triggers of anemia in G6PD-deficient people) has become legendary. Pythagoras founded a religious cult in which the avoidance of beans was an important belief. Opponents of his cult once captured Pythagoras by chasing him toward a bean field, which they knew he would not cross.

7.3.3.3 Ovalocytosis Ovalocytosis is an inherited condition in which a majority of red blood cells (erythrocytes) are oval instead of their usual circular, biconcave shape. The condition is usually rare, but it is most often found in Southeast Asia (Malaysia, Indonesia, the Philippines), New Guinea, and Melanesia. Symptoms vary greatly and are most often mild, but they can include enlargement of the spleen and hemolytic anemia (rupturing of the erythrocytes) in severe cases. Studies in New Guinea and elsewhere have found that the malarial parasites (*Plasmodium*) cannot complete their life cycle in the abnormally shaped erythrocytes, so the condition is probably protective against malaria in a manner comparable to sickle cell anemia.

7.3.4 Population genetics of malaria resistance

Polymorphism is the term used to describe a condition in which two or more alleles of the same gene are known in a given population at frequencies higher than the mutation rate. (This last restriction means that the alleles were inherited and could not all simply be the result of new mutation.) Polymorphism is a characteristic of the population, not of individuals; an individual may bear only one, or at most two, of the many alleles present in the population. Some alleles of polymorphic genes have harmful effects when homozygous, but they persist in populations because the same alleles also confer some important benefit (such as malaria resistance) when heterozygous. If the polymorphism persists for many generations, it is likely to be a **balanced polymorphism**. Balanced polymorphism arises when the homozygous genotypes suffer from some selective disadvantage or reduction in fitness, while the heterozygotes have the maximum fitness. For example, in a country in which malaria is present, Hb^AHb^A homozygotes have lower fitness because they are susceptible to malaria, and most Hb^SHb^S homozygotes die young from sickle cell anemia. The Hb^AHb^S heterozygotes have maximal fitness because they are malaria-resistant and because they have enough normal red blood cells for them not to suffer from fatal sickle cell anemia. Under conditions like these, natural selection brings about and perpetuates a situation in which both alleles persist.

The selection by malaria for genetic traits that offer resistance to it is at least as old as the open, swampy conditions (ideal for the breeding of mosquitoes) brought about by agriculture in warm climates. Evidence for this exists in the form of human bones found at a Neolithic archaeological site along the coast of Israel. Cultural remains found at this site show that it was an early farming community, one of the first in the area. Pollen analysis shows the presence of many plants characteristic of swampy areas. Some of the bones show characteristic increases in porosity (due to the increased production of red blood cells in the bone marrow) indicative of thalassemia.

7.3.5 Other diseases as selective factors

Hereditary diseases that confer some advantage in the heterozygous state are not confined to those that protect against malaria. In European populations of past centuries, tuberculosis, an infection caused by a bacterium called *Mycobacterium tuberculosis*, was an important force of selection, especially in crowded cities, from the Middle Ages to the early twentieth century. One scientist has proposed that people heterozygous for the alleles that cause cystic fibrosis (an inherited lung disorder discussed in Section 3.3.2) were protected against tuberculosis (and possibly also cholera and typhoid); they therefore survived epidemics of these diseases in greater numbers than did people without cystic fibrosis alleles. As the heterozygotes increased in number, some of them married one another, and, on average, one out of four of their children became afflicted with cystic fibrosis. Other studies have suggested that the bubonic plague, a fatal infection caused by the bacterium *Yersinia pestis*, was also an important agent of selection throughout Europe since the Middle Ages. Selection by bubonic plague has been hypothesized to explain the increased frequency of Tay–Sachs disease (and other lipid storage diseases) in Ashkenazi Jewish populations. Also, a mutation called CCR5-Δ32, present in frequencies as high as 10% in some European populations, helps cell membranes resist penetration by certain viruses, including HIV (the virus that causes AIDS; see Chapter 15). Several researchers have suggested that relatively high frequencies of this allele in certain populations are attributable to past centuries of selection by other viral diseases, such as smallpox.

What about the geographic variation in blood groups and other genetic traits? There is evidence that at least some of this variation may also result from the natural selection brought about by various medical conditions. In a smallpox epidemic in Bihar province, India, researchers found that those who died were more often of blood group A, while survivors were more often of blood group B. In similar fashion, cholera selects against blood group O and favors blood group B. (Note that these studies demonstrated a difference in fitness, but did not explain the mechanism.) Other studies have shown statistical correlation of various blood types with other diseases: Blood group O is correlated with an increased risk of duodenal ulcers and ovarian cancers, and blood group A with a slightly increased risk of stomach cancer. Associations of particular blood groups with cancers of the duodenum and the colon have also been suggested. Such statistical associations do not necessarily indicate a cause-and-effect relationship between the associated factors.

Fatal diseases are among the most striking agents of natural selection, but there are many other selective forces. We examine some of these other forces of natural selection in the next section.

THOUGHT QUESTIONS

1. How is an average life expectancy measured? Why is the average life expectancy of a population more affected by the deaths of children (e.g., from malaria) than by the deaths of elderly people?
2. All heterozygous carriers of the allele for G6PD deficiency are female. What does this tell you about the location of the G6PD gene? (You may need to review Chapter 3 to answer this question.)

7.4 NATURAL SELECTION BY PHYSICAL FACTORS CAUSES MORE POPULATION VARIATION

There are other agents of natural selection in addition to diseases. Among them are climatic factors such as temperature or sunlight, as well as climatic variation that makes food scarcer at some times of the year or from one year to another. Like the genetically based traits that confer protection against disease, genetic variation between populations has arisen in response to these other selective factors. In this section we look at some of these other factors and how they have selected in different geographic regions for differences in the genetically regulated aspects of physiology and of body shape and size.

7.4.1 Human variation in physiology and physique

During part of the Korean War (1950–1953), American soldiers were exposed to the fierce, frigid conditions of the Manchurian winter. Many soldiers were treated for frostbite. Most of the Euro-American (Caucasian) soldiers responded well to the medical treatment that was given, but a disproportionate number of African American soldiers did not, and many of them lost fingers and toes as a result. Disturbed by these findings, the U.S. Army ordered tests on resistance to environmental extremes in soldiers of different racial backgrounds.

In one series of tests, army recruits were required to perform strenuous tasks (such as chopping wood) under a variety of climatic conditions. In a hot, humid climate, the African American soldiers were able to continue working the longest and performed the best as a group; Asian American and Native American soldiers performed nearly as well as the African Americans, and Euro-American recruits lost excessive fluids through sweating and became easily fatigued and dehydrated. Under dry, desert conditions, the Asian American and Native American soldiers did best, the African Americans were second best, and again the Euro-American soldiers became dehydrated. Under extremes of cold, it was the Euro-American soldiers who did best, followed closely by the Native American and Asian American soldiers; the African Americans shivered the most and some became too cold to continue. These tests demonstrated definite differences between groups in bodily resistance to physiological stress under a variety of environmental extremes. The significance of these differences was enhanced by the fact that, in other respects, the recruits represented a fairly homogeneous population: 18- to 25-year-old males who had all been screened by the Army as being physically fit and free from disease and who had passed the same army physical and mental exams.

Other physiologists outside the Army conducted tests in which adult male volunteers immersed their arms in ice water almost to the shoulders. African Americans in general shivered the most and suffered the most rapid loss of body heat, as measured by a decline in body temperature. Euro-Americans and Asian Americans lasted longer without shivering, but they, too, eventually suffered loss of body heat. Only the Inuit (Eskimo) volunteers were able to keep their arms immersed indefinitely without any discomfort and without shivering. Subsequent studies that replicated these results made the additional finding that diet is also a factor: Inuit volunteers who ate high-protein, high-fat diets (traditional for the Inuit) did far better than other Inuit who had become acculturated to American dietary habits. It would be a mistake, however, to extrapolate findings from studies such as these beyond the groups used for the tests (adult males in good health) without further investigation. Many traits vary with age or sex or both.

7.4.1.1 Bergmann's rule Genetically based differences in physiology that correlate with climate are the basis for a number of ecogeographic rules. Adaptations can also work indirectly, through variables such as body physique. Biologists have long noticed certain general patterns of geographic variation among mammals and birds. In one such pattern, called **Bergmann's rule**, body sizes tend to be larger in cold parts of the range and smaller in warm parts. This can be explained by the relationship of body size to mechanisms of heat generation and heat loss. For example, an animal twice as long in all directions as another animal has eight times the volume of muscle tissue generating heat ($2 \times 2 \times 2 = 8$) as the smaller animal but only four times the surface area over which heat is lost ($2 \times 2 = 4$). Thus, the larger animal is twice as efficient as the smaller ($8/4 = 2$) in conserving heat under cold conditions. A survey of human variation confirms that the largest average body masses are found among people living in cold places (like Siberia), while most tropical peoples within all racial groups are of small body mass, even when their limbs are long. These relationships of body size to climate result from natural selection acting on genetic variation within populations over long periods. It does not mean that a person of a certain genotype will grow larger if they move to a cold climate.

7.4.1.2 Allen's rule Another broad, general phenotypic pattern in most geographically variable species of mammals and birds is **Allen's rule**: protruding parts like arms, legs, ears, and tails are longer and thinner in the warm parts of the range and shorter and thicker in cold regions. This rule is usually explained as an adaptation that conserves heat in cold places by reducing surface area and dissipates heat more effectively in warm places by increasing surface area. Human populations generally follow this rule: Inuit people have shorter, thicker limbs, while most tropical Africans have longer, thinner limbs (Figure 7.11). There are exceptions, however: A number of forest-dwelling populations along the Equator are much smaller than Allen's rule would predict, although they are usually

Figure 7.11 **Bergmann's and Allen's rules illustrated by comparisons between arctic and tropical body forms.**

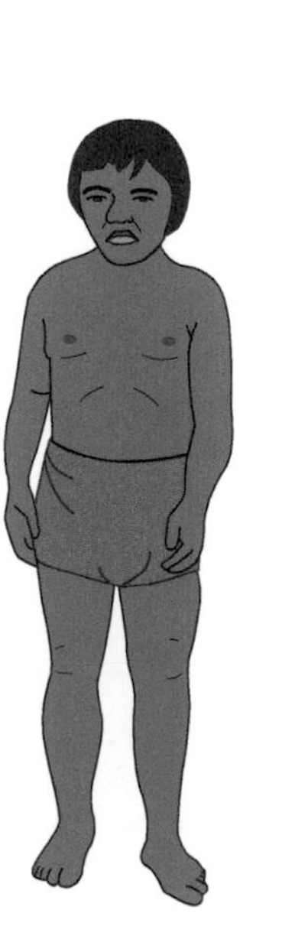

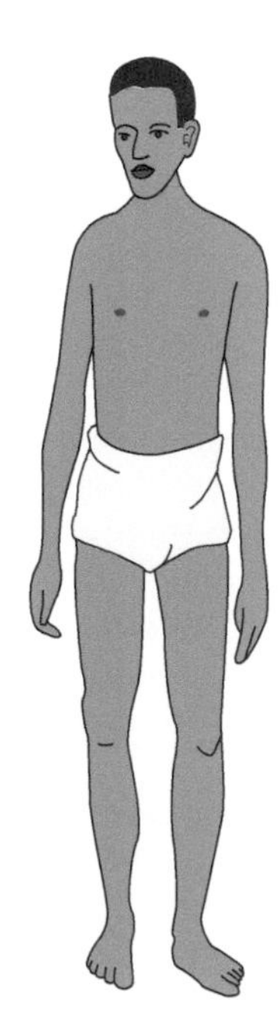

thin-legged. Also, the tallest (Tutsi) and shortest (Mbuti) people on Earth live near one another in the Democratic Republic of Congo, showing that climate is not the only instrument of natural selection influencing limb length or overall height within populations.

7.4.1.3 Diabetes and thrifty genes Diabetes, a potentially life-threatening illness in many populations, may be an indirect result of one or more of the so-called "thrifty genes" that protected certain people from starvation in past centuries. Ancestral Polynesians, for example, endured uncertain journeys over vast stretches of Pacific Ocean waters. Uncertain food supplies during such voyages selected for people who could withstand longer and longer periods of starvation and still remain active. The postulated "thrifty genes" may have caused excess food, when it was available, to be converted into body fat that could be used for energy in times of famine. The result was a population that was stocky in build and resistant to starvation in periods when food supplies were low but that was also more susceptible to diabetes under modern conditions, when physical exhaustion is rare and food is always available. People with diabetes fed "ordinary" diets have excess sugar in their blood, much of which is converted to fat and stored. Although diabetes is itself an unhealthy condition, the storage of fat may have been, under conditions like those described for the early Polynesians, an adaptive trait. Perhaps diabetes is an unfortunate modern consequence of having one or more alleles originally selected for their ability to convert sugar to body fat.

A similar history of selection for "thrifty genes" (not necessarily the same ones) might also explain the late twentieth-century upsurge of diabetes in certain Native American populations, notably the Pima of Arizona, who have the world's highest rate of type 2 diabetes, and also the nearby Navajo. The risks that selected for "thrifty genes" in the past were more significant in barren environments than in places in which the food supply was more assured. However, the commercial introduction of sugar-rich foods and a change from an active to a sedentary lifestyle have both raised the risks of diabetes, which are higher for sedentary people eating carbohydrate-rich diets. Because of these environmental changes, the genes that were once advantageous have, in some cases, turned into a liability, putting people of these genotypes at greater risk of diabetes. The Pima of Mexico are genetically similar to the Pima of Arizona, but only those in Arizona have such high rates of diabetes (presumably from obesity and unhealthy diets), while the diabetes rates in Mexico are comparable to the rates in surrounding non-Pima populations. The Navajo and Pima have discovered that a return to frequent long-distance foot racing (a traditional activity they had nearly abandoned) has kept their populations healthier and has significantly lowered the incidence of diabetes in the runners. Not enough time has yet elapsed for the allele frequencies of the "thrifty genes" to have again changed in this population, but the partial return to an earlier lifestyle has changed the environmental stresses and decreased the incidence of diabetes.

7.4.2 Natural selection, skin color, and disease resistance

The skin is the largest organ of the body and a major surface across which the body makes contact with the forces of natural selection in its environment. Human populations vary widely in skin color. Could these differences in skin color be adaptive?

7.4.2.1 Geographic variation in skin color Skin color is one of the most visible human characteristics and the one to which Americans have always paid the most attention when identifying race. Long-standing patterns of geographic variation are easier to understand if we ignore the population movements of the years since 1500 CE and consider only those populations still living where they did before that time.

Europe has for centuries been inhabited by light-skinned peoples, Africa and tropical Southern Asia by dark-skinned peoples, and the drier, desert regions of Asia and the Americas by people with reddish or yellowish complexions. What is even more remarkable is that we find geographic variation along the same pattern within most continents, and in fact greater variation within the larger population groups than between such groups. For example, among the group of populations spread continuously from Europe across Western Asia to India, we find the lightest skin colors (also eye and hair colors) in Scandinavia and Scotland, progressively darker average colors (and darker hair) closer to the Mediterranean Sea, further darkening as we move through the Middle East and across Iran to Pakistan and India, and the darkest at the southern tip of India and on the island of Sri Lanka. A similar gradient (a cline) for skin color can be found among Asians, from northern Japan south through China into the Philippines and Indonesia.

Why would it be adaptive for people to be light-skinned in Europe but dark in Africa, Sri Lanka, and New Guinea? Notice that there are some very dark-skinned people outside Africa, and they generally have few other physical or genetic characteristics in common with Africans other than their dark skin colors. The natives of Sri Lanka, for example, have very straight hair and blood group frequencies totally different from those of Africa. One clue to this puzzle is that all very dark-skinned peoples have lived for millennia in tropical latitudes.

7.4.2.2 Sunlight as an agent of selection Tropical regions receive on a year-round basis more direct sunlight than do temperate regions. In fact, the amount of sunlight received at ground level decreases with increases in latitude and corresponds more closely to belts of latitude than to variations in temperature. This is especially true for light in the ultraviolet region of the sun's spectrum.

If we exclude places where few people live, Europe receives the least sunlight of all the inhabited regions of the world. This is, first and foremost, a function of latitude. Europe includes populated regions of higher latitudes than on any other continent: London and 14 other European capitals are located north of latitude 50 degrees, while North America and Asia above this latitude contain few large cities and a great deal of sparsely inhabited land. Europe also has a frequent cloud cover that screens out even more of the Sun's rays. As a result of both high latitude and cloud cover, people in Europe receive much less exposure to ultraviolet light than most other people.

That sunlight levels select for body coloration is described by a third ecogeographic rule, **Gloger's rule**. While Bergmann's and Allen's rules, described earlier, take only temperature into account, Gloger's rule takes into account sunlight and humidity as well. Under Gloger's rule, most geographically variable species of birds and mammals have pale-colored or white populations in cold, moist regions, dark-colored or black populations in warm, moist regions, and reddish and yellowish colors in arid regions. We do not know all the reasons for this variation. Camouflage has been suggested as a cause, but vitamin D synthesis also plays an important role.

Vitamin D is needed for the proper formation of bone. Children who do not receive adequate vitamin D during growth suffer from a condition called rickets, a disease of bone formation that may result in weakness and curvature of the bones (especially those of the legs) and in crippling bone deformities if left untreated. Sunlight is necessary for vitamin D synthesis. Many foods are rich in vitamin D, such as egg yolks and whole milk, but most vitamin D found in foods is in a biologically inactive form. The final step of vitamin D biosynthesis takes place just beneath the skin, with the aid of the ultraviolet rays of natural sunlight. This is why vitamin D is sometimes called the "sunshine vitamin." To get adequate amounts of vitamin D, a population must have both adequate intake of the vitamin in the diet and adequate exposure to sunlight. European populations have the lightest skin colors (and they get lighter the farther north you go) as an adaptation that allows maximum sunlight penetration into the skin. Europeans also have many cultural adaptations related to vitamin D intake, such as the eating of cheeses and other fat-rich milk products containing vitamin D. These habits are aided by an allele, rare outside Europe, that permits lactose digestion beyond infancy and into adulthood. Also, Northern Europeans place great value on outdoor activity at all times of the year, including such occasional extremes as nude dashes into the snow after the traditional sauna.

In Northern Europe, people with dark skin could be at a very high risk of vitamin D deficiency because melanin pigment blocks out a large proportion of the Sun's ultraviolet rays. In times past, very few dark-skinned people lived in Northern Europe, even as immigrants. This has changed since World War II, when synthetic vitamin D became widely available. Because this prepared vitamin D is already in its active form, sunlight is no longer needed for its activation. Dark-skinned people can now live and remain healthy in northern latitudes without developing deficiency diseases.

At latitudes closer to the Equator another problem exists: The same wavelengths of ultraviolet that are needed in the final step of vitamin D synthesis are also cancer-causing. Skin cancer (malignant melanoma; see Section 10.6.4) is generally a disease of those white-skinned people who are overexposed to the Sun's direct rays. Populations of all racial groups living closer to the Equator have been selected over the millennia to have darker skins. Those individuals who had lighter skins in the past more often got skin cancer and died, in many cases before their reproductive years had ended. In the tropics, the sunshine is more than enough for all people to get adequate ultraviolet rays for vitamin D synthesis, even though melanin pigment absorbs much of the ultraviolet light.

There are no known genes that code specifically for skin color (except that the allele for albinism, present in all human populations, prevents synthesis of all melanin pigment and results in pale white skin). Apart from environmental effects (such as sun tanning), there are probably dozens of genes that produce enzymes that influence the synthesis of melanin and other skin pigments—so many that it is difficult to study any one of them apart from the others. Nevertheless, natural selection has favored different levels of pigmentation in different geographic regions.

7.4.2.3 Nutritional source of vitamin D in the far north For the reasons given in the preceding sections, populations living in the high latitudes are generally light-skinned and populations that are adapted to living in tropical latitudes are generally dark-skinned. There is one very interesting exception: the Inuit populations of Arctic regions, sometimes known as Eskimos. (These people have always called themselves Inuit; the name "Eskimo" was a pejorative name used by their enemies.) The Inuit are not very light-skinned, nor do they expose themselves much to sunlight. Most Inuit people live in places so cold that the exposure of bare skin poses a greater danger than any benefit of ultraviolet rays could overcome, and most Inuit are fully protected by clothing that offers hardly any exposure to the Sun. So how do they get enough vitamin D? The Inuit have discovered their own way of staying healthy. One of the world's richest sources of vitamin D is fish livers, especially those of cold-water fishes. (Cod liver oil is a very rich source of both A and D vitamins.) Moreover, the vitamin D in fish oils is fully synthesized and needs no sunlight to activate it. So, instead of having pale skins and traditions of exposing their skins to sunlight, the Inuit have traditions of catching cold-water fish (Figure 7.12) and eating them whole, liver and all. These traditions have allowed them to stay healthy in a climate that is too cold and too sunless for most other populations.

In all of these examples, a population that has lived in a particular geographic area for long periods has become adapted to the temperature, humidity, sunlight, and other conditions of their environment. The evidence presented in this chapter suggests that natural selection is largely responsible for these adaptations.

THOUGHT QUESTIONS

1. If people differ in their resistance to extreme cold or heat, does this mean that the difference is genetic? What would you need to know to answer this question? How could an experiment be arranged to test this?
2. Blood type O is statistically associated with duodenal ulcers, one of many such correlations between a blood type and a disease. Does a correlation demonstrate a cause? Does a correlation imply a mechanism of some kind? Does a correlation suggest new hypotheses? How can scientists learn more about whether there is a causal connection between the blood type and the disease?

Figure 7.12 Traditional Inuit fishing. The Inuit get most of their vitamin D from eating whole fish, including the liver.

CONCLUDING REMARKS

Throughout the history of biology, scientists have developed various ways of describing groups of people. Some of these groupings have been known as races. Some concepts of race have attempted to find biological explanations for the racial groupings already established by various societies. Morphological concepts of race divided humans on the basis of their physical appearance. Biologists and anthropologists of the past gathered descriptive data about the physical characteristics of different populations and assumed that each group was distinct and unchanging. More recently, biologists have abandoned these concepts, in part because of the racism that has flowed from them, but also because these ideas no longer fit the data that we now have. The population genetics theory of human variation views human populations as varying continuously, with no group being uniquely different from any other.

Biological differences among human populations are products of evolution. Like any other species, humans can evolve only when genetic variation is present in a population. When a population encounters some agent of natural selection, such as disease or climate, people with certain genotypes survive in greater numbers and leave more offspring than those with other genotypes. Over long periods, this process results in the adaptation of a population to its environment, with allele frequencies differing from one population to the next. This evolution continues today, although the increased mobility of people and technological alterations of the environment are slowly making populations less distinct than in past centuries. Populations vary in the frequencies of traits; they do not carry any unique traits. There is no biological phenotype, genotype, or DNA sequence that can assign an individual to a race or to a population. Although our biological concepts about race and other human variation have changed over time, racism will continue to exist if one group of people is held to be more valuable than another.

CHAPTER SUMMARY

- Human **populations** vary geographically. Phenotypic and genotypic variation within populations usually exceeds variation between them.
- Phenotypic variation within a population can be either **continuous** or **discontinuous**. Continuously variable traits, in which all intermediate values are possible, can be described in terms of average values for each population. Discontinuously variable traits, such as those that are either present or absent, can be described in terms of phenotypic frequencies or **allele frequencies**.
- Differences among populations have historically been described in terms of culturally defined or morphological **races**.
- **Population genetics** allows us to describe groups of populations that differ from one another by certain characteristic allele frequencies.
- Most allele frequencies vary gradually and continuously among populations, without abrupt boundaries. Continuous variation from one population to another is best described in terms of geographic gradients, also called **clines**. Clines can be plotted on maps for average values of continuously variable traits such as height, or for population frequencies of discontinuously variable traits such as particular blood groups or alleles or DNA sequences.
- When more than one allele of a gene persists in a population, this is called a **polymorphism**.
- The **Hardy–Weinberg equilibrium** describes the conditions under which allele frequencies remain constant in a population.
- Populations that were at one time small may have allele frequencies that have been shaped in part by **genetic drift**.
- Aside from genetic drift, most geographic variation among human populations has resulted from natural selection producing adaptation of the population to the environment.
- Disease is an important force of natural selection. **Malaria**, a widespread parasitic infection, has selected in different regions for high frequencies of alleles associated with **sickle cell anemia**, **thalassemia**, and **G6PD deficiency**, all of which protect heterozygous individuals against malaria. Malaria and other diseases result in **balanced polymorphism** whenever the heterozygous genotype enjoys maximum fitness.
- Temperature selects for geographic variation in the alleles influencing body size and body shape.
- Ultraviolet light at different latitudes selects for geographic variation in the population frequencies of alleles influencing skin color. Alleles producing pale skin are selectively favored at high latitudes as an adaptation to absorb more ultraviolet light and prevent vitamin D deficiency. Alleles producing dark skin are favored near the Equator as a protection against skin cancer from too much ultraviolet exposure.

BIBLIOGRAPHIC REFERENCES

Bibliographic references to material in this chapter can be found online at biologytrending.routledge.com/chapter7

GLOSSARY: KEY TERMS TO KNOW

These key terms are defined at the end of each chapter as an aid for student review.

Allele frequency The frequency of an allele in a population, or the fraction of gametes that carry a particular allele.

Allen's rule In any warm-blooded species, populations living in warmer climates tend to have longer and thinner protruding parts (legs, ears, tails, etc.) while the same parts tend to be shorter and thicker in cold climates.

Anemia An inability of the blood to efficiently transport oxygen around the body.

Antigen A molecule or other substance recognized by the immune system and capable of provoking an immune response.

Balanced polymorphism A situation in which different alleles of a gene persist in a population because of the superior fitness of the heterozygous condition.

Bergmann's rule In any warm-blooded species, populations living in colder climates tend to have larger body sizes, compared to smaller body sizes in warmer climates.

Biological determinism See *Determinism.*

Bottleneck effect A type of genetic drift that occurs when a population is temporarily small.

Cline A gradual geographic variation of a trait within a species.

Codominant Alleles capable of producing different phenotypic effects simultaneously.

Continuous variation Variation in which in-between values are always possible, such as a length of 23.15 cm between the values 23.1 and 23.2.

Determinism (genetic determinism) The belief that an individual's characteristics are wholly determined by its heredity (i.e., its genes).

Discontinuous variation "Either/or" variation in which intermediate conditions usually do not exist, as in the presence or absence of a disease.

Founder effect A type of genetic drift in which the allele frequencies of a population reflect the restricted variation present among a small number of founders of that population.

G6PD deficiency (favism) A deficiency of the enzyme glucose-6-phosphate dehydrogenase (G6PD), producing a blood-rupturing (hemolytic) anemia in response to certain drugs or when certain beans, especially fava beans, are eaten.

Gene pool The sum total of all alleles contained in a population.

Genetic determinism See *Determinism.*

Genetic drift Changes in allele frequencies in populations of small to moderate size as the result of random processes.

Gloger's rule In any warm-blooded species, populations living in warm, moist climates tend to be darkly colored or black; populations living in warm, arid climates tend to have red, yellow, brown, or tan colors; and populations living in cold, moist climates tend to be pale or white in color.

Hardy–Weinberg equilibrium A genetic equilibrium formed in large, random-mating populations in which selection, migration, and mutation do not occur or are balanced.

Hardy–Weinberg principle In a large, random-mating population in which selection, migration, and unbalanced mutation do not occur, allele frequencies tend to remain stable from each generation to the next.

Hemoglobin The oxygen-carrying protein in red blood cells.

Hereditarianism See *Determinism.*

Malaria A parasitic infection transmitted by mosquitoes in which a protist of the genus *Plasmodium* infects blood cells.

Mean (mean value) The mathematical sum of many values divided by the number of values.

Morphological (typological) race concept A definition of each race by its physical characteristics, based on the assumption that each characteristic is unvarying and reflects an ideal type or form shared by all members of the group.

Morphological species concept A now-discarded concept that defined each species according to its morphological (physical) features.

Multiregional model A model that views the human species as divided into various regional populations that exchange genes with one another frequently enough so that they all evolve together.

Normal distribution A mathematical description of random variation about a mean value.

Polymorphism The persistence of several alleles in a population at levels too high to be explained by mutation alone.

Population A group of organisms capable of interbreeding among themselves and often sharing a common descent as well; a group of individuals within a species living at a particular time and place.

Population genetics The study of genes and allele frequencies in populations.

Race A geographic subdivision of a species distinguished from other subdivisions by the frequencies of a number of alleles; a genetically distinct group of populations possessing less genetic variability than the species as a whole. This concept is called the *population genetics race concept* and is distinguished from other, older race concepts by defining race as a characteristic that can only apply to populations and not to individuals. Important older meanings include the *socially constructed race concept* (a definition of an oppressed group and the individuals in that group by their oppressors, using whatever cultural or biological distinctions the oppressors wish to use) and the *morphological (typological) race concept,* defined separately earlier.

Racism A belief that one race is superior to others.

Sickle cell anemia A genetic disorder in which hemoglobin A is replaced by hemoglobin S, with resulting deformed (sickle-shaped) red blood cells having a reduced oxygen-carrying capacity but also a resistance to malaria.

Subspecies A geographical subdivision of a species, characterized by less genetic variation within the subspecies than in the species as a whole.

Thalassemia A form of anemia, common in many Mediterranean countries, resulting from shortened forms of the β chain of hemoglobin molecules and protecting the bearers from malaria.

CONNECTIONS TO OTHER CHAPTERS

Chapter 1: Every study of human variation is conducted in a cultural context. Studies of human variation have ethical implications, including those arising from inappropriate use of the results.
Chapter 3: Many human variations have a genetic basis; such alleles arose ultimately from mutations.
Chapter 4: Genome sequences can now be used to trace the history of human populations.
Chapter 5: Human population variations reflect evolutionary processes, including mutation, natural selection, and genetic drift, all of which continue to work in modern populations.
Chapter 8: Different populations sometimes have different ways of meeting their nutritional requirements.
Chapter 9: Nearly all human populations are growing, and some are growing much faster than others. Population growth and migrations change various allele frequencies.
Chapter 10: Some types of cancer are more frequent in some human populations and less frequent in others.
Chapter 15: HIV and AIDS have unequal impact on different human populations.
Chapter 16: Infectious diseases have always influenced human populations. Newly emerging infections will have greater impact on some human populations than on others.
Chapter 18: Human variation is an example of biodiversity at the population level.
Chapter 19: Because of damage to Earth's ozone layer, people are being exposed to increased ultraviolet radiation, which may select over time for a shift in allele frequencies leading to a darkening of skin pigmentation in human populations.

PRACTICE QUESTIONS

1. How many different genotypes can code for the blood group B phenotype? What are they? Are they heterozygous or homozygous?
2. How many different genotypes can code for the blood group AB phenotype? What are they? Are they heterozygous or homozygous?
3. How many different genotypes can code for the blood group O phenotype? What are they? Are they heterozygous or homozygous?
4. How many different genotypes can code for the Rh^- phenotype? Are they heterozygous or homozygous?
5. How many different genotypes can code for the Rh+ phenotype? Are they heterozygous or homozygous?
6. If the allele frequency of Hb^S in a population is 0.1, how many people in that population will be heterozygous Hb^AHb^S? (Review the Hardy–Weinberg equation.)
7. How many different host species does the *Plasmodium* parasite need to complete its life cycle?
8. Why do people who are heterozygous for sickle cell anemia have less severe anemia than people who are homozygous Hb^SHb^S?
9. How does the bottleneck effect alter the allele frequencies of a population?
10. What is a balanced polymorphism? Give an example.

CHAPTER 8

Nutrition, Circulation, and Health

ISSUES

- Do all humans have the same dietary requirements?
- How are human diets related to good health? How are they related to chronic diseases?
- What is malnutrition? What are its causes and consequences?
- What social factors contribute to obesity or heart disease?
- Why do some people deliberately starve themselves?

BIOLOGICAL CONCEPTS

- Chemical and physical basis of biology (water, carbohydrates, lipids, proteins, enzymes, molecular structure, polar and nonpolar molecules)
- Energy and metabolism (energy conversion and storage, chemical bond energy, ATP, calories, glycolysis, Krebs cycle, oxidation–reduction reactions)
- Acidity and pH
- Cell membranes (diffusion, active transport)
- Organ systems (digestive system, circulatory system)
- Evolution (lactose intolerance, etc.)
- Homeostasis
- Health and disease (macronutrient malnutrition, micronutrient malnutrition, fiber, eating disorders, ecological factors)
- Species interactions (mutualism)

CHAPTER OUTLINE

DOI: 10.1201/9781003391159-8

When she weighed 65 kg (140 pounds), Melanie thought of herself as fat and ugly. Her menstrual periods stopped when her weight dropped to 45 kg (100 pounds). Now that she weighs 40 kg (90 pounds), all her friends tell her she is too skinny, but she is sure they are wrong because she still thinks of herself as chubby. She wants to lose even more weight. Melanie has an eating disorder known as anorexia nervosa. Her body is not getting the nutrition it needs. She could die if the situation remains untreated.

Melanie's father went in last week for a routine checkup. Although he was feeling fine, the doctor told him he had high blood pressure and needed to control his food intake. Unless he lowers his intake of saturated fats, he faces an increased risk of having a heart attack. He must now learn to eat a low-fat, low-salt, high-fiber diet, which will lower his chances for getting heart disease, the number-one cause of death in most industrialized countries.

All of us need food, but our dietary requirements vary according to our body size, age, sex, level of activity, and previous state of health. In addition, there are variations caused by hereditary differences in body constitution, metabolic rates, and other factors. The world's populations have found many different ways of meeting these nutritional needs. Different diets arose in different parts of the world because different kinds of plants grew best in each climate and in each type of soil, and each culture has its own preferences and prohibitions that limit their uses of the available foods in their environment—no culture makes use of all foodstuffs available to them.

We begin this chapter with a description of the molecules that make up all living things, including the foods we eat, and the chemical behavior of those molecules. We then examine the body's use of food, human dietary requirements, correlations between diet and the incidence of chronic diseases, and the effects of malnutrition that can result from eating too little or from eating the wrong foods. Malnutrition is one of the major health problems of the world, particularly among the poor and in areas of turmoil. Malnutrition can also result from eating disorders among people with access to sufficient foods.

8.1 THE CHEMISTRY BEHIND ALL OF BIOLOGY

8.1.1 Elements and chemical bonds

All biological systems are made of material composed of chemical elements arranged into compounds. Of the 92 naturally occurring chemical elements, only 4—carbon, oxygen, hydrogen, and nitrogen—make up the overwhelming bulk of most organisms, including about 96% of the human body. In addition, proteins also contain some sulfur, nucleic acids contain phosphorus (see Section 2.3), and our bones contain calcium and phosphorus. A few other elements (sodium, potassium, chlorine, magnesium) are dissolved in most body fluids, as is calcium. The remaining elements that occur in organisms are "trace elements" that together make less than 0.1% of our bodies, although some are important to our health in trace amounts (iodine, for example, is important for proper functioning of the thyroid gland).

Elements can combine to form compounds in two ways. Elements like sodium and potassium can lose electrons to become positively charged particles called **ions**, while chlorine can gain electrons and form negatively charged ions. Positive and negative ions then attract one another to form ionic compounds like salt (NaCl), most of which dissolve in water and therefore occur in most biological systems as dissolved ions (e.g., Na^+ and Cl^-).

The other way that elements can combine is by sharing electrons to form **covalent bonds**. If we exclude bones and body fluids, our bodies are made almost entirely of covalently bonded compounds. Most of these are the organic compounds to be described soon, held together by covalently bonded carbon.

8.1.2 Water

About 70%–90% of most organisms consists of water, H_2O or H-O-H (one oxygen atom forming two covalent bonds, one with each of two hydrogen atoms). Water also covers about 80% of Earth's surface in the form of the lakes, streams, and oceans that sustain life. Water molecules are held together by **polar covalent bonds**, meaning that the electrons are unequally shared: They spend the bulk of their time around the oxygen, which therefore has a partial negative charge (often denoted δ^-), and very little time around the hydrogens, which are therefore partially positively charged (δ^+). The partial positive and partial negative charges of adjacent water molecules form a type of bond called a **hydrogen bond** (Figure 8.1). Although hydrogen bonds are considered "weak bonds," the complete network of all of the hydrogen bonds between different water molecules adds up to hold the molecules together rather strongly, and this attraction accounts for water's ability to resist temperature changes (its high "specific heat") and for the high amount of heat energy needed for water to boil or evaporate. It is these features of water that keep lakes and oceans liquid all the time, making aquatic life possible year-round.

Water dissolves most ionic compounds, surrounding each ion with a shell of water molecules. Positive ions attract the negatively charged oxygens of the water molecules, while negative ions attract the positively charged hydrogens.

8.1.3 Organic compounds

Covalently bonded carbon compounds are called **organic compounds**, made almost entirely of just four elements: hydrogen, oxygen, nitrogen, and carbon. In these compounds, hydrogen atoms (H) always form bonds to one other atom, oxygen (O) always forms two, nitrogen (N) three, and carbon (C) four. (To remember these numbers in order, just say "HONC," pronounced like the sound that geese make.) In addition, phosphorus (P) always forms five bonds and sulfur (S) forms either six or two.

The bonds between carbon and hydrogen atoms, or between two carbon atoms, share their electrons rather equally. Compounds made mostly of hydrogen and carbon are therefore nonpolar and do not dissolve in water. On the other hand, bonds between oxygen and hydrogen, or between nitrogen and hydrogen, share their electrons very unequally and are therefore polar. Compounds with many polar bonds tend to be water-soluble, as is true of many sugars.

The structure of proteins, lipids, and carbohydrates are described in the next section. The structure of nucleic acids is described in Section 2.3.

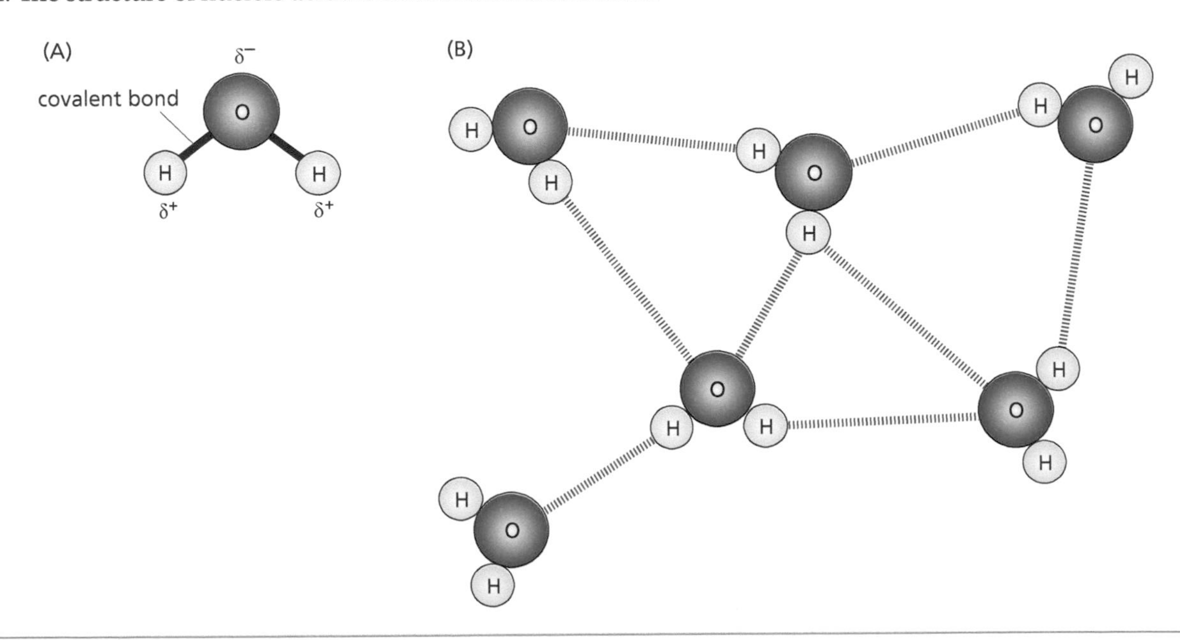

Figure 8.1 **Chemical structure of water. (A)** Detail view of a water molecule with partially positive (δ^+) hydrogen atoms (H) and partially negative (δ^-) oxygen atoms (O) connected by polar covalent bonds. **(B)** Water molecules are held together by hydrogen bonds (dotted lines) between the partially positive and partially negative portions of adjacent molecules.

8.2 ALL HUMANS HAVE DIETARY REQUIREMENTS FOR GOOD HEALTH

Foremost among our dietary needs is a need for energy, as measured in kilocalories (kcal). A **kilocalorie** is the amount of energy required to raise the temperature of a kilogram of water 1 degree Celsius. The "calories" that dieters count are actually kilocalories. Your body's need for caloric energy depends on many factors, such as body weight, level of activity, and sex (Table 8.1). A person who is completely inactive (e.g., in a hospital bed) requires a minimal number of calories, and this amount can be converted into a **basal metabolic rate**, the rate at which an inactive person uses energy when awake.

Caloric intake is the most important measure of dietary sufficiency. In most industrialized countries, most people are adequately nourished or overnourished. In the industrialized countries, particularly in the United States, many people are overweight. Many overweight people (and quite a few who are *not* overweight) have tried to modify their food intake by dieting, either for weight loss or because of a greater interest in health. In a general sense, a diet is the sum of all foods eaten by a person or by a population. In popular usage, being "on a diet" means something more restrictive: a conscious choice of foods to bring about some desired outcome. Diets that bring about weight loss do so by reducing caloric intake. Other diets, such as "heart-healthy" diets, aim to promote long-term health by eating or avoiding particular types of foods. Dieting can be carried to an extreme, however, and diets that are nutritionally very unbalanced can be harmful.

On a worldwide basis, inadequate caloric intake is the most widespread nutritional problem. Starvation kills millions of people each year, most of them children. Starvation and malnutrition are most noticeable in the nonindustrialized, or least developed, countries but are also present in impoverished areas, both rural and urban, within many developed nations. Many other nutritional problems, such as vitamin deficiencies, exist in undernourished people; most of these other problems are hard to treat if the caloric intake remains inadequate.

Most of what we call food can be classified chemically into three types of major constituents and several minor constituents. The major constituents, called **macronutrients**, include carbohydrates, proteins, and lipids (fats); the minor

Table 8.1 Calculating your body's caloric needs

A. Basal metabolic rate First, find the number of kilocalories required to maintain basal metabolism, that is, to keep you alive when you are inactive and lying down: Average adult woman: 21.6 kcal/kg body weight per day, or 9.8 kcal per pound per day Average adult man: 24.0 kcal/kg body weight per day, or 10.9 kcal per pound per day This works out to about 1225 kcal daily for a 125-pound woman or 1850 kcal for a 170-pound man (1 pound = 0.454 kg).
B. Level of activity Multiply the figure obtained earlier by a factor depending on your normal level of activity: 1.35 for sedentary activity (e.g., telephone sales, TV viewing) 1.45 for light activity (e.g., college studies, office work with occasional errands, light housekeeping) 1.55 for moderate activity (e.g., nursing, vigorous housekeeping, waiting on tables, light carpentry) 1.65 for heavy activity (e.g., pick-and-shovel work, bricklaying, full-time competitive athletics)
C. Other factors Figures obtained earlier need to be increased by as much as 10% for any of the following conditions: Growth (children 15 years old and younger) Pregnancy Recovery from a major illness or injury
D. Individual differences The figures calculated earlier are only guidelines or averages. Your individual need may either be greater or smaller. If you maintain a steady caloric intake on a day-to-day basis and you gain weight, your caloric intake is greater than your caloric needs. Conversely, if you lose weight, your intake is less than your caloric needs.

constituents, called **micronutrients**, include vitamins and minerals. Food is used to fuel all the activities of life, and macronutrients are the major sources for energy. In addition, people require fiber and micronutrients, which are not energy sources, but which have other vital functions. We will first examine the biology of these components of foods to help us understand why a balanced diet is necessary to maintain good health; we will also see the consequences of increased or decreased intake of specific foods.

8.2.1 Carbohydrates

Most people around the world derive the majority of their calories from **carbohydrates**, which include starches and sugars. Plants store energy as carbohydrates, so they are a good dietary source for carbohydrates. Cereal grains such as wheat, rice, oats, and corn are the most nutritious sources of carbohydrates because they also contain important vitamins, protein, and fiber. Breads, pastas, and other foods made from cereal grains retain all their nutritional value as long as the whole grain is used. Fruits and fruit products (including juices) generally contain sugars such as fructose or sucrose, together with important vitamins, minerals, and fibers. However, refined sucrose (table sugar) lacks these other nutrients and can also contribute to tooth decay (Box 8.1). Most vegetables contain carbohydrates but are even more important as sources of vitamins, minerals, and fiber.

Carbohydrates are molecules formed principally of three types of atoms: carbon, hydrogen, and oxygen (Figure 8.2). A single carbohydrate unit is called a monosaccharide, or simple sugar. Simple sugars differ from each other by their numbers of carbon atoms and the placement of their chemical bonds. More complex carbohydrates are built by linking these monosaccharides together in groups of two (disaccharides) or of many (polysaccharides). Starch is a common polysaccharide composed of repeated units of the sugar glucose (Figure 8.2).

Most simple carbohydrates are soluble in water because of an important similarity in the types of bonds in carbohydrates and in water. In the strongest type of chemical bond, a covalent bond, two atoms are held together by the sharing of electrons between the two atoms. These bonds can be either **polar** or **nonpolar** (Figure 8.3). Polar bonds have an unequal distribution of electrons and thus of electrical charge, whereas nonpolar bonds have a much more equal distribution of electrical charge. Water (H_2O) is one of the most polar liquids, with electrons unequally shared between hydrogen and oxygen atoms. Carbohydrates have many polar covalent bonds between hydrogen and oxygen atoms and also between carbon and oxygen atoms. Because of the high proportion of oxygen atoms and of polar bonds (Figure 8.2), carbohydrates tend to be soluble in water.

The human body's daily need for carbohydrates is measured in terms of total caloric intake, as indicated in Table 8.1. With regard to caloric content, all carbohydrates, both sugars and starches, are the same, providing 4 kilocalories per gram (kcal/g). From an energy standpoint, it makes little difference if the carbohydrates are eaten in the form of sugar or starch or whether the sugar is fructose or sucrose. There is, however, a difference in the rate of absorption: Starches generally take a few hours to be digested into absorbable sugars, while dietary sugars are capable of being absorbed within minutes. A meal containing both sugars and starches will therefore maintain the body's energy level (or blood glucose) more evenly over a longer period.

Box 8.1 How does sugar contribute to tooth decay?

The sugar that we add to coffee or cereal is known chemically as sucrose. There are many other sugars: fructose (fruit sugar), lactose (milk sugar), and dextrose (a synonym for glucose). Many bacteria live in our mouths and use these dietary sugars for their metabolic energy. One type of mouth bacteria makes a gluelike substance that attaches them to the tooth surface, and to make this substance, they require sucrose. Once the bacteria are glued to the tooth, they can use other sugars (including sorbitol, the sugar in "sugarless gum") as energy sources. When bacteria extract energy from sugars, acids are produced, and these acids dissolve tooth enamel, resulting in cavities. Without sucrose, the bacteria cannot make the glue and the acids are not trapped so closely against the enamel surface.

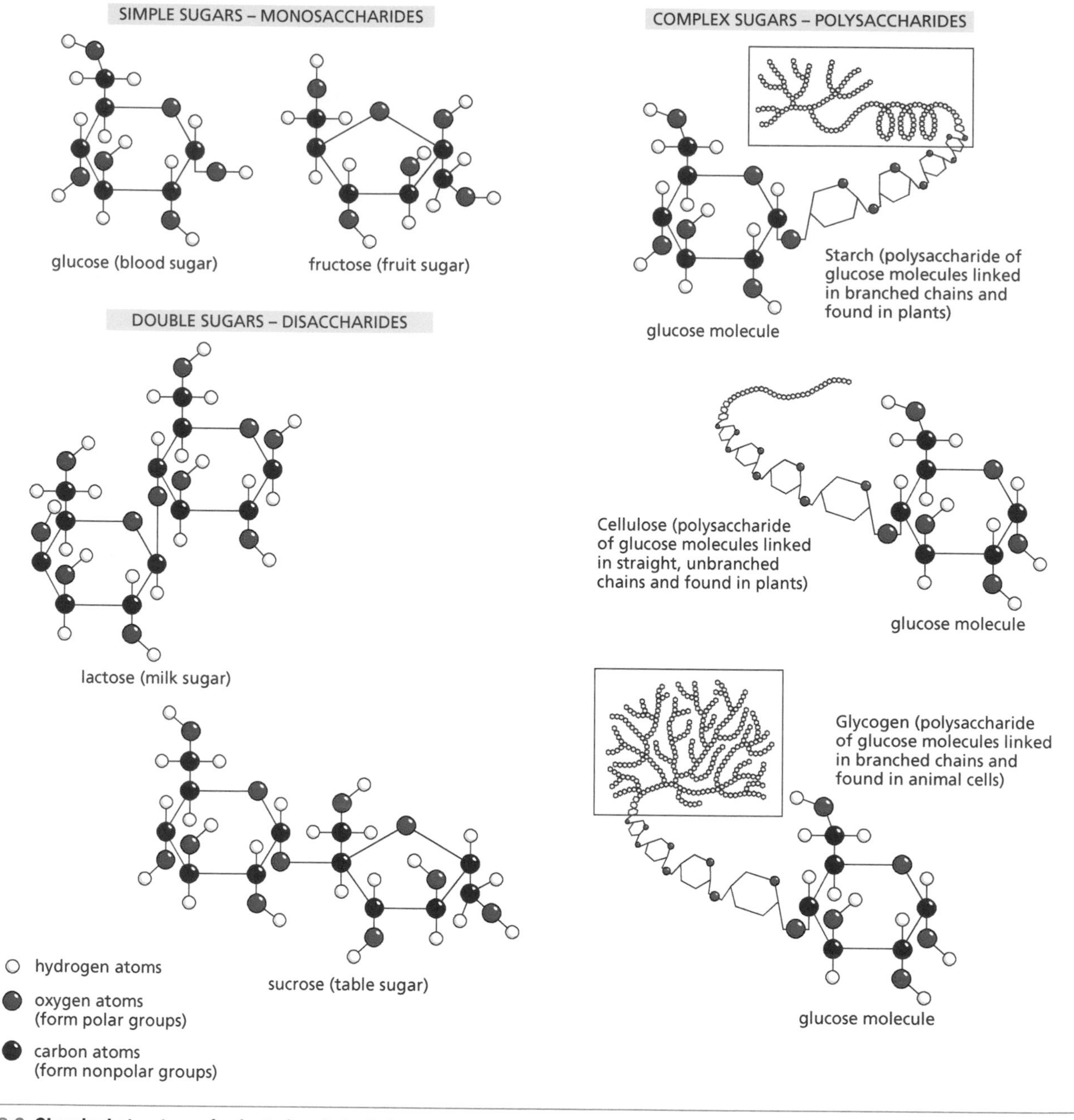

Figure 8.2 Chemical structure of selected carbohydrates.

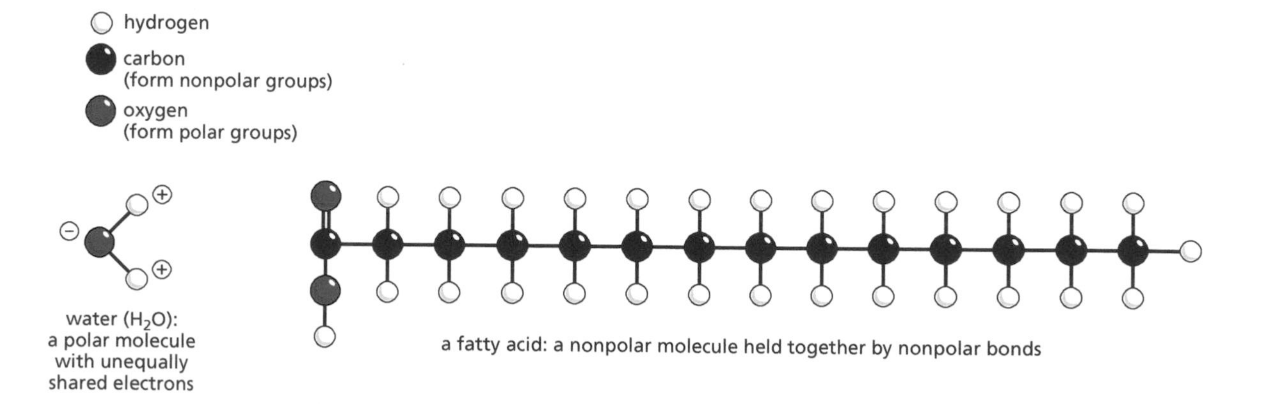

Figure 8.3 Polar molecules and nonpolar molecules.

In most populations, an increase in the consumption of carbohydrate-rich foods (especially whole grains) is desirable. Many diets in the nonindustrialized countries supply inadequate calories, and carbohydrates provide the most efficient and most economical means of improving these diets. Fewer kilocalories of labor, or fewer dollars, are needed to produce a kilocalorie of carbohydrate food than a kilocalorie of most fat-rich or protein-rich foods. In most industrialized countries, the replacement of dietary fats by complex carbohydrates, especially from whole-grain sources, would have many indirect health benefits, including a reduction in risks for heart attacks and certain forms of cancer.

8.2.2 Lipids

Lipids are organic compounds that do not dissolve in water because they are made mostly of hydrogen and carbon atoms linked together by nonpolar covalent bonds. Portions of molecules that are composed of only hydrogen and carbon are referred to as hydrocarbons. Carbon is a special atom because it can make four covalent bonds. Many hydrocarbons consist of linear chains of carbon atoms bonded to each other, with hydrogens occupying all bonding positions not taken by carbon–carbon bonds (Figure 8.3). Common lipids are fats, oils, waxes, phospholipids, and steroids. Dietary lipids are mostly **triglycerides**, molecules in which glycerol (a three-carbon molecule) is linked to three long chains of carbons and hydrogens called **fatty acids** (Figure 8.4). Triglycerides that are solid at room

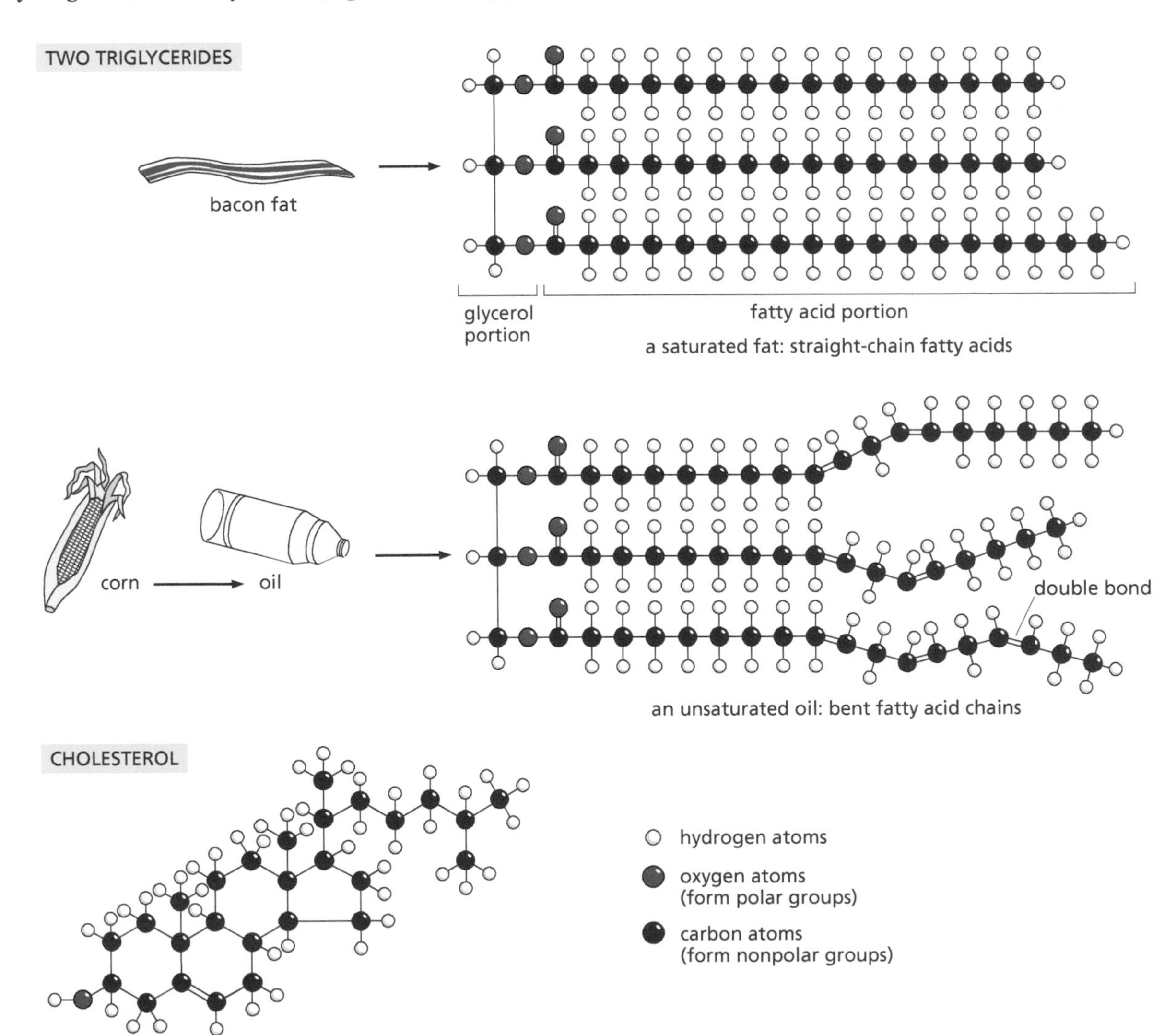

Figure 8.4 **Chemical structure of three types of lipids.** Notice that the general lack of OH groups (or other polar groups in which electric charges can partly separate) makes most parts of these molecules nonpolar and tending to separate away from water.

temperature are commonly called **fats**; those that are liquid at room temperature are commonly called oils. As sources of caloric energy, fats and oils contain almost 9 kcal/g, which is over twice as much as carbohydrates. A small amount of lipid is a dietary necessity, in part because the fat-soluble vitamins (especially A and D) cannot be absorbed without it. Lipids are also a source of fatty acids, which are the nonpolar portion of the phospholipid molecules that form the cell membrane. Two particular fatty acids (linoleic and arachidonic acids) are required from dietary sources because they cannot be made by the body, but they are required only in very small amounts (about 3 g or 1 tablespoonful per person per day). Most nonstarving people have an adequate intake of lipids. In the United States, many people consume too much lipid. The body tends to store excess lipid (and some excess carbohydrate) as fatty deposits within numerous adipose (fat-storing) cells.

8.2.2.1 Fatty acids, cholesterol, and cell membranes **Saturated fats** are fats whose fatty acids have only single bonds (Figure 8.4). Most saturated fats are derived from animal sources (or from a few tropical plants such as palm and coconut), and most are solid at room temperature. **Unsaturated fats**, often derived from plant sources, have double bonds as well as single bonds in their fatty acid chains, causing the chains to bend (Figure 8.4). Those containing only one double bond are sometimes called monounsaturated, while those with multiple double bonds are polyunsaturated. Both types are usually liquid at room temperature because the bends made by the double bonds prevent the molecules from packing together tightly and solidifying. In contrast, the straight chains of saturated fats can pack together more closely and thus solidify at room temperature.

Lipids are important in cell membranes. Fatty acids from dietary fats become incorporated into cell membranes as part of molecules called **phospholipids**. The fatty acid portion is nonpolar, as we have seen, but the other end of a phospholipid is polar. In water, phospholipids orient spontaneously to form bilayer membranes where the phospholipid polar heads face the watery surfaces and the nonpolar fatty acid tails are protected from the water by forming the interior of the bilayer (**Figure 8.5**). Membrane proteins are embedded in this phospholipid bilayer.

Cells remove the fatty acid chains from dietary triglycerides and incorporate the chains into membrane phospholipids. When unsaturated fatty acids are incorporated into the phospholipid cell membrane, the bends prevent their tight packing in the membrane, keeping the membrane more fluid. The phospholipid molecules need to be fluid to allow the embedded proteins to function. If a person's diet is high in saturated fats, more saturated fatty acids will be incorporated into the membrane, making it less fluid; this can reduce the functioning of membrane proteins, such as those involved in nutrient absorption into cells. It is hypothesized that, when dietary lipids cannot be properly absorbed into cells, they may tend to build up on blood vessel walls, contributing to heart disease.

Another important lipid is **cholesterol**, a fat-soluble molecule that is an important constituent of animal cell membranes (Figure 8.4 and notice the presence of hydrocarbon rings instead of chains in the chemical structure). Cholesterol serves as a "fluidity buffer," keeping membranes from becoming either too fluid or too solid and thus helping the cell and the organism to function properly. Cholesterol is also the precursor of several important hormones.

We need cholesterol in small quantities, but our bodies can usually synthesize this amount, so little or none is needed from food. Plant cell membranes do not contain cholesterol, so plant products are always cholesterol-free, although some (like coconut oil) contain saturated fatty acids that are easily converted into cholesterol by the body. All dietary fatty acids are broken down into one of the major starting materials of cholesterol synthesis; cholesterol synthesis is thus increased by nearly all fatty foods, even if they are advertised as "cholesterol free."

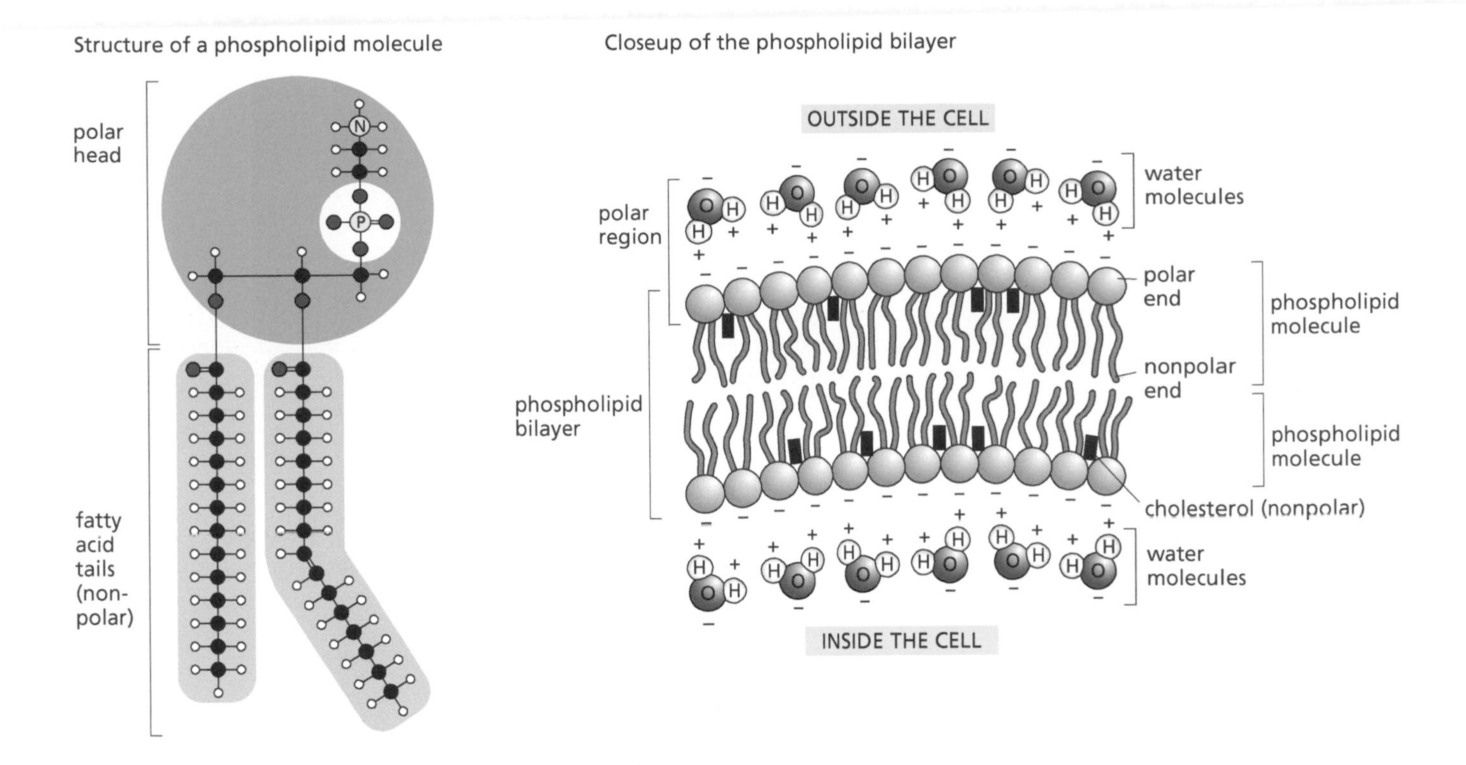

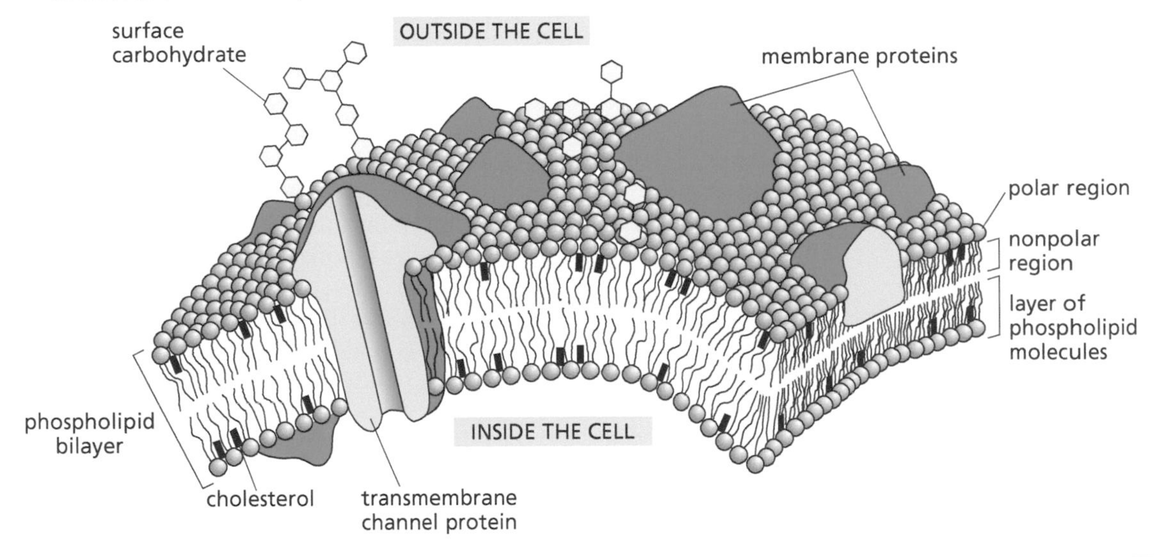

Figure 8.5 **The structure of a phospholipid.** The membrane shown is an animal cell membrane and therefore contains cholesterol.

Because the body makes about 75%–80% of its own cholesterol, and makes it from dietary fats, most of the cholesterol circulating in the bloodstream comes from dietary fats (especially saturated fats), not from dietary cholesterol. Excess cholesterol, like excess amounts of other lipids, can build up on blood vessel walls and increase your risk of disease. Most foods that contain cholesterol are also high in saturated fats, so avoiding either also helps you to avoid the other. Eggs are exceptional in having a lot of cholesterol with few other fats.

8.2.3 Proteins

Proteins have the most diverse set of functions of any of the major classes of biological molecules. The body uses **proteins** for tissue growth and repair, including

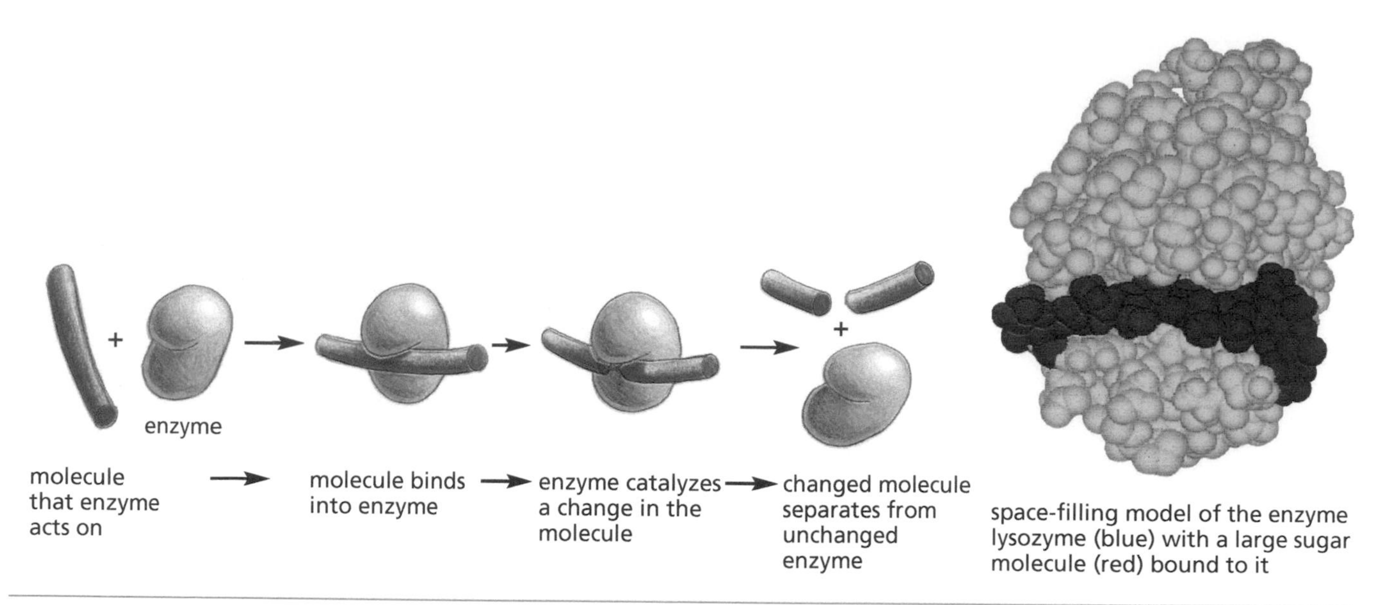

Figure 8.6 **Enzymes: Biological catalysts.**

the healing of wounds, replacement of skin and mucous membranes, and manufacture of antibodies (see Section 14.2). Proteins are important components of all cell membranes. Membrane proteins can function to transport other molecules into cells, as receptors for external signals, or in connecting cells to each other in a tissue, among other functions. Many proteins of the cellular interior provide structure, motility, and contractility to muscles and other cells. Other proteins, such as collagen and elastin, are located outside cells and give connective tissues their strength and thus help to support the entire body. Keratin, another protein found outside cells, is essential for healthy skin and is the main constituent of hair and fingernails.

A much larger assortment of proteins function as **enzymes**, which are factors that promote or speed up chemical reactions in cells without themselves being used up in the reactions (Figure 8.6). This speeding up of reactions is known in chemistry as catalysis, and enzymes are therefore biological catalysts. Because they are not used up in the reactions that they catalyze, enzymes can be reused over and over again, and they are therefore needed only in small quantities. Nearly all enzymes are proteins. Some enzymes (such as those used in digestion) function outside cells (extracellularly); many others function inside cells (intracellularly).

Although enzymes and certain other body proteins are needed only in small quantities, our muscles, blood, skin, and connective tissues need proteins in large amounts. Tendons and several other body parts are made of proteins that are relatively stable once they have been formed, but blood, skin, bone tissue, bone marrow, and many internal membrane surfaces all undergo constant reworking, repair, and replacement, requiring new protein supplies throughout life. Protein requirements are even higher, per unit of body weight, in growing infants and children, in pregnant or lactating women, and in persons recovering from a major illness or injury.

8.2.3.1 Dietary amino acids Proteins are built from chains of smaller molecules called **amino acids** (Figure 8.7). The digestive system breaks down the proteins in food into individual amino acids. After they have been absorbed by the body, these amino acids can then be used to build the body's own proteins. How a protein functions depends to a large extent on its three-dimensional shape after the linear sequence of amino acids has folded. The way in which a protein folds, and whether it is stable in the watery cytoplasm of the cell or in the nonpolar cell membrane, is largely determined by the arrangement of the polar and nonpolar

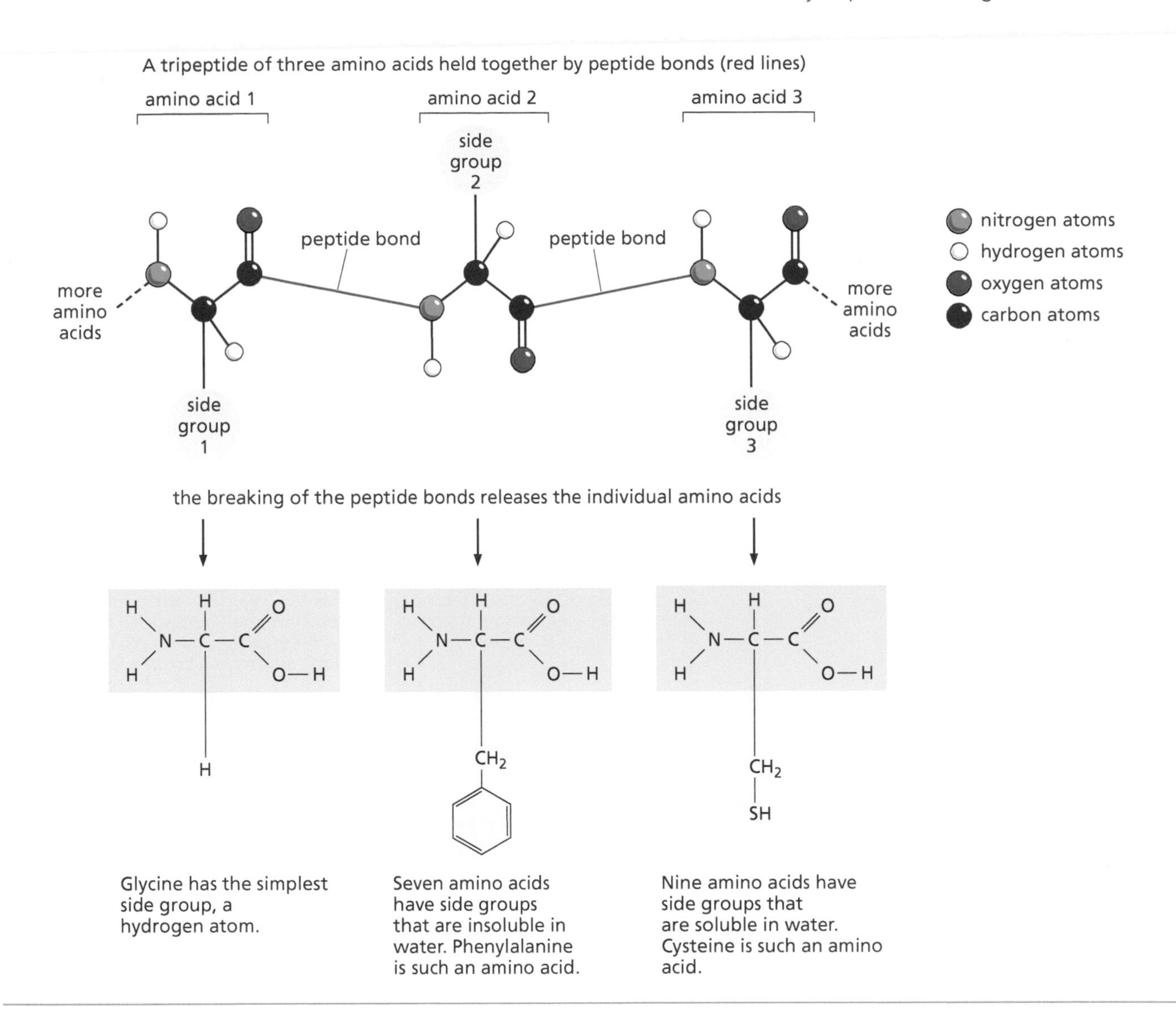

Figure 8.7 Chemical structure of part of a protein.

side groups of its amino acids (**Figure 8.8**). Enzymes are especially sensitive to small changes in shape at their "active sites," the places into which other molecules must fit. Changes in a single amino acid can often have profound effects on the way a protein folds, and therefore on its function as an enzyme.

Because proteins are synthesized by adding one amino acid at a time to the end of a growing chain (see Section 3.1.3), if one type of amino acid is missing from the cell, the synthesis of any protein needing that amino acid stops. An amino acid that is present in small quantities and is used up before other amino acids is called a **limiting amino acid**.

Proteins are necessary in the diet. The daily requirement is 0.8 g per kilogram of body weight, which is about 45 g for a 125-pound (57-kg) woman, or 64 g for a 175-pound (80-kg) man. Each species has its own capacities for making certain of the amino acids and therefore has its own dietary requirements for those it cannot make. Of the 20 standard amino acids, 8 (lysine, leucine, isoleucine, valine, methionine, threonine, phenylalanine, and tryptophan) are considered "essential" in the human diet because the body cannot synthesize them, and a ninth amino acid (histidine) is essential in human infants. The human body can make the remaining amino acids from these essential amino acids.

8.2.3.2 Complete and incomplete proteins

Most animal proteins are **complete proteins** in that they contain all the amino acids essential in the primary structure

the ordered sequence of amino acids in a protein (each colored circle represents an amino acid with a particular side group)

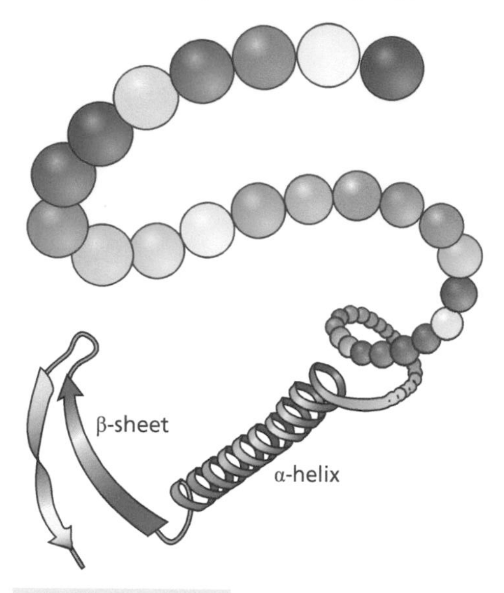

secondary structure

amino acids near each other in the primary structure fold into structures called α-helices and β-sheets

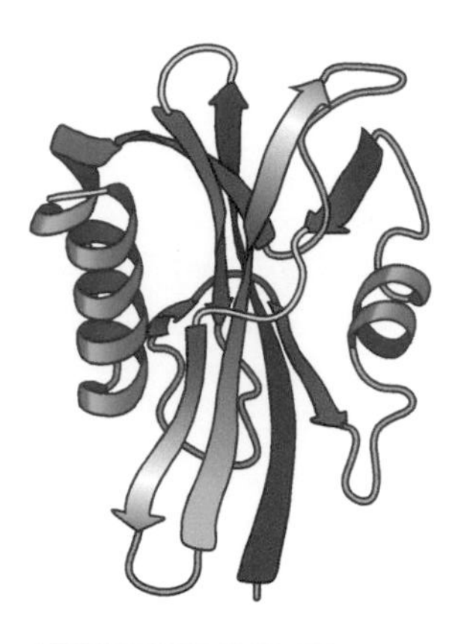

tertiary structure

the overall three-dimensional structure of a protein (secondary structure elements arranged in relation to each other)

quaternary structure

association of more than one protein chain to form a functional protein (the figure shows the four protein chains that make the oxygen-carrying protein hemoglobin)

Figure 8.8 **Different levels of protein structure: The primary structure is the linear sequence of amino acid sequence.** The secondary structure consists of portions of the amino acid chain twisted into the form of an "alpha helix" or folded back-and-forth into a "beta pleated sheet." Many different amino acid sequences can form these secondary structure elements. The tertiary structure is formed when the alpha helices and beta sheets interact to form more complex shapes, dictated in part by electrical attraction between positively charged and negatively charged amino acid side groups and in part by concealing the water-avoiding ("hydrophobic") portions in the interior while exposing the water-soluble ("hydrophilic") portions on the surface. The quaternary structure, not present in every protein, is an assembly of several folded protein chains to make the functional molecule, as shown here for hemoglobin, which consists of four subunits.

human diet. Soy protein is also complete, but most plant proteins lack at least one essential amino acid needed by humans. When an incomplete protein is eaten, the body uses all the amino acids until one of them, the limiting amino acid, becomes depleted. After the limiting amino acid is used up, the body uses the remaining amino acids to produce energy instead of making proteins because dietary protein cannot be stored for later use in the way that carbohydrates and lipids can.

To get around the problem of incomplete plant proteins in the human diet, we can eat them in combinations in which one plant protein supplies an essential amino acid missing in another one. The Iroquois and many other Native Americans commonly obtained complete protein by combining beans, squash, and corn in their diets. Most bean proteins, for example, are deficient in the amino acids valine, cysteine, and methionine, while corn proteins are deficient in the amino acids lysine and tryptophan. Alone, neither one of these proteins is nutritionally complete for humans, but in combination (as in corn tortillas with a bean filling, or succotash, a mixture of beans and corn cooked together), the two plant sources provide a nutritionally complete assortment of amino acids because each has the essential amino acids that the other lacks.

8.2.3.3 Vegetarian diets The amino acid inadequacy of plant proteins poses special problems for vegetarian (meat-avoiding) diets. Vegetarian diets are generally rich in carbohydrates, fiber, vitamins, and minerals, but they may be deficient in certain amino acids unless care is taken to combine several plant proteins at once. Some vegetarians avoid meat but consume fish or milk or eggs; these "ovolacto vegetarians" can usually meet their protein needs without much difficulty, especially if they combine proteins from both plant and animal sources in the same meal. Strict vegetarians, also called vegans, who do not eat food from any animal source (including milk or eggs), need to carefully combine plant proteins sources to supply their bodies with nutritionally complete protein. For example, legumes (beans, peas, and peanuts) can be combined with whole grains (such as rice, corn, or wheat). Nuts and seeds contain proteins that can supply amino acids missing from other plant-based proteins, and some vegetables also contain individual amino acids that can serve the same function.

Choline is another essential nutrient lacking from many vegan diets. The best sources of choline are egg yolks and liver, but vegans who avoid these animal foods can get adequate choline from peanuts and other legumes, soy, broccoli, and Brussels sprouts.

Because animal cells store energy principally as fat, proteins obtained from animals are accompanied by fat. Plant cells, in contrast, store energy in the form of complex carbohydrates such as starch, and plant proteins are thus accompanied by very little fat.

Plant-rich diets have other advantages. A given amount of arable land can support a larger human population if that land is used for raising crops for human consumption, including sources of plant proteins, than if the same land is used to raise food for animals that humans can eat. It takes 5–16 pounds of grain protein to produce 1 pound of meat protein. In well-fed countries with plenty of land, such as Australia or the United States, large tracts can be used for grazing or for the raising of crops primarily for animal consumption. However, poor countries of high population density can ill afford to feed crops to animals. Most of the world's poor eat little meat and get most of their protein from vegetable sources, or in some cases from fish. All whole-grain cereals contain some protein. If this protein is eaten with beans or other legumes, a high-quality protein source is created that is much less expensive than meat and contains far less saturated fat.

The consequences of inadequate protein intake are discussed in a later section.

8.2.4 Fiber

Not all nutrients are required as sources of calories. One example is **fiber**, material that the body cannot digest or absorb. Human diets should include both soluble fiber (pectin, gums, mucilages) and insoluble fiber (mostly cellulose), and both types are present in most fruits, vegetables, legumes, and whole grains. Many of these fibers are complex carbohydrate molecules, of which cellulose (Figure 8.2) is an example. An increase in dietary fiber reduces the incidence of several cancers, especially those of the intestine, but scientists are not sure about the exact mechanism of this effect. One intriguing possibility is that protection against these cancers depends on the rate of movement of food through the intestine and that fiber maintains the optimal rate of food movement. Higher rates cleanse the intestine of potentially toxic chemicals, while lower rates allow these chemicals to remain in one place long enough to undergo fermentation by bacteria into cancer-causing substances (carcinogens; see Section 10.6.4). Another possibility is that harmful carcinogens are frequently present inside the intestine for whatever reason, but a mucus secretion protects the intestinal lining from them, and the insoluble fiber rubbing against the intestinal lining stimulates the lining to secrete more of this protective mucus.

Soluble fiber such as oat bran may reduce the level of serum cholesterol and the risk of heart disease. The mechanism for this effect is not known with certainty, but one hypothesis is that certain soluble fibers bind strongly to bile, which is synthesized from cholesterol and secreted into the intestinal tract to aid in fat digestion. Without the soluble fiber, the bile would be reabsorbed by the intestinal lining and reused, but the soluble fiber prevents this reabsorption and ensures that the bile is eliminated with the stools. Without recycled bile, new bile must be synthesized from cholesterol, which the body withdraws from the blood, lowering blood cholesterol levels. Diets that are high in fiber are statistically associated with lower rates of coronary heart disease and stroke.

8.2.5 Vitamins

Plants are good sources of micronutrients—vitamins and minerals—because plants need these substances and use them for their own metabolism.

Vitamins are complex nutrients needed only in very small quantities because most of them work together with enzymes. Recall that enzymes need to fold into just the right shape in order to function properly. Some enzymes require a non-protein portion, called a **coenzyme**, in order to fold into the right shape and to function as a catalyst. Most vitamins are coenzymes. Like enzymes, these coenzymes are needed only in very small quantities because they are not used up in chemical reactions, but instead can be used repeatedly for many thousands of reaction cycles.

There are over a dozen vitamins categorized in two groups: water-soluble and fat-soluble (Table 8.2). They may be obtained either from pills or from food. The reasons for preferring vitamins in food are as follows:

- They are much less expensive this way.
- Foods rich in vitamins are also rich in other important substances, including minerals, fiber, and protein, nutrients that have other important health benefits. We do not know the complete nutritional requirements of any organism more complex than bacteria, and undoubtedly our food contains many unknown but needed nutrients. These other nutrients, known and unknown, are not obtained from vitamin pills.
- Some vitamins are more easily absorbed by the body in the combinations with other ingredients that exist in food than they are in the combinations that exist in vitamin pills.
- Purified vitamins can be toxic if taken in excessive amounts, an unlikely danger with vitamins contained in foods.

8.2.5.1 Vitamin overdoses and deficiencies The amounts of vitamins recommended for maintaining good health are called recommended dietary allowances (RDAs). Most of these amounts are the same for most healthy adults, but menstruation, pregnancy, and lactation can alter some values in women. Nutritional requirements also differ for growing children and for people recovering from a major illness or injury.

Either too much or too little of a vitamin can result in disease. Vitamin overdoses are possible, but are more likely with fat-soluble vitamins. Water-soluble vitamins, including vitamin C and the B group of vitamins, do not accumulate in the body. When you eat more than you need, the excess is simply excreted in the urine. They must therefore be consumed regularly. Because they are not stored, these vitamins cannot easily build up to toxic overdoses, especially if you get them from foods. It is, however, possible to overdose on water-soluble B vitamins taken in pill form or as concentrated liquids, particularly vitamin B_6. Vitamin B_6 (pyridoxine) is a coenzyme for many of the enzymes of amino acid synthesis; it therefore helps to build proteins and is sometimes used by body-builders. Daily doses of 500 mg or more can be dangerously toxic to the nervous system and liver. Fat-soluble vitamins, including A, D, E, and K (see Table 8.2), accumulate in the body's fat tissues and can build up over time. Overdoses of these vitamins, especially vitamins A and D, can be toxic.

Vitamin deficiencies are more common than overdoses (on a worldwide scale) and are less frequent in the industrialized countries. Disorders of fat absorption often cause deficiencies in fat-soluble vitamins because these vitamins are transported and absorbed along with dietary fats. People with such disorders may have plenty of the vitamins in their blood, but their cells are unable to absorb them. Diseases result from deficiencies of each of the vitamins; in fact, research on the cause of these diseases led to the discovery of vitamins.

8.2.5.2 Vitamin B_1 Vitamin B_1 (thiamine) was the first vitamin to be chemically characterized. While stationed on the island of Java in the 1890s, the Dutch physician Christiaan Eijkman noticed that polyneuritis, a neurological

Table 8.2 Vitamins and minerals in human health

	Importance for good health	Good food source
Water-soluble vitamins		
Vitamin B_1 (Thiamine)	▪ Helps to break down pyruvate ▪ Maintains healthy nerves, muscles, and blood vessels ▪ Prevents beriberi	Meat, whole grains, legumes
Vitamin B_2 (Riboflavin)	▪ Important in wound healing and in metabolism of carbohydrates ▪ Prevents dryness of skin, nose, mouth, and tongue	Yeast, liver, kidney
Vitamin B_3 (Niacin)	▪ Maintains healthy nerves and skin ▪ Prevents pellagra	Legumes, fish, whole grains
Vitamin B_6 (Pyridoxine)	▪ Coenzyme used in amino acid metabolism ▪ Prevents microcytic anemia	Whole grains (except rice), yeast, liver, mackerel, avocado, banana, meat, vegetables, eggs
Vitamin B_{12} (Cyanocobalamin)	▪ Required for DNA synthesis and cell division ▪ Prevents pernicious anemia (incomplete red blood cell development)	Meat, liver, eggs, dairy products, whole grains
Folic acid	▪ Used in synthesis of hemoglobin, DNA, and RNA ▪ Prevents megaloblastic anemia and spina bifida	Asparagus, liver, kidney, fresh greens, vegetables, yeast
Pantothenic acid	▪ Needed to make coenzyme A for carbohydrate and lipid metabolism	Liver, eggs, legumes, dairy products, whole grains
Biotin	▪ Used in fatty acid synthesis and other reactions using CO_2	Eggs, liver, tomatoes, yeast
Vitamin C (Ascorbic acid)	▪ Antioxidant ▪ Used in synthesis of collagen (in connective tissues) and epinephrine (in nerve cells) ▪ Promotes wound healing ▪ Protects mucous membranes ▪ Prevents scurvy	Fresh fruit (especially citrus and strawberries), fresh vegetables, liver, raw meat
Fat-soluble vitamins		
Vitamin A (Retinol)	Antioxidant Precursor of visual pigments Prevents night blindness and xerophthalmia	Yellow and dark green vegetables, some fruits, fish oils, creamy dairy products
Vitamin D (Calciferol)	Promotes calcium absorption and bone formation Prevents rickets and osteomalacia	Eggs, liver, fish, cheese, butter
Vitamin E (Tocopherol)	Antioxidant Protects cell membranes against organic peroxides Maintains health of reproductive system	Whole grains, nuts, legumes, vegetable oils
Vitamin K	Essential for blood clotting Prevents hemorrhage	Green leafy vegetables
Minerals		
Electrolytes (Na^+, K^+, Cl^-)	Maintain balance of fluids in body Maintain cell membrane potentials	Raisins, prunes K^+ also in dates
Calcium	Part of crystal structure of bones and teeth Maintains muscle and nerve membranes	Dairy products, peas, canned fish with bones (sardines, salmon), vegetables
Phosphorus	Part of crystal structure of bones and teeth	Dairy products, corn, broccoli, peas, potatoes, prunes
Magnesium	Maintains muscle and nerve membranes	Meat, milk, fish, green vegetables
Iron	Part of hemoglobin Used in energy-producing reactions	Meat, egg yolks, whole grains, beans, vegetables
Iodine	Maintains thyroid gland Prevents goiter	Fish and other seafood products
Fluorine	Strengthens crystal structure of tooth enamel	Drinking water, tea
Zinc	Promotes bone growth and wound healing	Seafood, meat, dairy products, whole grains, eggs
Copper	Cofactor for enzymes used to build proteins, including collagen, elastin, and hair	Nuts, raisins, shellfish, liver
Selenium	Statistically associated with lower death rates from heart disease, stroke, and cancer	Vegetables, meat, grains, seafood

Figure 8.9 Structure of a cereal grain seed.

A grain is highly nutritious as can be seen in the schematic diagram. Grains turned into flour lose the bran and embryo. In addition, flour is often treated to make it white, losing further nutritional value. Enriched white flour is a modern phenomenon whereby millers, by law, have to artificially add some of the nutrients back into flour.

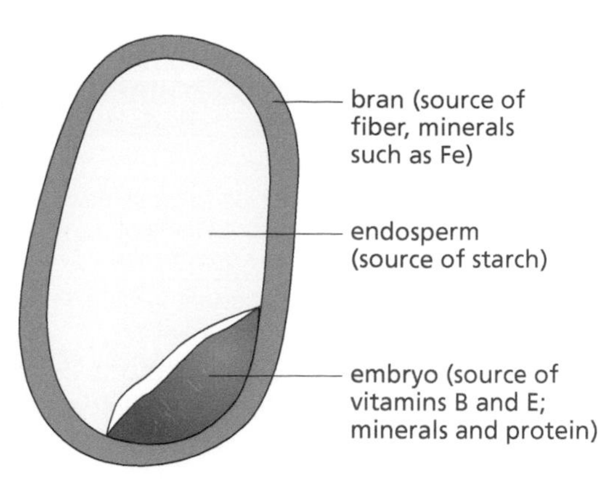

disease in chickens, had many symptoms similar to those of a human disease called beriberi. Both diseases caused muscle weakness and leg paralysis, resulting in an inability to stand up; both diseases were fatal if they persisted. Eijkman noticed that the chickens got polyneuritis only when they were fed on polished white rice, but the disease cleared up when unpolished rice was added to their feed or when whole brown rice was used. Thiamine was later isolated from the outer layer (bran) of the unpolished rice grains (Figure 8.9) and was found to be effective in both treating and preventing beriberi in humans. Because thiamine is a vital (necessary) substance and is also an amine (a chemical containing an $-NH_2$ group), it was called a "vital amine." This term was later shortened to "vitamine" and then "vitamin," and the name was applied to the entire class of substances needed only in small quantities. Eijkman's discovery, therefore, led to the concept of vitamins, and he earned a Nobel Prize for this work in 1929. Beriberi occurs primarily in people whose dietary carbohydrates come from a single, highly refined source such as white rice or white (unenriched) flour. Many countries now have laws requiring the addition of thiamine (and other B vitamins) to refined flour. For this reason, beriberi is now rare in the industrialized world, although it does occur in severe alcoholics whose dietary intake is inadequate.

8.2.5.3 Other B vitamins Other water-soluble vitamins are described in Table 8.2. Like thiamine, many other vitamins also owe their discovery to research on diseases that are collectively referred to as **vitamin deficiency diseases.** Vitamin B_6 deficiency (microcytic anemia) is seen frequently in people whose diets consist mostly of rice. A deficiency of niacin (vitamin B_3) causes pellagra, a disease of the skin and nervous system. The body can make its own niacin from the amino acid tryptophan, which occurs in many proteins. However, populations in which corn is the only protein source are often subject to pellagra because corn is particularly deficient in both niacin and available tryptophan. Hominy grits are made by a process that makes tryptophan (and thus niacin) available from corn protein. Folic acid (folate) is a B vitamin required during the early weeks of pregnancy (before most women even know that they are pregnant) to prevent such neural tube defects as spina bifida and anencephaly, and possibly also autism. Women of childbearing age are recommended to take a multivitamin that includes enough folate to prevent a deficiency in early pregnancy.

8.2.5.4 Vitamin C A deficiency of vitamin C causes scurvy, a disease once common in sailors at sea and among prisoners. Vitamin C is required as a cofactor in the synthesis of collagen, which is a major structural protein in the connective tissues of the body; defective collagen synthesis is what causes the symptoms of scurvy. A British naval surgeon discovered in the 1600s that limes and other fresh fruits (which are rich in vitamin C) would both prevent and cure scurvy. British ships began carrying limes and became so well known for this practice that British sailors came to be called "limeys." An inflammation of the mucous membranes, as in a cold, increases the body's need for vitamin C. Vitamin C may therefore decrease the severity of the symptoms of such an infection, but it cannot on this account be considered a "cure" or a prevention for the common cold, as has sometimes been claimed. Vitamin C is best obtained from fresh fruits, especially citrus fruits and strawberries, as it is less well absorbed when consumed in vitamin pills.

People who take large doses of vitamin C can suffer the symptoms of scurvy when they stop taking the vitamin. In addition, megadoses can produce hemolytic anemia (red blood cell deficiency caused by the rupture of red blood cells) in people with the G6PD metabolic deficiency (see Section 7.3.3) found in African American, Asian, and Sephardic Jewish populations. In addition, individuals who are genetically predisposed to gout find that high doses of vitamin C can sometimes bring on the condition by raising blood levels of uric acid. Vitamin C megadoses can produce deficiencies of another vitamin, B_{12}, in people who are iron deficient. Even in healthy people, megadoses of vitamin C can irritate the bowel sufficiently to result in diarrhea. Therefore, it is not advisable to consume such high doses, despite claims by some (most famously Nobel Laureate Linus Pauling) that they can prevent many adverse medical conditions.

8.2.5.5 Antioxidant vitamins Vitamin A (retinol) is essential in the synthesis of the light-sensitive chemical (retinal) used in vision. Vitamin A is also an **antioxidant**, meaning that it protects body tissues from certain types of chemical reactions called **oxidation** reactions, which can damage biological molecules. A molecule becomes oxidized when electrons are transferred from it to another molecule. Chemicals that bring about oxidation by taking up electrons are called oxidizing agents; among the most highly reactive oxidizing agents are a group of chemicals called **free radicals**, which have one or more unpaired electrons and a strong tendency to remove electrons from other molecules. Free radicals can thus damage many cellular molecules and are hypothesized to play a role in initiating some cancers and to play a role in the aging process (see Chapters 10 and 12). Free radicals are present in many substances, including cigarette smoke, car exhaust fumes, and meat that has been roasted over an open flame. Vitamin A and other antioxidants protect the body by destroying free radicals.

Vitamin A can be obtained from animal sources such as dairy products or fish. Many vegetables also contain a vitamin A precursor, the orange-yellowish pigment beta carotene, which is split after ingestion to produce two molecules of vitamin A. High consumption rates of whole foods rich in beta carotene are statistically associated with lower rates of lung cancer, but the causal link between the two is unclear. Laboratory studies on beta carotene, mostly in rodents, have shown that it suppresses or retards the growth of chemically induced cancers of the skin, breast, bladder, esophagus, pancreas, and colon.

Vitamin E (tocopherol) is another antioxidant vitamin that is especially important in breaking down a group of strong oxidizing agents called peroxides. Vitamin E also helps to prevent spontaneous abortions and stillbirths in pregnant rats, and for this reason it has acquired a reputation as an antisterility vitamin. However, health claims related to the effects of this vitamin on sexual function remain unproved, and overdose of vitamin E in pill form can result in low blood sugar, headache, muscle weakness, and other problems. Vitamin E occurs in several forms, of which alpha-tocopherol is the most potent. Some of the best food sources of vitamin E are nuts; seeds; and green, leafy vegetables. Because the vitamin E in these foods is destroyed by freezing and also by cooking, it is important to consume some fresh, raw foods.

8.2.5.6 Other fat-soluble vitamins Vitamin D (calciferol) is discussed at greater length in Section 7.4.2. It is essential to the body's use of calcium in bone formation and is therefore added as a supplement to most milk sold in the United States. Vitamin K is essential to blood clotting because it serves as a cofactor in reactions that produce blood-clotting factors from their inactive precursors. Most people get adequate amounts of vitamin K from the bacteria that live in their intestines. However, newborn infants, whose intestines have not yet been colonized by bacteria, and persons whose intestinal bacteria have been killed off by antibiotics, need more dietary vitamin K until gut bacteria have become established or reestablished.

8.2.6 Minerals

Minerals are inorganic (non-carbon-containing) ions and atoms necessary for proper physiological functioning. The ions of sodium (Na^+), potassium (K^+), and chloride (Cl^-) are the principal electrolytes (charged particles) of the body. Differences in the concentration of ions on opposite sides of a cell membrane constitute both a concentration gradient and an electrical gradient, which together are called a membrane potential. Like chemical bonds, membrane potentials are a means by which cells store energy in a usable form. One example of a membrane potential is the electrical potential of nerve cell membranes (see Section 11.1.2) created by the distribution of sodium and potassium ions. Because the body's electrically excitable cells (nerve and muscle cells) respond to changes in these membrane potentials, maintenance of these electrolytes within a very narrow concentration range is very important. If the number of ions, particularly of sodium ions, gets too high, the body compensates by retaining water that would otherwise be excreted in the urine. Because blood pressure is related to the volume of fluid in the circulatory system, too high a concentration of sodium in the body tissues results in high blood pressure (hypertension). This is an otherwise symptomless condition that increases the risks for vascular (blood vessel) diseases such as stroke and coronary artery disease. Overuse of salt (sodium chloride) makes this condition worse, but it is seldom the original cause. Many people in the United States consume too much sodium and not enough potassium. Potassium must be present in the proper amounts; either too much or too little can lead to heart failure and death. Raisins, prunes, dates, and bananas are good sources of potassium.

Other minerals important for human health are iron, calcium, fluoride, and a group of trace minerals.

8.2.6.1 Iron Soluble iron is needed for the formation of blood hemoglobin and as a cofactor for many enzymes. A deficiency of iron in cells causes an anemia that is more common in older people with poor dietary habits and in menstruating women. Iron deficiency anemia is in fact the single most common nutritional deficiency in most industrialized countries, including the United States. Menstruating women need almost twice as much iron as men, and pregnant women need even more for the proper synthesis of hemoglobin in the fetus's blood. Vitamin C increases the cellular absorption of iron; therefore, if vitamin C supplies are inadequate, cellular uptake of iron is also inadequate. Some people may have iron deficiencies resulting from low levels of the proteins that transport iron in and out of cells. Meats and legumes (beans, peas, peanuts, etc.) are the best dietary sources of iron.

8.2.6.2 Calcium Calcium (Ca^{2+}) is needed as an intracellular messenger for many processes, including muscle contraction (see Section11.2.3). In addition, the crystal structure of bones and teeth is composed of calcium combined with phosphate and other minerals. Vitamin D promotes calcium absorption, so most people suffering from vitamin D deficiency have symptoms of calcium deficiency as well. High dietary levels of protein can sometimes contribute to an increase in the rate of calcium excretion by the kidneys. Dairy products are usually the best source of calcium for most people, especially children; peas and many other vegetables are good sources, too.

Many older women suffer from low bone density and bone brittleness (osteoporosis). Although bone and teeth seem to be very unchanging because of their solidity, they are actually living tissues that are constantly exchanging molecules with the surrounding fluids. There is a balance between the calcium in bone and the calcium in blood; in osteoporosis the balance shifts, and calcium dissolves out of bone. Although low blood calcium levels are involved, the problem is not so simple that it can be solved by increasing the dietary intake of calcium later in life. Estrogenic hormones are important, and so is vitamin D, which promotes calcium absorption, but the exact processes are poorly understood. Supplementary doses of both calcium and vitamin D are recommended for postmenopausal women, although most bone loss within the first five years after menopause is caused by estrogen withdrawal, not by any nutritional deficiency. Higher levels of exercise in women aged 18–25 can increase their bone density and forestall the development of osteoporosis later in life. Vegetarians have an increased risk of osteoporosis at an early age unless they are very careful to eat a complete diet.

8.2.6.3 Fluoride Fluoride (the ion F^-) is important in the growth of strong teeth during the childhood years. Insufficient fluoride results in a greater incidence of tooth decay. Drinking water is the most important dietary source of fluoride. In some areas, the natural sources of drinking water contain high levels of fluoride. The observation that people in these areas had lower incidences of cavities led to a search for possible factors. Epidemiological research (research on diseases in whole populations) showed that nearby areas had similar diets and climate but higher rates of cavities. Analysis of the water showed higher fluoride levels in the low-cavity areas than in areas with a higher incidence of cavities. Many municipalities now add fluoride (in carefully measured amounts) to the drinking water supply as a preventive measure against tooth decay. Fluoride is also available as drops for breastfed infants and others who do not have access to fluoridated water. It is important to note that fluoride is toxic in very high doses. This is why fluoride-free toothpaste is recommended for very young children. If high doses are accidentally ingested, milk can neutralize the fluoride.

8.2.6.4 Trace minerals Most of the remaining mineral nutrients are sometimes called trace minerals because they are needed by the body only in very small quantities. Deficiencies of these trace minerals were more common in the past, when vegetables were grown locally in soils deficient in one or another trace minerals and when domestic animals grazing on plants growing in the same soils were the main supply of animal food. In the industrial world, such nutritional deficiencies are much less likely because our food supply comes from numerous sources grown in a variety of different soils and climates.

In addition to the trace minerals listed in Table 8.2, chromium and manganese are needed in carbohydrate metabolism; cobalt is an important part of the vitamin B_{12} molecule; molybdenum and nickel are required in the metabolism of nucleic acids; and silicon, tin, and vanadium are needed in trace amounts for proper growth, including the development of bone and connective tissue. Diets adequate in other nutrients usually supply sufficient amounts of these trace minerals.

Because of regional variations in the mineral content of soils, the mineral content of foods that grow in those soils also varies. Therefore, mineral deficiencies often vary geographically. Zinc deficiency, for example, is common in the Middle East. Iodine deficiency is found primarily in certain inland locations, such as the high Andes, the Himalayas, and parts of Central Africa.

8.2.7 Newly recognized micronutrients

In addition to the micronutrients associated with various deficiency diseases, other micronutrients are associated with improved human health and reduced cancer risks. Like vitamins, these organic micronutrients are required in small quantities, and many of them function as antioxidants. The health benefits of the additional micronutrients have been discovered in just the last few decades. The name "phytochemical" is sometimes applied to a diverse group of plant-derived chemicals found in foods that are associated with good health. Phytochemicals include lycopenes found in tomatoes, alkyl sulfides found in onions and garlic, certain flavonoids in foods such as red wine and green tea, curcumin in turmeric (a spice used in many curries), and many other compounds. Instead of being associated with the prevention of a deficiency disease like scurvy or beriberi, these micronutrients are associated with lower long-term risks of heart disease and cancer. Maintaining good health involves much more than the mere avoidance of deficiency diseases. Much of the evidence for the health benefits of these nutrients comes from epidemiological studies. For this reason, the mechanism by which any of these nutrients acts is in most cases unknown.

For several reasons, it is much more difficult to establish recommended dietary levels for any of these new micronutrients. For traditional vitamins, a recommended dietary allowance has generally been based on the amount needed by experimental animals to avoid developing a deficiency disease, but these new micronutrients are not associated with any known deficiency disease, so it is difficult to assess whether an experimental animal has an adequate or an inadequate amount. Another problem is that these micronutrients may have functional overlap or work together. For those that function as antioxidants, it is unclear whether an increase in one can compensate for a decrease in another, which would make it almost impossible to set a recommended daily amount for any one of them in isolation from the others. For those that work best in combinations, it is difficult to study any one of them in isolation. Given these uncertainties, it makes more sense to seek these nutrients in the natural combinations that occur in foods, rather than one at a time in pill form.

Most micronutrient needs can be met economically, even in poor countries, from grain and vegetable sources. Grains contain most B vitamins and also vitamin E and several important minerals, including zinc. Fresh vegetables contain additional vitamins (including A and C) and several important minerals, including calcium and iron. Fresh fruits provide additional vitamin C. Legumes provide calcium and iron in addition to proteins. Fruits, vegetables, grains, legumes, and even some spices are sources of a variety of phytochemicals. In general, all micronutrients can be supplied from plant sources, except for vitamin B_{12}, the one important vitamin that cannot be supplied from plant sources alone.

THOUGHT QUESTIONS

1. If two people drink soft drinks every day, but one chews sugarless gum and the other chews regular gum, would you expect they would have a difference in the number of cavities at their next dental check-up? Would it matter whether the soft drinks were sugar-free or not? (Refer to Box 8.1 for help with this question.)
2. Why are sugars absorbed from the intestine faster than starches?
3. How is it possible for different species (such as rats and people) to have different vitamin requirements? What does this mean on a biochemical level?
4. How is it possible for some studies to show that calcium supplements forestall osteoporosis whereas other studies do not?
5. In Mexico before European contact, the diet consisted mostly of corn, beans, chili peppers, and squash; these same foods are still the major elements of most Mexican diets, especially in rural areas. Can you think of any biological reasons why this diet has proved so stable?

8.3 DIGESTION PROCESSES FOOD INTO CHEMICAL SUBSTANCES THAT THE BODY CAN ABSORB

All organisms need energy to carry out life processes. Plants get this energy from sunlight through photosynthesis (see Section 17.1.2). Most other organisms, including humans, get their energy from the foods that they eat. Our food also supplies the building blocks we need to construct our own proteins and other body materials. However, most of the foods we eat cannot be used in the forms in which they are eaten. To be useful to our bodies, food must first be converted into substances that the body can absorb. Digestion is the process that breaks down food into these absorbable products. Digestion also functions to eliminate undigestible waste.

8.3.1 Chemical and mechanical processes in digestion

Digestion has two aspects: chemical digestion and mechanical digestion. **Chemical digestion** breaks foods down chemically using enzymes, which help speed up chemical reactions, as explained earlier in this chapter. In contrast to the enzymes that help synthesize molecules, the enzymes of digestion help to break down molecules. Chemical digestion works on the surfaces of food fragments. **Mechanical digestion** exposes new surface areas to chemical digestion by breaking fragments into smaller fragments and by removing partly digested surface material.

8.3.2 The digestive system

The digestive system is one of the organ systems in the body. An organ is a group of tissues that are integrated structurally and functionally. An organ system is a group of organs that perform different parts of the same process. Thus, the digestive system is a group of organs that function together to digest food. The body plan of the human digestive system is a common one: from roundworms (phylum Nematoda) to humans, most animals have digestive systems made of a continuous hollow tube, called a gut, with an entrance at one end and an exit at the other. As we go through the next sections, locate the human digestive organs on Figure 8.10.

8.3.2.1 The mouth Food is taken in through the mouth, and mechanical digestion begins in the mouth when the food is chewed. Chemical digestion of

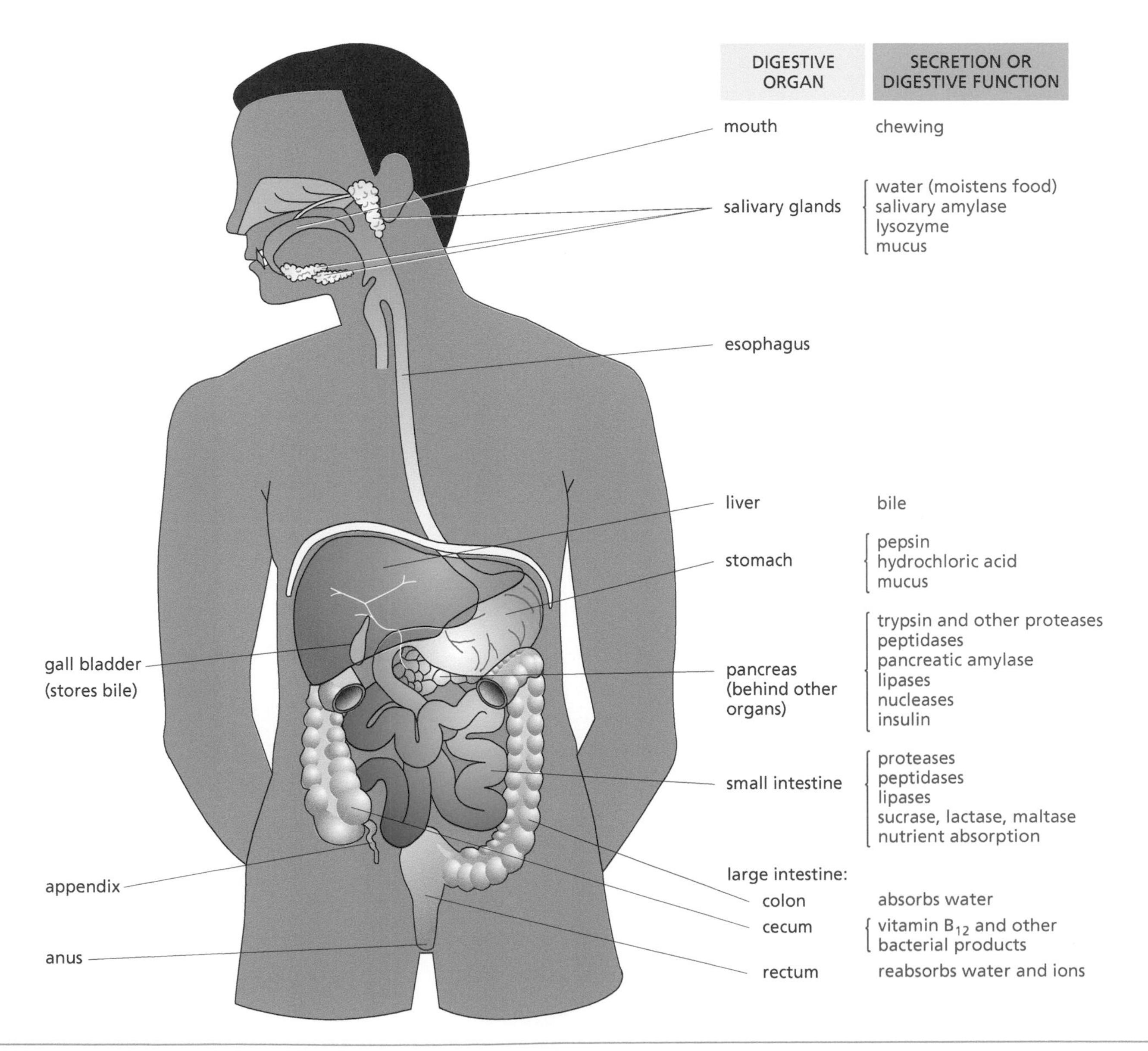

Figure 8.10 **The human digestive system.**

starches (carbohydrates) also begins in the mouth with the enzyme called salivary amylase. This enzyme, present in saliva, breaks down starches into smaller units (sugars). Starches are usually not in the mouth for long enough to be completely digested, however, and their digestion is completed later. Another salivary enzyme called lysozyme (see Figure 8.6) catalyzes the breakdown of bacterial cell walls, so that licking a wound (as all mammals do) helps sterilize it.

8.3.2.2 The stomach Once food is swallowed, it passes quickly through the esophagus and into the stomach. The stomach performs mechanical digestion through rhythmic contractions that knead the food back and forth, mixing it thoroughly, rubbing food particles against one another, and exposing new surface areas. The main activity in the stomach is the digestion of protein, accomplished with the aid of the enzyme pepsin, which breaks large protein molecules up into smaller fragments (called peptides). Like many other protein-digesting enzymes, pepsin is secreted in an inactive form, which protects the glands that secrete the enzyme from digesting themselves. The inactive form is converted into active pepsin by other digestive enzymes. Pepsin works best in an acidic solution, which the stomach provides by secreting hydrochloric acid (HCl). Acidity is measured by a scale called the pH scale; the lower the pH, the more acidic the solution (**Figure 8.11**). Fluids in the stomach are among the most acidic in biological

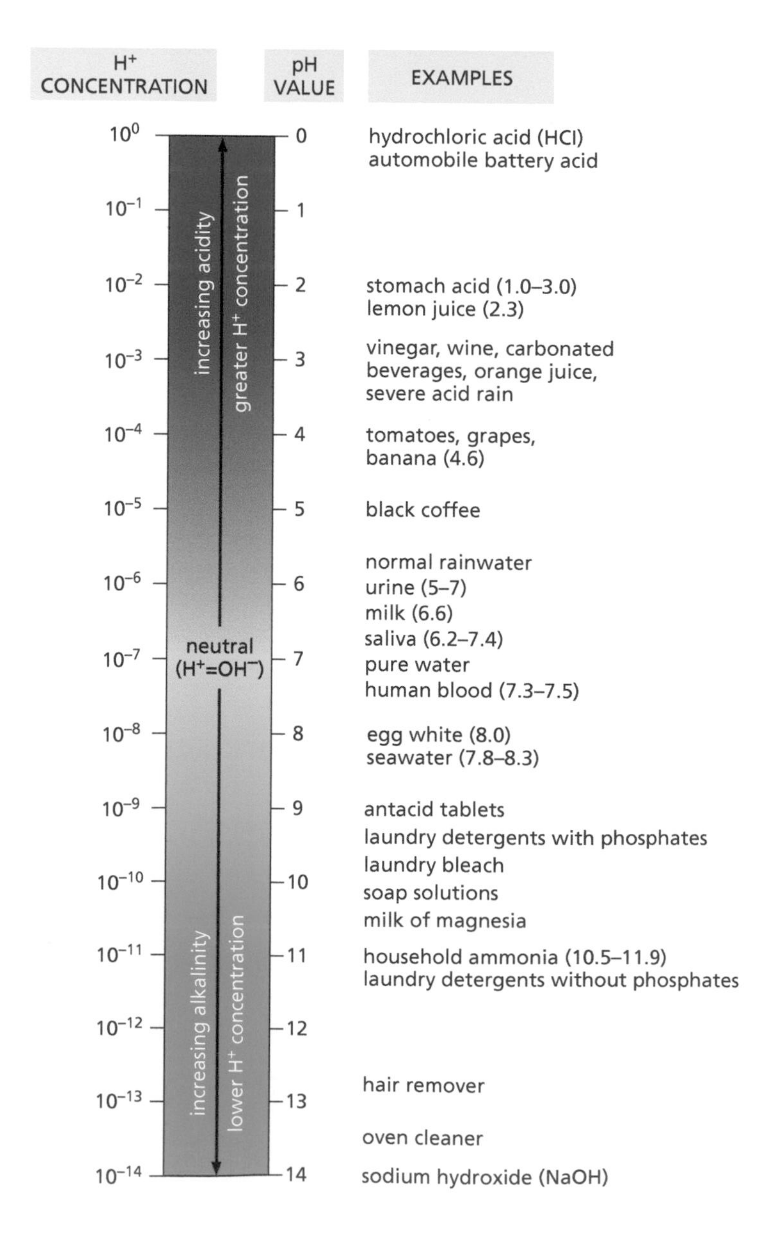

Figure 8.11 The pH scale. The pH of a solution tells us the concentration of hydrogen ions (H^+) in the solution. The H^+ concentration can be expressed as moles of H^+ per liter of solution, as shown to the left of the bar. It is more common, however, to express the concentration as pH, shown to the right of the bar. The pH scale is a reciprocal scale: the lower the pH, the higher the concentration of H^+ (and the more acidic the solution) and the higher the pH, the lower the concentration of H^+ (and the more basic the solution). The pH scale is also a logarithmic scale; that is, each number differs by a factor of 10 from the next number. A solution with a pH of 2 thus has 10 times more H^+ ions than a solution with a pH of 3.

systems, with a typical pH of 2. The stomach also secretes a mucus that protects the stomach lining (which is partly protein) from the pepsin and acid.

8.3.2.3 The small intestine: Processing of fat The lower end of the stomach empties into the **duodenum**, the initial part of the small intestine, where the pH is no longer acidic. The term "small" refers to the diameter, which is about 3 cm; the small intestine is actually very long (20 feet, or 6 m). The duodenum receives the secretions of the liver, called bile. In the watery environment of the intestine, fats tend to come out of solution and form large globules that coalesce to form even larger globules whenever they collide. Bile breaks up fat globules into smaller droplets and keeps these small droplets separate. Recall that fats are insoluble in water because their chemical bonds are nonpolar, while those of water are polar (Figure 8.3). One portion of each bile molecule is polar and, consequently, stable in water; another portion is nonpolar and is thus unstable in water but stable in fat. The nonpolar portions of the bile molecules dissolve in the fat droplets, leaving the polar portions of these molecules exposed on the surface, in contact with the watery intestinal fluids. The polar coating helps the fat droplets to mix with the water and also prevents the small droplets formed by mechanical action from coming back together to form large globules (**Figure 8.12**). This maintains the larger surface area of many small fat droplets and increases the efficiency of chemical digestion, which happens only on the surface.

Bile is secreted by the liver at a constant, slow rate but is used in large amounts when fats or oils are present in the intestine. Bile from the liver accumulates in the gallbladder until it is needed and is then released all at once under the stimulus of a digestive hormone. A hormone is a chemical messenger that is produced by one organ and secreted into blood or lymph vessels to travel throughout the body and cause a specific physiological change in one or more target organs. There are many kinds of hormones with very different functions; for example, several reproductive hormones are discussed in Section 9.2.1. The intestinal lining secretes a digestive hormone whenever fats are present in the intestine; it acts on the gallbladder—its target organ—to stimulate the release of bile into the intestine.

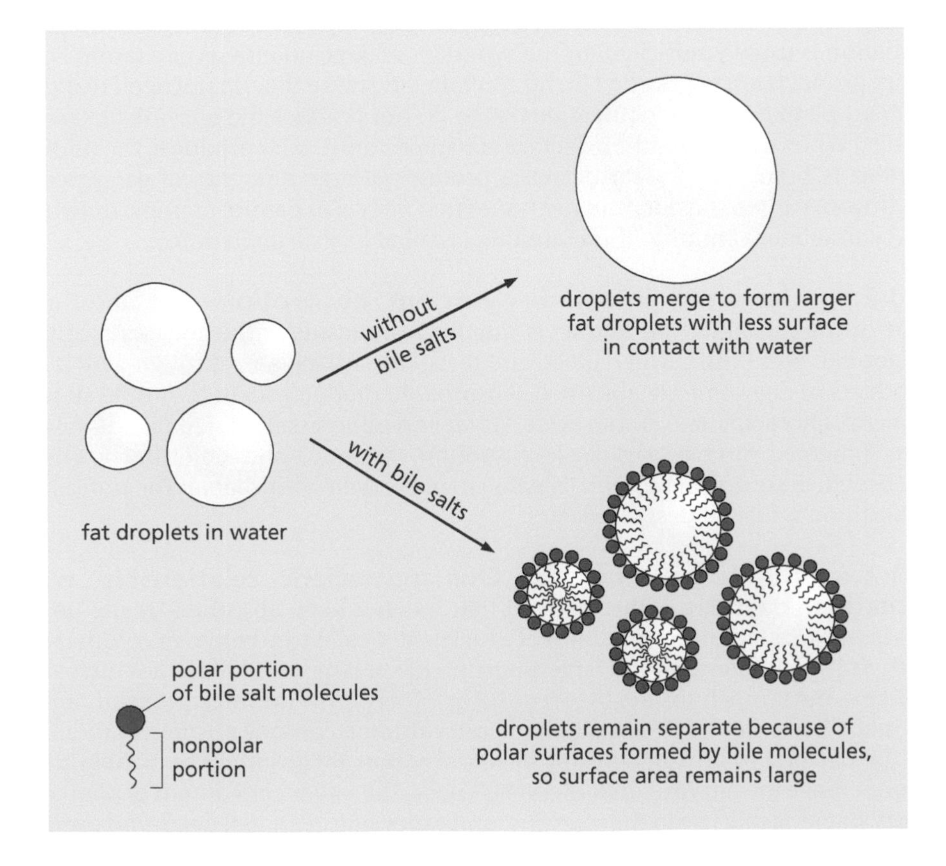

Figure 8.12 The action of bile salts in breaking up fats.

8.3.2.4 The small intestine: Digestive enzymes In the next portion of the small intestine, called the **jejunum**, chemical digestion is completed by enzymes secreted by the pancreas and by the intestine's own lining. Enzymes are often named by combining the suffix "-ase" with the name of the molecule on which the enzyme works: Proteases break down proteins, lipases break down lipids, and so forth. The intestinal enzymes include the following:

- Proteases, protein-digesting enzymes such as trypsin and chymotrypsin, secreted by the pancreas. Other proteases are secreted by the small intestine. Like the pepsin in the stomach, these enzymes break the chemical bonds between certain amino acids, thus breaking the proteins into smaller chains of amino acids called peptides. Each protease is specific and breaks only the bonds between certain specific amino acids.
- Peptidases, enzymes that complete the final stages of protein digestion by breaking peptides down into individual amino acids. Both the pancreas and intestinal lining secrete peptidases.
- Pancreatic amylase, an enzyme secreted by the pancreas. This enzyme continues the job, begun in the mouth, of breaking starches down into sugars.
- Lipases, fat-digesting enzymes, secreted by both the pancreas and the intestinal lining. These enzymes break down fats and oils into glycerol and fatty acids, molecules small enough to be absorbed.
- Sugar-digesting enzymes such as sucrase and lactase, secreted by the small intestine, which break down larger sugars (sucrose or lactose) into simple sugars such as glucose and fructose.

The production of certain digestive enzymes can vary among human populations. We saw in Section 7.4.2 that Northern Europe receives less ultraviolet radiation than other regions of the world, and populations living in Northern Europe therefore have less sunlight to help them synthesize vitamin D. To supply this vitamin, most Europeans consume dairy products rich in vitamin D, and these dairy products also contain significant amounts of lactose, the sugar in milk. Thus, natural selection acted on European populations to favor those individuals who possessed the enzyme lactase, needed to digest lactose. Outside of Europe, ultraviolet radiation is usually sufficient for the synthesis of large quantities of vitamin D, so dairy products are not needed in the adult diet. Because they do not need to digest lactose, people in these populations often do not produce the enzyme lactase as adults. When a person without lactase consumes most dairy products, the unused lactose is fermented by gut bacteria, producing large amounts of gas (mainly hydrogen, carbon dioxide, and methane) that results in painful cramps, diarrhea, and sometimes vomiting. This condition is called lactose intolerance.

8.3.2.5 The small intestine: Nutrient absorption The absorptive part of the intestine (the **ileum**) is lined on the inside with thousands of tiny finger-like tufts (**villi**), which greatly increase the surface area through which the products of digestion are absorbed. Absorbable products include simple sugars, glycerol, fatty acids, and amino acids. Water and mineral salts, including dissolved ions (charged atomic particles) of sodium, calcium, and chloride, are also absorbed; these do not require digestion to make them absorbable. The process of absorption is further described later.

8.3.2.6 The large intestine: Our mutualistic relationship with intestinal bacteria The material that has not been absorbed by the ileum passes into the large intestine, most of which is also called the **colon**. In comparison with the small intestine, the large intestine has a larger diameter (2.5 inches or 6–7 cm), but is much shorter (4 feet or 1.2 m). This part of the intestine is inhabited by many bacteria, which constitute the **microbiome**. Various nutrients produced by the bacteria or broken down by bacterial action are absorbed here. Mammals cannot make the enzymes that digest cellulose, the major constituent of plant cell walls. Bacteria that live in the intestine, and especially in a small dead-end portion (called the cecum), must do it for them. The cecum is especially important (and also much larger) in plant-eating mammals such as horses and rabbits, which consume large amounts of cellulose. Because we are mammals, humans cannot make the enzymes that degrade cellulose, and we also do not have the right species of intestinal bacteria to digest cellulose for us, so we cannot digest cellulose at all. We do, however, get some necessary nutrients from our intestinal bacteria. Humans cannot make vitamin K or biotin, needed for the synthesis of blood-clotting factors and for fatty acid synthesis, but our gut bacteria can synthesize them, and we can then absorb these micronutrients.

Gut bacteria live in a form of symbiosis with vertebrate organisms. **Symbiosis** means simply that two organisms live together; **mutualism** is the form of symbiosis in which the two species are beneficial to each other. Mutualistic gut bacteria derive nutrients from the food taken in by their human host. In exchange, they synthesize vitamins that humans need and break down many complex molecules into simpler components that are more easily absorbed from the intestine. The symbiosis may be disrupted by factors such as antibiotics, which kill the bacteria. Certain foods, such as yogurt, contain many beneficial bacteria that colonize the large intestine. Capsules containing bacteria are often marketed as "probiotics" and sold in health food stores and pharmacies, but there is no consensus among medical professionals concerning the health benefits of any one such formulation over another. Moreover, different people are hosts to different types of bacteria (depending on geographic location, food preferences, and closeness to pets and farm animals), and it is unclear that the bacterial flora inside the intestines of one healthy person would necessarily benefit other people living in different circumstances. Clinical studies of microbiomes (e.g., Zmora et al. 2018) support the following:

- Microbiomes differ greatly from person to person, depending on food preferences and many other factors.
- People with eating disorders such as obesity generally have microbiomes that differ from their healthy neighbors, but transplanting fecal bacteria is seldom an adequate solution for any such disorder. In particular, it can never be determined whether the microbiome is a consequence of the disorder or a factor that contributed to its cause.
- Microbiomes cannot be reliably sampled from stool samples.
- When people swallow "probiotic" capsules, the bacteria in them may or may not have any lasting effect, depending on individual differences that are poorly understood. Some people's intestines are readily colonized by the bacteria in "probiotic" capsules, while other people's intestines resist such colonization (and stool samples cannot reliably monitor whether or not the microbiome has changed).

Without the advice and supervision of a medical professional, the self-administration of "probiotic" formulations is an uncertain treatment at best.

8.3.2.7 The large intestine: Water absorption

The remainder of the large intestine consists of a straight portion called the rectum, which leads to a final opening called the anus. In the colon and rectum, water is absorbed, mostly by diffusion, from the material passing through the gut, giving this material a firmer consistency. Much of this material, which is called feces, is undigested food, but more than half is intestinal bacteria, which are rapidly replaced by bacterial cell division in the intestine. It is partly these bacteria and partly the bile pigments that give feces their characteristic brownish color.

8.3.3 Cellular respiration: Conversion of macronutrients into cellular energy

Carbohydrates, fats, and proteins are all macronutrients, the principal sources of calories for the body. After large macromolecules are broken down into simple subunits in the digestive system, the subunits are absorbed into the cells. They are

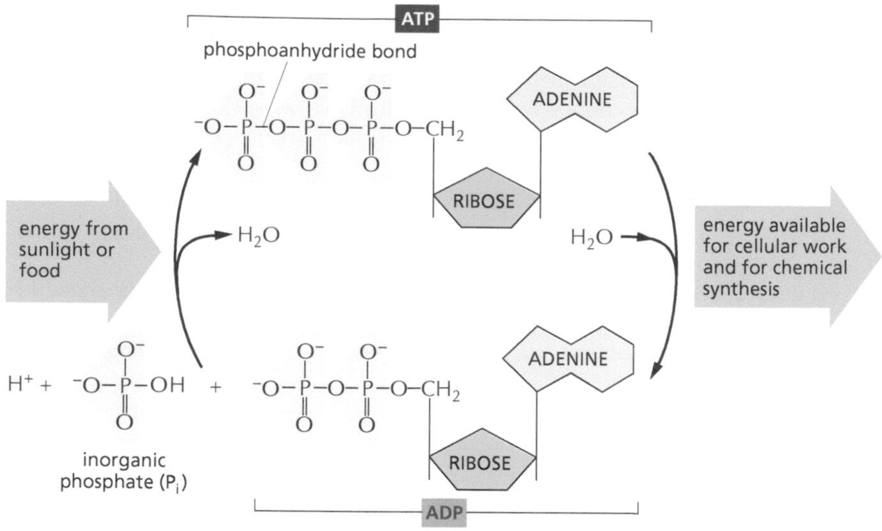

Figure 8.13 ATP as an energy carrier. Energy from sunlight (in plants) or from food (in animals and other organisms) is used to synthesize ATP, storing chemical energy in its bonds. The breaking of these bonds to form adenosine diphosphate (ADP) and inorganic phosphate releases energy for use by the cell.

absorbed first by the cells lining the digestive tract and are then transported via the blood to all of the other cells of the body. Once inside a cell, these simple subunits can begin a series of reactions that result in the storing of energy in the form of a molecule called **adenosine triphosphate (ATP)** (Figure 8.13). ATP is one of the principal molecules in which chemical energy is stored for later use by the cell. Food, therefore, provides the ability to make the ATP our cells use for all their work.

8.3.3.1 Absorption Food cannot be converted into energy until it is absorbed. Absorption takes place through the membranes of the cells lining the intestine. The polar chemical structure of most products of digestion means that they cannot enter the cell directly, because the interior of a cell membrane is nonpolar. The membrane thus acts as a gate that controls what enters (or leaves) the cell. Chemicals are absorbed by one of four mechanisms (Figure 8.14); these mechanisms also bring chemicals into cells elsewhere throughout the body.

- Some small molecules enter the cells by **diffusion**, a process that requires no added energy and is therefore sometimes called "passive diffusion" (see Figure 8.14A). Diffusion works only if a **concentration gradient** exists; each substance diffuses from a place where it is more concentrated to a place where it is less concentrated. Small, uncharged molecules such as water (H_2O) or oxygen (O_2) diffuse directly through the cell membrane. Charged molecules (ions) cannot cross the membrane but may diffuse through **membrane channels**, protein-lined "holes" in the membrane that are highly specific for particular ions that can pass through them.
- Polar molecules larger than water are often moved into cells with the help of proteins called carrier proteins that extend across the cell membrane. When this transport proceeds along a concentration gradient, it is called "facilitated diffusion"; see Figure 8.14B.
- Other molecules are absorbed *against* their concentration gradient: that is, from an area of lower concentration to an area of higher concentration of that type of molecule. This process requires an input of energy (usually from the breakdown of ATP) and is called **active transport** (see Figure 8.14C). Membrane proteins bind to the molecule being transported and use energy to carry it across the membrane.
- Large particles can be taken into the cell by a process called **endocytosis** (see Figure 8.14D), in which the plasma membrane is pulled in toward the interior of the cell. The plasma membrane first forms a pit that may contain large particles; the margins of the pit then draw closed, and the pit pinches off to form a vesicle inside the cell. This bulk process transports many molecules at once, either suspended in liquid or attached to membrane proteins called **receptors**.

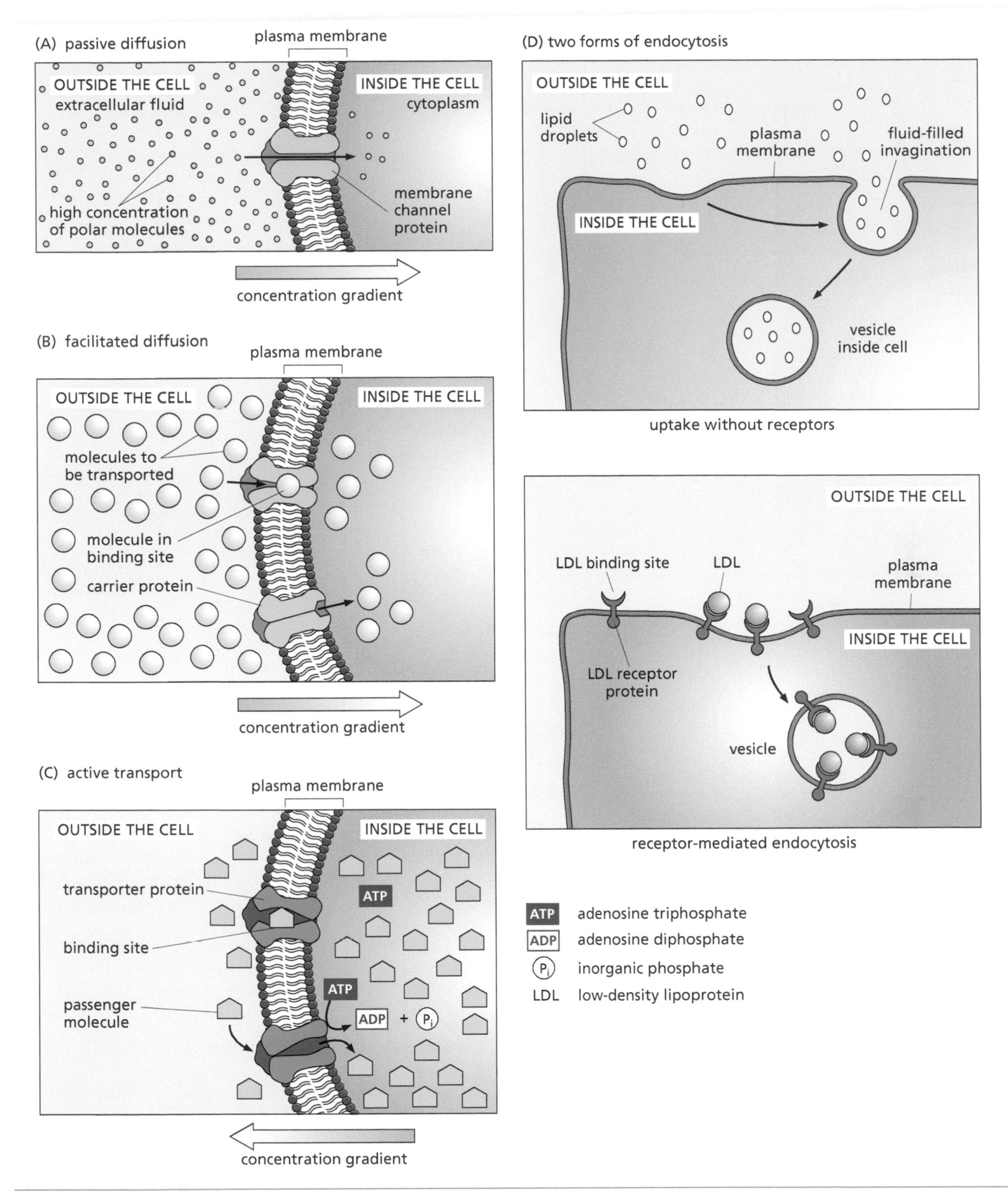

Figure 8.14 Membrane transport mechanisms.

8.3.3.2 Energy-releasing pathways The small molecules that are absorbed into cells are then broken down into even smaller molecules, as can be followed in Figure 8.15.

- The long carbon chains of fatty acids are broken down in stages. In each stage, two carbons at a time are broken off from the long carbon chains to make an acetyl group (CH_3COO^-). These acetyl groups are each put onto a

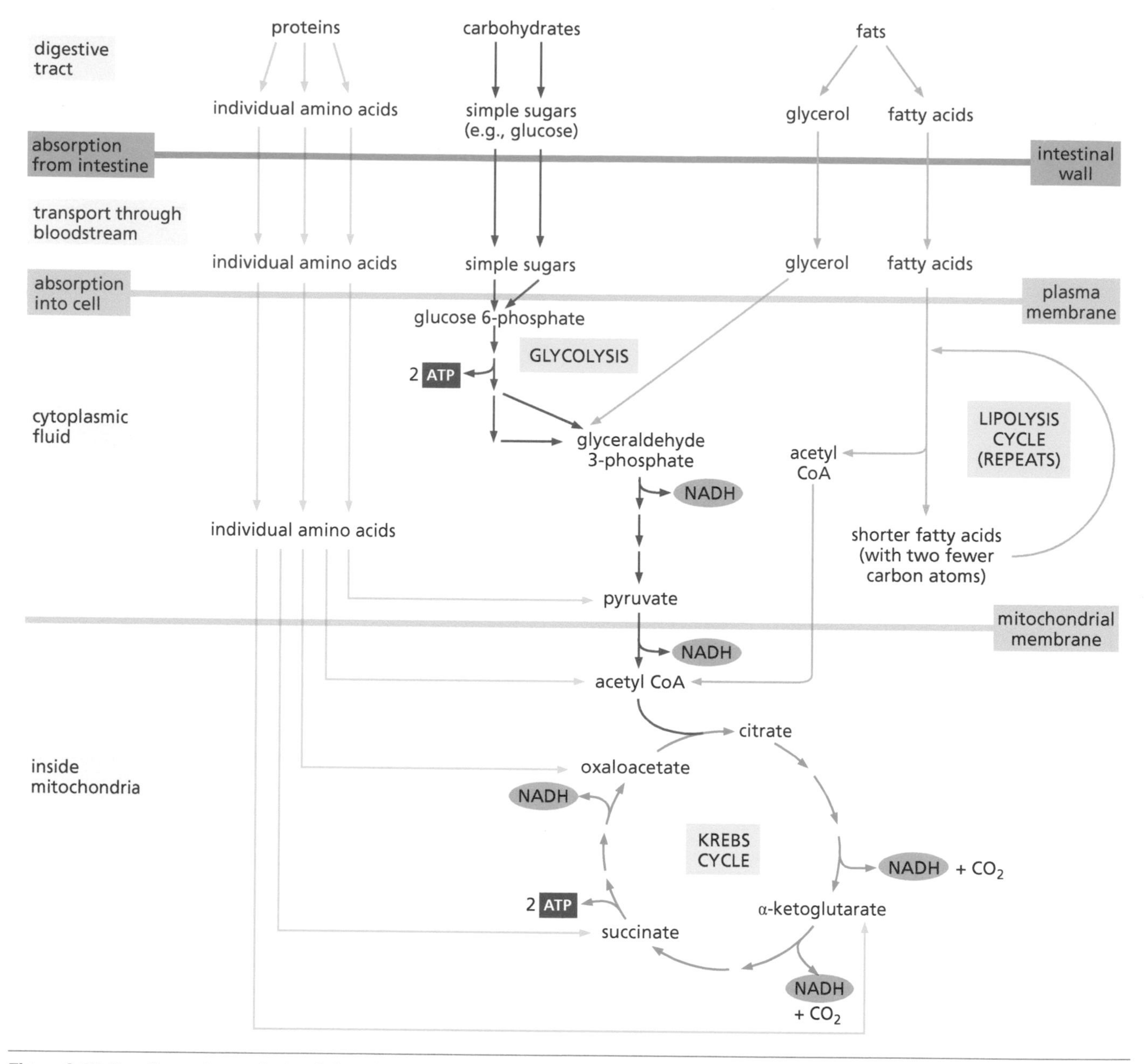

Figure 8.15 How the major products of digestion are broken down in a series of energy-yielding reactions.

carrier molecule called coenzyme A, so that the complex is called acetyl coenzyme A (acetyl CoA).

- Sugars such as glucose are converted into the three-carbon molecule **pyruvate** (CH_3COCOO^-) by a process called **glycolysis** (see Figure 8.15). Pyruvate is then converted into acetyl CoA, which enters the Krebs cycle, described later. During glycolysis, limited amounts of the energy-rich molecule ATP are synthesized. Glycolysis is one of the oldest known metabolic pathways and is found in most living things.
- Amino acids can be used as building blocks to make new body proteins or can be broken down and used for energy. Amino acids that are used for energy are first transformed into either pyruvate, acetyl CoA, or one of the molecules in the Krebs cycle.

Pyruvate and acetyl CoA produced from the breakdown of proteins, carbohydrates, or fats are then transported from the cytoplasm into organelles called mitochondria (see Section 6.2.4), where further energy is extracted from them in a cycle of reactions called the **Krebs cycle**, named after Hans Krebs, the scientist

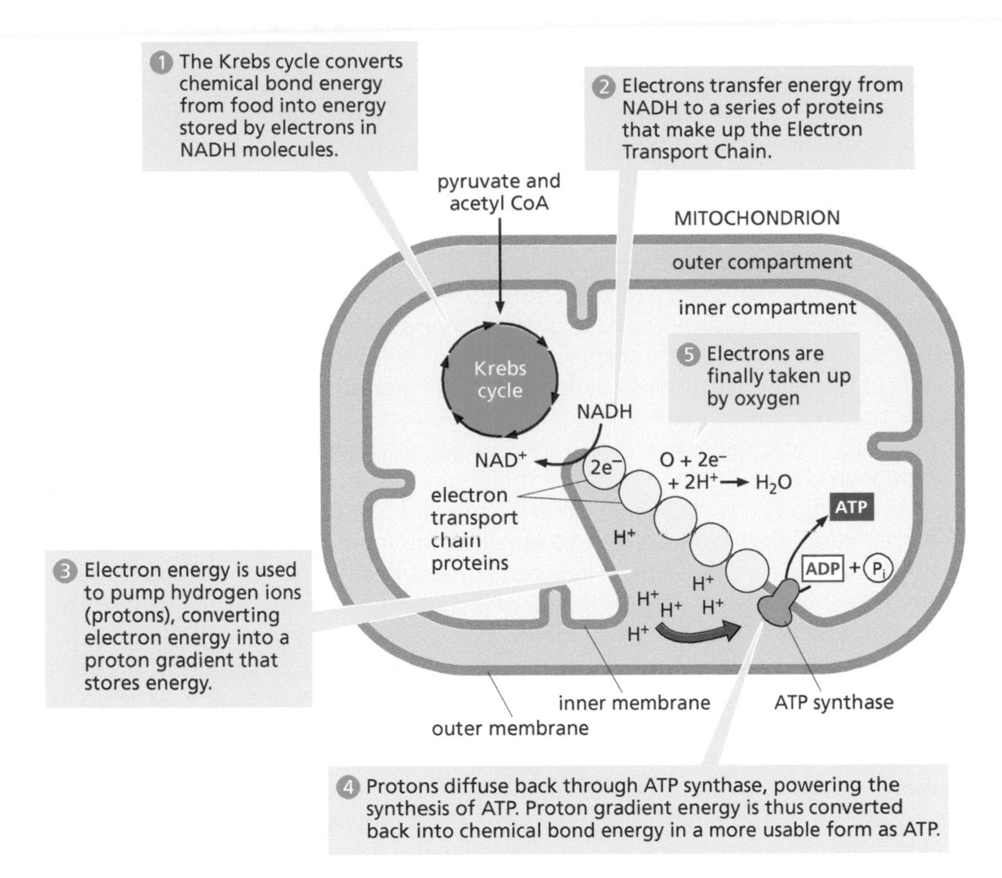

Figure 8.16 Energy-producing processes that take place inside the mitochondria. Electrons are shown as e^- and protons (hydrogen ions) as H^+.

who discovered it. The Krebs cycle is also called the citric acid cycle or tricarboxylic acid cycle. In each cycle, two ATP molecules are synthesized. The enzymes for this process are proteins located in the inner compartment of the mitochondria, called the matrix (Figure 8.16). One molecule of a compound called oxaloacetate combines with the two carbons carried by acetyl CoA, forming a new molecule called citrate; a series of reactions gradually removes electrons from this molecule, thereby extracting energy at each of several biochemical steps.

Each time a molecule gives up an electron, it is **oxidized**, and the molecule that receives the electron undergoes **reduction**. Therefore, reduction and oxidation go hand in hand, and these electron transfer reactions are given the name "**redox**" reactions. In the case of organic redox reactions, electrons are usually transferred together with protons in the form of hydrogen atoms. In the Krebs cycle, hydrogens removed from the molecules derived from citrate are accepted by a molecule called NAD^+, forming NADH. NAD^+ is the oxidized form of this molecule, while NADH is the reduced form. NAD^+/NADH is called a coenzyme because it is needed as part of the reactions catalyzed by the enzymes of the Krebs cycle. It serves as an "electron shuttle," since it receives electrons from the Krebs cycle (and one step of glycolysis) and then passes them on to the first in a series of mitochondrial membrane protein complexes, called the **electron transport chain (ETC)** (Figure 8.16).

The molecules of the ETC are ordered such that the next molecule in the chain holds on more tightly to the electron than the previous one. An electron being tightly held by a molecule is associated with chemical stability, and greater stability corresponds to a lower energy level. Therefore, the ordered redox reactions of the ETC result in a release of energy. Certain protein complexes that are part of the ETC can use this released energy to pump a proton (H^+) from one side of the mitochondrial membrane to the other, an example of active transport. These H^+ ions accumulate in the outer compartment of the mitochondria, called the intermembrane space (Figure 8.16). An unequal distribution of protons on opposite sides of the membrane (called a **proton gradient**) is thus formed. Because protons are charged, the protein gradient is an electrochemical gradient, which can store energy for later use by the cell. Some of this stored energy is used by an enzyme called ATP synthase to produce ATP. ATP synthase is an amazing protein complex that acts like a mechanical turbine: protons flow across the mitochondrial membrane through a channel in the ATP synthase complex from high to low concentration; this flow of protons turns a "rotor" that is also part of the complex, and this turning causes shape changes in the rotor subunits that force together ADP and phosphate to generate ATP. For an animation showing this complex in action, please go to biologytrending.routledge.com/chapter8

At the end of the ETC, oxygen is the final electron acceptor, which is why we need to breathe in oxygen from the atmosphere. Compared to the two molecules of ATP generated by one round of glycolysis or one round of the Krebs cycle, the reactions of the electron transport chain and ATP synthase can generate much more of this energy—up to 34 molecules of ATP per molecule of glucose. In addition, regulatory mechanisms block the Krebs cycle when no oxygen is present. Therefore, under anaerobic conditions (no oxygen), the yield from one molecule of glucose will be only the two ATP molecules produced by glycolysis, whereas glycolysis, the Krebs cycle, and the ETC can yield 38 total ATP molecules under aerobic conditions (with oxygen present). Some organisms can survive with the low energy yield available from anaerobic metabolism, but humans can do so for only very short periods of time. (Think of the amount of time you can sprint at top speed before you become "out of breath.")

Several steps in the Krebs cycle and in electron transport require vitamins as coenzymes. As an example, consider the role of the vitamin thiamine in the breakdown of the molecule pyruvate. An enzyme that contains thiamine as one of its constituent parts combines with the pyruvate molecule, releases CO_2, then emerges from a later reaction in its original form. Because the thiamine is not used up, it can participate in the reaction again and again. For this reason, only minute amounts of thiamine are needed to facilitate the breakdown of large quantities of pyruvate formed in carbohydrate metabolism. Vitamins B_2 and B_3 form parts of the larger molecules that carry electrons from the Krebs cycle to the ETC (Figure 8.16).

Digestion thus culminates with chemical changes at the cellular level. The process of turning foods into ATP begins when we eat and ends with electrons being taken up by oxygen. Each cell breaks down sugars and other organic molecules, converting their chemical bond energy to ATP. This process is called **cellular respiration**, but it should be noted that "respiration" in this case means chemical energy transfer rather than gas exchange. Ultimately, many of the carbon atoms from the food molecules are combined with oxygen as carbon dioxide (CO_2), which the body exhales. (For example, CO_2 is a by-product of several reactions of the Krebs cycle, as shown in Figure 8.15.) The hydrogen atoms from the food are split: Their protons form electrochemical gradients across the mitochondrial membrane, and their electrons are passed through the ETC; these protons and electrons recombine in the end with oxygen to form water (H_2O). At each intermediate step, some energy is extracted that is eventually converted into ATP. Other kinds of organisms extract energy by transporting carbon and hydrogen to molecules other than oxygen, enabling them to live in the absence of oxygen, but much more energy can be extracted from food when oxygen is used as the final acceptor. Eating causes the cellular ATP production process to swing into high gear, producing ATP for immediate use and for storage against future needs.

THOUGHT QUESTIONS

1. Why is it important for food to spend the proper length of time in the stomach? What would happen if food left the stomach too soon?
2. The contents of the digestive tract are pushed along rather slowly by rhythmic muscular action (peristalsis). How does this relate to the body's need for a long intestinal tract?
3. What would be the consequences of a mutation that prevented a person's body from making a membrane protein necessary for the active transport of sugar from the intestine?
4. Why do food substances need to be digested into smaller molecules before they can be absorbed from the intestine?

8.4 ABSORBED NUTRIENTS CIRCULATE THROUGH THE BODY

After nutrient molecules have been absorbed in the small intestine, they circulate throughout the body. Other materials, including dissolved ions, oxygen, and cells of the immune system, also circulate throughout the body.

8.4.1 Circulatory system

The circulation of materials is carried out by the circulatory system. In all vertebrates, this system consists of blood, contained within a series of blood vessels, and the heart, a muscular pump that keeps the blood circulating.

The blood consists of a fluid material (the plasma), in which cells and platelets are suspended. The cells include the red blood cells (erythrocytes), which contain the oxygen-carrying molecule hemoglobin (see Section 7.3.2), and several types of white blood cells, which form the immune system (see Sections 14.2 and 15.1.1). The platelets are cell fragments that release materials important in blood clotting. A soluble protein called fibrinogen is also important in clotting and is one of several soluble proteins that circulate within the plasma. When a wound brings fibrinogen into contact with the air, it can turn into an insoluble tangle in which cells become trapped. This tangle and its trapped cells form a blood clot that blocks further blood loss and initiates the process of wound healing.

The blood circulates through a series of blood vessels of differing diameters. The vessels leading away from the heart are called **arteries**; they branch into narrower and narrower vessels. The vessels leading back toward the heart are called **veins**. The veins are arranged as a series of tributaries that flow into larger vessels and eventually back to the heart. The thinnest vessels, called capillaries, carry blood from the smallest arteries to the smallest veins (Figure 8.17). Capillary walls

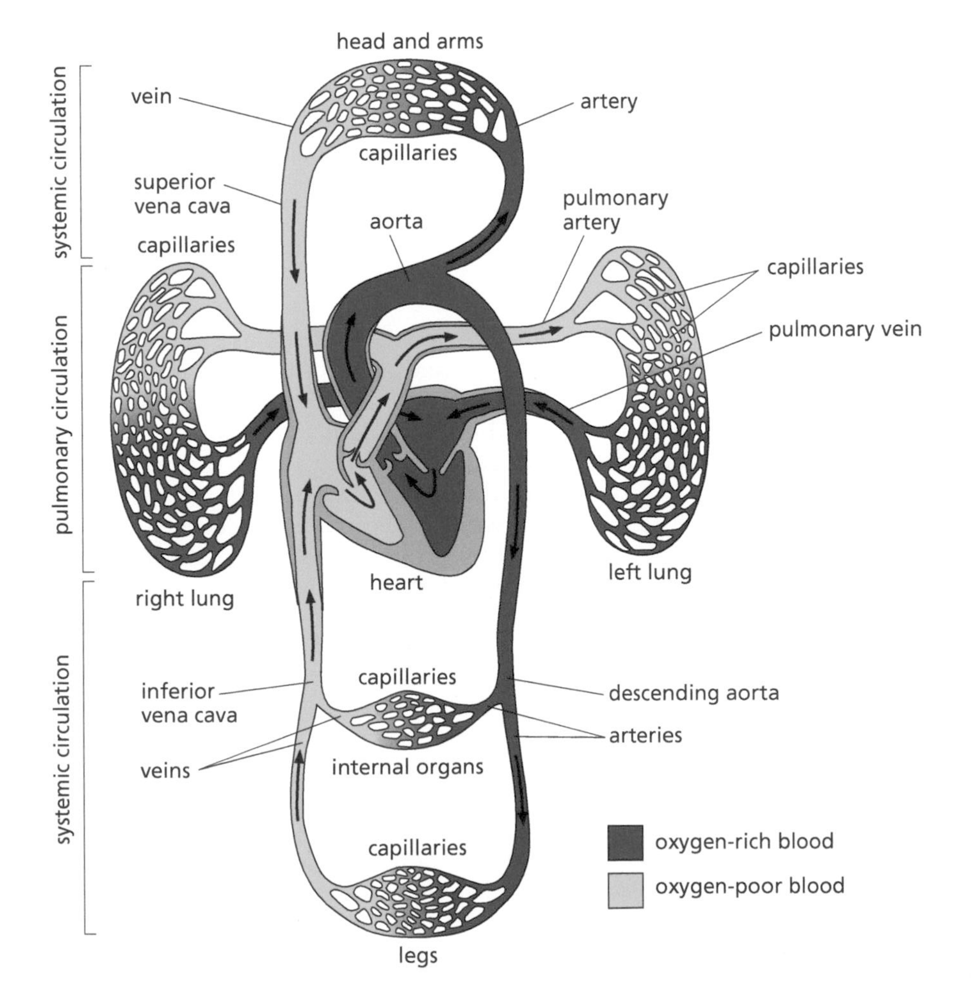

Figure 8.17 The human circulatory system.

are made of a single layer of cells. Materials diffuse from capillaries into tissues and from tissues into capillaries across the cells of the capillary walls. The very large capillary surface area permits diffusion on a large scale. Capillaries throughout the body provide cells with essential nutrients, so no cell of any tissue can be very far from a capillary. A key step in the development of cancer is the ability of a tumor to stimulate growth of new capillaries to supply it with nutrients (see Section 10.5.4).

Nutrients absorbed in the gut enter the capillaries of the gut lining and flow to the liver via the hepatic portal vein. If the blood contains more glucose than the body needs immediately, the excess is converted into the storage molecule **glycogen** (Figure 8.2), which is a polysaccharide. Glycogen storage takes place in most body cells, but the largest amount is stored in the liver. As the body uses up glucose from the blood, the liver cells convert glycogen back into glucose and release it into the bloodstream as needed, a mechanism that ensures a dependable but moderately low concentration of glucose in the blood. The storage of glycogen and the efficient use of glucose both require the hormone **insulin**, secreted by special clumps of cells within the pancreas. Persons in whom these cells have degenerated cannot produce enough insulin; their condition is known as type 1 **diabetes**, formerly called insulin-dependent diabetes mellitus (IDDM) or juvenile diabetes. The symptoms of diabetes can be controlled by supplying insulin or by controlling weight and diet, but there is no known cure for the disease itself. Researchers are working on ways to replace the damaged insulin-producing cells in the pancreas (possibly using stem cells; see Section 10.1) in individuals with type 1 diabetes.

Blood from the liver and the body's other organs flows through veins to the heart. Most veins have thin, flexible walls, and the blood within them flows at a relatively low fluid pressure. (Arteries, in contrast, have thick walls that can withstand the high, pulsating fluid pressure of the blood within them.) The blood within veins is propelled by the massaging action of nearby muscles and other organs. Valves within the veins keep the blood flowing in one direction, preventing it from flowing backwards.

In addition to distributing nutrients, the circulatory system also distributes oxygen to the body and transports carbon dioxide (a waste product of cellular respiration) from body cells to the lungs (which then exhale it). The way in which blood circulates through the heart keeps the oxygen and carbon dioxide from mixing.

8.4.2 The heart

The heart is a muscular organ whose rhythmic contractions keep the blood circulating throughout the body. In all mammals, the heart contains four chambers (**Figure 8.18**). Oxygen-poor blood from the body's various organs enters the right atrium and is pumped into the right ventricle. Contraction of the right ventricle propels the blood out through the pulmonary arteries and into the lungs. Here, the oxygen in the lung's tiny pockets, or alveoli (see Figure 13.1), diffuses into the blood, while carbon dioxide diffuses out and is exhaled. Oxygen-rich blood from the lungs returns to the left side of the heart, where it enters the left atrium. Contraction of the left atrium pushes the blood into the left ventricle, the heart's largest chamber. Contraction of the left ventricle propels the blood throughout the arteries and into the body's various organs; for this reason, the left ventricle has the thickest walls of the four heart chambers. Oxygen diffuses from the blood into the body's many cells across the thin capillary walls, and cellular wastes, including carbon dioxide, diffuse from the cells into the blood. The veins collect this oxygen-poor blood from the body's various organs and carry it back to the heart, where the pattern of circulation repeats (Figure 8.17).

The heart maintains its own rhythmic pattern of contractions. A heart cut from a living animal and placed in a salt solution will continue to beat for many hours. The rhythm is maintained even in the absence of any nerve input, showing that the heart's rhythm originates in the heart itself.

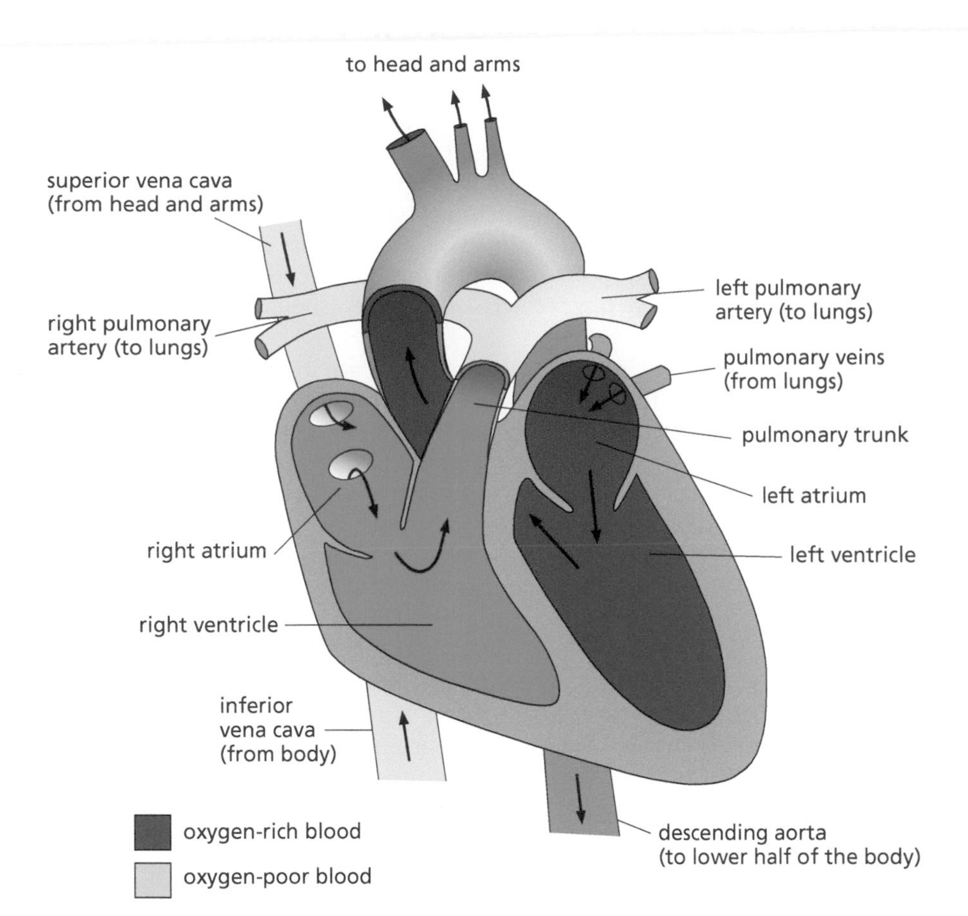

Figure 8.18 The human heart.

8.4.3 Cardiovascular disease

Cardiovascular disease includes both heart disease and diseases of the blood vessels. In the United States each year 500,000 people die of heart disease, making it the number-one cause of death. Another 1.5 million have nonfatal heart attacks. Each year almost another 500,000 die of strokes, a blood vessel disease. Men have more cardiovascular disease earlier than do women (2:1 male:female ratio overall before age 65), but cardiovascular disease is nevertheless the major killer of both men and women. Risk factors for cardiovascular disease include smoking, obesity, high-fat diets, lack of exercise, hypertension, atherosclerosis, high cholesterol levels, stress, and genetic predisposition. Proper diet, exercise, weight loss, and stress reduction are the major ways of reducing the risks.

8.4.3.1 Too much dietary fat The study of disease factors in large populations is called **epidemiology**. Epidemiological studies point to a connection between certain types of dietary fats and cardiovascular disease (heart disease and strokes). The United States, Australia, and New Zealand—all meat-producing countries—have a high consumption rate of meat products per person and also high incidences of cardiovascular diseases. Most meats are high in saturated fats. People in Mediterranean countries consume much of their lipid in the form of olive oil, an unsaturated fat, and their cardiovascular disease rates are lower. Heart attacks are very rare among the Inuit (Eskimos), whose diet contains large amounts of cold-water fish, a good source of a type of fatty acid called omega-3 fatty acid that has been shown to guard against the production of chemicals that damage cell membranes (Swanson, Block, and Mousa 2012). The Japanese also tend to have low consumption rates of saturated fats and low rates of cardiovascular disease.

Epidemiology also provides clues as to whether the association between dietary fats and cardiovascular disease is more closely related to diet or to genetically inherited traits: Japanese people in Japan have much lower rates of heart disease or stroke than do Japanese living in Hawaii or California, whose rates are similar to those of their non-Japanese neighbors. These findings (and similar ones on other immigrant groups) all point to diet, not heredity, as the major difference responsible for the different disease rates between populations.

Because saturated fats have been linked to a greater risk of cardiovascular disease, many experts recommend that saturated fats be replaced with unsaturated fats in most diets. Both monounsaturated fats (such as olive oil) and polyunsaturated fats (found in most fish and many plant sources) are desirable, especially if those fats replace saturated fats. Many experts recommend that the quantities of all dietary fats be reduced to lower the risk of cardiovascular disease.

8.4.3.2 Atherosclerosis Dietary fat can increase the risks for cardiovascular disease in several ways. One way is that excess dietary fat can result in fat deposits that build up in the arteries, causing **atherosclerosis** (Figure 8.19), a type of cardiovascular disease that can lead to heart attacks. The fat deposits obstruct the blood vessels, making the passages narrower; eventually these deposits may harden by accumulating calcium compounds that make the vessels more rigid, a process called "calcification." Atherosclerosis contributes to hypertension (high blood pressure), although a person can have hypertension without having atherosclerosis.

8.4.3.3 LDLs and HDLs In contrast with carbohydrate molecules, lipid molecules contain few oxygen and nitrogen atoms and have mostly nonpolar bonds in which electrons are shared equally around carbon and hydrogen atoms (Figure 8.3). Because the bonds in lipids are nonpolar, lipids are not water soluble. Blood plasma is mainly water, so lipids must be transported through the blood from one part of the body to another by transport particles such as **low-density lipoproteins (LDLs)** and **high-density lipoproteins (HDLs)**. These transport particles are proteins that bind lipids in such a way that they can move through body fluids.

People eating identical diets may not have the same serum cholesterol level. This difference seems to have a genetic component and relates in part to each person's ability to make LDLs and HDLs. No one has very much free (unbound) cholesterol because cholesterol is very nonpolar, so what is called serum cholesterol is actually the total of all the cholesterol bound to HDLs and LDLs. HDLs transport cholesterol, phospholipids, and triglycerides out of tissues to the liver, where they are used in the synthesis of bile acids, while LDLs transport cholesterol and lipids into tissues and cells. The HDL/LDL ratio is the ratio of outbound to inbound lipids. A high HDL/LDL ratio thus indicates that the proportion of "good cholesterol" (lipids on their way out) is higher than the proportion of "bad cholesterol" (lipids on their way into cells). A very low ratio, meaning a preponderance of inbound lipids, correlates strongly with an increased risk of

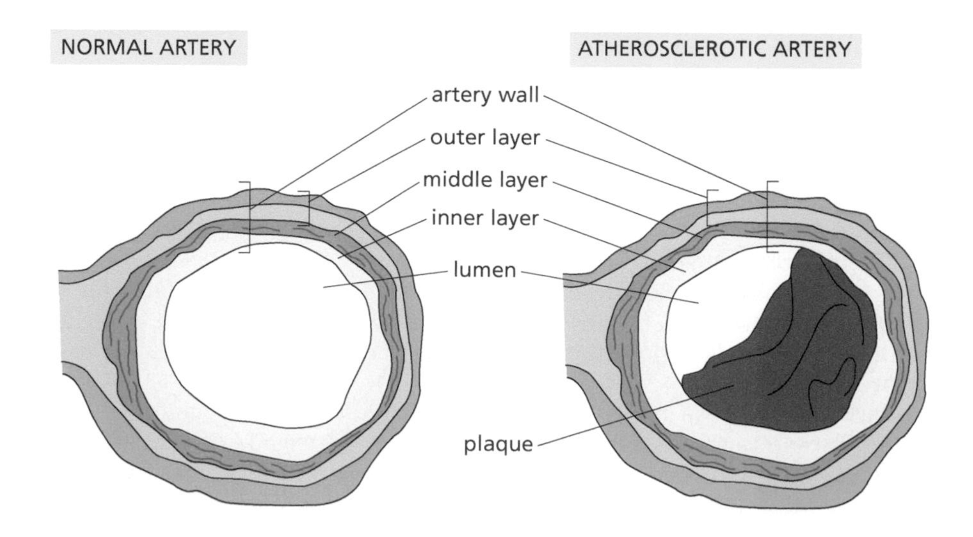

Figure 8.19 Atherosclerotic plaque reducing the effective diameter of an artery. Although the outer diameter of the vessel has not changed, the inner diameter (lumen) through which the blood flows has become smaller by the formation of lipid deposits on the inside of the vessel, making the blood pressure higher. These deposits may also become calcified, further increasing blood pressure.

atherosclerosis in the arteries of the heart (coronary arteries), a condition that can precipitate a heart attack.

Some people are genetically prone to high cholesterol levels because their cells lack LDL receptors on their surfaces. When cells need cholesterol, they get the cholesterol from the LDLs in the bloodstream by binding these particles to LDL receptors and internalizing the receptors and LDLs by endocytosis (Figure 8.14D). If the LDL receptors are missing or nonfunctional, the cells manufacture their own cholesterol even when LDL levels are already high, because they cannot take up LDL from the blood. Thus, diet is not the only factor leading to atherosclerosis; problems with lipid transport and lipid uptake are also factors.

THOUGHT QUESTIONS

1. Cancer is not as frequent a cause of death as is heart disease, yet cancer research seems to get more publicity than research on heart disease. Examine some newspapers and magazines to see whether this is a correct impression. If it is, what factors might contribute to this difference?
2. The human circulatory system is a continuous, closed system; in other words, none of the blood vessels are open-ended. Why must this be so?

8.5 MALNUTRITION AND DIETARY FADS CONTRIBUTE TO POOR HEALTH

The term **malnutrition** literally means "bad nutrition." Malnutrition can result either from eating too little or from eating the wrong foods, and it is one of the major health problems of the world, particularly among the poor and in areas of turmoil. Malnutrition also exists in people with access to sufficient foods, sometimes as a result of eating disorders.

8.5.1 Fad diets versus healthy eating

A **diet** is really the sum total of everything we eat, whether we are trying to alter our food habits or not. (Many people use the term "diet" only when they are trying to alter their food habits, especially when trying to lose weight.) Nutrition experts will always recommend balanced diets, containing adequate amounts of all important macronutrients and micronutrients, over imbalanced diets lacking one or more basic nutrients. Also, numerous studies have repeatedly found that rapid weight loss on an unbalanced diet is usually followed by gaining back all the weight that was lost, if not more.

8.5.1.1 Fad diets Many dietary fads severely restrict one or another particular group of nutrients. One such diet restricts carbohydrate intake to near zero in order to stimulate the body to make its own glucose out of fats. Although the diet works for a short time as a rapid weight-loss regimen, it is very unhealthy when followed for even a few months, causing a decrease in muscle mass that makes future weight loss much more difficult. Most people who have followed this diet quickly gain back the weight they lost, if not more, when they stop.

Diets that count only carbohydrates or that focus on the "glycemic index" that measures blood glucose levels following a meal may have their place in the control of diabetes (under proper medical supervision) but cannot be followed uncritically. First, there are important differences between high-fiber carbohydrate sources (fruits, vegetables, and whole grains) and refined carbohydrates such as table sugar and white flour: Those with high fiber help you feel full (and feel full longer) on fewer calories while lowering cancer risks (especially for cancers of the rectum and colon). The refined carbohydrates in sodas, pastries, candies, white rice, and most breads and pastas provide excess calories with few vitamins or other micronutrients and are thus a major contributor to the current obesity epidemic. Moreover, the fats allowed on most "carb"-counting diets contain over twice the caloric content of carbohydrates. Most overweight

North Americans eat far too many carbohydrates, but counting carbohydrates alone does not lead to a healthy diet unless fats are also controlled and high fiber intake is maintained.

One type of diet that has been promoted in recent years is the so-called "Paleolithic" diet. Based on the belief that Paleolithic hunters ate healthier food, diets of this kind promote meat and fresh vegetables at the expense of anything derived from dairy or cereal agriculture. Although one can eat a healthy diet conforming to these rules, the red meats encouraged on "Paleolithic" diets are unhealthy in large quantities, while the whole-grain cereals that these diets discourage can be very healthy. Among protein sources, red meats such as beef are usually accompanied by large amounts of saturated fats, while poultry has much less fat; fish has still less fat and also contains certain beneficial fatty acids (the omega-3 fatty acids). Most "Paleolithic" diets do not distinguish among these protein sources and also discourage such healthy proteins as low-fat milk, cottage cheese, and yogurt. A large-scale longitudinal study of over 120,000 health professionals found that an increase in red meat consumption of about one serving per day was associated with an increase in mortality from both cardiovascular disease (about 20%) and cancer (about 14%) and that substituting one serving per day of other protein sources instead of red meat would result in a 7%–19% decrease in mortality risk (Pan et al. 2012).

8.5.1.2 Healthy eating A recent study of food patterns in 195 countries reveals that not all diets are healthy. The three most common problems are excessive salt intake, inadequate eating of fruits and vegetables, and inadequate eating of whole-grain cereals. For good health, most nutritionists recommend balanced diets that limit salt and control calories.

A **balanced diet** is one that contains adequate amounts of all the important nutrients: fats, proteins, carbohydrates, and micronutrients, as well as fiber. Balanced diets are both healthier and easier to maintain long-term. A healthy body is easier to maintain with a diet that you can keep for the rest of your life. Such a diet should include adequate (but not excessive) amounts of healthy fats, such as those from fish, nuts, or olive oil. It should also include lean protein (from fish, poultry, legumes, or low-fat dairy products) while minimizing the intake of saturated fats from beef and pork. Plant foods (fruits, vegetables, and whole grains) are usually more filling than animal foods and have many more vitamins, minerals, and other micronutrients. A plant-rich, balanced diet will usually be much more satisfying and often lower in calories than the restrictive food choices promoted by many fad diets. Most nutritionists therefore recommend fiber-rich, balanced diets as the best for long-term health, including for permanent weight loss.

Eating a balanced diet does not require counting anything, but if you want to pay attention to numbers, the best numbers to count are dietary calories (kilocalories). Table 8.1 gives general guidelines as to the daily number of calories you should be eating to maintain a steady weight. Most packaged foods have calories specified on the label, and caloric values for fresh foods can easily be found on government websites. Perhaps the most important use of caloric information is in making informed decisions between different brands or different foods of the same type. Even if you cannot accurately count calories for the foods prepared by other people, knowing which alternatives are generally higher or lower in calories will help you in making informed decisions.

One recent clinical study of weight loss in generally healthy people compared a group that was instructed to eat a balanced diet but limit their carbohydrate intake, while another group was instructed to eat a balanced diet but limit their fat intake. Weight loss was comparable in the two groups; both groups lost around 5 kg, or 10 pounds, over 12 weeks. Balanced diets are thus effective in weight loss regardless of whether carbohydrate intake or fat intake is minimized.

Nutritionists who have studied the long-term health effects of different diets have generally found plant-rich balanced diets to be the best. Among these diets are the so-called Mediterranean diets that are based on whole grains, low-fat protein sources (plant proteins and fish in preference to beef and pork), and fresh fruits and vegetables in abundance. Not all such diets come from Southern Europe; many rural Asian and West African diets are broadly comparable and equally healthy. Small amounts of wine, sometimes described as part of these diets, are less essential. The National Heart, Lung, and Blood Institute (2018), one of the National Institutes of Health, has promoted Dietary Approaches to Stop Hypertension (DASH), a diet very similar to those of the Mediterranean type. Clinical trials of various diets (see Willett and Skerrett 2017; Mayo Clinic 2019) have shown that those promoting the best health outcomes are similar to the Mediterranean diet and DASH in limiting calories and in promoting healthy fats (containing omega-3 fatty acids or olive oil) over saturated fats. Other studies of such diets (Martinez-Lapiscina et al. 2013; Singh et al. 2014) have documented more specific benefits to better brain health.

In many cases, food choices can be made healthier by making simple substitutions as often as possible: choosing whole grains over refined carbohydrates, choosing plant-based protein or fish over beef and pork, choosing healthy fats over saturated fats, and choosing low-salt and low-calorie alternatives wherever possible. Train yourself to make such healthy choices as a matter of routine, and you have a much better chance to enjoy a longer, healthy life with greatly reduced risks of cardiovascular disease, obesity, Alzheimer's disease, and cancer.

8.5.2 Eating disorders

8.5.2.1 Obesity In the United States, the Centers for Disease Control and Prevention (CDC) estimates that 40% of adults are obese. **Obesity** is defined as a body weight 20% or more above the ideal for the particular subject's sex and height, which translates to a body mass index (BMI) of 30 or more (Box 8.2). Obesity increases the risk for high blood pressure, heart disease, stroke, diabetes, and several cancers (including those of the colon, breast, and prostate). The CDC has declared that obesity is one of the greatest public health threats in the United States.

Numerous factors contribute to obesity. There are some genetic contributions to a person's relative proportion of body fat or carbohydrate used for energy production and to the likelihood of energy surplus being stored as fat versus lean tissue. However, environmental factors are much stronger than genetic factors for most people. Factors that encourage obesity include decreased exercise, increased calorie intake, and increased consumption of fatty foods. In many cases, the

Box 8.2: Obesity and the body mass index (BMI)

The current definitions of overweight and obesity are based on the BodyMass Index (BMI), calculated as described here. A person with a BMI of 25 or higher is considered overweight, and a person with a BMI of 30 or above is considered obese. Based on these criteria, the Centers for Disease Control and Prevention (CDC) estimates that over 60% of adults in the United States are overweight or obese, and they have declared obesity to be a "public health epidemic." The CDC (at www.cdc.gov/nccdphp/dnpa/obesity) lists a total of 20 disease conditions to which overweight or obesity contributes.

TO CALCULATE YOUR BODY MASS INDEX

Divide your weight by your height, then divide by your height again.

The formula is $BMI = w/h^2$.

- If you measured your weight in kilograms and your height in meters, the result is your BMI.
- If you used kilograms and centimeters, multiply the result by 10,000 to get your BMI.
- If you used pounds and inches, multiply the result by 703.7 to get your BMI.

According to the CDC, the incidence of overweight and obesity has been increasing in the United States.

	Percentage of adults in United States		
	Overweight but not obese (BMI 25–29.9)	Obese (BMI 30 or above)	Total overweight (BMI 25 or above)
1980	33%	15%	48%
1999–2000	34%	30%	64%

amount of excess weight is proportional to the number of meals eaten away from home, where portions are large and fat content is high.

The marketing of food (including its advertising and packaging) encourages unhealthy behavior and overeating. High-calorie or high-fat foods are advertised heavily. Many foods, especially snack foods, are packaged in individual serving units that are convenient for a person on the go (often in containers that fit into the cup holders of automobiles), and these individual serving units can be much larger than any serving recommended by nutritionists, sometimes as much as two or three times as large. Fast food is often marketed in "super-size" portions, and customers are encouraged to have side orders of fried foods and beverages consisting of calorie-rich and vitamin-poor sodas and shakes. Eating in the car or while working or watching TV are forms of "unconscious" eating that can prevent a person from being aware of how much they have eaten.

Of course, no person can achieve results on a diet that they cannot stay on, and different people find different diets easier or harder to stick to. Nutritionists generally recommend balanced diets as both the healthiest and the easiest to maintain long-term.

8.5.2.2 Anorexia and bulimia In the middle and upper classes of the industrialized nations, some people suffer from a condition called **anorexia nervosa**. This condition is much more common in women than in men, at a ratio of 9:1. Anorexic individuals suffer from a mistaken perception of their body size. They imagine themselves to be heavier than they really are and desire to be thinner as a result, a feature that clearly distinguishes anorexia from all other forms of undernourishment (Figure 8.20). Anorexics also respond poorly to body cues of

Figure 8.20 **One of the earliest signs of anorexia is misperception of one's own body.**

hunger and satiety. The misperception of hunger, satiety, and body size are early symptoms that precede the most noticeable feature of the disease, which Dr. Hilde Bruch has called a "relentless pursuit of thinness," a self-imposed undernourishment that borders on starvation. At the same time, there is usually an absorbing or obsessive interest in food, which may include talking or reading about food, preparing food, collecting recipes, or serving food to others, all while avoiding actually eating.

One of the surest signs of the disease in anorexic women is that they usually stop menstruating because of a lack of the cholesterol needed for synthesis of the hormones that regulate the menstrual cycle (see Section 9.2.1). Other symptoms include changes in brain activity. The brain must be constantly supplied with glucose to provide cellular energy for nerve cell function; when it is not, many mental functions may be impaired. These impairments may manifest themselves in anorexic persons as deception (hiding things, keeping secrets) and a distrust of others, but only late in the process, after starvation has already set in. Untreated anorexia is usually fatal.

Anorexia is most common among Caucasian women between the ages of 15 and 30 years with an average or above-average level of education. The disease also occurs in Japan and in parts of Southeast Asia, but only in the uppermost social strata. Anorexia is virtually unknown among people living in poverty or in undernourished populations anywhere. One researcher has proposed that anorexia was originally an adaptation among nomadic people fleeing regions of famine, but the disease never seems to occur during documented times of famine or food scarcity. Even in countries where it occurs, it seems to vanish during economic hard times, such as the Depression of the 1930s.

Anorexia was once extremely rare. Its marked increase in the United States since World War II has been attributed by several experts to a general standard of beauty that has increasingly glorified thinness. Women in professions such as modeling and ballet dancing are particularly likely to develop anorexia. Also, female athletes in sports where competition is organized by weight classes (rowing, for example) are at high risk for developing a "female athlete triad," consisting of eating disorders such as anorexia, combined with loss of menstruation and a loss of bone mass that may cause osteoporosis later in life. The loss of menstruation resulting from overexercise has been attributed to the depletion of the body's estrogen.

Many anorexics also suffer from a related condition called **bulimia**, although bulimia can also occur independently of anorexia. Bulimia is characterized by occasional binge eating of everything in sight, usually including large quantities of high-calorie "forbidden" foods, in total disregard of any concept of a balanced diet. Immediately after a binge, bulimics typically force themselves to vomit or else purge themselves with an overdose of a laxative. Both the binge and the purge are usually done in secret; bulimics (like anorexics) become extremely skillful at hiding their condition from others. Persistent bulimia can lead to ulcers and other problems of the digestive tract and also to chemical erosion of the teeth from the frequent contact of the teeth with the acidic secretions of the stomach. Bulimia occurs in both sexes, but more often in women. Bulimia is especially common among educated women who have easy access to unrestricted amounts of food, a situation common on many college campuses.

8.5.2.3 Type 2 diabetes The CDC estimates that about 34.2 million Americans, or 10.5% of the U.S. population, suffer from **diabetes** (2018 statistics). Of this total, less than 10% is type 1 diabetes, an autoimmune disorder in which the body's immune system attacks the insulin-producing cells of the pancreas. Between 90% and 95% of diabetes cases are insulin-resistant or type 2 diabetes, an eating disorder that results in most cases from chronic obesity. This type of diabetes was once called "adult-onset" diabetes because it often took decades to develop, but it is now frequently diagnosed in children as well. Many other countries have lower prevalence rates, but these rates are increasing nearly everywhere at an alarming rate, especially among the poor. The World Health Organization reports that the worldwide prevalence of diabetes among adults nearly doubled from 4.7% in 1980 to 8.5% in 2014, with the greatest increases in poor and middle-income countries.

Diets low in fiber and high in calories, especially from refined carbohydrates, are considered to be the major risk factor for this disease. People living with diabetes are at increased risk for blindness and other vision loss, heart disease, stroke, kidney failure, and poor circulation, which often leads to the amputation of toes, feet, or legs. In 2015, diabetes was the seventh leading cause of death in the United States. Studies have shown that type 2 diabetes can usually be controlled with diet and exercise. Weight loss is key, but an increase in dietary fiber and a restriction in refined carbohydrates are often more important.

8.5.3 Starvation

Around the world, more people die each year of starvation than of any single disease or other preventable cause. Death by starvation occurs primarily in poor countries, but it also occurs in pockets of poverty within wealthier nations. The immediate cause of death by starvation is usually inadequate caloric intake. In populations with very low caloric intake, even a small or temporary interruption in the food supply can lead to mass starvation: just one bad crop, or political disruption that prevents the planting or harvesting of the crop, can have severe consequences.

Inadequate caloric intake can also result in protein deficiencies that take several forms, including both low total protein and low levels of particular amino acids. If protein intake is inadequate for either reason, a protein deficiency called kwashiorkor develops, most often when carbohydrate intake is adequate but protein intake is not.

In all organisms, cells and proteins are constantly being broken down, and in a healthy, adequately nourished body, they are constantly being replenished. When protein intake is insufficient for this replenishment to occur, there is considerable loss of muscle tissue, and the death of many cells releases numerous dissolved ions into the surrounding tissue. These ions retain water and contribute to tissue swelling (edema) that makes the loss of muscle tissue harder to see. Children suffering from kwashiorkor have swollen abdomens, a fact often noticed in photographs from protein-deficient areas. Their large bellies conceal the fact that these children are actually starving to death.

If protein intake is inadequate and carbohydrate intake is also inadequate, the combined deficiency produces a condition called marasmus, in which the body slowly digests its own tissues and wastes away. When carbohydrate is not available as an energy source, amino acids are used to produce energy. Because amino acids are not stored except in the form of the body's structural and functional proteins, the use of amino acids for metabolic energy degrades these proteins. Once the body's muscle mass falls below a certain minimum, marasmus is always fatal.

8.5.4 Ecological factors contributing to poor diets

Malnutrition is recognized as a worldwide disease with regional differences in its cause but with similar outcomes everywhere. Here we take Africa as an example, although many other examples exist. Many populations in Africa experience either chronic or periodic protein deficiency; Africa has therefore been described as a protein-starved continent. Fresh vegetables are widely available, so diets are high in fiber and most vitamins. Vegetables (including legumes) can fill nearly all of a population's nutritional needs as long as supplies are adequate and as long as some form of nutritionally complete protein is obtainable from animal sources (including fish) or from a combination of plant proteins. There are lakes and rivers where fishing provides adequate protein resources, and there are cattle-herding groups, such as the Masai and Fulani, who can meet their protein needs from their animals. Across most of the continent, however, animal protein sources are not readily available, and the supplies of grains that could offer protein by amino acid balancing are not always adequate.

Grain supplies are inadequate owing to a combination of ecological and social factors. Tsetse flies, which spread blood-borne diseases, have made much of the land uninhabitable to most domestic animals. This not only limits the availability of meat proteins but also limits the supply of draft animals that can pull plows and till the soil. Tractors and fuel are too expensive for most farmers in Africa, and many places are either too dry or too wet to support agriculture. The world's largest desert, the Sahara, occupies most of Africa's northern half; other deserts exist in Somalia and Namibia. Most of these deserts, including the vast Sahara, are growing larger each year as animals such as sheep and goats overgraze and destroy the plants on the desert fringes, a process known as desertification (see Section 18.3.2). In other parts of Africa, high rainfall leaches important minerals from the soil, leaving soils deficient in the minerals necessary for plant growth. Rice, wheat, and most other grains grow rather poorly in many African soils. Some success has been achieved with millet and corn (maize), but raising a sufficient quantity and variety of grains to provide complete protein is difficult. In most places in Africa, populations can maintain adequate nutrition in years when there are ample harvests and efficient distribution. Unfortunately, these two conditions are not always met. Protein deficiencies are made worse by political and military upheavals that drive people from their farms or that prevent the planting, harvesting, and distribution of crops. Protein starvation (marasmus and kwashiorkor) is all too common, particularly among children.

8.5.5 Effects of poverty and war on health

Climatic, economic, and political factors all contribute to the unequal production and unequal distribution of food across the planet. Poverty exists in all nations of the world. Many populations are marginally nourished and are therefore more vulnerable to a year of drought or a bad harvest, but malnutrition exists in every part of the globe, especially among the poor. Even in regions in which nourishment is usually adequate, people can become undernourished because of the disruptions of life associated with war.

8.5.5.1 Wartime starvation The effects of starvation on humans have been studied retrospectively in people who were, at an earlier time, subjected to starvation by war. Such a study on the short-term and long-term effects of starvation was conducted by using the birth records of infants born in the Netherlands in the winter of 1944–1945, when many people in certain cities experienced starvation (Stein et al. 1975). The food shortages varied from city to city, but the other effects of war were equal across the country. When food rations dropped below a threshold level, fertility in the population decreased, and the decrease was greatest among people in the lower classes. Fetuses carried by women who experienced starvation in the first trimester had a higher rate of abnormal development of the central nervous system. Maternal undernourishment in that period also carried forward to premature births, very low birth weight, and an increase in the infant death rate immediately after birth. Maternal undernourishment in the third trimester produced the greatest increase in the infant death rate in the three months after birth. Brain cells were depleted in infants who died. Among survivors, however, undernourishment in infancy was not correlated with long-term effects on mental development when males were tested at age 18. Other retrospective studies have found similar results.

8.5.5.2 Long-term effects of childhood undernutrition There are several indications that undernutrition during childhood decreases brain development and capacity. Infants who died of kwashiorkor or marasmus had less DNA, protein, and lipid in their brain cells compared with infants who died at similar ages of causes not related to nutrition. Evidence suggests that both the severity and the length of the period of malnutrition affect intellectual development. Iron deficiency, particularly in the early years of life, can also impair mental functions.

Malnutrition can also result from children's failure to eat properly because of a depressed mental state. Malnutrition of this type is called "failure to thrive." Infants must learn that their needs will be met, and this learned capacity is termed "basic trust." When circumstances are such that a child is not regularly fed and nurtured, it fails to develop trust. Loss of a caregiver or a diminution or loss of nurturing care can create depression in infants, who then lose trust and become malnourished. Medical intervention cannot rescue the child unless a trusting relationship is established. Adults recover their physical and mental capacities if they are rescued from starvation. For children, however, nutritional replacement will not lead to full recovery unless emotional and psychological support is also provided. The recovery of children from famine thus requires much more than food.

8.5.6 Micronutrient malnutrition

Malnutrition can still exist when total caloric intake is sufficient or even when it is excessive. Because the roles of micronutrients include the proper functioning of enzymes, micronutrient deficiencies prevent the proper utilization of other foods. As we saw earlier, mineral deficiencies can result when diets are restricted to those foods grown in mineral-deficient soils. Diets that include only a few types of foods, particularly processed "junk foods," may be lacking in necessary micronutrients. It is for this reason that Japanese children are encouraged to "eat at least 100 different foods each week." Limited incomes, limited mobility, and limited availability of fresh foods all make this goal difficult to achieve and generally make nutritional problems worse. Micronutrient malnutrition may be a danger that is overlooked among people who lose interest in eating, including elderly people, people with chronic diseases including cancer, and people who have mental illnesses such as depression. Some elderly people with dementia may also forget what they have or have not eaten.

THOUGHT QUESTIONS

1. Think about the definition of obesity as a weight 20% or more above the ideal weight for a person's sex and height. What is meant by "ideal"? Who sets these "ideals"? Are there cultural differences in what is considered "ideal"? How do cultural definitions of "ideal" relate to biological definitions?
2. Why do you suppose anorexia occurs only among well-fed populations and never among the poor or in times of famine? What does this imply about the possible causes of the condition?
3. Will efforts aimed at improving nutrition in people with various forms of malnourishment be readily accepted by the people they are meant to help? What factors determine acceptance? Give examples if you can.
4. If sufficient food can be produced, will every person have good nutrition? Why or why not?
5. Why might starvation have greater long-term effects in children than in adults?

CONCLUDING REMARKS

During the first half of the twentieth century, vitamin deficiencies were major public health concerns. Vitamin and mineral deficiencies can lead to various deficiency diseases, birth defects, or neurological damage. Nutritional deficiencies and deficiency diseases have declined since then in the industrial world as the result of better eating habits, more varied diets, and vitamin-fortified foods. Greater interest now centers on whether certain foods can promote good health and reduce the risks of chronic diseases. Since the 1960s, public health officials and nutritionists have increasingly turned their attention to heart disease, stroke, cancer, and chronic health problems such as obesity and high blood pressure that can influence the risks for these fatal diseases. Heart disease is the number-one cause of death in most industrialized countries and is significantly associated with a diet that is too high in fat, particularly saturated fats.

Malnutrition still exists, however, and it can take many forms. Inadequate caloric intake can result from crop failures, from poverty, or from eating disorders. Poverty and war usually worsen nutrition and contribute to stress among civilians. Adequate nutrition depends on getting all the necessary nutrients without taking in too much of any one of them. Much of the world's population gets inadequate food and inadequate protein. In many industrialized countries, problems are caused instead by high-fat diets with inadequate fiber. The best way to avoid both types of problems is to eat a varied and balanced diet while keeping fat intake low. Sensible eating habits such as these are important for good health. If we want to promote health for ourselves, our families, and the rest of humankind, then a very important step is ensuring that each person has an adequate and balanced diet.

CHAPTER SUMMARY

- Only four chemical elements, H, O, N, and C, make up the vast majority of all molecules in cells.
- Certain elements and molecules can donate or accept electrons, becoming charged **ions** that can engage in ionic interactions. Ionic compounds usually dissolve in water.
- Most molecules are held together by **covalent bonds**, where electrons are shared between atoms; in **nonpolar** covalent bonds, those electrons are shared equally, whereas in **polar** covalent bonds, one atom attracts the electrons more strongly, causing the two atoms in the bond to have partial charges.
- Water is a polar molecule with partial negative charge on the oxygen atom and partial positive charges on the two hydrogen atoms. Hydrogen bonds between these partial charges on adjacent water molecules are responsible for water's ability to resist temperature change.
- Digestion is a process in which materials consumed as food are broken down into substances that the body can absorb and then use. Digestion consists of **chemical digestion** with the aid of **enzymes** and **mechanical digestion**.
- The **macronutrients** are the major sources for energy, measured in **kilocalories**, and including **carbohydrates**, **lipids** (fats and oils), and **proteins**. Lipids are also needed for the formation of cell membranes, and proteins are needed as enzymes, receptors, and membrane transporters. **Complete proteins** have all the amino acids needed in human nutrition. Proteins are digested and used until one of the needed amino acids, called the **limiting amino acid**, is depleted, after which the remaining protein is used as an energy source in the manner of carbohydrates.
- Molecules or parts of molecules may be either polar and thus stable in contact with water or else nonpolar and thus water-avoiding.
- Digested food enters cells by passive **diffusion**, facilitated diffusion, **active transport**, or **endocytosis**. Diffusion always acts along a **concentration gradient**, from a place of higher concentration to a place of lower concentration; active transport always operates against such a concentration gradient. Concentration gradients can serve as a form of energy storage.
- Cellular energy is derived in the mitochondria by **oxidation** reactions in the **Krebs cycle** and electron transport to form **ATP**.
- Material that the body cannot absorb constitutes **fiber**.
- **Micronutrients** (**vitamins** and **minerals**, including electrolytes) are needed for good health.
- **Epidemiology** is the statistical study of diseases and risks in large populations. Data from epidemiology can help provide evidence for certain health risks and benefits, even when the mechanisms of action are not known.
- Good nutrition reduces such chronic conditions as high blood pressure and **obesity** and lowers the risks for **cardiovascular disease** and certain cancers.
- **Malnutrition** can result from inadequate food, inadequate protein, or vitamin and mineral deficiencies. Eating disorders such as **anorexia** and **bulimia** can lead to malnutrition.

BIBLIOGRAPHIC REFERENCES

Bibliographic references to material in this chapter can be found online at biologytrending.routledge.com/chapter8

GLOSSARY: KEY TERMS TO KNOW

These key terms are defined at the end of each chapter as an aid for student review.

Active transport Use of energy to transport a substance from an area where it is in low concentration to an area where it is in higher concentration. In cells, active transport is performed by membrane proteins called transporters.

Adenosine TriPhosphate (ATP) A molecule that provides energy for cellular processes.

Amino acid A small molecule containing both an amino group ($-NH_2$) and a carboxylic acid group (–COOH) that is capable of being joined with other similar molecules to form long protein chains.

Anorexia nervosa A psychological eating disorder characterized by self-imposed starvation.

Antioxidant A substance that prevents oxidation of a molecule by an oxidizing agent.

Artery A blood vessel that carries blood away from the heart.

Atherosclerosis Deposits of fat and cellular debris, which may become calcified, on the interior walls of arteries.

ATP See *Adenosine triphosphate.*

Balanced diet A varied diet containing ample (but not excessive) amounts of all nutrients.

Basal metabolic rate The rate at which the body uses energy when awake but lying completely at rest.

Bulimia A psychological eating disorder characterized by an overeating binge, followed by self-induced vomiting or laxative abuse.

Calorie The amount of energy required to raise the temperature of a gram of water by 1 degree Celsius. See also *Kilocalorie.*

Carbohydrates Polar molecules used by organisms as energy sources and consisting of carbon, hydrogen, and oxygen, with hydrogen and oxygen atoms in a 2 1 ratio.

Cardiovascular disease Any degenerative (age-related) disease of the heart or blood vessels, including atherosclerosis, arteriosclerosis, heart attack, and stroke.

Cellular respiration The process of energy use in cells by the chemical breakdown of macronutrients.

Channels See *Membrane channels.*

Chemical digestion The use of enzymes and chemical reactions to break down food into simpler molecules that cells can absorb.

Cholesterol A lipid with a multiringed structure, found in the membranes of most animal cells.

Coenzyme A nonprotein substance needed for an enzyme to function.

Colon The large intestine, excluding the cecum.

Complete protein A protein that contains all the amino acids considered essential for human nutrition.

Concentration gradient A situation in which the concentration of a substance is different in different locations or on opposite sides of a membrane.

Covalent bonds Chemical bonds formed by the sharing of electrons, either equally (nonpolar covalent bonds) or unequally (polar covalent bonds).

Diabetes Either of two metabolic disorders in which blood glucose levels are chronically high. Type 1 diabetes results from a lack of insulin production by the pancreas; type 2 diabetes (often accompanied by obesity) results from the body's inadequate response to insulin.

Diet An overall pattern of food consumption.

Diffusion A process in which molecules move randomly from an area of high concentration to an area of lower concentration until they are equally distributed.

Duodenum The initial portion of the small intestine, which receives the secretions of the liver and pancreas.

Epidemiology The study of the frequency and patterns of disease in populations.

Endocytosis Bringing a particle into a cell by surrounding it with cell membrane.

Enzyme A chemical substance (usually a protein, but sometimes an RNA) that speeds up a chemical reaction without getting used up in the reaction; a biological catalyst.

Fats Lipids that are generally solid at room temperature.

Fatty acids Long-chain, nonpolar organic acids, released by the digestion of fats or phospholipids.

Fiber Food components that pass through the digestive tract without being absorbed.

Free radicals Very reactive atoms or molecules with unpaired electrons.

Glycogen A carbohydrate consisting of many glucose units linked together, used as a storage molecule in animals and certain microorganisms.

Glycolysis The breakdown of carbohydrates into pyruvate.

High-Density Lipoproteins (HDLs) Proteins that carry lipids (like cholesterol) away from tissues via the bloodstream; often referred to as "good cholesterol."

Hydrogen bond A weak bond between a hydrogen atom with a partial positive charge and another atom (in water, an oxygen atom) with a partial negative charge.

Ileum The final third of the small intestine in which most absorption of digestive products takes place.

Insulin A hormone, produced by the pancreas, important in the metabolism and cellular uptake of carbohydrates.

Ion An electrically charged atom, one that is either positively charged because it has lost one or more electrons or negatively charged because it has gained one or more electrons.

Jejunum The second and longest portion of the small intestine, where most digestion is completed.

Kilocalorie (kcal) The amount of energy required to raise the temperature of 1000 grams (1 kilogram, equal to 1 liter) of water by 1 degree Celsius. Dietary "calories" are actually kilocalories.

Krebs cycle A series of biochemical reactions that break apart pyruvate and use the chemical bond energy to make ATP and NADH from ADP and NAD.

Low-Density Lipoproteins (LDLs) Proteins that carry lipids (like cholesterol) to tissues via the bloodstream; often referred to as "bad cholesterol."

Limiting amino acid An amino acid present in small amounts that, when used up, prevents the further synthesis of proteins requiring that amino acid.

Lipids Nonpolar molecules formed primarily of carbon and hydrogen, occurring in cell membranes and also used as energy sources.

Macronutrients Carbohydrates, lipids, and proteins, collectively.

Malnutrition Poor or inadequate nutrition.

Mechanical digestion Breaking food into smaller particles by physical means such as chewing and churning, exposing new surfaces for chemical digestion.

Membrane channels Proteins capable of transporting molecules across membranes.

Microbiome Collectively, the microbial flora inside the digestive tract or other body region.

Micronutrients Collectively, vitamins and minerals, nutrients needed in much smaller quantities than macronutrients.

Minerals Inorganic (noncarbon) atoms needed to regulate chemical processes in the body.

Mutualism An interaction between species in which both species benefit from the interaction.

Nonpolar Having electrical charges equally distributed (or nearly so) across chemical bonds; nonpolar molecules do not mix well with polar solvents such as water.

Obesity A condition in which ideal body weight is exceeded by at least 20%.

Organic compounds Carbon compounds held together by covalent bonds.

Oxidation Removal of electrons from an atom or molecule.

Oxidize To remove electrons from an atom or molecule.

Phospholipids Molecules containing long, nonpolar hydrocarbon chains attached at one end to a polar phosphate group.

Polar Having chemical bonds with electrons shared unevenly so that one part of the bond has more negative charge than the other. Water is polar and therefore dissolves other polar substances.

Polar covalent bonds Chemical bonds in which electrons are shared unequally between two elements.

Proteins Molecules built of amino acids linked together in long chains that fold up to produce complex shapes, functioning often as enzymes or as structural materials in or around cells.

Proton gradient A form of potential (stored) energy created by a separation or unequal distribution of protons (hydrogen ions).

Pyruvate A three-carbon molecule, CH_3COCOO^-, formed mainly from the breakdown of sugars in glycolysis.

Receptor A protein or other molecule that binds with a specific drug or other chemical substance and responds to the binding by initiating some cellular activity.

Redox Electron transfer reactions involving both oxidation and reduction.

Reduction Addition of electrons (negative charges) to an atom or molecule.

Saturated fats Lipids with no double bonds between their carbon atoms.

Symbiosis Any type of interaction between two species living together; mutualism and parasitism are two types of symbiosis.

Triglycerides Lipids formed from glycerol and three fatty acid units at a time.

Unsaturated fats Lipids with one or more double bonds between their carbon atoms.

Vein A blood vessel carrying blood toward the heart.

Villi Fingerlike processes such as those lining the inside of the small intestine.

Vitamin Carbon-containing molecules needed in small amounts to facilitate certain chemical reactions in the body.

Vitamin deficiency disease A disease caused by insufficient amounts of a vitamin and cured, in most cases, by adding the vitamin to the diet.

CONNECTIONS TO OTHER CHAPTERS

Chapter 3: Some genetic differences exist in the body's ability to digest certain substances (as in phenylketonuria) or to use certain nutrients after their absorption (as in diabetes).

Chapter 7: Nutritional requirements may differ among human populations for various inherited and environmental reasons.

Chapter 9: Unchecked population growth puts more people at risk for malnutrition.

Chapter 10: Certain cancers have frequencies that vary according to diet: High-fat diets promote certain cancers, whereas high-fiber diets lower many cancer risks. Researchers are investigating if stem cells may be used to regenerate insulin-producing cells to treat type 1 diabetes.

Chapter 14: Poor nutritional status impairs the immune system.

Chapter 15: People with HIV infection stay less sick if they have good nutrition, but appetite is often suppressed (and nutrition suffers) as AIDS progresses.

Chapter 17: Crop improvements may help to alleviate starvation in many populations.

Chapter 18: Biodiversity is threatened by the need to clear more land for farming to feed the world's population. Developing and conserving better and more varied crop plants will feed more people and support greater biodiversity at the same time.

PRACTICE QUESTIONS

1. Explain the differences between chemical digestion and mechanical digestion.
2. How many times greater is the hydrogen ion concentration in stomach acid than in blood?
3. What makes a molecule polar? What makes a molecule nonpolar? Can some molecules be both polar and nonpolar?
4. What kinds of food molecules can supply precursors to the Krebs cycle?
5. What type of transport is used to bring glucose into a cell?
6. What hormone regulates the uptake of glucose by cells?
7. What type of transport brings lipids into cells? How are nonpolar lipids able to travel through the blood to get to cells?
8. How does atherosclerosis result in an increase in blood pressure?
9. What is the HDL/LDL ratio and why is it important?
10. What are the functions of cholesterol in the human body?
11. Which food molecules produce the most ATP?
12. Where in the cell is ATP produced?
13. What are some of the functions of the water-soluble vitamins in the body?
14. How is an electrolyte different from other mineral micronutrients? What are some of the functions of each in the body?
15. Which can lead to health problems: vitamin deficiencies or vitamin megadoses?

CHAPTER 9

Populations and Reproduction

ISSUES

- Is the Earth overpopulated?
- How fast are human populations growing?
- Why might a population explosion be detrimental to the social good?
- What methods are available to people who want to control their reproduction?
- Can we restrict reproduction without violating people's rights?
- Can we diminish population growth and its impact?

BIOLOGICAL CONCEPTS

- Population ecology (populations, population density, growth rates, carrying capacity, population regulation)
- Biosphere (human influences, overpopulation and its effects, resource uses, habitat alteration)
- Reproductive biology (reproductive anatomy, reproductive physiology, reproductive cycles, hormonal controls)

CHAPTER OUTLINE

DOI: 10.1201/9781003391159-9

Imagine a world where people must share a room with 4–12 others. A room of one's own is a rare luxury. In fact, people who have any housing at all consider themselves fortunate, because so many people have none. Drinking water is in short supply each summer, and overworked sewer systems are breaking down all the time; many millions have no sewer system at all. Jobs are scarce, and well-paying jobs are almost unheard of. Beggars crowd every street, and each garbage can is searched through time after time by starving people looking for something to sustain them.

Some parts of the world already experience these conditions. Some experts predict that a future like this may be in store for all of us unless something is done soon, and on a massive scale, to control population growth.

The Earth has recently undergone the most rapid population increase in all of human history. From 2.5 billion people in 1950, the total world population more than doubled to 6 billion in 1999 (Figure 9.1). As of 2020, the United Nations estimates that the world's population increases each year by about 86 million people, a rate which would double the world's population again in about 64 years. The UN also estimates that the growth rate of the world's population has begun to slow down a bit, and they predict that the total world population will stabilize at about 11 billion people (plus or minus 1.5 billion) by the year 2100.

In this chapter we consider the factors that control the size and the rate of growth of populations, including the biological controls on populations that operate independently of any conscious planning. The ecological principles that we discuss include models of population growth, limits to growth, and some of the consequences of excessive growth. Although we focus mainly on human populations, these ecological principles apply equally well to populations of other organisms.

As with many of the other issues in this book, population growth cannot be looked at as a purely biological issue. There are political, religious, and ethical dimensions to population growth and its control, and so we consider some of these factors as well. In addition, even in the face of a global population increase, there has also been increasing research on the biology of infertility, in part because an increasing number of couples who are not able to conceive their own children seek these reproductive therapies. Both global population growth and individual fertility and infertility raise ethical questions. We have seen in Chapter 1 that the boundaries between the individual good and the social good are one of the subjects of ethics. Biology can inform ethical debate by assessing the biological risks to the individual and to society in different scenarios.

9.1 DEMOGRAPHY HELPS TO PREDICT FUTURE POPULATION SIZE

The study of the biological factors that affect the sizes of populations is called **population ecology**; the study of human populations in numerical terms is known as **demography**. Recall that a **population** is defined (Section 5.4.1) as a set of potentially interbreeding individuals in a certain geographical location at a certain time. Our ability to understand population growth depends on our ability to make predictions based on both population ecology and demography. Because populations are large aggregates, we need mathematical models to make these predictions and to study the factors that might influence and possibly curb population growth. These mathematical models were initially developed from the study of bacterial populations, but they pertain to the growth of all other species, including humans.

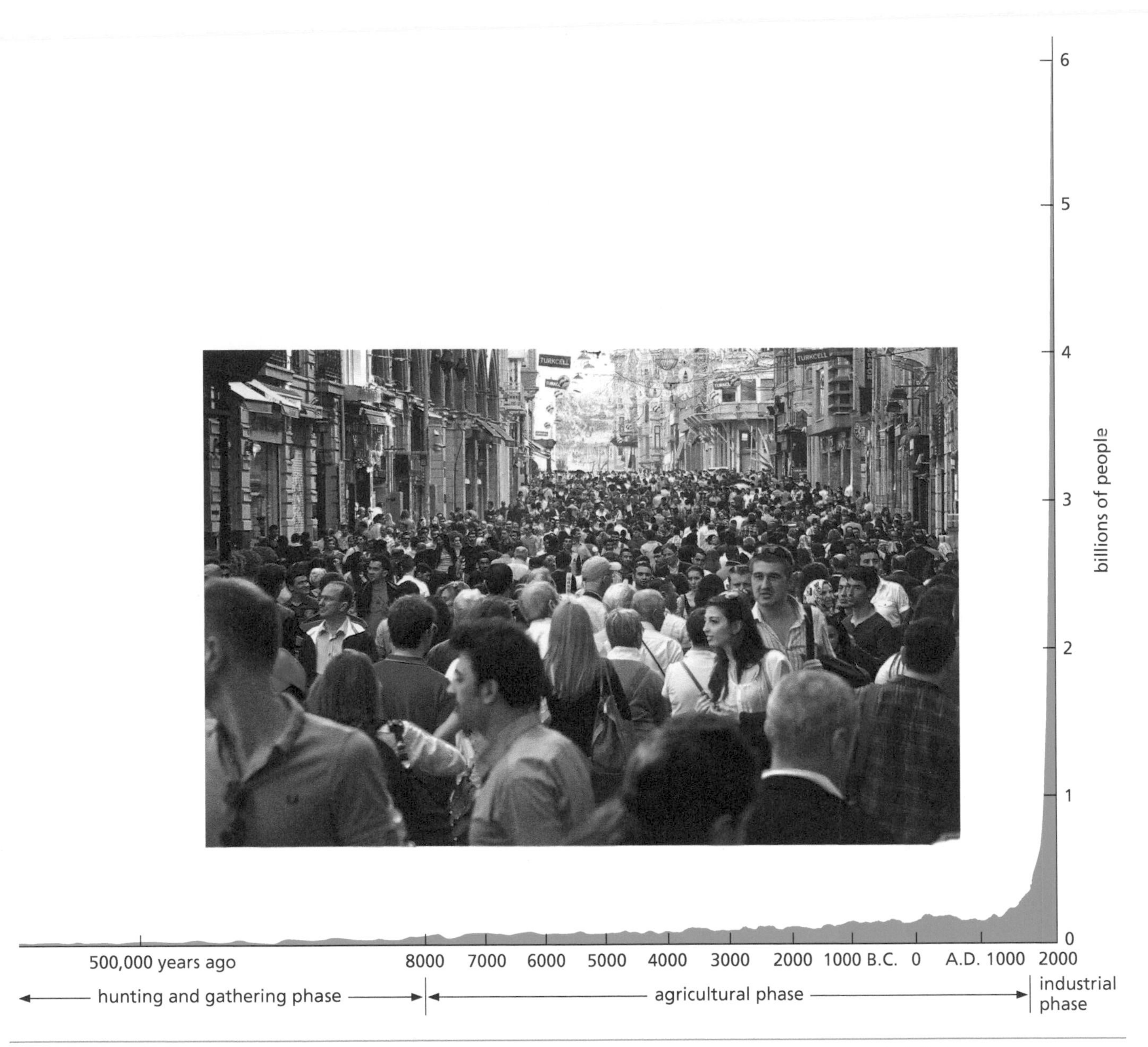

Figure 9.1 Graph showing the growth of the world's human population. The inset shows a street scene in Istanbul, Turkey.

Mathematical models of population growth begin with the gathering of census data. A **census** is, at minimum, a head count of all the individuals living in a specified area, usually within recognized political boundaries. Early censuses of human populations were often inaccurate, and the opposition of the censused populations (who did not want to be taxed) only compounded the inaccuracy. Moreover, these early censuses were hardly ever repeated over the same stable boundaries at a later time, the only conditions under which population growth could be accurately assessed.

The U.S. implemented the first nationwide census of any modern nation in 1790 for the purpose of achieving proportional representation of the different states in Congress. (This remains one of the major functions of the Unites States census today.) In the first few decades of the nineteenth century, most European nations began to census their populations and have continued to do so at regular intervals.

9.1.1 Population growth

The rate of change of the size of a population depends on its birth rate, its death rate, and the relation between the two. For a given time period, the **birth rate**, *B*, of a population is found by dividing the number of births during that period by the number of people already in the population, N:

$$B(\text{each year}) = \frac{\text{Number of births per year}}{N}$$

The birth rate, like any other rate, is always a fraction in which one number is divided by another. To illustrate with some actual numbers, if there are 10,000 people in a population (N = 10,000) and 200 babies are born that year, then the birth rate is B = 200/10,000 = 2/100 = 0.02 per year or 2% per year. The **death rate**, D (also called the mortality rate), is found in a similar manner to the birth rate and represents the fraction of the population that dies within the time interval in question:

$$D = \frac{\text{Number of deaths per year}}{N}$$

Notice that N, the population size, appears in the equations for both the birth rate and the death rate because both are expressed in terms of the size of the population, that is, both are fractions of N.

At the end of the year, the population will have changed because N will have increased by the number of births and decreased by the number of deaths. To express this mathematically, the previous equations can be rearranged and then combined.

By rearranging the equation for the birth rate, we get

$$\text{number of births per year} = BN$$

and by rearranging the equation for the death rate, we get

$$\text{number of deaths per year} = DN$$

Thus, to find the change in N per year, we add the births and subtract the deaths, to get

$$\text{change in } N \text{ per year} = BN - DN$$

We can rewrite this equation using the notation standard in mathematics for rates of change: If the change in N, called dN, is divided by the change in time, dT, we have

$$dN/dT = BN - DN$$

or

$$dN/dT = (B - D)N$$

The quantity (B - D), is called the **growth rate**, and is symbolized as r. It is the difference between the birth rate and the death rate.

$$dN/dT = rN$$

A population increases if its birth rate, B, exceeds its death rate, D, and in this case r is positive. If $B = D$, then $r = 0$, and the population is stable, neither increasing nor decreasing. A population whose death rate exceeds its birth rate decreases and r is negative.

The growth rates of 130 different nations are shown in Figure 9.2A. Growth rate does not correlate with population density (numbers of people per square mile, Figure 9.2B). Some nations with high growth rates have low population densities (for example, Afghanistan and Mali), although density can be assumed to be increasing wherever the growth rate is positive.

The preceding discussion assumed a "closed" population, unaffected by people migrating into or out of it. This is a safe assumption for the world as a whole, but for a single country or region we must also add terms for both immigration (people entering) and emigration (people leaving). The United States, for example, currently has a birth rate very close to the death rate, but the population continues to grow because more people move to the United States each year from

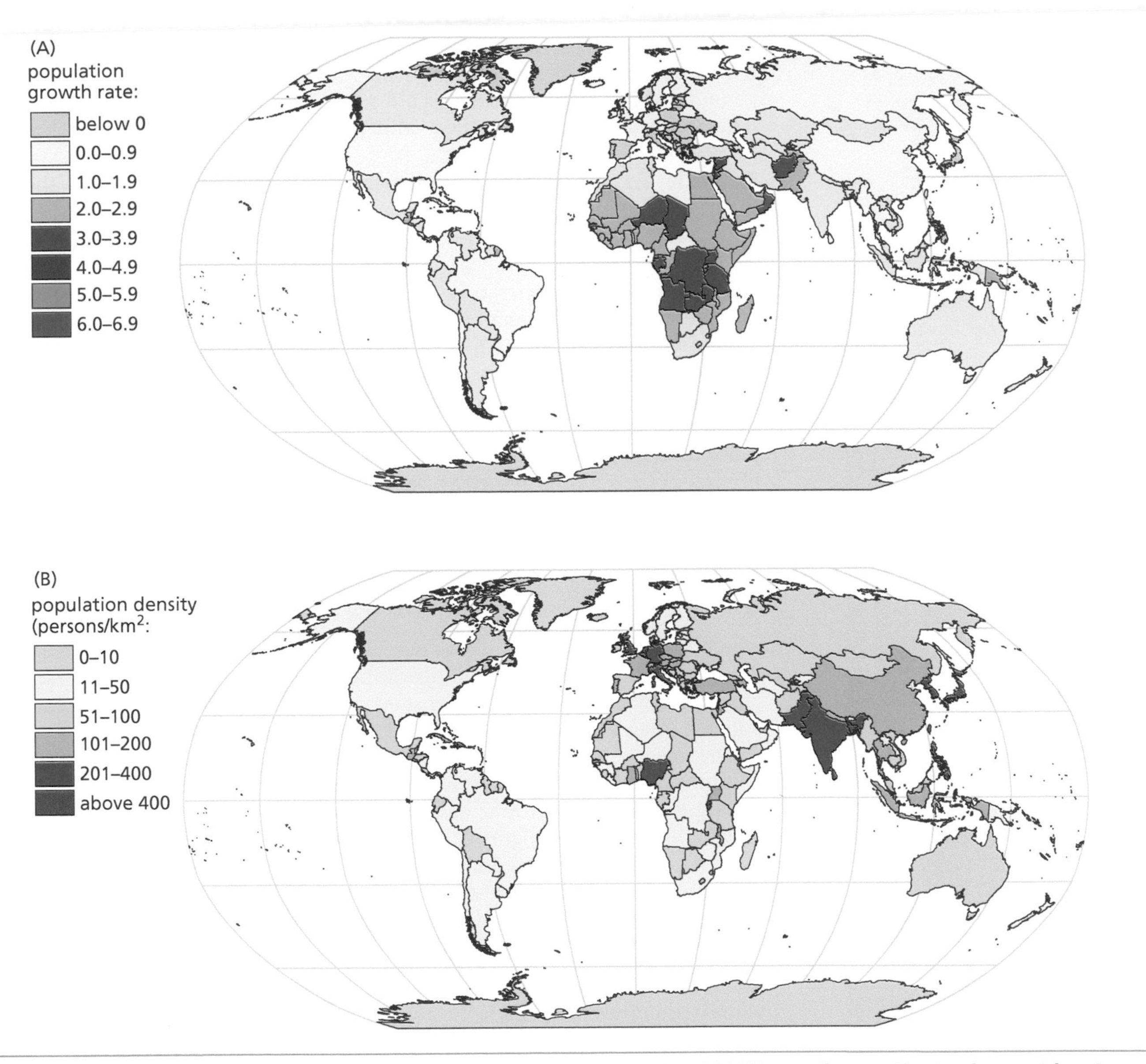

Figure 9.2 (A) Population growth rates (r) from 2010 to 2015 (UN estimates published 2018); negative growth rates for countries at war are usually temporary, (B) Population densities around the world. (CIA World Fact Book 2018.)

other countries than move away from the United States. In fact, one-fifth or more of the yearly population growth in the United States comes from immigration. To include migration rates in the calculation of r, we must write

$$r = B - D + i - m$$

where i is the rate of immigration (the number of immigrants in a year divided by the population size) and m is the similarly defined rate of emigration.

In most nations today, population growth results mostly from the excess of births over deaths rather than from the excess of immigration over emigration. In the 1990s, the U.S. population seemed to have reached a balance between birth and death rates. More recently, however, the birth rate (about 1.8% annually, or one birth every 7 seconds) has increased to about twice the death rate (about 0.9% annually, or one death every 14 seconds), and there also continues to be an excess of immigration over emigration. The net change is a population increase of one person approximately every 11 seconds in the United States. Many factors, some of them perhaps temporary, have contributed to the recent changes. For example, many "baby boomers" (people born from about 1945 to 1960) began having additional children many years after they first became parents, and there have also been increases in the number of people remarrying and starting second families. On the other hand, generations born since 1990 have been delaying marriage longer and having fewer children as a result (U.S. Census Bureau, 2018), in accordance with similar trends in most other countries.

9.1.1.1 Exponential (geometric) growth Although the definition of the term r may include immigration and emigration, the overall equation for the growth rate has not changed and remains

$$dN / dT = rN$$

The type of growth described by this equation is an example of geometric growth. A geometric series is one in which each number is multiplied by a constant to produce the next number in the series. For example, 1, 2, 4, 8, 16, ..., is a geometric series in which each number is multiplied by 2 to give the next number. The growth rate equation is a geometric series because each new value of N results from multiplying the previous value by $1 + r$. (Another familiar example of geometric increase is the growth of money by compound interest.) After an initial start, geometric growth always results in a rapid increase as the growth curve turns sharply upward.

Geometric growth is also called **exponential growth** because the expression for population growth can also be written as

$$N = N_0 e^{rT}$$

where N is the population size at time T, N_0 is the initial population size (at time $T = 0$), e is the base of natural logarithms (approximately 2.71828), and r is the growth rate. In this form, the constant by which each number is multiplied (r) appears as an exponent, hence the term exponential growth. The same equation can be used to describe populations of bacteria, fish, or humans, although different time units (minutes, months, or years) may be used in each case. Graphing this equation gives a growth curve such as the blue line in Figure 9.3. (Logistic growth, also shown in Figure 9.3, is explained later.)

For exponentially (geometrically) growing populations, we can calculate the **doubling time**, the length of time it will take for the population to double. Mathematically, doubling time is expressed as

$$\text{doubling time} = 0.69315 / r$$

where the number 0.69315 is the natural logarithm of 2.

In 2019, the United Nations calculated that the growth rate for the world's population was 1.1% per year, which will cause a doubling every

$$0.69315 / 0.011 = 63.0 \text{ years}$$

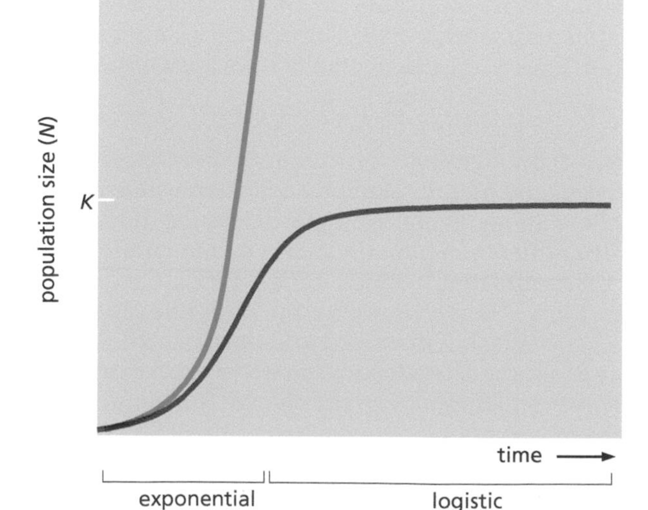

Figure 9.3 Exponential growth compared with logistic growth. The steeper (more vertical) the slope, the more rapid the rate of growth. Graphing exponential growth always gives a characteristic J-shaped curve (blue).

As we saw earlier, the value of r varies from place to place (Figure 9.2A). The fastest-growing nations have growth rates at or above 4%, which will cause them to double their population every 17.3 years or more rapidly. Many more nations are growing at about 3% per year and will thus double their population every 23.1 years.

9.1.2 Malthus's analysis of population growth

The world's human population has long been on the increase, but the rate of change was slow before modern times (Figure 9.1). During the seventeenth and eighteenth centuries, several European countries began to experience a great upswing in their populations, adding to the many motivations for sending forth expeditions to find and settle new lands. The need to clothe growing populations encouraged innovations that brought greater efficiency to textile production, marking the onset of the industrial revolution.

During the early part of the industrial revolution, philosophers and economists began to pay attention to the phenomenon of population growth. David Hume and Benjamin Franklin each described population increase as a blessing for civilization. The first person to emphasize the negative consequences of population growth was Thomas Robert Malthus. In his *Essay on the Principle of Population* (1798), Malthus explained the following dilemma:

1. A population tends to increase *geometrically* if its growth is unchecked. (As we saw earlier, a geometric series is one in which each number is *multiplied* by a constant to obtain the next number.)
2. The available food supply, in Malthus's view, increased only *arithmetically*. (An arithmetic series, like 3, 4, 5, 6, 7, ..., is one in which a constant, in this case 1, is *added* to each number to obtain the next number.)
3. Because the population increases faster than the food supply, the increasing population compounds human misery and poverty, especially among the lower classes.

Malthus divided the factors controlling the population increase into two broad categories that he called preventive checks and positive checks. **Preventive checks** were those that could prevent births from occurring. These were usually voluntary measures, operating on an individual level. They included delayed marriage (also called "moral restraint"), reduced family size, and several practices that Malthus condemned as "vice," including birth control and homosexuality. The **positive checks** on population were those that would operate automatically after births had taken place, whenever the preventive checks were not sufficient. Malthus identified as positive checks overcrowding, poverty, epidemic diseases, rising crime rates, warfare, starvation, and famine. Malthus opposed the social welfare legislation of his time because he thought that any measure to improve the condition of the poor would only encourage them to reproduce faster, further outstripping their food supply and compounding their own misery. One outgrowth of this type of thought was the nineteenth-century theory attributing warfare to the economic needs confronting the population of each nation, the so-called economic theory of war.

Malthus noticed that, even in European countries with stable populations, the *rate* of population growth temporarily rose in the years following a famine or plague until the population was restored to its pre-disaster level. This phenomenon also takes place after many wars and economic hard times (witness the "baby boom" in the years following World War II). Such events show that a population's potential for increase is much greater than is usually realized. The actual rate of population growth is kept lower than the potential rate of growth by positive and preventive checks.

Throughout the nineteenth and early twentieth centuries, technological progress in agriculture, especially in mechanized farming and in the use of chemical fertilizers, increased crop yields among the wealthy nations, providing a plentiful food supply for the expanding population. Also, European population crises were in many cases relieved by large-scale emigration to other continents. Malthus and his gloomy predictions of starvation and misery were largely forgotten.

Following World War II, attempts to deal with poverty, disease, and food shortages around the world ran into the harsh reality that burgeoning populations were exacerbating all these problems. Public health improvements (in public sanitation, in mosquito control, in vaccination against infectious diseases, and in the delivery of medical care) were diminishing the death rates, especially among the young. The result was in many cases a staggering population growth.

The population growth affected different countries unequally. Many nations of the developing world (those countries which had not become industrialized) were trying to diversify their economies, modernize their industries, and improve their living standards. Improvements in health care and sanitation became more rapid after about 1950 and led to more births and fewer deaths; soon these nations found their financial and other resources stretched thin as their populations continued to increase. Economic development, in other words, was being slowed by population growth, and many leaders of the developing nations became increasingly concerned. Without great wealth, the housing and other needs of the expanding populations could not be met. The positive checks that Malthus had foreseen were operating. The developing world was coming face to face with a population crisis (Figure 9.1).

The population crisis in the industrially developed countries of North America, Europe, and Japan was taking a different form: wealth was diverted into providing additional housing, roads, sewers, and needed services. However, this development cannot be sustained forever, because it depends on the use of nonrenewable resources such as fossil fuels and soil (see Chapters 17 and 18) and on resources and products imported from less developed countries, leaving less for them. The diversion of nonrenewable resources cannot indefinitely support stable or growing populations. By diverting resources, wealthy nations can postpone, but not avoid, the effects of global population growth.

9.1.3 Growth within limits

Malthus's assumption of an arithmetical increase in the food supply has been questioned. Although the limited data available to Malthus were consistent with an arithmetical increase, there is no biological or other theory that explains why this should be so. However, most biologists do agree with Malthus's point that the growth in food supplies is slower than the rise in population. One reason this is so is because there is a loss of energy at each stage in which one type of plant or animal becomes food for another, a topic that we develop more fully in Chapter 17. Because food and other resources increase more slowly than population, any population growing exponentially will outstrip its food supply and other resources, including the available space in its habitat. Clearly, no population can continue growing exponentially. After growing exponentially for a while, a population usually follows a pattern called **logistic growth**.

9.1.3.1 Logistic growth Logistic growth has been demonstrated in all types of experimental populations, and we know of no biological reason why this would not also apply to human populations.

Logistic growth can be modeled mathematically by the equation

$$dN/dT = rN(K-N)/K$$

In this equation, K is a new quantity called the **carrying capacity** of the environment. Like N, K is a number, not a rate. K refers to the maximum size of the population that can be sustained indefinitely by the environment. As population size approaches this carrying capacity, the population growth (symbolized by dN/dT) slows down, and when $N = K$ the population growth is zero (**Figures 9.3** and **9.4**). When the population reaches whatever carrying capacity is imposed by the environment (Figure 9.4C), the birth and death rates are balanced. If the population size overshoots K, then a population crash follows in which deaths outnumber births and the population size declines to the carrying capacity.

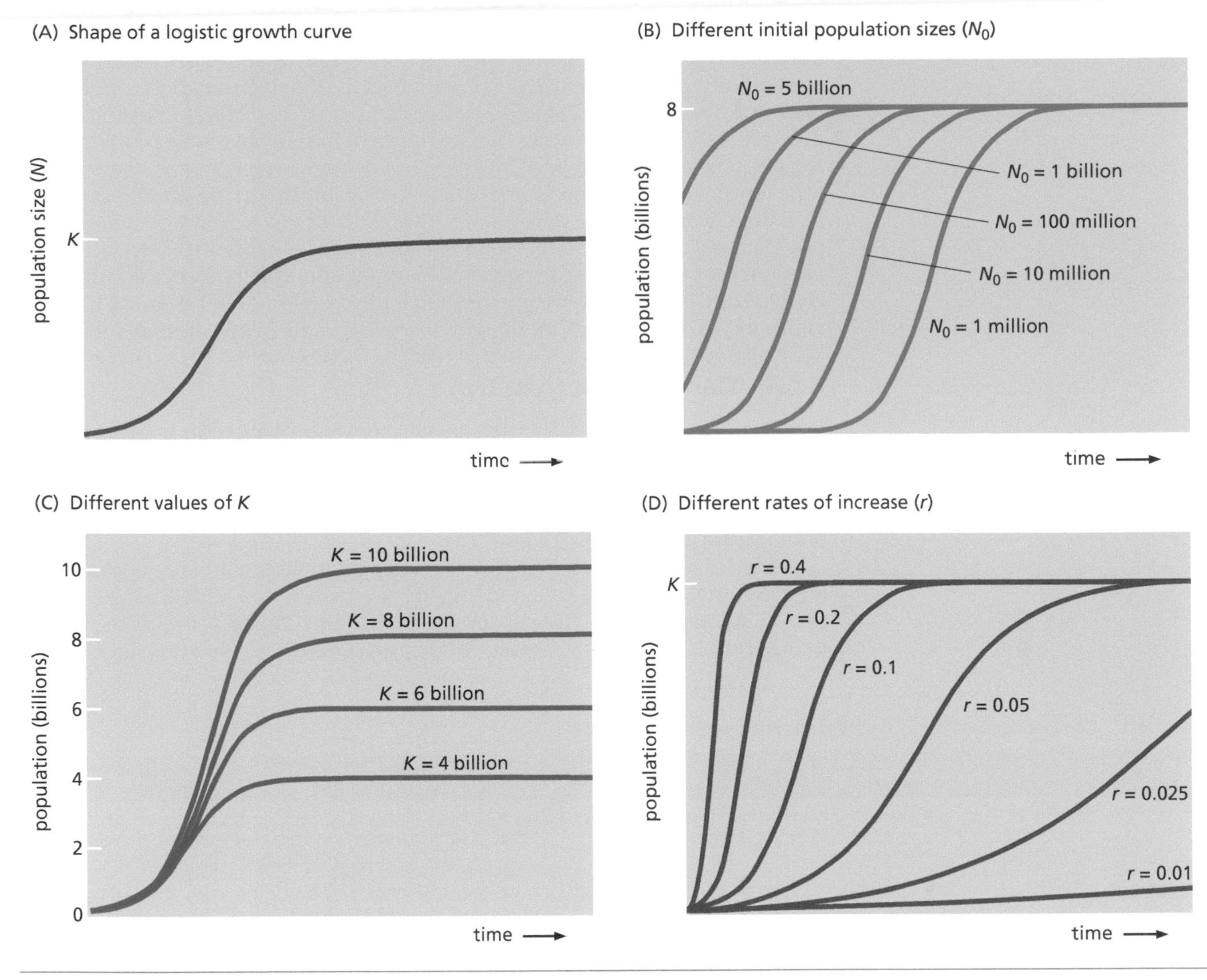

Figure 9.4 A variety of logistic growth curves.

K is related to the amount of space in an environment and the other resources available, including the amount of energy that is in a form that can be used by living organisms. For animal or plant species living in an unchanging environment, *K* is generally constant. For our own species, *K* varies with changes in technology, especially technology that can obtain more usable energy from the same environment—for example, by increases in the efficiency of energy use or of food production. A certain amount of land may support a particular population size of hunters and gatherers at a low carrying capacity (a low value of *K*). The development of agriculture generally results in an increased carrying capacity, and the industrial revolution (including the use of tractors and chemical fertilizers in agriculture) increases it still further.

Humans can avoid an environmentally imposed population crash by limiting the birth rate before the population reaches its carrying capacity; in addition, we need to consider the effects of human populations on the populations of other species. When two species interact, either one may modify the effective carrying capacity of the environment for the population size of the other. Population ecologists define competition as a type of interaction in which a species diminishes the population size of another. An increasing human population diminishes the sustainable population size of all species with which we compete, and it is driving many of these other species toward extinction (see Section 18.2).

9.1.3.2 *K*-selection and *r*-selection

Biological species can have very different types of population structures and life cycles. Two basic types of population growth are those that are limited by ***K*-selection** and those that are limited by ***r*-selection** (Table 9.1). Each is favored by natural selection, but in different sets of circumstances. In environments where favorable conditions can rapidly change or disappear, high mortality rates are common, and natural selection favors prodigious reproduction of offspring (high *r*, therefore *r*-selection) to compensate for this devastating mortality. In more stable environments, population sizes stay close to carrying capacity (*K*), and natural selection favors efficient use of resources, especially energy. In *K*-selection, the advantage generally goes to whoever can most efficiently convert food resources into new adults of the next generation. Humans are an example of a *K*-selected species.

9.1.4 Demographic transition

Because humans are a *K*-selected species, human population increases in the past have coincided with major advances in technology that have allowed the carrying capacity (*K*) to reach new levels. The development of agriculture made it possible for human populations to increase well above the size permitted by hunting and gathering (see Figure 9.1).

Most of our understanding of the changes that accompany a population increase come from studying the growth of a population with changes in economic development. Our current model of this process describes it as a **demographic transition**, characterized by an orderly succession of stages.

Table 9.1 Differences between *K*-selected and *r*-selected species

	K-selection	*r*-selection
Ecological conditions	▪ Favorable conditions dependable and change slowly (if at all) over time	▪ Unstable ▪ Very favorable and very unfavorable conditions appear and disappear erratically and unpredictably
Population size	▪ Stable at or near the carrying capacity (*K*)	▪ Fluctuates greatly over time, both above and below the carrying capacity ▪ Devastating mass mortality is frequent and often unpredictable
Values of *r*	▪ Low to moderate *r* ▪ Population growth is slower even when conditions are favorable	▪ High *r* leads to very rapid population growth under favorable conditions, which compensates for mass mortality under unfavorable conditions
Reproduction	▪ Always sexual	▪ Either asexual or sexual
Body size at reproduction	▪ Natural selection favors reproduction at a large body size	▪ Natural selection favors reproduction at a small body size and a young age
Reproductive rate	▪ Low ▪ Few offspring produced at once	▪ High ▪ Many eggs, seeds, or other reproductive stages produced at once
Frequency of reproduction	▪ Repeated many times throughout life	▪ Usually confined to a single occasion
Offspring size	▪ Offspring are individually large, and each represents a large proportion of its parents' reproductive output	▪ Released (as eggs or immature stages) at a small size and widely scattered ▪ Each represents only a tiny fraction of its parents' reproductive output
Parental investment	▪ Generally high ▪ May include provisioning of food for offspring and extensive parental care	▪ Little or no parental care or investment
Dispersal	▪ Dispersal of the population to new locations is generally slow ▪ Newly founded populations grow more slowly	▪ High capacity for dispersal (spread of the population) to new habitats and locations ▪ Most new locations are unsuitable, but the occasional favorable one permits a rapid explosion in numbers
Mortality pattern	▪ Most individuals live their full life span and die at an advanced age	▪ Mortality is extremely high among eggs, seeds, or larvae ▪ Most deaths occur very early in life
Examples	▪ Humans, cows, elephants	▪ Carrion-feeding beetles, tapeworms, weeds

9.1.4.1 Stages of a demographic transition As shown in Figure 9.5, the first stage of a demographic transition is a stable population in which a high death rate is balanced by a high birth rate. Population growth is, therefore, zero. Over the centuries, traditional societies with high death rates (especially from infant mortality and childhood infections) developed customs that encouraged high birth rates. In other words, high mortality rates encourage high birth rates; this is the pre-industrialized stage. Demographers estimate that overall mortality rates held steady or declined slowly in pre-industrial Europe, with occasional but temporary upsurges during wars and epidemics such as the great bubonic plague (the black death) that decimated European populations in the 1300s.

The second stage of the process is a period of exponential growth in which the population size climbs to a new high. This is brought about by technological changes (such as the agricultural revolution or the industrial revolution) that result in falling mortality rates, or death control. In Malthusian terms, the improvements (in sanitation, health care, and nutrition, for example) result in the alleviation of positive checks and thus a lowering of *D*. However, it always takes at least a few generations for the cultural values and customs to change so as to permit a matching decline in birth rates. In the meantime, the memory of high death rates in the recent past continues to encourage high birth rates. In some cases, the birth rate may even rise as the result of better nutrition and the improved physiological condition and reproductive health of prospective mothers. The combination of high birth rates and lower death rates causes the population growth that characterizes the middle of the demographic transition.

The third stage of the transition is marked by a decline in birth rates as the population adjusts to new conditions and as the incentives for high birth rates are removed. Population growth follows the logistic growth equation during this stage. As the birth rate declines to match the new lower mortality rate, the population once again stabilizes, but at a larger population size (and a larger *K*) than before. The demographic transition is complete.

England's demographic transition began in the early 1700s and took some 250 to 300 years to complete. Other industrialized countries (United States, Canada, Japan, and much of Europe) took closer to 200 years to complete their demographic transitions. The remaining countries of the world began their demographic transitions only during the twentieth century, and did so much more suddenly, often going from high traditional mortality rates to low modern rates in a single generation. All countries in the world will eventually experience advances

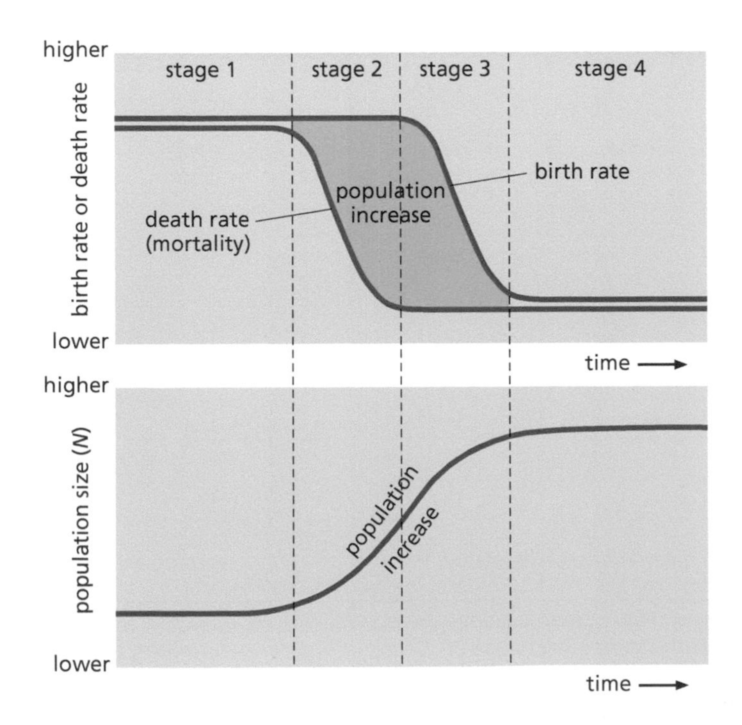

Figure 9.5 **Idealized stages of a demographic transition.** Some authorities recognize only three stages by combining the middle two.

in sanitation and medical care, even if these are imported. The resulting reduction in death rates will bring about a demographic transition, even in the absence of economic development. The only exceptions may lie in countries where repeated famine and warfare (both positive checks) keep the death rates high.

The United States may have completed a demographic transition a few years ago, when birth and death rates became approximately equal, a condition known as **zero population growth**. However, as we noted earlier, the U.S. population is once again increasing, and it remains to be seen whether the long-range trend will more closely resemble the zero-growth phase or the more recent increase.

Many parts of Europe have already reached stable populations, with birth rates close to death rates. As of 2019, over 20 countries now have stable or declining populations, including Latvia, Poland, Ukraine, and Japan. Meanwhile, most countries in Africa and many in Asia and the Pacific continue to have rapidly growing populations.

9.1.4.2 Age structure of populations The models we have examined so far treat all members of a population as being the same. However, we know that the probability of death and the probability of reproduction both vary with age. Thus, the true rate of population growth may vary depending upon the ages of individuals within the population. Grouping individuals by age gives the **age structure** of the population. The age structure can best be shown by a population pyramid, or **age pyramid**, such as those in Figure 9.6. Each horizontal layer on

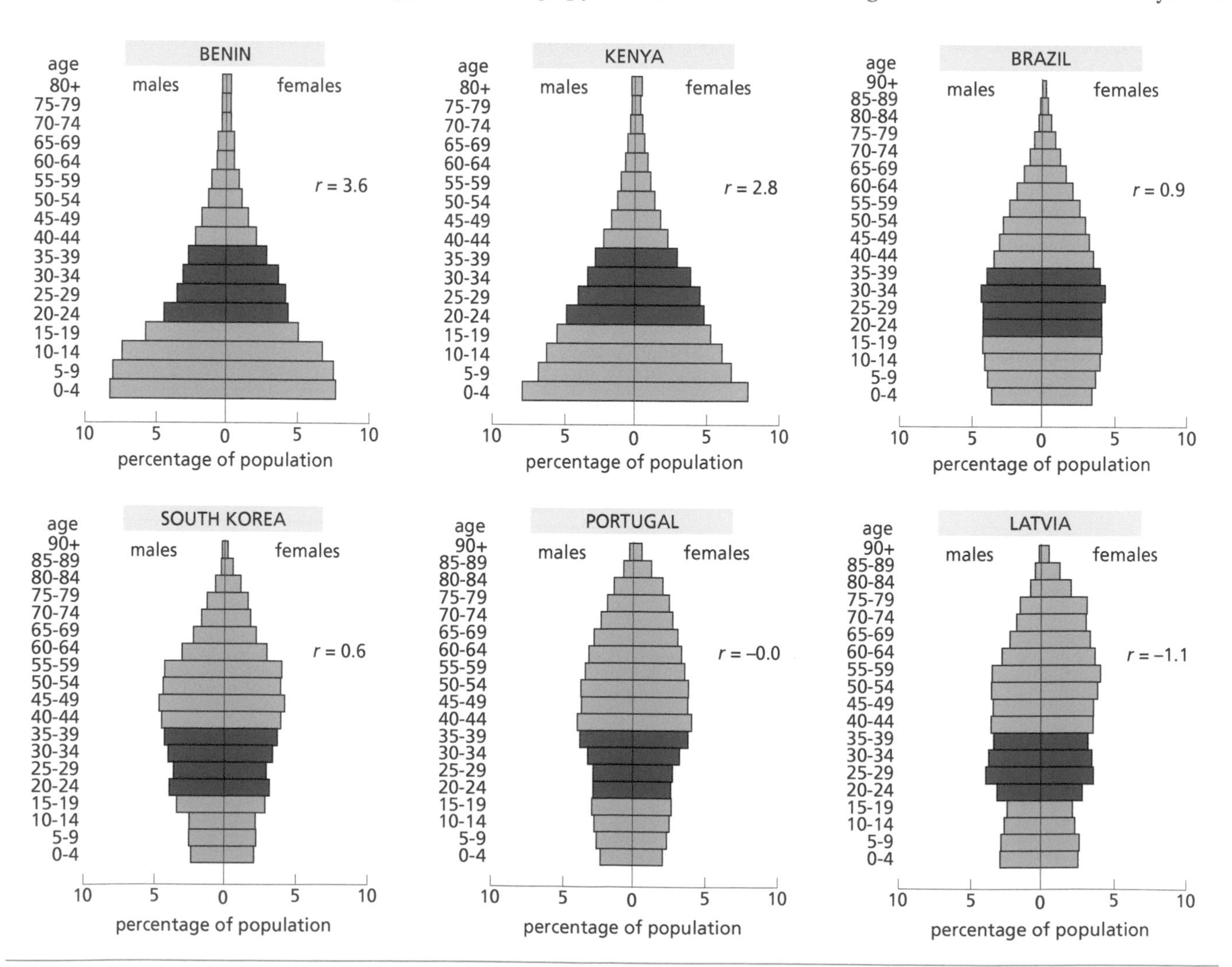

Figure 9.6 **Age pyramids for two rapidly growing populations (high *r*), two populations with slower growth rates, one rather stable population (zero *r*), and a declining population (negative *r*).** Age groups from 20 to 39 are darker in order to emphasize that most reproduction occurs in these age groups.

such a diagram represents the percentage of the population in a particular age group, with the youngest age groups on the bottom. Altogether, the age pyramid shows the distribution of individuals among the various age groups. To get a feeling for the scale of the age pyramids in Figure 9.6, notice that the 0–4 age group in Benin or Kenya constitutes about 15% of the population. Most age pyramids are divided by a vertical midline, with male age distribution shown on the left and female age distribution on the right. Notice in Figure 9.6 that the 0–4 age group has approximately equal numbers of males and females in all countries, but the 80+ age group has many more women than men in stable or declining populations. Among human populations, a pyramid with sloping sides and a wide base (many children) characterizes an expanding population. A shape maintaining more or less the same width throughout (except for the oldest few age classes) indicates a stable population. A stable age distribution is reached when the pyramid keeps the same shape as each age group grows older so that the numbers in each age group are replaced by an equal number advancing from the next younger group.

Sometimes there are age bulges, as with the post-World War II baby boom, when the birth rate increased temporarily. The United States and some other countries experienced a second baby boom (or a baby boom "echo") in the late twentieth century as members of the earlier baby boom reached their prime childbearing years. In these countries, schools had to accommodate the largest generation of school-age children that has ever lived.

Predictions of future values of r can sometimes be made on the basis of age structure. Clearly, a population of 10,000 individuals with 4000 females of reproductive age has much more potential for increase than one with only 400 females of reproductive age. Calculations of the potential for future increase are often carried out by multiplying the number of females in each age group by the number of children that each of those females is likely to bear, then adding up these products for all age groups.

9.1.4.3 Life expectancy In most of the industrialized nations, the control of many infectious diseases since the late 1800s and improvements in sanitation have resulted in decreased infant mortality. A few twentieth-century changes, such as increases in smoking (up to the 1980s, but declining since), auto accidents, and hand guns, have increased death rates, but these have generally been offset by much greater declines in the death rates from famines and infectious diseases. One consequence of declining death rates, especially among the young, is a greater **life expectancy**, meaning the average maximum age that people attain in life, or the age to which a person born in a particular place and at a particular time can expect to live (Figure 9.7). Life expectancy is calculated on a statistical basis,

Figure 9.7 Human life expectancy at birth in various countries.

taking into account the probability at any age of a person's proceeding on to the next age group. Therefore, both decreased mortality and increased longevity (the maximum age achieved by some individuals in the population) contribute to an increased life expectancy, with the largest increases resulting from reductions in childhood mortality. Life expectancy in the United States has risen from somewhere near 50 years in colonial times to about 78.5 years in 2016. Worldwide, the average life expectancy is around 72 years for people born in 2016. Twenty European countries, plus Singapore, Canada, Australia, and New Zealand, all have life expectancies of 80 years or higher, while Japan has the world's highest life expectancy, 84.2 years (World Health Organization, 2016). The age group over 80 is the most rapidly growing segment of the population worldwide. This "graying" of most populations results in a decrease in the proportion of the population in the most fertile age group and an increase in the proportion of older people, many of whom are dependent on the younger generation for their care (see Chapter 12). Numerous sociological changes follow the shift toward more people of advanced age and fewer children—more emphasis on medical care and less emphasis on schools, for example.

9.1.4.4 Demographic momentum Even after the birth rate falls to the level of the mortality rate, the population may continue to increase for another generation or two because of a **demographic momentum**. The momentum is caused by an age structure opposite to the one just described for a graying population. A population in which a large fraction of individuals is not yet of reproductive age (Figure 9.6), while only a small fraction is past the age of reproduction, has a future reproductive capacity larger than one in which a higher percentage of the population is beyond reproductive age. Because the younger age groups are more likely to reproduce in the next 20 years and less likely to die, a temporary increase in the birth rate (and a temporary decrease in the death rate) can easily be predicted. The population will continue to grow until the age distribution readjusts itself more evenly, that is, until a stable age distribution is reached. With a stable age distribution, the birth and death rates will no longer change, unless some external factor disrupts the stable situation.

9.1.4.5 World population estimates Demographers estimate that the world's population was about 50 million in 7000 BCE and increased to about 250 million by the time of Christ. The earliest date for which there are reliable population estimates is 1650, at which time the world's population stood at 500 million. By 1804, the world's population stood at an estimated 1 billion. Adding the next billion took 123 years (to 1927), but the third billion was added in only 33 years (by 1960). The world's population has increased even more rapidly since 1960, adding a fourth billion in only 14 years (by 1974), a fifth billion in 13 years (by 1987), and a sixth billion in just 12 years (by 1999) (Figure 9.1).

The United Nations has published tables with detailed predictions for the further growth of the world's population under different sets of assumptions. According to the model that demographic experts consider most likely, the world's population will grow to about 9.4 billion in 2050 and will stabilize at around 11 billion by the year 2200. Other models, with different sets of assumptions, predict population values in the year 2050 as low as 7.9 billion or as high as 10.9 billion. Each of these assumes logistic growth, except for the model in which population growth continues exponentially at its present rate—that model predicts a population size of 296 billion in 2200, nearly 50 times the present value! Few, if any, biologists think that the latter is in any way sustainable; population growth will have to level off from its present rate.

The logistic curve reaches its maximum slope (most rapid population increase) when the population is at half the carrying capacity ($K/2$). Because the growth of the world's human population seems to have passed its maximum rate of increase and is beginning to slow down, several experts have estimated that the world's population has already passed $K/2$ and will eventually level off at a carrying capacity of double that amount, or somewhere between 10 and 14 billion people, by the year 2100 or shortly thereafter.

Because people do not generally find raising the death rate to be an acceptable method of lowering population growth, decreasing the birth rate is the only other option. In the next section, we study reproductive biology and the factors that contribute to fertility. In the final section of the chapter, we see how fertility can be controlled to lower the birth rate.

THOUGHT QUESTIONS

1. What changes (biological, social, or economic) could increase the carrying capacity (K) of the entire world or of one nation? Should we be more interested in controlling r or in modifying K?
2. Suppose that you count the number of people in a given town every year for five years and you discover that N has stayed about the same during that time period; say, at 10,000. You also know that no one has moved into the town or away from the town in that time. What can you deduce about B and D? Confirm this for yourself by solving the equations given earlier in the chapter. Can you tell from this information how many people were born in the town in any of the five years?
3. Study the age pyramids in Figure 9.6. Does the age distribution of the females or that of the males have a greater impact on future population size? Why?
4. What values of r are typical during the several stages of a demographic transition? See Figure 9.5, and review the way in which r is defined.
5. Which is likely to produce a greater increase in the number of people, a small population with a high r or a large population with a small r? Try some calculations with any values of N and r that you would like to examine. Use the simplest model that seems appropriate, then ask what changes the more complex models would bring.
6. From the data presented in this chapter, can we estimate the value of K (the carrying capacity) for the human population on planet Earth? Can we make a minimum or maximum estimate? Have we already reached the carrying capacity of the planet?
7. Which would you think would contribute more to an increase in life expectancy, decreased infant mortality or increased longevity?
8. Among the "preventive checks," Malthus listed several forms of "vice," including the following:
   a. Heterosexual intercourse outside of marriage ("promiscuous intercourse"; "violations of the marriage bed");
   b. Sexual attraction to one's own sex (homosexuality), to animals (zoophilia), or to inanimate objects (fetishism);
   c. "Improper arts to conceal the consequences of irregular connections" (i.e., birth control).

   Malthus listed all these practices as preventive checks. Which of them would actually function to limit population growth? Which might be interpreted as preventive checks under certain assumptions that Malthus might have made? Do you think these assumptions are realistic? Do you think Malthus condemned these practices because of their effects on population or because of his own moral views?
9. What factors would need to be included in an equation to estimate what birth rate would produce zero population growth?
10. Which countries currently have the highest population growth rates? Are they wealthy or poor? Are they influential? Where do they stand in the current world order?

9.2 HUMAN REPRODUCTIVE BIOLOGY HELPS US TO UNDERSTAND FERTILITY AND INFERTILITY

In sexual reproduction, a male haploid gamete unites with a female haploid gamete. The two types of gametes are produced by individuals of different sexes. We begin this section by comparing the anatomy and physiology of the two sexes in humans.

9.2.1 Reproductive anatomy and physiology

Reproductive anatomy includes those structures that allow for hormonal secretion, gamete production, sexual intercourse, gestation of the fetus, and nourishment of the young infant. Males and females differ in appearance and in the type of gamete that they produce. These differences result largely from the actions of **hormones** that are present in both sexes but in differing amounts and with different consequences.

9.2.1.1 Sex determination The reproductive system, in the earliest stages of its development, is sexually indifferent, meaning that there are no indications as to the future sex of the embryo. The future gamete-producing reproductive organ (or gonad) is not yet male or female, but is just an indifferent gonad. As described in Section 3.2, several genes, including the *SRY* gene, begin to make products at this time. If the products of the genes *SRY*, *SOX9*, *DMRT1*, and *DMRT2* are all present and functional, then the developing gonad becomes a male gonad, or **testis**. The embryonic testis then begins to secrete the hormone **testosterone** and another hormone called anti-paramesonephric hormone (APH). In adulthood, the testis will also produce sperm. A developing gonad that does not become a testis becomes an **ovary** instead. The ovary will secrete the hormone **estrogen** throughout life and will produce eggs during adulthood. The developing testis and its hormonal secretions are essential to the development of male internal and external reproductive structures, but the corresponding female structures develop without requiring estrogen. Thus, a genetic male lacking function for any of the genes *SOX9*, *DMRT1*, or *DMRT2* will develop a female phenotype, even in the presence of *SRY*. For this reason, embryologists say that the default sex of the embryo is female and that male development can only occur when it is induced.

9.2.1.2 Maturation and puberty

Hormones are small molecules that are used for chemical communication throughout the body (see Section 10.4.2). Beginning during embryonic development, the ovaries secrete estradiol (the major estrogen) and the testes secrete testosterone and APH. The reproductive organs are formed during embryonic development under these hormonal influences, but they remain immature until **puberty**. At puberty, a group of hormones produce many changes in the body. At an age that varies greatly around an average of about 12 years, increasing levels of the pituitary hormone follicle-stimulating hormone (FSH) stimulate the final maturation of the ovaries in females to begin producing eggs and the testes in males to begin producing sperm. The ovaries and testes, in turn, step up their secretion of other hormones (estrogen and testosterone), stimulating (among other changes) the development of **secondary sexual characteristics**. Such characteristics include the widening of the hips, growth of breasts, and redistribution of body fat in females; the growth of facial hair and the deepening of the voice in males; and the growth of long bones and pubic hair in both sexes. The term "secondary sexual characteristics" means that these features, while characteristic of mature men and women, do not have a direct role in the production of gametes.

The age at which puberty occurs has declined steadily throughout the twentieth century in all countries in which such records have been kept. In Norway, which has kept records the longest, the average age at menarche (first menstruation) was just over 17 years in the 1840s but had declined to about 13.4 years by the 1950s. Data from other Western European countries are consistent with this, and all countries show similar and steadily declining trends. One possible explanation is that puberty requires a critical weight (about 47 kg or 106 lb for females) and that improvements in childhood nutrition have allowed this critical weight to be achieved at earlier ages over successive decades. Consistent with this hypothesis is the trend in the United States over the past century, where a similar decline occurred but where body weights have always tended to be above European averages. The average age at menarche has always been about 0.5–1.0 year younger in the United States than in Western European countries and is generally younger among African Americans than among U.S. whites. Beginning in the 1990s, this trend has begun to level off in many countries at an average age of between 12.0 and 12.5 years at menarche. At least some of the declining age of puberty may also be related to environmental pollutants with estrogen-like effects (Fisher and Eugster, 2014; also see Section 19.2.2), a hypothesis that would be quite difficult to test.

Male puberty is marked by the onset of sperm production (spermatogenesis). The pituitary hormone FSH initiates sperm production, and another pituitary hormone, luteinizing hormone (LH), induces other cells of the testis to secrete testosterone. Sperm production requires this testosterone for its completion, along with a small amount of estrogen. Female puberty is marked by the start of the menstrual cycle, as explained later.

After puberty, FSH, estrogen, and testosterone continue to have other roles throughout the lifetime of the individual. In males, some of the testosterone is converted into another hormone called dihydrotestosterone (DHT), and small amounts of estrogen appear to be essential for spermatogenesis and other processes. In females, FSH, estrogen, and progesterone are all important in the menstrual cycle, and small amounts of testosterone are thought to be important in generating the sexual appetite or libido.

9.2.1.3 Male reproductive organs and sperm production Sperm are the male gametes, formed by the process of meiosis (see Section 2.2.2), in which each gamete receives half the adult number of chromosomes—one chromosome from each pair. Sperm have very little cytoplasm surrounding their nucleus, but they have a long tail that moves rapidly back and forth to propel the sperm along the reproductive tract of the female after intercourse (Figure 9.8).

Sperm are produced in the testes of the male (Figure 9.9). The hormone testosterone is secreted by the interstitial cells that are crowded into the spaces between the sperm-producing tubules in the testes. (Testosterone is also made in smaller amounts by the brain in both sexes and by the ovaries in females.) The sperm accumulate in a series of wrinkled ducts that form the epididymis. The epididymis also secretes a fluid (the seminal fluid) that carries the sperm through a long duct, called the vas deferens, through which the sperm leave the testes. The left vas deferens and right vas deferens merge within the prostate gland, where they join the urethra carrying urine from the urinary bladder. The secretions of the prostate gland and the seminal vesicles add certain nutrients (including the sugar fructose) that help the sperm to swim more vigorously. Near the base of the penis, the bulbourethral gland helps squirt the seminal fluid and sperm through the length of the penis during ejaculation. Between ejaculations, sperm and seminal fluid are held in the ejaculatory ducts.

9.2.1.4 Female reproductive organs and ovulation The female reproductive organs (Figure 9.10) include the uterus and a pair of ovaries. The female gamete, or egg, also known as an **ovum**, is formed within the ovary by meiosis. The first meiotic division occurs before the egg is released from the ovary, but the second meiotic division is paused at metaphase (when the chromosomes are aligned at the middle of the cell, attached to the spindle; see Section 2.2.2) until after the egg is fertilized. During egg formation, the two cell divisions of meiosis occur in such a way that the cytoplasm is divided between the daughter cells in a very unequal manner. Only one of the four resultant cells becomes a

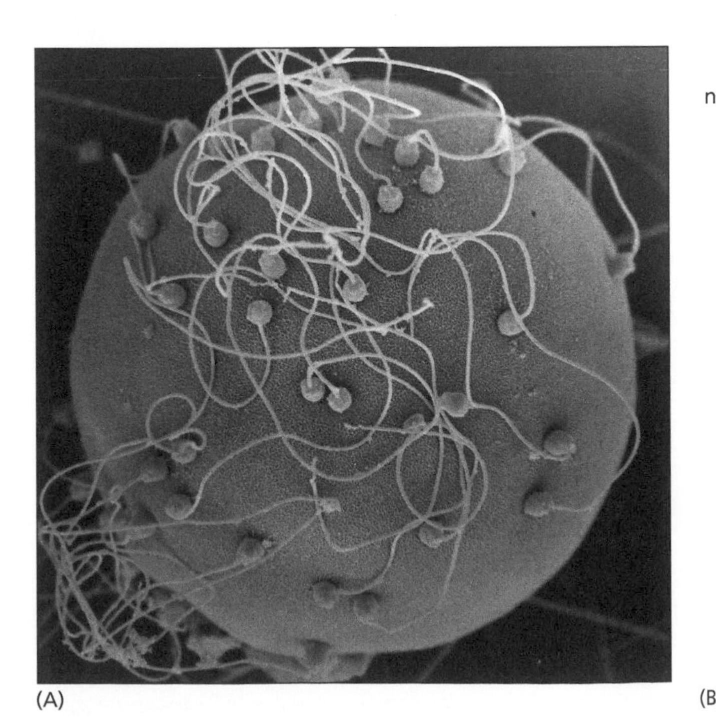
(A)

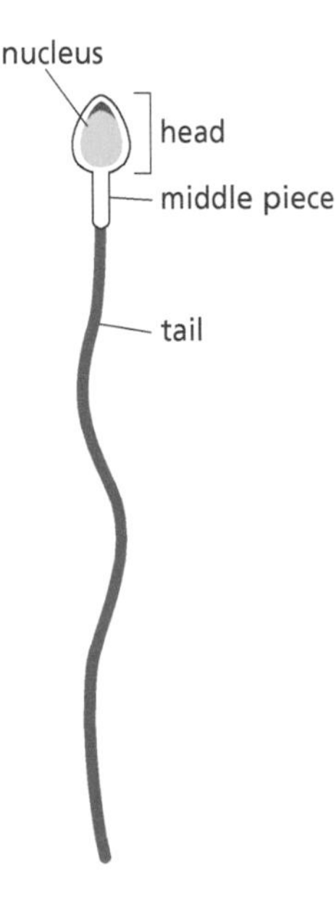

(B)

Figure 9.8 **(A) An egg surrounded by sperm, (B) schematic diagram of a human sperm.**

STRUCTURES IN THE MALE PELVIC AREA
urinary bladder
pubic bone
prostate gland
urethra
seminal vesicle
rectum
penis
ejaculatory duct
erectile tissue
bulbourethral gland
glans penis
foreskin
anus
vas deferens
epididymis
testis
scrotum

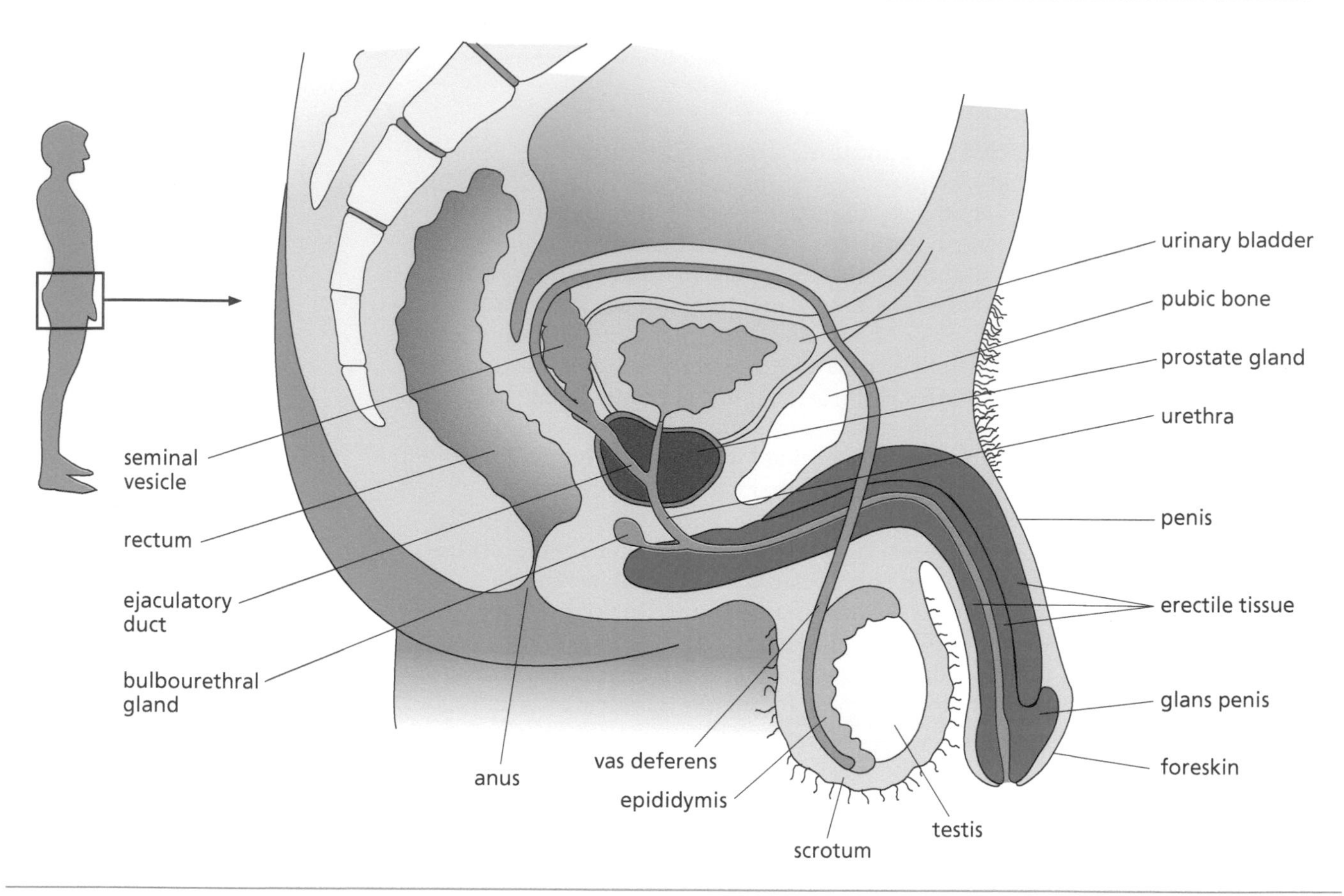

Figure 9.9 The human male reproductive system.

STRUCTURES IN THE FEMALE PELVIC AREA
midline view
anterior view
oviduct
ovary
uterine wall
uterine cavity
pubic bone
urinary bladder
urethra
cervix
cervix
vagina
rectum
external reproductive organs
clitoris
labium minor
labium major
vagina
anus

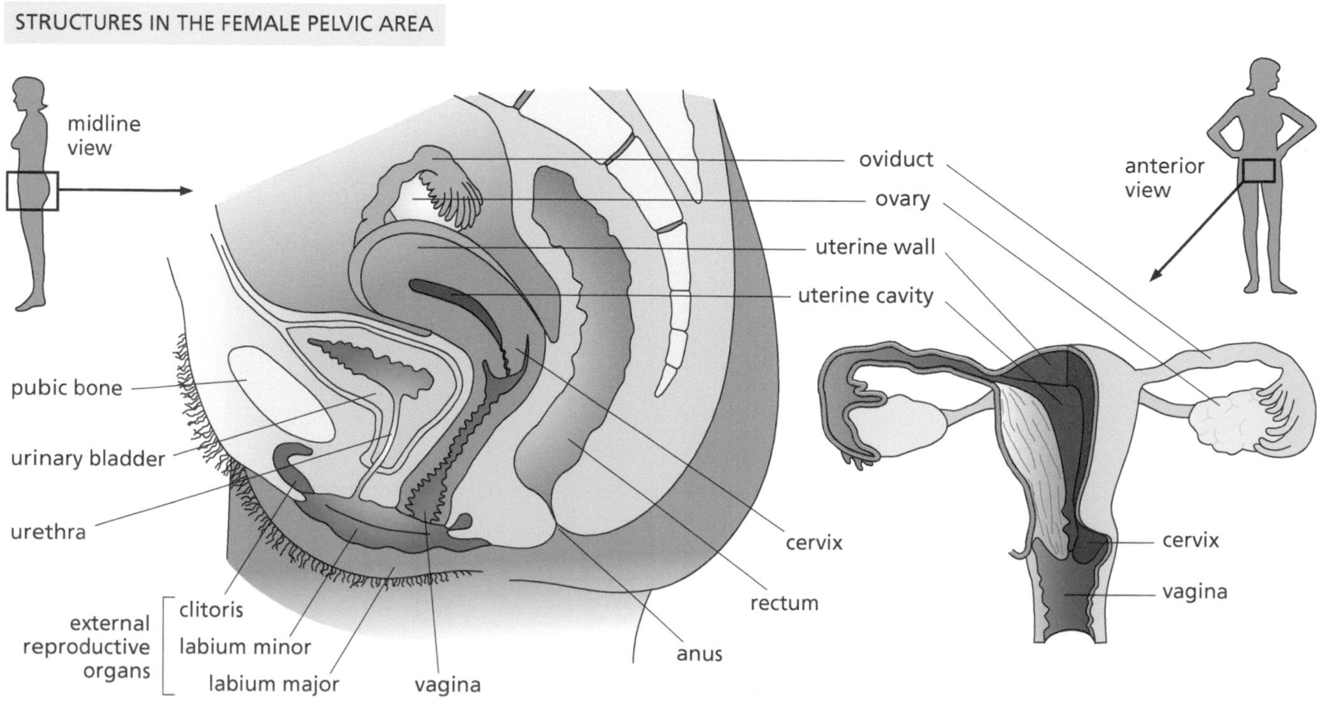

Figure 9.10 The human female reproductive system.

mature egg with a large amount of cytoplasm to nourish the zygote after fertilization. The other three products of meiosis become small cells with very little cytoplasm that are called "polar bodies." The polar bodies are usually lost, but in rare circumstances they can be fertilized and develop into embryos. The nonreproductive cells surrounding the egg enlarge to form an ovarian follicle. The rupture of this follicle and the release of its egg are called **ovulation**.

9.2.1.5 The menstrual cycle

Egg production and ovulation in mammals is a hormonally regulated cycle. In humans, this cycle lasts about 28 days and is called the **menstrual cycle**.

The cycle of egg production and changes in the uterus are controlled by two ovarian hormones, estrogen and progesterone, and by two hormones (FSH and LH) secreted by the pituitary gland at the base of the brain (Figure 9.11).

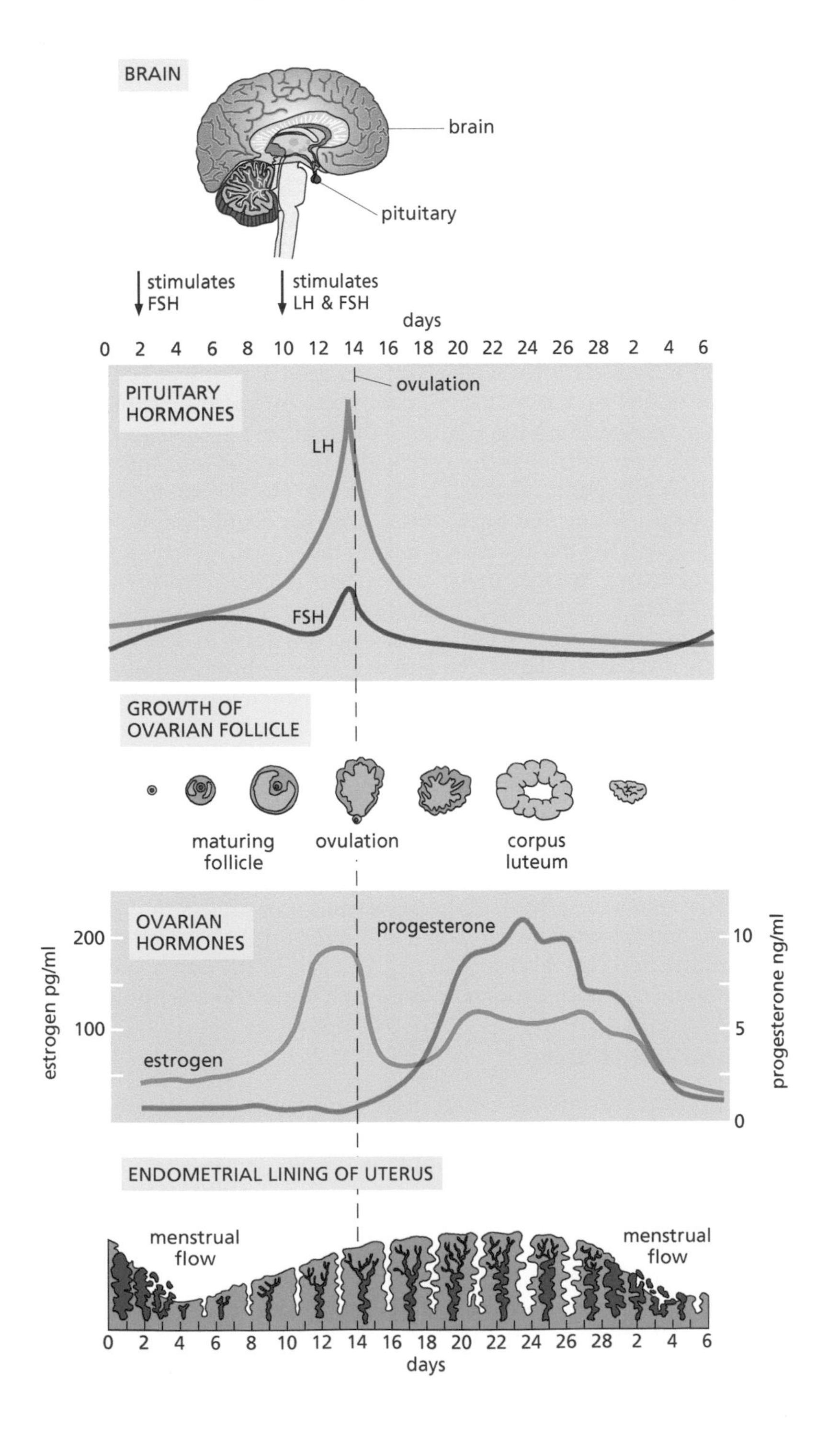

Figure 9.11 Reproductive cycles in the human female.

At the start of each cycle, which begins with menstruation, the pituitary secretes small amounts of FSH. The FSH stimulates two processes within the ovary: growth of an ovarian follicle and production of the hormone estrogen, which reaches a peak concentration during the second week of the cycle. The estrogen stimulates the thickening of the lining of the uterus and the release of a second pituitary hormone, LH, that induces the release of the egg (ovulation), after which the tissue that surrounded the egg is left behind to form a scar tissue called the **corpus luteum**. The corpus luteum then grows. As it matures, it begins to secrete the hormone **progesterone**, which maintains the uterine lining in a thickened and receptive condition, ready for the implantation of an embryo should the egg be fertilized. If fertilization does not occur and no implantation takes place, the corpus luteum degenerates and the supply of progesterone drops sharply, causing the uterine lining to break down. The egg and the uterine lining are then sloughed off in the form of menstrual bleeding. The absence of progesterone also releases the pituitary to begin secreting FSH once again, initiating a new cycle.

During the menstrual cycle, changes in the concentration of each hormone stimulate the production of the next hormone in the sequence. Hormones from the ovaries and hormones from the pituitary regulate each other. In several cases, the presence of a later hormone has an inhibiting effect on the secretion of the previous hormone. This regulation of a previous step of a cycle by a later step of the cycle is called a **feedback mechanism**. In this instance, feedback prevents the overproduction of any hormone and stops the production of a hormone once it has done its job.

9.2.1.6 Fertilization and implantation

Neither the sperm nor the egg lives very long. After its release from the follicle, the egg travels along the uterine (Fallopian) tubes or oviducts, and it is here that the egg is fertilized if sperm are present.

When the male ejaculates during sexual intercourse, approximately 300 million sperm are inserted into the vagina of the female. Of these, only about 3000 will successfully swim through the cervix and the uterus and into the oviducts, following hormonal signals released by the egg. Several sperm need to contact the egg to dissolve the coating that surrounds it. After the coating has dissolved, allowing a sperm to reach the plasma membrane of the egg, the sperm and egg plasma membranes fuse and the egg draws the sperm nucleus inside, a process first clearly explained by Ernest E. Just (Figure 1.4C). The coating closes again, preventing the entry of any more sperm nuclei. The egg completes its second meiotic division (releasing a second polar body), and its haploid chromosomes join with the haploid chromosomes from the sperm, a process called **fertilization**.

If an egg is fertilized by a sperm, the resultant zygote continues traveling along the oviduct for 4–5 days, during which time it undergoes several cell divisions and becomes an embryo. When it reaches the uterus, at a stage of approximately 64–200 cells, called the **blastocyst**, the embryo adheres to the inner (endometrial) lining of the uterine wall, a process called **implantation** (Figure 9.12). If the embryo does not implant, it does not develop further and is shed from the uterus. Implantation triggers the growth of the **placenta**, consisting of intertwined tissues of the embryo and the endometrium of the mother, through which the developing embryo is nourished (see also Figure 13.10). Most of the organs of the growing embryo form during the first month, after which the embryo is known as a **fetus**.

9.2.2 Impaired fertility

We have described the normal events of human reproduction, in which egg and sperm from a fertile female and a fertile male combine to make an embryo. Not all individuals are fertile, however. A couple is usually considered infertile if they have been unable to conceive after a year of trying. Infertility can result from lack of production of sperm or eggs, low motility of sperm, shortened viability of egg or sperm, or anatomical abnormalities preventing the sperm from contacting the egg. A number of diseases, most of them rare, can cause permanent sterility. These include several chromosomal anomalies (see Section 3.2.3) and other genetic disorders; developmental anomalies of the reproductive system, including failure of the testes to descend; and some cancers, including those of the reproductive

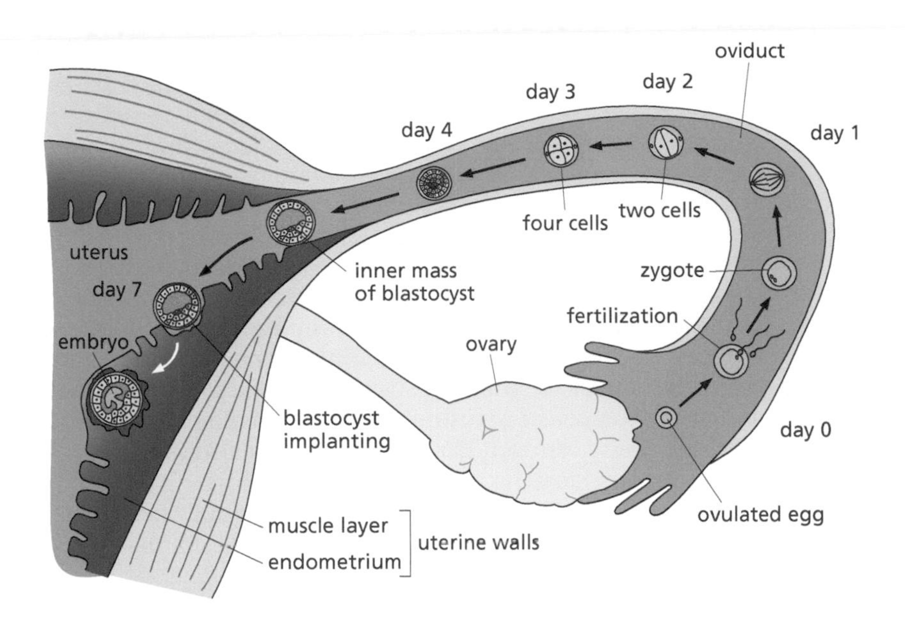

Figure 9.12 Fertilization in the oviduct and implantation in the uterus.

organs. Many sexually transmitted diseases such as gonorrhea and syphilis will cause a loss of fertility that can be permanent. Another large number of infectious diseases will lower fertility temporarily, but the fertility will return to previous levels or nearly so once the disease is cured.

A large number of other factors can also affect fertility in men and women. These are summarized in Table 9.2.

Table 9.2 Factors that can impair fertility

Factor	Description of effects
High blood pressure and other circulatory disorders	A major cause of erectile dysfunction in men (failure of the penis to enlarge and stiffen during sexual activity)
Opiate drugs (including morphine and heroin)	*In women*: loss of ovulation and of menstrual cycles *In men*: erectile dysfunction, loss of sexual interest (libido), and low sperm counts
Alcohol	*In women*: suppresses ovulation and menstrual cycling (but the effect is inconsistent and affects different women to different degrees) *In men*: can lead to erectile dysfunction, loss of libido, sperm abnormalities, and a conversion of testosterone to estrogen that can lead to breast enlargement and sterility
Tobacco	Decreased fertility in both sexes and earlier menopause in women
Marijuana	*In both sexes*: decline in sexual desire, lowering of fertility, and hormonal changes *In men*: breast enlargement and decreased sperm count
Cocaine (chronic use)	*In men*: abnormal sperm and decreased sperm counts and erectile dysfunction *In women*: loss or disturbance of the menstrual cycles and an 11-fold increase in uterine tube infertility
Drugs used to control high blood pressure and many antipsychotic drugs and antidepressants	*In men*: erectile dysfunction, decreased libido, inability to achieve orgasm or to ejaculate (inadequate study of women)
Many anticancer drugs	Inhibit both menstrual cycles and spermatogenesis and can lead to permanent sterility in both sexes
Stress (probably acting by stimulating the body's production of natural opiate-like compounds)	*In women*: suppresses ovulation and leads to loss or reduction of menstrual cycles *In men*: lowered sperm counts, decreased sperm mobility, increased sperm malformation, decreased testosterone levels, erectile dysfunction, and loss of sexual interest (libido)
Depression	Loss of sexual activity and interest
Anorexia in addition to physical overexertion, especially in female athletes and ballet dancers	*In women*: loss of menstruation and failure to ovulate
Heavy metals (mercury, lead, etc.)	Can cause permanent sterility in both sexes
Polychlorinated biphenyls (PCBs), dioxin, and many organic pesticides (including DDT)	Estrogen-like effects impair male fertility, and males exposed to these chemicals before puberty have very high rates of malformed internal and external reproductive structures, including testicular disorders

A further factor is age. The most fertile years for both men and women are between the ages of 20 and 24. Many people choose to delay having children until later than that. After age 35, female fertility (the percentage of women who become pregnant when attempting to become pregnant) is less than half what it was at its maximum. The rate of spontaneous miscarriage is around 20% for women in their thirties, increasing to over 30% for women aged 40–44. Male fertility also decreases with age, gradually from age 40 to 55, and then more steeply after age 55.

Since World War II, studies in various countries have documented a general decline in male sperm counts and a rise in the number of sperm abnormalities. From 1938 to 1990, the average sperm count for European men declined to about half its former value, from 113 million to 66 million sperm per milliliter. Although the reasons are unclear, many researchers suspect that the cause may be related to an environmental presence of industrial chemicals and pesticides that have estrogen-like effects. These estrogen-like chemicals may also play a role in the decreasing average age of menarche, mentioned previously.

In the next section, we describe some of the methods that have been developed for overcoming infertility.

9.2.3 Assisted reproduction

At the population level, there are many motivations for decreasing the birth rate, but on an individual level the urge to have children is very strong. Many couples who cannot conceive a child seek help in becoming pregnant. Usually the first step is a complete medical assessment of both partners to see if a medical cause can be found; in approximately 20%–30% of cases no cause can be identified. In parallel with research on birth control technologies, much research has gone into the development of several new technologies for assisted reproduction. Much of the research on infertility has been done in countries such as India and China, which also have active population control programs, a reminder that individuals and societies may have conflicting goals.

If a couple remains infertile even after medical treatment, or if no medical treatment is appropriate, a variety of remedies can be suggested. In some cases, gametes from both partners are used to produce a child that is genetically theirs. In other cases, gametes are taken from one partner but not the other. Sperm donation is relatively easy and is more commonly practiced, but egg donation requires surgery and is thus less common. (Both sperm banks and egg banks have become profitable businesses in a number of countries.) A possibility always to be considered is adopting a child who is genetically unrelated to both parents.

In this section, we examine a number of options for assisted reproduction.

9.2.3.1 Assisted gamete production Various treatments are available to increase fertility in either men or women (e.g., by using hormones to stimulate egg production and ovulation). Usually only one or a small number of eggs is released in a monthly cycle. Hormonal treatments for infertility may induce the release of many more eggs than normal and result in (fraternal, not identical) twins, triplets, and higher multiples if more than one egg is fertilized in the same ovulation cycle. The recent increase in the frequency of multiple births has been largely due to these treatments. (These treatments have no effect on the frequency of identical twins, which result from a fertilized embryo splitting at a very early stage, with each portion subsequently developing into distinct but genetically identical individuals.)

Treatments to increase ovulation are often used together with other forms of assisted reproduction. In males with low sperm counts, sperm may be harvested and then concentrated by centrifugation, then used in the process of artificial insemination, described next.

9.2.3.2 Artificial insemination **Artificial insemination** means introducing sperm into a female's reproductive tract through means other than sexual intercourse. The sperm could be derived from a woman's husband or from another man. The procedure is fairly simple (and is routinely performed on cattle and certain other domesticated species, as are in vitro fertilization and surrogate pregnancy). However, the legal rights and responsibilities of a sperm donor other than the recipient's husband are unclear in a number of jurisdictions.

9.2.3.3 In vitro fertilization and embryo transfer An in vitro process is one that takes place in laboratory glassware rather than inside the body (in vivo). In the case of **in vitro fertilization** (IVF), eggs are harvested from a woman (usually after hormonal treatment to stimulate egg development) and fertilized in a glass or plastic dish, using sperm contributed either by her husband or by another man (Figure 9.13). Fertilized eggs are then allowed to develop to the blastocyst stage, after which one or more of these embryos are implanted in a woman's uterus and allowed to develop to term. (The woman receiving the implanted embryo is usually the same woman who donated the eggs, but in some cases the recipient could be a surrogate.) Using sperm donated by someone other than the woman's husband raises the same kinds of legal issues (including custody and financial responsibility issues) as does artificial insemination.

IVF was first successfully practiced on humans in 1978. The Society for Assisted Reproductive Technology (SART) reported that IVF was responsible for 61,740 U.S. births in 2012, or just over 1.5% of the 4 million U.S. births for that year. Several newer modifications of this technique are also

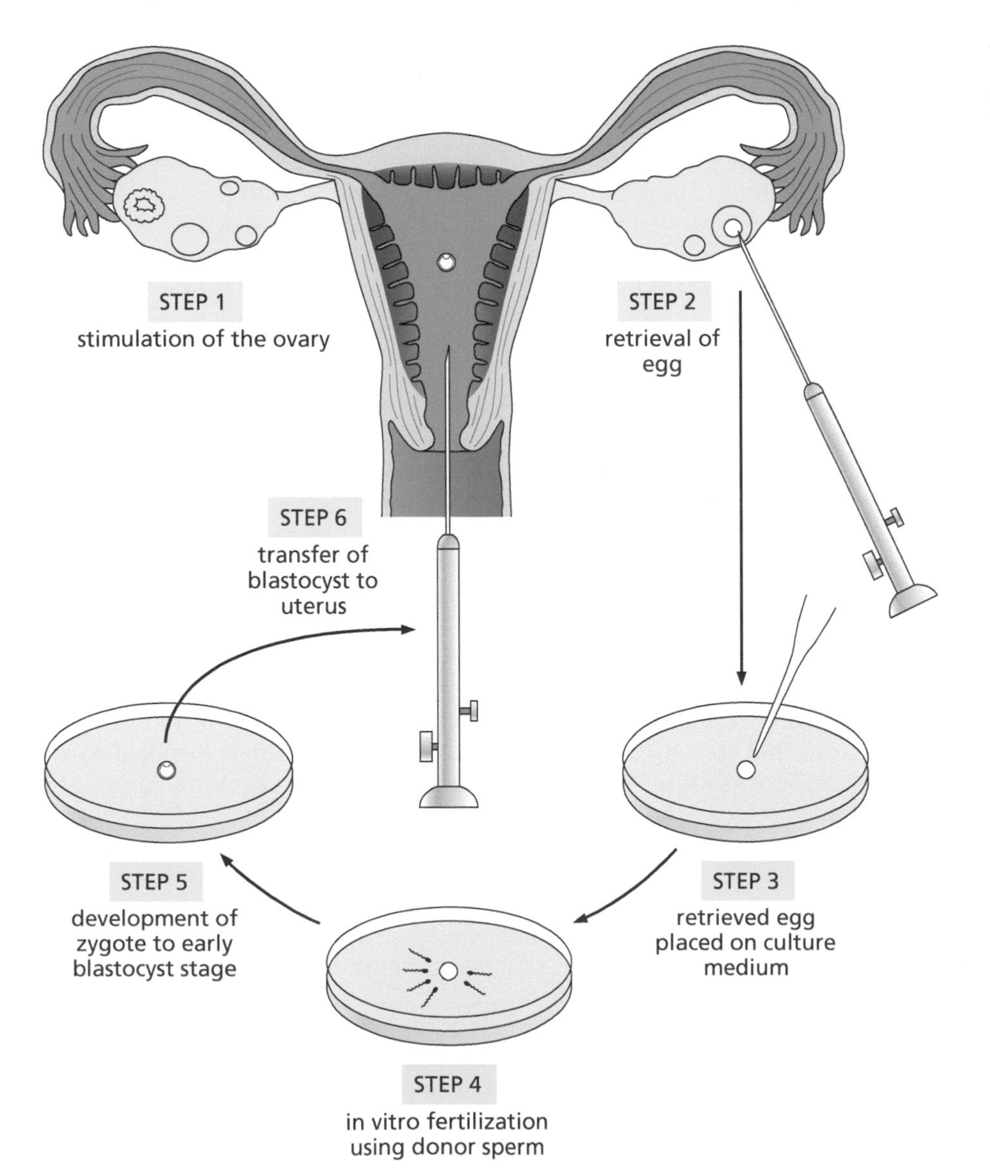

Figure 9.13 **Outline of the procedure used for in vitro fertilization.**

occasionally practiced. In one such technique, called zygote intrafallopian transfer (ZIFT), the zygote is transferred soon after fertilization rather than waiting until the blastocyst stage. In gamete intrafallopian transfer (GIFT), the eggs and sperm are inserted into the uterine tubes and fertilization takes place there rather than in vitro. In cases of low sperm count or low sperm motility, sperm heads or whole sperm may be injected into egg cells instead of allowing fertilization to take place by itself.

Persons who possess alleles that they do not wish to pass on to their offspring may seek IVF and embryo testing before implantation or artificial insemination using donated sperm. Genetic testing of the zygote or early embryo can be carried out prior to the use of any of the previously described techniques for implantation. Some researchers have been experimenting with techniques that test eight-cell embryos for genetic disorders. Because the testing process is usually destructive, only one of the eight cells is separated and tested, leaving the other seven cells available for implantation if desired; the seven-cell embryo develops normally, just as if all eight cells had been used. Embryos obtained by IVF can thus be tested before they are implanted into a woman's uterus to complete the pregnancy. This technique, sometimes called blastomere analysis before implantation (BABI) or preimplantation genetic diagnosis (PGD), has already been used to test human embryos in vitro for cystic fibrosis, allowing the selection of only those embryos that are free of the disease. Selected embryos can then be implanted, and the couple can be free of the fear that their child will be born with cystic fibrosis, a disease that is usually fatal by midlife. Many physicians and medical ethicists propose that embryo selection should only be used to allow parents to avoid passing on a serious genetic disease to their offspring. However, the majority of US fertility clinics also allow parents to select the sex of an embryo conceived by IVF. In the United States, there is not a strong cultural preference to have children of one sex over the other, but this is not the case in some other countries. Increased sex selection could lead to an imbalance in the numbers of men and women in populations where it occurs. Additional ethical concerns arise from the possibility of selecting yet other traits, as more genetic information becomes available regarding the genes that affect various human characteristics (Bayefsky, 2018). Therefore, selection of characteristics based on preimplantation genetics testing is often described as a "slippery slope."

9.2.3.4 Surrogate pregnancy Surrogate pregnancy is the use of another woman's womb to carry a baby to term on behalf of a woman who cannot undergo the pregnancy herself, usually for medical reasons. In most cases, the baby is conceived by IVF, using egg and sperm cells donated by a couple unable to conceive themselves. The resulting embryo, which is the genetic offspring of the donor couple, is then implanted into another woman who agrees to act as a surrogate mother, usually for a fee. In addition, medical expenses are generally paid by the donor couple. The legal status and rights of the surrogate mother are subject to many ethical and legal questions. Surrogacy contracts have been outlawed or held invalid in a number of jurisdictions that view the birth mother (i.e., the surrogate) as the legal parent who is therefore "selling" her baby if she receives any payment. Among the ethical issues raised are the exploitation of poor women by wealthy couples. Financial need is often a factor (one of many) in a woman's decision to become a surrogate. Other issues include the amount of compensation that can ethically or legally be given to the surrogate and the ways in which this situation is distinguished from "baby-selling." A final issue concerns the available alternative of adoption, which is generally less expensive and raises fewer legal and ethical objections.

The drive to reproduce is a strong force among all species, including humans. Not all humans want to have children, but for many it is something they desire greatly. When some people seek out assistive reproductive technologies to overcome infertility, it shows their motivation to have children who are genetically their own. In contrast, many people wish to limit the number of children that they have. In the next section, we discuss the ways in which the timing and number of births can be controlled.

THOUGHT QUESTIONS

1. Are there hormones that are only found in males? Only in females?
2. Most forms of assisted reproduction are expensive and are likely to remain so, and success rates are generally low. They are considered elective procedures and are therefore usually not covered by insurance.
    a. Is it discriminatory to provide assisted reproduction to those who can afford to pay for it, when so many other people cannot?
    b. Given the low success rate, should people be encouraged to give up after repeated failures? How many failures?
    c. Should an upper age limit be imposed on assisted reproduction? Should different limits be put according to age on the number of attempts at assisted reproduction?
3. If an embryo is gestated in the uterus of a surrogate mother who did not contribute the egg, what rights should the surrogate have? What rights should the egg donor (and her partner) have? What legal rights do these parties have where you live? Should these parties have the right to negotiate and agree to a certain division of rights by contract? Why or why not?
4. In most cases of assisted reproduction, multiple eggs are fertilized, only some of which are then implanted and brought to term. What should be done with the rest? If they are stored (usually in a frozen condition), who has the right to decide their fate? Does that include the right to destroy them, to refuse to pay for their continued storage, or to decide where to implant them? Can one member of a couple exercise any of these rights if the other member objects? What if the couple divorces or their relationship changes in another way?
5. Do reproductive health clinics offer assistive reproductive technologies to couples who are infertile due to reversible lifestyle decisions (such as cocaine or marijuana use)? Do you think they should?

9.3 CAN WE DIMINISH POPULATION GROWTH AND ITS IMPACT?

As Malthus realized, many factors influence population growth. Improving nutrition and health care generally increases a population by increasing fertility, decreasing infant mortality, and decreasing the death rate. Deaths from accidents are decreased by safety measures, and medical advances are decreasing the rate of death from many illnesses. To achieve a stable population size without increasing the death rate, the main option available to people is a voluntarily reduction of the birth rate.

We should make note of a distinction at this point: **Population control** is usually understood to operate on the level of populations, while **birth control** methods generally operate by preventing births one at a time. A birth control method is not a successful population control method unless it is widely adopted.

Birth control is not new. It has always been practiced among human populations. Ancient texts in China, India, and Egypt mention abortifacients (i.e., drugs that induce abortions). In Egypt, the Ebers Papyrus (1550 BCE) describes a medicated tampon made with ground acacia seed. Fermentation of the seed in the female reproductive tract produces lactic acid, which is toxic to sperm. Many more birth control options are available today.

Birth control methods work by controlling fertility, so the term "fertility control" is often used as a synonym for birth control. Because these methods allow the spacing and timing of the birth of children, they are also called family planning methods. Many of these methods prevent pregnancy by interfering with the reproductive anatomy or physiology of either the female or her male partner. The various methods of birth control form a spectrum of possibilities (**Figure 9.14**). By the timing of their action, they can be arranged into five distinct groups, which are highlighted in the gray boxes in the figure.

For each of the methods listed, a series of questions can be asked:

1. How does it work?
2. How effective is it in preventing pregnancy?
3. What costs or risks are involved?
4. What kinds of objections have been raised against it?
5. Does it have any benefits apart from birth control (e.g., in the prevention of sexually transmitted diseases)?

Method	effectiveness*	relative cost to provide
I. PREVENTING GAMETE RELEASE		
I.A. Sterilization Methods		
1. irreversible sterilization: (castration, hysterectomy, or ovariectomy)	100%	high
2. semipermanent sterilization: tubal ligation	99.6%	high
vasectomy	99.6%	medium
I.B. Hormonal Methods		
3. estrogen alone	general use 98%; experienced users 99.5%	medium
4. estrogen plus progesterone	general use 98%; experienced users 99.5%	medium
5. progestin only (injectible or 'minipill')	general use 97.5%; experienced users 99%	medium
6. male contraception	99%	medium
7. prolonged breast feeding	varies with nutritional status of mother	none
II. PREVENTING FERTILIZATION		
II.A. Abstinence Methods		
8. long-term sexual abstinence: (celibacy, delayed marriage)	100%	none
9. timed abstinence ('rhythm' methods)	general use 76%; experienced users 98%	none
10. withdrawal during intercourse (coitus interruptus)	general use 77%; experienced users 84%	none
II.B. Barrier Methods		
11. condom	general use 90%; experienced users 98%	low
12. vaginal diaphragm with spermicide	general use 81%; experienced users 98%	low
13. cervical cap with spermicide	general use 87%; experienced users 98%	medium
14. sponge with spermicide	general use 80%; experienced users 98%	low
II.C. Spermicidal Methods		
15. spermicidal creams, foams, and jellies alone	general use 82%; experienced users 97%	low
III. PREVENTING IMPLANTATION		
16. postcoital pills	not known	medium
17. intrauterine devices (IUDs)	general use 95%; experienced users 98.5%	medium
IV. INTERRUPTING GESTATION		
18. abortion	100%	high

* A 99% rate of effectiveness means that 1% of couples using that method will become pregnant in a year of use.

Figure 9.14 **Methods of birth control.**

9.3.1 Birth control acting before fertilization

Of all birth control methods, those that act to prevent pregnancy before fertilization (conception) are often called **contraceptive** measures.

9.3.1.1 Preventing gamete release by sterilization Sterilization, the elimination of reproductive capacity, usually involves surgery and is usually permanent, although some methods are potentially reversible. One way to achieve male sterilization and allow hormone secretion to continue is by vasectomy, the surgical cutting and tying off of the sperm duct (the vas deferens, Figure 9.9). Males with vasectomies continue to produce both testicular hormones and sperm for some time, but the sperm cannot reach the penis for release.

Among female sterilization methods, tubal ligation (tying off of the oviduct) is the only one performed primarily as a birth control measure. Tubal ligation is analogous to male vasectomy; eggs and hormones are still produced, but the eggs are blocked from traveling to the uterus. Surgical removal of the uterus (hysterectomy) is performed for medical reasons other than birth control (e.g., in the case of uterine cancer), but the removal of the uterus results in permanent sterility because the uterus is where the developing embryo grows. Surgical removal of the ovaries, the organs that produce the female gametes, also results in sterility, but this is usually avoided for the same reason as male castration is avoided: The gonads produce many hormones, and their removal has widespread effects on the individual.

All these sterilization methods involve surgery, which makes them expensive to implement on a very large scale, especially in poor or medically underserved areas. As with all forms of surgery, there are risks, such as those of infection or from the use of anesthesia. However, both costs and risks are experienced on a one-time basis only and do not recur. All sterilization methods are completely effective as birth control methods without any further action on the part of the individual.

One of the greatest objections to all these methods is that they are permanent. Many people, even people who want birth control, do not want to become permanently sterile. In the United States, about 60% of men who undergo vasectomy later regret that they did so. Vasectomy and tubal ligation can in some cases be reversed, but success depends on microsurgical techniques, and reported success rates vary greatly. A new method for male sterilization has been developed in China that is reversible under local anesthesia (Zhao, 1990). In this method, a polyurethane elastomer is injected into the sperm duct, where it solidifies to form a plug that effectively blocks the passage of sperm.

9.3.1.2 Preventing gamete release by hormonal methods

Several birth control methods depend upon alterations of the female reproductive cycle. Hormonal birth control methods take advantage of the feedback mechanisms shown in Figure 9.11. For example, estrogen and progesterone both inhibit the secretion of FSH, so that supplying these hormones (or a similar compound) prevents ovarian follicles from reaching maturity and releasing their eggs. The hormones can be given as birth control pills, as injections, as implants just under the skin, or as patches on the skin. Regardless of the method of drug delivery, all hormonal methods work by preventing the egg from maturing and being released. Because hormones have many effects throughout the body, hormones used in birth control can have many side effects, including the possibility of blood clots. For this reason, medical supervision is recommended, and hormonal methods require a prescription in most countries.

Early birth control pills contained estrogen alone, but progesterone was later added (producing the combination birth control pills, also called "combination oral contraceptive" or COC) to reduce the levels of estrogen and also its side effects. The continuous levels of these hormones prevent the usual hormonal cycling from taking place. The expense and the requirement of obtaining a prescription limit the use of birth control pills in many populations. Some developing countries have made birth control pills available *without* a prescription to encourage their more widespread use, but cost is still a problem. Birth control pills have become the most commonly used contraceptive method among the middle and upper classes in many countries. Most birth control pills sold today are of the combination type.

Newer types of birth control pills, developed in the 1980s, use progesterone-like compounds (progestins) only. These include the "minipill," which suppresses ovulation only half the time, but instead works primarily by thickening the cervical mucus. Minipills have a relatively high failure rate, largely because they need to be taken on a strict time schedule (even a three-hour delay can reduce their effectiveness). For this reason, subdermal implants of progestins have been used in many cases. Implants of a drug called levonorgestrel (trade name Norplant®) can be inserted beneath the skin in the underarm region, where they slowly release steady dosages of the drug. The early version, called Norplant I, quite commonly resulted in irregular menstrual bleeding, but this side effect is observed much less commonly with the newer version, Norplant II. Levonorgestrel is also the active ingredient in certain "morning-after" pills, discussed later.

Hormonal control of male fertility has been repeatedly suggested. One of the earliest such methods to be developed was a male contraceptive pill containing a drug called gossypol, derived from cottonseed. This drug was first developed around 1970 in China and is said to be about 99% effective. Gossypol interferes with sperm production through a mechanism that does not affect the hormone testosterone. Tests of gossypol have reported some toxic side effects, however, so an effort is now being made to develop a synthetic substitute.

Testosterone is known to inhibit LH and FSH production by the pituitary, which in turn can inhibit sperm production. However, complete suppression of sperm production by this method requires weekly testosterone injections, and these injections may have undesired side effects. It has also been discovered that the suppression of pituitary function is actually brought about by the small fraction of the testosterone that is converted into estrogen. As a consequence, several researchers have begun exploring the possibility that estrogen- or progesterone-like drugs might be capable of functioning as a male contraceptive. The drug Depo-Provera®, similar to progesterone, has been used for this purpose in Europe for several years, and in California it has been used to suppress testicular function in convicted rapists. (In this use, the drug does inhibit spermatogenesis, but it has no effect on violent behavior.)

One proposed method, developed in India, immunizes men against FSH, a hormone needed for sperm production. This method has not been shown to be effective in birth control, but it does seem to prevent fat gain in both sexes.

9.3.1.3 Preventing gamete release by extended breastfeeding An older hormonal method of birth control is the practice of extended breastfeeding (delayed weaning), which is common in many traditional African societies. Children in Africa are almost always breastfed, and many are not weaned until they are 4–6 years old. While a woman is breastfeeding, she is producing hormones that stimulate milk production. These same hormones also inhibit the rise and fall of the hormones produced by the ovary, thus interrupting the menstrual cycle and preventing egg maturation. Women who use this method of birth control do not wean their youngest child from the breast until they feel they are ready to have another child. This method of birth spacing is a widespread and seemingly effective practice in many parts of Africa, although studies have shown that it is unreliable among women living in the industrialized world at high caloric intake levels. For example, prolonged breastfeeding has a contraceptive effect among the !Kung San (Bushmen) of South Africa and Namibia. These women have a low caloric intake and walk 4–6 miles a day, conditions that seldom occur among North American women. Because the ability of breastfeeding to delay the return of the menstrual cycle is greatest among women who are physically active but who have a low caloric intake, improvements in maternal nutrition may actually decrease the effectiveness of delayed weaning as a method of birth control. Birth spacing by prolonged breastfeeding is also most effective when babies suck vigorously and often. Any contribution to infant nutrition other than breast milk (e.g., by bottled milk or cereal) reduces the effect. In most third-world countries, the use of bottled milk reduces the effectiveness of birth spacing by delayed weaning. The closer spacing of births leads to an increase in birth rates at the population level, possibly offsetting the effects of birth control programs. Even so, one expert surmises that, "In developing countries today, breast-feeding probably prevents more births than modern contraceptive methods."

9.3.1.4 Preventing fertilization by sexual abstinence

As a means of preventing fertilization, abstinence has the distinct advantage of being available to all people free of charge. However, the success of abstinence methods depends upon the determination of the people using them. Total voluntary abstinence (celibacy) has long been practiced as part of a regimen of religious devotion, but only by small numbers of people. Delayed marriage (with no other sexual activity) greatly reduces the birth rate, especially given that the years before age 30 are the most fertile period for a majority of women. Increasingly, in the United States and many other countries, more women are delaying the age at which they marry, and birth rates are falling as a result.

The so-called **rhythm method** is a form of partial abstinence, based upon the fact that a woman is fertile only for several days after ovulation, the time when the egg is in the oviduct or uterus; sexual intercourse at other times will generally not result in conception. Several versions of the rhythm method are practiced. The simplest version is a calendar method, in which a couple abstains from day 10 to day 17 (or, to be more certain, from day 7 to day 17) of a 28-day cycle in which the onset of menstrual bleeding is counted as day 1 (Figure 9.11). Another version, with a higher success rate, is called the Billings method. In this method, the woman feels the mucus just inside her vagina with her fingers. Estrogen causes this mucus to become more slippery and elastic just before ovulation, when estrogen reaches a peak; after that, the secretion becomes scant and dry. A couple using this method abstains from intercourse from the onset of slippery mucus until 4 days after the peak day, but may have intercourse on the remaining dry days.

If practiced correctly, the rhythm method is highly effective, but its effectiveness depends on several conditions, including the regularity of a woman's menstrual cycles (this varies individually), the ability of the couple to keep a calendar and count the days without making a mistake, and the willingness of the couple to refrain from sex (or else to practice another method of birth control) during the woman's fertile period. The effectiveness of the rhythm method can be increased by monitoring the woman's vaginal temperature, since a rise in temperature indicates the time of ovulation more precisely.

Sexual intercourse is also called coitus. Coitus interruptus is the withdrawal of the penis before ejaculation occurs. Some couples have used this method effectively, but the majority find it unsatisfying or difficult to follow. Since some sperm can be released before ejaculation, coitus interruptus is not reliable. On a population-wide scale, it is generally not as effective as other methods.

Abstinence methods are no-cost, which may seem advantageous, but because of their high failure rates, this advantage is often elusive. When the medical costs of an unwanted pregnancy are included, abstinence methods become among the most expensive methods available (Figure 9.15).

9.3.1.5 Preventing fertilization by barrier methods

Barrier methods are those that impose a barrier to the passage of sperm. Most **condoms** are designed for males and cover the penis, but a condom that is worn by women has also been developed. The male condom is the oldest of the barrier methods. First developed in England, traditional condoms were constructed of animal membranes (usually sheep intestine) and were therefore considered a luxury item. The development of rubber and then latex made condoms more widely available and also more reliable. Condoms have the added advantage of protecting against AIDS and other sexually transmitted diseases (see Chapters 15 and 16).

Other barrier methods include vaginal inserts (worn by women), such as cervical caps, vaginal diaphragms, and sponges. Vaginal diaphragms must initially be individually fitted by a physician or other trained medical worker and must be inserted correctly into the vagina before intercourse and left in place for several hours thereafter. When properly placed, vaginal diaphragms block the movement of sperm from the vagina to the uterus, thereby preventing their joining with the egg. (One study in Brazil reported a higher-than-usual success rate if the diaphragm was left in place nearly all the time, but this result remains to be confirmed

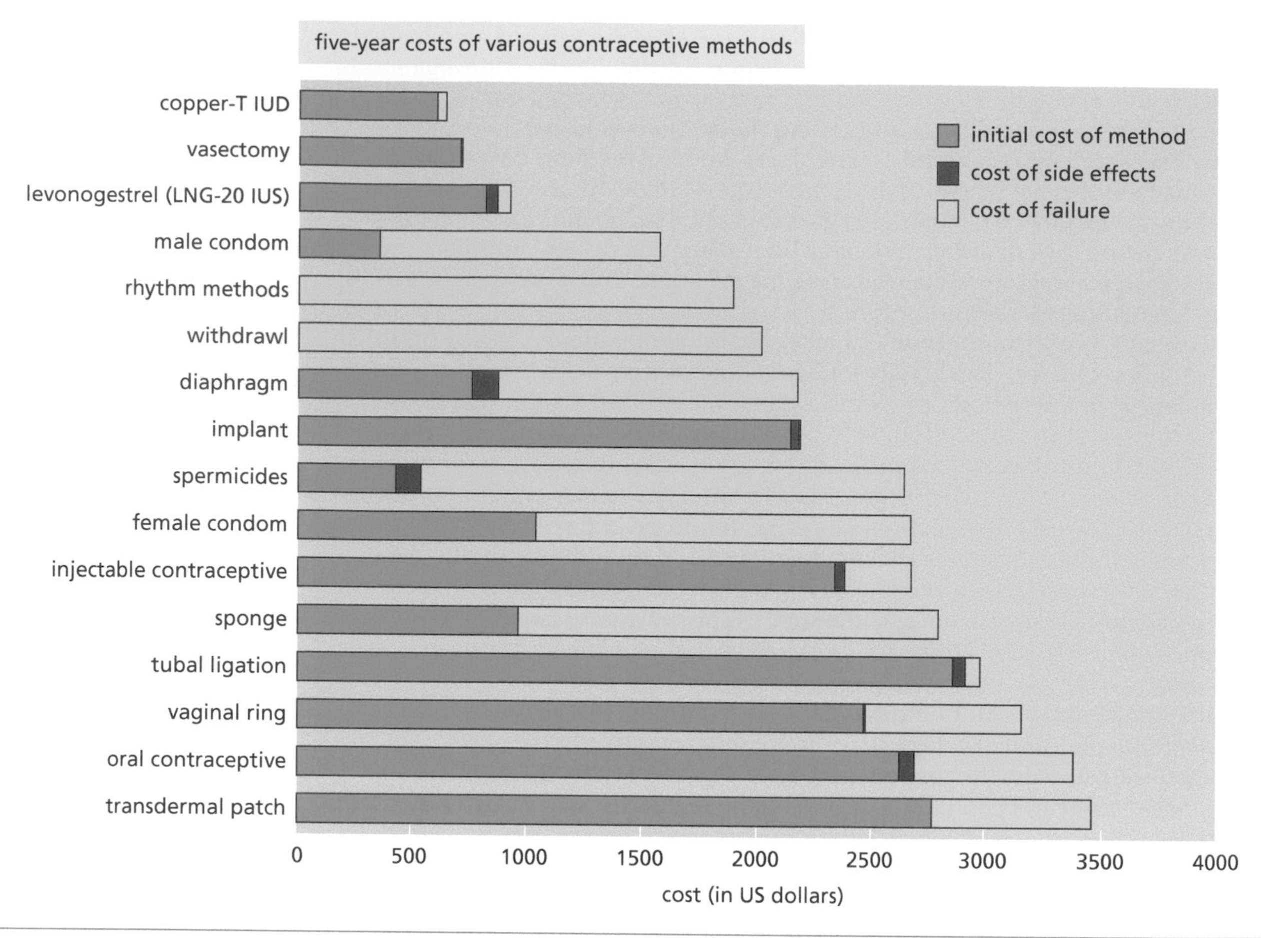

Figure 9.15 Relative costs (over five years of use) of various birth control methods under the type of medical insurance plans common in the United States. For many methods, the largest costs are those associated with unwanted pregnancies resulting from the method's failure.

in other populations.) Cervical caps are made to fit over the narrow portion (cervix) of the uterus, where they also block sperm.

Spermicidal agents, chemicals that can kill sperm, in the form of creams, foams, jellies, or suppositories, are often used together with a barrier method; the combination of barrier plus spermicide is much more effective than either method used alone. One of the newest methods is a sponge impregnated with spermicidal fluid. (Spermicides should not, however, be used with many types of condoms; the spermicide partly dissolves the condom, making it ineffective as a barrier to sperm or to sexually transmitted diseases.)

Barrier methods used with spermicides have extremely low failure rates when used by people familiar with their proper use; most pregnancies occurring with barrier methods are the result of improper use. Barrier methods are widely used in many countries.

9.3.2 Birth control acting after fertilization

Several birth control methods act after the egg has already been fertilized by a sperm.

9.3.2.1 Preventing implantation Hormones or hormone analogues can be used to prevent implantation of a fertilized egg if taken within 72 hours after intercourse. These postcoital or "morning-after" pills can thus be used for emergency contraception. One such drug is levonorgestrel, sold as Plan B and other brand names. It is available without prescription in the United States and in many other countries. The sooner it is taken after intercourse, the more effective it is at preventing implantation and pregnancy. However, once implantation has occurred, this drug has no effect—it does not cause an implanted embryo to abort.

Another drug for preventing implantation is called mifepristone (or RU-486), originally developed in France. Mifepristone blocks the action of the hormone progesterone, which is necessary for maintenance of a pregnancy. Although it requires a prescription, mifepristone can be used as a morning-after pill to prevent implantation from occurring during the first 5 days after intercourse. Its use after implantation is discussed later in this chapter.

9.3.2.2 Intrauterine devices An **intrauterine device** (IUD) is a small piece of plastic or wire, in one of several shapes (e.g., coiled or T-shaped), that is inserted by a physician or other health care professional into a woman's uterus, where it remains until removed by a health care professional. IUDs prevent pregnancy by preventing implantation, although the exact mechanism by which this occurs is not known. (Desert Bedouins have long practiced a similar method of birth control on their camels by inserting stones into the uteri of female camels to prevent pregnancy and removing the stone when breeding was again wanted.) A major advantage to this method is that, once inserted, the IUD works on its own with no need of further action on the part of the woman. On a worldwide basis, IUDs are used more than any other birth control method. This is largely due to the widespread use of IUDs in China, the world's most populous country, and the method is widely used throughout Asia and Europe. In the United States, the use of IUDs trails far behind the use of birth control pills and even sterilizations. The use of IUDs has met with great success in Colorado, and IUDs are largely responsible for a strongly declining rate in teenage pregnancy in that state. The steroid-releasing Progestasert and a version of the copper-T are the most-used IUDs in the United States.

Women who use IUDs have a higher rate of satisfaction than with any other form of birth control, including pills. Today's IUDs are effective, and their failure rate is low. Earlier IUDs were not as safe; one early model, the Dalkon Shield, caused infections and severe bleeding problems in many women, resulting in hysterectomies, large lawsuits, and the corporate bankruptcy of its manufacturer. Many people have shown renewed interest in IUDs in recent decades because of the development of newer, safer types.

9.3.2.3 Abortion The termination of a pregnancy, including the cleaning out of the uterine lining and the expulsion of the embryo or fetus, is called **abortion**. The traditional surgical method of dilation and curettage (enlarging the cervix, then scraping out the uterine interior with a spoonlike instrument) has now been supplemented by newer techniques such as vacuum aspiration (using a machine that uses suction to clean out the uterine contents). There are medical risks to the woman, including excessive bleeding, the chances of infection, and uterine injury, which can result in sterility. The sum total of risks to the life and health of the woman is less than the sum total of risks associated with completing the pregnancy and giving birth. The medical risks are especially low if the abortion is done early, during the first trimester, meaning the first three-month portion of a nine-month pregnancy. Second-trimester abortions using saline injections can also be done safely in most cases. At whatever time an abortion is performed, the risks are much higher if it is done by an untrained person.

Worldwide, the highest abortion rates are in the countries of the former Soviet Union, where other methods of birth control are not readily available. In 1990, an estimated 11% of all Russian women aged 15–44 had undergone an abortion. In contrast, the Netherlands has an abortion rate of less than 1%, one of the lowest among the countries for which data are available. Surveys have also identified the Netherlands as the country with the highest rate of virginity among women entering into their first marriage. Both of these findings have been attributed to a societal attitude that encourages open discussions of sexuality and sexual matters.

The drug mifepristone, previously mentioned as a morning-after pill, can also be used to induce an abortion during the first or second trimester of pregnancy. The success rate is between 93% and 98%, meaning that 2%–7% of women using mifepristone will require a surgical procedure to finish the abortion and clean out the uterus. Common side effects include uterine cramping and bleeding.

Another drug that can be used to induce abortions is methotrexate, originally developed as an anticancer drug and also used against certain autoimmune disorders. Success rates are comparable to mifepristone. Either mifepristone or methotrexate should be followed by misoprostol in order to clear out the contents of the uterus. A 2011 survey found that about one-fourth of abortions in the United States were medically induced rather than surgical.

Abortion is seldom a preferred method of birth control. Even among people who have no ethical objections to its use, abortion is typically considered as a last resort, to be used only if an earlier-acting method fails in a particular case. Abortions performed by untrained persons carry a high risk of injury, subsequent infertility, or death. Safe surgical abortions, performed by trained personnel under antiseptic conditions, are often expensive. The availability of drugs like mifepristone has made medical abortion a cost-effective, safe alternative. Most population planners have advocated that safe abortions be made a widely available option as a backup when the other, less costly methods have failed.

9.3.2.4 Infanticide Though not technically a means of birth control, **infanticide** has long been practiced as a means of population control in many parts of the world, especially in times of famine. There are records of infanticide from Medieval Europe, and the practice was still widespread in China, India, and other parts of Asia well into the twentieth century. In most cases, the infant is not directly killed, but is instead allowed to die through lack of care. In societies in which infanticide is practiced, female infanticide is more common than male infanticide. In China, infanticide is now officially outlawed, but boys outnumber girls in many areas. The government's strict population control policy allowing only one child per couple (Figure 9.16) has been relaxed in recent years, but social scientists suspect that female infanticide is still widely practiced by couples who want a boy but instead have a girl.

9.3.3 Cultural and ethical opposition to birth control

No single method of population control is best for all societies. Abortions and sterilizations, for example, require medically trained personnel. They are more expensive and more labor-intensive than other methods and are therefore unlikely

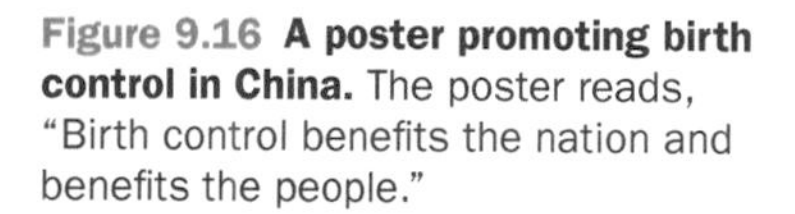

Figure 9.16 A poster promoting birth control in China. The poster reads, "Birth control benefits the nation and benefits the people."

to become the methods most widely used even among populations that have no objections to them. All methods need to be adapted to the customs of the people using them, and education in the use of certain methods may meet with resistance of various kinds. For instance, women in many Muslim societies are generally forbidden to discuss reproductive matters with anybody outside their families, including health care workers.

The attitudes of the Catholic Church toward birth control have varied over the centuries; official opposition to most forms of birth control is historically recent. A considerable debate about birth control took place within the Catholic hierarchy in the 1960s, resulting in two papal encyclicals, *Populorum progressio* (1967) and *Humanae vitae* (1968). The first of these acknowledges the population problem and the need for family planning in underdeveloped areas; the second denounces abortion, sterilization, and all forms of birth control except for the rhythm method. Surveys in many countries show that a large majority of Catholics use various forms of birth control despite the Church's official position. Attempts to spread birth control information have often been opposed by the Church, especially in Latin America, but Church teachings have not stopped Italy and Portugal, both of which are over 81% Catholic, from achieving stable (nongrowing) populations, with some of the lowest birth rates in Europe.

Other religious groups have generally been more tolerant of contraceptive methods. Abortion, however, is opposed by Catholics, Protestant fundamentalists, Orthodox Jews, and Muslims. In its simplest terms, the principal argument voiced by these groups is that a fetus is a living human being and that killing it is an act of murder. People who wish to keep abortion legally available have argued several major points, including a woman's right to choose, a child's right to be wanted by his or her parents, and the need to control the world's population. This is a highly charged issue.

9.3.3.1 The abortion debate Laws on abortion vary greatly from place to place and sometimes from one time period to another. In the United States, many states outlawed abortions until the Supreme Court ruling in *Roe v. Wade* (1973). Under this court opinion, women in the United States had a legally recognized right to an abortion under certain conditions. Abortion during the first or second trimester of pregnancy was therefore legal in the United States in most cases from 1973 to 2022. In 2022, however, the Supreme Court (in *Dobbs v. Jackson Women's Health Organization*) overturned *Roe v. Wade* in a ruling that allows individual states to legislate on abortions as they may choose. Anticipating this decision, 22 states had already written laws restricting most abortions if *Roe v. Wade* was overturned. As of early 2023, 14 states have outlawed most or all abortions, with some exceptions for cases involving rape or incest. Even in the 33 states where most abortions remain legal prior to 20 weeks of gestation, several impose various restrictions based on the qualifications of personnel performing the procedure, the availability of certain specialized equipment, or additional restrictions for pregnant teenagers. But the situation is still fluid, because over a dozen state laws are being challenged in various courts. Wherever abortions become illegal, pregnant women who do not wish to give birth will be forced to travel to other states seeking legal abortions or else seek illegal and unsafe alternatives nearby. Those who do give birth may decide to keep their babies or to place them up for adoption. Various people have suggested that federal legislation be written to make one uniform policy across the United States, but the prospects for any such legislation seem quite uncertain.

Internationally, abortions are generally legal in most European and Asian countries (except for those with large Muslim majorities) and also in Canada, Cuba, and Uruguay. Abortions are generally illegal in most parts of Africa. Abortions are illegal in most Latin American countries, but enforcement of anti-abortion laws varies greatly, and most women seeking abortions can now usually obtain them, especially with drugs such as mifepristone. Michelle Oberman's book, *Her Body, Our Laws* (2018) documents the effects of these laws in both Latin America and the United States.

Many of the questions usually raised in the course of the abortion debate revolve around matters of definition: When does "life" begin? Is a fetus a "person"? Is it a "human being"?

9.3.3.2 Biological definitions of life The usual definitions of "life" mention properties such as cellular structure, motility, metabolism, homeostasis, the ability to respond to stimuli, and the presence of genetic material that is inherited (see Section 1.1.4). By these criteria, a fetus has the characteristics of life as long as it remains in the womb, but it will quickly lose many of these characteristics if removed. Although the fetus is "alive," it is alive in the same way that the fertilized egg is alive; its "aliveness" is essentially the same as that of an unfertilized egg, a sperm, or an appendix or gallbladder.

Opponents of abortion usually argue that the fetus is "alive" in a way that is different from any of these other cells or organs and that abortion is therefore a form of murder. This different concept of a fetus being "alive" usually centers on its "potential" to become a human individual. A pro-choice counterargument might be that an organ such as the appendix is also "alive" and that an abortion is therefore the moral equivalent of an appendectomy. In addition, there are now experimental techniques that can return differentiated cells like those of the appendix or gallbladderto a state resembling stem cells that have the potential to become all types of cells in the body (see Section 10.1); so, the idea of "potential" is also somewhat ambiguous.

9.3.3.3 Legal definitions of personhood Is the fetus a person in a way that the appendix is not? "Personhood" is a social or legal concept, not a biological one, and it is differently defined in each culture or legal system. In the legal system followed in the English-speaking world, a "person" is defined as a legal entity having certain legally recognized rights and duties. In this tradition, corporations, estates, and government bodies have the legal rights of "persons." The "personhood" of a fetus is therefore a matter of legal definition and not of biology, and, because legislators can define personhood in various ways, the legal rights of a fetus can vary from one jurisdiction to another. Other cultures have their own ways of defining the rights or personhood of a fetus or newborn:

1. In Japanese tradition, a baby is not considered a person until it utters its first cry, so killing it is not considered homicide.
2. In parts of West Africa, a child is not considered human until it is a week old.
3. The Ayatal aborigines of Formosa had no punishment for killing a child before it was given a name at age two or three years.
4. Natives of the Pacific Island of Truk considered deformed infants to be ghosts and either burned or drowned them.

These examples illustrate the point that different cultures and different legal systems can reach remarkably different conclusions. Notice again that these are really not scientific questions, capable of being decided entirely by observational data.

9.3.3.4 Biological definitions of humanness To think about whether a fetus is human, we must first decide what we mean by "human." If we mean anything that possesses human genetic material in the form of DNA, then a fetus is definitely human, but so are white blood cells, haploid sperm and egg cells, and the many organs such as the appendix that are removed and discarded each year as "medical waste." If we distinguish the fetus from these others as being "potentially" capable of forming a human life capable of independent existence, then we need to examine what we mean by "potential" and when an "independent existence" begins.

Can biology help us define "humanness"? One book, *The Facts of Life*, by Morowitz and Trefil (1992) examines several possible definitions and argues that all of them point to an "acquisition of humanness" at around 25 weeks of age, the time at which most of the connections between nerve cells in the cerebral cortex are made. Electrical waves in the brain (electroencephalograms [EEGs]) (see Figure 11.20) begin at about this time, providing evidence that the fetus can respond to certain stimuli and can be described as having "experiences." Only after the 25th week are the lungs sufficiently well-endowed with a vascular blood supply to allow the baby's tissues to receive enough oxygen. Before the 25th week, the lungs tend to collapse when empty, and their internal mucous linings tend to stick together, preventing the lung from refilling. Morowitz and Trefil also place the upper limit of abortion safety at around the 25th week, which they say is also close to the current technological limit on the minimum age of "viability." The survival of premature babies is 80% or higher beginning with the 26th week of gestation, but 50% or less in the 25th week or earlier because of the underdeveloped state of the lungs and brain. A medical study published in 1993 by M.C. Allen and colleagues confirms low rates of survival before 24 weeks of age. According to this study, 69% of infants born at 25 weeks of age, but only 21% of infants born at 24 weeks of age, survive without severe abnormalities. According to the U.S. Centers for Disease Control and Prevention, only 1% of abortions are performed past week 21, and only a small fraction of these beyond the 24th week.

9.3.3.5 Ethical considerations Using the ethical principles discussed in Section 1.4.3, a deontologist who opposed abortions would simply argue that it is a wrongful act regardless of any type of medical or other evidence. Such a person would regard the matter as totally outside the bounds of science, because no possible observation or experimental evidence could change the wrongness of what they regard as a wrongful act. A utilitarian would weigh the possible consequences of an abortion (including the monetary costs and the medical risks) against the consequences of the birth if carried to term. Among the latter consequences are medical risks to the mother's health from the delivery, medical expenses, and the costs to the mother of raising the child (or the costs to society if, for some reason, the mother does not raise the child properly). From a utilitarian viewpoint, the mother's wishes, abilities, and financial circumstances are all important in the evaluation; to the deontologist, they are all irrelevant. Much of the frustration of the whole abortion debate is that deontologists and utilitarians talk in such different terms and use such different arguments that neither has much hope of convincing the other of anything.

In the meantime, a small number of antiabortion extremists have resorted to bombing clinics and terrorizing or shooting doctors who perform abortions. Many hospitals have discontinued performing abortions to avoid being the targets of such intimidation. One result is the increasing concentration of abortions performed at clinics that do little else, making them easier targets for these extremists. Other results are that fewer doctors are being trained to perform abortions, while many other doctors who might otherwise perform abortions are refusing to do so for reasons of personal safety. These developments further reduce the pool of doctors and hospitals willing to help someone seeking an abortion. In some states it has become increasingly difficult for a woman seeking an abortion to find a doctor willing to perform one.

9.3.4 Population control movements

Organizations to promote birth control and control population growth were first formed in the nineteenth century. In England, several of these organizations called themselves "Malthusian," even though Malthus, a curate in the Anglican Church, was opposed to nearly all of the birth control methods available in his day. Knowledge about reproduction and birth control was not widely available before the mid-twentieth century. An 1832 book on the subject by an American physician, Charles Knowlton, was banned as immoral in the United States and elsewhere. Two British reformers, Charles Bradlaugh and Annie Besant, and an American, Margaret Sanger, worked through the late 1800s and early 1900s to make information and contraceptives more widely available. Besant strongly influenced Mahatma Gandhi and later migrated to India. Sanger always viewed birth control from the perspective of giving individual women more control over their own lives. The availability of birth control in the United States owes much to her tireless campaigns. She also traveled widely, spreading the message of birth control to India, China, and Japan.

The influence of Gandhi, Besant, and Sanger led India to become the first modern nation to institute a government-funded campaign to control its population. Beginning in 1951, India implemented population control measures that featured easy access to both contraception and abortion and an information campaign to encourage their widespread voluntary use. During the 1960s, nine other nations, including China, Egypt, and Pakistan, also implemented population control programs.

The most ambitious population control program in history was adopted in China in 1962, at a time when their population was around 700 million and was growing at just above a 2% annual rate. The campaign for "only one child for one family" (Figure 9.16) was waged with special vigor. Parents who had only one child were given various benefits (such as pregnancy expenses paid for the first child only, free contraceptives, better housing, and educational benefits), while parents who bore more than one child were fined and sometimes imprisoned. The goal of this campaign was not just to limit population growth but to reduce the population to its 1962 level of 700 million as quickly as possible. Because of a large demographic momentum (that is, an age structure with many children), China's officials realized that they would have to cut the birth rate to somewhat *below* the mortality rate for a time in order to achieve a stable population. China's population in 2000 was around 1.3 billion, but the annual growth rate has been cut to 0.71%. In 2015, China revised its policy to allow two children per family without penalty, and in 2021 they announced plans to allow as many as three children per family in order to prevent many of the problems of a rapidly aging society (BBC, 2021).

9.3.5 The education of women

One of the most effective methods of reducing the number of children borne by each woman is to educate women. Studies in many parts of the world have shown that the population birth rate falls with each rise in the education level of women, even in the absence of any program aimed specifically at birth control (**Figure 9.17**). Around the world, the countries with the lowest rates of female literacy also have very high population growth rates, whereas those that have had higher female literacy rates for several decades all have lower growth rates. Niger and Chad have the lowest adult female literacy rates in the world (both below

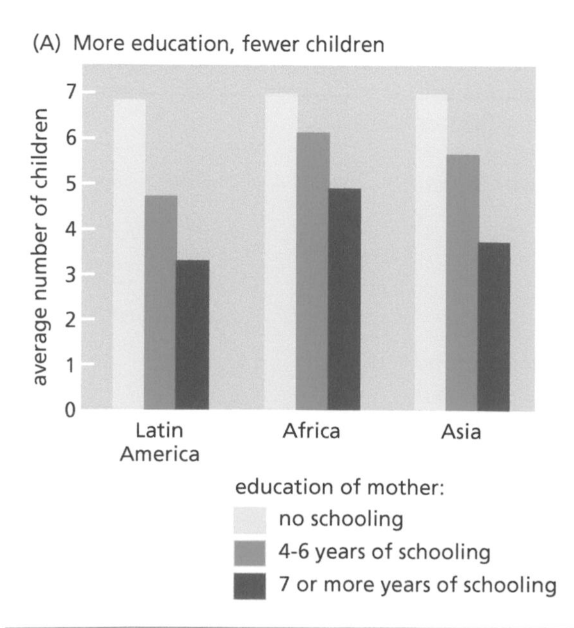

Figure 9.17 The education of women reduces the average number of children per family. Data for the graph are from the World Bank.

14%), and both have population growth rates of 3.2% or more. Afghanistan and Yemen, with adult female literacy rates near 17%, also have growth rates above 2.5%. By contrast, nonindustrialized countries with female literacy rates of 90% or higher include Sri Lanka, Chile, Uruguay, Cuba, and Thailand, all with population growth rates below 1.0%.

In nonindustrialized countries, women with a seventh-grade education or higher tend to marry later than other women (four years later, on average, and these are among the most fertile years). They also use voluntary means of birth control more often, have fewer (and healthier) children, and suffer far less often from either maternal or infant mortality in childbirth. The empowering of women (by giving them more education and more control over their reproductive lives) also raises the educational level of their children and results in more rapid economic development (at lower cost) than many other programs aimed more specifically at development.

In the United States, educational efforts are also an important part of most efforts to reduce pregnancy rates in teenagers. The rate of teenage pregnancy is lowest among women with the most years of schooling and highest among those with only a grade-school education or less.

Although most population control programs are carried out by national governments on their own populations, several programs are international in scope, including those run by the World Health Organization (a branch of the United Nations) and by the U.S. Agency for International Development (USAID). The United Nations has sponsored many conferences on population. Some past conferences emphasized the environmental impacts of population growth, but the 1994 conference in Cairo, Egypt, shifted the attention to the education of women. Women's rights advocates from a variety of countries stressed the need to improve both the education and legal status of women. Many people cautioned that overzealous government-sponsored programs aimed at population control could restrict the reproductive freedom of individual women as much as the earlier lack of birth control information. Although these people generally see the need to reduce the rate of population growth, they are deeply suspicious of programs that coerce individual women or restrict their freedom, and they are especially suspicious of programs urged upon nonindustrialized nations by male-dominated institutions in the industrial world. Instead, they favor programs to educate women and improve their legal status and reproductive choices. They point out that birth rates have diminished whenever the education, legal rights, and reproductive choices of women have improved.

9.3.6 Controlling population impact

Demographic transition brings about a marked increase in human population. In most parts of the world, the excess population tends to migrate to the cities, producing an overcrowding that strains the resources of those urban areas. In Europe and Japan, this process occurred gradually over a period of several hundred years (roughly 1600–1900), giving cities a chance to adjust to their changing conditions. Many cities have accommodated high population densities (meaning large numbers of people per square mile) without widespread misery. Since World War II, urbanization outside the industrialized nations has taken place much more rapidly than it did in Europe. Rapid, unplanned growth has strained most urban services to the point at which many of the newly arrived migrants have inadequate housing. Crowded slum areas or shanty towns often lack safe drinking water and may also suffer from chronic water shortage; sanitation and waste disposal are also frequent problems. It is often not crowding in itself that results in these problems, but crowding without sufficient facilities to support the population. Crime often increases and may become difficult to control, although many other factors besides population contribute to crime rates. Unemployment and economic hardship, when they occur, may compound these problems. The hardships of urban crowding usually fall disproportionately on the poor.

Pollution (Section 19.2) tends to increase approximately in proportion to population, most obviously because of corresponding increases in household garbage and water waste. Densely populated areas are dependent on food, water, and fuel coming in from a much wider radius; consequently, their environmental impact is felt far beyond their political borders. As population increases, more forests are cleared for agricultural use and more trees are cut down to build houses. The destruction of habitat for other organisms, particularly of forests, is one result (see Section 18.3). The loss of arable land (through topsoil erosion, desertification, and other processes) is accelerated by population growth, as is the depletion of nonrenewable resources such as minerals or fossil fuels.

9.3.6.1 Effects of consumption patterns

The impact on the environment is not, however, solely a function of the number of people. The amount of the world's resources that each person consumes is not equal around the globe. On average, the amount of resources consumed by a person in the United States is 54 times that consumed by a person in a developing country. The impact of this consumption is magnified still further by the fact that much of this consumption is of nonrenewable resources. In addition, resources that might be renewable are often consumed or discarded in ways that make them nonrenewable. The enormous size of municipal solid waste disposal sites in industrialized countries is a testament to these consumption patterns. These landfills are among the largest structures ever built by humans, and the materials within them are unavailable for reuse or biodegradation.

If the rate of energy use does not exceed the rate at which energy is captured from the sun in photosynthesis (see Section 17.1.2), then the energy use is considered **sustainable**. In many industrialized countries, however, present patterns of energy consumption are already unsustainable because they remove far more energy from global ecosystems than they produce.

Discussions of the world's population crisis frequently become linked to discussions of the environmental crisis. Many people, especially outside the industrial world, believe that the population crisis is only a small part of a greater environmental crisis. This environmental crisis, they say, is made worse by the industrial world's overconsumption more than by the developing world's population increase. Frances Moore Lappé is one of several American writers holding such views. Some analysts even question whether the industrial world's concerns over population are misdirected (and possibly racist). Developing countries, they say, could well support far larger populations than they do now if it were not for the export of so many of their resources to support the patterns of overconsumption that have become so typical of the industrial world. If the industrial countries, they say, were to give up their lavish patterns of consumption, then the developing world could well support a larger human population (at a larger carrying capacity) than it does in current circumstances.

Others take what may be called a neo-Malthusian position. Paul Ehrlich, for example, views most other problems as consequences of overpopulation. If the population were smaller, he argues, most environmental problems would diminish or even disappear. Because some countries (mostly in Europe) have already limited their population growth, the greatest efforts should be directed at those nations (mostly in the developing world; Figure 9.2) that have the highest population growth rates.

Overpopulation and overconsumption need not be viewed as opposing viewpoints. Population growth and profligate consumption are both widely recognized problems, and each makes the other worse. Some people see one of these as the bigger problem; some people see the other. Efforts directed at addressing either problem can only help to ameliorate both.

9.3.6.2 Limits on carrying capacity

Many scientists tell us that we will soon reach or even exceed the carrying capacity of the planet. In fact, this is one point on which people concerned with overpopulation and those concerned with overconsumption agree, although they postulate different causes for this condition. One of the few dissenters, economist Julian Simon, observes that the technological revolutions of past centuries have repeatedly brought about demographic shifts, each of which has increased the carrying capacity. He predicts that future technological revolutions will continue to enlarge the planet's carrying capacity indefinitely. Nearly all other scientists and writers who have contemplated the subject of population believe instead that the planet's carrying capacity has a limit.

Can the global carrying capacity be increased further? The answer is not known with certainty, but it depends in part on whether we assume the Earth's natural resources to be renewable and unlimited (Julian Simon's view) or limited and nonrenewable (the majority viewpoint). Those who accept the limits imposed by nonrenewable resources will be driven to the conclusion that carrying capacity cannot be increased very much. In fact, if we maintain our present patterns of consumption, we may not even be able to sustain the present population levels forever.

Human populations, like all other populations, are biological entities, requiring energy flow to survive. Populations are therefore subject to the laws of physics (energy is neither created nor destroyed), and populations cannot exceed the limits imposed by the availability of energy. When they approach K, populations are controlled by biological factors, such as starvation and disease. So, the question, "Should populations be controlled?" is academic because populations will be controlled by the forces of biology and physics, regardless of our answer. The relevant questions are, "Should we exercise preventive efforts at population control?" and, if we should, "How should we do so?"

THOUGHT QUESTIONS

1. Is an increase in population the only factor that puts resources in limited supply? Does population growth affect the availability of housing or medical care in the same way that it affects the availability of drinking water, sanitation controls, and food?
2. Find out whether your college makes birth control information and birth control itself available to the student population. Are certain methods favored over others? Why?
3. Why are most forms of birth control aimed at women rather than men? Why are most population control campaigns aimed at women? In societies in which women have little or no control over their lives, are they likely to be able to carry out family planning? Will family planning give them greater control?
4. Do you think it is proper to view birth control information as a freedom-of-speech issue? Do you think birth control methods should be taught in the public schools? Why or why not?
5. What social benefits are likely if population growth is controlled? What social and ethical problems need to be considered? What individual rights are at risk? In your opinion, what is the best way for a government to control its country's population growth without restricting the reproductive freedom of its women?
6. In what countries is abortion legal? Where is abortion illegal? Can you suggest reasons for these differences?
7. Can biology have any useful role in the abortion debate? What role? Would any biological data be persuasive to a person who opposed abortion on the basis of deontological principles? Would data be persuasive to someone using utilitarian ethics?
8. Most stem cells (useful in cancer research or cancer therapies) come from discarded embryos and aborted fetuses. Fertilized eggs are the ultimate stem cells. However, U.S. law prohibits the National Institutes of Health (NIH) from funding any experiments that deliberately create or destroy a human embryo. How does this restriction on research relate to the abortion debate in the United States? Do you agree with the restriction? If you do not, how would you change the law?
9. In Practice Question 15 at the end of this chapter, what considerations have been ignored? Do you think they may safely be ignored?

CONCLUDING REMARKS

Individual decision making in family planning is sometimes at odds with government decisions aimed at population control and also sometimes with various religious teachings. There are also many other reasons why people might resent strangers urging them to modify their most personal behaviors in one way or another. All of these factors, moreover, vary from place to place, and the lessons learned in one country or population cannot necessarily be applied uncritically to other populations elsewhere. However, any attempts to implement change based on biological data will necessarily take place in a context of many, often competing, social values. Many scientists think that moving to sustainable consumption levels may partly alleviate the problems of population growth, but only temporarily.

In addition, motivating people to decrease their consumption may be just as difficult as motivating them to have fewer children. Even well-planned efforts to address social, economic, or environmental problems may prove to be inadequate as resources are stretched to the breaking point in the face of increasing population pressure. "Whatever your cause," says one slogan, "it's a lost cause unless we can control population."

CHAPTER SUMMARY

- The principles of **population ecology** are applicable to all species.
- A new population shows rapid **exponential growth** at first, but its **growth rate** (***r***) levels off when it approaches the **carrying capacity** (***K***) of its environment, a phenomenon called **logistic growth**.
- The effects of population growth beyond the carrying capacity are numerous: Consumption of resources increases, pollution increases, and any inefficiency in the utilization of resources results in starvation and death.
- Humans and other ***K*-selected** species are characterized by stable populations at or near carrying capacity (*K*) and by most individuals living a long time. In contrast, ***r*-selected** species show more rapid growth, higher mortality early in life, and unstable population sizes.
- Human populations have grown markedly after each major advance in technology. Each major population increase has taken the form of a **demographic transition**, beginning with declining mortality and ending when the **birth rate** (***B***) declines to match the **death rate** (***D***).
- Many factors promote infertility, including those that interfere with the **menstrual cycle** in females or with sperm quality in males.
- **Ovulation** and many other aspects of sexual development and of fertility are controlled by **hormones**.
- Understanding reproductive biology can help us to treat infertility and also to control the birth rate.
- Many methods of **birth control** are available. They differ in their biological mechanisms, their costs, their medical risks, and their acceptance by different groups of people.
- Many studies have found that improvements in the education and legal status of women lower the birth rate in a cost-effective manner and bring other benefits besides.

BIBLIOGRAPHIC REFERENCES

Bibliographic references to material in this chapter can be found online at biologytrending.routledge.com/chapter9

GLOSSARY: KEY TERMS TO KNOW

These key terms are defined at the end of each chapter as an aid for student review.

Abortion Expulsion or removal of a fetus from the womb prematurely.

Age pyramid A diagram that represents the age distribution of a population by a stack of rectangles, each proportional in size to the percentage of individuals in a particular age group.

Age structure The distribution of members of a population into different age groups.

Artificial insemination Insertion of sperm into a female reproductive tract other than directly from a male.

Barrier method A birth control method in which a barrier is inserted across the path of sperm to prevent the sperm from reaching the egg.

Birth control Any measure intended to prevent unwanted births or to reduce the birth rate.

Birth rate (*B*) The number of births in a given time period divided by the population size at the beginning of that period.

Blastocyst An early mammalian embryo at the stage when it is ready to implant into the uterine wall.

Carrying capacity (*K*) The maximum population size that can persist in a given environment.

Census Any enumeration (counting) of the members of a population.

Condom A latex or other barrier to the passage of sperm worn as a covering over the penis.

Contraceptive Any method that prevents conception (fertilization).

Corpus luteum Progesterone-secreting scar tissue formed within the ovary by a follicle after the egg has been released.

Death rate (*D*) The number of deaths in a given time period divided by the population size at the beginning of that period.

Demographic momentum A temporary population increase that can be predicted in a population that has more pre-reproductive members and fewer post-reproductive members than a population with a stable age structure would have.

Demographic transition An orderly series of changes in population structure in which the death rate decreases before a similar change occurs in the birth rate, resulting in a population increase during the transition period.

Demography The mathematical study of populations.

Doubling time The time required for the number of individuals to double in a population.

Estrogen A hormone that stimulates the development of female sex organs prior to reproductive age and the growth of an ovarian follicle each month during the reproductive years.

Exponential growth A form of geometric growth without any limit, according to the equation $dN/dT = r N$.

Feedback mechanism or system Any process in which a later step modifies or regulates an earlier step in the process.

Fertilization The combining of an egg and a sperm to form a diploid zygote.

Fetus A developing embryo after all of its organs have formed; in humans, from about eight weeks postconception to birth.

Growth rate (*r*) The population increase during a specified time interval (usually a year) divided by the population size at the beginning of that time interval.

Hormone A chemical signal transported through the blood, producing physiological responses in target tissues elsewhere in the body.

Implantation The attachment of an early embryo to the wall of the uterus, where it later forms a placenta.

Infanticide The killing of an infant shortly after birth.

Intrauterine device (IUD) Anything inserted into the uterus to prevent implantation or pregnancy.

In vitro fertilization Fertilization that takes place outside the body in laboratory glassware.

***K*-selection** Natural selection favoring a stable population size near the carrying capacity, generally found in predictably stable environments.

Life expectancy The average length of time that individuals with certain characteristics are expected to live, or the average duration of life for the entire population.

Logistic growth Growth that begins exponentially but then levels off to a stable population size (*K*), according to the equation $dN/dT = rN (K - N)/K$.

Menstrual cycle A female reproductive cycle characterized by periodic loss of blood and uterine tissue approximately two weeks after ovulation.

Ovary The organ that produces eggs in females.

Ovulation Release of an egg from the ovary.

Ovum An egg cell.

Placenta A structure appearing in the development of most mammals, composed of tissue derived from both the embryo and the mother's uterine lining, by means of which the embryo is nourished during its development in the uterus.

Population A group of organisms capable of interbreeding among themselves and usually sharing common descent as well; a group of individuals within a species living in a particular time and place.

Population control All measures that reduce or limit the rate of population growth.

Population ecology The study of populations and the forces that control them.

Positive checks Involuntary controls that limit or reduce population growth, such as famine, war, and epidemic disease.

Preventive checks Voluntary measures that reduce population growth, including voluntary abstinence from sexual activity.

Progesterone A hormone that maintains the uterine lining in its enlarged, blood-rich condition, ready for implantation of an early embryo.

Puberty A series of hormonal changes and their consequences associated with the onset of sexual maturity.

***r*-selection** Natural selection favoring rapid and prolific reproduction, usually in environments where favorable conditions can quickly disappear and mortality can therefore be devastatingly high.

Rhythm method A birth control method of timed abstinence in which the couple avoids having sex during the time when ovulation is most likely.

Secondary sexual characteristics Features characteristic of one sex but not essential in reproduction; examples include female breasts and male beards in humans and antlers in male deer.

Sex determination The genetic or other control of development as a male or female individual.

Sustainable Any practice that could continue indefinitely without depleting any material whose supply is limited.

Testis (plural, testes) The sperm-producing organs in males.

Testosterone A steroid hormone that produces male primary and secondary sexual characteristics.

Zero population growth A condition in which a population no longer changes size because its birth rate and death rate are equal.

CONNECTIONS TO OTHER CHAPTERS

Chapter 1: Abortion and other forms of birth control raise important ethical issues.
Chapter 3: New reproductive technologies can be used to enhance fertility.
Chapter 5: Population size responds to evolutionary forces, such as competition from other species.
Chapter 8: Undernutrition and malnutrition increase the death rate and also lower rates of fertility.
Chapter 12: More and more people are living longer and experiencing Alzheimer's disease and other conditions that prevail among the elderly.
Chapter 14: Overcrowded conditions cause stress in many species.
Chapter 15: AIDS has greatly increased the death rate in many countries.
Chapter 16: Emerging infectious diseases increase infant mortality and decrease life expectancy.
Chapter 17: Plants are the main biological energy producers on which the life of humans and other consumer organisms depends. Populations cannot be sustained above levels that can be supported by producer organisms.
Chapter 19: Increasing population size usually makes pollution problems worse. Needs for water and sewage treatment increase with population size.

PRACTICE QUESTIONS

1. What is the birth rate *B* in a nation with 4 million men and 3.9 million women in which 200,000 children are born in one year? What is the birth rate *B* in a nation with 3 million men and 4.9 million women in which 200,000 children are born in one year?
2. What is the death rate *D* in a nation of 100 million people in which 500,000 people died in a year and 200,000 of those who died were children below the age of two years? What is the death rate *D* in a nation of 100 million people in which 500,000 people died in a year and 50,000 of those who died were children below the age of two years? What differences would you expect between these two cases?
3. What is the growth rate *r* in a nation of 500 million people if the birth rate is 2% and the death rate is 1% (assuming no immigration or emigration)? What is the growth rate *r* in a nation of 50 million people if the birth rate is 2% and the death rate is 1% (assuming no immigration or emigration)?
4. How many people will be added in one year to a population of 500 million people with a growth rate of 2%? How many will be added if the growth rate is 4%? How many will be added at growth rates of 2% or 4% if the initial population was 5 billion?
5. What is the doubling time of a population of 500 million people if the growth rate is 2%? What is the doubling time of a population of 5 billion people if the growth rate is 2%?
6. If a population is at its carrying capacity *K* of 500 million and its birth rate is 3%, what is its death rate (assuming no immigration or emigration)? If a population is at its carrying capacity *K* of 5 billion and its birth rate is 3%, what is its death rate (again assuming no immigration or emigration)?
7. A population of 4.5 million has a birth rate of 0.067 (or 6.7%) and a death rate of 0.024 (or 2.4%). Find the growth rate (*r*) and the number of years that it will take for the population to double.
8. For the population in the previous question, find:
   a. The population increase this year
   b. The size of the population after a year of increase
   c. The population increase in the second year
   d. The population size after two years of increase
9. Which will increase more rapidly this year: a population of 3.3 million with a growth rate of $r = 2.0\%$ or a population of 5 million with a growth rate of 1.1%?
10. Where in the body is testosterone produced? Where in the body is estrogen produced?
11. During the menstrual cycle, what hormones are secreted by the ovaries? Other hormones are involved in addition to the ovarian hormones. Where in the body are these other hormones produced?
12. How do hormonal contraceptive methods induce infertility?
13. How do barrier contraceptives prevent fertilization?
14. A particular birth control method is 98% effective. In a nation of 8 million people, if 400,000 women use this method, how many of them will become pregnant this year?
15. Suppose you are working for a family-planning program in a nation whose population is growing rapidly. Among several available methods of birth control are the following:
    Method A costs $1,600 per woman, lasts an average of 20 years, and is 99% effective.
    Method B costs $20 per month for each woman and is 78% effective.
    Method C costs $200 per year for each woman and is 88% effective.
    Which method is most cost-effective (reduces pregnancies by the largest amount per $1,000 spent)?

CHAPTER 10

Stem Cells, Cell Division, and Cancer

ISSUES

- What are stem cells and why are they so important?
- What are the ethical issues related to stem cells?
- Why is cancer so important?
- What are the causes of cancer? Are they more genetic or more environmental?
- What are the treatments for cancer?
- Can we prevent cancer?

BIOLOGICAL CONCEPTS

- The cell cycle and its regulation (receptors, growth factors)
- Levels of organization (cells, tissues)
- Gene expression and regulation (cell differentiation in embryos and adult animals, potentiality, stem cells)
- Embryonic development
- Interaction of genotype and environment (carcinogens, mutagens, diet)
- Health and disease (homeostasis, risk)
- Molecular biology of cancer (oncogenes, tumor suppressor genes)

CHAPTER OUTLINE

DOI: 10.1201/9781003391159-10

In this chapter, we consider two topics of great significance for human health: (1) stem cells, which can give rise to all types of cells in the body and hold therapeutic potential for many diseases, and (2) cancer, which the World Health Organization ranks as the first or second leading cause of death for people under age 70 in most countries of the world. Humans and all other multicellular organisms are made of cells that divide by mitosis to generate all the cells in a complete organism—more than 2 trillion cells in a human newborn and more than 35 trillion in an adult! During this process, specialized cell types with distinct forms and functions develop due to expression of particular subsets of genes from the organism's genome. Signaling molecules are produced that bind to and activate specific receptors on certain groups of cells within a developing organism, initiating signaling cascades that cause certain genes to be expressed (see Section 3.1 for steps in gene expression) and others to be silenced (not expressed). Early in development, all the cells are so-called "**stem cells**," which have the potential to divide indefinitely and to become all cell types in a complete organism; however, as cells become specialized, they lose these abilities. Fully specialized cells no longer divide and cannot become other types of cells except under artificial or abnormal circumstances. The regenerative capacity of stem cells has important potential for the treatment of human diseases that are caused by deficiencies in particular cell types.

In a fully formed organism, cell number is carefully controlled, with many proteins collaborating in pathways that regulate cell division, cell death, and many other behaviors. Genetic mutations (DNA sequence changes; see Section 3.1.4) may inactivate some of these regulatory mechanisms, leading to the out-of-control cell growth that is cancer. Cancer cells share many features of stem cells that are abnormal for the specialized cell types from which they originate. Thus, the same features that can be beneficial when they allow stem cells to generate and replace missing or damaged cells can also be the cause of a dreaded disease in the form of cancer. In this chapter, we first consider how cell growth, division, and behavior are regulated in normal cells, along with the mechanisms that allow stem cells to give rise to specialized cells. We then discuss what goes wrong in cancer and how it develops. We will also consider ethical issues related to stem cells and how we can most effectively treat and reduce our risks for various cancers.

10.1 MULTICELLULAR ORGANISMS ARE ORGANIZED GROUPS OF CELLS AND TISSUES THAT DEVELOP FROM UNDIFFERENTIATED STEM CELLS

As we introduced in Section 6.2, compartmentalization of individual eukaryotic cells into organelles and of whole organisms into individual cells promotes efficiency in the execution of life processes. In most multicellular organisms, cells are organized into **tissues**, with the cells of different tissues specialized to perform specific functions. For specialization to be beneficial, however, the behavior of one type of cell must be integrated with the behavior of other cells, just as the behaviors of individual organelles within a cell must be coordinated. Tissues are further integrated into organs and organ systems, in which two or more types of tissues coordinate to perform more complex functions, such as digestion or circulation (see Chapter 8), reproduction (see Chapter 9), external sensing (see Chapter 11), respiration, or excretion (see Chapter 13). A multicellular organism can thus be considered a complex ecosystem. Throughout evolution, specialization based on cooperation worked well and was favored by natural selection. All multicellular organisms—fungi, plants, and animals—have continued this basic strategy.

10.1.1 Cooperation and homeostasis

The proper functioning of the whole organism depends on the continued integration and cooperation of all the cells. When this integration is functioning properly, that is, when the ecosystem of cells is stable, we may consider that organism to be in a state of health. According to French physiologist Claude Bernard (1813–1878), cells are responsible for maintaining a "milieu intérieur" (an internal environment) within each cell and within the body as a whole. Good health is defined as the maintenance of conditions within this internal environment that fluctuate only minimally, a process that Bernard named **homeostasis.** This does not mean that during homeostasis there are no changes within the organism. Just the opposite is true: Molecules and cells are constantly being made and broken down, but these changes occur around a balance point. Homeostasis is the ability to return to that balance point. Disruption of homeostasis produces illness. As we will see later in this chapter, **cancer** is a disruption of the cellular homeostasis in which cell division is no longer in balance with cell death.

10.1.2 Cellular differentiation and tissue formation

Every multicellular organism that reproduces sexually begins as a single cell called a zygote, or fertilized egg (see Section 2.2.3). (There are also a few multicellular organisms that can reproduce asexually by budding, such as the freshwater invertebrates of the genus *Hydra.*) The immediate cellular descendants of the zygote, produced by mitosis, are able to form all the different cell types within the body. This capability is referred to as "totipotency." The long list of possible fates for **totipotent** cells includes skin cells, muscle cells, glandular cells, bone cells, liver cells, and so forth. In placental mammals, like humans, totipotent cells can also give rise to the placenta, which is required to nourish the embryo (see Section 9.2.1).

Within a developing multicellular organism, as cells proliferate, they also become different. **Differentiation** takes place in steps. At each successive differentiation step, the range of possible future identities for that cell lineage is progressively narrowed, until it narrows to a single cell type. Once a cell lineage has differentiated in the direction of muscle cells, for example, all progeny cells are committed to being muscle cells. A few groups of cells, however, remain undifferentiated, and these are called **stem cells.** How cells "know" what type of tissue to become has long been one of the major questions in biology. This involves cells producing and responding to signaling molecules, a process which we discuss in the next section.

Beyond the first few divisions of a human embryo, cells can no longer give rise to the placenta (which is an organ that is separate from the embryo; see Section 9.2.1) but can give rise to all embryonic cell types. These **embryonic stem cells** are referred to as **pluripotent.** At still later stages of development and in adult organisms, populations of stem cells remain, but their capabilities are more restricted than those of embryonic stem cells; these stem cells are called **multipotent** and can only develop into certain subsets of cells. For example, hematopoietic stem cells exist in the bone marrow throughout a person's life; they can generate all types of blood cells, but not the cells of other tissues.

Like cell division, differentiation is tightly regulated by the control of gene expression. The light-sensing rod cells in the eyes, the smooth muscle cells in the intestines, and the insulin-secreting β-cells in the pancreas are very different in both structure and function, yet they contain the same set of genes; the difference is that these three cell types express different *subsets* of that same genome (**Figure 10.1**). Thus, differentiation is largely about turning off certain genes while keeping others on or activating some genes at a higher level. Biologists refer to this as "regulation of gene expression," and mechanisms for this regulation are discussed later in this chapter. Much of what we know about cell differentiation has come from embryology, the study of the development of an organism from a zygote, and from the study of stem cells. Studies of normal differentiation have also taught us much about the abnormal conditions that exist in cancer. As it turns

Figure 10.1 Virtually all somatic cells contain the complete genome, but different genes are expressed by the cells of different tissues.

	genetic material	example of genes switched 'on'		
		rhodopsin	smooth muscle actin	insulin
 rod light sensor	complete genome - all genes	✓		
 smooth muscle cell	complete genome - all genes		✓	
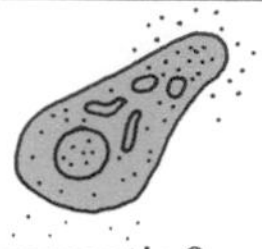 pancreatic β-cell	complete genome - all genes			✓

out, cancer cells share many similarities with stem cells, and the path by which a normal cell progresses to become a cancer cell partly involves a process of "de-differentiation."

10.2 SIGNALING BETWEEN CELLS IS THE BASIS OF COOPERATION AND SPECIALIZATION

Maintaining cooperation and homeostasis in a multicellular organism requires communication, or transfer of signals, among the various cells. Cells receive many different signals from their surroundings that are transmitted across the cell membrane to cause changes in their behavior, their shape, and their internal composition. Two crucial processes that are regulated by external signals are cell division and cell differentiation. Normal cells do not divide unless they receive signals to do so, and stem cells responding to signals proceed along particular differentiation pathways to generate specific cell types. In contrast, cancer cells can divide without the normal signals and do not obey normal differentiation patterns. Before considering how external signals regulate cell division and differentiation and how these processes are altered in cancer, we will first describe the general mechanisms by which cells receive and respond to signals.

10.2.1 Cells receive signals from the outside that are transmitted across the cell membrane to cause internal changes

Most signals are small molecules released (secreted) by other cells. These signals may affect other cells in the same area, in a process called **paracrine** signaling (**Figure 10.2A**), or they may enter the bloodstream to travel to other parts of the body and affect cells at a distance, which is known as **endocrine** signaling (**Figure 10.2B**). Endocrine signals are called "hormones," and several reproductive

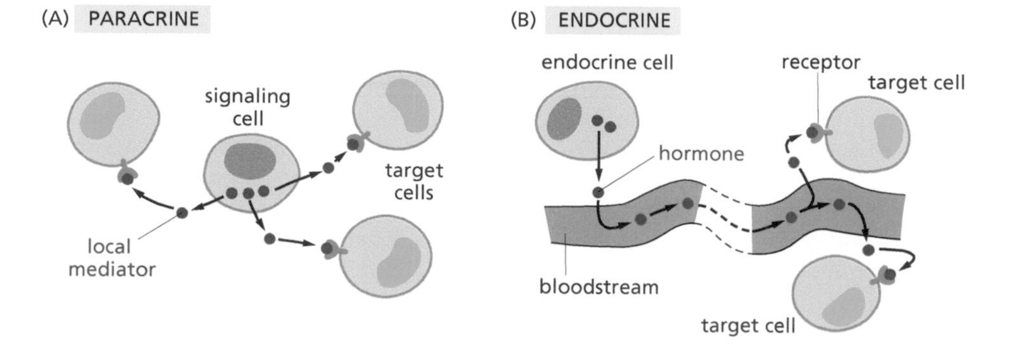

Figure 10.2 Signals can travel short or long distances to reach target cells. **(A)** Paracrine signals are released by one cell and affect neighboring cells. **(B)** Endocrine signals enter the bloodstream and travel throughout the body to reach distant target cells.

hormones are described in Chapter 9. Signals in one class, the steroid hormones (including the sex hormones testosterone and estrogen), can cross the cell membrane and cause immediate changes in gene expression inside the cells (**Figure 10.3A**). They can traverse the membrane because of their nonpolar chemical character, similar to that of the fatty acid tails of the phospholipids that make up the cell membrane (see Figure 8.5). In contrast, most other signal molecules are polar, or charged, and are thus blocked from entry into cells. Therefore, these signals must cause changes from outside the cell. How do they do this? The multistep process of communication across the cell membrane, shown in **Figure 10.3B**, is called "signal transduction." Because many steps are involved, with each leading to the next, such a signaling pathway is often called a "signaling cascade."

In the first step of signal transduction, the signal molecule binds to the extracellular (outside) portion of a specific receptor protein (Figure 10.3B, step 1). The term *specific* means that a given receptor can bind to only one particular type of molecule: The signal and its receptor match each other in terms of their shape and their ability to form chemical bonding interactions. Binding of a signal to the extracellular portion of its specific receptor changes the shape of the intracellular portion, the part of the receptor molecule that is inside the cell. This allows the intracellular part of the receptor to interact with other signaling proteins and, in turn, to cause changes in those proteins as well, thus propagating the signal

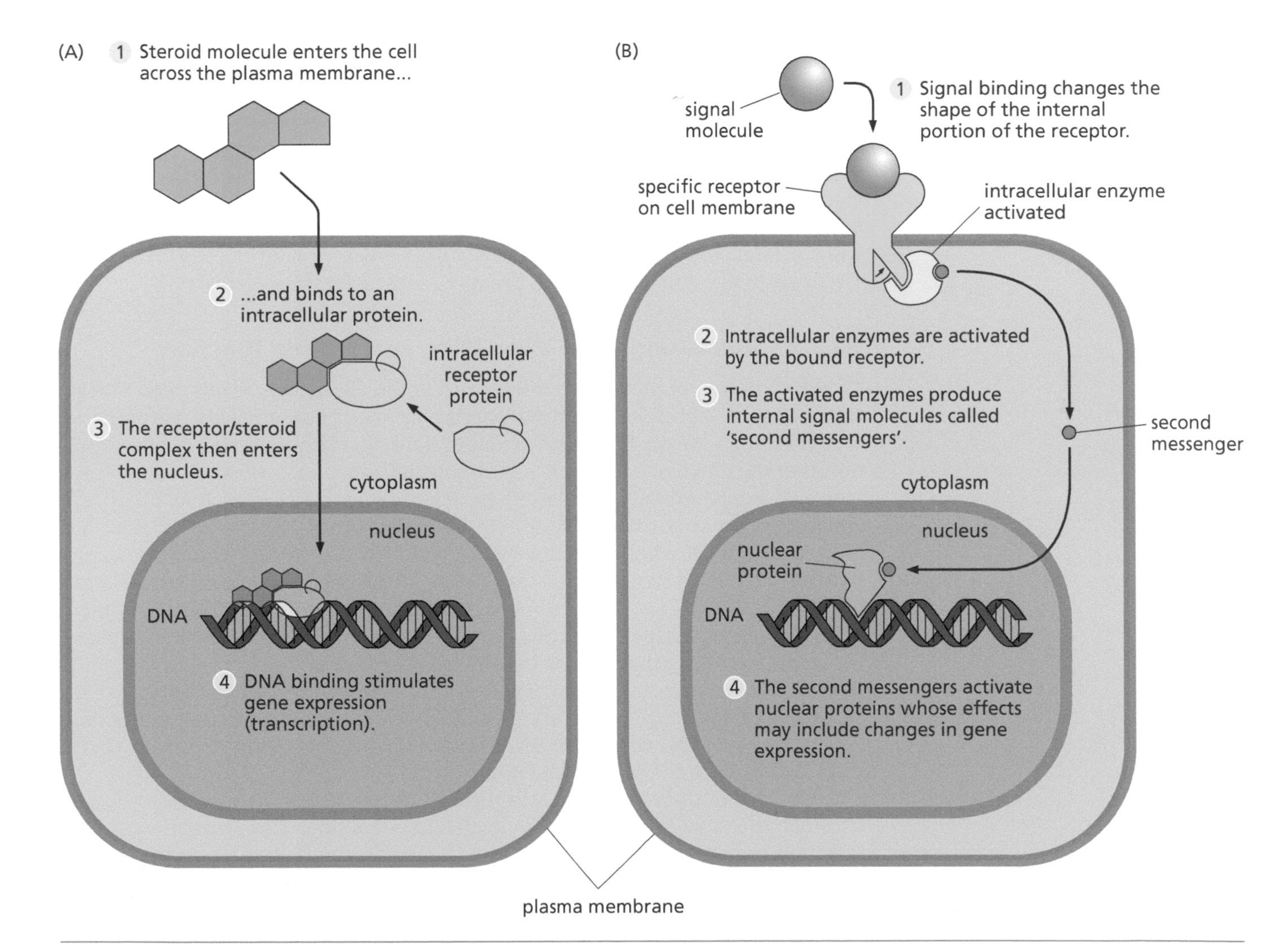

Figure 10.3 Signal reception and signal transduction. (A) Steroid hormones cross the cell membrane to bind internal receptors that regulate gene transcription. **(B)** Other signals bind cell-surface receptors to activate a signal transduction cascade inside the cell. (1) Signal binding changes the shape of the internal portion of the receptor. (2) Internal signaling proteins are activated by the bound receptor. (3) Some internal signaling proteins are enzymes that produce internal signals called second messengers. (4) Second messengers activate proteins that cause the final outcome of a signal, such as changes in gene expression. For an animated version of this figure, go to biologytrending.routledge.com/chapter10

through the multiple steps of a signaling cascade (Figure 10.3B, step 2). When a cell surface receptor is bound by a signal and thus “activated,” some intracellular proteins, including enzymes, are altered as a result. Once activated, these enzymes can catalyze chemical reactions that result in the production of internal signaling molecules called **second messengers** (Figure 10.3B, step 3). These second messengers go on to transmit signals to still other proteins, some of which will cause changes in the cell that constitute the end result of the multistep signaling pathway (Figure 10.3B, step 4). An external signal that carries the message for a cell to divide is typically called a **growth factor**. When a growth factor activates its specific cell-surface receptor, it triggers a signaling cascade, as was just described, that ultimately results in the cell initiating a sequence of events that lead to its division into two cells. The growth factor itself remains outside the cell, but *information* is transmitted across the cellular membrane and into the cell nucleus, stimulating cell division.

As previously mentioned, the defining feature of cancer is excessive and uncontrolled cell division. Therefore, it should be apparent that either too much of a growth factor or excessive activity of a growth factor receptor or other members of the signaling cascade could contribute to cancer. As one example, too many copies of a cell-surface growth factor receptor called HER2 are present in certain types of breast cancer. A second example is an intracellular signaling protein called Ras, which is part of many growth factor signaling cascades; Ras is mutated and hyperactive in about 30% of all human cancers, such that the National Cancer Institute has made a special initiative aimed at developing drugs to target this pathway. Ras was the first gene identified as being mutated in cancer cells (Tabin et al., 1982). In some cases, cancer cells themselves begin to produce their own growth factors and then respond to them, eliminating the need for neighboring cells to secrete the signaling molecules and increasing the local concentration of the signals. One signal, epidermal growth factor (EGF) is commonly overproduced in this way by a wide range of different types of cancer cells. As we will see later, knowing the specific signaling defects in cancer cells has allowed scientists to develop cancer drugs that precisely target those defects.

10.2.2 Regulating gene expression and protein function

To maintain cooperation and homeostasis, multicellular organisms need to regulate both gene expression and protein function. When biologists say that a process is **regulated**, it means that there are natural mechanisms that cause either more or less of the process to occur in a given time period, based on some sort of informational input. All cells regulate their activities by controlling how much of a given gene product (usually a protein, but sometimes an RNA) is present at a given time and/or whether or not that gene product is active at that time. In prokaryotic cells, both steps of gene expression—transcription of DNA to mRNA and translation of mRNA to protein—occur in one compartment, and translation can begin before transcription has finished. Therefore, the primary way that prokaryotes regulate gene expression is by controlling when and how rapidly transcription occurs. Eukaryotic cells have many more options for regulation because the two steps in gene expression are compartmentalized, with transcription taking place in the nucleus, where the DNA is located, and translation happening in the cytoplasm, where ribosomes are found.

Prokaryotic and eukaryotic cells regulate transcription in the same basic way, although the details are more complicated in eukaryotes. The regulation of transcription is summarized in **Figure 10.4**. Transcription begins when the enzyme RNA polymerase (see Section 3.1.2) binds to a special DNA sequence known as a **promoter** sequence (**Figure 10.4A**). In eukaryotic cells, the RNA polymerase is accompanied by a large set of other proteins called “general transcription factors” that are also required for initiation of transcription. Each eukaryotic gene has its own promoter, whereas in prokaryotes, multiple genes that function in the same process are often organized next to each other in the genome and transcribed from a single promoter. Proteins needed only in small amounts are controlled by promoters that only weakly bind RNA polymerase, so that initiation of

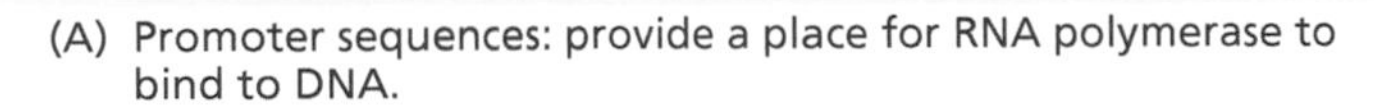

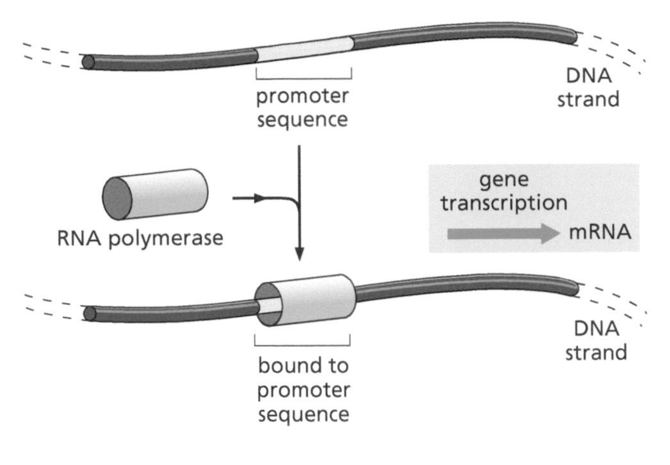

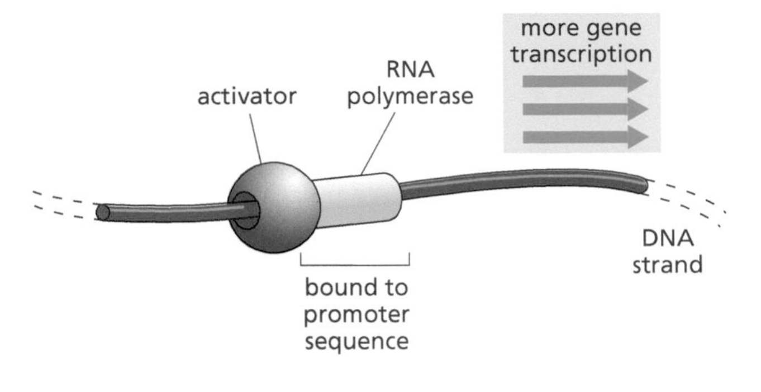

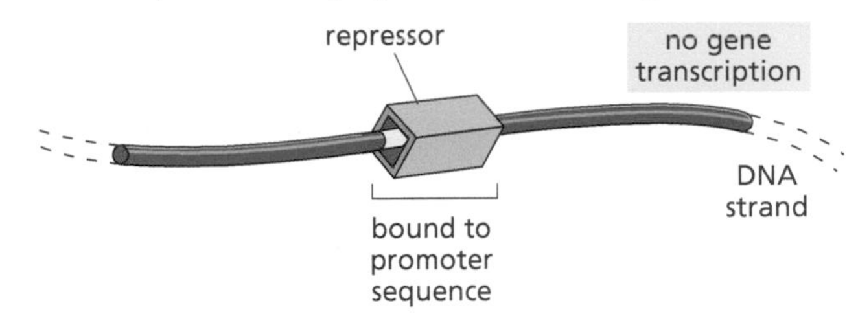

Figure 10.4 Regulation of transcription. (A) RNA polymerase binds to genes at promoter sequences to begin transcription. **(B)** When activator proteins bind to DNA, they help RNA polymerase bind to the promoter and increase the number of RNA transcripts produced. **(C)** When repressor proteins bind to DNA, they block RNA polymerase and prevent transcription.

transcription occurs infrequently. A protein needed in very large amounts is typically controlled by a promoter that strongly binds RNA polymerase, so that transcription initiation occurs more often.

On the DNA near a promoter (or, in eukaryotes, sometimes as many as thousands of nucleotides away from the promoter), there are regulatory gene sequences that can either activate or repress transcription by changing how RNA polymerase binds to the promoter. Proteins called activators and repressors bind to these sequences. If an activator binds to a regulatory sequence, RNA polymerase binds more strongly to the promoter, and more copies of mRNA are transcribed from that gene (Figure 10.4B). On the other hand, if a repressor binds, RNA polymerase is blocked from accessing the promoter, and transcription is halted (Figure 10.4C).

Activators and repressors themselves are also regulated (Figure 10.5). There are many possible ways in which this can occur. For example, an activator might require a signaling molecule to bind to it, causing a shape change that allows the activator to bind to DNA and increase transcription (Figure 10.5A); alternatively, an enzyme might add a chemical modification that prevents an activator from binding DNA, thus decreasing the rate of transcription (Figure 10.5B). Similarly, an inhibitory signaling molecule might bind to a repressor and change its shape, thus preventing the repressor from binding to DNA and allowing transcription to continue (Figure 10.5C), or an enzyme might add a chemical modification to a repressor that in turn allows it to bind DNA, thus repressing transcription (Figure 10.5D). Note that all types of regulation are possible: It is just as likely that an activator signal binding or chemical modification can either activate or inactivate any type of regulatory protein. Furthermore, mutations that alter the structures of activator or repressor proteins themselves can render these regulators nonfunctional or, in rarer cases, hyperactive. We will see that such mutations in proteins that regulate transcription of genes related to cell division or cell death can promote uncontrolled cell division, contributing to cancer.

We have just seen the general strategy by which transcription is regulated (Figures 10.4 and 10.5). As previously mentioned, this is the most prevalent

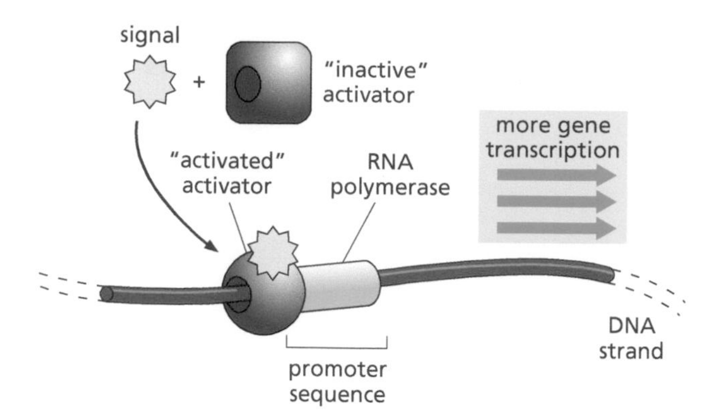

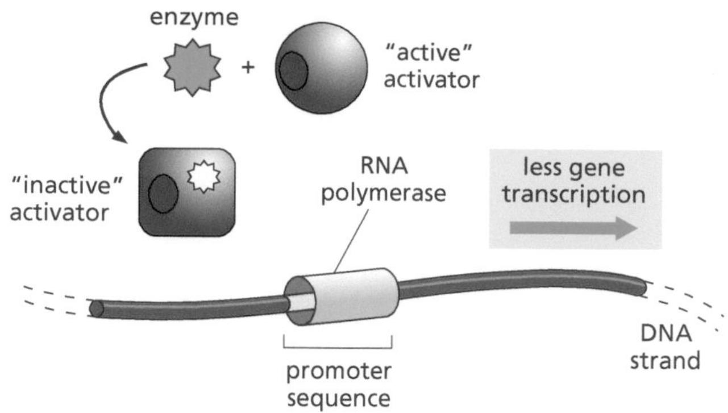

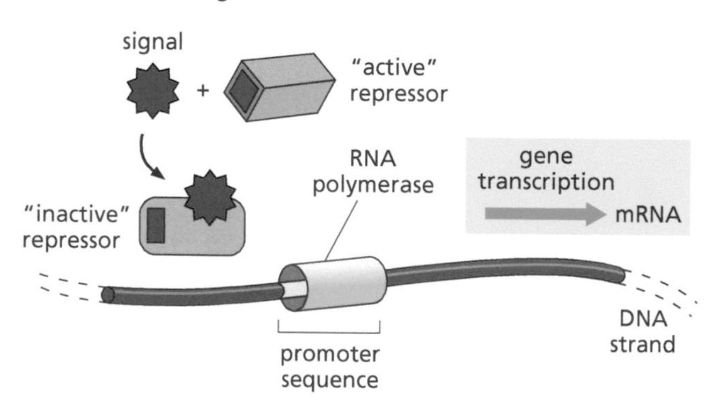

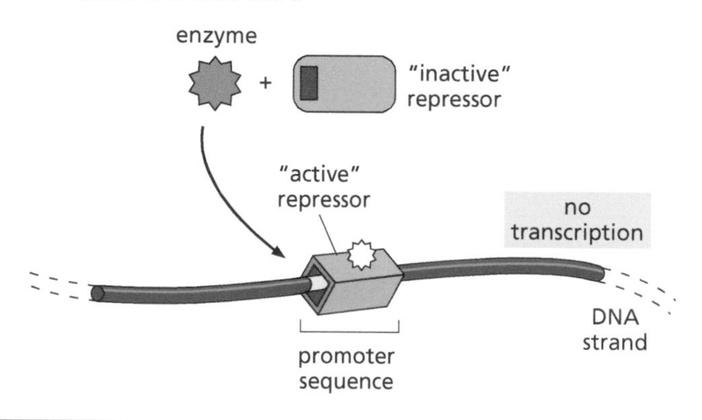

Figure 10.5 The activity of transcriptional activators and repressors is also regulated. The DNA binding activity of transcriptional activators and repressors can be altered by the binding of signaling molecules to these regulators or chemical modification of the regulators. Any change that promotes activator binding or reduces repressor binding will increase gene transcription **(A, C)**; conversely, any change that reduces activator binding or promotes repressor binding will reduce gene transcription **(B, D)**.

type of regulation in prokaryotic cells. Eukaryotic cells also regulate transcription (Figure 10.6, step 1), but they can additionally regulate gene expression at a number of other post-transcriptional steps (**Figure 10.6**, steps 2 through 6). As noted earlier, in eukaryotic cells, the mRNA synthesized during transcription must leave the nucleus before it can be translated into proteins because the ribosomes are in the cytoplasm. Many mRNAs must be chemically modified before they can leave the nucleus (Figure 10.6, step 2). One example of an mRNA modification is the removal of the noncoding regions (introns) by the process of RNA splicing, as described in Section 4.4.1. This is crucial, since an unspliced mRNA does not have the correct sequence to code for a functional protein. The two ends of the mRNA are also modified, with a chemical "cap" (a modified G nucleotide, 7-mG) being added to one end and a long series of "A" nucleotides being added to the other, forming a poly-A "tail." These end modifications stimulate efficient translation once an mRNA reaches the cytoplasm. The fact that mRNAs cannot be "exported" from the nucleus to the cytoplasm until splicing and end modifications have occurred is a regulatory mechanism that ensures that only mature mRNAs that are fully ready for translation can reach the translation machinery.

Once the processed mRNA is in the cytoplasm, the rate of translation can also be regulated: mRNAs contain sequences that bind to proteins that can influence how often a ribosome initiates translation on that particular mRNA. Rapid translation produces more copies of a protein, while slow translation produces fewer copies (Figure 10.6, step 3).

Further regulation often occurs after translation. The amino acid sequence first produced in translation may not be the sequence of the final protein. Some

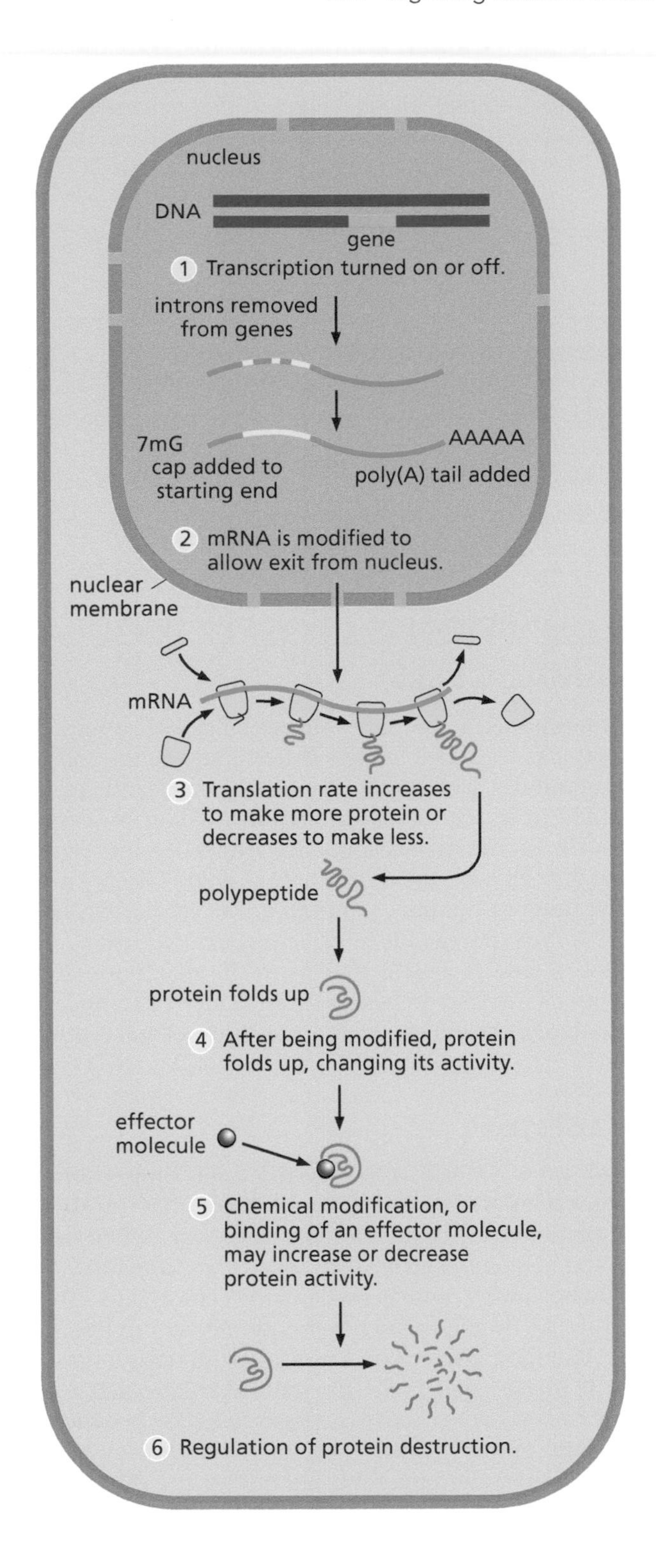

Figure 10.6 Regulation of gene expression can occur at many steps in eukaryotic cells. 1, Regulation of the rate of transcription; 2, mRNA modification prior to transport out of the nucleus; 3, Regulation of the rate of transcription; 4, Protein folding; 5, Chemical modification of a protein or binding of an effector molecule; 6, Regulation of protein destruction.

amino acids may need to be removed, or chemical groups may need to be added to certain amino acids, before the protein can fold properly into its functional shape. Without this processing, functional proteins are not produced (Figure 10.6, step 4).

Finally, as introduced previously in the context of transcriptional regulators, the activity of any mature protein can be regulated in several ways. Other molecules, called effector molecules, can bind to a protein and change its shape, either slowing down or speeding up the activity of the protein; also, chemical modifications can be added to and removed from a folded protein to regulate its activity (Figure 10.6, step 5). Lastly, a protein can be targeted for destruction at certain times but not at others, causing that protein to accumulate in the cell only when it is needed (Figure 10.6, step 6). The last regulatory mechanism, regulated protein destruction, is especially important for regulation of the cell division cycle, as we will see in the next section.

THOUGHT QUESTIONS

1. Why is it advantageous for differentiation to involve changes in gene expression rather than changes in the genetic material of cells?
2. Many animals, including insects, fishes, amphibians, and reptiles, do not maintain a constant internal temperature but allow their internal temperature to change with the external temperature. Are these animals in homeostasis? Why or why not? (Many of these animals can regulate their temperature behaviorally by moving to different locations.)
3. Why do eukaryotic cells have so many more ways of regulating gene expression and protein function compared to prokaryotic cells?
4. What can cause a protein to change its shape, and why are protein shape changes important?

10.3 CELL DIVISION IS CLOSELY REGULATED IN NORMAL CELLS

Cell division (the production of more cells, often called cell *proliferation*) is one cellular process that is regulated by signals from outside the cell. In Section 2.2.1, we learned about mitosis, which is the division of a cell's nucleus and the genetic material contained in it. Mitosis is one step of a larger process of cell division called the **cell cycle**. Scientists have found that the molecular signals that regulate cell division and the cell cycle are remarkably similar in highly diverse eukaryotic organisms, from fungi to humans. (Note that this similarity does not extend to prokaryotes and is thus shared only by cells with nuclei.) Just as abnormal growth factor signaling can lead to cancer, defects in the machinery that regulates the actual process of cell division are also seen in tumor cells, and, in fact, studying cancer cells has taught scientists a lot about the normal features of cell cycle regulation.

10.3.1 The cell cycle

Normal cells only grow a small fraction of the time. They continually make new proteins and other cellular chemicals to replace ones that have been used or damaged, but most of the time they do not increase in size. When cells do grow, they soon reach the size at which their surface area-to-volume ratio makes them inefficient. Instead of becoming more inefficient, the cells divide. When we talk about how fast cells grow, we usually mean how frequently they divide, not how fast they enlarge. Cell proliferation is therefore a more precise term for this "growth."

The series of phases that make up the eukaryotic cell cycle is shown in Figure 10.7. The cell cycle begins with a phase called G_1 in which protein synthesis is increased. Then, if the cell receives signals, it enters the synthesis, or S, phase, marked by DNA synthesis and the replication (copying through complementary base pairing) of both DNA strands (see Figure 2.18). When DNA synthesis is complete, the cell enters the G_2 phase in which preparations are made for mitosis. Mitosis itself (see Section 2.2.1) constitutes the M phase, at the end of the cell cycle, when the duplicated DNA is evenly split between two nuclei. Division of the cytoplasm, or **cytokinesis**, usually follows close on the heels of the M phase, yielding two so-called daughter cells.

After cytokinesis, one or both of the resulting daughter cells can re-enter the cell cycle and divide again. Most of the time, however, both daughter cells enter a resting stage, called G_0, that constitutes a pause between cell cycles. During the resting stage, other cellular metabolic processes proceed, but the cell does not re-enter the cell cycle to divide again unless it is signaled to do so. The duration of the cell cycle (G_1 through M) is fairly constant within a species, but the duration of G_0 varies greatly. For single-celled eukaryotic organisms, such as yeast, which

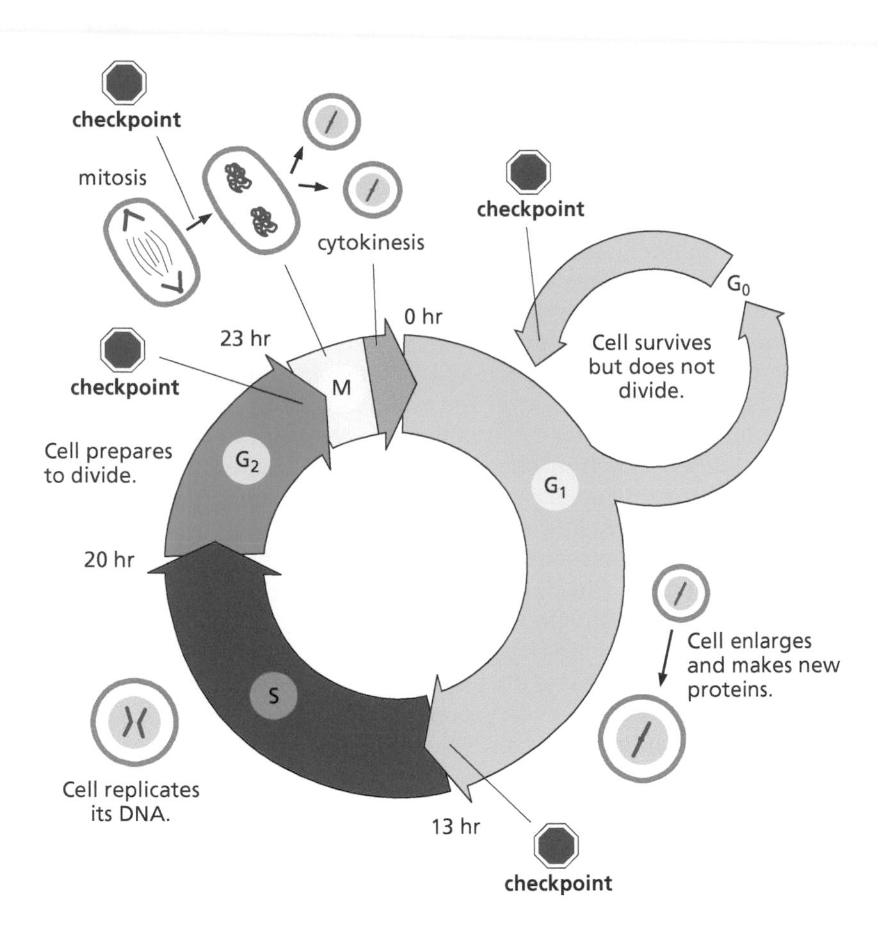

Figure 10.7 The cell cycle. Hours shown are the approximate lengths of time for each phase in a cell with a 24-hour cell cycle, typical of many eukaryotic cells.

have been key model organisms for studying the cell cycle, the length of time in the resting stage depends on the availability of nutrients. The length of G_0 in multicellular organisms varies with the developmental stage. When an animal or plant is developing as an embryo, the rate of cell proliferation can be very rapid, and cells spend little or no time in G_0. In most species, most types of cells spend more time in G_0 once adulthood is reached and will only re-enter the cell cycle when they receive an external signal to do so. For example, tissue damage might cause release of a signal that triggers cell division in neighboring cells within the tissue to repair the damage. Replacement cells are obviously needed if there has been an injury, but even in uninjured tissues some cells die and others divide to replace them, so there is typically a low baseline rate of cell division. This rate depends on the tissue: In skin, where surface cells slough off every day, the baseline rate of division is much higher than in an organ like the adult heart or brain.

10.3.2 Regulation of cell division

The cell cycle (and thus cell division) is a tightly regulated process in all types of eukaryotic organisms, both single-celled and multicellular. For two viable daughter cells to be produced, the stages of the cell cycle and the events of mitosis must occur in the correct order, and each step must be completed properly before the next step begins. Consider, for example, what would happen if a cell did not go through G_1 (the growth period) before it divided: Multiple rounds of omitting G_1 would lead to smaller and smaller cells, eventually leading to a nonviable size. Alternatively, imagine what would happen if mitosis began before DNA synthesis was complete, or if anaphase of mitosis (when duplicated chromosomes separate from each other) began before all the chromosomes were attached to the cytoskeletal filaments of the mitotic spindle: Either case would result in daughter cells with the wrong sets of genetic material. Because it is so important for the events of the cell cycle to occur in the correct order and at the proper time, regulatory mechanisms called **checkpoints** exist that serve as barriers, or brakes, to block cell progression until all conditions are favorable to proceed to the next step (Figure 10.7).

The primary regulators of the cell division cycle are **cyclin-dependent kinases** (CDKs). A kinase is an enzyme (catalytic protein) that can add phosphate groups to certain amino acids in other proteins, a process called **phosphorylation**. As their name suggests, the activity of CDKs requires a partner protein called a **cyclin** (**Figure 10.8A**). Cyclins earned their name because scientists studying sea urchin embryos noticed that the amounts of certain proteins increased and decreased in a characteristic manner with each round of cell division: that is, the levels of these proteins *cycled* up and down with each cell division cycle. Different cyclin proteins reach their peak levels at different times in the cell cycle, and it is the activity of the partner CDK that each cyclin activates that triggers progression into the next phase of the cell cycle (**Figure 10.8B**). For example, when the amount of cyclin D in a cell reaches a peak level, the cyclin D-CDK becomes active and phosphorylates other proteins that eventually cause the cell to enter S phase. On the other, hand, when cyclin B reaches its peak, the cell enters mitosis (M phase). Although the CDK proteins are always present in the cell, they are only active when their partner cyclins are also present.

The cell cycle checkpoints mentioned earlier influence the timing of the rises and falls in CDK activity. Checkpoint proteins bind to a cyclin-CDK complex and block its enzymatic activity, preventing target proteins from being phosphorylated and activated. This action pauses the cell cycle to allow for problems to be fixed before the cell moves on to the next stage. For example, the cell might pause to allow all of its chromosomes to become attached to the mitotic spindle or to give DNA repair proteins time to fix DNA damage so that mutations are not passed on to the daughter cells. Thus, if cyclin-CDK complexes are the "gas pedals" of the cell cycle, checkpoint proteins are the "brakes." In relation to cancer, both improper activation of the "gas pedals" and inactivation of the "brakes" contribute to the excessive cell division that produces tumors.

For cell division to take place in normal tissues, a number of conditions must be met: (1) As we have seen, signals must communicate the need for the cell to

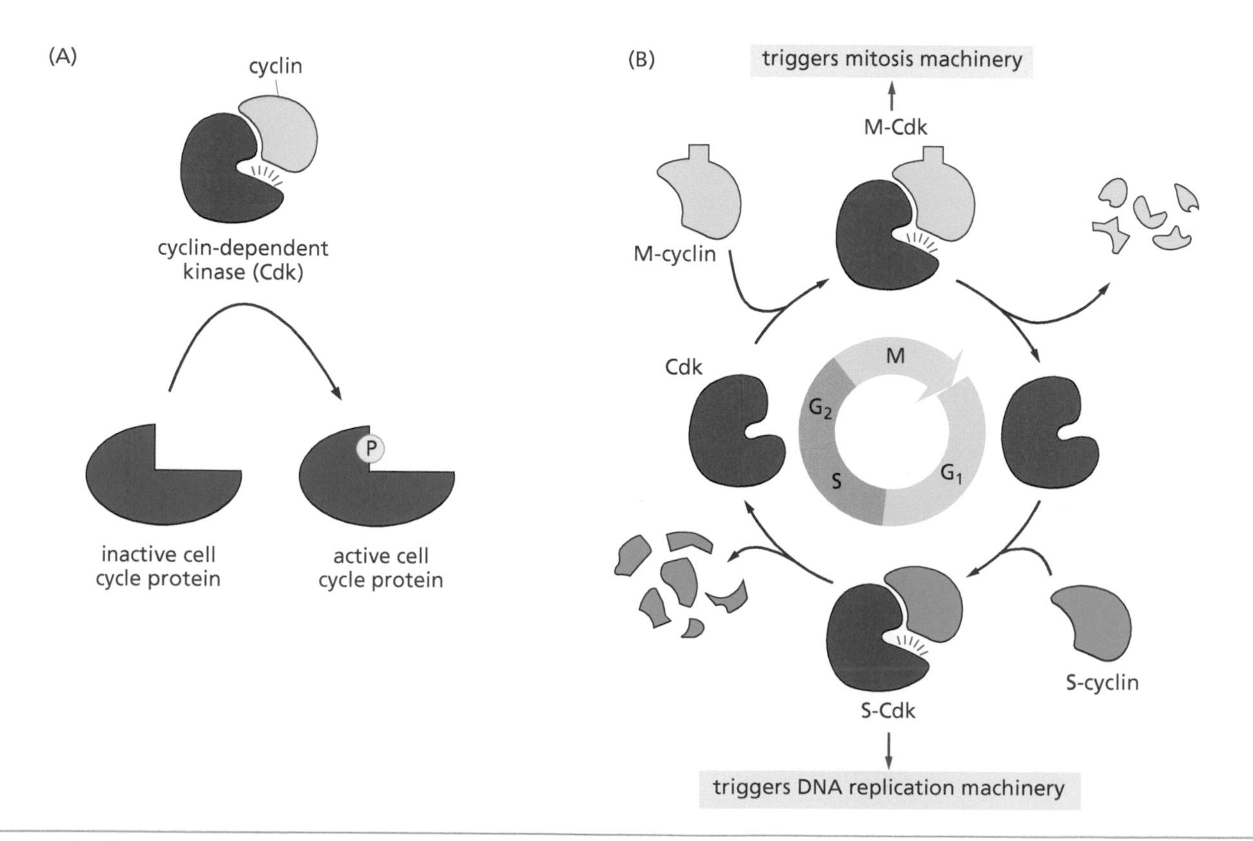

Figure 10.8 Cyclin-dependent kinases (CDKs) drive cell cycle progression. **(A)** CDKs are activated by binding to cyclins; phosphorylation of target proteins triggers cell cycle transitions. **(B)** Different cyclins accumulate and are destroyed at different times in the cell cycle, causing activation of the appropriate CDKs at the right times to trigger cell cycle events.

exit G_0 and re-enter the cell cycle; (2) there must be available space for the new cell in the tissue; and (3) the dividing cell must be attached to a surface (except in the case of blood cells, which naturally exist in liquid suspension). In adult organisms, cells do not divide unless an existing cell has died or been damaged, opening a space for a new cell. Contact with neighboring cells suppresses cell division in normal cells, a condition called **contact inhibition** (Figure 10.9A). Most cells (apart from blood cells) have an additional requirement called **anchorage dependence**: They divide only when they are attached to a surface (Figure 10.9B). In multicellular organisms, cells attach to complex organic molecules outside the cells (the **extracellular matrix** in some tissues or the basement membrane in others). When cells are isolated and grown in a laboratory **tissue culture**, they attach to the plastic dish or to a coating on the dish, forming a relatively flat layer (Figure 10.9A). A normal cell may be prevented from dividing if it loses its ability to adhere to such an external structure or if the structure changes in a way that prevents adherence.

All three criteria listed here must be met to allow a normal cell to divide, but cancer cells ignore all of these rules. Thus, in culture, cancer cells have a rounded, unattached appearance and form mounds of cells stacked one on top of each other (Figure 10.9A and C). Such mounds or heaps are called **tumors** when they form inside organisms.

10.3.3 Limits to cell division

Normal cells of most tissues have a limit to the number of times that they can divide. After a certain number of divisions, the cells stop dividing and enter a terminal "aged" state (eventually dying), even when optimal conditions exist. In the 1960s, Leonard Hayflick discovered that normal cells would only divide about 40–60 times before ceasing division in a process called **senescence**. This number of cell divisions came to be called the "Hayflick limit" and spurred further research in the field of cell aging, including the search for a biological "clock" that might keep track of the number of cell divisions.

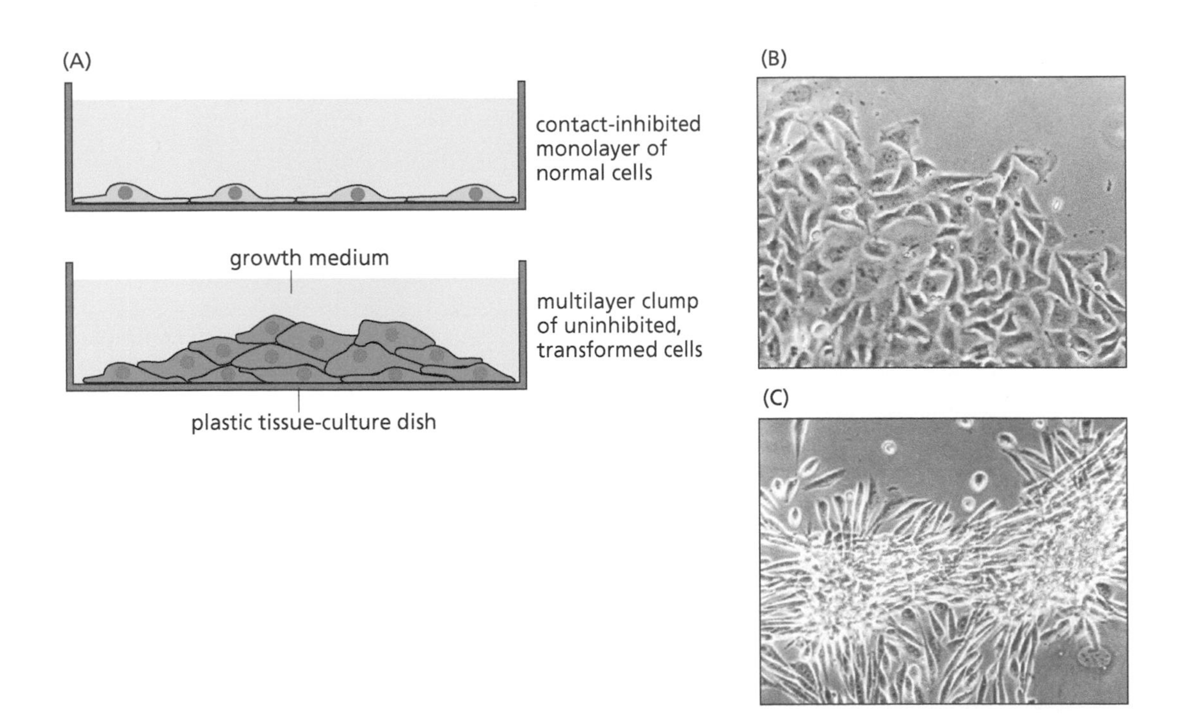

Figure 10.9 **Contact inhibition and anchorage dependence. (A)** Normal cells in a culture dish stop dividing when they have formed a complete layer, a property called contact inhibition; cancer cells continue to grow and form a multilayer clump. **(B)** Most normal cells need to be attached to a surface (culture dish or tissue) in order to divide, a property called anchorage dependence. **(C)** Cancer cells lose their attachment (rounding up) but continue to divide in an anchorage-independent manner.

In 2009, a trio of American scientists, Elizabeth Blackburn, Carol Greider, and Jack Szostak, were awarded a Nobel Prize for their studies of the molecular nature of this cellular lifespan clock (Szostak and Blackburn, 1982; Greider and Blackburn, 1985). They and others found that, in most eukaryotic cells, structures at the end of chromosomes called **telomeres** shorten with each round of cell division; once a cell's telomeres reach a threshold level of shortness, the cell no longer divides (**Figure 10.10**). Telomeres consist of thousands of repeats of a short DNA sequence (TTAGGG in humans). This repeat-containing DNA associates with a set of proteins that form a sort of "cap" that protects the chromosome ends from damage; to describe this protective function, telomeres are often compared to the plastic caps on the end of shoelaces that keep them from fraying. During DNA replication (see Section 2.3.3), one strand of a DNA molecule cannot be copied all the way to the end because of limitations of the replication enzymes. This is what causes the telomeres to shorten with each round of cell division. An enzyme called **telomerase** can extend telomere ends to counteract this shortening, thus preventing cell "aging" (Figure 10.10). Only cells in developing embryos and small populations of stem cells in organisms after birth normally express telomerase and thus avoid the aging process.

Cancer cells also escape the aging process, becoming "**immortal**," which means that they will go on proliferating indefinitely. Reactivation of telomerase expression is a major way that cancer cells achieve immortality. The immortal nature of cancer cells is reflected in the title of the book *The Immortal Life of Henrietta Lacks* by Rebecca Skloot (2010), which tells the story of an African American woman who died from an especially aggressive form of cervical cancer in 1951. Cells from the tumor that was surgically removed from Mrs. Lacks, named HeLa cells, have been proliferating in laboratories all over the world since her death. In the 1950s, patient consent regulations were not in place, so it was not until the 1970s that Mrs. Lacks's family learned of the fate of their deceased relative's cells, raising significant ethical issues.

10.3.4 Programmed cell death

In addition to the production of new cells being tightly regulated, cell death can also be a regulated process. Maintaining tissue homeostasis and a healthy organism requires the proper balance of cell division and cell death. One type of cell death, **necrosis**, occurs when a tissue is physically injured. As described

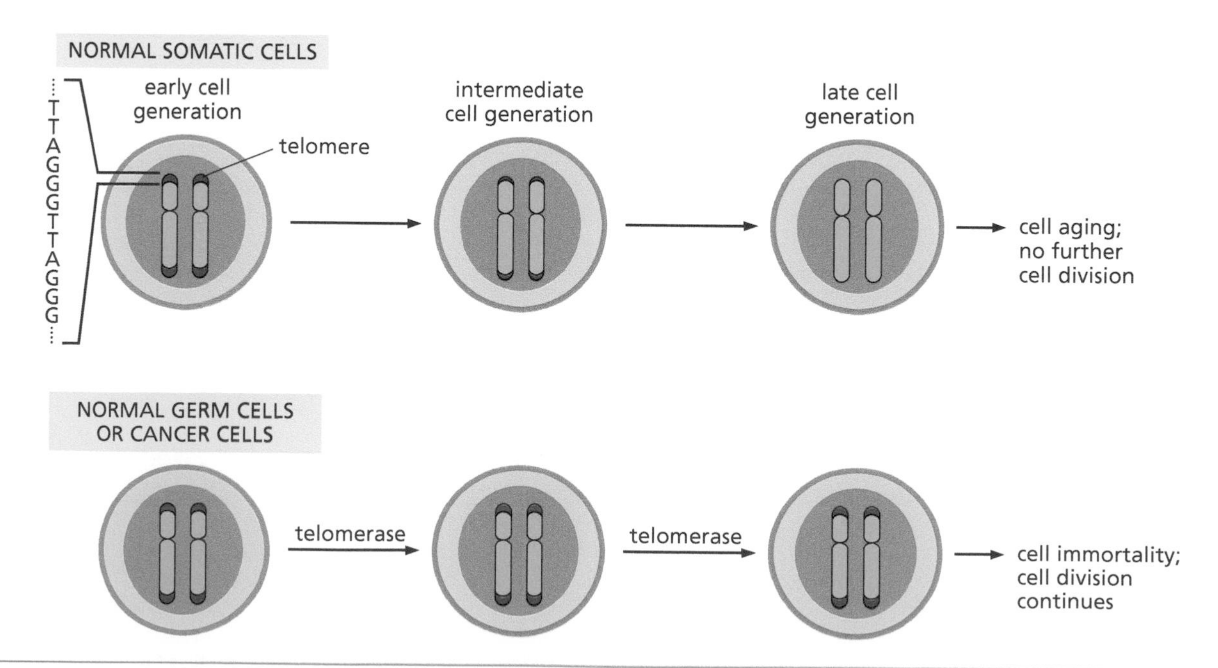

Figure 10.10 Telomeres are repeated DNA sequences (TTAGGG in humans) at the ends of eukaryotic chromosomes that shorten with each cell division. In normal somatic cells, telomere shortening with each round of cell division leads to cell aging and, eventually, lack of further division. In cancer cells, the enzyme telomerase is active and restores the telomeres, allowing the cells to continue dividing without limit.

previously, in such a case, signals are released that stimulate division of neighboring cells to replace the damaged cells. However, what if a cell undergoes internal damage, such as damage to its DNA or infection with a virus? In this case, the damaged cell itself can initiate an internal signaling cascade that causes it to die (essentially to "commit suicide") through a specialized process called **apoptosis**, or "programmed" cell death. It is called "programmed" because proteins encoded in the genome initiate the events that lead to cell death, rather than it being caused by external damage. When a cell dies by necrosis due to physical damage, its membrane usually breaks open, spilling the cellular contents into the tissue and causing inflammation. In contrast, a cell dying by apoptosis collapses inward and does not affect other cells in the tissue. Apoptosis can be initiated from within a cell without the need for any external signals, or it can be triggered by the binding of a signal molecule to a cell-surface receptor called a "death receptor." An example of the former would be when a cell sustains irreparable DNA damage that is recognized by "surveillance proteins" in the cell. An example of the latter would be a cell that is infected with a virus. A virus-infected cell will display pieces of the virus on its surface for recognition by cells of the immune system; some of these cells (called cytotoxic T cells) have surface molecules that bind to and activate death receptors on the infected cells (see Section 14.2.3). In either case (activation of internal surveillance proteins or binding of death receptors), a series of enzymes will be activated, which will cause the many changes associated with apoptosis. Most significantly, the cell's DNA will be cut up and its membrane structure will be altered such that it shrinks and shrivels up, detaching from other cells in the tissue. Specialized cells of the immune system, called **phagocytes** (literally "cell eaters") clean up the remains of cells that have died by apoptosis, internalizing the contents by wrapping a membrane around the debris and bringing the contents into internal compartments called "vacuoles," where the remains are then digested. Figure 10.11 contrasts cell death by necrosis vs. apoptosis.

Apoptosis is an essential process in normal tissues. During mammalian embryonic development, the digits of the hands and feet are initially webbed, and apoptosis removes the intervening cells, shaping the fingers and toes. In the developing brain, more neurons are formed than are ultimately needed, and a process called neural pruning eliminates the extra brain cells via apoptosis.

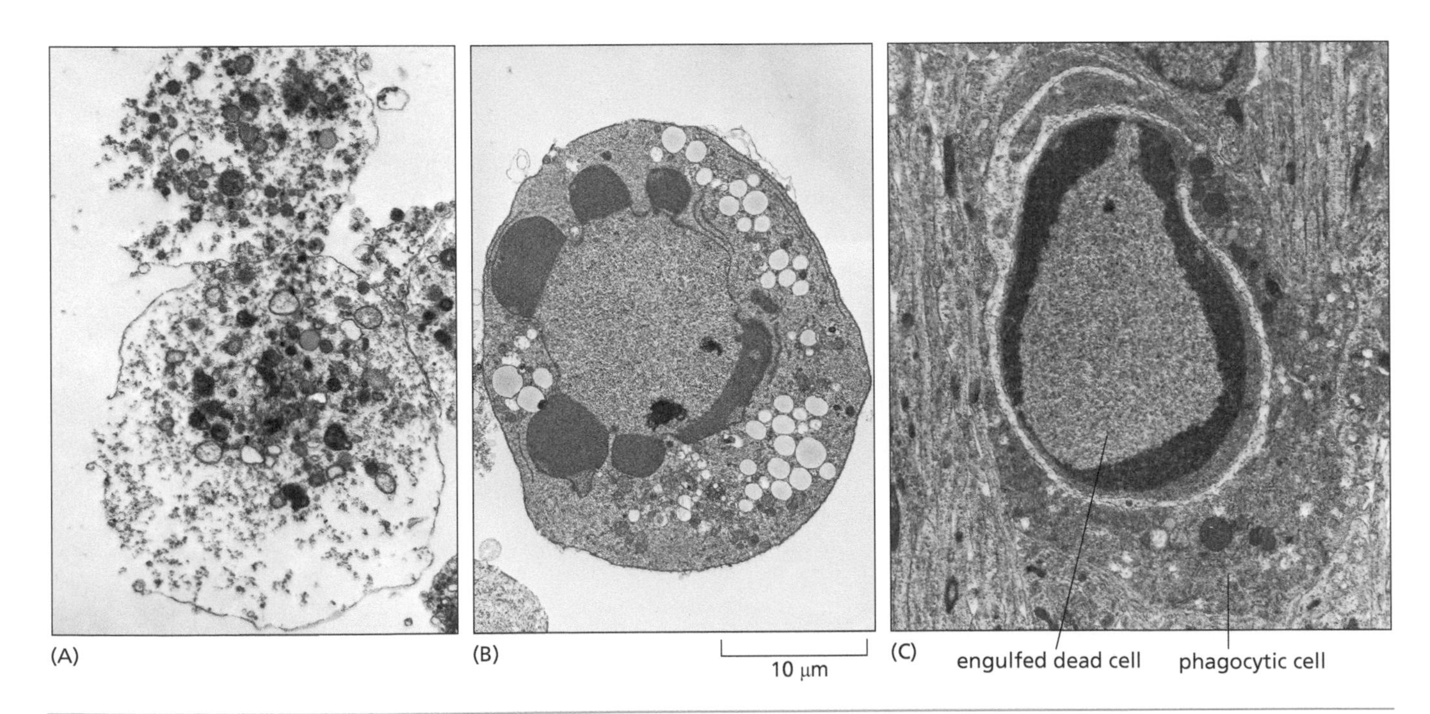

Figure 10.11 Types of cell death: Apoptosis vs. necrosis. (A) Cells that die by necrosis break open, spilling their contents into the tissue and causing inflammation. **(B)** In apoptosis, the nucleus fragments and the cell shrinks inward without bursting. **(C)** Immune cells called phagocytes engulf and digest apoptotic cells, clearing them from the tissue.

Interestingly, defects in this process have been associated with autism (Neniskyte and Gross, 2017). As mentioned previously, cells infected with viruses can be triggered to undergo apoptosis, usually when stimulated by an immune cell that can recognize the infected cell based on its expression of "foreign" (e.g., virus-induced) proteins; thus, apoptosis is a natural defense mechanism.

Lastly, apoptosis represents a powerful barrier to cells becoming cancerous. As we will learn later in this chapter, cancer occurs when a cell sustains multiple changes (mutations) to its DNA. Thus, DNA damage can promote cancer. If cells with damaged DNA die by apoptosis rather than continuing to proliferate with accumulated mutations, this risk is minimized. For this reason, defects in the process of apoptosis (due to mutations in genes that encode proteins that function in apoptosis) will promote cancer.

THOUGHT QUESTIONS

1. Why is it important for the transitions between different phases of the cell division cycle to be tightly regulated? How are these transitions regulated?
2. How would the relative proportions of cells in G_0 differ among a developing organism, an adult organism, and a tissue that has sustained damage in an adult organism?
3. In what way might research on cancer also lead to a better understanding of the aging process?
4. How is programmed cell death (apoptosis) beneficial to an organism and in what cases might it cause problems?

10.4 STEM CELLS CAN DIVIDE INDEFINITELY AND PRODUCE CELLS THAT UNDERGO DIFFERENTIATION

When a stem cell divides, it gives rise to one daughter cell that retains the stem cell identity and a second daughter cell that begins a pathway of differentiation (**Figure 10.12**). This second daughter cell will continue to divide a few more times (producing the "transit-amplifying cells" shown in Figure 10.12), but with each

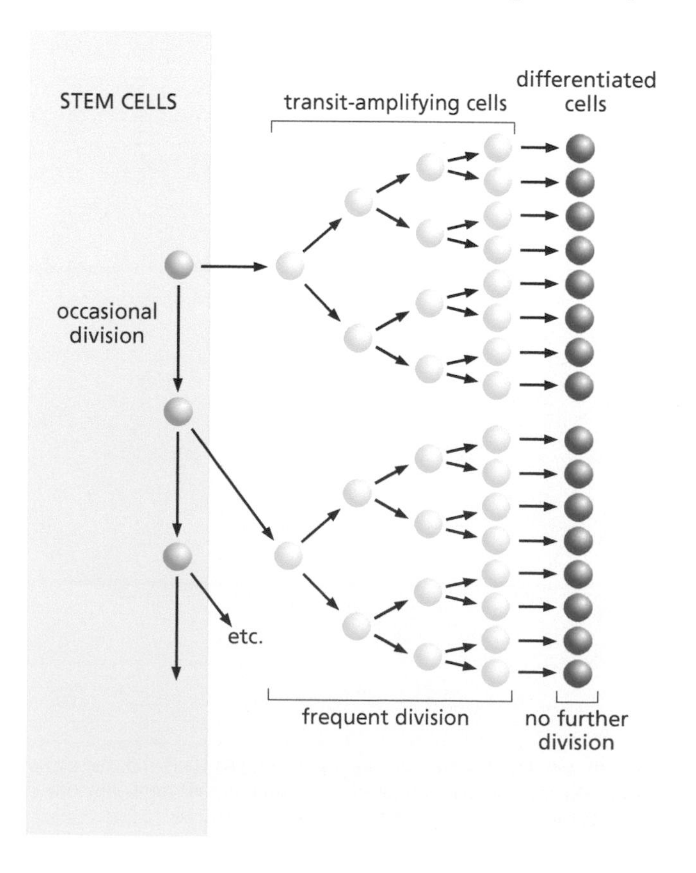

Figure 10.12 **When a stem cell divides, one daughter cell remains as an undifferentiated stem cell and the other begins the process of differentiating into a specialized, nondividing cell, passing through a stage of multiple cell divisions as a transit-amplifying cell.**

division, the daughter cells will be more specialized (more differentiated), and, eventually, terminally differentiated cells will be produced that no longer divide. In contrast, the daughter cell that remains a stem cell will retain the potential to divide indefinitely, although it will likely still only divide infrequently, except in developing organisms that require many new cells to construct all the structures of the body (Figure 10.12). As described at the beginning of this chapter, the list of possible types that a cell may become is called its **potentiality**. The zygote has maximum potentiality because it gives rise to all cell types. We will now consider the process by which this occurs and the various types of stem cells, which differ in their potentiality. We will also consider ways in which stem cells may be used to treat human diseases and ethical issues associated with their use.

10.4.1 Cellular differentiation and tissue formation

The potentiality of cells has been investigated by transplanting cells from the embryos of experimental animals. Up until the eight-cell stage in a mammalian embryo, cells are totipotent; that is, each cell could develop into a complete organism. As the cells continue to divide, they first form a hollow ball called a **blastula** (see Figure 6.16, p. 216). Cells then begin to differentiate and form tissue layers (ectoderm, mesoderm, and endoderm), a process that begins at a landmark called the dorsal lip. The differentiation process takes place with the help of certain parts of the embryos called **organizers**, which secrete signals that promote differentiation in nearby cells (Figure 10.13). As a result, cells in each layer are restricted to becoming certain types of tissues. At the stage where the embryo begins to form differentiated cell layers, it is called a **gastrula**. These early steps in embryonic development are shown in Figure 10.14A.

A group of cells removed from the ectodermal layer of the embryo at the gastrula stage and transplanted elsewhere on the same embryo can form various tissue types, but only types that are ectodermal. Their potentiality is still quite broad, but not as broad as that of the zygote. As shown in Figure 10.14B, each of the gastrula cell layers is destined to become certain types of cells. Cells transplanted later in development have a further narrowed potentiality. An ectodermal cell is restricted to one of two groups, epidermal cells or cells of the nervous system (Figure 10.14B). Finally, at a still later stage, the fate of these cells is completely **determined**, so that eye lens cells, for example, can form only eye lens tissue (Figure 10.14B).

We have seen that, as a cell becomes differentiated, its potentiality becomes restricted and that the cell somehow seems to "know" what these restrictions are. In the 1960s, a British cell biologist named John Gurdon asked the question of whether these restricted potentialities result from a loss of genes as a cell differentiates. To do this, he exposed some frog eggs (phylum Chordata, class Amphibia) to ultraviolet radiation. Because ultraviolet radiation is absorbed by DNA, a

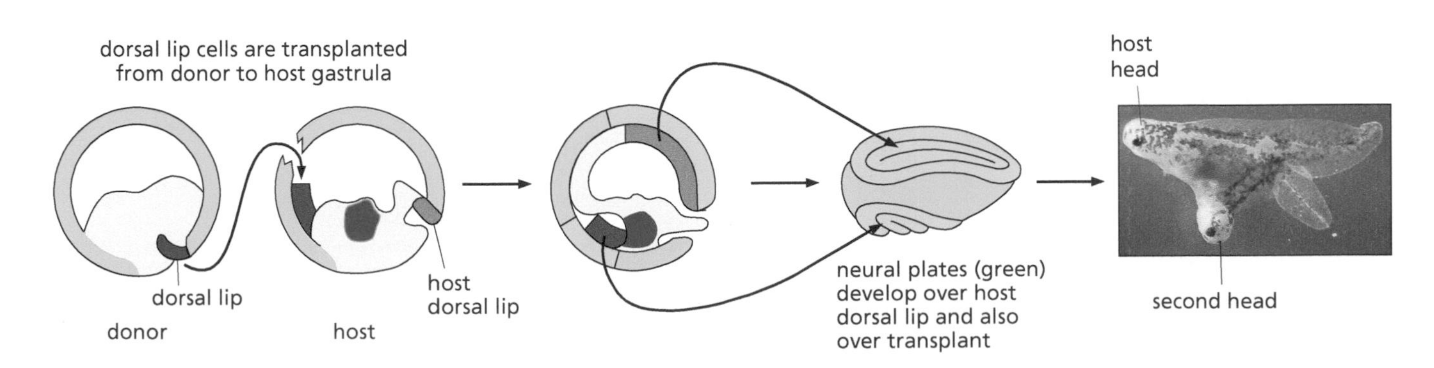

Figure 10.13 Experimentally induced differentiation. By manipulating cells from the gastrula stage of frog embryos, Hans Spemann and Hilde Mangold showed that an area called the dorsal lip served as an "organizer" to induce the formation of a nervous system. At the locations of both host and transplanted dorsal lip cells, neural plates developed and went on to form brains and other head structures. Subsequent experiments showed that the transplanted donor cells were not actually part of the induced neural plate. Rather, the dorsal lip secretes "organizer" signals that cause the overlying host cells to differentiate. Spemann was awarded the Nobel Prize for these experiments in 1935.

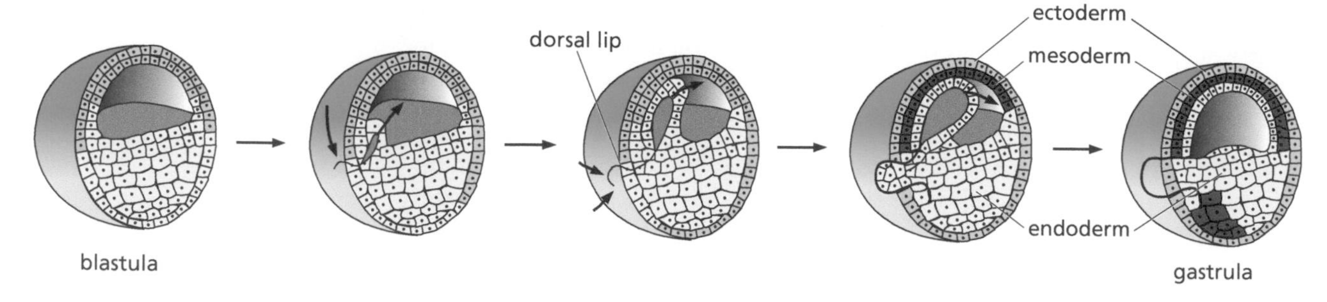

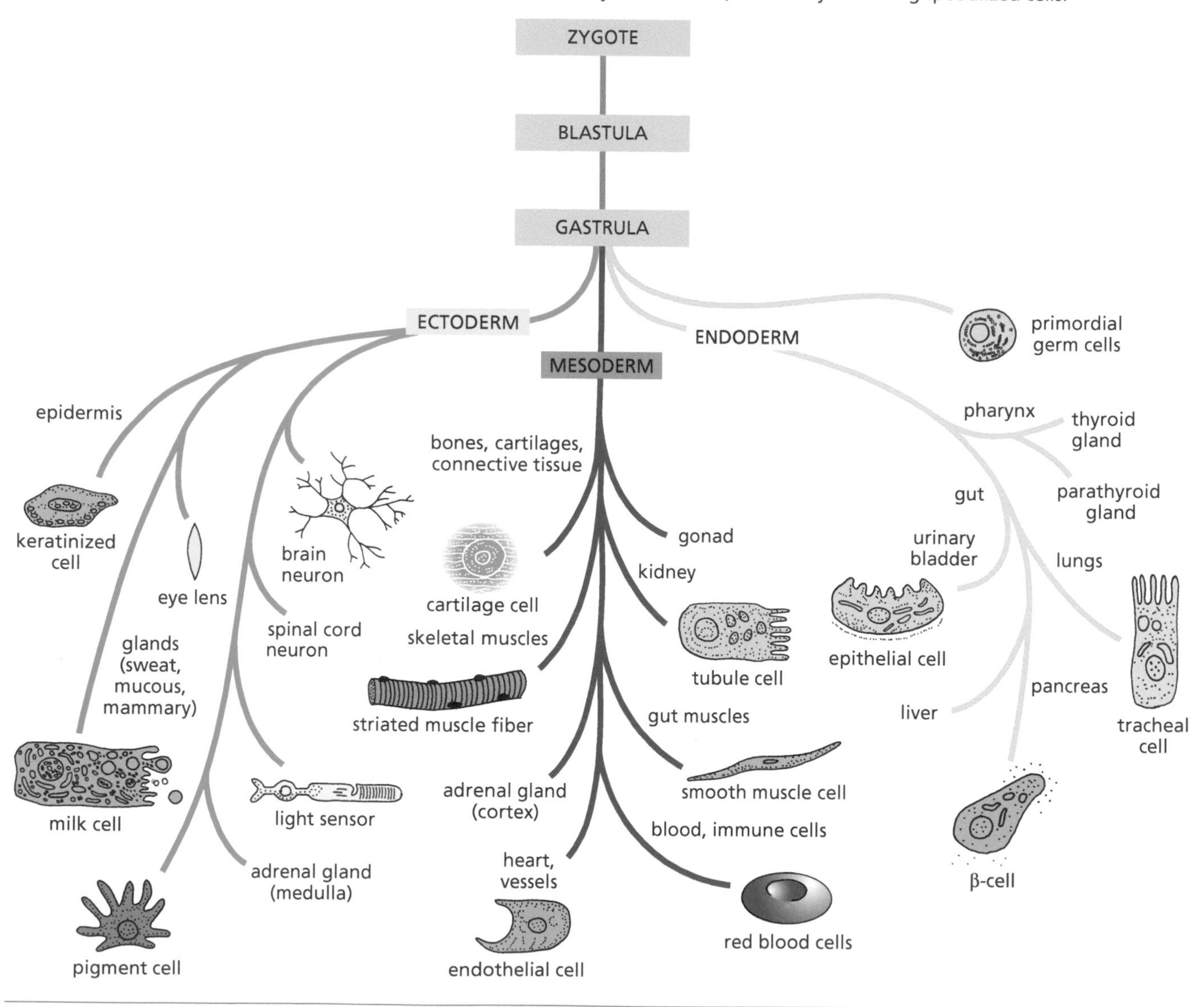

Figure 10.14 Differentiation of various cell types during embryonic development. (A) Cell layers form as a blastula develops into a gastrula, as shown here in a frog embryo. **(B)** As cells in the ectoderm, mesoderm, and endoderm divide, they differentiate, eventually becoming many different specialized cell types.

sufficient dose of ultraviolet radiation can be used to destroy the egg nucleus without damaging its cytoplasm. Gurdon then carefully inserted into each of these eggs the nucleus of a differentiated cell type, such as a skin cell. The resulting cell thus had cytoplasm from an egg but a nucleus from a differentiated cell. Gurdon was able to show that this cell, like a zygote, was able to produce an entire tadpole (**Figure 10.15**). This landmark experiment proved that the various types of cells of

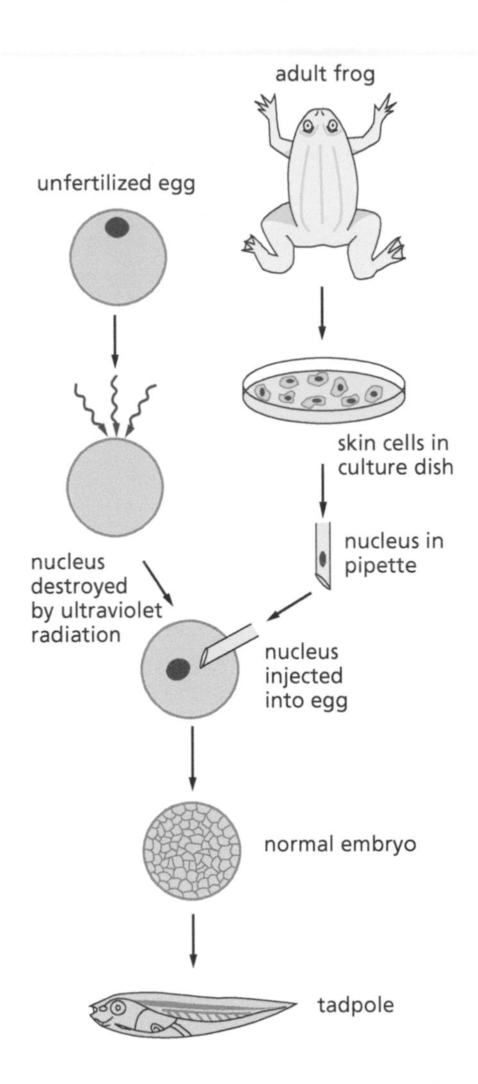

Figure 10.15 Gurdon's experiment demonstrating that a differentiated cell contains all the genes needed for the development of a complete organism. The nucleus of a frog egg was destroyed by ultraviolet irradiation and was replaced by the nucleus from the fully differentiated skin cell of another frog. The egg with its transplanted nucleus was allowed to grow, and it developed into a normal tadpole.

the body do not differ in the genes that they contain. Each nucleus usually keeps its full genome, and it is thus not the loss of genes that restricts potentiality, but rather which genes are expressed (transcribed and translated) and which are not expressed (silenced) in a given cell type (Figure 10.1).

10.4.2 Spatial distribution of signaling molecules drives differentiation

As cells in a developing organism lose potentiality and become more specialized, how do they "know" which genes to turn on and off? The answer lies in a complicated pattern of spatial signals to which different cells are exposed. At each round of cell division during development, a given cell receives chemical signals that determine whether it will differentiate and what type of cell it will form. From the very beginning, at the single-cell stage (zygote), some type of asymmetry exists that drives a cascade of differentiation, like a single domino setting off a branching chain reaction. In some organisms, the unfertilized egg is already asymmetrical due to the process by which it matures. For example, in fruit flies (*Drosophilidae*), cells called nurse cells surround one side of a developing egg in a structure called an egg chamber; these nurse cells (and other cells called follicle cells) transport certain proteins and mRNAs (which will later give rise to proteins via translation) into specific domains within the egg that will allow the two cells produced in the very first cell division to be different from one another (**Figure 10.16A**). These distinct domains within the egg define *axes of polarity* that will distinguish the head (anterior) from the tail (posterior) and the front (ventral side) from the back (dorsal side) in the developing embryo (Sander, 1975; Anderson and Nüsslein-Volhard, 1984). In some other organisms, such as frogs, a different mechanism results in the same basic outcome: the point where the sperm passes through the egg's

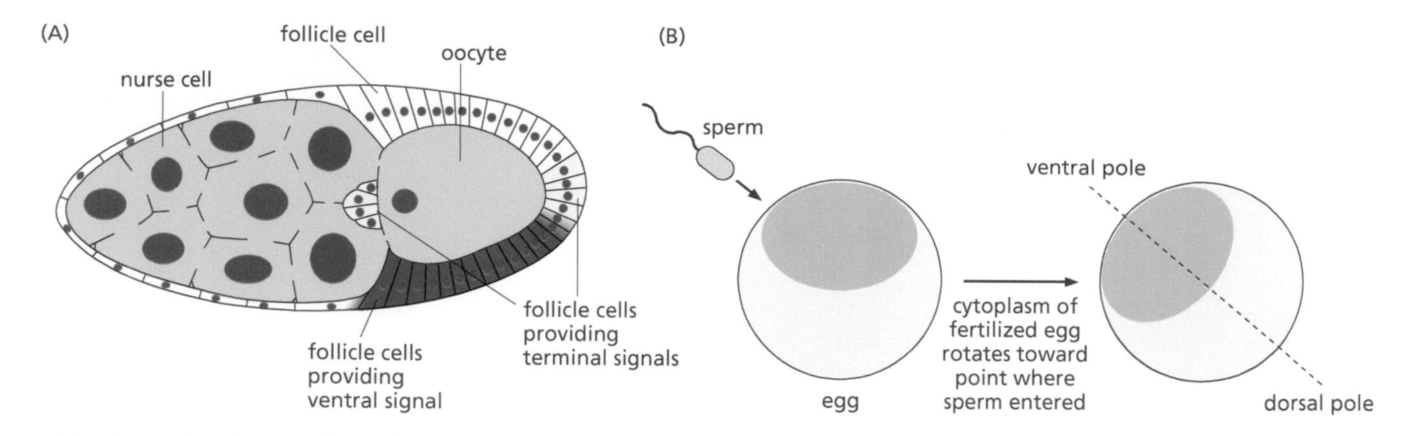

Figure 10.16 Establishing asymmetry (polarity) in a developing embryo. (A) Some eggs, such as those of fruit flies, have asymmetry even before fertilization. **(B)** In other organisms, including frogs, symmetry is established based on the position where the sperm enters upon fertilization.

membrane upon fertilization triggers reorganization of the egg's cytoskeleton and generates an axis of polarity that specifies the ventral vs. the dorsal side of the embryo (Figure 10.16B) (Gerhart et al., 1984). The mechanisms that generate polarity in mammalian embryos are more complicated, but the same requirement for asymmetry must be met at a very early stage in development.

Once an initial state of asymmetry is established in the zygote, it sets up a chain reaction that leads to the cascade of differentiation. As rounds of cell division proceed, the specific signals received by each cell turn on certain genes and silence others through signal transduction pathways like those illustrated in Figure 10.3. With many such rounds of signaling in the dividing cells, very different patterns of gene expression are established in the cells that go on to become distinct tissues. These disparate patterns of gene expression are significant for cancer because the specific type of cell that begins to proliferate out of control has a distinct pattern of gene expression that will influence the behavior of the cancer cells, including how they may respond to different treatments. Understanding the differences between cells and identifying the specific type of cell that initiates a tumor (the tumor **cell of origin**) is important in determining the best treatment plan for a particular cancer.

In the development of normal tissues, cells often do not originate in their end location; tissue formation relies on many migrating cells that travel through the organism until they find their "proper" location, where they adhere and join the tissue. These cells are generally partly differentiated at the time of their migration and become fully differentiated when exposed to growth factors in their new microenvironment. Such molecular "addresses" can take the form of membrane receptors that bind specifically to molecules expressed only on certain types of tissues. Abnormalities in cellular adhesion and cell migration are pertinent to the spread of cancer to other tissues, a process called **metastasis**, which is responsible for 90% of all deaths from cancer. We will describe metastasis in greater detail later in this chapter.

10.4.3 Stem cells and their therapeutic uses

As previously described, stem cells are (1) in an undifferentiated state, (2) able to differentiate into more committed cell types, and (3) able to renew themselves by cell division. They can be derived either from embryonic cells or from adult tissue. Stem cells can be induced to grow into many different cell types in tissue culture if given the proper signals. This has led to the idea that they might be used therapeutically when replacement cells are needed. The idea is that a person's own stem cells might be removed, isolated, grown in tissue culture, and stimulated with specific signaling molecules to cause them to differentiate into the needed replacement cell types as a cure for certain diseases. For example, insulin-producing pancreatic β-cells might be generated as a treatment for diabetes (see Section 8.4.1), or new neurons might be produced to treat Parkinson's disease (see Section 11.1.4). Stem cells may thus hold the key to a new era of regenerative medicine (Millman and Pagliuca, 2017; Parmar, Grealish and Henchcliffe, 2020).

10.4.3.1 Embryonic stem cells We have just seen how embryonic stem cells give rise to all of the tissues and organs of a new organism. Embryonic stem cells were first isolated in 1981 from early mouse embryos called **blastocysts** (Evans and Kaufman, 1981). In 1998, this process was copied in the laboratory using human blastocysts (Thomson et al., 1998). A blastocyst is a hollow, ball-shaped cluster of about 60–200 mammalian cells from an early stage in development—the stage when implantation to the uterine wall takes place (see Figure 9.12). Blastocysts maintained in culture in the laboratory for six months or more and without showing any signs of differentiating are referred to as "embryonic stem cell lines." They can be stored frozen in liquid nitrogen for later use. Under the influence of a growth medium containing specific signaling molecules (those normally produced by the "organizers" during embryonic development), they can be induced to differentiate into various types of cells.

Research is under way to determine whether embryonic stem cells transplanted into adult organisms can successfully replace damaged or degenerated tissues. Success has been demonstrated in mice, where mouse embryonic stem cells were directed to differentiate in vitro into the neurons normally defective in Parkinson's disease. When transferred to the mouse brain, the stem cell–derived neurons improved motor function (Cui et al., 2009). Human embryonic stem cells have similarly been induced to form neurons in the laboratory, but no clinical trials have yet attempted to transplant these neuronal cells into human patients, in part due to concerns that residual, undifferentiated cells could form tumors. Several clinical trials have been performed, however, in which human fetal tissue that presumably contained embryonic stem cells was transplanted into human Parkinson's patients. One Swedish study conducted in the 1990s reported remarkable benefits to some patients, but these results were questioned when two National Institutes of Health (NIH)–funded studies in the United States failed to replicate these results and also reported some significant negative side effects. Different transplantation procedures may possibly explain these different results; a new European Union–funded study is currently investigating this possibility. Additional clinical trials are also testing the implantation of induced pluripotent stem cells (iPS cells; see later). Although progress has been slow, the potential benefits of stem cell therapy continue to lend great hope to the human sufferers of many diseases, including Parkinson's disease.

Despite its tantalizing therapeutic potential, human embryonic stem cell research is a highly controversial area because it involves the destruction of a human embryo. Virtually all the existing human embryonic stem cell lines were derived from excess embryos from in vitro fertilization clinics. As we saw in Chapter 9, infertile couples can sometimes successfully conceive a child when their eggs and sperm are mixed in a dish, incubated in the laboratory, and the embryo implanted into the woman's uterus. In all cases, more embryos are created than are ever implanted. The extra embryos sometimes remain frozen and, with the couple's informed consent, can be used for research; in other cases, they are discarded. In 2001, President George W. Bush authorized the use of 64 already existing human embryonic stem cell lines for further research but declared that no U.S. federal funding could go toward developing any new lines. Many scientists were concerned by this restriction because the existing cell lines were not well characterized, and they feared that banning the development of additional stem cell lines could hamper progress on potentially lifesaving and life-transforming regenerative medicine techniques. While this ban was in place, development of new embryonic stem cell lines continued in other countries and in U.S. labs supported by nonfederal funding sources. In a bold move, the state of California passed legislation authorizing a $3 billion bond measure to fund embryonic stem cell research in 2004. Just five years later, the Bush-era ban was overturned by President Barack Obama, restoring U.S. federal funding for a broad spectrum of embryonic stem cell investigations. New restrictions, imposed in 2019, limit the use of human fetal tissue from elective abortions in federally funded research.

Most progress toward human therapy has remained incremental, but a milestone advance was reported in October 2014: Embryonic stem cell transplants into the eyes of legally blind patients suffering from macular degeneration were shown to partially restore their sight (Schwartz et al., 2014). Meanwhile, Doug Melton, a Harvard developmental biologist, decided to direct entire his research efforts toward a cure for diabetes when both of his children were diagnosed with type 1 diabetes (see Section 8.4.1). After years of work, Dr. Melton's team has succeeded in coaxing embryonic stem cells to become pancreatic β-cells with the right combination of signaling molecules, but they are still working out the best way to deliver the cells to patients and avoid their destruction by the immune system, which is a central feature of the pathology of type 1 diabetes.

10.4.3.2 Adult stem cells Stem cells do not only exist in embryos but are also found in some adult tissues and are important for normal tissue function. Very few types of cells are permanent: Some cells die and must be replaced by cell division and differentiation throughout the lifetime of the organism (Table 10.1). **Adult stem cells** are partly differentiated cells present in some tissues of adult organisms whose normal function is to divide and replace cells that are lost through routine physiological processes. These stem cells are located in several areas where cells are continually being lost: skin, gut lining, uterine cervix, bone marrow, and many glands. The division and differentiation of adult stem cells are tightly coordinated, so that cells lost from specific tissues are replaced by the correct number and type of cells.

One of the first locations in the body in which adult stem cells were identified was the bone marrow, the porous interior of the major bones. Throughout a person's life, bone marrow stem cells give rise to both red blood cells, which carry oxygen (see Section 7.3.2), and white blood cells, which are part of the immune system (see Section 14.2). New blood cells are produced to replace blood cells that have been lost through injury or have reached the end of their lifespan. Both the types and numbers of new cells produced are tightly regulated. If regulation of this process breaks down, the result can be leukemia or myeloma, which are blood cell cancers that develop in the bone marrow. Because the bone marrow contains stem cells, bone marrow transplants can reestablish blood cells and a complete immune system in individuals lacking them. Transplantation of bone marrow

Table 10.1 Average life spans of human differentiated cell types

Cell type	Life span (days)
Intestinal lining	1.3
Stomach lining	2.9
Tongue surface	3.5
Uterine cervix	5.7
Stomach mucus	6.4
Cornea	7
Epidermis: abdomen	7
Epidermis: cheek	10
Lung alveolus	21
Lung bronchus	167
Kidney	170
Bladder lining	333
Liver	450
Adrenal cortex	750
Brain nerve	27,375+ (75+ years)

from one person to another requires the precise matching of a set of inherited cell-surface proteins; otherwise, the recipient's body may reject the transplanted cells, which can lead to a life-threatening complication called graft-versus-host disease. In some cases, a person's own bone marrow cells can be removed and later transplanted back, for example, after the person's own immune cells have been killed by cancer therapy. This works very well because a person's own cells will not trigger graft-versus-host disease.

Although adult stem cells are not as undifferentiated as embryonic stem cells, they are still capable of forming many types of cells in addition to the type found in the tissue from which they were derived. In tissue culture, bone marrow stem cells can develop into other types of cells, including muscle or nerves, given the right sets of signaling molecules. Transplants of these adult stem cells are being investigated for their future potential to regenerate other types of tissue in addition to blood cells. Experiments in tissue culture suggest that it may be possible to bring adult stem cells back to full potentiality, that is, the ability to differentiate into any and all kinds of cells. This is not yet known for certain, however.

10.4.3.3 Induced pluripotent stem cells Because of the ethical issues regarding embryonic stem cells and the possible limitations of adult stem cells (including difficulties obtaining them from some adult tissues), scientists have sought ways to generate cells with stem-like properties. In 2006, a Japanese biologist, Shinya Yamanaka, succeeded in "reprogramming" skin cells from adult mice, returning the cells to a pluripotent state by engineering them to express just four transcription activator proteins normally expressed in embryonic stem cells but not in skin cells. The following year, Yamanaka and colleagues repeated the accomplishment using human skin cells (Takahashi et al., 2007). The reprogrammed cells are called "induced pluripotent stem cells," or iPS cells. Like embryonic stem cells, if given the proper signals, iPS cells can develop into a wide range of cell types. A major benefit of this strategy for regenerative medicine is that, theoretically, any cell type needed could be generated from a patient's own skin cells, which can be easily obtained and not provoke any immune reactions. In 2010, Yamanaka and British biologist John Gurdon (mentioned earlier in this chapter) were jointly awarded the Nobel Prize in Physiology or Medicine for their discoveries regarding the reversible nature of the differentiation process.

It is not yet possible to know for sure if iPS cells have a potentiality as broad as embryonic stem cells, as the signals required to specify all cell types have not yet been identified. However, in 2009, mice were born from iPS cells using a technique called tetraploid complementation. These embryos had a low rate of implantation in recipient mothers, and some had physical abnormalities, but some did survive and go on to reproduce, suggesting that at least some of the induced pluripotential stem cells were able to give rise to all necessary cell types. Such an experiment would not be considered ethical in humans.

10.4.4 Ethical and scientific questions

The ethical questions raised by stem cell research center mostly around the derivation of the embryonic stem cells. Among the opponents of stem cell research are deontologists who argue, often from a religious perspective, that embryos should be given the moral status of a human being and not be destroyed even in the interest of scientific research. Other people hold that humanness develops very late in gestation (see the discussion in Section 9.3.3), making the use of a blastocyst ethically acceptable. Many proponents of stem cell research share this view, including many scientists, patients, and people involved in the biotechnology industry.

Interestingly, healing and the promotion of health are part of all religious traditions. Most people see the use of adult stem cells for therapy as a potential good, and utilitarian ethical approaches would weigh the benefits of any such therapy against any harm done in the procurement of the stem cells. We do not yet know whether adult stem cells will have the full potentiality of embryonic stem cells, so many people are reluctant to restrict the research only to adult stem cells. Some, though certainly not all, people of various religions have accepted in vitro fertilization as a method of infertility treatment, and people embark on such treatments knowing that more embryos will be created than are used. Ethicists from most major religions have accepted the use of the excess embryos for stem cell research. The question of whether it is acceptable to create an embryo expressly as a source of stem cells, either though in vitro fertilization or therapeutic cloning, again usually depends on a person's view of whether or not an embryo is considered to have full human status from the point of conception. This varies among religions and religious sects. Several ethicists also oppose therapeutic cloning on the grounds that the embryo is being used "as a means only."

Apart from religious or moral objections, other people urge caution in pursuing stem cell therapies because there are so many still unanswered scientific questions about these processes and whether there might be unintended consequences. For example, some scientists wonder whether transplanted stem cells could become cancerous cells in the recipient. In the next sections, we will examine the characteristics of cancer cells and their similarities to stem cells that make some scientists cautious.

? THOUGHT QUESTIONS

1. Why is it essential to have some type of asymmetry in the zygote or very early embryo? How can such asymmetry be established?
2. Explain the following statement: "Both embryonic development and transformation of normal cells into cancer cells rely primarily on changing patterns of gene expression."
3. What are the arguments for continuing to develop new embryonic stem cell lines when adult stem cells and iPS cells can also give rise to various tissues? What are the ethical arguments for and against embryonic stem cell research?
4. Apart from ethical concerns that may favor iPS cells over embryonic stem cells, why else could iPS cells be a particularly good source of cells to provide a certain cell type needed by an individual patient?

10.5 CANCER RESULTS WHEN CELL DIVISION IS UNCONTROLLED

Now that we have seen how signal transduction, cell division, and differentiation are controlled in normal cells, we can examine these processes in cancer cells. Cancer is not just excessive proliferation of otherwise normal cells; cancer cells have escaped from many types of controls that operate in normal cells and exhibit a wide range of abnormal behaviors. Cancers can arise in any tissue whose cells are dividing, and all multicellular organisms can develop cancer. In this section we primarily discuss human cancers, although much of what follows also applies to cancer in other species, and animal cancer models are crucial for understanding human cancers and developing new ways of treating them.

10.5.1 Properties of cancer cells

Cancer can be described as "a genetic disease that arises over a person's lifetime." This is because cancer begins when a *single cell* (the **cell of origin**) accumulates enough mutations (DNA changes) in its genome to break down the normal regulatory mechanisms that enforce its "good behavior" in the tissue (or cell community) of its origin. Thus, development of cancer is an evolutionary process that happens within the population of cells that makes up an individual, instead of a population of individuals. This change from being a normal, "law-abiding" cell to being a "rogue" cancer cell is called **transformation**.

Transformed cells grow and divide more rapidly than normal cells, even in the absence of any cell division signals; they also exhibit many other characteristics that differ from those of normal cells (Table 10.2). Cancer cells continue to divide indefinitely and are therefore called "immortal." As in the case of the HeLa cells mentioned earlier, many cancer cell lines have been maintained in tissue culture for decades. In many cases, cancer cells are less differentiated than the cells from which they arose. The membrane transport systems of transformed cells carry

Table 10.2 Characteristics of normal cells and transformed (cancer) cells

Cellular behavior	Normal	Transformed
Limit to the number of cell divisions	Finite	Infinite (immortal)
Differentiation	Present	Inhibited
Transport of nutrients across cell membrane	Slower	Faster
Growth factors required for division	Yes	No
Contact inhibition	Present	Absent
Requirement for attachment	Present	Absent
Adhesiveness	High	Low
Secretion of protein-degrading enzymes	Low	High
Genetic material	Stable	Unstable

nutrient molecules into the cell at a higher rate. In the body, this gives transformed cells a competitive advantage over normal cells. Transformed cells are not inhibited by contact with other cells. In tissue culture, their growth does not stop when they have formed one-cell-thick monolayers, but instead continues, forming piles of cells growing over and on top of each other (Figure 10.9A). Cancer cells also grow this way inside organisms, forming piles of cells called **tumors**. Transformed cells also grow without the need to be attached (Figure 10.9B); in fact, this is the characteristic that best predicts whether a cell growing in culture will form a tumor if put into an animal. Changes back and forth from attached growth to unattached growth are believed to spread some tumors to new locations.

Cells become transformed when they are dividing; therefore, cells that are terminally differentiated and will never again divide, such as nerve cells in the brain and muscle cells of the heart, cannot become transformed. Instead, cancers arise from the transformation of stem cells or their immediate descendants that may have begun but not completed the differentiation process. Recall that when a stem cell divides normally, one daughter cell retains the stem cell identity and remains undifferentiated, while the other divides multiple times to generate a population of cells called "transit-amplifying cells," which will ultimately differentiate (Figure 10.12). Moreover, in an adult organism, normal stem cells divide relatively rarely, only as frequently as required to replace cells that have died. A transformed stem cell, however, will divide more frequently and may give rise to cells that do not follow the normal rules of differentiation, instead remaining in an undifferentiated state and continuing to proliferate. The partially differentiated products of stem cell division, the transit-amplifying cells, may also be the cells of origin for some cancers if they acquire mutations that prevent their further differentiation and give them certain other features of cancer cells, such as excessive proliferation, decreased apoptosis, and immortality, or failure to undergo senescence (aging). Regardless of whether the cell of origin for a tumor is a stem cell or a more differentiated cell, it must achieve immortality to become fully transformed, which involves acquiring the ability to extend the ends of its chromosomes (telomeres) to avoid senescence. This usually involves reactivation of expression of telomerase, but some cancer cells find other ways of doing this.

Interestingly, not all cells in a tumor are equivalent; a small subset of tumor cells has the potential to continue propagating the cancer or to found new tumors, while the rest of the tumor cells do not. The former cell population is referred to as "cancer stem cells." The unique properties of cancer stem cells are not yet understood, but some researchers seek to specifically eradicate cancer stem cells in the hope that this may be a much more effective treatment strategy than trying to kill cancer cells in general. One concept of experimental therapy for some cancers is to give drugs that promote differentiation to the terminal, nondividing state. In the other direction, one of the possible problems resulting from stem cell transplants is that the normal transformation-protective process may not operate in these cells and the cells meant to be therapeutic may divide without control and thus give rise to cancer.

10.5.2 The genetic basis of cancer

Some genetic diseases are caused by mutation of a single gene, and a person who inherits two mutated copies of the gene (or sometimes just one, depending on the disease) will be nearly guaranteed to have the disease. An example of such an inherited disease is cystic fibrosis, which is the most common inherited disease among Caucasians. In contrast, most mutations that cause cancer arise in somatic (body) cells during a person's lifetime instead of being passed down from a person's parents. A person can inherit a predisposition to developing cancer, which we will discuss later. However, first, we will consider the types of genes involved in cancer.

The normal growth regulatory genes fall into two categories: genes that encode proteins that promote cell division (**proto-oncogenes**) and genes whose protein products normally inhibit cell division (**tumor suppressor genes**). The proto-oncogenes can be thought of as the "gas pedal" of a car and the tumor suppressor genes as the "brakes." In addition to directly promoting cell division, proto-oncogenes can also encode proteins that inhibit cell death or differentiation (both processes that counteract cell proliferation), and tumor suppressor genes can encode proteins that promote these processes. Mutations in either type of regulatory gene increase the probability that cancer will arise.

10.5.2.1 Oncogenes and proto-oncogenes Proto-oncogenes include normal genes that encode components of the signal transduction pathways that promote cell division, such as growth factors, receptors, and internal signaling proteins, as well as the cell cycle proteins such as cyclins that respond to those signals (Figure 10.17). When a proto-oncogene becomes mutated such that its gene product is hyperactive (like a stuck gas pedal), it becomes a cancer-promoting

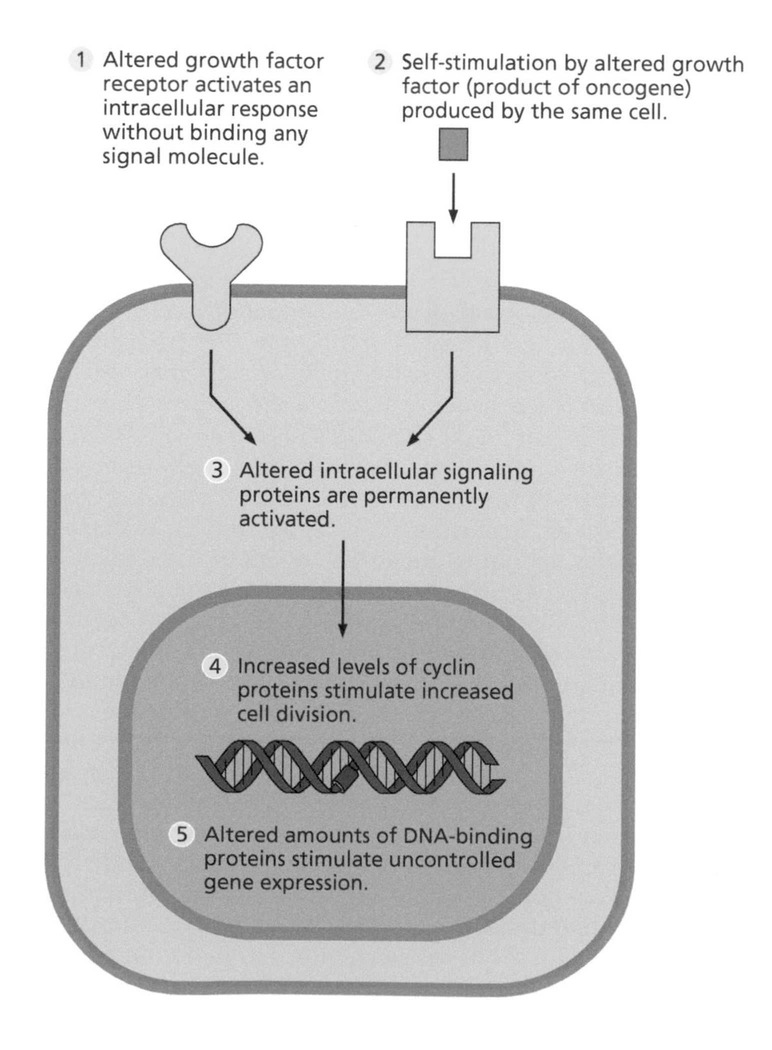

Figure 10.17 Five different ways in which the mutated protein products of oncogenes can cause continuous signaling of cell division in transformed cells.

oncogene. American cell biologists J. Michael Bishop and Harold Varmus received the 1989 Nobel Prize for their discovery of the relationship between proto-oncogenes and oncogenes, leading to a new era in our knowledge of both cell division and cancer (Stehelin et al., 1976a,b).

Oncogenes can differ from proto-oncogenes in any of three basic ways: the timing of their expression, the level of their expression, or the structure of their protein products (Figure 10.17). For example, the expression of a proto-oncogene might be silenced by a transcription repressor protein unless conditions are right for cell division, whereas the oncogene might be expressed all the time due to a mutation in the repressor binding site. Another type of mutation results in an increased number of copies of a proto-oncogene in the genome (gene amplification), which in turn leads to more protein being expressed and therefore more activity. Mutations to gene regulatory sequences can also result in increased expression. Finally, a mutation that alters the protein product of an oncogene by as little as a single amino acid compared to the protein encoded by the corresponding proto-oncogene can sometimes be enough to hyperactivate the protein, eliminating normal regulatory mechanisms. This is true for the most common mutation in the Ras protein, an internal signaling protein that was one of the first oncogene products discovered and is implicated in about 25% of all human cancers. Whereas the normal Ras protein can turn itself off, the single amino acid change in the oncogenic form of Ras keeps it stuck in the "on" state, constantly sending the signal for a cell to divide.

While the most common cancer-causing mutation in the *Ras* gene is very subtle (a change of a single letter in the DNA), other types of mutations can also generate oncogenes. Genes moved to another chromosome or another part of a chromosome due to abnormalities in crossing over (see Section 2.2) can also create oncogenes by placing proto-oncogenes in new locations within the genome. In its new location, a gene may no longer be under the control of the regulatory elements that operated at its previous location in the genome. In rarer circumstances, an abnormal DNA crossover may cause the coding sequences of two genes to fuse, creating a new, hybrid gene that encodes a novel protein with hyperactive growth-promoting properties. This is the case for the *bcr-abl* gene fusion that is implicated in a particular type of leukemia. Interestingly, the drug Gleevec (see Figure 4.17), designed specifically to inhibit the hybrid Bcr-Abl protein, was one of the first successes of so-called "designer" cancer drugs, as discussed later in this chapter.

10.5.2.2 Tumor suppressor genes

The protein products of tumor suppressor genes normally repress cell division or promote cell death (apoptosis). Mutations that inactivate these genes will promote cell transformation. In most cases, *both* copies of a tumor suppressor gene must be inactivated, resulting in the complete loss of function of that gene. This is different from the case for oncogenes. Because the types of mutations that convert a proto-oncogene to an oncogene result in a gain of function (hyperactivity), a single mutated copy is sufficient to promote cancer. In genetic terms, tumor suppressor mutations are recessive (two mutant copies needed for the effect to be observed), while oncogene mutations are dominant (only one mutant copy needed to affect the outcome) (**Figure 10.18**).

One crucial tumor suppressor gene, *p53*, is mutated in as many as 55% of human cancers. The p53 protein has been given the nickname "guardian of the genome." When something goes wrong inside a cell, such as the DNA being damaged, the normal p53 protein delays cell cycle progression by activating a cell cycle checkpoint (see Figure 10.7); this gives the cell time to try to fix the problem (that is, to repair the damaged DNA). If the problem is severe enough that the damage cannot be repaired, p53 eventually causes the damaged cell to die via apoptosis. When the *p53* gene is mutated, the altered p53 protein does not halt cell division, and apoptosis is not triggered; therefore, cells with damaged DNA continue to live and divide, passing on their accumulated mutations to their progeny cells. Some of these cells may be so damaged that they will not survive, but others will have acquired additional mutations that promote the evolutionary

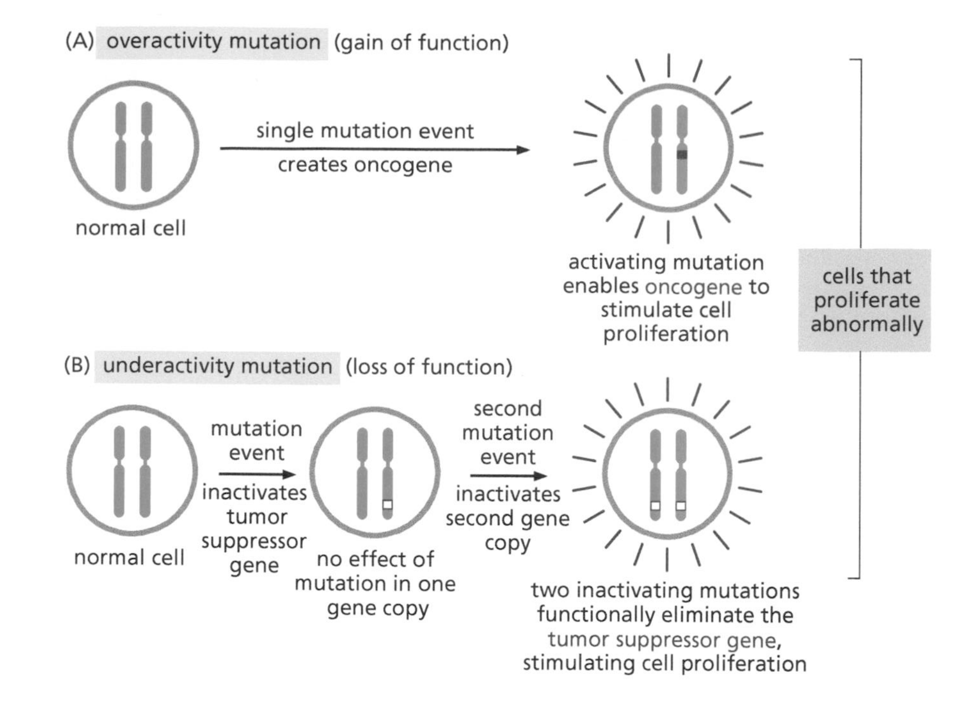

Figure 10.18 Only one oncogene mutation but two tumor suppressor mutations are required to promote cancer. Thus, oncogenes show dominant genetic behavior, while tumor suppressor mutations show recessive genetic behavior.

process of cancer progression. Many cancer therapies rely on damaging cells such that apoptosis is triggered. Because the p53 protein is so important for signaling apoptosis to occur, the loss of p53 function in many cancers represents a challenge for cancer treatment. In fact, p53 is so central to cancer biology that the status of the *p53* gene in a patient's tumor cells is often ascertained by DNA sequencing to inform treatment plans.

10.5.3 Accumulation of many mutations

Fortunately for us, transformation of cells requires a combination of mutations in several proto-oncogenes and tumor suppressor genes rather than a single mutation. If single mutations were able to cause cancer, the overall likelihood of developing cancer would be much higher, and the rate of incidence for new cancers would be the same for individuals of every age. That is clearly not the case for most cancers; rather, rates of most cancers increase exponentially with age (**Figure 10.19**). Based on mathematical modeling of cancer incidence as a function of age, it is estimated that five or six cancer-promoting mutations must occur *in a single cell* before it becomes transformed to a cancer cell. There are a few exceptions that can be explained by the unique biology of the affected tissues. For example, retinoblastoma, a tumor of the eye, occurs most often in children five years and younger and

Figure 10.19 Cancer risk increases strongly with age.

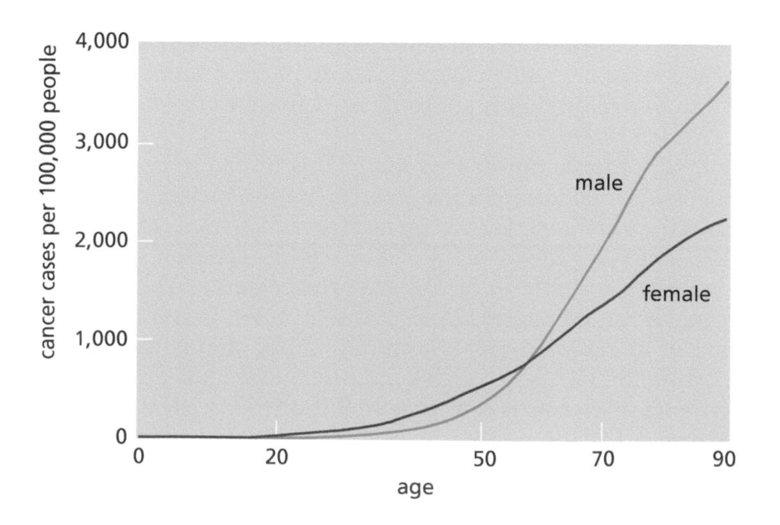

only very rarely in adults. This is because the cells of the eye's retina become terminally differentiated and stop dividing after this point. Another type of cancer with an unusual age dependence is testicular cancer. It is most common in males aged 15–40 years. This pattern is likely explained by the effects of male sex hormones, such as testosterone, on the progression of this particular cancer.

Most cancers arise in somatic cells (body cells), rather than in gametes (eggs or sperm), and are referred to as **sporadic** rather than heritable cancers. Somatic mutations are passed along to the progeny cells in that individual but are not passed on to the individual's offspring. Some types of cancer do run in families, but an individual does not actually inherit cancer per se; rather, they inherit a predisposition to develop cancer. In these hereditary, or familial, cancers, a mutated cancer-causing gene is passed on to sons or daughters through the gametes, and the mutated gene is therefore present in all cells of the body. This means that all cells are one step closer to accumulating the set of mutations needed to become fully transformed and to progress to cancer. Hereditary cancers are usually associated with inheritance of one inactivated copy of a tumor suppressor gene. This increases the likelihood that both copies of that gene will become inactivated in some cell in the body and that the function of that tumor suppressor gene will be lost entirely, thus directing that cell down the pathway of transformation. It is rare that an individual will inherit two inactive copies of a tumor suppressor gene, as the functions of many of these genes are required for normal embryonic development. One common familial cancer is breast cancer associated with inheritance of a mutated copy of either the *BRCA1* or *BRCA2* gene. Inherited mutations in these genes are more common in families of Ashkenazi Jewish descent than in other populations, so such individuals should be especially vigilant in screening for breast cancer. Another less common example is families that pass down a mutated copy of the *p53* gene. This is associated with a familial cancer susceptibility called Li-Fraumeni syndrome, in which family members have an increased risk of developing many different types of cancer.

What causes the mutations that generate oncogenes from proto-oncogenes or inactivate tumor suppressor genes? One source of mutations comes from the natural process of copying DNA every time a cell divides. The enzyme that copies DNA, DNA polymerase, makes a mistake about every 10^5 nucleotides. Fortunately, most mistakes are quickly fixed, either by the inherent "proofreading" activity of DNA polymerase itself or by other "spell-checking" proteins that can identify improper base-pairing (**Figure 10.20**). These mechanisms lower the error rate of DNA replication to only about one mistake in 10^{10} nucleotides copied. However, given that an estimated 10^{16} cell divisions occur in a human lifetime, this low rate of mutations from DNA replication cannot be ignored. An even more important source of mutations comes from environmental exposure to chemicals and radiation that can react

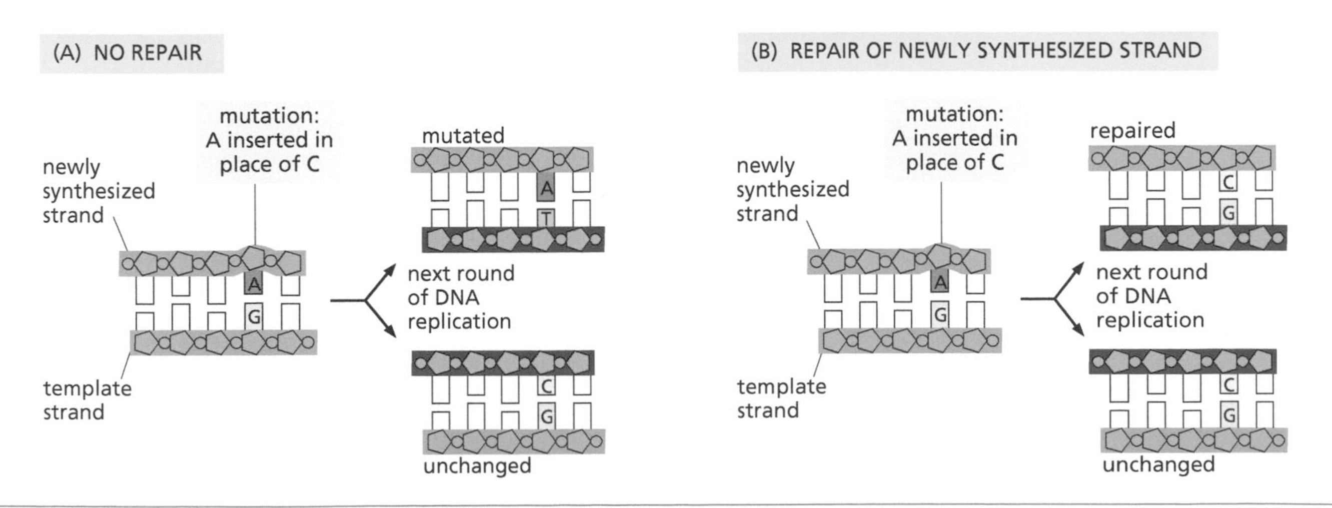

Figure 10.20 DNA mismatch mutations and their repair. (A) If a mismatch occurs during replication, it can lead to a permanent mutation if not corrected on the new strand. **(B)** A mismatch repair protein (spell-checking protein) checks for mismatched bases on the new strand as synthesis proceeds. If a mismatch is detected by the shape of improperly bonded nucleotides between the strands, part of the new strand is trimmed back before synthesis can restart and allow the correct base to be inserted.

chemically with DNA bases or cause physical damage to DNA. Such chemicals and radiation that can alter DNA are referred to as **mutagens**. The cell's "spell-checking" enzymes and other types of DNA repair proteins can help to minimize damage from mutagens, but some mistakes slip through the cracks. The genes that encode proteins that repair single-nucleotide changes and other types of DNA damage are one class of tumor suppressor gene. Importantly, if both copies of a DNA repair gene become inactivated due to mutations, the overall rate of accumulation of mutations will increase, and the risk of cancer will be elevated. The *BRCA1* and *BRCA2* genes that are implicated in familial breast cancer are two examples of DNA repair proteins with important tumor suppressor function.

Mutations caused by DNA replication errors or mutagens occur randomly throughout the genome. To contribute to transformation, uncorrected mutations must be in a proto-oncogene or a tumor suppressor gene. Because these genes represent a tiny fraction of the whole genome, the probability that one of them will receive an uncorrected mutation is low. Cells that suffer a significant level of DNA damage will be signaled to die by apoptosis (if the function of the p53 protein is intact). Additionally, for cancer to arise, recall that mutations in at least five to six growth regulatory genes must all accumulate in the same cell. Therefore, the barriers to cancer development are robust; nonetheless, the cancer rate statistics indicate that these barriers are surmounted at some point in many people's lifetimes.

10.5.4 Progression to cancer

We see the characteristics of transformed cells when we study them at the cellular level. However, cancer occurs in whole, multicellular organisms, not in isolated cells. As mentioned earlier, an organism can be considered an ecological system of many billions of cells. The interacting growth and differentiation signals keep the cellular ecosystem stable. After cells are transformed, progression to a tumor depends on many ecological factors. Mutated progeny cells may be killed (often by cells of the immune system), or they may not outgrow the normal cells and thus never progress to a tumor. Alternatively, the transformed cell and its progeny may continue to divide, taking up space and nutrients required by their neighbors, passing on their mutations to each new progeny cell, and usually accumulating even more mutations. Normal cells begin to die off, not because they are killed outright by cancer cells, but because they are deprived of space and nutrients. With the decline in the number of normal cells comes a reduction of their normal function, and the organism begins to show signs of illness. The exact symptoms depend on the type of cancer and the types of normal cells that are lost.

Transformed cells within organs may form solid tumors within those organs (Figure 10.21, middle right). For a tumor to be visible by X-ray, the original transformed cell must divide repeatedly until there are about 10^8 cells in the tumor. For a tumor to be large enough to be felt (about 1 cm in diameter), approximately 10^9 cells are needed. Long before tumors are this size, they have begun to influence their environment. One major influence they exert is to stimulate the growth of new blood vessels, a process called **angiogenesis**. The tumor cells secrete signals called angiogenic growth factors that induce nearby blood vessels to develop new branches that grow into the tumor. This is crucial for continued tumor growth, as all cells require a constant supply of oxygen from the blood, and the distance limit for oxygen diffusion through tissues is only about 0.2 mm. Thus, tumors contain other types of cells besides cancer cells, including the cells that make up blood vessels as well as connective tissue cells and various types of immune cells. In recent years, much research has focused on the interactions between cancer cells and other, normal cell types that are intermixed with cancer cells in tumors. Together, the various types of cells in a tumor define the tumor microenvironment, the features of which can significantly affect the rate of tumor growth and its progression to increased stages of disease. Dampening pro-growth signals in the tumor microenvironment is one strategy for developing new cancer drugs. For example, one class of cancer drugs inhibits the process of angiogenesis, with the goal of restricting a tumor's blood supply and essentially suffocating it.

A tumor is said to be **benign** if it is contained in one location and has not broken through the **basement membrane** to which normal cells are attached. Benign tumors, as their name suggests, often cause no health problems for the individual. Benign tumors can become large enough to interrupt the functioning of normal tissues, but their removal by surgery is generally successful because they have not intermingled with normal tissue. Tumor cells that invade normal tissues, rather than just pushing them out of the way, are said to be **malignant** (Figure 10.21, bottom). For cells to be invasive, they must produce protein-degrading enzymes such as collagenase, an enzyme that dissolves the collagen connective tissue that holds groups of cells together. The term "cancer" is generally reserved for malignant tumors.

Because malignant tumors produce enzymes that allow them to invade other tissue, they often spread to new locations in a process known as **metastasis**. In this process, one or more of the transformed cells lose their attachment to the other cells of the tumor, break through the basement membrane, and spread via the circulation to other areas of the body (Figure 10.21, bottom right). In the new location they regain attachment and continue to divide, forming new tumors. The new tumors are of the same type as the original tumor and thus when viewed with a microscope are seen to be different from the cells around them. Cancers that have begun to metastasize are far more serious and more resistant to treatment

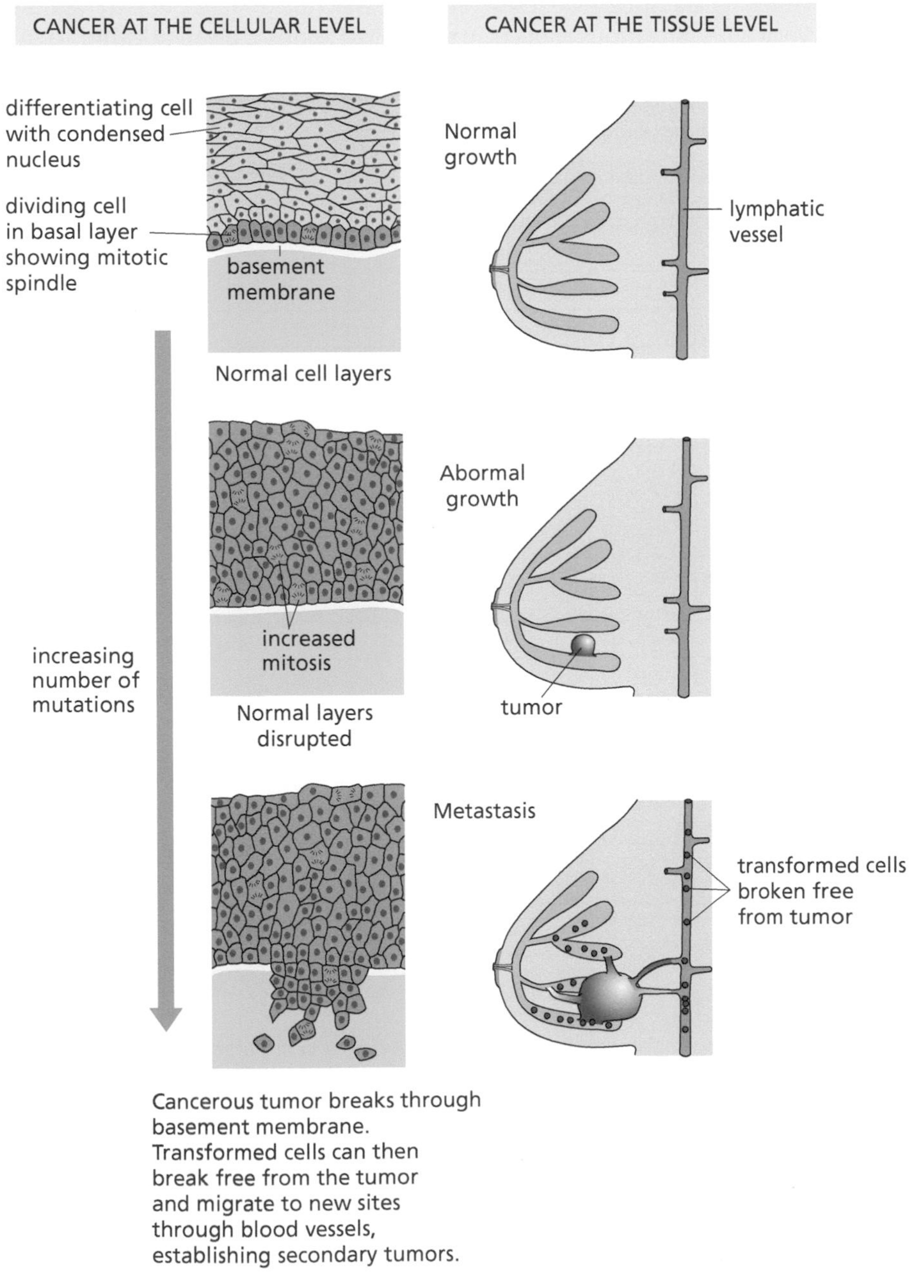

CANCER AT THE CELLULAR LEVEL

CANCER AT THE TISSUE LEVEL

differentiating cell with condensed nucleus

dividing cell in basal layer showing mitotic spindle

basement membrane

Normal cell layers

Normal growth

lymphatic vessel

increasing number of mutations

increased mitosis

Normal layers disrupted

Abormal growth

tumor

Metastasis

transformed cells broken free from tumor

Cancerous tumor breaks through basement membrane. Transformed cells can then break free from the tumor and migrate to new sites through blood vessels, establishing secondary tumors.

Figure 10.21 The growth of a tumor in the human breast.

than those that have not, because no amount of surgery can eliminate all the cancerous cells that have spread.

THOUGHT QUESTIONS

1. How do stem cells differ from transformed cells? How do stem cells differ from muscle cells or blood cells?
2. In what ways are benign and malignant tumors the same and in what ways are they different?
3. What does the statement "Cancer is a disease of the genes, but it is not an inherited disease" mean?
4. How might the discovery of genes mutated in cancer allow scientists to learn more about how normal cells regulate such processes as proliferation, apoptosis, and differentiation?

10.6 CANCERS HAVE COMPLEX CAUSES AND MULTIPLE RISK FACTORS

The study of disease at the population level constitutes the science of **epidemiology**. The basic epidemiological data for various forms of cancer have been compiled for the United States since 1950. Data on new cases (incidences), deaths, and five-year survival rates for various cancers in the United States is given in Table 10.3.

Table 10.3 Cancer cases, deaths, and five-year survival rates in the United States, arranged in decreasing order of the number of deaths

Type of cancer	New cases*	Deaths*	Five-year survival rate (%)ǂ
Lung	228,150	142,670	19
Colon and rectum	145,600	51,020	65
Pancreas	56,770	45,750	9
Breast	271,270†	42,260†	90
Liver and intrahepatic bile duct	42,030	31,780	18
Prostate	174,650	31,620	98
Leukemia	61,780	22,840	65^
Lymphoma	82,310	20,970	74, 88^
Brain and other nervous system	23,820	17,760	Varies widely by type and age at diagnosis
Urinary bladder	80,470	17,670	77
Kidney and renal pelvis	73,820	14,770	75
Ovary	22,530	13,980	47
Myeloma	32,110	12,960	52
Uterus, endometrium	61,880	12,160	81
Oral cavity and pharynx	53,000	10,860	65
Skin, melanoma	96,480	7,230	92
Soft tissue sarcoma	12,750	5,270	81
Uterus, cervix	13,170	4,250	66

Totals estimated by the National Cancer Institute from incidence rates found in the Surveillance, Epidemiology, and End Results [SEER] program.

* 2019 American Cancer Society Facts and Figures. Numbers are estimates for 2019 based on data from 2001 to 2015.

† Includes 2670 cases and 500 deaths in men.

ǂ Percentage of people diagnosed with cancer who are still alive after five years, compared with percentage in a comparable, cancer-free population. Data are for all clinical stages at diagnosis; cancers that have already spread regionally at the time of diagnosis have somewhat lower survival rates, and those that have metastasized have much lower survival rates. Unless otherwise indicated, data are from 2019 American Cancer Society Facts and Figures.

^ 2019 Leukemia and Lymphoma Society. Survival rates for leukemia are much higher in younger people. Five-year survival rate for non-Hodgkin lymphoma, 74%; for Hodgkin lymphoma, 88%.

Epidemiology uses descriptive statistics to find patterns in the incidence of diseases. Those patterns indicate possible risk factors that can suggest hypotheses that can be further tested in other ways. In general, the causes of adult cancers seem to be mainly environmental, not hereditary. Evidence to support this conclusion comes from epidemiological data for the United States and several European countries, showing in each case a marked increase in cancer death rates throughout the twentieth century. Most of this increase in cancer pertains to a single type, cancer of the lung (**Figure 10.22A**). This increase coincided with advancing industrialization and other changes in the environment, but very little change in the gene pool, suggesting environmental or lifestyle causes were primarily responsible. In the case of lung cancer, the death rate showed a striking parallel with increases in the consumption of cigarettes, with a lag of about 15–20 years (**Figure 10.22D**). Women adopted the habit of cigarette smoking later than men, and their increase in lung cancer deaths also occurred later (**Figures 10.22B** and **C**). Since 1960, cigarette consumption by males has decreased, and in 1990, the lung cancer death rate among males began to decrease (Figure 10.22B). Cancers of the pancreas and large intestine increased more slowly over the same period, while stomach and rectal cancers declined. The rapid change and irregular pattern of change both conform much better to the hypothesis of environmental causes than the alternative hypothesis of a genetic cause or causes. Interestingly, although media coverage of cancer has dramatically increased in the past decades, cancer death rates in the United States have been declining for the past 25 years, due in large part to improved treatment and implementation of regular screening as normal part of medical care. For some cancers, such as breast cancer, this decrease in mortality has occurred despite coincident increases in cancer incidence. Part of the increased incidence in breast cancer may be due to environmental factors, but increased screening (and thus identification of more cases that may not be life threatening) may also play a role.

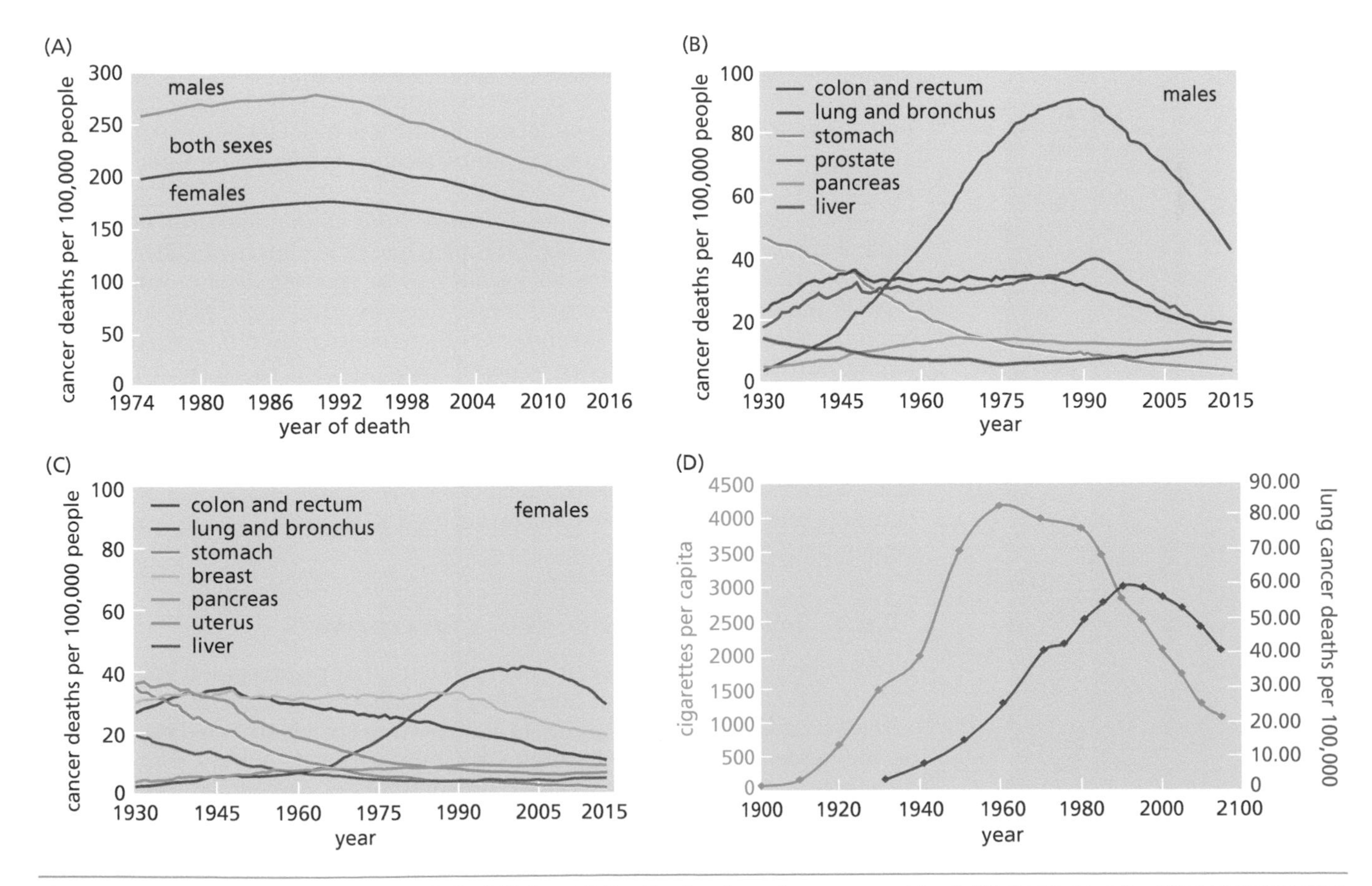

Figure 10.22 Deaths from cancers in the United States. (A) Death rates from all cancers combined. **(B)** Death rates for different cancer sites in males. **(C)** Death rates for different cancer sites in females. **(D)** Lung cancer death rate and cigarette consumption.

Human cancers are named according to the type of cell from which the cancer is derived (the cell of origin). A cancer that arises in epithelial cells (cells forming sheet-like tissues that line body cavities or glandular tissues) is called a **carcinoma**; a cancer that arises in connective tissue is called a **sarcoma**. Blood cell cancers such as leukemias and lymphomas make up a third class of cancers, and cancers that arise in cells of the nervous system define the fourth major class. Carcinomas account for most variety of cancers, largely because cells in epithelial tissues naturally turn over more frequently than cells in other tissues, and more cell division means more opportunities for accumulation of mutations. Cells in many epithelial tissues also come into contact with more environmental mutagens; for example, lung epithelia are exposed to substances in the air we breathe, and intestinal epithelia are exposed to chemicals in the food we eat. In addition to the major classes of tumors, there are also tumor subtypes: A liposarcoma, for example, is a sarcoma derived from fat cells, and the cells of origin of squamous cell carcinomas are skin cells called keratinocytes.

Cancers are rare in children, adolescents, and young adults, but certain cancers do occur in these populations. In particular, blood cancers (leukemias) occur much more frequently in children than do other types of cancers, making up about 26% of all childhood cancers. It is important to note, however, that adults still make up the vast majority of people who develop leukemia. Approximately 75% of childhood leukemias are acute lymphocytic leukemias that arise from stem cells in the bone marrow. Although these are very aggressive, they have a good cure rate because children still have many normal cells to take over after therapy.

About 85% of adult cancers are carcinomas, including cancers of the lungs, breast, colon, rectum, pancreas, skin, prostate, and uterus (Table 10.3). The incidence of these cancers (and many others) increases with age, so that cancers become more and more significant as causes of death with advancing age (Figure 10.19). Most of these adult cancers are believed to be caused mainly by environmental or lifestyle factors. Based on epidemiological evidence, the following factors have been suspected of causing at least one type of cancer or of increasing the rate at which at least some cancers occur: genes, increasing age, viruses, ionizing radiation, ultraviolet radiation, diet, stress, mental state, weak immune systems, unsafe sexual behavior, hormones, alcohol, tobacco, and some chemical substances.

In this section we examine the evidence from epidemiological studies and animal studies that have suggested the many possible causes of cancer. We will see how these seemingly disparate causes may be working by very similar pathways at the cellular and molecular levels. Keep in mind that when we speak of "causes" of cancer we often mean factors that are associated in epidemiological studies with increased incidence in populations. As such, these factors are more properly called "risk factors," not causes. A multitude of factors contributes to whether any individual gets cancer. We generally cannot say that one thing "caused" a particular cancer. As Clark Heath of the American Cancer Society has said, "Cancer cases are clinically nonspecific—you can't look at a leukemia case clinically and say, 'Ah, this a radiation-caused leukemia'" (*Scientific American*, September 1996, p. 86).

10.6.1 Inherited predispositions for cancers

As described previously, cancer is not inherited, but a predisposition for some cancers can be. Some of the genes implicated in familial cancers and the types of cancers associated with inheriting mutant alleles of these genes are listed in **Table 10.4**. As mentioned earlier, retinoblastoma is a rare cancer of the eye that primarily affects young children (**Figure 10.23A**). The presence of a growing tumor in the eye causes a white spot in photographs, and a cell phone app has been developed to help detect retinoblastoma at an early stage when it can be treated without loss of vision. Most people who develop retinoblastoma inherit one defective allele of the gene for a checkpoint protein called pRb. This protein blocks the expression of certain genes (including some cyclins) that are required for passage through a point in

Table 10.4 Examples of genes mutated in various familial cancers

Gene	Protein function	Type of cancer	Associated risk*
RB	G_1 cell cycle checkpoint	Retinoblastoma	90%†
XPA, XPB, XPC, XPD, XPE, XPF, XPG and *XPV*	DNA repair (UV damage)	Various skin cancers, including melanoma	~100%#
APC	Promotes cell differentiation	Colon	~100%#
BRCA1	DNA repair (via recombination)	Breast, ovarian	72% (breast) ‡, 44% (ovarian) ‡
BRCA2	DNA repair (via recombination)	Breast, ovarian	69% (breast) ‡, 17% (ovarian) ‡

* Data indicate the approximate percentages of people with inherited mutations in the given genes who will develop the associated cancers.

† 2019, American Cancer Society.

2019, National Institutes of Health, Genetics Home Reference.

‡ 2017, National Cancer Institute; data are for risk of developing cancer by age 80.

the G_1 phase of the cell cycle where a cell commits to another round of division (the "restriction point"). When pRb receives a signal initiated by growth factors, it becomes phosphorylated and inactivated; the cell can then pass the restriction point and continue through the cell cycle. Because of this important role, pRb is often referred to as the cell cycle "gatekeeper." Of the people who inherit a defective allele, 80%–90% develop retinoblastoma. The rapid cell division in the developing eye during early childhood seems to make it quite likely that the second functional

(A)

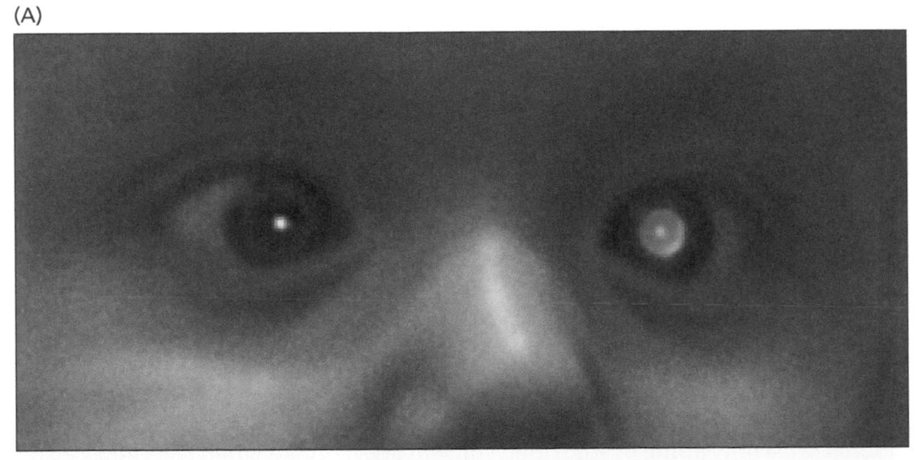

(B)

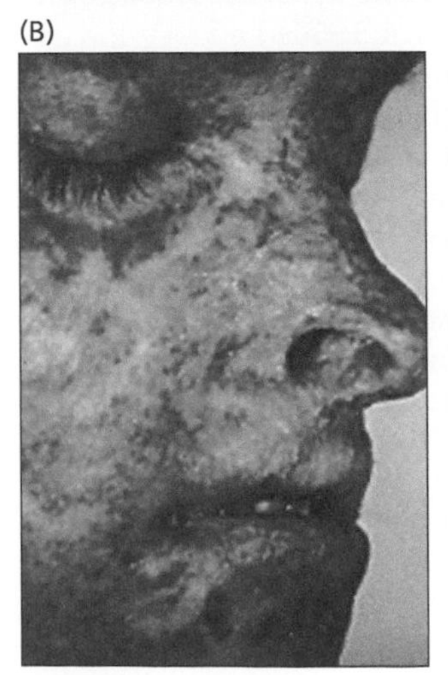

(C)

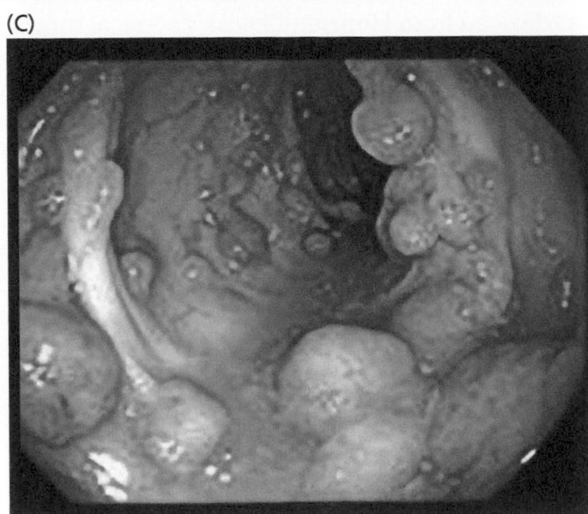

Figure 10.23 **Familial cancers. (A)** Retinoblastoma. **(B)** Xeroderma pigmentosum. **(C)** Familial adenomatous polyposis (precursor to colon cancer).

RB1 gene is lost, setting a cell on the pathway to becoming cancerous. Surgery, often paired with additional therapies such as radiation and chemotherapy, can cure up to 95% of retinoblastoma cases. Notably, however, people who have familial retinoblastoma are at higher risk for other types of cancer later in life, reflecting the central role of the pRb protein in cell cycle control.

Another rare cancer is xeroderma pigmentosum (XP), a condition that results in very high rates of skin cancer due to a defect in one of several proteins that function in a pathway that repairs DNA damage caused primarily by ultraviolet radiation from the sun. In the case of XP, affected individuals inherit two defective alleles of one of the genes that functions in the repair pathway. Children who are known to have inherited XP alleles are counseled to take extreme precautions to avoid sun exposure (**Figure 10.23B**). With such precautions, it is possible for some XP patients to live a relatively normal lifespan, but patients also exhibit neurological problems that may decrease their lifespan independent of threats from cancer.

The *APC* gene encodes a tumor suppressor protein that promotes cell differentiation to a nondividing state rather than continued cell proliferation. Inheritance of one defective copy of the *APC* gene causes a condition called familial adenomatous polyposis (FAP), which manifests as very high numbers of benign tumors, or polyps, in the colon, or large intestine (**Figure 10.23C**). While the polyps themselves are not cancerous, there is a very high chance (nearly 100%) that additional mutations will occur that cause cells in one or more polyps to become transformed, leading to malignant colon cancer. The presence of intestinal polyps can be identified by colonoscopy, where a camera mounted on a flexible tube is inserted into the colon through the anus. In the past 20 years, in the United States, it has become standard medical procedure to use colonoscopy to screen all people over age 50 every 10 years for polyps and cancerous lesions in the colon. Persons known to be at high risk based on family history are screened as frequently as every two years. During a colonoscopy, isolated polyps can often be removed by a simple clipping procedure, thus ensuring that they do not progress into malignant tumors. In patients with FAP, however, the number of polyps is usually so great by the time they reach their mid-thirties that the polyps cannot all be removed in this way, and many of such patients will eventually develop colon cancer. The *APC* gene also functions in tissues other than the colon, and individuals with FAP do have increased rates of some other cancers, but the risk is much lower than for colon cancer. The reason for this discrepancy is unknown. In addition, while retinoblastoma and xeroderma pigmentosum are quite rare, colon cancer is one of the most common cancers. Many people who develop colon cancer have not inherited a mutant *APC* allele and therefore have sporadic, rather than heritable, colon cancer. However, loss-of-function mutations in *APC* do seem to occur at an early stage in the development of sporadic colon cancers. The mutations associated with colon cancer progression are especially well characterized because of the availability of various stages of tumor tissue removed from patients during colonoscopies. (Note that any research using tissue samples from patients must now abide by informed consent laws, unlike when the HeLa cell line was derived from Henrietta Lacks's cervical tumor in the 1950s.)

Retinoblastoma, XP, and FAP occur in almost all people who inherit the relevant mutations, and the associated cancers often occur quite early in life. In contrast, mutations in the *BRCA1* and *BRCA2* genes (named for breast cancer 1 and 2) confer a breast cancer risk that is less extreme: By age 80, about 72% of women who carry a mutated *BRCA1* gene and 69% who carry a mutated *BRCA2* gene will develop breast cancer, whereas only about 12% of noncarriers will develop cancer in their lifetimes. Despite the prevalence of news stories about the "breast cancer genes," only 5%–10% of all breast cancers are associated with a genetic predisposition. Environmental factors play a much larger role in the overall prevalence of breast cancer, as is also the case for colon cancer.

10.6.2 Increasing age

Far more cancers seem to be caused by environmental rather than genetic factors, even after considering genetic predispositions for some cancers. The more

Table 10.5 Age-specific probabilities of developing invasive breast cancer for U.S. women*

Current age	Probability of developing invasive breast cancer in the next 10 years	
	%	1 in
20	0.1	1479
30	0.5	209
40	1.5	65
50	2.4	42
60	3.5	28
70	4.1	25
80	3.0	33
Lifetime risk	**12.8**	**8**

2019 American Cancer Society Facts and Figures.

* Note: For men, the incidence of invasive breast cancer is approximately 1/100th of that for women.

common type of breast cancer, a late-onset disease of postmenopausal women, is not linked to inheritance. The strongest risk factor for these breast cancers is age. The incidence of all cancers increases with age, presumably because there has been more time for environmental exposures to produce accumulated mutations. In fact, one of the reasons for the present-day higher incidence of cancers (and chronic diseases such as heart disease) is that people are living much longer because mortality from infectious diseases is lower.

From data on the incidence of cancers, the probability of acquiring cancer at different ages can be calculated. Table 10.5 gives data for breast cancer. As can be seen, the probability increases with age. Because age is such a strong factor in all health studies, epidemiologists must "control for age"; that is, they must either compare groups of the same ages or use mathematical formulas to "age-adjust" the data.

Cancer data are often shown as the probability of acquiring cancer by age 75. Examples are shown in Figure 10.24. Here, data of the type shown in Table 10.5 have been added up to give the "lifetime probability" of acquiring cancer by age 75. Note that these numbers do not mean, for example, that a 75-year-old white woman's chance of acquiring breast cancer while she is 75 is 10%. The chances for each age group are the numbers shown in Table 10.5, which are much lower.

Data of the type shown in Figure 10.24 are useful in comparing the probabilities for different types of cancers. They are also useful in comparing the probabilities for different segments of the population. In the United States, health statistics

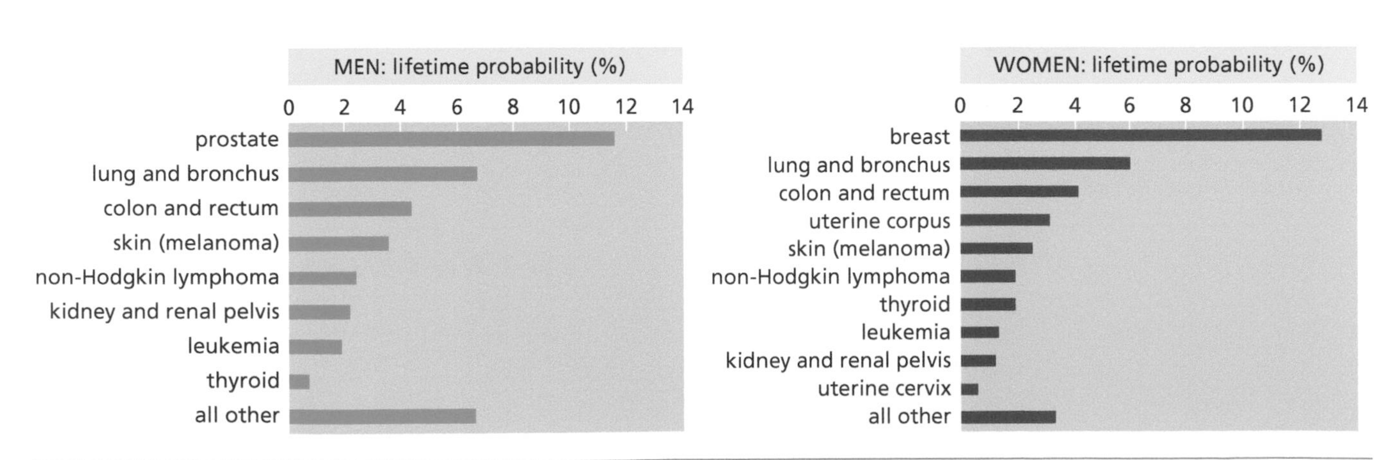

Figure 10.24 Lifetime probabilities of acquiring malignant cancers up to age 75 (U.S., 2014–2016).

are summarized by sex and by race. These are data for populations and do not mean that any individual's probability is the number shown. Individual risk is increased or decreased by all of the factors mentioned in this section.

10.6.3 Viruses

Several cancers are known to be associated with viruses and other infectious agents. In 1911, American pathologist Peyton Rous showed that a tumor of connective tissues (a sarcoma) in chickens was caused by a virus that was later named Rous sarcoma virus. (Rous received a Nobel Prize for this work, but not until 1966.) Chickens infected with this virus develop sarcomas at a high rate. Viruses seem to be associated with cancer in at least two different ways. First, a viral infection may cause a decrease in the activity of the immune system. Decreased immunity increases the likelihood that a transformed cell will progress to cancer, highlighting the role of the immune system in eliminating cancer cells, which have some abnormal features that can make them seem "foreign." An example of this type is Kaposi sarcoma, which occurs in people with AIDS (see Section 15.1.2). Second, some viruses carry genes that, when inserted into the host DNA, cause the host cell to become transformed into a cancerous cell. These genes are, therefore, oncogenes. The viruses do not cause the mutation; rather, they carry an entire mutated gene into the cell they infect. The 1989 Nobel Prize awarded to Bishop and Varmus honored their discovery of the fact that the oncogene in Rous sarcoma virus was actually a cellular gene that had been "picked up" by the virus and mutated to an oncogenic form that could then trigger transformation when reintroduced into cells via infection (Stehelin et al., 1976b). This relationship between cellular proto-oncogenes and viral oncogenes is illustrated in **Figure 10.25**.

The incidence of liver cancer is high in non-industrialized countries. A very high proportion of the people who develop liver cancer have previously had hepatitis B, a viral infection of the liver. Some people, after recovering from the acute symptoms of hepatitis, remain infected carriers of the virus. Over 250 million people worldwide are carriers of hepatitis B virus; carriers have a 100-fold higher risk of developing liver cancer, although this may occur up to 40 years after hepatitis infection. Hepatitis C virus is also associated with liver cancer, particularly in Japan. Worldwide, as many as 80% of liver cancers are caused by viral infections. The means by which the hepatitis viruses cause liver cancer are not as clear as for the cancer-causing mechanism of the Rous sarcoma virus, but research in this area is ongoing.

Cancers of the reproductive organs, especially cancer of the uterine cervix, are statistically related to both male and female sexual behavior. Incidence rates for cervical cancer are higher among women who were younger at the time of their first instance of sexual intercourse or who have had multiple sexual partners.

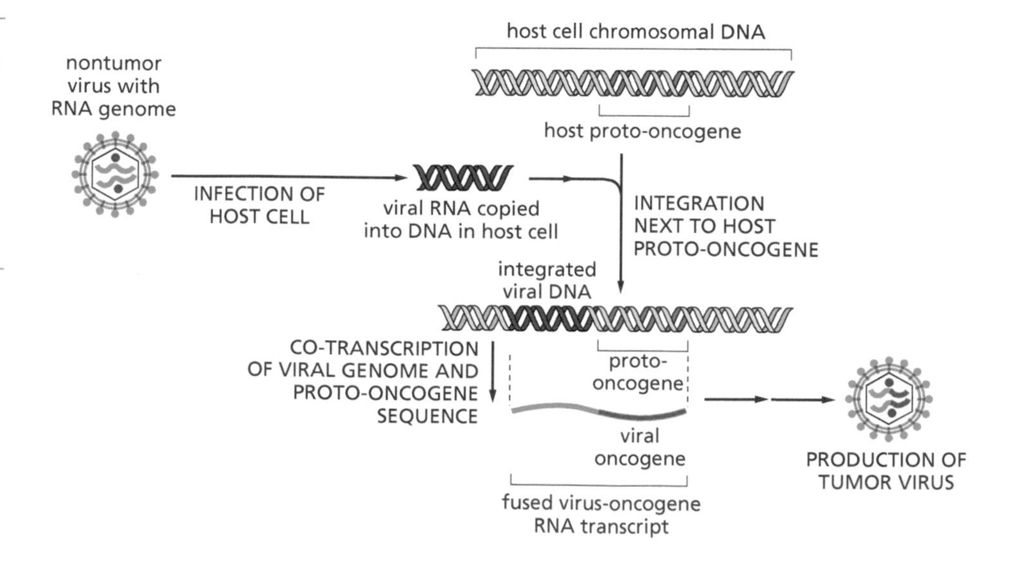

Figure 10.25 Tumor viruses carry oncogenes. Often, these oncogenes are derived from proto-oncogenes in the host cell DNA into which the viral genome has integrated. When RNA for the production of new virus particles is transcribed, the oncogene may be included in the transcript and thus becomes part of new viral genomes.

The rates of cervical cancers are also high in those countries in which women tend to have few sexual partners and to marry as virgins but where cultural tradition often encourages men to seek multiple sexual partners. This epidemiological evidence argues that male promiscuity is an important risk factor for cervical cancer even though it is women who develop the disease. Sexually transmitted human papillomaviruses (HPVs) often cause genital warts, but some types also cause cancer of the cervix, which is thus a sexually transmitted form of cancer. The HPV genome contains two oncogenes whose protein products inactivate two of the cell's key tumor suppressors (the previously mentioned pRb and p53 proteins); unlike for the case of the Rous sarcoma virus, these oncogenes are virus-specific and are not related to normal cellular genes. HPVs are associated with high proportions of cancers of the uterine cervix (90%), anus (90%), vagina and vulva (70%), penis (60%), and the oropharynx at the back of the throat (70%) (CDC, 2020). The first vaccine against HPV, called Gardasil®9, was approved for use in the United States in 2006, targeting preadolescents through young adults. A study that analyzed cervical tumors diagnosed during the period from 2008 to 2014 found a statistically significant decrease in HPV infection, indicating that the rate of HPV-induced cancer decreased with the onset of widespread administration of the HPV vaccine (McClung et al., 2019). Even individuals who had not been vaccinated showed a decrease in HPV incidence, illustrating the concept of "herd immunity" (see Section 16.1.3).

The epidemiological evidence for the association of each of these viruses with cancer has been verified by animal and tissue culture experimentation. Although the incidence of virally induced tumors is low in the United States, such tumors account for 20% of all cancers worldwide.

10.6.4 Physical and chemical carcinogens

A large and growing number of external agents are known to cause cancer; such agents are called **carcinogens**. Evidence that carcinogens cause cancer comes from studies in which animals are exposed to them. Evidence also comes from epidemiological studies of occupationally exposed persons, such as industrial workers who handle the agents. Still other carcinogens are discovered by recognizing "epidemiological clusters" such as unusually high numbers of people living in the area surrounding an industrial plant or waste disposal site who develop certain types of cancers. There are two types of carcinogens: those that induce DNA mutations (**mutagens**, also called **tumor initiators**) and those that promote the progression of transformed cells into cancer by causing DNA damage (**tumor promoters**). Some mutagens are physical agents—energy sources with high enough power to damage DNA. Other mutagens are chemical agents; these also damage DNA through chemical reactions with the DNA bases or the DNA backbone (see Section 2.3.2).

10.6.4.1 Physical carcinogens (radiation) Some carcinogens are physical agents, particularly certain types of energy sources. Exposure to ultraviolet (UV) radiation—most often from sunlight or tanning beds—is the primary risk factor for developing a range of skin cancers. UV radiation causes abnormal chemical bonds to form between certain DNA bases in the same DNA strand, most often between two neighboring "T" bases (Figure 10.26). This interferes with the normal base-pairing between those "T" bases and the complementary "A" bases on the opposite DNA strand, which can cause the introduction of mutations when the DNA is replicated. Cells have DNA repair mechanisms that can fix this type of damage, but the extent of damage can overwhelm the repair machinery, leading to mutations being passed on when skin cells divide. UV radiation can also produce reactive oxygen species (ROSs) in cells, which can damage DNA in other ways, as explained later.

Skin cancer rates have been steadily rising in the United States in recent years; currently, over 3 million people are diagnosed with some form of skin cancer in the United States each year. Most of these are squamous or basal cell carcinomas, which are among the less aggressive and more treatable skin cancers. In contrast,

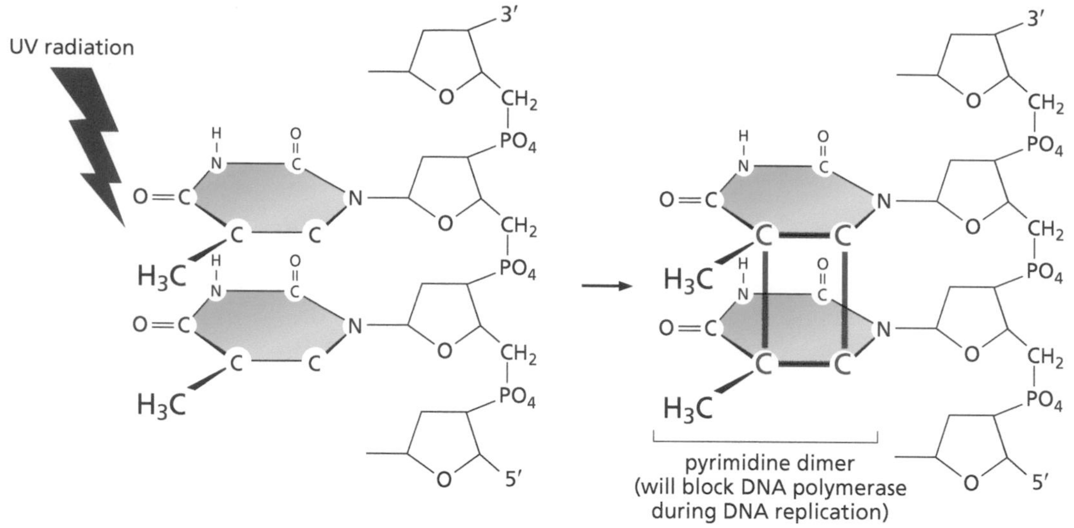

Figure 10.26 UV irradiation induces abnormal covalent bonds between neighboring nucleotides.

malignant melanoma is a cancer of the pigment cells (melanocytes) that is especially dangerous because melanoma cells readily spread around the body (metastasize), often before a person even realizes that they have a skin lesion. According to the American Cancer Society (2021d), 106,110 Americans were expected to be diagnosed with malignant melanoma in 2021, and 7180 were expected to die from it. Melanoma kills more women in their twenties than does breast cancer. Light-skinned people are more susceptible to skin cancers and should be especially vigilant about protecting themselves from UV radiation using sunscreen and physical barriers to sun exposure. Ninety percent of skin cancers are attributable to UV radiation, particularly UVB rays. In contrast to UVA rays, UVB rays have shorter wavelengths and cannot penetrate as deeply into the skin; instead, they are mostly absorbed by the surface layers of the skin, where they cause DNA damage. The adult incidence of skin cancer is correlated with the number of sunburns that a person received as a child. Epidemiological evidence shows that even exposures to natural levels of UV radiation increase the incidence rates for skin cancers, which are much higher in the southern half of the United States than in the northern half. In contrast to most skin cancers, one rare and especially deadly form of melanoma is not directly linked to UV exposure but is more determined by genetics. Dark-skinned people are more likely to develop this type of melanoma, particularly on the skin of their palms or the bottoms of their feet, and it is often diagnosed late because people of color and their physicians are less attuned to being aware of skin cancer in darker-skinned populations.

Ionizing radiation, such as that produced by radioactive substances, was clearly shown to be carcinogenic by studies on the Japanese survivors of the 1945 bombings of Hiroshima and Nagasaki. Leukemia was the first type of cancer to show increased prevalence in survivors of the bombings, exhibiting an uptick just five years later, but increased rates of a variety of other cancers have also been linked to the wartime radiation exposure as survivors have been followed through the years. The French (Polish-born) chemist Marie Curie (1867–1934), a two-time winner of the Nobel Prize and pioneer in the study of radioactive elements, died of a leukemia induced by her frequent handling of these elements. X-ray machines also produce ionizing radiation, so the level of exposure is carefully controlled to minimize the exposure. Medical and dental diagnostic X-rays do not increase cancer incidence. The only medical uses of radiation associated with increased cancer risk are the very high radiation doses used in cancer therapy itself. Although these treatments are necessary to save a patient, they do induce DNA damage, from which new cancers may arise a decade or more later. Radiation treatments, like all medical treatments, are based on estimates of risk-benefit ratios. When someone is very sick and would die without treatment, the probable benefit from the treatment may make a higher level of risk acceptable to the patient and her or his physician.

Ionizing radiation associated with cancer is also caused by radon, a radioactive element that occurs naturally in certain types of rocks. When these rocks are uncovered or a well is dug into them, radon gas may be released. A person with long-term exposure to radon gas may develop lung cancer; however, radon accounts for only about 10%–12% of lung cancers, whereas smoking accounts for at least 80%. Proper ventilation essentially eliminates radon risk.

Any time that charged particles move, whether they are electrons in a power line or ions moving across cell membranes, electric and magnetic energy fields are created. Therefore, all living things are sources of electrical and magnetic energy because of the ions moving through them. Low-frequency electric and magnetic fields from power lines or household appliances and radio-frequency electromagnetic radiation from cell phones or microwaves are too low in energy to cause ionizing damage to DNA. Although media stories have sensationalized the potential risk from electric and magnetic fields produced by these sources, carefully controlled epidemiological studies have found no correlation between exposure to them and increased cancer rates. In fact, the ambient level of energy from a cell phone is less than one one-hundredth of the electromagnetic radiation given off by the person holding the phone.

10.6.4.2 Chemical carcinogens: Tumor initiators and tumor promoters Other carcinogens are chemicals. As mentioned earlier, these make up two categories: **tumor initiators** and **tumor promoters**. Tumor initiators are agents that begin the process of transformation by causing permanent damage in the DNA; in other words, they are **mutagens**. Some chemicals are themselves harmless but become carcinogens when acted upon by metabolic enzymes in cells. Since the 1970s, the Ames test, described in Box 10.1, has been used as a simple method for screening chemicals for mutagenic activity. Although it is performed in bacteria, the Ames test incorporates a mechanism to allow for metabolic activation that mimics what would occur in human cells. When interpreting results from the Ames test, it is important to consider what concentration caused mutations and whether a person would ever be exposed to that concentration in normal life; for example, certain compounds found naturally

Box 10.1 The Ames test

Tens of thousands of known chemical substances have never been tested as possible carcinogens in animals. Animal testing is expensive and slow; it would take many, many decades (and many billions of research dollars) to test all these substances. Clearly, we need a quick screening method that tells us which substances are more likely to be carcinogenic; these substances can be tested first, while the testing of less likely carcinogens can wait.

The Ames test, devised by cell biologist Bruce Ames of Cornell University in the 1970s, is a screening method that detects mutagens capable of causing particular types of mutations in a culture of *Salmonella* bacteria, as shown in the diagram. The bacteria used are from a strain called *his*$^-$, which are unable to synthesize histidine, an amino acid required for the manufacture of bacterial proteins and hence for bacterial growth. Most bacteria are *his*$^+$, meaning that they can make their own histidine from other materials. In the Ames test, *his*$^-$ bacteria are grown in a medium containing just a small amount of histidine, which allows just enough growth for mutations to have a chance to occur. Soon, however, the histidine is used up, and the bacteria die unless they have mutated from *his*$^-$ to *his*$^+$ and thus have become able to make their own histidine. The rate of spontaneous mutation is very low. If a chemical is added to the culture medium and many more bacterial colonies grow than in a culture without this addition, the chemical can be assumed to have caused the mutations (i.e., to be a mutagen). Counts of the numbers of colonies also identify stronger and weaker mutagens.

Remember that the Ames test was designed as a *screening method* for carcinogens. A positive result in the Ames test does not guarantee that a chemical will be carcinogenic to animals or humans, but it identifies chemicals that should be tested further, first in mammalian mutagenesis assays (in tissue culture) and then in animal studies if the tissue culture tests are also positive. Reasons why a chemical that is mutagenic in bacteria might not be carcinogenic to animals include differences in the rate of uptake of the chemical by cells, differences in bacterial vs. mammalian cell metabolism, and differences in DNA repair. Bruce Ames, the originator of the Ames test, has also pointed out that nearly *any* substance is

(*Continued*)

mutagenic in a sufficiently high dose. Therefore, the effective dose to which cells in a whole animal are likely to be exposed must also be considered. Another limitation of the Ames test is that it cannot identify animal carcinogens that are tumor promoters, not tumor initiators, since these exert their cancer-causing effects without causing mutations.

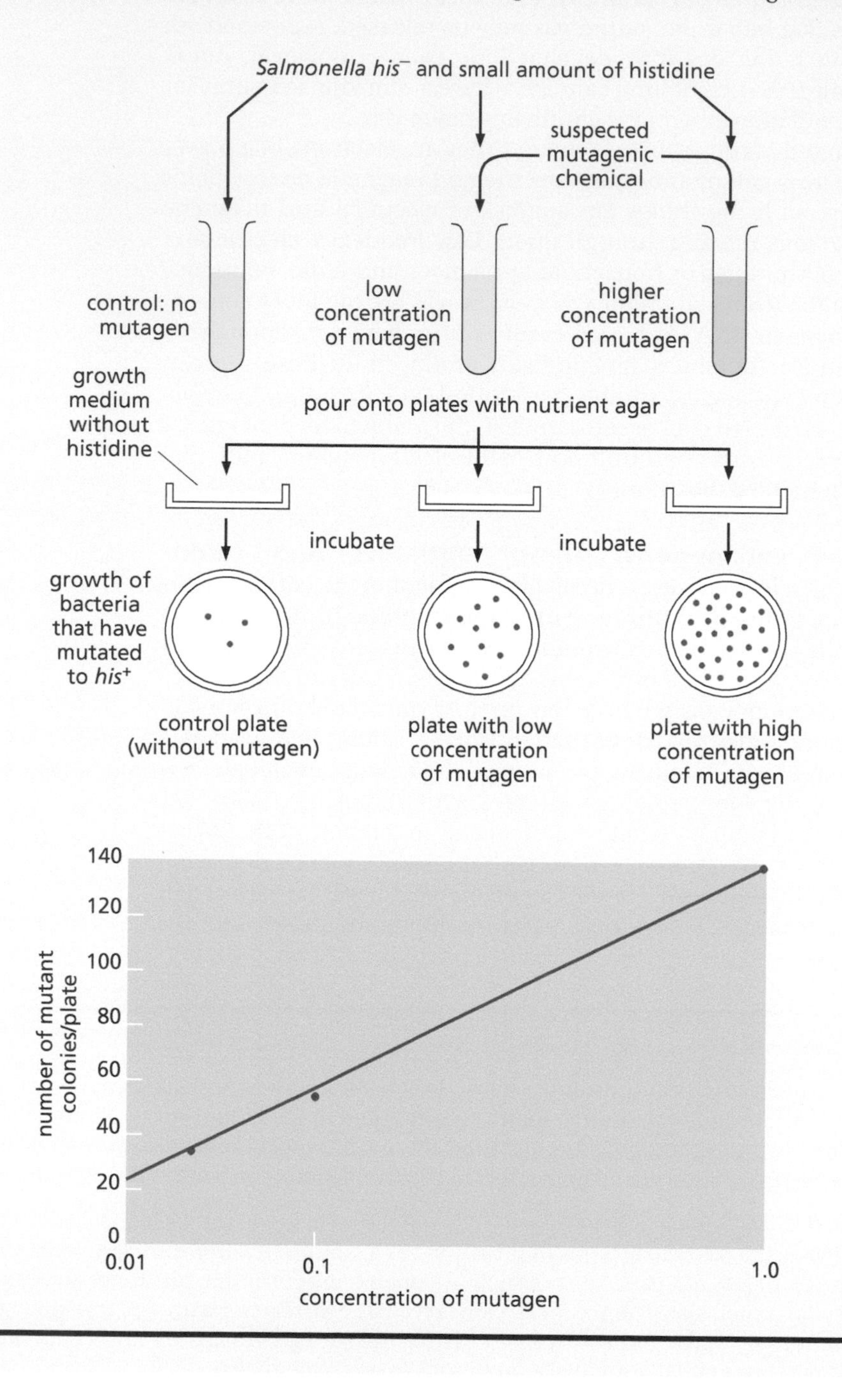

in celery give positive results in the Ames test, but that does not mean that eating celery will give you cancer because those compounds naturally occur at concentrations well below any level of concern. Compounds that warrant concern are those that cause mutations at low concentrations and might occur in contaminated drinking water or in polluted air, or compounds that could constitute realistic exposure concentrations for certain populations such as people in certain occupations.

In a cell whose DNA has been damaged by exposure to a mutagen, or tumor initiator, transformation can be completed by subsequent exposure to a tumor promoter. Tumor promoters by themselves do not cause mutation; therefore, they do not produce positive results in the Ames test. Instead, tumor promoters induce cell division, either directly, by affecting cell signaling, or, more commonly, by causing tissue damage or inflammation that triggers cell division to repair the tissue. If a dividing cell contains a mutation from an earlier exposure to a tumor initiator, its chance of acquiring additional mutations that lead to complete transformation are increased. Because the DNA damage from an initiator is permanent, a tumor promoter can have its effect years after exposure to the initiator. Tumor promoters include alcohol, phorbol esters (often used as tumor promoters in laboratory studies), natural hormones such as estrogen, dioxins (by-products of industrial bleaching processes), saccharin, and asbestos.

Tobacco smoke contains dozens of known carcinogens, including nitrosamines, formaldehyde, arsenic, nickel, cadmium, and several compounds called polycyclic aromatic hydrocarbons (PAHs). These are also present in smokeless tobacco (chewing tobacco and snuff). Many of the carcinogens in tobacco undergo metabolic activation to generate mutagens, and some are tumor promoters. A substance that is both a tumor initiator and a tumor promoter is referred to as a "complete carcinogen." Tobacco fits this description, and exposure to the chemicals in tobacco smoke, including secondhand smoke, is the largest single risk factor for cancer in the industrialized world. People who begin to smoke when they are teenagers or in college are more than 10 times more likely to develop lung cancer than people who have never smoked. Both smoked and smokeless tobacco also greatly increase the risk of cancers of the mouth and throat (oral cavity and pharynx) due to direct exposure of these tissues to the carcinogens when these products are used. Although a person's risk of cancer decreases after he or she stops smoking, the risk never drops as low as that for people who have never smoked. The danger of secondhand smoke is evident from the fact that nonsmoking women whose husbands smoke have higher cancer rates than nonsmoking women with nonsmoking husbands.

Many industrial chemicals are carcinogens, including vinyl chloride (used in the making of many plastics), formaldehyde, asbestos, nickel, arsenic, benzene, chromium, cadmium, and polychlorinated biphenyls (PCBs) (**Table 10.6**). People can become exposed to these chemicals through factory work or industrial contamination of the water supply. Some chemicals in herbicides and insecticides used in agriculture are also carcinogens, and people can similarly be exposed to these through farm work or through the water supply. Even when a company is known to have released carcinogens into the environment, it is nearly impossible to know whether any individual case of cancer was caused by environmental exposure to those chemicals. Instead, probable cause can be attributed to environmental contamination when a statistically significant increase in cancer cases is observed in a discrete area near the proposed site of release (a cancer "cluster"). This invariably involves long, expensive lawsuits. Consequently, there have been relatively few instances in the United States when a company has been found responsible for a cancer cluster and has been required to compensate victims or their families. Two notable cases were made into Hollywood films: the case against W.R. Grace in Woburn, Massachusetts, which was the subject of the film *A Civil Action*, and the case against Pacific Gas and Electric in Hinkley, California, which was the subject of *Erin Brockovich*. In the former, the chemicals in question were trichloroethylene and perchloroethylene, and in the latter, it was chromium-6. Recently, concerns have emerged over release of ethylene oxide gas into the air by two medical technology companies in Georgia. Ethylene oxide is used for sterilization, as well as for the synthesis of other chemicals. In late 2016, the U.S. Environmental Protection Agency raised its risk assessment on this chemical, concluding that it is indeed a human carcinogen, whereas it was previously categorized as a "possible carcinogen." It is estimated that additional antipollution measures being undertaken at the plants in question will reduce the cancer risk from ethylene oxide exposure to 2 cases in 1 million people from the current estimate of 100 per 1 million people. It can be argued that even the higher risk is quite low compared to the risk associated with some other exposures that are more under a person's individual control (such as smoking or not wearing sunscreen),

Table 10.6 Carcinogens in the workplace

Carcinogen	Cancer type	Exposure of general population	Examples of workers frequently exposed or exposure sources
CHEMICAL AGENTS			
Arsenic	Lung, skin	Rare	Insecticide and herbicide sprayers Oil refi refinery workers
Asbestos	Lung, other sites	Uncommon	Brake lining Shipyard; Insulation and demolition workers
Benzene	Bone marrow	Common	Painters Distillers and petrochemical workers Dye users Furniture finishers Rubber workers
Diesel exhaust	Lung	Common	Railroad and bus garage workers Truck operators Miners
Formaldehyde	Nose, pharynx	Rare	Hospital laboratory workers Manufacture of wood products, paper, textiles, garments, and metal products
Heavy metals (cadmium, uranium, nickel)	Prostate	Rare	Metal workers
Synthetic mineral fibers	Lung	Uncommon	Wall and pipe insulation Duct wrapping
Hair dyes	Bladder	Uncommon	Hairdressers and barbers (inadequate evidence for customers)
Mineral oils	Skin	Common	Metal machining
Nonarsenical pesticides	Lung	Common	Sprayers Agricultural workers
Painting materials	Lung	Uncommon	Professional painters
Polychlorinated biphenyls (PCBs)	Liver, skin	Uncommon	Heat transfer and hydraulic fluids and lubricants Inks Adhesives Insecticides
Soot	Skin	Uncommon	Chimney sweeps and cleaners Bricklayers Insulators Firefighters Heating-unit service workers
Vinyl chloride	Liver	Uncommon	Plastic workers
PHYSICAL AGENTS			
Ionizing radiation	Bone marrow, several others	Common	Sunlight Nuclear materials Medicinal products and procedures
Radon	Lung	Uncommon	Mines Underground structures

Data modified from *Scientific American*, September 1996.

but this provides little comfort to a person who believes his or her cancer was caused by industrial pollution.

The popular weed killer Roundup® contains glyphosate, an herbicide implicated in causing non-Hodgkin lymphoma and several other cancers and is listed as a possible carcinogen by the WHO and the International Agency for Research on Cancer, among others. Its manufacturer, Monsanto (now owned by Bayer), has been sued by multiple users of the herbicide who later developed cancer, and $2 billion has been awarded (through 2019) in various legal cases.

10.6.5 Dietary factors

The American Cancer Society states that "for the majority of Americans who do not use tobacco products, dietary choices and physical activity are the most important modifiable determinants of cancer risk." Evidence that dietary factors contribute to the development of cancer is best established for cancers of the digestive tract, including the colon and rectum. The evidence comes from laboratory studies of animals exposed to experimentally controlled diets, from clinical studies on human patients, and from epidemiological studies of large populations.

10.6.5.1 Dietary fiber and fats Diets high in fiber and low in fats are associated with a lower incidence of cancers of the intestinal tract (including the colon and rectum) and also pancreatic and breast cancer. In countries where fiber consumption is high and fat consumption is very low, as in most of equatorial Africa, incidence rates of colon and rectal cancer are only a fraction of what they are in the industrialized world. Australia, New Zealand, and the United States, where diets are lower in fiber and higher in fats, have high rates of colon and rectal cancers. In fact, diet is much more strongly correlated with cancer incidence than is industrial pollution. Studies comparing the cancer rates in Iceland and New Zealand, where diets are similar to those of Americans but where there is far less industrialization, have shown that the cancer incidence is the same in these countries as it is in the United States. In contrast, cancer rates overall, and for many specific types of cancer, are much lower in Japan, which, like the United States, is an industrialized nation, but one in which dietary fat intake is very low. The incidence of cancers among Seventh-Day Adventists, who are vegetarian and do not smoke or drink, is much lower than the incidence in their neighbors, despite both groups' living in the same conditions and being exposed to the same environmental pollutants. Several studies have also shown that eating fresh vegetables, particularly those rich in vitamins A, C, E, and beta-carotene (a vitamin A precursor), reduces the incidence of many cancers. Epidemiological data on people who migrate from one country to another further support the idea that diet and lifestyle factors have a more significant role in cancer risk than genetic factors for most cancers. One study compared cancer incidences for ethnic Japanese people living in Osaka, Japan, vs. those who had migrated to California and Hawaii. The Japanese typically have a low risk of breast cancer but a higher risk of stomach cancer compared to Americans; the ethnic Japanese living in California and Hawaii were found to have cancer incidence profiles more similar to that of their fellow Americans than to those in Japan, suggesting that their adoption of American dietary and lifestyle habits altered their patterns of cancer risk (Dunn, 1975).

10.6.5.2 Salty and pickled foods Cancer of the stomach follows a different epidemiological pattern correlated with a different set of dietary factors. This cancer is most frequent in Japan and in certain Latin American countries, where it seems to be correlated with the eating of very salty foods and pickled vegetables. The incidence of this cancer in Japanese immigrants to Hawaii and California decreases after a generation or two, while that for cancers of the colon, rectum, and breast increases. Among Japanese Americans in Hawaii, the incidence of stomach cancers correlates closely with the retention of other aspects of Japanese culture: that segment of the Japanese American population who maintains more of their traditional culture have higher rates of stomach cancer than those who adopt more Western cultural practices. This evidence suggests that diet has a larger role than genetics in the incidence of stomach cancer (Dunn, 1975).

10.6.5.3 Alcohol Ethyl alcohol has been identified as a risk factor for cancer by many studies, but the increased cancer risk is largely confined to people who also smoke. The risk of developing a cancer of the mouth or throat, for

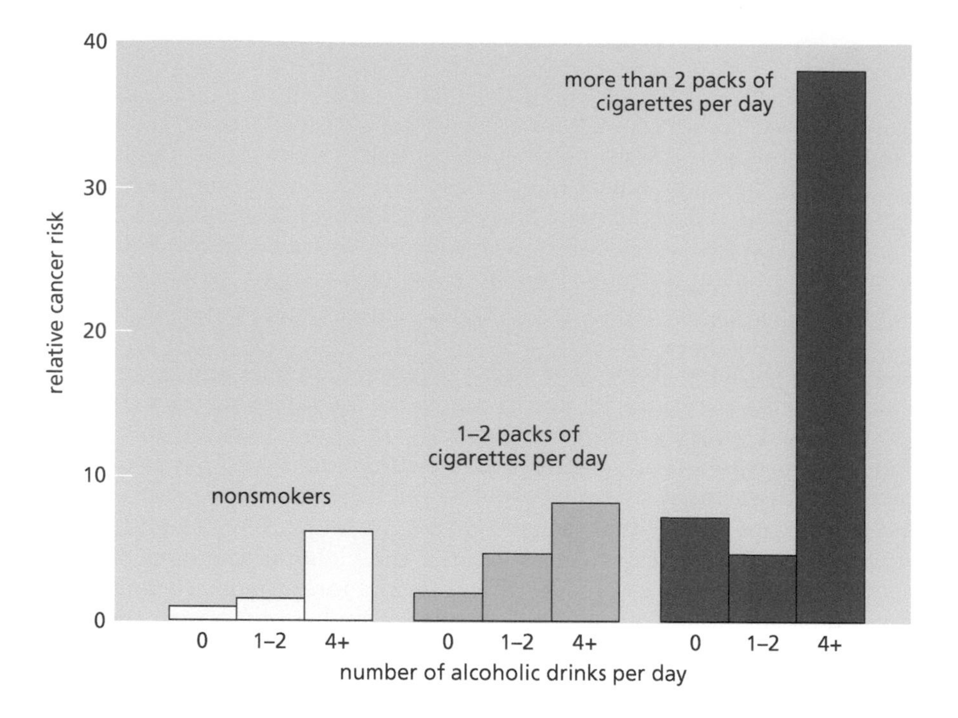

Figure 10.27 Synergism between alcohol and cigarettes in producing cancers of the mouth and throat.

example, is much higher in people who both smoke and drink (**Figure 10.27**). This is an example of a **synergistic effect**, meaning that the increased risk due to two causes is much more than the additive combination of their effects taken separately. Alcohol and tobacco are also synergistic in producing other forms of cancer. The reason for their synergy is that tobacco is a tumor initiator and alcohol is a tumor promoter.

10.6.6 Internal resistance to cancer

A good deal of evidence shows that people vary in their resistance to cancer. People with the same exposure to all known risks do not get cancer at the same rate. People also vary in their recovery rates once they get cancer. Individual variation in hormones, stress, mental outlook, and immune function may be involved.

10.6.6.1 Hormones Hormones are implicated in several types of cancers, including uterine, ovarian, breast, prostate and testicular cancers. In addition to these cancers that have a clear relationship to female and male sex hormones, cancers for which obesity is a key risk factor also have links to hormones. Obesity causes chronic inflammation, which leads to elevated levels of the hormones insulin and estrogen (in both men and women). These in turn stimulate cell division, thus promoting cancer development.

The relationship between hormones (primarily estrogen) and cancer risk has been studied most extensively for breast cancer, but the results are complex. The risk of some breast cancers can be reduced by ovariectomy (removal of the ovaries) or by taking the estrogen-inhibiting drug tamoxifen. A five-year course of tamoxifen treatment is commonly prescribed for women whose breast cancer has been eradicated by surgery, radiation, and/or chemotherapy to prevent recurrence of the cancer. Breastfeeding causes hormonal changes that protect women from both breast and ovarian cancer; experts recommend that women breastfeed for at least six months to receive these benefits. In addition, children who were breastfed as infants have a lower risk of obesity, which results in them also having a lower risk of some cancers. A woman's reproductive history also influences her risk of breast cancer due to hormonal changes that occur during pregnancy; women who have a child before age 35 receive some long-term protection against breast cancer—although the risk of developing breast cancer is elevated for 10 years immediately after a first birth—and each successive birth adds a small amount of additional long-term protection.

Women who take oral contraceptives (birth control pills) have a slightly increased risk of breast cancer, but a *decreased* risk of ovarian cancer. Birth control pills contain two hormones, estrogen and progesterone (see Section 9.3.1). The slightly increased rate of breast cancer is associated with the increased dose of estrogen, while the decreased risk of ovarian cancer is due to the blockage of ovulation (release of eggs from the ovaries) by the combination of hormones used in birth control pills. Hormone replacement therapy to lessen symptoms of menopause is also linked to an increase in breast cancer and a smaller increase in ovarian cancer. A comprehensive report that reanalyzed data from more than 50 studies found a slight elevation in risk of breast cancer in women who took birth control pills (Collaborative Group on Hormonal Factors in Breast Cancer, 1996). For every 10,000 women who are currently using the pill and who started using it between the ages of 25 and 29, 48.7 are expected to develop breast cancer in the next 10 years. Among women of the same age who have never used the pill, 44 out of 10,000 are expected to develop breast cancer in the next 10 years. Therefore the "attributable risk" (the number that can be attributed to oral contraceptives) is 48.7 minus 44, or 4.7 cases per 10,000 women. Data such as these are often reported as "relative risk," calculated as 48.7 divided by 44, or 1.16. Relative risk is generally what is reported to the public and would be stated as follows: Women using oral contraceptives (who started at age 25–29) are 1.16 times as likely (or 16% more likely) to develop breast cancer in 10 years. Most experts focus on women with known genetic susceptibility to breast cancer (those with an inherited mutation in *BRCA1* or *BRCA2*) as the population who should consider the additional risk posed by the pill when deciding on a method of birth control. The additional breast cancer risk from postmenopausal hormone replacement therapy is about double that due to oral contraceptives. Because there are other, safer ways of treating menopausal symptoms, many physicians counsel women against choosing hormone replacement.

The news media have paid significant attention to the possible breast cancer risks associated with plant estrogens that occur naturally in some foods, especially soybeans, as well as to so-called endocrine disruptors, which are chemicals with structures similar to those of human hormones that can have hormonal effects in the body at certain concentrations (Calaf et al., 2020;). Because consumption of soy products is also known to have significant health benefits, and because Asian countries where soy products are frequently consumed have low levels of breast cancer, many experts believe the benefits of eating soy outweigh any possible risks (Messina, 2016). Bisphenol-A (BPA) is an estrogen-like endocrine disruptor that is found in many plastics. It binds rather weakly to the estrogen receptor, so only high concentrations have hormone-disrupting effects. However, it can also affect cells through other mechanisms, including causing DNA damage (at lower concentrations), promoting inflammation, and interfering with intracellular signaling pathways. There has been a strong consumer backlash against BPA, but the true risk remains unclear.

Imbalances between estrogens and androgens (male hormones) are implicated in testicular cancer. In many cases, such hormonal imbalances may be genetically determined. Ongoing research is also examining the hypothesis that endocrine disruptors in the environment may promote development of testicular cancer. The rates of testicular cancer have been increasing worldwide over the past few decades, especially in developed countries. These epidemiological data support the existence of an environmental component to testicular cancer risk (Fénichel and Chevalier, 2019). In contrast to testicular cancer, rates of prostate cancer have remained relatively steady in recent decades. Research has not identified a clear hormonal influence on the risk of developing prostate cancer; however, androgens are required for continued growth and development of prostate tumors, and differences in hormone levels among patients do seem to affect treatment outcomes, although the relationships are not yet clear.

10.6.6.2 Stress There are two types of stress: acute stress, which is short term and related to a specific event (such as having to give a presentation in class), and chronic stress, which is long term and related to prolonged difficult life circumstances (such as poverty, a dysfunctional family, abuse, etc.). Chronic stress causes sustained changes in the levels of certain hormones that affect the nervous system, which are called neuroendocrine hormones. These include cortisol (commonly referred to as the "stress hormone") and catecholamines such as dopamine, norepinephrine, and epinephrine (formerly called adrenaline). Altered levels of these hormones can have wide-ranging effects on health. Most studies examining stress and cancer have found no solid correlation between chronic stress and cancer incidence; however, a 2017 report did find a slight increase in prostate cancer risk in men younger than 65 who reported chronic workplace stress versus those who did not (Blanc-Lapierre et al., 2017). No physiological data (e.g., levels of stress hormones in blood) were examined to classify the men who were considered to be chronically stressed. In contrast to the relatively weak evidence for stress promoting cancer initiation, there is considerable evidence that cancer is more likely to progress more rapidly in individuals who are experiencing chronic stress relative to those who have lower stress levels. In particular, stress seems to promote cancer metastasis, which is associated with much worse clinical outcomes than nonmetastatic cancer. Thus, many cancer treatment facilities now offer support services such as counseling, meditation, yoga, and massage to lower stress levels in patients. Among women who have been treated for breast cancer, survival rates are significantly higher for those women who participate in social support groups (see Section 14.4.5).

10.6.6.3 Immune function Evidence is increasing that people with weakened immune systems develop cancers more frequently. This factor is striking in conditions that severely damage the immune system, such as AIDS (see Chapter 15), but is also present in people whose immune systems are weakened less drastically from other causes, including chronic stress, sleep disorders, and so forth. Far more cells are mutated and transformed than ever develop into cancers. A healthy and active immune system eliminates most of these cells as they arise in a process called **immunosurveillance**. Any weakening of the immune system increases the number of transformed cells that grow and proliferate, and a higher rate of cancer is one of the results. People's immune systems also weaken with age, which is consistent with the finding that the incidence of new cancers increases with age.

Animal studies have shown that natural killer (NK) cells, a component of the innate immune system (see Section 14.2), are especially important for combating cancer. NK cells recognize abnormal cell-surface features shared by many cancer cells (but not normal cells) and inject so-called "cytotoxic granules" that contain proteins that kill the cancer cells. Genetically engineered mice that have defects in NK function were shown to develop spontaneous tumors at an increased frequency than normal mice (Guerra et al., 2008).

The adaptive (specific) immune system (see Section 14.2.3) is also important for carrying out immunosurveillance and eliminating transformed cells that would otherwise proliferate to form tumors. Cancer arises from a person's own cells, so cancer cells are not "foreign" in the sense that a virus or a bacterium is. However, because cancer cells contain numerous genetic mutations, they express proteins that are different from those that were present in the body early in life when the immune system was being "trained" not to attack the body's own cells (see Section 14.2.3). These new (novel) features of the mutated proteins expressed by cancer cells therefore function as antigens to attract the attention of antigen-specific B and T cells, activating immune responses that eliminate the cancer cells expressing the novel proteins. The novel proteins in cancer cells that stimulate a specific immune response need not be the same mutated proteins that are responsible for the cells' transformation (that is, the proteins produced by oncogenes or inactivated tumor suppressor genes). In fact, one key feature of transformed cells is that they acquire new mutations at a much higher rate than normal cells (often because of loss-of-function mutations in DNA repair genes). Many of these mutations are so-called "passenger" mutations; they don't "drive" the cancerous progression but are just "along for the ride." Nevertheless, passenger mutations can code for novel proteins that target cancer cells for elimination by the adaptive immune system. The novel proteins produced by a given cancer can differ in the degree to which they activate B and T cells—that is, in their "antigenicity." Thus, some tumors might grow very slowly because they express mutated proteins that are highly antigenic, causing the cancer cells to constantly be attacked by immune cells, whereas others may express mutated proteins that are less antigenic and can "fly under the radar" of the body's immunosurveillance system. These differences in antigenicity of mutated proteins produced by cancer cells can have important implications for therapies that exploit the body's immune system (immunotherapies), which are discussed later in this chapter.

10.6.7 Social, economic, and racial factors

Socioeconomic status refers to a person's income and education level. Together with race, socioeconomic status influences incidence rates, and especially survival rates, for various cancers. With regard to race, some differences likely relate to genetic susceptibilities: For example, African American women seem especially susceptible to developing a particularly aggressive form of breast cancer (triple-negative breast cancer, so-named because the cancer cells do not express three cell-surface receptors common on breast cells). Other racial differences relate instead to socioeconomic forces more likely to be experienced by particular racial groups.

In the United States, data for the years 2011–2015 showed higher incidence rates for black men for all cancers combined compared to all races or to whites (American Cancer Society, 2019). Black women had lower rates of cancer incidence, but both black men and women had higher rates of death from cancer (mortality). The mortality rates for blacks are higher, in part, because they are less likely to receive routine medical exams that would detect the common cancers in their earliest and most treatable stages. These include screening measures such as mammograms, colonoscopies, Pap smears, and prostate-specific antigen (PSA) tests, discussed elsewhere in this chapter.

From the 1950s to the 1980s, socioeconomic status was *directly* correlated with cancer incidence and mortality rates; in other words, people with *higher* income and education were *more* likely to develop and die of cancer than those with lower income and education. At the end of the 1980s, that trend reversed; it now shows an *inverse* correlation, such that *lower* socioeconomic status is now correlated with *increased* cancer incidence and mortality rates for all races (Singh and Jemal, 2017). People of all races with lower socioeconomic status are more likely to exhibit behaviors that increase cancer risk, such as smoking, physical inactivity, and poor diets. Differences in socioeconomic status are associated with a number of disparities that can influence cancer rates and outcomes: access to affordable health insurance (and therefore access to regular, preventive care); access to transportation to be able to receive care; ability to take time off work (paid sick leave) or find childcare for medical appointments; access to healthy foods; and education surrounding habits that can reduce cancer risk. In the United States, the Medicaid expansion that occurred in many states as a result of the Affordable Care Act passed by Congress in 2014 made health insurance more attainable for people with low incomes. While this helped all races, it was especially effective in narrowing the gap in the percentages of whites versus nonwhites who have health insurance. This is important progress toward ensuring health care access that can decrease cancer mortality rates; however, significantly higher percentages of nonwhites remain uninsured compared to whites, especially for Hispanic and Native American individuals. Apart from health care access and the lifestyle factors already mentioned, chronic stress related to socioeconomic status may also play a role in cancer incidence rates and outcomes. Several recent studies have indicated that many nonwhite people who are not disadvantaged economically or educationally nevertheless experience higher levels of stress than their white counterparts due to real or perceived encounters with racial discrimination or stereotyping.

As we have seen, many factors may contribute to cancer incidence rates, including genetic predisposition, lifestyle, and exposure to environmental carcinogens.

Table 10.7 Summary of various causes of cancer worldwide

Cause	Relative percentage of cancer deaths[*]
Smoking and tobacco use	22
Diet	30
Alcohol	6
Food additives (salt)	1
Sedentary lifestyle	3
Radiation	2[†]
Pollutants (air, water)	2
Viruses	20
Chemical carcinogens	Variable[‡]
Genetic susceptibility	<10

[*] Derived from statistical analysis of epidemiological data. A figure of 30 means that 30% of all cancer deaths worldwide are attributable to that particular cause.

[†] Over 90% of skin cancers are caused by UV radiation.

[‡] The percentage of deaths in the general population attributable to carcinogen exposure is low, but can be locally very high, for example, in industries with high or prolonged exposure.

Estimates of the relative contributions of various biological causes is given in Table 10.7. These are rough averages for global cancer incidence. The percentage of cancers attributable to viruses is about 15% worldwide but is much lower in the United States. The percentage attributable to diet also varies from one part of the world to another, as does the percentage attributable to chemical carcinogens in the environment. With the exception of the low percentage of cancers that may be attributable to genetic predisposition, changes that can be made by individuals or by societies have the potential to significantly decrease incidence rates for most cancers, a topic we explore further in the next section.

THOUGHT QUESTIONS

1. Not everyone who smokes gets cancer. Does this mean that smoking is not a risk factor for cancer?
2. What is the difference between a risk factor and a cause? Can we say what caused cancer in a given individual?
3. Why do different individuals respond differently to cancer risk factors?
4. Does an increase in the percentage of deaths due to cancer necessarily mean that cancer rates have increased? What else could explain such findings? How could you go about determining which of the possible explanations best fits the data?
5. Recently the genetic defect associated with an inherited form of colon cancer was identified as a defect in a DNA-checking protein. This finding was reported in the lay press as the discovery of "the colon cancer gene." Is this name misleading? To what extent can a defective repair mechanism be considered to be the same thing as a cause of a cancer?
6. Tobacco smoke contains chemicals that are tumor initiators and other chemicals that are tumor promoters. How does this combination contribute to the carcinogenicity of tobacco smoke?
7. Tobacco and alcohol act synergistically in increasing cancer risks. Can you explain this in terms of what is happening inside cells?

10.7 WE CAN TREAT MANY CANCERS AND LOWER OUR RISKS FOR MANY MORE

An understanding of the mechanisms that produce cancer has greatly increased our understanding of how to prevent and treat it. Cancer therapies have improved significantly over the past few decades and continue to do so, resulting in increased survival rates for most cancers. In the period from 1991 to 2018, overall cancer death rates in the United States decreased 31% (American Cancer Society, 2021b). In this section we examine medicine's current strategies for treatment and prevention.

10.7.1 Traditional cancer therapies: Surgery, radiation, and chemotherapy

The traditional treatments for cancer comprise a triad of approaches that have been used since the 1940s: surgery, radiation, and chemotherapy. Surgery aims to remove visible tumors, but it cannot eliminate cancer cells circulating in the bloodstream that have the potential to spawn metastatic tumors or existing metastases that are too small to detect. Therefore, surgery is often combined with radiation or chemotherapy. With either radiation therapy or chemotherapy, the strategy is the same: Cancer cells are dividing cells; therefore, agents that interfere with cell division should stop tumor growth or even cause tumors to shrink due to apoptosis. Radiation causes breaks in the DNA of dividing cells that are so large that the cell cannot repair them and usually cannot live with the damage. Chemotherapeutic drugs prevent DNA synthesis at several steps. Some of the drugs inhibit the synthesis of the nucleotides needed to build DNA; some substitute for certain nucleotides in newly synthesized DNA, preventing its further replication; and some inhibit an enzyme needed to unwind and rewind the double helix during its replication. Other chemotherapeutic drugs, some of which are natural plant products, prevent RNA synthesis or damage the mitotic spindle, thus blocking mitosis.

An important drawback to radiation and chemotherapy treatments is that, because they damage DNA, they increase the risk of secondary cancers derived from damaged cells that survive the treatments. Another drawback is that both radiation and chemotherapy are nonspecific: They both kill any type of dividing cell without distinguishing cancer cells from healthy cells. Two examples of healthy dividing cells are hair follicle cells and immune cells. A high proportion of cells in hair follicles are dividing, so hair loss frequently accompanies chemotherapy treatments. A large percentage of cells of the immune system are also dividing and so are killed. Not all hair follicle and immune cells are killed because not all were dividing at the time of treatment, so they can repopulate. Hair grows back, and people regain their full immune function. During the time when people's immune systems are compromised, they need to avoid exposure to infectious diseases. Both radiation and chemotherapy may destroy memory B and T cells, which "remember" which diseases the person has been exposed to or vaccinated against (see Section 14.2.4). If these memory cells are killed, even a person who has regained the ability to form new immune responses has lost previous immunities and therefore may need to be revaccinated.

A further risk from chemotherapeutic drugs is that they put a selective pressure on the population of transformed cells. As a result, any cells that become resistant to the drug quickly outcompete the drug-susceptible cells. Because cancer cells mutate quickly, they often acquire mutations that confer resistance to specific drugs. Several drugs, each of which works by a different mechanism of action, are often used in combination to minimize the development of drug resistance. However, cancer cells may also develop mutations that prevent them from transporting drug molecules across their membranes or start expressing cell-surface "pump" proteins that can pump a range of different drugs out of the cell. This phenomenon of "multidrug resistance" is a major problem in cancer treatment.

Despite the drawbacks and risks, surgery, radiation, and chemotherapy have been very effective, increasing the survival rates of many types of cancers. Moreover, in the vast majority of patients, melanoma, Hodgkin lymphoma, and breast, prostate, testicular, cervical, and thyroid cancers can now be "cured" by these treatments, as measured by five-year survival rates. Leukemia also has very high "cure" rates in patients under 55, and especially in children. It is impossible to know whether cancer cells have been eliminated from a person's body, so many physicians and scientists hesitate to use the word "cure" when talking about cancer outcomes. More often, they say that a person's cancer is "in remission," which means that cancer cells cannot be detected, but there is a chance that it could recur later. Recurrence rates vary widely depending on multiple factors, including the type of cancer, the cancer stage at diagnosis, and the treatment received. In some cases, even patients who have been in remission for more than five years are monitored in the hope than any recurrence can be detected early. For some cancers, especially those diagnosed at later stages, surgery, radiation, and chemotherapy cannot achieve complete remission; they may not fully eliminate solid tumors or circulating tumor cells, but they cause tumors to shrink and the numbers of circulating tumor cells to be significantly reduced. In such cases, cancer is treated as a chronic disease, with patients continually receiving treatment to try to prevent further progression of their cancers without achieving true remission.

10.7.2 Targeted cancer drugs

Research on cancer treatments continues at a rapid pace. New chemotherapeutic agents, both natural and artificial, continue to be sought and developed. Some of these are similar to "traditional" chemotherapies, in that they target features common to all dividing cells. However, there is greater emphasis on therapies that exploit specific features of cancer cells that are not shared by normal cells, which minimizes side effects due to generalized cell death and maximizes killing of cancer cells. Such agents are often referred to as "designer drugs" and are the basis of a new trend in cancer therapy called "precision medicine." In precision medicine, a patient is not just diagnosed with a particular type of cancer, but the specific features of the cancer cells are assessed at the genetic and molecular level to identify possible routes of targeted therapy. For example, DNA may be isolated from cells taken from a tumor that has been removed surgically (or from a small piece of a tumor taken in a biopsy), and the sequences of certain genes known to be commonly mutated in that cancer (i.e., driver mutations in oncogenes and tumor suppressor genes) will be determined to identify mutations. In most cases, genetic testing of tumors focuses on a panel of specific driver genes, but given the rapid advances in next-generation DNA sequencing (see Section 4.4.2), the entire genomes of tumors are sometimes sequenced. Other types of tests may detect the expression level of particular proteins known to be overexpressed in certain types of cancers.

One of the first precision drugs approved for use in patients targets a specific chromosomal rearrangement that occurs in more than 90% of cases of chronic myelogenous leukemia (CML). This chromosomal mutation fuses the coding sequences of two signaling proteins, Bcr and Abl, creating a fusion protein that has hyperactive growth-promoting activity, as mentioned earlier in this chapter. Scientists determined the three-dimensional structure of the Bcr-Abl oncoprotein and designed a drug that could bind specifically to this protein and block its function (see Figure 4.17). The result was the drug Gleevec®, which was approved for use in the United States in 2001. About 90%–95% of CML patients survive five years or more after diagnosis when treated with Gleevec (Druker et al., 2006). CML accounts for only 15% of leukemia cases, but the Bcr-Abl oncogene is also implicated in a small percentage of cases of another leukemia (acute lymphoblastic leukemia [ALL]), and Gleevec is highly effective for those patients as well.

As described earlier in this chapter, the hormone estrogen drives the proliferation of many breast cancers. Therefore, drugs that block the estrogen signaling pathway can be effective in treating so-called estrogen-dependent breast cancers. About 80% of breast cancers express the estrogen receptor. These can be treated with a class of drugs called selective estrogen receptor modulators (SERMs). The SERMs bind to the estrogen receptor and block its ability to promote breast cancer cell division. Estrogen has additional functions in cells apart from promoting cell division; these include maintenance of bone density and cholesterol regulation. Importantly, while SERMs block the proliferative effect of estrogen, they do not block its other, desirable effects. Tamoxifen is one example of a SERM that has proved effective in reducing mortality from breast cancer and in decreasing its onset in women at high risk due to genetic predisposition or age (over 60). Another class of drugs, the aromatase inhibitors, block an enzyme required to make estrogen. These are frequently used to treat estrogen-dependent breast cancer in postmenopausal women, but they do not have the selective action of the SERMs, so patients taking them must be monitored and sometimes treated for bone loss and elevated cholesterol.

Another type of targeted therapy aims to deliver toxins selectively to tumor cells but not normal cells in a "seek and destroy" tactic. One such approach, called photodynamic therapy, is used for esophageal cancer, for a type of lung cancer that affects the large breathing tubes (bronchi), and for precancerous skin lesions. In this treatment, the patient is given an intravenous injection of a photosensitizing dye, which accumulates in tumor cells, while being expelled by normal cells. The patient's tumor is then illuminated using fiber optics. Light reacts with the dye in the tumor cells to create a toxin that kills the cells. Photodynamic therapy is limited to tumors that are accessible to the fiber optic light required to activate the toxin. Other toxin-based therapies are delivered in nanoparticles that can be injected into the bloodstream. When it was first developed in the mid-1990s, nanoparticle-mediated cancer therapy relied on the fact that the extra blood vessels that develop to feed tumors have "leaky walls"; when injected into the bloodstream, nanoparticles bound to chemotherapy drugs can exit the leaky blood vessels into tumors but are too large to exit intact vessels in normal tissues. In more recent years, nanoparticles have been developed that use targeting molecules such as antibodies that bind specifically to proteins found preferentially on cancer cells to increase the concentration of the drug-carrying nanoparticles in tumors. Another approach uses nanoparticles made of metals such as gold or silver; once these have been given time to concentrate inside tumors, a laser pulse or radiation is administered, which causes the metal nanoparticles to heat up and kill tumor cells. Many cancer researchers are working to develop additional nanoparticle-mediated therapies, but one paper published in 2019 cautions that care must be taken to ensure that nanoparticles do not inadvertently create larger holes in blood vessels that will promote migration of tumor cells into the vessels, thereby stimulating metastasis.

Because we now know so much about the changes that occur in cells during transformation and cancer progression, scientists have pursued many different routes to exploit the varied characteristics of cancer cells that we discussed earlier in the chapter. These include the following: drugs that inhibit the enzyme telomerase, which confers "immortality" on cancer cells; angiogenesis inhibitors that target production of new blood vessels required to provide nutrients and oxygen to tumors; inhibitors of many proteins that function in growth factor signaling pathways or drive cell cycle progression; and inhibitors of the proteasome, which regulates many cellular pathways via protein degradation. Unfortunately, none of these has proved to be a "magic bullet"; in other words, there does not seem to be one specific feature of cancer that can be targeted that will result in its eradication. Rather, the best outcomes are typically achieved by combining multiple angles of attack, and different strategies work better for different types of cancers and even different patients with the same basic type of cancer.

One additional behavior of cancer cells that could be a very important clinical target is the process of metastasis. Since many cancer deaths are due to metastasis, a treatment that inhibits the formation of second-site tumors could make a very large impact on cancer mortality rates. One study (Frankowsky et al., 2018) described a compound the authors called "metarrestin" that suppressed metastasis in mouse models of pancreatic cancer, and clinical trials testing the safety and efficacy of this drug in humans have already begun.

After extensive laboratory research, new therapies are tested only on patients who have given their informed consent (see Section 1.5.3) to be part of the studies; for the first stage of clinical trials patients accepted to participate are usually the very sickest who have exhausted all other options, since unexpected, possibly dangerous side effects can occur, and it would not be ethical to subject healthier patients to those potential dangers. Once a new therapy has been deemed safe, other types of patients are accepted into later rounds of clinical trials that are designed to test its efficacy and, particularly, to determine whether it works better than the treatments currently used for a given cancer. The gold standard of clinical trials is a double-blind trial, where neither the patient nor the physician knows which patients are receiving the therapy and which are receiving a mock therapy, or placebo.

Because cancer is greatly feared and not always curable, some people put their hopes in unproven remedies. For a number of years, a compound called laetrile (also known as amygdalin) achieved a large and devoted following as an alternative therapy. Laetrile is a bitter compound found in some fruit pits, raw nuts, and other plant parts; when taken into the body, it is converted into cyanide, which was supposedly effective at killing cancer cells. The supporters of laetrile became so influential that, in the early 1980s, the National Cancer Institute conducted careful clinical trials. The clinical trial results showed no positive effects of laetrile, but even years later, it is still produced in Mexico (without batch standardization) and is sought by some patients as a last resort.

10.7.3 Immunotherapies

In the past few decades, much progress has been made toward understanding how the human immune system responds to cancer, and these discoveries have yielded novel therapies that are achieving significant clinical results (Table 10.8). As discussed earlier, cancer cells represent unique targets of the immune system because they are "self," but they may also have features that distinguish them from normal cells. A variety of animal and human studies have shown that, during the course of cancer progression, an unrelenting "tug of war" plays out between the cancer cells and a person's immune system. Cancer cells may express novel, mutated proteins that function as antigens to attract the attention of immune cells, resulting in elimination of some of the cancer cells; however, additional mutations typically occur in the cancer cells that cause changes in gene expression that allow new subpopulations of cancer cells to escape immunosurveillance. Thus, there is a continual back-and-forth between the cancer cells and the immune system, with one temporarily gaining ground and then the other fighting back. The goal of cancer immunotherapies is to tip the balance in favor of the patient's immune system.

One type of immunotherapy is "passive immunization." In this approach, patients are given immune products such as antibodies or antigen-specific immune cells that stimulate destruction of the cancer cells. One of the first and most successful antibody-based passive immunization treatments is a drug called Herceptin, which consists of an antibody that recognizes a cell-surface receptor called HER2 that is highly expressed on about 25%–30% of breast cancers. Herceptin works in several ways: Its binding to the HER2 receptor can block signaling pathways that stimulate breast cancer cell proliferation; it can also cause the HER2 protein to be taken into the cell and degraded (thus also blocking signaling); and, further, it can attract NK cells that kill cancer cells by injecting them with cytotoxic granules.

Another type of passive immunization, called adoptive cell transfer, involves giving a patient immune cells—either from a donor or expanded cell populations derived from the patient themselves. One type of adoptive cell transfer has been

Table 10.8 Examples of common chemotherapy drugs and their modes of action

Drug	Mode of action
Cisplatin	Damages DNA by adding chemical groups to DNA bases (alkylating agent)
5-Fluorouracil	Incorporated into DNA in place of normal nucleotides (antimetabolite)
Methotrexate	Incorporated into DNA in place of normal nucleotides (antimetabolite)
Hydroxyurea	Inhibits nucleotide synthesis
Doxorubicin	Interferes with DNA replication through several mechanisms
Bleomycin	Interferes with DNA replication through several mechanisms
Etoposide	Inhibits an enzyme needed for unwinding DNA strands during replication
Camptothecin*	Inhibits an enzyme needed for unwinding DNA strands during replication
Paclitaxel (Taxol)*	Inhibits mitotic spindle function by binding to microtubules
Vinblastine and vincristine*	Inhibit mitotic spindle function by binding to microtubules

* Originally isolated from plants (plant alkaloids).

used for many years in bone marrow transplants to replace blood cell-generating bone marrow in patients with blood cancers whose own bone marrow has been destroyed by radiation treatments. In recent years, it has become evident that the best outcomes are achieved when the surface proteins on the bone marrow donor cells differ very slightly from the patient's so that the donor immune cells recognize remaining cancer cells as foreign and destroy them. (Donor cells that are too different from the patient can cause a lethal immune response, so patients must be very carefully matched with donors.) Another passive immunization therapy that has been approved for clinical use more recently is called chimeric antigen receptor T-cell therapy (CAR-T). In this strategy, a patient's own T cells are harvested from their blood and genetically engineered in the laboratory to express a particular receptor protein (a chimeric antigen receptor) that will recognize a molecule found on the surface of the patient's cancer cells. These engineered T cells are then infused into the patient's bloodstream, where they seek out cancer cells expressing their target antigen and kill them. CAR-T therapies have been approved for several blood cancers, and more are being developed. One final example of passive immunization involves administration of tumor-infiltrating lymphocytes (TILs, or T and B cells harvested from excised tumors) that have been allowed to proliferate in tissue culture, often stimulated by activating immune signal molecules. These expanded TIL populations are introduced back into the patient as "reinforcement troops" to kill remaining tumor cells. Studies have shown that patients with tumors that naturally have a higher number of TILs have better prognoses. TIL therapy has been especially effective for metastatic melanoma, allowing some patients to achieve complete remission and significantly improving three-year survival rates.

In addition to the passive immunization strategies just described, several active immunization approaches to cancer therapy have been developed in recent years. These aim to stimulate the body's own immune response, rather than delivering immune cells or molecules. One such strategy is called immune checkpoint blockade. The immune system has natural mechanisms that limit the activity of stimulated T cells so that out-of-control immune responses are not produced that might attack normal cells. In the case of cancer, these so-called immune "checkpoint" mechanisms can limit the efficacy of antitumor immune responses. Therefore, molecules that interfere with the immune checkpoint responses can maximize the cancer-killing potential of a patient's own T cells. Antibodies that bind and block the function of three key immune checkpoint proteins, PD-1, PDL-1, and CTLA-4, are currently in use to treat many different types of cancer and are being studied for additional applications. Former U.S. President Jimmy Carter is one high-profile success story for immune checkpoint inhibitors. He had melanoma that had spread to his liver and his brain; after three months of checkpoint inhibitor therapy, no remaining cancer could be detected anywhere in his body. Such dramatic results are seen in only about 20% of patients, but they provide hope to many patients with advanced cancers.

Cancer vaccines are another type of active immunization treatment. In contrast to vaccines against viruses or bacteria, which are administered to prevent infections, cancer vaccines are designed to prime the immune system to fight existing tumors. Only one cancer vaccine is currently approved for clinical use, but others are in development. The approved vaccine is called Provenge, or sipuleucel-T, and it is used to treat advanced prostate cancer. In this treatment, dendritic cells, which are a type of immune cell that process and "present" antigens to other immune cells, are isolated from a patient's blood and cultured in vitro in the presence of a protein called PAP that is nearly universally expressed by prostate cancer cells. When these dendritic cells are infused into the patient, they stimulate the patient's B cells to produce antibodies against the prostate cancer-specific antigen and activate T cells that can kill the cancer cells. Because most prostate cancers express PAP, this vaccine can be used in most prostate cancer patients. One small clinical study reported in 2017 developed personalized vaccines for six melanoma patients that consisted of small segments of up to 20 novel proteins (neoantigens) expressed by the melanoma cells but not by normal cells; the neoantigen mixture for each patient was generated by sequencing the complete genomes of the patient's cancer cells and their normal cells, identifying cancer-specific mutations likely to be antigenic, and expressing protein fragments corresponding to those mutated genes (Ott et al., 2017). Melanoma was chosen for this study because it is caused by exposure to UV light, which is highly mutagenic, so melanoma cells tend to have higher overall numbers of mutations than other cancers, including many "passenger" mutations that do not contribute to malignant behaviors but cause the expression of novel proteins not present in normal cells. Of the six patients in the study, four achieved complete remission with the vaccine alone, while the other two went into remission after receiving the vaccine plus an immune checkpoint inhibitor. While these results are very promising, scaling up production of individualized vaccines for larger numbers of patients will present great financial and technical challenges.

10.7.4 The cost of cancer treatment

The process of developing a new cancer drug is estimated to take 15 years or more, including the time spent on laboratory research to identify and optimize a drug and as many as 8 years of clinical trials prior to approval by regulatory agencies (in the United States, the Food and Drug Administration). This is very expensive. Drug companies typically cite a figure of $2.7 billion for the whole process; an independent study claimed the real figure may be closer to $650 million, but this is still a high price. Once a drug is marketed, patients and their insurance companies also pay high prices, as drug companies seek to recoup their investments and earn a profit. For many patients, the cost of cancer therapy can lead to bankruptcy or other serious financial problems. For a new cancer drug to be approved for treatment of a given cancer, it must not only be safe and effective it also must produce better outcomes than the current standard of care for that cancer. However, the definition of a "better outcome" is often measured in months of extended survival rather than years. For example, the prostate cancer vaccine mentioned earlier in this chapter was shown to extend the time for which 50% of trial participants survived by only about 4 months compared to the placebo; the cost of this treatment is approximately $93,000 per patient. Extending a patient's life by a matter of months may enable that person to attend a child's wedding or see the birth of a grandchild; however, it is important to consider whether the benefits of a short extension of life outweigh the financial costs of treatment. This is a difficult ethical question, and its emotional content makes it extremely difficult to enact governmental policies that aim to place any sort of limits on late-stage cancer care, even when there is very little chance that a patient will survive. Some cancer therapies are extremely expensive but have a better chance of producing lasting remission. For example, one type of CAR-T therapy was shown to produce complete remission for two years or more in more than 50% of patients; the cost of CAR-T for a single patient is currently $1.5 million. As cancer treatments become ever more specialized and the numbers of patients affected by cancer remain high, the cost of cancer care will be an issue that cannot be ignored.

10.7.5 Cancer detection and predisposition

Early detection greatly increases the probability of successful cancer treatment. Monthly breast self-examination is very effective at finding tumors while they are treatable. Diagnostic breast X-rays (mammograms) and ultrasound (sonograms) detect tumors that are too small to be felt but are less effective in younger women than in postmenopausal women whose breast tissue is less dense. Microscopic examination of tissue from the cervix taken during a medical examination, called a Pap smear, is effective at early detection of cervical cancer. Testicular cancer, although rare, is the most frequent cancer in men between the ages of 15 and 34; as for breast tumors, monthly self-examinations to detect lumps in the testes are an important method of early detection. As mentioned earlier, regular colonoscopies in people older than 50 (or 45, according to a recommendation issued by the U.S. Preventive Services Task Force in May 2021) have also contributed to decreased mortality from colon cancer.

An emphasis on early detection has contributed to the increased survival rate for certain cancers. For other cancers, it is not so clear what is meant by an increase in survival rate and how this relates to a "cure." When we try to evaluate the meaning of statistical statements like "survival rates," we need to know a lot about how the numbers were gathered and what definitions are being used for certain terms. What is often reported as survival rate refers to the proportion of persons with cancer still living after five years compared with the proportion of surviving persons without cancer. For example, it is estimated that it takes nine years for a breast cancer to develop, spread, and kill a person. If past methods of detection led to discovery of the cancer seven years after its inception, very few people would have been alive five years after its discovery. Now, with better detection methods and better public education for breast self-examination, the five-year survival rate looks much improved. Does this mean that therapies have improved, or does it simply mean that people are now finding the cancers at two years into their development rather than at seven years? It is not always easy to distinguish advances made through better or earlier diagnosis from advances made in the treatment of cancers once they have reached comparable stages of development.

In addition to the tests mentioned earlier, two types of laboratory cancer tests currently exist. One type is for the early detection of existing cancers. The other type is genetic testing for cancer predisposition.

An example of the first type is the PSA test for prostate cancer. This test measures the level of PSA in a person's blood. This protein is elevated when prostate cancer begins to develop. Thirty percent of people with elevated PSA are found to have cancer when the test is followed by a biopsy (a tissue sample examined by microscope). The PSA test can detect cancer up to five years before there are other symptoms. The clinical question then becomes what to do about it. Cancers detected by PSA tests are generally still localized and can therefore be successfully removed surgically. However, prostate cancer is very slow growing, and it has been said that most men die with prostate cancer, not because of it. One-third of men have some form of prostate cancer by the time they are over the age of 50, but only 3% die from it. For many individuals, after a biopsy has confirmed the presence of prostate cancer, the recommended treatment is therefore to do nothing. However, a recent study (Holmberg et al., 2002) has shown that men with newly diagnosed prostate cancer who have their prostate surgically removed have a 48% reduction in their risk of dying from this disease (4.6% mortality) compared to a similar group of men who underwent "watchful waiting" (8.9% mortality). There are similar uncertainties related to some types of breast cancer. Many breast tumors develop very slowly and have a low probability of spreading to other sites, so, especially in women over age 75, minimal treatment may be the best option.

The second type of laboratory test does not detect cancer, but instead identifies DNA sequences that are statistically correlated with an increased probability of someday acquiring the disease. Tests of this type are available for *BRCA1* and *BRCA2* (breast cancer predisposition mutations), DNA mismatch repair genes (predisposing to some kinds of colon and uterine cancers), *APC* (predisposing to colon cancer), *p53* (predisposing to a range of tumors), and a few other tumor suppressor genes and oncogenes. The interpretation of these gene-specific tests can be difficult; even if a DNA sequence difference is identified (compared to the "normal" gene sequence), it is not always clear what effect the change might be expected to have on the protein product of the gene.

Tests for genetic predispositions for cancer are controversial for other reasons as well. They are very expensive and do not give much more useful information than is gained from knowing your family medical history. A negative test does not mean that a person will not get cancer; cancers arise spontaneously in people with no family history of the disease, so the recommended preventive measures discussed later should still be followed. A positive test is also not a guarantee that a person will get cancer. The probability is increased, but we cannot always tell by how much. The increase in risk is different for different mutations, but none increases the probability to 100%. The genetic mutation cannot be repaired, so there is little that a person with increased risk can or should do beyond what is recommended for everyone (regular check-ups and the lifestyle choices summarized later). Still, because some of these cancers are difficult to detect early, being aware of a predisposition for them can ensure that physical examinations are done even more thoroughly than usual, and perhaps more often.

Some women with increased breast cancer risk due to mutations in *BRCA1* or *BRCA2*, including actress Angelina Jolie, choose to have "prophylactic mastectomies," that is, removal of their breasts before there is any evidence of disease. A recent study was widely publicized as showing that women who had their breasts removed reduced their risk of dying by 90%. While this statement is not untrue, it is only part of the story and is an example of reporting "relative risk" instead of "absolute risk." In the study, 639 women had their breasts removed. Based on calculations made from the number of deaths among their sisters who faced the same increased susceptibility but did not have their breasts removed, it was estimated that 20 of the 639 women would have died. In actual, only 2 died, so the relative risk was decreased by 90% $[(20 - 2)/20 \times 100 = 90\%]$. When absolute risk is considered, it can also be correctly stated that 97% of the women had their breasts removed unnecessarily $[(639 - 20)/639 \times 100 = 97\%]$. The difficulty for any woman faced with such a choice is that there is no way to predict whether she will be one of the 18 saved by the procedure or one of the 619 who did not need it.

10.7.6 Cancer management

Some people feel that more research dollars should be spent on cancer management, not just cancer treatment. Cancer management includes the development of drugs or strategies to minimize the side effects of cancer treatments. Examples include cold capping, a procedure in which a cold pack is applied to the scalp during chemotherapy so as to slow cell division in hair follicle cells, thus decreasing hair loss. Another possibility is the development of antinausea drugs. Use of medicinal marijuana to overcome nausea and restore appetite in chemotherapy patients is becoming increasingly common (see Section 13.2.2). As of 2020, 33 U.S. states and the District of Columbia have legalized medicinal use of marijuana, and 14 others have approved restricted medicinal use; however, a 2018 study reported that, while 73% of oncologists (cancer specialists) believed marijuana could provide benefits to their patients, only 46% felt comfortable recommending it to patients (Braun et al., 2018). Because of the federal ban on marijuana, there has been little funding for proper scientific studies of its efficacy; however, this is beginning to change as more states legalize even recreational marijuana and the stigma surrounding this drug is reduced.

Cancer management also includes providing psychological and emotional support through group or individual therapy and complementary therapies such as massage, acupuncture, yoga, meditation, and nutritional counseling. The aim of these approaches is to improve the quality of life—to treat the person, not just the disease. Women who were in support groups after recurrent breast cancer lived longer than those who were not. Survival rates have been found in some cases to be influenced by mental attitude (see Section 14.4.5). Patients who were optimistic, who were aggressive, or who were "determined fighters" had statistically longer survival rates and higher cure rates than those who were pessimistic or who resigned themselves early to their fate. Complementary therapies aim to keep the body and mind as strong as possible while an individual endures the often grueling treatments required to fight their cancer.

10.7.7 Cancer prevention

As we learn more about the causes of cancer, it seems that preventing cancer may be far easier and less expensive than curing it.

Smoking remains a major cause of cancer (**Figure 10.28**). The American Cancer Society estimates that about 80% of lung cancer deaths are from smoking (American Cancer Society, 2021c). Exposure to secondhand smoke (for example, living in a home with someone who smokes) causes death from lung cancer in over 7,000 nonsmokers a year in the United States. Tobacco smoke contains both tumor initiators and tumor promoters, and it suppresses the immune system.

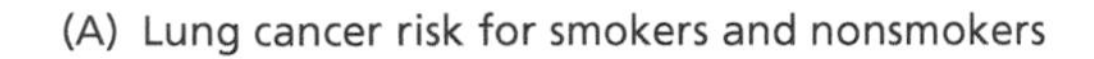

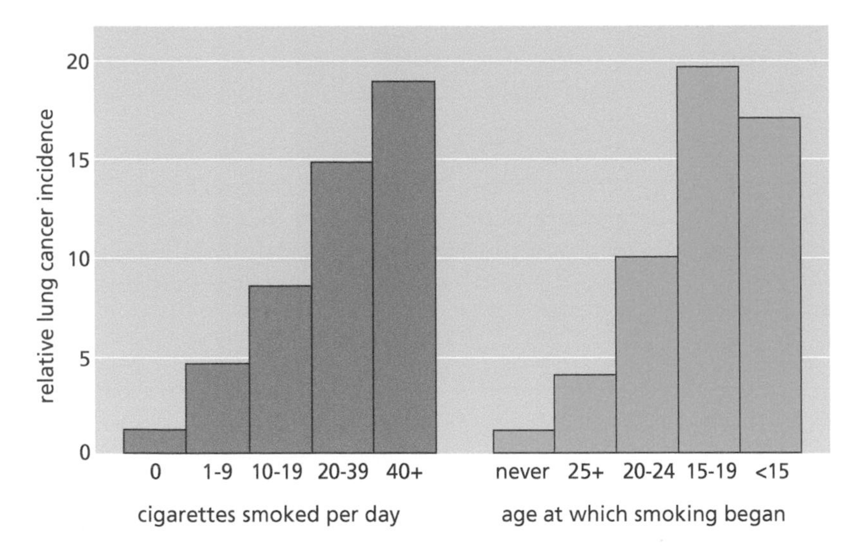

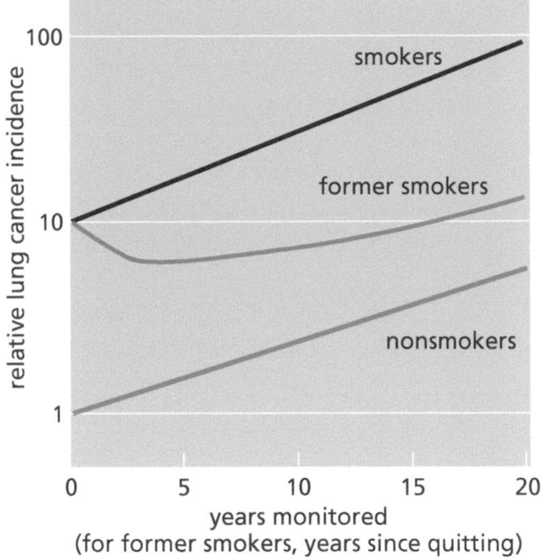

Figure 10.28 The effects of smoking on the incidence of cancer.

Some dietary regimens have been associated with a decreased risk of cancer: lower total calorie and lower fat intake, for example. Foods containing antioxidants may also help, as these compounds protect cells from the DNA-damaging effects of mutagens; such antioxidants include beta-carotene and vitamins A, C, and E (see Section 8.2.5), which can be found in a wide range of fruits and vegetables (especially brightly colored ones), nuts, and even dark chocolate. Fresh vegetables of the plant family Cruciferae (including cabbage, radish, turnip, broccoli, cauliflower, kale, kohlrabi, mustard greens, and brussels sprouts) are especially rich in these and other protective compounds, and these are recommended by the National Cancer Institute for that reason. Green tea also contains antioxidants and other compounds that may help protect against cancer. Clinical studies have shown that some vitamin supplements are not as effective in preventing cancer as the same vitamins obtained from foods, probably because other food ingredients are also at work or because the vitamins in the supplements are not absorbed as well as from foods. High-fat diets and obesity are important risk factors in colorectal cancer and in postmenopausal breast cancer. There are also evidence that high-fiber diets lower the risks of colon and rectal cancers, mainly by decreasing calorie intake and helping people maintain a healthy weight.

Given the available evidence, the most important actions you can take to lower your cancer risks are the following.

1. *Don't smoke*! Also, avoid secondhand smoke from poorly ventilated rooms where others smoke. *These are the single greatest steps you can take to reduce your cancer risk, far outweighing all other possible measures.*
2. Limit alcohol consumption, the American Cancer Society recommends no more than two alcoholic drinks per day for men and no more than one per day for women.
3. Follow a diet low in fats, high in fiber, and high in antioxidants, and strive to maintain a healthy weight through both diet and regular exercise.
4. Avoid occupational exposures to potential carcinogens; minimize exposure through the appropriate use of safety equipment.
5. Avoid exposure to radioactive substances and X-rays above necessary minimum levels, and avoid needless exposure to UV radiation from the sun or tanning booths.
6. As you age, be sure to get checkups at regular intervals, including screening that detects the common cancers in their earliest and most easily treated stages. If you are a woman, learn to practice breast self-examination and, if you are a man, testicular self-examination.

THOUGHT QUESTIONS

1. Explain how the theory of evolution accounts for the development of cancer cells that are resistant to chemotherapy.
2. What characteristics would you look for in an ideal chemotherapeutic drug for the treatment of cancer?
3. Secondhand smoke is a cancer risk. Does this biological reality change the ethical debate about smoking in restaurants or smoking in the workplace? What rights are in conflict on either side?
4. Consider the following statement: When someone's chances for survival are predicted to be very low, any and all treatments are justified. In other words, any treatment is good as long as it is not harmful. Do you agree? Try to apply this thinking to such unproven remedies as laetrile.
5. Is it ever ethically permissible to give up on treatment? Are there things other than treatments that can be done for a dying person? Do you think people who have terminal cancer should be told of their condition? Try to justify your answers. What ethical assumptions underlie your argument?

CONCLUDING REMARKS

Research aimed at understanding the normal signals that cause stem cells to differentiate into specialized cell types and those that regulate cell division and cell death have provided much insight into what goes wrong in cancer cells; similarly, studies of cancer cells have increased understanding of normal cell regulation.

Discovery of the signaling molecules that direct differentiation of stem cells into insulin-secreting pancreatic β-cells provides hope for people with diabetes, who could be helped by transplantation of healthy β-cells. The use of embryonic stem cells as the starting point for such therapies remains controversial, primarily due to ethical and religious objections, so research continues to explore other types of cells that might be used, including partially differentiated pancreatic precursor cells and induced pluripotent stem cells. Similar research is examining ways to generate specific types of specialized cells to treat a wide range of other diseases that involve problems with specific tissues.

Scientists have identified many mutations that occur in cancer, both those that generate oncogenes from proto-oncogenes and those that inactivate tumor suppressor genes. These have enabled researchers to develop targeted drugs that exploit specific genetic features of a given patient's cancer cells to kill them more efficiently than the traditional chemotherapy drugs that rely on inhibiting DNA replication or mitosis, while also sparing normal cells that do not express the mutated proteins that are targeted by the drugs. However, most of these "designer" drugs, as well as the newer immunotherapies (and stem cell therapies for other diseases), are very expensive.

Research directed toward understanding the basic biology of cancer continues. Some people feel that treatment cannot be rationally designed unless the underlying biology is known, while others feel that such basic understanding is not important. The former group point to the development of new treatments such as tamoxifen and Herceptin. The latter group use arguments such as the following: We still do not know the basic biology underlying the disease polio, but development of a vaccine for its prevention has eliminated our need to know. The real dilemma, a problem in the allocation of resources, is a difficult issue. How much money should we spend on treating cancer patients, how much on improving methods of treatment, how much on laboratory research to discover more information on the causes of cancer, and how much on cancer prevention activities? How much funding should be aimed at particular types of cancers, such as breast cancer as compared with colon cancer? There are no clear answers here because we cannot accurately predict how well or how soon funds spent on certain activities (especially research) will translate into a reduction of cancer incidence rates or cancer deaths. Cancer prevention is clearly very cost-effective, but the cost-effectiveness of the other alternatives may be very difficult to assess.

On an individual level, we can reduce our exposure to cancer risk factors by many choices that we make in our lives. However, not all types of exposure among those listed earlier are matters of personal choice. Dumping of carcinogens on the land and in the water of poor people with little or no political power has become a global environmental issue. Air pollution affects people at great distances from the source. Therefore, as part of any effort to prevent cancer, people need to work together to prevent or remove environmental hazards from workplaces and communities.

CHAPTER SUMMARY

- Multicellular organisms originate from a single cell. Many rounds of cell division by mitosis give rise to all the cells in a complete organism.
- As cells of a multicellular organism divide, they also **differentiate**, becoming more and more fully **determined**, that is, restricted in their **potentiality** to form different kinds of cells that in most species are organized into **tissues**.
- Cell differentiation results from differential **gene expression**, which is triggered by molecular signals secreted by certain cells and received by others. These signal molecules are bound by receptors, and this information is then transferred into the cell nucleus by a multistep process called a signaling cascade, which often involves internal signal molecules called **second messengers**.
- Normal cells occasionally enter the **cell cycle** and divide, but only when they receive signals to do so; signals that trigger cell division are called **growth factors**. Normal cells stop dividing when enough cells are present because of such phenomena as **contact inhibition** and other processes that maintain **homeostasis** of cell number.
- **Stem cells** retain the ability to differentiate into various kinds of cells and, if given the correct signals, they have the potential to generate specific cell types to treat many diseases.
- Cancer cells are cells that have undergone **transformation**, sustaining multiple mutations that disrupt the normal regulation of cell division, cell death, differentiation, and other behaviors. They share several properties with stem cells but do not obey the same regulatory signals.
- A few **cancers** have genetic predispositions, but most are caused by environmental factors. These factors include exposures to certain viruses and infective agents, dietary and other behavioral factors, and exposure to a long list of **carcinogens** including ionizing radiation (radioactivity), ultraviolet radiation, tobacco smoke, and a variety of chemicals.
- **Mutagens** damage DNA, and other chemicals can then increase the risk that these mutated cells will develop into cancer. Mutagens and tumor promoters have **synergistic effects**, that is, the increased risk of exposure to both is greater than the sum of the risks of the two separately.
- Cancerous **tumors** can be removed surgically, but other forms of cancer therapy target any cells that are dividing, destroying many healthy cells along with the cancer. New, more specific chemotherapies are being developed, including therapies based on boosting the body's immune system.
- We can best reduce our risks for cancer by avoiding tobacco smoke and other carcinogens (including ultraviolet and ionizing radiations) and by eating a diet low in fats, high in fiber, and rich in antioxidants.

BIBLIOGRAPHIC REFERENCES

Bibliographic references to material in this chapter can be found online at biologytrending.routledge.com/chapter10

GLOSSARY: KEY TERMS TO KNOW

These key terms are defined at the end of each chapter as an aid for student review.

Adult stem cell An undifferentiated cell that retains into adulthood the ability to divide and differentiate; bone marrow cells capable of differentiating into blood cells are an example.

Anchorage dependence The inability of most cells to divide unless attached to a surface, a restriction lost in most cancer cells.

Angiogenesis The stimulation of nearby blood vessels to grow into a structure and supply it with blood.

Apoptosis Programmed cell death (or cellular suicide) that begins with breakage of a cell's DNA, followed by cell shrinkage and detachment from its surroundings.

Basement membrane A membrane to which cells are attached at their base, especially in epithelial (sheetlike) tissues.

Benign tumor A tumor that has not broken through its basement membrane.

Blastocyst An early mammalian embryo at the stage when it is ready to implant into the uterine wall.

Blastula An early embryonic stage consisting of a hollow ball of cells.

Cancer A group of diseases characterized by DNA mutations in growth control genes and in which some cells divide without regard to growth control signals.

Carcinoma A cancerous growth of epithelial (sheetlike or glandular) tissue.

Carcinogen A physical, chemical, or viral agent that induces cancer; its action is called carcinogenesis.

Cell cycle The process by which a cell duplicates its DNA and then divides into two cells.

Cell of origin The initial cell whose transformation begins a cancerous growth.

Checkpoints Molecules that do not allow the cell cycle to proceed until certain repairs are made or processes are completed.

Contact inhibition The inability of a normal cell to divide if it is surrounded by other cells.

Cyclin A protein whose abundance varies cyclically, increasing at a particular time in the cell cycle that it appears to control.

Cyclin-dependent kinase An enzyme that controls progress through the cell cycle by adding phosphate groups to other proteins in response to the abundance of cyclins.

Cytokinesis Division of the cytoplasm into two distinct cells following mitosis, or following either of the divisions of meiosis.

Determined A state of development in which the future identity of a cell's progeny is predictable.

Differentiation The process of becoming different; a restriction of the set of future possibilities of a cell's progeny.

Embryonic stem cell An undifferentiated cell in an embryo that is able to divide and differentiate.

Endocrine A form of secretion in which a product is secreted into the bloodstream to be carried to a target elsewhere.

Epidemiology The study of the frequency and patterns of disease in populations.

Extracellular matrix Material produced by cells but located outside any cell. Connective tissues have large amounts of extracellular matrix.

Gastrula An early embryonic stage in most animals, containing two to three distinct cell layers, formed from a blastula by the invagination (tucking in) of some cells to form an inner cell layer (endoderm) that is usually separated from an outer cell layer (ectoderm) by an intervening middle layer (mesoderm, absent in Cnidaria).

Growth factor A messenger molecule that stimulates a cell to divide.

Homeostasis The ability of a complex system (such as a living organism) to maintain conditions within narrow limits. Also, the resulting state of dynamic equilibrium, in which changes in one direction are counteracted by other changes that bring the system back to its original state

Immortal A property of transformed cells that relieves them from having a limit on the number of times they can divide.

Immunosurveillance Elimination of abnormal cells by the immune system, thus reducing the chance of those cells developing into cancer.

Malignant A tumor that has grown through the basement membrane or extracellular matrix.

Metastasis The ability of transformed cells to leave the original tumor, travel through the body, and adhere to and form new tumors in other locations.

Multipotent Capable of forming a restricted variety of cellular progeny of several different types.

Mutagen An agent that causes mutations in DNA.

Necrosis Death of a damaged or diseased cell by breakage of its outer membrane and release of its contents.

Oncogene A mutated growth control gene that leads to the transformation of a cell, which may then lead to cancer.

Organizer An embryonic tissue whose chemical secretions induce the differentiation of other cells.

Paracrine A form of secretion in which a product is secreted locally, affecting only nearby cells.

Phagocytes Cells that engulf and digest other cells or the remains of dead cells.

Phosphorylation Addition of a phosphate group to a protein, usually causing activation.

Pluripotent Capable of forming a wide variety of cellular progeny, such as all ectodermal cell types.

Potentiality The range of possible futures for a cell's progeny.

Promoter A DNA sequence where RNA polymerase binds and where transcription of a gene therefore begins.

Proto-oncogene A normal gene from which an oncogene is derived; it encodes a product that regulates cell division.

Regulated Causing a process to occur either more or less in a given time period, based on some sort of informational input.

Sarcoma A cancerous growth originating in connective tissue.

Second messengers Molecules within the cytoplasm of a cell that carry information from membrane receptors to other locations in the cell.

Senescence Progressive loss of physical functioning as a result of age. Also, the aging of individual cells as a consequence of the telomeres of chromosomes reaching a threshold shortness and resulting in no further cell division.

Sporadic Arising in somatic cells, and therefore not inherited.

Stem cell An undifferentiated cell that retains the ability to divide and differentiate.

Synergistic effect A physiological response to two drugs given simultaneously that is greater than the sum of the effects of the same two drugs given separately.

Telomerase An enzyme that lengthens the telomeres of embryonic cells or cancer cells to counteract their shortening during cell division.

Telomere A repetitive DNA sequence at the end of a chromosome whose small, progressive shortening during each cell division limits the number of times a cell can divide.

Tissue A group of similar cells and their extracellular products, built together (structurally integrated) and working together (functionally integrated).

Tissue culture A growth of cells and tissues in a laboratory, artificially maintained outside any organism.

Totipotent Capable of forming cellular progeny of all different types, as in embryonic stem cells.

Transformation In multicellular organisms, a multistage process in which normal cells acquire the phenotypes of cancer cells, including immortality and lack of contact inhibition or anchorage dependence. (NOTE: Bacterial transformation, discussed in Chapters 2 and 4, is unrelated.)

Tumor A solid mass of transformed cells that may also contain induced normal cells such as blood vessels.

Tumor initiator An agent that begins the process of transformation by causing permanent changes in the DNA; mutagens and radiation are tumor initiators.

Tumor promoter An agent that completes the process of cell transformation after the process is started by a tumor initiator; tumor promoters are not mutagenic by themselves but cause partly transformed cells to go into cell division.

Tumor suppressor genes Genes whose protein products normally inhibit cell division.

CONNECTIONS TO OTHER CHAPTERS

Chapter 2: Cancers are caused by mutated growth-control genes or by DNA damage during chromosome crossovers.
Chapter 3: Cancers can result from changes that bring normal genes to abnormal locations in the genome.
Chapter 7: Several cancers have different incidence rates in different human populations.
Chapter 8: High-fat diets with low fiber content increase the risks for several cancers.
Chapter 14: Good mental outlook and immunological health can improve cancer survival rates.
Chapter 15: Certain otherwise rare cancers occur more frequently in AIDS patients.
Chapter 17: Many cancer-fighting drugs are plant products.
Chapter 18: Rainforest plants have been sources for new cancer-fighting drugs.
Chapter 19: Pollution increases the incidence rates of several forms of cancer.

PRACTICE QUESTIONS

1. What are the four phases of the cell cycle and what happens in the cell during each phase? Is G_0 part of the cell cycle?
2. In which cells is cell division inhibited by contact with other cells: normal cells or cancer cells?
3. Compartmentalizing an organism by dividing it into cells increases which of the following: the volume or the surface area?
4. Which of the following develop by differentiation from the zygote: tracheal cells in the lungs, muscle cells, or cells of the eye? Which develop from the gastrula? Which develop from the endoderm?
5. How long does an average human tongue cell live? How long do human liver cells live on average? Human nerve cells?
6. When the telomere region of the chromosomes becomes too short, which of the following happens?
   a. The cell can no longer divide.
   b. The cell becomes cancerous.
   c. The cell can no longer differentiate.
7. Do the protein products of proto-oncogenes induce cells to divide or do they prevent cells from dividing? What about the protein products of tumor suppressor genes?
8. How many cells need to be transformed for cancer to develop? Does every transformed cell result in cancer? Why or why not?
9. Which of the following causes more cases of cancer: heredity, smoking, or viruses?
10. What processes are induced in a cell by binding of a growth factor to its receptor? Does a growth factor induce these processes in every cell?
11. Are the same genes transcribed and translated in every cell of the body?
12. Does cellular differentiation take place in adult animals or only in embryos?

CHAPTER 11

The Nervous System and Senses

ISSUES

- What happens when the nervous system malfunctions?
- Can mental illness be understood and treated biochemically?
- How can changes in brain chemistry interfere with message transmission?
- How can many dissimilar malfunctions have similar causes?

BIOLOGICAL CONCEPTS

- Evolution (comparative anatomy, specialization and adaptation, form and function)
- Hierarchy of organization (ions, neurons, brain, behavior)
- Matter and energy (membranes, ion potentials, action potentials)
- Detection of environmental stimuli (receptors, neurotransmitters, sense organs)
- Movement (muscular system)
- Homeostasis (feedback mechanisms, nervous system)
- Health and disease
- Behavior (sleep, learning)

CHAPTER OUTLINE

DOI: 10.1201/9781003391159-11

A man in his sixties finds it difficult to walk and shuffles his feet along the floor, just an inch or two at a time. Last week he tried to walk by lifting his feet, but he stumbled after only two quick but awkward steps. A young woman has lost all motivation to work, to keep herself clean, or just to go on living. "What for?" she asks, "none of it matters." A man who is usually friendly has episodes when he seems fearful and suspicious of others. During these episodes, he hears voices telling him that "they" (he's not sure who) are out to get him, though he has broken no laws. These three people have three very different diseases: Parkinsonism, depression, and schizophrenia. Could it be that all their assorted symptoms are caused by chemical imbalances in the brain?

The brain coordinates the activities of the rest of the body, including moving, sleeping, eating, and breathing. Changes in the brain can thus produce diseases such as Parkinsonism, whose symptoms include uncoordinated movements.

Does the brain also produce activities that we associate with the mind? Does biochemical activity in the brain produce perceptions, emotions, moods, and personality? Does it produce thoughts, dreams, and hopes? A branch of biology called neurobiology is the scientific study of the brain and nervous system. A central theory of neurobiology is that the mind and the brain are one and the same. As a framework that guides (and limits) research, the assumption that mind equals brain is a good example of a research paradigm (see Chapter 1). There is probably no theory in biology that is more controversial, both among biologists and between biologists and the public.

There is now considerable evidence in support of this theory. Electrical and biochemical activities have been measured in the brain during dreams and thought. Some diseases, such as Alzheimer's disease (Chapter 12), in which there is brain degeneration, are accompanied by changes in personality. Mental illnesses such as depression are associated with changes in brain chemistry and can, in many cases, be treated with drugs.

In this chapter we examine the workings of the brain and the nervous system and consider the extent to which the mind is another name for brain activity. The nervous system is adapted for sending and receiving messages. You will learn how these messages travel electrically along cell membranes and are carried from cell to cell by chemicals called neurotransmitters. You will also learn how the brain is organized, how sense organs handle incoming messages, how muscles react by contracting, how the brain processes and stores messages, and how mental processes in both health and disease derive from activity at the chemical and cellular levels.

11.1 THE NERVOUS SYSTEM CARRIES MESSAGES THROUGHOUT THE BODY

Central to the science of neurobiology is the theory that all functions and dysfunctions of activity, thought, and behavior, including mental activities and mental illnesses, result from the actions and interactions of cells and chemicals of the nervous system. The diseases described in this chapter all involve malfunctions related to chemicals in the brain; several are also characterized by the degeneration of particular groups of cells.

In 1817, Dr. James Parkinson first described a disease marked by muscle tremors, including a distinctive "pill-rolling" movement of the thumb and forefinger. The disease is also characterized by a walking gait in which the body above the waist leans forward while the feet shuffle slowly in small steps and are barely lifted from the ground. People with **Parkinsonism** have difficulty in initiating voluntary movements. This difficulty is not a true paralysis because, by stopping other activities and concentrating on their voluntary movements, Parkinson patients can temporarily improve their performance.

For voluntary movements of any kind to take place, the brain must initiate and send messages to the muscles. Parkinsonism presents some of the clearest evidence of a malfunction in this message-sending system. However, before we can understand this malfunction in greater detail, we must first describe the cells of the nervous system and how the sending of messages usually works.

11.1.1 The nervous system and neurons

Nearly all animals (and only animals) have nerve cells arranged in some type of nervous system. The simplest nervous systems are netlike arrangements of interconnected cells with no control center; hydras and other cnidarians have such nerve nets. The nervous systems of flatworms are a bit more organized, with a concentration of sense organs (eyes, hearing organs) at the front end of the animal. A pair of enlarged control centers called "cerebral ganglia," the beginnings of a simple brain, coordinate the input from these sense organs. Many worms, especially parasitic worms, have greatly simplified nervous systems. The nervous system of the roundworm *Caenorhabditis elegans* has fewer than two dozen individual nerve cells, yet *C. elegans* has been an important model system for neuroscientists, due to its simplicity and the ease with which its genetic information can be manipulated. More complex animals need a more extensive nervous system to coordinate the movements of legs and other appendages, and those with more elaborate sense organs need a larger central nervous system to receive and interpret messages from these sense organs. Especially among mollusks and arthropods, animals with better sensory capabilities and locomotor skills have larger brains to act as centers of coordination and control.

The largest brains are found among the vertebrates, or backboned animals, of the phylum Chordata. In these animals, the brain continues rearward into a long spinal cord, and brain and spinal cord together make up the **central nervous system (CNS)**. The rest of the nervous system is called the peripheral nervous system, and it includes the remainder of the nerves, called **peripheral nerves**, and the special sense organs such as the eyes and ears (**Figure 11.1**).

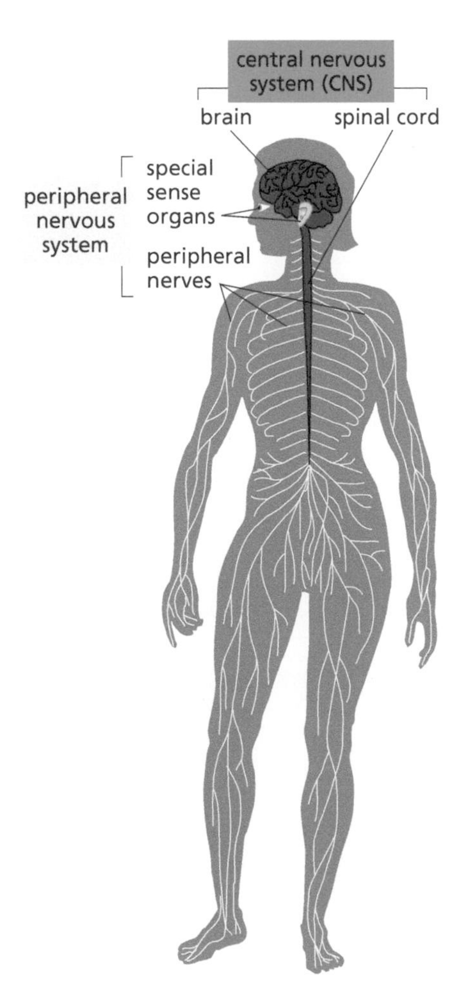

Figure 11.1 Major divisions of the nervous system.

The functions of the nervous system, including those of the brain, are carried out largely by nerve cells, also called **neurons**, cells that carry nerve impulses along their membrane surfaces. Each neuron contains a cell body that includes a nucleus and surrounding cytoplasm. Extending out from this cell body are the branches called **dendrites**, which conduct nerve impulses toward the cell body. Each neuron also has another extension, called an **axon**, that conducts impulses away from the cell body (Figure 11.2A). Some neurons are oriented so that their axons conduct impulses from a sense organ inward toward the central nervous system. Other neurons conduct impulses away from the CNS and terminate either at a muscle or a gland. Nerve impulses pass from neuron to neuron (or from neuron to muscle) across a gap known as a **synapse** (Figure 11.2A).

Many axons, but not all, are surrounded by a series of special cells whose rolled-up plasma membranes form a structure called the **myelin sheath**

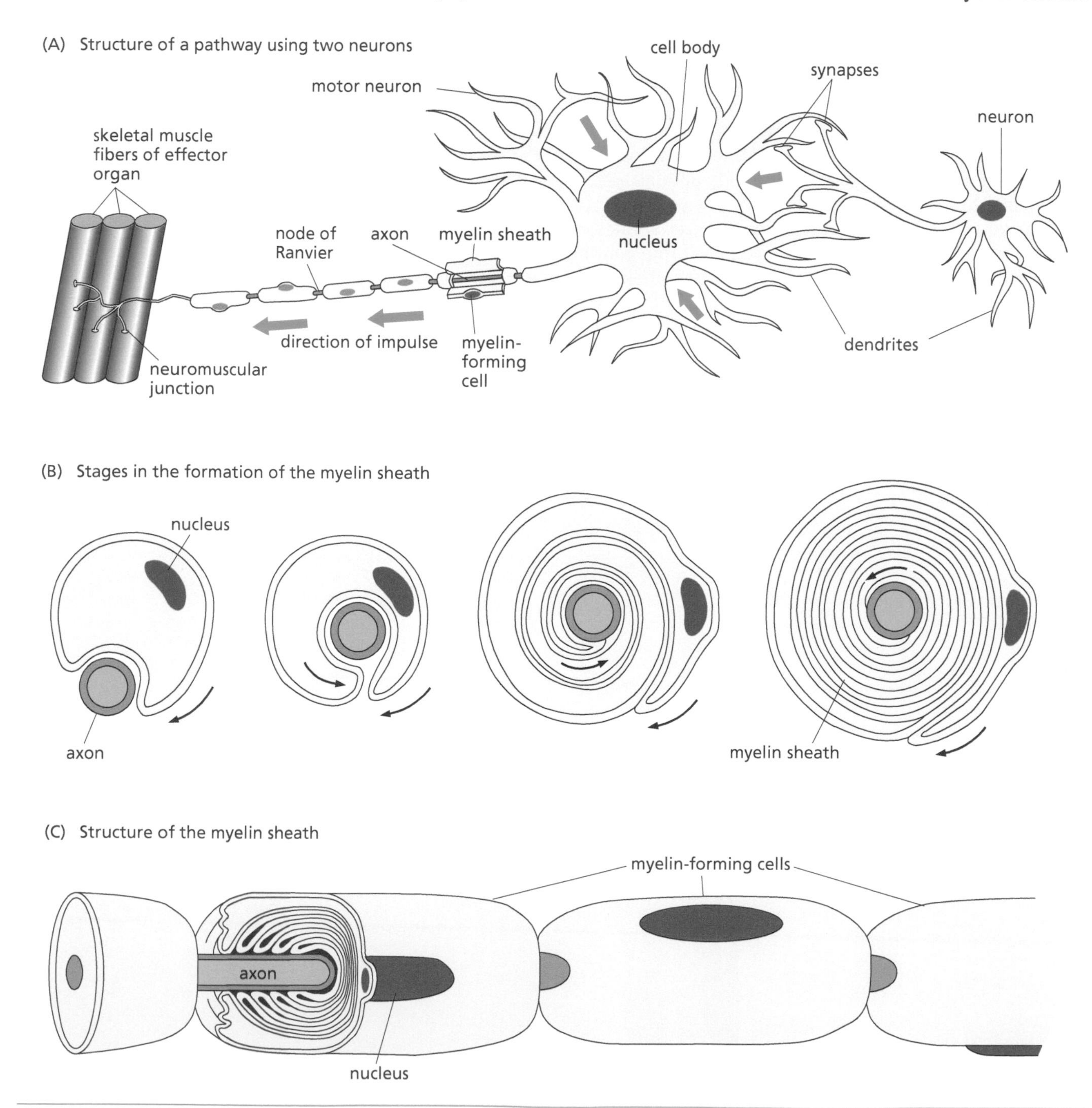

Figure 11.2 The neuron and its myelin sheath.

(**Figure 11.2B,C**). The myelinated axon looks a bit like a string of sausages because the sheath is thinner where the adjacent cells of the myelin sheath meet one another, a structure called the **node of Ranvier**. The myelin sheath acts as an insulator that prevents nerve impulses from spreading sideways from one axon to another, and it keeps the impulses traveling along the length of the axon. When myelin is disrupted by disease, transmission of nerve impulses becomes disordered. In the disease called multiple sclerosis (MS), the myelin sheath is destroyed by cells of the body's own immune system. The resulting disturbance of nerve conduction results in weakness or trembling of the arms or legs and in hazy or double vision, among other symptoms.

Aggregations of neurons or their parts have distinctive names in the nervous system. Bundles of axons are called **nerves** throughout the peripheral nervous system and tracts within the brain and spinal cord. Clumps of cell bodies are called **ganglia** throughout the peripheral nervous system and nuclei within the CNS, except for one group of brain structures called the basal ganglia.

In addition to neurons, the nervous system contains other types of cells called **neuroglia**. One group of neuroglia are the small **microglia**, which serve as an immune defense of CNS tissue by reacting to the presence of foreign or diseased material in the brain and spinal cord. Activated microglia can detect, surround, and consume diseased material by phagocytosis, a process more fully described in Section 14.2.2. The other types of neuroglia supply nutrition, structural support, and electrical insulation to the neurons, usually through cellular extensions that wrap around the neurons. Other extensions from the neuroglia wrap around small blood vessels, especially along the brain surface. Through these extensions, these cells carry nourishment from the small blood vessels to the neurons and waste products from the neurons to the bloodstream. Neuroglia cells with few branches are called **oligodendrocytes**; those with many branches and starlike shapes are called **astrocytes**. Astrocytes can also help detect and destroy certain nonfunctioning proteins, so changes in astrocyte activity are thought to play a role in conditions like Alzheimer's disease (see Chapter 12).

11.1.2 Nerve impulses: How messages travel along neurons

Much of what we know about nerve impulses was learned from studying the giant axons of the squid (**Figure 11.3A**), a member of a group called the Cephalopoda (which includes the octopus and chambered nautilus), which is a class of the phylum Mollusca (see Figure 5.7). The animals of this class need to react both quickly and forcefully to stimuli that could signal danger; they all have giant axons as part of their quick-response system. These axons are many times the diameter of typical axons in humans and so are ideal for studying nerve impulses, as we will soon describe.

11.1.2.1 Electrical potentials One way to study nerve impulses within a single neuron is by using a very sensitive voltmeter attached to needle-like probes (electrodes) that conduct electricity. Voltmeters read differences in the amount of electric charge in contact with its two electrodes. A giant axon has a large enough diameter that one of the electrodes can be placed inside the axon and the other electrode outside (**Figure 11.3B**). When placed this way, voltmeter readings show a difference in charge between the inside and outside of the axon, a difference called an **electrical potential**, measured in units called millivolts (mV). Unlike electrical potentials in nonliving systems (such as electrical wiring), which are due to electrons, electrical potentials in living systems are due to differences in the concentration of charged particles called ions. Electrical potentials in the nervous system result from differences in concentration of sodium ions (Na^+) and potassium ions (K^+) inside and outside the neuron. These differences are chemical gradients: There are more sodium ions outside the cell than inside and more potassium ions inside than outside (**Figure 11.3C**). There is also an electrical potential because there are more positive charges outside the cell than inside. Together, these two differences make an electrochemical gradient. The excess of positive charge outside the cell can be measured as an electrical potential of

Figure 11.3 Resting potentials in the giant axons of the squid.

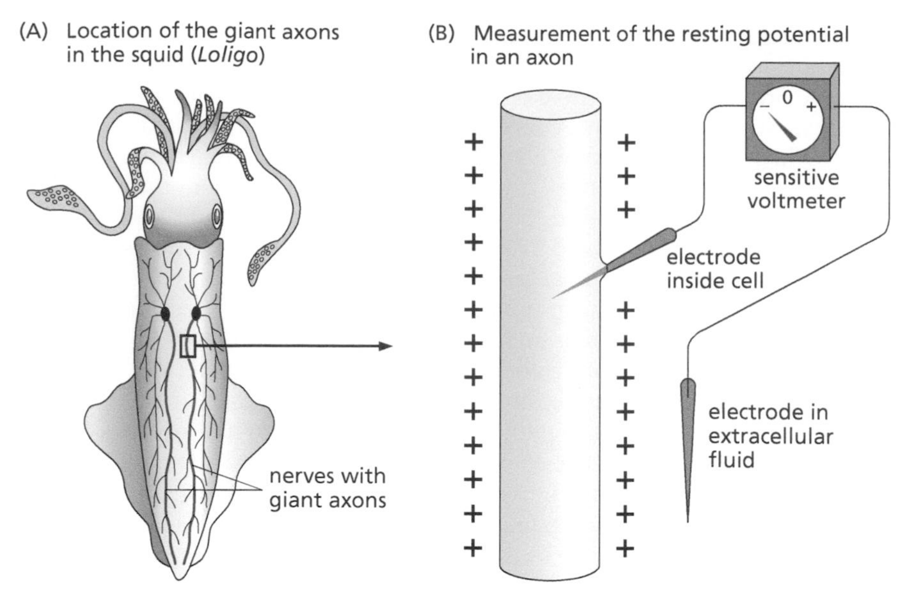

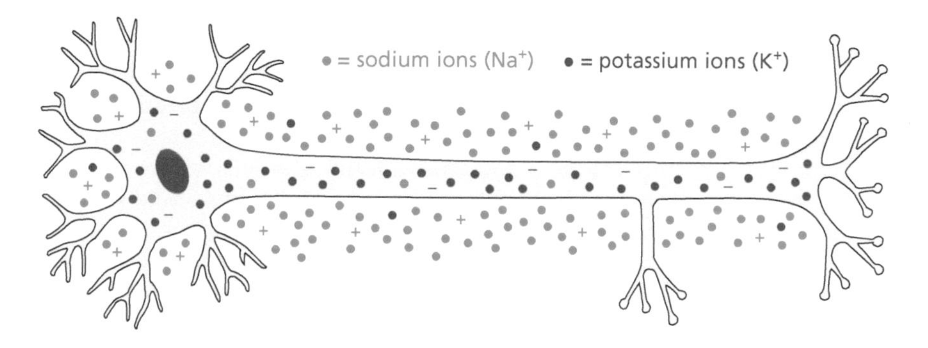

about –70 mV across the cell membrane of the neuron, meaning that the inside of the cell is negatively charged in comparison with the outside. (The magnitude of this potential is different in different types of neurons.) When a neuron is not conducting an impulse, the electrical potential across its cell membrane remains constant and is called a **resting potential** (Figure 11.3B,C).

11.1.2.2 Measuring nerve impulses We say that a membrane is **polarized** when there is an electrical potential across it. When the electrical potential decreases, we say it has **depolarized**. A nerve impulse is a wave of depolarization traveling along the cell membrane. The easiest way to demonstrate this is by using a voltmeter with electrodes applied to two points along the outside of the axon. The voltmeter compares the charge at the second site outside the axon with the charge at the first site. When the neuron is not conducting an impulse, the concentration of positive ions is the same at the two sites, so the two electrodes detect the same charge, and the voltmeter needle reads zero. When a nerve impulse arrives at the first site, it depolarizes because some positive ions move inside the axon. The concentration of positive ions, and thus the charge, is less at the first electrode than at the second, so the voltmeter reads negative. Later, successive sites down the axon depolarize, while the concentration of positive ions at the first site is restored. When the concentration of positive ions, and thus the charge, is higher at the first electrode than at the second, the voltmeter reads positive. As a nerve impulse passes, the needle deflects first one way and then the other, as shown in Figure 11.4.

Along the length of an axon, the nerve impulse travels at a relatively rapid rate that is faster in large-diameter axons and also faster with the presence of a myelin sheath. The impulse is made faster in most myelinated axons by "jumping" from one node of Ranvier to the next, a process called "saltatory conduction."

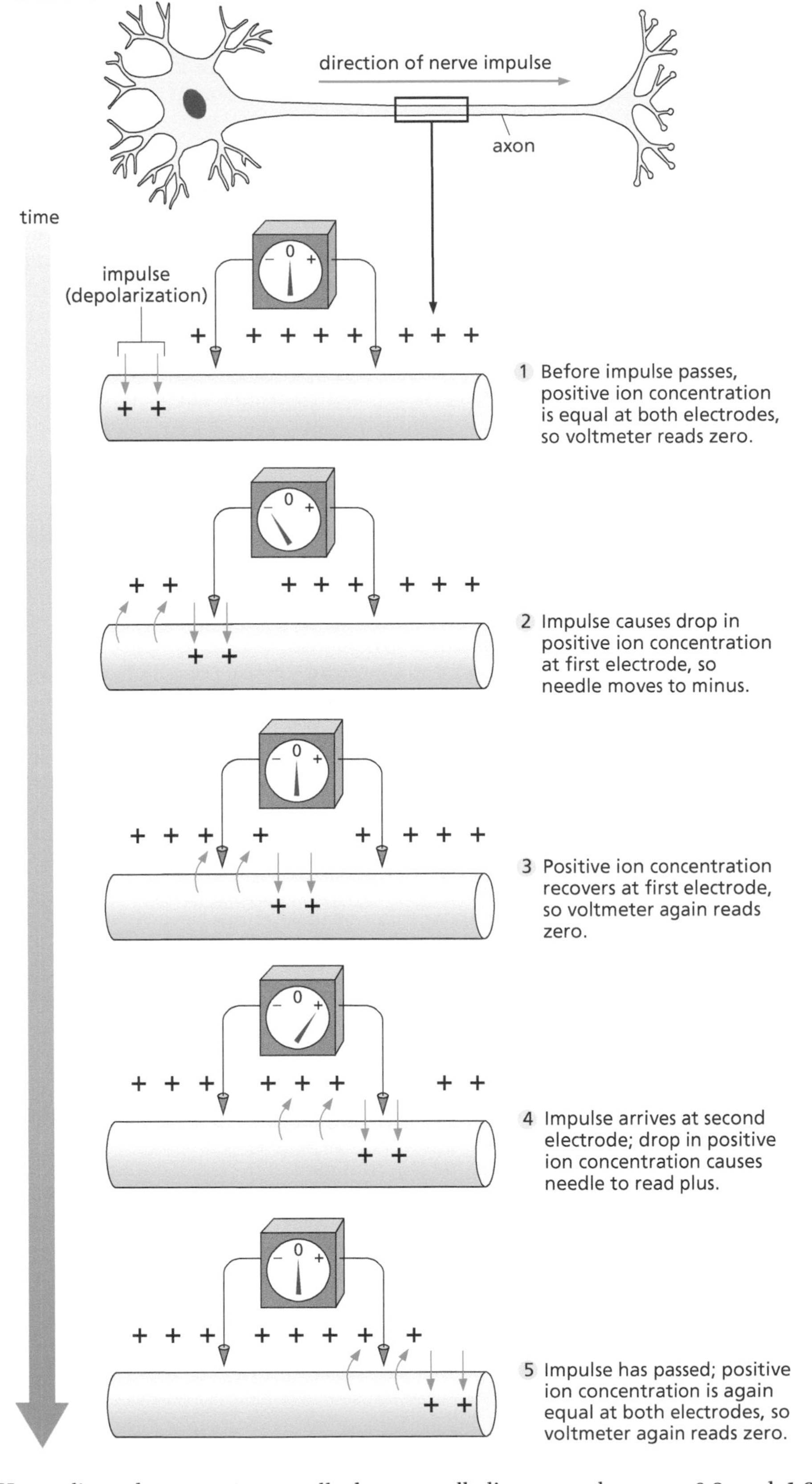

Figure 11.4 Detection of a nerve impulse by using a sensitive voltmeter. For an animated version of this figure, please go to biologytrending.routledge.com/chapter11

Unmyelinated neurons generally have small diameters, between 0.3 and 1.3 micrometers (1 μm = 10^{-6} m) and conduction velocities of 0.5–2.3 meters per second (abbreviated m/sec). Myelinated neurons have larger diameters (3–20 μm) and faster conduction velocities, varying from 3 m/sec in the smallest fibers to 120 m/sec in the largest. The giant squid axons, so named because of their large diameters, have a very high conduction velocity.

Neurons are stimulated to depolarize in a small portion of their cell membranes; however, not every localized depolarization leads to a sustained impulse. Depolarizations that are strong enough will trigger a nerve impulse that travels the length of the axon. We will now look at the propagation of a sustained nerve impulse in more detail.

11.1.2.3 Action potentials Recall that, at its resting potential, a neuron has an excess of sodium ions outside the cell and an excess of potassium ions inside (Figure 11.3). Such neurons are electrically excitable; that is, they can be stimulated by changes in electrical charge. When a nerve cell membrane is sufficiently stimulated, channels through the membrane open and let sodium ions flow in, locally depolarizing the membrane for a short time. This can be measured as a change in voltage by placing electrodes as in Figure 11.3B, and voltage changes over time can be recorded in graphs. A slight depolarization of the membrane is transmitted a very short distance to adjacent portions of the cell membrane; these effects decrease rapidly with distance and may thus disappear quickly. However, a greater depolarization that reaches or exceeds some **threshold** triggers nearby membrane channels to open. The rapid inflow of sodium ions results in a characteristic type of electrical discharge known as an **action potential**. The action potential appears as a "spike" on a graph of potential versus time, as shown in Figure 11.5. From a resting potential of −70 mV (Step 1), a portion of the axon membrane depolarizes due to the opening of a small number of sodium channels (Step 2). If the depolarization passes the threshold value of −50 mV (for this neuron), more sodium channels open, inducing a depolarization to zero, then a reversal of charge to +50 mV (Step 3). Depolarization ends as the sodium channels close; the membrane becomes repolarized by the opening of other channels—the potassium channels—and by the outward flow of potassium ions (Step 4). The electrical potential becomes negative once again and returns to a resting potential as the potassium channels close. Because there are now more sodium ions inside the axon than is normal and more potassium ions outside, the original ion distribution must be restored. This is accomplished by **sodium-potassium pumps**, membrane proteins that engage in the active transport of sodium ions and potassium ions in opposite directions across the cell membrane. (Active transport and ion gradients are described in Section 8.3.3.) For every two potassium ions transported into the neuron, three sodium ions are transported across the membrane to the outside. Because each sodium ion and each potassium ion carry a single positive charge, this also moves three positive charges out of the cell for every two that are moved in. Active transport thus restores the sodium and potassium concentration gradients and additionally contributes a few millivolts to the resting potential (Step 5). For an animated view of this process, please go to biologytrending.routledge.com/chapter11

The action potential spike in one location causes the adjacent portion of the cell membrane to depolarize, repeating Step 2 and producing another action potential equal in strength to the first. In this way, an action potential rapidly spreads along the entire neuron membrane with no reduction in its size or intensity, with each spike being the trigger for the next depolarization. It is this traveling of successive action potentials along the neuron that we recognize as a nerve impulse. Action potentials maintain their direction of travel because once sodium channels have opened and then shut, they are prevented from reopening for a short period of time.

11.1.3 Neurotransmitters: How messages travel between neurons

Even though neurons are electrically excitable, the transfer of information from cell to cell (across a synapse) is generally chemical, not electrical. Any chemical substance that can stimulate or inhibit an action potential is called a **neurotransmitter**. Most synapses between nerve cells are chemical synapses, in

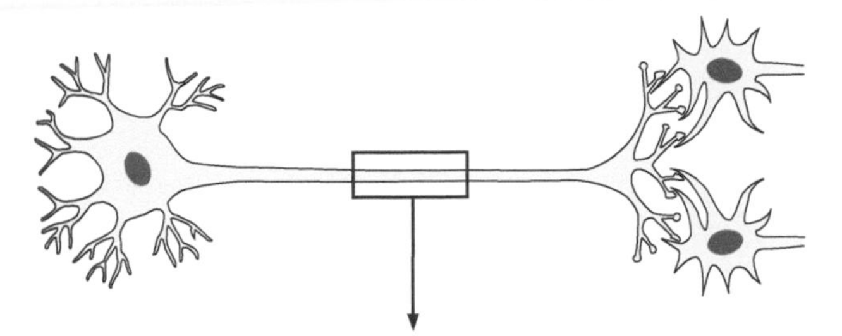

time

1 Resting potential when no impulse is present

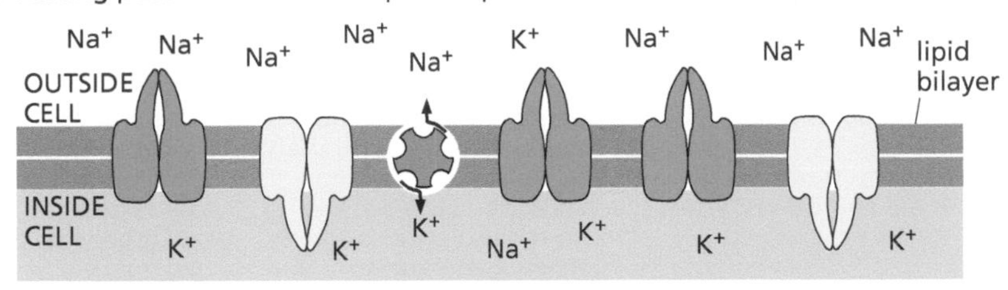

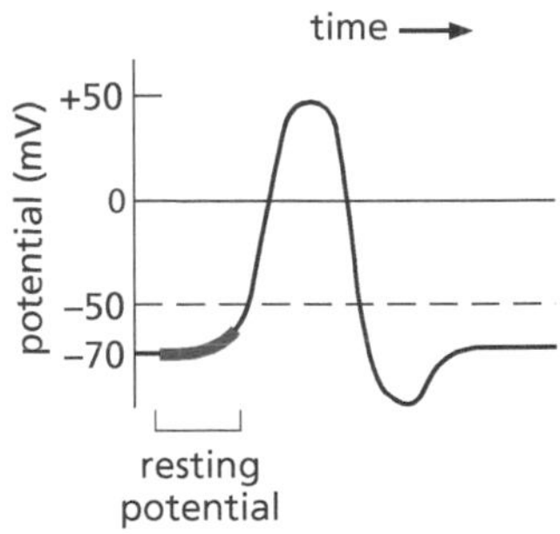

2 Stimulus causes a few channels to open; sodium ions leak in, depolarizing the membrane

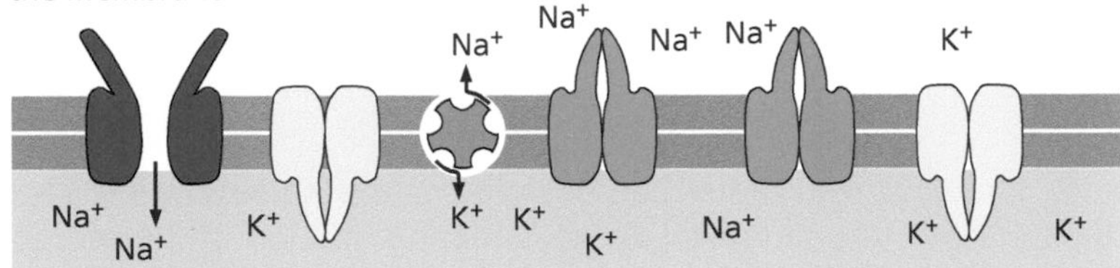

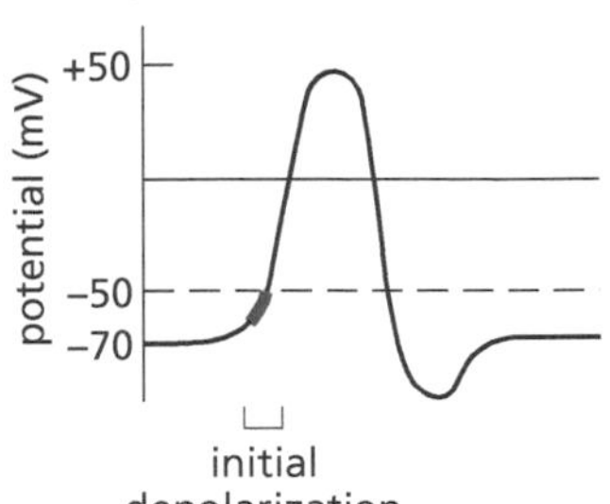

3 At –50 mV, all sodium channels open (*bright red*); sodium rushes in, causing charge to reverse and form an action potential

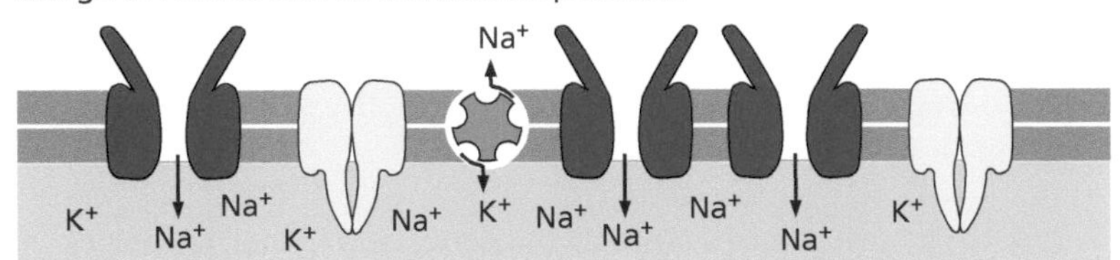

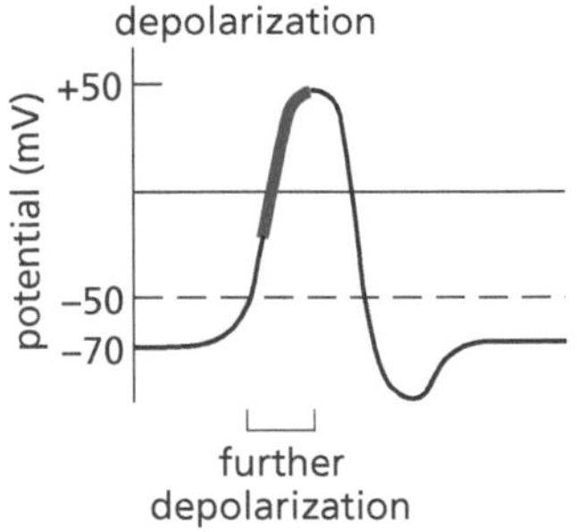

4 Sodium channels close as potassium channels open (*solid green*); potassium ions rush out, repolarizing the membrane

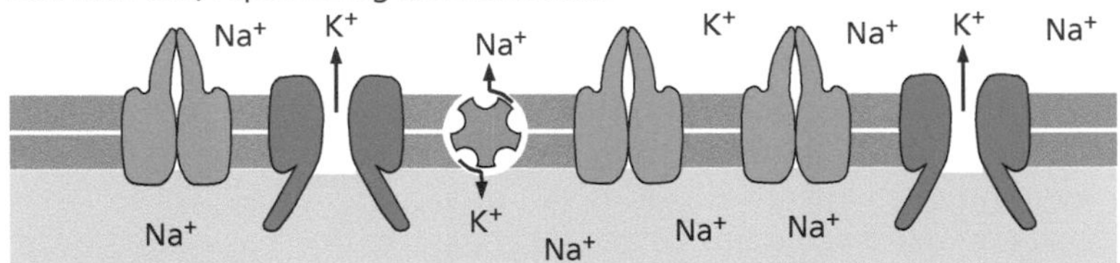

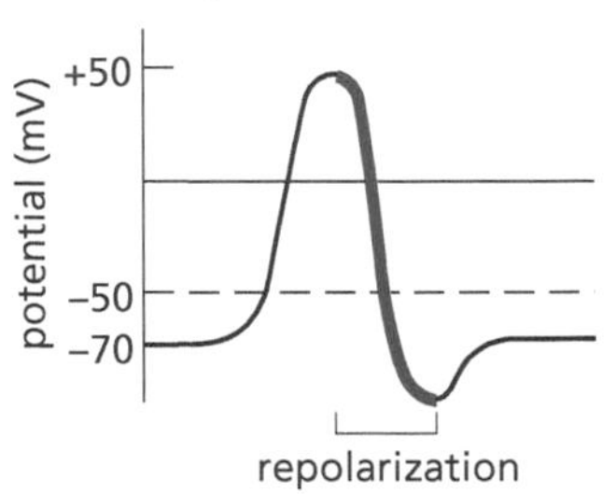

5 Closing of potassium channels restores resting potential; sodium–potassium pump (*purple*) redistributes ions

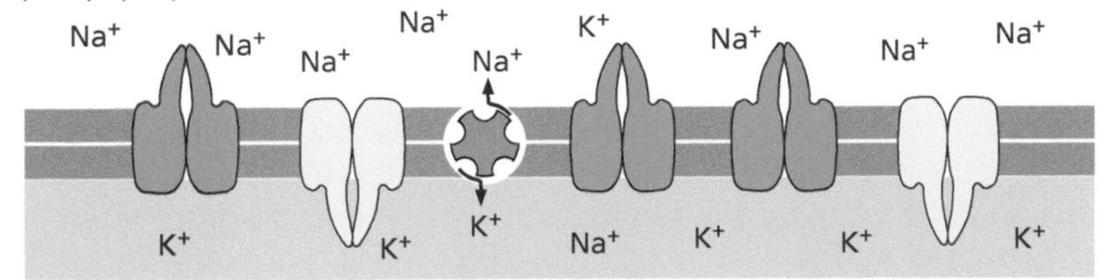

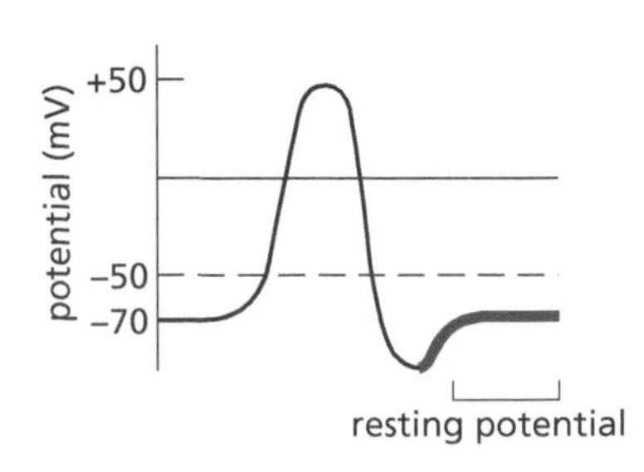

Figure 11.5 Generation of an action potential in a nerve cell. The action potential corresponds to steps 2 through 4 and extends from one resting potential to the next. For an animated version of this figure, please go to biologytrending.routledge.com/chapter11

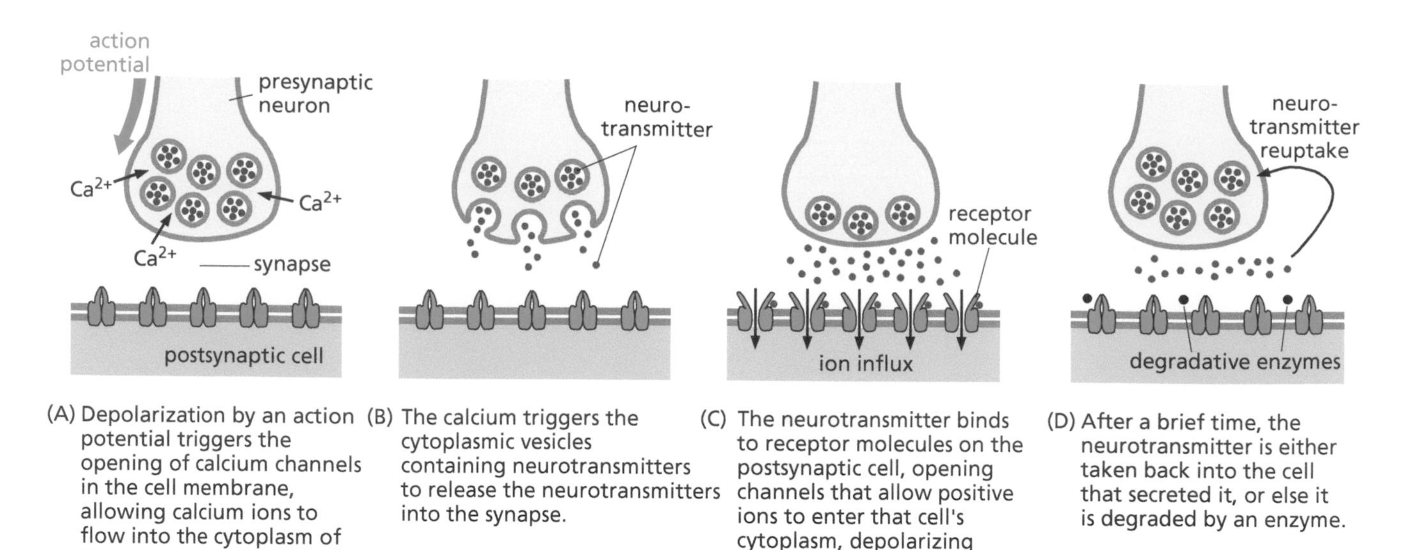

(A) Depolarization by an action potential triggers the opening of calcium channels in the cell membrane, allowing calcium ions to flow into the cytoplasm of the presynaptic neuron.

(B) The calcium triggers the cytoplasmic vesicles containing neurotransmitters to release the neurotransmitters into the synapse.

(C) The neurotransmitter binds to receptor molecules on the postsynaptic cell, opening channels that allow positive ions to enter that cell's cytoplasm, depolarizing its membrane.

(D) After a brief time, the neurotransmitter is either taken back into the cell that secreted it, or else it is degraded by an enzyme.

Figure 11.6 How a chemical synapse works. For an animated version of this figure, please go to biologytrending.routledge.com/chapter11

which a chemical neurotransmitter crosses the narrow space separating two adjacent cells. In a chemical synapse, depolarization of the membrane in the presynaptic or transmitting neuron (**Figure 11.6A**) causes it to release neurotransmitter molecules (**Figure 11.6B**). The neurotransmitter released into the synapse binds to receptors on the next cell (the postsynaptic cell), opening channels in its membrane. Ions flow through these channels, thus altering the electrical potential of the postsynaptic or receiving cell (**Figure 11.6C**). The neurotransmitter thus carries the impulse across the synapse but never enters the postsynaptic cell (**Figure 11.6D**).

11.1.3.1 Experiments demonstrating neurotransmitters The concept of a chemical neurotransmitter was first suggested by the British physiologist Henry H. Dale but was first demonstrated by his German American colleague Otto Loewi, with whom Dale shared the Nobel Prize in 1936. Dale and Loewi already knew that the vagus nerve, running down from the brain into the abdominal cavity, sends branches to the heart that slow down the heartbeat. They also knew that the heart of a frog separated from the rest of the animal and maintained in a physiological salt solution (a solution containing various ions at concentrations close to those in the intact organism) would keep beating for hours. Loewi dissected out the beating hearts of two frogs and placed them in salt solutions. One of his preparations contained nothing but the heart, but the other contained a carefully preserved vagus nerve. When Loewi stimulated the vagus nerve, the heart connected to this nerve slowed down. Loewi then used a dropper to take some of the fluid surrounding this heart and transfer it to the other heart, which no longer had any nerves leading to it. The second heart also slowed down, showing that some chemical had been present in the first preparation that could be transferred by dropper to the second, for this was the only connection between the two. The chemical was isolated and was identified as **acetylcholine**, the first neurotransmitter to be studied experimentally. Loewi's preparation also allowed the testing of various drugs or other substances that could block or otherwise modify the effect of acetylcholine. These drugs, in turn, could often be used to study whether a particular synapse used acetylcholine as a neurotransmitter.

11.1.3.2 Types of neurotransmitters Many chemicals are now known to be able to function as neurotransmitters; those that have been positively

Table 11.1 Neurotransmitters

Neurotransmitters
Amine neurotransmitters
Epinephrine
Norepinephrine
Dopamine
Serotonin
Histamine
Anandamide
Amino acid neurotransmitters
Aspartic acid (aspartate)
Glutamic acid (glutamate)
Gamma-aminobutyric acid (GABA)
Glycine
Protein or peptide neurotransmitters (neuropeptides)
Somatostatin
α-Endorphin
β-Endorphin
Leu-enkephalin
Met-enkephalin
Substance P
Gas neurotransmitter
Nitric oxide (NO)
Ester neurotransmitter
Acetylcholine

identified are listed in Table 11.1, grouped by their chemical structure. Most neurons in the peripheral nervous system use the neurotransmitter acetylcholine; a few use **norepinephrine**. But neurons in the brain use many different neurotransmitters, including all of those shown in Table 11.1. Each particular neuron secretes one type of neurotransmitter primarily or exclusively. Each cell that responds to a neurotransmitter has a specific receptor for that neurotransmitter, in the same way that other types of molecular signals are received by specific receptors, generally residing on the plasma membrane (see Section 10.2.1). The term "neurotransmitter" is used specifically for signaling between neurons, but the basic mechanisms of cell signaling are the same as for other signaling pathways.

11.1.3.3 Removal of neurotransmitters from the synapse After a neurotransmitter evokes its response, there are mechanisms that stop further transmission by removing loose (unbound) neurotransmitter molecules from the synapse (Figure 11.6D). Many neurotransmitters are reabsorbed by the cell that secreted them, permitting the same molecules to be recycled and reused. This process, called **reuptake**, is typical of many synapses in the brain. Also, several neurotransmitters can be chemically degraded by enzymes. The amine neurotransmitters can be degraded by the enzyme monoamine oxidase (MAO). Another enzyme, cholinesterase, breaks down acetylcholine molecules in synapses outside the brain. Any process interfering with the chemical breakdown or reuptake of neurotransmitters lets the neurotransmitter stay in the synapse and thus excessively stimulate the postsynaptic cell; likewise, any enhancement of chemical breakdown or reuptake may decrease neurotransmission. As we shall see, changes in neurotransmission can result in disease.

11.1.4 Dopamine pathways in the brain: Parkinsonism and Huntington's disease

One of the amine neurotransmitters (Table 11.1) is **dopamine**. Dopamine transmits messages across synapses between neurons in the brain. Some of the neurons that are stimulated by dopamine act, in turn, on peripheral neurons that stimulate voluntary muscle cells. Dopamine does not act directly on muscle cells; rather, it acts within the brain to smooth and coordinate signals to the muscles.

Two diseases have helped us to understand what can happen when dopamine secretion is out of balance: In Parkinsonism, there is too little dopamine, and in Huntington's disease, there is too much.

11.1.4.1 Parkinsonism: Too little dopamine

When autopsies are performed on the brains of Parkinson patients, a consistent finding is the degeneration of a bundle of darkly pigmented neurons (called the substantia nigra). Experimental staining shows that these neurons secrete the neurotransmitter dopamine. Among the cells stimulated by this dopamine are those of the basal ganglia. The degeneration of these neurons in Parkinson patients thus deprives the basal ganglia of dopamine. The neurons of the basal ganglia stimulate and coordinate muscle movements by acting on the acetylcholine-secreting neurons that trigger muscle contraction. If sufficient dopamine is not present, the cells of the basal ganglia do not function normally, and the person suffers from muscle tremors and has difficulty walking.

Evidence for the hypothesis that dopamine underproduction has a large role in Parkinsonism comes from the effectiveness of the drug L-DOPA (levodopa) in temporarily alleviating many of the symptoms of Parkinsonism. Because DOPA is a precursor of dopamine, it is hypothesized that supplying L-DOPA (a synthetic form of DOPA) can increase dopamine production and thus relieve the symptoms of dopamine deficiency. Unfortunately, this form of treatment increases dopamine production everywhere, not just to the portion of the brain that needs it. The excess of dopamine in other places may cause serious side effects, such as schizophrenia-like symptoms. Similar caveats apply to the Neupro® transdermal patch, approved in 2007, which supplies rotigotine, a drug that mimics the effects of dopamine on receptors in the brain.

One promising form of therapy that has already been tested is the implantation of fetal tissue into the brains of Parkinson patients. The hypothesis is that the fetal tissue will grow and replace the missing or damaged dopamine-secreting cells. Adult tissue is unsuitable for this purpose because the brain loses much of its capacity to regenerate new neurons at an early age; even tissue from newborn babies or infants has much less regenerative capacity than fetal tissue. However, the use of fetal tissue raises objections from opponents of abortion because the tissue is obtained in most cases from aborted fetuses (see Section 10.4.4). U.S. government regulation of the use of fetal tissue has fluctuated politically in the United States. From 1988 to 1993, there was a ban on federal funding of studies using fetal tissue. Because of the strong therapeutic potential of fetal tissue, many people in the medical community and patient advocate groups welcomed the lifting of this ban in 1993. Most recently, regulations revised in 2019 allow existing, federally funded projects to continue to their termination dates, but stipulate a new ethics review process for the evaluation of any new projects involving fetal tissue. It is not yet clear what impact this new process may have on future research.

11.1.4.2 Huntington's disease: Too much dopamine

If Parkinsonism is a disease in which voluntary movements are made difficult by a lack of dopamine, then too much dopamine might cause excessive movements. This is precisely what happens in Huntington's disease, a degenerative neurological disease whose genetic basis was discussed in Section 3.3.2. Huntington's disease is marked by uncontrollable spasms or twitches of many muscles, usually beginning between ages 40 and 50. As the disease progresses, the spasms become more pronounced, and the patient gradually loses control of all motor functions and of mental processes. A slow death occurs within a few years of the disease's onset. Autopsies of Huntington's patients reveal that some of the brain cells in the basal ganglia have been destroyed. Because these cells normally inhibit the production and release of dopamine, one major effect of their destruction (and one sign of the disease) is an overproduction of dopamine. Drugs known to inhibit the action of dopamine will temporarily reduce the symptoms of Huntington's disease.

11.1.4.3 Feedback systems How does the destruction of some of the cells in the basal ganglia result in an overproduction of dopamine? One answer may lie in another neurotransmitter, gamma-aminobutyric acid (GABA). GABA functions in many neuronal pathways by inhibiting a neuron that has fired from firing again unless another stimulus, larger than the first, is received. Because the GABA-secreting neuron inhibits a neuron at an earlier point in the pathway, the information is "feeding back" and is thus described as a **feedback system**. Feedback occurs in many biological systems whenever a later step regulates an earlier step in any process. One common type of feedback is a feedback inhibition (or "negative" feedback), in which a later step inhibits an earlier step. Feedback systems of this kind function to keep certain variables such as body temperature within narrow limits, turning metabolic processes higher when the body is cold or taking measures (such as sweating) to dissipate heat when the body is warm. We saw in Section 9.2.1 that levels of certain hormones are regulated by feedback from other hormones. In the nervous system, feedback can often be seen literally as neurons that feed information back to the earlier neurons that stimulated them.

GABA normally has this type of feedback effect on the dopamine-secreting neurons mentioned earlier. Because it acts to inhibit dopamine, it is an example of feedback inhibition. The destruction of the GABA-secreting neurons in Huntington's disease removes this inhibition. Because the dopamine-secreting neurons are no longer inhibited, they begin to overproduce dopamine (Figure 11.7). Too much dopamine leads to overstimulation of

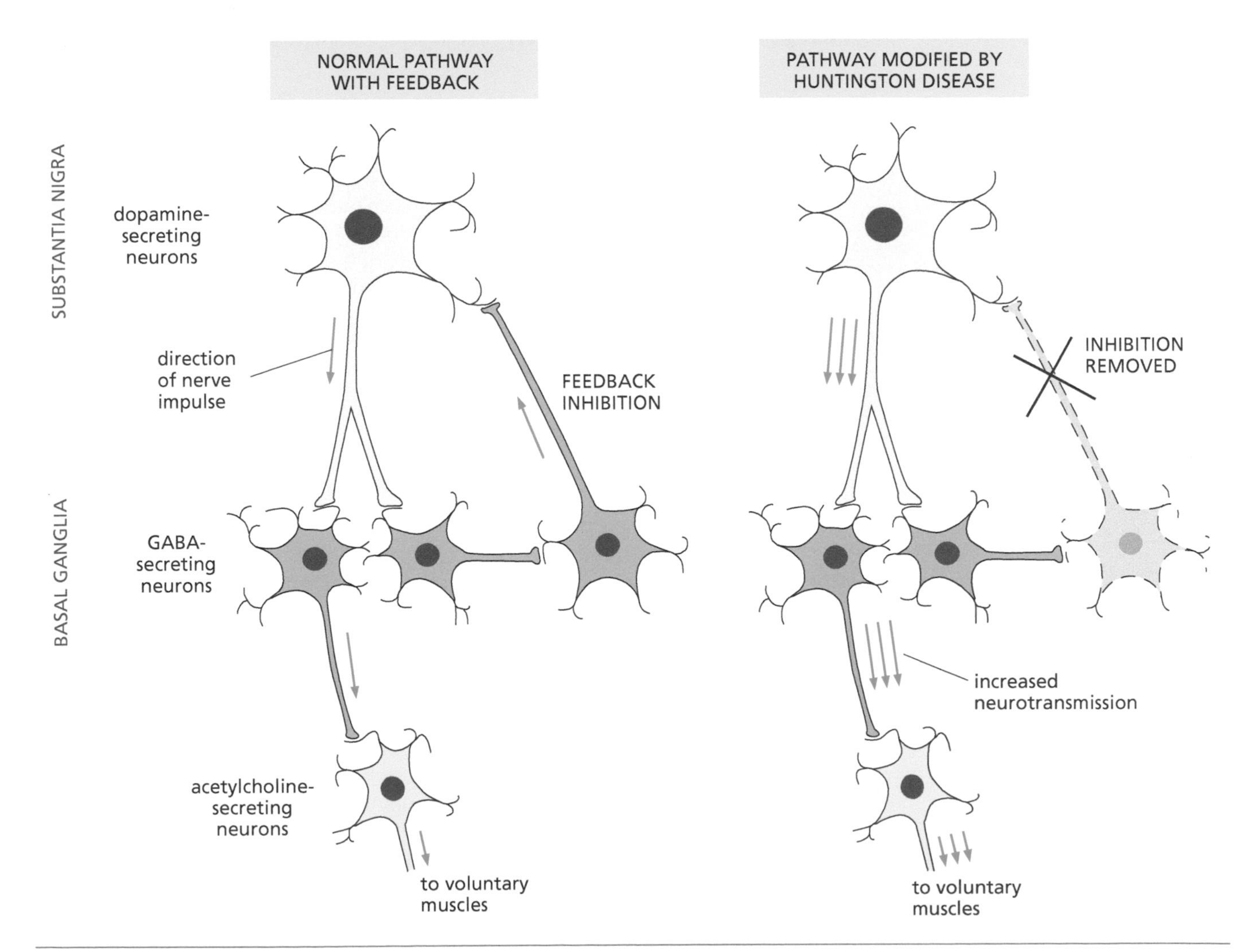

Figure 11.7 Feedback inhibition usually prevents dopamine overproduction, but impairment of the feedback pathway in Huntington's disease allows dopamine overproduction.

other GABA-secreting neurons, ones that trigger the acetylcholine-secreting neurons, stimulating excessive muscle contractions.

THOUGHT QUESTIONS

1. To test the clinical benefits of a technique such as the surgical implantation of fetal cells into the brains of Parkinson's patients, a double-blind comparison needs to be made with a group of control subjects. The control subjects in such a study are generally given sham surgery, meaning that a hole is drilled in their skull and a probe inserted into their brains, but no fetal cells are introduced. Because of a strong placebo effect, patients receiving this treatment often improve at least temporarily. Do you think it is ethical to conduct tests in this way? Do you think it is ethical to adopt such a procedure without this kind of clinical testing? Construct both utilitarian and deontological arguments for your position.
2. Suppose a scientist proposes the hypothesis that, in a select group of hard-to-reach cells, a particular amino acid functions as a neurotransmitter. How would you test this hypothesis? What kinds of drugs or poisons would you look for to help you study the properties of the hypothesized neurotransmitter?
3. Explain how Loewi's experiment demonstrated that neurotransmitters are chemicals, not electric current.
4. What are the arguments for and against the use of fetal tissue for the treatment of Parkinsonism? Can both utilitarian and deontological arguments be used for each position?
5. If a neuron is stimulated halfway along its axon, will an impulse travel in both directions on the axon?

11.2 MESSAGES ARE ROUTED TO AND FROM THE BRAIN

Some types of neuron messages are internal. As seen in the feedback system illustrated in Figure 11.7, neurons can carry information about the internal workings of the organism. They can also regulate internal processes, both within the brain (see the previous discussion) and in other parts of the body (see Chapter 14). Many other messages processed by the brain originate outside the organism. The nervous system is one of the major body systems through which an individual receives information about the surrounding environment. External stimuli cause specialized cells to trigger depolarizations (changes in membrane potentials) that travel to other neurons by means of neurotransmitters, and eventually to the brain, where huge numbers of incoming messages are processed. Some, but not all, of these messages become conscious perceptions.

11.2.1 Message input: Sense organs

The brain receives sensory input from the outside world through a variety of sense organs, all containing many specialized **sensory neurons**. Different types of sensory neurons respond to different types of stimuli. Although **sensory reception** is a function of these sense organs, **sensory perception**, which involves interpretation, is largely a function of the brain.

11.2.1.1 The skin: Reception of touch, temperature, and pain The main functions of skin, our body's largest organ, are protective: It protects us from temperature variations, harmful chemicals, water loss, and pathogens that might cause disease. Our skin also protects us by sending messages to the rest of the body from its various sensory nerve endings. Five types of specialized sensory endings are known, corresponding to five different skin senses: light touch, deep pressure, warmth, cold, and pain (Figure 11.8). When these are stimulated, their axons depolarize. The altered membrane potential produced in sensory cells is called a **generator potential**. Unlike an action potential, which operates on an all-or-none principle, the magnitude of a generator potential depends on the strength of the stimulus.

The different sensory receptors can be distinguished using a microscope but are much too small to be stimulated individually in an experiment. So how do we

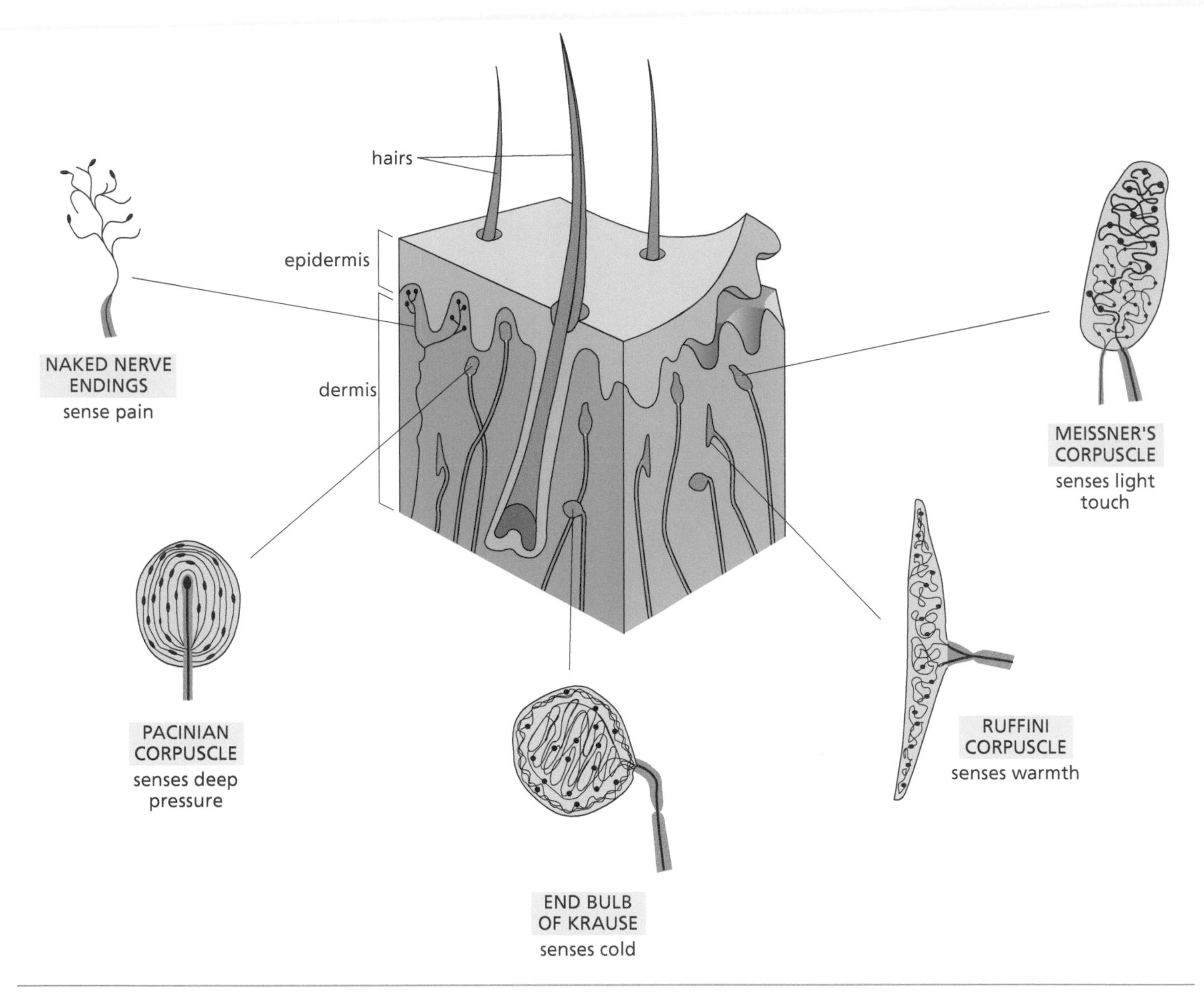

Figure 11.8 Nerve endings in the skin.

know which type of nerve ending serves which function? Our knowledge is largely based on the spatial distribution of the nerve endings over the body's surface. For example, the places most sensitive to touch (such as the lips and the fingertips) have the highest densities of the nerve endings thought to sense light touch, while the places most sensitive to cold (such as the cornea of the eye or the tip of the penis) have the highest densities of the nerve endings thought to sense cold.

Several of these sensory receptors can report anomalous sensations when they are overstimulated. For example, temperatures a few degrees lower than the surroundings are perceived as cold, but much colder temperatures may be perceived as heat or burning, and temperatures still lower are simply perceived as pain. In fact, a sufficiently large stimulus of any sort, even a sound much too loud or a light much too bright, is perceived by the brain as pain. The nerve endings that are specifically pain receptors are naked nerve endings, which are easily overstimulated to produce sensations of pain. In contrast, the deep pressure receptors are surrounded by onion-like layers, ensuring that the nerve ending within can only be stimulated to produce changes in membrane potential if there is enough pressure to deform the shape of these layered capsules deep inside the skin.

Our sense of touch requires direct pressure on the skin. Fishes have an additional sense organ for touch (called the "lateral line system") that is sensitive to very small changes in water pressure caused by nearby obstacles or by the swimming movements of other fish (Figure 11.9).

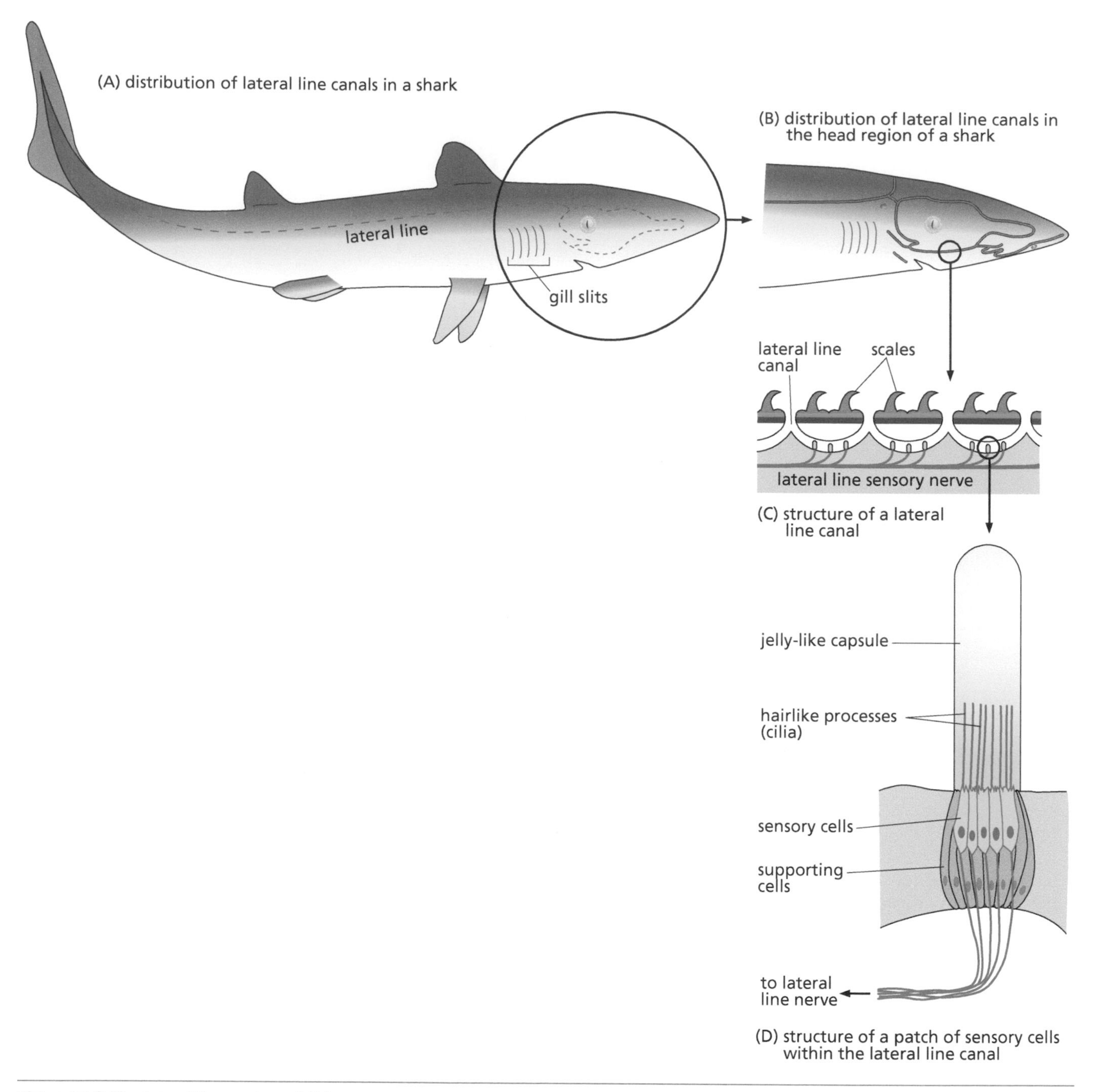

Figure 11.9 The lateral line system of fishes.

11.2.1.2 The eye: Reception of light The organ of vision in all vertebrates is the eye (**Figure 11.10A**), which is surrounded by two outer protective layers, the choroid, and the sclera. Light reaches the interior of the eye through a transparent front layer, the cornea. The light is concentrated by a transparent lens, which changes shape to bring into focus objects at varying distances from the eye. A suspensory ligament holds the lens in place, while a series of smooth muscles change the shape of the lens. The iris diaphragm, which gives each of us our individual eye color, controls the amount of light that enters the lens through the pupil, an opening that enlarges in response to dim light and becomes smaller in response to bright light.

The lens focuses light on the **retina**, the sensory sheet along the rear of the eye that responds to light. In a nearsighted person, the retina is too far from the lens, so the lens is not able to focus light on the retina. In a farsighted person, the

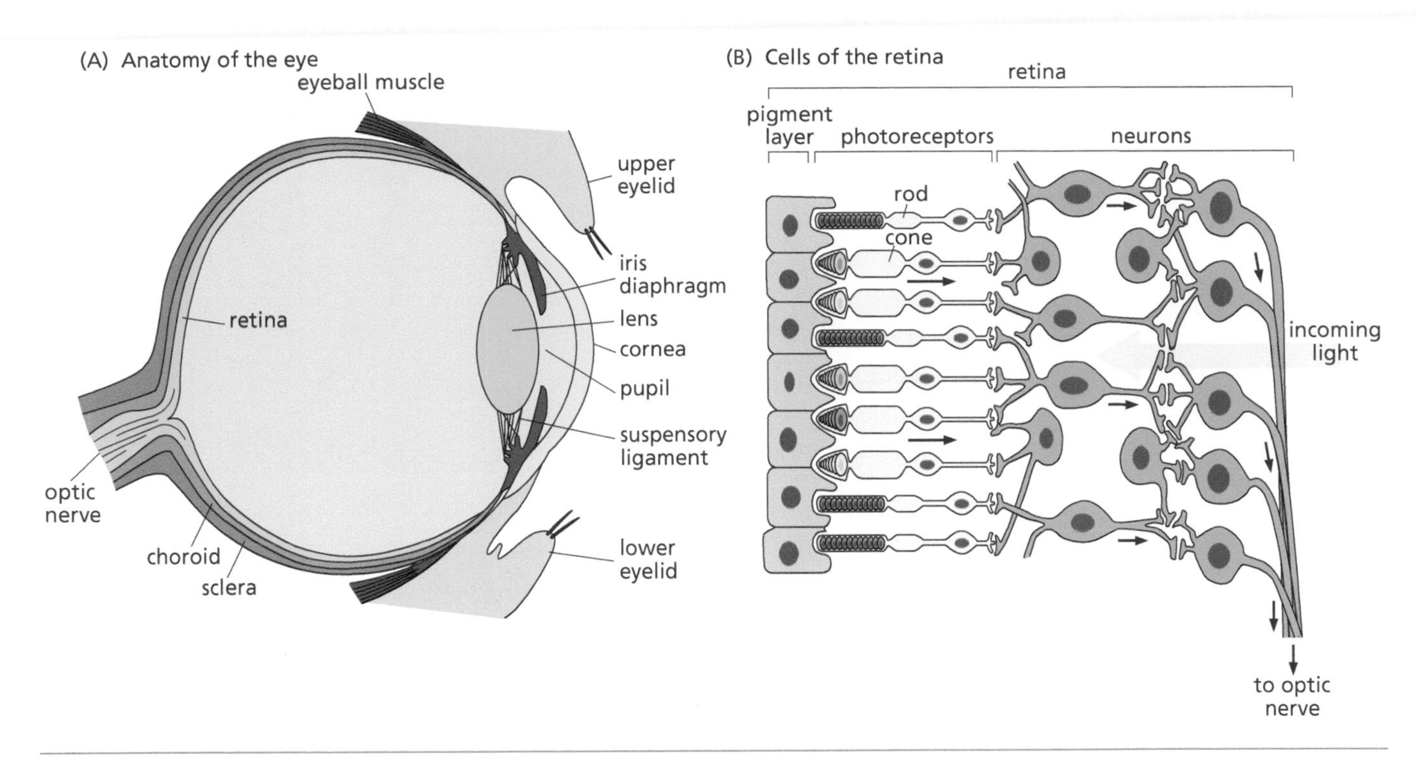

Figure 11.10 **The human eye.**

opposite is true—the lens is too close to the retina. For both nearsighted and farsighted people, glasses or contact lenses bring the focal point of the light to the retina. Until about age 40, the lens itself can change shape to accommodate for the vision of either near or distant objects. However, as a person ages, the lens becomes less flexible, and accommodation may no longer work well. For this reason, many people who had good vision earlier in life may need to wear glasses as they age.

The retina contains photoreceptor (light-receiving) cells called rods and cones, and light that falls on the retina causes changes in the chemical structure of pigment molecules in these cells. These pigment changes alter the membrane potential in the rods and cones, inducing impulses. The impulses travel along the retina to the optic nerve, which carries the information to the areas of the brain responsible for vision (**Figure 11.10B**). Rod cells respond in dim light. Cone cells require brighter light to trigger impulses, and different cone cells respond to different wavelengths of light. The brain receives messages from cones sensitive to different wavelengths and processes these messages, creating our perception of color.

Many details of eye structure differ across the animal kingdom (**Figure 11.11**; see also Section 5.3.2). Vertebrate eyes have a retina that conducts a nerve impulse from back to front, while various mollusks have retinas that conduct impulses from front to back. Insect eyes are constructed on a totally different optical principle, and they are sensitive to a wider range of wavelengths than are vertebrate eyes, especially in the ultraviolet range.

11.2.1.3 The ear: Reception of sound and balance The ear is a highly elaborate structure. In humans and other mammals, it is arranged in outer, middle, and inner parts (**Figure 11.12A**). The outer ear, in which sound waves travel as vibrations in air, consists of an external flap, the pinna, and a tube called the ear canal. The pinna acts as a funnel to focus the sound waves into the tube leading to the eardrum (tympanic membrane), which marks the boundary between the outer ear and middle ear. The middle ear is connected to the back of the throat by the Eustachian tube, which allows pressure to equalize on the two sides of the eardrum. If the Eustachian tube swells shut, as can happen in

Figure 11.11 Differences in the structure of eyes in different groups of animals.

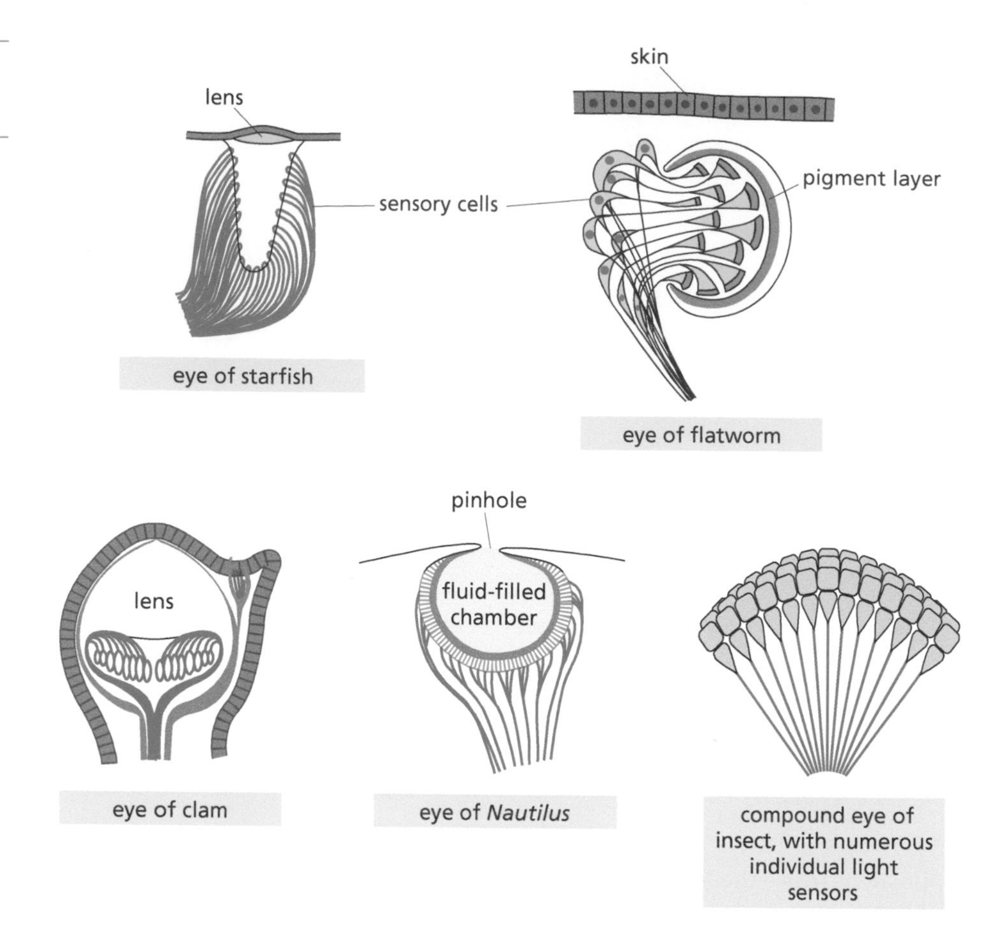

middle-ear infections or changes in altitude, pressure builds up inside the middle ear preventing the vibrations of the ear bones. This can lead to partial hearing loss, which is usually a temporary loss that is alleviated when the tube opens.

Sound waves in the middle ear travel to the inner ear as vibrations through a series of small bones, the hammer (malleus), anvil (incus), and stirrup (stapes). The inner ear is divided into a **cochlea**, responsible for sensing sound, and the semicircular canals, responsible for sensing gravity and balance. Both portions of the inner ear are filled with fluid, and the sensory nerve endings respond to the movement of fluids. The fluid in the coiled cochlea picks up the vibrations from the bones of the middle ear. The cochlea contains a sensory membrane with hair-like cells that respond to these vibrations (**Figure 11.12B**, **C**, and **D**). Very loud noises (as from industrial sources or rock concerts) can damage these hair cells and result in a permanent loss of hearing. Hair cells at different positions along the cochlea trigger action potentials in response to different pitches or frequencies of sound.

The range of pitch detectable by different species varies greatly from that detectable by humans. Whales detect extremely low-pitched sound waves, which travel very far; this allows them to communicate across great distances. Dogs can detect much higher-pitched sounds than we can, and bats higher still. The high-pitched squeaks of bats are reflected off objects, and the bats use these echoes for echolocation, permitting them to avoid obstacles at night and to catch insects in mid-flight.

The other portion of the inner ear includes the three semicircular canals, oriented at right angles to one another. Accelerations due to gravity and to body movements result in the movement of the fluid within these canals, and this information is detected by a patch of sensory cells near the end of each canal. People with damage to the nerve connected to these patches have difficulty in standing up and maintaining balance.

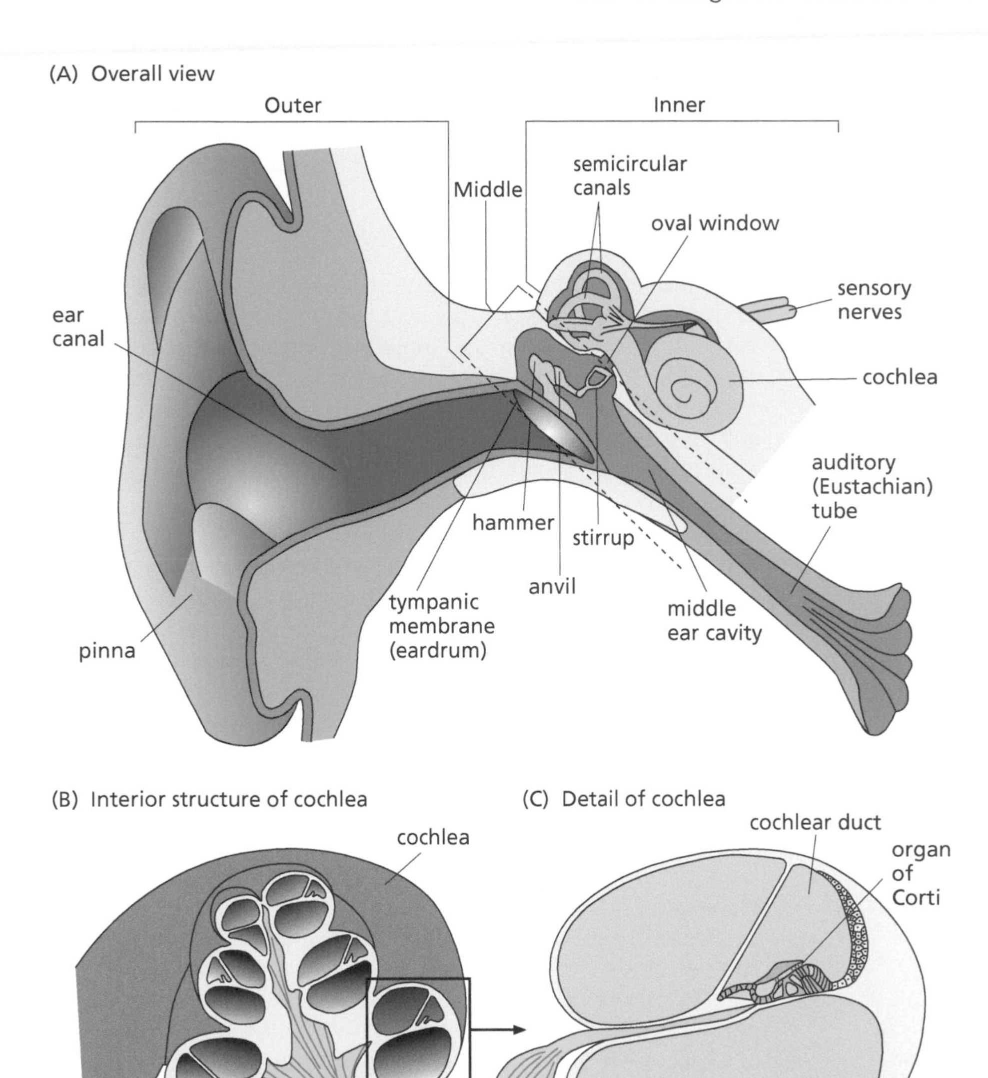

(D) Three rows of outer hair cells and one row of inner hair cells from the inner ear of a chinchilla (left) and a closeup view of outer hair cells (right).

Figure 11.12 The human ear.

11.2.1.4 The tongue: Reception of chemical signals as taste

Humans sense taste through a series of taste buds (Figure 11.13A) located along the tongue (Figure 11.13B) and the roof of the mouth. There are five different basic tastes in humans, usually described as sweet, salty, sour, bitter, and "umami" (the Japanese word for "yummy"). The taste buds for these different tastes all look the same but differ in their sensitivity to different chemicals: The sweetness receptors trigger action potentials in response to sugars, the salt receptors do so in response to sodium ions (Na^+), the sour receptors in response to the hydrogen ions (H^+) present in acids, and the bitter receptors in response to various other chemicals, including bitter-tasting plant compounds called alkaloids. The umami receptors respond to the glutamate ion, the active form of an amino acid present in many proteins, especially those found in meat products, soy, and fish. A few taste buds are present in unexpected locations, such as the respiratory passages, where they trigger sneezing and coughing rather than taste sensations.

11.2.1.5 The nose: Reception of chemical signals as smell

Although we are sensitive to thousands of different odors, the mechanism of smell is poorly understood. There is even disagreement as to the number of basic odors, with some experts naming as few as 7 while others list 20 or 30. Nerve endings sensitive to smell are most abundant in the lining of the nose (Figure 11.13C), although most vertebrates also have a similar sense organ opening into the roof of the mouth. These nerve endings trigger action potentials in response to chemicals just as the taste buds do. We perceive these chemical messages as smells, not tastes, because the nerves from the nasal lining go to a different area of the brain than do those from the tongue. However, taste and smell perception interact to a large extent. Therefore, food often does not taste very appealing when one has a respiratory infection that compromises smell.

Virtually all organisms, including single-celled prokaryotes and eukaryotes, can respond to chemical signals by means of receptors. Only animals have nervous systems, however, so only animals respond to chemical signals by producing nerve impulses. The sense of smell is well developed among the vertebrates, most of which are more acutely sensitive to odors than humans are.

11.2.2 Message processing in the brain

The brain receives information from various parts of the body through peripheral nerves and is central to the function of the entire nervous system. The brain is the hub where most activity of the nervous system takes place, where most decisions are reached, and where most bodily activities are both directed and coordinated. Nearly all activities that the body performs are carried out in response to signals

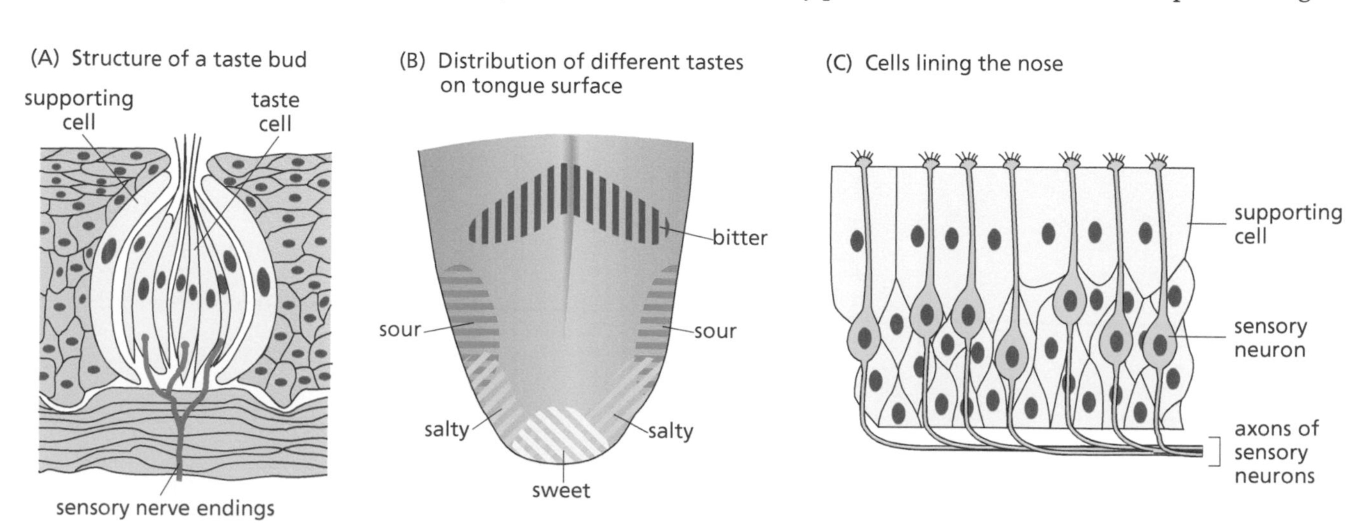

Figure 11.13 **Organs of taste and smell.**

originating in the brain and sent to the rest of the body through the peripheral nerves. We are aware of many of these signals; such signals are often said to represent "conscious" activity, although it is difficult to define consciousness in a manner that permits rigorous experimental investigation.

11.2.2.1 Studying the brain Our knowledge of the brain and its functions derives from many types of investigation. Many studies use experimental animals. Brain function can be studied in experimental animals by implanting a recording electrode in the brain to measure electrical activity during various brain activities. Another method uses an electrode through which brain activity can be stimulated in experimental animals for observation of their behavioral responses. Yet another approach involves destroying a portion of the brain (usually with an electric current) and studying the resulting changes in behavior.

Invasive techniques cannot ethically be used for studying brain function in humans. Electroencephalograms (EEGs), in which electrodes are pasted to the scalp, measure electrical activity within the brain (see Section 11.3.4). Another technique, positron emission tomography (PET), creates a computer-integrated picture of various brain activities, such as glucose metabolism and blood flow, within the living brain.

The brain can also be studied anatomically by dissections of the brain and its parts and also by microscopic examination of the brain and its cells. What are some of the major anatomical features of the brain?

11.2.2.2 Brain anatomy The brain is an immense network of neurons, yet it weighs less than 3 pounds (about 1.3 kg) in adult humans. The folds and constrictions in the partly developed brains of embryos allow us to recognize several major brain divisions. These same divisions are also recognizable in the brains of adult animals of various species. The major regions of the brain are the forebrain, midbrain, and hindbrain (**Figure 11.14**), each composed of millions of neurons and each containing a central cavity. These divisions are similar in various vertebrate species, but their proportions differ (Figure 11.14). Comparison of the brains of various species reveals a great deal about evolution and adaptation. The midbrain and hindbrain make up a larger proportion of the total in primitive vertebrates, while the forebrain, especially the cerebral hemispheres, is larger in mammals, especially primates. In general, animals with more complex behavior patterns have larger brains with larger and more highly folded forebrain areas. The increasing size and complexity of the forebrain are especially apparent in humans and other species that rely primarily on learned behavior. Other changes in proportions are related to the senses that each species uses: Brain regions concerned with vision are larger in those species that rely upon the sense of sight, whereas species that rely on smell or hearing have larger brain regions devoted to those functions.

Next, we look in more detail at the anatomy of the human brain (**Figure 11.15**).

11.2.2.3 Forebrain The **forebrain** includes the cerebrum, olfactory bulbs, hippocampus, and diencephalon. In early vertebrates, the forebrain was concerned with the sense of smell, and much of it still is through the olfactory bulbs, which process nerve impulses from sensory cells in the nose. Smells can influence both hormonal secretions and emotional responses such as anger, fear, and sexual response, which are therefore also processed in the forebrain, specifically in the diencephalon. In many cases, these responses can take place on an unconscious level, meaning that we may not even be aware that the changes are taking place. The hippocampus, concerned with certain types of learning, is located off center in the forebrain.

The **cerebrum** is divided into two halves, the right and left **cerebral hemispheres**. Thoughts and actions originate in the cerebral hemispheres, as do most of our "higher" functions of intellectual thought and reasoning. The cerebral hemispheres of mammals are much larger than those of other vertebrates. This is particularly true in humans and closely related primates, in which the cerebral hemispheres make up the largest part of the brain. Conscious activity and higher

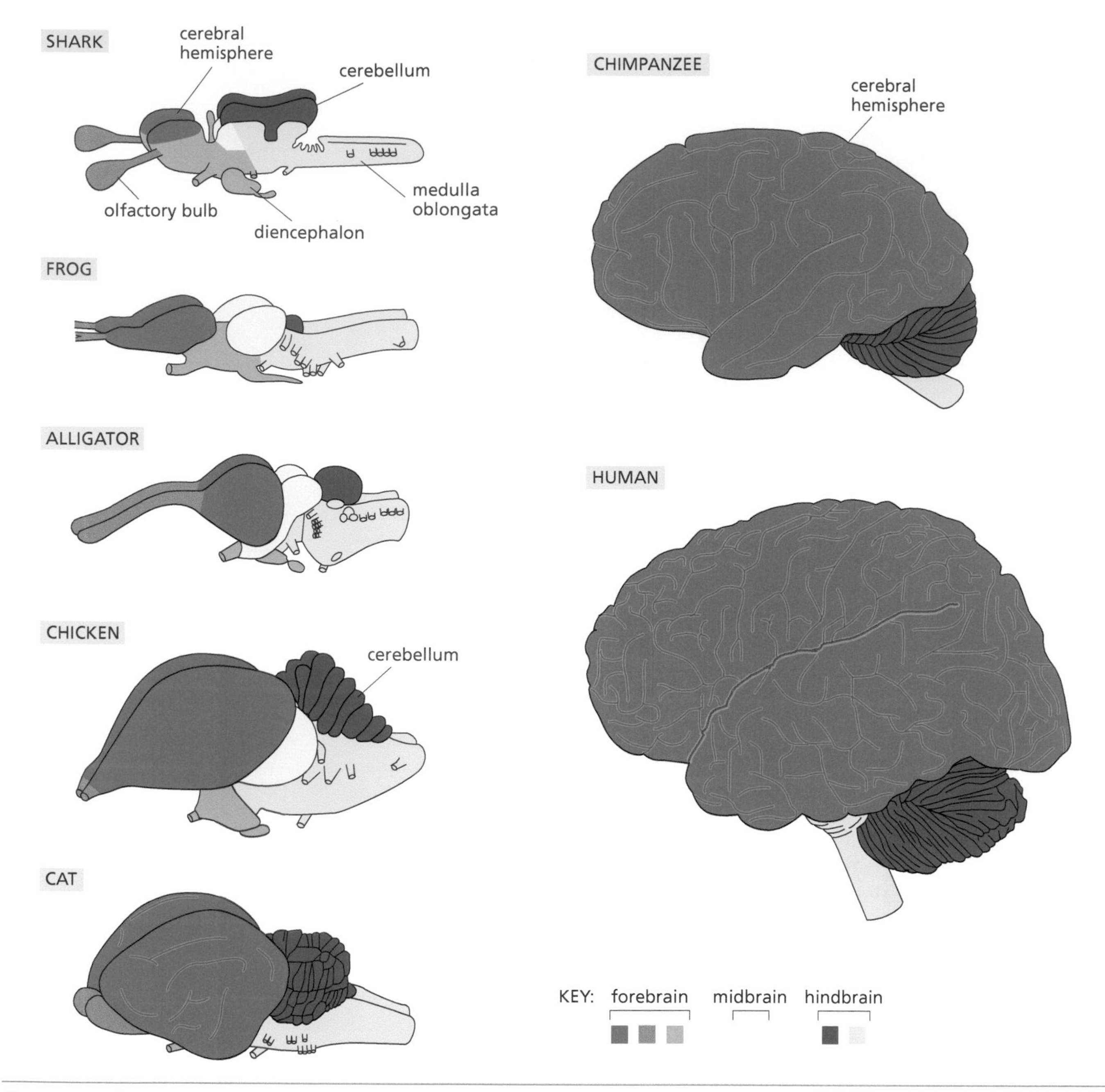

Figure 11.14 Correspondences among the brains of several vertebrate species (not to scale). In the chimpanzee and human brains, the olfactory bulbs and midbrain are concealed beneath the expanded cerebrum.

thought originate in the highly folded surface layer of the cerebral hemispheres known as the **cerebral cortex**. Below the cortex lie many series of neuron interconnections that constitute the cerebral white matter. One large part of this white matter is the **corpus callosum**, a series of fiber tracts that connect the right and left cerebral hemispheres across the midline, allowing messages received on just one side of the brain (e.g., from unilateral sensory input) to be shared with both hemispheres. Deeper still lie several clumps of neuron cell bodies known as basal ganglia. Several of these structures are involved in the disorders described elsewhere in this chapter.

11.2.2.4 Midbrain The **midbrain** contains the superior colliculi, the ventral tegmental area, and the reticular formation. The **superior colliculi** coordinate sensory information to help locate objects spatially. The ventral tegmental area

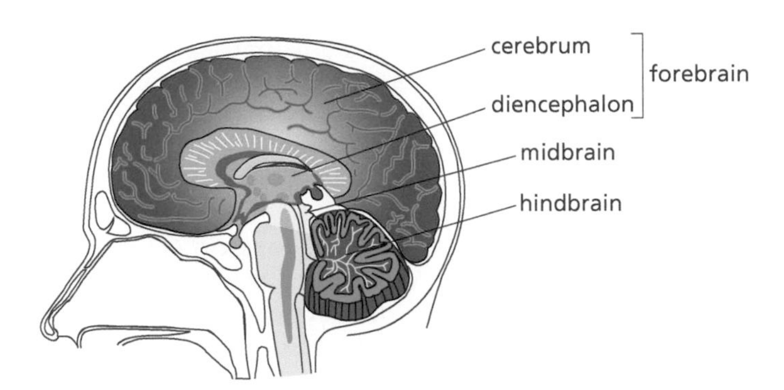

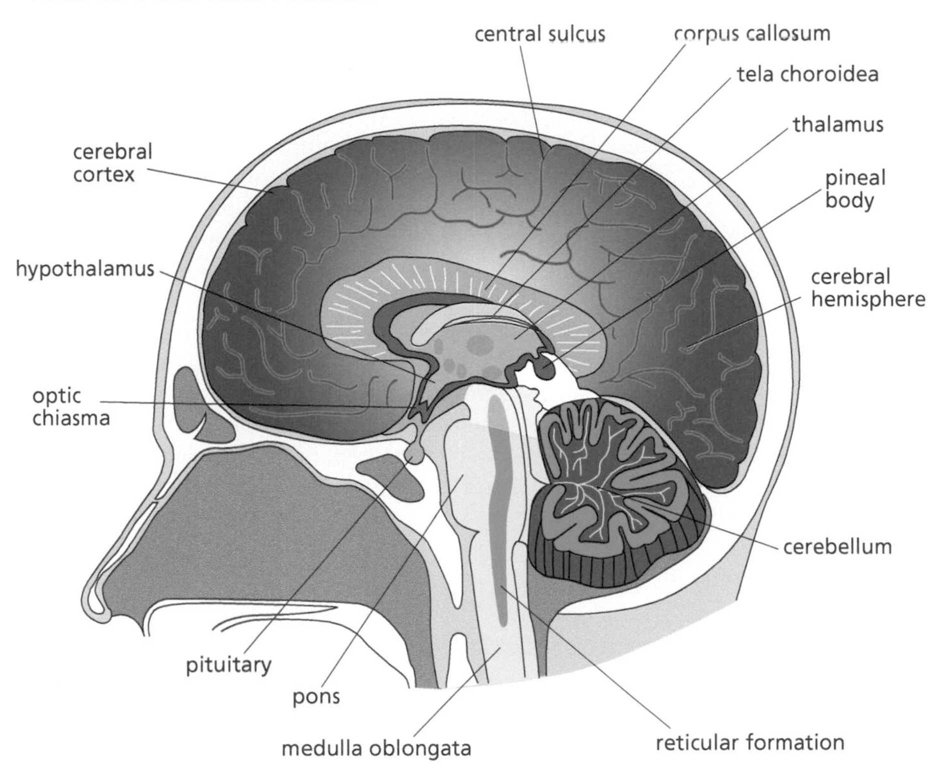

Figure 11.15 Section through the human brain. Only structures located in the midline plane are visible in this view; off-center structures (such as the olfactory bulbs) cannot be seen.

contains the brain's positive reward centers (see Section 13.3.2). The **reticular formation**, which also extends into part of the hindbrain, is important in keeping us awake and alert.

11.2.2.5 Hindbrain The **hindbrain** includes the **cerebellum** and the **medulla oblongata**. Functionally, the cerebellum is concerned primarily with balance, processing neuron impulses from the semicircular canals of the inner ear. The cerebellum also coordinates complex muscle movements. The medulla is concerned with such involuntary functions as breathing, functions that must continue even in sleep.

11.2.2.6 Blood–brain barrier The brain has few internal blood vessels; most of the arteries and veins that supply the brain run along the brain surface only. The cells deep in the interior must therefore receive most of their nutrition through the **cerebrospinal fluid**, which fills the brain's interior cavities. There is no direct flow of fluid from the blood to the cerebrospinal fluid. Instead, the cerebrospinal fluid communicates with the blood supply across a thin membrane, the **tela choroidea**, which serves as part of a **blood–brain barrier**. Nutrients and other small molecules cross this membrane or move through the neuroglia (nutritive cells) mentioned earlier, but many types of molecules, particularly larger molecules, cannot cross the blood–brain barrier. This barrier thus functions to prevent many molecules that have access to the rest of the body from entering the brain. The integrity of the blood–brain barrier is important to maintaining healthy brain function; several brain diseases, including Alzheimer's disease (Chapter 12), have been hypothesized to begin with leakiness in the blood–brain barrier.

11.2.2.7 Epilepsy: Abnormal message processing Messages come to the brain from the peripheral nervous system and the sense organs. These signals trigger action potentials in a series of interconnected neurons in different parts of the brain. It is this message processing that produces our perceptions of the world.

In some disorders, neurons in the brain trigger action potentials spontaneously, without an outside signal. **Epilepsy** is a disorder marked by brain seizures, usually mild, characterized by uncontrolled electrical activity in the cerebral cortex. Both the cerebral cortex and the hippocampus contain many feedback pathways involving GABA-secreting neurons. Recall that GABA inhibits a neuron that has fired from firing again (Figure 11.7). An impairment of one or more of these GABA feedback pathways is hypothesized to allow a stimulated neuron to keep firing, perhaps causing an epileptic seizure. Some of the drugs that block GABA synthesis or GABA receptors can bring on epileptic seizures, thus lending support to the hypothesis. What is not fully explained by this hypothesis is why the seizures are temporary, why long periods intervene between them, and why certain events precipitate the onset of seizures.

In epilepsy, the processing of messages by the brain is abnormal. Some people with epilepsy experience "auras." These take many forms depending on the part of the brain in which the spontaneous neuronal signals occur. The person might have a visual hallucination or smell a bad smell when no actual source of such a sight or smell is present. A specific thought may be triggered, or a vague feeling of a place being familiar, even if it is not. Another of the symptoms of uncontrolled brain activity in epilepsy is muscle seizures, in which muscles are continually stimulated to contract. To understand how changes in brain activity could affect muscle activity, we next examine the pathways of normal muscle contraction.

11.2.3 Message output: Muscle contraction

In addition to message processing by neurons forming synapses to other neurons within the brain and nervous system, the brain also coordinates the functioning of other structures within the body. Neurons, as well as forming synapses with other neurons, can also form synapses with cells of other types, such as muscle cells and gland cells. Glandular secretions are regulated by the autonomic nervous system, a part of the peripheral nervous system that will be described in Section 14.3.

Voluntary movements are produced when the brain sends nerve impulses to the body's skeletal muscles. Because the muscles are attached to the skeleton, contraction of the muscles brings about movement of the skeleton and of the body as a whole. Most movements would be jerky and uncontrolled if only a single muscle were involved; controlled, steady movements usually require the simultaneous contraction of several muscles that pull in different directions to smooth and steady the movement. Coordination of messages to thousands of cells is a function of the brain. Earlier we saw that diseases (Parkinsonism and Huntington's disease) can result when the brain does not function properly. Now we will examine in more detail how neurons induce muscle contraction under normal conditions.

11.2.3.1 Muscle contraction at the molecular level Neurons whose axons form synapses with muscle cells rather than with other neurons are called **motor neurons** (**Figure 11.16A**). Muscle cell membranes, like the cell membranes of neurons, carry an electrical potential and are electrically excitable. The muscle cell electrical potential can be depolarized after the cell receives a neurotransmitter signal from the motor neuron. Motor neurons

(A) Overall structure of skeletal muscle, showing cross-striations

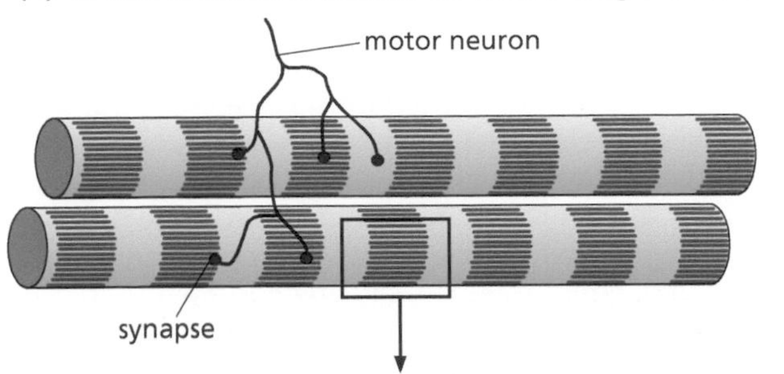

(B) Enlargement showing arrangement of actin and myosin filaments in a relaxed portion of skeletal muscle

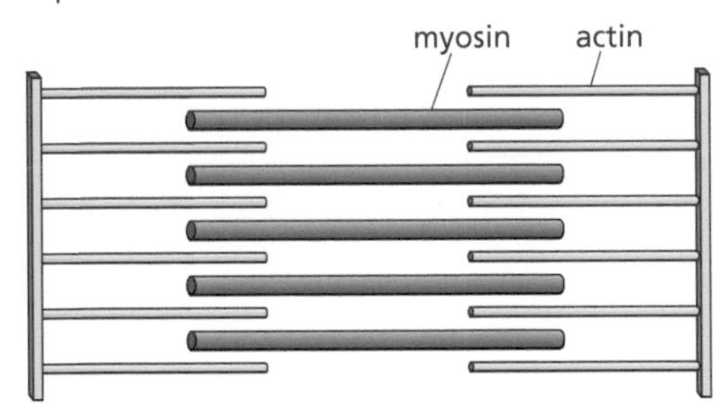

(C) The same portion of muscle, contracted

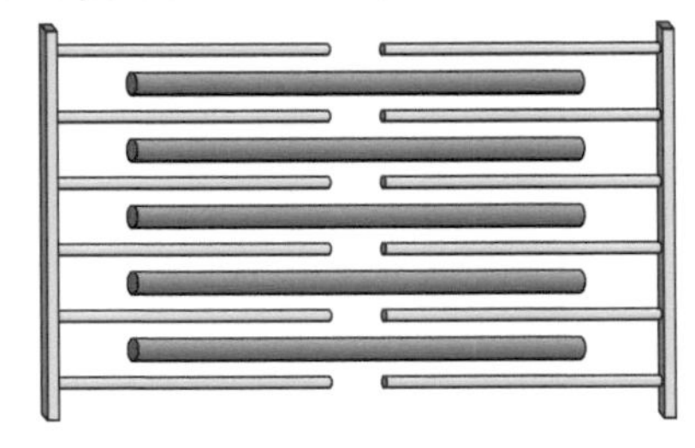

Figure 11.16 **Structure of a portion of skeletal muscle.**

generally secrete the neurotransmitter acetylcholine, and muscle cell membranes have specific receptors for this neurotransmitter.

When a motor neuron releases acetylcholine into the synapse with a muscle cell, the acetylcholine binds to receptors on the muscle cell membrane, causing the membrane to depolarize. This depolarization of the muscle cell membrane causes a release of calcium ions from compartments within the cell. These calcium ions trigger reactions that allow cross bridges to form between the two major muscle proteins, **actin** and **myosin**. The cross bridges pull the actin filaments, increasing their overlap with the myosin filaments and producing the muscle contraction (Figure 11.16B and C). Many filaments of actin are pulled over many filaments of myosin to produce a forcible contraction of the muscle fiber as a whole.

Actin and myosin may be arranged in orderly bands (see Figure 11.16A), giving certain types of muscle fiber a striated (cross-banded) appearance common to both skeletal and heart muscle tissue. Another type, smooth muscle tissue, found in blood vessels and internal organs, lacks this cross-banding because the actin and myosin fibers are arranged at random intervals.

11.2.3.2 Muscle relaxation Muscle cell contraction is a neuron-stimulated process; muscle cell relaxation is not. When the muscle cell membrane is no longer receiving depolarization signals, calcium ions are removed from the cell cytoplasm. Removal of calcium breaks the cross-bridges between actin and myosin, allowing them to return automatically to their original positions relative to each other (Figure 11.16B). They do not receive another neuron signal that pulls them back apart. Removal of calcium ions does require energy, however. This energy is supplied as adenosine triphosphate (ATP). When an organism dies, no more ATP is made, so calcium cannot be removed, and the muscle cells stay contracted. This produces rigor mortis.

Within the whole organism, controlled muscle contraction may bend an arm. Relaxation of those same muscles does not return the arm to its original position; for that movement, another set of muscles must contract. Skeletal muscles are therefore arranged in opposing sets, with flexor muscles bending limbs and extensor muscles straightening them (**Figure 11.17**).

As we saw earlier (Figure 11.6), neurotransmitters released into a synapse are normally taken back up by the neuron that secreted them or are degraded chemically. Motor neurons release acetylcholine into the synapse, where it is normally broken down by the enzyme cholinesterase, causing contraction to cease. Some insecticides work by blocking the action of cholinesterase at these synapses. In this case, any synapse stimulated by acetylcholine remains continually stimulated (because the acetylcholine is never broken down), and the insect receiving the insecticide dies with most of its muscles in a state of rigid contraction. Because most animal nervous systems use acetylcholine as a neurotransmitter, such pesticides are toxic to all animals, great and small, including insects, pets, and humans. The toxic dose, however, depends on body size, so an amount that might kill most insects would only make your pet sick and might not noticeably affect you at all.

Many messages to and from the brain pass through neuron synapses connecting the peripheral and central nervous systems. We have seen that sense organs in the peripheral nervous system send messages to the brain. The brain also sends messages to the peripheral nervous system, for example, through motor neurons that induce muscle contraction. Many other messages, and message storage, pass through synapses within the brain itself, a subject we take up in the next section.

THOUGHT QUESTIONS

1. The function of a portion of the brain of an experimental animal can be investigated by destroying it and studying the defect produced. What are this technique's limitations? Could you study how a piano or an automobile works by inserting a probe, destroying some local region, and studying the resulting defects? Could you study the function of a radio in this way, or a computer? Which of these is more comparable to the brain in complexity?
2. What does it mean to say that there are five basic tastes, not four or six? Compare this with the several different types of receptors in the skin.
3. Why would your being on a roller coaster or in a stunt airplane confuse your inner ear with misleading stimuli?
4. What causes muscle contraction to stop? What happens to an organism if muscle contraction does not stop?
5. Why does the body have so many different kinds of neurotransmitters?

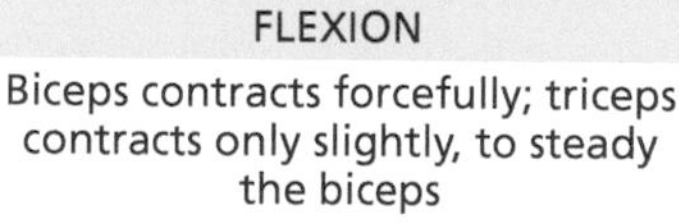

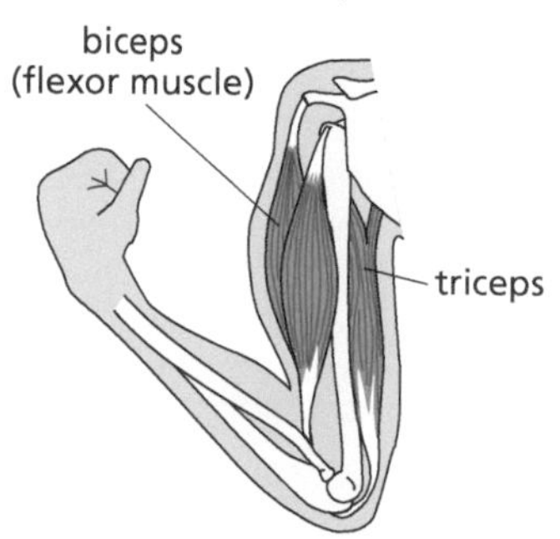

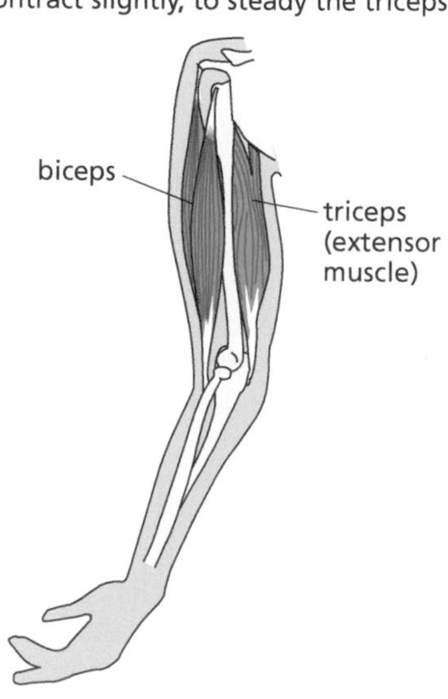

Figure 11.17 The major muscles that flex and extend the arm. Most muscles act together with other muscles that steady their action by pulling in the opposite direction.

11.3 THE BRAIN STORES AND REHEARSES MESSAGES

Neuron activity within the brain produces many of the functions that we associate with the mind. Some messages are stored as memory for future retrieval. Some of these modify our future behavior, a process that constitutes learning. At times, but especially when we sleep, the brain may rehearse sending and interpreting certain messages without remembering them. Brain activity produces personality, emotions, thoughts, and dreams, and, as we will see, mental illnesses can result from abnormal brain activity.

11.3.1 Learning: Storing brain activity

When we change in response to changes in the world around us, this is known as **adaptation** (a separate meaning, unrelated to the adaptations that arise from the operation of natural selection). Adaptation can take place through many physiological, immunological, and neurological mechanisms and can be either conscious or unconscious. **Learning**, which consists of lasting changes in behavior or knowledge in response to experience, is an important form of adaptation brought about by the nervous system. Different types of learning are distinguished by the types of information that are learned and by ways in which the nervous system processes and stores the information. **Declarative learning** is mostly conscious remembrance of persons, places, things, and concepts, requiring the actions of neurons in the hippocampus and certain parts of the cerebral cortex. Memory of how to do things, **procedural learning**, does not require the hippocampus or the temporal lobe of the cerebral cortex, and is not necessarily conscious. People with hippocampal damage can still learn how to do new things, although they will not consciously recall that they can do them.

11.3.1.1 Procedural learning Procedural learning can be very simple—in fact, simpler animals such as mollusks and insects are capable of procedural learning because it does not require the forebrain structures that evolved in vertebrates. In human development, procedural learning becomes possible earlier than declarative learning. Infants at first learn procedurally: how to eat, how to move, how to respond to gravity.

The three simplest kinds of procedural learning are habituation, sensitization, and classical conditioning. These simple kinds of learning are distinguished from declarative learning in that they can be involuntary: The learner changes his or her behavior without showing any awareness of the learning process. Most animals, no matter how minimal their nervous systems, can learn in these simple ways.

When a stimulus is presented repeatedly, an animal may learn to no longer respond to it. This is known as **habituation**. Habituation occurs both behaviorally and at the level of the neuron. Single neurons can stop making action potentials owing to changes in their receptors or to an increase in the threshold for generating action potentials. When a stimulus does not result in harm, an organism may learn to change its behavior and not expend energy in responding to the stimulus (**Figure 11.18A**). Humans habituate to all kinds of signals: If you move to a new location, you see and hear many things that people who have lived there for a while have learned not to notice any more. After a time, you no longer see or hear them either. If something unusual happens, the habituation can be overcome. Habituation is thus context-specific to some extent.

Sensitization (**Figure 11.18B**) is the opposite of habituation. An organism becomes sensitized when an intense and aversive stimulus, such as a loud gunshot, increases subsequent responses to other stimuli. On the cellular level, for a prolonged time, neurons become more capable of generating an action potential, a change known as long-term potentiation. Several hypotheses have been suggested to account for long-term potentiation. One recent hypothesis involves a type of receptor known as an *N*-methyl-D-aspartate receptor (NMDA receptor). These receptors respond to the secretion of glutamate (a neurotransmitter) by

Figure 11.18 Three kinds of procedural learning.

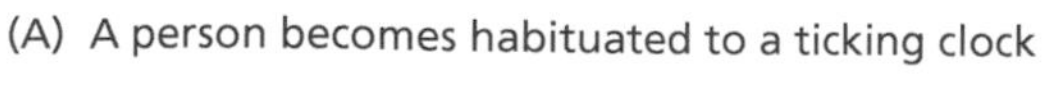

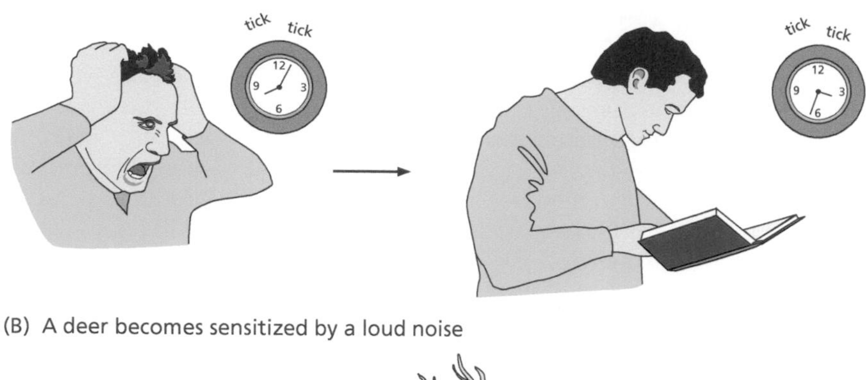

(B) A deer becomes sensitized by a loud noise

bang

crack

deer is startled by loud noise

deer is now more sensitive to other sounds

(C) A dog undergoes Pavlovian conditioning

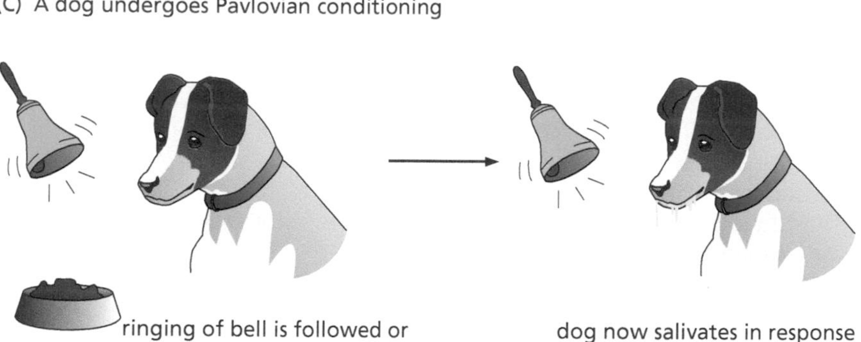

producing nitric oxide. The nitric oxide then diffuses back to the glutamate-secreting (presynaptic) cell, where it enhances the future release of glutamate. This type of feedback is called a positive feedback loop because a condition stimulates more of itself and often leads to an extreme condition. In this case, glutamate secretion enhances future glutamate secretion. The extreme overstimulation can lead to glutamate poisoning (characterized by convulsions), a condition made worse by the food additive monosodium glutamate (MSG).

A third type of simple procedural learning is called **classical conditioning** (Figure 11.18C), a change in which an organism learns to associate a stimulus with a particular response. Classical conditioning is also called Pavlovian conditioning because it was first demonstrated by the Russian physiologist Ivan Pavlov around 1900. Dogs salivate when they see or smell food. If a bell is rung each time food is presented, a dog learns after a very few repetitions to salivate when it hears the bell ring, even if no food is present.

Humans can learn through classical conditioning, very often without being consciously aware of it. Fears sometimes become associated with various objects, colors, or smells when we are young because those stimuli were present when we were hurt in some way, or because they remind us of other unpleasant stimuli. One person we know grew up with a decades-long aversion to gelatin desserts.

It seems that at the kindergarten he attended, such desserts were brought in on a tray, piled in cubes whose wiggling movements reminded him of certain caterpillars. Caterpillars are something that most children would avoid eating, so the shaking movements in this case conditioned an aversion to the desserts. Adults too can become conditioned. In one case, several people became nauseated every time they saw the carpeting in a hospital. It turned out that the carpet was the same color as the chemotherapeutic drugs that these people had taken for treatment of their cancers, drugs that had made them sick to their stomachs. The people had become conditioned, associating the color of the carpet with the cause of their nausea.

11.3.1.2 Declarative learning

Declarative learning includes recognition of a familiar person or object, or learning to associate a sound or name with it, like associating a dog with a barking sound or with the word "dog." In contrast to procedural learning, more complex types of learning require the activity of the hippocampus and cerebral cortex. There is evidence that complexity of experience actually contributes to the size of the cortex. Rats raised in "enriched environments"—in large cages with other rats and with "toys"—develop a thicker and more elaborate cortex. In humans, at about the age of two, declarative learning begins. At this age, the number of brain cells reaches its maximum and a critical level of complexity of connections is achieved. Although the exact age varies from child to child, the capacity for declarative learning always develops later than the capacity for procedural learning.

11.3.2 Memory formation and consolidation

To become a **memory**, a piece of information must be acquired, stored, and retrieved. Many acquired pieces of information can be retrieved for only a short period. For example, you may be able to recall having heard a particular sound if someone asks you about the sound within a few minutes of the time you heard it, but not after that. Information that is quickly forgotten is said to be part of short-term memory. Short-term memory is mostly chemical, having to do with temporary changes in neurotransmitters and their receptors. Long-term storage of information requires actual structural changes (the formation of new synapses) within many parts of the brain.

11.3.2.1 Long-term memory

One part of the brain that is essential to change a sensory input from a short-term into a long-term memory is the hippocampus (**Figure 11.19**). People who have suffered damage to the hippocampus are unable to form new long-term memories. They can recall things from before the time of the damage, indicating that the storage sites themselves are not in the hippocampus. Their recall of newly acquired information or experiences is limited to the short term, after which it is forgotten. Normal long-term retention of a memory may therefore be said to require a "lack of forgetting." This is especially true of memories that can be consciously known.

The hippocampus turns short-term memories into long-term ones by making new cellular connections with other parts of the brain. Neurons form new synapses, connecting several neurons in loops with each cell synapsing on the next cell in the loop. The stimulation of any one of these neurons results in information transfer around the whole loop. Stimulation of these assemblies of cells may need to continue for years before a memory is permanently stored.

As time passes after the acquisition, **memory consolidation** occurs. While long-term memories are forming, and even after they have been stored, they are organized and restructured based on even more recent experiences. Existing knowledge is constantly being reordered in the light of new knowledge. Memory consolidation relies on information processing, one innate aspect of which is the capacity to generalize from specific experiences. If we live in a city and have walked in a forest only once, we mentally picture all forests as being like the one we walked in. Further experience, either "in person" or acquired through seeing pictures or reading stories, enables us to reformulate the initial generalization that

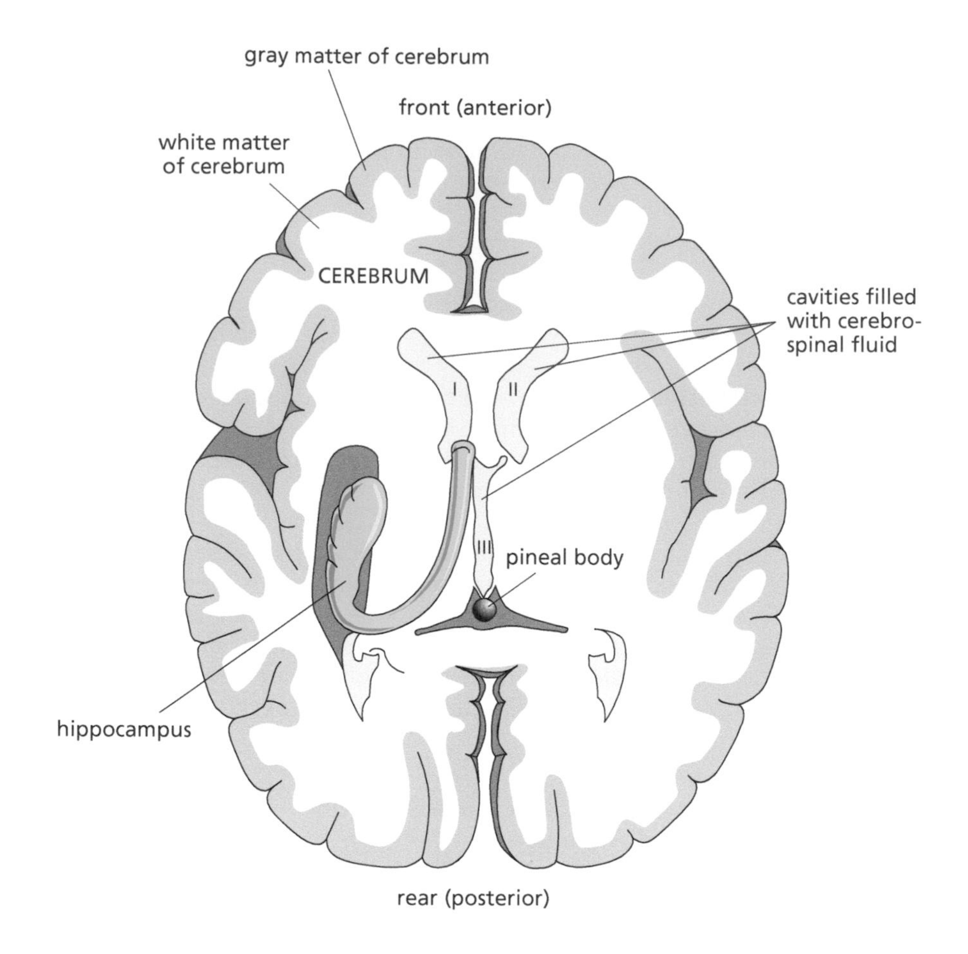

Figure 11.19 Horizontal section through the brain, showing the hippocampus, an important structure in the formation of certain types of memory. (Only one hippocampus is shown; the other is symmetrically located.)

we made, replacing it with another generalization. In addition, we can still recall some very specific aspects of the particular forest that we first walked in.

Emotional states can sometimes modify the process of memory consolidation: We are more apt to remember something that we will normally not consider worth remembering if we associate it with an event that had great emotional meaning for us (either positive or negative). People who were old enough at the time of the assassination of John F. Kennedy, the tearing down of the Berlin Wall, or the September 11, 2001, attacks on the World Trade Center and Pentagon can remember vivid details of where they were and what they were doing. The same is generally true of events with great personal meaning, such as weddings, deaths, or natural disasters such as earthquakes or floods.

11.3.2.2 Abstraction and generalization Memory consolidation also relies on the capacity to conceptualize, an extension of the ability to generalize. Researchers can demonstrate the capacity for generalization by using what are called "oddity problems," typified by the childhood learning game "one of these things is not like the others." Monkeys are shown three or four objects, all alike except one, and they are rewarded for picking the different one. If a set of objects consists of three toy trucks and a car, they must learn to pick the car. Presented with a totally different set, two oranges and an apple, they must learn to pick the apple. After many such sets, each set different, have been presented, the monkeys develop a concept of "oddness" and pick the single object immediately. This type of declarative learning requires activity in the temporal lobe of the cerebral cortex. Much of human learning and memory is processed by the cerebral cortex, consolidating memories, and forming concepts, although other parts of the brain are also involved.

Memory formation requires the hippocampus, but memory consolidation requires more parts of the brain, particularly the cerebral cortex. The cortex is also important for memory retrieval. In a series of experiments in the 1940s, a neurosurgeon named Wilder Penfield electrically stimulated the cerebral cortex of conscious patients who were undergoing brain surgery for various neurological diseases. Such stimuli cause a person to recall a memory so vividly that they feel they are reliving the experience.

11.3.3 Biological rhythms: Time-of-day messages

In addition to the conscious processes of declarative learning, the brain sends itself many unconscious messages. Our bodies respond all the time to messages that tell us what time of day it is. In response to these messages, we establish a biological rhythm that governs our pattern of sleep and wakefulness.

11.3.3.1 Circadian rhythms and their control There are many kinds of biological rhythms, biological processes that repeat at somewhat predictable intervals. Those of approximately 24 hours' duration are called **circadian rhythms** (Latin *circa*, "about," and *die*, "day"). Various biological functions can be monitored on a 24-hour basis, and most of them show some recurrent circadian rhythm. Even individual cells experience circadian rhythms in the rise and fall of certain protein levels.

Where do circadian rhythms originate? If the rhythms originate internally, how do they keep tuned to the 24-hour cycle of the world around us, and how can they adjust to different time zones, seasons, and work shifts? If, in contrast, the rhythms originate externally, how do external cues regulate the body's cycles, and can these external cues be easily manipulated? The first serious attempts to answer these questions began with isolation experiments, such as those conducted in Mammoth Cave in Kentucky. In these experiments, volunteers who were kept from all sources of natural light were allowed to set their own daily routines. Nearly all individuals maintained fairly constant circadian rhythms in their sleep–wake cycles and also in body temperature, activity, and a variety of physiological measurements. Circadian rhythms were all maintained despite the constant environment and were rather uniform for each individual. Most significantly, the circadian rhythms maintained in these isolation experiments were generally slightly more than 24 hours, between 25 and 26 hours for most individuals. The maintenance of such a rhythm after its drift from synchrony with the day–night cycle of the outside world showed that internal rhythm-keeping mechanisms existed.

How, then, does the external world exert its influence, causing most of us to maintain a 24-hour daily rhythm? Our current understanding is that the external environment provides us with certain time-related clues. The most important of these clues is the natural rise and fall of light intensity throughout the day; other important clues include our own activity and our bombardment with external stimuli (including social stimuli) during daylight hours.

A group of neurons located in the **hypothalamus** (Figure 11.15) are important to circadian rhythms. Destruction of these neurons abolishes all circadian rhythms. Under experimental conditions of continuous darkness or continuously dim lighting, these neurons maintain an internal circadian rhythm with a cycle of slightly more than 24 hours in length. The rhythm is also regulated by the **pineal body**, a structure about the size of a pencil eraser, located on the roof of the forebrain (Figures 11.15 and 11.19). Under natural conditions, the light received by the pineal body adjusts the rhythm to follow a 24-hour cycle, with a peak of activity in the late morning and with greatly reduced activity throughout the hours of darkness. During times of darkness, the pineal body secretes a hormone called **melatonin**, which is not secreted during times of illumination. By illuminating various parts of the body separately from the rest, biologists have experimentally determined that the pineal body is sensitive to the light that it receives right through the skull and brain! If a laser (a highly focused beam of light) is focused on the pineal body through the head and is turned on and off in a 24-hour rhythm, all parts of the body follow the established circadian rhythms, even when the rest of the body is in darkness. If, instead, the head is kept in darkness while other parts of the body are illuminated, the effect is the same as if the body and head were both in total darkness.

Gradual or slight disturbances in a person's 24-hour rhythm can result from short-distance travel, seasonal changes, and the semiannual change of clocks at the beginning and end of daylight-saving time. The effects of these changes are minimal in most cases. More drastic effects are felt in the phenomenon known as jet lag, the disturbance of our 24-hour rhythms as a result of travel through several successive time zones. The major symptom of jet lag is fatigue, plus a desire to sleep or remain awake at inappropriate times for a few days until the body readjusts to the new cycle. The rigors of travel add to the fatigue, but travel north to south within a time zone is much less fatiguing than travel east to west across time zones. People who have traveled long distances by air are also statistically more susceptible to infection. While some of this may be due to other conditions of air travel, the increase in susceptibility is greater among people who have flown east to west than among those who have flown south to north, suggesting that interrupted circadian rhythms are involved. People whose bodies are not able to adjust to any particular rhythm because of irregular work schedules also show an increase in fatigue-related events such as the number of accidents.

11.3.3.2 Sleep One mental state that shows a strong circadian rhythm is sleep. At the end of each day, we usually have a strong urge to sleep; we can postpone this urge, but only to a limited extent. People deprived of sleep do not function well when awake (they make more mistakes, for example), and people who awake from a "good night's sleep" feel refreshed and alert.

As we have seen, the electrical activities of single cells take place across their membranes. The cumulative effect of ions flowing across the membranes of many cells can be seen in an electroencephalogram (EEG), a graph of electric activity obtained from electrodes pasted onto the scalp (Figure 11.20A). By using an EEG, we can detect different patterns of ion flow in the brain. EEGs show several levels or stages of sleep. Stage 1 is characterized by regular respiration, slowed heart rate, drifting mental imagery (similar to daydreaming), and low-voltage EEG wave patterns of about 7–10 Hz. (Hz stands for Hertz, a measure of frequency equivalent to cycles, or waves, per second.) Subjects aroused from stage 1 sleep will often say, "I wasn't sleeping." Sleep stages 2, 3, and 4 also have characteristic EEG patterns (Figure 11.20B). A further sleep stage is characterized by rapid eye movements (**REMs**), noticeable as movements of the eyeball beneath the closed eyelids

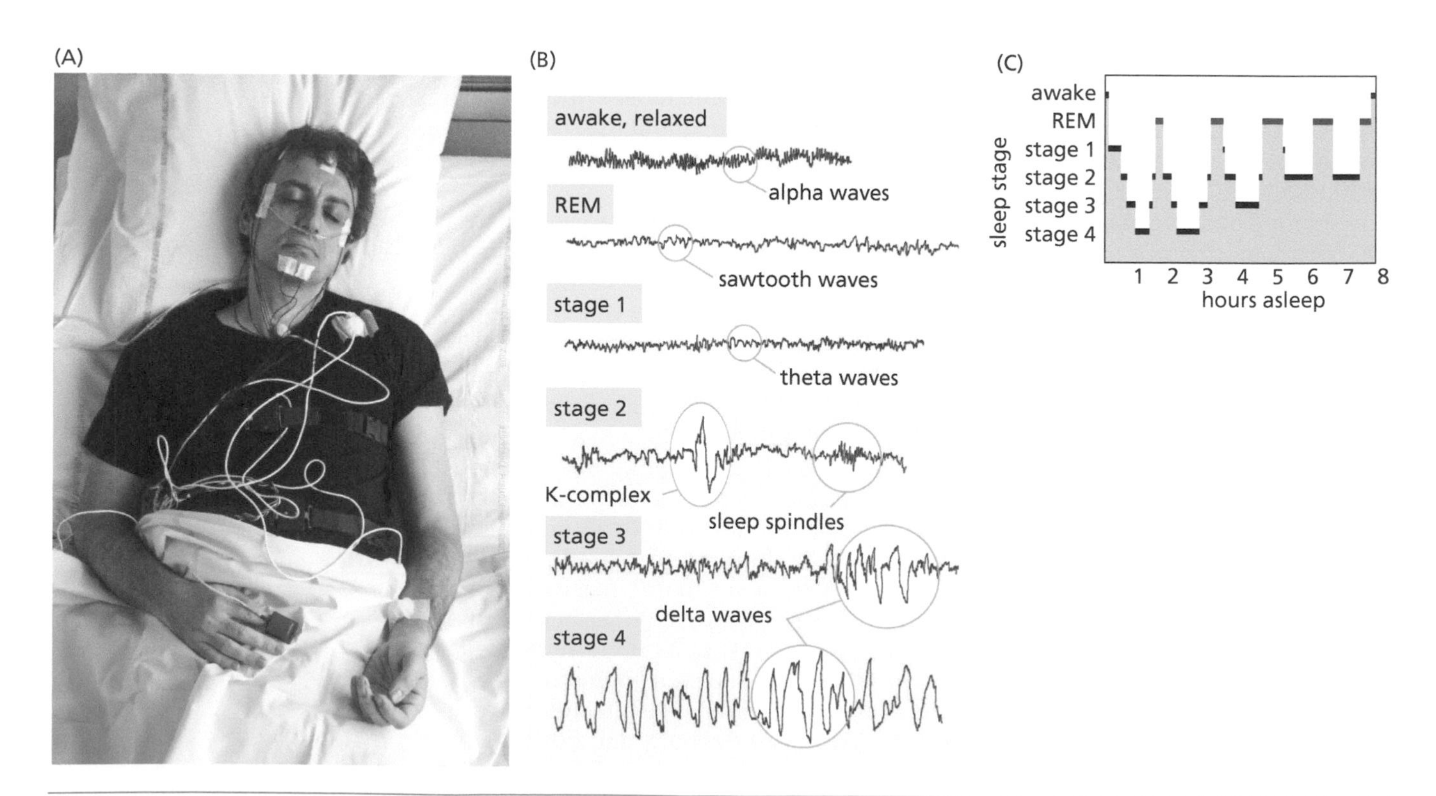

Figure 11.20 Electrical activity in the brain during different stages of sleep.

of sleeping subjects, including cats and dogs as well as humans. About 80% of human subjects awakened during REM sleep tell of some dream that they were having, often in vivid detail, while subjects awakened during other sleep phases do not remember any dreams.

During a typical night's sleep, an adult goes through about four or five sleep cycles. As shown by the steps in Figure 11.20C, each cycle goes through stages 1, 2, 3, and 4, in that order, then in reverse order through stages 3, 2 and 1, followed by a REM period, after which the next cycle begins. The proportion of REM sleep increases with each successive cycle, as shown by the increasing width of the REM bars in the later sleep cycles in Figure 11.20C. The later sleep cycles may also skip some of the stages, for example, proceeding only to stage 3 before reversing back to 2, 1, and REM, and at a still later cycle, proceeding only to stage 2 before returning to 1 and REM. If subjects are allowed to awaken by themselves (with no alarm clock or other external stimulus), they usually awaken near the end of a cycle—that is, following a REM stage or during stage 1 (see Figure 11.20C). Although this pattern is fairly typical, sleep cycle patterns vary widely with the person and the circumstances.

Evidence continues to mount that REM sleep is extremely important. Volunteers who are deprived of REM sleep for a day or two begin to daydream and have their thoughts wander. They also make more mistakes and have more accidents, even if their total amount of sleep and their amounts of all the other sleep stages are normal or above normal. If finally allowed to sleep as long as they wish, REM-deprived subjects sleep longer than usual, and a higher-than-usual proportion of their sleep is REM sleep. People deprived of any of the non-REM stages do not show any abnormal symptoms.

Some drugs can alter the natural occurrence of sleep. Caffeine, amphetamine, and other stimulants (see Chapter 13) may interfere with the onset of sleep, although sensitivity to this effect varies with the person. Alcohol and barbiturate drugs bring on sleep more readily, especially in persons already tired, but this drug-induced sleep is less restful because it has longer stage 3 and stage 4 intervals and shorter REM episodes. Muscle relaxants have no effect on sleep except to overcome muscle tension, which may sometimes inhibit the onset of sleep. (Most drugs taken as aids to sleep are either barbiturates or muscle relaxants.) There is no totally safe sleeping pill, and all drugs that alter sleep patterns usually have other effects as well. In recent years, the sleep hormone melatonin has become popular as a natural supplement to combat insomnia or help people adjust to jet lag. However, most medical professionals do not recommend the habitual use of supplemental melatonin; rather, it is preferable to support the body's own production of melatonin by such practices as dimming the lights and minimizing screen exposure before bedtime.

The role of neurotransmitters in the control of sleep and wakefulness is unclear. The rate at which neurotransmitters or certain other chemicals are produced or degraded during sleep can be studied by labeling these chemicals radioactively. For example, studies in which animals are fed radioactive tryptophan (a chemical precursor from which the neurotransmitter serotonin is synthesized) show that the rate at which tryptophan is converted into serotonin increases during sleep. Likewise, studies have found that serotonin breaks down more rapidly during sleep. Such experiments have enabled us to locate areas of neurotransmitter activity during sleep and wakefulness. In these studies, as in other biological studies with radioactively labeled substances, the radioactive dosage is kept low to minimize the risks to the experimental subjects and experimenter.

An important part of the brain governing sleep and wakefulness is a group of neurons called the **reticular activating system**, which radiates outward and upward from the reticular formation of the brain stem (Figure 11.21). These neurons send "alertness" signals through the diencephalon to widely scattered parts of the cerebral hemispheres. These signals seem to accompany most types of sensory input but do not seem to vary depending on the type of stimulus. Their message seems to be simply "pay attention!" The reticular activating system is usually more active by day and quiescent at night. Low-level activity allows sleep, but higher activity maintains wakefulness. Fortunately, it is also possible for the

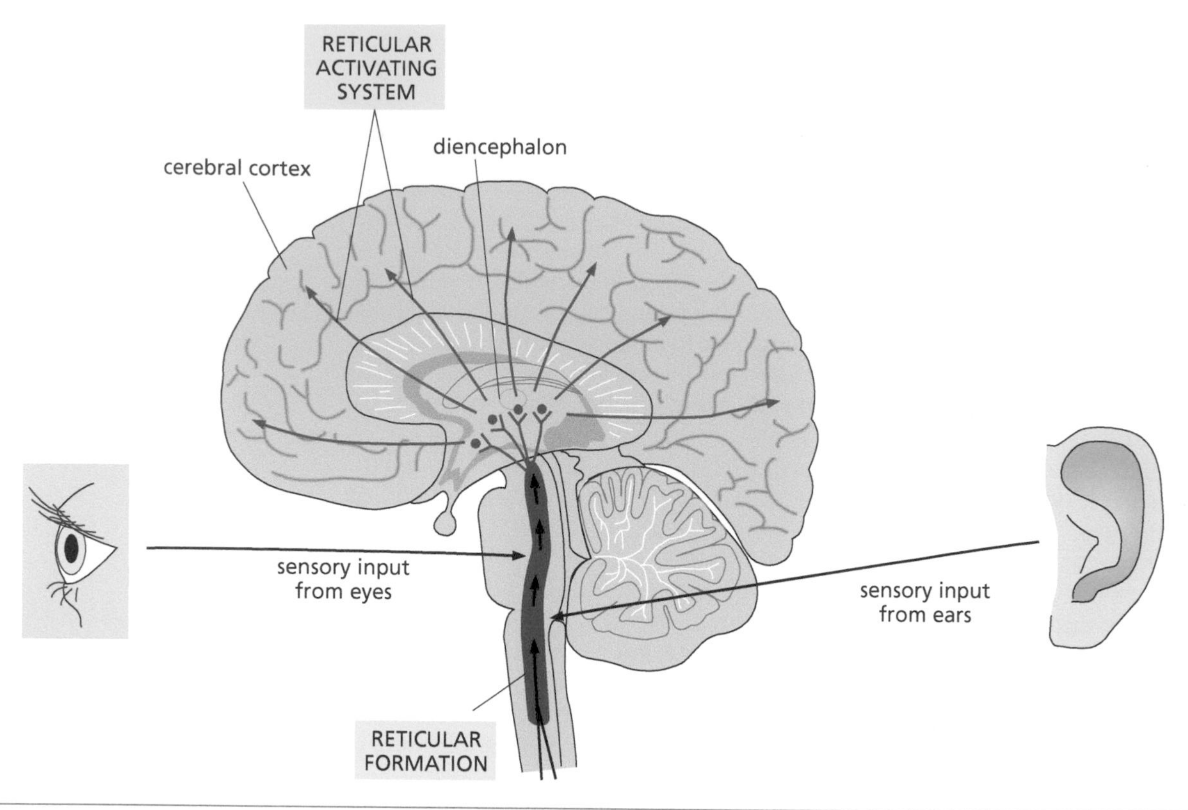

Figure 11.21 The reticular activating system.

reticular activating system to awaken us to an emergency in the middle of the night or to deviate in other ways from its usual 24-hour rhythm, as conditions demand.

11.3.4 Dreams: Practice in sending messages

Dreams are an important phenomenon of sleep. Philosophers and poets, fortune-tellers and psychiatrists have each had their ways of interpreting dreams. Modern-day dream researchers have used several techniques to study dreams. One method is to have the subject keep a "dream diary" in which they record as much of a dream as they can remember upon awakening. Among the findings using this technique are the following: Most dreams are visual, including those of people who became blind after age five or six, but excluding those of people who were blind from birth or infancy. Other senses (hearing, touch, smell, taste) also are mentioned in dream reports, but less often than vision. (These other senses predominate, however, in the dreams of people blind since birth or infancy.) Familiar persons, places, and types of events appear in most dreams, but not always congruously: The dreamer sometimes experiences familiar activities in the wrong setting, and places that are supposedly far away or unfamiliar often look very familiar in the dream. There are frequent changes of scene, of mood, or of persons present. Stimuli in the dreamer's environment are often incorporated in the dream imagery itself. Actual sounds, smells, flashes of light, and other sensations from the sleeper's environment may be worked into the dream.

Perhaps one of the most important and most constant features of our dreams is that we remember so little about them during our waking hours. There seems to be a good reason for this. The EEGs recorded during REM sleep show spontaneous electrical activity in the brain. This is our brain's way of rehearsing its message-sending functions, and it might be related to the strengthening (through practice) of existing synaptic connections or the establishing of new ones.

Our brains are usually programmed to interpret any such electrical activity as a coherent picture of the world around us. When we are awake, our brains usually record these coherent pictures so that we can remember them at some later time. In dreams, the integrating mechanisms are still at work, so a coherent or semi-coherent picture of the world is drawn, but the mechanism whereby these pictures are remembered for later use is suppressed, at least most of the time. When people have disturbing nightmares, what is usually most disturbing about them is that they are remembered as if they had been real. Of all people, those who suffer from schizophrenia have the greatest difficulty in distinguishing reality from dream activity.

11.4 MENTAL ILLNESS AND NEUROTRANSMITTERS IN THE BRAIN

We have described brain functioning during normal mental states. Several mental illnesses are associated with chemical imbalances in the brain, such as abnormalities of neurotransmitters.

11.4.1 Depression and serotonin

Depression is a disorder marked by feelings of total helplessness, despair, and frequent thoughts of suicide. Many people suffering from depression attempt suicide, and some succeed. The smallest task, such as getting out of bed in the morning, can seem overwhelming. Everyone has unhappy or pessimistic feelings from time to time, but a person suffering from depression has these feelings nearly all the time and to a severe degree. Depression is about twice as prevalent among women as among men. About 7.1% of all adults in the United States have experienced a major depressive episode (N.I.M.H., 2021), defined as severe depression lasting at least two weeks; significantly larger numbers of teens and adults suffer milder and often chronic forms of depression.

Patients suffering from depression have smaller amounts of several neurotransmitters than do other people. In particular, the brains of depressed patients who commit suicide are found at autopsy to contain low concentrations of the neurotransmitter **serotonin**.

Several drugs are effective in treating the symptoms of depression. These drugs act in either of two ways: Some drugs act by blocking the action of the enzyme MAO, which degrades many neurotransmitters, including serotonin, mostly in synapses. Other drugs, including Prozac®, Elavil®, and Tofranil®, inhibit the reuptake of serotonin and several other neurotransmitters by the presynaptic neurons that secreted them. In either case, the effect is the same: The serotonin remains active for a longer time after it is secreted and results in a greater (or more lasting) stimulus being passed on to the next neuron. Drugs that inhibit the body's production of serotonin can reverse the effects of antidepressant drugs, but drugs that inhibit the production of several other neurotransmitters do not have this effect. Use of antidepressant drugs is quite common: During 2015–2018, 13.2% of American adults were taking an antidepressant drug (Brody and Gu, 2020).

How might decreased amounts of a neurotransmitter bring about the symptoms of depression? One possible explanation is that, when something good happens, most people receive a pleasurable stimulus in the form of a stimulation to the brain's positive reward centers, and this stimulation makes them more likely to repeat whatever behavior led to the reinforcement. However, the reinforcement mechanism is not working in patients who are depressed, so they are never rewarded, nor do they learn to repeat whatever behavior led them to the sensation.

A normal bloodstream chemical, acetyl-L-carnitine, occurs in greatly reduced levels among people suffering from major depression and among rodents who show similar symptoms, according to a study by Nasca et al. (2018). The availability of a quantitative blood test for this substance may help in the diagnosis of depression, in the monitoring of treatment for depression, and in the testing of drugs used to treat the condition.

Patients suffering from depression (or, less often, from bipolar disorder) are often reported to have disturbances in their circadian rhythms. Some clinicians have proposed treating these disorders by voluntarily manipulating the sleep-wake cycle, and a study by Martiny et al. (2012) using randomized controls found that circadian cycle changes have positive benefits for treating depression. However, such studies are few, and so these proposals have not been widely adopted.

11.4.2 Schizophrenia and dopamine

Schizophrenia is a disorder characterized by frequent delusions and auditory or visual hallucinations. Schizophrenia seems to result from an excess of the neurotransmitter dopamine; drugs that stimulate or mimic dopamine (such as amphetamines or L-DOPA) make schizophrenic symptoms worse, and drugs that block dopamine (such as chlorpromazine and haloperidol) generally lessen symptoms. Further evidence comes from schizophrenic side effects observed when drugs are used to treat Parkinsonism and Parkinsonian side effects observed when drugs are used to treat schizophrenia. In addition to dopamine, serotonin may also be involved in schizophrenia.

THOUGHT QUESTIONS

1. Suppose that you were investigating why depression occurs more often in women than in men. How might you test whether differences in upbringing and other cultural influences were at work? Would it be useful to study the prevalence of depression in different cultures? What methodological, ethical, or social problems would such a study face? Would you examine many people or a few? What additional difficulties would you face if depression were defined differently in each culture? Could you investigate depression by studying an outcome such as suicide? What new ethical or social problems might arise from the results of such a study?
2. Is it ethical to give a drug that changes someone's personality? If a drug such as Prozac is given to a patient and the patient kills someone, can the doctor be held responsible?
3. If a person's social situation may be contributing to a problem such as depression, is it proper for a doctor to treat the condition with drugs without also addressing the social situation?

CONCLUDING REMARKS

Many of the functions that people have historically attributed to the mind have been shown by neurobiology to originate in activity in the brain. Neurons transmitting action potentials and stimulating other neurons across synapses can account for mental activities such as sensations, dreams, learning, and memories. Mental illness can also be seen to have a neurochemical basis, at least in part. However, whether these brain and nervous system activities actually *are* the mind, or whether they are the mechanisms by which the mind becomes material, is a question that philosophers will continue to debate. Certainly, a person is not an isolated collection of biochemicals, but exists in a social relationship to other people and is necessarily linked to the rest of the world. So, although much progress has been made in drug treatments of mental illness by thinking of the mind and the brain as synonymous, it is a concept that could be carried too far if it leads us to ignore the importance of the whole person and their relation to others and to the environment.

CHAPTER SUMMARY

- The work of the brain is carried out by nerve cells (**neurons**).
- Neurons maintain differences in ion concentration across their cell membrane and thus a difference in electrical charge known as the **resting potential**.
- If a **threshold** of depolarization is exceeded, neurons carry nerve impulses in the form of successive **action potentials** down the length of their **axons**. **Sodium–potassium pumps** then restore the ion distribution, which reestablishes the resting potential.

- Neurons communicate with one another across spaces called **synapses** by means of chemical **neurotransmitters** such as **acetylcholine, norepinephrine**, serotonin, **dopamine**, and GABA. Decreased dopamine neurotransmission may lead to Parkinsonism, while excessive dopamine neurotransmission may produce the uncontrolled movements that characterize Huntington's disease.
- Neurons control the activity of other neurons by **feedback systems.** Feedback can be either inhibitory (feedback inhibition, negative feedback) or stimulatory (positive feedback).
- Sense organs in the peripheral nervous system are responsible for receiving external stimuli such as visual images, sounds, tastes, smells, touch, pressure, heat, cold, and pain.
- Sensory messages are carried to the central nervous system and processed in the brain; such processing produces our perceptions of the stimuli.
- The brain also sends out messages such as those that stimulate muscle contraction.
- Learning requires activity in the brain, but not necessarily consciousness of the activity.
- **Procedural learning** includes **habituation**, **sensitization**, and **classical conditioning**, none of which require conscious awareness or activity in the hippocampus or the temporal lobe of the **cerebral cortex**.
- **Declarative learning** is conscious learning, requiring the hippocampus and cerebral cortex for stimulus processing, **memory** consolidation, and the formation of generalizations and abstractions.
- Many biological functions follow a **circadian rhythm** that is set internally and fine-tuned by light acting on the **pineal body** within the brain.
- Sleep follows definite stages, each with a distinctive EEG pattern. Dreams coincide with periods of rapid eye movements (**REMs**) and result from the brain's own practice at sending messages.
- Abnormal neurotransmitter concentrations are associated with some mental illnesses.

BIBLIOGRAPHIC REFERENCES

Bibliographic references to material in this chapter can be found online at biologytrending.routledge.com/chapter11

GLOSSARY: KEY TERMS TO KNOW

These key terms are defined at the end of each chapter as an aid for student review.

Acetylcholine The most common neurotransmitter, present wherever neurons stimulate muscle contraction and also at a majority of the synapses between neurons.

Actin A contractile protein found in many eukaryotic cells, especially muscle cells, forming part of the cytoskeleton and responsible (with myosin) for muscle contraction.

Action potential (spike) A large reversal of polarization in a nerve cell membrane, resulting in a nerve impulse.

Adaptation In nervous systems, a physiological change in response to a stimulus that prepares the body to better withstand or react more vigorously to similar stimuli.

Astrocytes Star-shaped neuroglial cells that nourish and protect brain cells, also capable of engulfing or destroying cellular debris or abnormal proteins in brain tissue.

Axon An extension of a nerve cell that carries an impulse away from the nerve cell body.

Blood-brain barrier A membrane system that separates the bloodstream from the cerebrospinal fluid while allowing certain small molecules to diffuse between the blood and the brain.

Central nervous system (CNS) The brain and spinal cord.

Cerebellum Part of the hindbrain that controls muscular coordination and balance.

Cerebral cortex The outer part of the cerebrum.

Cerebral hemispheres The two halves of the cerebrum.

Cerebrospinal fluid The fluid contained within the cavities of the brain and spinal cord, from which nutrients and oxygen diffuse to the neurons.

Cerebrum The part of the forebrain controlling conscious activity and thought; it is the major part of the brain in humans.

Circadian rhythm A biological change whose pattern repeats approximately every 24 hours.

Classical conditioning A form of learning in which one stimulus (the conditioned stimulus) that repeatedly precedes or accompanies another (the unconditioned stimulus) becomes capable of evoking the response originally elicited only by the unconditioned stimulus.

Cochlea The coiled part of the inner ear, responsible for sensing sounds.

Corpus callosum A series of fiber tracts that connect the right and left cerebral hemispheres across the midline.

Declarative learning Conscious remembrance of persons, places, things, and concepts, requiring the action of the hippocampus and the temporal regions of the brain.

Dendrites Nerve cell processes that receive signals and respond by conducting impulses toward the nerve cell body.

Depolarization The disappearance of a separation of unequal electric charges.

Depression A mental disorder characterized by low levels of serotonin and other neurotransmitters, by lack of motivation, and, in severe cases, by suicidal thoughts and actions.

Dopamine A neurotransmitter synthesized from the amino acid tyrosine and terminating in an amino group.

Electrical potential A form of stored energy resulting from the separation of positive and negative electric charges.

Epilepsy A brain disorder characterized by uncontrollable muscle seizures and other symptoms.

Feedback mechanism or system Any process in which a later step modifies or regulates an earlier step in the process.

Forebrain The front portion of the brain, containing the paired olfactory bulbs, olfactory lobes, cerebral hemispheres, and several unpaired portions including the hypothalamus.

Ganglion (plural, ganglia) A clump or cluster of nerve cell bodies, usually in the peripheral nervous system.

Generator potential The membrane potential generated by sensory cells in response to the sensory information they receive.

Habituation A form of learning in which an organism learns not to react to a stimulus that is repeated without consequence.

Hindbrain The rear portion of the brain, containing the cerebellum and medulla.

Hypothalamus A structure at the base of the forebrain that regulates body temperature and controls the release of various pituitary hormones.

Learning The modification of behavior or of memory on the basis of experience.

Medulla The innermost part of any organ, such as the medulla oblongata, a portion of the hindbrain that controls breathing and other involuntary activities that continue even during sleep.

Medulla oblongata See *Medulla.*

Melatonin A hormone produced by the pineal body during darkness; its changing levels of concentration entrain the body to follow circadian rhythms.

Memory The ability to recall past learning.

Memory consolidation The formation of long-term memory, involving action by the hippocampus.

Microglia A type of small neuroglial cell that creeps through brain tissue with amoeboid motion, engulfing bacteria or cellular debris.

Midbrain The middle portion of the brain, containing most of the reticular formation.

Motor neuron A neuron that conducts impulses away from the central nervous system.

Myelin sheath A lipid-rich covering that surrounds and insulates many neurons.

Myosin A contractile protein found principally in muscle cells.

Nerve A bundle of axons outside the central nervous system.

Neuroglia Cells of the nervous system other than neurons.

Neurons Specialized cells that conduct nerve impulses along their surface.

Neurotransmitter Any chemical that transmits a nerve impulse from one cell to another.

Node of Ranvier The thinning of the myelin sheath where two adjacent myelin cells meet.

Norepinephrine A neurotransmitter that transmits impulses to postganglionic neurons of the sympathetic nervous system.

Oligodendrocyte A type of neuroglial cell with few branches, generally nutritive in function.

Parkinsonism A neurological disorder involving degeneration of certain darkly pigmented brain cells (the substantia nigra), an insufficiency of dopamine, muscle tremors, and difficulty walking or initiating voluntary movement.

Peripheral nerves Nerves outside the brain and spinal cord.

Pineal body A structure on the roof of the diencephalon of vertebrate brains that maintains circadian rhythms.

Polarized Having opposite ends or surfaces differing from one another in electrical charge.

Procedural learning Learning how to do things, a process that does not require the hippocampus and is not necessarily conscious.

REM A sleep stage characterized by Rapid Eye Movements beneath the closed eyelids and associated with dreaming episodes.

Resting potential The difference in electric charge maintained by a nerve cell membrane in the absence of a nerve impulse.

Reticular activating system A system of neurons, radiating upward from the midbrain, maintaining the body's alertness and readiness to respond to stimuli.

Reticular formation A center in the midbrain and part of the hindbrain from which the reticular activating system radiates upward.

Retina The light-sensitive membrane coating the rear of the eye.

Reuptake Absorption of a neurotransmitter by the cell that secreted it.

Schizophrenia A disorder that results in an inability to distinguish real from imaginary situations or stimuli and characterized by frequent auditory or other hallucinations.

Sensitization A form of learning in which an intense and often aversive stimulus increases subsequent responses to other stimuli.

Sensory neuron A neuron that conducts impulses toward the central nervous system.

Sensory perception The interpretation of sensory messages as images, sounds, etc.

Sensory reception The translation of sensory information into nerve impulses.

Serotonin A neurotransmitter, 5-hydroxytryptamine, synthesized from the amino acid tryptophan, terminating in an amino group.

Sodium–potassium pump A group of membrane proteins that can actively transport sodium ions from the inside to the outside of a cell, such as a nerve cell, while actively transporting potassium ions in the opposite direction.

Superior colliculi Midbrain structures that coordinate sensory information to help locate objects spatially.

Synapse A meeting of cells in which a nerve cell stimulates another cell by secreting a neurotransmitter; the postsynaptic cell must have receptors to which the neurotransmitter binds.

Tela choroidea A thin-walled roof of one of the brain cavities, across which nutrients and wastes are exchanged between the blood and the cerebrospinal fluid.

Threshold The minimum level of a stimulus that is capable of producing an action potential, or the minimum level of a drug below which no physiological response can be detected.

CONNECTIONS TO OTHER CHAPTERS

Chapter 1: Equating the mind and the brain is an example of a research paradigm.
Chapter 1: Ethical issues are raised by the use of fetal tissue for research and for therapy.
Chapter 3: Susceptibility to some brain disorders has a genetic basis.
Chapter 5: Similarities in the brains of different species have resulted from evolution.

Chapter 8: Neurotransmitters are made from dietary amino acids.
Chapter 10: Neurotransmitters bind to and stimulate specific receptors on the surface of target cells.
Chapter 12: Alzheimer's disease involves the death of neurons in the brain.
Chapter 13: Psychoactive drugs alter the functions of the brain, in some cases by affecting neurotransmission.
Chapter 14: Brain activity can influence our general health. Sleep deprivation, for example, can impair the immune system.
Chapter 18: Many drugs that influence brain activity are obtained from tropical rainforest plants.
Chapter 19: Brain disorders may arise from environmental pollution.

PRACTICE QUESTIONS

1. In a neuron that is not conducting an impulse, there is an excess of _____ ions inside the cell and an excess of _____ ions just outside the cell membrane.
2. An action potential is caused by a large number of _____ ions moving across the cell membrane toward the _____.
3. Which neurotransmitter(s):
   a. Is a gas?
   b. Is broken down by cholinesterase?
   c. Is overproduced in Huntington's disease?
   d. Was the first to be experimentally investigated?
   e. Do depressed patients often have in insufficient amounts?
   f. Are amines?
4. _____ is a neurotransmitter but not an amine; the enzyme _____ breaks it down in the synapse.
5. Name three parts of the forebrain.
6. For each of the following, name the sense organ that contains it:
   a. Anvil bone
   b. Iris diaphragm
   c. Cones
   d. Naked nerve endings
   e. Tympanic membrane
   f. Cochlea
   g. An area sensitive to acidic compounds
7. Each of the following is an example of what form of learning? Please be as specific as you can.
   a. Learning to associate horses with the word "horse."
   b. Being more alert to other stimuli when the fire alarm rings.
   c. Learning not to pay attention to traffic noises in one's neighborhood.
   d. Learning to recognize a TV show by its theme music.
8. How does the removal of GABA produce more dopamine?
9. What is the effect of monoamine oxidase inhibitors on neurotransmission and why?
10. Name the types of sensory input that induce generator potentials in neurons of each of the following sense organs: skin, eye, ear, tongue, and nose.
11. What are the functions of the blood–brain barrier?
12. What are the major ways in which brain anatomy has changed during evolution?
13. What are the major functions of each of the three divisions of the brain: forebrain, midbrain, and hindbrain?

CHAPTER 12

Aging and Alzheimer's Disease

ISSUES

- What is aging at the level of cells and molecules?
- What determines the average lifespan of a species?
- Why does aging occur, and can it be slowed or reversed?
- Why do plants usually live longer than animals?
- Is there an upper limit to the human lifespan?
- What is Alzheimer's disease, and how is it related to aging?

BIOLOGICAL CONCEPTS

- Determinants of lifespan
- Aging and natural selection
- Stem cells as preservers of the genome
- DNA damage and repair
- Telomeres and cellular senescence
- Protein aggregation and disease

CHAPTER OUTLINE

DOI: 10.1201/9781003391159-12

Throughout history, across most human cultures, there have been stories about people searching for the fountain of youth or the elixir of life. Similarly, the promise of some version of eternal life is common to many of the world's religions. These universal human themes reflect the inescapable fact that life is always accompanied by aging and, eventually, death. We can all recognize the signs of aging in ourselves and our family members; consequently, many commercial products target our natural desire to slow or reverse the aging process—but why do we age? What changes occur at the cellular and molecular levels as organisms age, and why do different organisms age at different rates, leading to vastly different lifespans? In this chapter, we consider the biology of the aging process, its underlying causes at the cellular and molecular levels, and current research into how the aging process might be slowed. We briefly address the many diseases that are associated with aging and then focus on one of the most common, Alzheimer's disease.

12.1 AGING IS A NATURAL PROCESS

The development of a single fertilized egg (a zygote) into a viable organism and the maturation of a juvenile into an adult are tightly controlled by genetic programs (see Chapter 10). Once they reach adulthood, all organisms age and ultimately die, but should aging be considered part of development? Traditionally, many developmental biologists have argued that aging is a default process that occurs because the laws of the universe dictate that everything proceeds toward greater disorder (**entropy**). However, more recent research has demonstrated that aging is associated with distinct processes at the molecular level that can be influenced by genetics as well as diet and cellular signaling pathways, changing our view of what aging is and how it might be altered by pharmaceutical or lifestyle interventions.

The **maximum lifespan** of a species is the oldest age that has been reached by an individual of that species. The oldest human for which credible birth and death records exist is thought to have been a French woman who lived 122 years and died in 1997. A theoretical model has determined that the upper limit on the human lifespan may fall around 125 years. The maximum lifespan varies greatly for different types of organisms: Mayflies live their entire adult lives in a single day, whereas Greenland sharks are thought to be able to live 400 years or more. Although impressive, this shark's lifespan pales in comparison to those of many trees, some of which can live over 5000 years!

While lifespan records may make the *Guinness Book of World Records*, we are often more interested in **life expectancy**, which is a statistical measure of the likely lifespan for a given population. This can vary widely even within the same species, depending on environmental conditions. For example, according to the United Nations (UN) Population Division, the average life expectancy of a person born in the United States in 2015 is 79.24 years, whereas in Sierra Leone, a war-ravaged country on the west coast of Africa, it is only 51.42 years. Globally, human life expectancy has significantly increased in the past century (Figure 12.1), due largely to the advent of modern medicine.

The large maximum lifespan range for different species (e.g., mayflies vs Greenland sharks) and varied average life expectancies for different populations of the same species (e.g., humans living in different areas of the world) show that both genetic and environmental factors influence lifespan. It follows that these factors influence the rate of aging. Geneticists working with model organisms such as nematode worms (*Caenorhabditis elegans*), fruit flies (*Drosophila melanogaster*), zebrafish (*Danio rerio*), and mice (*Mus musculus*) have identified genetic variations that increase or decrease lifespan. A gene that was found to be altered in one long-lived strain of fruit flies was named *Methuselah (Mth)*, after the Old Testament patriarch who was said to have lived 969 years. This gene encodes a

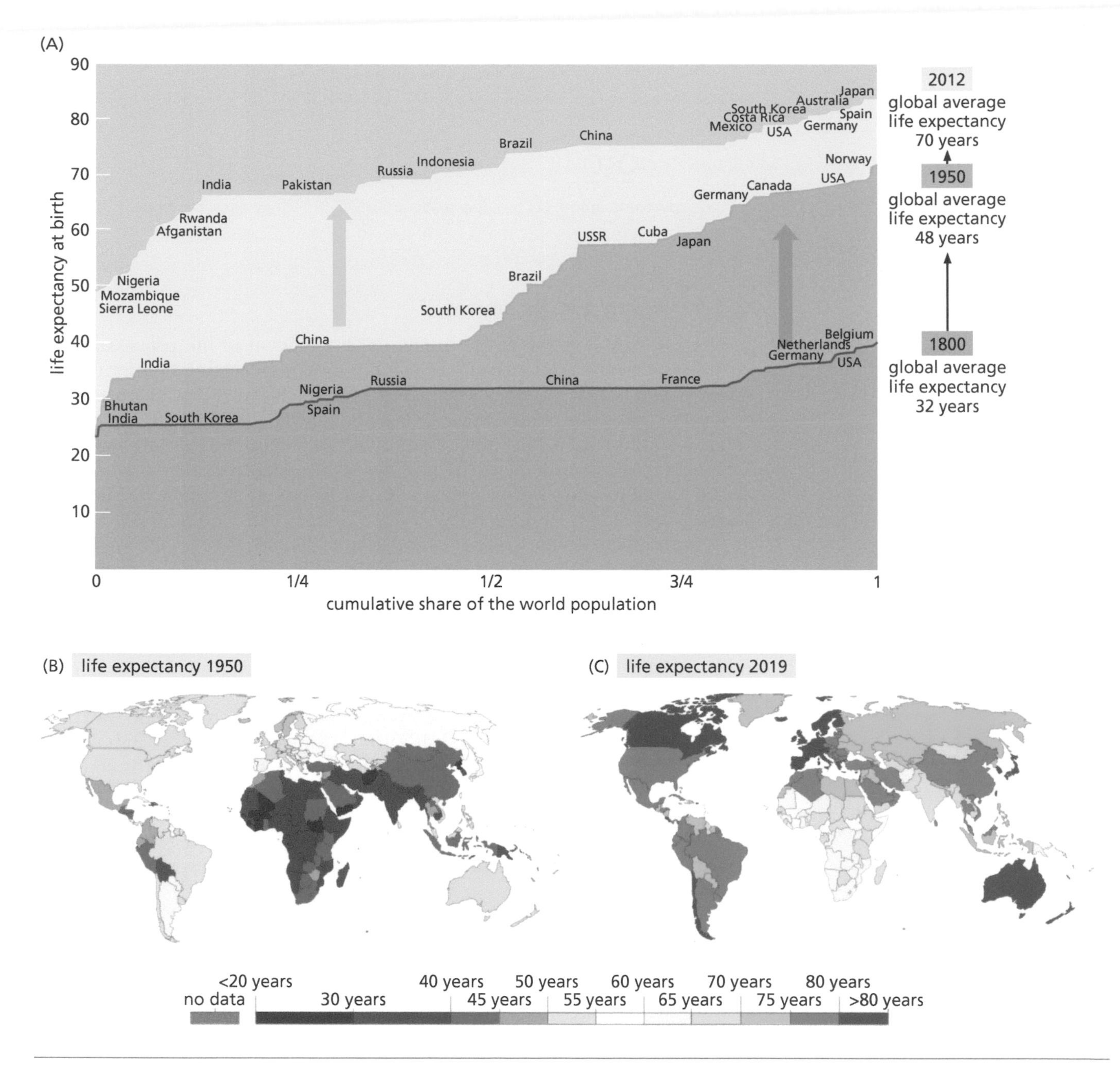

Figure 12.1 Life expectancy has improved over the last hundred years in nearly every country of the world, with occasional setbacks during wars and other crises. **(A)** Data for each country plotted for 1800, 1950, and 2012, showing that life expectancy in 1950 varied from around 24 to 73 years in different countries but increased by 2012 to a variation from 46 to 84 years. **(B)** Data by country for 1950. **(C)** Data by country for 2019.

cell-surface receptor protein that responds to various stresses, such as starvation, heat, and free radicals, and also to insulin signaling, which impacts metabolism. The mutant version of the *Mth* gene that extends the fruit fly lifespan reduces the function of the encoded protein, but complete elimination of the gene is lethal during embryonic development. Thus, it seems that there is a happy medium level of stress response that promotes longevity, but too little response is fatal and a higher level of response leads to aging. Interestingly, the genome of the long-lived white shark was recently sequenced, and it shows evidence of genetic adaptations associated with genome stability (e.g., genes involved in DNA repair or DNA damage responses) and also wound healing. Similarly, in plants, the stem cells, which give rise to all other types of cells, have been shown to be more resistant to DNA damage than those of animals, providing a possible clue as to why many plants have such exceptionally long lifespans.

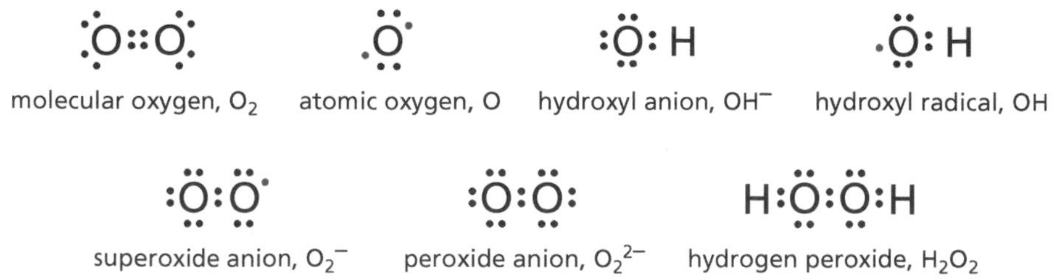

Figure 12.2 Some reactive oxygen species (ROS). Those that have unpaired electrons (shown in red) also qualify as "free radicals."

molecular oxygen, O_2 atomic oxygen, O hydroxyl anion, OH^- hydroxyl radical, OH

superoxide anion, O_2^- peroxide anion, O_2^{2-} hydrogen peroxide, H_2O_2

12.1.1 Aging and its causes

Aging is broadly defined as the functional breakdown of life processes in an organism over time. Biologists often use the term **senescence** to refer to aging (see Section 10.3.3). Changes associated with senescence can be observed at the level of the whole organism as well as at the level of individual cells and the molecules within them. Recently, researchers have identified a set of "hallmarks" of the aging process that relate the root causes of aging to a range of changes in cellular function and also suggest means by which aging might be slowed or reversed.

Aging is primarily due to the accumulation of damage to cellular components, including DNA, proteins, and organelles, especially mitochondria, which are crucial for providing energy to cells. This damage is in part caused by **reactive oxygen species** (ROS), chemically reactive molecules that are derived from reactions involving molecular oxygen. Examples include hydroxyl radicals, superoxides, and peroxides (Figure 12.2), many of which are also called **free radicals** because they have unpaired electrons. These can be generated exogenously (from the outside) by exposure to various chemicals in the environment and in food. But many ROS are also produced endogenously (from within) by natural metabolic processes, including the electron transport chain in mitochondria that is central to cellular respiration (see Section 8.3.3). Cells have enzymes that can detoxify ROS, and ROS can activate cellular signaling pathways that protect against ROS-mediated damage by promoting expression of these enzymes. However, as we age, these signaling pathways become less efficient, causing the balance to tip in favor of cellular damage.

12.1.2 Aging at the cellular and molecular levels

When individual cells become senescent, they stop dividing, and aging at the organismal level is associated with accumulation of senescent cells. Since the mid-1960s, it has been known that most mammalian cells cultured in vitro will undergo a finite number of divisions—about 60—then become senescent (see Section 10.3.3). Cancer cells, in contrast, have undergone changes that make them immortal, capable of continually dividing without reaching senescence (see Section 10.5). Senescence in eukaryotic cells is now thought to be controlled by **telomeres**, structures at the end of chromosomes that shorten with each round of cell division (see Section 10.3.3). Recent large-scale studies have shown that the average length of human telomeres decreases with age. Telomere shortening is also associated with certain rare diseases that cause premature aging (**progeria**), including loss of hair, loss of skin elasticity, loss of subcutaneous fat, and hardening of the arteries (arteriosclerosis), usually leading to cardiovascular disease and premature death. One form of progeria, Hutchinson-Gilford syndrome (Figure 12.3), is associated with a mutation in the gene for the protein lamin-A, one of several proteins that forms filaments that support the membrane that encloses the cell nucleus. Werner syndrome is a different type of progeria whose symptoms do not usually appear until age 15–25. An unrelated autoimmune disease called ataxia-telangiectasia features premature senescence of the immune system and also shortened telomeres. Since these diseases are all accompanied by shortened telomeres, it might seem that telomerase, the enzyme that extends

Figure 12.3 A child with Hutchinson–Gilford progeria syndrome, causing premature aging. Children with progeria usually die in their teens from cardiovascular disease more typical of the elderly. Sam Berns, shown here, lived to the age of 17. His mother, Leslie Gordon, also shown, is a medical researcher at Brown University and medical director of the Progeria Research Foundation.

telomeres, might lengthen life. However, telomere lengthening is a double-edged sword, since it is also a feature of cancer (see Section 10.5.1).

Apart from telomere shortening, several other cellular and molecular changes are associated with aging (Table 12.1). One of these is the accumulation of mutations, both in the nuclear and mitochondrial genomes; these in turn lead to compromised function of the proteins encoded by the mutated genes. There are also so-called **epigenetic changes** that do not affect the DNA sequence but do cause changes in gene expression. Examples of epigenetic changes include modifications to DNA packaging proteins and the overall structure of chromatin, as well as direct chemical modification of DNA bases.

Cells have quality control mechanisms that eliminate proteins that do not fold properly, and these protein quality control responses also diminish with aging. The accumulation of misfolded proteins contributes to a number of human neurodegenerative diseases, such as Alzheimer's disease, which is discussed later in this chapter.

The aging process is also associated with decreased efficiency of cellular energy metabolism and of the signaling pathways that sense nutrients. These changes contribute to metabolic diseases like diabetes that are more common in older people. Interestingly, caloric restriction (eating less) has been shown to extend the lifespan in many organisms, and some recent studies suggest that a carefully controlled, occasional fasting regimen may promote health as humans age. Caloric restriction diminishes flux through nutrient-sensitive cellular signaling pathways. However, no fasting regimen should be attempted without consulting a physician, since it may not be appropriate for some people; moreover, severe calorie restriction, such as occurs in people who suffer from anorexia nervosa, can be life-threatening (see Section 8.5.2).

Finally, as organisms age, the stem cells that replenish tissues (see Section 10.4) become depleted, causing an overall decline in function. Special adaptations that enable plants to protect the genetic integrity of their stem cells likely contributes to the longevity of many plants.

Table 12.1 Some phenomena related to aging

Cellular and molecular changes	Changes in body function
Cells stop dividing (enter senescence)	Bone mass and density decrease
Telomeres shorten	Muscle loss (sarcopenia)
Mutations accumulate in DNA	Connective tissues become less elastic
Epigenetic changes occur in chromatin	Skin wrinkles and becomes less elastic
Protein quality control decreases, misfolded proteins accumulate	Basal metabolic rate decreases
Nutrient-sensing signaling pathways become deregulated	Organ reserve decreases
Stem cell populations become depleted	Memory loss
	Decreased immune function
	Chronic inflammation

12.1.3 Changes in body function with aging

Some familiar signs of human aging include graying and thinning hair, wrinkled skin, brittle bones, vision changes, weight gain, decreased cardiovascular function, and worsening memory (Table 12.1), and many of these changes also occur in other mammalian species. These diverse "symptoms" of aging can be linked to a few fundamental changes that occur at the cellular and molecular level as we age. Many tissues of the body lose cells with age due to depletion of stem cell populations that are needed to replenish them. This causes an overall decrease in organ function, especially in their **reserve capacity**. The reserve capacity is the ability of an organ to function at a higher level than is required to sustain life, and therefore to respond to stress conditions that require elevated function. As an example, both the heart of a 20-year-old and that of a 60-year-old may function well during normal activity, but the 20-year-old would be able to exercise safely at a higher heart rate than the 60-year-old. The stem cells that give rise to hair follicle cells and pigment cells also become depleted, leading to hair thinning and graying.

Bones consist of a collagen matrix with embedded mineral crystals along with cells, lipids, and water. Throughout a person's life, cells called osteoblasts produce new bone, while other cells, called osteoclasts, resorb old bone. This cycle of bone renewal becomes less efficient with age, leading to an overall decrease in bone size and density. This is why older people are more prone to bone breakage, especially with falls, and also why people tend to shrink somewhat in height as they age.

Some symptoms of aging relate to the loss of elasticity in various tissues. Connective tissues lose the protein elastin and show increased crosslinking of the protein collagen; these changes are associated with skin wrinkling and decreased elasticity in blood vessels, which compromises cardiovascular function. Similar changes in the walls of the digestive tract decrease the force of contraction and can cause digestive problems. The lenses of the eyes also become more rigid, which causes presbyopia, or old-age farsightedness (the inability to focus clearly on close objects), requiring many people with previously good vision to begin wearing eyeglasses as they age.

Another typical feature of advancing age is a decrease in a person's **basal metabolic rate** (or BMR, the rate that your body consumes energy while at rest), which commonly results in increased body weight. Many factors may contribute to this change in BMR with age, including hormonal changes, altered sensitivity to nutritional signals, damage to mitochondria, and decreased physical activity. Muscle tissue has a higher energy requirement than fat, and older adults tend to lose muscle as result of decreased physical activity and hormonal changes. This muscle loss is called **sarcopenia**. For men in particular, decreasing testosterone levels with age contribute to sarcopenia. The good news, however, is that exercise and a healthy diet can minimize both muscle and bone loss with age and also maintain BMR.

Finally, aging is associated with memory loss and diminished immune function, and at least one version of a gene associated with increased longevity also correlates with a much lower risk for cognitive decline in old age (Barzilai et al. 2006). Even in healthy individuals, brain size decreases with age, but it seems to be due more to cell shrinkage and decreased neural connections rather than cell death. However, age is also the primary risk factor for many neurodegenerative diseases, such as Alzheimer's, Huntington's, and Parkinson's diseases. One of these, Alzheimer's disease, is discussed in detail later in this chapter. Changes in the immune system with aging are discussed next.

12.1.4 Aging and the immune system

Elderly people are at elevated risk for both contracting and dying from infectious diseases. This is because the function of the immune system decreases with aging in a process called **immunosenescence**. The immune system is highly complex, consisting of many different types of cells, numerous signaling pathways mediated by soluble molecules secreted into the bloodstream, and various lymphoid organs, including the thymus, spleen, bone marrow, and lymph nodes (see Chapter 14). All of these exhibit changes as we age, resulting in an overall decrease in the capacity of the immune system to combat pathogens. Ongoing research is investigating how immune signaling molecules, or **cytokines**, may be used to "jump-start" immune function in the elderly.

Somewhat paradoxically, aging is also associated with chronic inflammation, which represents low-level activation of the immune system in various parts of the body when no infection is present. This phenomenon has been referred to as "inflammaging," and it causes pain and organ damage. Chronic inflammation is also a risk factor for many disease conditions, such as heart disease, diabetes, and cancer, and it can contribute to these conditions even in younger people.

Antibodies are proteins produced by the B cells of the immune system; their job is to recognize foreign molecules in the body (such as those associated with bacteria, viruses, and parasites) and to trigger immune responses to eliminate them. The immune system has complex mechanisms to eliminate B cells that produce antibodies that recognize "self" molecules (or **autoantibodies**), since these can cause **autoimmune diseases**. Interestingly, older people often express significantly higher levels of certain autoantibodies than their younger counterparts, but autoimmune diseases in the elderly are rare. Debris from dead cells that is not properly cleared from circulation may trigger production of autoantibodies. Such debris is normally cleared by immune cells called **macrophages**, and these cells have been shown to decline in function with age. In addition, senescent cells secrete inflammatory cytokines that also contribute to chronic inflammation. Recently, the so-called "anti-inflammatory" diet has gained popularity as a possible means of preventing or treating a range of negative health conditions. This diet emphasizes foods that are rich in antioxidants and omega-3 fatty acids, such as fruits, vegetables, nuts, seeds, and lean proteins like fish.

12.1.5 Aging and cancers

Although certain cancers such as leukemia and retinoblastoma occur in children, the vast majority of cancers occur later in life. In fact, age is the most significant risk factor for developing cancer. Figure 10.19 shows the age-based risk of all types of cancer. Overall, cancer is infrequent in the first three decades of life and begins to rise sharply around age 40. The reason why the risk of cancer increases with age—and does so in an exponential rather than a linear manner—is that cancer starts with a single cell that has accumulated multiple genetic changes (mutations) that would be insufficient to cause cancer on their own but that, together, cause the cell's regulatory mechanisms to fail, leading to out-of-control cell growth and division (see Chapter 10). Based on mathematical modeling of age-based cancer risk data, cancer researchers have determined that human cells require at least five to six separate mutations to become cancerous. Because the cells in an older person's body have had more time to accumulate mutations than those in a younger person—whether due to random errors in DNA replication, spontaneous chemical reactions, exposure to chemicals in the environment or diet, etc.—a person's risk of cancer increases with age. It's also important to realize that cancer is often considered a "modern disease" simply because, for generations, human life expectancies were much lower due to the prevalence of infectious diseases, malnutrition, accidents, etc., so most people did not reach an age at which they were likely to develop cancer.

THOUGHT QUESTIONS

1. Which is more important: the maximum lifespan, the maximum age of good physical health, or the maximum age of good mental functioning?
2. Which is more important in maintaining good health longer: good genes or a good environment? Does either one of these impose a maximum limit? Does either one simply impose risks to be avoided or hurdles to be overcome?
3. What role does the immune system play in maintaining good health over time? What role does it play in lengthening lifespan independent of health?

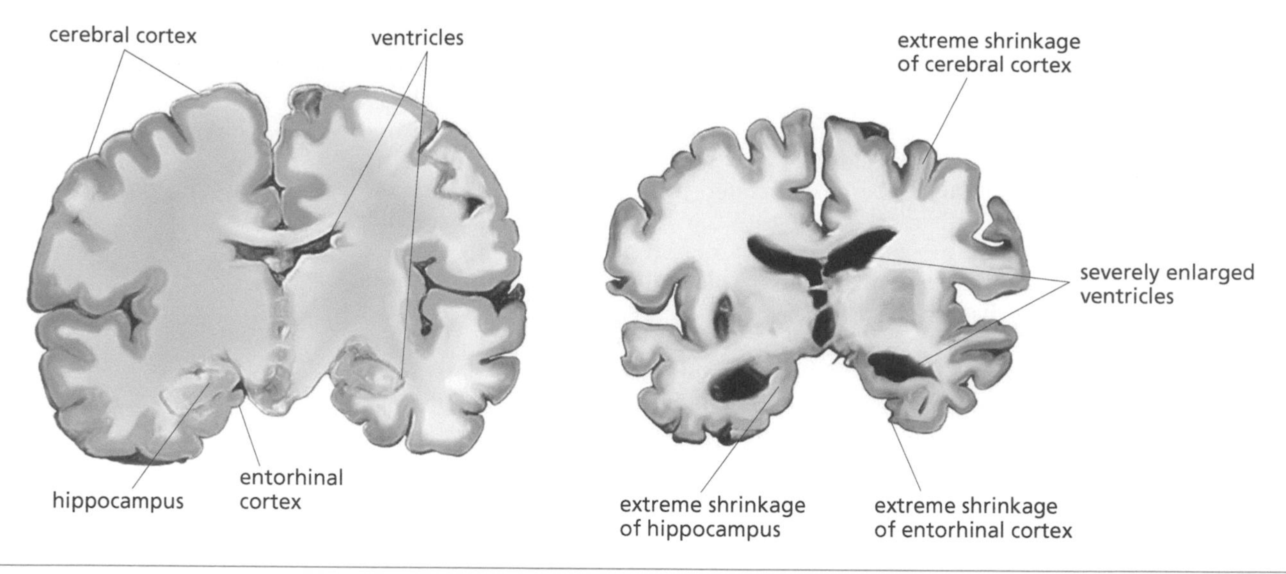

Figure 12.4 **Changes in the brains of Alzheimer's patients.** Transverse brain sections, showing a largely healthy brain (preclinical, left) and a brain with late-stage Alzheimer's disease (right).

12.2 ALZHEIMER'S DISEASE

In 1901, Dr. Alois Alzheimer encountered a patient with a previously unknown mental condition. Her responses to Dr. Alzheimer's questions often made no sense, and she frequently stopped in mid-sentence, unable to continue. When she died in 1906, Dr. Alzheimer did an autopsy and studied her brain, which showed a drastic shrinkage of brain tissue (Figure 12.4). He also discovered the plaques of amyloid protein and the neurofibrillary tangles of tau protein in the brain that are now considered the pathological hallmarks of **Alzheimer's disease** (Figures 12.5–12.7).

The devastating brain disorder that Dr. Alzheimer described is a physical illness that causes radical changes in the brain. Even before healthy neurons degenerate and die, brain function diminishes, and persons with the disease experience a steady decline in memory and in the ability to use their brain to perform previously familiar tasks such as dressing themselves, feeding themselves, and swallowing, leaving most patients to eventually require institutional care. Currently, there is no cure for Alzheimer's disease, and the severe deterioration of brain function ultimately leads to death.

12.2.1 General description and course of the disease

Alzheimer's disease is a form of **dementia**, a loss of brain function that impairs thought and especially memory. Although speech and movement may be

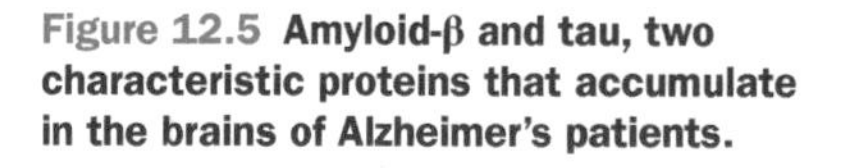

Figure 12.5 **Amyloid-β and tau, two characteristic proteins that accumulate in the brains of Alzheimer's patients.**

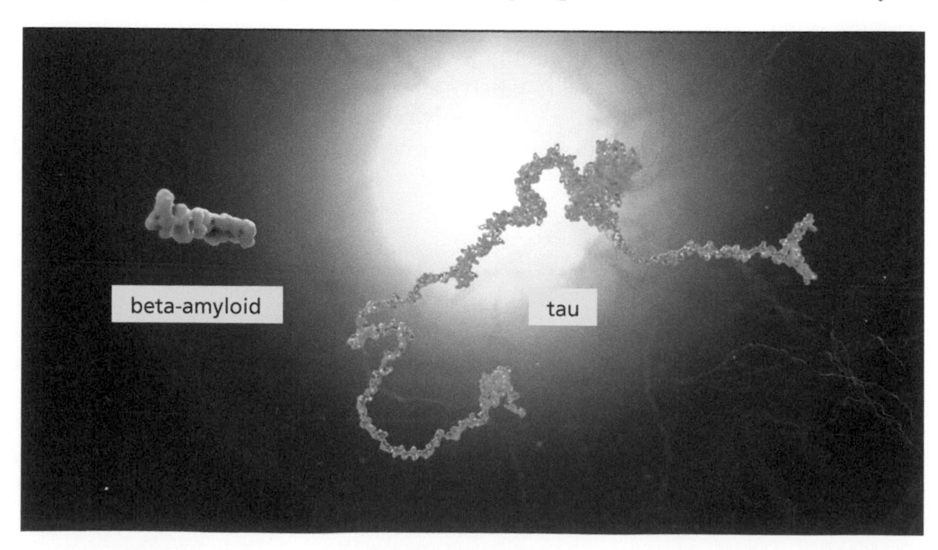

Figure 12.6 If amyloid-β proteins accumulate in the brain outside neurons, they clump together and form dense plaques, as shown here on the right. Notice that the neuron on the left also has an accumulation of tau protein (blue) that has formed tangled neurofibrils.

minimally affected at first, the formation of new memories is severely impaired, as are associative learning and logical reasoning. It is a **progressive** disease, meaning that it can only worsen, or stop worsening for a time, but it can never get better. The Alzheimer's Association estimates that 5.4 million people in the United States are currently living with Alzheimer's disease, making it the most prevalent type of brain function loss, accounting for 60%–80% of all dementia cases. Globally, over 46 million people were suffering from this disease in 2015 (according to Alzheimer's Disease International, 2015), and this number is doubling about every 20 years as the world's population ages. In the United States, most other causes of mortality experienced a decline during the period 2000–2008, but Alzheimer's disease experienced a 66% increase. Even these estimates may be low, because Alzheimer's disease and other dementias are underrecognized and therefore underreported across all cultures.

All dementias, including Alzheimer's disease, affect a person's memory, mood, and behavior. Alzheimer's disease has many symptoms, including mental confusion, behavioral and personality changes, difficulty speaking or performing familiar tasks, failure to recognize familiar places or people, impaired judgment, impaired recognition of what is real or imagined, and problems with abstract thinking. Alzheimer's symptoms are often subtle at first. They start with slight memory loss, subtle changes in behavior, and confusion; as the disease progresses, memory problems persist and worsen. People with Alzheimer's disease often find themselves lost in familiar surroundings, forget conversations, repeat themselves, routinely misplace things (often putting them in illogical locations),

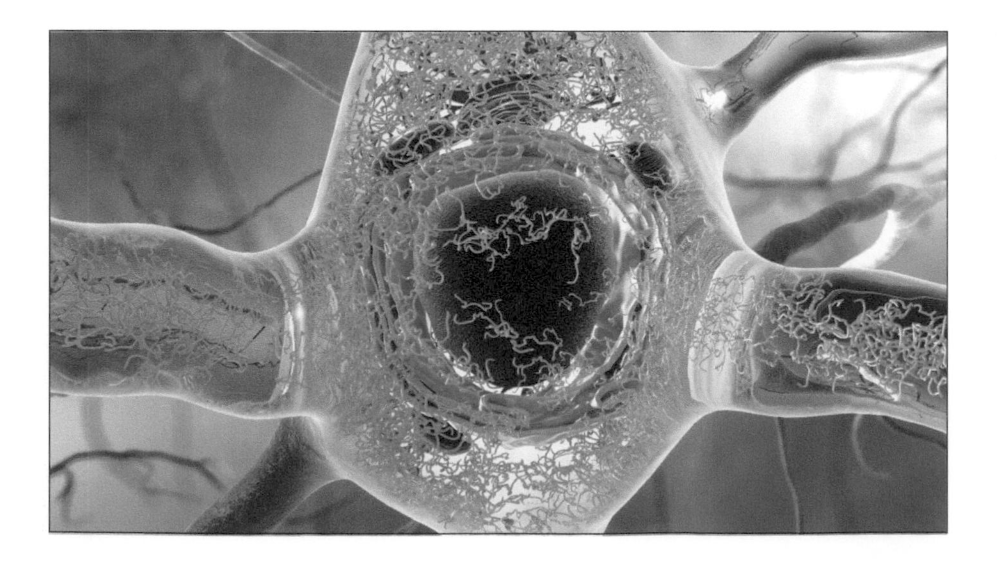

Figure 12.7 In the brains of Alzheimer's patients, tau protein accumulates inside neurons and causes neurofibrils to form the tangles shown here in blue.

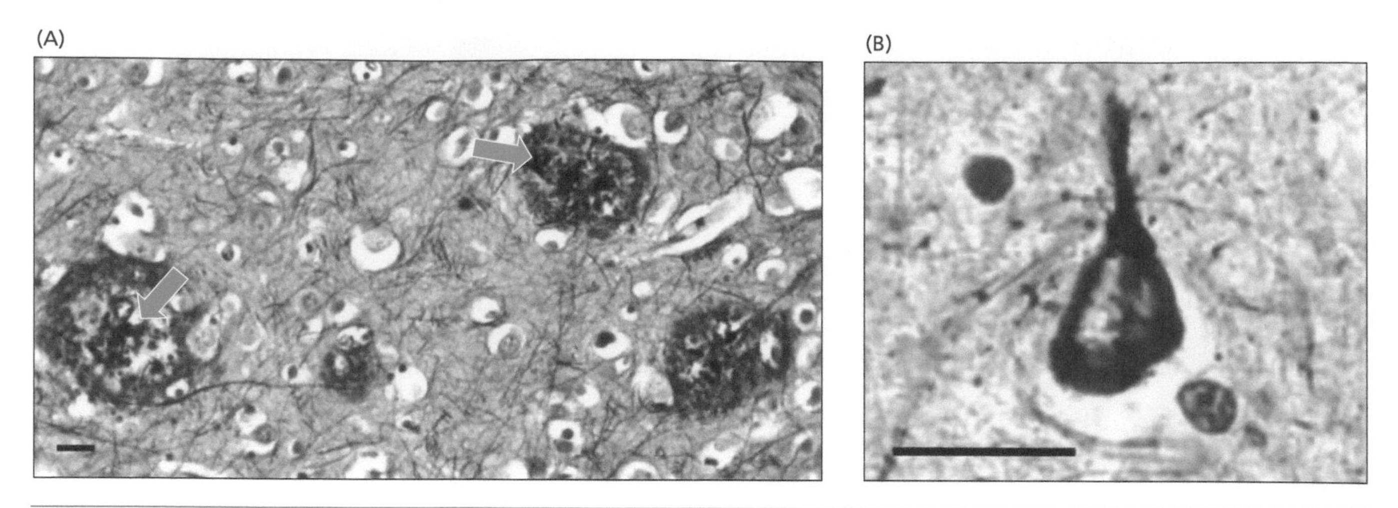

Figure 12.8 **Sections of brain tissue in an Alzheimer's patient, showing dark-staining amyloid plaques (arrows, A) and dead nerve cell bodies (B).**

become unable to maintain a schedule or keep appointments, and eventually forget the names of family members and everyday objects. One early sign of the condition is the inability to balance a checkbook or properly manage finances. Disorientation is another early sign, including the inability to locate familiar places around town or in the home or the inability to find the bathroom in a place one has been to many times before, such as a friend's home. Personality changes can also be an early sign of Alzheimer's disease. Persons suffering from the condition are often moody, restless, and sometimes mean. They may exhibit distrust in others, increased stubbornness, social withdrawal, depression, anxiety, aggressiveness, and compulsive behaviors. Even in the early stages of the disease, these behavioral changes can have devastating effects. Many Alzheimer's patients eventually forget how to chew and swallow, which often causes death by choking, by aspiration-induced pneumonia, by starvation, or by other indirect causes.

12.2.2 Changes in the brain

On a molecular level, the hallmark of Alzheimer's disease is the buildup in the brain of an extracellular protein called **amyloid-β** (Aβ) or β-amyloid. The Aβ clumps together to form deposits known as amyloid plaques, the principal microscopic marker of Alzheimer's disease (Figure 12.8). Clinical studies show that higher plasma levels of Aβ are associated with more rapid cognitive decline. Compared to healthy neurons (Figure 12.9), the brain tissues of Alzheimer's

Figure 12.9 **Appearance of a healthy neuron.**

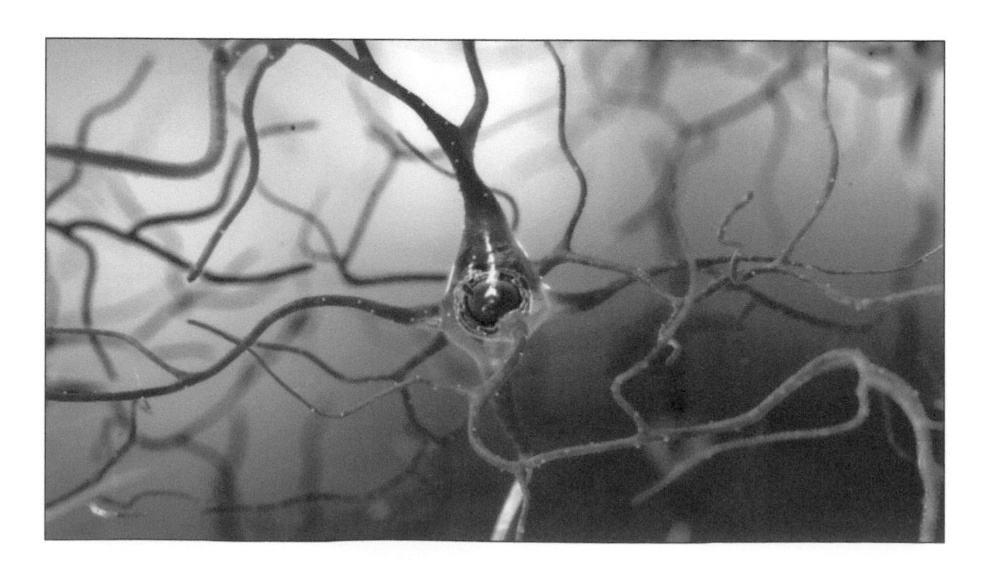

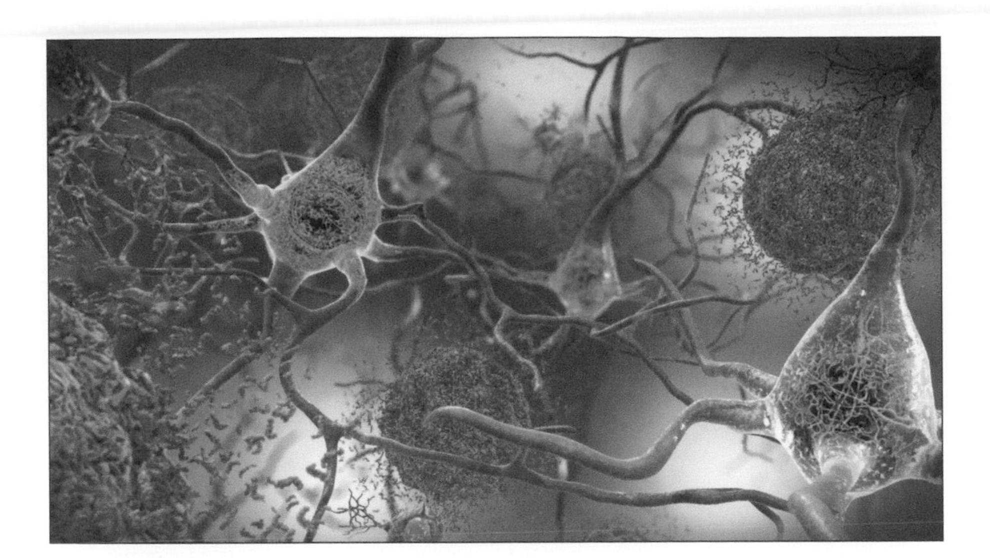

Figure 12.10 Amyloid-β plaques (light brown) and neurofibrillary tangles of tau protein (blue, inside the neurons) in the brain of an Alzheimer's patient.

patients show increased levels of both Aβ and tau proteins (**Figure 12.10**), and the levels of these proteins can differentiate well between demented and nondemented patient populations.

Figure 12.11 shows the hypothesized cascade of events leading up to the development of Alzheimer's disease. A protein called amyloid precursor protein (APP) is normally cleaved by the enzyme α-secretase into healthy products. If, however, APP is instead cleaved by the enzyme β-secretase (followed by γ-secretase), it produces the Aβ protein, which can build up and lead to the formation of amyloid plaques (**Figure 12.12**) and diseased neurons (**Figure 12.13**). It is still unclear what triggers the aberrant processing of APP in individuals who develop Alzheimer's disease; however, the postmortem examination of these patients invariably shows a buildup of amyloid plaques in the brain. These amyloid deposits are most prevalent in the hippocampus and the parts of the cortex that are involved in memory. Efforts to cure or prevent Alzheimer's disease have often focused their attention on either breaking down amyloid-β or preventing its buildup in the first place. However, amyloid plaques can build up in the brain

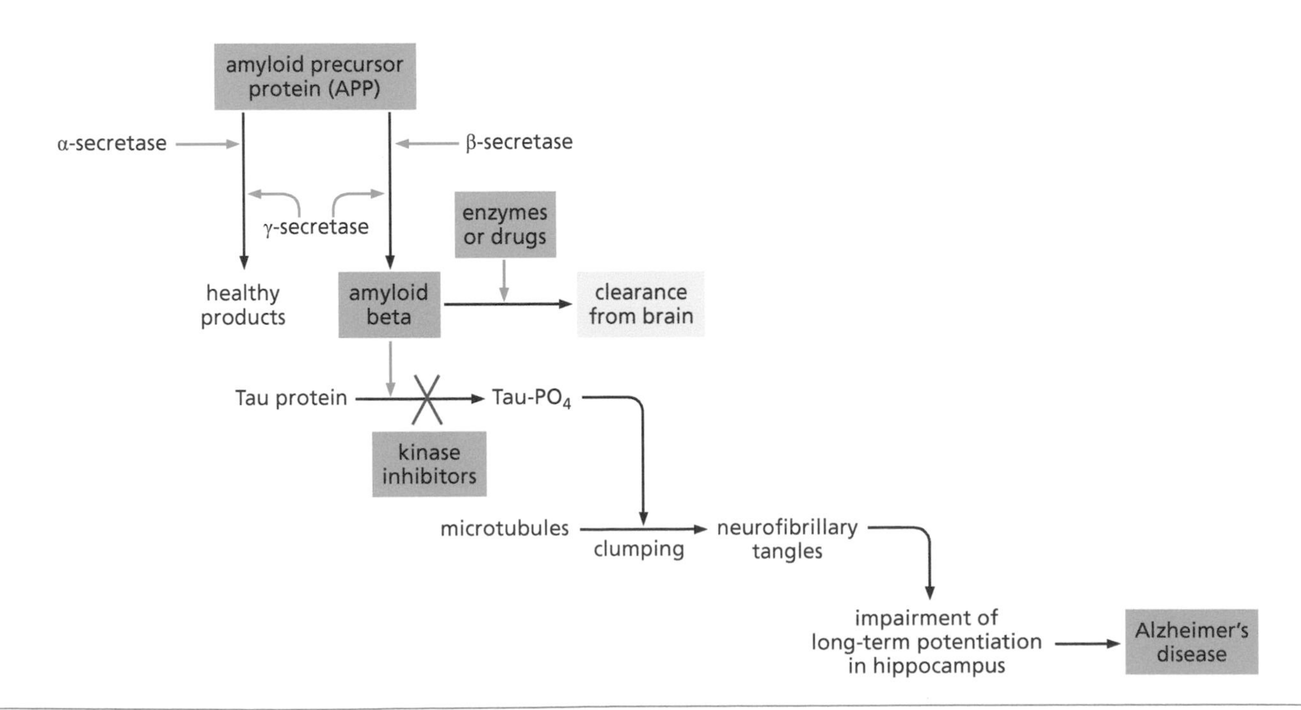

Figure 12.11 Hypothesized pathway for the control of Alzheimer's disease.

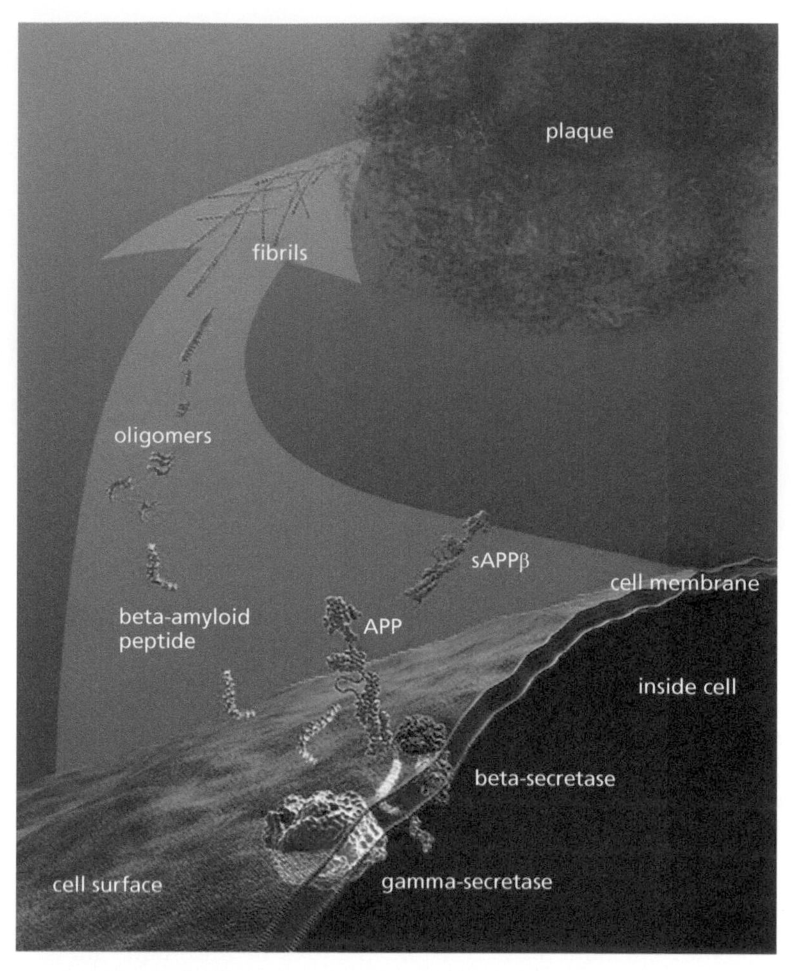

Figure 12.12 The process that leads to the formation of amyloid plaques.

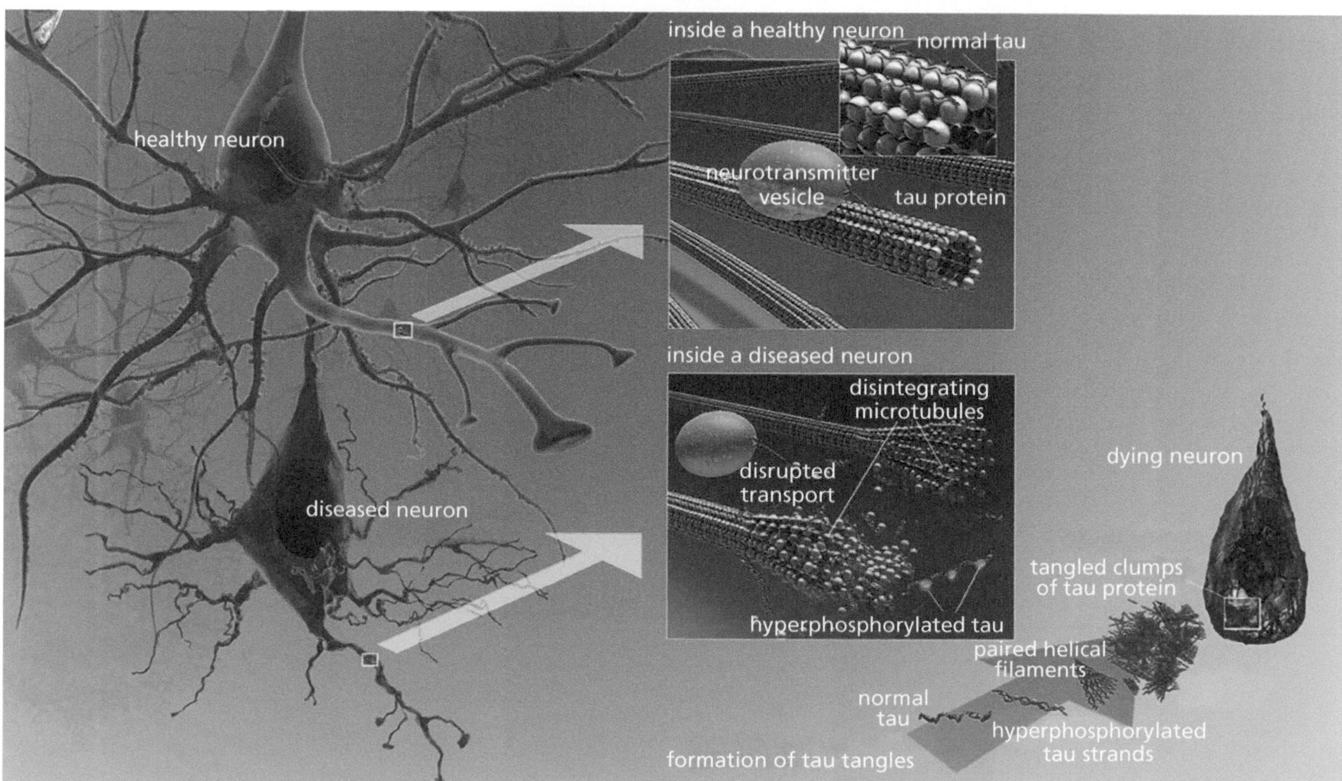

Figure 12.13 Comparison of a healthy neuron and a neuron affected by Alzheimer's disease, showing the disruption of microtubules and formation of neurofibrillary tangles (tau tangles).

without much cognitive impairment, and symptoms are few unless and until tangles of tau protein are also present (Hanger, Anderton, and Noble, 2009) (Figures 12.7 and 12.10).

Some researchers (Hyman, 2011) distinguish the early stages of Alzheimer's disease as amyloid-dependent (characterized by amyloid buildup) and the later stages as amyloid-independent (characterized by tau tangles). Many early-stage patients have a "cognitive reserve" that allows them to remain mentally alert even in the presence of increased Aβ and amyloid plaques, but no such protective effect has been reported for the later stages once tau tangles accumulate or neurons begin to die. Once neuronal death sets in, the course of the disease becomes irreversible. Therefore, clinical trials for anti-amyloid therapies are now focusing on the earlier stages of the disease, to try to intervene before the toxic accumulation of tau. Because abnormal proteins like Aβ can appear long before they cause harm, sometimes even decades earlier, increased attention is now focusing on **biomarkers** (clinically useful indicators) for the earliest stages of the disease, when treatments are likely to be most effective.

The ultimate cause of neuron death in Alzheimer's disease is not known, but one strong contributing factor is the abnormal processing of the APP protein mentioned earlier. The proteins presenilin-1 and presenilin-2 are the catalytic subunits of γ-secretase, one of the enzymes that causes the alternative cleavage of APP to form the Aβ that accumulates in amyloid plaques. Mutations in APP or in either type of presenilin result in greatly enhanced production of an especially damaging form of Aβ known as Aβ-42, which leads to the early-onset form of Alzheimer's disease.

Once Aβ is formed, several enzymes are capable of breaking it down. One such enzyme is the insulin degrading enzyme (IDE). IDE also degrades several other proteins, including insulin and amylin, which are both secreted by pancreatic β-cells. The deposition of Aβ in Alzheimer's disease bears much in common with the deposition of amylin in type 2 diabetes, and the two diseases have much in common, including similar epidemiological and biochemical profiles, as well as the degradation of their very similar proteins by the same enzymes. Enhancement of IDE activity decreases extracellular levels of Aβ and also effectively reduces brain levels of Aβ (Qiu et al., 1998); a treatment that could increase IDE activity could thus be useful in controlling amyloid plaques and potentially slowing the progression of Alzheimer's disease. Various other molecular mechanisms exist for the degradation and clearance of Aβ from blood or brain tissue. For example, some researchers have reported that Aβ can be degraded by parkin, a protein whose malfunctions cause Parkinson's disease (Burns et al., 2009; Rosen et al., 2010).

Methods of measuring the rate of Aβ formation and degradation have been developed, and these rates can be used to assess both the disease process and the efficacy of possible therapies. Such measurements have been used to show that early-onset Alzheimer's disease results largely from Aβ overproduction, while the much more common late-onset Alzheimer's disease is characterized by decreased clearance of Aβ from brain tissue (Mawuenyega et al., 2010). Some researchers now suspect that small oligomers (composed of several Aβ molecules bound together) may be more damaging than the plaques themselves, which are made of larger aggregates of the protein. Aβ oligomers can be detected in blood plasma, and their levels correlate well with other clinical biomarkers and with mental decline. Some researchers have suggested that drugs that target these oligomers might slow the progression of the disease (Tolar et al., 2021). There is also good evidence that the buildup of amyloid-β in blood serum damages the blood–brain barrier and makes it leaky, allowing brain-damaging amyloid-β to leak into the brain (Biron et al., 2011). Working under this hypothesis, researchers are trying to develop and test drugs (including antibody-based drugs) that may clear Aβ from the blood and brain and possibly prevent the damage.

When inflammation occurs in brain tissue, small brain cells called **microglia** (**Figure 12.14**; see also Section 11.1.1) identify and remove the damaged tissue. Intriguingly, genetic studies of late-onset Alzheimer's disease have identified several genes that encode proteins only found in microglia, suggesting that

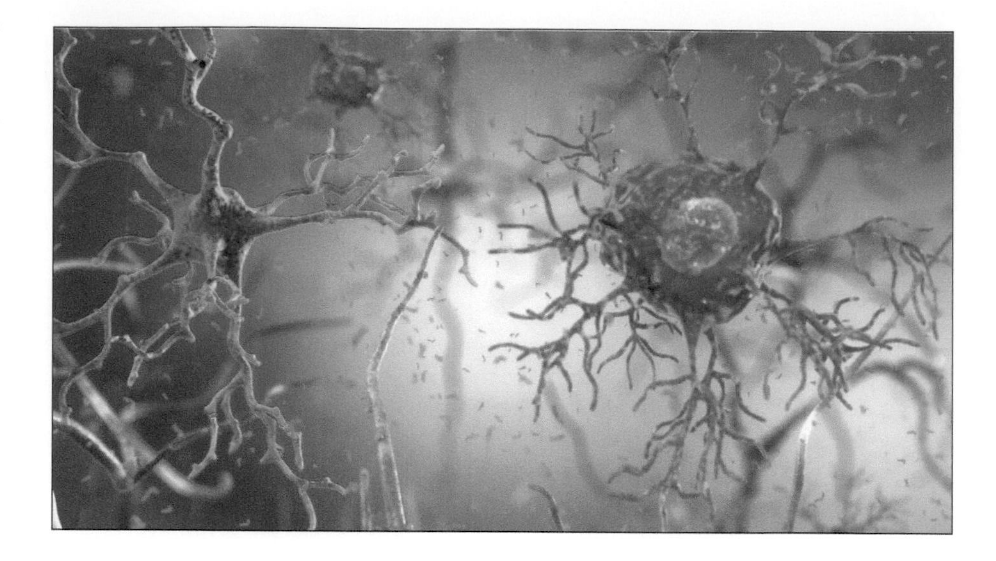

Figure 12.14 Astrocyte (right) and microglial cell (left), two types of glial cells in the brain.

dysfunction of microglia may contribute to risk for the disease. However, the damage in Alzheimer's disease is often so great as to overwhelm the microglia, leaving other brain cells, called **astrocytes**, to react by clearing some of the debris from brain tissue (Figure 12.15). In particular, the astrocytes are attracted to amyloid plaques and are effective in degrading Aβ within brain tissue. An enzyme called matrix metalloproteinase-9, produced by astrocytes, has been shown to degrade both soluble Aβ and complex amyloid plaques that have already formed. Also, a signaling molecule called interleukin-3 programs astrocytes to switch from inflammatory responses to protective responses within the brain, limiting the progression of the disease (McAlpine et al., 2021).

The internal support structure for brain cells also depends on the normal functioning of the protein **tau**. Tau stabilizes microtubules, which are part of the cell's cytoskeleton (see Section 6.2.4). In people with Alzheimer's disease, threads of tau protein become abnormally modified by phosphate groups (phosphorylated), which changes the shape of tau, causing it to become twisted. The twisted strands of phosphorylated tau form tangles of neurofibrillary material inside neurons. Many researchers believe that these tau tangles may seriously damage neurons and cause them to die. Evidence now shows that, even with many amyloid plaques in the brain, neither cognitive decline nor Alzheimer's disease occurs unless tau tangles are also present. Indeed, Alzheimer-like memory loss in mice can be prevented by reducing the presence of tau without altering the levels of Aβ. This immediately suggests the possibility that drugs targeting the tau protein may be effective in treating Alzheimer's disease, and several such drugs are currently being tested.

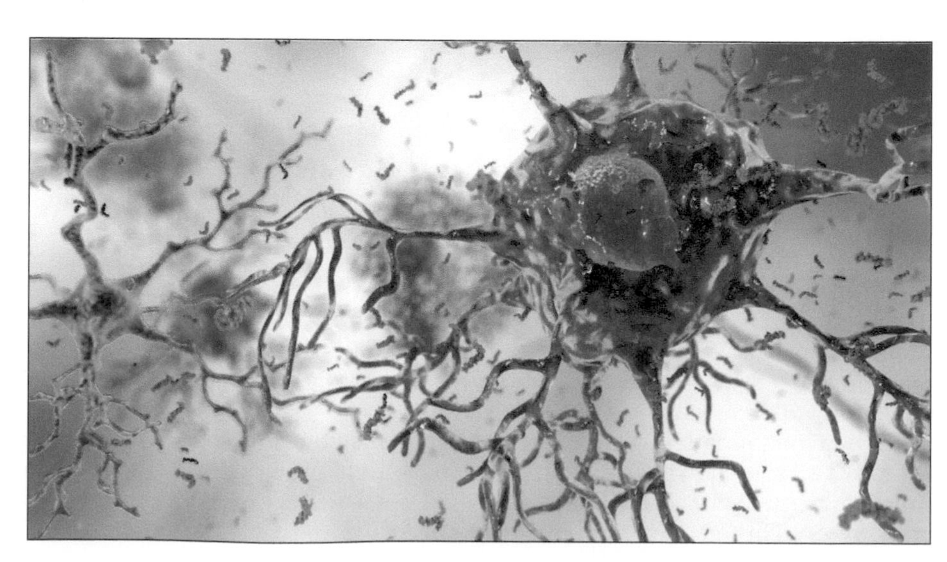

Figure 12.15 The microglial cell on the left functions as part of the immune system, patrolling the brain tissue and destroying damaged proteins. When Aβ (shown in light brown) accumulates in large amounts, it overwhelms the microglia, leaving the astrocytes, like the one on the right, to clear Aβ from the brain. Some drugs increase astrocyte activity in an attempt to rid the brain of accumulated Aβ.

12.2.3 Neuroimaging and diagnostic tests

The brains of Alzheimer's patients exhibit shrinkage of brain tissue, as noted earlier. Cellular hallmarks of the disease include the buildup of amyloid plaques (Figure 12.6) outside neurons and neurofibrillary tangles (tau tangles, Figure 12.7) inside neurons, abnormalities that postmortem staining of microscopic sections can verify. People who die of other causes can have evidence of early-stage Alzheimer's disease in their brains. In a study of 150 patients clinically diagnosed with Alzheimer's disease before death, autopsies confirmed 87% of the diagnoses, with other neurological diseases accounting for most of the incorrect diagnoses of Alzheimer's disease. Because other conditions can masquerade as Alzheimer's disease, the Fisher Center for Alzheimer's Research Foundation states that "a definitive diagnosis of Alzheimer's is possible only by examining brain tissue after death." Nevertheless, tentative diagnoses are made with increasing frequency using computerized axial tomography (CT or CAT) scans to assess brain shrinkage. These scans use X-rays that pass through the body at various angles, and a computer uses this information to create cross-sectional images, or slices, of the body part scanned.

In addition to CT scans, other types of neuroimaging are often used in diagnosis. Magnetic resonance imaging (MRI) uses radio waves and a strong magnetic field to produce detailed images of the brain. Positron emission tomography (PET) can reveal areas of the brain that may be less active and thus detect the existence and density of both amyloid plaques and tau tangles. Such images can usefully distinguish between patients with Alzheimer's disease, mild cognitive impairment, and no cognitive impairment. Images of the brain may also enable doctors to pinpoint any visible abnormalities, such as clots, bleeding, or tumors, that may be causing symptoms of impaired memory. Shrinkage of the brain may also be evident in advanced cases, but usually not until serious memory impairment has already occurred.

To help distinguish Alzheimer's disease from other causes of memory loss in living patients, doctors typically rely on a variety of tests because no single test is definitive. A thorough diagnosis requires a medical history, a physical examination, a neurological examination, neuropsychological tests, and brain imaging. Neuropsychological testing may include functional tests of memory and other tests of thinking and memory skills. Neurological tests are often conducted to assess reflexes, eye movements, speech, muscle tone, and voluntary movements that can help distinguish Alzheimer's disease from other neurological conditions. Blood tests may be done to help doctors rule out other potential causes of dementia, such as thyroid disorders or vitamin deficiencies. Spinal taps may be done to study the levels of molecular biomarkers of Alzheimer's disease, such as Aβ and tau, in cerebrospinal fluid. Despite all available tests, autopsies may still be performed when needed to confirm diagnoses of Alzheimer's disease made during life.

12.2.4 Early-onset Alzheimer's disease

Although most cases of Alzheimer's disease develop in people over age 65 (normal onset or late onset), an estimated 5% or less of cases develop in people below this age, usually with an onset in their forties or fifties. This early-onset form of Alzheimer's disease is especially devastating because those affected are often still actively engaged in careers or still have children living at home. Estimates of the number of people in the United States who have early-onset Alzheimer's disease vary from approximately 200,000 (Mayo Clinic) to about 500,000 and increasing (Alzheimer's Association).

Inherited mutations play a clear, causative role in early-onset Alzheimer's disease. The early-onset form is associated with mutations in at least three different genes: *PSEN1*, which encodes the protein presenilin-1; *PSEN2*, which encodes presenilin-2; and *APP*, which encodes the amyloid precursor protein (Tanzi, 1999).

The gene that encodes APP (the unprocessed form of Aβ) is located on chromosome 21, which is present in an extra copy (trisomy) in people with Down syndrome (see Section 3.2.3). People with Down syndrome have a greatly increased risk of early-onset Alzheimer's disease, and nearly all of them will develop the condition if they live long enough. A few Down syndrome individuals were identified who had an incomplete third copy of chromosome 21 that lacked the *APP* gene, and these individuals did *not* get Alzheimer's disease. Also, a rare duplication of the *APP* gene on chromosome 21 produces early-onset Alzheimer's disease without Down syndrome. These facts clearly implicate the *APP* gene as one cause of early-onset Alzheimer's disease. Other genes on chromosome 21 may also contribute to an increased risk.

12.2.5 Genes and Alzheimer's disease

In addition to the genes related to early-onset Alzheimer's disease, several genes that increase the risk of late-onset (or normal-onset) Alzheimer's disease have also been identified. Many of these are related to lipid or cholesterol metabolism. The activities of the β- and γ-secretase enzymes are known to be affected by the amount of cholesterol in the cell membrane, and the enzyme cholesterol acetyltransferase has emerged as a potential therapeutic target for the disease. Researchers have studied families that have an especially high incidence of Alzheimer's disease and have found that such families often have a particular allele of the gene *APOE*, which encodes the protein apolipoprotein E. The APOE-e4 allele, which occurs in about 25% of people, raises the risk of developing Alzheimer's disease after age 60, compared with the risks for the more common APOE-e3 or -e2 alleles. Apolipoprotein E transports cholesterol and other lipids through the blood.

Other genes that have been implicated in late-onset Alzheimer's disease include *TREM2* and *SORL1*. *TREM2* encodes a receptor protein that appears on the surface of several types of myeloid cells, including the microglia that patrol brain tissues and react to inflammation by increased activity in destroying certain abnormal proteins, including Aβ. *SORL1* (also called *SORLA*) interacts with the APP molecule and can therefore modify the production of Aβ. Additional genes (*CD33, CR1, ABCA7, SHIP1*) have also been implicated in affecting Alzheimer's disease risks. Many such genes have small effects, but these small effects may accumulate to cause the condition.

12.2.6 Causes and risks for the disease

The causes of Alzheimer's disease are heterogeneous and not well understood. Currently, scientists believe that a combination of genetic and environmental factors (including lifestyle factors) may lead to its onset. No single factor has been identified as the direct cause of the disease.

Epidemiology, the study of diseases at the population level, can often provide clues about the causes and risk factors for various diseases. Age is the most important risk factor for Alzheimer's disease. At age 65, only 10% of people suffer from the condition; however, the risk doubles with every 10 years of life, and, by age 85, the Alzheimer's Association estimates that 50% of persons suffer from this disease.

Women are more likely to develop the disease than men are, in part because women generally live longer than men and make up a larger fraction of advanced age groups. Hormones also appear to play a role. Some studies suggest that women who undergo postmenopausal estrogen hormone replacement therapy have a lower risk of Alzheimer's disease than those who do not, but other studies have failed to confirm this finding. Some evidence suggests that a woman's reproductive history can influence her risk of the disease, but the connection is currently unclear and in dispute. Elevated levels of the hormone adiponectin, produced by visceral (deep abdominal) fat deposits, is also a risk factor for Alzheimer's disease and other forms of dementia, as is obesity, which causes numerous hormonal changes.

Catholic nuns who live in convents are a unique population for epidemiologists to study because their lifestyles, diets, and daily routines are much more uniform than in most other groups of people. One study (Riley et al., 2005) followed a group of nuns whose written applications to enter their religious orders had been preserved. Using these records to evaluate the language ability of the nuns at the time of their entrance into the convent (as one measure of cognitive function), researchers found that those whose applications showed superior language ability had a reduced chance of developing Alzheimer's disease many decades later, compared with applicants whose early writing samples were judged lower in terms of cognitive score (Snowdon, Greiner, and Markesbery, 2000). The clinical significance of these findings is unclear, but one possibility is that people with higher intelligence have a "cognitive reserve" built up that compensates for some degree of memory loss until a greater amount of damage to brain tissue has occurred. People with higher intelligence generally do better on cognitive tests, so it may take more nerve damage to the brain before any dementia can be detected by such tests.

Many of the same factors that put someone at risk of heart disease also increase the likelihood that a person will develop Alzheimer's disease; these factors include high blood pressure, high cholesterol, and poorly controlled diabetes. For this reason, most physicians recommend that Alzheimer's patients attempt to control their blood pressure, exercise regularly, and follow a heart-healthy diet. More recent studies have specifically identified the Mediterranean diet (see Section 8.5.1) as the one that most strongly reduces the risk for Alzheimer's disease (Martinez-Lapiscina et al., 2013; Singh et al., 2014).

The well-known benefits of regular exercise may be mediated by a recently discovered hormone called irisin (Aydin, 2014; Munoz et al., 2018). Despite some skepticism (Albrecht, 2015), a study by Islam (2021) confirms the cognitive benefits of this hormone among mice used to model the effects of Alzheimer's disease.

12.2.7 Drugs to treat Alzheimer's disease

Although no known drug can cure Alzheimer's disease, several have been used to treat its symptoms. These include cholinesterase inhibitors, anti-inflammatory drugs, cholesterol-lowering drugs, antibodies, and kinase enzyme inhibitors.

Because of the realization that Alzheimer's patients suffer from a deficiency in the neurotransmitter acetylcholine, the early medications used to improve symptoms were generally cholinesterase inhibitors. These inhibitors raise acetylcholine levels in the brain by blocking its degradation by the enzyme cholinesterase. However, most cholinesterase inhibitors are helpful only in a certain percentage of patients; they also tend to lose their efficacy over time, and they do nothing to arrest the progression of the disease. They also have significant side effects, such as nausea, vomiting, or diarrhea, that are severe enough to force many patients to stop taking these medications.

Another type of drug used to treat Alzheimer's disease works by controlling inflammation in the brain. There is some evidence that nonsteroidal anti-inflammatory drugs (NSAIDs) lower the risks of Alzheimer's disease, but the effects seem to vary, and the causal relation between inflammation and the progression of Alzheimer's disease is not well understood. Many of these drugs work by inhibiting enzymes such as cyclooxygenase-2 (Cox-2); Naproxen (trade name: Aleve®) is representative of these Cox-2 inhibitors. Studies have shown that some drugs given to control inflammation will lower the risks for Alzheimer's disease, but not all anti-inflammatory drugs have proven helpful in Alzheimer's patients. Cholesterol-lowering drugs are sometimes prescribed for Alzheimer's patients in the belief that any improvement in cerebral blood supply would be beneficial. Such an approach is even more widely used against vascular dementia, the second most common form of dementia. In many instances, vascular dementia and true Alzheimer's disease are hard to distinguish and may often occur together. Epidemiological evidence does show higher rates of Alzheimer's disease in people with circulatory problems, and the pharmacological or dietary control of high cholesterol or high blood pressure does appear to lower Alzheimer's rates by as much as 50%.

To actually reverse or cure Alzheimer's disease, treatments would likely need to target and eliminate the amyloid-β deposits and tau tangles that are the molecular hallmarks of the disease or to prevent their buildup in the first place. Researchers have tried using antibodies (see Chapter 14) that recognize Aβ to promote Aβ clearance from the brain. One drug based on this principle, aducanumab (trade name Aduhelm®), is a monoclonal antibody that binds to Aβ and labels it for destruction by the immune system (Sevigny et al., 2016). In June 2021, the Food and Drug Administration (FDA) approved aducanumab for clinical use under its "accelerated approval" pathway (US FDA, 2021), which allows approval for therapies targeting serious or life-threatening diseases that are *reasonably likely* to provide a clinical benefit to patients, rather than the more stringent clinical benefit criteria applied under the normal approval process. Several members of the FDA's scientific advisory committee argued strongly against approval and in fact resigned from the board following the decision because they thought the evidence for clinical efficacy was insufficient and the cost of the drug was also very high (Jimenez, 2021). Under the rules of the accelerated approval pathway, the drug maker (Biogen) must provide clearer evidence of clinical benefit within nine years or risk reversal of the approval.

In January 2023, the FDA approved another monoclonal antibody drug, lecanemab (trade name Leqembi®), that also targets and binds to Aβ (van Dyck et al., 2023). This drug also received "accelerated approval" (US FDA, 2023). Patients receiving lecanemab experienced "greater reductions in brain amyloid burden" than with a placebo (van Dyck et al., 2023) and had slower memory decline. Both lecaneab and aducanumab target the early stages of Alzheimer's disease to slow disease progression.

An effective cure for Alzheimer's disease would require *both* (1) an effective treatment to prevent the cellular damage in its earliest stages and (2) a series of biomarkers (e.g., Eyigoz et al., 2020) to detect these earliest (preclinical) stages, before any substantial damage is done. The efficacy of many drugs, like aducanumab, depends upon using biomarkers to identify patients early enough for effective treatment. Many drugs that are eventually approved to prevent cellular damage were initially rejected because they were tested on patients too late in the course of their disease, after the damage had already been done. By the time that symptoms permit a diagnosis, the damage has become so great that no medication can possibly reverse the course of the disease, and critics of the drug can claim that the drug is ineffective in patients already diagnosed.

Other methods of Aβ clearance are also being studied. A team at University College, London, has investigated a serum amyloid P component (SAP) that cross-links many kinds of amyloid deposits, including those involved in Alzheimer's disease (Bodin, 2010). Antibodies to SAP can remove this protein from amyloid deposits. This team has also developed a drug, CPHPC, which can clear SAP from blood serum, cerebrospinal fluid, and brain tissue (Kolstoe et al., 2009). Clinical studies of SAP and CPHPC in Alzheimer's patients are in their early stages. Other drugs are being developed to inhibit the β-secretase and γ-secretase enzymes.

Drugs targeting the tau protein include antibodies against tau and also drugs that might prevent tau phosphorylation. Enzymes that phosphorylate proteins are called protein kinases (see Section 10.3.2), and phosphorylation of the tau protein is an important event in the development of Alzheimer's disease. Many researchers are now testing protein kinase inhibitors as possible anti-Alzheimer's drugs that might reduce the formation of tau tangles. The FDA has already approved one such drug, sorafenib, as an anticancer drug, since protein kinases are also important in cancer biology. It is often easier to get a drug approved for clinical use if it is already approved for some other condition, because its safety record and method of action are already known.

As of 2023, only one drug, memantine (trade name Namenda®), has received FDA approval for the treatment of moderate to severe Alzheimer's disease (Stanford Medicine, 2023). This drug relieves symptoms only, and it does so by reducing the amount of glutamate in the brain. Once the disease has become moderate to severe, no currently approved medication can slow disease progression.

12.2.8 Living with Alzheimer's disease

A cure for Alzheimer's disease may be years or decades away. For the many families already living with this condition, any such cure will be too late. What they hope for more immediately are coping strategies, therapies, and other interventions that can alleviate the suffering as much as possible. The many victims of Alzheimer's disease include not only the patients themselves but also their family members, who are usually their principal caregivers. The burden on caregivers is particularly heavy because Alzheimer's patients become increasingly needy as

Table 12.2 Useful resources for families dealing with Alzheimer's disease (most websites listed here have links to many additional resources)

Resource	Website
General information	
Alzheimer's Association	www.alz.org
National Institute on Aging	www.nia.nih.gov
Puzzles To Remember	www.PuzzlesToRemember.org
The Alzheimer's Caregiver	www.TheAlzheimersCaregiver.com
Alzheimers & Dementia Weekly	www.alzheimersweekly.com
Support groups for caregivers	
Caregiver Action Network	www.caregiveraction.org
Family Caregiver Alliance	www.caregiver.org
Caregiver resources	www.caregiver.org/resources/alzheimers-disease-caregiving
Kid Caregivers	www.kidcaregivers.com
Recommended activities for Alzheimer's patients and caregivers	
	http://www.kidcaregivers.com/2018/05/two-minute-activities-for-dementia.html
Puzzles made to help Alzheimer's patients	
	https://www.springbok-puzzles.com/alzheimer-s-puzzles-s/1849.htm
Finding resources for families with Alzheimer's disease	
D.H.H.S. Eldercare Locator	https://eldercare.acl.gov/Public/Index.aspx
Senior Guidance	https://www.seniorguidance.org

the disease progresses, often requiring care just to navigate around the house, to dress, to bathe or shower, and often to feed themselves or go to the toilet, and this task usually falls to family caregivers. Caregivers, as a consequence, are at a greatly elevated risk for stress-related illnesses, for depression and for certain types of injuries sustained in caring for the Alzheimer's patient. In many cases, the emotional toll is compounded by the fact that the patient may no longer recognize the caregiver by name or express any appreciation for or awareness of the care. In addition to the many support groups for Alzheimer's patients, there are also a number of support groups for the caregivers and other family members who help care for Alzheimer's patients (Table 12.2). Maintaining physical activity as much as possible helps slow the progression of the disease (Friedland et al., 2001). This can be as simple as going for a walk with a caregiver around the neighborhood or some other familiar place.

Activities for the brain, such as reading or playing cards, will also calm down Alzheimer's patients, alleviate negative symptoms, and slow the progression of the disease (Friedland et al., 2001; Postal, 2010). Working simple jigsaw puzzles has been shown to be particularly effective in calming patients, bringing on delight not just when the puzzle is complete but even when they recognize an image in the puzzle design. Table 12.2 has links to many simple suggestions for caregivers, including recommended activities for Alzheimer's patients.

THOUGHT QUESTIONS

1. Should Alzheimer's disease be defined on the basis of amyloid-β, tau tangles, or memory tests? How would this decision affect research on the disease? How would it affect clinical practice?
2. Several of the symptoms of Alzheimer's disease, such as the ability to recall things or to finish sentences, seem to come and go or to fluctuate from day to day. Can you think of a possible mechanism to account for this? How would you test for such a hypothesized mechanism? How would you test for any possible relation to sleep, stress, or dehydration?
3. If early-onset Alzheimer's disease and late-onset Alzheimer's disease have different risk factors and possibly different causes, would it be helpful to call them two different diseases? If Alzheimer's disease shares many risk factors with diabetes, would it be helpful to call it type 3 diabetes, as some have suggested?
4. If a drug like aducanumab is only available at a very high cost, should private insurance or Medicare be required to cover it? What would you say to the many people who could be demanding it because they were desperate for a treatment? What if a particular case were so far advanced that the drug would be unlikely to have any benefit?
5. After the FDA approved the drug aducanumab, two major hospitals (the Cleveland Clinic in Ohio and Mount Sinai Hospital in New York) announced that they would not recommend it to their patients. Do you agree with this decision? Do you agree with the FDA's approval of the drug? What assumptions, and what ethical principles, are involved?

12.3 SOCIAL CONSEQUENCES OF AN AGING POPULATION

Advances in the control of infectious diseases and other causes of premature death allow more and more people to live longer lives. As a result, the age structure of most populations is changing: The proportion of children in the population is diminishing (see Section 9.1.4), while the proportion of adults over age 60 is increasing. For the most part, this is a welcome change, but one that causes us to consider the social consequences of an aging population. Alzheimer's disease was once rare, but its increasing prevalence is one consequence of an aging population. Alzheimer's disease has become a leading cause of death (second only to cardiovascular disease) in many countries, including France, Spain, and Sweden. In the United States, it ranks sixth, and its incidence is increasing while most other causes of death are decreasing.

Because of aging populations, medicine and allied fields must focus increasingly on maintaining the health of the elderly. Society must learn to accommodate the needs of the many adults who are caring simultaneously for children and for elderly parents. The needs of children may remain undiminished, but their numbers are shrinking, while the numbers of the elderly continue to increase. Society's resources will therefore require reallocation, with fewer new schools needed but more hospitals and nursing facilities caring for the elderly. Businesses will find themselves selling more goods and services to older adults, while an increasing number of families will span three generations or even four. Some of these social consequences can easily be foreseen, but many other changes will also occur in our ever-changing society.

THOUGHT QUESTIONS

1. In addition to the possibilities mentioned earlier, in what other ways is society likely to change as a result of changing proportions of elderly people or as a result of changing proportions of healthy versus dependent people among the elderly?
2. Why was Alzheimer's disease not even recognized until the twentieth century? Does this fact reflect the lack of attention paid to diseases of the elderly in the past? Are there other diseases of the elderly awaiting discovery?

CONCLUDING REMARKS

Aging is a process that affects everyone who lives long enough. It involves many changes at the cellular level, including changes in gene expression, telomere shortening, loss of elasticity in certain tissues, and a slowing down of immune function. As more treatments are found to the causes of early death, and as more people learn to stay healthier throughout their lives, more and more people are living long enough to experience the effects of aging. Also, many elderly people are living longer, and a greater proportion of most societies is coming to consist of elderly people. More and more people in middle age are caring for elderly parents, and an increasing proportion of health care dollars is being spent on care of the elderly. As explained in Section 9.1, these changes accompany the shift from a rapidly growing population to a more stable one. One of the great challenges faced by societies in North America and Europe, and increasingly elsewhere, is their need to adjust to the changing proportions of people needing care and providing care.

CHAPTER SUMMARY

- Aging is a natural process that involves accumulated damage to cells and tissues. Some of this damage is caused by **free radicals** and other chemical agents.
- To some extent, the **maximum lifespan** is limited by the length of **telomeres** that "cap" ends of DNA sequences in each cell.

- The **reserve capacity** of cells or tissues is their ability to function at levels above those required to maintain life. In general, this reserve capacity diminishes with age, leading to diminished ability to function under stress.
- Immune system function diminishes with age and may contribute to **senescence**.
- **Alzheimer's disease** results from the altered folding of a membrane protein, forming a molecule called **amyloid-β**. Extracellular clumps of this protein then accumulate inside the brain.
- A later development is the phosphorylation of the **tau** protein to form tangles inside brain cells. The accumulation of tau tangles correlates with mental decline.
- In most countries, the proportion of the population living to old age is increasing, and the diseases of old age, including Alzheimer's disease, are becoming more and more prevalent.

BIBLIOGRAPHIC REFERENCES

Bibliographic references to material in this chapter can be found online at biologytrending.routledge.com/chapter12

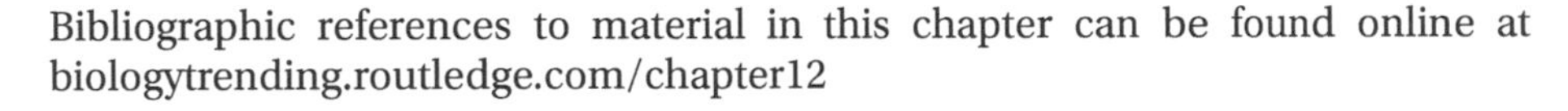

GLOSSARY: KEY TERMS TO KNOW

These key terms are defined at the end of each chapter as an aid for student review.

Alzheimer's disease A progressive form of dementia, most often affecting the elderly, characterized by buildup in the brain of amyloid-β clumps ("plaques") and tangles of tau protein and by memory loss.

Amyloid-β A protein that forms abnormal deposits ("plaques") in the brains of patients with Alzheimer's disease.

Antibodies Proteins secreted by B-lymphocytes during an immune response that bind to and disable the specific type of molecule that induced their secretion, thus helping to protect the body's health.

Astrocytes Star-shaped neuroglial cells that nourish and protect brain cells, also capable of engulfing or destroying cellular debris or abnormal proteins in brain tissue.

Autoantibodies Antibodies against one's own proteins.

Autoimmune disease Any disease in which the immune system abnormally secretes antibodies that attack the body's own healthy tissues.

Basal metabolic rate The rate at which the body uses energy when awake but lying completely at rest.

Biomarker Any molecule or other change that can easily be measured in patients to monitor the course of a disease or its treatment.

Cytokine A signaling molecule in the body's immune system.

Dementia A type of brain disease marked by mental deterioration, memory loss, and decreasing ability to carry out such everyday functions as feeding oneself or bathing.

Entropy A measure of randomness (disorderliness) in physical systems.

Epidemiology The study of the frequency and patterns of diseases in populations.

Epigenetic changes Changes in gene function that do not involve changes in the genes themselves.

Free radicals Very reactive atoms or molecules with unpaired electrons.

Immunosenescence The decline of immune function as a consequence of the aging process.

Life expectancy The average length of time that individuals with certain characteristics are expected to live, or the average duration of life for the entire population.

Lifespan See *Maximum lifespan.*

Macrophage A type of phagocytic cell that can engulf and destroy foreign cells or cellular debris in various tissues.

Maximum lifespan The maximum number of years that an individual can hope to live.

Microglia A type of small neuroglial cell that creeps through brain tissue with amoeboid motion, engulfing bacteria or cellular debris.

Progeria Premature aging (senescence) in a young person.

Progressive disease A disease that can worsen with time or sometimes not worsen but can never improve.

Reactive oxygen species Atoms or molecules that owe their reactivity to the presence of oxygen.

Reserve capacity The ability of any tissue or organ to function above its normal level.

Sarcopenia Loss of physical strength due to loss of muscle tissue.

Senescence Progressive loss of physical functioning as a result of age. Also, the aging of individual cells as a consequence of the telomeres of chromosomes reaching a threshold shortness and resulting in no further cell division.

Tau A protein that normally stabilizes microtubules in cells but that can form abnormal tangles that impair brain function in Alzheimer's patients.

Telomere A repetitive DNA sequence at the end of a chromosome whose small, progressive shortening during each cell division limits the number of times a cell can divide.

CONNECTIONS TO OTHER CHAPTERS

Chapter 1: Allocation of health care (and government funding of health care) to various age groups raises many ethical issues.

Chapter 3: Various human genes have effects on the aging process and the development of Alzheimer's disease.

Chapter 8: Good, heart-healthy nutrition is key to healthy aging in general and to the avoidance of Alzheimer's disease in particular.

Chapter 9: As people are living longer, populations around the world are increasing, and most of the increase occurs in the fraction of the population that is elderly.

Chapter 10: The shortening of telomeres with each cell division limits the number of times cells can divide, and thus limits the lifespan, but enzymes that prevent shortening can lead to cancer.

Chapter 11: Nervous system function declines generally with age, and this decline is rapid and progressive as Alzheimer's disease advances.

Chapter 13: Certain drugs can improve Alzheimer's symptoms, offering temporary relief. No drug has yet been shown to arrest or reverse the course of the disease more than temporarily.

Chapter 14: Inflammation in the brain is a factor in Alzheimer's disease, but some antibodies can help clear amyloid plaques from brain tissue.

PRACTICE QUESTIONS

1. What molecule is β-amyloid (Aβ) made from? What enzymes are involved?
2. What effect does aging have on the immune system? On collagen? Why does each of these matter? Which matters more and why?
3. What chemical group is added to the tau protein? What effect does this have?
4. In what part of the brain are the effects of Alzheimer's disease most prominent? What does this part of the brain govern?
5. Why might biomarkers for brain diseases be hard to find? What kind of biomarkers should physicians look for?
6. What role do telomeres play in aging?
7. Why might dietary modifications help in treating Alzheimer's disease? Why might dietary modification be beneficial for the elderly in general?
8. What role do microglia play in the brain? Why is this important for understanding Alzheimer's disease? What role do astrocytes play? Why is their role important?

CHAPTER 13

Drugs and Addiction

ISSUES

- What is a drug?
- How do drugs interact with one another?
- What is addiction?
- Why are some drugs addictive?
- Is drug abuse the same as addiction?
- Does drug abuse lead to addiction?
- What are the social effects of drug abuse?
- Are all forms of drug addiction treated as crimes?
- Should pharmaceutical companies bear responsibility for the opioid crisis?
- How are people's attitudes to marijuana changing? Why are many states changing their marijuana laws?

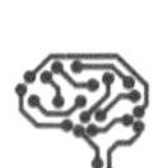

BIOLOGICAL CONCEPTS

- Molecules (structure, diffusion)
- Membranes and cell-surface receptors (agonists, antagonists)
- Energy and metabolism
- Organ systems (respiratory system, excretory system, placental circulation)
- Homeostasis (drug tolerance, drug metabolism, excretion)
- Central nervous system (brain)
- Behavior (reinforcement, addiction, behavior modification)
- Health and disease (public health issues)

CHAPTER OUTLINE

DOI: 10.1201/9781003391159-13

Most of us are accustomed to the idea of taking drugs when we are sick. Many of these drugs are prescribed to fight bacterial infections, to fight cancer, or to regulate the body's physiological processes. Many other drugs are taken without medical supervision and for a wide variety of purposes. Many of these drugs are legal; some are not. The United States is the number-one drug-producing and the number-one drug-using country in the world. There is said to be a "drug problem" that has led to a "war on drugs," yet many in society are uncertain about the answers to some very basic questions: What is a drug? What are the various legitimate uses of drugs? What is addiction, and what makes a drug addictive? In this chapter we examine some recent research in biology and in related fields relevant to these questions. We also examine the basic biological principles that underlie our understanding of how drugs work. An understanding of the respiratory, circulatory, and excretory systems is needed to see how drugs enter and become distributed around the body, how long they stay in the body, and how they are eventually removed.

13.1 DRUGS ARE CHEMICALS THAT ALTER BIOLOGICAL PROCESSES

The term **drug** can have many meanings, depending on the context. To biologists, a drug is any chemical substance that alters the function of a living organism other than by supplying energy or needed nutrients. In a medical context, a drug may be thought of as any agent used to treat or prevent disease. Those who work in the field of drug addiction define a **psychoactive drug** as any chemical substance that alters consciousness, mood, or perception. As with all definitions in science, these are open-ended; no one definition can cover every possible situation. Each of these definitions expresses a slightly different concept, and each is correct within its contextual field.

This chapter emphasizes psychoactive drugs—that is, those that alter consciousness, mood, or perception. All psychoactive drugs alter biological functions and are thus considered to be drugs on the basis of the first definition. Many, but not all, are also drugs by the medical definition.

13.1.1 Drugs and their activity

To understand how psychoactive drugs work, we need to know some general principles that apply to all types of drugs. The study of drugs, their properties, and their effects is called **pharmacology**. Pharmacological principles explain how drugs can be both effective and dangerous, how both concentration and time affect the activity of drugs, and how interactions between drugs can change drug activity.

The activity of any drug varies with its **dose**, meaning the amount given at one time. There is an effective dose, which is the amount that is "effective" in producing the desired change; in medicine, the effective dose is also called a therapeutic dose. For almost all drugs there is also a toxic dose—the amount at which the drug produces harmful effects—and a lethal dose—the amount that kills the organism. The more commonly used term, overdose, includes both toxic and lethal doses. Some drugs have a wide margin of safety, meaning that the toxic dose is many hundreds or thousands of times as much as the therapeutic dose. For other drugs, the toxic dose, or even the lethal dose, may be very close to the effective dose.

The exact amount that is effective, toxic, or lethal differs with the chemical structure of the drug. It also differs from one individual to the next depending on body size and many other physiological variables. We discuss some of these variables later in the chapter.

The drug concentration achieved at any given location in the body depends on several factors: How the drug enters the body, how it is carried around the body, how it is transformed by the body's cells, and how it is removed from the body.

13.1.2 Routes of drug entry into the body

Most drugs enter the body through the digestive system or the respiratory system; a few are injected directly into the bloodstream. Drugs that are taken orally and swallowed must be able to withstand the acidic environment of the stomach and then be absorbable by the cells of the intestinal lining in the same ways in which food substances are taken up (see Section 8.3). Drugs that enter through the respiratory system face different obstacles.

13.1.2.1 The respiratory system The lungs are the body organ by which vertebrate land animals take up oxygen and give off carbon dioxide, a waste product of their metabolism (Figure 13.1). This process is controlled by the respiratory system and has two parts: the mechanics of breathing and the exchange of gases.

Breathing is accomplished not by the lungs themselves, but by the diaphragm, a muscular layer below the lungs. When the muscles of the diaphragm contract, they pull the diaphragm downward, expanding the chest cavity. The laws of physics tell us that the pressure of a gas depends on the number of gas molecules in a given volume; therefore, when the volume of the chest cavity becomes larger, the pressure inside it decreases. The air pressure inside the lungs is then less than the air pressure outside the body. Physics also tells us that gases flow from areas of higher pressure to areas of lower pressure, so the enlargement of the chest cavity causes air to enter the lungs (inhalation). (Hiccups result when the downward movement of the diaphragm is more rapid than normal.) When the diaphragm relaxes, it returns to its higher position, decreasing the volume of the chest cavity and pushing air back out (exhalation) (Figure 13.1A).

Air passes into the lungs from the trachea, or windpipe, through branching airways called bronchi into thin-walled sacs called **alveoli**. Oxygen and carbon dioxide are exchanged between the alveoli and the blood vessels of very small diameter (capillaries) that surround them (Figure 13.1B). The gases move by passive diffusion (see Figure 8.14A) across the membranes of the cells of the alveoli and the cells of the capillaries. Oxygen is in higher concentration in the inhaled air in the alveoli than it is in the capillaries, so oxygen diffuses *into* the blood. Carbon dioxide is in higher concentration in the blood than it is in the inhaled air, so it diffuses *out of* the blood into the alveoli, to be exhaled. The rate of diffusion depends on the difference in concentration and on the amount of surface area over which the diffusion can occur. The huge number of capillaries and the compartmentalized, saclike structure of the alveoli provide an enormous surface area (Figure 13.1C), making diffusion very rapid (the effects of compartmentalization on surface area are discussed in Section 6.2.1).

Gaseous drugs, including many anesthetics, can enter the lungs, diffuse across cell membranes, and enter the blood by the same rapid mechanism as does oxygen. Toxic inhalants, including various solvents and propellants, can enter in the same way. Particulate matter, such as the chemicals in smoke, can adhere to the inner surfaces of the alveoli, causing damage to those surfaces while delivering drugs to the bloodstream (see later).

13.1.2.2 Route of entry and effective drug dose The way in which a drug enters the body often affects its resulting concentration in body tissues. As an example, consider cocaine, a product of the coca plant, *Erythroxylon coca*, native to the Andes Mountains. Cocaine exists in many forms that differ in both the concentration of the drug and in its molecular form. Purification of the drug makes it easier for the user to receive a toxic or lethal dose. The addictive potential (see later), while always great, also increases with each purification.

Coca leaves were and are chewed by indigenous South Americans, especially those living at high altitudes. The concentrations absorbed from the gut when coca leaves are chewed are quite low because the cocaine is present in an ionized form. Ionized molecules carry a charge and thus do not readily diffuse across the nonpolar portion of the plasma membranes of the cells lining the gut (see Section 8.3.3).

In contrast, when cocaine is purified into a powder, although the cocaine itself is still in an ionized form, sniffing the powder increases both the rate of absorption

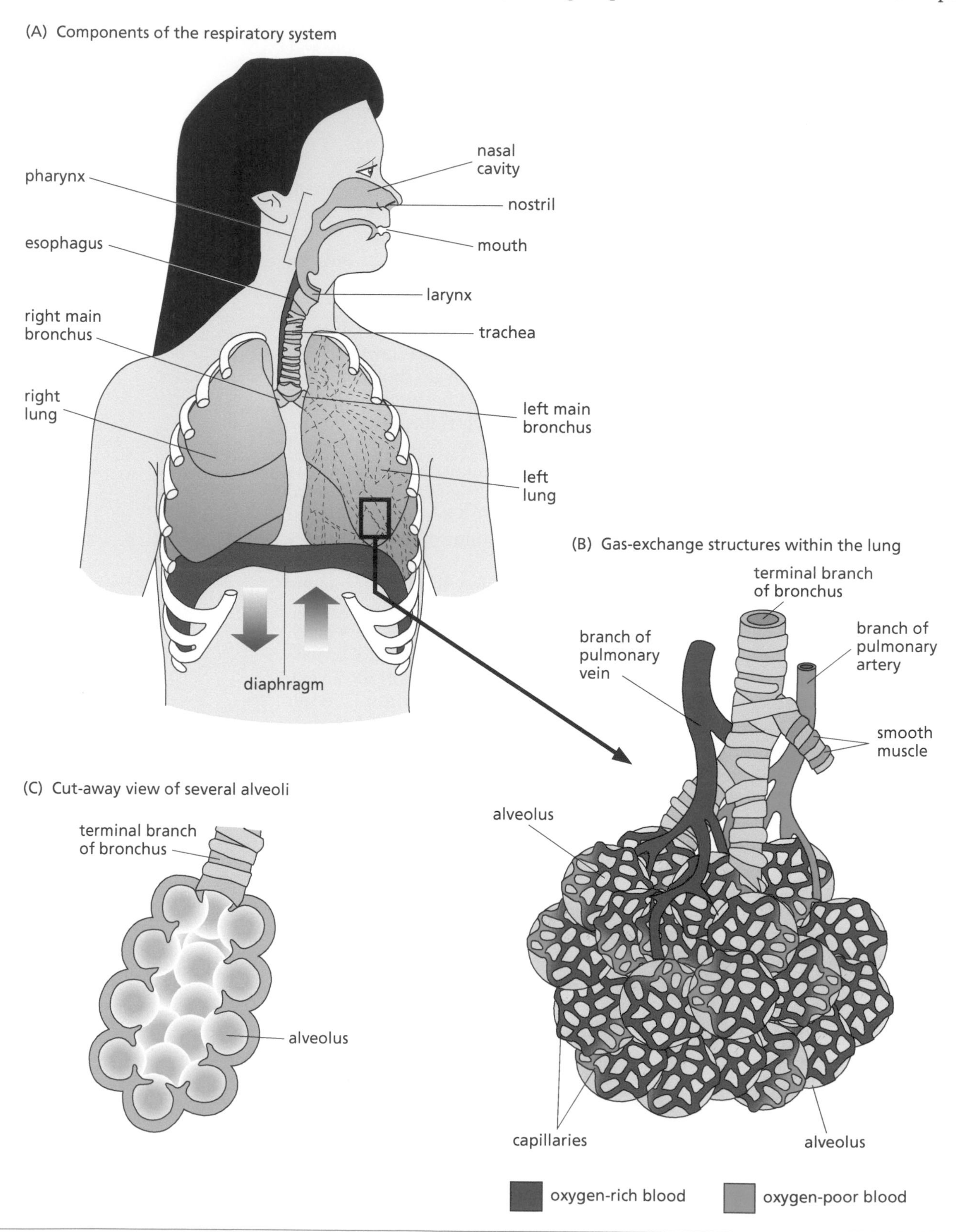

Figure 13.1 The human respiratory system.

and the amount absorbed into the capillaries under the lining of the nasal passages. Cocaine can be further purified, resulting in the uncharged form called "free-base" cocaine and the still more highly purified form known as "crack." Uncharged molecules are less polar than ionized molecules and are therefore more rapidly absorbed across cell membranes. In addition, free-base and crack cocaine are no longer powders. They are smoked rather than sniffed, moving the drug beyond the nasal passages and into the lungs. Smoking these drugs further increases the rate of absorption because the surface area of the lungs over which the drug is absorbed is much larger than the surface area of the nasal passages.

Whether cocaine is sniffed or smoked, it is inhaled not as a gas, but as small particles. All forms of smoke, including smoke from crack cocaine, cigarettes, marijuana, and the burning of fossil fuels (industrial smoke and automobile exhaust), are particulate. For a chemical to be absorbed from these particles, the particle must first adhere to the lung tissue. Various chemicals can be highly concentrated within a particle, even if the concentration is low when calculated as amount of chemical per volume of air. Thus, an adherent particle may cause substantial damage to the lung tissue, inhibiting the normal functioning of the tissue and damaging the body's health. Similarly, sniffing particles, such as cocaine powder, can cause local damage to the cells that line the nasal passages.

Other drugs may be administered by injection, that is, with a needle. In a drug injection, the needle can be placed in many locations, including into a muscle (intramuscular), under the skin (subcutaneous), or directly into a vein (intravenous, or IV). Each of these routes may be appropriate for different therapeutic drugs. For example, insulin, a drug given to control blood sugar fluctuation in diabetes, is usually given intramuscularly, so that it will be released into the blood over a period of time rather than all at once; insulin cannot be given orally because it is a protein that would be digested in the stomach (see Section 8.3.2).

Uptake from an intravenous injection is faster than it is from other routes because entry is not dependent on absorption. When a drug must be absorbed from the muscles, the gut, the nasal passages, or the lungs, only a portion of the drug is taken up; intravenous injection puts the drug directly into the bloodstream. Accidental overdoses can more easily arise from injection than from other routes of administration. The biological consequences of drug abuse became more severe after the invention of the hypodermic needle and syringe in 1853. In addition, all forms of drug injection can be dangerous if contaminated needles are shared because these practices can transmit infections such as hepatitis and HIV from one user to another (see Chapter 15).

13.1.3 Distribution of drugs throughout the body

Although the concentration of a drug is initially highest at the local site of entry, the blood vessels (veins and arteries, see Section 8.4.1) quickly distribute substances throughout the body. Transport through veins and arteries is one aspect of drug distribution; transport into the tissues is another. Drugs and other molecules can leave the bloodstream and enter the body's tissues via the smallest blood vessels, the capillaries. In most areas of the body, almost every tissue cell is within a few cells of a capillary. Areas of the body that have many blood capillaries, such as the lungs, liver, kidneys, and heart, receive a higher dose of most drugs and receive it faster than those areas with fewer capillaries (e.g., skin, muscle, and fat).

Various parts of the circulatory system let different molecules pass through. To act on the nerve cells of the central nervous system (CNS), psychoactive drugs must traverse the blood-brain barrier (see p. 453). Drugs that do not easily cross this blood-brain barrier remain in the bloodstream while they are in the brain, even though they can leave the bloodstream in other locations of the body. Most antibiotics, for example, do not cross the blood-brain barrier, making it difficult to treat bacterial infections in the brain. In contrast, ethyl alcohol is particularly effective at crossing the blood-brain barrier and in causing both short-term and long-term damage to brain cells. (Chemists recognize many types of alcohol, but the only one that we consider here is ethyl alcohol, the type in alcoholic beverages.)

13.1.4 Elimination of drugs from the body

In contrast to food molecules, which may be stored, most drugs begin to be broken down or removed from the body as soon as they enter it. One way in which drugs are eliminated from the body is in the urine produced by the excretory system. Another way is for the active form of a drug to be chemically altered, in a process called drug metabolism, so that it is no longer active. Drug metabolism can occur in a central location such as the liver, or in scattered locations, as in the inactivation of neurotransmitters within synapses between nerve cells (see Section 11.1.3). In this section we look at these two mechanisms by which drugs can leave the body.

13.1.4.1 Metabolic elimination Many drugs are chemically altered by the body into substances that no longer produce the drug's effects, although some of these breakdown products may have effects of their own. One example of a drug that is eliminated by drug metabolism is ethyl alcohol (ethanol), which is metabolized by the cells of the liver. The metabolic breakdown of ethyl alcohol involves several steps, but the overall rate is limited by the amount of the proton carrier NAD available to be reduced to NADH (see Section 8.3.3) and by the levels of the enzyme alcohol dehydrogenase. The liver can metabolize 7–10 mL (about 0.25–0.33 fluid ounces) of alcohol per hour; if the rate of intake exceeds the rate of metabolism, intoxication results. As Table 13.1 shows, the intake of most forms of alcoholic beverages can easily exceed this limit.

One step in the metabolic breakdown of alcohol involves the production of acetaldehyde, which is immediately broken down by another enzyme, acetaldehyde dehydrogenase. Some people become sick from small amounts of alcohol because they lack this second enzyme, leading to a buildup of acetaldehyde, a toxic chemical. In some populations, the proportion of people who lack acetaldehyde dehydrogenase is quite high; for example, as many as 50% of all people in Japan and China lack this enzyme. Disulfiram (Antabuse), a therapeutic drug used in the treatment of alcoholism, works by inhibiting acetaldehyde dehydrogenase in people with normal levels of this enzyme. People who take this drug and then drink alcohol become very sick, so they are both incentivized to and conditioned to avoid alcohol. The breakdown of alcohol to acetaldehyde is also partially responsible for the link between alcohol consumption and cancer (see Section 10.6.5): Acetaldehyde has been shown to be mutagenic in animal studies, and acetaldehyde derived from ethanol breakdown is thought to promote esophageal cancer in humans.

13.1.4.2 The excretory system In addition to being altered by enzymes, drugs are also eliminated by the same system that rids the body of normal waste products, such as the nitrogen compounds derived from the breakdown of proteins. The process of removing these waste materials from the body is called **excretion**. The major route for the excretion of drugs and other substances is through the urine. The excretion of urine is vital to maintaining the proper concentrations of many types of ions in the blood and tissues. For example, calcium (Ca^{2+}) is an important second messenger in muscle contraction, and potassium (K^+), sodium (Na^+), and chloride (Cl^-) each function in maintaining

Table 13.1 Beverages containing equivalent amounts of ethyl alcohol (ethanol)

Beverage	Alcohol content (%) (varies somewhat)	Amount of beverage equivalent to "one drink"	Quantity of ethanol
Beer and ale	4	12 oz. can (350 mL)	1/2 oz. (15 mL)
Table wine	12–15	4 oz. glass (100 mL)	1/2 oz. (15 mL)
Dessert wine (sherry or fortified wine)	20	Two-thirds of a glass (2.5 oz. or 70 mL)	1/2 oz. (15 mL)
Distilled liquor (whiskey, vodka, etc.)	40	1.25 oz. (35 mL)	1/2 oz. (15 mL)
Liqueur	40	1.25 oz. (35 mL)	1/2 oz. (15 mL)

charge gradients across membranes and in the propagation of nerve action potentials (see Section 11.1.2). Levels of these ions in the blood are monitored and controlled in the kidneys, as we will soon describe.

Urine is produced in three steps—filtration, reabsorption, and secretion—all of which occur in the kidneys. The kidneys are made up of many complex, tubular structures called **nephrons** (Figure 13.2). The first step in urine production,

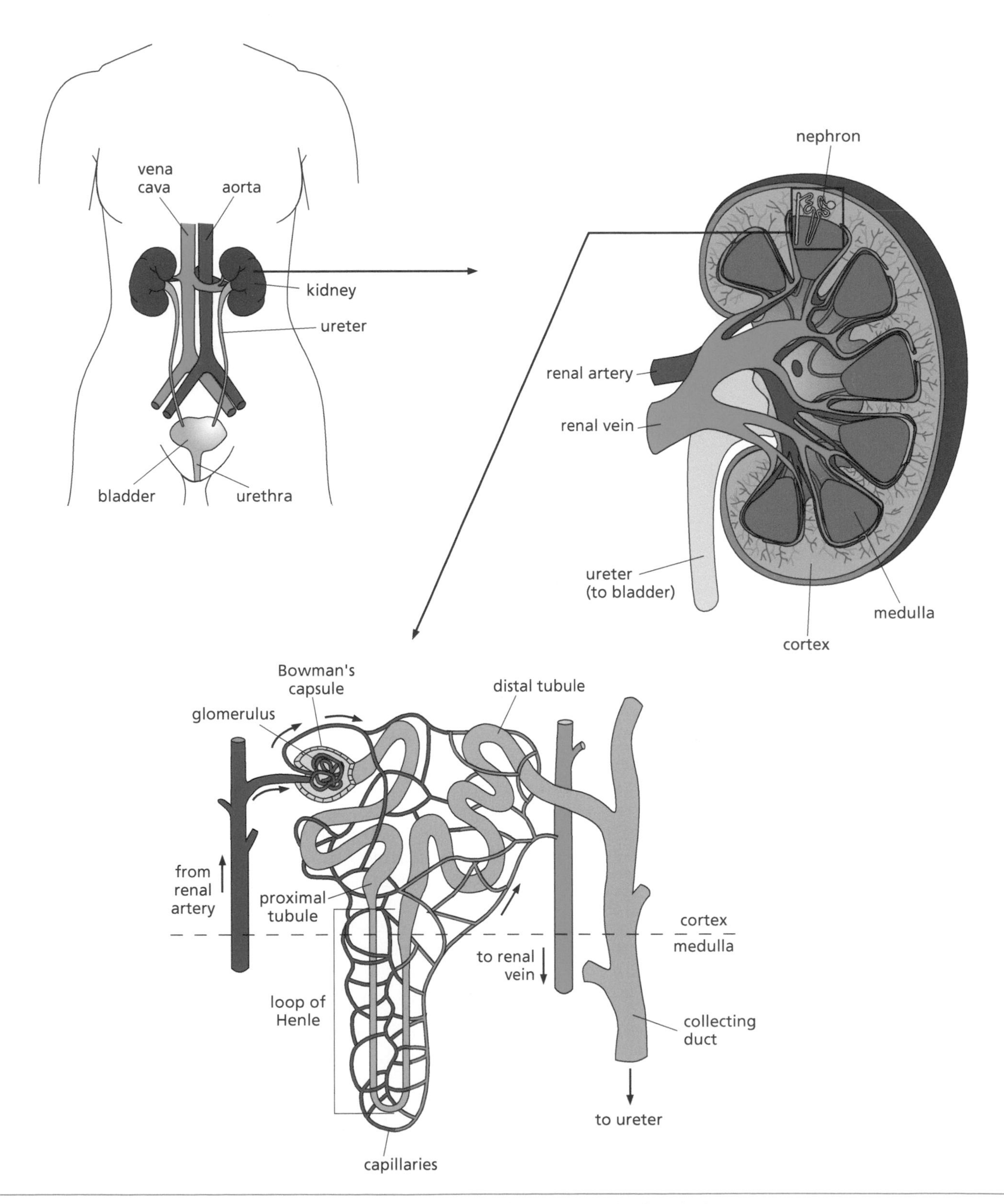

Figure 13.2 The major organs of the human excretory system responsible for the production of urine.

filtration, takes place in substructures within the nephrons called glomeruli. Each **glomerulus** consists of a meshwork of capillaries surrounded by a sheath called Bowman's capsule. The circulatory system brings blood in via the renal (kidney) artery, and low-molecular-weight substances, including ions, amino acids, glucose, urea, and water, are filtered out of the blood and into Bowman's capsule, which empties into the proximal tubule of the kidney. Larger molecules such as proteins are retained in the blood.

The second step, reabsorption, takes place in other substructures of the nephrons, the proximal tubules and the loops of Henle, both of which are surrounded by capillaries (Figure 13.2). Here the ions, amino acids, and glucose that the body can reuse are reabsorbed, meaning that they move from the tubules back into the blood. Reabsorption is by active transport (see Section 8.3.3) carried out by transporter proteins in the membranes of the nephron cells. Much of the water is also reabsorbed.

In the medulla of the kidney, various ions (especially Na^+) are taken up from the descending portion of the loop of Henle, while lesser amounts are returned to the ascending portion of this loop. The process is controlled in part by the hormone antidiuretic hormone (ADH, also called vasopressin), which regulates the permeability of the ascending part of the loop of Henle and also smooth muscle tension in the collecting ducts, thus controlling the amount of water and dissolved ions held within the kidney.

The third step in urine formation, secretion, takes place farther along, in the distal tubules of the nephrons (Figure 13.2). Higher-molecular-weight substances, including drugs and toxins, are secreted from the blood into the tubules. After these three stages, the liquid in the tubules is called urine. Urine goes to the collecting ducts and then from the kidneys to the bladder, where it is stored until it is excreted (Figure 13.2). Many drugs, or their breakdown products (metabolites), enter the nephrons at either the filtration or secretion steps and are excreted in the urine.

Drugs are also excreted in smaller amounts via saliva, sweat, and the air exhaled from the lungs. In nursing mothers, drugs may also be excreted into breast milk, with obvious consequences for the infant if the excreted drugs are still active, which they often can be.

13.1.4.3 Drug half-lives The concentration of a drug in its active form can be measured, usually in the blood. The length of time required for the drug concentration to be decreased by half is called the **half-life** of the drug (**Figure 13.3A**). Cocaine, for example, has a half-life of 5–15 minutes, meaning that, regardless of the mode of intake of the drug or the initial starting dose, half of it will be gone in 5–15 minutes. A drug with a half-life of 10 minutes remains active for a shorter time than a drug with a half-life of 10 hours (**Figure 13.3B**), even if they start out at the same initial concentration in the blood. Drugs that are inactivated in a simple one-step process often have a shorter half-life than those whose inactivation pathways are more complex. For example, the half-life of alcohol is very short in most people, while compounds present in marijuana remain in the blood for up to a week. Taking more of a drug before the original dose is completely eliminated increases the concentration of the drug to a level higher than the amount actually taken in the second dose (**Figure 13.3C**).

13.1.5 Drug receptors and drug action on cells

The activity of a drug depends on its dosage and its concentration in tissue. In many cases, drug activity also depends on the existence of specific receptors for the drug.

13.1.5.1 Specific receptors for drugs After a drug has reached the site of its action, it acts on individual cells at that site. The ability of a drug to act on a particular cell most often depends on the presence of a cellular **receptor** for that drug (see Section 10.2.1). Picture a receptor as being like a glove. For a drug to bind to that receptor, the drug must have the shape of a hand. A small hand fits

(A) Regardless of initial dose, a drug's concentration is reduced by half in one half-life.

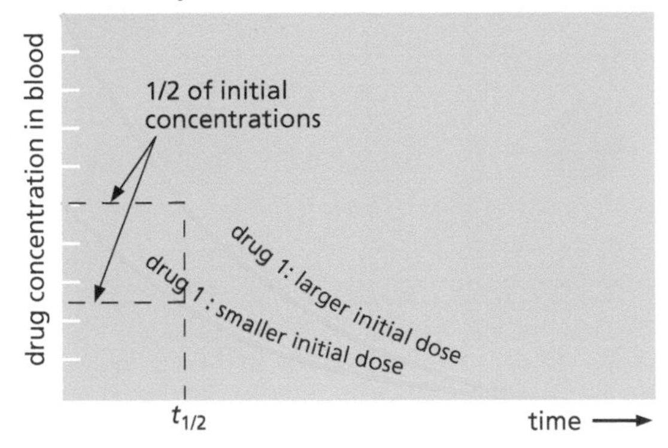

(B) Drugs with longer half-lives stay active for longer.

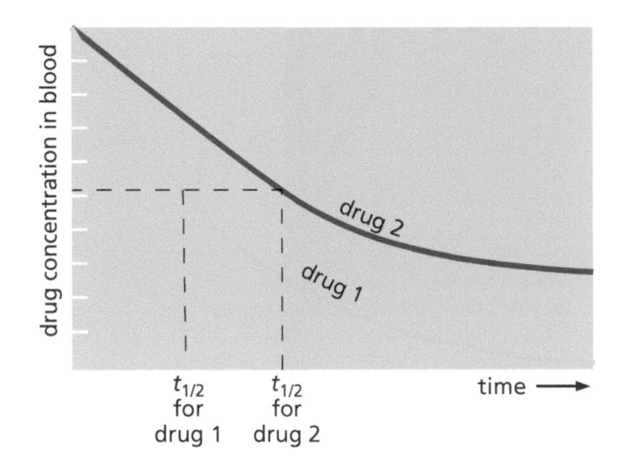

(C) Taking a second dose of drug before the first dose is completely eliminated raises the concentration, even if the two doses taken are equal.

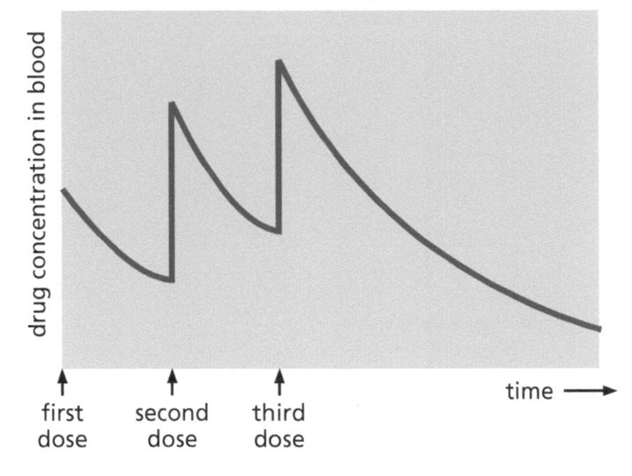

Figure 13.3 Half-lives of various drugs. The half-life of a drug ($t_{1/2}$) is the time required for its active concentration to be reduced by half.

into a large glove, but, for any glove, there is some hand size that fits best. A large hand does not fit into a small glove. Those substances that have the best fit to a receptor have the greatest activity. Those that do not fit have no activity.

Many chemical molecules can have right-handed or left-handed shapes depending on the directions in which their bonds are arranged. That is, two molecules composed of identical atoms can differ in their shape. Right-handed molecules are said to be mirror images (or "enantiomers" in chemical terms) of left-handed molecules. As an aid in visualizing this, hold your hands flat. Place them in front of you, with the fingers of both hands pointing up and with the thumbs of both hands pointing toward the left (so that the palm of your left hand is towards you and the palm of your right hand is away from you). Look at your right hand in a mirror. Does the image of your right hand in the mirror look like your left hand without the mirror? Now put your right hand into the right glove of a fitted pair, then try to put it into the left glove. Your right hand fits into only one glove of the pair. Similarly, biological receptors are specific for right-handed or left-handed molecules, not both. This "handedness" of molecules is illustrated in Figure 13.4A. Even if you rotate one enantiomer, it will never be able to be superimposed on the other (Figure 13.4B). Because the binding of a drug to its receptor requires very specific interactions between atoms, usually only one enantiomer can bind to a given receptor. The other enantiomer may be inactive or, in some cases, may interact with other receptors to have harmful effects. One tragic example of this is the case of thalidomide, a drug that was prescribed to pregnant women in the late 1950s to treat the symptoms of morning sickness. One enantiomer, the so-called "R" version, had the desired sedative activity, but the other,

(A) (B)

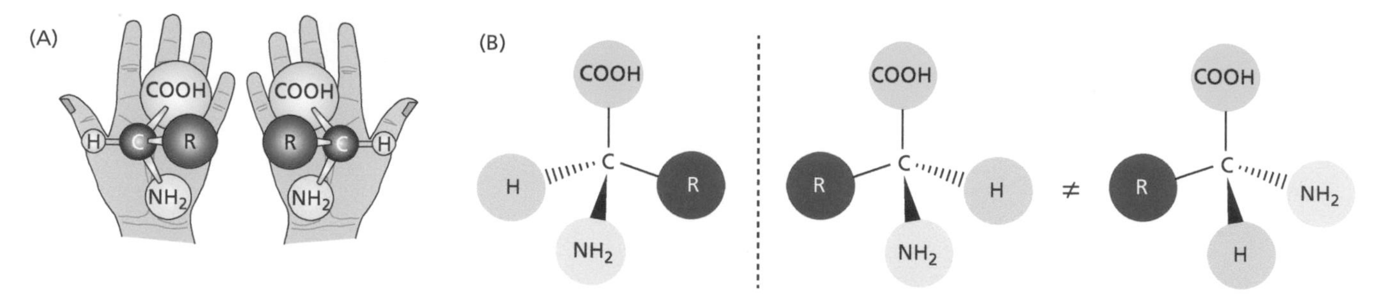

Enantiomers have the same atoms linked by the same bonds, but they differ in their spatial arrangement such that they are mirror images, like left and right hands.

There is no way you can turn one enantiomer such that it can be superimposed on the other, so they are distinct molecules and cannot bind to the same cellular receptor proteins. In the figure, the dotted bonds are pointing backward into the page while the wedge-shaped bonds are sticking upward from the page.

(C)

The two enantiomers of the drug thalidomide. The left one (called "R") is a sedative that was once used to treat morning sickness during pregnancy, but the right one ("S") causes birth defects.

Figure 13.4 Enantiomers are mirror-image molecules that often have very different biological effects.

called "S," induced birth defects in the women's babies, including malformations of the arms and legs as well as other symptoms, some of which were lethal. Approximately 10,000 infants were born with thalidomide-induced deformities before use of the drug was halted. One enantiomer of thalidomide can spontaneously convert into the other, so it was not possible to prepare a safe version of the drug consisting only of the "R" enantiomer. More recently, thalidomide has found a safe use as treatment for certain cancers (Latif et al., 2012); it only causes harm during a particular stage of embryonic development, not to adult cancer patients.

13.1.5.2 Tissue locations of receptors and locations of drug actions

Receptors can be located on the cell surface (plasma membrane receptors) or within the cytoplasm of the cell. Not all types of cells have all types of receptors, and the number of copies of a particular receptor molecule on a cell may change over time. The only cells that can respond to a particular drug are those with receptors for that drug. Thus, whether a drug is active in a particular tissue depends both on the ability of the drug to get to the tissue and on the presence of receptors for that drug on the cells of that tissue.

The binding of neurotransmitters to receptors on postsynaptic cells (see Section 11.1.3) is equivalent to drug-receptor binding. In fact, many psychoactive drugs act as either agonists or antagonists of neurotransmitters. An **agonist** is a substance that elicits a particular response or stimulates a receptor. For example, an agonist of a neurotransmitter is any other substance that produces the same response as that neurotransmitter (Figure 13.5A and B). An **antagonist** is a substance that inhibits a response (Figure 13.5C). The summative effects caused in all of the receptor-bearing cells produce functional effects on the body as a whole.

13.1.6 Side effects and drug interactions

Because drugs are not taken by isolated cells, many effects are likely to be produced, especially for drugs that circulate throughout the body. Although a drug is

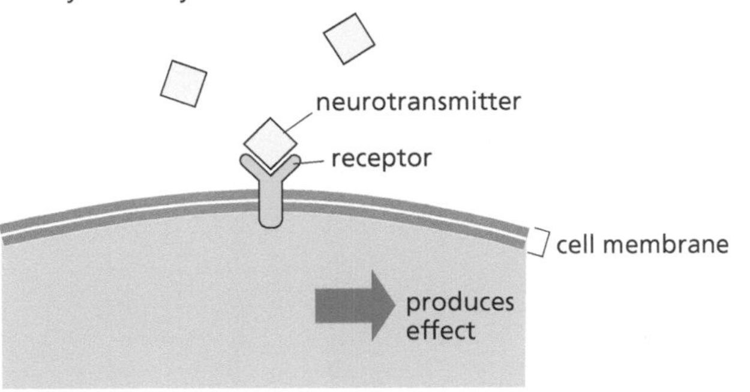

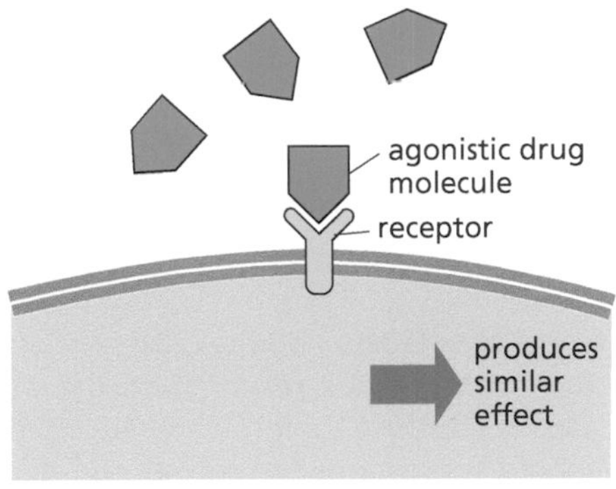

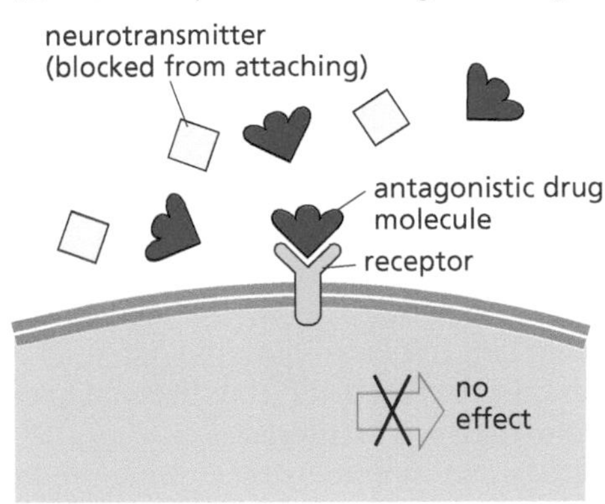

Figure 13.5 **Drug agonists and antagonists.**

usually intended to produce a specific effect, drugs almost always have effects other than the intended ones; these other effects are called **side effects**. Side effects may be weak or strong and can vary from person to person. Side effects can also vary from beneficial or harmless to harmful or even lethal. Calling something a side effect simply means that it is not the main effect or the reason for which the drug was used.

13.1.6.1 Types of drug–drug interactions When two or more drugs are taken, they may interact in various ways. The simplest interaction is called an **additive effect**. When two drugs have an additive effect, the response to taking them together equals the sum of the responses produced by each drug individually (Figure 13.6A). In a **synergistic interaction**, one drug increases the action of the other so that the total response is *greater* than the sum of the responses to each drug separately (Figure 13.6B). An **antagonistic interaction** is one in which one drug inhibits the action of another so that the total response is *less* than the response to the two drugs individually (Figure 13.6C).

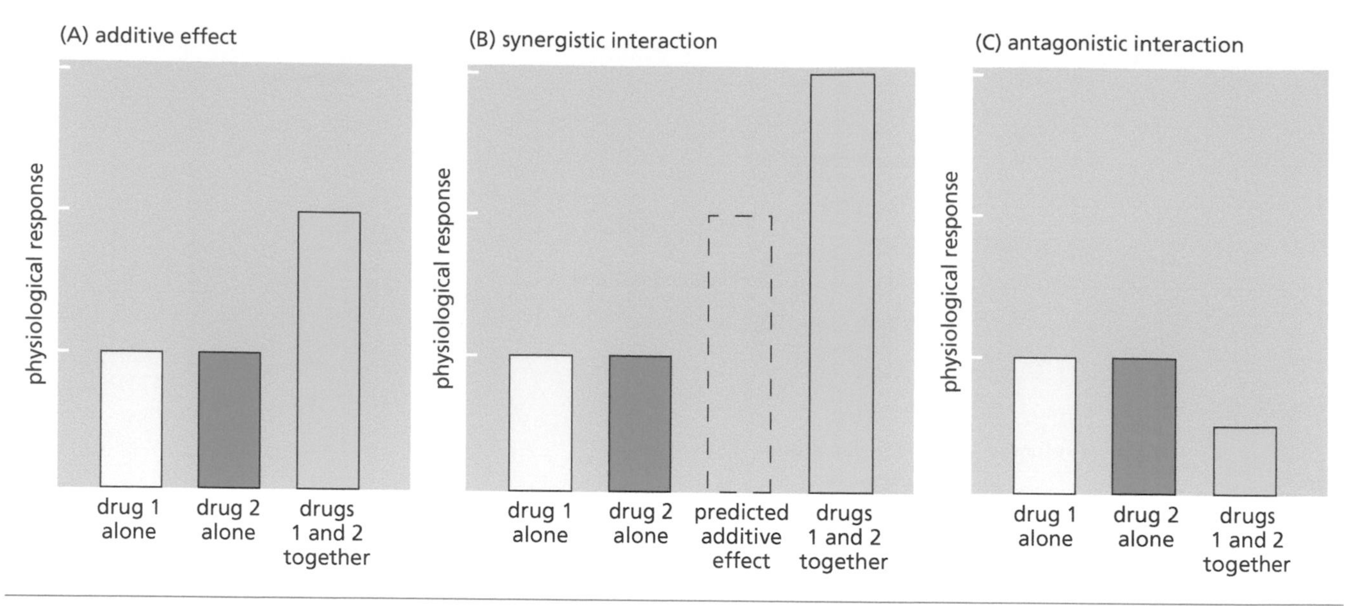

Figure 13.6 How drugs interact.

13.1.6.2 Mechanisms of interactions One example of drug interactions is when one drug changes the threshold for response to the other drug. The **threshold** is a value (in this case, a drug dose) below which no effect is detectable. One drug may lower the threshold for response to the second drug, making the body responsive to a lower concentration of the second drug. This could happen, for example, if one drug increased the number of receptors or the affinity of the cell receptors for the other drug. In contrast, one drug may antagonize the action of a second drug by raising the threshold for response to the second drug or decreasing the number or affinity of receptors for it. Again, this is similar to the actions of neurotransmitters (see Section 11.1.3).

Differences in drug half-lives have implications for drug dosage and interactions between drugs. Because drugs with long half-lives remain in the body for a long time, a second dose adds to the remaining fraction of the first dose to produce a concentration higher than would result from either dose separately (Figure 13.3C). Drugs with long half-lives also have the potential to interact with other drugs taken a long time after the first drug. Thus, it is not necessary that two drugs be taken at the same time for them to interact.

Another way in which drugs can interact is by altering the body's ability to metabolize other drugs. Barbiturates are eliminated from the body by enzymes contained in liver cells, and the use of barbiturates causes more of these enzymes to be made. Because these same enzymes eliminate other drugs, barbiturate use lowers the body concentrations of other medications, including steroid hormones. Body concentrations of estradiol, a steroid in birth control pills, is decreased by barbiturates, so taking birth control pills along with using barbiturates may result in an unwanted pregnancy; the same is true of certain antibiotics, which is one reason why it is important to read the warning labels on prescription drugs.

13.1.6.3 Frequency of drug–drug interactions Drug-drug interactions are an increasingly important problem because many people take medication on a lifetime basis for the control of chronic conditions such as high blood pressure. One study on an elderly population demonstrated that the probability of having some adverse interactions from medications was 75% if they were taking five different medicines and 100% if they were taking eight or more medications on a regular basis. However, potentially harmful drug interactions are not solely a problem for elderly people. There can also be harmful interactions between various street drugs, between street drugs and medications, or between any of these and alcohol.

13.1.7 Drug safety

People obtain drugs in various ways. Some are sold by prescription only, a process that ensures at least some medical supervision. Other drugs, called "over the counter" (OTC) remedies, are sold freely in stores. In many countries, a government agency supervises the formulation, but not the sale, of these OTC drugs. Plant products containing drugs are less widely regulated. In the United States, "herbal remedies" and "food supplements" can be advertised and sold with essentially no regulation, and the same is true of beverages containing caffeine. Alcohol and tobacco products are largely unregulated for adults, although various laws restrict their sale to minors. In addition, there are various "street drugs," sold illegally but not legally available. The definition of drugs thus covers a very wide spectrum. As we discuss later in this chapter, the position of marijuana along this spectrum has changed in recent years in many places, from being completely illegal to being approved for medicinal use only or also being approved for recreational use.

Many variables can influence the safety of drug use. Drugs that are prescribed by a physician are usually safe if taken as prescribed, but harmful side effects and drug interactions can still occur, even if directions are followed. The fact that a medicine is available only by prescription usually means that there is potential harm from unsupervised use. Most jurisdictions regulate the content of prescription medications to ensure that the stated ingredients are formulated to benefit most users with only minimal risk (if taken as directed) and that the formulation is appropriately labeled.

13.1.7.1 Over-the-counter drugs OTC drugs are those that can legally be sold without a prescription. Although people tend to view OTC drugs as "safe" because a prescription is not required, there are actually thousands of deaths from OTC drugs each year and over a million cases of drug poisoning, generally from overdose. Acetaminophen (e.g., Tylenol® or Excedrin®) can cause liver damage from repeated use, and aspirin can cause increased internal bleeding such as from stomach ulcers. Aspirin is second only to barbiturates as the drug most frequently used in suicides and suicide attempts.

Some OTC preparations contain combinations of multiple drugs. Most of the OTC sleeping pills are combinations of aspirin and antihistamines. Cold remedies often have many ingredients, and the liquid versions contain high concentrations (up to 25%) of alcohol. These combination drugs are marketed as an easy way of treating all possible symptoms of common problems like a cold or the flu. There is the unstated assumption in such advertising that the taking of unneeded drugs is harmless, but a larger number of active drug ingredients in a medication raises the chance that harmful side effects will occur in some portion of consumers who take it.

When used as intended, OTC drugs are generally safe, but their use in the wrong dose or for purposes other than those designated on the package can be harmful. People in the United States often have cultural expectations that problems can be fixed rapidly and with little effort on the part of the patient. This cultural expectation, encouraged by advertising, is certainly a factor in the tremendous use of OTC drugs. Many OTC drugs have a potential for abuse, particularly if they are psychoactive.

There are approximately 20,000 prescription drug formulations and over 300,000 OTC drugs approved for sale in the United States, which amounted to nearly $500 billion in revenue in 2019. New drugs seeking entry into the U.S. market must undergo an extensive review process by the Food and Drug Administration (FDA). The drug must be proved to be effective in its intended use and must be demonstrated to be free of adverse side effects. Many of the OTC drugs on the market have not undergone this process because they were "grandfathered" into the list of approved drugs—that is, exempted from testing because of their many years of prior use. Although official terminology lists these drugs as Generally Regarded As Safe (GRAS), many studies have shown that common OTC drugs, including aspirin, have side effects that cause harm in so many cases that they would not be able to meet the more stringent standards applied to new drugs.

13.1.7.2 Drug contaminants and additives Although prescription drugs and OTC drugs are not always "safe" unless taken as directed (and, even then, are not without side effects), consumers can at least be assured that these products contain the drugs they claim to contain and that the dosages are standardized from one batch to the next. Although no medication is sold as an unmixed, pure compound, the purchaser can be assured that harmful impurities are not present. None of these assurances exist, however, for herbal remedies, food additives, or street drugs. Street drugs can contain many impurities left over from chemical synthesis, and other substances are sometimes added deliberately. Strychnine is sometimes added to LSD ("white acid"), supposedly to sensitize the nerves to the LSD. It is also sometimes added to marijuana without the knowledge of the purchaser. By acting on the nerve cells, strychnine causes abnormal muscle contractions (convulsions) and is therefore sometimes used as rat poison. There is no specific antidote for strychnine, so it cannot be counteracted once it is taken, and it can cause permanent nerve damage in the brain and elsewhere.

13.1.7.3 Herbal medicines Herbal medicines have a long history of use in many cultures, and a majority of the medicines that we now use were originally derived from plant sources. For example, aspirin was first prepared from the bark of willow trees; prior to its commercial availability in the late 1890s, people would chew on the bark to receive the medicinal effect. In recent decades, many people have been returning to the use of herbal medications, sometimes favoring them as an alternative to the medicines sold by drug companies. At least some of this trend stems from a belief that such "natural" remedies are better or safer than synthetic drugs. Although many of these herbs are very active as medications, it is not accurate to think of them as always being "safe." Unprocessed herbs have unpredictable variations in the quantities of both the desired ingredients and the potentially harmful ones. Moreover, many of the most potent psychoactive drugs are plant products. For example, valerian is a plant that contains a chemical called chatinine, which is a tranquilizer, and Scotch broom contains a strong sedative, cytisine. Other herbal products, flowers, and some spices (nutmeg, sassafras, mace, saffron, crocus, and parsley) are strong stimulants of the CNS, while opium, strychnine, snake venom, and botulism toxin are all deadly poisons but are perfectly natural.

Chemically, the active ingredient in an herbal medication is identical to the same active ingredient if it is in a medicine produced by a drug company. The difference is that herbal medications are present in an unpurified form; that is, the active ingredient is present along with many other chemicals whose effects are not known and many of which are unidentified. A potential problem with herbal medications is the inconsistency of dose, both in the amount present in a particular plant and in the amount present in a particular extract. For example, many herbal remedies are taken like tea, with the active ingredients extracted in hot water. The length of brewing time and the temperature of the water greatly affect the amount of drug extracted.

Unlike OTC drugs, which are regulated by law to contain predictable doses of labeled medications and to contain no harmful ingredients, herbal remedies are unregulated in most places. A customer buying an herbal remedy cannot be assured that harmful substances are absent or that the product contains any active ingredient at all. Moreover, there is no guarantee that the product in question has been tested for safety or efficacy. Thus, people should be careful when considering herbal remedies and should do so in consultation with their physician.

In most countries, prescription medications and OTC medications are regulated by a government agency. In the United States, this is the FDA. However, a regulatory loophole permits substances to escape FDA oversight if they are marketed as "food supplements." Unfortunately, the term food supplement is so ill-defined that nearly any substance can be so designated by the company that sells it, even if it fits the biological definition of a drug. Many herbal formulations are marketed this way in the United States, including echinacea, *Ginkgo biloba*, saw palmetto, and St. John's wort. Ephedra, the dried leaves of the plant *Ephedra sinica*, contains the stimulant ephedrine, which has been used for a variety of purposes, including treatment of asthma, promotion of weight loss, and increased athletic performance. The San Diego–based company Metabolife once earned millions marketing an ephedra-containing weight loss supplement but came under investigation in the early 2000s because several deaths were linked to its use. The company was found guilty of withholding information about the risks of ephedra and ultimately went bankrupt. This case led the FDA to ban sales of ephedra-containing dietary supplements in the United States as of 2004. Without regulation, companies can market drugs as food supplements without testing them for dosage, efficacy, harmful side effects, or interactions with other drugs; they can also make health claims that have never been substantiated by rigorous scientific research (although they are prohibited from making therapeutic medicinal claims). For these reasons, people taking such substances should realize that they are dealing with untested risks.

THOUGHT QUESTIONS

1. The physiological effects of a drug usually depend on its concentration in body tissues. If two people, one of whom weighs 60 kg (about 130 pounds) and the other 90 kg (about 200 pounds), take the same *amount* of a drug (one beer each, for example, or two aspirins), will the concentration reached in their body tissues be the same? If other factors were equal, which person do you think would be more strongly affected by the drug?
2. How would you set up an experiment to monitor the rate at which a drug is delivered to various body tissues? Consider both animal and human test subjects.
3. Why do left-handed and right-handed forms of the same molecule frequently differ in their ability to act as drugs? When a difference of this kind affects the activity of a drug molecule, what does it indicate?
4. How long will it take for the blood concentration of a drug whose half-life is 20 hours to be reduced to one-eighth of its peak concentration?
5. Devise a model of the mechanism of action of two drugs with their receptors that would account for additivity between the two drugs. How would you modify this model to account for antagonism? How would you modify it to account for synergism?

13.2 PSYCHOACTIVE DRUGS AFFECT THE MIND

The brain coordinates the activities of the rest of the body (moving, sleeping, eating, breathing), but does it also produce activities that we associate with the mind? As we saw in Chapter 11, there is now considerable evidence in support of this theory. Electrical and biochemical activities have been measured in the brain during dreams and thought. Some diseases that involve brain degeneration, such as Alzheimer's disease, are frequently accompanied by changes in personality (see Chapter 12). Mental illnesses such as depression are associated with changes in brain neurotransmitters and can be treated with drugs (see Section 11.4). Additional evidence for the "mind equals brain" theory comes from research on psychoactive drugs. Psychoactive drugs act directly on the nerve cells of the brain or the CNS to produce changes in consciousness, mood, or perception in highly drug-specific ways. Both sensation and perception can be altered, sometimes permanently. Sensation refers to the information obtained through the senses (sight, hearing, taste, touch, smell), while perception refers to how this information is interpreted. Some drugs can induce paranoia, which is the perception of danger without sensory information that supports the existence of real danger. Drugs can also induce hallucinations, which are chemically induced changes in sensation.

Some psychoactive drugs (opiates, marijuana, and nicotine) work through specific receptors, and some (like amphetamines and hallucinogens) work because they are structurally similar to a neurotransmitter and bind to receptors for that neurotransmitter. Drugs that do not work solely via receptors (e.g., caffeine and alcohol) have a more generalized effect because they act on many types of cells.

13.2.1 Opiates and opiate receptors

Opiates are **narcotics**, which are drugs that cause either a drowsy stupor or sleep. Most narcotics also cause some degree of **euphoria** (literally, "good feeling"), and most are highly addictive. In low doses, most narcotics can be used as painkillers (analgesics). High doses of narcotics can produce coma and death. Most common narcotics, including heroin, morphine, and codeine, are derived from the opium poppy (*Papaver somniferum*) and thus are known as opiates. The painkilling and euphoria-inducing properties of opiates have been known for thousands of years in Asia, as has their ability to cause addiction. Pain reduction results from changes in the release of the neurotransmitters acetylcholine, norepinephrine, and dopamine and a pain-related substance called substance P. In people who are not in pain, opiates produce euphoria by acting on neurons in certain locations in the brain. Beginning in the mid-1990s, some prescription opioids were heavily marketed for pain control by pharmaceutical companies under misleading claims that they were not likely to be addictive. This claim was patently disproven, as increased opioid prescriptions resulted in an opioid addiction crisis that is still ongoing in the United States. The opioid crisis is addressed further later in this chapter.

Opiates and similar chemicals (opioids) act on cells via one or more specific opiate receptors to produce their various effects. Synthetic antagonist drugs have been made that block the binding of opiates to their receptors. These narcotic antagonists, of which naloxone (Narcan®) and naltrexone are examples, are useful in the treatment of opiate overdoses. Some other synthetic drugs act as both agonists and antagonists: They bind to opiate receptors, thus blocking the effects of narcotics like morphine (antagonistic action), but their own binding produces some euphoric effect (agonistic action), sometimes leading to their abuse. Buprenorphine is an example of the latter; it does produce a partial euphoric effect, but one that is minimal compared to that of other opiates. The combination drug Suboxone® contains both buprenorphine and naloxone. It is used as a tool in addiction recovery. The buprenorphine lessens symptoms of opiate withdrawal, while the naloxone decreases the chance of addiction or overdose. Suboxone® can be used safely as a component of addiction treatment, but it can still be abused and should be used under close medical supervision.

Opiate receptors are found on the membranes of many other types of cells in addition to neurons. Opiates consequently relax various muscles, including those of the colon. Consequently, morphine is an ingredient in some prescription anti-diarrheal medications (Paregoric), and severe constipation is an effect of long-term opiate use.

Why would opium poppies and other plants produce chemical substances that have these many effects? The likely reason is that poisoning animals that feed upon the plant will deter future feeding and thus benefit the plant by increasing its survival. Remember also that toxic doses are proportional to body size, so a dose that might sicken a 91-kg (200-lb) human would likely be lethal to a 20-g (0.7-oz.) rodent.

Several neurobiologists hypothesized that opiate receptors would never have evolved unless they served some adaptive function beyond allowing for learned addictive behaviors. This type of thinking led to the hypothesis that normal physiological molecules must exist that could bind to opiate receptors. The search for such molecules led to the discovery of endorphins and enkephalins, sometimes called endogenous opiates because they are produced within the body (endogenous), rather than being taken in from the outside (exogenous). These are peptides (short chains of amino acids) that have a molecular shape very similar to that of a portion of certain opiate molecules. The current hypothesis is that the endogenous opiates act as inhibitory neurotransmitters, decreasing the activity of the neurons that normally signal pain and stress. These same endogenous opiates have other effects throughout the body, including actions on the immune system, which are discussed in Chapter 14. Some people experience a so-called "runner's high" from intense aerobic exercise, which is a feeling of exhilaration and reduction in stress that is accompanied with decreased sensation of pain. It was initially hypothesized that exercise-induced release of endorphins was responsible for this effect. However, it has been shown that endorphins cannot cross the blood–brain barrier, and people have experienced the runner's high even when administered chemical blockers of opioid receptors. A more recent hypothesis proposes that the runner's high is instead caused by release of endocannabinoids, which are endogenous signals that bind to the same receptor as Δ_9-tetrahydrocannabinol (THC), the primary psychoactive component of marijuana. Marijuana and its effects are considered in the next section.

13.2.2 Marijuana and THC receptors

Marijuana is an extract obtained from plants of the genus *Cannabis*, which also produces the hemp fiber used in ropes. A number of psychoactive drugs are contained in marijuana, the most active of which is THC. THC binds to two receptors called cannabinoid receptor type-1 (CB_1R) and cannabinoid receptor type 2 (CB_2R). CB_1R is very abundant in the brain, particularly in the parts that influence mood, but it is also expressed in the peripheral nervous system and other tissues. CB_2R is less prevalent in the central and peripheral nervous systems but is expressed in a wide range of other tissues. Given this wide range of expression, it is evident that THC may potentially affect many body systems. Similar to the endorphins, the endogenous molecules that bind to opiate receptors, endocannabinoids are endogenous molecules that bind to cannabinoid receptors. Two endocannabinoids were discovered in the early 1990s, anandamide and 2-arachidonoylglycerol (2-AG) (Hanus et al., 1993). Anandamide is also a component of cocoa powder, and cocoa contains yet other compounds that stimulate production of endogenous anandamide. People who refer to their love of chocolate as an addiction may thus be close to the truth.

Binding of THC to cannabinoid receptors in the brain produces an altered sense of time, an enhanced feeling of closeness to other people, and elevates the intensity of sensory stimuli. In higher doses, marijuana can also cause hallucinations. Stimulation of the THC receptors causes a release of norepinephrine by the nerves in the median forebrain bundle, producing euphoric effects. There are also THC receptors on the cells of the hypothalamus (see Figure 11.15), a secretory part of the brain that regulates the steroid sex hormones. Long-term marijuana use decreases testosterone levels in males and alters the menstrual cycle in females. Of particular concern with regard to adolescents and young adults is the effect of THC on the prefrontal cortex, an area of the brain involved decision making and impulse control. This part of the brain does not fully mature until a person's mid-twenties, and studies using magnetic resonance imaging (MRI) have provided evidence that regular use of marijuana in adolescents and young adults causes structural changes in the brain. It is not clear whether those changes are permanent or might be reversed with subsequent abstinence from the drug.

Marijuana has long been recognized as a potentially dangerous and addictive drug, and its use has been illegal in most countries. However, significant attention has been paid to the potential medical uses of marijuana since the mid-1990s, and legal restrictions on its medical use have been relaxed in many places. More recently, some countries and U.S. states have legalized the recreational use of marijuana, with age restrictions. In addition, many claims have been made about the health benefits of cannabidiol (CBD), a non-THC (nonpsychoactive) component of marijuana. There has been a marketing boom for products that contain CBD, although many of the claims made about its beneficial effects have yet to be rigorously proven. These changes in attitudes toward marijuana and CBD are discussed at greater length later in the chapter.

13.2.3 Nicotine and nicotinic receptors

Cigarette smoke contains over 1000 drugs, a large number of which are carcinogens (see Section 10.6.4). The primary psychoactive drug among them is nicotine. In the brain, nicotine acts to stimulate the cerebral cortex, possibly by a direct effect on the cortical neurons, which have a series of nicotinic receptors. Nicotine also acts by stimulating nicotinic receptors on the neurons in the sympathetic ganglia, releasing the neurotransmitters acetylcholine, epinephrine, and norepinephrine. In the brain, norepinephrine produces increased awareness. In the rest of the body, these neurotransmitters produce a variety of physiological effects: increased heart rate and blood pressure, constriction of blood vessels, and changes in carbohydrate and fat metabolism. Because of these various effects on the body, nicotine promotes cardiovascular disease and also carries reproductive risks. In addition to nicotine, cigarette smoke contains some as-yet-unidentified substance that decreases expression of the enzyme monoamine oxidase (MAO) in the brain. MAO breaks down dopamine, the so-called "pleasure hormone," so this effect contributes to the addictiveness of cigarettes. The highly addictive nature of nicotine and the serious health effects of cigarettes, as well as also other nicotine delivery devices, such as e-cigarettes, are discussed further later in this chapter.

13.2.4 Amphetamines: Agonists of norepinephrine

Methamphetamine and other amphetamines are examples of a type of drug called CNS stimulants. All amphetamines are derivatives of ephedrine, a drug originally obtained from the mah huang plant (*E. sinica*), that was mentioned earlier as a now-banned weight loss supplement. CNS stimulants increase behavioral activity by increasing the function of the reticular activating system. This is the portion of the brain that maintains a baseline level of neuronal activity in the brain as a whole, thus establishing a foundational set-point for wakefulness and awareness (see Section 11.3.3). CNS stimulants influence judgment and have side effects on organs outside the brain, including the heart and diaphragm; effects on the diaphragm in turn influence breathing (Figure 13.1). Consequently, they are dangerous drugs, accounting for 40% of all drug-related trips to the emergency room and 50% of all sudden deaths due to drugs.

Amphetamines mimic the effects of the neurotransmitter norepinephrine by binding to norepinephrine receptors. They can also indirectly increase norepinephrine activity by blocking its reuptake from the synapse and by inhibiting MAO, the enzyme that normally breaks down norepinephrine. Either mechanism results in more norepinephrine remaining in the synapse to act on the receiving (postsynaptic) neuron (see Figure 11.6). Prolonged use of high doses of amphetamines can induce a form of psychosis that includes aggressiveness, delusions, and hallucinations, possibly because of an oversupply of an enzyme involved in norepinephrine synthesis.

Methylphenidate (Ritalin®) is a controversial drug in this group. In most human subjects, it has mild amphetamine-like effects similar to those described earlier. However, the drug has quite the opposite effect (called a "paradoxical effect") in children who have attention-deficit hyperactive disorder (ADHD, formerly called ADD): In such individuals, it reduces their hyperactivity. The reason for this paradoxical effect seems to be related to the fact that hyperactive children actually have a lower-than-normal function in the reticular activating system, the area of the brain that keeps the rest of the brain alert. Such children may need to be constantly moving around to arouse their reticular activating system. Chemical stimulation of this area by Ritalin obviates the children's need for movement to maintain alertness. Ritalin is sometimes abused by teens and adults who do not need it to treat ADHD but use it to achieve a "high."

Another currently abused drug in this group is "ecstasy" (technically methylenedioxymethamphetamine or MDMA), a drug associated with many all-night "rave" dance parties because it suppresses feelings of fatigue. Ecstasy triggers the release of all stores of the neurotransmitter serotonin from brain neurons. It also blocks the reuptake of serotonin, thus producing prolonged alterations of sensory perceptions. In addition, one of the normal targets of serotonin is the hypothalamus, the part of the brain that regulates body temperature and thirst. Prolonged triggering of the hypothalamus can dehydrate the body and raise body temperature to damaging—even fatal—levels.

13.2.5 LSD: An agonist of serotonin

Lysergic acid diethylamide (LSD) is derived from the fungus *Claviceps purpurea*, which grows on rye. In contrast to ecstasy, which triggers release of serotonin itself, LSD and the related compound psilocybin (from mushrooms found in Central and South America) are structurally similar to serotonin. These drugs therefore act as serotonin agonists and activate the nerve cells that normally respond to serotonin, leading to altered sensory perception and hallucinations. "Altered perception" means a change in the awareness of stimuli that actually exist. Colors may become brighter or sounds clearer. Perceptions of the sizes of objects and of speed or time may be altered. Hallucination, in contrast, is the perception of things for which no outside physical stimuli have been received. Hallucinations can be visual, auditory, olfactory, or cognitive. Heavy use of these serotonin agonists leads to permanent brain damage, with symptoms ranging from impairments of memory, attention span, and abstract thinking to severe, long-lasting psychotic episodes.

13.2.6 Dissociative drugs: Glutamate antagonists

Dissociative drugs are another type of hallucinogen. They function as antagonists of the amino acid neurotransmitter glutamate, which stimulates N-methyl-D-aspartate (NMDA) receptors. This class of drugs includes PCP (phencyclidine), ketamine, and dextromethorphan (DXM). They alter pain perception, emotions, and a person's response to their environment. PCP was originally developed as a surgical anesthetic, but its medical use was discontinued due to its neurotoxic effects, and it is now only sold illegally as a street drug. Ketamine is mostly used as an anesthetic in animals and is also subject to abuse, mainly as a nightclub drug; however, it has also recently shown promise as a treatment for severe depression in patients who do not respond to other antidepressant drugs (Murrough et al., 2013). DXM is a common ingredient in cough syrups; it is harmless in small amounts, but its hallucinogenic potential may be exploited by individuals (usually teenagers) seeking dissociative experiences.

13.2.7 Caffeine: A general cellular stimulant

Not all psychoactive drugs produce their effects via action on CNS neurotransmitters and their receptors. Some can cross the cell membrane and produce effects within cells. One example of such a drug is caffeine, the most widely used CNS stimulant worldwide. Caffeine does bind to adenosine receptors on the cell surface, acting as an antagonist of those receptors and affecting the release of a number of neurotransmitters. In addition, it also enters cells and works from within to increase their rate of metabolism. Caffeine thus has a general stimulatory effect on cells throughout the body, including neurons in the brain. Like other CNS stimulants, caffeine increases the general level of awareness via action on the neurons of the reticular formation of the brain. In higher doses, or in more susceptible individuals, it can also produce insomnia (inability to sleep), anxiety, and irritability. It increases the heart rate, the respiratory rate, and the rate of excretion of urine by the kidney. It dilates peripheral blood vessels but constricts the blood vessels of the CNS, which produces headaches in some people at high concentrations.

Caffeine is derived from several types of plants, including the beans of the coffee plant (*Coffea arabica*), the leaves of the tea plant (*Thea sinensis* or *Camellia sinensis*), the seeds of the cocoa plant (*Theobroma cacao*, from which we get chocolate), and nuts from the kola tree (*Cola acuminata*, an African tree, the source of cola beverages). Table 13.2 shows the amounts of caffeine in different beverages and nonprescription medications.

Table 13.2 Caffeine content of beverages and nonprescription medications

Form of intake	Approximate caffeine dose (mg)
HOT BEVERAGES, PER 6-OZ. CUP (170 mL)	
Coffee, brewed	100–180
Coffee, instant	100–120
Tea, brewed from bag or leaves	35–90
Cocoa	5–50
Coffee, decaffeinated	2–4
CARBONATED BEVERAGES, PER 12-OZ. CAN (350 mL)	
Cola drinks and others	35–60
NONPRESCRIPTION (OTC) DRUGS, PER TABLET	
Stimulants	
Vivarin®, Caffedrine®	200
NoDoz®	100
Analgesics (pain relievers)	
Excedrin® extra strength	65
Midol® maximum strength	60
Anacin®, Bromo Seltzer®, Cope®, Emprin®	32

13.2.8 Alcohol: A CNS depressant

Ethyl alcohol belongs to a category of drugs called CNS depressants because they reduce the functioning of the CNS by inhibiting the transmission of signals in the reticular activating system. Also included in this category are barbiturates and tranquilizers, including Valium®, Xanax®, and Rohypnol®, the so-called "date rape" drug. All these drugs lower the general level of awareness and are therefore also called sedatives or hypnotics. Because they all affect the reticular activating system, when two or more CNS depressants are taken together, the effect is stronger (either additively or synergistically) than when either is used alone, and their combined use can be deadly. The actions of barbiturates and tranquilizers are mediated through receptors on neurons in the brain, whereas alcohol produces a more generalized effect because it acts by making all cell membranes more fluid.

Alcohol is soluble in both water and fat and is thus readily able to pass through the plasma membrane of the cells forming the blood-brain barrier. The portions of the brain affected depend on the dose: the higher the dose, the deeper into the brain the alcohol penetrates. Even low doses of alcohol can impair a person's response time, with devastating (often fatal) consequences if that person is driving a motor vehicle or boat. In the United States, alcohol is involved in over 60% of all motor vehicle fatalities and in over half of all drownings. The likelihood of sexual abuse is also greatly increased if alcohol has been consumed first.

Higher doses of alcohol suppress the reticular activating system's stimulation of those portions of the brain involved in involuntary processes, such as the brain stem and the medulla oblongata. Depression of the medulla can result in the cessation of breathing (respiratory arrest) and death. Each year, thousands of college students engage in "binge drinking"—consuming large quantities of alcohol in a short time. Binge drinking often results in coma ("passing out") and can result in death, including death from the alcohol directly, as well as death from falls or other accidents while intoxicated.

The effects of alcohol on behavior can be predicted by the blood alcohol level (Table 13.3). Blood alcohol level is measured as the number of grams of alcohol in each 100 mL of blood. Thus 1 g of alcohol per 100 mL equals a 1% blood alcohol content and 100 mg equals 0.1%. Since alcohol is evenly distributed throughout the body, the actual blood alcohol content that results from drinking a given amount of alcohol varies with the blood volume of the person, which is approximately proportional to the muscle weight of the person. A blood alcohol level of 0.1% is the level at which a person can be charged with driving while intoxicated

Table 13.3 Brain and behavioral effects of alcohol consumption

The greater the amounts of alcohol consumed, the deeper within the brain the alcohol penetrates after it is absorbed across the blood–brain barrier. Blood alcohol content is given for a 68 kg (150 lb) person; in general, persons weighing less will experience higher blood alcohol levels after consuming the same amounts.

Blood alcohol content (%)	Number of drinks*	Effects
0.01–0.04	1–2	▪ Slightly impaired judgment ▪ Lessening of inhibitions and restraints ▪ Alteration of mood
0.05–0.06	3–4	▪ Disrupted judgment ▪ Impaired muscle coordination ▪ Lessening of mental function
0.07–0.10	5–6	▪ Deeper areas of cortex affected ▪ Slower reaction time ▪ Exaggerated emotions ▪ Talkativeness or social withdrawal ▪ Mental impairment ▪ Visual impairment
0.11–0.16	7–8	▪ Cerebellum affected ▪ Staggering ▪ Slurred speech ▪ Blurred vision ▪ Greater impairment of judgment, coordination, and mental function
0.17–0.20	9–10	▪ Midbrain affected ▪ Inability to walk or do simple tasks ▪ Double vision ▪ Outbursts of emotion
0.21–0.39	11–15	▪ Lower brain affected ▪ Stupor and confusion ▪ Increased potential for violence ▪ Noncomprehension of events
0.40–0.50	16–25	▪ Activity of lower brain centers severely depressed ▪ Loss of consciousness ▪ Shock
0.51 or more	26 or more	▪ Failure of brain to regulate heart and breathing ▪ Coma and death

*Based on equivalent amounts of alcohol: 12 oz. of beer, 4 oz. of wine, or 1.25 oz. of liquor (Table 13.1).

(DWI) or operating [a motor vehicle] under the influence (OUI) of alcohol in most U.S. states, and several states have amended their laws to make the limit even lower. At a blood alcohol content of 0.05%, the probability of being involved in an automobile accident is 2–3 times higher than for someone who has not been drinking.

THOUGHT QUESTIONS

1. Does the finding that many drugs act directly on the cells of the brain to alter perception, mood, and consciousness necessarily mean that there is no "mind" or "spirit" apart from the physical entity of the brain?
2. Why is the use of certain drugs, such as nicotine, caffeine, and alcohol, more widely accepted than the use of other drugs, including marijuana? Are these differences based on the real levels of risk associated with these drugs?
3. Alcohol concentrations in the body are measured as "blood alcohol" levels. After alcohol is consumed, and before any of it is eliminated from the body, does all of it remain in the blood? Where else does it go?
4. In many countries the drinking age is the same as the driving age. In the United States, the driving age is lower than the drinking age. What kinds of impacts has this had in various societies?
5. How many drinks is someone likely to consume on a binge? Look at Table 13.3. What effects are likely to result from this number of drinks?

13.3 MOST PSYCHOACTIVE DRUGS ARE ADDICTIVE

Some drugs that have uses as medicines, and many others that do not, are also used socially. All cultures have used at least some drugs, particularly psychoactive drugs, for nonmedical purposes that can be described as social. For example, in many cultures, wine is a common accompaniment to food, and wine is also used in many religious ceremonies.

Excessive or harmful social use of a drug is considered **drug abuse** or **substance abuse**. Drug abuse is a major problem that affects many thousands of people, their families, and many other people with whom they interact. Most drugs that are abused socially are psychoactive drugs. These drugs have direct effects on the user as well as indirect effects on other people owing to the user's altered behavior.

If psychoactive drugs have been used by all cultures, why not approve their use? In most cultures in which the social use of psychoactive drugs has been endorsed by tradition, the uses have been highly ritualized or ceremonial, and the decision of how much drug to use is not left up to the individual. Psychoactive drugs such as alcohol impair a number of higher-order mental functions (Table 13.3). One of these is judgment, the very mental function needed for a person to be able to distinguish between use and abuse. In addition, most, but not all, drugs that are abused are addictive.

Addiction has been defined as a compulsive physiological and psychological need for a substance, implying that there is both a biological basis and a mental basis for addiction. However, as more psychologists have accepted the neurobiology paradigm that "mind equals brain" and that all brain functions are biochemically based, the distinction between physiological and psychological addiction has become increasingly blurred. The above definition of addiction carries with it the assumption that all addictive drugs are psychoactive. The number of people in the United States who use addictive drugs is given in **Figure 13.7**. Note that the data in Figure 13.7 report use of a given drug in the past month; therefore, not all users shown in the graph are necessarily addicted, but regular use of these substances does carry a significant risk of addiction.

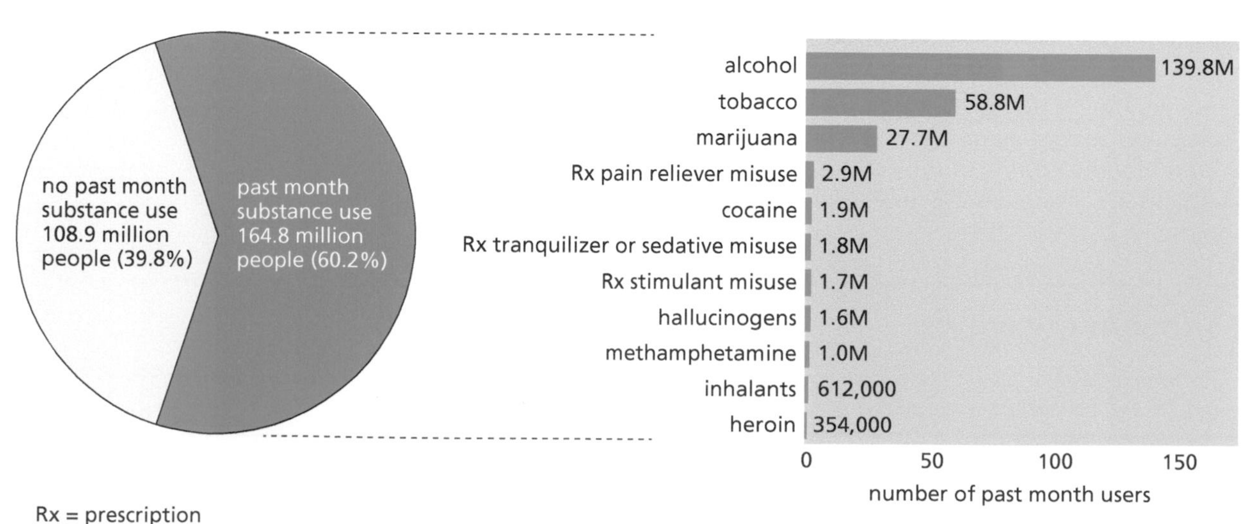

Figure 13.7 Users of addictive drugs in the United States in 2018. People are listed as users if they used the drug at least once in the past month. This figure does not include the most popular addictive drug, caffeine; a 2014 report estimated that 85% of Americans (approximately 271 million) used caffeine on a daily basis.

13.3.1 Dependence and withdrawal

Addictive drugs cause a physiological **dependence**, meaning that the person can no longer function normally without the drug. Once a person has become dependent on a drug, cessation of drug taking causes a person to feel physically unwell. These biological symptoms constitute the physical manifestation of **withdrawal**. The sensations felt during withdrawal tend to be the opposite of the sensations produced by taking the drug. For example, a person dependent on depressant drugs may become anxious and agitated during withdrawal, while people who have become dependent on painkillers feel pain when the drug is withdrawn, long after the original, biological source of the pain has been eliminated. The length of time required for the development of dependence varies with the drug, as does the severity of withdrawal. Dependence on morphine and related drugs develops very quickly; withdrawal begins within 48 hours of the last dose and lasts for about 10 days. People are very ill during this time, but withdrawal is rarely fatal. Withdrawal from alcohol dependence is physically much more severe and can sometimes be fatal. Anyone who regularly drinks coffee and has abstained for some period of time is likely familiar with the much milder but still obvious effects of caffeine withdrawal, the most common symptom of which is a headache but can also include cognitive and mood disruptions.

Another aspect of addiction is psychological dependence. This is characterized by a strong urge to use the drug that is difficult to ignore. Cocaine, caffeine, nicotine, and marijuana all produce greater psychological dependence than physiological dependence. Withdrawal from these drugs is dominated by symptoms that affect the physiology of the brain rather than the physiology of the entire body, and because of this they were originally thought to be nonaddictive. In light of the neurobiology paradigm, it is now known that all four are addictive. In fact, if we define the level of addiction as the degree of difficulty of getting through withdrawal without returning to the drug, nicotine must be considered as one of the most highly addictive drugs known.

One aspect of psychological drug dependence in humans involves the context—the places in which the drugs are taken and the people sharing the experience. "Conditioned withdrawal syndrome" refers to the fact that visual cues associated with drug taking (for example, seeing one of these people or places) can bring on the physiological symptoms of withdrawal in drug-dependent people even when the body is not actually in withdrawal, possibly leading them to seek the drug again. For this reason, many drug recovery programs recommend that recovering addicts stay away from specific people and locations associated with previous drug use.

13.3.2 Brain reward centers and drug-seeking behaviors

Neuroscientists have discovered both negative and positive reward, or reinforcement, systems in the brain. Certain things make us feel good (positive reward, or positive reinforcement); others make us feel bad, and their removal or avoidance constitutes a negative reward, or negative reinforcement. Negative reinforcement is associated with avoidance of a negative consequence or removal of an unpleasant stimulus. Stimulation of the nerves in the positive reinforcement system leads to a repetition of the behavior. Basic biological functions like eating and sexual activity are repeated because they activate these nerves. Throughout our lives we learn other experiences that stimulate these centers. Our ability to enjoy certain experiences, such as the feeling of "a job well done" or the feelings evoked by a beautiful painting or a sunset, is "learned" to a large extent; our families, religions, cultures, and other influences operate from birth to teach us to view certain experiences as positive and certain experiences as negative.

An important distinction should be made between "pleasure" and "satisfaction." A person who has enjoyed a good meal or a restful night's sleep feels *satisfied* and may say, "That was great, but I've had enough and do not need any more." The brain centers implicated in addictive behaviors produce a very different sensation, *pleasure*, characterized as "That was great, and I want even more." Pleasure-seeking behavior may lead to addiction, but behavior that brings satisfaction does not.

13.3.2.1 Two hypotheses of drug addiction One hypothesis of drug addiction is that people use drugs to escape from some kind of pain, either physical or psychological. This hypothesis suggests that drug-taking behavior results from the attempt to inhibit or avoid the negative consequences associated with that pain and then, upon becoming dependent, to avoid the unpleasant symptoms of withdrawal. A newer hypothesis suggests that drug addiction results from stimulation of the positive reinforcement centers. Of the psychoactive drugs, the ones that cause addiction most rapidly are those that stimulate the positive reward system in the brain. They directly stimulate these centers, neurochemically producing the positive sensation. This hypothesis is supported by the findings that people may become addicted well before physiological dependence (and therefore the possibility of withdrawal) has begun. The two hypotheses may be valid for different drugs or different individuals, meaning that there may be more than one mechanism for addiction.

13.3.2.2 Evidence from behavioral experiments with rats Much of the research on drug-seeking behavior has been performed using rats. Rats have been fitted with tubes so that they can give themselves drugs, either into their bodies or directly into specific parts of their brains. These rats quickly learn to self-administer addictive drugs, and they increase the frequency of self-administration if allowed. They do not, however, self-administer nonaddictive drugs. Although physiological dependence on the addictive drugs does develop, the experiments suggest that dependence is the result of addiction rather than its cause.

A part of the brain stem known as the ventral tegmental area (VTA) (Figure 13.8) is thought to be the positive reinforcement center, or "pleasure center." This has been demonstrated in several types of experiments. If electrodes are placed into the VTA, rats will activate the electrodes, electrically stimulating that area of the brain. They quickly learn to repeat the behavior, giving themselves repeated electrical stimulation, even preferring it over food. The same type of experiment has been done on rhesus monkeys, with the same results: They refuse food if their choice is between eating and stimulating their VTAs electrically (even after starvation). Electrodes placed in other areas do not produce repeated self-stimulation, and destruction of the VTA of the brain stops the behavior.

In another type of experiment on rats, tubes are placed into an area of the brain so that, by pushing a lever, the rat can administer drugs directly to the area. By their action on neurons or their receptors, psychoactive drugs stimulate nerve impulses in the brain, producing the same effects elicited by electrical stimulation of those neurons. The negative reinforcement centers are in the periventricular areas of the brain, while the positive reinforcement center, as previously stated, is in the VTA. The nerves of the VTA make synaptic connections to the nerves of another brain area, the nucleus accumbens (NA), which is involved in the processing or interpretation of the signal (Figure 13.8). Self-administration of drug to the negative reward center is not reinforcing and is therefore not repeated, while self-administration to the VTA or the NA is.

When rats are able to take drugs by a more normal route, rather than directly into their brains, the results are similar: They repeat the drug taking if the drug is addictive and do not when the drug is nonaddictive. Measurements of changes in electrical activity in different parts of the CNS have shown that activity increases in the VTA (and generally nowhere else) after the administration of an addictive drug.

The VTA of the brain includes the reticular formation, the central part of the reticular activating system described earlier. Many of the drugs that act on the reticular activating system also act on the positive reinforcement area of the VTA. Amphetamines indirectly stimulate the neurons of the VTA, elevating mood. For this reason, they have been successful in the treatment of depression. Cocaine acts on brain cells of the VTA that secrete dopamine. Most researchers think that the euphoria produced by cocaine is due to its effects on the dopamine-secreting cells because the euphoria can be stopped with drugs that block dopamine receptors (Figure 13.8). Opiates, marijuana, caffeine, and alcohol all produce ventral tegmental, self-reinforcing effects.

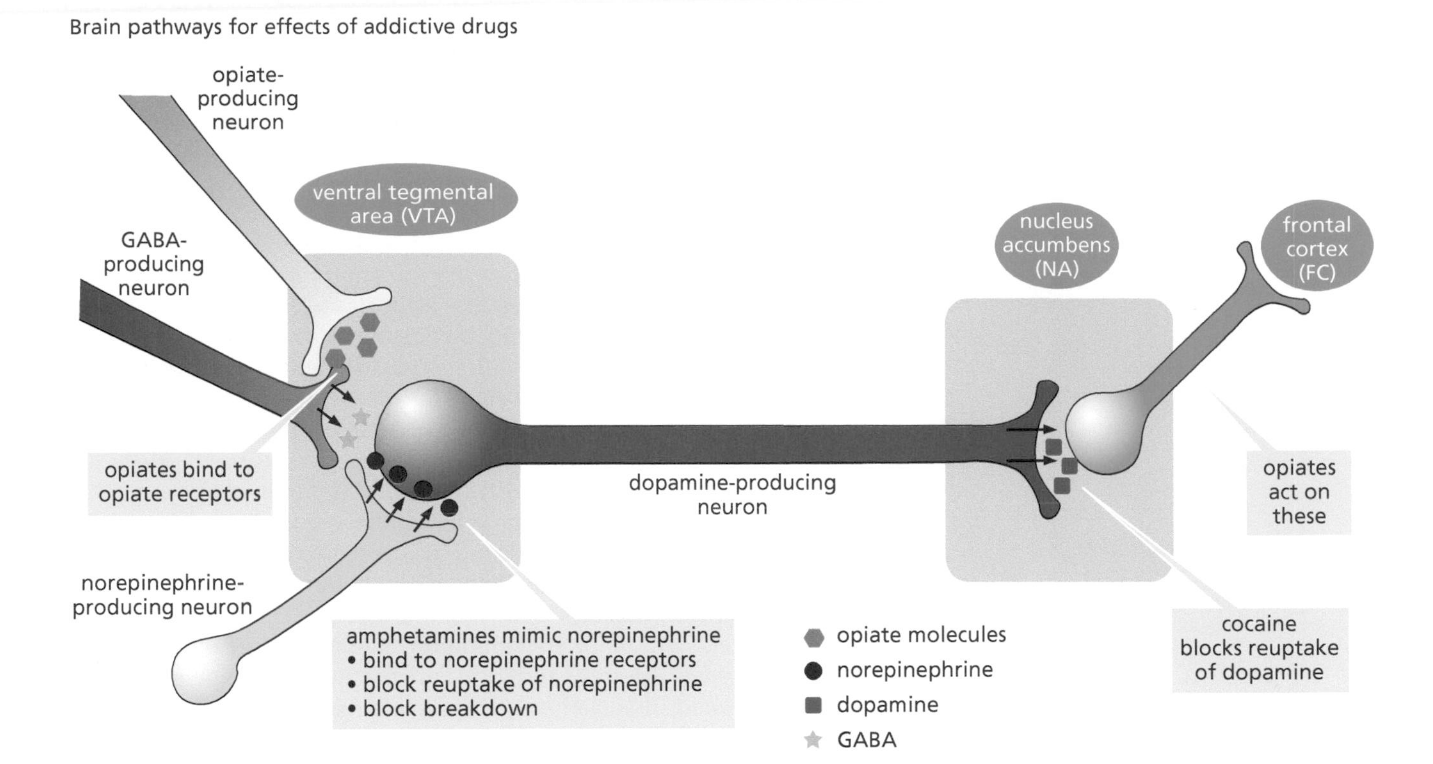

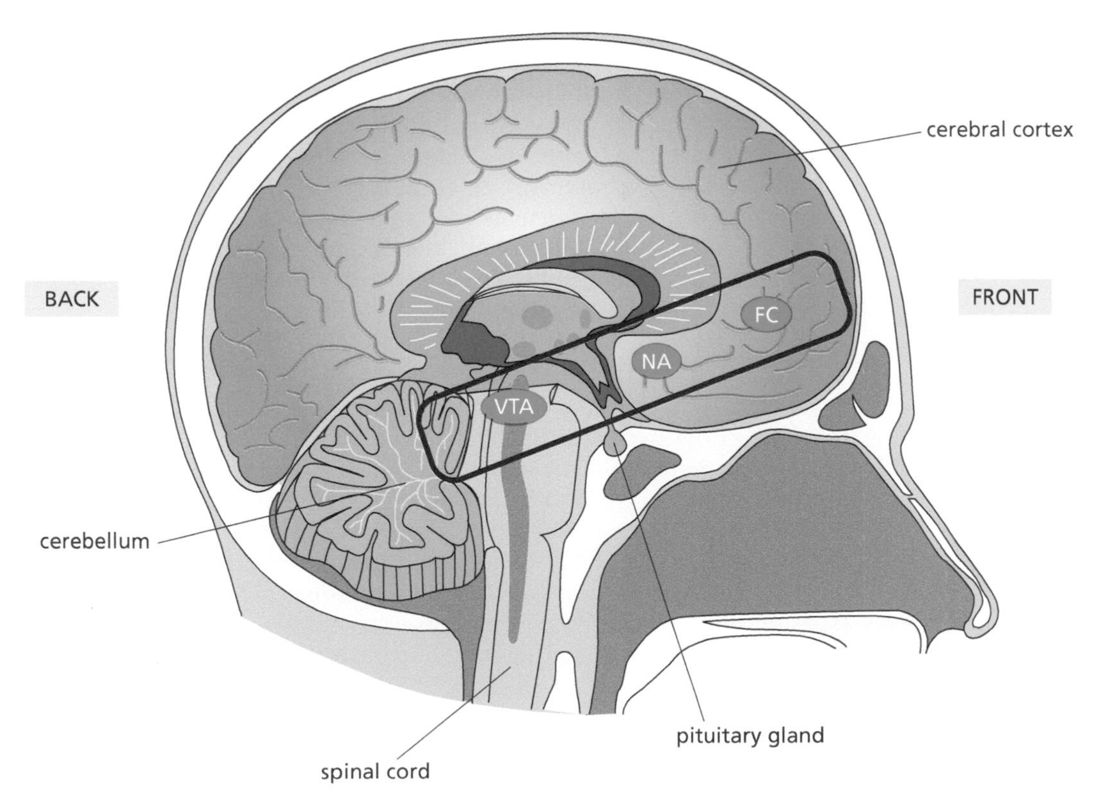

Figure 13.8 Positive reinforcement areas of the brain and the locations of action of addictive drugs on synapses in the cells of these areas.

Not all psychoactive drugs are addictive. Hallucinogens, for example, do not produce repeated self-administration by rats and therefore are not considered addictive under this hypothesis. Nicotine does not initially produce self-administration in rats; in fact, initially it is strongly aversive. However, after a rat has been exposed to nicotine several times, it begins to self-administer the drug, and this becomes a strongly persistent behavior that is hard to stop.

13.3.2.3 Conditioned learning in drug addiction Addiction is both a biological response and a learned behavioral response in which the behavior being learned is the drug-seeking and drug-taking behavior. In this type of learning (called operant conditioning), behavior is learned as a result of its consequences. Drug seeking and drug taking can be learned this way if taking the drug is usually associated with stimulation of the positive reinforcement centers in the brain. As mentioned previously, contextual cues are important. As is true in classical conditioning, one stimulus can become associated with another (see Section 11.3.1). Being in certain places or being with people with whom a person has taken drugs provide strong learned cues that bring on a physical sensation of craving for the drug. Thinking of those places or longing for those people brings on a craving for the drug and can also bring on the sensations produced by the drug itself. Seeing or thinking about aspects of the drug taking itself is often sufficient to bring on these feelings. A person dependent on cocaine reported that the sight of someone wearing a gold watch would bring on the sensations of a cocaine high because gold watches were among the items he had often stolen to purchase his cocaine. The contribution of the "drug culture" to the reinforcement of drug-taking behavior is enormous: Many of the symbols and rituals associated with the culture become contextual cues. Attempts to block and reverse drug dependence must consider these learned associations, which are called "conditioned place preferences." The rehabilitation of drug addicts is usually very difficult because of the strength of these learned associations. However, the chances of successful rehabilitation are increased if the addict can be helped to develop an aversion to the old behaviors while learning to substitute new behaviors.

13.3.2.4 Addictive behaviors Some pleasure-seeking behaviors can produce addiction without exogenous drugs. Compulsive gambling, for example, is an addictive behavior now recognized as a clinical diagnosis, gambling disorder, by the American Psychiatric Association in their latest manual of mental disorders (DSM-5, 2013). Neurobiology research on this disorder shows a remarkable similarity to drug addiction in its effects on the same brain reward centers, neural pathways, neurotransmitters, and membrane receptors (Potenza, 2013; Heinz et al., 2019). In particular, the VTA is affected in the manner as for drug addiction and shows the same type of conditioned learning.

13.3.2.5 Effects of long-term use One of the effects of long-term use of psychoactive drugs is that they erase the ability of the ventral tegmental nerves to respond to the normal positive signals of satisfaction: Appreciation of a good meal, enjoyment of the company of friends, and happiness from helping others may all disappear. We tend to interpret positive experiences as "pleasure," so the positive reinforcement centers have sometimes been called the "pleasure centers." Drugs compete with the normal neurotransmitters in these brain centers. Long-term use of addictive drugs decreases the number of receptors on the nerve cells (see the next paragraph) so that these centers are only triggered in the drug-abusing person by taking drugs and no longer by the experiences that used to be pleasurable.

13.3.3 Drug tolerance

One of the biological effects of addictive drugs is that they produce **tolerance**, which is also called "homeostatic compensation" (i.e., the body adjusting to new conditions). This means that the same dose of drug exerts a decreased effect when administered repeatedly and, therefore, greater amounts of the drug must be taken to produce the same effect. Some well-meaning people have suggested that addicted people be given "all the drugs they want" to keep them off the street. The simple biological reason why such an approach would not work is that drug tolerance develops. A person needs more and more drug to produce the original

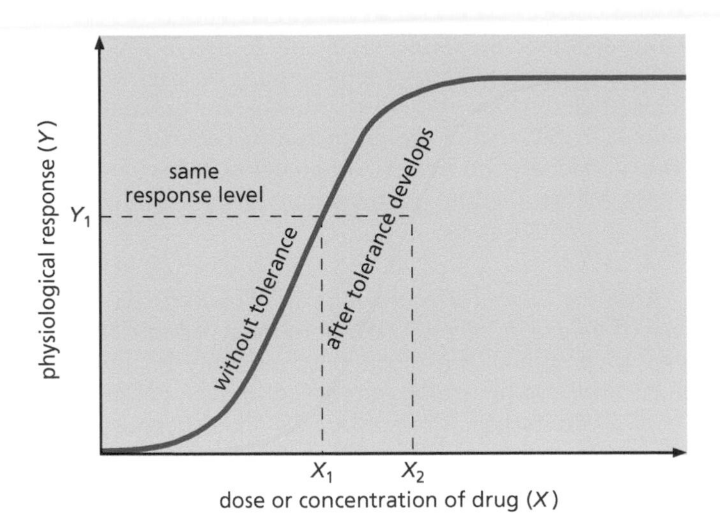

Figure 13.9 Effect of drug tolerance on the dose–response curve. A higher dose (X_2) is required to produce the same response after drug tolerance develops.

psychoactive effect. Higher doses affect other body systems and begin to produce negative mental states such as hostility and paranoia. The increasing doses may often become toxic or lethal.

Drug tolerance can be shown graphically as a shift of the drug's dose–response curve to the right (**Figure 13.9**). There are two broad categories of mechanisms that produce tolerance: metabolic and cellular. Metabolic tolerance is the body's production of an increased amount of the enzymes that break down the drug. Tolerance to barbiturates develops at least in part because the drug stimulates the synthesis of the liver enzymes responsible for its elimination, so the rate of elimination increases with repeated use.

Cellular tolerance results from changes in the receptors for the drug, principally the receptors on the nerve cells in the brain. Tolerance results from either a drug-induced decrease in the number of receptors or an increase in the threshold needed to trigger a response. Heroin produces these changes in the brain within a week or two of daily use. Some drugs can cause permanent damage to receptors and, therefore, tolerance to these drugs becomes permanent. For most drugs, however, tolerance is not permanent but disappears gradually with time. The time period varies greatly from one drug to another.

So far, we have emphasized the effects of psychoactive drugs on the brain. However, these drugs also have effects throughout the body, as we see in the next section.

 THOUGHT QUESTIONS

1. If a person has never learned to feel pleasure or satisfaction from daily activities, will they be more attracted to the artificial pleasure offered by drugs? Can this suggest anything to us about drug prevention strategies? One slogan says that "Hugs are better than drugs." Does this slogan correlate with neurobiological findings? Can this idea be applied in drug prevention programs?
2. Because psychoactive drugs work directly on the brain cells, are individuals exempted from responsibility for their own drug use? (Can a drug user justify their behavior by saying "the drug made me do it"?) Apply this reasoning to alcohol and tobacco as well as illegal drugs.

13.4 DRUG ABUSE IMPAIRS HEALTH

In addition to addiction, the use of psychoactive drugs can impair the health of individual drug users in other ways. It can also affect their gametes, and thus the development of their children.

13.4.1 Drug effects on the health of users

Drug abuse negatively affects the health of individuals and society. In the United States, deaths from drug overdoses rose steadily from under 20,000 in 1999 to over 70,000 in 2017. In recent years, the vast majority of these overdose deaths have been caused by synthetic narcotics, especially fentanyl, which is 50–100 times more potent than morphine, from which heroin is derived. While synthetic and natural opioids (including heroin) are responsible for the majority of drug overdoses (about 68% in 2017), other drugs, including cocaine, methamphetamines, tranquilizers (e.g., Valium and Xanax), and alcohol, also cause significant numbers of overdose deaths. In addition, many more deaths result from health conditions caused by chronic drug use.

13.4.1.1 Alcohol Among the most harmful of drugs, particularly in view of the frequency of its use, is alcohol. While other types of CNS depressants are designated as a controlled substance that can only be obtained legally via prescription, alcohol can be readily purchased in most countries as long as the purchaser is of a certain age. The reasons for the difference are certainly not biological, because ethyl alcohol is a powerful depressant comparable to the others in terms of both short-term and long-term risks to health.

Most of the biological effects of alcohol on the body are **acute effects**, reversible in a matter of hours or days. Over time, however, permanent damage results from the **chronic effects**. Alcohol-induced biochemical imbalances permanently damage tissues such as the brain, liver, and muscle tissue (including the heart), resulting in dementia, cirrhosis and other liver diseases, and cardiovascular disease, respectively. Tissue damage may in turn contribute to altered uptake of and decreased sensitivity to medically necessary drugs, including antibiotics, antidiabetic drugs, and other medications. Alcohol in the gut also destroys certain vitamins and interferes with the absorption of others. This is why vitamin deficiencies rarely seen in industrial countries occur in those countries among alcoholics.

Alcohol depresses the immune system, leaving alcoholics very susceptible to infectious disease, including tuberculosis. Because alcoholics frequently substitute alcohol for food, they are often malnourished either in total calories or in micronutrients, which can further depress the immune system (see Chapter 14). Alcohol consumption is also associated with an increased risk for cancer, heart disease, and high blood pressure. The Centers for Disease Control and Prevention (CDC) in the United States defines excessive alcohol use as more than 8 drinks per week for women and more than 15 drinks per week for men. Many individuals who are not alcoholics consume these quantities of alcohol regularly, and the full extent of the health effects of such consumption are not well characterized; however, abstaining from or strongly limiting alcohol intake is likely to reduce one's risk of many common health conditions.

13.4.1.2 Caffeine Caffeine, which is even more widely used than alcohol, can have significant negative effects on health if consumed in large quantities. The consumption of more than 10 cups of coffee a day increases chromosome damage (which can lead to birth defects), respiratory difficulties, and heart and circulatory problems.

13.4.1.3 Tobacco and nicotine Tobacco is carcinogenic in any form. Smoking tobacco is strongly associated with lung cancer, whereas chewing tobacco is associated with cancer in the mouth, a form of cancer that often metastasizes (spreads) to other parts of the body. Tobacco and alcohol together produce a synergistic increase in the number of cancer deaths, beyond what would produce without the other (see Figure 10.27, p. 412). Nicotine, the psychoactive component of tobacco, suppresses the immune system; consequently, smokers have a high incidence of other lung diseases, including emphysema, COVID-19, and respiratory tract infections such as pneumonia and bronchitis.

The negative health effects of smoking are not limited to smokers themselves. Passive smoke (smoke produced by someone else's smoking activity) has been shown to increase cancer risk in adults and to increase deaths from sudden infant death syndrome; it also increases the incidence of pneumonia and bronchitis in the first year of an infant's life. Because an infant's immune system is only partly developed, passive smoke can be much more damaging to infants than to adults.

In addition to direct effects on immune cells, tobacco smoke (including passive smoke) paralyzes the cilia on the lining of the respiratory tract. When we breathe, dust, bacteria, and other particles enter our respiratory tract along with the air. Many of these particles are trapped on a sticky fluid (mucus) that coats the respiratory lining. The cells lining the upper respiratory tract have many hairlike projections (cilia) on their surfaces. These cilia beat rhythmically in a coordinated way; the beating of the cilia moves the mucus and its trapped material upward, out of the respiratory tract. Tobacco smoke stops the beating of the cilia; inhaled bacteria and viruses consequently work their way down into the lungs instead of being eliminated. People who smoke therefore have a much higher incidence of respiratory tract infections and other infectious diseases of the lungs than do nonsmokers. The particulate matter from smoke, which can include cancer-causing chemicals in highly concentrated form, also works its way into the lungs and begins the process of transformation of normal cells into cancer cells (see Section 10.6.4). Smokers therefore have much higher rates of lung cancer than nonsmokers.

For many years, the tobacco industry in the United States suppressed evidence of the very strong link between cigarette smoking and cancer, along with other negative health impacts of tobacco use. In 1998, the major U.S. tobacco companies agreed to a settlement of lawsuits filed by the attorneys general of the majority of U.S. states, referred to as the Tobacco Master Settlement Agreement. The companies agreed to pay $246 billion to states over 25 years, with the intention that the money would fund smoking cessation, education, and other public health programs. The companies also promised not to pursue marketing strategies directed at youths. Adult cigarette smoking has steadily declined in the United States from a high point of 42.4% of adults smoking in 1965 to 14% in 2017. Youth smoking rates (partly influenced by changes in cigarette prices) increased to a peak of 36.4% of youths reporting smoking in 1997 to 8.8% by 2017. Since the 1990s, many laws were enacted banning smoking in public places such as restaurants and bars. Smoking was also banned in many workplaces, universities, and public transportation systems.

Given the highly addictive nature of nicotine, several companies saw a niche for a new nicotine delivery vehicle: e-cigarettes, in which volatile liquids are heated to the point where they give off vapors containing nicotine and various other products. Since e-cigarettes, patented in 2003, were introduced to the North American market in 2007, the use of e-cigarettes, or vaping, has steadily increased, especially among people aged 25 and under. Several companies marketed vaping products containing flavors that appeal to younger users.

Because hot vapors are inhaled, e-cigarette use can dry out and damage lung membranes. Vaping is associated with an increased risk of several respiratory diseases, including chronic bronchitis and emphysema (two forms of chronic obstructive pulmonary disease [COPD]), and also asthma. In 2019, the many reported cases of deadly lung damage caused by vaping prompted public health agencies to act, and, in 2020, a federal ban on the sale of all flavored vaping products was enacted in the United States (although nonflavored products and those containing menthol remain available for purchase by adults in most states). There are still many uncertainties surrounding the risks of vaping, including those that may be caused by the nicotine itself or by other chemicals in the vape liquids. A further complication is the rise of vaping products containing THC, the primary psychoactive ingredient in marijuana (CDC, 2020). Several studies identified vitamin E acetate, a thickening agent used in some vaping products, especially those containing THC, as the main culprit in the acute lung injuries associated with the 2019 outbreak (Blount et al., 2020).

13.4.1.4 Marijuana In the United States, federal law categorizes marijuana as a Schedule I drug, which is defined as an illegal substance with no accepted medical use that has a high potential to cause addiction. Other Schedule I drugs include heroin, cocaine, LSD, PCP, and ecstasy. It is known that drug tolerance does develop from regular marijuana use, as does physiological dependence in some heavy users. However, since the mid-1990s, there has been growing interest in the possible medicinal uses of marijuana; more recently, its acceptance for recreational (nonmedical) use has also been growing.

California was the first state to legalize medical use of marijuana in 1996, and now 36 states (as of 2021) allow its use under certain conditions, although the federal ban on all marijuana use remains in place. In 2018, the U.S. FDA approved CBD, a nonpsychoactive component of marijuana, as a treatment for two severe forms of epilepsy. Many states that have not legalized medical use of marijuana itself do allow prescription-based use of CBD to treat epilepsy; some also allow use of CBD for other medical conditions, but such laws in these states vary widely. Apart from the FDA-sanctioned use of CBD to treat epilepsy, medicinal uses for marijuana for which there is good scientific evidence include the stimulation of appetite and suppression of nausea in cancer patients undergoing chemotherapy, the stimulation of appetite in AIDS patients, reduction of anxiety in patients with posttraumatic stress syndrome, and reduction of pain, especially that caused by nerve damage (Nat. Acad. Sci., 2017). Marijuana and CBD for pain reduction have attracted particular attention as safer, less addictive alternatives to opioid medications. Marijuana and CBD have also been suggested to have potential therapeutic benefits for many other conditions, including glaucoma, multiple sclerosis, anorexia, irritable bowel syndrome, and neurodegenerative diseases such as Parkinson's, Alzheimer's, and Huntington's diseases. Cannabinoids have even shown promising antitumor effects, especially in brain and breast cancers. The degree to which existing clinical evidence validates these claims varies. All of the potential therapeutic effects are thought to involve the body's endocannabinoid system, which is poorly understood. Historically, research concerning the physiological effects of marijuana was limited by the lack of funding and the extensive regulatory requirements that had to be satisfied due to its classification as a Schedule I drug.

Hemp is a term for *Cannabis* that contains less than 0.3% THC; it contains a variety of other nonpsychoactive cannabinoids. Hemp was previously regulated the same way as marijuana, but in December 2018, passage of a Farm Bill in the United States legalized agricultural production of hemp for the purpose of CBD extraction. Almost overnight, this spawned a new industry that has grown exponentially—an estimated 700% in 2019 alone. CBD is offered in many forms, including oils, creams, teas, pills, and gummy candies, and the health-related marketing claims made by the companies that sell them are extensive. Currently, clinical research lags far behind the marketing claims made by purveyors of CBD (Eisenstein, 2019), and there is no real regulation of the industry to verify the composition of the products. Possible side effects are also not well characterized. Thus, use of any CBD product carries the same risks as the use of nutritional supplements, discussed earlier in this chapter.

While research into the physiology of the endocannabinoid system and the possible therapeutic benefits of cannabinoids is at a quite early stage, we do know that there are harmful effects of marijuana, particularly when it is smoked (Volkow et al., 2014). Much of the particulate matter in marijuana smoke stays in the lungs and builds up to form tar (Taskin, 2013). Marijuana smoke produces more tar per weight of plant material than does tobacco smoke, and the tar is equally carcinogenic. It also inhibits the immune cells that clear debris from the lungs and protect against airborne infectious bacteria and viruses. Given the lung-related effects of smoking marijuana, alternative methods of delivery are preferred for its therapeutic use; these may include ingestion in food products, absorption of liquids placed under the tongue, or application to the skin via creams.

Numerous studies in recent years have highlighted both the acute (short-term) and chronic (long-term) effects of marijuana (Nat. Acad. of Sciences, 2017; Joy, Watson, and Benson, 1999; Sohn, 2019). In addition to the lung-related risks of smoking marijuana, all forms of the drug alter the production of reproductive hormones, decreasing the production of sperm in men and ovulation in women. Men who use marijuana over long periods of time often develop fatty enlargement of the breasts (gynecomastia). Mental illnesses (psychoses) are diagnosed five times as often in marijuana users as in nonusers (Sohn, 2019). Several studies have found associations between marijuana use and a decline in cognitive abilities, particularly when it is used in adolescence and early adulthood (up to age 25), when the brain is still developing (Meier et al., 2012; Nader and Sanchez, 2018; Burggren et al., 2019).

Two of the largest and most comprehensive long-term studies of marijuana use come from New Zealand (Poulton, Moffitt, and Silva, 2015; Poulton et al., 2020). Both are prospective longitudinal cohort studies, meaning that groups of marijuana users vs nonusers were distinguished at the start (prospectively) and were followed and assessed for several decades. These studies found that chronic use beginning in adolescence correlated with a 6- to 8-point lowering of mid-adulthood IQ, and other studies have suggested that marijuana use early in life can hasten memory loss as people age. Chronic marijuana use has also been found to correlate with lowered academic performance (e.g., high school graduation rates) and higher rates of motor vehicle accidents and deaths (Poulton et al., 2020). Perlson (2020) has criticized these studies because of the presence of many confounding variables (like income or social status), but Poulton et al. (2020) insist that they have controlled statistically for possible confounding effects, and the results have also been supported by similar studies in Germany, Australia, and the Netherlands.

Despite the health risks just mentioned, proven medicinal benefits of marijuana have led most states to legalize its medical use, as previously stated. In addition, social attitudes toward the recreational use of marijuana have also been evolving. In recent years, a number of U.S. states have either decriminalized the recreational use of marijuana (substituting a fine rather than jail time for possession of small amounts of the drug) or actually legalized its use and sale for nonmedical purposes (despite the federal ban still existing). Colorado and Washington were the first states to legalize recreational marijuana in 2012. As of mid-2020, 11 states and the District of Columbia had legalized recreational use of marijuana (with some restrictions), and 9 of these states had set up legal markets for its taxable sale. Sixteen additional states have implemented a decriminalization policy. Outside the United States, marijuana use is still illegal in most of Asia and Africa, while many countries in Europe and Latin America have begun to relax enforcement of antimarijuana laws or to allow some medical marijuana uses, and adult use is now legal in Uruguay, Canada, and South Africa.

In the past, a major argument against marijuana legalization was that it could function as a "gateway drug," increasing the chance that a person will try more dangerous drugs, such as heroin or cocaine. However, epidemiological research has shown that alcohol and tobacco, both legal, are more likely to play the role of gateway substances than is marijuana. Arguments in favor of marijuana legalization include the following: that it is already widely used and, overall, has less negative public health consequences than alcohol and tobacco; that enforcement of marijuana laws has overwhelmed the criminal justice system and has disproportionally affected nonwhite, often low-income individuals, who are otherwise law-abiding; and that it represents a significant potential for tax revenues to help states facing budget shortages. Most studies have shown that recreational marijuana legalization has led to a small increase in the number of adult users but has not increased teen use.

One issue related to marijuana legalization is the need for accurate tests to detect impaired drivers. Under certain conditions, marijuana use does produce an intoxication that impairs driving ability and results in an increased risk of auto accidents. However, this intoxication is poorly understood and does not correlate with commonly measured blood concentrations of THC or its metabolic breakdown products (Gilman et al., 2022). While police officers can readily administer breathalyzer tests to determine if a driver is intoxicated with alcohol, similarly reliable tests for marijuana are not yet available; an article by Gilman et al. (2022) discusses one possible approach to developing such a test.

13.4.1.5 Designer drugs The term "designer drugs" refers to those drugs that are slight structural alterations of existing drugs. Designer drugs are often made to circumvent laws that ban particular drugs by name. New designer drugs may be legal until laws are rewritten to cover them, but they can be dangerous, nevertheless.

MPTP and MPPP are two designer derivatives of Demerol (meperidine), itself a derivative of opium. These two have psychoactive properties similar to those of other opiates but are also potent neurotoxins (nerve cell poisons). They destroy the nerve cells in the area of the brain that controls movement, causing movement defects similar to those found in Parkinson's disease, a condition that otherwise mostly affects people over age 50 (see Section 11.1.4). In 1985, 400 cases of Parkinsonism in young people were found to be due to MPTP. After these cases, MPTP and MPPP were made illegal in the United States.

13.4.1.6 Prescription painkillers In 2016, an estimated 100 million people in the United States suffered from pain; of these, 9–12 million experienced long-lasting, chronic pain. The medical management of pain has long been influenced by social attitudes regarding drug addiction. Historically, because of the dangers of addiction, there was a reluctance to prescribe opioid painkillers such as morphine. Beginning in the 1990s, however, there was a shift toward a greater willingness to prescribe such medicines to alleviate the pain experienced by patients, including those with terminal illnesses but also those with short-term pain caused by injuries or accidents. Around this time, the medical establishment designated pain as the "fifth vital sign" (along with the four traditional vital signs of temperature, heart rate, blood pressure, and respiratory rate), giving it much greater prominence in patient care. Pharmaceutical companies developed new formulations of opioid medications that were designed to release their active ingredient slowly over a period of time. The most prominent of these drugs is oxycodone (trade name OxyContin®, produced by Purdue Pharma). Drug manufacturers claimed that these new formulations were safe and unlikely to result in addiction when used for pain management in persons who never previously abused drugs. Purdue Pharma and other companies aggressively marketed these drugs, offering financial incentives to physicians who wrote more prescriptions for them. Prescriptions for opioid medications increased nearly 300% from 1991 to 2013. By the early 2000s, it was clear that the new-generation opioid drugs were not nearly as safe as was claimed. Many patients who were not drug addicts previously did become addicted, and drug overdose deaths from opioids increased dramatically. The initial wave of overdose deaths from prescription opioids triggered two subsequent waves of overdose deaths: the first from heroin, beginning in 2010, and the second from synthetic opioids such as fentanyl, beginning in 2013. Patients who became addicted to prescription opioids often resorted to "doctor shopping," a pattern of visiting multiple providers to obtain additional prescriptions. Prescription drug monitoring programs were developed to prevent this behavior, which caused many addicted patients to turn to the illegal drug market. The opioid crisis was declared a national emergency in the United States in 2018, and the attorneys general of many states pursued lawsuits against drug companies that manufactured opioid pain medicines. In 2019, Purdue Pharma agreed to a more than $10 billion settlement agreement, which resulted in its bankruptcy and subsequent break-up, acknowledging that they mislead the public regarding the addiction risk of OxyContin®. Total drug overdose deaths in the United States decreased for the first time from 2017 to 2018 but again increased in 2019 to a new high, fueled mainly by deaths from fentanyl and other synthetic opioids manufactured and sold illegally. Thus, the opioid epidemic remains an extremely serious public health problem.

13.4.1.7 Steroids and other performance-enhancing drugs Although most commonly abused drugs are psychoactive drugs, there are some exceptions. Anabolic steroids, such as testosterone, are abused because of their hormonal effects on physical development. Certain athletes of both sexes (body builders, weight lifters, swimmers, runners, football players, and others) have used these drugs because they cause an increase in muscle mass, resulting in a bulkier and more powerful physique. These drugs are not addictive, but their many dangerous side effects include damage to the reproductive organs and the circulatory system, especially the heart, as well as increased hair in some places and premature baldness in others. Deaths have occurred among amateur and professional athletes from the abuse of steroids. In addition to steroids, a number of other substances, including human growth hormone (HGH) and the red blood cell development factor erythropoietin (EPO), have been abused by athletes seeking performance-enhancing effects. These performance enhancers are mostly hormones that occur naturally in the body, but only in very small amounts;

in contrast, the larger doses used by those seeking to enhance their athletic performance have significant side effects on many different body systems. The fact that these "drugs" are naturally present in humans can complicate the job of regulatory agencies seeking to enforce fairness in sports, as some individuals in the population will naturally have larger amounts of them than others, making it difficult to prove the intentional taking of a performance-enhancing drug.

13.4.2 Drug effects on embryonic and fetal development

Aside from the biological effects on the person taking a drug, there are many effects on developing embryos. Many drugs can affect fetuses in utero. Additional harm can result from damaged gametes from either a mother or a father who has used drugs.

The placenta in pregnant women is very rich in capillaries. The mother's blood does not circulate directly into the fetus, but maternal blood vessels in the placenta are close to the fetal blood supply (Figure 13.10). This close association of

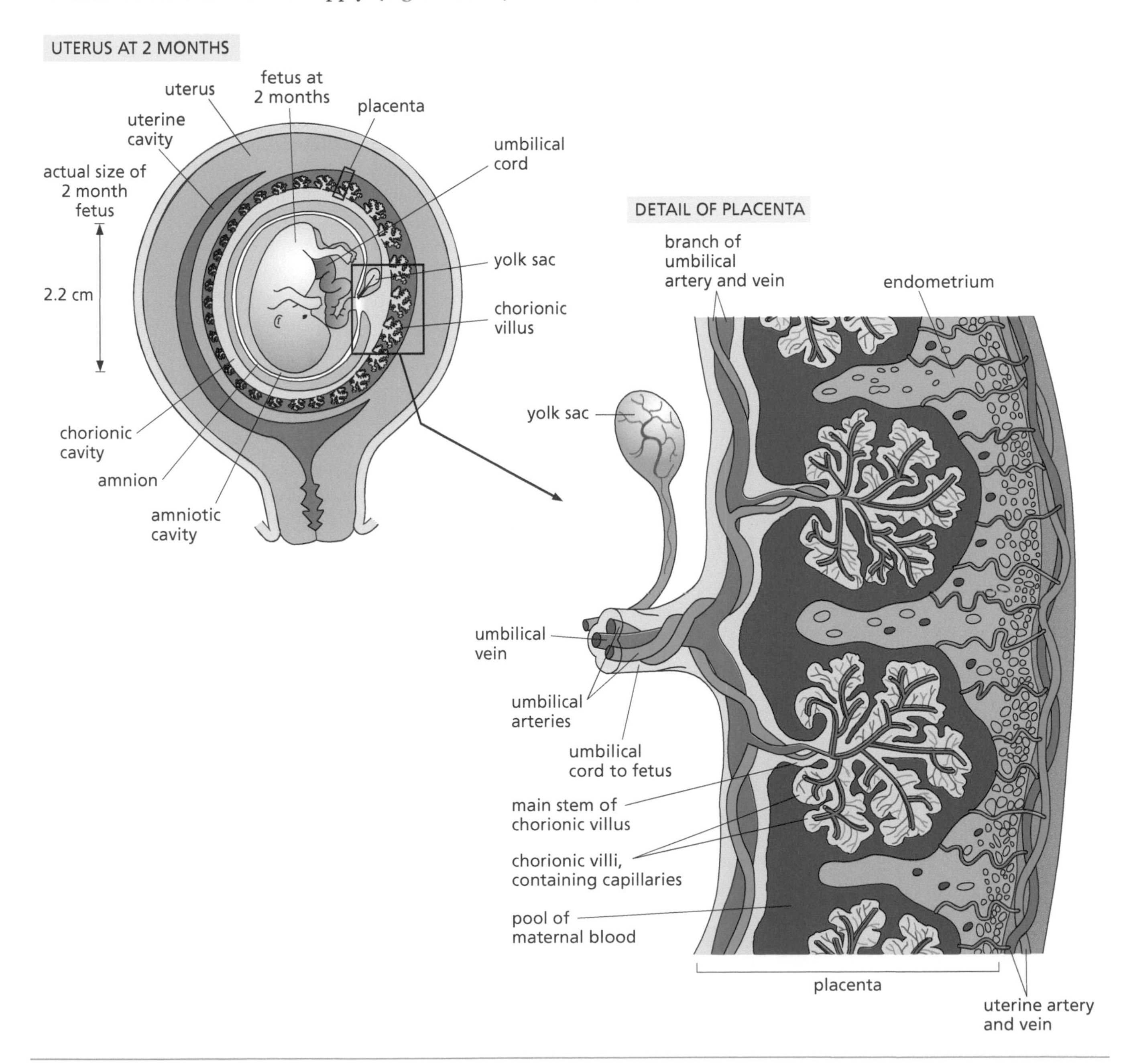

Figure 13.10 Maternal and fetal circulation in the placenta. Fetal blood flows through vessels in the umbilical cord to the chorionic villi, which are in close contact with maternal blood.

maternal and fetal blood systems effectively delivers nutrients and other molecules from the mother's blood into the fetal capillaries of the chorionic villi. Thus, in a pregnant woman, drugs transported throughout her body are also distributed to the fetus. In many cases, doses that are toxic or lethal for the fetus are much lower than the doses that are harmful to adult tissues. Drugs with only a small effect on the mother may therefore profoundly and permanently damage the fetus.

13.4.2.1 Caffeine Many studies using a variety of research approaches have investigated potential negative effects of caffeine on pregnancy. Negative outcomes reported to be associated with caffeine intake have included miscarriages, stillbirths, and premature births; skeletal and brain defects; and behavioral effects. Some of these studies have used experimental laboratory approaches in mice or rats; in general, rodent studies that have reported birth defects have exposed the animals to extremely high doses of caffeine that would never occur in humans. Therefore, their relevance for human pregnancies is negligible.

Studies of caffeine consumption in humans have reached inconsistent and inconclusive results for a number of reasons. Among these reasons are inconsistencies in the number of people and the types of beverages studied (e.g., coffee, tea, caffeinated sodas and energy drinks, or just one of these types), inconsistencies in the different brewing methods and cup sizes used for coffee and tea, the lack of control for noncaffeine ingredients in caffeine-containing beverages, and a variety of other methodological differences.

Experimental studies cannot ethically be performed on humans. Instead, researchers use two types of epidemiological studies referred to as retrospective or prospective. In a retrospective study, participants are interviewed and data are collected on their prior activities after an outcome (e.g., a birth or spontaneous termination of a pregnancy) has already occurred. A major drawback of retrospective studies is that they rely on participants' recall, which may not be accurate and is especially unreliable over long time periods. In addition, participants who have experienced a negative outcome (like a miscarriage) may be more likely to attribute that outcome to something they did and perhaps to overestimate their intake of a given substance, while participants who experienced a positive outcome (like a healthy childbirth) may be inclined to underestimate their intake. For these various reasons, prospective studies, which enroll participants prior to any outcome and collect data on their behaviors in real time, are considered more reliable. For example, a prospective study might follow low caffeine users and high caffeine users over time and determine which group experiences more of a particular disease or other outcome.

An additional type of study called a meta-analysis has also been performed to try to overcome the ambiguities of individual studies. In a meta-analysis, no new data are collected, but the results of many existing studies are examined together to try to discern the most robust trends. Based on such meta-analyses, the American College of Obstetricians and Gynecologists currently asserts that caffeine consumption of less than 200 mg (about one 8-oz. cup of coffee) per day can be considered safe and is unlikely to be associated with any increase in miscarriage or premature birth; nonetheless, many women choose to completely eliminate or strongly limit caffeine during pregnancy out of an abundance of caution.

13.4.2.2 Nicotine Nicotine damages the placenta, increasing the likelihood of miscarriages, premature births, and damage to the fetus. Nicotine crosses the placenta very quickly and remains in the fetal circulation longer than it does in the mother's bloodstream (Figure 13.10). Nicotine causes oxygen deprivation in the fetus, as do the carbon monoxide and cyanide also contained in cigarette smoke. Because oxygen is required in the production of adenosine triphosphate (ATP), cells deprived of oxygen have less ATP and are less able to perform the cell synthesis functions necessary to produce new cells in the growing fetus, especially in the brain. Oxygen deprivation is made worse by nicotine-induced damage to the blood vessels, including those of the placenta. Because of these serious risks, pregnant women or women hoping to become pregnant who smoke or consume nicotine in another form are strongly advised to stop, which may require seeking medical or behavioral help for their nicotine addiction.

13.4.2.3 Alcohol Alcohol that is consumed by a pregnant woman will be quickly distributed into the blood of the fetus at the same concentration as is present in the mother's blood, causing a range of severe and permanent mental and physical birth defects referred to as fetal alcohol spectrum disorders (FASDs) or fetal alcohol syndrome. Precise statistics for the rates of FASD are difficult to determine. One complicating factor is that the period during which the fetus is most sensitive to damage from alcohol is the first month, often before a woman knows she is pregnant. Some studies have estimated FASD to occur in as little as 0.5–2 cases per 1,000 births in the United States (i.e., 0.05%–0.2%), while others have estimated the prevalence to be 1% or even as high as 5% in certain communities. There is also evidence that alcohol abuse by women before conception correlates with decreased fetal growth, even when the mother abstains during pregnancy itself. There are few data on the fetal effects of heavy alcohol consumption by fathers, in part because they can be difficult to separate from maternal effects. However, several experimental rodent studies have suggested that paternal alcohol consumption can reduce the growth rate of the fetus. In addition, alcohol can reduce testosterone production, which can decrease the number and quality of sperm produced, thus contributing to infertility. Men seeking to conceive children are therefore advised to moderate their alcohol intake.

13.4.2.4 Drug combinations Alcohol, caffeine, and nicotine all increase the blood levels of the neurotransmitter acetylcholine, lowering placental blood flow. The effects increase with the dose of the drug and also with its duration in the body. Because the fetus lacks the enzymes for breaking down either alcohol or caffeine, the concentrations of these drugs stay higher longer in the fetus than in the maternal circulation. Because there is a higher incidence of smoking in people who abuse alcohol, the interactions of these drugs are also significant. Marijuana also crosses the placenta and is correlated with low birth weight and prematurity. Barbiturates readily cross the placenta, and the use of barbiturates by pregnant women can cause birth defects. The combination of marijuana or barbiturates with any of the other drugs mentioned increases the risks to the fetus.

13.4.2.5 Drug persistence after birth Drugs passed on to the fetus in utero may remain in the child for a long time, particularly if the enzymes needed to metabolize these drugs are not present. PCP, known as "angel dust," can still be present in the blood of a 5-year-old child of a PCP-using mother. Because children's brains and immune systems continue to develop after they are born, toxic drugs may continue to interfere with the development of the brain and the immune system in young children long after birth.

13.4.3 Drug abuse: Public health and social attitudes

Recall that drug abuse has been defined as drug use that negatively affects the health of individuals or society. We have examined the effects of drugs on cells and on individuals, but an ecological perspective on biology teaches us that actions in cells and organisms generally have consequences for populations.

There are different conceptions of how to protect society from the consequences of drug use by some members of society. One theory views drug addiction as a crime and drug abuse as a law enforcement problem. U.S. efforts to halt the trade in dangerous drugs have relied almost exclusively on this approach, which emphasizes arrests and interdictions instead of prevention or rehabilitation. Billions of government dollars have been spent on the "war on drugs," yet drug use continues to flourish and those who profit from the drug trade grow richer with each passing year. As mentioned earlier in this chapter, shifting dollars spent on marijuana sales from black market dealers to legal, taxed businesses has been one argument for the legalization of recreational marijuana. In addition, a large percentage of violent crime is related to the illegal drug trade.

Another concept, referred to as "harm reduction," is based on a view of addiction as a disease and drug abuse as a public health problem. Many European countries have followed harm reduction strategies, particularly in dealing with such drugs as heroin. The drug abuse rates (and crime rates) of these countries are much lower than those of the United States, which has followed the crime concept for the most part. In Great Britain, for example, heroin can legally be prescribed to those who are addicted. The approach seems to work in several ways: First, harm to the addicts from overdose or impure formulation is minimized; second, there is no incentive for the addict to commit crimes to pay for drugs or to recruit others into becoming addicts. (Many drug dealers are addicts who recruit others so that they will have a steady supply of customers and profits to support their own habit.) British rates of heroin use have dropped under the harm reduction approach, while rates of heroin use in the United States have risen. The illegal heroin trade has withered in Great Britain because it is no longer profitable, and new cases of addiction are rare because the drug is available only to persons registered as already addicted. Most promising of all is the fact that about 25% of British heroin addicts spontaneously give up the habit on their own. A 2019 report by the European Monitoring Centre for Drugs and Drug Addiction (https://www.emcdda.europa.eu/countries/drug-reports/2019/united-kingdom/drug-use_en) shows that, during the period 2006–2017, the number of patients seeking treatment for heroin addiction declined to about one-fourth of its former level while those seeking treatment for amphetamine addiction declined by about half. Also, amphetamine use for the 16- to 34-year-old age group in the UK declined to about one-third of its former level. Under the guidance of its health officials, Britain's National Health Service continues to follow this harm reduction approach.

The approach in several other European countries, including Germany, Switzerland, and the Netherlands, allows drug users freedom from arrest if they follow a few simple rules (staying in certain locations, for example). Instead of spending money mostly on enforcing drug laws (as in the United States), more government funds are spent in these countries on public health campaigns aimed at education, prevention, and rehabilitation. Although the unlicensed selling of drugs is illegal in these countries, the criminal justice system is used very little in attempts to minimize the harm done by addictive drugs, either to the addicts themselves or to society as a whole.

Education is at the heart of most measures aimed at preventing drug abuse, including most harm reduction strategies. For example, the risk of marijuana use as perceived by 12th-graders in the United States shows a clear inverse relationship with actual marijuana use in that population (Figure 13.11). Where prevention of addiction has been attempted, it is cheaper and more successful than rehabilitation. European countries that emphasize education and prevention have much lower drug abuse rates than the rest of the industrialized world.

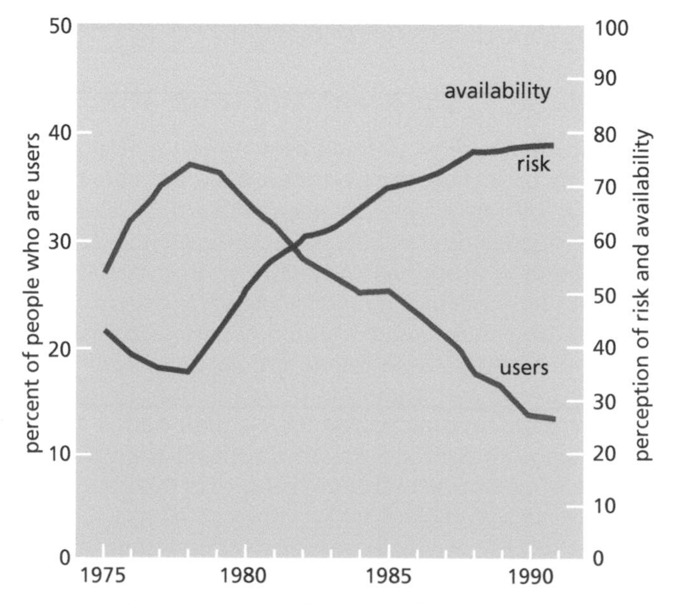

Figure 13.11 Influence of perception of risk on the use of marijuana. The availability of marijuana in the United States remained uniform over the years covered by this graph, but use decreased in inverse proportion to the perception of risk.

Some critics of U.S. drug policy on both the political left and right have occasionally suggested that all drugs should be legalized, taxed, and treated as public health problems the way that we treat alcohol and tobacco, with heavy reliance on education and prevention programs. Apart from the recent trend of marijuana legalization, this idea has gained little traction due to the significant dangers of many drugs and the difficulty of breaking the addiction cycle. However, the large increase in drug overdose deaths caused by the opioid crisis has led some U.S. cities and towns to consider the idea of supervised drug injection sites. These are locations where drug users can obtain clean needles to prevent the spread of blood-borne diseases like HIV and can inject drugs intravenously in the presence of trained medical personnel who can respond with treatment in the case of an overdose. Such facilities have existed in Canada, Australia, and Europe for a number of years but received significant pushback in the United States; in late 2021, the first two supervised injection sites did open in New York City, and debate about opening sites in other cities continues.

The harm done by alcohol shows that a drug need not be illegal to cause considerable social harm. In addition to its negative effects on the user's health, alcohol abuse also causes harm to others. Alcohol use currently causes deaths in motor vehicle accidents, boating accidents, drownings, and many other causes of accidental injury including industrial accidents. It also is responsible for much employee absenteeism, job loss, and school failure. Alcohol is frequently a factor in acquaintance rape (also called "date rape"), child neglect, child abuse, spouse abuse, divorce, and suicide. Alcohol and other psychoactive drugs can also do great harm in safety-sensitive occupations such as commercial transportation (airline pilots, air traffic controllers, railroad engineers), power plant operations, the nuclear industry, and much of the military. We tend to hear about the increases in criminal activity and gangster influence when alcohol use was illegal during the prohibition on alcoholic beverages in the period from 1920 to 1933 in the United States, but it is also true that the incidence of alcohol-related accidents and disease was greatly *decreased* during Prohibition.

THOUGHT QUESTIONS

1. Divide a piece of paper into three columns. In one column list all the characteristics you can think of to describe addictive drugs; in the second, list the characteristics of psychoactive drugs; in the third column list the characteristics of drugs of abuse. Also list specific drugs in each column. Are all addictive substances drugs of abuse? Are all drugs of abuse addictive? Are all psychoactive drugs addictive? Are all psychoactive substances drugs of abuse?
2. Many factors contribute to making a drug dangerous. In what ways are street drugs more dangerous than a chemically similar drug obtained from a licensed manufacturer?
3. Does the American cultural expectation of a quick fix for life's pains contribute to our tremendous use of legal drugs? Does the widespread and somewhat casual use of legal drugs contribute to drug abuse?
4. Read the package inserts for some drugs that you have purchased over the counter. Do these inserts indicate that there are any potential negative effects? What kinds of warnings do the package labels contain? Study the advertisements for these medications. Do the advertisements and the labels give you the same impression of the products?
5. Use the information given in this chapter to determine whether you agree or disagree with the increasing number of state laws permitting marijuana use, either for medicinal or for recreational purposes. What type of ethical reasoning are you using? What facts about marijuana are you using in your argument? What do we not yet know about marijuana that might cause you to change your position?
6. Select a substance sold as a "food supplement" and find out what research (if any) has ever been conducted to test (a) whether it has any of the effects claimed for it, (b) whether it is safe (specify at what dose and under what conditions), (c) whether it has harmful side effects, and (d) whether it interacts with other drugs. If any research studies have been conducted, did they use adequately large samples? Was there a control group? Was the supplement compared with a placebo in a double-blind study?
7. Is information about the biological effects of drugs an effective prevention or a deterrent against drug use? See whether you can find any published information on the effectiveness of various educational programs.
8. What criteria could be used to distinguish between the use and abuse of caffeine? Can these same criteria be applied to other drugs?
9. What ethical considerations govern the use of animals in testing drugs for safety and effectiveness? What about the use of human volunteers?

CONCLUDING REMARKS

Where is the boundary between the rights of the individual and the rights of the group in relation to drug use? Many people are inclined to leave this matter up to the individual if there is no "harm to society." However, in many, if not all, forms of drug abuse, there is clearly a harm to society. Two obvious effects of drugs on populations are the bodily harm done to others by persons under the influence of a psychoactive drug and the commission of crimes to pay for the drugs. In some instances, once the societal effects of individual drug abuse have been documented, laws have been passed to limit these effects. The U.S. public has accepted laws designed to protect the nondrug-using citizen from the harmful effects of others' use (or abuse) but has not accepted laws that are perceived as infringement on individual rights. For example, laws designed to protect others from exposure to secondhand smoke have been successful at limiting where or when people may smoke, while attempts to pass laws prohibiting individuals from smoking at all have not been successful. A similar approach is being tried on college campuses in educational efforts to raise awareness about the secondhand effects of binge drinking, which include higher risks for rape and for sexually transmitted diseases including AIDS.

In Chapter 1 we outlined steps for arriving at policy decisions on societal issues that are influenced by science. Have these methods been followed on the issues presented by drug use and abuse? Nicotine addiction causes far more deaths (from lung cancer) than all other drugs combined, yet it is legal. Alcohol abuse ruins more families and careers than do illegal drugs and causes more fatal accidents, yet it is legal in most places. Marijuana users, on the other hand, cause far less harm to others, yet the substance is illegal in the United States (at the federal level and in many states) and in many other countries. Caffeine is not regarded by most people as a drug and is readily available even to children, yet it is certainly addictive. The inconsistencies go on and on. Clearly, decisions as to which drugs should be legal and which should be illegal are not always made on scientific criteria.

CHAPTER SUMMARY

- A **drug** is a chemical substance that produces one or more biological effects, and usually several.
- The effects of any drug depend on the molecular structure of the drug, on its **dose** or concentration, and in many cases on the bodily location of specific receptors for the drug.
- All drugs, whether used medically or nonmedically, must enter the body by some route, usually orally or via the respiratory system. The drug must then be distributed around the body, usually by the circulatory system, and eventually eliminated either by metabolic breakdown or by **excretion**. The length of time that a drug remains metabolically active in the body is measured by its **half-life**.
- Drugs have both **acute effects** and **chronic effects**.
- Drugs have their effects on cells either by direct action on the cell membranes or by stimulating **receptor** molecules to alter one or more cellular functions. Actions on cells in different tissues produce different physiological effects.
- There are always other effects in addition to that for which the drug was taken, and these **side effects** are every bit as real as the intended effect.
- Drugs can interact either **additively**, **synergistically**, or **antagonistically**.
- **Psychoactive drugs** cross the blood–brain barrier and act directly on the nerve cells of the central nervous system. Those that produce **addiction** do so by stimulating activity in the positive reinforcement center of the brain. Addictive drugs induce **dependence** and **tolerance** to the drug, as well as **withdrawal** symptoms that appear when the drug is stopped. The possible side effects of psychoactive drugs

include permanent damage to the brain cells and interference with normal physiological functioning of other organ systems, impairing the health of drug users.

- Many drugs can cross the placental barrier and cause damage to a fetus in utero; fetal alcohol syndrome is an example. Other drugs can be transmitted to infants through breast milk.
- **Drug abuse**, also called **substance abuse**, has a tremendous cost to society.

BIBLIOGRAPHIC REFERENCES

Bibliographic references to material in this chapter can be found online at biologytrending.routledge.com/chapter13

GLOSSARY: KEY TERMS TO KNOW

These key terms are defined at the end of each chapter as an aid for student review.

Acute effect An effect that ceases soon after its cause is removed.

Addiction A strong physiological and psychological dependence.

Additive effect A physiological response produced by two drugs given together that is the same as the sum of the effects of each drug given separately.

Agonist A drug that stimulates a particular receptor or that has a stated effect.

Alveoli Small pouches or cavities, especially the air-filled pouches in which gas exchange occurs in the lungs.

Antagonist A drug that inhibits another or that inhibits a particular receptor.

Antagonistic interaction A combined effect in which two drugs together produce less of a physiological response than either drug given separately.

Chronic effects Lasting or lifelong effects.

Dependence Inability to carry out normal physiological functions without a particular drug.

Dose The amount of a drug given at one time.

Drug Any chemical substance that alters the function of a living organism other than by supplying energy or needed nutrients.

Drug abuse (substance abuse) Excessive use of a drug or use which causes harm to the individual or to society.

Euphoria A feeling of elation and well-being, especially one unrelated to the true state of affairs.

Excretion The production of waste products, especially by the kidney, and their subsequent removal from the body.

Glomerulus A small clump of thin-walled capillaries within the kidneys that filters blood into a surrounding capsule (Bowman's capsule).

Half-life For a drug, the time that it takes for the level of the drug in the body to be reduced by half.

Narcotic Any drug capable of producing sleep or loss of consciousness.

Nephron One of the many tubules within the kidney that filters blood and produces urine.

Pharmacology The study of drugs and their effects.

Psychoactive drug Any chemical substance that alters consciousness, mood, or perception.

Receptor A protein or other molecule that binds a specific drug or other chemical substance and responds to the binding by initiating some cellular activity.

Side effect A drug effect other than the one for which the drug was intended.

Substance abuse See *Drug abuse*.

Synergistic interaction A combination of two causes that lead to an effect greater than the sum of the effects that would have been produced by the two causes independently; for example, a combined effect in which two drugs together produce a greater physiological response than the sum of the effects of each drug given separately.

Threshold The minimum level of a drug below which no physiological response can be detected.

Tolerance An acquired condition in which progressively greater amounts of a drug are required to produce the same physiological effect when the drug is taken repeatedly.

Withdrawal Physiological changes or unpleasant symptoms associated with the cessation of drug taking.

CONNECTIONS TO OTHER CHAPTERS

Chapter 1: "The mind is the same thing as the brain" is an example of a research paradigm.
Chapter 1: Attempts to limit the effects of drug abuse raise numerous ethical issues.
Chapter 8: Drugs interfere with nutrient pathways at many levels.
Chapter 9: Drug use can affect the physiological regulation of sex hormones. Drug use can also affect sexual behavior, including sexual abuse and unprotected intercourse.
Chapter 10: Drugs such as tobacco and marijuana contain many cancer-causing agents. Alcohol and tobacco act synergistically as causes of cancer.
Chapter 11: Drugs may interfere with the normal processes of the brain.
Chapter 14: The brain and the endogenous opiates interact with the immune system.
Chapter 15: Drug use by injection is a major risk factor in the transmission of HIV infection and AIDS.
Chapter 17: Most drugs are plant products or are derived from plant products.

PRACTICE QUESTIONS

1. If the toxic dose of caffeine is about 16 g for a person weighing 80 kg (about 180 lb), how many cups of coffee, consumed at one time, would it take to realize this dose?
2. Which of the following drugs are addictive?

Cocaine	Heroin	Alcohol	Methadone
Morphine	Codeine	Tobacco	Meperidine

3. How many equivalent "drinks" are there in a bottle of wine, which is typically 750 mL? How many "drinks" are there in a quart of liquor? (1 ounce = 28.4 mL; 1 quart = 32 ounces).
4. Of the many drugs described in this chapter, which causes the largest number of deaths from people under its influence?
5. Name three drugs that can enter the body through the respiratory system.
6. If you take 100 mg of a drug, how much remains metabolically active in your body after one half-life?
7. If the half-life of the drug in Question 6 is one hour and you start with 100 mg of the drug, how much drug is left after 2 hours?
8. If you take 100 mg of a drug that has a half-life of 1 hour and after 1 hour you take another 100 mg of the drug, how much of the drug will be in your body after the second dose?
9. Of all addictive drugs, which is the most widely used in the United States?
10. If a drug is described as an opiate agonist, what does that mean?
11. If another drug is described as an estrogen antagonist, what does that mean?
12. If two drugs interact synergistically, what does that mean?
13. If the body develops a tolerance to a painkilling drug, what does that mean?

CHAPTER 14

Mind, Body, and Immunity

ISSUES

- What are the causes of disease? Can we promote health, or is it simply the absence of disease?
- Why does one person get sick and not another?
- How do our mental and emotional states affect our physical health?
- How do our body systems communicate with our emotions?
- Can stress make you sick?
- How has psychoneuroimmunology reworked our definitions of disease and its prevention?

BIOLOGICAL CONCEPTS

- Dynamic equilibrium (detection of environmental stimuli, homeostasis)
- Organ systems (immune system, lymphatic circulation, autonomic nervous system, endocrine system)
- Health (specific immunity, inflammation, tissue healing, vaccination, stress response, relaxation response, placebo effect)

CHAPTER OUTLINE

DOI: 10.1201/9781003391159-14

What do we mean by health, and how do we achieve it? What makes one person healthy and someone else chronically ill? Certainly, there are many answers to these questions, and we have touched on some of them in previous chapters. A person's genetic heritage plays a role in some diseases (Chapter 3), as does their diet (Chapter 8). How and where a person lives is important because they influence the person's exposure to infectious microorganisms (Chapter 16) and to hazardous chemicals (Chapter 19). Our immune system helps to remove damaged tissue and repair the body, preventing some diseases before we know we have been exposed and bringing us back to health after we have been sick. Genetics, nutrition, and exposure to chemicals and microorganisms all affect the functioning of the immune system. It is often the case, however, that some people in a particular area will get sick, while other people in the same area with about the same exposure, diet, and genetic background do not. The COVID-19 pandemic has been a prime example of this phenomenon. In Chapter 11, we saw that many so-called mental illnesses have mechanisms based in brain biochemistry. In this chapter, we will examine the theory that a person's mental and emotional states are factors in physical health or disease and that the mind exerts its effect on the body because it interacts with the immune system.

14.1 THE MIND AND THE BODY INTERACT

In the preceding chapters of this book, we have discussed some biological fields in which theories have been debated, tested, and modified for 150 years—a long time in the science of biology. In this chapter we will be discussing the workings of the immune system and its interactions with the mind and body. This fairly new subject area, **psychoneuroimmunology**, is built upon a *central organizing theory*: the premise that the mind, through the action of the nerves, affects the functioning of the immune system (the body's defense apparatus against harmful or malfunctioning molecules) and therefore affects human health. Each part of the name contributes to the overall meaning: *psycho*, from the Greek word *psyche*, meaning "the mind"; *neuro*, referring to the nerves and the brain; and *immunology*, the study of the immune system. Psychoneuroimmunology is the study of how the nervous and immune systems, which were previously assumed to be independent of one another, interact with and influence each other. Psychoneuroimmunology is thus a new paradigm (see Section 1.2.1) since it embodies both a new theory and a new field of study.

Central to the paradigm are large and important issues with which people have struggled for millennia: What do we mean by **health**? Why do we get sick? How we answer these questions will help to define what areas of investigation are valid with regard to the cure and prevention of disease and the promotion of health.

No theories arise spontaneously, and neither did psychoneuroimmunology. Its roots lie in ancient observations that personality, emotional state, and attitudes influence when and if people get sick and how sick they get. Doctors in China, India, and (later) Greece rejected supernatural forces (the gods, evil spirits, or magic) in favor of natural (biological) forces as the explanations for both health and disease. Each of these traditions maintained that there is a life force or life spirit, called *qi* (or *ch'i*) by the Chinese, *prana* by the Indians, and *pneuma* by the Greeks, that is present in humans and other organisms for the duration of their lives. A person is healthy when the life force is balanced and unhealthy when it is out of balance. A person's mental and emotional states alter the balance or imbalance of the life force.

The Greek physician and teacher Hippocrates taught that diseases had natural causes and that those causes should be discernible and knowable entirely by rational thought. Hippocrates taught that the body contained four fluids, or *humors* (from the same Greek word that gives us *humid*). Each of the four humors corresponded to a personality type. Each personality type also predisposed people to a further excess of the corresponding humor, creating diseases, also classified into four types according to which humor was present in excess. A person with an excess of black bile, for example, would have a "hot" personality and would be prone to "hot" diseases accompanied by fevers. People who had an even balance of all four humors enjoyed good health and were thus said to be in "good humor."

As European science developed in the 1600s, the criteria for "knowability" changed. In 1616 an English physician, William Harvey, described the circulation of the blood on the basis of dissection and observation, not purely on rational thought. At about the same time, Galileo invented the thermometer, and this was used by another Italian, Santorio, to demonstrate that people said to have an excess of black bile were no hotter than other people. Not only did these two discoveries falsify the notion of the four humors, they ushered in an era in which hypothesis testing was added to the standards of "knowing" about human health. Although the scientific method has vastly increased our knowledge of health and disease, a negative aspect has been that things that could not be seen or in some way quantified have come to be viewed as irrelevant to the explanation of health and disease.

Christianity viewed humans as unchanging reflections of God and therefore beyond the scope of study by the scientific method. The French philosopher René Descartes offered a way around this problem by positing that the mind was separate from the body: The mind was the seat of the spirit and hence belonged to the realm of the church. The body was purely physical and hence suitable for scientific study. This split of mind from body (or Cartesian dualism as it is sometimes called) had a profound effect on the study of biology and medicine. As medicine strived to become more of a science, the split became wider. It is just this split, however, that psychoneuroimmunology rejects as artificial because it does not fit the evidence.

In seeking to rejoin the mind to the body, scientists working in this field test three major hypotheses: that the immune system maintains health, that the mind and emotions can affect the functioning of the immune system, and that mental states can, therefore, affect health. There is an underlying assumption here, as there was in Chapter 11, that all of the functions of the "mind" (thoughts, emotions, hopes, and dreams) can be studied in terms of brain biochemistry. Some take this assumption one step further, postulating that the mind is more than just the brain; it is the integrated, inseparable network that includes the nervous system, the endocrine system, and the immune system. Other scientists in this field would be more likely to say that the psyche (or mind or spirit) affects the brain and the body in ways that can be studied by biology. Think about these distinctions as you proceed through this chapter.

In order to understand how the central theory of psychoneuroimmunology can be tested, we need to learn about the biology of the immune system, the nervous system, and the endocrine system. These three systems in the body have communication as their primary function.

THOUGHT QUESTIONS

1. Think about the difference between the following two statements: (a) The mind *is* the neuroendocrine-immune system and (b) The mind can be studied by studying the neuroendocrine-immune system. Are both scientific statements? Would both allow the formulation of falsifiable hypotheses?
2. Can a person accept the data produced by psychoneuroimmunologists without taking a stand on which of the assumptions stated in thought question 1 is correct?
3. Can a person accept the scientific findings of neuroscience without accepting the central theory that the mind and the brain are one? Can this central theory ever be proved beyond doubt?

14.2 THE IMMUNE SYSTEM MAINTAINS HEALTH

A healthy multicelled organism can be viewed as an ecosystem of cells in which the parts are in dynamic equilibrium, or **homeostasis**. The immune system is the sense organ that detects whether this homeostasis exists and attempts to bring the organism back to this state if it does not exist. We can talk about homeostasis in many physiological contexts: temperature regulation in organisms that maintain a constant body temperature, for example. Immunological homeostasis is a more general state, suggesting that the cells have a way of asking, "Are we all together?" "Are we in harmony?" The immune system can be viewed as a communication network that carries on a "conversation" throughout the organism, checking to make sure that all the parts are contributing. In this view, the central function of the immune system is the maintenance of "self," meaning the aggregate of cells forming a cooperative unit that we recognize as a multicellular organism. As we see in this chapter, the immune network is aided in this task by its ability to exchange chemical messages with the nervous and endocrine systems.

14.2.1 Cells of the immune system and the lymphatic circulation

Cells of the immune system are primarily **white blood cells**. They are called that because under the microscope they look clear or white by comparison to the oxygen-carrying red blood cells. The immune system consists largely of these mobile cells that travel throughout the body and often are not confined to specific locations. The white blood cells comprising the immune system that are discussed in the ensuing sections of this chapter are shown in Figure 14.1.

White blood cells develop in the bone marrow, spleen, and thymus and are then transported throughout the body by the bloodstream, particularly to the areas of the body that contact the external environment: the skin, the nasal passages and lungs, and the intestinal lining. They leave the blood and "crawl" through the spaces between the cells in tissues; later they are transported via a second circulatory system called the **lymphatic circulation** to the lymph nodes. Other immune cells may differentiate in the lining of the small intestine and in the lower layers of the skin. The skin and the gut lining are thus important organs of the immune system. These structures of the immune system are shown in Figure 14.2.

The lymphatic circulatory system drains liquid from tissues. The spaces between cells in all tissues (interstitial compartments or tissue spaces) are filled with a water-based liquid called **interstitial fluid**. (Recall from Chapter 8 that all cells must be constantly in contact with water both inside and outside the cell, as

(A)

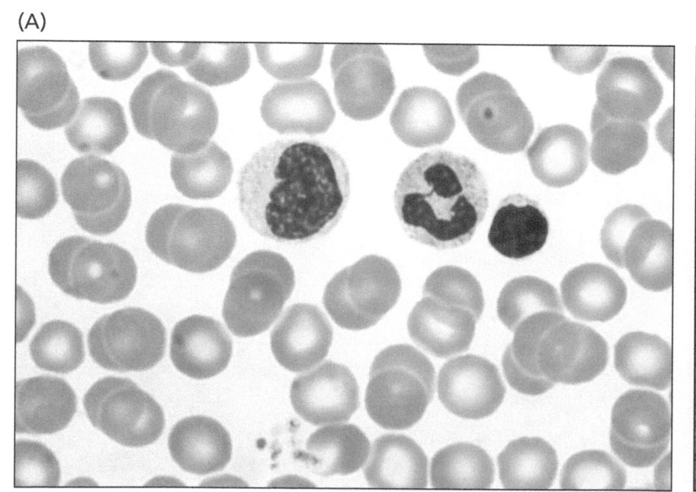

(B)

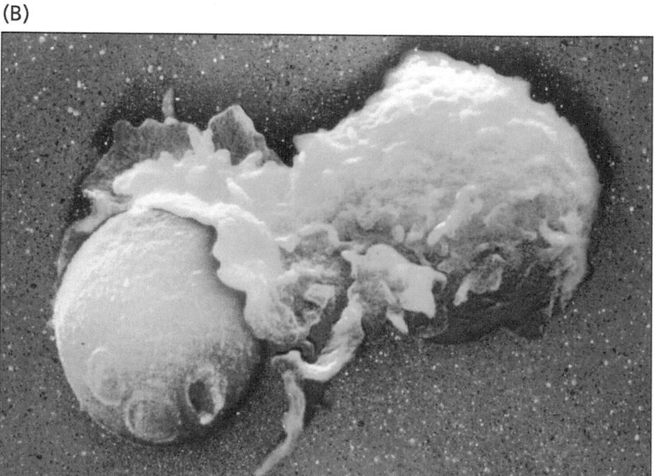

Figure 14.1 **White blood cells that provide innate and specific immunity. (A)** Three types of white blood cells (left to right: monocyte, neutrophil, lymphocyte), surrounded by smaller red blood cells (erythrocytes). Monocytes and neutrophils are phagocytic, functioning in innate immunity. Lymphocytes, which function in specific immunity, are of two types that cannot be distinguished by appearance alone: B lymphocytes secrete antibodies, while T lymphocytes kill virally infected cells and cancer cells. **(B)** Phagocytic leukocyte (right) engulfing a yeast cell.

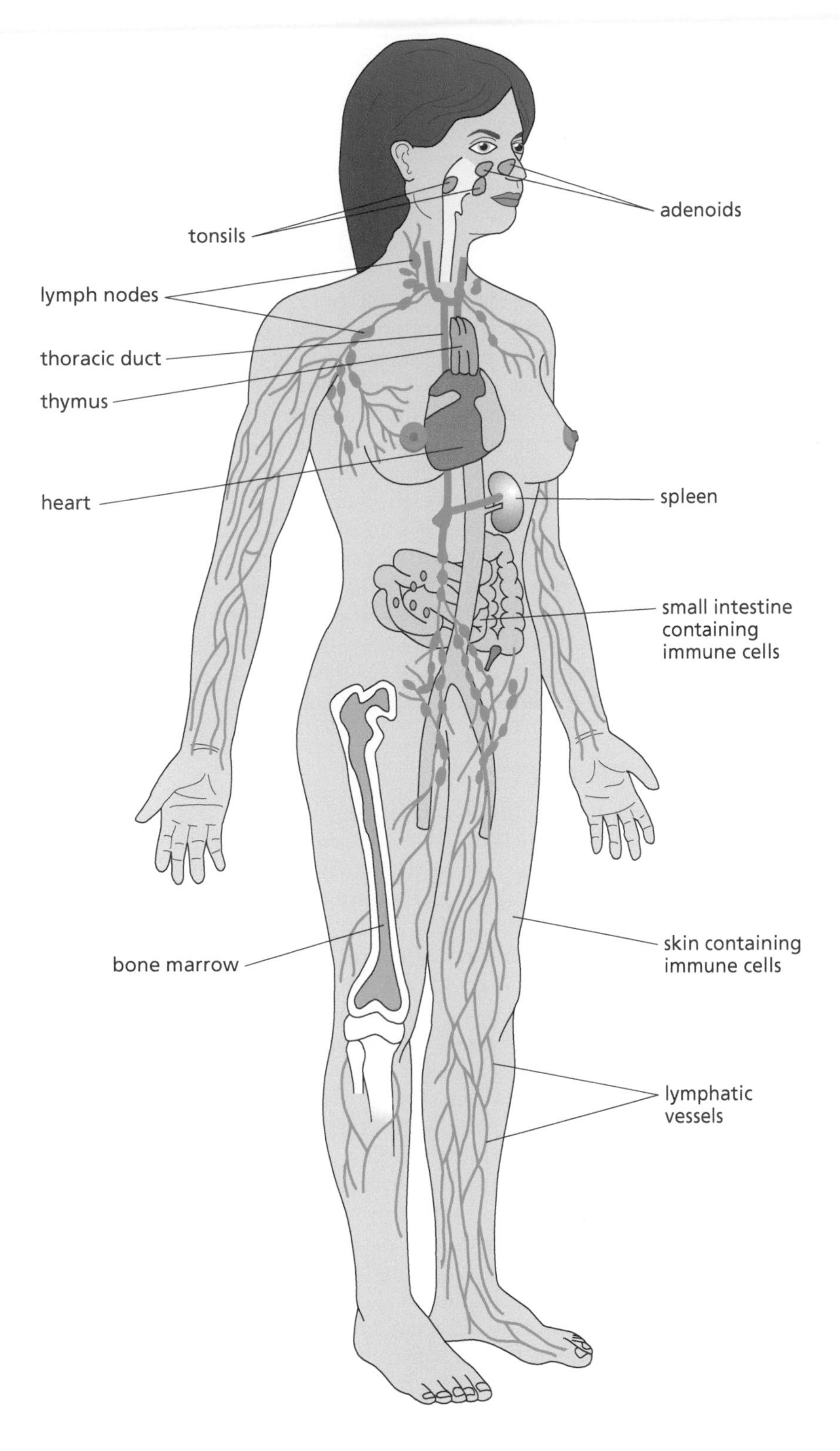

Figure 14.2 Macroscopically visible parts of the immune system. The parts of the lymphatic circulation are in blue, and the major lymphatic tissues (thymus, bone marrow, tonsils, etc.) are colored orange.

it is the tendency of lipid molecules to avoid water that keeps cell membranes intact.) Immune cells, but not red blood cells, move in this liquid, cleaning up any dead or damaged cells. This fluid diffuses from the intercellular spaces into lymphatic capillaries. Once inside the lymphatic capillaries, the fluid is called **lymph**. From the lymphatic capillaries, lymph drains into larger collecting vessels, the lymphatic vessels. Lymph is chiefly returned to the blood via the thoracic duct, which empties into a large vein near the heart (Figure 14.2). There is no pump to move fluids through the lymphatic circulation the way the heart moves fluids through the blood. Muscle contractions and movements of the individual provide what little push this system gets.

Cells and molecules that are not contributing to the homeostasis of the cellular ecosystem are called *non-self.* Thus, dead, damaged, or cancerous cells, as well as molecules from outside the organism, are non-self. Immune cells spend much of their time checking tissues for these non-self molecules. The lymphatic circulation not only helps remove wastes and damaged cells from tissues, but it also solves the problem of how the immune system can monitor all of the cells and molecules in all of the tissues of the body. This is accomplished by the lymphatic circulation bringing molecules to centralized locations (lymph nodes, tonsils, and adenoids, Figure 14.2) to be checked by other immune cells. In an active immune response, 10 times the normal number of white blood cells enter the nodes, causing the nodes to "swell." Our lymph nodes are what we commonly refer to as "swollen glands" when we are sick. The fact that they get larger during sickness indicates that an immune response is occurring; thus, swollen glands are generally a sign that a return to health is underway (although lymph nodes that remain chronically swollen sometimes indicate other problems and should be checked by a physician).

Tonsils and adenoids also enlarge with additional cells during an immune response. In the mid-twentieth century, the function of tonsils and adenoids as tissues of the immune system was poorly understood. Tonsils and adenoids were routinely removed from children who had repeated respiratory or middle ear infections because the swelling of these tissues during infections can make children's breathing difficult or block the tube that connects the back of the throat to the middle ear (see Figure 11.12). Fortunately, there are backup systems in the immune tissues so that removal rarely had serious consequences, but today tonsils and adenoids are left in place unless the blockage is extreme.

14.2.2 Innate immunity

One important function of the immune system is to promote growth and repair after injury, whether the injury is due to microorganisms or to physical damage. This process is a capacity we are born with; hence the part of the immune system that carries out this function is called the **innate immune system**. The mobilization of innate immune cells to eliminate microorganisms or damaged cells and to repair wounds is called **inflammation**. Small molecules called **cytokines** are also involved in inflammation, helping with cell-to-cell communication. In the first century CE, the Roman physician Celsus described the "four cardinal signs of inflammation": *rubor* (redness), *calor* (heat), *dolor* (pain), and *tumor* (swelling). If you have ever had a scraped knee or a splinter in your finger, you no doubt have experienced these cardinal signs. The injured area becomes red, hot, sore, and slightly swollen; all symptoms are caused by changes in local blood vessels.

The various processes of inflammation that lead to healing are illustrated in Figure 14.3. Inflammation is carried out by various white blood cells. Principal among these are the macrophages and neutrophils that can engulf bacteria and damaged tissue by surrounding them with pseudopods, bringing them inside the cell, and digesting them, a process called **phagocytosis**. Cell damage from a wound also changes the acidity (pH; see Figure 8.11) of the local fluids, activating various chemicals. Some of these chemicals attract macrophages and neutrophils to the area (Figure 14.3A). Other chemical factors constrict the blood vessels beyond the site of the wound, causing blood to build up in the capillaries close to the wound. These changes in blood flow result in the redness, heat, and swelling of inflammation. The localized swelling puts pressure on nerve endings, causing pain, the fourth characteristic of inflammation. Still other chemical factors, particularly histamine, increase the permeability of the capillaries in the local area. The neutrophils and macrophages are able to crawl through the capillary walls and into the tissue. There, the neutrophils remove the bacteria and the macrophages remove the damaged tissue. Macrophages also secrete chemicals called "growth factors" that are the first step in wound healing. These stimulate cell division, providing offspring cells to replace the damaged cells, for example, the skin cells lost because of the wound (Figure 14.3B).

During inflammation, macrophages also secrete several cytokines that induce fever (Figure 14.3B), the symptom we often recognize as a symptom of having an infection. These cytokines induce fever by acting on part of the brain called the

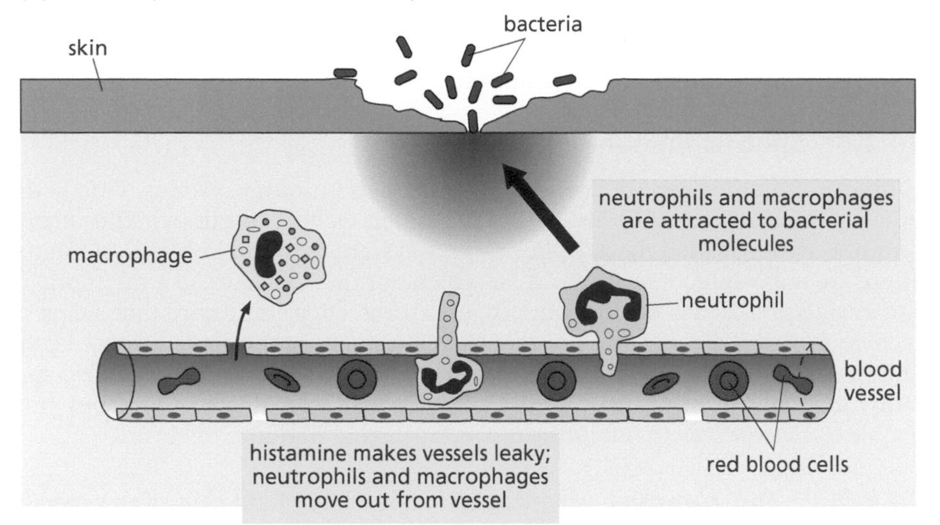

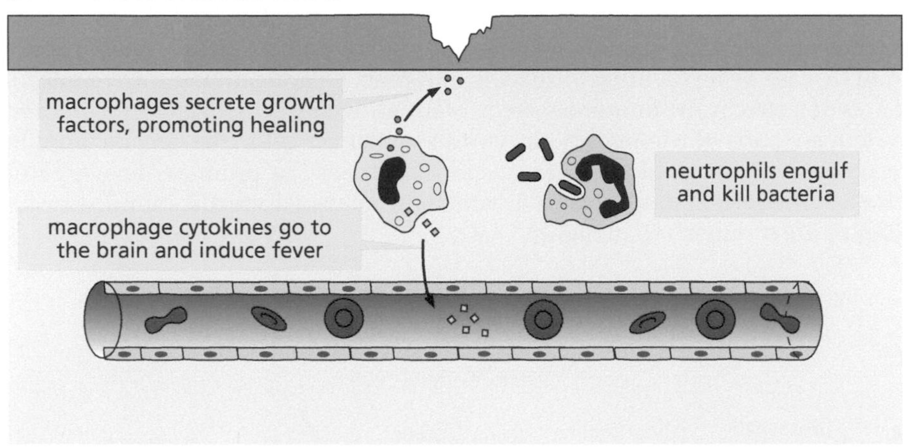

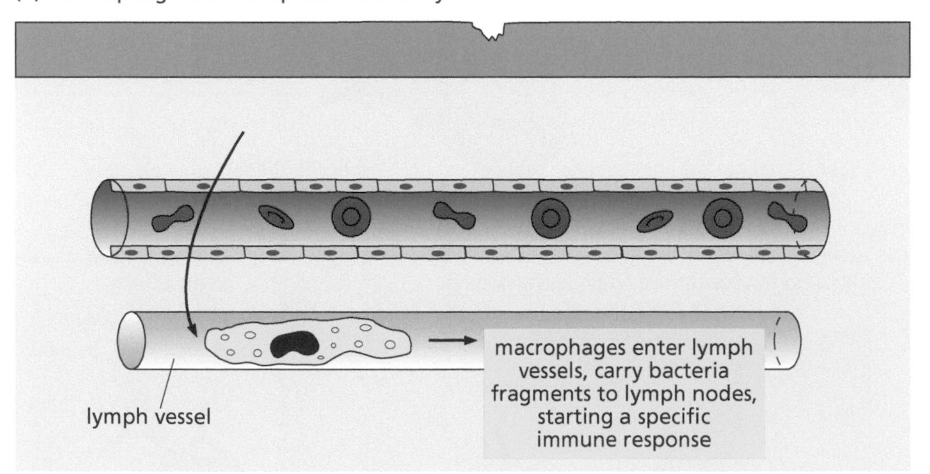

Figure 14.3 The stages of inflammation.

hypothalamus (see Figure 11.15), the first of many examples in which products of the immune system act on the cells of the nervous system. The increased body temperature inhibits the growth of bacteria and also enhances the immune response to the bacteria. Human body temperature is normally 98.6°F (37°C). Small increases enhance immunity; however, increases above 105°F (40.6°C) may result in convulsions or death.

Finally, the macrophages enter the lymphatic vessels and carry bacteria (or other non-self molecules) to the lymph nodes (Figure 14.3C). Here, the macrophages show the bacterial molecules to other immune cells, triggering another type of immunity called specific immunity.

14.2.3 Specific immunity

Humans and other vertebrates actually have two immune systems. One is the innate immune system we are born with, which we just discussed. The second immune system, called the specific immune system, is found only in vertebrates. There are many infectious diseases for which our first exposure to a microorganism results in disease but also builds up a protective immunity that helps us recover from the disease and protects us against getting the same disease again. This protection against future exposures to the same microorganism is called **specific** (or **acquired**) **immunity**. We are not born with specific immunity, but must acquire it by exposure to microorganisms during our lifetime.

14.2.3.1 The removal of antigen The white blood cells of the specific immune system are called **lymphocytes** (Figure 14.1). There are two types: B lymphocytes (**B cells**), which make blood proteins called antibodies (explained in the next paragraph), and T lymphocytes (**T cells**), some of which kill infected cells directly and some of which help other immune responses. These cells protect us from disease by responding to specific **antigens**, an antigen being any molecule that is detected by the immune system. Many antigens recognized by the immune system are parts of whole bacteria, viruses, or cancer cells. The specific immune response removes or blocks these to prevent disease or promote recovery from disease. The several mechanisms by which the activated specific immune system does so are summarized in Figure 14.4.

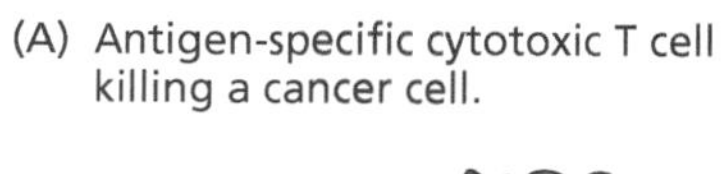
(A) Antigen-specific cytotoxic T cell killing a cancer cell.

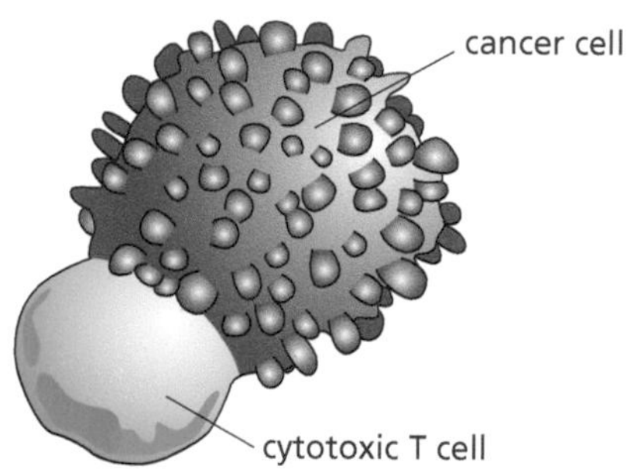

(B) Antibody and complement forming holes in a bacterial membrane.

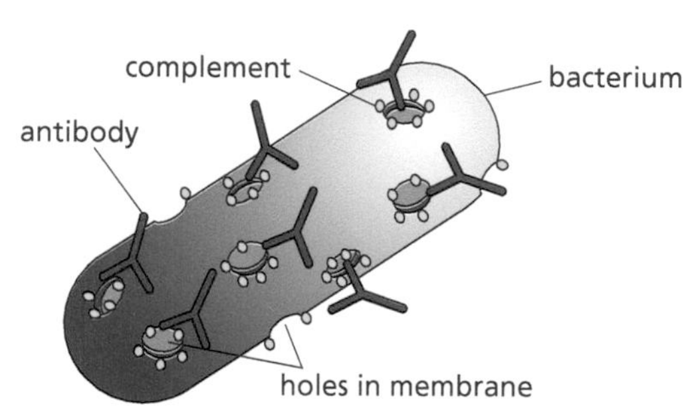

(C) Neutrophil phagocytosing an antibody-coated bacterium.

antibody-coated bacterium
neutrophil

(D) Antibody blocking of bacterial adherence to host tissue, allowing the bacterium to be washed away.

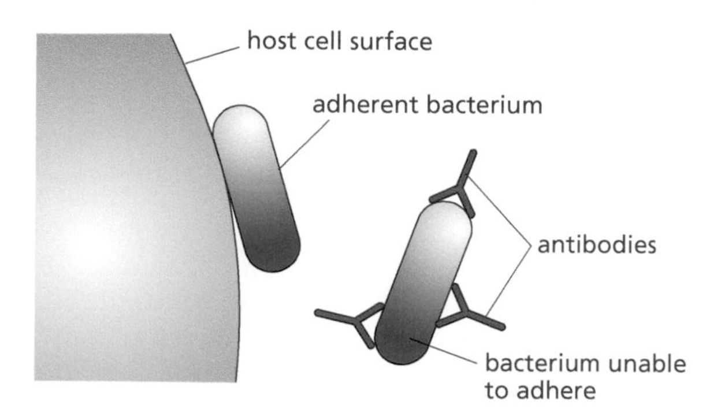

(E) Antibody inhibition of toxin activity.

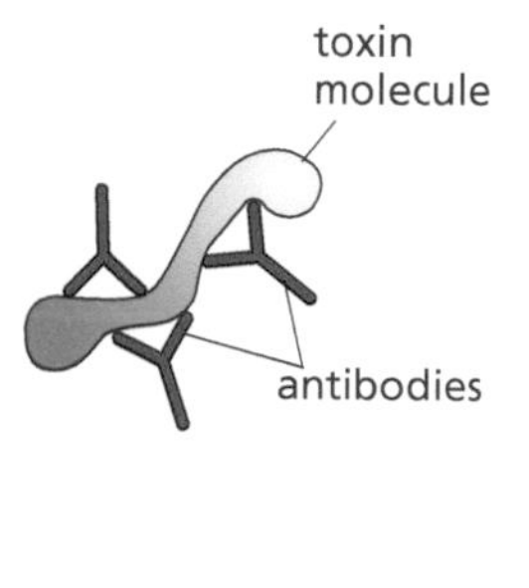

Figure 14.4 Ways in which the immune system can eliminate non-self antigens.

A type of T cell called cytotoxic T cells can directly kill cancer cells (see Section 10.3.4) or cells infected with a virus (**Figure 14.4A**). In contrast, B cells do not kill antigens directly. Instead, they make and secrete proteins called **antibodies** that are specific for a particular antigen, which then circulate in the body fluids. Antibodies are proteins present in the blood, lymph, and other body fluids, and they bind to specific antigens. Once antibody is bound to the bacteria or virus, the antibody can then combine with other blood proteins called **complement**. The antibody–complement combination kills bacteria by making holes in their cell membranes and can also inactivate viruses that are not yet inside cells (**Figure 14.4B**).

In addition, antibodies and complement can coat bacteria, allowing the bacteria to be engulfed and killed by the white blood cells called **neutrophils**. Unlike lymphocytes, neutrophils cannot bind to most bacteria directly. Instead, they have receptors that bind to one end of antibody molecules. As a result, once a bacterium is coated with antibodies, the other ends of the antibody molecules can be bound by a neutrophil, which will then take up the bacterium, kill it, and digest it (**Figure 14.4C**).

Some antibodies work not by killing microorganisms but by preventing their binding to the host. Most microorganisms cannot initiate disease without attaching to the host's cells; this is especially true of respiratory viruses and oral bacteria. Antibodies bound to these organisms block their adherence, preventing disease. Antibodies present in the mucous linings of the respiratory tract and the throat and mouth are especially important in blocking the adherence of organisms trying to gain entry through those routes, causing them to pass harmlessly through the body and to be excreted as waste (**Figure 14.4D**).

Other antigens, such as toxins, are soluble molecules secreted by bacteria, not parts of bacterial cells, and it is the toxins, rather than the whole bacteria, that induces the disease. Toxins are inactivated by having specific antibodies bind to them (**Figure 14.4E**). Many of our most successful vaccines actually stimulate the production of antitoxins, that is, antibodies against toxins. For example, the lethal results of diphtheria, a disease that formerly produced epidemics that killed many hundreds of thousands of people, result from the action of a bacterial toxin. Immunity conferred by diphtheria vaccine produces an antitoxin that prevents the disease and has virtually eradicated diphtheria from areas of the world where people are vaccinated against it.

14.2.3.2 Antigen discrimination The specific immune system has three unique characteristics that allow the development of protective immunity:

1. It can distinguish one molecule from another.
2. It can further distinguish "self" molecules from non-self.
3. It shows *memory* of having seen particular molecules in the past.

Each of these is explained by a singular characteristic of lymphocytes. Each individual lymphocyte can bind to only one specific antigen. The population of lymphocytes as a whole, however, can recognize a vast diversity of antigens. This combination of individual cell specificity and population diversity allows the immune system to distinguish one antigen from another.

14.2.3.3 Antigen receptor specificity Each B or T cell has receptors that can bind to only one specific antigen, yet the immune system as a whole is able to recognize over 10^{11} (100,000,000,000) different antigens. If each different receptor resulted from a different gene, as is typical for many proteins, we would come up against a large problem: The number of different antigen receptors is more than 1 million times greater than the total number of genes in humans! However, this huge array of antigen receptors is made from just a few genes. Something happens to these antigen receptor genes that is not known to happen in any other genes. During the differentiation of each B and T cell, the DNA in the antigen receptor gene is rearranged, and some strings of nucleotides are cut out. This is different from the alternative ways of processing mRNA (Chapter 4, p. 131) that can result in different proteins being synthesized at different times in a cell. Here it is the DNA sequence that is changed, not the mRNA. Thus, the rearrangements are permanent in that cell and its offspring cells. Each mature T cell or B cell is therefore able to synthesize a unique protein that functions as a receptor for only one specific antigen.

The DNA is rearranged randomly in each developing T or B cell so the antigen receptor proteins made on one lymphocyte differ from those on another. The whole population of T and B cells thus contains cells that can bind to over 10^{11} different antigens. Not only are the rearrangements random, but they happen *independently* of the person's being exposed to any particular antigen. *Even before* being infected by a particular type of virus, for example, a person has a small number of lymphocytes that can bind to that virus.

14.2.3.4 Self/non-self discrimination Immature T cells develop in the bone marrow, then travel through the blood to an organ called the thymus (Figure 14.2), where they complete their differentiation into T cells. Self-reactive T cells, meaning those capable of reacting against the body's own molecules, are eliminated, as are self-reactive B cells (**Figure 14.5**). This critical process of immune selection usually protects you from autoimmunity (discussed later in this chapter), where your immune system reacts against the tissues of your own body. The immune system's capability of discriminating self from non-self thus results from the selection of a population of responding cells from those that arose randomly and is a characteristic of the *population* of cells, but not of any single cell.

14.2.4 Immunological memory

A person is not born with immunity to specific antigens but is born with the capacity to acquire it. This capacity is the result of the two processes just explained: production of vast numbers of different antigen receptors by DNA rearrangements, followed by selection of a population of lymphocytes that are not self-reactive. Lymphocytes are released into the bloodstream, where some circulate at all times, ready to go into action when needed. These lymphocytes are carried by the bloodstream to the lymph nodes, tonsils, and adenoids (Figure 14.2), where they encounter foreign (non-self) antigens brought from the tissues by the lymphatic circulation. Most of the B cells cannot bind to the antigen and so continue on their way. However, a small number of B cells can bind, and when they do so, their binding triggers them to begin to divide, producing a group of identical B cells, all able to bind to the same antigen (**Figure 14.6**). This group of cells then differentiates into two types: One type of activated B cell produces the antibodies that help destroy the antigens as we saw previously (see Figure 14.4),

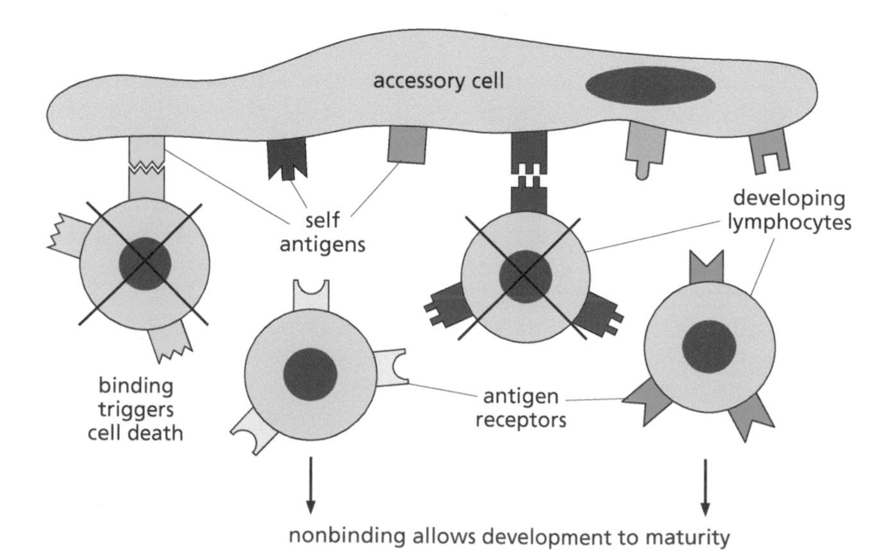

Figure 14.5 Immune selection. Developing B and T lymphocytes, each with a unique antigen receptor on its cell membrane, encounter characteristic "self" antigens on accessory cells. Any developing lymphocyte whose antigen receptor binds to a self antigen on an accessory cell dies. Lymphocytes whose receptors do not bind develop to maturity and are available to bind to any foreign antigens that match their antigen receptors at a later time.

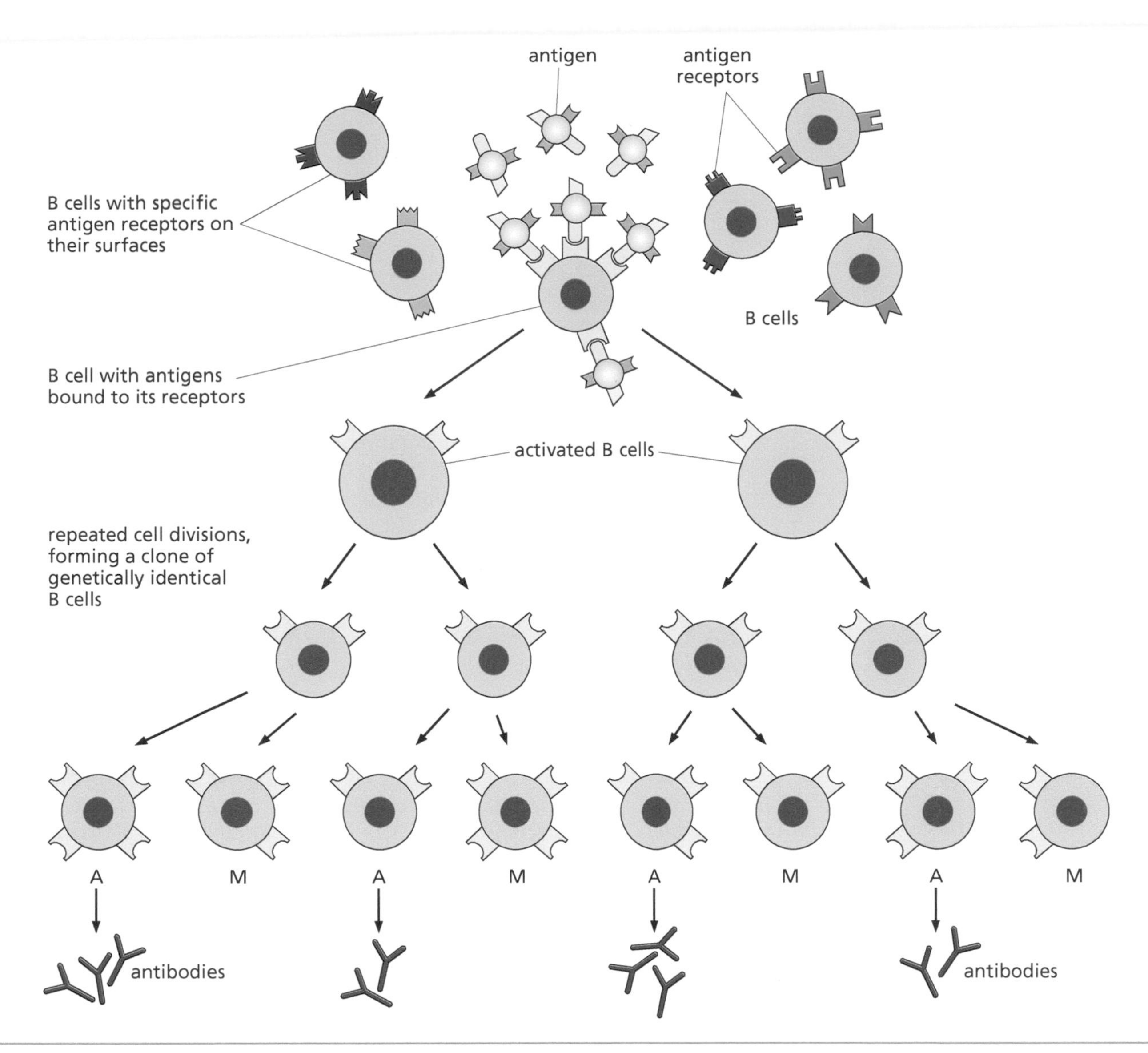

Figure 14.6 **B-cell activation. When a B cell binds to an antigen, it becomes activated and starts to divide repeatedly, forming a clone of genetically identical B cells.** Some of these cells become antibody-secreting cells (A), while others become memory cells (M). Both types of B cells have receptors that bind to the same antigen that was originally encountered, and this is the basis of antigen-specific immunity.

while the other type are long-lived **memory cells** that help with later attacks by the same antigen.

T cells are similarly activated by exposure to antigens that match their cell-surface antigen receptors. If you are exposed to a particular virus (such as influenza virus), for example, a group of cytotoxic T cells that recognize the virus will develop and kill off the virus. In addition, a group of memory T cells will be ready to make a fast response on the next exposure to the same virus.

Antigen recognition and cell division take some time, so when a person first encounters a particular infectious agent (pathogen), the pathogen is able to make that person sick before the immune system fights it off (Figure 14.7). In other words, on the first exposure, the immune system may not block the pathogen fast enough to prevent the disease, but then, as specific immunity develops, it is able to stop the infection, thereby ending the disease. Some of the specific B and T cells produced in that first encounter remain in the body as memory B and T cells, sometimes for the lifetime of the individual. The second time the person is exposed to that same pathogen, there are more memory cells than on the first encounter, so the body responds more quickly and more strongly. In most cases this heightened response is enough to prevent the illness from occurring on the second exposure.

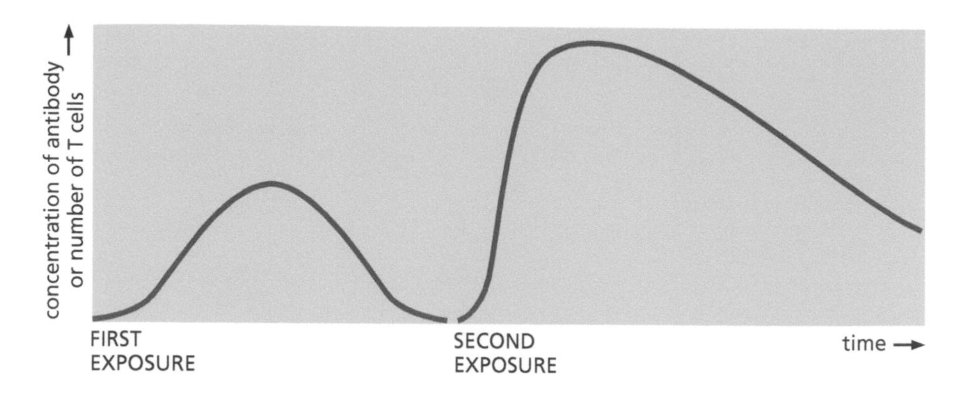

Figure 14.7 The activity of the immune system changes in response to a second exposure to an infectious microorganism. The secondary response is much stronger than the first and also much more rapid.

B and T cells also encounter all of the body's molecules in the lymph nodes, but remember, in a fully functional immune system, that the self-reactive B or T cells were eliminated before they matured. If the body's molecules become altered due to mutation, B and T cells may exist that can bind the altered molecules; this mechanism can eliminate some cells that have become transformed and is therefore an important defense against cancer (see Chapter 10).

14.2.4.1 Immunization Remember that we are not born with specific immunity, but with the potential to develop it. We develop specific immunity only to those things to which we are exposed in our individual lifetimes, and so one person's "immune repertoire" will not be the same as another's.

This is the basis for **immunization**, also called **vaccination**. When you receive a vaccination, you are given molecules from various disease-causing bacteria or viruses, but in a form that will not cause disease. Your immune system responds to this artificial challenge and establishes a group of memory cells, which stand ready to protect you from later exposure to the real bacteria or virus. The vaccine, in other words, substitutes for the first exposure shown in Figure 14.7. If you are later exposed to live, virulent forms of the pathogen used in the vaccine, you will have the rapid and strong response typical of the second exposure in Figure 14.7, protecting you from disease. Immunization, working *with* the disease-preventive powers of the body, has become one of our most effective ways of preventing many infectious diseases. Immunization does not guarantee that a person will not get a disease, as different people's immune systems will mount immune responses of varying strengths upon vaccination. In general, however, if a vaccinated person does become infected with a pathogen, the resulting disease is likely to be much less severe than if they were unvaccinated.

Because the activation of each immune response is antigen-specific, several immune responses can be occurring at one time. Because each antigen-specific response is essentially independent of any others, responding to one bacterial or viral infection does not prevent responses to others. This also means that it is safe to receive more than one vaccination at a time. Further, for pathogens that mutate quickly, such as influenza and several other viruses, new vaccine formulations are required on a regular basis (as often as annually) to respond to changes in the immunogenic molecules of the pathogen (see Section 16.1.1 for the concepts of antigenic shift and antigenic drift).

14.2.5 Passive immunity

If we have to *acquire* our own specific immunity, why aren't we killed by our first exposure to bacteria when we are very young? First, newborns do temporarily have some specific immunity transferred from their mothers; some antibodies cross the placenta, enter the fetal circulation, and can protect an infant for six months or so after birth. Other antibodies can be passed from mother to child in breast milk, particularly in the first week or two of breastfeeding (**Figure 14.8**). Remember that these antibodies protect the infant against antigens for which the mother has acquired immunity, which may or may not be the antigens to which

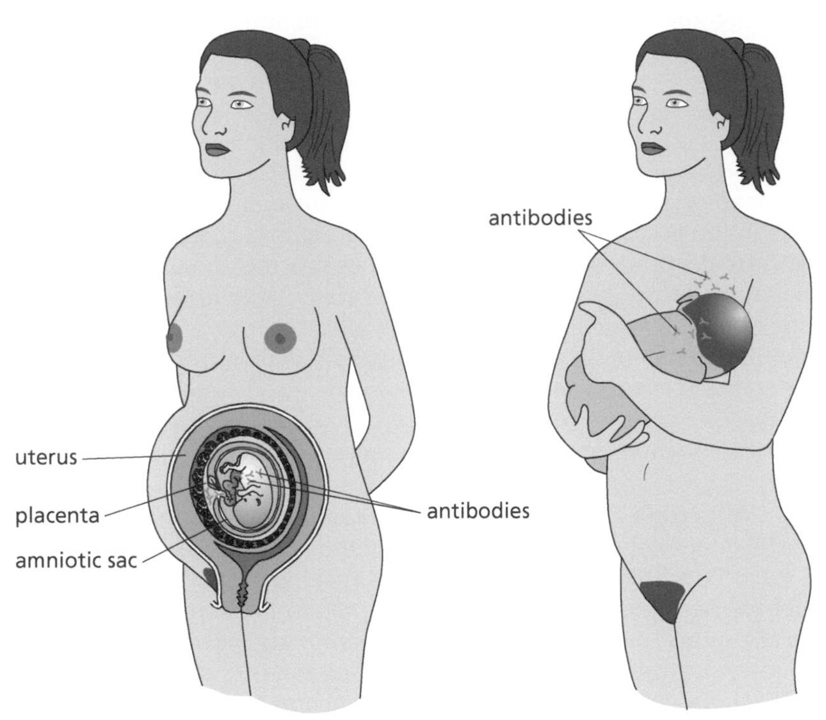

Figure 14.8 **Passive immunity transferred from mother to child.**

the infant is exposed. This is a type of **passive immunity**, meaning antigen-specific immunity transferred from another individual. Passive immunity is only temporary. Antibody proteins, like all proteins in the body, are constantly being degraded, and the infant has not acquired the mother's antigen-specific B cells to produce more. Passive immunity can also be transferred from one adult to another by transfusing blood containing specific antibodies or by giving an antibody-containing fluid derived from blood called "gamma globulin".

In addition to passive immunity transferred from its mother, a baby is born with innate immunity. The parts of the innate immune system operate even on our first exposure to some, but not all, new infections. The blood proteins called complement are able to bind to and kill some types of bacteria or to inactivate some types of viruses without the aid of antibodies. Viruses nonspecifically induce lymphocytes to secrete a cytokine (called **interferon**) that prevents replication of other virus strains, as well as the virus strain that induced its secretion. Cells capable of engulfing particles (macrophages and neutrophils) can engulf some types of bacteria and fungi without antibodies and thus seem to be analogous (and also homologous) to the nonspecific immune system of invertebrates. Because innate immunity does not involve antigen-specific T cells or B cells, it does not show memory and is not stronger on the second exposure to the same antigen.

14.2.6 Harmful immune responses and immunosuppression

Almost every biological system is changeable to some degree; healthy organisms are constantly adapting and adjusting to their environments. The immune system is probably among the most changeable of all the body's systems. Some or all parts of the immune system can either be inhibited or strengthened by both the internal and external environments of the organism.

The widespread metaphorical view of the immune system as our defender against disease may lead us to assume that the immune system is always protective. However, there are several situations when it is not. The abnormal reactions of the immune system are generally against a specific antigen or a small number of antigens, while other antibodies and immune cells function normally. In other cases, the whole immune system can be suppressed.

14.2.6.1 Autoimmune diseases **Autoimmune diseases** result when the immune system begins to make a self-targeted immune response, resulting in both antibodies and cytotoxic T cells that react with antigens in the body's own tissues. The immune cells and/or antibodies then try to rid the body of these antigens as if they were non-self, resulting in damage to the body's own tissues. Although the mechanisms that produce tissue damage are known for some autoimmune diseases, the factors that trigger autoimmunity are unknown.

Multiple sclerosis is an autoimmune disease in which some cytotoxic T cells develop that are specific for an antigen on the insulating myelin sheath around nerve cells in the brain (see Section 11.1.1 and Figure 11.2). These cytotoxic T cells migrate to the brain, where they kill the cells bearing this antigen. The ensuing damage to the nerve sheaths causes a variety of neurological problems, depending on exactly which nerves have been affected.

In insulin-dependent diabetes mellitus (type 1 diabetes), self-reactive T and B cells develop that recognize and destroy the cells in the pancreas that produce the hormone insulin. Because insulin controls the cellular uptake of glucose, its absence produces severe consequences throughout the body.

14.2.6.2 Allergies An **allergy** is an inflammatory immune response to a substance that does not usually pose a threat to the body. People who suffer from allergies do so because their immune systems react atypically to some antigens from which the host does not need protection (pollen or dust mites, for example). The atypical response produces a special type of antibody called IgE that is specific for these antigens, which are called **allergens**. IgE binds to cells of the immune system called **mast cells**. When the person later encounters the same allergen, the allergen binds to the IgE on the mast cells, triggering the explosive release of histamine. Histamine is one of the chemicals that plays a positive role in the first stages of inflammation, making blood vessels leaky to allow the entrance of neutrophils and macrophages into the tissue (Figure 14.4A). In an allergic reaction, however, larger amounts of histamine are suddenly released (**Figure 14.9**), producing the various symptoms of allergy. Whether an allergic response produces runny eyes, sneezing, or shortness of breath (such as in asthma) depends on the tissue in which the mast cells were triggered (the eyes, nasal lining, or the lungs). Because the symptoms are produced by histamine, antihistamine medications stop the symptoms by blocking the binding of histamine to cells in the blood vessels. Antihistamines, however, do not prevent the immune response or the release of histamine by the mast cells.

Other allergies such as skin rashes that occur in response to contact with poison ivy, latex, or the dyes and other chemicals in cosmetics or clothing are mediated by T cells, not by IgE antibodies and mast cells. Consequently, antihistamines do not block these reactions, although they may alleviate the associated itching. Severe T-cell-mediated allergic responses are treated with steroid drugs, which we discuss later.

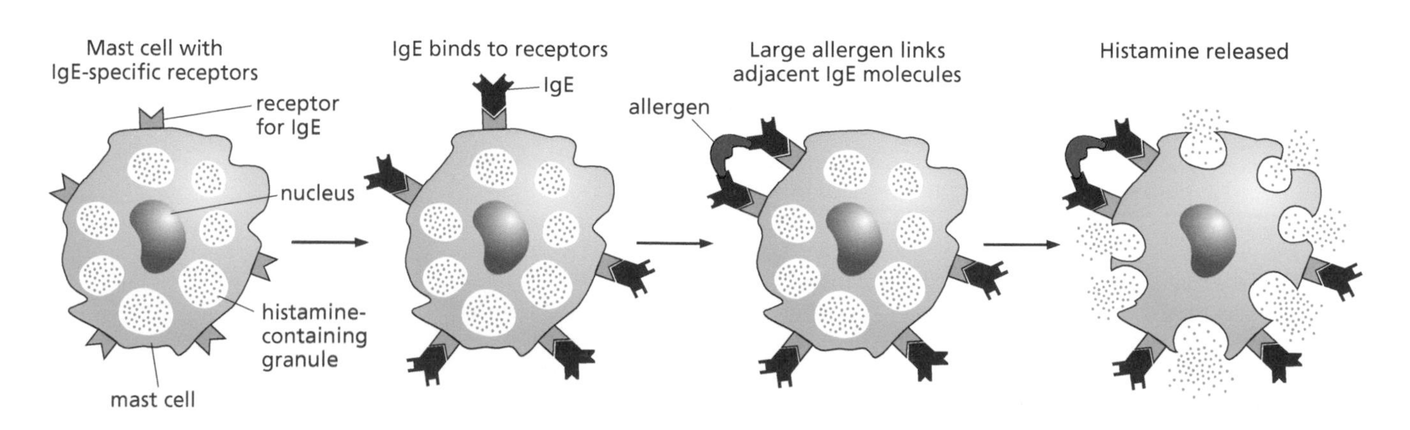

Figure 14.9 Allergic release of histamine.

Because each allergy is an antigen-specific immune response, it shows memory and a greater response on the next exposure, which is why people's allergies can worsen over time. Although there are thousands of different substances that produce allergy in some people, each person with allergies is usually bothered by only a few. The severity of allergy to one substance will not predict the severity of allergy to some other substance, since each is a separate, antigen-specific response. We are not yet able to predict who will become allergic or what they will become allergic to, but there does seem to be some inherited component, because allergies do run in families.

14.2.6.3 Transplant rejection Organ transplantation is a very effective (although very expensive) way of replacing damaged organs. To be successful, the organ donor and recipient must "match" in a series of molecules called human leukocyte antigens (HLAs) and also known as **transplantation antigens**. The HLA proteins are cell-surface proteins that "present" various antigens to the cells of the immune system so that those immune cells can screen them for being "self" or "non-self." These HLA proteins are encoded by about a half-dozen genes, most of which have several hundred known alleles. A person's full set of HLA antigens will be expressed on their cells. Because there are so many alleles, the chances of an unrelated donor having the same set of antigens are slim. However, all of the genes are located close together on one chromosome, so they are generally inherited as a unit. Two siblings thus have a 25% chance of having the same set of antigens.

If donor tissue has the same transplantation antigen type as the recipient's own tissue, the transplant is accepted. However, if the donor organ is of a different transplantation antigen type, the recipient's immune system will mount an immune response against it. Both T cells and antibodies will arise and kill the transplant. This is why organ banks attempt to find the closest possible match when an organ becomes available. A perfect match is very rare, so an organ recipient must take immunosuppressive drugs for the rest of their life to minimize the chance of organ rejection. These drugs will also make the person more susceptible to infections, as explained later. For bone marrow transplants used to treat blood cancers (Section 10.4.3), if the donor tissue has a small amount of difference in HLA antigens, the donor cells can mount an immune response that helps to eliminate the cancer cells; however, if the incompatibility is too great, a life-threatening response can occur. Therapeutic cloning using induced pluripotent stem cells derived from the same person they are used to treat (discussed in the online supplement to Chapter 10) overcomes the problem of HLA antigen incompatibility, since the antigens of the therapeutic tissue are identical to those of the patient's other cells.

14.2.6.4 Immunological tolerance When immune responses to a specific antigen are inhibited, it is called immunological **tolerance**. Such tolerance can be induced. For example, people who suffer from allergies can often be desensitized—that is, made nonresponsive—to those particular antigens. The desensitization procedure consists of giving the person repeated small doses of the substance to which they are allergic, usually in the form of "allergy shots." The procedure must be carried out very carefully because giving the wrong dose, either too much or too little, will make the allergy worse, not better. Because both allergy and its desensitization are antigen-specific, the procedure must be carried out for each separate allergen. Induced tolerance to one allergen does not change the ability of the individual to react to other allergens or to truly harmful antigens. Once established, tolerance is long lasting and prevents that allergic response for years or decades.

14.2.6.5 Immunosuppression Other factors can inhibit the functioning of all or parts of the immune system to all antigens and are thus said to produce **immunosuppression**. Because immunosuppression inhibits the workings of innate immunity or of all B cells or all T cells, rather than just antigen-specific groups, it is generally correlated with an increase in disease.

Many environmental pollutants and other chemicals suppress the whole immune system. Other immunosuppressive factors include alcohol, cocaine, and heroin. Chronic use of these leads to increased incidence of infectious disease and, in the case of alcohol, increased incidence of cancer (Chapter 10). Particular foods have not been found to be immunosuppressive, but too much food (overnutrition) has been. People who are obese have a higher incidence of infection-related sickness and death. Chronic protein undernourishment, even of a moderate nature, also impairs the ability of the immune system to fight off infectious diseases. The high rate of mortality in undernourished infants is partly due to immunosuppression, leading to a high incidence of infections. Deficiencies of many of the micronutrients, including magnesium, selenium, zinc, copper, vitamin A, and vitamin C (see Section 8.2), impair various aspects of the immune system. Many elderly people become deficient in one or more of these micronutrients and consequently become immunosuppressed. Many of the infectious diseases of the elderly can be minimized in frequency and severity if nutrition is adequate.

On the other hand, in the case of autoimmunity, where the immune system has turned against itself, or in organ transplantation, suppression of immune reactivity may be needed to prevent tissue damage. Because immunosuppression is nonspecific, it also inhibits protective immunity, thereby increasing susceptibility to infectious disease; therefore, the risk–benefit ratio of immunosuppressive drugs must be carefully assessed in each case.

Psychological factors, such as stress, change the internal environment and are among the factors that can induce immunosuppression. Psychological factors can also positively affect the immune system; in this case, the effect is called immune potentiation. We will examine psychologically produced immunosuppression and immune potentiation in more detail after we examine the workings of the neuroendocrine system.

THOUGHT QUESTIONS

1. Since the beginnings of immunology just before 1900, its language has often been very military. The immune cells are said to "protect us from invaders" or to "kill off foreign antigens." Locate an immunology textbook or an article on immunology from the popular press. Can you find examples of military language in these accounts?
2. We need words to convey what we imagine the immune system to be doing, based on experimentation and hypothesis testing. Do you think that the new imagery of the immune system as a communications system has taken hold at this time because we are in the "information age" or because new discoveries brought about a need for new terminology? To what extent are scientific terms metaphors for reality and to what extent are they models? Can the words we choose cloud our view or prevent us from being open-minded about new hypotheses?
3. Make a list of the thoughts that come to your mind when the word *self* is used in a nonimmunological context. Make a second list for your thoughts about what *self* means in immunology. Is there any overlap between your two lists?
4. When people get a bacterial or viral infection, they often get a fever. Why? Why do people sometimes get a fever after a vaccination?

14.3 THE NEUROENDOCRINE SYSTEM CONSISTS OF NEURONS AND ENDOCRINE GLANDS

The **endocrine glands** are a series of organs that secrete chemical products directly into the bloodstream. (In contrast, glands that release their secretions into the digestive tract or through the skin are called **exocrine glands.**) These secreted chemicals alter the function of target organs and thus can be said to carry messages from one organ to another. The chemicals secreted by endocrine glands are called **hormones.** Because hormones are distributed throughout the body by the bloodstream, the target or receiving cells can be far removed from the endocrine glands.

It was once common for the endocrine glands to be described as an endocrine system because the actions of these glands were thought to be separate from the other known communication system of the body, the nervous system. In the past few decades, however, biologists have discovered that several hormones originally thought to be secreted only by cells of the endocrine glands are also secreted by brain cells. Other endocrine secretions are chemically related to the neurotransmitter substances originally thought to be secreted only by the cells of the nervous system. Today, the endocrine and the nervous systems are considered to be so completely intertwined that many scientists now refer to them collectively as the **neuroendocrine system**.

All vertebrate nervous systems have a central nervous system (CNS), consisting of a brain and spinal cord, and a peripheral nervous system (Chapter 11). The peripheral nervous system connects the CNS to the more distant parts of the organism (the sensory receptors, muscles, glands, and organs) and is itself composed of the **somatic nervous system** and the **autonomic nervous system**.

14.3.1 The autonomic nervous system

The autonomic nervous system has been considered, at least by Western scientists, to be largely involuntary, carrying signals to and from the gut, blood vessels, heart, and various glands, and thus regulating the internal environment of the body. *Autonomic* literally means "self-governing" or "self-regulating," reflecting the fact that the autonomic nervous system can work by itself without any input from the centers of conscious awareness in the brain. The autonomic nervous system can thus regulate body functions even while we are distracted or asleep. In contrast, the somatic nervous system, largely under conscious, voluntary control, carries signals to and from the skeletal (voluntary) muscles, skin, and tendons.

14.3.1.1 The sympathetic and parasympathetic divisions The autonomic nervous system consists of two functionally separate divisions, called the sympathetic and the parasympathetic nervous systems. As is shown by Figure 14.10, the two divisions of the autonomic nervous system have opposite effects. In general, the neurons of the **sympathetic nervous system** and its principal neurotransmitter, norepinephrine, ready the organism for heightened activity, while neurons of the **parasympathetic nervous system** and its neurotransmitter, acetylcholine, do the opposite. Tissues and organs throughout the body receive nerve endings and neurotransmitters from both of these divisions of the autonomic nervous system.

During rest or when the organism is not receiving much sensory input, the parasympathetic system is predominant, diverting much of the body's work towards digestion and restorative processes. While most voluntary muscles are inactive, the involuntary rhythmic muscle contractions (peristalsis) of the stomach and intestine are stimulated, and so are the secretions of saliva, bile, and stomach enzymes. The pupils of the eye and the bronchi of the lungs meanwhile constrict, and the heart rate slows. In contrast, during moments of physical exertion or emergency, the hypothalamus at the base of the brain signals the sympathetic system to dominate, stimulating the pupils and bronchi to dilate, the storage molecule glycogen to break down into glucose to be metabolized for energy, the heartbeat to accelerate, and the digestive processes to slow down. When a sympathetic response is no longer needed (and the stimuli that produced it are no longer present), the parasympathetic system predominates once again. In extreme cases, this switch can be rapid, producing a rebound effect, as when a person feels woozy or faint after an emergency situation is over.

14.3.1.2 Fight or flight Imagine that you are crossing a street. You hear a loud noise! You turn your head suddenly, and a large truck is heading right at you! Your heart begins to pound faster, your sweating increases, your voluntary muscles are stimulated (and their threshold for action is lowered), your breathing speeds up, and your digestive organs stop digesting your last meal. (You may even feel nauseated in extreme cases.) All these are the results of stimulation of different organs and tissues by the sympathetic nervous system and its neurotransmitter, norepinephrine (also called noradrenaline). In general terms, the sympathetic nervous system prepares the body for reactions that require large amounts of energy and oxygen for voluntary muscle contraction, including the "fight-or-flight" response, so-called because the individual is primed to fight hard or to get away fast.

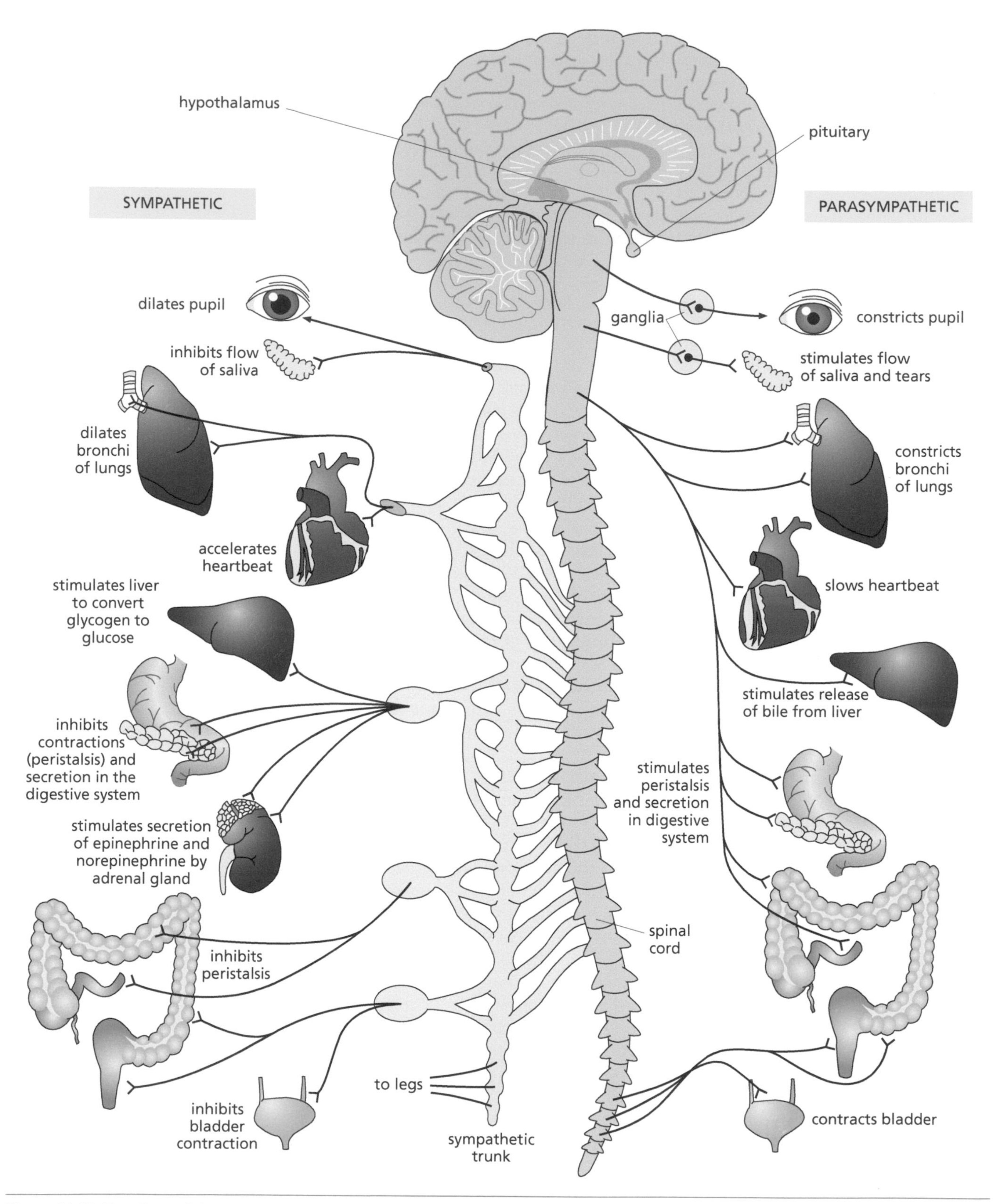

Figure 14.10 Functions of the sympathetic and parasympathetic nervous systems.

14.3.1.3 Rest and ruminate Now imagine having finished a sumptuous candlelight dinner with your favorite food, elegant service, and quiet music playing softly in the background. You are relaxing in a comfortable chair or sofa with your favorite drink in your hand and wonderful company nearby. Just thinking about this scene (or re-reading this paragraph to yourself slowly and calmly) can relax you, cause your heartbeat and your breathing to slow down, your sweating to stop, your voluntary muscles to relax (and to raise their threshold for action), and your blood to be diverted to digestive organs, which are now digesting the sumptuous meal. These are all the effects of the parasympathetic nervous system and its neurotransmitter, acetylcholine. The parasympathetic division prepares the body to "rest and ruminate," activities that use less oxygen while replenishing the body's store of energy supplies.

As you may have been able to demonstrate to yourself as you imagined the scenes just described, the actual frightening or relaxing situation need not be present. The fight-or-flight response can be triggered by just thinking of tense situations or by watching a frightening movie. Similarly, a rest-and-ruminate response can be brought about just by relaxing comfortably and imagining a relaxing, pleasurable situation. The triggers for these responses can thus originate completely within the brain of the person imagining them.

14.3.2 The stress and relaxation responses

Research on the fight-or-flight response by a Canadian (Austrian-born) physiologist, Dr. Hans Selye, showed that it is the first step of a larger series of physiological reactions. These reactions are produced by chemicals secreted by the nervous system and also by the endocrine and immune systems. The process begins when some stimulus or force, called a **stressor**, causes the body to deviate at least temporarily from its normal state of balance. The body's response to this deviation from homeostasis is called **stress**, or the **stress response**. Stress consists of physiological and immunological changes that allow our bodies to fight off or remove ourselves from stressors and return to homeostasis, and thus stress can be a useful response. However, when stress persists too long, it can become harmful, causing disease or even death.

14.3.2.1 Alarm **Alarm**, the first stage of the stress response, includes the fight-or-flight response. The hypothalamus stimulates the sympathetic neurons to secrete norepinephrine, stimulating an endocrine gland called the adrenal gland to secrete epinephrine (also called adrenaline). Epinephrine and norepinephrine together bring about the physiological changes known as fight or flight. In addition, sympathetic neurons release norepinephrine directly into the lymphoid organs (the spleen, thymus, and lymph nodes), stimulating these organs to release their store of lymphocytes into the bloodstream. As the lymphocytes are released, the organs in which they were stored decrease in size. The hypothalamus also secretes a hormone called adrenocorticotropic hormone (ACTH), which stimulates the adrenal gland to produce corticotropin-releasing hormone (CRH), which in turn stimulates steroid hormones, particularly cortisol (the so-called "stress hormone"), to be released from the cells of the outer layer of the adrenal gland. The alarm phase of the stress response is shown in pink in the upper part of Figure 14.11.

14.3.2.2 Resistance If the stress continues, the body enters the second stage, **resistance**, in which resources are mobilized to overcome the stressor and regain homeostasis. Epinephrine acts on the heart to increase the heart rate (the number of contractions per minute) and cardiac output (the strength of the contractions). Norepinephrine increases the flow of blood to the heart and muscles for possible increased activity, while constricting other vessels, diminishing the flow of blood to the gut, skin, and kidneys. If the stressor is a disease or injury, inflammation begins and innate immune cells are chemically attracted to the inflamed area. New stores of steroid hormones are synthesized, keeping blood levels of these hormones elevated. Chronic elevation of cortisol begins to reduce antibody and cytotoxic T-cell activity. Cortisol also suppresses inflammation. The resistance phase of the stress response is shown in red in the middle part of Figure 14.11.

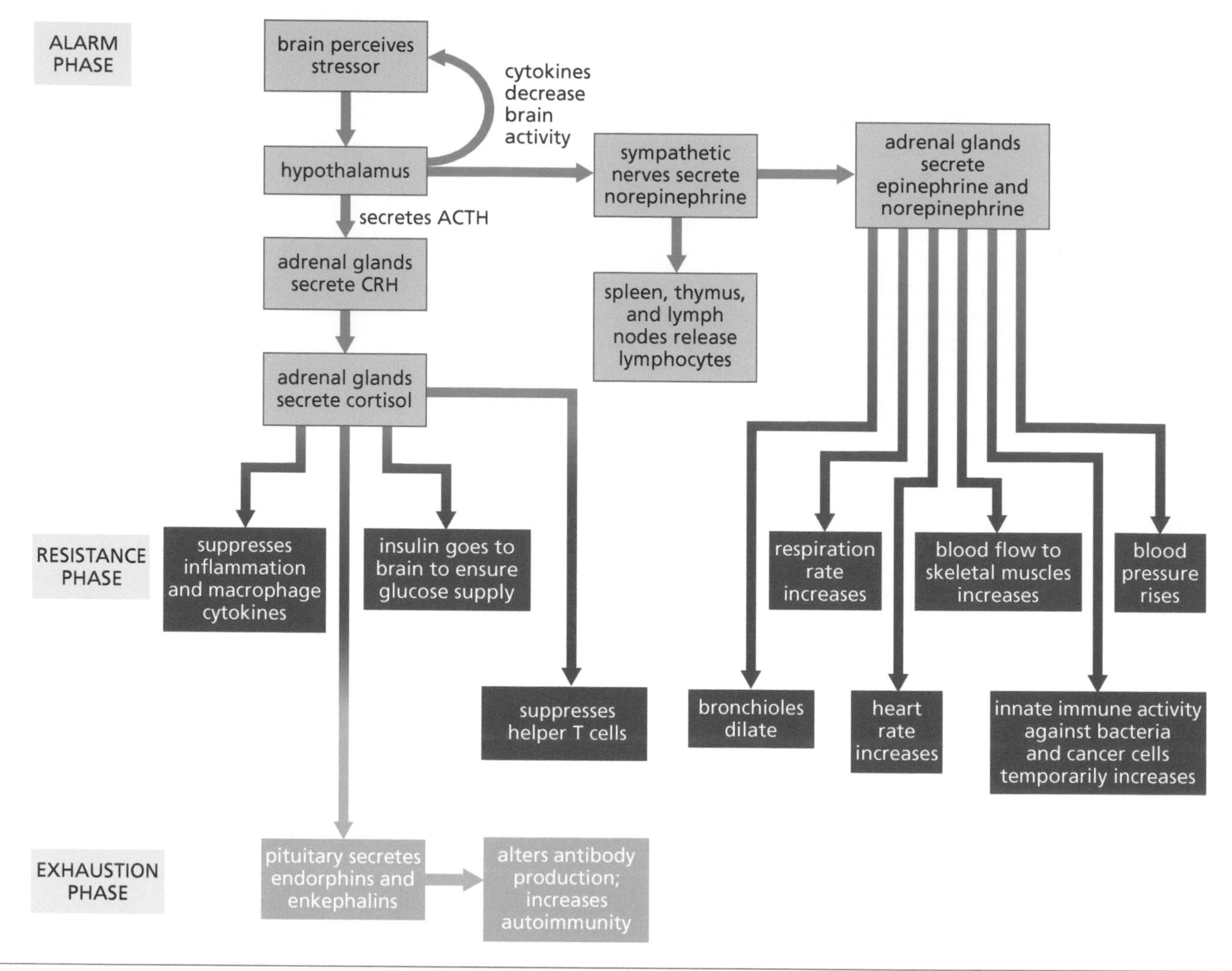

Figure 14.11 The phases of the stress response. Stressors induce a variety of physiological changes, including immunological changes, which are mediated by the sympathetic nervous system and adrenal hormones. CRH is corticotrophin-releasing hormone.

14.3.2.3 Exhaustion If the stressor is not successfully overcome, the stress response reaches its third phase, **exhaustion**. The steroids made in the resistance phase are used up and the body is unable to make more. During the exhaustion phase, the action of the sympathetic nerves tapers off, while the endocrine organs take over. Adrenal hormones stimulate another endocrine gland, the pituitary gland, to secrete chemicals (endorphins and enkephalins) that are structurally related to opioid drugs (see Section 13.2.1) and that alter the activity of neurons and of immune cells.

The pituitary was once called "the master gland" because its hormone secretions control the activity of many other endocrine glands, but it is now known that the pituitary itself is under the control of the brain via the hypothalamus. Chemicals secreted by the hypothalamus also act on other cells within the brain itself, changing the activity levels of neurons, and this change affects behavior, heat production, and many other functions.

Some stress responses are due to actual physical danger, but stress responses can also result from the demands of work or school (e.g., deadlines or examinations), the actions of the people in our lives, or from other external events. Researchers have surveyed people experiencing certain types of events and have found many types of events that produce stress. A stress response occurs in nearly everyone experiencing the death of a spouse and in 73% of people undergoing a divorce. Nearly any change, whether happy or sad, also produces stress, indicated by the fact that 63% of people experiencing marital separation but also 45% of those experiencing marital reconciliation exhibited a stress response, as did 50%

of those getting married and 39% of those welcoming a new child into the family. All stressors, whether real or perceived, can have the same biological consequences: the physiological changes that characterize the stress response.

14.3.2.4 The relaxation response

When actual physical danger has passed, the stress response will abate. During a stress response, the adrenal and pituitary glands mediate the response of the sympathetic nervous system. This response will gradually be reversed by the actions of the parasympathetic nervous system. This reversal is called the **relaxation response**.

It is sometimes more difficult to turn off our mentally induced stresses. Since the stress response can be mentally induced, it has been hypothesized that the relaxation response should be mentally inducible as well. Some cultures and some religions have been more open to this idea than others. Many practices that are aimed at evoking this relaxation response, including yoga, transcendental meditation, and others, have their origins in Asian traditions. Hindu yogis have learned how to consciously control the actions of their autonomic nervous systems and to bring about levels of activity even below the normal levels for the resting state. Measurements made by Western scientists have shown that these yogis are able to lower their blood pressure, breathing, oxygen consumption, heart rate, and metabolic rates.

Other cultures, including many Western cultures, have been less open to the idea that the relaxation response can be controlled. The traditional definition of the autonomic nervous system emphasized that it governed involuntary functions over which we do not have conscious control, and this had the unintended effect of discouraging research on any possible interactions of the autonomic nervous system with our emotions and other conscious body states. Recently, however, some of the methods for conscious control of autonomic processes are being borrowed from other cultures. Some athletes have learned meditation, while others have learned to invoke specific mental imagery in order to put one's body as well as mind in a certain state of relaxed determination to succeed in sport. Western medicine has begun to use similar techniques to help cancer patients in fighting their cancers, as described further in a later section on mental imaging.

Less conscious strategies can also produce the relaxation response. Studies using measurements of blood pressure and other physiological indicators have shown that contact with pets can reduce stress and bring about relaxation. Older people who keep pets have been shown to live longer than those who do not, even when comparison is made between people of comparable health status initially, and the effect is even more pronounced for cardiovascular mortality specifically (Kramer et al., 2019). Heart attack victims have reduced mortality and longer survival if they care for a pet, and they are much less likely to have a second heart attack (Mubanga et al., 2019). Studies such as these show a statistical correlation between two factors. However, from such data by themselves, we cannot assign a causal relation between the two; that is, we cannot say definitively that caring for pets caused the increased longevity or improved health, just that the two often go hand in hand.

THOUGHT QUESTIONS

1. Is adrenaline (epinephrine) a hormone or a neurotransmitter?
2. How does thinking about an annoying or threatening event produce stress?

14.4 THE NEUROENDOCRINE SYSTEM INTERACTS WITH THE IMMUNE SYSTEM

We have discussed how the nervous system and the endocrine system communicate with each other. In recent decades, it has also become apparent that both of these systems also communicate with the immune system. The emerging concept is that the nervous, endocrine, and immune systems do not exist as separate entities, but are rather one interacting communications network. Because, as we have seen, the immune system protects against disease, communication between the brain and the immune system suggests a mechanism by which the mind can affect health.

Psychoneuroimmunology has an appealing central model, which suggests that the mind and body are intertwined. But models or theories that are appealing or that fit with our common sense do not always stand up to scientific scrutiny. Is there any scientific evidence that supports the hypothesis that the nervous, endocrine, and immune systems are interconnected and that the mind, therefore, affects health? Yes. We examine the evidence in this section. Furthermore, we extrapolate findings to discuss the possibility that a person can learn to control their immune system.

14.4.1 Evidence for one interconnecting network

The first line of evidence comes from the discovery that the immune, endocrine, and nervous systems use the same cytokines for cellular communication. In addition, there is both structural and functional evidence gathered from in vitro studies ("in glass," that is, studies in laboratory glassware) and in vivo studies ("in life," that is, studies in animals and humans).

14.4.1.1 Shared signals Many different types of signaling molecules are produced by cells and function in communication with other cells. Signals initially discovered in particular body systems have historically been given different names: signals produced by neurons are called neurotransmitters, those produced by endocrine tissues are called hormones, and those that modulate immune function are called cytokines. The ability of any cell to respond to a particular signaling molecule is dependent on whether the cell has receptors for that signal. Therefore, to consider the idea that signals secreted by the nervous, endocrine, or immune systems might have effects on the other systems, one must ask if there are receptors for them on cells of the other systems. The answer is yes, there are. For example, consider endorphins, which are endogenous opioids and also neurotransmitters. American pharmacologists Candace Pert and Solomon Snyder, who discovered endorphin receptors on nerve tissue in 1973, discovered in the early 1980s that there were similar receptors for endorphins on immune cells. Since that time, receptors have been found for numerous other signals that interconnect the nervous, immune, and endocrine systems. B lymphocytes have receptors for many different neurotransmitters, including norepinephrine and enkephalins. The finding that the number of these neurotransmitter receptors per lymphocyte increases during immune activation suggests that they serve an important role in the immune response. Lymphocytes also possess receptors for several endocrine hormones induced under conditions of stress (ACTH and steroid hormones) or pain (β-endorphin). Neurons in the hypothalamus have receptors for the cytokines secreted by immune cells, and it is the effect of these cytokines on the hypothalamus that induces fever. The neuroglia cells that feed the neurons in the brain (see Section 11.1.1) also have receptors for immune cytokines, as do some endocrine cells. Thus, a picture emerges in which these various classes of signaling molecules may in fact be less distinct and have more overlapping functions than was previously believed.

14.4.1.2 Nerve endings in immune organs Evidence of another sort came from studies performed by American scientist David Felten on the nerve supply of the organs of the immune system. These studies used a technique called immunohistochemistry. In this method, immune organs from animals are frozen and sliced very thin (*histology*, *histo-*) and then are stained with antibodies (*immuno-*) bound to enzymes (*chemistry*). This procedure takes advantage of the exquisite antigen specificity of the immune system. Antibodies that recognize some molecule you want to detect (usually a protein) are produced in a laboratory animal or in cultured cells. The antibodies used by Felten's group recognize and bind to a protein that is an enzyme used in the synthesis of norepinephrine, but

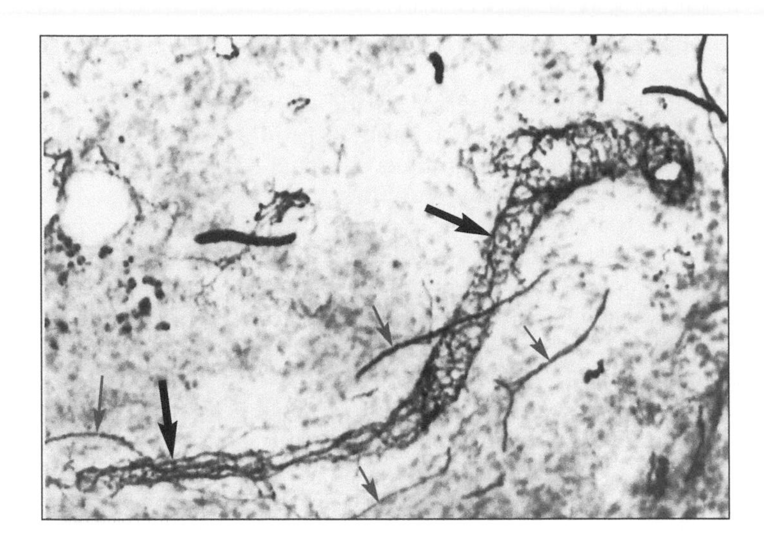

Figure 14.12 Immunohistochemical staining showing the presence of the neurotransmitter norepinephrine in the rat spleen. Large black arrows: blood vessels. Small red arrows: nerve fibers synthesizing norepinephrine.

do not bind to other proteins. Thus, the antibodies give specificity to the technique, just as they do during an immune response in the body. The detection enzyme associated with the antibodies produces a color change in the location where the antibodies have bound, thus allowing identification of sites where the target molecule, the norepinephrine synthesis enzyme, is present. Some typical results are shown in Figure 14.12.

While most nerve cells terminate in synapses with other nerve cells or on muscle cells (Section 11.1.1), Felten's studies showed nerve cells terminating and releasing norepinephrine in proximity to the cells of the immune system. Neurons of the sympathetic nervous system were found to terminate in the immune organs, such as the spleen, thymus, lymph nodes, bone marrow, and lymphoid tissue in the gut. The sympathetic nerve cells release norepinephrine, and the immune cells in these organs have receptors for norepinephrine.

14.4.1.3 Analysis of cytokine function in animals Demonstrating that a receptor is present, or even that both the receptor and its cytokine are present together in a tissue, does not by itself show that any effect follows the binding of the cytokine to its receptor. For that, experiments known as functional assays (techniques used to measure a response) need to be performed. Such studies are often carried out using animals as experimental models.

Using functional assays, some hormones have indeed been shown to have effects on the immune system. During an acute bacterial infection, the secretion of adrenal and pituitary hormones increases. Functional studies showed that the effect is not a direct one. If pituitary cells are stimulated with bacterial molecules in vitro, they do not secrete these hormones. When immune cells are exposed to these bacterial products, however, they secrete cytokines. If immune cell cytokines are administered to animals, the level of pituitary hormone in the blood increases. Thus, the cytokines produced by the immune cells are a necessary intermediate for stimulating the pituitary gland to secrete hormone.

Neuroendocrine cytokines have also been shown to have effects on the immune system. Among the cytokines secreted by the brain cells are the enkephalins and endorphins. When enkephalins are given to living rats, immune responses are altered, including antibody responses and T-cell-mediated responses. Interestingly, low doses of enkephalin increase antibody production, while high doses suppress it. Low doses of enkephalins also increase some destructive aspects of the immune response, including both allergic and autoimmune responses, while high doses of enkephalins depress those responses.

Just as low doses of enkephalins may actually strengthen aspects of the immune system, so too does short-lasting stress. Engulfment of bacteria by white blood cells in mice (see Figure 14.4C) is increased by short-term stress brought on by conflict (see Figure 14.11). The ability of immune cells to kill tumor cells can be increased by restraining rats on a single day so as to increase their stress response.

However, several days of restraint-induced stress causes a decrease in tumor-killing activity.

The steroid hormones secreted by the adrenal cortex during the stress response can have inhibitory effects on the immune system. Steroid hormones such as corticosterone given to animals decrease the numbers of cells in lymphoid organs and suppress secretion of immune cytokines (Figure 14.11). Daily fluctuations (circadian rhythms, see Section 11.3.3) in the plasma levels of corticosterone also correlate with the circadian rhythm of the numbers of B and T cells circulating in the blood. Further, steroids increase the susceptibility of the animals to disease and activate latent infections (disease caused by infectious organisms that were previously present without causing any symptoms but are now stimulated to do so).

A compound chemically similar to the stress hormones, hydrocortisone, is used medicinally to block inflammation. Remember that normal inflammation is the healing phase of the immune response (Figure 14.3). In insect bites, severe poison ivy, athletic injuries, and rheumatoid arthritis, the swelling and pain are the result of the inflammatory response. In these situations, the annoying or harmful symptoms are actually an indication that the immune system is at work. Thus, a person who chooses to take an anti-inflammatory drug is choosing to suppress the healing processes of the immune system in order to suppress the negative symptoms of inflammation. The symptoms may be so severe that immunosuppression is needed, but long-term immunosuppression by corticosteroids will likely have adverse consequences on other aspects of health.

14.4.2 The placebo effect

In clinical trials testing new drugs, a common experimental design is for one group of people to receive the test drug and for another group (the **control group**) to receive a **placebo**, a preparation that is similarly colored and flavored but that does not contain the test ingredient. For the drug to be considered effective, there must be a statistically significant difference in outcome between the group receiving the experimental drug and the control group receiving the placebo.

Studies like this were initially designed to demonstrate whether particular drugs were effective. What they have also shown, over and over again, is that the people who receive the placebo in such tests have a significant change from the baseline values of whatever parameters are being measured, an effect known as the **placebo effect** (Figure 14.13). They also experience many "side effects," although they have not received a drug. In experiments on pain perception, people who are given placebos instead of painkillers very often experience a reduction in pain. Similarly, people who have purchased street drugs will often feel the

Figure 14.13 The placebo effect.

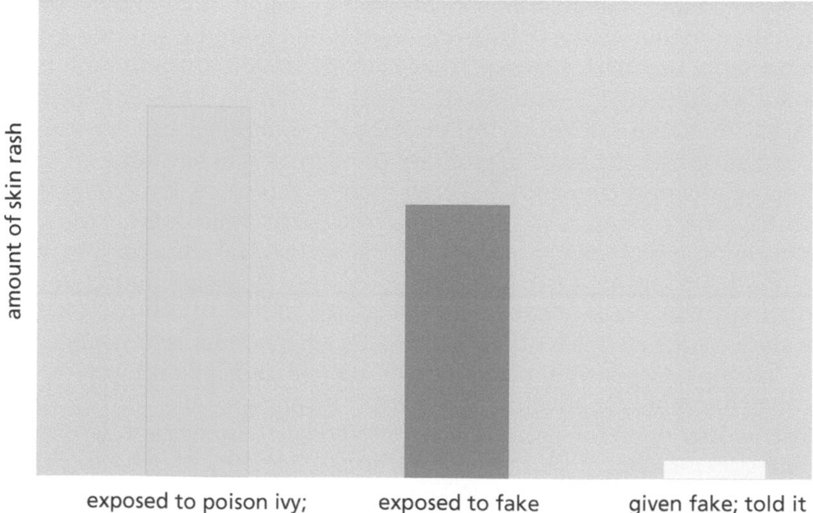

reaction they seek even when the drugs are in such low concentration that no real effect could be produced.

Such placebo effects have often been considered an annoyance in research; that is, they make it more difficult to demonstrate the "real" effects of test compounds. The existence of such effects, however, is additional evidence that the mind can bring about physiological changes in the body in addition to the effects of the stress response and the relaxation response.

14.4.3 Effects of stress on health

To what extent does the stress response influence human health? The answer is, to a great extent. Stress (i.e., the stress response) is an important risk factor in heart disease, and there is considerable evidence that people exposed to chronically high levels of stress are statistically more likely to become ill with infectious diseases, to remain ill for longer periods, and to suffer more severe consequences, even death.

The biochemical events of stress can be triggered by psychological factors, and prolonged stress can suppress the immune system. There have been many demonstrations of immune suppression in people undergoing various types of stress. It follows that long-term psychological stressors might produce conditions in which disease can develop. In one experimental design, blood samples were taken from young, basically healthy medical students during an exam period, and various immune parameters were compared to baseline levels measured in blood samples from the same students one month prior to the exam period (Glaser et al., 1993). In this type of experiment, each person serves as his or her own control, minimizing differences due to factors other than the tension of exam situations. Exam periods brought on an increase in adrenal gland hormones and a decrease in immune cell activity.

There were also decreases in the number of another type of lymphocyte called helper T cells. Helper T cells do not eliminate antigen; instead, they secrete cytokines that "help" boost the strength of the responses of cytotoxic T cells and B cells to antigens (see Section 15.1.1). Without these helper T cells, the immune response of the cytotoxic T cells and B cells is often not strong enough to prevent disease. Thus, when stress decreases the activity of helper T cells, this results in an overall decrease in both B- and T-cell responsiveness. In these studies, such short-term stress was correlated with an increase in disease, primarily upper respiratory tract infections. Many college health centers report increases in student admissions for infectious diseases during exam periods.

Another study examined the immune function and health status of men who had separated or divorced within the previous year (Kiecolt-Glaser et al., 1988). The experimental design was different from the one used with the medical students. One group of people, divorced men, was compared to people in a control group of married men. Not only were the immune systems of the divorced men found to be impaired, they experienced a greater number of illnesses than did the controls. Comparisons were also made between those men who had not initiated the separation or divorce and those who had. Those who had not initiated the break were significantly more immunosuppressed and had more illnesses than those who did. Studies of this type employ statistical methods for determining if the differences between groups are greater than could have been predicted by chance only.

Other researchers have found that elderly people who had been caring for a spouse with Alzheimer's disease demonstrated a decrease in three different measures of immune function (Kiecolt-Glaser et al., 1991). The elderly people undergoing prolonged stress got sick more often than those in the control group (people of similar age and health status but who were not caring for a spouse with Alzheimer's disease). The immune systems of these caregivers stayed depressed after the death of the spouse.

Several studies have showed decreased immune function in people with clinical depression. A meta-analysis integrating the results of a large number of prospective studies showed that people who had experienced prior depression and then developed cancer were more likely to die from their cancer than those who had not been depressed (Pinquart and Duberstein, 2010). A **prospective experimental design** is one in which a group of people are examined at the start of the study (using either physical or psychological exams) and their outcomes are monitored at various later times. One strength of a prospective study is that no one knows ahead of time who will be sick and who will not; baseline data are taken before the outcomes are known. One weakness is that the percentage of people in any particular group who will get a particular disease may be very low, so the number of people in the study must be very large. Many people will leave the study for unrelated reasons. Many other factors can influence the outcome; to some extent this can be corrected by statistical methods, but only those factors that have been identified can be factored out by statistical methods.

What mechanisms could bring about cancer after clinical depression? The immune system normally serves as an important barrier to tumor formation, targeting abnormal cells for destruction before they can develop into cancer via a process called immunosurveillance (see Section 10.6.6). In the depression study, immune activity was compromised in the depressed individuals, and other functions were also impaired, including levels of an enzyme that repairs damaged DNA. Breaks and mistakes in copying DNA occur rather frequently, but normally several "proofreading" mechanisms check the DNA and repair most of the mistakes; mistakes that remain uncorrected result in mutations that may promote the development of cancer (see Section 10.5.3 and Figure 10.20). The suppressed immune systems of the depressed individuals would be less successful at immunosurveillance, thus increasing the likelihood of cancer developing.

14.4.3.1 Individual variation in the stress response There are many types of stressors. The effects of stressors on health are highly variable, since they are modified by many additional factors. For example, of the people exposed to infectious mononucleosis, prevalent in college-age populations, not everyone becomes sick, and of those who do, some become sicker than others. It makes a difference whether the stress occurs before or after the immune challenge started. The severity and duration of the stress are also important. Genetic factors have some role; in animal studies, different strains of mice (each inbred to minimize genetic variation within the strain) respond differently to stressors. Psychological factors are just as important. If an animal is able to establish coping behaviors, the effect of the stress period on the immune system will be lessened.

Personality profiles and life events have some bearing on disease susceptibility and disease progression in humans as well. Testing methods have been developed for quantifying the psychosocial impacts of life events. The use of such methods has indicated that certain life events can increase the probability that cancer will develop, although the results have been highly variable from study to study.

We cannot predict how much immunosuppression will be sufficient to result in disease in a given person. Both the degree and duration of suppression that result in disease will likely be different for different people. Some studies suggest that coping styles can help regulate the degree of impact that stressful events will have on individual health. When psychological tests are given to matched sets of cancer patients (with the same kind of cancer and in comparable stages of the disease), those with more optimistic or aggressive personalities show higher survival rates than those who are more easily resigned to what they perceive to be their fate. Other factors, such as environmental pollutants, drugs, alcohol, and malnutrition, may also weaken the immune system. If a person's immune system is already weakened by one or more of these factors, the additional immunosuppressive effects of stress are more likely to result in disease.

14.4.4 Conditioned learning in the immune system

If brain cells can interact with immune cells and psychological factors can influence the onset and progression of disease, could a person learn how to control his or her own immune response? As odd as this idea may sound, evidence in support of it is accumulating.

Classical, or Pavlovian, conditioning is a type of unconscious learning in which an organism learns to associate one stimulus with another. After such conditioning, presentation of the second stimulus will bring on the physical effects of the first stimulus (see Section 11.3.1). Robert Ader, a psychiatrist, and Nicholas Cohen, an immunologist, worked together to try to explain why some of Ader's mice had been dying unexpectedly in his studies on a drug called cyclophosphamide. This drug suppresses the immune system and in fact is given to recipients of organ transplants so that their immune systems do not reject their transplants. Ader's mice had been receiving cyclophosphamide along with saccharin in their drinking water. Later, the mice received only the saccharin, but their immune systems again became suppressed as they had when given the cyclophosphamide, even though saccharin itself has no effect on the immune system. What Ader and Cohen demonstrated in several controlled studies is that the dying mice had been conditioned. After the mice learned to associate the immunosuppressant chemical cyclophosphamide with the saccharin, the immunosuppressant effect could be produced by giving them only the saccharin water without cyclophosphamide (Ader and Cohen, 1993).

These experiments demonstrating conditioned immunosuppression have been repeated with several other paired stimuli. Further, immunosuppression triggered by such conditioning has been shown to improve the health of mice with autoimmune disease; preliminary results suggest that conditioned immunosuppression is also useful in the treatment of people with autoimmune disease.

Experiments have similarly demonstrated conditioning enhancement of an immune response, as well as conditioned immunosuppression. If animals are challenged with an antigen paired with another stimulus, classical conditioning that boosts immunity can be demonstrated (**Figure 14.14**). In a normal immune response, a second exposure to the same antigen would produce an increase in specific immunity to that antigen. In a typical conditioning experiment, the animals are given their first exposure to an antigen paired with a conditioned stimulus such as saccharin. Saccharin itself does not induce antibody for the antigen (**Figure 14.14A**). When saccharin is given at the same time as the first exposure to antigen, animals make antibody to the antigen (**Figure 14.14B**). After conditioning, when the animals are exposed to saccharin, they react as though they were being exposed for a second time to the antigen. Saccharin induces an increased secretion of antibody for the antigen, without a second exposure to antigen (**Figure 14.14C**).

14.4.5 Voluntary control of the immune system

Other work has shown that people can learn to voluntarily regulate many of the physiological processes mediated by the autonomic nervous system. People can learn to regulate the temperature of their hands, their blood pressure, their heart rate, and their galvanic skin resistance (resistance to electrical conductivity, which is a measure of the amount of sweat on the skin). Since the immune system communicates with the autonomic system, these findings raised the hypothesis that parameters of the immune system may also be subject to voluntary control. Several studies have shown that this is possible using different self-regulation procedures, including relaxation, mental imaging, biofeedback, and emotional support. Voluntary potentiation of several immune parameters has been shown, including white blood cell engulfment of bacteria, antibody production, lymphocyte reactivity, and natural killer cell activity. For example, women with metastatic breast cancer, all of whom were receiving medical treatment of their cancers, survived longer if they were part of support groups than if they were not (Spiegel et al., 1989).

14.4.5.1 Mental imaging Several studies have shown that mental activities, including the formation of mental images, can affect the cells of the immune system. One such study (Rider and Achterberg, 1989) randomly assigned subjects to two groups. Images of neutrophils, accompanied by background music, were shown to one group, while images of lymphocytes were shown to the

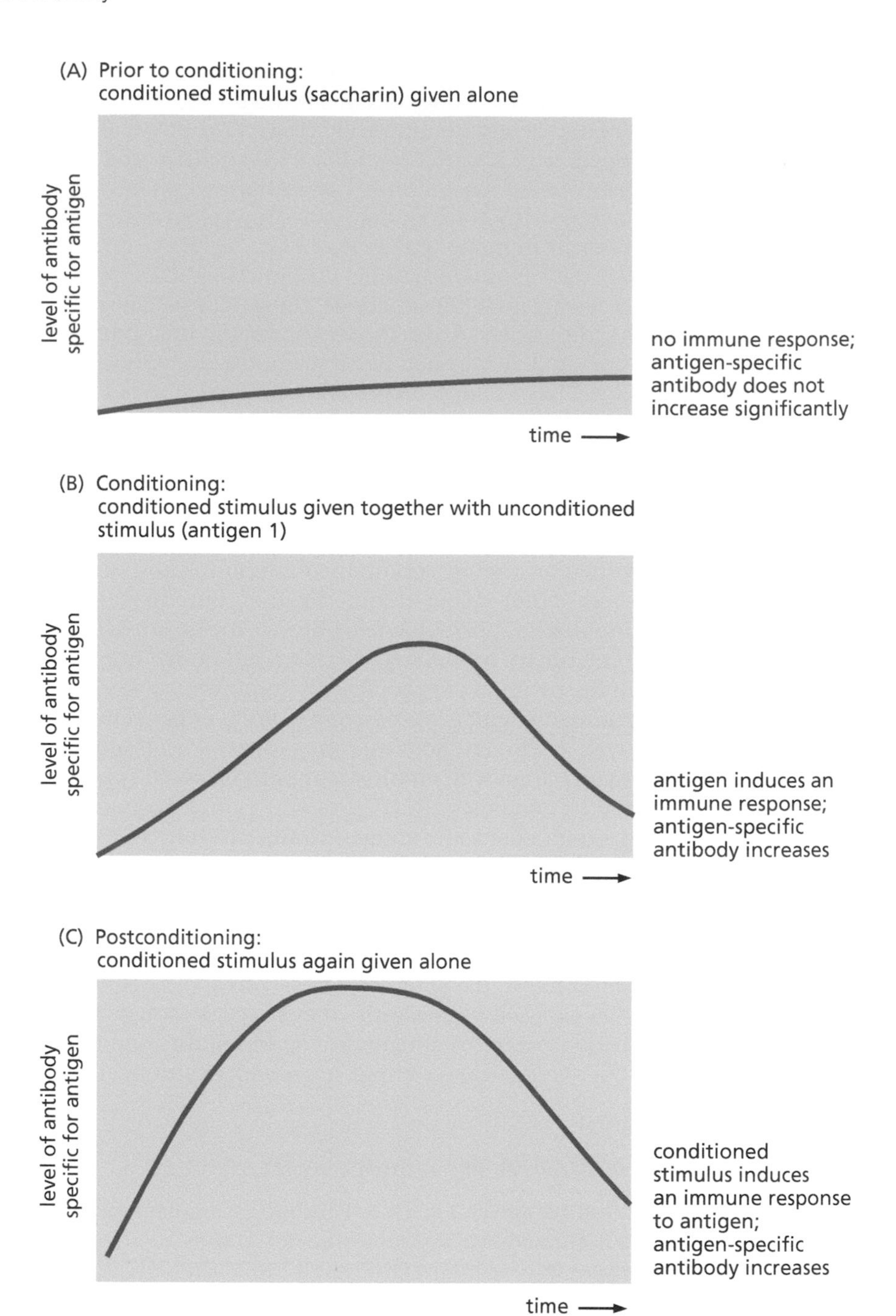

Figure 14.14 Classical conditioning of the immune response.

other group. After six weeks of training, the first group was able to significantly decrease the level of circulating neutrophils (but not lymphocytes) in their blood, while the second group was able to significantly decrease the level of circulating lymphocytes but not neutrophils. A study by Hall, Minnes, and Olness (1993) found that subjects were able to increase the adherence activity of their neutrophils through training sessions in forming mental images of neutrophil activity. In a follow-up study (Hall et al., 1996), one group of subjects was able to increase the adherence of their neutrophils by means of a relaxation response, while another group was able to decrease neutrophil adherence by forming mental images. As these studies show, mental images and other brain activities can have measurable effects on the cells of the immune system.

14.4.5.2 Biofeedback In **biofeedback**, measurement devices are placed on people so that they can monitor the results of their self-regulation. Biofeedback has proved to be effective for some people in the management of chronic pain and migraine headaches. Although it is too soon for there to be conclusive data on whether voluntary regulation of immunity will translate into improved health, preliminary studies suggest that it will. As one example, people with HIV infection and AIDS have remained healthier when they have used these techniques.

14.4.5.3 Studies on populations If the mind and the emotions can influence the immune system, and the immune system helps fight off many diseases, how far can the disease-fighting process be controlled by the mind? In recent years, statistical evidence has been accumulating from studies in both the United States and China that shows that dying patients can exercise control over their disease processes to the extent that they can actually influence the time of their death.

Large-scale studies are often done on entire populations by studying death certificates and comparable records. When the date of death is examined for a large number of patients in a population at large, several interesting regularities appear. For instance, the overall mortality rate is lower for a period of several days to either side of each person's birthday, and this is compensated for by an increasing mortality rate about a week or two later. Such data make it appear that dying patients are eager to survive to reach their birthdays and that they can postpone the inevitable by as much as a week or two. Other studies have shown similar statistical effects demonstrating the ability of people to postpone the time of their death until after holidays or family events (e.g., weddings) of special importance to them. In China, this effect even has a name: the Harvest Moon phenomenon. The Harvest Moon Festival is a traditional family celebration in which the oldest and most respected woman in each family is expected to prepare a large feast to celebrate with her entire family. Studies of death certificates show that older Chinese women have a reduced mortality rate around the time of this festival (Phillips and Smith, 1990) and that the timing of the effect follows the calendar variations in the date of this lunar festival. It is difficult to determine whether this effect is primarily the result of the activities in which the matriarch engages, the increased esteem or importance that she receives, her desire not to disappoint others, or simply the anticipation of the big event. Studies on Jewish populations have shown a similar decline in mortality around the time of Passover.

A complex interaction between expectation and mortality has also surfaced in another study in China. Traditional Chinese astrology divides the calendar into 12-year cycles. Each year is represented by a different animal, and people born in that year are said to be under control of that animal's influence, with which certain diseases are associated. In this study, elderly Chinese patients were surveyed to see whether or not their disease matched the predictions of Chinese astrology. Patients with a disease that matched their astrological year were then compared with patients having the same disease but a different astrological year. The patients whose diseases matched their astrological year experienced higher mortality, and this effect was proportional to the patient's belief in traditional Chinese astrology. Presumably, patients who believed that they had the disease that was fated for them in the stars more willingly gave up the struggle and resigned themselves to an earlier death. Although these phenomena are well documented, the mechanism(s) by which they are produced are not known.

THOUGHT QUESTIONS

1. Given that chronic or severe stress generally weakens the immune system, how might such an apparently harmful relationship have evolved?
2. Is stress always harmful?
3. List the types of experiments done in psychoneuroimmunology. Do the results of any of these falsify the hypothesis that the mind and the body interact? Do any of these results prove that the mind and the body interact?
4. Is the Cartesian concept of the mind the same thing as the brain? Is *mind* simply the name we give to the *workings* (or functions) of the brain?

CONCLUDING REMARKS

Since the mid-nineteenth century, Western medicine has tended to view disease as having external causes. The ascendancy of this view can be traced back to the work of Louis Pasteur and Robert Koch who championed the theory of "specific etiology" as part of the germ theory of disease. In this theory, each disease has one specific and identifiable cause. This theory led to much highly successful research that associated single species of microorganisms (bacteria, viruses, and parasites) with specific diseases. Such research ultimately produced vaccines and antibiotics, which have successfully controlled many infectious diseases. However, many of the diseases that are still without effective cures today are chronic diseases (cancer and heart disease, for example) that do not appear to have simple, single causes. A new concept of disease and of health may therefore be needed to find therapies and preventive measures for these diseases.

Psychoneuroimmunology is redefining our concepts of health and disease. Scientists in this field are using new technologies to reexamine some old concepts of disease causation. Working at the same time as Pasteur, another French scientist, Claude Bernard, questioned what it meant for a microorganism to "cause" a disease. He observed that there were very great differences in individual response to microorganisms; some people got sick and even died, while other people who were also exposed did not get sick. Bernard, a physiologist, developed an alternative theory, that of the *milieu intérieur*, or inner environment, as being an equal determinant in whether or not a person became sick. The past century of research in immunology and, more recently, in psychoneuroimmunology suggests that even the diseases for which an infectious agent is known are not caused by the microorganism alone, but rather by the outcome of a complex process in which the microorganism disturbs homeostasis while host mechanisms attempt to restore it. Bernard's theory is compatible with a view of the neuro-endocrine-immune communication network as a sensory organ by which deviations from homeostasis are detected and corrected.

Psychoneuroimmunology also borrows from Chinese traditional medicine and ayurvedic medicine in India, as well as other Asian traditions that view health as the balance of life forces. Like these, the psychoneuroimmunology paradigm uses a *functional* model of the body, a model that regards the body as an entity that is in a constant state of change. Health is the state in which these forces are in balance, in homeostasis; disease is the state in which they are not. African traditions in which a person's health and well-being are seen to be influenced by the social environment in which the person lives coincide with the psychoneuroimmunology view that mental and emotional factors can affect health and disease. Within the psychoneuroimmunology paradigm, scientists are testing hypotheses suggested by these ancient traditions. Widening one's point of view and being open to new ideas are integral parts of science. New hypotheses are formed by the melding of ideas and then must be followed by the hard, and often slow, work of hypothesis testing.

CHAPTER SUMMARY

- The immune system is a system that works to detect and correct deviations from **homeostasis** within the organism. It is composed of **white blood cells** that travel throughout the body in the blood and in the **lymphatic circulation**.
- **Health** is the ability to maintain or return to homeostasis.
- **Innate immunity** is present from birth and does not depend on exposure to develop. Innate immunity rids the body of some pathogens and initiates **inflammation** and wound **healing**.
- **Specific immunity** is acquired by the individual after exposure to specific **antigens** such as bacteria or viruses.
- By forming specific groups of **B cells** and **T cells**, the immune system retains a memory of the encounter so that it can react faster and more strongly to subsequent exposures to that antigen. This is the basis for **immunization**: An artificial first exposure gives protection against later natural exposure to the same disease.

- Products of the specific immune response (**antibodies** and cytotoxic T cells) rid the body of the specific antigen that induced their production.
- **Passive immunity** is specific immunity developed in one person and transferred to another, for example, by transferring antibody to them.
- Specific immunity can be selectively turned off. This is called development of **tolerance**.
- The immune system interacts with the **autonomic nervous system** and the **neuroendocrine system** in an integrated and multidirectional way. Communication among these systems is mediated by chemicals called **cytokines**.
- Factors that interrupt this communication network will prevent the restoration of homeostasis within the organism, producing disease.
- Mental states can affect the functioning of the immune system and can either increase or decrease its disease-fighting activity, a theory known as **psychoneuroimmunology**.
- The **stress response**, through the action of the sympathetic nervous system and stress hormones, can produce **immunosuppression**, while the **relaxation response**, through the action of the parasympathetic nervous system, can reverse this process.
- A **placebo** is a compound without physiologic effect that is given, for example, to the control group in a clinical test. Mental expectations can result in people experiencing physiological symptoms in response to the placebo, a result known as the **placebo effect**.

BIBLIOGRAPHIC REFERENCES

Bibliographic references to material in this chapter can be found online at biologytrending.routledge.com/chapter14

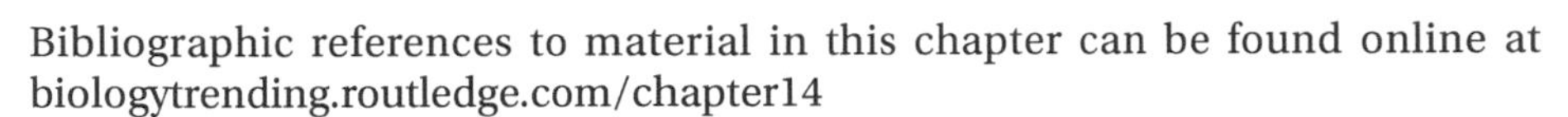

GLOSSARY: KEY TERMS TO KNOW

These key terms are defined at the end of each chapter as an aid for student review.

Acquired immunity See *Specific immunity*.

Alarm The first phase of the stress response, in which the body quickly secretes hormones such as cortisol and norepinephrine, preparing the body for "fight or flight."

Allergen Anything provoking an allergic response by the immune system.

Allergy An inflammatory immune response to a substance that does not usually pose a threat to the body.

Antibodies Proteins secreted by B lymphocytes during an immune response that bind to and disable the specific type of molecule that induced their secretion, thus helping to protect the body's health.

Antigen Any molecule or part of a cell that is detected by the immune system.

Autoimmune disease Any disease in which the immune system attacks "self" cells not normally recognized by healthy immune systems.

Autonomic nervous system Part of the peripheral nervous system that regulates "involuntary" physiological processes of the body; consists of the sympathetic and parasympathetic divisions.

B lymphocyte (B cell) A type of lymphocyte that makes antibodies.

Biofeedback Monitoring one's own physiological activity as a means of learning how to modify this activity (e.g., to reduce stress).

Complement Blood proteins that, in combination with antibodies, can destroy some bacteria and viruses.

Control group In an experiment, a group used for comparison. For example, if animals are experimentally exposed to a drug, then a control group might consist of similar animals not exposed to the drug but treated the same in every other way.

Cytokines Molecular signals that produce effects on cells and functions of the immune system.

Endocrine glands Glands that secrete their products, called hormones, into the bloodstream rather than into a duct.

Exhaustion The third and final phase of the stress response, in which the sympathetic nervous system ceases to respond and the pituitary gland secretes endorphins and enkephalins that lessen the pain.

Exocrine glands Glands that secrete their product into a duct.

Health The ability of an organism to maintain homeostasis or to return to homeostasis after a disease or injury.

HLA antigens See *Transplantation antigens*.

Homeostasis The ability of a complex system (such as a living organism) to maintain conditions within narrow limits. Also, the resulting state of dynamic equilibrium, in which changes in one direction are offset by other changes that bring the system back to its original state.

Hormones Chemical signals secreted into the bloodstream that produce physiological responses elsewhere in the body.

Immunization (vaccination) Artificial exposure to an antigen that evokes a protective immune response against a potential disease-causing antigen similar in structure to the antigen in the vaccine.

Immunosuppression Decreasing the strength of future immune functions in any manner that is not antigen-specific.

Inflammation A physiological response to cellular injury that includes capillary dilation, redness, heat, and immunological activity that stimulates healing and repair.

Innate immune system Host defenses that exist prior to that individual's exposure to an antigen and are not antigen-specific.

Interferon A cytokine secreted by lymphocytes that prevents viral replication.

Interstitial fluid A water-based fluid filling in spaces between cells, especially in connective tissues, and containing immune system cells such as lymphocytes.

Lymph A fluid containing white blood cells but no red blood cells.

Lymphatic circulation An open circulatory system in vertebrate animals that gathers intracellular fluid and returns it to the blood along with cells of the immune system.

Lymphocytes White blood cells that have specific receptors for antigens and are therefore capable of forming an antigen-specific immune response.

Mast cells Cells whose release of histamine causes inflammation.

Memory cells Cells of the immune system that retain the ability to respond rapidly to an antigen that the body has encountered before.

Neuroendocrine system The nervous system and the endocrine system considered as an interactive whole.

Neutrophils White blood cells that can surround and digest bacteria by phagocytosis.

Parasympathetic nervous system A division of the autonomic nervous system that brings about the relaxation response and secretes acetylcholine as the final neurotransmitter.

Passive immunity Antigen-specific immunity acquired in an organism by transfer of antibodies or specific immune cells from another organism.

Phagocytosis A process in which one cell surrounds, engulfs, and digests another.

Placebo A drug formulation lacking the active ingredient being tested.

Placebo effect Physiological response to a placebo that does not result from the chemistry of the placebo but that often produces the response expected by the subject.

Prospective experimental design An experimental design in which subjects are chosen beforehand and data are subsequently gathered on events as they happen.

Psychoneuroimmunology A theory that postulates that the mind and body are a single entity interconnected through interactions of the nervous, endocrine, and immune systems.

Relaxation response A stimulation of the parasympathetic nervous system, sometimes voluntary and self-induced, in which the stress response is ended, blood pressure and breathing are reduced, the threshold of excitation of nerve cells becomes higher, and digestive activity is stimulated.

Resistance The second phase of the stress response, in which the body's resources are mobilized to regain homeostasis and overcome the stressor by increasing breathing and blood flow and by mobilizing innate immune activity and temporarily causing inflammation.

Somatic nervous system Part of the peripheral nervous system that carries sensory information to the brain, and signals from the brain to the muscles to allow voluntary movement.

Specific (acquired) immunity An acquired, antigen-specific ability to react to a previously encountered antigen.

Stress (stress response) A physiological response or state of heightened activity brought about by the sympathetic nervous system and maintained for a longer time by the endocrine and immune systems.

Stressor Any stimulus or condition that brings on a stress response.

Sympathetic nervous system A division of the autonomic nervous system that brings about the fight-or-flight response and secretes norepinephrine (sometimes epinephrine) as its final neurotransmitter.

T lymphocyte (T cell) A type of white blood cell that helps bring about an antigen-specific immune response without releasing antibodies.

Tolerance In immunology, an acquired lack of response to a specific antigen after repeated contact with that antigen.

Transplantation (HLA) antigens Antigens that cause a host-graft reaction that attacks transplanted tissue from another individual, causing tissue rejection.

Vaccination See *Immunization*.

White blood cells (leukocytes) The several types of blood cells that perform various protective (immune) functions but do not possess hemoglobin and do not carry oxygen.

CONNECTIONS TO OTHER CHAPTERS

Chapter 1: Psychoneuroimmunology is a good example of a new paradigm.
Chapter 2: The great variety of antigen receptor proteins, which can each recognize one of the huge variety of antigens in the world, is based on the rearrangements of just a few genes.
Chapter 5: The immune system has evolved.
Chapter 7: The ability to form an immune response to any particular antigen depends on cell surface proteins. The allele frequencies of the genes that code for these proteins vary among populations, making some populations more susceptible than others to a particular disease.
Chapter 8: Poor nutrition suppresses the immune system.
Chapter 10: Suppressed immune function increases the risk of cancer. Also, some new cancer therapies rely on immune system stimulation.
Chapter 11: The brain can affect many immune functions.
Chapter 12: Immune system functions decline rapidly in old age.
Chapter 13: Many drugs suppress the immune system, and other drugs mimic some of the activities of the autonomic nervous system.
Chapter 15: Immunosuppression is characteristic of AIDS.
Chapter 19: Pollution can suppress immune function.

PRACTICE QUESTIONS

1. What type of white blood cell secretes antibodies?
2. What type of white blood cell can kill cancer cells?
3. What type of white blood cell can engulf bacteria and kill them?
4. What type of cell releases histamine?
5. What part of the body is acted on by macrophage cytokines to induce fever?
6. Which part of the autonomic nervous system mediates the flight-or-flight response?
7. Which part of the autonomic nervous system mediates the relaxation response?
8. Which of the following are parts of the immune system: bone marrow, spleen, skin, intestines, bladder, tonsils, eyes?
9. What types of scientific evidence suggest that the brain and nervous system interact with the immune system?
10. How do the terms *stress* and *stressor* differ?
11. What are the stages of the stress response, and what characterizes each stage?

CHAPTER 15

HIV and AIDS

ISSUES

- How were AIDS and HIV discovered? What other diseases helped or hindered scientists and doctors in the discovery?
- What do HIV tests tell us?
- Will there be a cure for AIDS? What about this disease makes a cure so difficult?
- Will there be a vaccine to prevent AIDS? If a vaccine is possible, will it solve all the problems associated with AIDS?
- Will studying HIV teach us all we need to know about the AIDS pandemic? What social factors contribute to the global spread of AIDS?

BIOLOGICAL CONCEPTS

- Health and disease (immune system, pathogens and response, receptors, Koch's postulates, routes of transmission, behavior)
- Biodiversity (virus structures and life cycles)
- Mutations and evolution of viruses
- Testing for pathogens
- Drug development

CHAPTER OUTLINE

DOI: 10.1201/9781003391159-15

AIDS is a disease caused by the virus HIV. HIV undermines the immune system, leaving the infected person vulnerable to other diseases. As we saw in Chapter 14, the immune system has several ways of protecting the body from disease. When people have AIDS, their immune systems no longer function properly, so they are at risk of becoming ill from infections that would barely affect a healthy person. There is currently no cure for AIDS, and, without proper treatment, many people with the disease suffer long and painful deaths. However, much progress has been made in developing drugs that can control replication of the HIV virus in infected people, allowing many who receive consistent drug therapy to live long and relatively healthy lives. AIDS was first identified in the United States in 1981, but it was soon recognized in countries around the world, making it a **pandemic** disease (see Chapter 16). It quickly became one of the most feared and widely discussed diseases of our time. According to the Joint United Nations Programme on HIV/AIDS (UNAIDS), about 76 million people became infected with HIV from 1980 through 2019, around 33 million died from AIDS or its consequences, and another 38 million people were living with the disease in 2019. Worldwide, more than 17 million children have lost one or both parents to this disease. HIV is spread from person to person in certain infected body fluids (blood, semen, vaginal and rectal fluids and breast milk, but not saliva, urine, or feces). Sexual contact is one of the main routes of transmission; thus, AIDS is a sexually transmitted disease (see Chapter 16). Throughout the HIV/AIDS pandemic, infected individuals have been stigmatized, often due to misconceptions concerning the disease, and this has caused infected individuals to experience significant psychological and emotional burdens, as well as interfering with robust testing and treatment programs. In this chapter we describe the biology of HIV and the path to scientific understanding of the virus; we also address the current status of the HIV pandemic, modern HIV therapies, and the social aspects of HIV/AIDS.

15.1 AIDS IS AN IMMUNE SYSTEM DEFICIENCY

The acronym **AIDS** stands for **Acquired Immunodeficiency Syndrome**. **Acquired** means that the illness is not genetically inherited as the result of a defective DNA message; rather, it is caused by an infectious agent (a virus). **Immunodeficiency** means that some part of the immune system is not functional, and **syndrome** means that a wide range of symptoms are associated with the disease. People whose immune systems are deficient can become seriously ill with infectious diseases or cancers.

The body has many ways of protecting itself from diseases. The skin protects the body's surface against entry of bacteria, viruses, or other microorganisms. The mouth, vagina, and many other potential entry points are coated with mucous secretions that continually wash away adherent bacteria, inhibit bacterial growth, and promote healing.

A more specific type of protection is afforded by the immune system. As discussed in Chapter 14, an organism's immune system distinguishes between molecules that are part of the organism (self) and ones that are not (non-self). Non-self molecules are carried by bacteria, viruses, and cancer cells. Many non-self molecules trigger an immune response that inactivates or destroys the invading pathogens or abnormal body cells (see Figure 14.4).

An **immunodeficiency** is an absence of one or more of the normal functions of the immune system. People who are immunodeficient get sick more often than people with healthy immune systems, and their illnesses last longer and are more severe.

How does someone become immunodeficient? Some of the many causes are inherited and some are environmental. One type of inherited immunodeficiency, severe combined immune deficiency (SCID), caused by a lack of the enzyme adenosine deaminase (ADA), is discussed in Section 4.1.2. Inherited immunodeficiencies are rare; much more common are those that are acquired as a result of environmental exposures. The functioning of the immune system can be depressed, for example, by alcohol, cocaine, or marijuana (see Chapter 13); psychological stress and depression (see Chapter 14); cigarette smoke and other pollutants; malnutrition (either total calorie deficit or micronutrient malnutrition; see Chapter 8); and cancer therapies (see Chapter 10). Clearly, AIDS is not the only kind of immunodeficiency, but it is among the most severe. Many immunodeficiencies are temporary and reversible: If the causative factor is removed, the immune system recovers. In contrast, AIDS is long-lasting; the immune system does not recover, and the disease is fatal unless it is carefully controlled with lifelong drug regimens.

15.1.1 AIDS is caused by a virus called HIV

AIDS is caused by the **Human Immunodeficiency Virus** (**HIV**). This virus specifically infects **helper** (**CD4**) **T cells**, a type of lymphocyte or white blood cell (see Section 14.2). These cells secrete a signaling molecule (cytokine) called interleukin-2, and both the cells and the interleukin-2 are necessary for the B-lymphocyte and **cytotoxic** (**CD8**) **T-cell** responses of the immune system (**Figure 15.1**). As you may recall from Section 14.2, B cells make antibodies, our main defense against bacteria and fungi. Cytotoxic T cells protect against viral infections and against

Helper T cells, B cells, and cytotoxic T cells have receptors for nonself molecules. Within each cell population, individual cells have receptors for different nonself molecules.

B lymphocytes and cytotoxic T lymphocytes require two signals to become functional immune cells: (1) binding specific nonself molecule to their receptors, causing receptors for IL-2 to be brought to the cell surface, and (2) binding of IL-2 to IL-2 receptors.

Helper T cells that have bound their specific nonself molecule are the source of IL-2; thus removal of helper T cells shuts down both B-cell and cytotoxic T-cell activity.

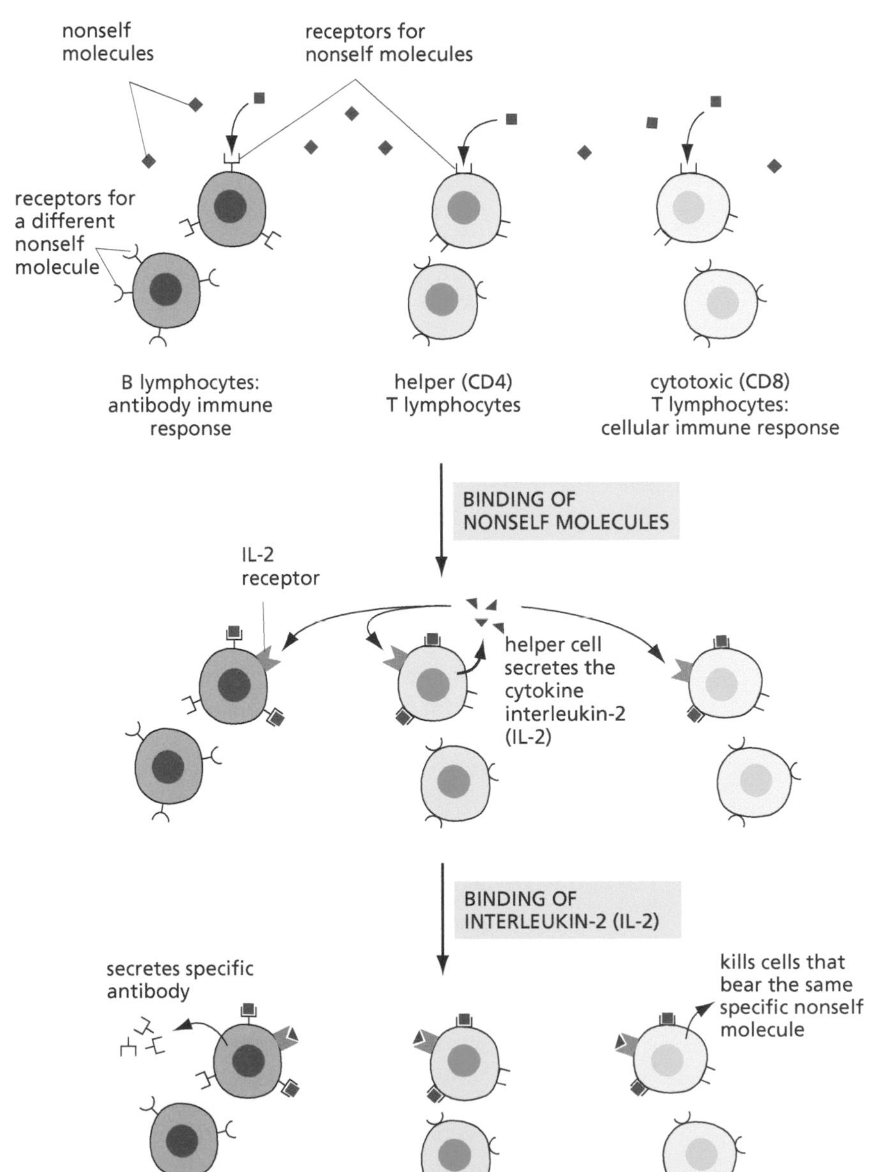

Figure 15.1 Interactions between lymphocytes and their cytokines.

cancer cells. When helper T cells are destroyed in AIDS, both the B-cell and cytotoxic T-cell arms of the immune system are lost, leaving the individual vulnerable to bacterial, fungal, and viral infections and also to cancer. In addition, the neutrophil and macrophage white blood cells of nonspecific immunity (see Figure 14.1) are also severely weakened. Macrophages, like helper T cells, can be directly targeted in AIDS, and, without antibodies, neutrophils cannot engulf bacteria. Therefore, an infection that would be minor in a person with a healthy immune system can quickly become life-threatening in a person with AIDS.

When the syndrome now known as AIDS first began to appear in the United States at the very end of 1980, the cause, and even the fact that it was an immunodeficiency, were not known. In this section we trace the steps (more fully documented by Shilts, 1987) that led to the identification of this immunodeficiency and its causative agent, HIV, while considering several questions:

- How were hypotheses developed?
- How were they tested?
- What types of evidence are necessary to call something the "cause" of a disease?
- Does such evidence rule out other hypotheses?

We then look at the virus itself and how it lives in human cells.

15.1.2 Discovery of the connection between HIV and AIDS

In June 1981, five cases of pneumonia in San Francisco were reported to be associated with a microorganism called *Pneumocystis carinii*. The report appeared in *Morbidity and Mortality Weekly Review* (*MMWR*), a publication of the Centers for Disease Control and Prevention (CDC), which tallies all cases of sickness (morbidity) and death (mortality) in the United States due to certain kinds of diseases called reportable diseases. They are called "reportable" diseases because a physician seeing a patient with one of these diseases is legally obliged to report it to the CDC. (The numbers of cases are reported; patients' names are not.) Reportable diseases listed in *MMWR* include most of the serious contagious diseases caused by microorganisms. Each issue of *MMWR* also contains articles written by alert clinicians who have observed patterns of disease that are unusual for a particular geographic area or season, are too frequent, or are occurring in an age group that does not usually get the disease. Tallies of reportable illnesses are often the first indication of an unusual spread of a known disease or the appearance of a new disease. The statistical study of information about the occurrence and spread of diseases in whole populations is called **epidemiology**.

Pneumonia, a disease characterized by fluid in the lungs, can be caused by many different bacteria and viruses. The five cases of *Pneumocystis* pneumonia reported in *MMWR* were quite unusual because *Pneumocystis* is a parasite that is neither a bacterium nor a virus. All five cases were from a single geographic area and occurred close together in time, thus representing what epidemiologists call a "case cluster."

Later in the summer of 1981, a dermatologist in New York City, Dr. Alvin Friedman-Kien, noticed an unusual cancer, called Kaposi sarcoma, among many of his young homosexual male patients, a finding that he reported in *MMWR*. Kaposi sarcoma, a cancer of the cells lining the walls of blood vessels, causes red or purple raised patches on the skin. Kaposi sarcoma was rare in the United States and had previously been found only in elderly men of Italian or Eastern European Jewish descent. The Kaposi sarcoma seen in the reported cluster was far more aggressive than that seen in elderly men, meaning that it spread much faster and was present in the internal organs, not just on the skin. Aggressive Kaposi sarcoma is, however, seen in kidney transplant patients, who take medication to suppress their immune systems, which would otherwise reject the transplanted tissue (see Section 14.2.6). This fact suggested the hypothesis that the Kaposi sarcoma becomes aggressive when the immune system is suppressed.

15.1.2.1 The immunodeficiency hypothesis

Were the Kaposi sarcoma cases in any way related to the unusual pneumonia cases? Did immunodeficiency

Figure 15.2 The AIDS quilt, commemorating those who have died of AIDS. In 1999, the quilt had grown to be 42,960 panels commemorating 83,279 names.

underlie both the unusual pneumonias and the unusual cancers? Both the *Pneumocystis* pneumonia patients and the patients with Kaposi sarcoma were found to have severely decreased numbers of helper T cells. As we have seen, helper T cells are central in the immune system; a person lacking helper T cells has a suppressed immune system (Figure 15.1). The evidence thus fit the hypothesis that the cancer and the pneumonia belonged to a syndrome resulting from the same underlying mechanism, namely immunodeficiency. This syndrome was given the name AIDS. Reported cases accumulated quickly: 87 in the first six months of 1981, 365 in the first six months of 1982, and 1,215 in the first six months of 1983. AIDS had been given a name, but its cause was still unknown. Funding for research was slow in coming, as was interest on the part of many scientists. Even after the cause was established, there was silence on the part of government agencies. Various projects, such as the AIDS quilt (started in 1987; Figure 15.2), were started by private individuals and nongovernmental organizations (NGOs) to increase public awareness and push for increased research.

15.1.2.2 Lifestyle hypotheses Epidemiologists gathered information from AIDS patients, trying to establish any common links between the cases: Had they all been exposed to the same chemical agent? Did they all live in the same geographical area or in the same household? Were the patients known to one another? For a while, because the first AIDS patients were homosexual men, the search was for common lifestyle factors, based on the questionable assumption that there is such a thing as a "homosexual lifestyle," one characterized by some drug or dietary factor that was shared by most or all homosexual men. Several researchers hypothesized that the immunodeficiency was due to an overload of the immune system by chronic exposure to non-self molecules via promiscuous sexual activity. Others doubted these hypotheses because the effects seemed to be specifically targeted to one type of cell, the helper T cell; they searched instead for infectious microorganisms that homed in on this type of cell and that might be transmitted by sexual contact. Because the new disease first appeared in a minority community (homosexual men), many in this community (not just affected individuals) were stigmatized due to fear and lack of knowledge about the disease. Both the general population and health care workers feared the disease, leading to a lack of adequate care for many affected individuals.

15.1.2.3 The viral hypothesis Was the infectious microorganism a bacterium, a fungus, a protist, or a virus? Support for the hypothesis of a viral agent came when some cases of AIDS were reported among hemophiliacs. People with hemophilia lack functional genes that code for certain blood proteins necessary for forming blood clots after an injury. Hemophilia can be life-threatening because the person can bleed to death from a minor cut or scrape. As protection, hemophiliacs are given blood-clotting proteins from other people. This works well, but only temporarily; transferred clotting agents, like all proteins in the body, are eventually broken down by protein-degrading enzymes. The clotting factors must therefore be supplied repeatedly to hemophiliacs. These clotting factors are obtained from blood pooled from many donors and filtered to remove bacteria and fungi. Viruses, however, can pass through the filters. Because hemophiliacs receiving a filtered blood product were contracting AIDS, it was reasoned that the infectious agent could be a virus.

One laboratory that was studying viruses at the time was the National Cancer Institute's Laboratory for Tumor Cell Biology, headed by Robert Gallo. The occurrence of AIDS in hemophiliacs convinced Gallo that the infectious agent must be a virus. Gallo's laboratory was studying **retroviruses**, a type of virus whose genetic information is RNA that is copied to make DNA. (Recall from Chapter 3 that the normal flow of information in cells is from DNA to RNA through the process of transcription, carried out by the enzyme RNA polymerase. The RNA genomes of retroviruses contain genetic instructions that encode a type of enzyme called **reverse transcriptase**, which performs the reverse process, copying RNA into DNA.) One retrovirus being studied was the human T-cell leukemia virus, HTLV-I. This retrovirus was known to cause a form of leukemia (a blood cell cancer) that, in some patients, was associated with a mild immunodeficiency. It could not be said, however, whether the immunodeficiency seen in the HTLV-I patients was caused by the virus or was the result of the cancer or some other factor. Nevertheless, because it was known that HTLV-I was transmitted from person to person by sexual contact and that it specifically attacked T cells, it fit the pattern seen for AIDS.

In 1983, Luc Montagnier and his co-workers at the Pasteur Institute in France found important new evidence in the tissues of a patient with chronically swollen lymph nodes, a condition common in the early stages of AIDS. (The lymph nodes are the structures that temporarily enlarge when the body is fighting an infection; people often refer to them as "swollen glands.") The scientists found the enzyme reverse transcriptase in the lymph tissues. Montagnier's group had not yet found the virus, just one of its enzymes, but the reverse transcriptase was strong evidence of the presence of a retrovirus.

The presence of any retrovirus can be detected by finding reverse transcriptase (as Montagnier's group had done), but the identification of a specific retrovirus requires testing of large quantities of viruses, which are obtained by growing them in laboratory culture. Viruses cannot replicate outside a host cell, but these host cells may be grown in the laboratory rather than in an animal. Gallo's laboratory had developed a method for growing human T cells in the laboratory and for growing HTLV-I in those cells. Antibodies were made to HTLV-I grown this way. These antibodies were used to show that the new retrovirus from AIDS patients was not HTLV-I, since the antibodies specific for the HTLV-1 virus did not react with samples from the AIDS patients.

When scientists tried to grow the new retrovirus from AIDS patients in human T cells in the laboratory, it killed the cells. Michael Popovic in Gallo's lab found a type of leukemia T cell in which the new retrovirus could be grown without killing the host cells. Once the method for growing quantities of the new virus had been developed, antibodies were made that were specific for it. These antibodies were then used by Gallo's group to test viruses isolated from three groups of people: a control group that consisted of healthy heterosexuals, a group of AIDS patients, and a group of patients with AIDS-related complex (ARC), a set of symptoms assumed to be an early stage of AIDS. The viruses isolated from AIDS patients and some ARC patients were identified as being the same as the new virus. None of the healthy subjects had this new virus.

Using the specific antibodies, scientists found the virus in 80–100% of AIDS patients, in varying percentages of people in certain defined risk groups, and only rarely in healthy individuals outside the risk groups. These results were strong evidence that this new retrovirus was associated with AIDS. The retrovirus found by Gallo and the retrovirus found earlier by Montagnier were determined to be two strains of the same virus. Each group had given their virus a different name, and each group wanted the name they had chosen to become the standard. The International Committee on the Taxonomy of Viruses studied the naming problem and decided in 1986 that neither name should be used. They instead assigned a new name to the virus, **human immunodeficiency virus** or **HIV**.

15.1.3 Establishing cause and effect

Just because a microorganism is associated with a disease does not necessarily mean that it causes the disease. How do we know that HIV is the actual cause of AIDS? There is a set of rules that have traditionally been used to pinpoint a microorganism as the cause of a particular disease. These rules were formulated in the late 1800s by Robert Koch, a German physician, and have come to be known as **Koch's postulates**:

- First, the microorganism suspected as the causative agent must be present in all (or nearly all) animals or people with the disease.
- Second, the microorganism must not be present in healthy animals.
- Third, the microorganism must be isolated from a diseased animal and grown in pure culture (that is, a culture containing no other microorganisms).
- Fourth, injection of the isolated microorganism into a healthy animal must reproduce the original disease in that animal, and the microorganism must be found growing in the infected animal and be reisolated from it in pure culture.

Koch used these rules to show that the bacterium *Bacillus anthracis* was the cause of anthrax, a fatal disease in sheep that was decimating European herds in the 1870s. He later used his postulates to identify a bacterium now known as *Mycobacterium tuberculosis* as the causative agent for human tuberculosis. In 1905, Koch received a Nobel Prize for these discoveries.

15.1.3.1 Limitations to Koch's postulates While Koch's rules can be stated in a straightforward manner, they are difficult to fulfill. Every animal is host to many bacteria and viruses, so finding one that is present only in diseased animals is not easy. Koch and his contemporaries developed bacterial culture media and techniques that enabled them to grow pure cultures of certain bacteria, but the growing of many other bacterial species (such as those killed by exposure to air) required technology that did not exist in Koch's time. Viruses can be still harder to culture, as they may replicate only in certain types of cells.

Despite the difficulties associated with meeting the requirements of Koch's postulates, they have been very useful. Many infectious diseases are controllable by vaccination, including rabies, poliomyelitis, whooping cough, tetanus, measles, and diphtheria. The infectious microorganisms responsible for these diseases and others were identified on the basis of Koch's postulates. An infectious agent that has been shown to cause a disease is called a **pathogen**.

There are other diseases for which these postulates have not yet been demonstrated. Most bacteria can grow outside cells, so they can be grown in "pure cultures" if the proper growth conditions can be found. However, viruses grow only inside a host cell, so that difficulties in growing the infectious agent in culture become even more acute when viral, rather than bacterial, diseases are studied. Despite these and other limitations, Koch's postulates are the standards by which scientists establish cause and effect for most *infectious* diseases, meaning those diseases that can spread by infection with a microorganism. (Some virologists have proposed updating Koch's postulates for the 21st century: see https://www.virology.ws/2010/01/22/kochs-postulates-in-the-21st-century).

15.1.3.2 HIV and Koch's postulates Because Koch's postulates are the accepted standard of proof for asserting cause and effect in infectious disease, to say that "HIV *causes* AIDS" carries with it the implication to many in the scientific community that Koch's postulates have been fulfilled. This implication was contested by the American virologist Peter Duesburg, who pointed out, for example, that most AIDS patients were identified by the presence in their blood of antibodies to the virus, not by the virus itself. (Newer, more sensitive techniques, such as the polymerase chain reaction (PCR) described in Chapter 3, can now be used to detect the virus directly, so this objection no longer applies; the virus has now been found in virtually all persons with AIDS.) The overwhelming majority of scientists who study AIDS now recognize HIV as its cause and reject Duesberg's original objections.

15.1.3.3 Criteria other than Koch's postulates Newer criteria have been suggested for establishing disease causality, particularly of viral diseases. For example, the criteria used by Blattner, Gallo, and Temin (1988) for stating that HIV is the sole cause of AIDS are as follows:

- HIV or antibody to HIV is found in the vast majority of persons with AIDS.
- HIV is found in a high percentage of people with ARC.
- HIV is a new virus, and AIDS is a new disease.
- Wherever HIV is found, AIDS develops; where there is no HIV, there is no AIDS.
- People who received transfusions of blood contaminated with HIV developed AIDS.
- HIV infects CD4 helper T lymphocytes, a cell type depleted in AIDS.
- On autopsy, HIV is found in the brains of people who have died of AIDS, and dementia (loss of brain cell function) is a symptom of AIDS.

15.1.3.4 Necessary causes and sufficient causes At present, the vast majority of scientists agree that HIV is a *necessary cause* of AIDS; that is, someone who is not infected with HIV will not get AIDS. However, not all scientists agree that HIV is the sole or *sufficient cause* of AIDS (that is, no other factors are required) because there is a great difference in the course of the infection among various persons with AIDS.

Saying that a particular virus causes a disease, especially a disease with such a diffuse group of symptoms as AIDS, really tells us very little by itself. All of the "how" questions remain. How does the virus infect? How do cellular effects progress to clinical symptoms? How can the disease be prevented or stopped? How is the infection transmitted? How contagious is it? How likely is it that HIV infection will become AIDS? How do people cope with such a disease? Most articles written by scientists, either for other scientists or for the public, are devoted to answering the first three questions, while articles written by nonscientists are much more concerned with the remainder. We examine each of these questions later in this chapter.

15.1.4 Viruses and HIV

Many human diseases are caused by viruses. AIDS is one such disease; measles, mumps, polio, herpes, and COVID-19 are also caused by viruses. In addition to this role in disease, viruses are interesting to biologists because they challenge our understanding of what it means for something to be alive (Section 1.1.4). **Viruses** are bits of either DNA or RNA contained within a protein coat that cannot reproduce by themselves but can replicate inside a cell (called the host cell) by using the biochemical machinery of the host. Biologists define a living organism as one that can reproduce itself, which viruses cannot; yet once inside a host, viruses can cause the host to replicate the virus, something that is not a characteristic of any known nonliving thing. What is the structure of a virus, and how do viruses accomplish this?

15.1.4.1 The viral life cycle A virus consists of nucleic acid, an outer protein shell, and, for some viruses, including HIV, a phospholipid bilayer membrane called the viral envelope that also contains some viral proteins. The genome of a particular virus is made of either DNA or RNA, and the nucleic acid is either single-stranded or double-stranded; these distinctions are used in viral classification. All retroviruses, including HIV, have single-stranded RNA genomes. Viral genomes vary in size: Some have only enough nucleic acid to code for 3–10 proteins, while others code for 100–200 or more proteins. HIV, the virus that causes

AIDS, is at the small end of this size range. Viruses are very small, even in comparison with bacterial cells. A single human cell is typically 10 μm (10 millionths of a meter) in diameter. Magnifying this human cell 100,000 times would make it a meter wide; at the same magnification a bacterium would be about the size of a football and a virus would be only the size of an M&M candy.

Some viruses can survive outside cells, and transmission via surfaces occurs for some viruses, but not for HIV. However, no virus can replicate unless it is inside a host cell. Human cells, animal cells, plant cells, and bacteria can all serve as hosts to viruses. For each virus, there are only certain species that can serve as its host, and within an individual of the host species only certain types of cells can be host cells. This is because, to enter a cell, the virus must attach to some molecule on the host cell surface. Each type of virus is able to bind only to specific host cell molecules, which are usually membrane proteins. The species of virus that can adhere to dog or cat cells, for example, usually cannot adhere to human cells, which explains why we usually cannot catch viral diseases from our pets.

After a virus attaches to a host cell, the viral nucleic acid, sometimes with some viral proteins, enters the cell's cytoplasm, usually with the help of energy derived from the host cell. In viruses whose nucleic acid is DNA, many copies of the virus are made using the host's molecular machinery for DNA replication. Viruses whose nucleic acid is RNA encode enzymes that can replicate their RNA genome directly or, in the case of retroviruses like HIV, first convert their RNA into DNA, using **reverse transcription**. The host's machinery is then used to make more viral particles. The final stage of the viral life cycle consists of the release of viruses from the host cell in one of two ways: The viruses may cause the host cell to break open (a process called lysis), thus destroying the host cell; alternatively, some viruses, mostly those with a phospholipid bilayer coat (called enveloped viruses), bud through the host cell membrane, with the host membrane yielding their outer envelope. The new viruses can then infect other cells and repeat their life cycle.

15.1.4.2 HIV structure and life cycle As mentioned earlier in Section 15.1.4, HIV is an enveloped virus whose genome consists of two copies of a single strand of RNA. The structure of HIV is shown in Figure 15.3. The virus is

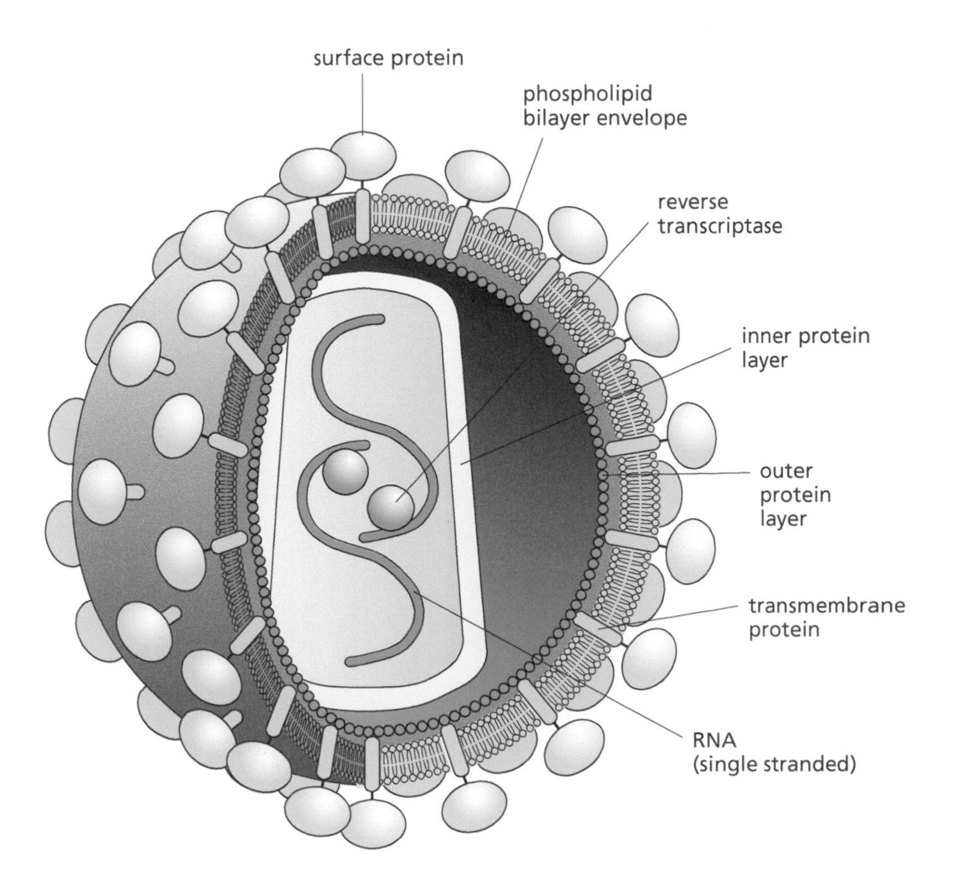

Figure 15.3 The structure of HIV.

surrounded by a phospholipid bilayer envelope that contains viral proteins important for attachment to and entry into specific host cells. Inside the viral envelope are two protein layers. Inside the inner protein layer is the viral genome.

The HIV life cycle is shown in Figure 15.4 and is typical for many retroviruses. Proteins in the viral envelope bind to the CD4 protein found on only a few types of human cells: helper T cells and some macrophages, a type of white blood cell

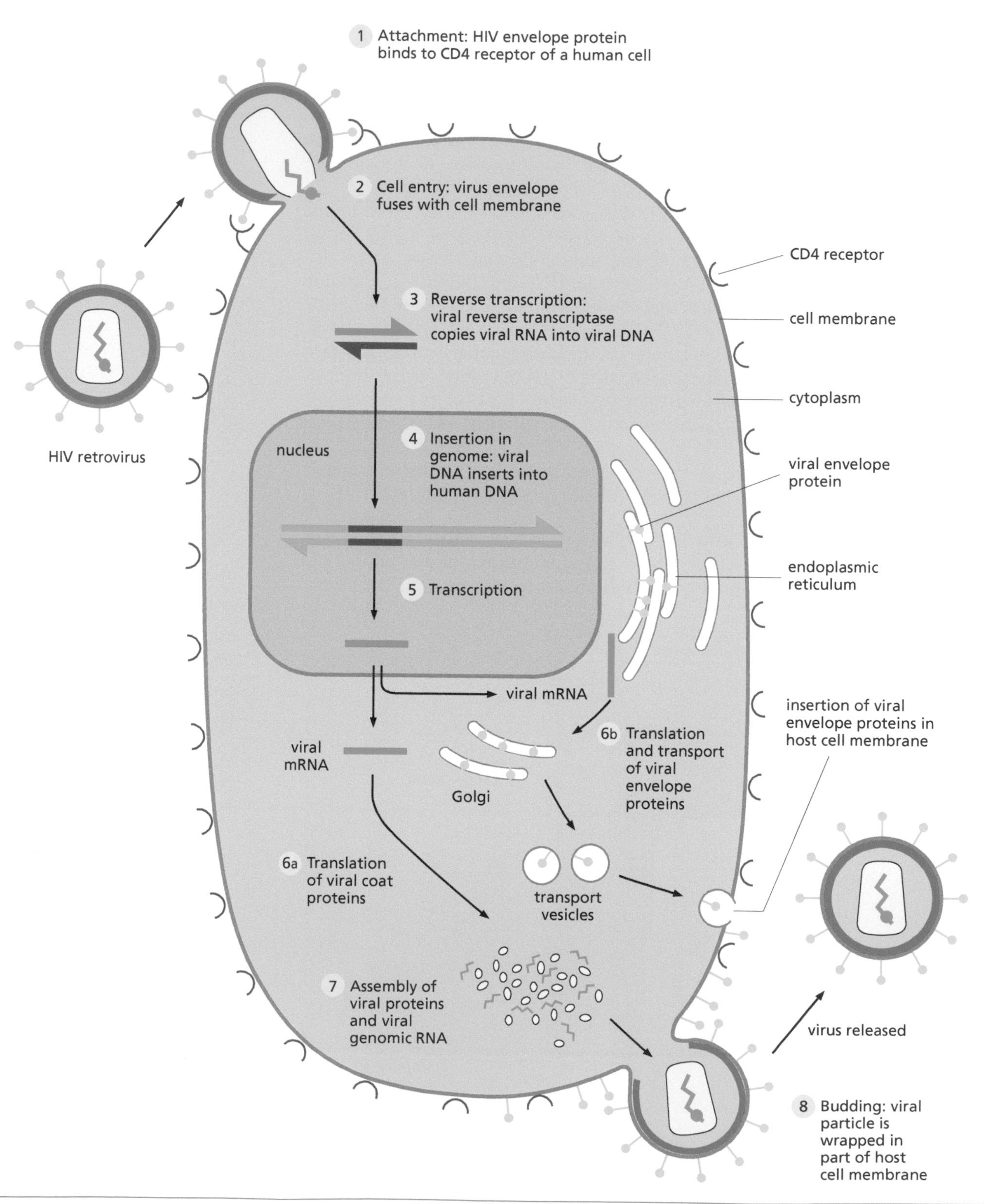

Figure 15.4 The retroviral life cycle.

important in inflammation and in engulfing pathogens through the process of phagocytosis (see Figure 14.1). Helper T cells, macrophages, and cells related to macrophages are thus the only cells that become directly infected by the HIV virus. Other species of mammals also have CD4 and helper T cells, but the structure of CD4 in each species is different, such that HIV attaches only to human CD4. Once attached to the host cell, HIV enters by fusion of its viral envelope with the host's plasma membrane. This fusion step requires other proteins in the host cell membrane.

After entry into the cell, the viral enzyme reverse transcriptase, which is packaged into the virus along with its RNA genome, becomes activated. This enzyme first uses the viral RNA as a template and synthesizes complementary DNA (**cDNA**). The first DNA strand is, in turn, the template for synthesis of the second strand of DNA, which is carried out by a separate enzymatic activity of the same viral enzyme. The now double-stranded DNA is incorporated into the host's DNA. The viral RNA is meanwhile broken down, or degraded. This process of integration of viral DNA into a host genome was described in Section 10.6.3 for viruses that carry oncogenes and therefore promote cancer (see Figure 10.25).

Once inserted into the host cell's DNA, the viral DNA can be transcribed by host RNA polymerase enzymes into many copies of viral messenger RNA (mRNA). Some of these mRNAs undergo several different patterns of RNA splicing (see Section 4.4.1) and are then translated by ribosomes in the host cell cytoplasm to yield the protein coats of new virus particles, viral envelope proteins, reverse transcriptase, and a small number of regulatory proteins. The alternative mRNA splicing allows more different proteins to be encoded by a viral genome than if the mRNA were unspliced or only spliced in one way. Full-length mRNAs that are not spliced are exported from the nucleus to the cytoplasm, where they are ultimately packaged into new viral particles as the viral RNA genome. This requires one of the viral regulatory proteins, as the host cell will not normally allow unspliced mRNA to reach the cytoplasm.

Like host cell membrane proteins, viral envelope proteins are made on the host cell's endoplasmic reticulum, transported through the Golgi apparatus, then carried via transport vesicles to the plasma membrane (see Section 6.2.4). Multiple copies of viral envelope proteins build up on the surface of the host cell (Figure 15.4). The viral genomic RNA and critical viral regulatory proteins join a portion of the host cell plasma membrane containing viral envelope proteins. The virus then buds out, carrying along a piece of the host cell membrane, which becomes the viral envelope. A photograph taken at very high magnification through an electron microscope shows HIV budding from a helper T cell (**Figure 15.5A**). Upon leaving the host cell, the virus is a mature, cell-free virus, ready to infect a new host cell. Many new viruses bud out of a single helper T cell (**Figure 15.5B**). HIV can also lyse (rupture) the T cell.

The HIV envelope and genome contain all of the molecular features that make the virus virulent (able to cause disease), infective (able to enter a cell), cell-specific (entering only certain types of cells), and cytopathic (able to kill or inactivate the host cell). As previously mentioned, the HIV genome also contains some regulatory genes, including those that encode proteins that affect the expression of other viral genes (transcriptional regulators; see Section 10.2.2). All of the HIV regulatory genes seem to be essential for the life cycle; thus, they are possible targets for drugs or vaccines because blocking any essential step would inhibit the whole cycle.

Two distinct HIV viruses are currently known, HIV-1 and HIV-2. Both viruses have the same basic structure, life cycle, and routes of transmission, but the rate of transmission of HIV-1 is 5–10 times higher than that for HIV-2, and HIV-1 accounts for the vast majority of infections worldwide. People infected with HIV-1 are 3–8 times more likely to have a decrease in helper T cells, lose immune function, and progress to AIDS than people infected with HIV-2. Both HIV-1 and HIV-2 exist as multiple viral strains characterized by different genomic sequences. Some of these sequence variations can affect how particular drugs function against the virus.

How does HIV bring about the disease called AIDS? We examine this question next.

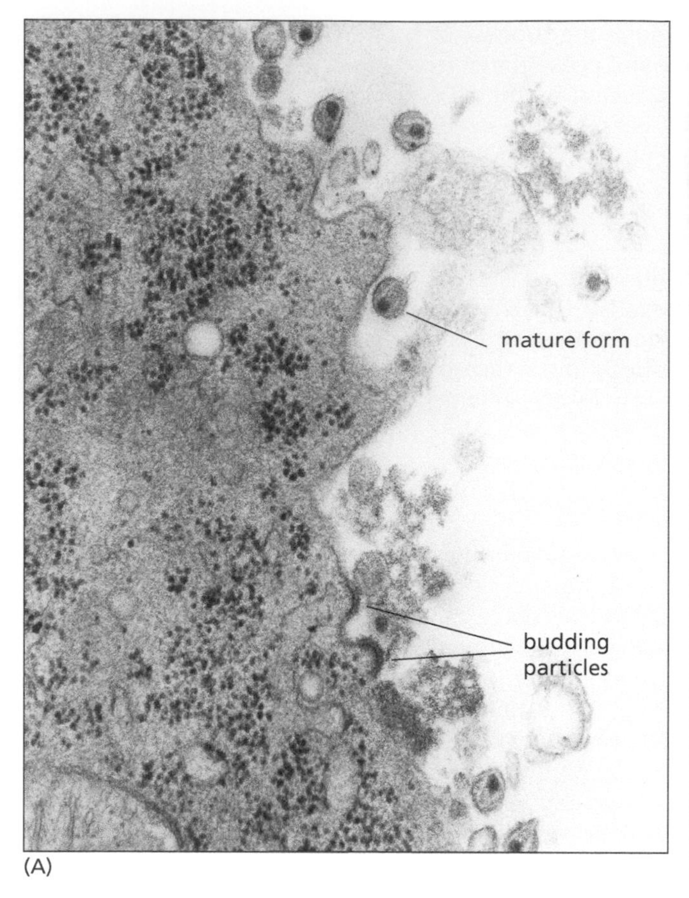

(A)

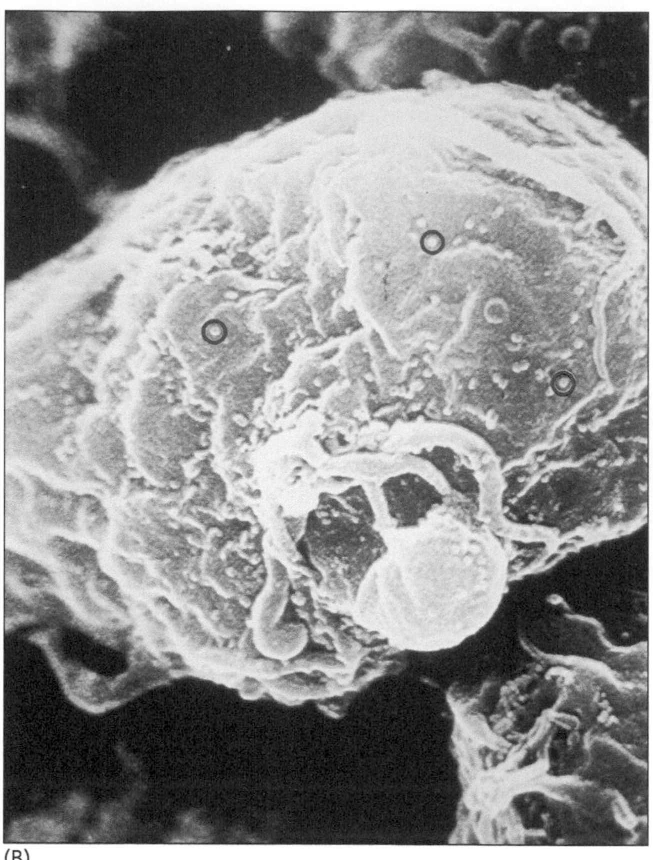

(B)

Figure 15.5 HIV budding from a helper T cell. (A) Budding and mature HIV. The mature viruses are free of the cell; the viral protein can be seen inside the viral envelope (a piece of the cell membrane that the viruses have taken with them as they budded out). **(B)** Many HIV particles bud from the same cell.

THOUGHT QUESTIONS

1. How might scientists decide whether or not two strains of virus are actually the same?
2. What are some of the reasons why evidence of the types called for by Koch's postulates may be impossible to obtain for some infectious diseases?
3. What are some differences between bacteria and viruses?
4. If an RNA strand from HIV contains the base sequence AAUGCA, what would be the base sequence on the first strand of DNA produced by reverse transcription? What would be the sequence of the second DNA strand transcribed from the first one? (You may need to review material from Chapter 2 to answer this question.)

15.2 HIV INFECTION PROGRESSES IN CERTAIN PATTERNS, OFTEN LEADING TO AIDS

We have already seen that HIV binds to cells in the immune system that have CD4 molecules on their surfaces. It then enters these cells, replicates, and goes on to infect more cells. HIV infection diminishes both the number and the activity of the CD4-bearing cells, thus reducing their ability to perform their disease-fighting functions. Because CD4-bearing helper T cells are central to both arms of the specific immunity (Figure 15.1), their elimination results in immune deficiency,

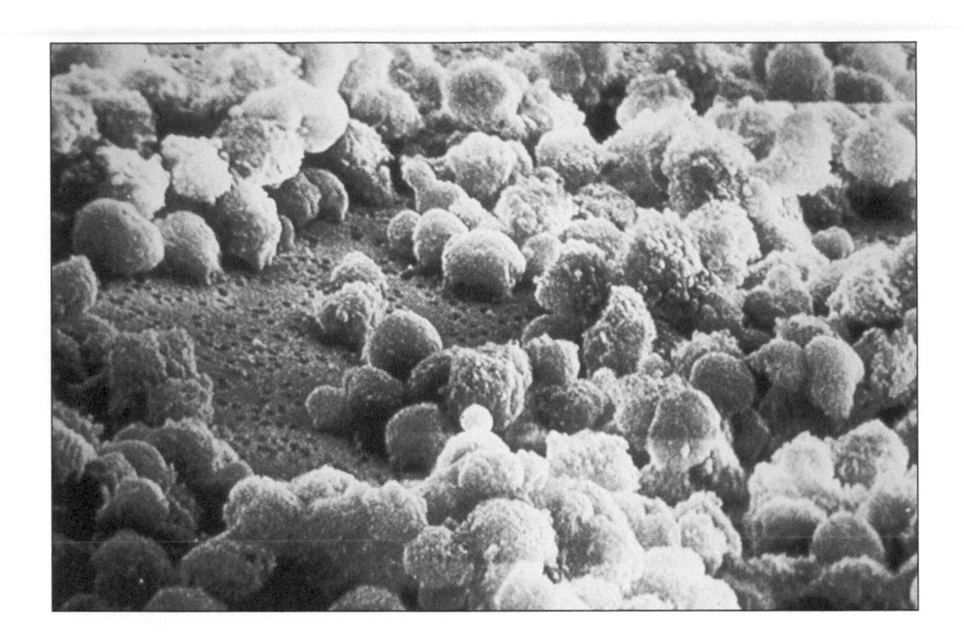

Figure 15.6 **Many HIV particles emerging from a helper T cell.** The dark circles are holes in the cell membrane left when the viruses bud out. These holes eventually kill the cell.

which in turn results in disease. This process can be described and studied at many levels. At the cellular level, exactly how does HIV eliminate helper T cells? At the organismal level (the person), how does infection progress to disease? At the population level, how is HIV transmitted? We look in this section at the cellular and organismal effects and in a later section at the population effects.

15.2.1 Events in infected helper T cells

How does HIV eliminate helper T cells? There are several different mechanisms.

1. **Direct killing.** As we have already seen, HIV can directly kill the cell it has entered by rupturing (lysing) it. Repeated budding out of replicated viruses also eventually kills the cell (Figure 15.6).
2. **Cell suicide (apoptosis).** HIV may also change a helper T cell so that, when it responds to another infection, it commits suicide instead of dividing. Healthy T cells, when activated by a pathogen, begin to synthesize DNA and divide. Under the same conditions, HIV-infected helper T cells undergo the process of apoptosis, or programmed cell death, where the DNA and nucleus break into fragments, the membrane becomes convoluted, and the cell shrinks inward without lysis (see Figure 10.11).
3. **Killing by cytotoxic T cells.** When helper T cells make viral proteins, these proteins are expressed (or "presented"), along with human leukocyte antigen (HLA) proteins, on the surface membranes of these T cells. This process marks these helper T cells as targets for killing by cytotoxic T cells (see Section 14.2.3), a process that normally eliminates many viral infections. With HIV, however, the cytotoxic T cells target the infected helper T cell for destruction, which further decreases the number of helper T cells and greatly impairs the immune response.
4. **Cell fusion.** HIV carries in its viral envelope a protein that helps it bind to the CD4 protein on the surface of the cell it infects. The infected host cell also expresses some of this viral envelope protein in its plasma membrane before new viruses bud out (Figure 15.4). The infected host cell can thus bind to the CD4 protein on the surface of other, uninfected helper T cells. The plasma membranes of the infected and uninfected host cells then fuse, bringing about cell fusion and spreading the virus to a new cell. This fusion can be repeated until a multinucleated "giant cell" is formed. Although these giant cells are still alive, they can no longer perform the immunological activities of normal helper T cells.

5. **Indirect inactivation**. Certain strains of HIV cause the production of the wrong signaling molecules (cytokines) or inhibit the production of the cytokines needed for T-cell growth. Without these cytokines, the helper T cells cannot divide to perform their normal disease-fighting processes or to replace cells lost to HIV-induced lysis and apoptosis.

Although these mechanisms can be demonstrated in laboratory experiments, it is not certain that they all occur in infected human hosts. We do not yet know which of these mechanisms causes the greatest loss of T cells or of T-cell functions. The result, however, is the same: The loss of healthy, active helper T cells results in immune deficiency. Many of these events also occur in HIV-infected CD4-bearing macrophages, thus impairing the antigen nonspecific, innate portion of the immune system as well.

15.2.2 Progression from HIV infection to AIDS

When a person becomes infected with HIV, the virus keeps spreading to more and more cells. When an HIV infection has progressed to AIDS, 1–10% of the helper T cells have become infected. How does infection of some cells progress to disease in a person?

15.2.2.1 Three stages of HIV infection The progression from HIV infection to AIDS follows three stages: the initial infection, an asymptomatic phase, and a third phase called disease progression (Figure 15.7). An infected person can transmit HIV to another person at any of the three stages but is most likely to do so in the first and third stages, when the numbers of cell-free viruses and infected cells in the body fluids are highest.

In the initial stage, virus levels in the blood are high. As more and more helper T cells are infected and killed, the helper T-cell count begins to drop. The initial infection may be accompanied by flulike symptoms—fever, swollen lymph nodes, and fatigue—which, because they are similar to the symptoms for many other diseases, are often not diagnosed as being an acute HIV infection. The initial infection also stimulates two types of immune response. Antibodies to HIV are produced by B cells, and there is an increase in

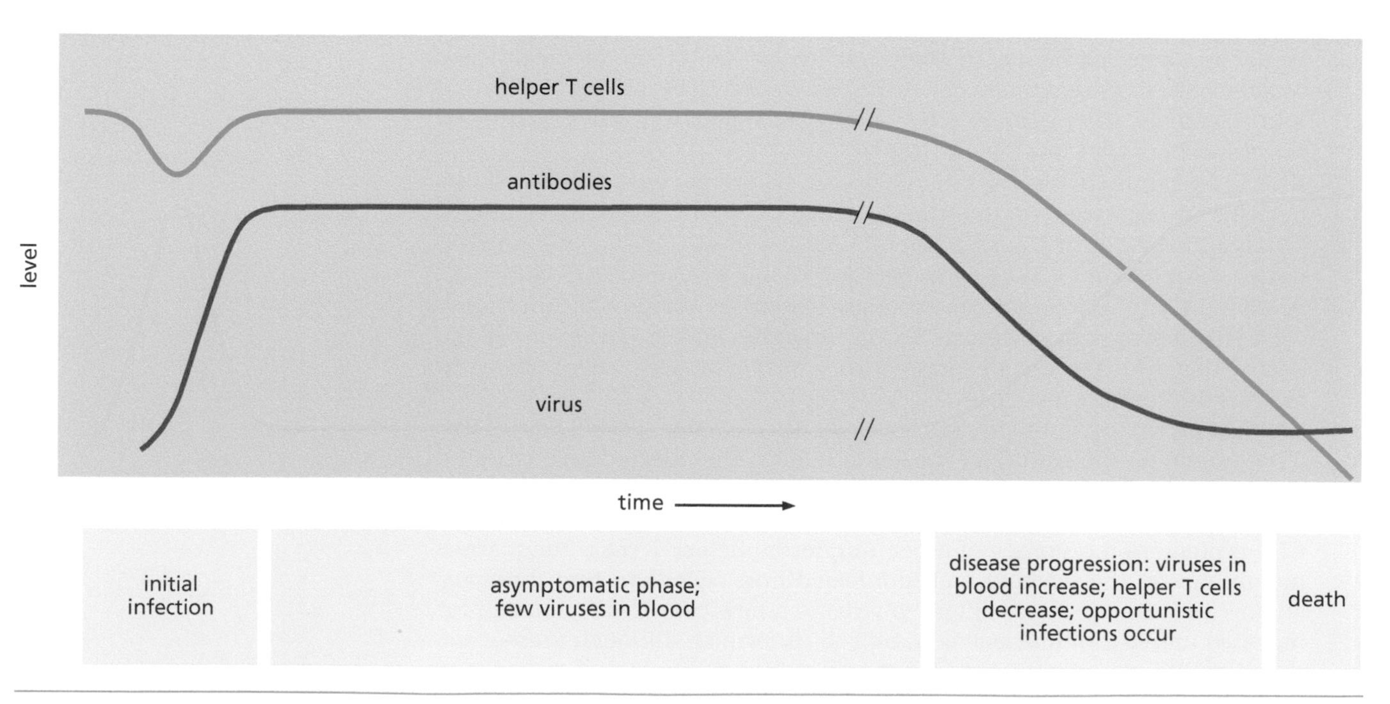

Figure 15.7 The course of HIV infection.

HIV-specific cytotoxic T cells that can kill cells containing HIV virus. These processes are initially able to contain the HIV, so the levels of virus in the blood decrease. New helper T cells develop to replace those that were killed, and helper T-cell counts return to normal.

With the decrease in virus in the blood and the return of helper T cells, the person enters the second, or asymptomatic, phase, which can last for a few months to many years, with 10 years being typical. The levels of virus in the blood decrease, but the viral population in the lymph nodes continues to grow by viral replication and infection of new helper T cells. By binding to the virus, the antibodies that were synthesized in the initial phase can neutralize the virus, preventing it from infecting more cells. During the asymptomatic phase, the immune system is still able to keep the infection under control, so the person does not feel ill.

In the third phase, the levels of virus in the blood increase once more, while the numbers of helper T cells in the blood decrease. It is uncertain what triggers the onset of the third phase, although malnutrition, stress, or other immunosuppressive factors seem to hasten the onset. Originally it was assumed that HIV infection inhibited the maturation of new helper T cells. It is now known that the maturation rate is actually normal or greater than normal, but the rate of cell death is so great that the overall helper T-cell population decreases, particularly as an HIV-infected person progresses to AIDS. The virus mutates to forms that no longer match the antibodies produced in the acute phase, which thus cannot bind to the virus and neutralize it. The mutation rate of the HIV genome is much higher than that of cells because reverse transcriptase makes mistakes much more often than cellular DNA polymerase. With fewer antibodies that can neutralize the virus, the host's own cytotoxic T cells turn against the infected helper T cells. A person is defined as having progressed to AIDS when their CD4 helper T-cell count (also called the T4 count) falls from a normal value of 1,000 cells per microliter of blood to less than 200. Death generally follows when the level of helper T cells declines still further in the third phase.

Other cells bearing CD4 may also be infected and inactivated by HIV. As mentioned previously, macrophages are white blood cells that engulf and remove pathogens and damaged cells or molecules and also secrete cytokines that strengthen immune responses. The elimination of macrophage cells contributes to the risk for opportunistic infections that may, in turn, lead to the patient's death.

15.2.2.2 Opportunistic infections and other symptoms There are microorganisms that are always present in a person or the environment but are kept in check by a healthy immune system, so that they seldom cause illness. Infections caused by these microorganisms in immunodeficient people are called **opportunistic infections** and are one of the primary symptoms of AIDS. These infections can be very severe and even fatal in a person with AIDS. Some of the pathogens that cause these infections are described further in Chapter 16. People with intact immune systems need not fear catching opportunistic infections from a person with AIDS. In the United States, typical opportunistic infections that accompany AIDS are *Pneumocystis* pneumonia, caused by *P. carinii*, and fungal infections with *Toxoplasma* or *Histoplasma*. Recall that the appearance of a cluster of cases of this rare *Pneumocystis* pneumonia was the first hint of AIDS. Another fungus called *Candida* (a yeast) causes mild infections of the mouth, esophagus, or vagina in the absence of AIDS, but *Candida* infections in people with AIDS are much more severe. The same is true of viral diseases, including shingles, cytomegalovirus eye infections, and herpes viruses. In Africa, the most common opportunistic infection accompanying AIDS is tuberculosis, a bacterial infection caused by *Mycobacterium tuberculosis*. Tuberculosis is a disease in which there are active periods and periods of remission; HIV infection increases the frequency of reactivation of tuberculosis and also the mortality rate. Worldwide, tuberculosis is the leading cause of death in HIV-infected people.

AIDS patients may also suffer from high fevers, night sweats, general weakness, mental deterioration (dementia), and severe weight loss, although these last two symptoms may not develop for a long time. Dementia may be related to the elimination of macrophage-like cells from the brain. In the gut, there is a type of CD4-bearing cell that has a role in the absorption of nutrients; elimination of these cells may be related to the weight loss.

15.2.2.3 Variations in disease progression Many people infected with HIV develop ARC, a set of symptoms milder than AIDS. Originally it was thought that ARC was a pre-AIDS condition and that everyone who had ARC would end up with AIDS. The CDC did not initially require the reporting of ARC, assuming that these cases would later be reported as AIDS cases, which resulted in underestimates of HIV infection rates. As time passed, researchers noticed that several people died while still showing only the symptoms of ARC, not AIDS. Distinctions are no longer made between ARC and other categories of HIV infection; they are all simply called HIV infection. Does everyone infected with HIV get AIDS? Does everyone with AIDS die from the disease? We do not have definitive answers to these questions. The speed with which HIV infection progresses to disease varies greatly. Some people have been infected with HIV for many years without developing AIDS. Several studies have shown that nonprogressive HIV infections are often characterized by a very small amount of the virus, but it is not clear whether this reflects a low infective dose initially or an immune system that has successfully kept the viral population low. A long-term study of HIV-infected homosexual men in San Francisco showed that, after 12 years, 65% had progressed to AIDS, but 35% had not. It may yet turn out that the progression from HIV infection to AIDS is not inevitable. Certainly, the avoidance of other immunosuppressive factors, including drugs, alcohol, and stress, can help to maintain health (see Chapters 13 and 14).

Researchers studied several dozen professional sex workers in West Africa who were infected with HIV-2, a less virulent strain of HIV than HIV-1 (Travers et al., 1995). Significant findings of this and other studies are that people infected with HIV-2 were less sick than people infected with the more virulent strain, HIV-1, and that HIV-2 infection seems to offer some degree of protection against HIV-l by delaying the progression to AIDS (Esbjörnsson et al., 2012). The sex workers did have high rates of infection for other sexually transmitted diseases, falsifying the hypothesis that the lower HIV-1 rates were simply the result of safer sex practices. A different group of professional sex workers in Kenya was identified by Strydom (2010) as protected from AIDS for up to 13 years despite repeated exposure to HIV. Large numbers of cytotoxic T lymphocytes (killer T cells) were found in their blood, presumably stimulated by their repeated exposure to HIV in the course of their work. However, many of these women became HIV-positive and developed AIDS after abstaining from their work for several months and then returning, showing that they had lost this protection.

Another group of people was found who remained uninfected with HIV despite numerous exposures. Many of these people turned out to have genetic mutations of the cell-surface protein CCR5, a "chemokine receptor" that greatly enhances the infection process when it accompanies CD4 molecules on host cell membranes (Samson et al., 1995). People who carry two copies of a naturally occurring CCR5 mutant allele express a truncated (shortened) protein that cannot function as a co-receptor; these individuals are thus largely resistant to HIV infection, and those with one copy of the mutation usually stay asymptomatic for longer than most other infected people. Several researchers hypothesize that natural selection by other pathogens (such as plague and smallpox) helped spread this mutant allele among many European populations in past centuries (Galvani and Slatkin, 2003; Galvani and Novembre, 2005). As described in Section 4.1.3, a Chinese geneticist has used CRISPR gene editing to inactivate the *CCR5* gene in a human baby, which will likely provide that individual with some level of resistance to HIV, but this genetic manipulation was considered unethical by most geneticists. It also seems probable that *CCR5* mutations may not offer protection against all strains of HIV.

15.2.3 Tests for HIV infection

How can people tell whether they are infected with HIV? The B lymphocytes of a person infected with HIV respond to the virus. This response, which is the basis of most testing for HIV, takes a couple of weeks or months and results in the production of antibodies to HIV in the blood. The development of specific antibodies is called "seroconversion," and once the antibodies have developed, the person is said to be HIV-positive (HIV^+).

There are two common tests for HIV, the enzyme-linked immunoassay (ELISA) test and the Western blot test, as described in Box 15.1. Both of these tests detect antibodies to HIV. As mentioned earlier, there are also now PCR-based tests to detect the virus itself (by detecting viral RNA), but these tests are very expensive, so most HIV tests are still based on the presence of antibodies to HIV.

For every diagnostic test there exist the possibilities of **false positives**, test results that are positive when the person does not really have the condition, and **false negatives**, test results that are negative when the person really does have the condition. The frequency of false negatives determines the **sensitivity** of the test; the frequency of false positives determines the **specificity** of the test. The reliability of a given test depends on both its sensitivity and its specificity. The more sensitive a test, the less often it will miss a truly positive case; the more specific a test, the fewer will be the cases that are truly negative but that are reported as positive. Every diagnostic test must be thoroughly validated using samples from thousands of individuals whose actual status is known before the test can be sold. These trials must be conducted blind; that is, the person doing the testing cannot know during the trials whether the test samples came from persons infected with HIV or not. Afterwards, the true infection status (known beforehand but concealed from the researchers) is compared with the status revealed by the test. In this way the frequency of false results can be quantified.

Box 15.1 ELISA and Western blot tests for detection of antibodies to HIV

The Enzyme-Linked Immuno Sorbent Assay (ELISA) and the Western blot tests use immunological techniques to detect HIV-specific antibodies in a person's blood and thus are called immunodiagnostic techniques. In the ELISA, laboratory-grown HIV are immobilized onto a surface, generally small wells made of a plastic designed to bind protein molecules tightly (1). The rest of the plastic is coated with other proteins to block any nonspecific binding of proteins used later in the assay. The immobilized virus particles then act as binding sites for specific antibodies: They are exposed to blood serum from the person being tested, and if that person's serum contains antibodies whose specific binding sites match molecules on the virus, the antibodies bind to the immobilized virus (2). There are many other antibody molecules in the person's blood that do not match any HIV molecule; these do not bind. The wells are then rinsed, removing any unattached antibody molecules. The viral molecules and any specific anti-HIV antibodies bound to them are so tightly attached that they do not wash away. Anti-HIV antibodies are then detected by a second antibody to which an enzyme has been attached (3). The binding sites on the enzyme-linked antibody match amino acid sequences on human antibody molecules; they bind, not to the plastic or to the HIV, but to any human antibodies present. Again, any unbound antibody is washed away, then enzyme substrate is added (4). Enzymes are proteins that catalyze biochemical reactions; the enzymes used in these tests cause a color change in the medium (5). A well in which the medium has changed color (from clear to blue in the photo) is thus a well in which the blood plasma used in the test contained antibody specific for HIV, an initial bit of evidence that the person is HIV-positive.

If an ELISA suggests the presence of anti-HIV antibodies, the Western blot test is done. The viral proteins are separated by using a technique called electrophoresis (1) (shown for DNA in Figure 3.16). The separated proteins are then transferred out of the gel onto special paper that has a high affinity for proteins (2). The paper is then exposed to blood serum from the person being tested. Specific antibodies bind, but in this case, they bind not to the whole virus but to some

(*Continued*)

individual viral protein. The serum may contain specific antibodies that bind to some viral proteins but may lack antibodies to other viral proteins (3). Unbound antibody is rinsed away and bound antibody is detected, as in the ELISA, with enzyme-linked antibody that binds to all human antibodies (4). If the enzyme-linked antibody finds human antibody to bind to, the enzyme makes dark bands on the paper where the blood contained antibody specific for that viral protein; where there is no specific antibody, no band appears (5). If antibody to specific proteins is present, the person has tested HIV-positive (also referred to as seropositive, because the test is done on serum, the liquid part of the blood after blood has been allowed to clot).

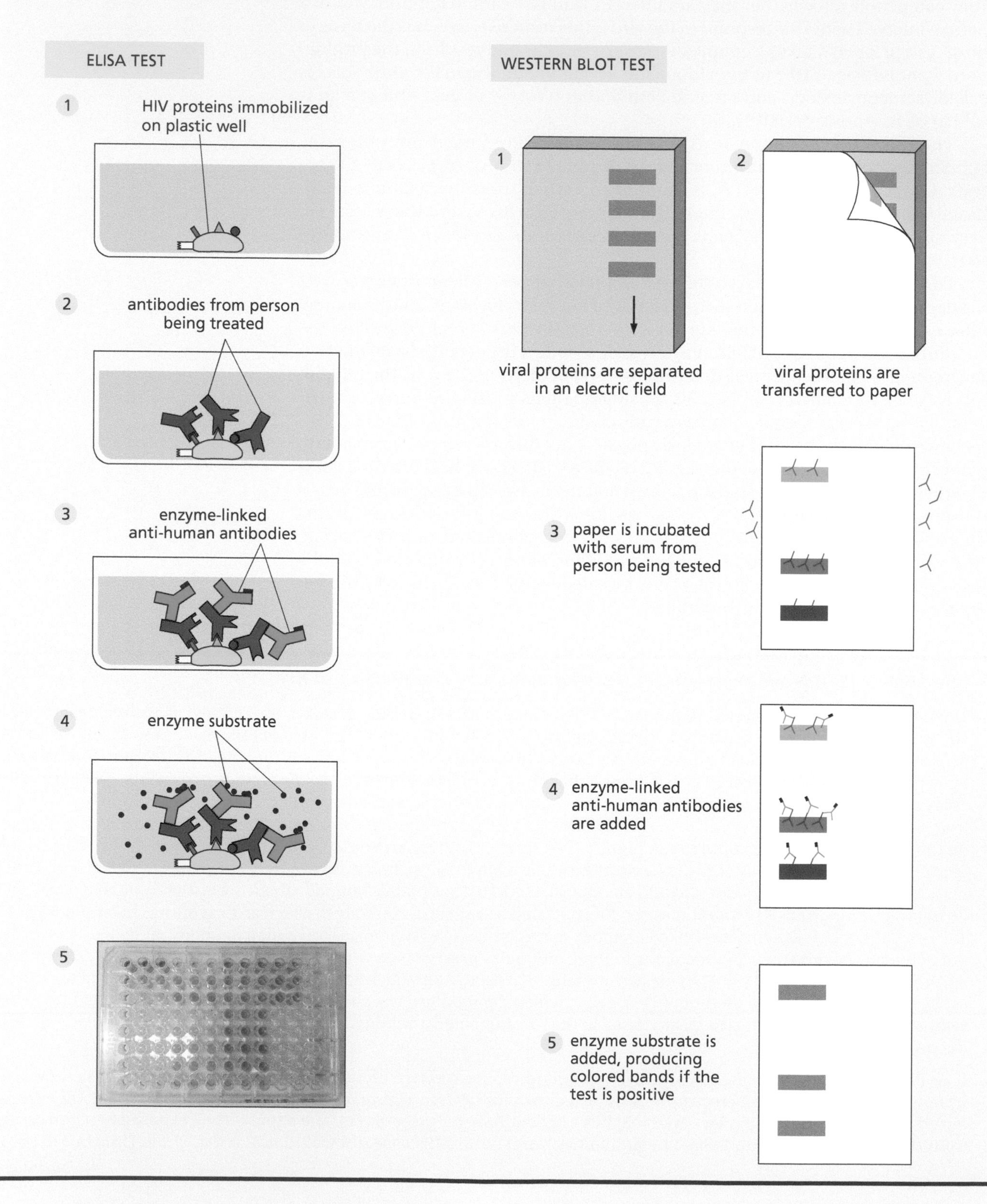

For HIV testing, the ELISA test is done first. The sensitivity of the ELISA test is high—less than 1% false negatives—but it is not very specific: There can be as many as 2–3% false positives. For this reason, when an ELISA result is positive, the result is rechecked with a Western blot test, which rarely gives false positives. Why not use the Western blot as the initial test? The reason is that Western blots are more costly and technically more difficult. Even the ELISA test is too costly for widespread use in many countries. The PCR-based tests that detect viral RNA are even more accurate than the Western blot and can detect the presence of virus just days to weeks after infection. However, they are more time-consuming and costly than the antibody-based tests. PCR tests are regularly used to screen donated blood for HIV, but individual patients usually first receive the ELISA test, followed by a Western blot test if the ELISA result is positive.

Should everyone be tested for HIV? One argument against testing everyone is the frequency of false results. The problem of false results is more severe when the true frequency of infection in the test population is low. A little mathematics will illustrate the point. The frequency of false positives in HIV ELISA tests is between 2% and 3%, while the frequency of false negatives is less than 1%, for an overall inaccuracy of about 3%, or 3 false tests out of every 100 tests done. The true frequency of HIV infection in the overall U.S. adult heterosexual population is 15 per 100,000 (**Table 15.1**). Therefore every 100,000 tests should reveal an average of 15 true cases and 3,000 false positives. The false positives translate to a failure rate of 99.5% for the test (3,000 false positives out of 3,015 positive test results). If, on the other hand, the true rate of infection were 1 in 3, then every 100,000 tests would produce an average of 33,000 true cases and 3,000 false positives, a failure rate of 8% for the test.

15.2.4 A vaccine against AIDS?

A highly successful strategy for the prevention of many infectious diseases has been vaccination, which is a controlled exposure of a person to molecules similar or identical to those carried by the pathogen (see Section 14.2.4). Exposure to a vaccine stimulates the immune system to make an immune response to the molecules, and vaccination is therefore also called immunization. The pathogen itself is not used so that the person is not given the disease. The vaccine may be another microorganism, closely related to the pathogen but nonvirulent to humans, as when vaccinia from cows was used to protect against smallpox. (Smallpox vaccination succeeded in eliminating this disease from the globe; therefore, smallpox vaccinations are no longer routinely given.) Alternatively, a vaccine may be the pathogen itself but treated so as to make it nonvirulent or to kill it. Older vaccines used whole microorganisms, but today molecules vital to the pathogen's life cycle or to its ability to cause disease are more frequently used instead. Many laboratories are attempting to develop vaccines that would prevent HIV infection

Table 15.1 Prevalence of HIV infection in 1998 (near the high point in the pandemic)

Group (Adults 15–49) in the United States	Prevalence	Number per 100,000	Percentage
General heterosexual	1 in 6,666	15	0.015
College students	1 in 500	200	0.2
Prison population	1 in 495	202	0.2
Male homosexuals*	1 in 5	20,000	20
Bisexuals, infrequent homosexuals	1 in 20	5,000	5
Injection drug users	1 in 3	33,300	33
Worldwide	1 in 109	917	0.9

A prevalence rate of 1 in 109, or 917 per 100,000, means that, in a population of 100,000, 917 people (or 1 in every 109) are expected (on average) to have the condition being discussed.

* Rate among men who have sex with men and who were attending clinics for sexually transmitted diseases.

(pre-exposure immunization) or would prevent the progression of HIV infection to AIDS (postexposure immunization).

There are many biological barriers to developing vaccines against AIDS. These roadblocks include genetic variation of the virus, a lack of knowledge about which immune responses are protective against HIV, and a limit on the availability of suitable animal models in which to test trial vaccines.

HIV nucleic acid sequences change very rapidly. As mentioned previously, reverse transcription is error-prone, with 1–5 mutations per round of reverse transcription. In part, this is because the RNA is single stranded; there is no complementary strand to use as a template for corrections. In addition, there are no correcting and editing enzymes like those that keep the mutation rate low in DNA replication. Because there are two single strands of RNA per virus particle, these strands can recombine, further adding to genetic diversity. The virus thus evolves rapidly within a single host. The amount of genetic change within the HIV in a single patient over a 10-year period of disease is estimated to equal millions of years of change in the human species. The enormous resulting genetic variation presents a major problem for the design of vaccines. Is it possible to develop one vaccine that could stimulate a protective immune response in every person vaccinated and that would continue to protect infected people as the viral nucleic acid sequences changed? The answer right now is "maybe": Maybe there are some sequences that do not change very much or for which changes have no effect on recognition by the immune system. The latter is possible because most antibodies actually recognize protein *shapes*, not specific sequences of amino acids. A change in nucleic acid sequence may cause one amino acid to be substituted for another in the protein during its synthesis, but some substitutions may not significantly alter the shape of the folded protein. If the shape does not change, the immune system will still recognize the altered protein.

Not all immune responses against HIV are fully protective, as can be seen by the fact that HIV-infected people develop antibody and CD8 cytotoxic T-cell immune responses to HIV but still eventually get AIDS. Proteins that function in the viral life cycle are targets for vaccine development, but it is not known whether these will stimulate protective responses. There has been some concern that stimulating an immune response by vaccination might actually trigger progression to AIDS in someone already infected with HIV, since activated CD4 T cells are more susceptible to HIV infection than resting T cells. However, most studies of HIV-positive individuals who have received vaccinations for other viruses (especially influenza) suggest that these vaccinations are safe and do provide some protection against viral pathogens, though not to the extent that they do in HIV-negative individuals with fully intact immune systems.

The effects of vaccination on each step in an immune response can be studied in vitro, but protection from disease can only be evaluated in an animal that gets the disease. The lack of animal models has been a significant problem for HIV vaccine development. Chimpanzees become infected with HIV, but they rarely develop a disease that replicates human AIDS. Macaques, another primate species, do develop an AIDS-like disease when infected with a version of HIV with sequence modifications in a single regulatory gene and thus have proved more useful for studies of disease progression. However, primate research is extremely expensive. So-called "humanized" mice that have been engineered to exhibit key features of the human immune system relevant for HIV have also yielded important research findings, but they too have limitations with regard to vaccine development research. Scientists could use human volunteers, but ethical considerations call for extreme caution in the testing of vaccines on human volunteers in a disease known to have a high percentage of fatalities and for which there is no known cure.

In 1994, several vaccines were tested in small-scale trials on humans in Europe and North America. These vaccines did not prove to be entirely protective: A few individuals contracted HIV after vaccination. They did not get HIV from the vaccine; rather, the vaccine failed to protect them from transmission by the routes described later in this chapter. The National Institutes of Health did not allow larger-scale tests to proceed in the United States. The World Health Organization (WHO) took a different stand and allowed vaccine tests to be conducted in both

Uganda and Thailand. Following some promising results in 2009 in Thailand, a new and much larger clinical trial began in South Africa in 2016. These vaccines are targeted against the HIV strains prevalent in Africa and Asia. Earlier research focused largely on the subtype common in Europe and North America, but this subtype is not the one common in other parts of the world. Most recently, in early 2021, a small trial conducted in the United States by researchers at the Scripps Research Institute (2021) in collaboration with the International AIDS Vaccine Initiative produced positive results for the initial stages of a B-cell immune response that may be broadly protective against many different strains of HIV (https://www.scripps.edu/news-and-events/press-room/2021/20210203-hiv-vaccine.html); this trial also demonstrated that the vaccine was safe. These researchers are using an mRNA-based vaccine similar to some of those developed for the SARS-CoV-2 virus that causes COVID-19. Their strategy will employ multiple rounds of vaccination with slightly different versions of the vaccine in an attempt to stimulate broad protection against the many HIV variants that are currently circulating worldwide. Developing a successful vaccine represents the "holy grail" for AIDS research because drug therapies, which we look at in the next section, are much more expensive.

Prevention of HIV infection by education and by vaccination are currently the best and most affordable options. In addition, HIV-negative individuals who are at risk for HIV infection can now be prescribed one of two pre-exposure prophylaxis (PrEP) drugs as a preventative measure (though not a perfect one) against infection (CDC, 2021b). Both of these PrEP agents inhibit the viral enzyme reverse transcriptase through a combination of two antiviral drugs; the first was approved by the U.S. Food and Drug Administration (FDA) in 2012, and the second in 2019. The idea behind PrEP is that replication of any viral particles to which an individual is exposed will be rapidly shut down by the drugs that block the enzyme that converts the viral RNA genome to double-stranded DNA (Figures 15.4 and 15.8),

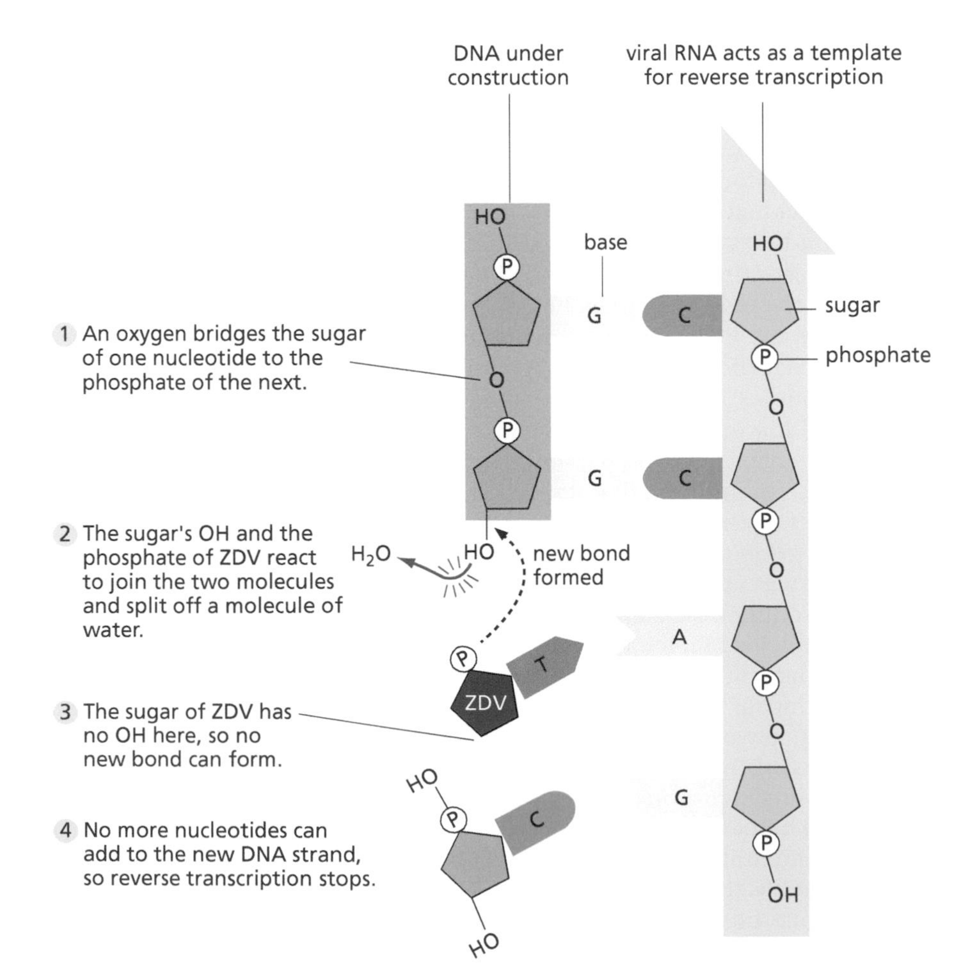

Figure 15.8 How the drug zidovudine (ZDV) interrupts the replication of the HIV virus.

thus halting any amplification of the infection and allowing the person's immune system to clear the virus. The downside of PrEP in comparison to a vaccine is that the PrEP drugs must be taken continuously, whereas a vaccine could generate lasting immunity.

15.2.5 Drug therapy for people with AIDS

In general, very few drugs are helpful against viral diseases. Antibiotics, which are highly effective against bacterial diseases, do not work against viruses. As detailed knowledge has become available about the few enzymes encoded by the HIV genome itself, new drugs have been developed to target these enzymes specifically. These are referred to as "antiviral" drugs.

15.2.5.1 Antiviral drugs The first HIV drugs, developed at the beginning of the AIDS pandemic in the 1980s, were *reverse transcriptase inhibitors*. These drugs are similar in chemical structure to nucleotides, the building blocks of DNA, so they can bind to reverse transcriptase at the same site where nucleotides bind (the enzyme's "active site"; see Section 8.2.3). They become incorporated into the growing DNA strand like normal nucleotides, but they lack the chemical group that is needed to form the link between a subsequent nucleotide in the chain. This terminates the reverse transcription of viral RNA to DNA (Figure 15.8), thus preventing any retrovirus from making the full-length DNA it needs to complete its infective cycle (Figure 15.4). One such drug is zidovudine or ZDV (trade name Retrovir®, formerly known as azidothymidine, or AZT). Didanosine (Videx®, formerly called dideoxyinosine or DDI) is another drug that inhibits reverse transcriptase. Other reverse transcriptase inhibitors that are not similar to nucleotides were developed later. These so-called "non-nukes" bind to reverse transcriptase at sites other than the enzyme's active site and change the shape of the enzyme in such a way that continued DNA synthesis is similarly blocked.

In addition to keeping the virus in check in infected individuals, reverse transcriptase inhibitors are commonly used to minimize transmission of the virus from an infected individual to others, either from pregnant women to their babies or between sexual partners. This is different from the PrEP strategy mentioned previously in Section 15.2.4 because, in this case, the drugs are given to individuals who are already infected to prevent transmission to uninfected individuals, rather than to uninfected individuals to prevent their own infection. According to 2017 UNAIDS data, implementation of antiretroviral therapy in infected individuals has reduced mother-to-child transmission of HIV to below 5% in several African countries (compared to 15%–45% without treatment, according to the WHO). A study by Rodger et al. (2019) followed 782 gay male couples in 14 European countries in which one partner was HIV-positive and the other was HIV-negative; although these couples engaged in high-risk behavior (anal sex without condoms) for a median study time of two years, the use of antiretroviral drugs reduced the transmission of HIV within these couples to zero.

Protease inhibitors were the second type of HIV antiviral drug developed; the first of these was approved for use in 1995. Later in the viral life cycle, after the reverse transcription step, an enzyme called a protease is required to trim newly translated proteins into their functional form. Blocking this step stops viral replication and infectivity. HIV protease is very different from human protease enzymes, reducing the effects of the drug on the human host.

Two even newer classes of anti-HIV drugs are the *fusion inhibitors* that prevent the virus from entering host cells and *integrase inhibitors* that prevent viral DNA from being incorporated into the host genome. After HIV binds to the CD4 molecule on helper T cells or macrophages, it enters the cells by triggering the fusion of its membrane with the host cell membrane. Fusion requires host cell proteins, and it is these proteins that are targeted by fusion inhibitors, blocking viral entry into the cell. The first of the fusion inhibitor drugs goes by the name T-20 (or Fuzeon®) and was approved for clinical use in 2003. The first integrase

Table 15.2 HIV antiviral drugs and their modes of action

Drug class	Mode of action	Examples*
Nucleoside/Nucleotide reverse transcriptase (RT) inhibitors	Bind to active site of RT enzyme	Zidovudine (Retrovir®) Didanosine (Videx®) Emtricitabine (Emtriva®) Tenofovir disoproxil (Viread®) Tenofovir alafenamide (Vemlidy®)
Non-nucleoside RT inhibitors	Bind to another site on RT enzyme	Nevirapine (Viramune®)
Protease inhibitors	Interfere with processing of HIV proteins	Saquinavir (Invirase/Fortovase®)
Fusion inhibitors	Block HIV from entering cells	Enfuvirtide (Fuzeon®)
Integrase inhibitors	Block insertion of HIV genome into host cell DNA	Bictegravir (component of Biktarvy®)
Gp120 attachment inhibitor	Blocks HIV surface protein from binding to CD4 T cells	Fostemsavir (Rukobia®)
CCR5 antagonist	Blocks CCR5 protein used as co-receptor for entry into human cells	Maraviroc (Selzentry®)
PrEP	Used to prevent HIV infection in healthy individuals	Emtricitabine + tenofovir disoproxil (Truvada®) Emtricitabine + tenofovir alafenamide (Vemlidy®)

* Trademarked names are in parentheses.

(Based in part on https://hivinfo.nih.gov/understanding-hiv/fact-sheets/what-start-choosing-hiv-regimen.)

inhibitor, raltegravir (Isentress®), was approved by the FDA in 2007, and other types of integrase inhibitors are still in development. Both of these newest types of drugs have been shown to work on patients whose HIV has become resistant to other antiviral drugs, and they are now frequently used in combination with other classes of drugs (see the next section). In addition, drugs with other modes of action continue to be developed and added to the HIV arsenal (Table 15.2).

15.2.5.2 Combination therapy With its rapidly changing genome, HIV has the potential to evolve resistance to any particular drug quickly. In combination therapy, the ideal approach is to combine two or more drugs that work by different mechanisms. If resistance to one mechanism of action evolves, the other drug(s) will still be effective. Combination therapy against HIV at first used two or more drugs with the same mechanism of action, that is, two or more reverse transcriptase inhibitors, because they were the only drugs available. Now, more than two drugs are included in what is called the "drug cocktail," and the combinations include several different classes of HIV antiviral drugs. These combinations have been very successful, reducing viral loads below detectable levels in a high percentage of HIV-infected people.

Drug combinations reduce the probability of HIV becoming resistant and rebounding, but these drugs must be taken on schedule. Early multidrug regimens involved taking as many as 16 pills at 6 different times during the day, making adherence to the schedule a big challenge. Current drug combinations involve fewer pills and are much simpler to administer because they only need to be taken once or twice a day. These drug regimens still have a variety of negative side effects, however, which makes therapy difficult for many patients. The drugs can also be enormously costly, formerly as high as $150,000 per year. The prices have declined in recent years but are still above $20,000 annually in many cases, and some treatments are not covered by insurance plans in the United States. For example, a three-drug, once-a-day pill called Biktarvy®, which contains an integrase inhibitor and two reverse transcriptase inhibitors (Table 15.2), is now available at a cost of $3,000 out-of-pocket for a 30-day supply. There is also significant controversy about the cost of drugs in the United States compared to other countries. For example, a different three-drug, one-a-day generic pill is now available in Africa for $75 per year but is not an option for patients in the United States because of patent restrictions. Despite all these challenges, many HIV-infected people—even some who have progressed to AIDS—have been restored to functional lives, and an HIV-positive diagnosis is somewhat less devastating than it once was.

It is not yet known how long a person would need to stay on therapy to be truly cured or if that is even possible. Because HIV is a retrovirus that integrates its genetic material into the genomes of host cells, so-called "latent" copies of the viral genome typically remain in patients in whom viral particles have been eliminated from the bloodstream by antiviral drug therapy, and those latent viruses have the potential to restart production of viral particles, leading to a resurgence of active infection. Reducing virus to "below detectable limits" is the goal of HIV therapy, but this cannot be considered a true "cure" for HIV or AIDS; some people's viral loads have returned when they stopped taking the drugs. Importantly, reaching the point at which virus cannot be detected vastly decreases the chance that a HIV-positive person will transmit the virus to others, so successful treatment is beneficial to society as a whole, not just to individuals dealing with HIV infection. Recently, a new approach that aims to inactivate latent copies of the HIV genome in host cells using CRISPR-Cas9 gene editing technology (see Chapter 4) has been pioneered in a humanized mouse model of HIV infection. This strategy, if it eventually proves successful in humans, and if it can be widely implemented, could represent a true cure for HIV, but this is likely still years away.

Another possible strategy for obtaining a cure for HIV involves transplantation of stem cells from a donor that has a genetic mutation in the CCR5 receptor so as to inhibit the ability of the virus to enter cells. So far, this approach has only been undertaken in HIV-infected patients who also have a blood cell (hematologic) cancer. Stem cell transplants are routinely used to treat people with hematologic cancers that have become resistant to other therapies (see Chapter 10). The procedure involves killing the patient's own blood cells with radiation, then introducing stem cells from a genetically matched donor to reconstitute the patient's blood cell population. In the case of HIV-infected patients, the fact that the stem cell donor is chosen to lack the CCR5 receptor means that the reconstituted blood cell population is no longer susceptible to HIV infection, and the virus may eventually be cleared from the body. As of 2022, this strategy has eliminated all detectable HIV virus from three patients, and two of these patients are still alive, while one eventually died of his cancer; in two cases the donor cells were from bone marrow, and in one they were from umbilical cord blood, which is easier to obtain. Stem cell transplants carry a high risk due to potential rejection, so this is unlikely to become a routine cure for HIV, but the fact that what has long been considered an uncurable disease has actually been cured in a small number of people has brought hope to many others living with HIV.

In summary, the advances in HIV antiviral drugs have certainly been a source of optimism, but further research into prevention of HIV infection (e.g., through vaccines) and elimination of latent copies of the virus (e.g., through gene editing) will be critical to ending the HIV/AIDS crisis. Meanwhile, much more needs to be done to provide access to lifesaving antiviral therapies for all infected patients and, on a population scale, to minimize the chance that an HIV-infected person will transmit the virus to others.

15.2.5.3 Prevention and treatment of opportunistic infections in persons with AIDS Nearly all AIDS deaths are caused by opportunistic infections. Attempting to prevent opportunistic infections in people with AIDS is thus highly important. During the phase of CD4 helper T-cell depletion, people are very susceptible to infectious diseases carried by people who are not infected by HIV. A cold or the flu can have grave consequences in an immunodeficient person. Bacteria picked up from food can be equally hazardous. In the developed world, therapeutic drugs are available for the treatment of many of the opportunistic infections, such as a combination of the drugs trimethoprim and sulfamethoxazole for *Pneumocystis* pneumonia. However, in many parts of the world, such drugs are unavailable because of their cost. The only other way to stop

AIDS deaths in populations is by preventing its transmission, which is the topic of the next section.

THOUGHT QUESTIONS

1. Calculate the failure rates (total false positives and false negatives) on your own for various sets of conditions using Table 15.1. Should everyone be tested when the true frequency of infection in a population is low? What if the frequency of infection is different in a specific population, such as a particular country or among people that have certain risk factors?
2. What features of HIV have made it especially difficult to develop a vaccine?
3. What physiological differences between individuals might affect their susceptibility to HIV infection or the progression of disease if they become infected?
4. Why do you think the use of antiviral drugs by HIV-positive people to prevent transmission to HIV-negative children or partners was approved before their use by uninfected people as a preventative measure (e.g., PrEP)?
5. How should continued testing of any promising new AIDS vaccines be performed? Who should be included in trials, and would your test have a control group? How would you ensure that the conduct of the test was ethical?
6. What are the financial considerations associated with HIV therapies and PrEP? Is the distribution of AIDS medications equitable?

15.3 KNOWLEDGE OF HIV TRANSMISSION CAN HELP YOU TO AVOID AIDS RISKS

The general term for the transfer of a pathogen from one individual to another is **transmission**. How is HIV transmitted from one person to another? HIV does not have any other animal hosts and does not remain infective in water or in air. It can pass from one person to another only in certain body fluids—blood, semen, and rectal and vaginal fluids. To enter another person, these fluids containing HIV or HIV-infected cells must rapidly come in contact with the bloodstream of the other person via breaks in the mucous membranes or skin. HIV must make rapid contact with cells bearing the CD4 molecule. Cell-free HIV does not remain infectious very long if the fluids are outside a person, for example, in a blood spill. Washing with ordinary soap and water kills cell-free HIV when it is outside a person, because soap dissolves the viral lipid envelope. A mother can transmit HIV to her unborn fetus or to her baby during delivery, as well as during breastfeeding after birth; antiretroviral therapy can, however, reduce these risks, as mentioned previously. HIV is not found in feces or urine. There are small numbers of HIV particles in the saliva or tears from 1% to 2% of HIV-infected people, but saliva contains antiviral activity, and HIV has never been known to be transmitted through saliva, including human bites, or via tears. How do the fluids containing HIV get passed? The percentage of new HIV diagnoses (incidence) in the United States transmitted by various routes (in 2018) is shown in Figure 15.9.

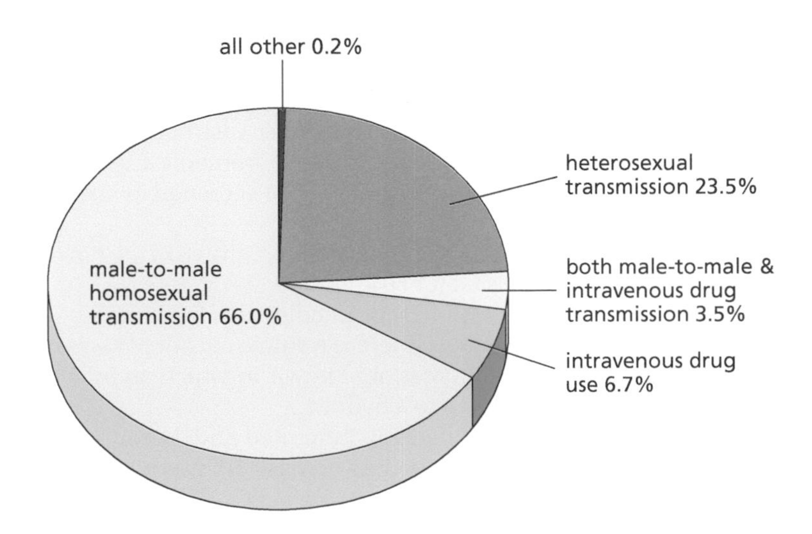

Figure 15.9 **Routes of HIV transmission in the United States, based on statistics for the year 2018 from the Centers for Disease Control and Prevention (https://www.cdc.gov/hiv/risk/estimates/riskbehaviors.html).** The percentages indicate the proportion of cases transmitted by each route.

15.3.1 Risk factors: Behaviors that may lead to infection

When scientists use the term **risk** of HIV infection, they mean the mathematical probability of transmission of the infection. Various behaviors or activities have been grouped into "risk categories," based on what is known about transmission routes (CDC, 2021a). High-risk behaviors are those that give a high probability of transmission from an infected person to an uninfected person. Notice that risk is now categorized by specific **risk behaviors** and not by population groups. The safest way to make choices about these behaviors is to assume that every person whose HIV status is unknown to us may be HIV infected and a potential source of HIV transmission.

15.3.1.1 Category I: High-risk behaviors

Transmission while engaging in a **high-risk behavior** is very likely. These behaviors, which account for over 97% of all HIV transmissions, are:

1. Behaviors in which the passage of blood, vaginal fluids, or semen is very likely, such as anal or vaginal intercourse with an infected person without the protection of a condom (unsafe sex).
2. Injection drug use in which needles or syringes are shared with an infected person.
3. An infected mother's going through pregnancy and giving birth without receiving antiretroviral treatment. (This route of transmission, once very significant, has been greatly reduced by the widespread availability of antiretroviral drugs.)

Anal intercourse is more risky than vaginal intercourse because semen contains an enzyme (collagenase) that breaks down the lining of the rectum and exposes blood vessels, a form of injury to which the vagina is much more resistant. In addition, vaginal intercourse is not as likely to result in HIV infection as anal intercourse because the cells of the rectal mucosa have the CD4 molecule on their surfaces and so can be infected with HIV, whereas the intact epithelial lining of the vagina is a significant barrier. The risk of male-to-female transmission of HIV infection during vaginal intercourse is greater than for female-to-male transmission.

15.3.1.2 Category II: Likely risk behaviors

HIV transmission has been documented for routes in this category, but with lower frequency:

1. Anal or vaginal intercourse using a condom (safer sex, not safe sex). Condoms do not make intercourse completely safe. Condoms fail as birth control for about 10% of couples who use them; this means they can also fail to prevent HIV transmission. Although the rate of failure seems to be low (less than 5%) for condoms *used properly*, many users do not exercise proper care in putting condoms on or taking them off, so estimates of failure rates *in general use* can be as high as 20%. Although reliable data of this kind are difficult to obtain, one study found that 17% of women whose husbands were HIV-positive became infected despite proper and consistent condom use.
2. Breastfeeding (transmission to a baby from an infected mother).
3. Receiving a blood transfusion or organ transplant. This risk was high prior to 1987, but is now very low in the United States because of careful screening of blood products and donated organs. The risk remains higher elsewhere, where blood donors and blood products are not screened as stringently (see later in this chapter).
4. Artificial insemination. As with blood transfusion, the risks are now low when donated semen has been tested for HIV.
5. Infection of health care professionals by needle-stick injuries.
6. Dental care by an infected worker. There is a single cluster of six cases involving only one dentist; no other cases are known in which an infected health care worker has transmitted HIV to a patient.
7. Deep kissing. A single case has been documented and, because both people had severe periodontal disease (bleeding gums), HIV was probably transmitted through blood, not saliva.

15.3.1.3 Category III: Low-risk behaviors

This category includes routes that are biologically plausible, but no cases have been confirmed.

1. Sharing toothbrushes or razors or other implements that may be contaminated with infected blood.
2. Being tattooed or body pierced to produce ornamental scars or for jewelry.
3. Receiving tears or saliva.
4. Oral sex. Although there is anecdotal evidence of transmission via oral sex, there are no cases in which transmission by this route is documented.

15.3.1.4 Category IV: No-risk behaviors

Transmission of HIV from engaging in the following behaviors with an infected person is considered not biologically possible:

1. Shaking hands
2. Sharing a toilet
3. Sharing eating utensils
4. Being sneezed on
5. Working in the same room
6. Handling the same pets
7. Close-mouthed kissing (kissing with no exchange of saliva or blood)

Also, exposure to mosquitoes or other biting insects is a no-risk behavior, as explained in Box 15.2.

15.3.2 Communicability

Another question people ask about HIV is, "How contagious is it?" meaning, "If I am exposed, how likely is it that I will become infected?" The term used by the medical community to mean the likelihood of transmission after exposure to HIV is **communicability**. ("Contagious" simply means "capable of being transmitted"; it does not refer to the probability of transmission.) The concept of communicability, or likelihood of transmission, is directly related to the concept of risk. High-risk behaviors (see Section 15.3.1.1, Category I) increase the probability of transmission.

Box 15.2 Can mosquitoes transmit AIDS?

Two frequently asked questions are, "Why don't mosquitoes transmit HIV?" and "How do we know that they don't?" Epidemiological evidence shows that the frequency of AIDS and of the number of unexplained cases is no higher in mosquito-infested areas of the United States than in other areas. In Africa, the people with AIDS are mostly babies and sexually active young adults; mosquitoes do not bite people in these groups more frequently than they bite other people. On all continents, children who are not yet sexually active often get mosquito bites, but they do not get AIDS unless they are infected from their mothers at birth. Laboratory experiments have shown that although HIV and HIV-infected cells may be taken up by mosquitoes who bite infected people, HIV is not transmitted to other people through mosquito bites. Several factors help to explain this: HIV cannot replicate in mosquitoes or survive long in their bodies (because mosquitoes are not a host for HIV), and the amount of blood ingested (3–4 μL, millionths of a liter) is too small to contain enough HIV or HIV-infected cells to infect a person. Saliva (including mosquito saliva) may also have substances that inhibit the virus. Furthermore, because of the many biological factors involved in the transmission of a disease by an insect, it is highly unlikely that a single mutation in either the mosquito or the virus would significantly alter this situation.

A study done at the Institute for Tropical Medicine in North Miami, Florida, proposed in 1986–1987 that the high rate of AIDS in the town of Belle Glade, Florida, could be attributed to the squalor and crowding of its people and to the mosquitoes breeding in nearby swampy lands, "where 100 insect bites a day are not unusual." The U.S. Centers for Disease Control and Prevention (CDC) studied this situation and concluded that the high incidence of AIDS in Belle Glade was attributable to sexual contact and shared needles, not insects.

Transmission of HIV by other blood-sucking animals such as bedbugs (which are insects) and ticks (which are more closely related to spiders) has also been ruled out. A tick that is endemic in the same parts of Africa where AIDS is common carries enough blood and live virus to make transmission theoretically possible. However, the possibility does not fit with the epidemiology: Children below the age of sexual activity do get bitten in significant numbers but do not get AIDS.

Table 15.3 Estimated efficiency of transmission of HIV by various routes

Route of transmission	Efficiency (incidence per 10,000 exposures)
Anal intercourse, receptive (unprotected)	138
Anal intercourse, insertive (unprotected)	11
Vaginal intercourse, receptive (unprotected)	8
Vaginal intercourse, insertive (unprotected)	4
Mother to child during pregnancy or childbirth (without use of antiretroviral drugs)	2,260
Intravenous injection with infected needle	63
Needlestick with infected needle	23

(Data from Patel et al (2014), AIDS. 2014 Jun 19; 28(10): 1509–1519. doi: 10.1097/QAD.0000000000000298.)

There are at least two ways to answer the question of communicability: One takes an epidemiological approach and the other a microbiological approach. The epidemiological approach compares the number of encounters with the number of infections throughout the population or within certain population subgroups. It is difficult to designate the probabilities for HIV transmission because there is a period of weeks or months before antibodies develop, and there is often a period of years between infection and disease symptoms, during which people may not know they are HIV-infected. The number of encounters with HIV is often not known for an individual, and is even less known for all the people constituting a population.

Probabilities are, however, more accurately known for some routes of transmission than for others. For example, there is a 22.6% chance that an infected mother will transmit the virus to the fetus if she is not treated with antiretroviral drugs. The efficiency of transmission of HIV by various routes is shown in **Table 15.3**. The microbiological approach to determining communicability is to quantify (measure numerically) what is called the infective dose—the number of pathogenic particles that must be transferred to result in an infection of an individual. This value is not known for HIV, although one estimate is that the transfer of 10,000–15,000 HIV particles can establish an infection. Very early in an HIV infection, in the weeks or months before antibodies develop, and also very late, when the CD4 helper T-cell count is low and the antibody concentration has dropped, the number of HIV particles in blood and genital fluids is much higher than at other stages (Figure 15.7). The number of HIV particles in general is higher in semen than in vaginal fluids, but each varies at different stages of infection. One study found 4.2 million HIV particles per milliliter of blood on average in people with AIDS, but this can vary tremendously.

From accidents in which health care workers have been exposed to infected blood, it is known that there is a higher probability of infection when a person has been splashed with large quantities of blood onto open sores in the skin and a lower probability when they have been pricked by a needle. For every 250 reported needlesticks, there has been one transmission (0.4% efficiency). The effect of blood volume can also be seen in Table 15.3, where the transfer of greater quantities of blood by intravenous injection or transfusion increases the efficiency of transmission.

Viral load is certainly a factor in determining the efficiency of HIV transmission, but only one of many factors. The precise infective dose is not known and probably varies from one person to another. For example, persons with open genital sores due to other sexually transmitted diseases such as syphilis or herpes are 10–20 times more likely to become infected than other people. Gonorrhea or chlamydia infections increase the probability of HIV transmission three-fold or four-fold. (These sexually transmitted diseases are discussed further in Section 16.2.1.)

In general, it may be said that HIV is much less communicable than many other viruses. For example, for hepatitis, another virus spread by contact with contaminated blood, there is a 6–30% chance of transmission by a needlestick in comparison with the 0.4% efficiency of HIV transmission by this route. However, the communicability of HIV varies with the risk behavior. Transmission of opportunistic microorganisms from a person not infected with HIV to an HIV-infected person is much more likely than transmission of HIV from an HIV-infected person to an HIV-uninfected person. Because people with AIDS have severely impaired immune systems, their risk of acquiring diseases from other people is very high.

15.3.3 Susceptibility versus high risk

What is the difference between the terms susceptibility and high risk? To examine this question, let us go back and look further at how knowledge of AIDS and HIV developed. The fact that early cases were reported in hemophiliacs suggested an infectious cause for AIDS. As more cases were reported, the affected individuals seemed to fall into five groups, which became known among epidemiologists as "the five H's": homosexual males, hemophiliacs, heroin addicts, Haitians, and hookers (prostitutes). From an epidemiological perspective, these categories served a useful purpose to describe groups within which cases were occurring most frequently. But what was useful scientifically had quite negative social consequences. The terms quickly became imprinted in the minds of scientists and the public, allowing complacency on the part of people who were not in these five groups and prejudice against those who were in the groups. To epidemiologists, the five H's were merely a convenient way to designate clusters of reported cases of a mysterious, new syndrome. Such identification did not in itself imply anything about cause and effect, about transmission, or about all of the people in the groups, but it was useful in suggesting hypotheses that could be tested. Because of the fear generated by AIDS, however, these groups of people became stigmatized and often treated badly by the general public and even by some health care professionals. The book *All the Young Men* (Burks and O'Leary, 2020) documents the neglect and mistreatment of AIDS patients early in the epidemic and the heroic efforts of one Arkansas woman to provide them with medical care, education, and advocacy.

To epidemiologists, the term **high-risk group**, as applied in connection with a particular disease, simply means that there is a higher frequency of the disease among members of that group (frequency equals the number of people with the infection divided by the number of people in the group). Use of the term implies nothing about the possible reasons for the increased frequency, which may stem from increased exposure, increased susceptibility, or both. *Exposure* to a disease means coming in contact with the disease agent. Increased exposure can sometimes be due to shared behaviors, but there are many other possible explanations. A disease may have a higher frequency in a certain group of people if all the people in that group came from the same geographic location so that they were exposed to the same toxic chemical or if they all ate food from the same source and were therefore exposed to the same foodborne pathogen. It does not mean that these people are more susceptible; anyone else exposed to the same factors would also have become sick. **Susceptibility** to a disease means the ability to contract that disease *if exposed*. Humans are susceptible to HIV, while most other animals are not. Susceptibility can vary from one person to another, and it can be genetically or environmentally influenced (malnutrition, for example, may make a person more susceptible to many infectious diseases).

Several misperceptions about AIDS resulted from the early identification of specific high-risk groups. First, some people not in the identified high-risk group assumed they were not susceptible. Some people assumed that every person within a high-risk group was equally likely to be infected (and that people outside these groups were unlikely to carry the disease). Haitians, in particular, suffered adverse consequences by being classified as "high risk." In efforts to screen blood donors before the cause of the disease was known and before appropriate tests were available for screening blood, all Haitians were barred from donating blood in the United States. In the resultant hysteria, some Haitians were evicted from their homes and lost their jobs, Haiti's tourist trade collapsed, and Haitian dictator

Jean-Claude Duvalier's state police rounded up and incarcerated homosexuals in Haiti. Haiti's ambassador to Washington wrote a letter published in the *New England Journal of Medicine* deploring the damage done by North American semantic carelessness. As he pointed out, being from a certain country does not contribute to disease in the way that socially acquired behaviors do (for example, having multiple sex partners or using intravenous drugs).

The fifth H, hookers, always seemed problematic because the other risk categories were predominantly or exclusively male. Why did so few women contract the disease at first? If women could contract the disease, why only sex workers? There was a period of time when women were thought of only as "carriers," even though some women were dying of AIDS themselves. Scientists now think that women are just as susceptible as men to HIV infection, but that the epidemic in the United States began among homosexual men and spread more slowly to women.

Once mechanisms of transmission are known, it becomes more appropriate to focus on high-risk behaviors than on high-risk groups. However, the frequency of infection within a discernible group of people can sometimes have a role in an individual's risk. People within a high-risk group are at risk to the extent that they engage in high-risk behaviors. Their risk may be increased to the extent that their partners in high-risk behaviors are also members of a group in which the frequency of infection is high. A higher population frequency of a disease increases risk by increasing the chance of encountering an infected person, not by altering any individual's susceptibility. (Remember that membership in a group, either a group with a high frequency of infected individuals or one with a low frequency, does not tell you whether a particular individual is or is not infected.)

As we have seen, HIV infection rates in the United States have always been highest in homosexual men. They continue to be so, with the overall rate of male infection in the United States being five times that for women in 2019; however, from 2015 to 2019 the infection rate among men decreased, while that for women remained stable. The lower frequency of HIV infection in females in the United States initially led many people to assume that women were less susceptible. Therefore, when more women began to fall ill, there was a further misconception that the virus must have mutated to change its infectivity. In reality, however, women and heterosexual men have always been susceptible to HIV infection, as amply demonstrated by the pattern of the infection in Africa, where men and women have been infected almost equally. The pattern of infection in the United States has changed over time, but it is possible to explain all of the changes on the basis of frequency of HIV in various subpopulations, not on the basis of changes in infectivity of the virus.

Transmission of HIV by vaginal intercourse is not as likely as it is by anal intercourse, but this does not mean that vaginal sex is safe, only that the number of infections per number of encounters is lower. It also does not mean that women are less susceptible than men. It seems that, if there are breaks in the vaginal epithelium (for example, as a result of other sexually transmitted diseases), women are just as likely to be infected as men. So, again, the risk is related to particular practices, not to differences in susceptibility, and these practices carry comparable risks for all groups of people. For example, data collected in both the United States and Africa seem to show that anal sex is just as risky for females as for males.

Epidemiology shows that there is a positive correlation in both sexes between the rate of HIV infection and the number of sexual partners; that is, persons with greater numbers of partners have higher rates of HIV infection. There is also a positive correlation between the infection rate and the frequency of previous infections with other kinds of sexually transmitted diseases. People who have contracted any other sexually transmitted disease have already engaged in behavior that puts them at risk for HIV infection. Because properly used condoms can, in general, greatly reduce transmission rates for other sexually transmitted diseases as well as for HIV, someone who has contracted a sexually transmitted disease has probably not used a condom during intercourse. Moreover, they have further increased their risk for HIV infection because the presence of open genital sores greatly increases the probability that contact with HIV will result in HIV infection.

The use of illegal drugs (injectable or not) also increases the rate of infection, particularly in women. In New York City, 32% of female crack cocaine users were HIV-positive, compared with 6% of other women. It has not been shown that crack is a cofactor (a factor that increases susceptibility), but the subculture in which crack is used is often one of a high incidence of sexual activity and of sexually transmitted diseases. On college campuses, where the "drug of choice" is frequently alcohol, the impaired judgment that accompanies alcohol (or other drug use) is a factor working against sexual abstinence or the practice of safer sex.

Another aspect to risk is a person's ability or inability to say "no" to high-risk behaviors. The ability of a person to say no is termed his or her refusal skills. Economic and cultural factors can put severe limitations on a person's refusal skills. In some cultures, for example, women may not be able to insist that their male sexual partners use a condom. Education about the risks of HIV and AIDS must do much more than provide people with information about transmission routes, as we will soon discuss.

15.3.4 Public health and public policy

Whereas medicine deals with individual cases of disease, public health deals with populations and seeks to minimize the levels of particular diseases in those populations. Many public health efforts require legislation, and most require funding, so they are most often conducted by governments or large organizations. Many nongovernmental organizations (NGOs) have been crucial in educating the public about AIDS and in caring for persons with AIDS and their families. Early in the epidemic, they were also crucial in pressuring governments for more funding for research on this disease.

15.3.4.1 History of public health responses to disease In each nation in which the AIDS epidemic has spread, the governmental response was molded by the unique history and social customs of that nation. Some nations sought to restrict the immigration of HIV-infected people; others did not. Some jurisdictions segregated certain types of AIDS patients, while others did not. Hospital care and medical insurance for AIDS patients varied greatly from one country to another. Some nations instituted needle-exchange programs for drug addicts; others did not.

The response of the U.S. government to AIDS has been influenced partly by the nature of transmission of the disease (that is, it is not transmitted by casual contact), partly by the political organization of the "AIDS community," and partly by changes in civil rights laws during the period since 1950. Before 1950, U.S. law gave only weak support to individual rights. The Supreme Court decision in *Brown v. Board of Education*, the resultant Civil Rights Act of 1964, and the Voting Rights Act of 1965 gave much greater strength to the rights of individuals in the face of discrimination by the many. The Rehabilitation Act of 1973 mandated that employers receiving federal funds cannot discriminate against someone with a handicapping condition who is otherwise qualified, and a disease that does not endanger others is considered a handicapping condition. This was upheld in *School Board of Nassau County v. Arline* (1987), in which a person with tuberculosis won the right to continue to work. These principles have been further extended in the United States by the Americans with Disabilities Act (1990). Under this act, persons with disabilities cannot be denied "full and equal enjoyment" of goods, services, and public accommodations. AIDS is considered a disability under this act. In 1998, the U.S. Supreme Court ruled that asymptomatic HIV infection can also be a disability. A disability is defined under this law as "a physical or mental impairment that substantially limits one or more of the major life activities of the individual." The court case involved an HIV-positive woman who was denied treatment by a dentist in his regular dental office. The court ruled that the woman was disabled because HIV infection limited her ability to have children (because unprotected intercourse would put her husband and a child they conceived at risk for infection). The dentist had wanted to offer the woman dental treatment in a hospital, but the court also ruled (5–4) that the Universal Precautions for handling body fluids (see Section 15.3.4.3) should make a dental office a safe environment for treatment.

A bioethical principle known as the harm principle provides a moral limit on the exercise of freedom of individuals when others may be injured. By this principle, a person with AIDS could morally be prevented from deliberately spreading the disease but could not be prevented from working or attending school or living in a particular place.

15.3.4.2 HIV testing and notification AIDS is a reportable disease in all 50 of the United States. Twenty-seven states require the notification of public health officials of HIV infection, including the person's name, so that cases can be followed. The issue of mandatory testing for HIV antibodies remains very controversial. Issues of confidentiality are involved, because many people fear (correctly or incorrectly) that they would be discriminated against if they were identified as being HIV-positive. At the same time, medical professionals would like to know the HIV status of their patients so that they can provide both better care for the infected person and better protection for health care workers against unintentional infection. Forty-four states permit notification of health care workers of patients' HIV status; two states (Arkansas and Missouri) require it.

Current laws generally prohibit testing a person without the person's consent. However, the U.S. government does test everyone who applies for immigration, the Peace Corps, the Job Corps, the military, or the Foreign Service and also tests the spouses of Foreign Service applicants. The 1992 International Conference on AIDS, which was to have been held in the United States, was moved to Amsterdam in protest against the U.S. policy of denying entry to any HIV-positive person; that policy has since been relaxed. As of 2012, 11 states screen all prisoners for HIV (Aguilar, 2012), but only 6 segregate those who are HIV-positive.

Another issue relates to the notification of sexual partners of HIV-infected persons. For other sexually transmitted diseases, public health officials notify and test the sexual partners of all infected persons, then the partners of those sexual partners, and so on. Some doctors have argued against this practice for HIV-infected women, who are often diagnosed during prenatal testing when they are pregnant. Many HIV-positive women (and their unborn fetuses) become victims of domestic violence when their HIV status is reported to their partner. Partner notification is permitted ("duty to warn"), but not required, in most states. Notification obviously depends on infected persons' accurate disclosure of the names of all their partners.

Around the world, many countries have passed strongly punitive laws against being HIV-positive, against having sex while HIV-positive, or simply against homosexuality. According to UNAIDS, such punitive laws needlessly force HIV-positive people into hiding and greatly interfere with public health campaigns that provide education, condoms, and antiretroviral drugs.

15.3.4.3 Guidelines for handling blood People likely to come in contact with blood—for example, dentists and surgeons as well as their auxiliary workers, sports coaches and trainers, and security personnel—are now required to follow a series of guidelines known in the United States as the *Universal Precautions for Blood-Borne Pathogens*. The term "universal" refers to the fact that all blood must be handled as though it were infected. The guidelines, which are intended to limit the transmission of any blood-borne pathogen, include wearing gloves when handling blood (**Figure 15.10**) and additional personal protective equipment when handling large quantities of blood. The guidelines also specify procedures for cleaning blood spills, reporting accidents and injuries in which workers have come into contact with blood, and educating workers about the risks in handling blood. The U.S. guidelines were developed by the CDC, and the Occupational Safety and Health Administration (OSHA) has been charged by

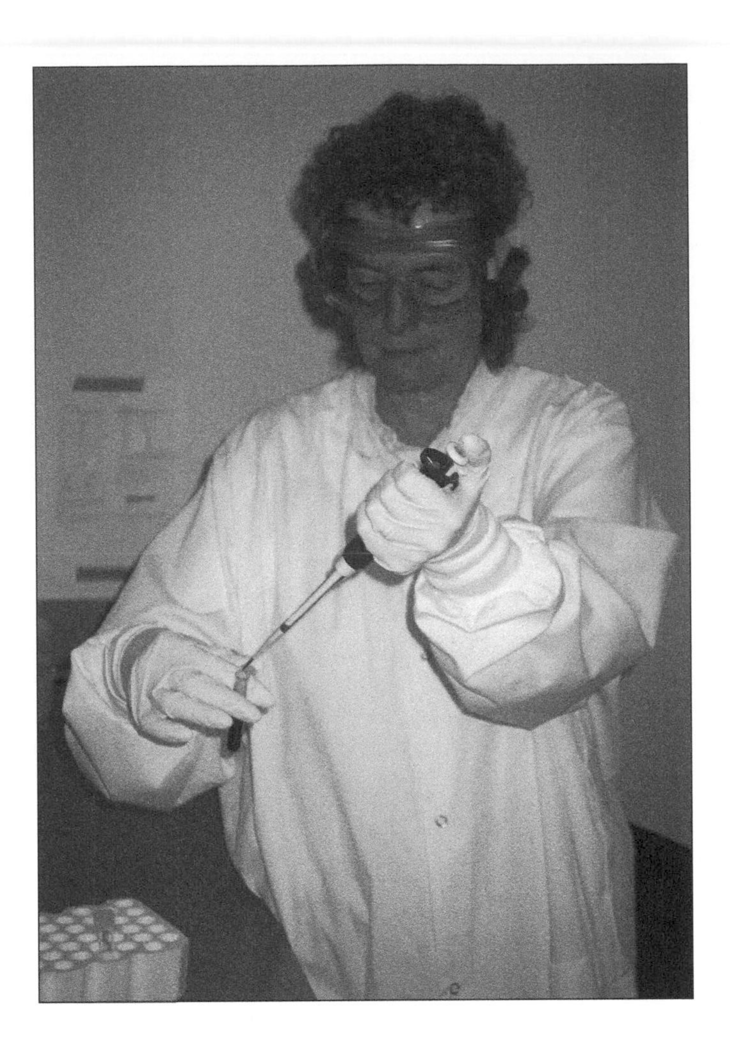

Figure 15.10 Health care workers, researchers, and others who handle human blood must follow the Universal Precautions for Blood-Borne Pathogens.

the federal government with monitoring the compliance of employers with these guidelines. Every college and university, for example, must have an infection control plan.

The risk of transmission of HIV through blood transfusion depends a good deal on the methods of blood donor selection. In the United States, blood donors are volunteers. Blood is screened for antibodies to HIV, but blood donors are not. The behaviors that transmit HIV are explained to blood donors, and they can anonymously tag their donated blood as having come from a person involved in high-risk behaviors. In the United States, the risk of transmission from a blood transfusion is currently estimated to be 1 in 450,000–650,000 (Table 15.3).

In countries in which donors are paid, the safety of the blood supply is much less assured than in countries in which donations are voluntary. In India, 30–50% of blood donors are professional donors who sell their blood an average of 3.5 times per week. Although the blood is screened for HIV, the virus can escape detection in the early stages of infection, and thousands of people in India have contracted HIV/AIDS from blood transfusions as a result (Mangle, 1993; Kumbhakar, 2021).

15.3.4.4 Access to health care

In the United States, the CDC develops the criteria that define AIDS. The criteria have changed as new information has become known. The wording of the definition is important because people with AIDS are eligible for some types of care from the government not afforded to the general population. AIDS was originally defined as a set of symptoms, including particular opportunistic infections. The CDC definition now includes all those persons who are HIV-positive and have a CD4 helper T-cell count below 200.

HIV-positive women have a poorer prognosis (predicted outcome) than HIV-positive men, both in the United States and elsewhere in the world. This is probably a result of women's generally having poorer access to medical care for the infections that accompany AIDS. The care of people with AIDS has put a strain on public health monies and personnel. In some countries of Africa, more than half of all public health expenditures are for AIDS.

15.3.4.5 Educational campaigns

In educational campaigns aimed at increasing AIDS awareness, some organizations distribute free condoms and promote their use, while others emphasize abstinence. The former surgeon general of the United States, C. Everett Koop, stated that the only safe sex is a faithfully monogamous relationship with a faithfully monogamous uninfected partner, and the next best thing is the use of a condom.

Education about HIV and its transmission changed the behavior of homosexual men so that the rates of infection within this group began to decline. This subgroup is generally well educated and has provided many model programs that have been copied in educational efforts to reach other groups. Because the factors guiding people's private behavior differ from one group to another (on the basis of language, income, geography, religion, and cultural background), educational campaigns need to be designed for each different locale and target group.

Information is not the same thing as education. Giving people information about how HIV is spread may not help unless the reasons underlying their high-risk behaviors are addressed. The motivations of people having consensual sex differ from the motivations of commercial sex workers (prostitutes) and street children having sex for survival. Many teenagers and young adults engage in sexual activity (often including high-risk activity), and those who do not are frequently subjected to very strong peer pressure to conform. Education often includes strategies for raising self-esteem and providing support for avoiding high-risk behaviors.

15.3.5 Worldwide patterns of infection

As stated in the opening of this chapter, HIV is a pandemic, a worldwide epidemic. The number of people living with HIV worldwide grew to 24 million people in 2000 and to 38 million by 2019, according to UNAIDS (2021). By 1998, AIDS had become the seventh leading cause of death worldwide, but the availability of antiretroviral drugs has greatly lessened the disease burden and mortality rates since then, even as the number of people living with HIV continues to grow. By 2016, AIDS was no longer among the top 10 causes of death worldwide, but it was still the fourth most common cause of death in low-income countries, according to UN statistics.

UNAIDS (2021) estimates that 38.0 million people worldwide were living with HIV infection in 2019 (**Figure 15.11**), including 25.6 million in sub-Saharan Africa (over 67% of the total), with about 1.7 million new HIV infections annually worldwide. AIDS-related deaths stood at about 690,000 worldwide in 2019, down from an annual rate of 1.7 million in 2004, with the decline largely attributed to the use of antiretroviral drugs. In 2017, about 59% of the people living with HIV infection were receiving antiretroviral therapy. Tuberculosis accounts for about one-third of AIDS-related deaths worldwide.

In 2019, the CDC estimated that 1.2 million people were living with HIV infection in the United States and its territories (https://www.hiv.gov/hiv-basics/overview/data-and-trends/statistics). The annual number of new cases has been declining since 2016, from about 38,000 in that year to 34,800 in 2019. The rate varies geographically, with over 50% of new diagnoses in 2019 occurring in 48 counties plus Washington, D.C., and San Juan, Puerto Rico; these locations are specifically targeted by a plan announced in 2019 by the U.S. Health Resources and Services Administration (HRSA, 2021). In 2019, 69% of new cases were attributable to male-to-male homosexual transmission, 7% to intravenous drug use, and 23% to heterosexual contact (also refer to Figure 15.9).

Transmission patterns vary from country to country. In North America and Western Europe, male-to-male homosexual transmission is the major transmission route. In other parts of the world, heterosexual transmission is much more

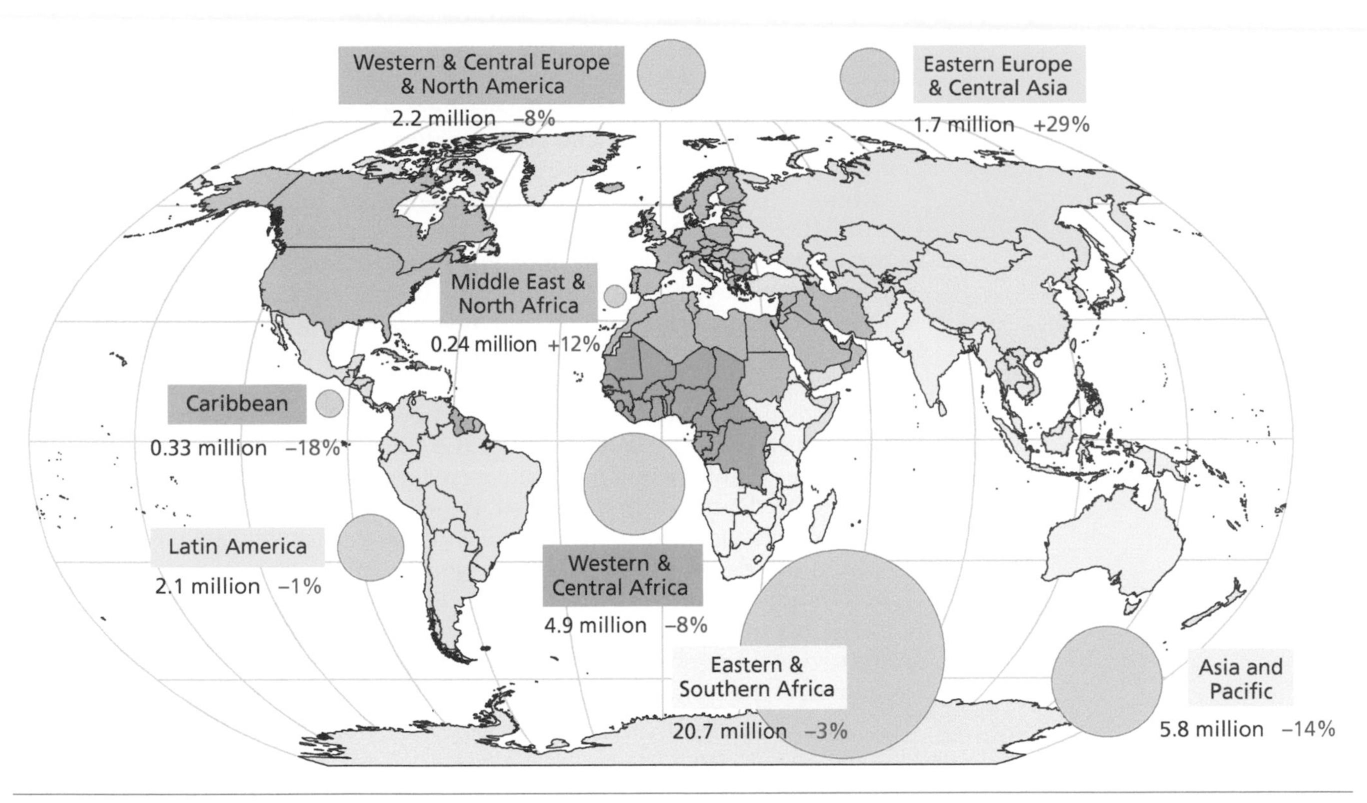

Figure 15.11 HIV/AIDS in the different regions of the world recognized by UNAIDS. Circles and numbers in black show *prevalence*, the number of people living with HIV/AIDS in 2019. Numbers in red show the trend (or change) in *incidence* (number of new cases per year) from 2000 to 2017.

common, although the reporting of data is less detailed and possibly less reliable in countries where homosexuality is illegal. In Russia, Eastern Europe, and Central Asia, injection drug use accounted for 39% of HIV transmission in 2017, according to United Nations statistics, and these drug users, together with their sexual partners, account for over half of all HIV infections in the region. This is also one of only two regions where HIV prevalence continues to increase.

Statistics on AIDS may also vary according to the way in which AIDS is defined. The WHO criteria for diagnosing someone with AIDS are very different from the CDC criteria. The WHO criteria, based on symptoms, not on HIV status or T-cell counts, are used in countries where monetary or technical considerations make testing for HIV infection impossible. Worldwide surveillance of numbers of AIDS cases is therefore not the same as surveillance of HIV infection, which must be estimated from the numbers of AIDS cases. Moreover, many people with AIDS are difficult to distinguish from people who are immunodeficient from other causes, such as undernourishment, and so may or may not be counted as AIDS cases. The numbers may thus underestimate the true AIDS incidence.

AIDS is now the leading cause of death in much of Africa. Africa also has the highest prevalence of AIDS (the number of infected people divided by the total population), and in most areas the rate is still increasing. In the five years from 1997 to 2002, the number of infected people increased 30% in Africa, from 23 million to 30 million, but the use of antiretroviral drugs has halted and even reversed this increase in many places. Eight African countries had a prevalence rate above 10% in 2019, including South Africa (prevalence rate 18.9%), Botswana (21.9%), Lesotho (25.0%), and Eswatini (Swaziland, 27.2%). Sickness of people living with HIV and deaths from AIDS are having a devastating effect on the economies of these countries, particularly because the prevalence of HIV is highest among working-age people. Labor-intensive industries such as food production and mining have been especially hard hit, leading to further hardships as food becomes scarce and mineral exports cannot pay for needed medicine. Education of the next generation is being strained at all levels. Universities have been greatly affected as sickness and death among faculty, staff, and students have soared.

HIV infection is not uniformly prevalent throughout Africa. Several countries in West Africa have lower rates owing to early and sustained implementation of prevention programs. Prevalence is also very low in North Africa. In Botswana, however, inaccurate beliefs and risky practices among traditional healers have contributed to the spread of the infection.

China, Russia, and India now each have more than 1 million people with HIV or AIDS. HIV infections are increasing most rapidly in Eastern Europe and Central Asia, especially among injection drug users and their sexual partners. In Africa the rate of infection is about equal in men and women, while in North America, South America, Europe, and Australia it is higher in men. In the United States, a majority of infected women are between the ages of 15 and 25, and 80% of newly infected women are either intravenous drug users or sexual partners of intravenous drug users. Starting in 1997, new AIDS cases in the United States began to decline for the first time since the epidemic began, although the numbers of cases among women still increased. In developed countries, the numbers of AIDS deaths have decreased as new drugs have been developed and made available. Because of the high cost of these drugs, however, most infected people in the developing world have not benefited from them. In 2001, the United Nations began the Global Fund to Fight AIDS, Tuberculosis and Malaria. In 2019, this fund invested $4 billion toward these efforts, including funding antiretroviral therapy for 20.1 million people worldwide, but this is still insufficient to meet the overall need.

THOUGHT QUESTIONS

1. What advice would you give to college students about the best ways to avoid becoming infected with HIV? How might you modify the advice for different groups of people?
2. What are the misconceptions surrounding HIV and AIDS? Have they changed over the years?
3. How are the Centers for Disease Control and Prevention's criteria for AIDS different from those of the World Health Organization? Why are they different?
4. HIV infection is associated with "risk behaviors." To what extent is risk the result of an individual's choice? To what extent do societal factors remove choice from individuals?

CONCLUDING REMARKS

What does the future hold? Successful drug therapies now exist that can keep HIV in check in infected individuals for very long periods of time—potentially a person's lifetime—but access to these treatments varies greatly between countries and certain populations. Even in the United States, socioeconomic factors and differences in medical insurance may limit access to these treatments for many people. Prevention remains the best hope for the control of HIV infection. The relatively new strategy of PrEP for individuals who are at high risk for HIV infection is having an impact, at least in more developed nations, but education surrounding high-risk behaviors is critical. Effective prevention programs require that research scientists, medical professionals, and educators work together, rather than in isolation. Common language must be found in which these groups can communicate with each other and with the people that they serve. Education cannot be unidirectional: Professionals educating the general public must also learn from the populations they serve. Social inequalities and differences of opportunity have played a large role in determining the populations most affected by HIV; addressing these foundational drivers of the HIV pandemic will require persistent effort. As the former head of the WHO AIDS office declared (Mann, 1992),

> *If we believe that the entire problem of AIDS is really only about a virus, then we really only need a virucide or a vaccine. Yet if AIDS is deeply, fundamentally about people and society and if societal inequity and discrimination fuel the spread of the pandemic then, to be effective against AIDS, we would have to address these issues.*

CHAPTER SUMMARY

- The antibodies and **cytotoxic T cells** of the immune system protect the body from disease, but they are produced only in the presence of the cytokine interleukin-2 secreted by the **CD4 helper T cells.**
- When some part of the immune system fails to work, an **immunodeficiency** results. Immunodeficiency leads to an increased probability and severity of sickness.
- **Human immunodeficiency virus (HIV)** is the virus that causes **acquired immunodeficiency syndrome (AIDS)** by destroying CD4 helper T cells and other immune cells displaying CD4 molecules.
- **Epidemiology**, **Koch's postulates**, and other types of evidence helped to establish HIV as the cause of AIDS.
- HIV is a retrovirus that uses **reverse transcription** to convert its RNA genome to DNA and then uses the host's cellular machinery to replicate itself. Microorganisms that are normally kept in check by healthy immune systems can cause serious and possibly lethal opportunistic infections when the immune system is compromised by AIDS.
- Most HIV testing detects antibodies to HIV; some tests detect virus itself. All tests have rates of **false-negative** results (related to the **sensitivity** of the test) and **false-positive** results (related to the **specificity** of the test) that must be considered when interpreting a test result.
- HIV **transmission** is via behaviors in which blood, semen, or vaginal fluids are passed from an infected person to an uninfected person. Essentially everyone is **susceptible** to HIV infection if they receive body fluids from an infected person. **High-risk behaviors** (those with the highest likelihood of transmission) include unprotected sexual intercourse and sharing injection drug needles.
- Since 1981, AIDS has spread to become a worldwide **pandemic**, burdening individuals and their loved ones and also public health resources.
- There are currently no cures, although many drugs have been developed that can both reduce HIV transmission and keep infected individuals from developing AIDS for long periods of time. Vaccines are currently being tested.
- Prevention is key, but it depends on many groups of people listening to each other, learning from each other, and working together.

BIBLIOGRAPHIC REFERENCES

Bibliographic references to material in this chapter can be found online at biologytrending.routledge.com/chapter15

GLOSSARY: KEY TERMS TO KNOW

These key terms are defined at the end of each chapter as an aid for student review.

Acquired Not present at birth, but developed subsequently, often as a response to some environmental circumstance.

Acquired immunodeficiency syndrome (AIDS) Impairment of most of the immune system from infection with the human immunodeficiency virus (HIV), often resulting in death from opportunistic infections or rare cancers.

CD4 See *Helper T cells.*

CD8 See *Cytotoxic T cells.*

cDNA (complementary DNA) A strand of DNA produced by reverse transcription from an RNA template using the retroviral enzyme reverse transcriptase.

Communicability The probability that a disease-causing microorganism will be transferred to another individual, either directly or indirectly.

Cytotoxic (CD8) T cells Lymphocyte cells of the immune system that react specifically to a non-self molecule, becoming activated to kill cells bearing that molecule.

Epidemiology The study of the frequency and patterns of disease in populations.

False negative A negative test result in a sample that actually has the condition being tested for; indicates a lack of sensitivity of the test.

False positive A positive test result in a sample that does not actually have the condition being tested for; indicates a lack of specificity of the test.

Helper (CD4) T cells Lymphocyte cells of the immune system that react specifically to a non-self antigen by secreting interleukin-2, a cytokine needed for full activity of B cells and CD8 T cells.

High-risk behaviors Behaviors or actions that greatly increase the probability of undesirable outcomes, such as the transmission of a disease.

High-risk group A subpopulation of people who share some behavioral, geographic, nutritional, or other characteristic and who have a higher frequency of a particular disease than the general population.

Human immunodeficiency virus (HIV) The virus that causes AIDS by infecting and inactivating cells of the immune system that bear a molecule called CD4.

Immunodeficiency A decreased activity of some part of the immune system as the result of genetic, infectious, or environmental factors.

Koch's postulates A set of test results that must be obtained in order to demonstrate that a particular pathogen is the cause of a particular infectious disease.

Likely risk behaviors Behaviors or actions that moderately increase the probability of undesirable outcomes, such as the transmission of a disease.

Low-risk behaviors Behaviors or actions that seldom increase the probability of undesirable outcomes, such as the transmission of a disease.

No-risk behaviors Behaviors or actions that do not increase the probability of undesirable outcomes, such as the transmission of a disease.

Opportunistic infection Infection in a host with suppressed immunity, resulting from microorganisms that are normally present in the host's environment but that do not cause disease in a host with a normal or healthy immune system.

Pandemic A worldwide epidemic or disease outbreak.

Pathogen An infectious agent that causes a disease.

Retrovirus An RNA virus that begins its reproduction by synthesizing DNA from its RNA.

Reverse transcriptase A viral enzyme that helps transcribe viral RNA into DNA.

Reverse transcription Transcription of complementary DNA from an RNA template.

Risk The probability of an event or condition such as a disease.

Risk behaviors Behaviors classified according to the probability of disease transmission.

Sensitivity The probability that a test will give a positive result if the condition being tested for is actually present; the lack of false negatives; also the smallest amount of a substance that a given test can detect.

Specificity The degree to which a test detects only the molecule it is meant to detect and does not detect other molecules; the lack of false positives.

Susceptibility The probability that a person who is exposed to a pathogen will become infected with that pathogen and get a disease.

Syndrome A condition that includes a combination of multiple symptoms.

Transmission Passing of an infectious agent such as a virus or bacterium from one host to another.

Virus A particle of nucleic acid (DNA or RNA) enclosed in a protein coat that cannot replicate itself but that can cause a cell to replicate it.

CONNECTIONS TO OTHER CHAPTERS

Chapter 1: The initial identification of AIDS and the search for its causes present several examples of scientific reasoning.

Chapter 3: HIV carries out reverse transcription, which produces DNA from an RNA template, the opposite of normal transcription.

Chapter 5: HIV probably evolved from a similar virus in apes, simian immunodeficiency virus (SIV). It has evolved at least two distinct strains (HIV-1 and HIV-2) and continues to evolve.

Chapter 8: A decline in appetite (and therefore a decline in nutritional status) often occurs in the late stages of AIDS.
Chapter 9: Condoms are useful for both contraception and preventing the transmission of sexually transmitted diseases, including AIDS.
Chapter 10: A healthy immune system kills many cancer cells. Kaposi sarcoma is an otherwise rare cancer that occurs more frequently in AIDS patients.
Chapter 11: Loss of brain cell function is a late-developing symptom of AIDS.
Chapter 12: Dementia can result as a symptom of AIDS.
Chapter 13: The abuse of injectable drugs is a major risk factor for AIDS.
Chapter 14: The immune system that normally protects us from infection is greatly impaired by HIV infection.
Chapter 16: HIV/AIDS represents one of many new infectious threats encountered in the past century.

PRACTICE QUESTIONS

1. What nucleic acid is in the genome of HIV?
2. Where does the phospholipid bilayer envelope of HIV originate?
3. What does the enzyme reverse transcriptase do? Is reverse transcriptase an enzyme used by the virus, the host cell, or both?
4. Why does HIV infect only cells that carry human CD4 molecules on their surface?
5. What cellular machinery of the host cell does HIV use to replicate itself?
6. What criteria are used for saying that a person is HIV-positive?
7. What criteria are used to say that a person has gone from being HIV-infected to having AIDS?
8. Define risk.
9. What behaviors produce the greatest risk of HIV transmission? Why?
10. How do reverse transcriptase inhibitor drugs stop HIV replication?
11. What are the modes of action of other types of HIV drugs?
12. What species is/are susceptible to HIV infection? What species is/are susceptible to AIDS?
13. Among humans, who is susceptible to HIV infection?
14. When body fluids such as blood are spilled, can cell-free HIV be killed by soap and water? Why or why not?
15. What are the advantages to an ELISA test for HIV infection and what are the disadvantages? What are the advantages and disadvantages of the Western blot test? The PCR test?

CHAPTER 16

New Infectious Threats

ISSUES

- What factors have led to changes in the global patterns of infectious disease?
- What can be done to reduce the spread of diseases?
- How may microorganisms be used as bioweapons?

BIOLOGICAL CONCEPTS

- Health and disease (pathogens and response, routes of transmission, behavior)
- Evolution (virulence, host–pathogen co-evolution)
- Diversity (viral, bacterial, protist, prion, fungi, parasite)

CHAPTER OUTLINE

DOI: 10.1201/9781003391159-16

In Chapter 15, we saw how a new disease, HIV/AIDS, quickly spread around the globe, but many other diseases have also seen major changes in patterns of infection worldwide. Some of these infectious diseases are caused by bacteria instead of viruses. Unlike viral diseases, bacterial diseases can generally be treated by antibiotics. For the last half of the twentieth century, people thought that antibiotics would make widespread bacterial disease a thing of the past. Over time, however, many species of bacteria have become resistant to antibiotics.

In this chapter, we will mainly focus on human diseases. Keep in mind, however, that all organisms can suffer from disease. Production of disease-resistant strains of plants has been a major goal of both traditional plant breeding and of genetic engineering of plants (see Chapter 17). Diseases of agriculturally and domestically important animal species have long been a focus of research in veterinary medicine. Diseases among wild animal and plant species are a research focus in ecology because of their importance to the health of wild populations and because of their interaction with human disease.

Although a particular organism can be demonstrated to be the causative agent of a disease (Chapter 15), there are many other factors that contribute to the spread of the disease in populations. The development of treatments for specific diseases in individuals depends on knowledge of the pathogen. However, development of public health strategies to prevent the spread of disease depends just as much on knowledge of the routes of transmission of the pathogen. One disease, smallpox, has been eliminated entirely. However, according to the Centers for Disease Control and Prevention (CDC), some 30 new infectious diseases, including AIDS, SARS, MERS, and COVID-19, have emerged in the past few decades as technology and lifestyles have changed. Other infectious diseases have been with us for millennia; they may cycle through long periods of quiescence, only to re-emerge as epidemics once more.

16.1 PATHOGENS ARE INFECTIOUS AGENTS THAT CAUSE DISEASE

In every phylum, the vast majority of species do not cause any disease, either in humans or in any other organisms. However, many phyla contain some species or strains within species that can cause disease. Some of these species are poisonous; that is, they produce chemicals that have adverse effects. Examples include insect or snake venoms as well as plant poisons. "Red tide" is another example. The organisms responsible for it are one-celled eukaryotes (dinoflagellates, described in Chapter 6) that produce toxins. When shellfish such as clams and oysters eat these protists, there is no harm to the shellfish, but the toxins become concentrated in the shellfish tissue and can cause acute sickness and even death in a person who eats the shellfish.

As harmful as these organisms can be, they do not cause infections. The term **infection** implies that an organism finds an ecological niche in which to grow within or on another species, causing it harm and spreading to other individual hosts. A **pathogen** is an organism or virus that can cause an infection and produce a **disease**, that is, a condition in which normal activity is prevented and which, at the extreme, may cause death. Diseases that are present persistently, usually at a low level, in a local area are called **endemic** diseases. Chicken pox, caused by the varicella-zoster virus, is an endemic disease in the United States and Europe, as is pneumococcal pneumonia. When an infectious disease increases in frequency and spreads to new geographic areas or to new populations, it becomes an **epidemic** disease. Outbreaks of **severe acute respiratory syndrome** (**SARS**) in

2003–2004 or of measles in 2018–2019 are examples of epidemics. When a disease becomes frequent worldwide, as with AIDS or COVID-19, it is a **pandemic** disease.

Many organisms colonize humans, but the vast majority are not pathogens. Our microbiome, for example, contains many bacteria that are necessary to our health and well-being, either because they produce some product we need (see Section 8.3.2) or because they occupy a niche preventing pathogens from infecting us. Examples of the latter include the normal bacteria of the mouth that help prevent the infections that cause tooth decay and gum disease, as well as the intestinal bacteria that help prevent some of the infections described later in this chapter. Some diets discussed in Chapter 8 are designed to encourage beneficial intestinal bacteria to flourish.

Many types of organisms are pathogens. Pathogens can be viruses, bacteria (kingdom Bacteria), protists of various types, or fungi (kingdom Mycota). Animals from several phyla, such as roundworms (phylum Nematoda) and tapeworms (phylum Platyhelminthes), also cause disease. These animals are also considered **parasites**, organisms that live within other species, using their host as a nutrient source. Pathogens from many different kingdoms and phyla are shown in Figure 16.1. Pathogens described in other chapters include the virus HIV (Chapter 15) and the protist *Plasmodium*, which causes malaria (Chapter 7).

16.1.1 Characteristics of pathogens

As mentioned earlier, pathogenic organisms exist within many phyla, but the vast majority of organisms are not pathogenic to humans. Pathogens must have some way of attaching to or entering the host, such as being breathed in via the respiratory tract or being taken in with food or water through the digestive system. Many pathogens have specific molecules that can bind to host molecules (as we saw for HIV binding to CD4 in Chapter 15). These attachment mechanisms are usually very specific and are one reason why a particular pathogen can only attack a particular host species.

Once inside, pathogens must adapt to life within the environment of the host. Pathogens that cause human disease must be able to live and replicate at human body temperature. Because humans are aerobic organisms (adapted to living in

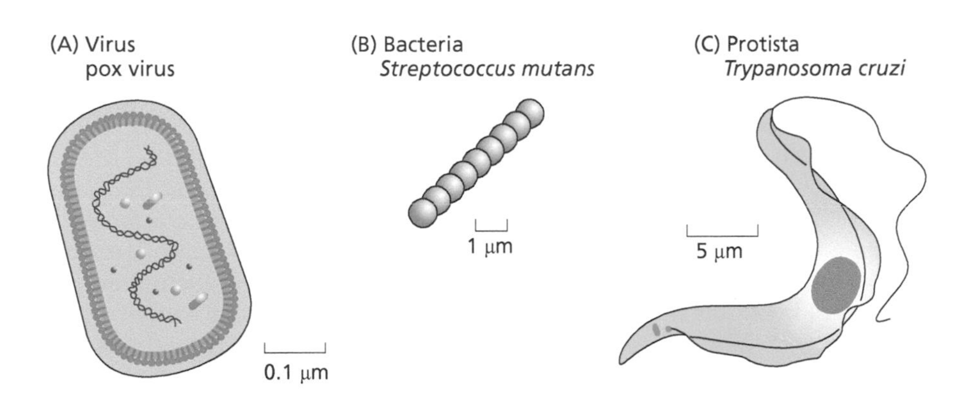

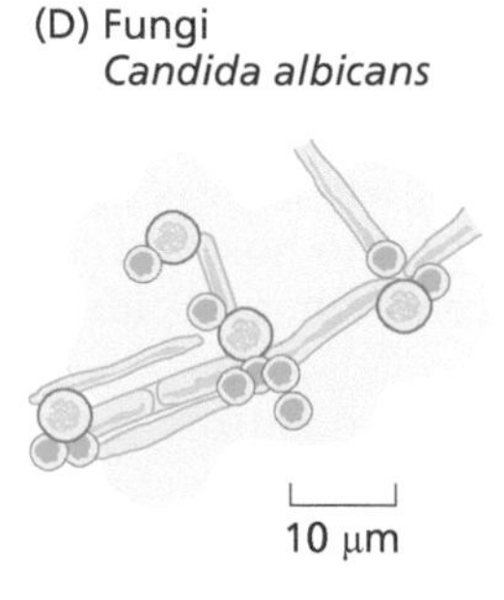

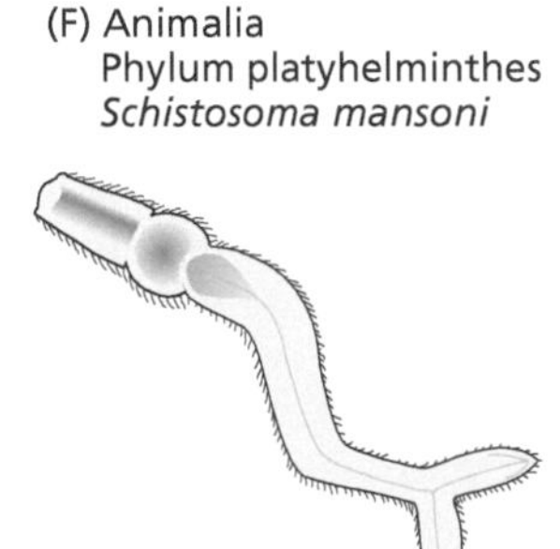

Figure 16.1 **Pathogens are found in many kingdoms and phyla of living things.**

atmospheric oxygen), many organisms that are pathogenic for humans are also aerobic. Pathogens that enter the gut must have a way of surviving in the stomach's low pH and resisting the various digestive enzymes.

Pathogens must also have some mechanism for at least temporarily escaping the immune system. Sometimes they shift to making new protein variants so antibodies that the host made to the first pathogen proteins do not work against the new versions (called **antigenic drift**). Such changes can lead to epidemics because there will be many newly susceptible hosts for the altered pathogen. Others can change their protein structure even more radically by combining new gene segments. Virus are adept at this process (called **antigenic shift**), particularly when a single host is coinfected with more than one species of virus. Viruses that undergo antigenic shift can lead to pandemics because almost all hosts will be newly susceptible after these large-scale changes in pathogen structure. Some pathogens hide by covering their surface protein antigens in a carbohydrate capsule or by living inside a human cell where they become inaccessible to the immune system. Still others can make enzymes that destroy parts of the immune system, such as enzymes that degrade antibodies.

Finally, for a disease to spread within a population, there must be a mechanism for the pathogen to exit from one host and be carried to the next. Some pathogens are excreted with feces, some are sneezed or coughed out of the respiratory tract, and others require intimate person-to-person contact.

A successful pathogen must therefore be adapted to gain entry to the host, to adhere and replicate within the host, to avoid or trick the immune response, and finally to spread from one host individual to another. Many nonpathogenic organisms, particularly bacteria, live on or in the host and share these adaptive characteristics. In addition to these features, pathogens also have some characteristic that makes them **virulent**, that is, able to cause disease. Some pathogens directly damage host tissue, such as skin, muscle, or bone. Others induce such a high fever that they cause brain damage in the host. The bacterium *Neisseria meningitidis*, which causes the brain infection called meningitis, is an example of the latter.

16.1.2 Evolution of virulence

Host species and their pathogens have co-evolved a delicate balance. As we saw in Section 7.3, infectious disease is a significant force for natural selection in human evolution. Adaptation to life within a human is, however, a significant selective pressure for the evolution of the pathogen.

Virulence is the ability of a pathogen to overcome host defenses, thereby causing serious illness or death. From an evolutionary standpoint, virulence poses a severe problem for the pathogen: If it kills its host, it deprives itself of a suitable habitat and food supply. Clearly, a pathogen that causes minimal harm to its host is assured a longer time span for itself and its offspring to continue living in the same place than a pathogen that kills its host. HIV, for example, is related to several viruses that infect higher primates such as monkeys and apes. These viruses are nonvirulent—they spread from one host to another without causing serious illness or death. Thus, they have been around long enough—thousands of years at the very least—for symbiotic relationships with their hosts to have evolved.

Evolutionary biologists who study bacteria and viruses believe that the evolution of virulence is related to the pathogen's fitness, meaning its capacity to leave offspring (see Section 5.1.4). When a new strain originates, it must compete with the older, nonvirulent strains. A virulent pathogen can proliferate rapidly within a host, but if it spreads from host to host at a slow rate, it might kill off its hosts before being able to colonize new ones; such a virulent strain will be less fit and will soon die out. A virulent strain that spreads rapidly from host to host will soon outcompete its nonvirulent relatives. If the process is rapid enough, an epidemic occurs.

In the long run, natural selection favors the evolution of host defenses against the pathogen, including both physiological and chemical defenses. These changes in the host reduce microbial virulence directly, and the adoption of host behaviors less conducive to the pathogen's spread also slows transmission. When transmission slows, less virulent strains once again become more fit than the virulent strains, and the cycle repeats.

You will recall that evolutionary change is dependent on genetic change. Virulent strains will be selected under circumstances where the rate of transmission is rapid. Public health efforts to change human behaviors and slow transmission will, in contrast, select for less virulent mutations. Thus, virulence and pathogenicity are not unchanging traits of an infecting organism; rather, they have to do with the dynamic relationship between the pathogen and its host. The host–pathogen relationship is constantly changing, thereby changing the severity and nature of infectious disease in the host population, leading to the emergence and re-emergence of infectious diseases.

16.1.3 Factors governing the spread of pathogens

There are four types of factors that influence the spread of pathogens. The first, as we have seen, is a combination of susceptible hosts with microorganisms virulent for that host. Host susceptibility and pathogen virulence are each, in part, genetically controlled, and hosts and pathogens exert selective pressures on each other.

Second is a factor called **herd immunity**, which is the proportion of nonsusceptible hosts within a population. The lower this proportion, the more likely the spread of disease. Herd immunity can be increased by vaccination campaigns. Even if not every individual is vaccinated, as long as a high percentage are, the spread of disease will be limited. Measles and polio, for example, are viral diseases for which there are very effective vaccines that increase herd immunity. Worldwide, measles caused an estimated 2.6 million deaths in 1980, but only 110,000 deaths in 2017, mostly among unvaccinated people.

Herd immunity lowers individual risk of disease, but individual risk rises when herd immunity is low. If enough people refuse vaccination, they lower herd immunity and increase the individual risk for themselves as well as for the entire population. In some areas where vaccination is common, many people within a single generation may forget that measles and polio are serious pathogens. In 2000, health officials declared that measles had been eradicated from the United States, and various people chose not to be vaccinated, sometimes for religious reasons, or because of distrust of the agencies supplying the vaccines, or because there is some very low risk of complications from vaccines. As a result, a 2014 measles outbreak began near Los Angeles, centered at first in Disneyland, before spreading to seven states. Most of the 140 people affected in California were unvaccinated for "philosophical" objections, including fear of vaccination. An even larger U.S. measles epidemic surfaced in 2019, with over 1200 cases in 31 states, about 75% of them in New York State. Medical authorities have long insisted that the dangers of being unvaccinated are hundreds to thousands of times greater than any risks from the vaccines themselves, but strong distrust of vaccines still exists among certain populations in the United States as well as worldwide, fueled in part by the Internet and social media. Measles is also a large problem in many other countries. In the first half of 2019, measles killed nearly 1,900 people in the Democratic Republic of Congo, more than the more publicized Ebola virus, and it sickened about 40 times as many people as Ebola. From 2018 to 2019, the World Health Organization reported large increases in the number of measles cases in most parts of the world, including a 900% increase in Africa!

A third factor in the spread of pathogens is weather. Disasters such as floods can spread some diseases, if, for example, floods spread sewage into drinking water supplies. Relief workers emphasize that the most urgent public health priority in response to a disaster is separating people from their waste to prevent the spread of disease. More long-term effects are starting to become apparent from climate change as a result of global warming (Chapter 19). This is already having an impact on the ecology of local regions, changing the niches that may serve as reservoirs for pathogens. This can include changes in water temperature, as many pathogens live in water. It can also include changes in the geographical spread and local size of animal populations. Some human diseases are caused by pathogens that can also live in another animal host. Rabies is an example. This viral disease can infect several other animals. Infection in domestic pets can be controlled by vaccination of the pets. But rabies spreads in populations of wild animals, and climate change affects the geographic distribution of these animals. Physiological stress from drought or starvation as a result of changes in their food supply can weaken animals' immunity, increasing the percentage of diseased animals. Other human diseases are spread by **vectors**, organisms that carry the pathogens without getting the disease themselves. As we have already seen for malaria (see Section 7.3), insects can be significant vectors for human pathogens. Climate change alters the range and population size of these vector species.

Finally, disease is spread by various means that carry pathogens from an infected person to a new susceptible host. These are referred to as the **routes of transmission**. Some routes of transmission are direct: that is, from one person to another by direct physical contact (handshakes, kissing, or sexual contact) or by aerosols (tiny droplets formed by coughing or sneezing). Other routes of transmission are indirect, requiring another species or some object for transmission. Indirect routes include insect carriers (vectors), food or water contamination, and needles or syringes.

Regardless of the route, the size of an epidemic is increased by two factors. One is the increased size of human populations. The other is the extent of travel: Because of globalization, disease is now more often spread from one population to another than it would have been in the past. In 1950 there were a total of 5 million international arrivals at all destinations around the world. By the year 2000, that number had grown to 800 million. The numbers of people who have migrated as a result of war or famine or economic dislocation are larger now than at any other time in world history.

In terms of disease, as in so many other ways, the world population is now one interconnected population. The health, social, and economic impacts of infectious disease in one part of the world quickly have impacts on the health of societies and economies around the world. In 1995, the World Health Organization, in recognition of this, drafted a resolution entitled *Communicable Diseases Prevention and Control: New, Emerging and Re-emerging Infectious Diseases* (WHA48.13). This resolution ushered in a new era of public health measures designed to increase disease surveillance (the detection of disease patterns in populations) and improve control programs. Applied research is leading to more effective and low-cost ways to prevent the spread of disease. These efforts have been aided by computer databases and the sharing of such databases by governments and non-governmental organizations around the world. The infrastructure for local health care providers to report disease cases to centralized public health systems has become better. In addition, geographical information system (GIS) computer software has allowed those data to be instantaneously mapped so disease trends can be spotted quickly.

In the remainder of this chapter, we will examine various infectious organisms that can spread through human populations and cause disease. We will also examine some of the factors that change the balance between pathogens and their hosts. Before we do so, we discuss the threat from pathogens that can be introduced into the human environment intentionally.

16.1.4 Intentional transmission turns disease into bioterrorism

In 2001, the world learned what can happen if a person decides to spread a disease intentionally. On a small scale, a deranged individual who knows they have an infectious disease may deliberately expose others in the hope of making them suffer. But the incident that began in October, 2001 was different from this. Someone prepared large quantities of a bacteria called *Bacillus anthracis* and distributed it widely through the U.S. postal system, spreading the disease anthrax to a total of 22 people and causing 5 deaths and great economic and social disruption. Intentional transmission to large numbers of people for the purpose of spreading disease or suffering is called **bioterrorism**.

The idea of bioterrorism is not new. During the bubonic plague in thirteenth-century Europe, some armies used catapults to hurl the dead bodies of plague victims over castle walls at their enemies. The Allies were testing anthrax for use as a weapon in World War II. But the anthrax scare of 2001–2002 was an act of bioterrorism on a larger scale.

16.1.4.1 Anthrax

Anthrax is ideally suited to bioterrorism. If it is grown in a pure culture under the right conditions, it will form spores (**Figure 16.2**). The DNA becomes condensed, and the cytoplasm becomes dehydrated. Bacterial spores are exquisitely resistant to killing by heat, by freezing, or by drying, which are the factors that kill most pathogens in the environment. Anthrax spores can survive in the soil at least for decades, and spores of some bacteria can even survive for centuries. Spores are very tiny, so their purified form is a powder that will easily spread at the slightest movement. Some strains of *B. anthracis* are more deadly than others; the one that was spread in the postal system in 2001–2002 was the deadliest combination of fine aerosol-forming spores and high infectivity.

When anthrax spores enter a human or animal host, they rehydrate and begin to divide. The bacteria cannot divide in the soil, only in a host animal. The bacteria then secrete a three-protein toxin. One toxin protein binds to host cell membranes and triggers the entry of the other two proteins into the cell cytoplasm. One of these proteins is an enzyme that degrades a host protein necessary for cell signaling. The consequences depend on where in the body the spores entered. In cutaneous anthrax, the spores enter through the skin, usually at the site of an existing break in the skin. This form results in local lesions, and, if treated with antibiotics, the person usually recovers. In another form, inhalation anthrax, spores are breathed into the lungs, and the disease results in swelling and hemorrhage in the lungs and a high death rate (including 45% of the bioterrorism exposure cases), even with antibiotic treatment.

Different isolates of *B. anthracis* can be identified, using the same techniques outlined in Section 4.2.2 for identifying individual humans (Figure 4.10). The *B. anthracis* genome contains repeat sequences, and different isolates have variable

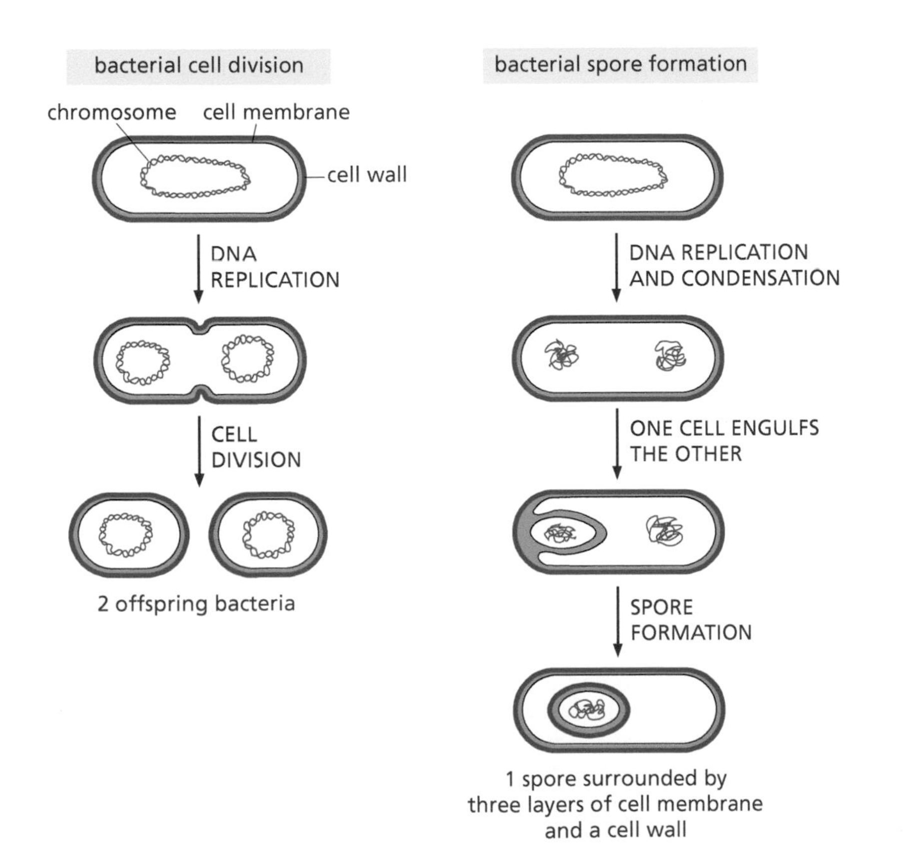

Figure 16.2 Spore formation compared with normal bacterial cell division.

numbers of these repeats. Analysis of these genetic markers allowed the strain used in the 2001 incident to be traced to a research lab in Ames, Iowa. One researcher who had worked in this laboratory was identified as a suspect but committed suicide before any involvement could be proved.

16.1.4.2 Smallpox Another organism that people fear as a potential bioterrorism agent is smallpox. **Smallpox** is caused by a virus and has a high fatality rate. It was already prevalent in India and China in 1000 BCE, and it appeared in Europe by the year 700. The Spanish carried it to Central America in 1520, leading to the death of 3.5 million Aztec Indians within two years. Similar epidemics occurred in South America in the 1530s. Throughout the seventeenth and eighteenth centuries, epidemics continued in Europe, killing 400,000 people a year. It is thought that smallpox has killed more people throughout history than all other infectious diseases combined (Behbehani, 1991).

From ancient times, it was known that the few people who lived through a smallpox infection were protected from future infections. Prevention by deliberate exposure to dried crusts from the pox was practiced in China for millennia and in the Middle East and West Africa at least as long ago as the 1600s. The practice was finally adopted in Britain and its North American colonies in the early 1700s. It worked frequently, but not always, and many people contracted the disease, rather than being protected. (We now know that this was because the crusts usually contained "attenuated" virus, a virus that was no longer virulent but sometimes contained live virus.) By the 1700s, it was also well known among dairy herdsmen and milkmaids that exposure to cowpox, a similar, but far less deadly, disease of cows, would protect against smallpox. In 1796, a British physician named Edward Jenner used this knowledge to immunize an 8-year-old boy with material from the cowpox lesion of a milkmaid. The child was later exposed to material from a smallpox patient and proved to be fully protected from the disease. Such experiments would never be approved by ethics review boards today, as children are not deemed capable of giving informed consent (see Section 1.5.3) and because there was no known treatment should exposure lead to disease. But in a way, our ethical stance today is allowed by the luxury of living in a world from which smallpox has been eradicated. Jenner's experiments began the practice that was named "vaccination," after the Latin word for "cow," *vacca*. Vaccines against smallpox were gradually developed that offered lifelong protection to virtually all people who were vaccinated. By the 1950s, it was known that smallpox can spread only from one person to another and that there are no vectors or other reservoir hosts for the virus. The combination of lifelong immunity and the lack of other reservoirs made it possible to eradicate the disease entirely. In 1958, V.M. Zhadanov of the USSR proposed such an eradication to the World Health Assembly of the United Nations. A stable freeze-dried vaccine was developed in the 1950s that could be carried without refrigeration into remote areas around the world. The eradication campaign began in earnest in 1967. The last cases in Europe occurred in Yugoslavia in 1972, and the last case in the world occurred in 1977 in Somalia. In October 1979, the World Health Organization declared the world free of the disease.

No smallpox cases have occurred in the United States since 1949; vaccination of children in the United States was discontinued in 1971 and of hospital employees in 1976. Elsewhere in the world, immunization has also been stopped. The small risk from reactions to the vaccine is now higher than the zero risk from the disease. Because we no longer immunize, herd immunity is lost.

But the smallpox virus itself has not been totally eliminated; virus stocks are stored in two places: the CDC in the United States and the Research Institute for Viral Preparations in Moscow. These were scheduled to be destroyed a few years ago, but by then, people had begun to be concerned that if anyone else had the virus illegally, there would be no stocks from which to reestablish the manufacture of vaccines. Consequently, they have not been destroyed. Smallpox is greatly feared for its potential for use in bioterrorism because its spread is rapid once a few people are infected and the fatality rate is so high. Because anyone born after 1970 in the United States is a susceptible host, concern is heightened, yet there remains no credible evidence that anyone has the virus or is thinking of using it. Therefore, the threat, although high in theory, remains only a potential threat, not an actual threat.

THOUGHT QUESTIONS

1. What are the differences between the terms "virulence" and "pathogen"?
2. Are the same species of organisms pathogens in every person?
3. How should the rights of individuals be balanced against the needs of society as a whole in deciding who should be vaccinated against a particular disease?
4. What methods are in place to control bioterrorism?
5. Should everyone be vaccinated against pathogens that are potential bioterrorism threats? Why or why not? Can people be vaccinated after an outbreak has started?
6. Could bubonic plague be effectively used in bioterrorism? Why or why not?

16.2 SOME DISEASES THAT SPREAD BY DIRECT CONTACT ARE INCREASING IN PREVALENCE

Some pathogens are spread directly from one person to another via direct contact, which includes sexual contact. As we saw in Chapter 15, AIDS is a sexually transmitted disease (STD), but there are many others. In the United States and worldwide, the number of new cases per year (**incidence**) of AIDS is lower than that of many other STDs (Figure 16.3). Each year, in the United States, 3 million people are infected with *Chlamydia*, and there are more than 300,000 cases of gonorrhea and 120,000 new cases of syphilis. The incidence of all of these STDs is rising among teenagers and young adults. In addition, about 31 million people in the United States carry type 2 **herpes simplex** (the most common genital herpes virus), and 500,000 new type 2 herpes cases are reported to the CDC annually. Except for AIDS and untreated syphilis, most STDs are not fatal, but they have other serious consequences, including (for different diseases) sterility, paralysis, arthritis, and chronic pain in adults, as well as severe disease in newborns when transmitted from the mother. We will first examine what some of these diseases are and then look at the factors that are leading to their increasing prevalence.

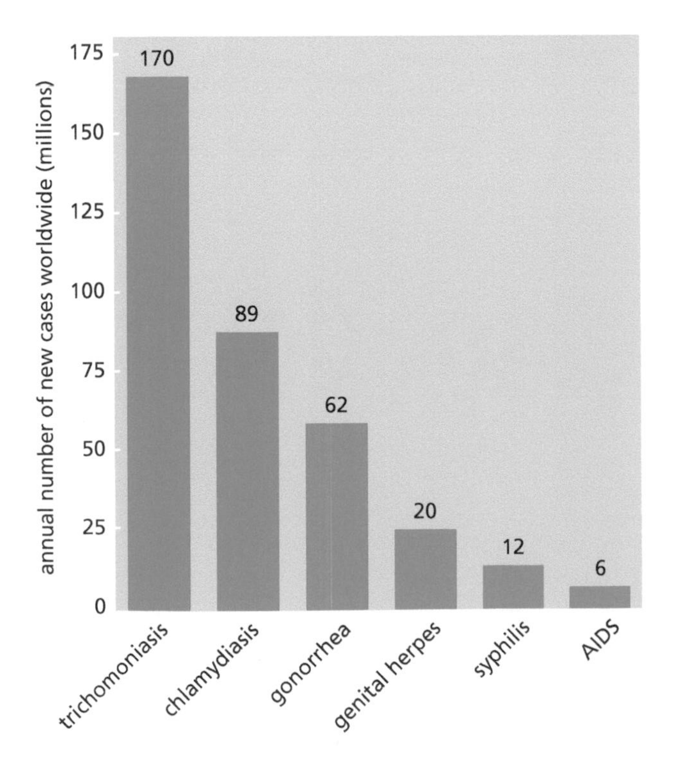

Figure 16.3 Worldwide incidence (new cases per year) of some sexually transmitted diseases (STDs). (Data are from the World Health Organization Report from 1996, with the exception of HIV/AIDS, which are data from 1998.)

Table 16.1 Summary of sexually transmitted diseases

Disease	Type of pathogen	Name of pathogen	U.S. incidence rate (cases per 100,000 population)
Chlamydia	Bacteria	*Chlamydia trachomatis*	529
Gonorrhea	Bacteria	*Neisseria gonorrhoeae*	172
Syphilis	Bacteria	*Treponema pallidum*	31.4
Mycoplasma genitalium	Bacteria	*Mycoplasma genitalium*	N.A.
Genital herpes	Virus	Herpes simplex virus-2	N.A.
AIDS	Virus	HIV	14.3
Trichomoniasis	Parasite	*Trichomonas vaginalis*	N.A.

N.A. means incidence rate not available.

(Compiled from CDC, Sexually Transmitted Diseases Surveillance 2017, and from 2016 HIV Surveillance Report.)

16.2.1 The major sexually transmitted diseases

The STDs shown in Figure 16.3 fall into several categories. Some are bacterial (chlamydia, gonorrhea, **syphilis**, and ***Mycoplasma genitalium***), some are viral (genital warts, genital herpes, and AIDS), and still others are parasitic protists (trichomoniasis), as is summarized in Table 16.1.

16.2.1.1 Chlamydia **Chlamydia** is an infection of the lower genital tract, rectum, or throat by the bacteria *Chlamydia trachomatis*. The bacterial life cycle alternates between a hardy, extracellular stage and an intracellular stage. The extracellular stage is engulfed by the host cell, forming a vesicle (see Step 2 in Figure 16.4), but, rather than being killed, it lives and replicates inside the vesicle.

A microbiology textbook from 1973 describes chlamydia as "not common except in individuals who are highly sexually promiscuous" (B.D. Davis, R. Dulbecco, et al., *Microbiology*, Hagerstown, MD: Harper & Row, 1973). That statement remains true; what has changed is the number of people who would be considered "promiscuous" by the standards of 1973. In the class of diseases for

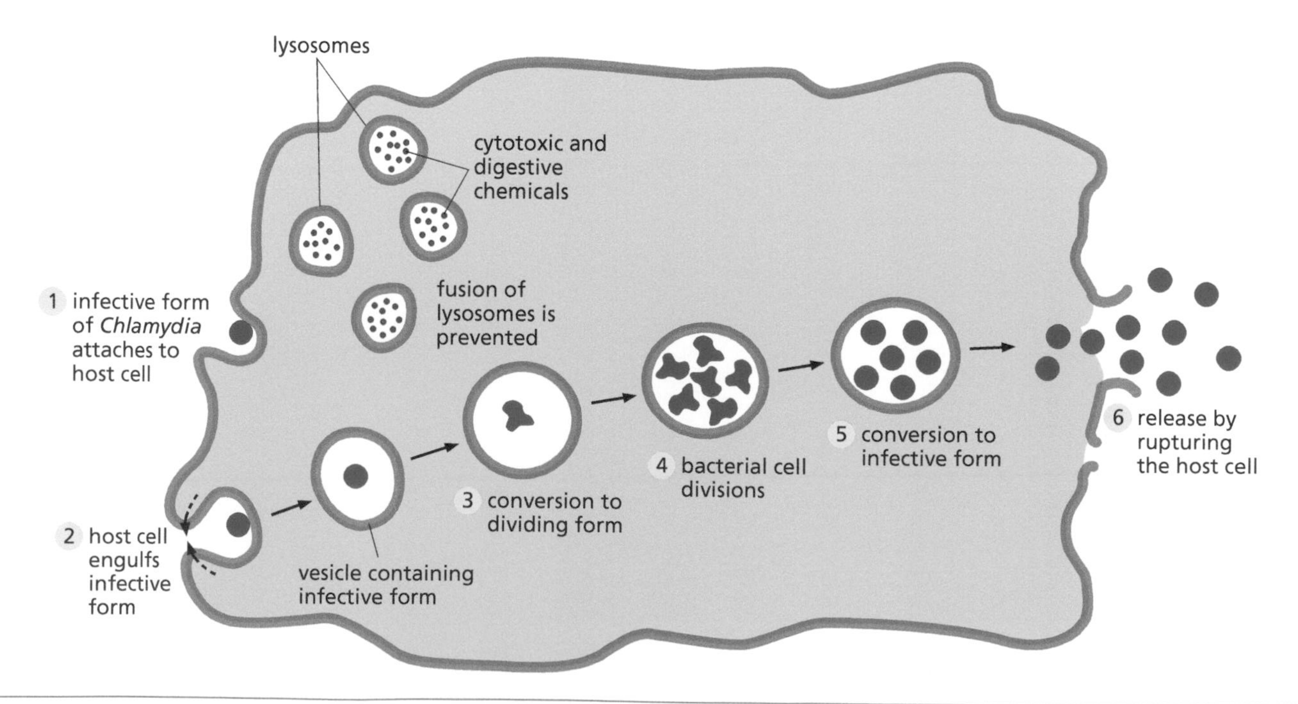

Figure 16.4 Life cycle of the bacterium *Chlamydia trachomatis*.

which reporting of cases to the CDC is required, chlamydia is now the most frequently reported infectious disease in the United States. It increased from 3.2 cases per 100,000 population in 1984 to 207 cases per 100,000 in 1997, then to 528.8 cases per 100,000 in 2017. Even these numbers may underestimate the rates because 75% of infected women and 50% of infected men show no symptoms and their cases may go unreported; estimates of the true rate of infection range as high as 3 million per year in the United States. Teenage girls have the highest rates of infection: 15- to 19-year-old girls account for 46% of the infections and 20- to 24-year-old women another 33%.

Chlamydia cells lack the genes that encode the machinery needed for adenosine triphosphate (ATP) synthesis and so are "energy parasites," robbing host cells of the ATP they produce. After several generations of division in a host cell, the noninfectious, intracellular form shrinks up and is released as the infectious extracellular form by rupture of the host cell (Figure 16.4).

Damage caused by the rupturing of cells results in pelvic inflammatory disease in 40% of women with untreated cases, and one-fifth of these women will become permanently infertile. Thus, *Chlamydia* infection is one of the major causes of female infertility. Pelvic inflammatory disease from *Chlamydia* infection can also result in lifelong severe pelvic pain and in life-threatening tubal pregnancy. In fact, it is the leading cause of pregnancy-related death among U.S. women. Men infected with *Chlamydia* can develop infections of the urethra or swelling and pain in the testicles. Unprotected sex, which allows transmission of any STD, also allows transmission of HIV. If a person is infected with *Chlamydia*, they are three to five times more likely to become infected with HIV after exposure.

In addition to the sexually transmitted genital infections, *C. trachomatis* can also cause trachoma, an eye infection that is the most common cause of blindness worldwide. It is a different strain of the same species of bacteria that is responsible for most eye infections, a strain that can be readily spread from one person to another by touch or by flies. The genital strain can, however, be passed from an infected mother to her baby at birth, causing eye infections and pneumonia in newborns and chronic respiratory infections in children.

16.2.1.2 Gonorrhea **Gonorrhea** is an infection of the urinary and genital tissues by a Gram-negative bacterium called *Neisseria gonorrhoeae*. It works its way past the mucous membranes and into the spaces between epithelial cells. There it induces an acute inflammatory response by the immune system, resulting in a discharge of pus from the urethra or the vagina. Despite the reaction of the immune system, protective immunity does not develop because the bacteria are capable of antigenic shift, that is, changing the structures of their cell-surface proteins. People can therefore be reinfected multiple times. In untreated males, scars may develop in the prostate and urethra, and, in untreated females, scarring of the fallopian tubes may result in permanent infertility. The rectum and the throat can also be infected, so the bacteria can be spread by anal sex (heterosexual or homosexual) and by oral sex in addition to vaginal sex. The infection can spread to the blood, heart, and brain and can also lead to painful arthritis in the joints. Infected mothers can pass the bacteria along to their babies at birth, and *N. gonorrhoeae* infections can result in blindness in newborns.

16.2.1.3 Syphilis This ancient disease is caused by the spiral bacterium *Treponema pallidum*. The infection results in open sores on the genitals, vagina, anus, or rectum and also on the lips and mouth. It is spread by direct contact with these sores. Therefore, vaginal, anal, or oral sex can spread the bacteria. If untreated, the initial sores progress to the secondary stage consisting of a rash and sometimes a fever, sore throat, patchy hair loss, weight loss, muscle aches, and tiredness. Obviously, many of these symptoms can also be associated with many other diseases besides syphilis, so a person may not be aware of their infection or it may be misdiagnosed. Infected people, whether they are aware of their infection or not, can easily pass the bacteria to their sexual partners. Outward symptoms may then disappear, but the infection spreads internally, damaging internal organs including the brain, blood vessels, liver, and bones and resulting eventually in loss of muscle coordination, numbness, paralysis, dementia, and even death. Syphilis that spreads to the nervous system is nearly always fatal.

Pregnant women infected with syphilis often have stillbirths or babies who die shortly after birth. Babies who live may be infected and, if untreated, the babies will show brain problems, including seizures and developmental delays.

16.2.1.4 *Mycoplasma genitalium*

This is an emerging bacterial infection (Manhart et al., 2011; CDC, 2021b), first identified in 1981. It causes inflammation of the urethra in men and more varied symptoms in women. It can also occur asymptomatically in the vagina, uterine cervix, or endometrium. *M. genitalium* lacks a cell wall, and because a large class of antibiotics (including the well-known penicillins) attack cell walls, these antibiotics are ineffective against this pathogen. A few antibiotics that work differently can kill *M. genitalium*, but some strains have developed antibiotic resistance to these drugs. It is currently uncertain whether *M. genitalium* can cause sterility. Interestingly, *M. genitalium* has one of the smallest genomes known. A complete artificial genome similar to *M. genitalium* was the first to be fully synthesized outside of a living cell, an accomplishment first achieved in 2008 (see Chapter 4).

16.2.1.5 Viral STDs

Genital warts caused by the **human papilloma virus (HPV)** and herpes infections caused by herpes simplex virus (HSV) type 1 or type 2 are the two most common viral STDs. HPV also causes cervical cancer, which can result in sterility and even death. There is now a vaccine to protect people (and their sex partners) from HPV, and public health campaigns have begun in many places to encourage preadolescents to receive this vaccine and establish herd immunity in the population against this infection. In several African countries, vaccination campaigns have succeeded in drastically reducing the incidence of cervical cancer and other HPV infections.

One-fifth of adolescents and adults in the United States are estimated to be infected with HSV type 2, which causes a genital infection. Herpes infections cause blisters, but the virus can also be released through skin that looks unbroken. HSV type 1 causes blisters in the mouth and lips, commonly referred to as cold sores, which shed virus and cause a genital infection during oral sex. HSV type 1 can also be spread by saliva. While HSV does not produce a disease as severe as the bacterial STDs, it can be painful, and if passed from an infected mother at birth, it can kill a newborn baby. In addition, herpes makes people more susceptible to HIV infection after exposure. If a person is infected with both HIV and HSV, their HIV becomes more easily transmissible.

16.2.1.6 Parasitic STDs

The most common of these, and the most common STD of any kind, is trichomoniasis, an infection caused by a microscopically sized flagellated protozoan called *Trichomonas vaginalis*. It is most common among 16- to 35-year-old women, especially those who have had multiple sex partners. The infection results in a vaginal discharge or vaginal itching, and it may also make intercourse painful. Men can also be infected and can transmit the parasite, although they generally do not show any symptoms.

16.2.2 Factors increasing prevalence

The **prevalence** of a disease is the number of people who have the disease at any given time. The prevalence of a disease is generally higher than the incidence because prevalence includes both new cases and cases that have not been cured. The global incidences of common STDs are summarized in Figure 16.3. The prevalence of each of these is higher than its incidence. The prevalence of all STDs is rising, and there are many factors contributing to this rise.

Antibiotics are medicines that kill or stop the growth of bacteria. Different antibiotics work by different mechanisms, but in each case, they target some molecule or pathway found only in bacteria. In this way, they exert little direct effect on the human host. Because chlamydia, gonorrhea, and syphilis are bacterial diseases, most cases can be successfully treated with antibiotics if treatment is started early enough. However, because so many people do not show symptoms from these infections, they often do not seek treatment and therefore remain chronically infected and infectious to others.

For an antibiotic to stop bacteria, it must be taken up by the bacteria. Regardless of whether the antibiotic is taken orally or given intravenously, it is delivered to the site of the infection by the circulatory system. Most bacteria live and grow outside of the cells of their host, making access of the antibiotic easier. *Chlamydia* infection is more difficult to treat with antibiotics because of its intracellular growth. Many antibiotics do not reach the insides of host cells, so antibiotics for *Chlamydia* must sometimes be continued for several weeks.

Antibiotics put a strong selective pressure on bacterial populations, selecting for mutated variants that are resistant to the antibiotic. Many strains of bacteria, including those that cause syphilis and gonorrhea, are now resistant to many antibiotics, severely limiting the options for treatment of a disease caused by these strains. For example, 29% of isolates of *N. gonorrhoeae* have become resistant to all of the penicillins or tetracyclines, or to both groups. Many are now also resistant to a newer family of antibiotics, the fluoroquinolones. Penicillin antibiotics block the synthesis of bacterial cell walls, but resistant bacteria are able to break the penicillin molecule, rendering it inactive. Tetracycline antibiotics inhibit bacterial protein synthesis by binding to bacterial ribosomes, which are sufficiently different from eukaryotic ribosomes such that the drugs do not affect the host's cells. Bacteria become resistant by acquiring the gene for a membrane protein that pumps the antibiotic back out of the bacterial cell. The fluoroquinolones, which include such drugs as ciprofloxacin, work by preventing the bacterial chromosome from replicating. A bacterium has a single chromosome in the form of a ring. To replicate, the double helix must unwind, but this presents problems for a circular chromosome. Consequently, bacteria have a special enzyme that breaks the chromosome, allowing its unwinding for replication. Fluoroquinolones, in principle, block this enzyme; however, some strains of *Neisseria* are now resistant. The incidence of gonorrhea in the United States had declined 75% from 1974 to the 1990s, but is now again on the rise, largely due to the increase of antibiotic resistance among *N. gonorrhoeae*. Genes for resistance can be carried on either the bacterial chromosome or on the separate small circles of DNA present in bacteria, the plasmids. Because plasmids are easily transferred from one bacterial cell to another and even from one bacterial species to another, the spread of plasmid-mediated resistance is especially rapid. These various mechanisms by which antibiotics stop bacteria and the paths by which bacteria develop antibiotic resistance are illustrated in **Figure 16.5**.

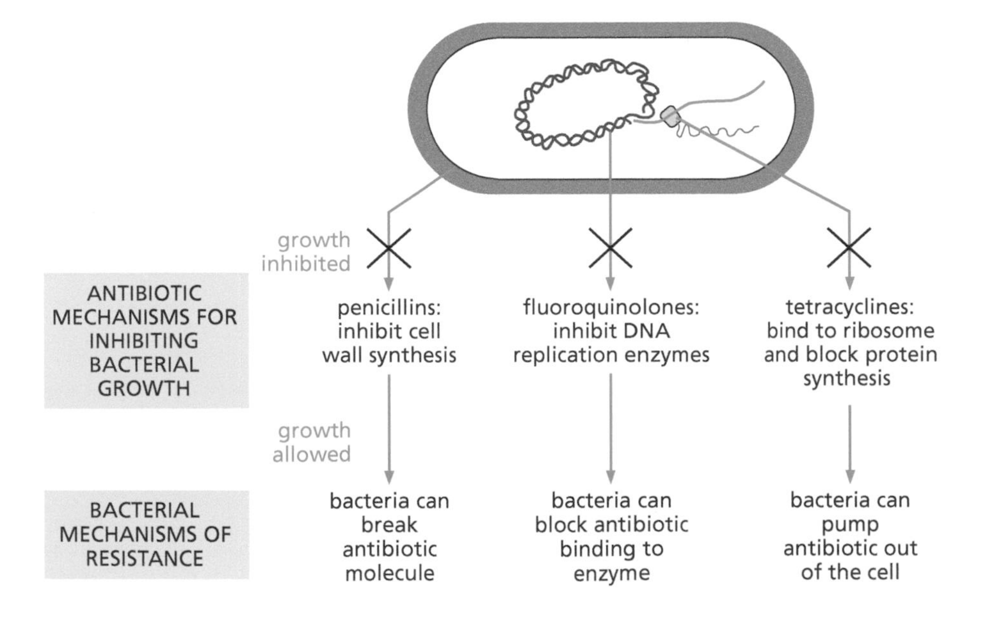

Figure 16.5 Sites and mechanisms of action of antibiotics and pathways to bacterial resistance to antibiotics.

Any given mixture of bacteria is a heterogeneous population containing cells with varying degrees of resistance. This is why it is important that antibiotic treatment be continued for the full number of doses prescribed. If it is not, the most resistant bacteria are much more likely to persist and multiply, causing a relapse of the infection and an overall increase in the antibiotic resistance of the bacterial population (**Figure 16.6**).

In contrast to the bacterial STDs, there is no drug to eradicate herpes or papilloma, which are viruses. Antiviral medications can shorten an outbreak but do not eliminate the virus. Currently, the best defense against HPV is preventative vaccination.

Only one antibiotic, metronidazole, is active against the protozoan *T. vaginalis*. *Trichomonas* can become resistant to metronidazole. There is no other drug that will work, leaving a higher dose or a longer time course as the only choices. Metronidazole, especially in high doses, has many adverse side effects. The numbers of white blood cells of the immune system can be temporarily lowered, leaving the person vulnerable to other infections (see Section 14.2). If alcohol is consumed while the medication is being taken, a violent reaction ensues. Also, metronidazole kills off so many of the protective bacteria in the body that the yeast called *Candida* that are normally present on the skin and in body cavities can overgrow and cause additional discomfort.

The increases in human population density and the changes in attitudes towards sexual activity that have occurred in the past few decades have increased the transmission of STDs in many parts of the world. One of the largest risk factors for STDs is having multiple sexual partners. People might not think of themselves

Figure 16.6 Evolution of resistance during antibiotic treatment.

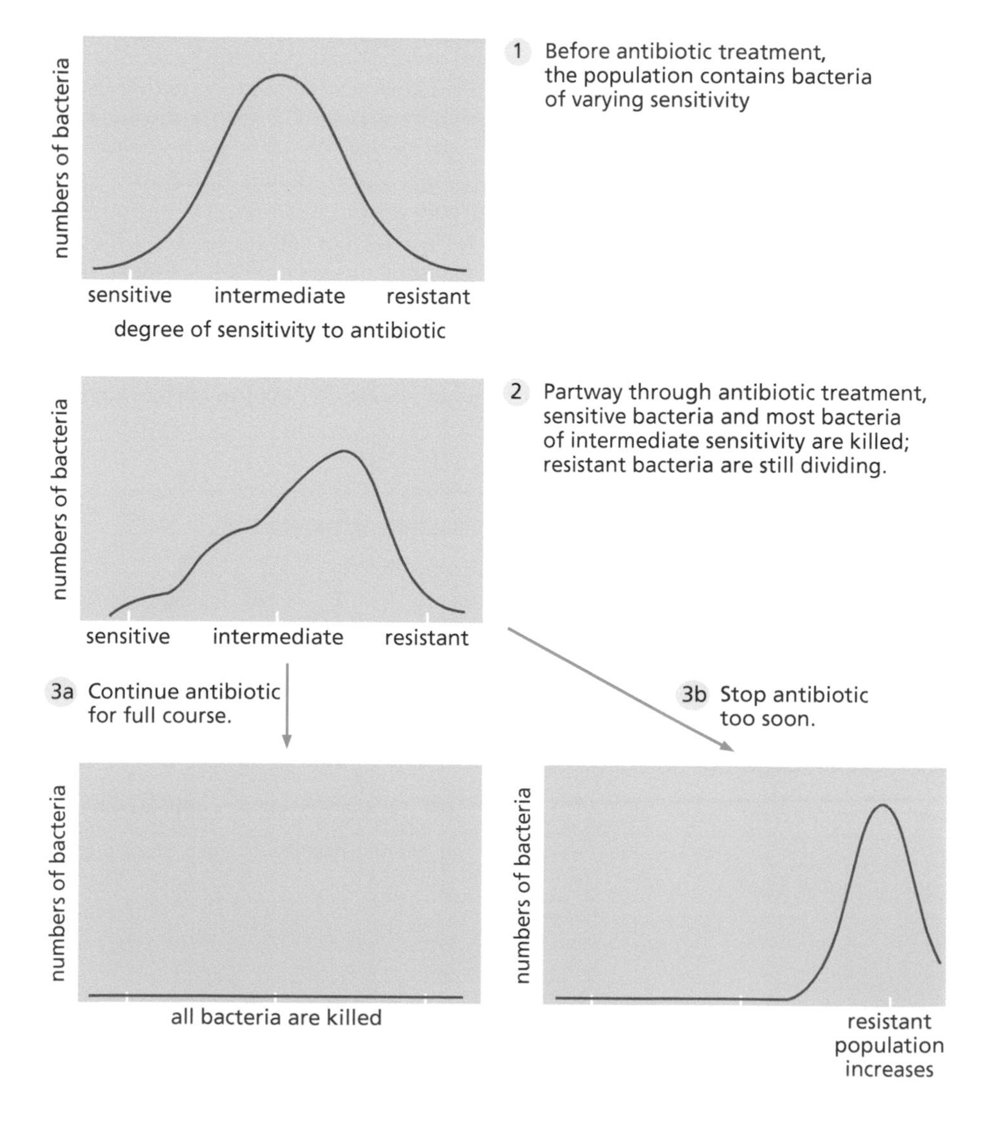

as having multiple sexual partners if they are monogamous for a time and then move on to a new monogamous relationship, but one's risk increases with each change in partners. The risk of transmission of an STD from a former partner to the next also increases. Many STD pathogens can remain infective for decades, and all sexual partners over this long period are therefore at risk.

Abstinence or monogamy with an uninfected partner will prevent STDs. Consistent use of condoms is quite effective, although not 100% (see Section 15.3.1). Seeking treatment early, when there is any suspicion of having contracted an STD, will help protect oneself and others; however, because of the risk from resistant pathogens, treatment is never as good an option as prevention. For any STD, it does little good to treat only the initial patient. That person will be asked for the names of all of their sexual contacts, so that each of them may also be treated. In the instances where people do not know their sexual contacts or do not know how to locate them for notification, further infections will continue. When people continue to be sexually active while they are taking antibiotics, the spread of antibiotic-resistant STDs is enhanced because bacteria will have been transmitted before it is obvious that the bacteria have become resistant.

16.2.3 Tuberculosis

Person-to-person transmission is required for any pathogen to succeed if it cannot live outside of a host organism and has no hosts other than humans. Not all such pathogens need to be transmitted by sexual contact, however. One example is **tuberculosis** (TB), caused by the bacterium *Mycobacterium tuberculosis*, which grows intracellularly. Infections with this bacterium can occur in many parts of the body, with different sets of symptoms, but the most common and the most infectious is pulmonary TB, an infection of the lungs. This infection leads to permanent damage to the lung tissue. It is spread by droplet contamination from coughing. The droplets (aerosols) are small enough that they hang in the air for a few hours, and the bacteria remain infectious during this time. Another person can therefore pick up the bacteria by breathing the droplets without being in direct contact with the infected person.

TB cases declined greatly in industrialized countries until the 1980s, but there was a large resurgence during the 1980s and 1990s. This is a primary example of a "re-emerging disease." Worldwide, there were 8 million new cases per year and 2 million deaths from TB in the year 2000 alone, about the same as the number of deaths from AIDS. Part of the increase in deaths from TB is due to its association with HIV infection in many parts of the world because TB is much more aggressive in immunocompromised persons. TB can be treated with antibiotics, but eradicating the bacteria is difficult because of its intracellular location. During the long treatment time, antibiotic resistance often develops. It is estimated that, even in susceptible strains, one bacterium in a million is resistant. That bacterium can be selected over time, so a person may at first get better and then relapse, requiring treatment with more than one antibiotic working by different mechanisms. This approach is analogous to the antiviral drug "cocktails" used to treat HIV (see Section 15.2.5). Population density and poverty that result in large numbers of people sharing a living space increase transmission of TB, as does the drug resistance that develops if a person does not finish the course of treatment, often due to lack of money or difficulty accessing health centers.

Over time, several drug-resistant strains of *M. tuberculosis* have emerged and become more widespread, and some strains are resistant to multiple drugs. It is thus welcome news that the Food and Drug Administration (FDA) approved in 2019 a new therapy for extensively drug-resistant TB, consisting of a new antibiotic plus two other existing drugs (FDA, 2019). The new therapy has a cure rate close to 90%, about three times higher than previous medications. The new antibiotic, pretomanid, was developed by the nonprofit TB Alliance because profit-making drug companies have little incentive to develop expensive drugs that mostly benefit poor people.

Both the health impact and the economic impact of TB are enormous, a fact that has been recognized by both the World Bank and the World

Health Organization. An international standard treatment has been developed, called directly observed treatment short-course (DOTS). Diagnosis is by direct microscopic observation of the bacterium in smears from the lung fluid coughed up by people with the disease. Patients are then given a short course of antibiotics but are directly observed during treatment to be sure that the drugs are taken. In the United States, individuals are required to receive treatment, and public health officials deliver the drugs to infected persons to ensure compliance. The World Bank included DOTS as one of the essential clinical services necessary for economic development. The World Health Organization has included TB as one of the three major diseases targeted by the Global Fund to Fight AIDS, TB, and Malaria. DOTS is considered one of the most effective public health strategies ever devised, although multidrug-resistant TB is still a problem. Because the disease is only transmitted from one person to another, the chain of transmission can be broken by decreasing the percentage of people with the disease.

16.2.4 Other diseases spread by contact

Several other infectious diseases that spread by contact include *Staphylococcus* infection and Ebola. *Staphylococcus aureus* is a bacterium that can spread through the bloodstream or connective tissues and can cause death. It spreads to other people by direct person-to-person contact (via pus from sores) or by contact with soiled bedsheets, rugs, furniture, or aerosol droplets from coughing and sneezing. Symptoms of infection can include blistering or abscesses of the skin, heart value problems, and pneumonia. *Staphylococcus* infections were formerly controlled with antibiotics such as methicillin, but methicillin-resistant *S. aureus* (MRSA) has evolved and has caused localized epidemics. In Holland, strict isolation and quarantine of *Staphylococcus* patients has greatly reduced the incidence of MRSA to only 1% of *Staphylococcus* infections. In Denmark, strict laws have restricted the overuse of antibiotics (especially in agriculture for cattle and pigs), and, as a result, MRSA declined from 18% of *Staphylococcus* infections to less than 1% (Rosdahl and Knudsen, 1991).

Ebola is a horrifying viral disease that causes a hemorrhagic (blood cell–rupturing) fever and internal bleeding that result in fatigue, weakness, and pain in many parts of the body. The mortality rate in various Ebola outbreaks has ranged from 25% to 90%. After first appearing along Africa's Ebola River in 1976, it caused minor sporadic outbreaks that killed between 100 and 280 people at a time in the Democratic Republic of Congo (formerly Zaire) and neighboring Uganda (CDC, 2021a). Tropical fruit bats are a reservoir for the virus. In the early 2000s, rebel groups in Guinea displaced many people and also cut down biodiverse forests to sell the timber. The fruit bats and human refugees were squeezed out of their habitats and often came into contact in crowded refugee camps. Burial customs (hugging dead relatives) greatly added to the danger of transmission. From 2014 to 2016, an outbreak of Ebola in Guinea spread to five nearby countries, infecting about 26,000 people and also gorillas and chimpanzees. It then spread to overcrowded urban slums, including the national capitals of Guinea, Sierra Leone, and Liberia, causing over 11,000 deaths before it subsided. An effective vaccine was meanwhile developed, but a new outbreak began in August 2018 in the eastern highlands of the Democratic Republic of Congo. It was contained at first, affecting only small villages, making it easier to trace and vaccinate all known contacts of infected persons. Unfortunately, antigovernment rebel groups became active in the area, and many people distrusted the government workers and the medical workers (mostly from international agencies) who came to help. Over 100 attacks occurred against Ebola clinics, and some international medical agencies withdrew from the conflict zone. The original vaccine is now in short supply, but several other vaccines are becoming available, and more are under development. Also, two new drugs that boost the immune response have been shown to reduce mortality rates for infected people by as much as 90%. Nevertheless, over 3,000 cases occurred during 2018–2019, with over 2,000 deaths, and another outbreak occurred briefly in Kivu Province, Democratic Republic of Congo, in 2021 (WHO, 2021a).

THOUGHT QUESTIONS

1. Which disease might it be more difficult to eradicate from a population, an STD or tuberculosis? Why?
2. What differences might there need to be in public health strategies in dealing with STDs or with tuberculosis?

16.3 FOODBORNE DISEASE PATTERNS REFLECT CHANGES IN FOOD DISTRIBUTION

Many other infectious diseases are more readily spread than STDs because they can be carried in ways other than by direct person-to-person contact. Many pathogens can survive and grow on the surfaces of foods, and eating or handling the foods can spread the bacteria to people. Such pathogens can then be spread from one person to another, but, more commonly, a disease cluster results from several people eating the same batch of contaminated food.

The major foodborne pathogens are summarized in Table 16.2. The incidence of many of these has increased in the past 25 years. *Campylobacter jejuni*, the leading cause of foodborne bacterial infection in the United States, was not even recognized as a cause of human illness until the late 1970s. Although these infections are common, severe sickness or death is much more likely in people who have weakened immune systems. People with AIDS must be very careful in the preparation of the foods they eat, as must people whose white blood cell count is temporarily low from chemotherapy treatment for cancer. One of the other significant causes of weakened immunity is age, so foodborne illnesses are becoming more prevalent as populations age. Ironically, people are living longer due to the control of many types of infectious diseases (smallpox, typhoid, diphtheria, and polio, to name a few), so they now live long enough to become susceptible to foodborne diseases. In 1900, less than 5% of the U.S. population was over the age of 65; in 2000 more than 15% was.

16.3.1 One example: Variant Creutzfeldt–Jakob disease

Evidence has been accumulating to consider one form of **Creutzfeldt-Jakob disease** (CJD) as a foodborne disease. CJD is a very rare brain disease that is invariably fatal. It can occur spontaneously or be initiated by exposure to a pathogen. In the latter case, the pathogen responsible does not fall into any of the categories of organisms previously mentioned. Rather, the agent is a **prion**, an altered (misfolded) shape of a normal protein that becomes self-replicating (Figure 16.7). Prions are therefore not organisms at all, but can spread from one person to another with the characteristics of an infection, so prion diseases are considered to be infectious diseases. Prions accumulate in the brain, causing its deterioration and, ultimately, death. Another form of CJD (sporadic) can arise by spontaneous misfolding of normal prion proteins; this becomes much more likely (nearly 100%) in individuals who inherit a mutation of the gene for the prion protein that

Table 16.2 The major foodborne pathogens in the United States

Bacteria	Incidence	Deaths per year	Food sources
Campylobacter jejuni	4,000,000	200–1000	Unpasteurized milk, undercooked poultry
Salmonella	2,000,000	500–2000	Raw or undercooked eggs, poultry, other meats
***Escherichia coli* 0157:H7**	20,000	100–200	Undercooked ground beef, fresh produce, unpasteurized milk or fruit juice
***Vibrio* species**	10,000	50–100	Raw seafood (mollusks, crustaceans, fish)

(Data from Swerdlow and Altekruse, in *Emerging Infections*, Scheld, Craig and Hughes, editors, 1998.)

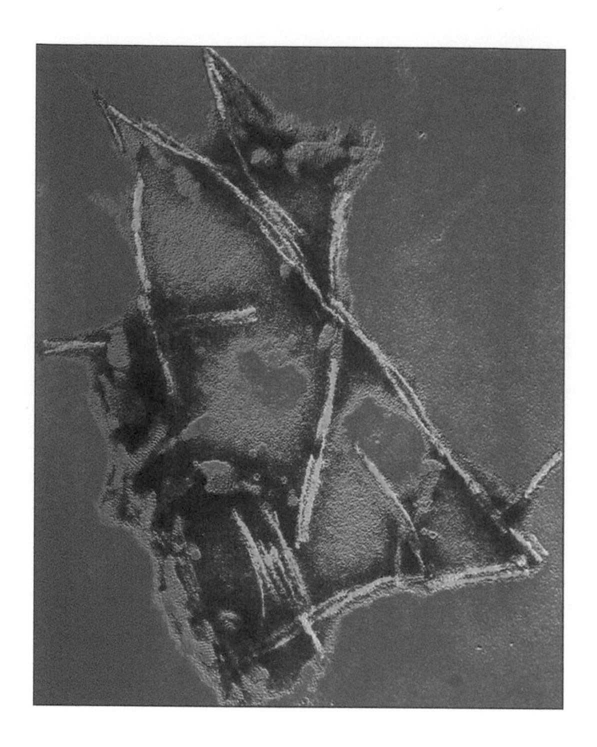

Figure 16.7 Prions are altered forms of proteins that self-replicate in huge numbers, forming crystalline protein rods.

makes it more likely to misfold. The latter case can be considered an inherited disease rather than a transmissible disease. However, both the sporadic and inherited forms of CJD are exceedingly rare, occurring with an incidence of about one in a million worldwide each year (with 85% of these cases being sporadic and 10–15% being inherited). Starting in 1995 in Britain, however, there was a sudden increase in the number of cases of CJD. The disease occurred in young, healthy people and progressed much more rapidly than the spontaneous or inherited CJD caused by mutations; it was thus given the name **variant CJD (vCJD)**.

It now appears that humans can contract vCJD by eating meat from beef cattle that have bovine spongiform encephalopathy (BSE). "Bovine" refers to cows, an "encephalopathy" is pathology in the lining of the brain, and "spongiform" refers to the spongy look of the affected brain tissue. BSE is more widely known as "mad cow disease." The BSE prion can be detected in the brain, spinal cord, retina, and bone marrow of infected cows, and it causes a brain disease that makes cows act very strangely ("mad") and leads eventually to their death. The prion most likely arose by spontaneous mutation in cattle, but it was likely spread by the practice of using meat and bone scraps to make animal feed that was fed back to cattle. Large-scale agricultural techniques probably contributed to its spread. By 1997, in Britain, 170,000 cows were infected, and efforts began to stop its spread among cows. In the same year, the U.S. FDA prohibited the use of mammalian protein in the manufacture of food for ruminants (cows, sheep, goats). In June 2000, the European Union Commission on Food Safety and Animal Welfare strengthened its measures to prohibit the use of meat from the spinal column or brains of cows, sheep, and goats as animal feed. (Sheep also carry the prion and develop a disease called scrapie, so-called because they exhibit neurological defects that cause them to scrape their bodies repeatedly against fences.)

Initially, no one knew that the prion could also be spread to humans. The incubation period between infection and symptoms is several years, making discovery of the connection much more difficult. As of May 2015, a total of 229 cases of vCJD had been confirmed, 77% of them in the United Kingdom (UK) (Maheshwari et al., 2015). Most individuals who were not residing in the UK when they were diagnosed nonetheless had previous connections to the UK, but BSE was documented in some other European Union (EU) countries. There are no cases of vCJD where there is no BSE. As evidence accumulated that the human prion is an infection acquired from eating meat from diseased cows, particularly brains or hamburger that contains nerve tissue, food safety regulations were strengthened. Meat from the spinal column of cows, sheep, and goats has been prohibited in human food in member countries of the EU since 2001. All cows over 30 months of age must be tested and are killed if they are found to carry the prion. Particular vigilance is required to prevent transmission of vCJD to humans through consumption of BSE-containing beef because normal cooking techniques do not inactivate the disease-causing prions; the disease form of the prion protein forms aggregates (similar to the amyloid deposits that occur in Alzheimer disease; see Section 12.2) that can only be destroyed by long periods of heating at extremely high temperatures.

16.3.2 Social and economic factors contributing to disease outbreaks

In many respects, food has become safer as sanitation, refrigeration, and better canning practices have developed. However, in many industrialized countries, eating customs have changed so that a much higher percentage of meals are eaten away from home. Indeed, 80% of outbreaks of foodborne illnesses are traced to sources outside the home. Furthermore, food production and distribution systems have changed enormously. People are eating many more fresh fruits and vegetables and meats, such as ground beef, than ever before. The fresh foods are arriving out of season from locations around the globe. A 1990 outbreak of 245 cases of *Salmonella* across 30 states was traced to a single source of cantaloupe imported from Central America. Similarly, a 1995 *Salmonella* outbreak was traced to one source of orange juice from within the United States that resulted in 63 severe infections across 21 states. The short amount of time between harvest and shipment to distant markets can mean that more bacteria are still alive when the food is purchased. Bacteria on the outside of a fruit can be transferred inside when the fruit is cut and then multiply if the food is held at room temperature.

Centralization of food processing has resulted in outbreaks being larger and more widespread when they occur. An outbreak of *Salmonella* in 1994 resulted from the transport of ice cream premix in a tanker truck that had not been thoroughly disinfected after carrying raw liquid eggs. After the ice cream was produced and distributed, 224,000 cases of *Salmonella* infection resulted.

Escherichia coli is a bacterium that is normally present in the guts of humans and many other animals. However, some strains of *E. coli* are very virulent, including strain O157:H7, which was responsible for a 1993 outbreak traced to a fast-food restaurant chain in Washington State. Trimmings from many beef cattle were included in one 2,000-pound lot of ground beef. This lot was then distributed among hamburger patties delivered to several restaurants in the chain; 500 illnesses, 151 hospitalizations, and three deaths resulted. In 2018, the same strain of *E. coli* was detected in romaine lettuce grown in Arizona, infecting 210 people across 36 states and causing five deaths. The contamination was traced to the water used in an irrigation system.

A type of *Salmonella* (called *Salmonella* serotype Enteritidis) has been found to infect the ovaries of chickens, from which it can be transmitted into the egg contents within intact eggshells. Farming practices have changed such that a typical hen house may now contain 100,000 hens and many houses may be serviced by the same machinery, making possible widespread contamination that was not possible when a hen house had 500 hens or when each farm had its own small hen house.

16.3.3 Improvements needed

More thorough testing of foods and food sources would help prevent and monitor outbreaks. In the 1990s, due to budget cuts within state public health agencies, 12 states in the United States no longer had any personnel monitoring foodborne illnesses. New detection methods using bacterial DNA amplification by polymerase chain reaction (PCR) (see Section 3.4.1.1) are very sensitive but are time-consuming and expensive. Testing is also complicated by the pooling of ground beef from animals or pooling of eggs from many chickens into a single transport truck. The level of bacteria on an infected animal may be low enough not to be detected once it is combined with many others; however, the bacteria can grow and contaminate the whole lot. The surface of animal carcasses can be disinfected by antimicrobial rinses and, if a few bacteria remain on the surface, they are readily killed by cooking. When beef is ground into hamburgers, however, any remaining surface contamination is brought inside the meat where it will not be killed unless the meat is thoroughly cooked all the way through. In both commercial food processing and at home, all food preparation surfaces and equipment must be thoroughly and repeatedly sterilized, and all foods must be maintained at adequately low storage temperatures and cooked at adequately high cooking temperatures long enough to kill all foodborne pathogens. Health officials in Europe have repeatedly emphasized the raising of food animals under sanitary conditions and have criticized the widespread U.S. practice of generously using antibiotics to compensate for poor sanitation. Agricultural use of antibiotics in animals that are not sick is not only ineffective in preventing foodborne illness, it also increases the chances that the bacteria will develop antibiotic-resistant strains, as in the case of MRSA, described earlier.

Fruits and vegetables are washed as they are harvested, but often with water from farm ponds, and this water is not chlorinated. Chlorination of water used in processing and for making the ice on which foods are transported would stop many cases of contamination. Disinfection of farm machinery and transport vehicles between uses happens frequently, but not always.

As our food supply has become globalized, the World Trade Organization has recognized the sanitary and food safety standards developed by the World Health Organization. The World Trade Organization has no authority over member governments to accept these standards, but those that follow the standards allow distribution only of products that meet the requirements of the standard.

The last step in the chain of food safety is what you do in your own home. Contamination of food or spreading of pathogens from contaminated food can be prevented by washing your hands and utensils thoroughly with soap and water. Refrigeration slows the multiplication of pathogens in stored food, and thorough cooking kills most bacteria.

THOUGHT QUESTIONS

1. Mayonnaise and chocolate mousse are often made with raw eggs. What special precautions should be taken with these foods if you wanted to serve them in a picnic on a hot day?

16.4 WATERBORNE DISEASES REFLECT CHANGES IN LIFESTYLE AND CLIMATE

Contaminated water is one of the major sources of infectious diseases. One-third of the people on Earth do not have access to clean water. Some waterborne diseases are very ancient, but outbreaks have become larger due to the lack of access to clean water that is associated with poverty and increased population density. Other waterborne diseases are very new and reflect changes in technology and lifestyle.

16.4.1 Cholera

Cholera has been recognized as a disease since the early 1800s. Five widespread cholera epidemics were documented before the bacterial cause of the disease was discovered. Robert Koch, the German physician responsible for Koch's postulates about infectious disease causation (see Section 15.1.3), isolated and described *Vibrio cholerae* as the causative agent of cholera in 1886. This was one of the important milestones in establishing the germ theory of disease during the 1880s. Prior to that time, people had recognized that the disease was spread by "bad" water (without knowing what was bad about it), and this recognition led to the sanitary movement that started municipal water treatment and sewage treatment. Water treatment is still one of the most effective ways of stopping cholera.

The seventh global cholera pandemic started in 1961. In 1970 the disease returned to West Africa, where no cases had been seen for over 100 years. In 1991 the disease also returned to Latin America, where again there had been no cases in a century. One of the reasons for its resurgence is the international transport of ballast water contaminated with bacteria, especially in warmer climates. (Ships often take on water as ballast weight when they are not full of cargo and then pump it out to reduce the ships' weight when they enter a different port to take on cargo.) Increases in the sizes of ships and the amount of water they carry as ballast increase the likelihood for contamination to spread great distances. The *V. cholerae* bacteria live in warm water and can be carried by warm water crustaceans (shrimp and crayfish), so another source for transmission is by eating undercooked crustaceans. As global water temperatures rise as a result of global warming (see Chapter 19), the geographic area where *Vibrio* can live is expanding.

V. cholerae causes an intestinal infection. The bacteria are completely specific for humans and will not infect the guts of any other animals. This is because infection requires adherence of the bacteria via a pilus: a hairlike extracellular structure whose end protein binds specifically to human gut mucosal cells. Once bound, the bacteria begin to reproduce and then produce a toxin that induces a watery diarrhea. The severe dehydration that results leads to death in 25% of infected people.

The toxin and the protein needed for adhesion are carried close together in the *Vibrio* genome. The expression of these genes is turned on and off together in response to environmental conditions. The virulence genes can be transferred to nonvirulent strains of *Vibrio* by a bacterial virus (bacteriophage). The bacteria are capable of large shifts in their outer molecules, a problem that long hindered vaccine development. One bacterial variant that caused a 1992 outbreak was not recognized by antibodies that would bind other *Vibrio* strains. In this new strain, a large part of the outer membrane had been replaced by a separate nonmembrane-associated polysaccharide capsule. Such a large deletion/replacement mutation has not been found in any other bacteria. Protective immunity typically develops when antibodies are formed that can bind to some molecule on the outer surface of a bacteria. If the entire outer molecule structure changes, immunity must develop all over again.

A vaccine against cholera is now available (Verma, Khanna, and Chawla, 2012), but the immunity it confers takes time to develop. The course of the disease is very rapid, often not allowing time for immunity to develop or for vaccinations to stop the initial spread. A toxin that causes diarrhea can quickly spread the infection to new hosts if the host population density is great enough. From the standpoint of the evolution of virulence, such a toxin is an advantage to the pathogen only when host density is high, so that the bacteria can spread to new hosts as it kills its first host. Cholera also places a selective pressure on humans. Those people with blood type O develop the most severe diarrhea; likely as a result of this, the frequency of blood type O has been found to be very low in areas of the world, such as the Bay of Bengal, where cholera is present at a low endemic rate and where the first widespread cholera epidemics may have originated.

Water treatment is an effective public health strategy to prevent the spread of cholera. As with most infectious diseases, prevention of disease by prevention of transmission is more effective, and certainly more cost-effective, than treating people after they are sick. It is even said that, for waterborne diseases, 100 plumbers can have a greater effect on disease rates than 100 doctors. In urban areas where water systems are not adequately maintained, water pressure can be intermittent, and a pressure drop in the water outflow pipes can pull sewage back into the water pipes if sewage pipes are adjacent to them, leading to large outbreaks. Communal bathing and laundry facilities can spread the bacteria, and even clean water can be easily contaminated if water is scooped from a storage container by someone's bare hands.

For cholera, the other effective public health strategy is oral rehydration therapy. This is very inexpensive and replaces the electrolytes and fluids lost during diarrhea. With rehydration, the mortality rate drops from 25% to less than 1%. However, oral rehydration therapy requires sterile fluids, which are not always available in poor countries.

Cholera is now considered a preventable disease and is almost unknown in areas of the world with good sanitation and adequate water treatment. But cholera is also an important re-emerging disease because hurricanes, floods, or earthquakes may disrupt water supplies and cause cholera outbreaks, as happened in Haiti following an earthquake in 2010. Civil war in Yemen also disrupted water and sanitation systems, causing a cholera epidemic that has continued out of control since 2016, infecting over 1.2 million people and causing over 2,500 fatalities.

16.4.2 Legionnaire's disease

In 1976, several members of the American Legion became seriously ill at a convention at a hotel in Philadelphia. Several people developed a severe pneumonia, and some of them died. Pneumonia is the accumulation of fluid in the lungs; it can be caused by many different bacterial species or viruses. It was quickly determined, however, that the Legionnaires were not ill with any known microbial source of pneumonia. The hotel closed, and the search began for an explanation. The cause was found very rapidly, in one of the great success stories of public health epidemiology. The cause turned out to be a previously unknown species of bacteria that was given the name *Legionella pneumophila,* and the illness it caused was termed **Legionnaire's disease**. It was so different from any known bacteria that it was placed in its own family, the Legionellaceae.

Scientists assume that the bacterium has been with us for a while, but probably not in high enough density to be a problem. We now know that *Legionella* thrives in the water-cooling circulation of air conditioning systems, and this modern convenience caused the emergence of the disease. The bacterium is not passed from one person to another but is only transmitted by contaminated water. Showerheads and fine mist from whirlpool baths are additional sources. Most people who are exposed do not become sick, but those with weakened immune systems due to age, smoking, and use of alcohol are more susceptible. (Many of the Legionnaires at this convention were older individuals, and quite a few of them either smoked or used alcohol or both.) The pneumonia is fatal in about 10% of cases. Initial flulike symptoms are followed by difficulty in breathing and also by mental changes including memory loss, disorientation, or even hallucination. Since 1976, several outbreaks have occurred, including outbreaks in Britain in 2001 and 2002. Some outbreaks have occurred on cruise ships, underscoring the need for disinfection of air conditioning systems and for location of ventilation system air intakes at a distance from the cooling towers of the air conditioning system.

16.4.3 The emerging waterborne parasites: *Giardia, Schistosoma,* and *Cryptosporidium*

Several waterborne infections have emerged that are caused by eukaryotic parasites. One such disease is **giardiasis**, a type of severe diarrhea caused by an intestinal parasite (*Giardia intestinalis*) that is resistant to many standard water treatments. *Giardia* is an unusual organism (**Figure 16.8**) and is interesting to evolutionary biologists. It is unique in having two nuclei, each containing a full genome. It also has a cytoskeleton, and the presence of the nuclei and cytoskeleton make it eukaryotic. However, it does not have any of the other membrane-bound organelles usually found in eukaryotic cells: no mitochondria, chloroplasts, endoplasmic reticulum, or Golgi apparatus. It may therefore represent a very early stage in the evolution of eukaryotic organisms.

Humans infected with this parasite are unlikely to be fascinated by its evolutionary oddity. The diarrhea it causes is severe and, if untreated, will last for several weeks, leading to significant weight loss and dehydration. It lives and

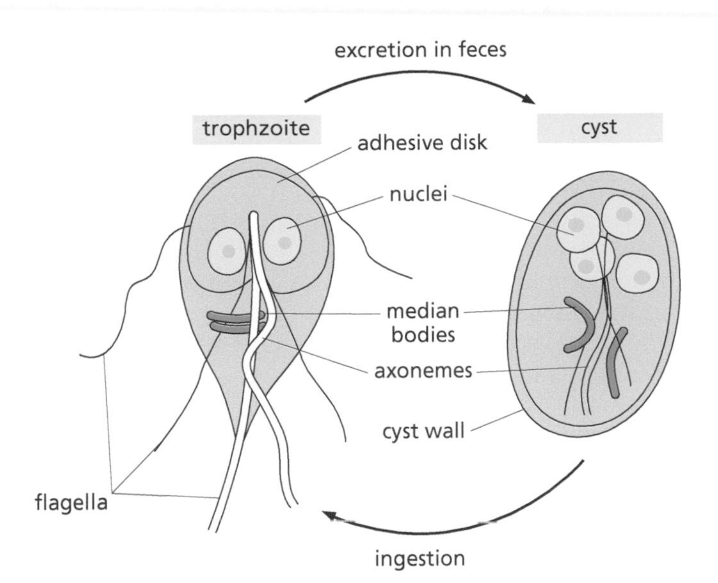

Figure 16.8 ***Giardia intestinalis*, a single-celled parasite with two nuclei and a cytoskeleton but with no other subcellular organelles.**

reproduces in the intestines of humans and other animals and is spread to water by fecal contamination. The parasite has an outer covering that enables it to stay alive for long periods of time in water, and hikers and others pick up the parasite by drinking water out of streams and rivers throughout the United States and worldwide. It is, in fact, one of the most common waterborne diseases in the United States. Filtering water will remove this parasite, but water treatment with iodine does not always kill it. *Giardia* also survives in swimming pools, hot tubs, and Jacuzzis, as it is relatively resistant to disinfection by chlorination. Hot tubs are a particular problem because the water temperature causes rapid evaporation of the chlorine. Now that this has been recognized, separate standards have been set for chlorination of hot tubs to maintain chlorine at levels effective against most bacteria, but *Giardia* may survive even at those levels. The parasite can also be spread directly by feces. It can be spread by fecal contamination of food, either direct contamination, or contamination by washing food with contaminated water, which emphasizes the connection between waterborne and foodborne routes of disease transmission. Infected people should help protect others by not swimming for several weeks after the symptoms have ended. Hand washing is very effective at preventing the spread of this disease. While many other waterborne parasites are tropical, *Giardia* occurs worldwide and is routinely found among people in the United States and Europe.

Also emergent in recent years is **cryptosporidiosis**, which is now one of the most common waterborne diseases in the United States and is also found throughout the rest of the world. It is caused by a parasite (*Cryptosporidium*) that spreads when a water source is contaminated, usually with the feces of infected animals or humans. In 1993 in Milwaukee, Wisconsin, there was an incident in which 400,000 people were affected.

The most serious infectious threats in many parts of the world are waterborne blood flukes (*Schistosoma*), flatworm parasites of the phylum Platyhelminthes (see Figure 16.1). Infection with these flukes causes the disease **schistosomiasis** (also called bilharziasis), leading to chronic disability and liver damage. The disease has been recognized since the time of the pharaohs, but the worm that causes it was first identified in 1851 by a German doctor named Theodore Bilharz. Two hundred million people are infected, making it the second most important tropical disease after malaria (see Section 7.3.1). Through the use of drugs like ivermectin, schistosomiasis has been successfully controlled in Asia, the Americas, North Africa, and the Middle East, only to emerge in new areas. Areas that did not previously harbor the blood flukes do now, largely as a result of large-scale water projects. Diama Dam on the Senegal River in West Africa led to endemic infections in Mauritania and Senegal, two countries in Africa where the disease was previously rare. Similarly, a dam on the Volta River in Ghana has created Lake Volta, the largest artificial lake in Africa. This lake has proved a fertile breeding ground for the worms, and villages along its shore have infection rates as high as 90%. Another important factor has been mass movements of refugees, which have recently spread the worms and the disease into Somalia and Djibouti.

16.4.4 Aerosol-dispersed viral diseases, including COVID-19

Many waterborne diseases are also viral. Hepatitis A is a viral disease of the liver. The virus can live in water, but the disease is also highly contagious and spreads from person to person by contact. It is most common where crowding and poor sanitation facilitate transmission.

Several viral diseases are commonly transmitted in the form of aerosols—tiny water droplets formed by coughing and sneezing—and also by contact with the countertops, doorknobs, and other hard surfaces on which such aerosol droplets may fall. Measles and influenza viruses are commonly transmitted this way.

H1N1 swine flu was first encountered during the trench warfare of World War I, causing 40 million deaths in 1918 alone. The disease then died out in humans, but it persisted in pigs until it reemerged in 2009. Like other influenza viruses, it is spread primarily by coughing and sneezing or by the tiny droplets produced by coughs and sneezes that persist on environmental surfaces. Frequent hand washing is generally the best protection.

16.4.4.1 Coronaviruses Aerosol transmission is common for viruses in the family Coronaviridae (Figure 16.9), a group of single-stranded RNA viruses that are surrounded by an envelope studded with spike-shaped proteins that give the appearance of a crown (Latin: *corona*). Coronaviruses are responsible for the common cold and also for a handful of more serious emerging infections. Because of the envelope that surrounds them, coronaviruses can easily be rendered harmless by diligent and frequent hand washing or by most common soaps, detergents, and disinfectants.

16.4.4.2 SARS Severe Acute Respiratory Syndrome (SARS) is a waterborne infection caused by a coronavirus named SARS-CoV-1, often transmitted in aerosol form. Horseshoe bats form a reservoir for the virus. In 2003, an outbreak began in China's "wet markets" in Guangzhou, where many species of both wild and tame animals are caged in close proximity, often with poor sanitation. The disease was reported in raccoons, ferrets, badgers, snakes, palm civets, and pigs; it then mutated to infect people as well. Increasing wealth in China has led to increased demand for exotic cuisine and also more pig farms close to bat-infested forests. Bats have proved to be an especially rich source of novel viruses. This is in part because many bats live close together, allowing easy transmission from one animal to another; recent studies have also shown that the immune systems of

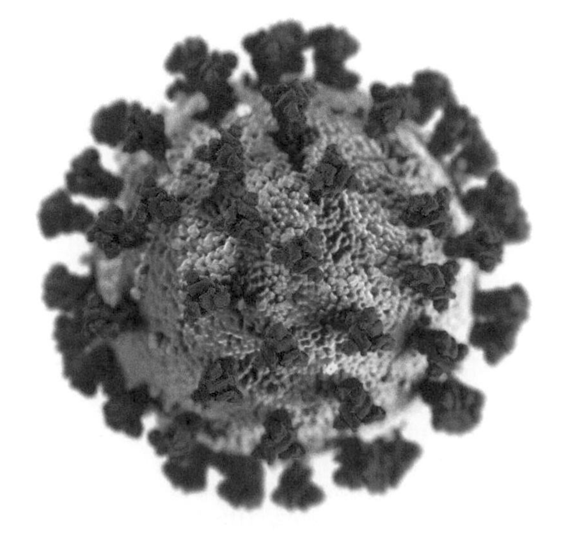

Figure 16.9 **Structure of SARS-CoV-2, the coronavirus responsible for the COVID-19 pandemic of 2020–2022.**

bats have some unique features that allow them to tolerate much higher viral loads than other animals without becoming diseased. In the case of SARS, an immigrant from Hong Kong to Toronto spread the illness to Canada, closing down two hospitals; as a result, many Asians in Canada were shunned, pointing to the social implications of disease epidemics.

16.4.4.3 Middle East Respiratory Syndrome Another new illness, **Middle East Respiratory Syndrome (MERS)**, first emerged in 2012 in Saudi Arabia. Like SARS, MERS is caused by a coronavirus that is commonly transmitted in aerosol form. The virus occurs in bats and is thought to have been transmitted to humans through contact with camels as an intermediate species. Most cases of MERS were from the Arabian Peninsula or neighboring countries or from local outbreaks initiated by travelers from the Arabian Peninsula.

16.4.4.4 COVID-19 In December 2019, a new viral disease emerged in Wuhan, China, that was named **COVID-19**. The virus that causes COVID-19, SARS-CoV-2, is related to the earlier SARS coronavirus. The outbreak was first noticed at a seafood market where live animals are also sold. As with several other viruses, bats are suspected as supplying the reservoir from which the virus first spread to humans, possibly via snakes or pangolins as intermediate hosts. There were also suggestions that the virus originated in a laboratory in Wuhan, but the vast majority of geneticists find no evidence in the virus's genome to support this hypothesis and instead find evidence of its likely animal origin. Direct person-to-person transmission was soon confirmed. The spread of the new coronavirus was amazingly rapid: Within about five weeks, more people (around 10,000) developed COVID-19 than was true of SARS during the entire 2003–2004 epidemic of that disease, and COVID-19 rapidly spread worldwide, becoming a pandemic. Travel from Wuhan to other parts of China was restricted, and many nations closed their borders to travelers who had visited China.

COVID-19 manifests mostly as flu- and pneumonia-like symptoms that make it difficult to distinguish from other diseases without a viral culture (Huang et al., 2020). Some other unusual symptoms are also associated with the disease, such as loss of smell and taste, circulatory problems (sometimes resulting in reddish discoloration of the toes, known as "COVID toes"), and cognitive problems ("brain fog"). SARS-CoV-2 is highly contagious and spreads more rapidly than SARS-CoV-1 or other coronaviruses, although the fatality rate is lower than for the earlier SARS epidemic. The actual fatality rate is difficult to determine because a significant number of people who become infected do not show any symptoms (although they can still transmit the virus to others). Elderly people and people with pre-existing health problems or weak immune systems are the most vulnerable, but many young, healthy individuals have also become severely ill from COVID-19 or have even died from it. Nearly 50,000 people were infected in the first eight weeks of the epidemic and over 1,000 died. As of May 2022, there were nearly half a billion cases of COVID-19 worldwide, including 6.3 million deaths; of these, the United States experienced over 80 million cases and over 1 million deaths. These large case numbers seriously strained health care systems, and many people also died from otherwise treatable conditions that were inadequately addressed or neglected during the pandemic. Furthermore, a significant portion of people infected with SARS-CoV-2, even those with relatively mild disease who were not hospitalized, report long-term symptoms persisting for weeks or months after their infection. This phenomenon is referred to as "long COVID" and includes symptoms such as fatigue, shortness of breath, and memory and attention problems (Chen et al., 2022).

In addition to the COVID-19 disease, whose fatality rates are greatest among the elderly and those with pre-existing heart or lung conditions or diabetes, the SARS-CoV-2 virus also causes a rare but potentially deadly inflammatory disease in children. About four to five weeks after exposure to the virus (often without previously showing any symptoms), some children develop a Pediatric Inflammatory Multisystem Syndrome (PIMS, also called Multisystem Inflammatory Syndrome in Children, or MIS-C), similar to Kawasaki disease or to toxic shock syndrome (Radia et al., 2021; Viner and Whittaker, 2020). Symptoms include high fever, abdominal pain, diarrhea, frequent bright red rashes, and severe inflammation to blood vessels and to multiple organ systems. Tests for antibodies to the SARS-CoV-2 virus show that PIMS is an overreaction of the immune system to the virus.

In response to the COVID-19 pandemic, travel was greatly restricted, both internationally and within many countries, and schools at all levels sent students home and switched to online activities only. People were told to practice "social distancing" by avoiding large gatherings and maintaining a minimum six-foot (two-meter) distance from others. Schools, restaurants, bars, gyms, and other gathering places were closed, and sporting events were cancelled. Billions of people were advised to stay home as much as possible, sometimes under legal penalty if they did not comply. Economic activity was disrupted, work was halted for many industries, and the world experienced its biggest economic decline (and unemployment increase) since the Great Depression of the 1930s. The social and economic disruption caused by COVID-19 was certainly the greatest of any disease since the "Spanish flu" pandemic of 1918–1919.

The number of COVID-19 infections in the United States reached an early peak in April 2020, a second peak over twice as high in late July, then a third peak more than twice again as high toward the end of the year (**Figure 16.10A**). The CDC recommended (1) frequent hand washing, (2) avoiding large gatherings and maintaining a social distance of six feet from other people, and (3) wearing face masks in all public places. State and local public health agencies echoed these recommendations, and disease levels temporarily abated in locations where these recommendations were widely followed. However, the recommendations were ignored in many other places, and, as a result, infection rates (and subsequently death rates) climbed during late 2020, as they did in several other countries (**Figure 16.10B**). Some schools and colleges reopened in fall 2020; many more offered remote classes. Local outbreaks caused many places that had opened to close again.

Vaccines against SARS-CoV-2 were developed and tested in countries around the world with unprecedented speed. Some of these vaccines use a rather novel method based on messenger RNA (mRNA). The vaccine contains some of the viral mRNA that codes for the "spike" protein on the surface of virus particles. Injection of the vaccine causes the body's cells to make some of this protein, causing the immune system to recognize it as a foreign substance and to make antibodies against it. The antibodies are thus primed to respond against any virus that displays this protein on its surface. Other SARS-CoV-2 vaccines use the more traditional method of an inactivated virus or a viral protein that provokes an immune response more directly (Corum and Zimmer, 2021).

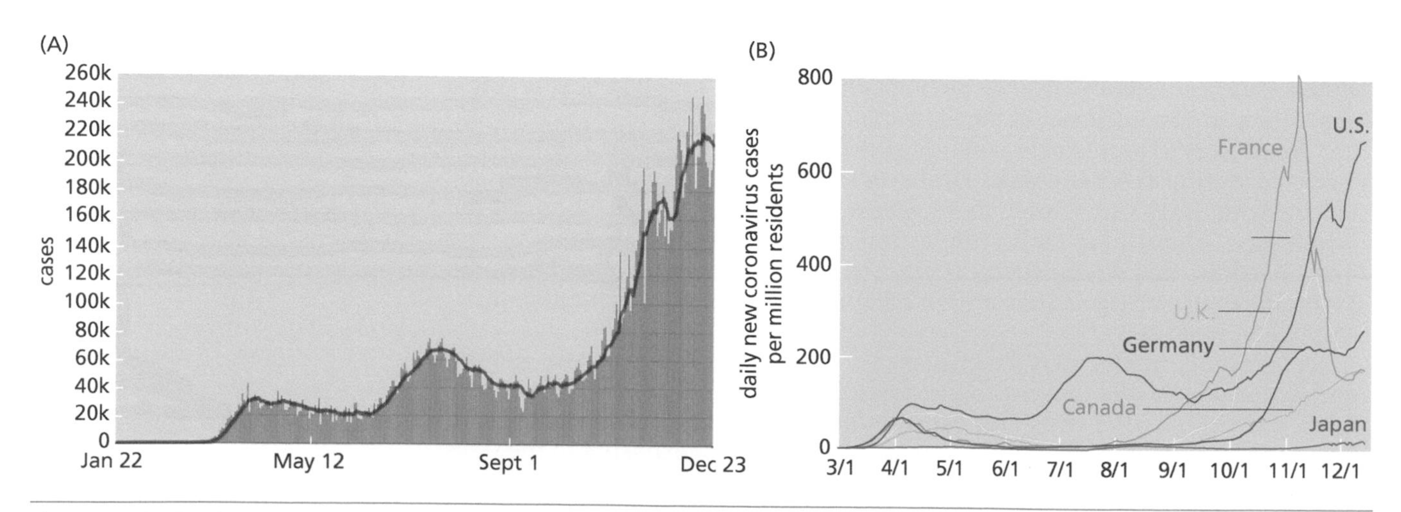

Figure 16.10 Timeline of COVID-19 infections during 2020. (A) In the United States. **(B)** In seven selected countries.

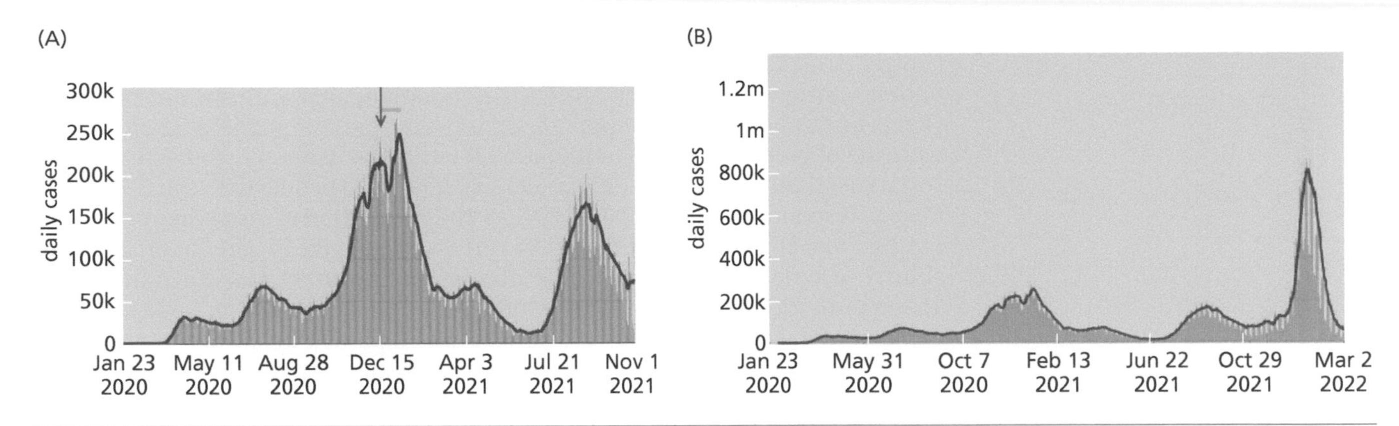

Figure 16.11 Timeline of COVID-19 infections in the United States. (A) January 2020 through November 1, 2021. The red line represents a rolling seven-day average. The arrow represents the time when the first vaccines were approved (December 15, 2020), but it took a few weeks for their distribution, as shown by the green bar. **(B)** January 2020 through March 2, 2022. The large spike in cases early in 2022 was largely due to the omicron variant.

In the United States, three vaccines became available in late 2020 and early 2021, and a similar number of other vaccines were developed and used in other countries. The number of COVID-19 infections and deaths fell dramatically once these vaccines became available (**Figure 16.11A**), but new strains of the virus emerged later. The "delta" and "omicron" variants of the virus spread much more rapidly than the earlier variants, and so they increased in prevalence, especially among unvaccinated people. The omicron variants in particular were also responsible for many "breakthrough" infections in vaccinated individuals (part of a very large spike in COVID-19 cases in the winter of 2021–2022; see **Figure 16.11B**), but vaccination still provided strong protection against hospitalization and death from COVID-19. The continued evolution of the virus spurred scientists to develop new versions of SARS-CoV-2 vaccines to provide more specific protection against a range of virus variants; this proved much easier to do for the mRNA vaccines than for traditional vaccine methods. It remains to be determined whether a yearly SARS-CoV-2 immunization targeted against the viral variants in current circulation will be recommended, as is the strategy for protection against influenza; however, the mutation rate for SARS-CoV-2 is significantly lower than for the influenza virus, so annual vaccination may not be necessary.

Adverse effects of COVID-19 vaccination have been reported but are uncommon (CDC, 2022). The more serious of these, such as myocarditis and other cardiovascular problems, have been attributed to the spike protein (Trougakos et al., 2022), which has been described as neurotoxic because "it impairs DNA repair mechanisms" (Seneff et al., 2022). Traditional vaccines contain this protein in carefully measured amounts, and a new vaccine, announced in 2022 (UW Medicine, 2022), minimizes the risk even further by using only a small portion of this protein. However, the vaccines that use mRNA continue to produce the entire protein for a long time in quantities that cannot be controlled (Seneff et al., 2022). It is too early to know the long-term effects of vaccination, and risk/benefit analyses that would include these effects are clearly desirable. Meanwhile, the CDC and other public health agencies continue to stress the need for repeated vaccinations.

Because COVID-19 appeared and spread so quickly, and because it had a wide range and severity of symptoms for a respiratory illness, doctors scrambled to find drugs that they hypothesized might be able to help patients already infected. Those that received the most attention were remdesivir, an antiviral nucleotide analog; dexamethasone, a steroid drug that dampens inflammation; and two antimalarial drugs, chloroquine and hydroxychloroquine, which, in cell culture systems, were able to interfere with the ability of the virus to fuse with cell membranes to gain entry into cells. Significant media hype surrounded these drugs, and various people declared them either to be miracle cures or to be completely ineffective. As more carefully controlled trials were conducted, the preponderance of data suggested the following conclusions: that Remdesivir shortens hospital stays for COVID-19 patients but does not affect mortality rates (Beigel et al., 2020); that dexamethasone is effective in reducing deaths among patients receiving oxygen or mechanical breathing support, but not otherwise (Recovery Collaborative Group, 2021); and that the antimalarial agents, which were never properly studied or prepared for antiviral use, are not effective and, in some cases, can actually be dangerous, especially for people with the G6PD deficiency described in Section 7.3.3 (FDA, 2020). In addition to these chemical drugs, monoclonal antibodies that target the SARS-CoV-2 spike protein have also been shown to be effective, an example of passive immunity (see Section 14.2.5). These monoclonal antibodies are produced in the laboratory in large-scale cell culture "bioreactors." They are introduced into a patient by intravenous infusion and work by interfering with the ability of newly replicated virus particles to attach to host cells and spread the infection within the body, augmenting the effect of antibodies produced by the infected person's own immune system.

In fact, all the COVID-19 drugs discussed so far must be given intravenously in a hospital or clinic setting. A significant breakthrough in COVID-19 treatment came in late 2021, when the U.S. FDA granted emergency use authorization (EUA) for two medications in pill form that can be taken by patients at home. The first of these, molnupiravir, interferes with replication of the SARS-CoV-2 viral genome such that it acquires so many mutations that it is rendered inactive; the second, Paxlovid, is a two-drug combination, one that directly inhibits the enzyme needed for viral replication and a second that acts to maintain higher levels of the antiviral drug in the body for longer periods of time. Both drugs decrease the rate of hospitalization of patients at high risk of COVID-19 complications—molnupiravir by about 30% and Paxlovid by as much as 90%. The former is marketed with a strict warning against taking it during pregnancy because animal studies revealed evidence of harm to fetuses; no such warning has been issued for Paxlovid, but it has not been studied extensively for possible negative effects in pregnancy.

A third drug in pill form that received considerable attention as a potential SARS-CoV-2 antiviral agent is ivermectin, originally developed in 1975 and widely used to fight parasitic roundworm infections in both humans and domestic animals (Crump and Omura, 2011). Ivermectin was shown to reduce the viral load in cultured cells infected with SARS-CoV-2 (Caly et al., 2020). Roundworm infections (including schistosomiasis, discussed earlier in this chapter) are common in tropical and subtropical climates, so doctors in those climates are very familiar with ivermectin's use in humans and with its proper dosing and possible side effects. Many researchers, particularly in tropical and subtropical countries, have conducted clinical trials on ivermectin. Some studies have reported reductions in rates of infection, hospitalization, and death in patients who were either already taking ivermectin prior to SARS-CoV-2 infection or began taking it soon after infection (Bryant et al., 2021). However, other studies have found no statistically significant protective effects from taking ivermectin, and there has been considerable controversy in the scientific community over the methodologies used in studies that demonstrate its efficacy in clinical trials.

In the United States, parasitic roundworm infections in humans are rare, and ivermectin is seldom prescribed. Although it can be prescribed to treat head lice and a few other conditions, ivermectin is used principally as a veterinary medication to treat parasitic roundworm infections in horses and cattle. The formulations used for these large animals contain much larger doses of ivermectin than can safely be prescribed for humans, and these formulations also contain other ingredients that have never been approved for human use, including some possibly harmful ones. People who tried to self-medicate with these veterinary formulations found themselves poisoned and often hospitalized, and the CDC issued cautionary statements against such unauthorized use. The U.S. FDA has not approved ivermectin for use against COVID-19, and Merck, which manufactures the drug, finds "no scientific basis for a potential therapeutic effect against COVID-19." The National Institutes of Health (NIH, 2021) further concludes that "there is insufficient evidence for the COVID-19 Treatment Guidelines Panel... to recommend either for or against the use of ivermectin for the treatment of COVID-19" and that better clinical trials are needed.

Some of the data needed to definitively assess the efficacy of ivermectin may come from countries that are actively administering ivermectin as a COVID-19 treatment. A very large study in Brazil studied the effects of ivermectin given as a prophylactic treatment to prevent COVID-19; the authors of this study concluded that ivermectin "significantly reduced COVID-19 infection, hospitalization, and mortality rates" (Kerr et al., 2022a). A follow-up study concluded that "regular use of ivermectin as prophylaxis for COVID-19 led up to 92% reduction in COVID-19 mortality rate in a dose-response manner" (Kerr et al., 2022b). ("Dose-response manner" means that larger doses of ivermectin had a stronger effect.) A meta-analysis based on 18 randomized controlled treatment trials of ivermectin in COVID-19 found large, statistically significant reductions in mortality, time to clinical recovery, and time to viral clearance, and concluded that "the many examples of ivermectin distribution campaigns leading to rapid population-wide decreases in morbidity and mortality indicate that an oral agent effective in all phases of COVID-19 has been identified" (Kory et al., 2021).

The Infectious Disease Society of America (IDSA, www.idsociety.org) has published guidelines on the treatment and management of COVID-19 patients (Bhimraj et al., 2023), and these guidelines are periodically updated. The guidelines are quite specific and summarize clinical studies on the efficacy of over 30 drugs.

Extensive testing and rapid contact tracing, along with vigorous vaccination campaigns and adherence to precautions such as masking and social distancing, have proven very effective in combatting the spread of SARS-CoV-2. However, few countries have been able to implement this whole suite of public health measures. Australia and New Zealand were especially successful in enforcing lockdowns and travel bans to keep case numbers very low throughout most of the pandemic and to minimize deaths. In India's most populous state of Uttar Pradesh, a vigorous campaign of effective and rapid testing, contact tracing, and isolation and treatment of the sick (described by WHO, 2021b) resulted in fewer than 200 active cases of COVID-19 remaining in the entire state of 230 million people, and the positivity test rate was reduced to less than 0.01% (Hindustan Times, October 11, 2021). China implemented very strict controls on its citizens, preventing them from leaving their homes for weeks and even months at a time except for testing when cases began to rise in certain areas. Overall, China's strict measures kept cases relatively low, but Chinese citizens began to resist the extreme lockdowns around two years into the pandemic. Other countries experienced pushback against public health mandates for quarantine, vaccination, and masking much sooner, and these became highly politicized in some instances, particularly in the United States. The less successful implementation of precautionary public health measures in the United States compared to some other countries resulted in many deaths that public health officials largely viewed as having been avoidable. As a comparison, the populations of Florida, Texas, and Australia all fall between 20 and 29 million people, but Australia experienced fewer than 9,000 deaths from COVID-19, while the death rates in Florida and Texas were approximately 10 times higher.

As new mutations occur, new variants of SARS-CoV-2 emerge and spread. In the summer of 2021, the delta variant spread rapidly and killed millions of people worldwide. Later in 2021, the omicron variant proved even more infective, and thus spread much more rapidly, but caused a much milder infection and thus fewer deaths. A very large spike in COVID-19 cases in the winter of 2021–2022, attributed almost entirely to the omicron variant, infected hundreds of millions of people worldwide before it subsided.

For diseases like COVID-19, several long-range fates are possible. The simplest and best possibility is that safe and effective vaccines could limit the spread whenever an outbreak occurs. Ideally, worldwide use of such a vaccine could completely eliminate the disease. This achievement has thus far only been reached for one disease (smallpox) and nearly achieved for just one more (polio). A second and more likely scenario is that herd immunity will develop over time, through a combination of virus exposure and vaccination. In this case, the disease may still sicken and kill people in small numbers or in local outbreaks, but it will have difficulty spreading and will soon dissipate wherever it occurs. Note that prior exposure or vaccination may not completely eliminate the risk of a person developing COVID-19, but these factors are likely to lessen the severity of the disease when it does occur, similar to what is seen for influenza (the flu). Experts have given varying estimates, from 60% to 80% or higher, of the percentage of the population that must become immune to COVID-19 (either through vaccination or having previously been infected with the virus) to reach the point of herd immunity. Still another possibility is for the disease to persist in many or most populations at low (endemic) levels, with deaths and transmission both limited by medications that control the symptoms (such as sneezing and coughing) and by practices (such as frequent hand washing and wearing of masks) that limit contact with the virus but do not attack the pathogen itself. Malaria and HIV/AIDS are two diseases that persist endemically in many populations around the world, and many experts predict that COVID-19 will follow a similar course.

Johns Hopkins University has developed an excellent online learning tool to provide easily understood information about COVID-19 and its spread (https://coronavirus.jhu.edu/covid-19-basics/understanding-covid-19). The Worldometer website maintains up-to-date information on the COVID-19 pandemic at https://www.cdc.gov/coronavirus/novel-coronavirus-2019.html. The CDC also has a website at https://www.cdc.gov/coronavirus/2019-ncov/cases-in-us.html.

THOUGHT QUESTIONS

1. In what ways does disease transmission by food differ from transmission by water?
2. In what ways do food and water routes of transmission interact?

16.5 ECOLOGICAL FACTORS ESPECIALLY AFFECT PATTERNS OF VECTOR-BORNE DISEASES

Vector-borne diseases are those in which the pathogen is carried from one host to another by some other animal that is not itself infected. Often the vector is an insect (phylum Arthropoda, class Insecta), but other sorts of animals, such as snails and ticks, can also serve as vectors. The pathogen can be a parasite as we saw for malaria (Box 7.3), but it can also be a bacterium or a virus.

16.5.1 Plague

An ancient example of a vector-borne bacterial disease is bubonic plague. **Plague** is a deadly infection caused by a bacterium, *Yersinia pestis*, and is transmitted principally by fleas. The disease is spread by many mammals, principally rodents, and transmitted to humans by fleas that have bitten an infected animal. Like most bacteria, *Y. pestis* does not have the ability to penetrate the natural barrier of the skin; it is the flea bite that provides the mechanism for transmitting the bacteria past this barrier. Direct transmission of the bacteria from one person to another can only occur if the material from an open sore is transferred to a cut in another person's skin.

Vector-borne diseases often have a secondary host, another species that can be infected by the pathogen, which thus acts as a reservoir of the pathogens in nature. For plague, the secondary hosts are rodents such as squirrels, chipmunks, and especially urban rats, or occasionally the carnivores (like cats) that bite these rodents. When there is a secondary host, disease eradication is difficult because both the vector and the secondary host reservoir must be controlled.

The most common form of plague begins with flulike symptoms and progresses to painfully swollen lymph nodes that can grow as large as chicken eggs. These swellings are called "buboes," after which this form of the disease is called the bubonic plague. It is from the word "bubo" that we get "boo-boo," the child's term for any painful sore.

Plague has been present in Asia throughout history. In the sixth century, it spread to Europe as a pandemic known as the Plague of Justinian. In the fourteenth century, an even larger plague pandemic, often called the Black Death, spread to Europe, where it wiped out one-third of the entire population. A third pandemic began around 1860 in China and then spread via rats aboard ships to port cities throughout the world. In the United States, the disease was first reported in 1900, and repeated outbreaks occurred during the period 1900–1925 from the Rocky Mountains to the Pacific, principally in California. Since then, the disease has been controlled in most countries by a combination of rodent control, flea control, and antibiotic treatment of the disease itself. Large outbreaks do, however, continue to occur in many places, especially in Africa. In the Western United States, the disease still occurs among wild rodents and is transmitted to humans by fleas (less often by animal bites).

From 2017 to 2018, a plague epidemic struck in Madagascar, affecting approximately 1,800 people. Most cases were of the lung-infecting (pneumonic) rather than the lymph node-infecting (bubonic) form of the disease; the pneumonic version is the more deadly of the two. The outbreak also affected nine other African and Indian Ocean countries and island territories. In the United States, various health officials fear that a new epidemic of the bubonic plague could easily occur in California, where large homeless populations and poor sanitation create breeding grounds for infected rats, wild rodents (squirrels, chipmunks, etc.), and the carnivores (including cats and coyotes) that may bite them.

16.5.2 West Nile virus and Eastern equine encephalitis

A vector-borne disease recently introduced in the United States and Canada is **West Nile virus**. This virus causes a brain inflammation that can be fatal. During 2002 there were over 100 deaths in the United States from a virus that was unknown in North America before 1999. It was first identified in New York City, and in three years' time it spread from the east to the west coast. Cases were confirmed in 32 states, and the disease has become the most common mosquito-borne infection in the United States (CDC, 2021c). The first human cases in Canada were in 2001. West Nile virus is thus considered an emerging disease in the United States and Canada, although it has existed in Africa for a long time. It is not known how it was carried to the United States, but it is believed to have been discovered soon after its arrival by epidemiologists noting the unusual deaths of birds. Because it was detected early on, its spread has been well documented, and the speed of its spread across the country has surprised public health epidemiologists. It is spread by mosquitoes of the genus *Culex*, different from the *Anopheles* mosquito that spreads malaria (see Section 7.3). West Nile virus cannot spread from person to person, either directly or by way of mosquitoes. The virus does not replicate to high numbers in people, so people are considered **dead-end hosts**: They get sick but they do not contribute to disease transmission (Figure 16.12). Even in areas where the virus is present, most mosquitoes do not carry enough virus to make a person sick, and most people do not become ill even if mosquitoes do inoculate them with virus during a mosquito bite. Of those who do become ill, over 95% recover. Horses are also "dead-end hosts," but appear to be much more susceptible, and 40% of infected horses die. Birds are the **primary host**, that is, the host in which viruses replicate and from which they are spread by mosquitoes to humans and horses. Over 110 species of birds are known to be hosts for the virus. The virus kills jays and crows, as noted in New York City at the start of the outbreak, but most birds most show no sign of infection. As we saw in Chapter 15, viruses have a life cycle at the cellular level (see Figure 15.4), but they may also have a much larger-scale life cycle in the environment (Figure 16.12). Controlling transmission thus depends on mosquito control.

Eastern equine encephalitis (EEE) is another viral disease that is transmitted by the same types of mosquitoes as West Nile virus. EEE is much rarer than West Nile, but it is much more deadly, killing approximately 30% of those affected. Individuals can protect themselves from either of these viruses by using insect repellent or by staying indoors at hours when mosquitoes are active (peak biting

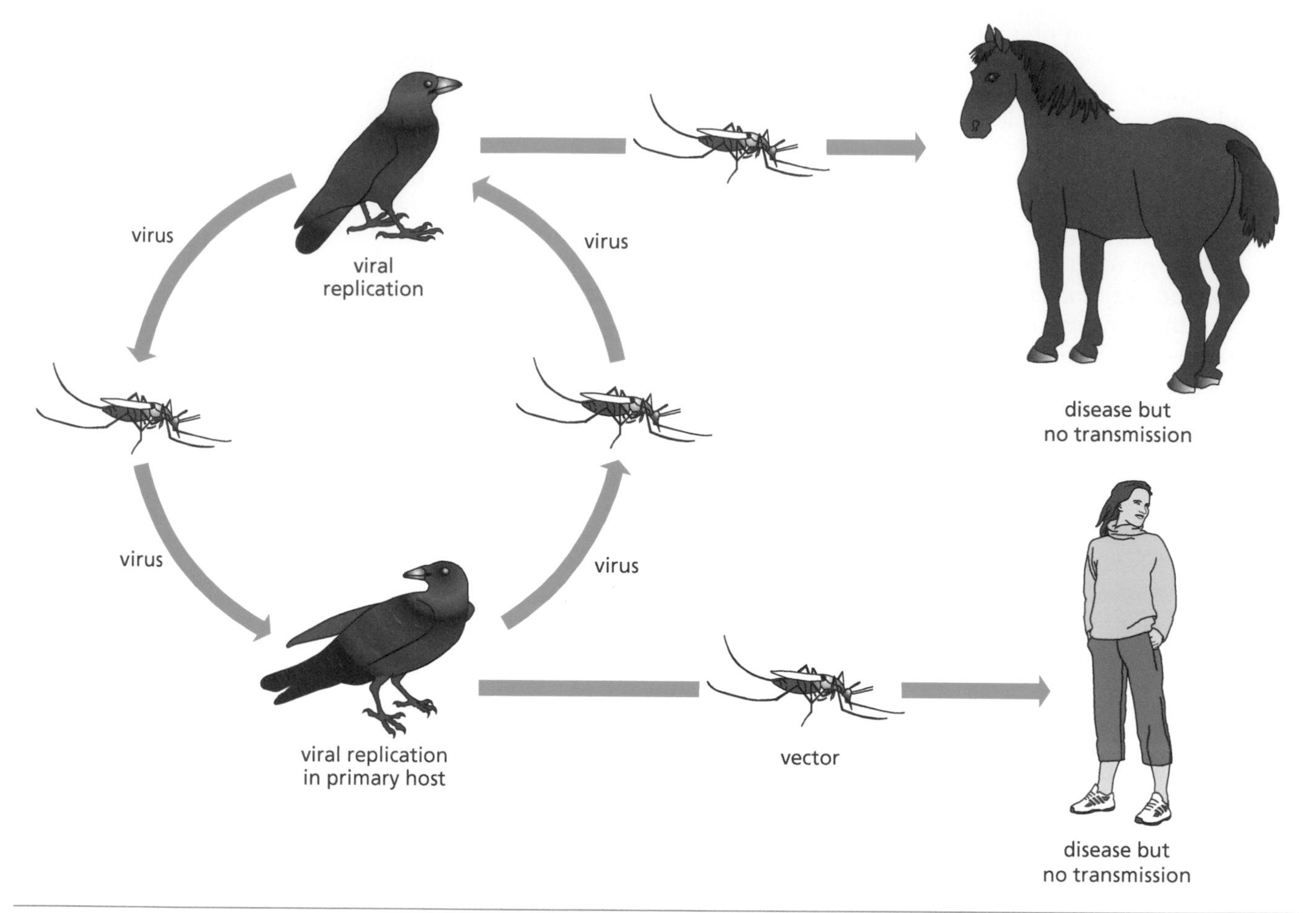

Figure 16.12 Life cycle of West Nile virus.

times are from dawn to dusk). Such measures will not prevent the spread of the virus in nature because humans are not the primary hosts.

As this example illustrates, control of transmission of vector-borne diseases depends largely on knowledge of the ecology of the vectors and primary hosts. Factors that increase the population size of nonhuman host species will likely increase the spread. Factors that separate humans from the primary host species or from the vector will control a disease outbreak but will not eliminate the pathogen from the environment.

16.5.3 Leishmaniasis

Leishmaniasis is a disease caused by *Leishmania*, a eukaryotic parasite related to the flagellated protist that causes African sleeping sickness. It is far more common and far more deadly worldwide than the last few diseases mentioned here. *Leishmania* is spread by over 30 different species of phlebotomine sandfly. There are many hosts, in addition to humans, including rodents and dogs. The sandfly can transmit the parasite by carrying blood from one person to another or from an animal to a person. The disease itself takes different forms, one involving fevers, weight loss, and anemia; the other resulting in disabling skin lesions.

The regions in which the *Leishmania* parasites are endemic have expanded significantly since 1993. This has led to a parallel increase in human cases, which number 1.5 to 2 million per year. One factor in this expansion has been large projects like dams and irrigation systems that have led previously unexposed people into new geographic areas to farm. The parasite was already endemic in these areas and was thereby provided with a new pool of susceptible human hosts. This was followed by massive human migrations from rural to urban areas, spreading the parasite into new locales. Factors contributing to these migrations include deforestation and war. Leishmaniasis is now endemic in 88 countries in Africa, Asia, Europe, and North and South America, wherever the climate is warm enough to support the life cycle of the sandfly vector. In Afghanistan alone, the World Health Organization estimates that 226,000 new cases occurred each year from 2003 to 2007.

Another factor has been the incidence of coinfection with HIV. Most people who are bitten by sandflies carrying *Leishmania* do not acquire the parasite; however, when a person is also HIV-infected, they are 100 to 1,000 times more likely to pick up the parasite and to develop leishmaniasis. Likewise, if they are already infected with the parasite, the onset of AIDS after an HIV infection is much more rapid. *Leishmania* and HIV coinfection is the greatest problem in Southwestern Europe (Spain, Portugal, Italy, and France), where over 70% of the coinfected people are intravenous drug users, so the parasite and HIV are being spread by blood in contaminated needles, rather than by the sandfly.

This vector-borne disease is thus a good illustration of the fact that biological factors, social factors, and economic factors all interact in the spread of disease. All will need to be taken into account when trying to control disease.

16.5.4 Tick-borne diseases

Lyme disease is a bacterial disease caused by a spirochete (corkscrew-shaped) bacterium, *Borrelia burgdorferi*, and several related species (Kurokawa et al., 2020). It is transmitted by the black-legged deer tick *Ixodes scapularis*. Symptoms include fever, headache, fatigue, and skin rashes; untreated cases can spread to affect the joints, heart, and nervous system. The disease is ancient and has been found in mummies but was never understood until a cluster of children in Lyme, Connecticut, first brought attention to the disease in the 1970s. From 1975 to 1995, prevalence of this disease increased about 25-fold. Subsequent research led to the discovery of its tick-borne transmission and the identification of the bacterium in 1981. Once diagnosed, the disease can be treated with antibiotics. Preventive measures include minimizing exposure to ticks by using pesticide sprays and clothing that tightly covers all areas of the body, especially in wooded areas, and also checking people and pets carefully for ticks after exposure to wooded areas. Ecological changes have allowed this disease to emerge as an increased public health threat in recent decades. Weasels, badgers, and opossums once controlled ticks through their grooming activities; one opossum destroys an estimated 6,000 ticks each week when grooming its fur. Small rodents, once controlled by coyotes, bobcats, and other wild predators, are now the greatest reservoir of Lyme disease, but the killing of the predators has allowed the rodent population to flourish. Fragmentation of forests through human activity has also resulted in more deer and white-footed mice, and fewer opossums and badgers.

A more recently emerging tick-borne disease is Powassan disease, a form of viral encephalitis that causes fever, severe headache, vomiting, confusion, loss of coordination, speech and memory problems, and eventually death. First identified in North America around 2009, the virus is actually endemic to the Russian Far East. It infects wild rodents such as squirrels and white-footed mice and is transmitted to humans by several species of ticks, principally of the genus *Ixodes*. There is currently no effective cure or treatment; the principal defense against the disease is the avoidance of ticks.

16.5.5 Other vector-borne diseases

Several other vector-borne diseases have emerged as new threats in recent decades. One example is **Chikungunya**, a viral disease transmitted by mosquitoes. From its discovery in 1952, the disease has caused repeated sporadic outbreaks in Asia and Africa. The 2005–2006 outbreak on the Indian Ocean islands of Réunion and Mauritius infected over 272,000 people, and a 2006 outbreak in India infected approximately 1.5 million. The most recent outbreak occurred in 2018 in Sudan, sickening nearly 14,000 people.

Avian influenza (bird flu) is caused by type A influenza virus. A large reservoir for this disease persists among ducks, geese, swans, and other waterfowl, all of which have immunity and can transmit the virus without getting sick. Chickens, however, have no such immunity—the virus kills them quickly and dies out if flocks are small and isolated. Great increases in chicken flocks, proximity to wild fowl, and international transport (increased 20-fold in the last 30 years) have all spread the virus further. In 1996, wild bird flu (strain H5N1) in Guangdong Province, China, spread from geese to other bird species and then to humans. The disease is now spread by migratory birds through their feces and by unsanitary disposal of waste among captive flocks, especially in Asia. Viral strains infecting birds were formerly distinct from those infecting humans, but viral recombination was able to occur in pigs because they are susceptible to both the bird flu virus and the human flu virus. This recombination, which gave the bird flu the ability to infect humans, probably occurred in the 1990s in China, which has over half the world's supply of pigs (about six times the U.S. supply).

Zika virus is a viral disease that first appeared in Zika Forest, Uganda, in 1947 (WHO, 2018); it has since spread throughout the tropics (except at high altitudes). It is transmitted mostly by mosquitoes (especially *Aedes aegypti*), but it can also be transmitted sexually. Because symptoms are mild in many people, precautions (like isolation of infected persons) are often ignored, allowing the disease to spread. In pregnant women, however, Zika, can spread to the developing fetus and cause severe neurological defects including underdeveloped brains (microcephaly). Zika-associated birth defects are estimated to occur in 5–10% of women infected with the virus during pregnancy. In addition, a small number (about 0.02%) of patients infected with Zika develop Guillain–Barré syndrome, an autoimmune response against neurons that causes muscle weakness and often temporary paralysis and can sometimes be fatal. To prevent transmission, the most important precaution is to avoid exposure to mosquitoes, especially when traveling to a region where Zika occurs. In 2015–2016, an outbreak spread from Brazil to other parts of the Americas, including the Caribbean (especially Puerto Rico), and also back to Africa and to Hawaii. From 2015 to 2019, all reported cases of Zika in the mainland United States were from travelers who visited Zika-affected areas, except for one laboratory worker who contracted Zika from a needle-stick injury.

Global warming strongly affects many vector-borne diseases. Harsh winters kill off many mosquitoes, ticks, and fleas, but milder winters allow larger populations of these vector species to persist and to transmit more disease.

THOUGHT QUESTIONS

1. What is the difference between the terms "host" and "vector"?
2. What factors influence the geographical distribution of vectors? Are these the same factors that influence the geographical distribution of hosts?

CONCLUDING REMARKS

Although this chapter has examined new infectious threats, we wish to emphasize that the epidemics and pandemics of today, although not to be ignored, are nothing like those of the past. Twenty million people died in an influenza epidemic during 1918–1919. Millions used to die every year from diseases of which we no longer even hear mention. Highly effective vaccines exist for prevention of many viral and bacterial diseases. Understanding the routes of pathogen transmission has led to more effective measures for hygiene control. The interconnectedness of today's world has both positive and negative consequences for disease transmission: Cooperation between nations and worldwide reporting of disease incidences have made it possible to control most outbreaks before they become epidemics; however, as the recent COVID-19 pandemic made clear, the rapid movement of people from place to place around the globe also makes us more vulnerable to new diseases. For most existing diseases, we know what to do to stop their spread. Of primary concern is extending access to clean food and water and to adequate nutrition to the areas of the world where these do not exist, the same areas where many diseases remain endemic. Control of pollution will also be important, because pollution weakens immunity, thereby increasing host susceptibility. Global warming and other forms of climate change also contribute to the increased spread of pathogens, especially for vector-borne diseases. We will examine both pollution and climate change in Chapter 19.

CHAPTER SUMMARY

- Many types of **pathogens** are **virulent**, meaning that they can cause disease.
- Hosts evolve mechanisms to protect themselves against pathogens, but pathogens evolve mechanisms to evade these protective adaptations.
- New infections emerge when pathogens evolve new adaptations or when environmental changes (including changes in host behavior) allow new opportunities for a pathogen to spread.
- Routes of transmission include direct contact, foodborne transmission, waterborne transmission, and transmission by a **vector** species.
- Sexually transmitted diseases and tuberculosis are spread by direct contact.
- Foodborne transmission can result from unsanitary conditions and also from the inadvisable feeding of certain animal wastes to other animals.
- Waterborne transmission can result from poor sanitation, especially when treatment systems are overwhelmed or disrupted in floods or other natural disasters, as in the 2010 cholera outbreak in Haiti.
- Many viral diseases, including SARS, MERS, and COVID-19, are spread mostly by aerosol droplets from coughing and sneezing. COVID-19 reached worldwide **pandemic** levels in 2020.
- Ecological disruption and climate change can strongly influence diseases transmitted by mosquitoes, ticks, and other vectors. Examples include plague, West Nile virus, leishmaniasis, Lyme disease, Chikungunya, and Zika.

BIBLIOGRAPHIC REFERENCES

Bibliographic references to material in this chapter can be found online at biologytrending.routledge.com/chapter16

GLOSSARY: KEY TERMS TO KNOW

These key terms are defined at the end of each chapter as an aid for student review.

Anthrax A lung or skin infection caused by *Bacillus anthracis* and its spores.

Antigenic drift Spontaneous single-nucleotide mutations that change the shape of antigenic molecules within a species, often reducing the effectiveness of previously acquired immunity within a host.

Antigenic shift Genetic reassortment of large sequences of nucleotides that result in major changes in the antigen shapes on the cells of organisms, making those organisms "invisible" to previously immune hosts.

Bioterrorism The use of biological agents to spread harm and also fear of greater harm in human populations.

Chikungunya A viral disease transmitted by mosquitoes in tropical regions.

Chlamydia A sexually transmitted disease caused by the bacteria *Chlamydia trachomatis.*

Cholera A waterborne infection caused by the bacteria *Vibrio cholerae.*

COVID-19 An infectious disease caused by the coronavirus SARS-CoV-2, responsible for a pandemic that began in 2019.

Creutzfeldt-Jakob disease A brain infection in humans caused by a prion.

Cryptosporidiosis A waterborne infection caused by the parasite *Cryptosporidium.*

Dead-end host A host in which a pathogen reaches the end of its life cycle and is not transmitted to subsequent hosts.

Disease Any condition of an organism in which normal biological function is lessened or impaired.

Ebola A virus endemic to Central Africa that caused an epidemic in West Africa from 2014 to 2016.

Endemic Persistently found in a specified location (e.g., a disease that maintains a low to moderate prevalence over a long time).

Epidemic An outbreak of a disease at much greater prevalence than usual.

Giardiasis A severe diarrhea caused by the waterborne intestinal parasite *Giardia intestinalis.*

Gonorrhea A sexually transmitted disease caused by the spherical bacteria *Gonococcus.*

H1N1 swine flu See *swine flu.*

Herd immunity Resistance of a population to a disease by the presence of many individuals who do not transmit it.

Herpes simplex A common sexually transmitted disease caused by the herpes simplex virus (HSV).

Human papilloma virus (HPV) A sexually transmitted virus that causes genital sores and cervical cancer.

Incidence The number or frequency of new cases of a disease or other condition. Compare to *prevalence.*

Infection The colonization and growth of a pathogen within a host organism.

Legionnaire's disease A waterborne infection caused by *Legionella* bacteria.

Leishmaniasis An infection caused by the flagellated parasite *Leishmania* and transmitted by sandfly vectors.

Lyme disease A tick-borne infection caused by the bacteria *Borrelia burgdorferi.*

Middle East respiratory syndrome (MERS) An infection caused by a coronavirus responsible for a 2012 epidemic.

Mycoplasma genitalium A bacteria responsible for a sexually transmitted infection.

Pandemic A worldwide epidemic or disease outbreak.

Parasite A species that lives in or on another species (the host) to which it causes harm.

Pathogen An infectious agent (organism, virus, or protein) that causes a disease.

Plague A highly fatal infectious disease caused by the bacteria *Yersinia pestis.*

Prevalence The number or frequency of existing cases of a disease or other condition at any particular time. Compare to *Incidence.*

Primary host The host in which a pathogen or parasite spends the majority of its life cycle, usually including the reproductive stages.

Prion A protein capable of producing an infection. Creutzfeldt-Jakob disease in humans and bovine spongiform encephalopathy (BSE or "mad cow disease") are examples of diseases caused by prions.

Route of transmission The means by which an infectious disease spreads from one host individual to another.

Schistosomiasis A waterborne infection caused by the parasitic flatworm *Schistosoma.*

Severe acute respiratory syndrome (SARS) An infection caused by the coronavirus SARS-CoV-1, responsible for a 2003–2004 epidemic.

Smallpox A viral disease, once deadly, that has been eradicated by worldwide vaccination.

Swine flu A viral disease spread from pigs to humans responsible for a 2009 epidemic.

Syphilis A sexually transmitted disease caused by the bacteria *Treponema pallidum.*

Tuberculosis An infectious lung disease caused by the bacteria *Mycoplasma tuberculosis.*

Variant CJD See *Creutzfeldt-Jakob disease.*

Vector An insect or other intermediary that transmits a disease organism.

Virulence The ability of a microorganism to cause a disease.

Virulent Capable of causing an infectious disease.

West Nile virus A virus that can be transmitted to humans by mosquito vectors from a host reservoir in multiple species of birds.

Zika virus A brain-damaging virus transmitted to humans by mosquitoes.

CONNECTIONS TO OTHER CHAPTERS

Chapter 3: Susceptibility of individual hosts varies genetically, as does virulence of pathogens.
Chapter 4: Complete genome sequences have been deciphered for several pathogenic species. In some cases, genes responsible for virulence have been identified.
Chapter 5: Virulence follows a long-term pattern in its evolution.
Chapter 7: Malaria and other infectious diseases have always been important mechanisms of natural selection in human evolution.
Chapter 8: Poor nutrition increases susceptibility to infectious disease.
Chapter 14: Immunocompromised people are more susceptible to infectious disease.
Chapter 15: People with AIDS become susceptible to many other infections.
Chapter 19: Pollution and global warming can both contribute to the spread of diseases.

PRACTICE QUESTIONS

1. What is the difference between the terms prevalence and incidence? Under what conditions will prevalence increase if incidence remains the same?
2. What is the difference between the incidence and the incidence *rate* of a disease? What additional information do you need to know to calculate an incidence rate that you do not need to know to make a table of disease incidence?
3. Why does the risk of antibiotic resistance rise if people do not finish taking all of their antibiotic pills?
4. What do the bacteria that cause chlamydia, tuberculosis, and Legionnaire's disease have in common?
5. Could bubonic plague be effectively used in bioterrorism? Why or why not?

CHAPTER 17

Plants to Feed the World

ISSUES

- What is plant science?
- How has plant science changed the world?
- Can plant science feed the world?
- Why is the "law of unintended consequences" so important?
- Are genetically engineered plants different from other plants?

BIOLOGICAL CONCEPTS

- Photosynthesis (autotrophs, light-dependent reactions, dark reactions)
- Osmosis
- Nitrogen cycle
- Plant structure and function (tissues, water transport, gas exchange)
- Ecosystems (trophic levels: producer, consumer, decomposer; soil; sustainable agriculture; pest control; integrated pest management)
- Limiting factors (fertilizers, irrigation)
- Plant genetic engineering

CHAPTER OUTLINE

DOI: 10.1201/9781003391159-17

Hunger, starvation, and malnutrition are endemic in many parts of the world (see Chapter 8). Rapid increases in the world's population (see Chapter 9) have intensified these problems. Because all the food that we eat comes either directly or indirectly from plants, any attempt to produce adequate food for the world's growing human population must focus on the production of food by plants. However, the amount of land under cultivation has its geographic limits, and extending these limits carries a very high biological cost in terms of the destruction of natural ecosystems (see Chapter 18). Another way to increase crop production is to increase the yields or the nutrient content of crop species through such techniques as soil improvement, use of fertilizers and advanced irrigation techniques, pest control, and the use of different plants or new genetic strains of plants. In this chapter, we concentrate on food production by plants because that is where most of our food comes from.

Plants are essential to all of life on Earth. In addition to food, plants also provide the oxygen that humans and other organisms breathe. Without plants, most other forms of life would soon die out. Plants are the world's richest energy source. On a worldwide basis, the amount of energy produced by plants is about 6×10^{17} kilocalories per year (abbreviated kcal/year), which is the equivalent of the energy contained in a sugar cube 5 km (or 3 miles) on each side, but, as the human population continues to increase, we are rapidly outgrowing this energy resource. This chapter discusses how plants make energy available to other organisms and how humans, through application of this knowledge, might learn to grow more crop plants and grow them more efficiently.

17.1 PLANTS CAPTURE THE SUN'S ENERGY AND MAKE MANY USEFUL PRODUCTS

As we saw in Chapter 6, plants are a kingdom of living organisms that have eukaryotic cells containing specialized structures for capturing energy from the sun's light. Plants are essential to life on Earth, because they capture light energy from the sun and convert much of this energy into forms that can be stored for later use. While building and maintaining their own bodies, plants make products that serve their own needs and also many needs of other organisms. How exactly do plants trap light energy and use it to feed entire ecosystems and to produce the many plant products that we depend upon? We begin with a discussion of these processes.

17.1.1 The importance of plants

An **ecosystem** includes all the species that live together and interact, plus all the physical resources (including water, soil, and atmosphere) with which they interact. The energy that passes through all ecosystems comes from the solar energy captured by plants and other photosynthetic organisms, and these photosynthetic organisms form the base of all food chains. Plants store many carbohydrates (and some oils) for their own use as energy sources for a later time. Humans and other animals eat many plant products as foods, that is, as sources of energy and nutrients (see Chapter 8). As we shall see, humans also find many other ways of benefiting from the use of plants.

17.1.2 Photosynthesis

Plants use the energy that they capture from the sun to make energy-rich carbohydrates by a process called **photosynthesis**. They store the carbohydrates for their own use as energy sources for a later time. When humans and other animals eat plants, they harvest some of this energy.

17.1.2.1 Energy producers and energy harvesters Organisms, such as plants, that can use light or other inorganic energy sources to make all of their own organic (carbon-containing) molecules from simpler molecules are called **autotrophs**. Most other organisms, such as animals, are **heterotrophs**, which means that they cannot make their own organic compounds from inorganic materials and are therefore absolutely dependent on the organic compounds made by plants and other autotrophs. Autotrophs store energy in the chemical bonds of energy-rich molecules such as carbohydrates. Autotrophs and heterotrophs alike use cellular respiration to release this energy and power life's processes.

Autotrophs are also called **producers** because they produce compounds usable by other organisms, including ourselves. Most heterotrophs are **consumers** that must obtain their energy by eating other organisms. Plants are the ultimate source of food energy not only for the primary consumers that eat plants directly, but also for the secondary and higher-order consumers that eat other consumers. When organisms die, their complex organic molecules are broken down by other heterotrophs called decomposers, which are mostly fungi and bacteria; the breakdown products can then be recycled and used by other living things.

Energy enters the biological world as sunlight and flows through producer, consumer, and decomposer organisms in turn (Figure 17.1). At each step, some energy is lost by being converted into forms that are not usable by organisms. Most of this unusable energy escapes as heat. Because of this loss of energy in the form of heat, the process would run down and stop altogether unless new energy were continually supplied. The new energy for the living world is sunlight. Plants are essential to the global energy flow because they are the principal means by which light energy is captured and changed into forms that other organisms can use.

17.1.2.2 Energy and pigments By the process of photosynthesis, plants gather energy from sunlight and use this energy to make carbohydrates such as

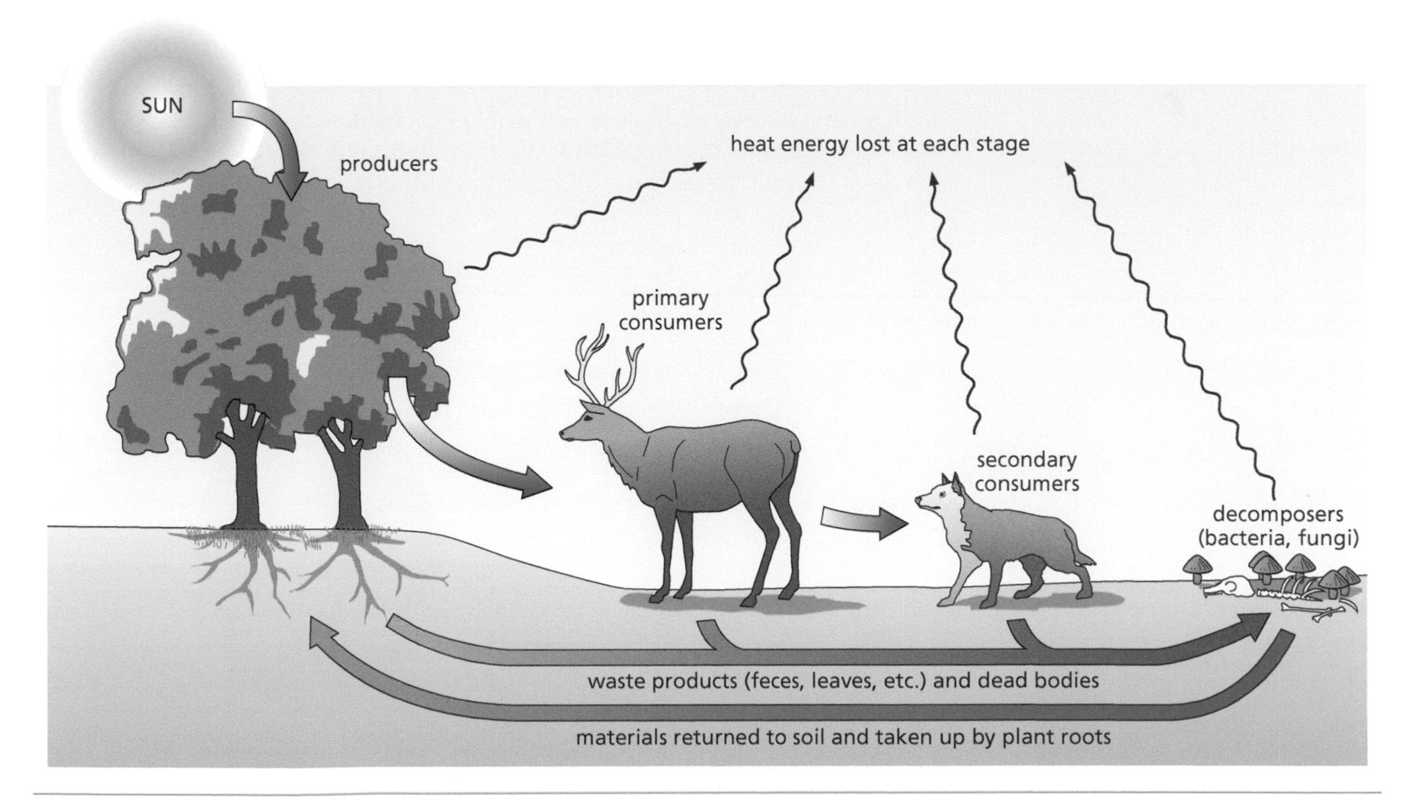

Figure 17.1 **Energy flow through a biological system.** Energy enters as sunlight. Producers convert the sunlight to chemical bond energy usable by other organisms, the consumers. Energy locked in the chemical bonds of dead producers and consumers is released by decomposer organisms. Energy is also lost as heat at each step, so more energy must continuously enter the system for the process to continue.

sugars and starches from atmospheric carbon dioxide and water. The overall process can be summarized by the following equation:

$$\underset{\text{carbon dioxide}}{6CO_2} + \underset{\text{water}}{12H_2O} \xrightarrow{\text{light energy}} \underset{\text{glucose (a sugar)}}{C_6H_{12}O_6} + \underset{\text{oxygen}}{6O_2} + \underset{\text{water}}{6H_2O}$$

Plants thus sustain the composition of the atmosphere by using up carbon dioxide and releasing oxygen in exchange. They also sustain animal (including human) life by supplying the very oxygen that we breathe. (Plants also use oxygen in cellular respiration, but they produce much more than they consume.)

The capture of light energy for photosynthesis takes place in certain light-sensitive molecules called pigments, of which **chlorophyll** is the most important. Each of these molecules absorbs some wavelengths of light and not others. Chlorophyll absorbs blue and red light but not green light. The colors that we see are those that are reflected rather than absorbed, which is why so much of the living world looks green (**Figure 17.2**). In addition to chlorophyll, plants and other

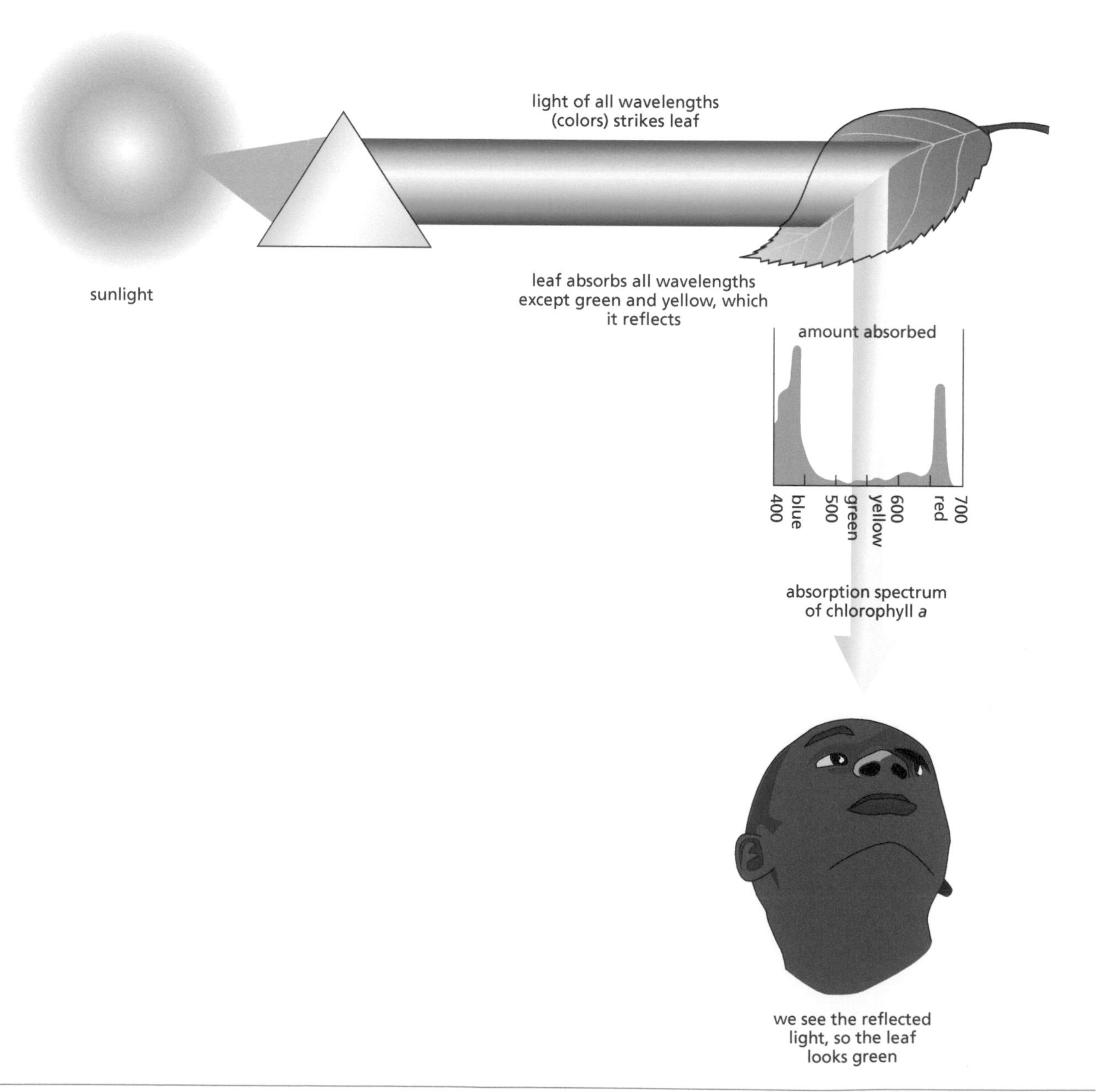

Figure 17.2 Perception of color.

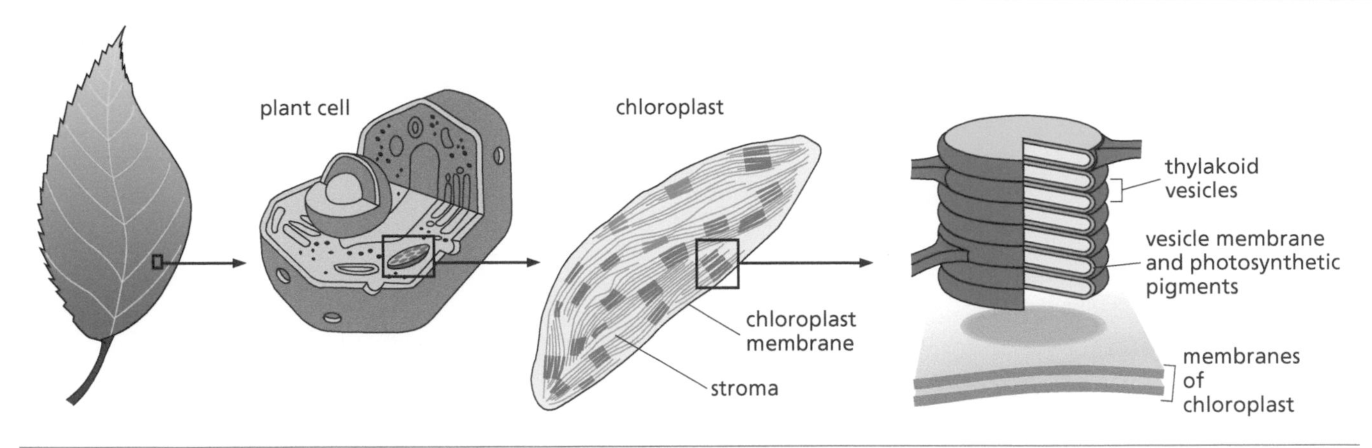

Figure 17.3 Location of photosynthesis in plant cells. Within plant cells there are chloroplasts. Each chloroplast contains stacks of thylakoids.

photosynthetic organisms possess various other pigments, such as carotenes and xanthophylls, which absorb light of other colors and pass the energy on to chlorophyll. These pigments are useful to the organism because they enable it to use light energy of different wavelengths. Because particular light-absorbing pigments are found only in certain groups of photosynthetic organisms, pigments are used to identify these groups and reconstruct their evolution. For example, similarity between the pigments of green algae, bryophytes, and vascular plants is strong evidence that bryophytes and vascular plants probably evolved from green algae (see Chapter 6).

Many of the broad-leaved trees of temperate regions stop making chlorophyll in the autumn. In the absence of chlorophyll, the other pigments in the leaves become apparent. These pigments absorb green and blue and reflect many other colors of light, resulting in the fantastic rainbow of leaf colors in the fall foliage (see Figure 1.1).

In all photosynthesizing organisms that have a eukaryotic cell structure (plants and various algae; see Chapter 6), the photosynthetic pigments are contained in organelles known as **chloroplasts**, which are located within the cells of the green parts, especially leaves. The fluid interior of these chloroplasts contains stacks of flattened membrane vesicles (**Figure 17.3**). Photosynthesis takes place along the membranes of these stacked vesicles, called thylakoids.

Although most photosynthesizing organisms are plants, there are also a number of photosynthesizing species of prokaryotes, including a few bacteria and all cyanobacteria. In these prokaryotic organisms, photosynthesis takes place along the cell's plasma membrane and in the cytoplasm because there are no specialized chloroplasts. Photosynthetic prokaryotes (bacteria and cyanobacteria) and eukaryotes (diatoms, dinoflagellates, and other "algae") fill our oceans and lakes with nutrients that can be used by other organisms.

17.1.2.3 Photosynthesis: Light reactions Plant photosynthesis takes place in two stages. The first stage consists of the reactions that take place in the membranes of the stacked vesicles; these require light energy and are therefore called "light reactions." You can follow the steps of the light reactions in **Figure 17.4**. In Step 1, photosynthetic pigments in the vesicle membrane capture light energy, represented by the gold rays. The captured energy is used to split water molecules into hydrogen and oxygen (Step 2). The oxygen is released to the atmosphere (Step 3), where it is essential to oxygen-dependent organisms, including humans. Each hydrogen atom is further split into a hydrogen ion (H^+) and an electron (e^-). A hydrogen atom contains a single proton and a single electron, so a hydrogen ion that has lost its electron is just a proton; therefore, hydrogen ions are often simply referred to as protons. These hydrogen ions (protons) are shown as gray circles in Figure 17.4. The hydrogen ions accumulate inside the stacked vesicle membranes (Step 4). The electrons are shown as the blue circles in Figure 17.4, and their movements are shown as blue arrows in

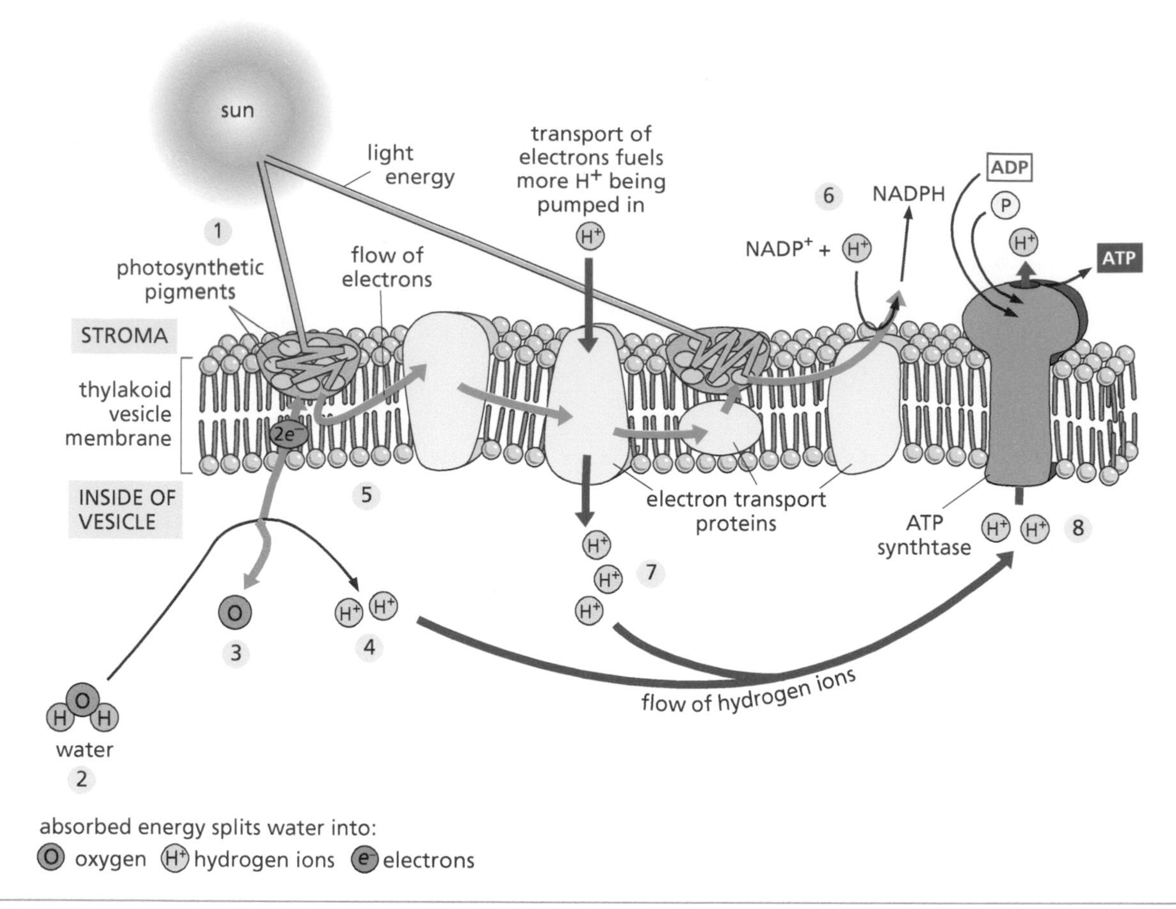

Figure 17.4 **The light reactions of photosynthesis.** The flows of energy, electrons, and hydrogen ions are shown, as are the syntheses of ATP and NADPH, both of which contain high-energy chemical bonds.

that figure. The electrons move through a series of proteins that are embedded in the vesicle membrane (Step 5) in what is called an "electron transport chain." It is often said that the electrons move "down the electron transport chain" because, with each transfer, they move to a lower energy state. At the end of the electron transport chain, the electrons are delivered to a molecule called $NADP^+$, forming NADPH (Step 6). $NADP^+$ is an energy-carrying molecule that is made in part from niacin, a vitamin described in Section 8.2.5. NADPH is one of two key products made by the light reactions of photosynthesis; we will next consider the second, which is the energy storage molecule adenosine triphosphate (ATP).

Some of the proteins that form the electron transport chain are also ion pumps. As the electrons move to a lower energy state, these pumps use some of the released energy to pump hydrogen ions into the interior space of the stacked thylakoid membranes (Step 7), thus forming an ion gradient. As with other ion gradients, such as those across the membranes of mitochondria (see Figure 8.16), the hydrogen ion gradient stores energy. The stored energy is then used to power the synthesis of the energy-rich molecule ATP by the ATP synthase complex, also as in mitochondria (Step 8). The plant cell uses ATP for most biological activities that require an energy source, just as we saw for animal cells in Chapter 8.

In summary, in the light reactions of photosynthesis, captured light energy is converted into the high-energy chemical bonds in NADPH and ATP. This stored, light-derived energy is later used in another set of reactions called the Calvin cycle that form the remainder of photosynthesis (Figure 17.5). Some light energy is also unavoidably transformed into heat.

17.1.2.4 Photosynthesis: Dark reactions

The ATP and NADPH created in the stacked membrane vesicles are not released from the chloroplasts; they move from these vesicles into the surrounding fluid of the chloroplasts, called

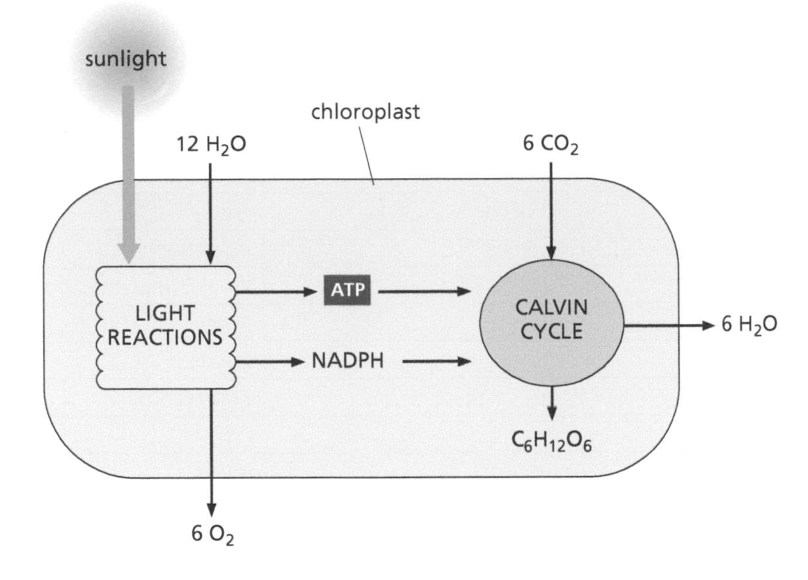

Figure 17.5 Summary of the reactions of photosynthesis.

the stroma (Figure 17.3). Here, the ATP and NADPH participate in the second of the two stages of photosynthesis. The ATP supplies energy and the NADPH provides electrons in the form of hydrogen atoms for the synthesis of glucose ($C_6H_{12}O_6$) from carbon dioxide (CO_2), the source of the carbon and oxygen atoms. Because the reactions that use ATP and NADPH do not directly require light, they are sometimes called the "light-independent" or "dark" reactions, but it is important to note that the dark reactions need a continuous supply of the products of the light reactions, so they will come to a halt in the absence of light. The net outcome of the dark reactions is that atmospheric carbon dioxide is "fixed," or incorporated, into plant organic material as the sugar glucose. Figure 17.5 summarizes the overall process of photosynthesis.

Glucose can be used immediately as an energy source in cellular respiration, or it can instead be converted into sucrose, fructose, starch, or other compounds for long-term energy storage. If the energy is later needed by the plant at a time when photosynthesis is not possible or in a nonphotosynthetic part of the plant, storage compounds in the plant can be converted back into glucose and broken down using cellular respiration to supply energy. Table sugar is sucrose, the storage product found in sugar cane and sugar beets. Starch is the storage product in potatoes and cereal grains. In addition to these carbohydrate storage molecules, many plants (including corn, palm, and most nuts) use oils (see Section 8.2.2) as storage products in their seeds; the energy stored in seeds is used when the seed germinates. Growth of the new plant depends on energy from the seeds until enough new leaves have been produced to carry on photosynthesis for themselves.

17.1.3 Agriculture and other plant uses

Agriculture, meaning the cultivation of plants as food for humans, is often regarded as the one critical achievement that led also to the establishment of human civilization. The cultivated plants most important in the development of civilization were the cereal grains (wheat, rice, corn, oats, and others). Today, wheat, rice, corn, and potatoes provide more of the world's food than all other crops combined. Of the almost 250,000 species of plants known, some 80,000 are edible by humans. Of these, however, only about 30 form the major crop plants.

People also use plants as sources of beverages, flavorings, fragrances, dyes, poisons, decorations, building materials, and medicines. Beverages made from plants include coffee, tea, cola, beer, wine, spirits, and many juices. Many plant parts are used as spices, fragrances, and flavorings, including barks such as

Table 17.1 A few of the many vascular flowering plants used as spices and fragrances

Plant taxon	Plant name	Plant use	Plant part used
Class Dicotyledonae			
Mint family	*Mentha* spp.	Mint	Leaves or essential oil
	Ocimum basilicum	Basil	Leaves
	Origanum vulgare	Oregano	Leaves
Pepper family	*Piper nigrum*	Pepper	Seeds
Nightshade family	*Capsicum frutescens*	Chili peppers, paprika	Fruit
Laurel family	*Cinnamomum* spp.	Cinnamon	Inner bark
Myrtle family	*Eugenia aromatica*	Cloves	Whole, unopened flower buds
Ginger family	*Zingiber officianalis*	Ginger	Roots
Mustard family	*Brassica nigra*	Mustard	Seeds
Rose family	*Rosa* spp.	Rose	Essential oil
Class Monocotyledonae			
Lily family	*Allium sativum*	Garlic	Bulbs
Iris family	*Crocus sativus*	Saffron	Petals
Orchid family	*Vanilla* spp.	Vanilla	Immature seed capsule

Notes: In many cases, the plant part used is dried and ground into a powder. The abbreviation spp. means that various species in the same genus are used.

cinnamon, seeds such as black pepper, roots such as ginger, and flowers or flower parts such as cloves, saffron, and vanilla (Table 17.1). Often, an essential oil or other ingredient is squeezed or extracted from the appropriate plant part and used in concentrated form. Some of these extracts are used as fragrances or food ingredients; others are used as animal poisons. For example, roots containing rotenone are used by native South Americans to help in capturing fish by temporarily paralyzing them. Rotenone is also used as a pesticide because it paralyzes and kills insects. Manioc, a tropical root, is the source of both an arrow poison and tapioca, a multipurpose food that also serves as a thickening agent.

Wood is a commercially important plant product used the world over as a fuel and as a building material in the form of lumber. The history of human civilization would have been very different without spears, axes, hoes, boats, houses, and many other objects made mostly of wood. Also, paper and paper products are made largely from wood.

Our modern arsenal of prescription medicines is derived largely from plant products, many of them tropical (Table 17.2). Many of these drugs are now

Table 17.2 A few of the many medicinal plants

Plant		Drug	
Latin name	**Common name**	**Name**	**Use**
Digitalis purpurea	Purple foxglove	Digitalis	Strengthens heart contractions
Rauwolfia serpentia	India snakeroot	Reserpine	Lowers blood pressure
Atropa belladonna	Deadly nightshade	Atropine	Blocks neurotransmitters, antispasmodic
		Belladonna	Blocks neurotransmitters, antispasmodic
Datura spp.	Jimson weed (thorn apple)	Scopolamine	Sedative, controls nausea
Papaver somniferum	Opium poppies	Codeine	Cough suppressant, painkiller
		Morphine	painkiller
Cinchona ledgeriana	Cinchona tree bark	Quinine	Malaria treatment
Catharanthus roseus	Madagascar rose periwinkle	Vinblastine	Cancer chemotherapy
		Vincristine	Cancer chemotherapy
Taxus brevifolia *Artemisia annua*	Pacific yew Sweet wormwood	Taxol Artemisin	Cancer chemotherapy Malaria treatment

manufactured synthetically but were originally derived from plants. Aspirin, for example, was originally derived from the bark of willow trees (*Salix* spp.). Willow bark, or tea-like infusions made from willow bark, were used to treat aches and pains for centuries by the Greeks and by Native Americans, among others. Many herbal medicines continue to be used in many parts of the world.

THOUGHT QUESTIONS

1. Consider the value judgments that may be hidden in the term *useful* (as in "useful products"). Are plants that are useful to humans more valuable than plants that are not? Is utility the only criterion for making something valuable? Is economic value the only way of measuring value? What might some of the other ways be? How would you measure them?
2. What are some ways in which large plants (e.g., trees) are used by other plants? What are some ways in which plants are used by animals? List as many possible uses as you can.
3. Do animals have any molecules that can absorb light energy? Where might you expect to find them?
4. The sun is a renewable, free energy source. Humans cannot photosynthesize, but in what other ways have they made use of the sun's energy? What other possible uses can be developed in the future?

17.2 PLANTS USE SPECIALIZED TISSUES AND TRANSPORT MECHANISMS

In most plants, photosynthesis is carried out principally in the leaves. Other plant parts are also specialized for particular functions. Roots are specialized for water absorption and fruits for reproduction and dispersal (see Section 6.4.2). The division of labor among different parts of the plant would not be possible without the specialization of plant tissues, nor would it be possible without efficient mechanisms for the transport of materials from one part of the plant to another. These adaptations are described in Section 6.4.2, but a brief review is presented here.

The majority of plants are vascular plants that possess separate organs such as roots, stems, and leaves. The roots absorb water and nutrients from the soil. The leaves carry out photosynthesis and produce sugars and other carbohydrates. Gas exchange in the leaves brings in CO_2 and releases both oxygen and water to the atmosphere. The vascular tissues, especially in the stems, conduct water upwards from the roots and distribute photosynthetic products to all parts of the plant. The vascular tissues are strong enough to support the weight of the plant and allow it to grow very tall—up to 80 m (250 ft.) in some tropical trees.

17.2.1 Tissue specialization in plants

In many algae and simple plants, each cell carries out all essential functions, including photosynthesis, absorption of nutrients, and excretion of waste products. More complex plants, like animals, have groups of similar cells organized into tissues, and groups of tissues organized into organs such as leaves and roots. For example, each leaf is an organ, while each cell layer within a leaf is a tissue. The simplest plants containing separate types of tissues are the mosses, liverworts, and hornworts, often grouped as Bryophyta and known as nonvascular plants.

The most familiar and ecologically dominant group of plants are the **vascular plants**, including all plants that have vascular (conducting) tissues that transport fluids, generally through tubular cells surrounded by rigid cell walls. Vascular plants contain two types of conducting tissue: xylem, which usually conducts water and minerals upward, and phloem, which conducts a water solution of photosynthetic products (mostly sugars) in both directions but more often downward. The existence of these vascular tissues allows the parts of the plant to specialize (Figure 17.6).

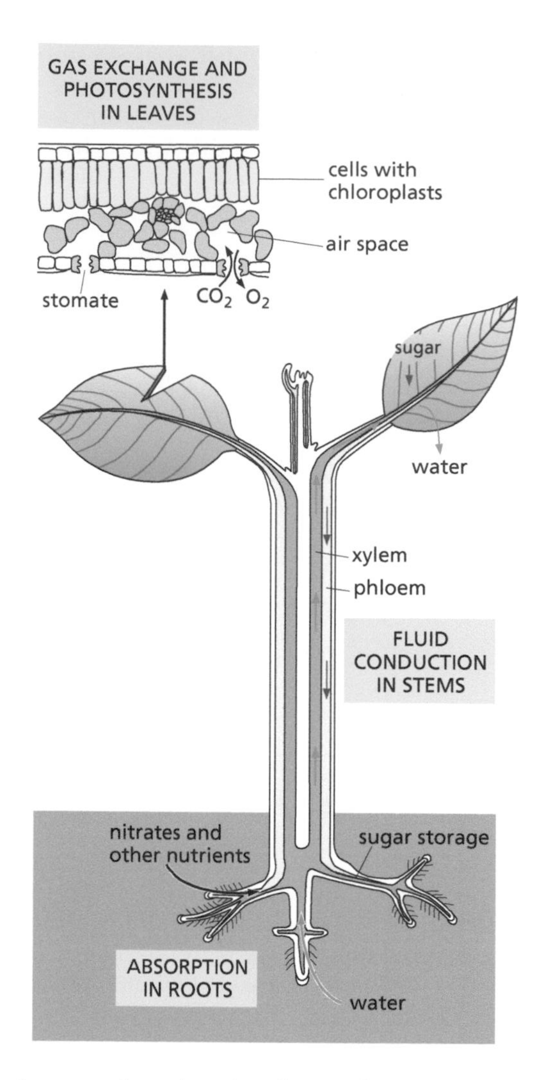

Figure 17.6 Different functions carried out by specialized parts of vascular plants. Detail at the top shows a cross-section through a leaf.

The roots grow underground, anchor the plant in the soil, and absorb water and dissolved nutrients. The vascular tissues conduct the water and nutrients from the roots through the xylem of the stem to the above-ground parts, where the water is needed for both photosynthesis and support of the upper parts of the plant. The roots can receive photosynthetic products through the phloem from the chlorophyll-containing tissues above. Roots thus need not contain chlorophyll or carry out photosynthesis themselves, so they are not green and do not require light.

Vascular tissues have rigid cell walls, and, as we see in the next section, water pressure makes plant tissues even stronger as well as helping plants to stand upright. Thus, in addition to transport, these adaptations in stems allow many vascular plants to grow tall without having the skeletons that support vertebrate animals.

17.2.2 Water transport in plants

Water moves over great distances within plant bodies. Plants would not be able to absorb nutrients, conduct photosynthesis, synthesize chemical compounds, or get rid of waste products if water were not able to move in and out of plant cells and through the plant as a whole.

17.2.2.1 Osmosis Like all cell membranes, plant cell membranes stay intact only because they are surrounded by water. Lipid bilayer membranes are impermeable to ions, yet are permeable to water. This allows water to move by a process called **osmosis** whenever opposite sides of the membrane contain solutions at different concentrations. We usually think of the concentration of a solution in terms of the number of dissolved ions or other particles in a given volume of water. We have seen that a membrane can store chemical potential energy in the form of differences in ionic concentration on the membrane's two sides. But we can also think of the solution in terms of the concentration of water, that is, the numbers of molecules of water per volume of solution: the higher the concentration of ions and other dissolved materials, the lower the concentration of water. All substances tend, according to the laws of physics, to diffuse from a region of higher concentration to a region of lower concentration. Osmosis is a special type of diffusion in which water passes through a membrane from a region of high water concentration (and low ion concentration) to a region of low water concentration (and high ion concentration), while the ions are prevented by their charge from moving through the membrane. There is no attraction or repulsion in the process of osmosis, just a flow of water molecules from an area where their number is higher to an area where their number is lower (Figure 17.7).

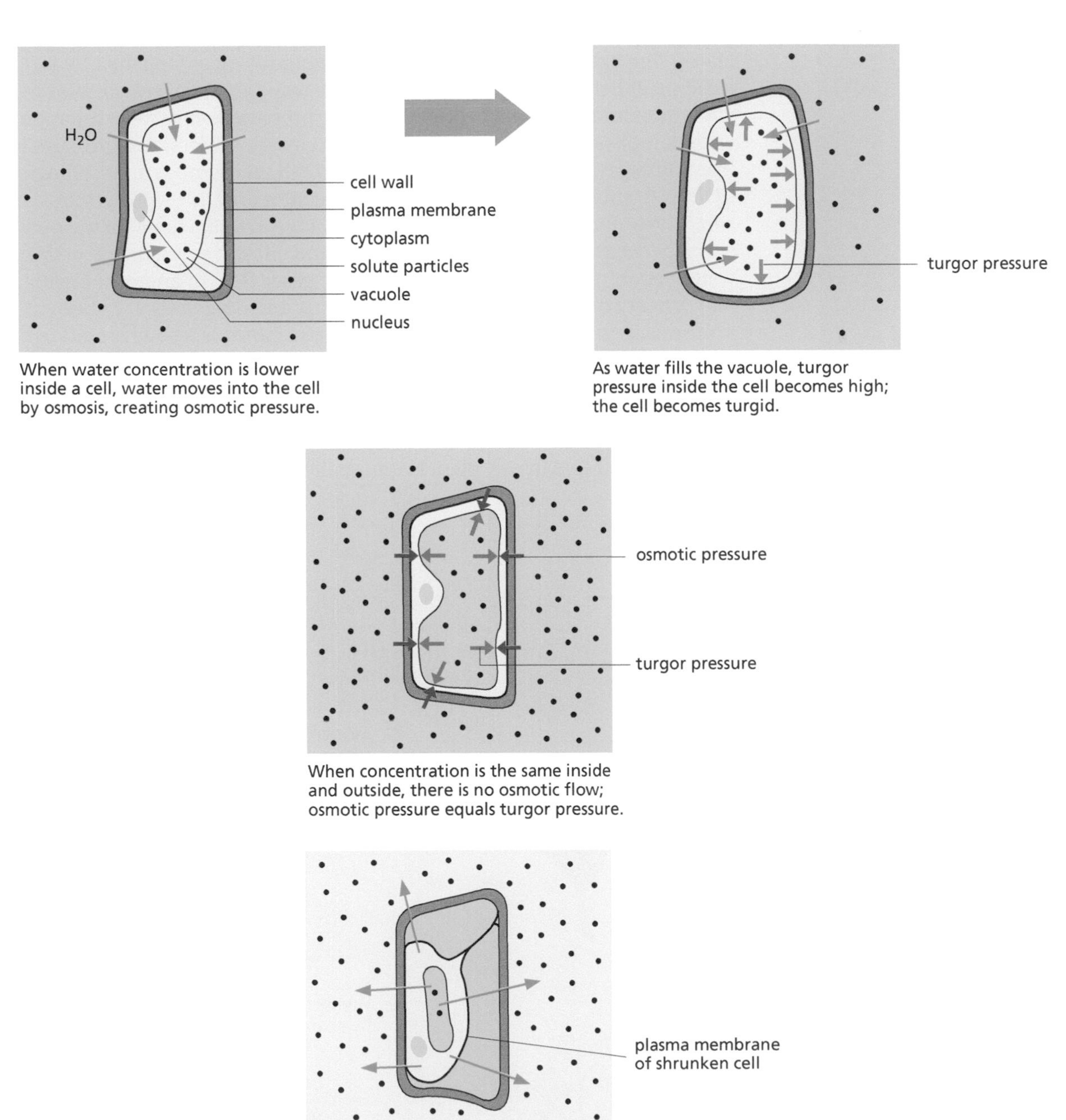

Figure 17.7 **Osmosis in plant cells.** Water moves across cell membranes toward the side with the lower water concentration and higher ion concentration.

Plant cells are surrounded by a rigid cell wall made of cellulose, which lies outside the plant cell membrane. When water flows into plant cells by osmosis, the cell membrane eventually pushes against this cell wall, producing a form of fluid pressure, called **turgor**, that makes plant tissues stiff (see Figure 17.7). It is thus water pressure, rather than a skeleton, that keeps nonwoody plants upright. In woody plants, cell walls are further stiffened with a **lignin**, a rigid material that provides additional support.

17.2.2.2 Water transport A plant moves water from its roots to all of its cells. The rate of water movement is greatest when the plant is in sunlight, a time when the water loss by evaporation from the leaves (called **transpiration**) is greatest. Transpiration lowers water pressure in the leaves, and the higher pressure from below moves the water upward.

Transpiration is controlled by openings called **stomates** in the undersides of leaves. The opening and closing of the stomates is accomplished by large "guard cells" on either side, which change shape in response to increasing and decreasing water pressure (**Figure 17.8**). When water pressure is low (which happens when water supplies are low), the guard cells are closed. When water pressure increases in adjoining cells, ion channels in the guard cell membranes open, allowing K^+ and Cl^- ions to enter the guard cells. Water molecules flow by osmosis, increasing pressure in the guard cells and resulting in a change in their shape. In the new shape, a space between the two guard cells opens and allows the exchange of CO_2, O_2, and water vapor. When the stomates are open, gases flow in and out: Carbon dioxide (CO_2), needed for photosynthesis, is taken up, and oxygen (O_2), a product of photosynthesis, is given off (see Figure 17.6). Water vapor is also given off because the water concentration of the air is lower than that of the xylem.

Transpiration also serves to cool the plant. Some of the light energy absorbed by plants is transformed into heat and cannot be used in photosynthesis. Plants avoid overheating by evaporating water from their leaves through transpiration. Most of the water absorbed by any plant is eventually lost through the leaves in this way.

Figure 17.8 Opening of a stomate. For an animated view of this process, please go to www.biologytrending.routledge.com/chapter17

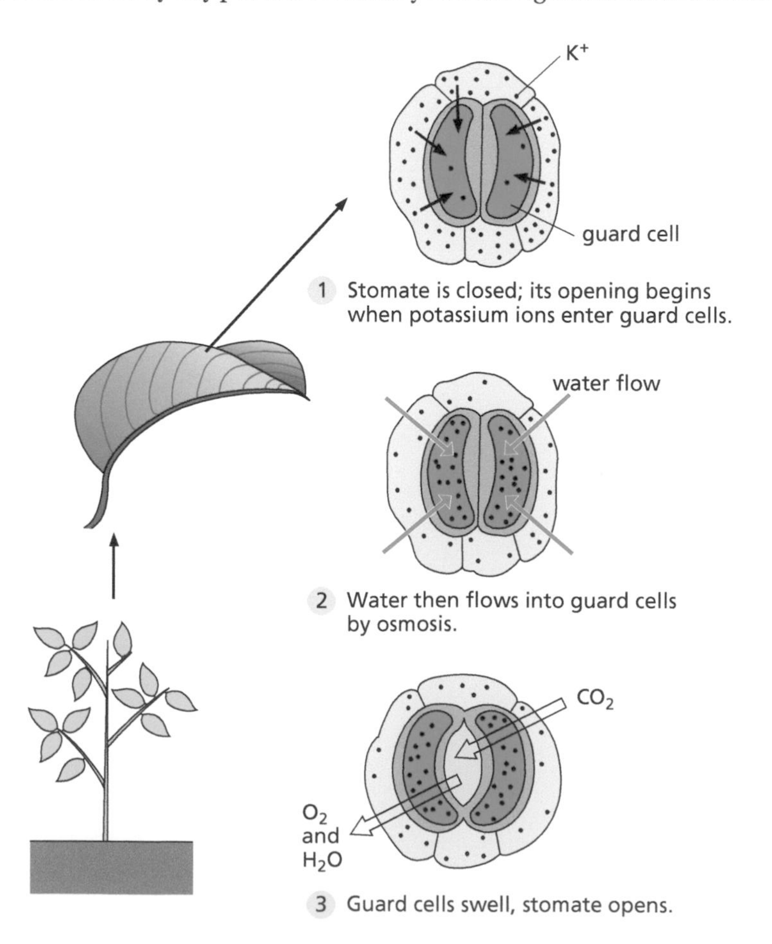

When this sensitive plant (*Mimosa pudica*) is touched, the leaflets fold together within a fraction of a second.

When an insect touches the sensitive hairs of this Venus fly trap, the leaf halves snap together in less than half a second, trapping the insect.

Figure 17.9 **Rapid and selective movement in plants.**

17.2.2.3 Rapid movement in plants While animals can use nerve cells to stimulate movement through muscle contraction, plants have neither muscles nor nerves and thus have only limited powers of movement. Some plants, however, can move their parts quite rapidly; their ability to regulate turgor pressure is the key to how plants such as the Venus fly trap (*Dionaea muscipula*) and the sensitive plant *Mimosa pudica* are capable of rapid movements (Figure 17.9). Touching one of these plants causes ion channels in the tissue touched to open and ions to flow out from the cells. Water flows out in response, and turgor drops quickly. The cytoplasm of a plant cell is connected to the cytoplasm of adjoining cells through small openings. The changes of ion concentration are thus passed along to the next cell, where ion channels are stimulated, passing along the change in turgor. When turgor drops in the cells where a *Mimosa* leaflet joins the stem, the leaflets collapse towards each other (see Figure 17.9). When changes in turgor have propagated to the base of the leaf stem, which may take more than one touch, the whole leaf droops. The plant will return to its former position when water flow reverses, reestablishing turgor pressure. Thus, in plants, movement can be accomplished without contractile muscle fibers.

THOUGHT QUESTIONS

1. Do plants have sense organs?
2. How can plants move without nerves and muscles?
3. How are plants able to stand upright?
4. Why can plants grow without soil but not without water?

17.3 NUTRIENT MATERIALS CIRCULATE THROUGH THE WORLD'S ECOSYSTEMS

Recall that living species and their physical surroundings make up ecosystems and that energy passes through ecosystems in only one direction and is ultimately lost as heat. Materials, however, are naturally recycled throughout ecosystems, and each chemical element has its own cyclical pattern.

17.3.1 Carbon and oxygen cycles

Carbon and oxygen circulate through biological systems in rather simple patterns. Atmospheric oxygen, for example, combines with organic compounds in all organisms during the cellular respiration described in Section 8.3.3. Oxygen is released to the atmosphere during photosynthesis.

The carbon cycle is also simple: Carbon is incorporated (or "fixed") into organic compounds during photosynthesis, and it is released to the atmosphere as CO_2 as an end product of both respiration and decomposition. Other chemical elements that plants need to make their essential biological molecules can be obtained from water, which serves as a source for hydrogen, oxygen, and many dissolved ions.

17.3.2 Nitrogen for plant products

Compared to carbon, nitrogen follows a more complex cycle, and plants have many uses for nitrogen. DNA, RNA, and amino acids all contain nitrogen, and the amino acids are also used to make proteins and many other compounds. Amino acids can be made from simpler compounds by the addition of an amino group ($-NH_2$), which can be supplied from soluble ammonium compounds containing the NH_4^+ ion. Amino groups can also be transferred from one amino acid to another. Amino acids are the starting materials for all the other nitrogen-containing compounds, including proteins, DNA and RNA, vitamins, plant hormones, and pigments. Plants also make bitter-tasting nitrogen compounds called alkaloids as a chemical defense against the insects and other animals that might feed on the plants; examples include caffeine, codeine, cocaine, nicotine, quinine, and scopolamine. All alkaloids are poisonous in high concentrations, but, in low concentrations, some have proven useful as medicines.

17.3.2.1 Nitrate: A limiting nutrient Although plants need a source of amino groups ($-NH_2$) to help make amino acids and proteins, most plants cannot absorb NH_4^+ ions directly. Instead, they get their nitrogen from the soil as dissolved nitrates (NO_3^- ions) or nitrites (NO_2^- ions), which they then convert into ammonia (NH_3) for immediate use. Ammonia quickly reacts with water to form ammonium hydroxide (NH_4OH), the source of the NH_4^+ ions needed for amino acid synthesis. However, a plant does not accumulate any more ammonia than it immediately uses because excess unused NH_4OH is very alkaline and damaging to most organic tissues.

Many plant species are limited in where they can live by the availability of dissolved nitrates. A nutrient whose absence makes a species unable to grow in a particular place or on a particular food source is called a **limiting nutrient**. If the amount of the limiting nutrient were increased (by the use of fertilizers, for example), more growth would take place, assuming that other nutrients were present in adequate amounts. For many plants in many places, nitrates are a limiting nutrient. In order to understand how plants get their nitrogen, we need to take a global view.

17.3.2.2 The nitrogen cycle Nitrogen moves through the world's ecosystems, changing from one molecular form to another in a cyclical flow called the **nitrogen cycle**. Plants and other living organisms are an important part of this cycle: Chemical end products released by one type of organism are used in biochemical reactions by other organisms. Some of these organisms can fix (incorporate) atmospheric nitrogen into molecular forms that other organisms can use, while other organisms release nitrogen as a gas into the atmosphere. In each ecosystem, living organisms are thus united with each other and with their physical surroundings (including the atmosphere) by the nitrogen cycle. The nitrogen cycle is a good example of a nutrient cycle in which materials are exchanged among producers, consumers, decomposers, and their surroundings, including the atmosphere.

The key stages of the nitrogen cycle are shown in Figure 17.10. First, follow the outer circle, starting with free nitrogen (N_2) in the atmosphere. Earth has an abundant source of nitrogen in nitrogen gas (N_2), which makes up about 78% of our atmosphere. Plants cannot use nitrogen directly from the atmosphere. They are therefore dependent upon certain prokaryotic microorganisms that have the ability to convert atmospheric nitrogen into ammonia (NH_3) in the process called nitrogen fixation. Several kinds of nitrogen-fixing organisms live in soil, including

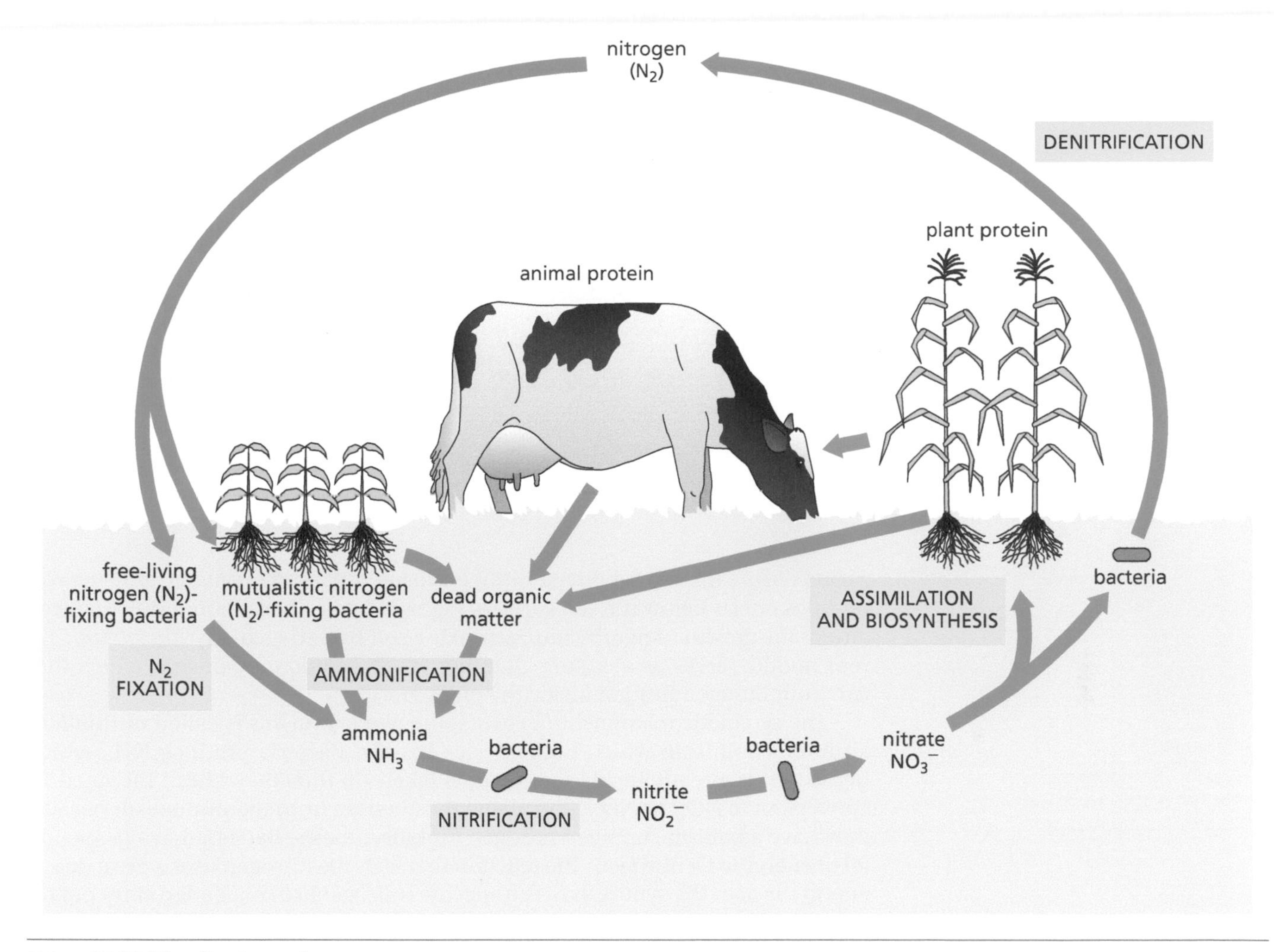

Figure 17.10 The nitrogen cycle.

cyanobacteria such as *Nostoc* and many nonphotosynthetic bacteria such as *Rhizobium*, *Azotobacter*, *Klebsiella*, and *Clostridium*.

Ammonium compounds produced by nitrogen-fixing bacteria in the soil can be converted into nitrites and then into nitrates by still other bacteria such as *Nitrosomonas* and *Nitrobacter* in a process called nitrification. Plants can absorb both nitrites and nitrates, then convert these ions back into NH_3 (only a little at a time) to be used in the biological synthesis of amino acids, proteins, and other nitrogen-containing compounds.

The nitrogen cycle is completed by still other soil bacteria that take excess nitrate and convert it into molecular nitrogen (N_2), which is then released into the atmosphere as a gas. In a balanced nitrogen cycle, the amount of nitrogen released as a gas equals the amount converted into nitrogen compounds.

Nitrogen also cycles through other organisms. Animals get their nitrogen by eating plants or other animals. When plants and animals die or give off waste products, their nitrogen is returned to the soil by still other species of bacteria that convert these waste products into ammonia. This process (called ammonification) is a second source of the ammonia needed for making nitrates. The cycling of nitrogen from ammonia to nitrites, nitrates, plant proteins, animal proteins, and back to ammonia forms a subcycle within the nitrogen cycle, as shown by the inner circle in Figure 17.10.

17.3.3 Mutualistic relationships

As we have seen, ammonia can enter the soil via free-living nitrogen-fixing bacteria or by the breakdown of dead organic material. A third source of soil ammonia is bacterial nitrogen fixation within the roots of some plants. Some species of

Figure 17.11 Underground nitrogen fixation in root nodules.

plants ensure the availability of nitrogen-fixing microorganisms by growing root nodules not far below the soil surface (**Figure 17.11**). These root nodules actively attract the growth of nearby nitrogen-fixing soil bacteria, chiefly *Rhizobium*. The root nodule serves as a culture chamber for these microorganisms, which then carry out nitrogen fixation inside the plant roots.

The symbiotic relationship between the two organisms is called **mutualism**, an interaction from which both species benefit. Bacteria produce NH_3, which, because it is already inside the plant, is taken up directly without the need for conversion to NO_2^- or NO_3^-. The plants thus benefit from the mutualism because they have a built-in supply of nitrogen for biosynthesis. Bacteria have to spend a lot of energy to fix nitrogen. The reaction that fixes the nitrogen uses a great deal of energy, as does the synthesis of the enzyme (nitrogenase) needed to carry out the reaction. The bacteria benefit from the mutualism because they get the energy for both enzyme synthesis and nitrogen fixation from the plant in the form of organic compounds.

Plants of the family Leguminosae, including beans, peanuts, peas, locusts, and alfalfa, are all capable of adding nitrogen compounds to the soil because of their mutualistic association with nitrogen-fixing *Rhizobium* bacteria. This is the basis for the use of "green manure," the planting and subsequent plowing under of a legume crop in a field that has been depleted of nitrates by the earlier growing of plants (such as corn and wheat) that absorb large quantities of nitrates from the soil. When the legume is plowed under, the nitrogen compounds that it has produced with the help of its mutualistic nitrogen-fixing bacteria are returned to the soil in a form that other bacteria can convert to nitrates. Green manure thus adds nitrogen compounds to the soil, reducing or eliminating the need for the addition of fertilizers containing nitrate or ammonia.

Most plants do not have mutualistic nitrogen-fixing bacteria living in their roots. They therefore need to absorb their nitrogen compounds from the soil, principally as nitrates. A few plant species, however, have evolved other means of obtaining nitrogen.

17.3.4 Plants living in nitrogen-poor soils

Plants that have evolved to live in nitrogen-poor soils have a number of different ways of coping with the scarcity. One rather unusual solution to the problem of obtaining nitrogen is to digest animal proteins. Few plants, however, can be so fortunate as to have an animal die and leave its carcass in the soil just within reach of their root system. How, then, are they to obtain animal protein? Carnivorous plants have evolved adaptations to trap and kill small animals (mostly insects) and derive nitrogen from the digested proteins. Most carnivorous plants live in such nitrogen-poor habitats as acid bogs, where moisture and insects are both

Figure 17.12 Plants that obtain nitrogen from animals.

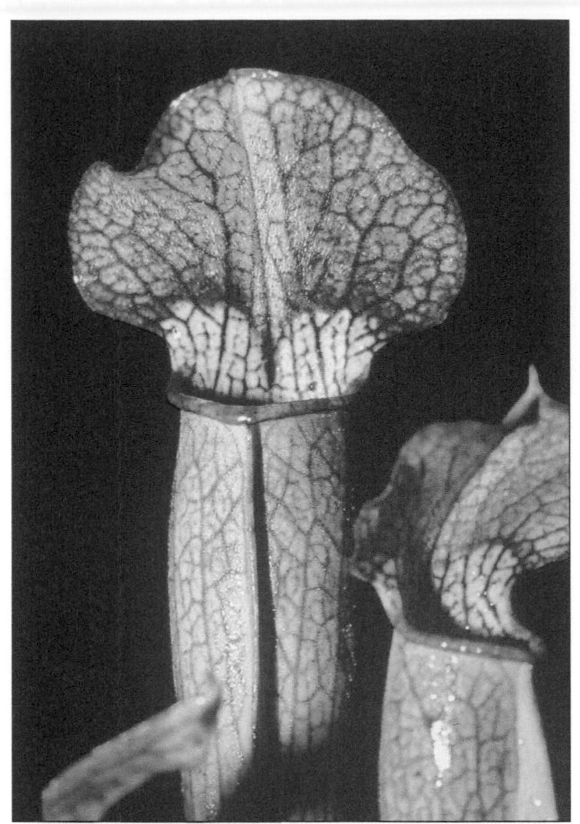

Pitcher plant (*Sarracenia oreophila*), showing a passive trap (pitfall type), with slippery inside surfaces and fluid pool at bottom containing digestive enzymes.

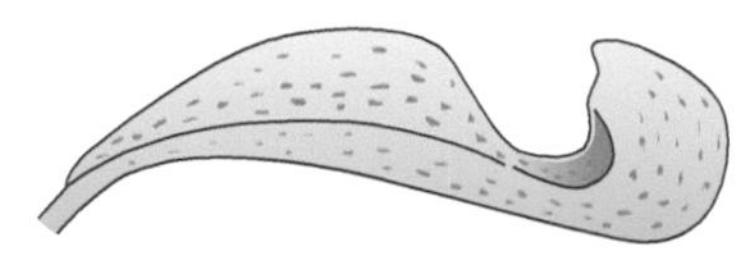

Sarracenia psittacina, a pitcher plant with two passive traps. The entrance works like a lobster trap because insects that have entered the chamber have difficulty finding the opening by which they came in. When they crawl into the long tube, the hairs inside form a second trap that allows them to crawl deeper in only one direction.

The sundew plant, *Drosera*, whose sticky hairs close over an insect, forming an active trap (flypaper type).

usually abundant. A variety of mechanisms have evolved by which these plants trap their prey, digest their proteins, and absorb the resulting amino acids. The traps can either be passive or active (Figure 17.12). Passive mechanisms can include pitfalls (as in pitcher plants), sticky materials that form a passive flypaper, and traps shaped like lobster pots, from which victims find it hard to exit once they have entered. Active mechanisms can include trapdoors, active flypapers (such as the sundew, whose hairs bend to enclose and further hold their victim), or pads that swing shut when an object touches them (such as the Venus flytrap, which uses a mechanism also shown in Figure 17.9).

THOUGHT QUESTIONS

1. Why do you think plants can take up NO_3^- or NO_2^- ions but not NH_4^+ ions?
2. Some plants have mutualistic associations with bacteria in root nodules. Do humans also have mutualistic associations with bacteria?
3. What is the definition of acidity? How does the uptake of NH_4^+ ions by some plants make the soil around them more acidic?

17.4 CROP YIELDS CAN BE INCREASED BY OVERCOMING VARIOUS LIMITING FACTORS

In order to feed the world's growing human population, we must increase crop production. The amount of land under cultivation is, however, limited, and most efforts to increase yields of crop species have therefore focused on such techniques as soil improvement, fertilizers, irrigation, and pest control. Each of these methods seeks to increase yield in a different way: by supplying a limiting nutrient, by supplying water (which is often limiting in the same way), or by controlling natural enemies of the crop species. Moreover, these improvements must be carried out sustainably, that is, without lasting harm to the ecosystem that supports the growth of crop plants. (In a later section, we will also examine the development of new strains of crop plants.)

17.4.1 Fertilizers

In those locations where a nutrient is a limiting factor for the growth of plants, crop yields can often be dramatically increased by adding the appropriate fertilizer to the soil. A **fertilizer** is a substance that supplies organic or inorganic nutrients needed by plants. In the largest number of locations, nitrogen is the limiting nutrient for many plant species, and nitrates are among the most important fertilizers used in agriculture. In many places, crop yields can be dramatically increased by adding nitrates to the soil. Phosphorus (P), in the form of phosphates (PO_4^{3-}), can also be an important limiting nutrient, especially for ornamental flowers or crops whose edible portion is a seed, a flower, or a fruit. Phosphorus often signals the plant to begin setting flowers, seeds, and fruit. Potassium (K^+) is often important as well. It is used by plants to control the opening and closing of stomates (see Figure 17.8), and it may be scarce (and therefore limiting) in certain soils. The hazards as well as the benefits of applying fertilizers need to be considered.

Fertilizers can be either organic, including various manures and composts, or inorganic. The nutrients in organic fertilizers are complex molecules that break down slowly. The slow breakdown gives these fertilizers the advantage of slow release, a process that provides a steady supply of soluble nutrients that is more likely to be absorbed by growing plants than to be washed away as are many inorganic fertilizers. Because organic fertilizers contribute to the long-range health of the soil, they are preferred by the many farmers (and consumers) that make up the backbone of the "organic food" movement. One disadvantage of organic fertilizers is their high bulk and cost of transportation, features that often limit their use to the immediate locality of their production.

17.4.1.1 Manures Waste products from domestic animals are used all over the world as fertilizers, and these animal manures are often readily available near where crops are grown. In China, human and animal wastes have been used as fertilizers for centuries. This practice provides needed nutrients to crops, but it also has the undesirable effect of spreading parasites (especially flukes of the animal phylum Platyhelminthes) and other infectious diseases. In some countries, animal wastes are treated to kill parasites or to reduce water content (and therefore transportation costs).

We have seen that plants can accumulate nitrogen compounds. Green manure consists of the remains of plants that are plowed under at the end of the season, providing organic fertilizers without need of transportation. For example, grasses in a field previously used as pasture may be plowed under. Some plants are grown only to be plowed under; more often, some part of the plant is harvested, and the remainder is plowed under. Lima beans, for example, may be mechanically harvested and the roots and stems then plowed under. Alfalfa is a legume commonly grown to provide both food for animals and then added nitrogen to the soil when it is plowed under.

Many farmers plant a certain fraction of their fields with plants that accumulate nitrogen; the location of the various plants is then rotated from year to year (a practice called crop rotation), giving each field a turn to accumulate nitrogen in the soil. The alternation of cereal grains with soybeans or alfalfa (both legumes) is a form of crop rotation that contributes to high agricultural productivity in many large regions of North America and Asia. Organic farmers use green manuring, sometimes in combination with the use of various animal and plant wastes as fertilizer. The benefit of crop rotation for maintaining the fertility of the soil was recognized in ancient Rome and several other ancient civilizations.

17.4.1.2 Composting Organic material (leaves, grass clippings, food waste) can be layered with manure or soil. The mixture provides a place for microorganisms to live, and they aid in the decomposition of the plant and manure layers. Many U.S. municipalities encourage composting, either by teaching people how to compost in their own backyards or by having centralized composting facilities to which people can bring their yard wastes. Even so, less than 1% of organic wastes are composted in the United States, compared to much higher rates in Europe. It is estimated that solid waste in U.S. landfills could be reduced by 20% if organic material were composted.

17.4.1.3 Chemical fertilizers Inorganic chemical fertilizers are usually nitrate or phosphate salts sold as powders or granules to be applied to fields, usually by mechanical equipment. When they were first introduced, chemical fertilizers were cheap and readily available. The benefits in increased crop production were immediately evident. Bountiful harvests created surplus crops, and the use of chemical fertilizers increased greatly in countries that could afford them.

Ecological problems associated with the prolonged use of inorganic chemical fertilizers became evident only after several decades of general use, in accordance with what some call the "law of unintended consequences." This has also been stated as the first law of ecology: Any change imposed upon nature has a number of effects, many of which are unpredictable, so "you can never change just one thing." For example, excessive use of inorganic fertilizers can kill off the soil organisms required for maintaining soil fertility, thus requiring the use of even more fertilizer for subsequent crops.

Compared with organic fertilizers, inorganic minerals are relatively expensive to produce or extract, and most of them must be transported over great distances. Cheap local supplies have in most cases been depleted, and many are now mined far from the places where they are used. For example, large quantities of nitrate minerals are mined in Chile for use as fertilizers in the Northern Hemisphere. Fossil fuels are used in the mining and transportation of fertilizers and by the tractors used to spread these fertilizers on fields. Most chemical fertilizers are costly to transport to where they are needed, and in most of the developing nations of the world they are prohibitively expensive. Organic fertilizers are often available more locally, and therefore need not be transported as far.

17.4.1.4 Sources of phosphorus Fertilizers are often needed to supply phosphorus. Important nutrients, including both calcium and phosphates, are slowly released from fish meals and bone meals, which are powdered, dried bone. The phosphates provided by these fertilizers are insoluble; they are converted to soluble form only slowly, often by the very plants that use them. If the practice is begun in soil that is nutritionally deficient, the slow solubility is a disadvantage at first; chemical fertilizers may therefore be necessary during the initial phase-in period. Fish or fish meals also contain additional phosphorus in the form of nucleic acids from the nuclei and organic phospholipids from the membranes of the cells of the dead animal. Native Americans throughout the eastern woodlands traditionally grew their corn in mounds, under which they customarily buried a fish that provided both phosphate and nitrogen as fertilizer for the corn.

The presence of phosphorus in fertilizers, or of any fertilizer in soluble form, is often associated with problems of runoff and stream pollution. Fertilizers are concentrated nutrient sources for all plants, not just for the crop plants for which they are intended. When fertilizers run off from agricultural fields into bodies of water, they supply nutrients to algae in the water. The growth of the algae may have previously been held in check by the lack of some limiting nutrient, such as phosphorus. Once that nutrient is present, the growth of these algae can be so rapid that they use up all the oxygen in the water, causing the death of many fish. The decomposition of dead fish releases more nutrients that further accelerate the growth of the algae. This process of nutrient enrichment, algal growth, and oxygen depletion is called **eutrophication**. Eutrophication led to the deaths of most fish, and the commercial fishing industry, in Lake Erie in the 1960s. International regulations that cut down on runoff pollution have allowed the lake to recover to a great extent. Eutrophication remains a problem in more than half of the lakes in the United States (**Figure 17.13**).

The problems of runoff and eutrophication can originate from animal wastes (e.g., from dairy or hog farms or from human sewage) or from inorganic chemical fertilizers, especially soluble nitrates and phosphates. Laundry detergents containing phosphates were also a major source of nutrients, but public awareness pressured manufacturers into changing their formulations, so that many detergents are now phosphate-free. The amount of rainfall and its seasonal distribution can greatly influence runoff, as can human factors such as pavement near the edges of bodies of water. Runoff can sometimes be minimized by applying fertilizers at the proper time of year, just before the nutrients are most needed by the plants and are most likely to be absorbed.

Figure 17.13 Eutrophication: Algal blooms accelerated by fertilizer runoff. The two portions of this lake in Ontario were separated by a plastic curtain, and phosphates were added to one side of the curtain only. An algal bloom occurred only on the far side, to which the phosphates had been added, making the water look more cloudy.

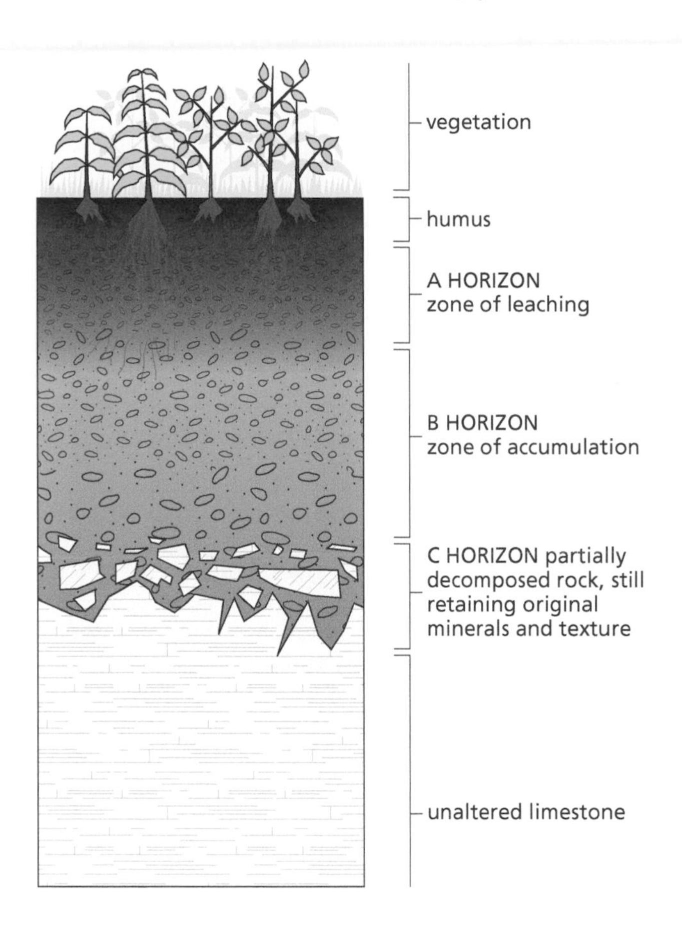

Figure 17.14 Soil formation from the weathering of rock (in this case, limestone) and from biological processes.

17.4.2 Soil improvement and conservation

Soil is the loose material derived from weathered rocks and supplemented with organic material from decaying organisms. This organic material supports the growth of plants and other species that, in turn, add to the soil-building process. The type of soil formed depends on the parent rock type, the climate, the removal of soluble materials by percolating groundwater, and the mechanical and chemical activities of living organisms such as soil microbes (both eukaryotic and bacterial), lichens, earthworms, soil nematodes, and plant roots. Soil is thus the product of both biological and geological processes (**Figure 17.14**).

17.4.2.1 Humus Organic material in the soil is broken down by bacteria and fungi, which function as decomposers. The partially decomposed organic matter is called **humus**, and it serves several important functions in the top layer of the soil. It binds the particles of weathered rock together to form small aggregates, which give the soil its structure. The small holes between particles help to hold water and oxygen and provide space for the growth of minute root hairs, extensions of single cells on the roots of plants. Although the larger roots provide the anchorage for the plant and in some plants can reach deeply into the ground in search of water, it is the root hairs that provide the enormous surface area for absorption of nutrients. Because root hairs are so delicate, they cannot push into soil, but must grow into already existing spaces in the soil structure.

Humus has an overall negative charge, which helps to hold positively charged nutrient ions such as potassium (K^+), calcium (Ca^{2+}), and ammonium (NH_4^+) in the topsoil, where they are available to plants' roots. Inorganic fertilizers can provide chemical nutrients, but they do not contribute to humus and soil texture as organic fertilizers can. The plowing under of organic matter helps to build humus. Rather than being left bare, fields can be planted with grasses during the season when cash crops do not grow, or with legumes that will be plowed under when it is time to sow the crops. This both prevents erosion and adds to humus. Humus can also be produced by composting; the decomposer bacteria from the soil turn the organic matter into humus over a period of weeks or months.

17.4.2.2 Soil as a nonrenewable resource

Topsoil is eroding faster than it is being formed on approximately one-third of the world's croplands. Soil that is lost can be regenerated by biological processes, but only very slowly—one inch of topsoil takes between 200 and 1,000 years to renew, depending on the climate and other factors. Groundwater pollution, excess fertilizer, and runoff from highway salt can kill soil microorganisms and stop the regeneration of topsoil. Excessive plowing followed by years of drought contributed to the loss of topsoil and formation of a "dust bowl" in parts of Oklahoma and neighboring states in the 1930s. Because soil cannot be regenerated within a few years, it must be treated in the same way as a nonrenewable resource. By treating soil as a nonrenewable resource, farming and other land uses are best done in ways that are sustainable and can thus continue indefinitely. **Sustainable** practices are those that lead to no net loss of a resource (Clark and Tilman, 2017; *Lancet*, 2017). Important practices in sustainable agriculture include crop rotation; use of organic, humus-building fertilizers; and the prevention of erosion.

17.4.3 Irrigation

Plants need water to supply the hydrogen ions used in photosynthesis. They also need water for transport of their nutrients and the products of their synthesis reactions. About 90% of the water absorbed by a plant is evaporated through its leaves during transpiration, dissipating excess heat. Also, the rigidity and strength of plants is based partly on the properties of water, as explained earlier.

Irrigation is the process of supplying added water, a vital substance for the growth of plants. In much of the world, crop yields are limited by a shortage of water. Water is thus a limiting resource for agriculture in most places where crops could be grown.

Irrigation is expensive in regions where water is scarce and where irrigation is therefore most needed. Traditional methods of spray irrigation lose much of the water to evaporation. Newer methods include drip irrigation, in which irrigation tubes with tiny holes are laid at or below ground level. The tiny holes, spaced every few inches, are designed to leak or drip water slowly into the soil, providing moisture but minimizing evaporation. All irrigation systems, whether drip or traditional, require a large initial investment in pipes or ditches, pumping stations, and the like, plus a supply of fresh water or desalinized water (seawater from which the salts have been removed). Freshwater rivers and lakes can be overused if too much water is diverted for agricultural irrigation. Freshwater sources can become the subject of political disputes between neighboring governments, such as those between Israel and Jordan, between Turkey and Syria, or between California and Arizona.

17.4.4 Hydroponics

Although plants cannot live without water, they can live without soil. Some plants grow naturally without soil. The growing of crop plants without soil is called **hydroponics** (Figure 17.15). In places where there is very little soil, or where the soil is unsuitable, hydroponics offers a possible alternative agricultural method. In a hydroponic system, the plants are grown with their roots immersed in tanks through which water carrying dissolved nutrients is allowed to flow. The water is recirculated (and therefore conserved), and its dissolved materials are frequently monitored and adjusted (Benton, 1983).

17.4.4.1 Advantages of hydroponics

A well-managed hydroponic system can produce greater yields than traditional soil-based systems. Because hydroponic systems are generally inside greenhouses, where they are protected from insects and other plant pests, the produce that results is free from the blemishes often caused by these pests. Much of the labor traditionally concentrated on soil care, including tilling, planting, fumigation, and irrigation, is eliminated. Water is provided to the plants directly and more efficiently, and it is recycled so

Figure 17.15 Hydroponics: Growing plants in water tanks, without soil. The water supply is often covered over, as here, to retard evaporation.

as to minimize its use. This is extremely important in arid climates, where sunshine may abound but where water is scarce. Instead of traditional application of fertilizers that can diffuse beyond the reach of plant roots, mineral nutrients are added directly to the water, where the excess not taken up by the plants remains available and can be recycled. In this way, the use of nutrient supplements is minimized, and runoff problems are also minimized. Disease and pest control can be handled easily by adding the necessary chemicals to the recycled hydroponic water, with far less danger that these substances will spread to local water supplies or to domestic animals and humans.

17.4.4.2 Disadvantages of hydroponics Many of the disadvantages of hydroponic systems have to do with cost: the initial construction costs, the costs of maintaining equipment and greenhouse facilities, the costs of nutrients, and the salaries of trained personnel for constant monitoring for nutritional problems (such as nutrient depletion) and waterborne diseases. Hydroponic systems may be in locations distant from the Equator, where the provision of artificial light and heat can add to the cost.

Plants are subject to disease, as are all living organisms; waterborne diseases are those in which the pathogen (disease-causing organism) can live in water. Even if a pathogen does not grow well in water, its ability to survive in water might enable it to be transmitted from plant to plant in a hydroponic system. Most hydroponic farms concentrate on a single plant species, a practice known as **monoculture** that brings a risk of the rapid spread of disease, whether the monoculture is hydroponic or land-based. The risk relates to the fact that many pathogens are able to infect only one or a few host species. When a pathogen encounters a monoculture, it can spread very quickly because of the proximity of other individuals of the same host species.

For these reasons, hydroponic systems are easier to justify economically for plants that are economically profitable in small quantities (e.g., pharmaceutical plants or certain vegetables) rather than for staple crops such as cereal grains. Hydroponics is potentially useful where either growing space or soil are limited, as in proposed space stations. On Earth, hydroponics works best in warm climates such as Israel or southern Italy, where light energy is abundant and artificial heating is unnecessary, or in countries such as Japan, where people are willing to pay extra for produce that is "picture perfect." In several college communities, including Worcester State University in Massachusetts and Maryville University in Missouri, the hydroponic approach is now used to grow lettuce and other vegetables for the dining halls.

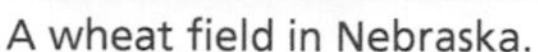
A wheat field in Nebraska.

A lettuce field in Ontario.

Figure 17.16 **Monoculture: The planting of a single species only.**

17.4.5 Chemical pest control

Each year, 30% or more of many crops are destroyed by insects and other crop pests. In addition to insects and their larvae, which damage the plants themselves, other crop pests, including various rodents, damage many crops in the field and also consume stored grains. Fungi also destroy up to 25% of stored crops. Nutrients that are consumed by pests are not available to the plant or to humans for food.

17.4.5.1 Monoculture and pests Some of the farming practices that have increased crop yields over the last few centuries have also increased the susceptibility of crops to damage by pests. For example, large, mechanized farms (farms that are heavily dependent on the use of machinery) are often plant monocultures that are thousands of hectares in size. (The hectare, or metric unit of land area, is an area 100 m by 100 m, equivalent to about 2.47 acres.) Monocultures are especially suitable for mechanized agriculture (Figure 17.16), and large-scale operations tend to be more economical than small-scale ones; as a result, monoculture is extremely profitable for a while. However, as we have seen, monocultures make it easier for a pest species to spread rapidly, especially if neighboring farms are also planted with the same crop. If neighboring fields were planted with different crops, the spread of pest species would be interrupted or made more difficult. Also, planting the same crop year after year enables pest species to survive from year to year in the form of eggs or larval stages. In contrast, planting different crops in rotation interrupts the life cycle of a pest that depends entirely on one particular crop species for its propagation.

17.4.5.2 Pesticides Any substance that kills organisms that we consider undesirable is called a **pesticide**. The ideal pesticide would (1) kill only the target species, (2) have no effect on nontarget species (including humans), (3) avoid the development of resistant genotypes in the target species, and (4) be capable of breaking down into harmless substances in the soil by natural processes in a reasonably short time period. It is hard to develop a pesticide that has all these characteristics. The ideal pesticide unfortunately does not exist.

Chemical pesticides have been used since ancient times. For example, the Romans dusted sulfur, a fungicide, on their grapes. The use of chemical pesticides, including arsenic and copper compounds, increased enormously during the nineteenth and twentieth centuries. These compounds have largely been replaced by pesticides derived from petroleum. Chemical pesticides can greatly reduce the amount of crop damage due to pests, and they continue to be used in many countries because they increase crop yields.

For much of the twentieth century, economic pressure encouraged the use of pesticides on crops and postharvest treatment with fungicides. The postharvest treatments have given many farm products longer shelf lives, allowing transportation across longer distances and a globalization of food production. As a result, people in the industrialized world have come to expect perfect, blemish-free produce at almost any time of the year, even for crops that do not grow at all in their local area.

A classic example of a chemical pesticide is DDT. First developed in the late 1930s, DDT was found to kill large numbers of different insect species. During World War II, DDT was sprayed on soldiers to control body lice and similar pests. Its effectiveness in insect control was followed by extensive spraying campaigns on crops after the war. Because it killed so many species of insects, DDT reduced crop damage and greatly increased the food supply in many countries. DDT was also sprayed onto the surface of bodies of water to control mosquitoes and other disease-carrying insects that have aquatic stages in their life cycles, thereby reducing human disease. Disease reduction and increased food supplies both contributed to population increases in many countries (see Chapter 9). The use of DDT marked the time of greatest optimism in the use of chemical pesticides. This era was brought to a close by the discovery of DDT's toxic effects on nontarget species, findings that were convincingly made public by Rachel Carson's book *Silent Spring* (1962). The term "nontarget species" is an apt one for a phenomenon that is an example of the operation of the law of unintended consequences: The target species is the one you intended to kill, and the nontarget species are those that are killed unintentionally. Often, unintentional effects are not detected immediately, not because of ill will, but because it can be difficult to predict where, when, and how the unintended effects will show up.

17.4.5.3 Negative consequences of pesticide use

There are many problems associated with the use of chemical pesticides such as DDT. The pesticides themselves are generally expensive; most of them are petroleum derivatives, and a great deal of energy is used in their extraction and further synthesis. Attempts to control pests with chemical pesticides have in several cases brought about increased levels of pest-related devastation several decades later. Pesticides like DDT are toxic to a wide variety of harmful and beneficial species alike. They may kill so many of the target species' natural enemies that the population size of the target species subsequently increases (after a time delay) above its earlier levels. Another problem with frequent pesticide use is that the target species develop pesticide-resistant mutations and are no longer killed by the spraying. Over 400 insect species, for example, are now DDT-resistant. Widespread use of the same pesticide year after year favors the evolution of pest populations with mutations that make them pesticide-resistant. Once they originate, these resistant strains of pest species spread rapidly because of selection by the pesticide itself.

As with fertilizers, there is also a runoff problem. A pesticide can find its way into groundwater supplies and from there into streams and lakes. Pesticides may contaminate drinking water supplies in this way, and they may also poison the fish in our lakes and streams.

Because of concerns over the effects of pesticide residues on aquatic ecosystems and human health, many consumers are willing to pay higher prices for food that is grown organically. In the United States and elsewhere, food can only use the term "organic" if it is grown without the use of inorganic fertilizers, chemical pesticides, or genetically modified organisms (GMOs) and if it is monitored by an independent third party to ensure compliance with these practices. When first introduced, organic foods could only be found in certain specialty stores or had to be purchased directly from the farm. Now, organic food sections are common in most major supermarkets.

17.4.5.4 Biomagnification

DDT and many other long-lasting insecticides that are applied to crops become concentrated in the bodies of the pests that eat those crops, and then further concentrated (and thus more toxic) in the bodies of animals that eat the pests, a principle known as **biomagnification**. In addition to pesticides, a wide range of other pollutants, including heavy metals such as

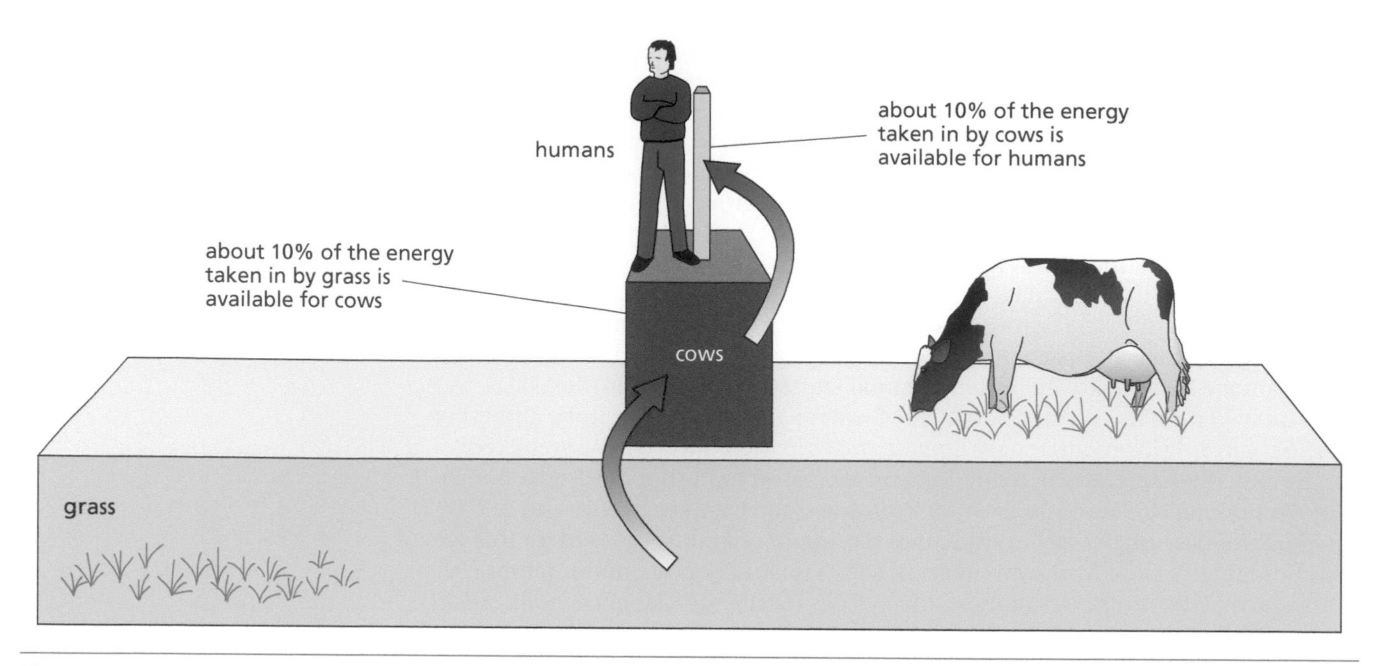

Figure 17.17 A simple food pyramid. The width of each column is proportional to the amount of energy available as food for the next higher step up the pyramid. At each energy conversion, 90% or more of the energy is lost, leaving only 10% available as food, as expressed by the widths of each step.

mercury, can become concentrated through biomagnification. To understand biomagnification, recall that producer organisms such as plants supply food energy to the primary consumers that eat them. The primary consumers, in their turn, supply food energy to secondary consumers, and so on. In each conversion, much of the food energy is lost in the form of heat, so that 10,000 kilocalories (kcal) of sunlight provided to the grass produces only 1000 kcal of grass energy to the cows that eat the grass, who in turn provide only 100 kcal of energy to the people that eat beef steak. These relations can be represented in a food pyramid (also called a trophic pyramid), as shown in Figure 17.17. Similar food pyramids could be drawn in proportion to biomass (quantity of biological tissues) or numbers of individual organisms rather than energy; most such diagrams would have basically the same pyramidal shape. The energy relations in a food pyramid are one reason why it is inefficient and wasteful to have intermediate steps between the producer plants and the top-level consumers, and thus why a certain supply of crops will support more people if the people eat the crops directly instead of eating animals that eat the crops. This basic principle of "eating low on the food chain" has major benefits for feeding the world's population by supplying plant foods directly to people instead of to the animals that people later consume. Feeding meat to people requires much more land area and more nonrenewable energy than feeding people with plants, and diets rich in plants are also much healthier.

When pesticides are used, an additional problem arises. Many pesticides are not broken down by decomposer organisms and thus persist in the environment. Also, many pesticides accumulate in biological tissues without being excreted. If a long-lasting pesticide is taken up by plants at the base of a food pyramid, the same amount of the pesticide is concentrated into a smaller and smaller amount of biological tissue with each successive conversion. The concentration of the pesticide increases with each successive step in the food pyramid as the result of biomagnification.

As a case in point, Figure 17.18 shows the biomagnification of the long-lasting pesticide DDT. In Long Island Sound, New York, the DDT concentration was measured as 0.000003 parts per million (ppm) in the water but was 25 ppm in the tissues of fish-eating birds such as ospreys and eagles, an increase in concentration of more than 8 million times.

DDT and other chemical pesticides that concentrate in fish and fish-eating birds interfere with calcium metabolism, causing a thinning of the shells of the

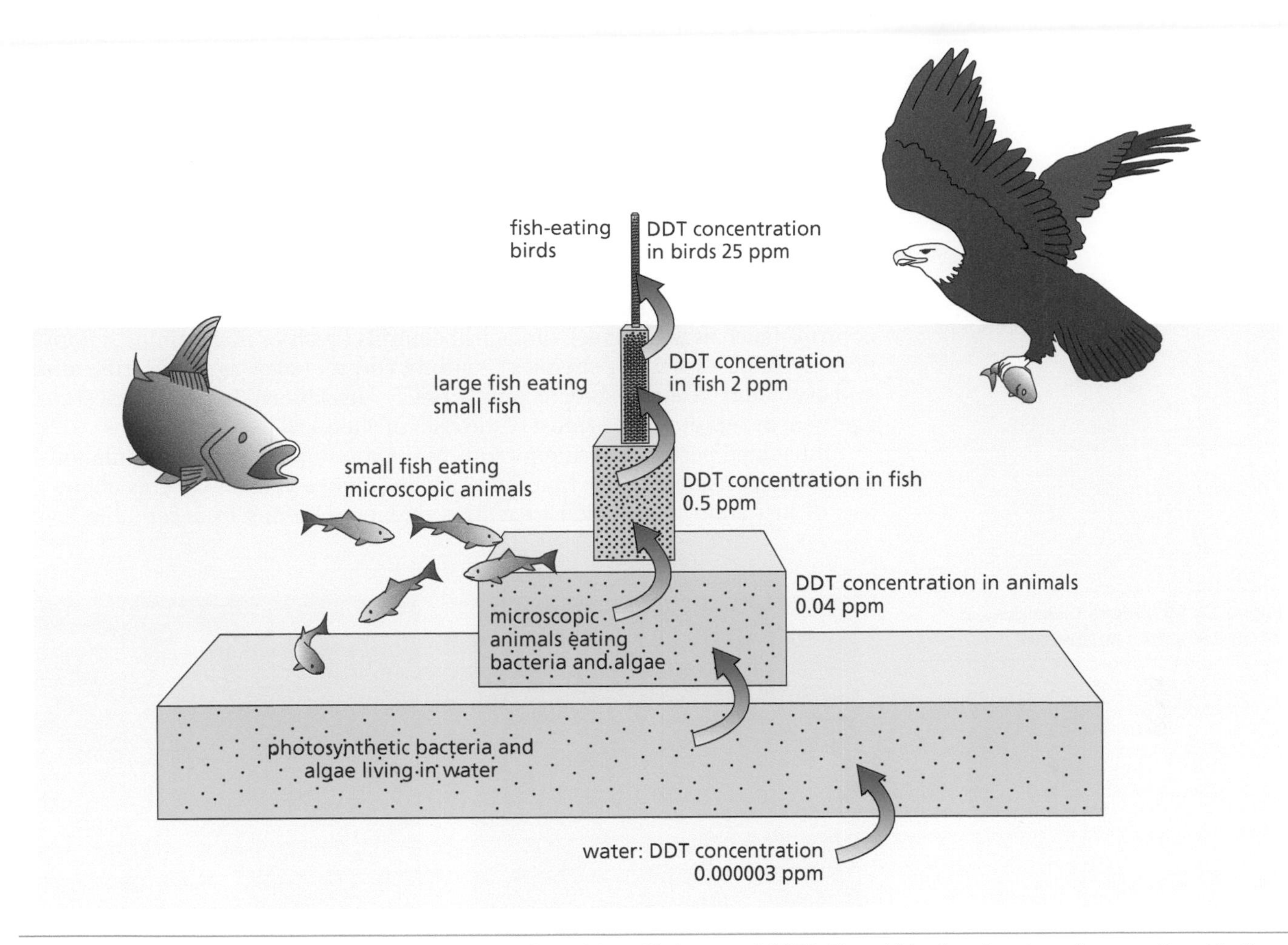

Figure 17.18 Biomagnification of DDT in Long Island Sound, New York, around 1970. The width of each column is proportional to the biomass.

eggs laid by the birds. The thin shells break before the chicks can develop, and the bird populations, unable to reproduce, decline. For birds that reproduce only once a year, laying only one or a few eggs at a time, it can take many decades after the pollution is removed for the species to recover its former numbers. For this and other reasons, DDT is now banned in most industrialized countries, though it is still used in some parts of the world.

17.4.5.5 Pesticides and neurotransmitters Many pesticides kill insect pests by interfering with the chemicals (neurotransmitters) that conduct impulses from one nerve cell to the next (see Section 11.1.3) and to muscle tissue. Blocking the enzymes that break down these chemicals induces continuous muscle contraction. Because some of the same chemicals are present in the nervous systems of vertebrates (including humans) as in insects, these pesticides can also impair nerve function in humans. Although the recommended concentrations properly used for pest control would not kill a person, they can cause permanent injury. The concentrated forms in which the pesticides are manufactured, transported, and stored are more dangerous and may be lethal. About a thousand cases of pesticide poisoning are reported every year in the United States, primarily among agricultural workers, and the actual incidence is probably much higher. Containers with warning labels are sometimes not in the language of the people using the pesticide. Safety standards, where they exist, are often not enforced. High concentrations can also result from biomagnification. Unfortunately, the banning of DDT use in many countries has resulted in the development and use of other chemicals that are even more toxic to nontarget species, including humans.

17.4.6 Integrated pest management

Integrated pest management (IPM Institute, 2021; EPA, 2021) is a newer approach to pest management, one that uses a combination of techniques (**Figure 17.19**). The term "management" means that pest populations are kept under control, so that they stay below the levels at which they cause economic harm. Total pest eradication is in most cases viewed as a goal that can only be achieved at an unacceptably high cost (including the cost to the environment or to society as a whole) or that cannot be achieved at any cost. The term "integrated" means that all available tools are used in a mix of strategies that includes chemical controls (such as pesticides), biological controls (such as maintaining a population of the pest's natural enemies), cultural control (such as public education), and regulatory control (such as public policy legislation). Integrated pest management avoids or reduces most of the risks of chemical pesticides.

Integrated pest management requires the monitoring of pest populations to assess the possible damage that they may do (**Figure 17.19A–C**). This allows the use of just enough pesticide to reduce pest populations to acceptable levels,

Figure 17.19 Various techniques of integrated pest management.

(A) Several small traps set to monitor the levels of pest populations.

(B) Closeup view of a trap that uses pheromones to attract target species for monitoring.

(C) Corn earworm, an insect whose presence in corn ears can be monitored visually.

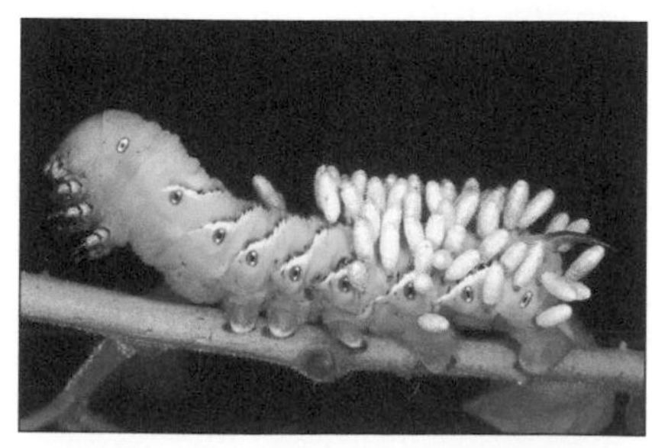

(D) Caterpillar pests can be controlled by parasitic braconid wasps, whose eggs develop into larvae that kill and eat the caterpillar. New wasps emerge from the pupae (cocoons) seen here.

saving expense and reducing runoff and possible harm to nontarget species. Because integrated pest management relies more on biological controls than previous techniques, it requires a good working knowledge of the ecology of the pest species, including knowledge of its natural enemies (**Figure 17.19D**).

17.4.6.1 Economic impact level An important feature of any integrated pest management program is the concept of an **economic impact level**, meaning a low level of the pest population considered economically acceptable. The economic impact level is the threshold level above which corrective action must be taken. Pest populations are constantly monitored, and as long as the populations stay below the economic impact level, they are left alone, and the cost of countermeasures is saved. The cost of corrective measures includes the cost of expendable materials such as pesticides, the cost of using and maintaining necessary equipment, the cost of labor, and the costs to the environment (including cleanup). Because it usually saves money overall, integrated pest management has become widely adopted. As compared with "calendar" spraying (spraying at a particular time of year, without any regard to the level of the pest population or the need to spray), integrated pest management saves costs in chemicals and equipment, but there are costs in monitoring pest populations and in using biological controls.

17.4.6.2 Introduction of predator species Planting crops in smaller, separated patches instead of larger, single-species blocks is one way in which the spread of pests can be controlled without the use of chemical pesticides. Planting seasons can sometimes be modified so as to interrupt the life cycles of the pests. The most important techniques in integrated pest management, however, are those that take advantage of the natural enemies that keep the pest species in check (Flint, 1998). If a predator that preys upon the pest can be identified, then measures that encourage the growth, development, and proliferation of the predator may be able to keep the pests in check. For example, the bacterium *Bacillus thuringiensis* (Bt) can attack the larvae of insect pests and prevent them from destructively feeding on crops; therefore, many plant growers use *B. thuringiensis* on their crops. There are many subtypes of *B. thuringiensis*, each producing a toxin specific for only certain types of insects, so the *B. thuringiensis* type needs to be matched with the pest. Integrated pest management can often cost less than the application of chemical pesticides. For example, a predator species need not be applied repeatedly because it will reproduce naturally on its own, particularly if conditions that support all life stages of the predator species are maintained. As another advantage, because only predators with a very specific and limited range of prey species should be selected as biological controls, there should be less damage to nontarget species.

As an example of integrated pest management, consider cotton production. Cotton has long been a crop of commercial importance in the southern United States, India, Egypt, and elsewhere. Cotton pests include the boll weevil and the pink bollworm. Spraying with chemical pesticides initially reduced the levels of these pests, but, by the 1960s, pesticide resistance had developed in both pest species. Despite increased spraying, pest populations continued to increase. Worse yet, the chemical sprays destroyed many of the pests' natural enemies, such as the spined soldier bug, and the destruction of the natural predators allowed other pest species, such as the tobacco budworm (previously unimportant as a pest of cotton), to become significant pests—in some cases more devastating than the original ones.

In both Texas and Peru, integrated pest management techniques have been used successfully to control cotton pests. Soldier bugs and other natural predators are collected, reared, and released on the cotton fields. Meanwhile, chemical spraying has been greatly reduced and, although not eliminated entirely, it is used only selectively. The planting season is timed so as to disrupt the life cycle of the pink bollworm moth; when the moths emerge, they can find no cotton plants on which to lay their eggs. Stalks and other unused parts of the plants are shredded and plowed under soon after each harvest, denying the pest insects places to hide and overwinter until the next growing season. In some places, corn and wheat are interplanted with the cotton to help the growth of populations of natural predators and to reduce the ability of the cotton pests to spread from one field to the next.

Alfalfa is another important crop for which integrated pest management techniques have been used successfully. As a nitrogen-rich legume, alfalfa is useful in crop rotation, and its high-quality protein is valued as an animal feed for many domestic animal species. In the United States, alfalfa ranks fourth (behind corn, cotton, and soybeans and ahead of wheat) in area under cultivation. The principal pests of alfalfa are two related species of alfalfa weevils (genus *Hypera*). At least nine natural enemies of these weevils have been identified, most of them wasps that parasitize either the weevil larvae or other stages of the weevil life cycle. The weevil's life cycle can also be disrupted by harvesting alfalfa early. Alfalfa pests were formerly controlled with chemical pesticides, but this practice was sharply curtailed when one such pesticide, heptachlor, showed up in the milk produced by cows that had eaten the treated alfalfa.

17.4.6.3 Use of pheromones Another integrated pest management technique is the spraying of pheromones, hormones that function in animal communication. The pheromones are targeted at disrupting reproductive activities, thus preventing the production of new larvae; it is usually the larvae that destroy crops by their feeding activities. An example of such a pheromone is gossyplure, used by female pink bollworm moths to attract their mates. Spraying this pheromone on cotton fields confuses the male moths and interferes with their ability to locate the females, resulting in a natural birth control that is very specific to the pink bollworm and that has no effect on other species. It is, moreover, a chemical to which the bollworm can never develop a natural resistance without impairing its own ability to mate. Pest control via pheromones depends on decreasing the pest population size. Pheromones do not directly affect the juvenile stages of the insect, which in many species is the stage that causes the actual crop damage.

THOUGHT QUESTIONS

1. The runoff of fertilizers from agricultural fields often produces algal blooms. Can you explain why this would be so? (What limits algal growth under normal conditions?) Would the problem be greater with organic fertilizers or with inorganic ones? Why do you think so?
2. Compare the volume occupied by the same weight of commercially available potting soil and sand. Compare the amount of water that can be held by equal weights of soil and sand. Now compare those results with the amount of water that can be held by soil from your area. Does the soil in your area contain a lot of humus? What ways can you think of to improve the quality of your soil?
3. Can you see any evidence of erosion in the area where you live? What natural processes or human activities might contribute to soil erosion?
4. Once the toxic effects of DDT on nontarget species received widespread publicity, agricultural use of the chemical was banned in the United States and in many European countries. The United Nations considered imposing a worldwide ban, but this effort was stopped by the insistence of many nonindustrialized nations that they needed the pesticide to help control both agricultural pests and mosquitoes. Do you think DDT should have been banned in countries like the United States? Do you think the ban should have been extended worldwide?
5. What do you think would happen if DDT were allowed to be used in limited amounts? How might the limits be enforced? What would be done if a farmer found that he or she could kill more pests (and thus increase crop yields) by using more DDT and causing potential future harm to the environment? How intrusive would enforcement agencies need to be? Is a total ban more practical than a limited ban? Would rationing work? Would an "agricultural prescription" system (similar to medical prescriptions for drugs) work?
6. Once pest resistance to a pesticide arises, it can spread quickly through any pest populations treated with the pesticide. Explain this fact using your knowledge of genetic mutations and natural selection.
7. Why is "eating low on the food chain" beneficial to agriculture? Why is it beneficial to the economies of both rich and poor nations? In what ways is it beneficial for human health?

17.5 CROP YIELDS CAN BE INCREASED FURTHER BY ALTERING PLANT GENOMES

Many plant characteristics, including the size, texture, and sweetness of the edible portion, are at least partially genetically determined. Also under genetic influence are many factors that determine the hardiness of crop plants, their drought resistance, their rate of growth under different soil conditions, their dependence on artificial fertilizers, and their resistance to various pests and plant diseases. Therefore, the yield, both in terms of the amount of crop per hectare (or per acre) and the amount of nutrition per unit of crop, can be increased by improving plant genomes. Other possible goals of genetic improvement include the development of new crop plants such as drought-resistant strains, pest-resistant strains, or strains with diminished nutrient requirements, capable of growth on marginal or poor soils.

The kinds of changes to a species that can be accomplished by traditional methods of selective breeding are limited by the genetic variation that exists within the species or its close relatives. Genetic engineering offers a newer method for customizing food crops by giving them genetic traits that they normally lack. We will discuss traditional methods first and then genetic engineering afterwards.

17.5.1 Altering plant genomes is not new

Selective breeding is also called **artificial selection**. As carried out by both animal and plant breeders, the practice was already well known in Charles Darwin's time and served as a model for his theory of natural selection (see Section 5.1.4). Darwin realized that great changes in agriculturally important plants and animals had been made within his own lifetime by British animal and plant breeders. These breeders chose the individuals of the species that best exemplified the trait they desired. They allowed these individuals to mate, while preventing mating between individuals that did not have the desired trait.

Artificial selection can be used to change almost any trait of a crop species in one direction or another. A closely related wild species may offer a desired trait, such as a nutritionally more complete protein, in which case the wild species may be crossed with the crop plant as a first step toward the production of a nutritionally superior strain. If this makes you wonder how the concept of crossing members of different species can be reconciled with the biological species definition given in Section 5.4.1, remember that the definition refers to populations (not individuals) that do not *naturally* interbreed. It is often possible to get individuals under domesticated conditions to do what is not natural for entire populations, for example, by dusting pollen artificially from a cultivated plant species onto a wild relative.

Figure 17.20 shows the results of 50 years of selection to produce corn plants with high or low oil content, or high or low protein content. However, attempts to change only one trait at a time can often result in the production of an inferior strain. For example, it does no good to select for corn plants with larger kernels or larger ears unless the stalks and root systems are capable of supporting them and unless the plants are sufficiently drought resistant and disease resistant to survive under field conditions. Modern breeding practices include the selection for

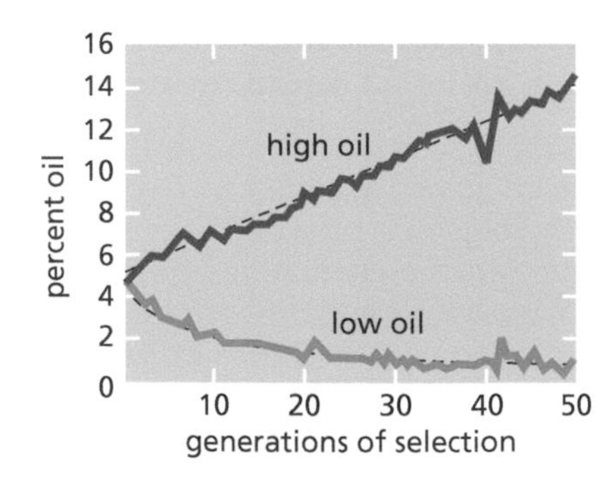

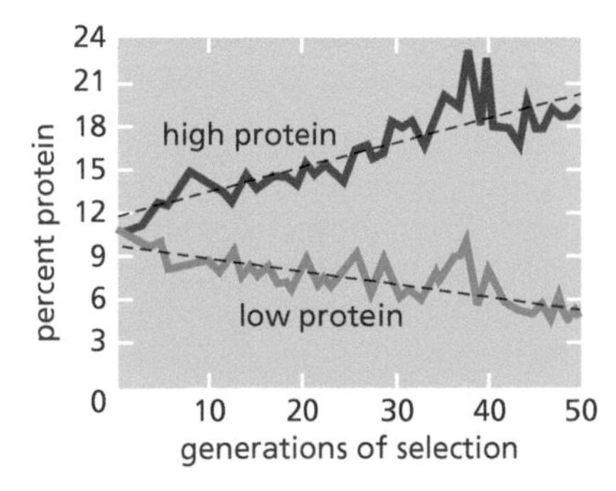

Figure 17.20 **The results of 50 years of selection on the oil content and protein content of corn (*Zea mays*).** By selectively breeding only those plants that had the highest or lowest protein or oil content in each generation, plant scientists have changed the inherited characteristics of each strain. See Dudley and Lambert (2004) for a 100-year follow-up.

several traits at once, resulting in harmonious combinations of traits that are well adapted to function together as a whole.

By selectively planting strains with desirable traits and by avoiding the use of genetic strains with less desirable traits, agricultural scientists in many nations have dramatically increased crop yields in the last few hundred years. The seeds of high-yielding strains command a high price. Around the world, nations that have achieved the most efficient agricultural production (high cash value yields of major crops per worker-day or per cash unit invested) have generally become wealthy, while those countries with the least efficient agricultural production are generally among the poorest. Thus, there is a high correlation between the affluence of a nation as a whole and the efficiency of its agricultural production. The development of new strains of crops, each suited to a particular climate and soil type, is among the most important components of agricultural efficiency, rivaling even the mechanization of agricultural work. Since about 1920, these strains have contributed to the increase in crop yields in industrialized countries. In the United States, for example, crop production doubled between 1940 and 1990 even though some land was taken out of cultivation.

17.5.1.1 The green revolution In the 1960s and 1970s, an effort was made to export many new and improved genetic strains of plants from North America and Europe to other parts of the world. This effort, loosely termed the "green revolution," was aimed at improving both the agricultural yields and the nutritional content of crops in the recipient countries. For example, some agricultural scientists developed a more nutritious variety of corn, high in lysine, an amino acid in which corn is usually deficient. High-lysine corn provides more complete protein for human nutrition (see Section 8.2.3). Yield was increased by the development of wheat and rice strains with short stems. They produce more grain on less stem and mature earlier so that more than one crop can be planted in a year. Many of these strains were developed in nonindustrialized countries under the auspices of international plant breeding institutes established there.

Greatly improved crop strains are, however, subject to the law of unintended consequences. For example, many of these new strains grew well with mechanized agriculture, but getting comparably increased yields in poorer countries meant the adoption not just of the new plant strains but also of irrigation and fertilizer use. Although production on farms was increased 50–100%, it never increased to the extent that it had on research stations. The results of the promised "green revolution" have been mixed, as is summarized in this quote:

> *Forty years after the first adoption cycles of the Green Revolution began in Mexico, and 15 years after they came to completion in Asia, we see that the world is not much better off. A similar percentage (10–15%) of the world population that was undernourished in the 1950s and 1960s is undernourished in the 1990s. The increases in agricultural production, while impressive, have kept just ahead of population growth, and not led to a more even distribution of food to all people. In addition, many problems plague sustainability in the high-input system, in the Third World just as well as in the developed countries.*
>
> (M.J. Chrispeels and D. Sadava, Plants, Genes and Agriculture, 1994)

17.5.2 Altering plant strains through genetic engineering

As we noted earlier, artificial selection is limited by the genetic variation that exists within the species or its close relatives. Genetic engineering offers a newer method for customizing crops by giving them genetic traits that they normally lack (Fox, 1992; US FDA, 2020). These may include better nutritional qualities, pest or herbicide resistance, the ability to live in nitrogen-poor soils and other marginal habitats, or the ability to fix atmospheric nitrogen and make their own nitrates (Kung and Wu, 1993; Grunewald and Bury, 2016). In this section, we examine some techniques of genetic engineering in plants and the uses to which plant genetic engineering is put. A later section provides insight into the controversy that surrounds this area of biology.

Genetic engineering is not simple. The genetic traits to be changed must first be identified. As in animals, many plant traits are not controlled by single genes and thus are not easily altered by genetic engineering. Most of the traits related to hardiness (drought resistance or cold resistance) are multigene traits. Other traits, for example, pest or herbicide resistance, have been successfully introduced by transferring single genes into plants, often a gene from another species.

Whether or not a genetically altered plant actually makes a desired protein depends on whether the desired gene has been inserted into a portion of the genome that is transcribed into messenger RNA (mRNA). This means that the gene must be located "downstream" from the binding sites ("promoter" sequences) that control gene expression (see Section 10.2.2). We also need to know how the altered plant will function at the ecological level, knowledge that we usually do not have until after the genetic engineering has been performed.

Plant genetic engineering follows the general concepts of genetic engineering that were described in Chapter 4. First, the gene of interest must be isolated and the gene must be coupled with (or relocated near) an easily identified marker such as a gene for antibiotic resistance. Second, the new gene must somehow be inserted into the plant genome. Third, because gene uptake is always a chance event with only a low to moderate chance of success, the organisms that successfully took up the new gene must be screened and separated from those that did not. Last of all, the genetically altered plant needs to be cloned so that many such plants can be produced.

17.5.2.1 Organizing the genes and markers to be inserted For most genetic engineering methods, the gene or genes that are targeted for insertion into a plant's genome must first be assembled into a continuous piece of DNA. This DNA contains the nucleotide sequences of the gene associated with the trait being modified. In addition, it must contain regulatory DNA sequences that will enable this gene to be transcribed and translated into protein. The regulatory DNA may also control the tissues of the plant in which the protein gene product will be made. Finally, the DNA assembly must carry what is called a "marker gene." As we will see shortly, the product of this marker gene will allow the separation of those cells that took up the assembled gene complex from those cells that did not. The parts of the DNA assembly are shown in Figure 17.21.

17.5.2.2 Methods of inserting new genes into plants Once the necessary DNA segments have been put together into one piece of DNA, the whole piece can be inserted into plant cells. Scientists can use at least three different methods to insert a new or altered gene into a plant genome: viruses, bacterial plasmids, and mechanical insertion.

Viruses and plasmids can both be used as "vectors" to carry the new DNA sequences into plant cells (Low et al., 2018). In both cases, restriction enzymes are used to make "sticky ends" on the assembled DNA piece (Figure 17.21) and on the viral or plasmid DNA, as described in Section 4.1. The complementary sticky ends of the "insert" DNA then join with those of the vector DNA to generate a DNA construct that can be introduced into plant cells. Viruses can naturally enter into the plant cells and incorporate into the host genome, whereas plasmids require other methods of introduction, as described later.

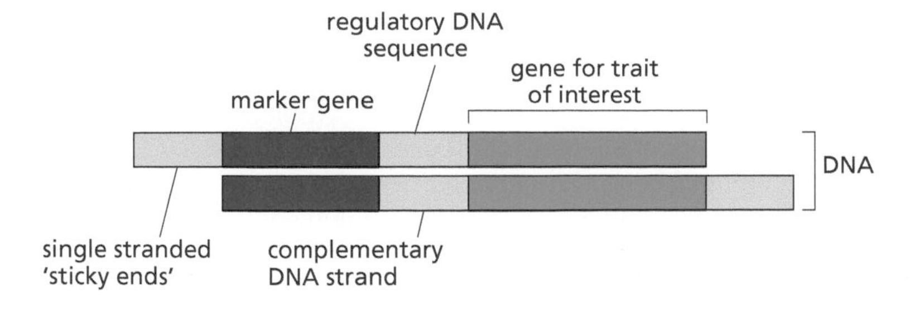

Figure 17.21 **Structure of a DNA assembly for use in genetic engineering.**

One virus used in plant genetic engineering is the tobacco mosaic virus. Like most viruses, the tobacco mosaic virus is restricted in its choice of hosts. Because the virus reproduces only in tobacco plants, it is very unlikely that it could accidentally spread new traits to other species. Tobacco mosaic virus has the advantage that both its biology and that of its host, the tobacco plant (*Nicotiana tabaccum*), have been intensively studied for decades. In cases in which it does not matter what plant is used in producing a particular compound, the tobacco plant is a logical choice, because methods for its cultivation and for the growth and insertion of tobacco mosaic virus are well known.

Because tobacco mosaic virus does not enter cells of species other than the tobacco plant, this virus is unsuitable for changing most crop species. For these species, genetic engineers have experimented principally with plasmids introduced into *Agrobacterium tumefaciens*, a bacterial species that causes tumors in many plant species (Nester, 2008; Nester, Gordon, and Kerr, 2005). *Agrobacterium*, like many bacteria, naturally carries small, circular plasmid DNA molecules. In the normal life cycle of *Agrobacterium*, plasmids can introduce bacterial genes into the cells of the host plants, and these genes cause tumors to form. The plasmids used in genetic engineering are modified so that they are still capable of introducing bacterial genes into the host plants but lack the gene that induces tumors. When a gene of interest has been inserted into a plasmid (Figure 17.21), the engineered plasmid can then be introduced into *Agrobacterium* cells through the process of transformation (see Section 4.1). The transformed bacteria are then incubated with pieces of the target plant, and the bacteria transfer the plasmid carrying the desired genes and regulatory DNA sequences into the plant cells, where they become incorporated into the plant genome (**Figure 17.22**). Many different plant species can serve as hosts to *Agrobacterium* and can incorporate the bacterial plasmids. However, many other important crop species, such as corn, cannot incorporate *Agrobacterium*. Several other known plasmids are useful in only one or two host species.

For plant species where host-specific virus or plasmid transfer methods cannot be used, mechanical methods of gene transfer have been developed. These methods are less specific and can therefore be used on nearly any plant species. In the particle-gun method, plant pieces are literally shot with tiny pellets (Figure 17.22) whose surfaces carry the DNA for the gene of interest and its regulatory DNA sequences (Figure 17.21). The force of the bombardment rams some of the pellets through the plant cell walls and membranes. The "naked DNA" carried by the pellets will sometimes incorporate into the DNA of the plant cells. In another nonspecific method called electroporation, temporary holes are made in plant cell membranes by disrupting them with electrical current. While these holes are open, DNA can enter the cytoplasm and then the nucleus, before the hole closes again.

17.5.2.3 Selection and cloning Regardless of the method of entry, the uptake and incorporation of new DNA do not happen in many plant cells that are treated. Scientists therefore need a method of selection (or screening) to identify which cells actually have the new gene. As mentioned earlier, this is done in most cases by incorporating another gene along with the gene for the trait being modified. This additional "marker" gene is usually one that confers resistance to some antibiotic, as shown in Figure 17.21. After the plant pieces have been subjected to some insertion method, they are then incubated in growth medium containing the antibiotic to which the marker gene confers resistance. Those cells that have not taken up the transferred genes are killed by the antibiotic, while those cells in which the new genes have been successfully incorporated will survive and grow. Once it has been determined which cells have incorporated the gene of interest, those cells can be used to grow complete plants.

Cloning (and therefore genetic engineering) is much easier to accomplish in plants than in animals because many plant species can regenerate a whole individual from just one or a few adult cells. Many genetically identical plants can thus be grown and produced asexually from a single cell containing the integrated DNA (see Figure 17.21). Cloning the plants in this way is usually done by growing

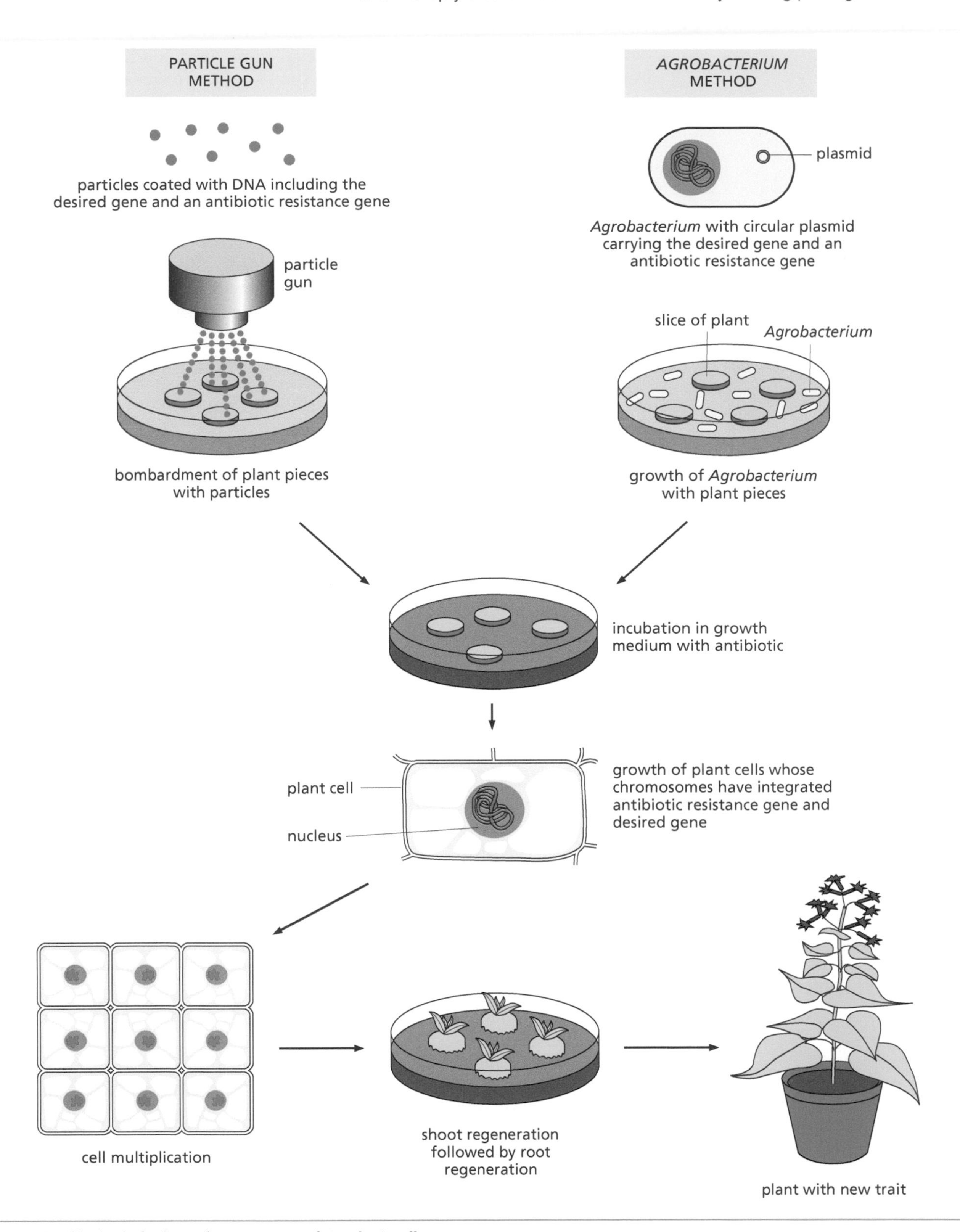

Figure 17.22 Methods for inserting new genes into plant cells.

the cells in small dishes in growth medium containing plant growth hormones, from which complete plants will develop without sexual reproduction or seed formation. Plants are particularly amenable to cloning, and methods have been developed for the cloning of many species of plants. Cloning techniques are not just restricted to growing genetically engineered plants; plants with desirable traits that have been developed by traditional plant breeding (artificial selection) can also be grown by such techniques. In fact, the cloning of traditionally developed plant strains is far more generally practiced and has had a far greater impact on agriculture around the world than any form of genetic engineering.

The advent of the CRISPR/Cas9 system of gene editing (see Section 4.1.3) has created even more options for the genetic modification of plants as well as other organisms (Norman, Aqeel, and He, 2016). Because of the high efficiency of CRISPR, mutants can be isolated without the need for antibiotic selection. Plasmid vectors containing DNA sequences that express both the CRISPR guide RNA and the CRISPR Cas9 enzyme are introduced into plant cells using one of the methods described earlier in Section 17.5.2.2. The treated plants or plant pieces are then allowed to develop to the point of producing seeds, and those seeds are used to produce a new generation of plants. The genomes of these progeny plants are then screened by polymerase chain reaction (PCR) or next-generation DNA sequencing (see Section 4.4.2) to identify those that contain the desired genetic modification. CRISPR methods are being tested for a wide range of crop plants, especially fruits. Apples that have been made resistant to browning via CRISPR gene editing are already available in the United States. It is important to note that mutations introduced by CRISPR/Cas9 editing cannot be identified as having occurred through genetic engineering, since they do not contain any foreign DNA, just a mutated form of the normal plant genome that could conceivably have arisen due to spontaneous mutation. This has significant implications for those who oppose GMOs, as we will soon discuss.

17.5.3 Uses of transgenic plants

Genetic engineering (also called "bioengineering") using *Agrobacterium* was first achieved in 1983. Since then, many plant species have been genetically modified using the techniques just described, producing genetically modified organisms (GMOs). In all, **transgenic plants** (meaning those with genes derived from another species) have been produced in over 20 species, including tomatoes, potatoes, carrots, alfalfa, corn, soybeans, peas, cotton, rice, apples, canola, and sugar beets (Nottingham, 1998; Teitel and Wilson, 1999). An estimated 92% of all corn grown in the United States is now genetically modified, along with 94% or more of soybeans and cotton (Center for Food Safety, 2021; USDA, 2020).

17.5.3.1 Transgenic plants with altered nutritional content Although genetic engineering holds the promise of producing plant strains with higher or more complete nutritional content, very few of the strains that have been developed have actually been changed in this way. One that has is the potato, in which the starch content has been increased by insertion of a bacterial gene. Potatoes are already 21–22% starch, but an even higher starch content makes the potatoes better for processing into potato chips and frozen French fries, the major ways in which potatoes are consumed in the United States. Another such crop is "golden rice," containing β-carotene, a vitamin A precursor, as described later.

The canola plant (*Brassica napus*), whose seeds are the source of canola oil, has been genetically engineered to produce lauric acid, a saturated fatty acid used in the food industry (for food additives such as coffee creamers and cake and candy coatings) and in the detergent industry. Lauric acid is not ordinarily made by *B. napus*. It is naturally synthesized by laurels and by tropical plants such as palm and coconut, but by inserting a gene from the California bay laurel tree, scientists have modified a canola strain to produce lauric acid.

17.5.3.2 Pest-resistant transgenic plants Sometimes genetic engineering is used to modify a trait that makes growing agricultural crops more efficient, thus increasing yield. Currently, the most widespread transgenic plants are those in which the gene for a toxin from the bacterium *Bacillus thuringiensis* (Bt) has been inserted. As mentioned earlier, the toxin is a natural insecticide, and *B. thuringiensis* is often sprayed onto plants as part of integrated pest management. Transgenic plants that contain the toxin gene make the toxin themselves, doing away with the need for repeated spraying of the bacteria onto crops. The toxin protects the crop plants against damaging pests but does no harm to the natural enemies of those pests. Cotton, potatoes, and corn with Bt toxin genes are all commercially available.

Resistance to nematodes has also been engineered into potatoes, tomatoes, and sugar beets. Nematodes are roundworms (kingdom Animalia, phylum Nematoda) that do a tremendous amount of damage to the roots of plants in many parts of the world, but particularly in tropical climates. Protease inhibitors and other genes conferring resistance to parasitic nematodes have been successfully transferred into several commercially important plant crops from other plant species (Ali et al, 2017).

In 2017, U.S. approval was granted to a gene-based insecticide called DvSnf7 dsRNA that was genetically engineered into corn plants to protect them against an insect pest. The gene makes use of RNA interference (RNAi), a regulatory mechanism that occurs naturally for some genes in most organisms but has been adapted relatively recently to target specific genes of interest. RNAi "silences" (turns off) the expression of an essential gene in the Western corn rootworm, a species called the "billion-dollar pest" for the damage that it causes to corn crops. The silencing occurs because the RNA produced from the DvSnf7 dsRNA gene is able to interact with and trigger the destruction of the mRNA transcribed from the rootworm target gene, thus preventing its translation into protein. The rootworm cannot survive with this essential gene silenced, and the corn is therefore protected by its transgene that has no effect on nontarget species.

17.5.3.3 Herbicide-resistant transgenic plants Another use for genetic engineering in plants is the introduction of genes for herbicide tolerance. If a crop plant is given a gene that allows it to resist (or tolerate) a particular herbicide, then the herbicide can be used as a weed-killer to control weed species that would otherwise compete with the crop for water and other limited resources. A bacterial gene conferring resistance to the popular wide-spectrum herbicide glyphosate (Roundup®) has been introduced into the most commonly planted strains of corn and soybeans, making them resistant to the herbicide. Fields can now be sprayed with the herbicide, killing the weeds but sparing the crops that carry the resistance gene. Over half of all soybeans grown in the United States now have this genetically engineered gene. The practice has been criticized, however, and is banned altogether in Europe, because glyphosate has entered the human food chain, where its long-term effects on human health are largely unknown. Glyphosate has been associated with increased incidences of non-Hodgkin lymphoma and other cancers in agricultural workers who use it, and its maker, Monsanto (since bought out by Bayer) agreed to pay a $2 billion settlement of various lawsuits brought by users of this chemical. Most marine mammals are likewise unable to break down glyphosate and other organophosphorus compounds, so they are susceptible to poisoning by the runoff of such compounds from agricultural land (Meyer et al., 2018). It is important to note here that the criticism is not directed at the transgene itself, but rather at the widespread use of the herbicide against which the transgene offers protection.

17.5.3.4 Plants with longer shelf lives In 1994, the Flavr-Savr™ tomato was introduced into markets in the United States as the world's first commercially available, genetically engineered fresh produce. These tomatoes were modified by another method of genetic engineering. Rather than inserting a gene from another species, here a tomato gene was turned off using RNAi. The gene that was targeted codes for a tomato enzyme that softens the tomatoes as they ripen. The engineered plants in which this gene was silenced thus secreted less of the softening enzyme, allowing the tomatoes to be vine ripened (instead of picked green), and thus to become more flavorful, while still being able to withstand transportation and storage (Martineau, 2001a,b). The Flavr-Savr™ tomato did achieve a longer shelf life but did not actually resist softening, so it was never commercially successful and was withdrawn after a few years.

17.5.3.5 Molecular farming One type of genetic engineering that uses tobacco plants is called molecular farming (or "pharming") (Lim, 2002). The goal here is not to make a better tobacco plant, but rather to use the tobacco plant as a biological factory to produce, say, a medicine, even small quantities of which would be valuable. Any medicine produced in this way must be purified to remove the nicotine and other tobacco plant molecules, but this may not be any more expensive than purifying the medicine from its original plant source. Further, the tobacco plant may grow in places where the original plant source will not, or it may grow faster. Various people hope that success in tobacco-based pharming might encourage tobacco farmers to grow more plants for pharmaceutical uses and fewer for cigarettes. Especially if the pharmaceutical plants became more profitable, tobacco farmers would have reason to switch away from growing the crop that is currently the leading cause of lung cancer (see Chapter 10) and other diseases such as emphysema.

A recent example of such molecular farming is the insertion into tobacco plants of a gene from cows that codes for lysozyme. Lysozyme is found naturally in the saliva of many animal species, where it has an important antibacterial function because of its ability to digest the cell walls of many bacterial species. Cow lysozyme produced by tobacco plants makes the tobacco leaves resistant to those bacteria. The lysozyme produced can then be used to treat seeds from many plant species. Disease-transmitting bacteria on seeds are a major agricultural problem. Treating seeds with dilute lysozyme produced by genetically engineered tobacco plants clears harmful bacteria from the seeds.

17.5.3.6 Other transgenic plants The bioengineered strains (GMOs) that we have described thus far are all now commercially available. Many more strains are in development but are not yet marketed. Plant strains have been engineered to make both soybeans and canola oil more nutritionally complete by the inclusion of methionine, an amino acid not normally produced in these plants. A strain of rice has been produced that contains genes for β-carotene, a substance metabolized by the body to produce vitamin A (see Section 8.2.5). Vitamin A deficiency is a significant cause of blindness, with 50,000 children a month affected worldwide by some estimates. The β-carotene gives the rice a yellow color; hence, its name golden rice. The hope is that such enhanced rice could be a factor in reducing vitamin A deficiency worldwide. Currently the level of β-carotene achieved in the rice is lower than the minimum daily requirements for the vitamin, but researchers hope to increase the amounts. The environmentalist organization Greenpeace has protested the development of golden rice because it is genetically modified (see later), but a letter signed in 2016 by over 100 Nobel Prize-winning scientists (Agre et al., 2016; Roberts, 2018) has affirmed the importance of this crop in overcoming an important human nutritional deficiency. Addressing the destruction of golden rice by Greenpeace activists, these Nobel laureates ask, "How many poor people in the world must die before we consider this [the Greenpeace action] a 'crime against humanity'?" In 2018, the United States, Canada, Australia, and New Zealand all approved the use of golden rice for food. The β-carotene-fortified rice has yet to be cultivated widely in the poor countries where it is hoped to do the most good, but Bangladesh is on track to be the first country to do so.

Plants resistant to viral and fungal infections are also being developed, as are plants with increased abilities to fix nitrogen or to be tolerant of high-salinity soils or drought conditions. In addition, so-called "edible vaccines" are being developed to provide protection against several pathogens, including measles, cholera, and several hepatitis viruses. In this case a food plant is engineered to express a protein from a pathogen to stimulate a person's immune system to respond to that pathogen when it is encountered. Various plants are being investigated for such purposes, including bananas, lettuce, and potatoes. While much research has been conducted in this area, most edible vaccines are still in the animal testing stages; a few small-scale human trials have been conducted for certain edible vaccines, but none have yet been approved for general use.

Lastly, plants have been engineered to produce nonfood products, such as polyhydroxybutyrate, which is used in the manufacture of plastics that can be broken down by decomposer organisms. Monsanto (now a division of Bayer) developed a transgene for a blue pigment that is inserted into cotton plants to make them produce blue cotton, decreasing the need for chemical dyeing to produce denim for blue jeans.

17.5.4 Risks and concerns

Genetically modified crops are widely grown in over two dozen countries (Fox, 1992). In terms of acreage devoted to genetically modified crops, the United States was the largest producer of GMO crops in 2017, followed by Brazil, Argentina, India, Canada, China, Paraguay, and South Africa. Most European countries prohibit or severely restrict the growing of genetically modified crops, although these countries import large quantities of GMOs, largely in the form of soybeans used for animal feed. The labeling of foods derived from genetically modified organisms is mandatory throughout most of Europe. Starting in 2022, foods sold in the United States that contain GMO products must be labeled as "bioengineered."

Whether genetically modified plants will become further commercialized, and whether those commercially available will gain widespread acceptance, will depend in part on overcoming opposition to genetic engineering (Levin and Strauss, 1991). Part of the opposition is biological, and part of it is ethical.

17.5.4.1 Ethical arguments Ethical questions include both deontological and utilitarian concerns. A few critics, notably Jeremy Rifkin, have consistently opposed all biotechnology, especially transgenic research. We are, these critics say, attempting to alter nature by going considerably beyond the bounds that nature intended. Rifkin goes so far as to question whether *any* transgenic research can be ethical. In the terminology explained in Section 1.4.3, Rifkin might be described as a deontologist who believes that any transplantation of a gene from one species to another is inherently unethical, a stance that prevents the experimental measurement of certain risks (among other things). One possible answer to such criticisms is to point out that genomes are being rearranged all the time in nature. Gene transfer between related plant species happens fairly often in plants, and cross-breeding has been transferring genes between domestic plant strains of the same species for centuries.

Other deontological objections include those from some vegetarians who are opposed to products that might contain a gene from an animal. Religious groups have other objections, such as Muslim fears that genetically engineered foods might contain genes transferred from pigs, which may not be eaten under Muslim dietary laws.

Various Christian and Jewish theologians view genetically modified plants as equivalent to plants modified by more traditional agricultural methods. This is the view generally taken by scientists also, although they are more likely to consider this to be a scientific, rather than an ethical, issue. For example, the Scientific Committee on Problems of the Environment (SCOPE), a committee of the International Council of Scientific Unions, has stated that "there are no convincing scientific grounds for distinguishing engineered organisms from natural ones" (H.A. Mooney and G. Bernardi, *Introduction of Genetically Modified Organisms into the Environment*, 1990). Also, "because organisms of either type could pose unforeseen hazards, some safety testing is desirable before large-scale propagation."

Utilitarian ethical arguments opposing genetically modified organisms center more on the premise that the risks of genetic engineering are poorly understood and quantitatively uncertain. Risks probably vary from one plant species to another, making it important for the questions to be raised (and the research conducted to answer them) again and again for each new application. European legislation on genetically modified crops generally takes this utilitarian, case-by-case approach, as in dealing with "gene silencing" by dsRNA (Arpaia et al., 2020). As another example, the use of the *Agrobacterium* plasmid has been criticized because the original form of this plasmid stimulates plant tumor formation.

Suppose that a genetically altered strain of a plant containing an *Agrobacterium* plasmid managed to reacquire the gene for plant tumor formation, either by mutation or (more likely) by genetic recombination with wild strains of *Agrobacterium*; plants carrying the plasmid might then grow tumors. The probability of such an event must be carefully estimated if reliable risk–benefit ratios are to be obtained and if defensive measures against such mutant strains are to be planned in advance. Potential economic benefits from genetically engineered crops are generally easy to estimate; risks often are much more uncertain. Many utilitarians are not opposed to genetic engineering in principle but argue that the risks are great and should be evaluated before we proceed.

17.5.4.2 Concerns about damage to the environment The risks (or potential dangers) of biotechnology do exist, and some of them have been alluded to earlier. Many biologists, particularly ecologists, argue that there are biological questions that need to be answered and that will not be addressed if we think only at the molecular level (Mooney and Bernardi, 1990; Marvier, 2001). When plants modified at the molecular level are introduced into the environment, we need to consider the functioning of the ecosystem as a whole. The Union of Concerned Scientists has published a summary of their concerns, which are in two areas: the possible escape of genetically altered strains as "superweeds" and the spread of plant viruses. They argue that because agricultural crops cannot be isolated from their surrounding ecosystems, transgenic plants pose risks that other genetically engineered species, such as bacteria grown in factory vats for the production of medicines, do not. They caution that transgenic plants could escape from cultivation and become weeds. They give as examples various plants, such as kudzu and purple loosestrife in the United States, that have been introduced into new locales and have overrun the environment, altering the habitats of other plants and of animals. Members of the Scientific Committee on Problems of the Environment (SCOPE) do not see this as an equivalent example; genetically engineered crops, they say, would most probably be introduced into areas where the nonengineered form of the crop had already been grown, so natural ecological balances should still apply. A larger risk, though, is whether engineered plants would cross-pollinate or otherwise transfer their new genes to weeds. Plant species frequently do cross-pollinate with weeds of related species. This might be particularly worrisome if herbicide-resistant plants passed their transgenes to weeds, making the weeds herbicide resistant. The Union of Concerned Scientists cautions that small field trials are not necessarily good predictors of full-scale agricultural conditions, particularly if the field trials have not been designed to examine ecological effects. They propose expanded testing protocols that should be followed before crop species are approved.

The issue of the spread of plant viruses, or of the creation of new viruses from recombination of the virus used to insert the transgenes with normal viruses already present in the plant, is also of concern. Although viruses are known to recombine, the risk is considered to be low, simply because viruses are not now frequently used for gene insertion into plants.

A further ecological concern centered on the Bt bacterial pesticide genes that have now been engineered into several crop plant species. Opposition to Bt-containing crops was enhanced by reports that pollen from such plants killed the larvae of monarch butterflies. The initial reports merely established the fact that, under laboratory conditions, enough Bt protein is produced by these plants to kill some butterflies. The report did not contain controls, however, such as a comparison to the number of butterflies killed by Bt or other pesticides sprayed onto plants. Although the initial report continues to receive much attention, further studies have found that, under field conditions, the effect on monarchs is negligible.

17.5.4.3 Concerns about human health Another issue raised by biologists is a food safety issue, but again it has to do with thinking in a more integrated way about how systems function. The issue here has to do with the antibiotic resistance genes that are engineered into plants along with the gene of interest (Figure 17.21). The unanswered question is whether these genes will contribute to humans and animals becoming unable to use the antibiotics due to bacterial resistance. Many disease-causing bacteria are already becoming resistant to antibiotics owing to their overuse in animal feeds and their improper use medically (for example, in the treatment of colds, which are viral diseases against which antibiotics have no effect). If transgenes in food add to this growing problem, it could help undo the control of infectious diseases that has been achieved since the early 1900s.

Much of the opposition of member nations within the European Union (EU) to genetically engineered foods centers on this last point. The EU banned Bt transgenic corn in 1996 because it also contains an inserted gene for ampicillin resistance. While this would not pose any direct threat to human health, the possibility exists that the resistance gene could be transferred to gut bacteria. Laboratory tests show that most, but not absolutely all, of the DNA ingested in food is destroyed by strong acids present during digestion. The possibility of resistance transfer is further diminished in any food that is cooked, since the DNA and its proteins would be destroyed in the cooking. Because ampicillin is used to treat many human diseases, and because resistance to it often confers resistance to many other penicillin-type antibiotics, the development of resistant bacteria in the gut, which could possibly transfer resistance to pathogenic bacteria, is a risk that many scientists feel should be more thoroughly investigated. The EU reversed its outright ban, but now requires labeling of any food that contains live GMOs, or has modified ingredients that are not equivalent to, or materials that are not present in, the original, or has substances that might be objected to on ethical grounds (such as animal genes that might be opposed by vegetarians).

In the United States, opposition to corn containing Bt has had a different basis. Much of the publicity has been about a particular corn variety called StarLink™. This corn contains both the Bt transgene and another gene that makes it tolerant to a commonly used herbicide. Because Bt is a pesticide, products containing it are regulated in the United States by the Environmental Protection Agency (EPA), not by the Food and Drug Administration (FDA). In tests done as part of the approval process, it was found that the protein product of the *Bt* gene (called Cry9C) could partially survive cooking and digestion. Initially there were also concerns about the possible capacity of the protein to induce allergies. Consequently, the corn was approved only for use in animal feeds. In September 2000, however, traces of the Bt protein were found in human food products. The Aventis company voluntarily agreed to recall the food products, cancel its EPA registration for the corn, and pay the farmers to remove their corn from the market. Although further testing has cast some doubt on both the protein's persistence and on its allergic potential, the corn is currently off the market and mired in lawsuits.

The potential of other transgenic proteins to cause problems for people with allergies remains a concern. This is one reason for the desire to have genetically modified foods accurately labeled. The StarLink™ corn experience has cast doubt on the feasibility of ever separating foods for human use from other uses, leading some to advocate instead for truth in labeling. Others see practical barriers to such labeling. Products refined from GMO crops include oil and corn syrup from corn and protein, oil, lecithin, and several vitamins from soybeans, all of which are now a common part of the food supply in the United States. The transgenes have to do with pest or herbicide resistance, not with the character of the food produced by the plant. Once a refined product like corn syrup has been extracted, there is usually no way to detect the genetic background of the plant source.

17.5.4.4 Social and agricultural concerns Another line of opposition to genetically engineered crops is that they do not contribute to efforts to develop sustainable agricultural practices. The Union of Concerned Scientists states that we should be developing sustainable practices that prevent environmental problems in the first place, rather than focusing on solving problems after they are created. The proponents of genetic engineering point to crops modified to be pesticide resistant as an example of ways in which plant engineering could cut down on pesticide use. On the other hand, herbicide-resistant crops have no benefit if they are not used in conjunction with the matching herbicide.

(Farmers who used Monsanto's RoundupReady® soybean seeds were required to sign an agreement with the company to use only Monsanto's Roundup® herbicide and faced heavy fines imposed by the company if they did not.) Several species of weeds have meanwhile developed their own natural resistance to the herbicide, rendering it much less effective, and its use has greatly declined as a result. As alluded to previously, the widespread use of Roundup® herbicide has come under further criticism because its main ingredient, glyphosate, has been implicated in cancer. Several agricultural workers have won large lawsuits seeking damage awards for cancers that they have attributed to the use of glyphosate herbicides.

Some see genetically engineered plants, in part because they are developed and marketed by large multinational corporations, as inherently contributing to monoculture practices, thereby also contributing to the loss of biological diversity, a topic that we examine in greater detail in Chapter 18.

The other half of a risk–benefit equation is the benefit side. Many see the benefits of engineered plants as potentially immense. Others see the profits as potentially immense, but the benefits to society as very small. The countries in which farmers will be able to afford these seeds are countries that are already awash in excess food. Examples are raised such as the engineering of lauric acid into canola, which will benefit North American farmers that grow canola, but this will be at the expense of tropical farmers who grow palm and coconut, the natural sources of lauric acid. Because engineering is seldom applied to the primary food crops used in poor countries, it is unlikely to be of any direct benefit in these countries. Moreover, the use of patented or trademarked plant strains, or the increasing use of plant strains that require mechanized agricultural methods, will probably make farmers in poor countries more dependent on imported seed supplies and on foreign debt. Many groups of scientists also caution that genetic engineering will not solve the world's food problems; they see these problems as being largely due to the unequal distribution of food, not to a lack of food production. They point to the example of the green revolution, which has tremendously increased production but has not done away with hunger.

The possible benefits of the genetic engineering of crops are very large; depending on the particular genes introduced, they include better drought resistance or pest resistance in the field, decreased need for nitrogen fertilizers, and higher nutritional content. The monetary costs may also be large, but they are very difficult to estimate because of the great uncertainties involved. The biological dangers are not fully known. Given the law of unintended consequences, difficulties should not be underestimated. The risks may prove to be minimal, or they may prove to be significant, and the dangers may prove to be easily controlled even in worst-case scenarios—we will never know unless we undertake the relevant investigations for each species. Only if we have investigated the possible dangers will we be able to assess the possible risks. That is one reason why it may be desirable to proceed with testing, why all applications should be closely monitored, and why many people await the evaluation of risks as well as benefits.

THOUGHT QUESTIONS

1. Artificial selection and genetic engineering are both ways of modifying plants. In what ways are the two methods similar? In what ways are they different?
2. Experimenters attempting to alter plant strains by selecting for one trait sometimes end up changing not only that trait but some other trait along with it. Use your knowledge of genetics to explain why this is so.
3. Do genetically engineered foods increase nutritional quality for consumers or are they of more benefit to mechanized farming and the food processing industries? Develop arguments on both sides.
4. Jeremy Rifkin and other critics of genetic engineering have argued that the escape of a genetically engineered strain from cultivation would be a chaotic event whose consequences are inherently unknowable. Do you agree or disagree? Is there any way of planning for such events? If a genetically altered strain of plants such as golden rice were found growing outside cultivated areas, what should our response be?
5. Most scientists view genetically engineered (GMO) foods as safe to eat. Why is there so much public concern about GMO food safety and the need to label such foods as "bioengineered"? Why are so many consumers attracted by foods advertised to be "GMO-free"?

CONCLUDING REMARKS

As the world's population continues to increase, methods are being developed to make agricultural production more efficient. Overcoming various limiting factors, such as nutrients or water, increases food production. Sustainable agricultural practices aim at ensuring that production will remain high far into the future. Plants need light energy for photosynthesis, and light energy is more abundant in countries close to the Equator, including many poor countries. Wherever light energy is naturally abundant, the need for fertilizers can be diminished by planting crops that harbor symbiotic nitrogen-fixing organisms in their roots. Alternatively, the genes for nitrogen fixation or for other desirable traits can be genetically engineered into plants that do not naturally possess them. Genetic engineering can also be used to make plants more nutritious and more resistant to drought and to pests.

Although there is considerable opposition to genetic engineering, most groups of scientists conclude that the risk–benefit ratio still argues in favor of continuing research on genetic engineering of food crops. Genetically engineered crop plants may increase food yields and nutritional value. The possible benefits of the genetic engineering of crops are thus immense, although the dollar costs may also be immense. Many scientists caution that increased food production by itself will not solve the world's food problems. They see these problems as being also due to the social, economic, and political forces that result in unequal distribution of food. Plant science will be an important part of any solution, but world food problems will not be solved by science alone.

CHAPTER SUMMARY

- Plants make carbohydrates by the process of **photosynthesis**, using water and atmospheric carbon dioxide as raw materials and sunlight as an energy source. Sunlight is absorbed by plant pigments such as **chlorophyll**, located in plant cell organelles called **chloroplasts**.
- Because plants can make all of their own organic compounds, they are **autotrophs**; organisms that cannot, including all animals, are **heterotrophs**, dependent on autotrophs for their food.
- **Vascular plants** absorb water through their roots and evaporate water through the **stomates** in their leaves. **Osmosis** generates water pressure, and this pressure contributes strength to most plant tissues; a lack of water pressure causes wilting.
- Proteins, nucleic acids, vitamins, and other plant products require nitrogen for their synthesis. Most plants get their nitrogen from dissolved nitrates, which limits the distribution of many plant species to soils that contain adequate nitrogen. Some plants form **mutualisms** with microorganisms that can fix atmospheric nitrogen and convert it into a form that the plants can use. The cycling of nitrogen through the biosphere and atmosphere is called the **nitrogen cycle**.
- Crop yields can be increased by supplying **limiting nutrients** through **fertilizers** or soil improvement, by supplying water, by controlling pests that compete with the plant or with humans for the energy produced by plants, and by altering the traits of the plants either by **artificial selection** or by genetic engineering.
- Pesticides and chemicals can become concentrated in biological tissues by **biomagnification**.
- **Monocultures** allow the rapid expansion of pest species.
- **Integrated pest management** can reduce our dependence on chemically produced pesticides and herbicides. Integrated pest management techniques begin with a strategy of setting an **economic impact level** for each pest, below which countermeasures will not be taken and money and labor will be saved.
- **Transgenic plants** may also be used to introduce desirable traits into crop plants (e.g., making them hardier in the field or making them more nutritious).

BIBLIOGRAPHIC REFERENCES

Bibliographic references to material in this chapter can be found online at biologytrending.routledge.com/chapter17

GLOSSARY: KEY TERMS TO KNOW

These key terms are defined at the end of each chapter as an aid for student review.

Artificial selection Consistent differences in the contribution of different genotypes to future generations, brought about by intentional human activity.

Autotroph An organism capable of making its own energy-rich compounds from inorganic compounds.

Biomagnification The increasing concentration of pollutants as one proceeds up the food pyramid from one trophic level to another.

Chlorophyll A green pigment molecule that traps light in the light reactions of photosynthesis.

Chloroplasts Cytoplasmic organelles that contain chlorophyll and carry out photosynthesis.

Consumers Organisms that consume food energy by eating other organisms.

Economic impact level The threshold population density of a pest species above which economic damage will occur.

Ecosystem A biological community interacting with its physical environment.

Eutrophication Depletion of oxygen in a body of water due to bacterial processing of excessive algal and plant mass arising from high nutrient levels derived from chemical run-off from the land.

Fertilizer Any substance artificially furnished to promote the growth of crops.

Heterotroph An organism not capable of manufacturing its own energy-rich organic compounds and therefore dependent on consuming such compounds as food.

Humus A dark-colored, nutrient-rich layer of soil in which organic matter is abundant; also called topsoil.

Hydroponics The practice of growing plants without soil.

Integrated pest management (IPM) An approach to the management of pest populations that emphasizes biological controls and frequent monitoring of pest populations.

Lignin A material that stiffens the cell walls of woody plants.

Limiting nutrient Any nutrient whose amounts constrain the growth of an organism or population; supplying greater amounts of this nutrient therefore allows a population of organisms to increase or grow more vigorously.

Monoculture Growth of only one species in a particular place, as in a field planted with only a single crop.

Mutualism An interaction between species in which both species benefit from the interaction.

Nitrogen cycle A cyclical series of chemical reactions occurring in nature in which nitrogen compounds are built up, broken down, and changed from one form to another with the help of living organisms.

Osmosis Diffusion of water molecules across a semipermeable membrane in response to a concentration gradient of some other molecule or ion.

Pesticide A chemical used to kill undesired (pest) organisms.

Photosynthesis A process by which plants and certain other organisms use energy captured from sunlight to build energy-rich organic compounds, especially carbohydrates.

Producers Plants and other photosynthetic autotrophs that produce food energy from sunlight.

Stomates Pores on the underside of leaves in most vascular plants through which gases are exchanged.

Sustainable Any practice that could continue indefinitely without depleting any material whose supply is limited.

Transgenic Containing a gene (transgene) from another species.

Transpiration Water loss from the leaves of plants by the escape of water vapor through the stomates.

Turgor Fluid pressure that causes swelling and stiffness in plant cells and other cells that have cell walls.

Vascular plants Plants containing tissues that efficiently conduct fluids from one part of the plant to another.

CONNECTIONS TO OTHER CHAPTERS

Chapter 1: Genetic engineering of crops raises ethical issues. Pest control and fertilizer uses have both costs and benefits.
Chapter 4: Genetic engineering techniques can be used on crop species as well as other species.
Chapter 5: Artificial selection can be compared to natural selection.
Chapter 6: Plants have adapted to their environments in the course of evolution.
Chapter 7: Some human populations have evolved that cannot digest certain plant crops that are useful nutrition for most humans.
Chapter 8: Feeding the world's growing population will be aided by increased crop yields and more nutritious crops.
Chapter 9: Undernutrition and malnutrition affect human health. These problems will only get worse if our populations continue to grow.
Chapter 10: Several anticancer drugs are plant products. The plasmid used in plant genetic engineering originally induced tumors in host plants.
Chapter 11: Plants do not possess nervous systems or contractile muscle fibers, in contrast with animals.
Chapter 13: Most drugs are plant products, and several are psychoactive in humans.
Chapter 18: Clearing more land for agriculture threatens biodiversity and destroys ecosystems.

PRACTICE QUESTIONS

1. What is the difference between an autotroph and a heterotroph?
2. Name at least:
   a. One microscopic heterotroph
   b. Two heterotrophs larger than your thumb
   c. Two autotrophs
3. The major chemical process in the light reactions of photosynthesis involves the splitting of _____ and the release of _____.
4. In the dark reactions of photosynthesis, _____ from the atmosphere is incorporated into organic molecules such as _____.
5. What form of energy enters photosynthesis? In what form is that energy stored at the end of photosynthesis? In what other forms is energy stored during photosynthesis?
6. What carbohydrates can be obtained by eating plants? What function(s) do these carbohydrates serve in the plant?
7. Name three types of compounds containing nitrogen that plants need to make. Do humans need nitrogen for these same compounds?
8. How do plants obtain nitrogen? How do humans obtain nitrogen? Can either plants or humans obtain nitrogen from the air?
9. What is a limiting nutrient? What are some examples of nutrients that can be limiting for the growth of plants? How can farmers supply limiting nutrients to plants?
10. Which part of a vascular plant is responsible for each of the following?
    a. Uptake of water
    b. The bulk of photosynthesis
    c. The major portion of fluid transport
    d. Holding the plant up
    e. Anchoring the plant in place
11. Are all vascular plants flowering plants? Are all flowering plants vascular plants?
12. What molecules move through membranes during osmosis?
13. What are the functions of humus?
14. Under integrated pest management, name:
    a. Two items that cost more time or money than in traditional forms of pest management
    b. To items that cost less time or money than in traditional forms of pest management
15. Name four ways of introducing a gene into a strain of plants in which it is not already present.

CHAPTER 18

Biodiversity and Threatened Habitats

ISSUES

- What is biodiversity? How is it measured?
- What do we lose if we lose biodiversity? Why should humans be concerned?
- How do humans contribute to loss of biodiversity or to habitat destruction?
- How do economic disparities among people influence habitat destruction?
- How do economic disparities among nations influence habitat destruction?
- How can societies limit the threats to habitats and to biodiversity?

BIOLOGICAL CONCEPTS

- Biodiversity (species diversity, species richness)
- Scales of time (geological time, human experience)
- Extinction
- Specialization and adaptation
- Community structure (mutualisms and other interactions between species)
- Biogeography (biomes)
- Ecosystems (habitats, habitat alteration, human influences, biosphere)
- Conservation biology (renewable and nonrenewable resources, sustainable and nonsustainable uses)

CHAPTER OUTLINE

DOI: 10.1201/9781003391159-18

In the shadow of trees over 60 m tall (more than 200 feet, or as high as a 17-story building), workers use bulldozers and other heavy equipment to clear a path 100 m wide (328 feet, over twice the width of an American football field) through a tropical forest. They are building a new road that will bring commerce and communications to the people of the region and will enable them to send their agricultural products, crafts, and minerals to markets in faraway countries. For each kilometer (0.62 mile) of roadway, they are destroying 10 hectares (about 24.7 acres) of tropical rainforest. The building of the road brings many high-paying jobs to the workers who build it, and the road will also open up new land for agriculture and human settlement, a process that will destroy even more forest. The trees are important in themselves, and also because they provide **habitat** (a set of environmental conditions that make up a place to live) for thousands of species.

The number and variety of species in a place are referred to as biological diversity or, simply, **biodiversity**. Biodiversity is measured most easily by the number of distinct species present. More broadly, biodiversity also includes genetic diversity within species and ecological diversity within habitats or ecosystems. In this chapter we consider the importance of biodiversity, the conditions that support biodiversity, some of the threats to biodiversity, and some ways in which humans can reduce these threats.

Recall from Chapter 5 that species are reproductively isolated groups of interbreeding natural populations. Most new species originate by a process of geographical speciation, in which reproductive isolation evolves during a period of geographic separation. The processes that give rise to new species increase biodiversity, while the processes that result in the extinction of species decrease biodiversity.

18.1 BIODIVERSITY RESULTS FROM ECOLOGICAL AND EVOLUTIONARY PROCESSES

There are nearly 1,500,000 species of organisms currently known to science. More than half of these (53.1%) are insects, and another 17.6% (approximately 250,000 species) are vascular plants, so that over 70% of all known species are either vascular plants or insects (**Figure 18.1**). Animals other than insects make up 281,000 species, or about 19.9% of the total, and the remaining 9.4% are fungi, algae, and various single-celled organisms that are visible only with the aid of a microscope (Wilson, 1988, 1992).

Our knowledge is far from complete. Various estimates put the total number of species—known and not yet identified—between 5 and 30 million. Such estimates are extrapolations from the few studies in which an effort has been made to identify every species in a given area. In one such study, between 4,000 and 5,000 species were found in a single gram of sand. In another study, 5,000 species were found in a gram of forest soil, and these 5,000 species were almost completely different from the species found in the gram of sand. The more thoroughly we look, the more species we find. Some recent studies suggest that bacteria and other prokaryotic organisms are particularly underrepresented in these surveys and that many hundreds of thousands of undescribed prokaryotic species await discovery.

18.1.1 Factors influencing the distribution of biodiversity

Despite our incomplete knowledge of the extent of biodiversity, we have begun to use what we do know to test hypotheses about the factors that contribute to its richness. Present levels of biodiversity are the result of many processes. Above all, the process of speciation increases biodiversity and extinction decreases biodiversity. Because these are both evolutionary processes, the study of biodiversity

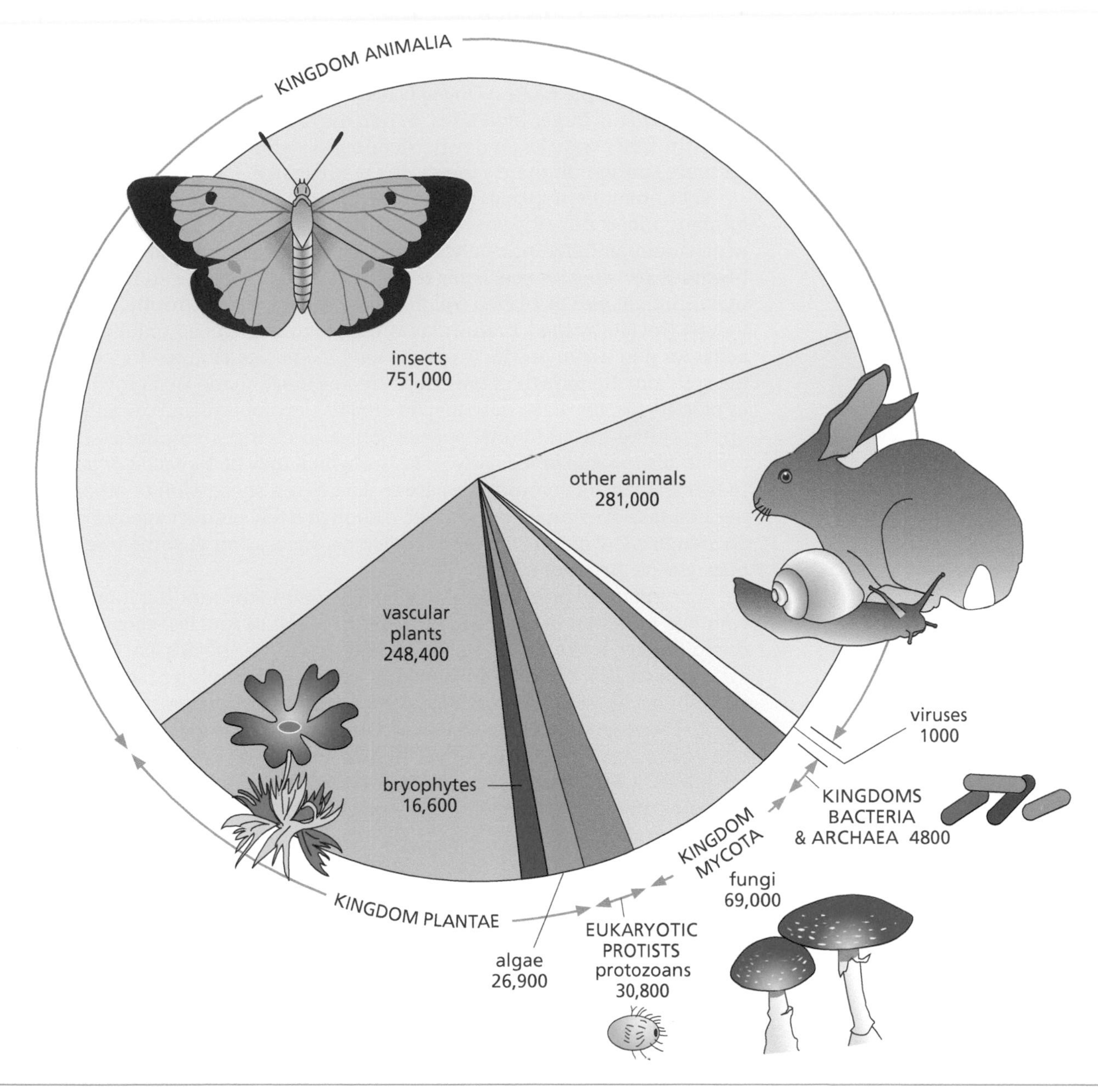

Figure 18.1 Numbers of species currently known in the major groups of organisms.

depends on an understanding of evolution. To understand biodiversity, we also need to become familiar with the ecological concepts of communities, ecosystems, climate, and energy flow. These concepts will help us to explain patterns in the distribution of biodiversity across the Earth's surface and also how our own species interacts with the millions of other species that inhabit this planet. By looking at present-day biodiversity in different places, we can test certain hypotheses that give a better understanding of evolutionary and ecological processes that influence biodiversity.

18.1.1.1 Communities and ecosystems Biological diversity implies more than just the number of species; the numerous habitats and the diverse ways of life within each habitat are also important. In any given place, many species live together and interact with one another in an organized **community**. Each species also has its own **niche**, meaning it has a unique way of life and role in the life of the community. Each species (or its products) provides part of the niche of many other species in its community. An increase or decrease in the population size of any one of the species is therefore likely to have consequences for all the others. Competition between species often results in partitioning of a niche into multiple niches: One species may compete better at smaller food sizes and the other at larger food sizes, or one may look for food higher up in the trees while the other stays lower down. Niche subdivision provides more opportunities for more species, ultimately increasing biodiversity.

Communities of species sometimes replace one another over a period of years to many thousands of years. After a devastating storm, landslide, or wildfire in which many animals and plants are wiped out, a process of **ecological succession** begins. A few species specialize in being the first colonists to take advantage of open sunlight and disturbed soil that follows such a catastrophe, and they create habitat (including food, hiding places, etc.) for other species to follow. In time, the activities and waste products of new animal species change the composition of the soil, and the growth of new trees creates more shade that crowds out shade-intolerant species. Each new species creates new food resources and other niche opportunities for additional species, but also changes conditions in a manner unsuitable to some of the early colonists, which may be replaced or pushed aside. In this way, entire communities are replaced by a succession of other communities that take over, one after another. Although a few pioneer species will colonize an area in a few months or years, ecological succession in some tropical habitats may take centuries or more.

A community plus the physical environment surrounding it and interacting with it is called an **ecosystem**. In addition to living species, ecosystems include the soil, water, rocks, and atmosphere in which those species live. The largest ecosystem of all, that of the entire planet, is called the **biosphere** (see Chapter 19).

When a new species originates (see Section 5.4.2), it must find its niche and fit into the ecosystem in which it lives. Otherwise, it fails to flourish in that ecosystem and quickly becomes extinct. If, however, a new species finds a niche that integrates it into the ecosystem, its population increases, and the ecosystem becomes more complex owing to its presence. Biodiversity is thus a measure of ecosystem complexity.

18.1.1.2 Energy and biodiversity

One of the major determinants of biodiversity is latitude. Tropical ecosystems contain a much richer diversity of species and genera than temperate-zone ecosystems. The richest diversity on land is in tropical rainforests, while the richest marine ecosystems are those of warm-water coral reefs (Figure 18.2). Some large taxonomic groups, including many entire families and several orders, are confined to tropical ecosystems, and

Figure 18.2 Coral reef diversity in the Red Sea, Egypt.

nearly every major group reaches its maximum diversity in the tropics. In marked contrast are the Arctic and Antarctic ecosystems, which are relatively sparse in terms of biodiversity and in ecological complexity.

Why should this be true? One of the great theoretical problems in evolutionary biology is why species diversity is greater in the tropics. One possible explanation is provided by the "energy-stability-area theory" of biodiversity, which begins with the observation that the species-rich tropics receives the greatest amounts and most continuous levels of solar energy. Each biological population requires a certain minimum amount of energy to maintain a population size capable of reproducing itself. Photosynthesizing plants capture the energy that they need directly from sunlight; most other species obtain their energy from food. For an equal amount of nutrients (an important proviso), greater quantities of biomass (mass of living things) grow in areas that are the hottest (receive the most solar energy) and the most humid (have a constant supply of water for photosynthesis). Also, in tropical regions, the climate varies little throughout the year and is more stable over the centuries or across geological time. Within each unit of area, there are many more niches than would be present in an equal-sized portion of the temperate zone, with each niche differing slightly from the next in the amount and type of energy that is available. Tropical species can specialize to fill these different stable niches in many ways, living at different heights in a vertically stratified forest, occupying different kinds of microhabitats, exploiting different food resources, or attracting different species of pollinators. Another explanation of this theory is that species living in temperate climates must withstand greater seasonal fluctuations in temperature and sunlight. A species that can tolerate both a Minnesota winter and a Minnesota summer could probably tolerate a variety of conditions across most of North America, and, indeed, hardly any species living in Minnesota is geographically restricted to that state. Temperate regions will therefore tend to support fewer wide-ranging species, while tropical regions will support many more species with geographically smaller ranges.

Genetic diversity is another important aspect of biodiversity. If a species is to be resilient in its response to shifts in habitat, it must have the genetic resources that might allow it to display new phenotypes. In many species, however, shrinking populations have reduced the amount of genetic diversity. Humans have also reduced the genetic diversity of many domestic species of animals and plants; one instance of this will be discussed shortly.

18.1.2 Interdependence of humans and biodiversity

The study of biological diversity is as old as our need for food, clothing, shelter, and medicines, because all of these things, as well as tools and weapons, are made from the millions of other species that inhabit our planet. At the same time, humans have tilled farms, grown crops, and built cities and roads. All of these activities have altered many ecosystems, some of them profoundly, and have thus influenced biodiversity. Human activities affect biodiversity, and biodiversity affects our lives in return.

18.1.2.1 The biological value of preserving species The preservation of biological diversity is important for many reasons, of which three broad types can be distinguished. First, our ignorance as to which species might be beneficial is a reason for simply preserving all species. We know that many plants have yielded important drugs; other plants have yielded important foods, dyes, paper, and rubber. Among the species now living but poorly known, some probably possess a wealth of new possibilities for such uses; therefore, we should preserve them all for the sake of those that may someday prove useful to humans. For example, cures for AIDS or cancer may lie hidden in the depths of the rainforest; in fact, several useful cancer medicines have already come from rainforest plants.

A second reason for preserving biodiversity pertains to the wild relatives of our domesticated species. The store of genetic variation, and therefore the possible number of genetic traits from which to choose, is greatly reduced in each of our domestic species and is much greater in their wild relatives. It is therefore in our long-range best interests to preserve the wild relatives of all domestic species and varieties so that newly discovered desirable properties (or properties that become desirable) can be bred into domestic stocks from their wild relatives.

For example, corn (*Zea mays*, also called maize) is one of the world's most valuable domesticated species of plants, but the domesticated variety is an annual plant that must be replanted each year at considerable labor and expense. In the 1970s, however, a wild relative named *Zea diploperennis* or teosinte ("mother of corn" in the Aztec language) was discovered growing in the Mexican state of Jalisco, confined to a small mountain tract (Iltis et al., 1979). The discovery was made just days before the land was scheduled to be cleared, which would have wiped the species out. *Z. diploperennis* was found to be resistant to a number of diseases that afflict domestic varieties. Best of all, unlike all other species and varieties of corn in the world, the newly discovered species grows as a perennial, meaning that an individual plant produces corn year after year without replanting. If some of these genetic traits could be introduced into domestic corn, either by breeding or by genetic engineering (see Section 17.5), the new strains could represent billions of dollars' worth of savings for the farmers of all corn-producing regions (Camara-Hernandez and Manglesdorf, 1981; Murray and Jessup, 2013). Had the Jalisco corn not been discovered in time, an important genetic reserve for this important domestic species would have been lost forever.

Another example concerns bananas. Of the many varieties of banana formerly cultivated, only one, the Cavendish variety, remains commercially profitable, and this one extremely inbred strain is highly susceptible to a fungus that wiped out many African banana crops in the 1960s. Preservation of the last few domesticated varieties and wild banana relatives may be the only hope of finding genetic traits that can resist this fungus. Apart from corn and bananas, similar arguments can be given for the preservation of the genetic resources of other species in zoos, botanical reserves, and gene banks, but the most cost-effective way to preserve these genetic resources is to promote the survival of the wild species or varieties in their natural habitats.

18.1.2.2 Preserving ecosystem stability A third reason for preserving biological diversity is that species affect one another. *There are no ecosystems that are made up of only one or a few species.* Recall that a community is a group of species whose needs are interdependent. Stable communities are stable in part because materials are recycled: Many producer, consumer, and decomposer species (see Section 17.1) are present. A small group of species is much less likely to form a complete and stable community than a larger one. Multiple species of each kind make the stability less likely to be disrupted. For example, multiple prey species provide a more stable food supply for predators because the predators can survive a scarcity of one prey species by switching to other species for their food.

Many communities are unstable in the sense that the removal of just one "keystone" species can cause the balance among dozens of other species to collapse, such that the disappearance of one species causes other extinctions and leads to other drastic changes. Some of these changes may even affect the physical environment, as when the removal of beavers causes dams not to be built and allows water to flow more freely, or when corals can no longer grow and provide reef habitat for a great diversity of fish and invertebrate species. Other communities, such as tropical rainforests, are thought to be more stable than these examples as a consequence of their large number and variety of species. In a typical rainforest there are hundreds of species of trees, with no single species constituting more than 5% or so of the total. There are also hundreds of bird species, thousands of insect species, and a large diversity of other animals and plants, some of which are illustrated later in this chapter.

The health of animals, including humans, is promoted by the variety of plants available for them to eat or to climb or to nest in. Likewise, the health and well-being of many of the plants depends on the variety of animals that can pollinate them, disperse their seeds, or fertilize the ground near their roots with their feces and other remains. While it is obvious that the survival of any of these species depends on the survival of the ecosystem as a whole, it is equally true that the stability of the community (and often of the entire ecosystem) depends on the survival of certain key species or groups of species.

Many evolutionary and ecological processes increase biodiversity. Life as we know it depends on the rich biodiversity of the world's major ecosystems, yet we are currently in an era when biodiversity is rapidly declining. What forces are producing this decline? Will the current decline in biodiversity match those of the great mass extinctions of the past? These are some of the issues addressed in the next section.

THOUGHT QUESTIONS

1. Why would the perennial growth habit in the corn from Jalisco be considered a valuable trait? Under what agricultural conditions (and in what nations) would this trait be especially valuable?
2. How easily could the genes for perennial growth be identified? How easily could these genes be introduced from the Jalisco corn into the domestic varieties? (You will probably need to review parts of Chapters 2, 4, and 17 to answer this question.)
3. What would be the effects of clearing 100 square kilometers (100 km^2) from a small rainforest 5000 km^2 in size (about 2,000 square miles, approximately the size of Delaware)? What would be the effect of clearing 1,000 km^2 of this same forest?

18.2 EXTINCTION REDUCES BIODIVERSITY

Evolutionary change often produces new species and thus increases biodiversity. But evolution can also lead to the disappearance of species, a phenomenon called **extinction**. The species that we see around us today are in fact only a small proportion of all those that have ever lived.

Species that have no living members are said to be extinct. Extinct species are known through the fossil record. As an example, consider the Age of Reptiles, or Mesozoic era, a time interval from approximately 200 to 65 million years ago (see Figure 5.8). Of all the species that lived during that time, none are still alive today—they are all considered extinct. These species did not die out all at once, but rather a few at a time, although many perished in a mass extinction at the end of the Cretaceous period. The reasons behind these extinctions differed from case to case and are imperfectly known for most species. In some cases, the fossil record shows that a competing group of species appeared on the scene shortly before the extinction occurred. In other cases, no specific cause can be identified.

18.2.1 Types of extinction

If all Mesozoic species are now extinct, how did life manage to persist? To answer this question, we must distinguish between two major types of extinction. First, we must recognize the concept of a **lineage**, which is an unbroken series of species arranged in ancestor-to-descendant sequence, with each later species having evolved from the one that immediately preceded it. If we had a complete record of the history of life on this planet, every lineage would extend back in time to the common origin of all earthly life. Working forward from earlier times, each lineage extends either to a species alive today or to one that has become extinct.

When an entire lineage has died out without issue, no living descendant species exist. We call this true extinction. Many groups of organisms have, according to current theories, undergone true extinction, meaning that no living species are descended from them—their lineages have ended. Among these groups are the trilobites and conodonts shown in Figure 18.3 and several Mesozoic reptile groups, including six of the seven suborders of dinosaurs as well as the marine ichthyosaurs and plesiosaurs.

When a species no longer exists, its lineage may continue in the form of descendant species. The ancestral horse, *Hyracotherium* (formerly called *Eohippus*), is extinct in one sense: There are none alive today; however, they do have living descendants, the modern horses. This type of change, called **pseudoextinction** or phyletic transformation, occurs when a species evolves into something recognizable as a different species. Many of the traits of *Hyracotherium*, and the genes that contributed to these traits, persist among modern horses. Likewise, many of the

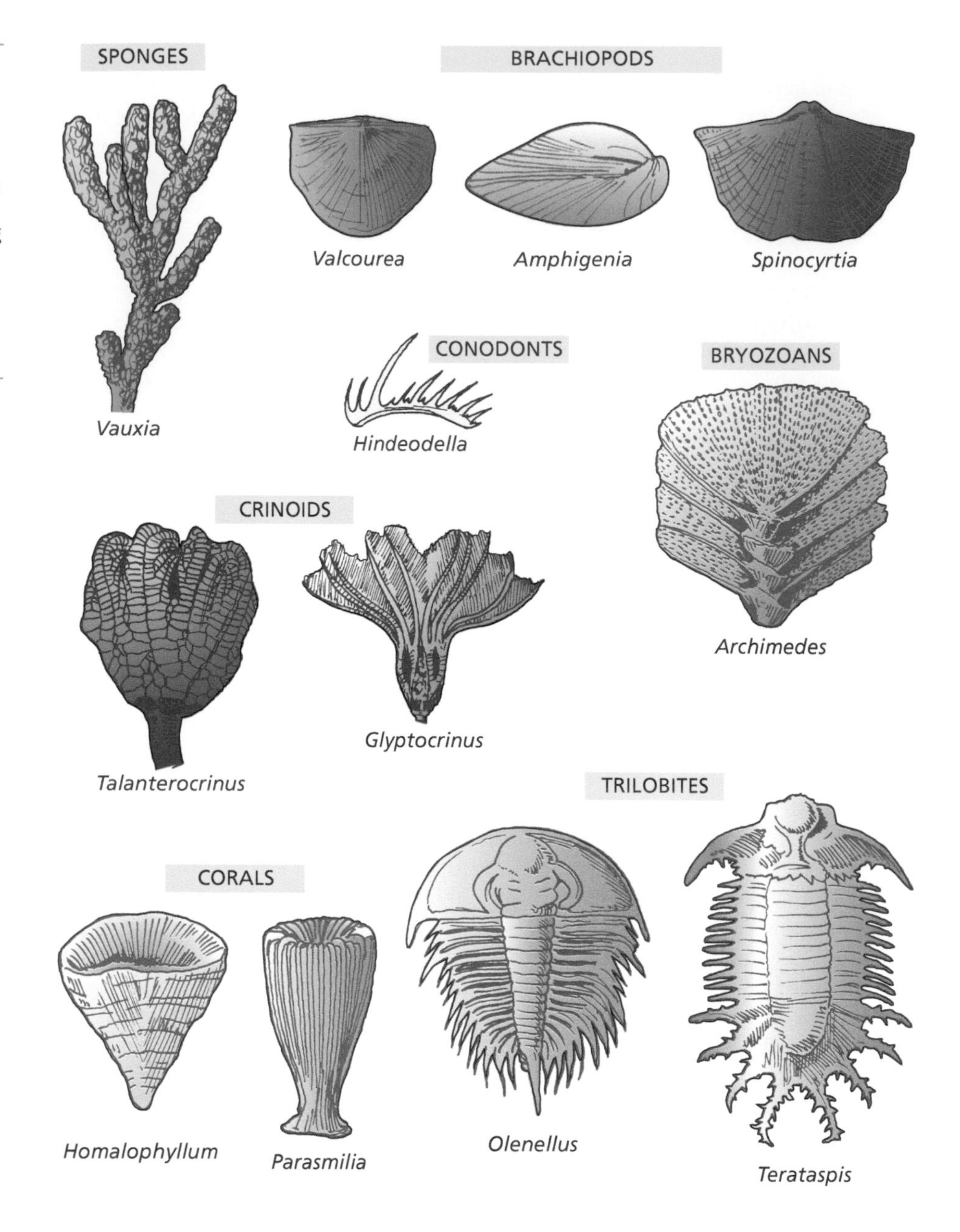

Figure 18.3 Extinct species of the Paleozoic era. The fossils shown here represent groups that were abundant during Paleozoic times. All these groups suffered considerable extinction at the end of the Permian period. The trilobites (belonging to the phylum Arthropoda) and conodonts (jaw structures belonging to the phylum Chordata) are truly extinct groups with no living descendants. The other groups have some living species, but the species alive today are not the same as the Paleozoic species.

genetic traits of one dinosaur group (the Theropoda) persist among their descendants, the modern birds. In many cases we do not know whether an extinct group has undergone true extinction or only pseudoextinction—the evidence does not permit us to make a clear choice. For example, the swamps of the Carboniferous time period were populated by many types of amphibians, and it is unclear which of these belong to truly extinct lineages and which are ancestral to modern frogs and salamanders and are thus only pseudoextinct. Increasing efforts to extract DNA from fossils now allows us to detect DNA sequences from Pleistocene cave bears in modern brown bears and Neanderthal DNA sequences in modern humans, showing that both Neanderthals and cave bears (who inhabited many of the same caves) are only pseudoextinct (Barlow et al., 2018).

18.2.2 Analyzing patterns of extinction

Has extinction occurred randomly? Does the probability of extinction remain constant over time and from place to place? Several late twentieth-century biologists have hypothesized that extinction occurs at random over vast time periods. This hypothesis has been tested in a mathematical model that compares actual extinctions with theoretical predictions based on a model of random extinctions. Comparison of actual data with a theoretical model is a good research strategy (see Section 1.1) because it allows us to identify both circumstances for which the model holds and circumstances for which it does not (and for which additional explanations are therefore needed).

Several studies have compared extinction in the fossil record of particular animal groups with the random model. Most of these comparisons counted all extinctions together, without distinguishing true extinction from pseudoextinction. The fossil record of mammals from the Tertiary period is one of several such groups that conforms to the random extinction model; that is, the actual extinctions match the rates predicted by the model. Many instances of nonrandom extinction have also been discovered. For example, many early invertebrate groups suffered most of their extinction early in their history rather than at a constant rate through time.

Comparison with mathematical models has revealed two major types of departure from randomness: situations in which fewer species become extinct than random models predict and also situations in which more species become extinct than those models would predict.

18.2.2.1 Living fossils We first examine situations in which the frequency of extinctions is reduced. "Living fossils" are species or genera that have survived for many millions of years without true extinction and with only minimal pseudoextinction, meaning that very little morphological change separates the living species from their fossil relatives. Several of these living fossils are described next, and some are shown in Figure 18.4; the time periods mentioned are shown in Figure 5.8:

Psilophyton, a primitive vascular plant (kingdom Plantae, phylum Psilophyta) closely resembling the earliest land plants of the Silurian period.

Ginkgo, a tree (kingdom Plantae, phylum Ginkgophyta) native to China that closely resembles its Mesozoic ancestors and is planted in many urban areas around the world because it tolerates urban pollution.

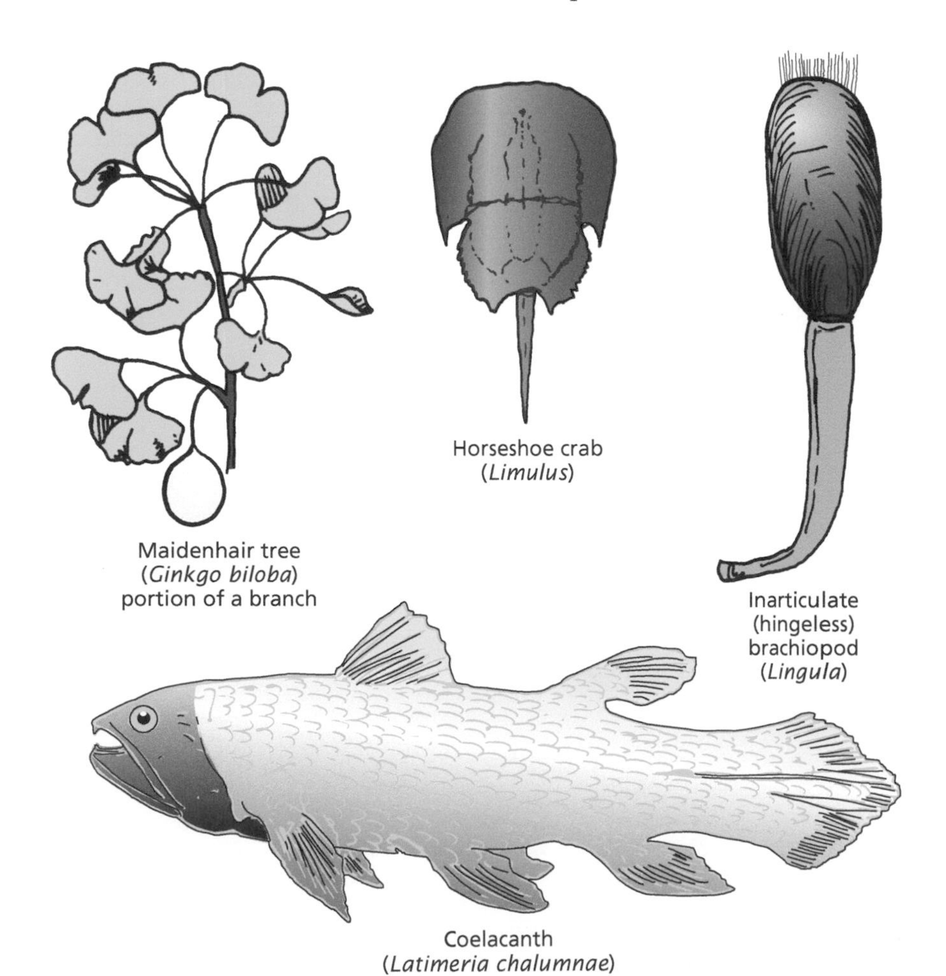

Figure 18.4 Living fossils that have avoided extinction over long periods of time.

Lingula, a type of brachiopod (kingdom Animalia, phylum Brachiopoda) with a wormlike body enclosed in a two-valved shell that has no hinge and with a feeding structure that strains suspended particles from the water.

Neopilina, a deep-water mollusk (kingdom Animalia, phylum Mollusca) with a low-domed conical shell resembling that of the extinct genus *Pilina*, one of the most primitive mollusks.

Limulus, the horseshoe crab (kingdom Animalia, phylum Arthropoda), which closely resembles its Paleozoic ancestors.

Latimeria, a large, rare Indian Ocean fish (kingdom Animalia, phylum Chordata, class Osteichthyes, order Crossopterygii) belonging to a group (the coelacanths) whose other members became extinct during the Mesozoic era.

What might make a taxon of organisms less likely to suffer extinction? These living fossils share several characteristics: They all have locally large populations (with a sufficient gene pool to maintain a large amount of genetic variation); they are all adapted to dependably persistent habitats (such as deep ocean waters); those that are animals do not depend on a narrow range of food species (some of them will eat anything within a certain size range); and they all have reproductive stages (pollen, spores, or larvae) that are dispersed mechanically by wind or ocean currents rather than by other species. If there is any secret to long-term survival, these species have surely stumbled upon it.

18.2.2.2 Mass extinctions Departing in the other direction from the random extinction model are examples in which many more species became extinct within a geologically short interval of time than would have been predicted by the model. We call these mass extinctions. There was one such event at the end of the Cretaceous period, which was also the end of the Mesozoic era. There was another, even larger, mass extinction at the end of the Permian period (Figure 18.5).

Mass extinctions devastate biodiversity. Over half of the families and 85% of the genera became extinct in the last 5 million years of the Permian period, and much higher percentages were lost in some classes and phyla (Erwin, 1990, 1993; Broadley et al., 2018). Although many more species and genera perished in this mass extinction than in any other, the Permian event has attracted much less attention than other mass extinctions because nearly all the species were unfamiliar types of

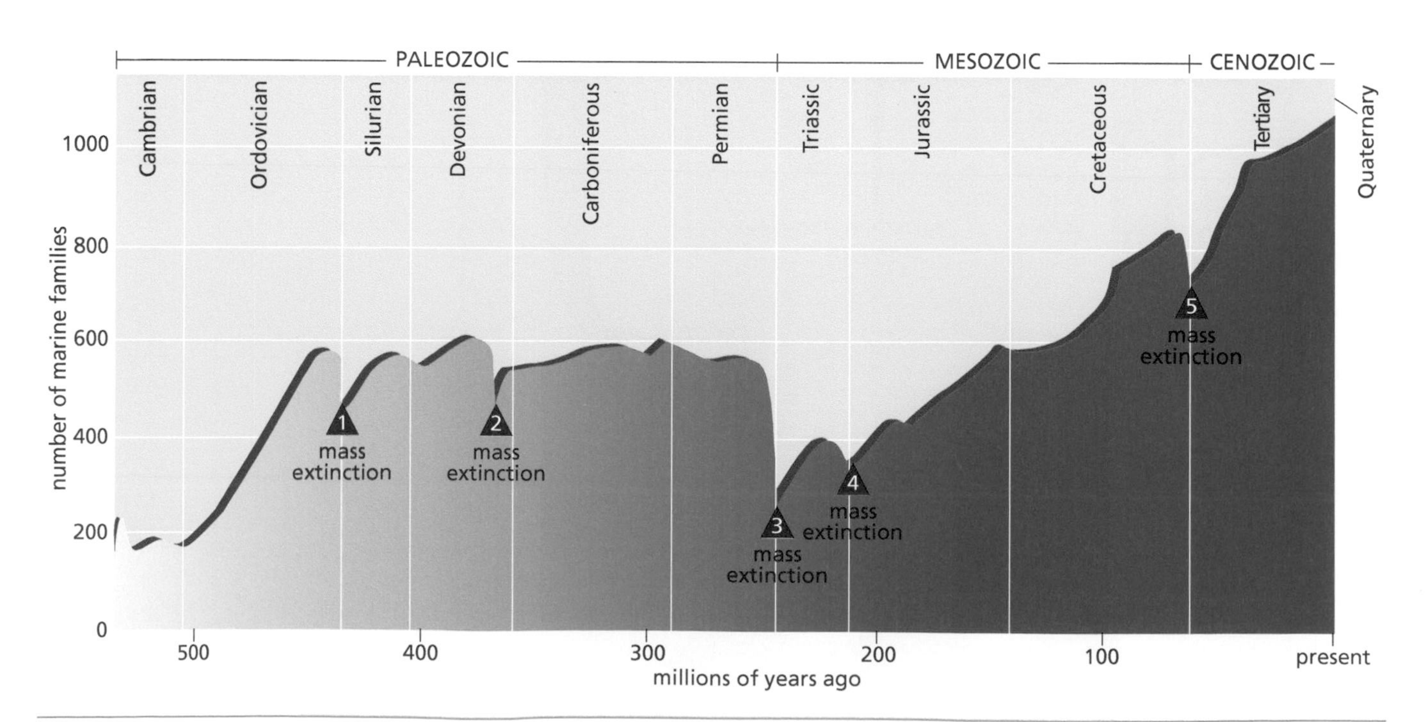

Figure 18.5 Changes through time in the number of families of marine organisms. Mass extinctions are indicated by numbered triangles.

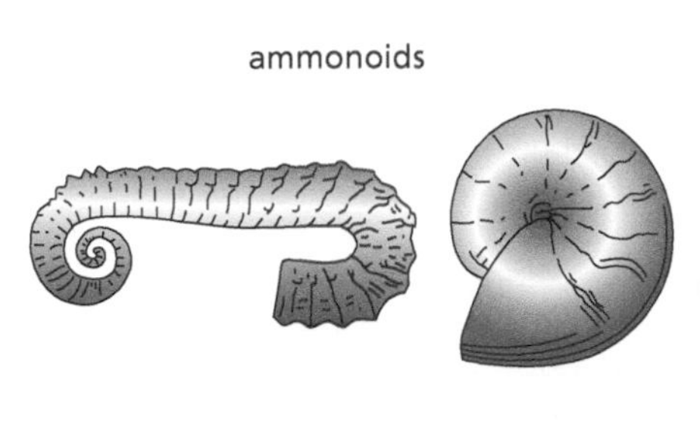

giant ammonoid

artist's view of various Mesozoic reptiles

Figure 18.6 Extinct species belonging to groups that died out completely at the end of the Cretaceous period.

organisms (for example, crinoids and brachiopods; Figure 18.3) that lived in underwater habitats such as the shallow inland seas that were abundant at that time.

The mass extinction at the end of the Cretaceous period has attracted the most attention (Archibald, 1996; Gulick, 2019) because many well-known animals became extinct at this time: dinosaurs (the reptilian orders Saurischia and Ornithischia); flying reptiles (order Pterosauria); several types of marine reptiles (orders Sauropterygia, Ichthyopterygia, and others); ammonoids (phylum Mollusca, class Cephalopoda, mostly with large, coiled shells); and several groups of fish, plants, and other organisms (Figure 18.6).

18.2.2.3 Possible causes of mass extinctions The fossil record shows at least five mass extinctions in which many families of marine organisms died out (Figure 18.5) (Donovan, 1989; Hallam, 1997). The rates of extinction happening today are as great as the rates during these mass extinctions. Many scientists have therefore concluded that a sixth great mass extinction is currently in progress. A large body of evidence points to human hunting and human habitat destruction as the causes responsible for the current mass extinction.

What could have caused such high rates of extinction in past geological times? There are several hypotheses, including the following: warming or cooling of the Earth, changes in seasonal fluctuations or ocean currents, and changing positions of the continents (plate tectonics). Biological hypotheses include ecological changes brought about by evolution of cooperation between insects and flowering plants or of bottom-feeding predators in the oceans. Some of the proposed mechanisms require a very brief period during which all extinctions suddenly took place; other mechanisms would be more likely to have taken place more gradually, over an extended period, or at different times on different continents. Some hypotheses fail to account for simultaneous extinctions on land and in the seas.

Each mass extinction may have had a different cause or multiple causes. The largest mass extinction of the geologic past was the end-Permian "Great Dying" that closed out the Paleozoic Era. Geologists who have studied rock formations of this age have noted an unusually large amount of basaltic lava flows in Siberia that went on for a million years or more. Hot lava of this composition would be expected to release many poisonous gases into the atmosphere, and some of these gases may have formed chlorine monoxide (ClO) and other highly reactive free radicals that would persist for many years and drift upwards into the stratosphere. Once in the stratosphere, such free radicals would be capable of destroying much of the Earth's ozone layer, as we shall discuss in Chapter 19. This, in turn, would cause an increase in the ultraviolet radiation reaching Earth's surface, with likely increases in both cancer rates and mutation rates among nearly all species. At about the same time, large areas of shallow-water seas ("epeiric" seas) were disappearing as the result of tectonic movements, and the extinction of many bottom-dwelling species (especially brachiopods, echinoderms, early mollusks, and the last of the trilobites) may have caused the further extinction of the predatory fishes and arthropods that fed upon them.

American paleontologists David Raup and John Sepkoski, who have studied extinction rates in a number of fossil groups, suggest that episodes of increased extinction have recurred periodically, approximately every 26 million years since the mid-Cretaceous period (Raup and Sepkoski, 1984; Raup, 1991). The late Cretaceous extinction of the dinosaurs and ammonoids was just one of the more drastic in a whole series of such recurrent extinction episodes. The possibility that mass extinctions may recur periodically has given rise to such hypotheses as that of a companion star with a long-period orbit deflecting other bodies from their normal orbits, causing some of them to fall to Earth as meteors, wreaking widespread devastation upon impact.

18.2.2.4 The asteroid impact hypothesis Of the various hypotheses attempting to account for the late Cretaceous extinctions, the one that has attracted the most attention is the asteroid impact hypothesis first suggested by Luis and Walter Alvarez (see Figure 1.4E). According to this hypothesis, the Earth collided with an asteroid with an estimated diameter of 10 km, or with several asteroids, the combined mass of which was comparable. The force of collision spewed large amounts of debris into the atmosphere, darkening the skies for several years before the finer particles settled. The reduced level of photosynthesis led to a massive decline in plant life of all kinds, and this led to massive starvation, first of herbivores and subsequently of carnivores. The mass extinction would have occurred very suddenly under this hypothesis (Broadley et al., 2018; Gulick et al., 2019).

One interesting test of the Alvarez hypothesis is based on the presence of the rare-earth element iridium (Ir). The Earth's crust contains very little of this element, but most asteroids contain a lot more. Debris thrown into the atmosphere by an asteroid collision would presumably contain large amounts of iridium, and atmospheric currents would carry this material all over the globe. A search of sedimentary deposits that span the boundary between the Cretaceous and Tertiary periods shows that there is a dramatic increase in the abundance of iridium briefly and precisely at this boundary (visible in Figure 1.4E). This iridium anomaly offers strong support for the Alvarez hypothesis.

An asteroid of this size would be expected to leave an immense crater, even if the asteroid itself was disintegrated by the impact. The intense heat of the impact would produce heat-shocked quartz in many types of rocks. Also, large blocks thrown aside by the impact would form secondary craters surrounding the main crater. To date, several such secondary craters have been found along Mexico's Yucatan Peninsula, and heat-shocked quartz has been found both in Mexico and in Haiti and as far away as North Dakota. A location called Chicxulub, along the Yucatan coast, has been confirmed as the primary impact site. A recent discovery in North Dakota (DaPalma et al., 2022) reveals a fossil deposit, dated at 65.76 ± 0.15 million years of age, containing glass fragments from the impact as well as multiple species killed by the impact event. Evidence of near-simultaneous sudden death of many species on a massive scale includes multiple fish specimens with glass spherules lodged in their gills. Also, marine and freshwater species are heaped together in a manner suggesting a sudden inundation and rapid burial, with no evidence of scavenging or decay.

A 2021 analysis of fossil pollen from carefully dated deposits in Colombia that span the Cretaceous-Tertiary boundary does indeed reveal a devastating removal of plants from this region about 66 million years ago, followed by a lengthy interval of at least 8 million years, during which very little pollen was present (Carvalho et al., 2021). The older plants that were killed off were mostly wind-pollinated cycads and other gymnosperms, spaced widely apart, with bare ground between them and relatively few places for insects to hide. The vegetation returned slowly, by a process of ecological succession (described earlier) that took an estimated 8 million years. The new plant species were mostly angiosperms (Anthophyta), and most of them were now insect-pollinated. Many of these newer angiosperms were trees, much taller than the plants that they replaced, and they grew closer together to form a continuous canopy, as described later in this chapter. Beneath these tall trees, new habitats were created for a great variety of smaller, shade-tolerant plants. The great biodiversity of tropical rainforests developed gradually at this time, providing both habitat and many new niches for many thousands of species of insects (including social insects); birds; and insect-eating mammals, frogs, and lizards. Today's tropical rainforests were made possible by the mass extinction that killed off many earlier plants and most of the reptiles that roamed around them.

18.2.2.5 Quaternary extinctions There were also many extinctions during the past 2 million years, an interval that includes the Pleistocene epoch and its several glacial episodes, plus the Recent epoch (beginning with the end of the last ice age, about 10,000 years ago). A number of species of large mammals, reptiles, and flightless birds became extinct during these 2 million years, and several hypotheses have been advanced as explanations. The most obvious hypothesis attributes the extinctions to changes in climate and the advance and retreat of Pleistocene glaciers. But paleontologists who have carefully examined the fossil record point out that this hypothesis fails to explain the timing of the extinctions, most of which did not coincide with the extremes of temperature.

A second hypothesis is that newly introduced species brought about extinction through increased competition. Widespread glaciation caused a decline in sea levels, which resulted in the emergence of land bridges, including those across Panama, the Bering Strait, and the English Channel. Many species were thus introduced from one landmass to another, and the newly introduced species competed with other species already present and established new predator-prey interactions. Many species could not adjust to the new conditions and became extinct as a result. This hypothesis has been used to explain many of the animal extinctions that followed the emergence of the land bridge across Panama, connecting South America (previously an island continent) with North America. Most of this group of extinctions, however, happened early in the Pleistocene, leaving another large group of later extinctions still to be explained. We examine these most recent extinctions next.

18.2.2.6 The human role in extinctions From the comparison of species that became extinct in the past 50,000 years with others that did not become extinct, an interesting pattern emerges: Only large, conspicuous species of mammals, reptiles, and flightless birds became extinct, while smaller animals (including rodents, bats, and small birds) or marine animals suffered very little extinction or else none at all. This pattern suggests yet another hypothesis: that the activities of humans, including both hunting and alterations of habitat, played a large role in the extinctions of the past 50,000 years (Martin and Klein, 1984).

A great deal of circumstantial evidence favors the hypothesis of extinction by human agency: in those places where the time of first human arrival can be dated (e.g., Madagascar, New Zealand, and certain Pacific islands), dense piles of animal bones accumulated beginning at the times of human arrival, and most of the extinctions took place soon afterwards, within several hundred years of the arrival of humans at each place (Figure 18.7). Moreover, the species that became extinct were those that humans would be apt to hunt, mostly large herbivores, whereas most species that would have been difficult to hunt, or too small to be worth

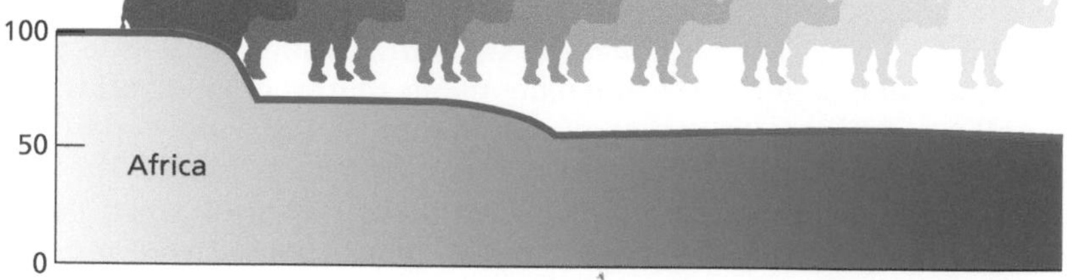
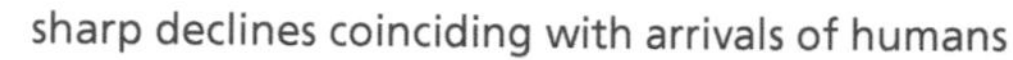
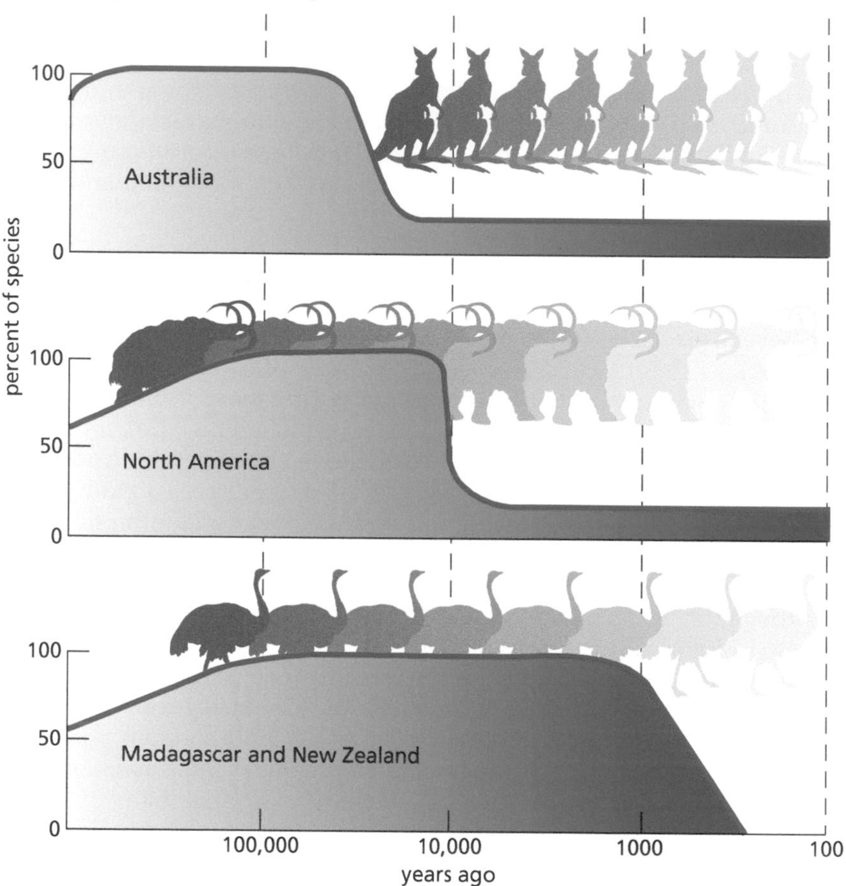

Figure 18.7 The extinction of many large mammals and flightless birds followed soon after human arrival in many parts of the world. In Africa, however, extinctions took place more gradually because humans had been present for a much longer period.

hunting, survived. On the island continent of Australia, for example, the arrival of humans some 50,000–60,000 years ago was soon followed by the extinction of 20 species of giant kangaroos, along with a marsupial lion, a marsupial wolf or tiger (*Thylacinus*, surviving into the 1800s on Tasmania), and the giant, cow-sized herbivore *Diprotodon*. On New Zealand and the Hawaiian Islands, the extinction of flightless birds began with human arrival and was nearly complete by the time of European discovery.

Several vertebrate species have become extinct within historical time: The dodo (a large, flightless bird) and the passenger pigeon are two famous examples. Many other species of birds and also many plants and insects also became extinct. Hawaii was home to some 50 species of land birds at the time of its European discovery in 1778. Humans and the animals that they have introduced since that time have caused the extinction of one-third of these species, and archaeologists have shown that the indigenous Hawaiians had hunted an additional 35–55 species to extinction before the arrival of Captain Cook. On New Zealand, the giant moa (an ostrich-like flightless bird) was hunted to extinction by the Maoris, the indigenous people of that island nation. Much the same thing happened on Madagascar, where an even larger flightless bird, the elephant bird, had become extinct long before European colonists arrived.

Most recently, and on a worldwide scale, it is now estimated that more than 100,000 plant and animal species became extinct during the decade of the 1980s and that the rate of extinction is accelerating (Ehrlich and Ehrlich, 1981; Nitecki, 1984; Gore, 1989; Eldredge, 1994; Grant, 1995). Most of the extinctions of the last few decades were on land and were scattered among many families, so they do not show on the graph of Figure 18.6, which counts only marine families. In the United States, many species of migratory songbirds have been greatly reduced in numbers and several have become extinct. Bachman's warbler, a bird species that was once common throughout the Southeastern United States, was last seen in 1988 (U.S. FWS, n.d.). The Ohio River and Lake Erie once had dense populations of 78 species of freshwater mollusks, of which 19 are now extinct and another 29 are rare. Biologists now believe that a single dam, Wilson Dam, on the Tennessee River, caused the extinction of 44 species of freshwater mollusks when it was built (Sherpa Guides, n.d.). In Africa's Lake Victoria, over 100 species of fishes became extinct after the introduction of the Nile perch, a large, predatory sport fish introduced to the lake in 1959 (Pringle, 2005). In 2021, the U.S. Fish and Wildlife Service declared that 11 bird species, 8 mollusks, and 4 other species have recently been driven to extinction in the United States (U.S. FWS, 2021; BBC News, 2021).

18.2.3 Species threatened with extinction today

A species threatened with extinction is called an **endangered species**. An example is the northern spotted owl (*Strix occidentalis*) of California, Oregon, and Washington (Portland Audubon, 2021). These owls nest only in old pine trees in "old-growth" forests, meaning forests with many species of trees of mixed ages and sizes, including very old trees (up to 200 years old or more) and also dead wood, both standing and fallen. A large area of old-growth forest is needed to support enough rodents, small birds, and other wildlife for each individual spotted owl to find enough food. The species is currently endangered because of logging of the Pacific coastal forests for timber.

Various governments and international organizations maintain official lists of endangered species. The United States, for example, maintains a list of endangered species in the United States as part of the Endangered Species Act. International lists are maintained by such organizations as the International Union for Conservation of Nature (www.iucn.org). These lists differ from one another, however, because they are based on different criteria.

18.2.3.1 Predictors of extinction How do we know whether a species is endangered? One indication that a species may be endangered is a reduction in its numbers. The extinction of a species is nearly always preceded by its becoming rare, and rarity may thus be the prelude to extinction. However, the only type of rarity qualifying as the harbinger of extinction, and therefore qualifying a species as endangered, is the kind in which the entire species is represented by only one or a few populations, all of which are small. Once the population size falls below a certain minimum, several factors can increase the risk of extinction. One of these is genetic drift (see Section 7.2.2), a pattern in which allele frequencies in small populations change erratically and not necessarily adaptively, which may hasten extinction. A second factor is that in very small populations there are more matings between related individuals than in large populations, and this inbreeding makes homozygous recessive traits more frequent in these populations than in non-inbred populations. Because many homozygous recessive traits are harmful, they reduce survival and thus cause the population to decline further and possibly to die out. A third factor is that environmental fluctuations (resulting from changing seasons, weather phenomena, and so forth) may favor different genotypes at different times. Large populations that contain many different genotypes are thus more likely to survive, but populations with less genetic diversity are more susceptible to extinction.

18.2.3.2 Extinction of a niche Other indications that a species is threatened with extinction are the disappearance (or impending disappearance) of its habitat or the disappearance of another species on which its niche depends.

A remarkable example of the latter is the tambalacoque tree (*Sideroxylon* or *Calvaria grandiflorum*) of the island of Mauritius in the Indian Ocean. None of the seeds of these long-lived trees had been observed to germinate, even when planted. A botanist who studied these seeds noticed that they have a very hard outer husk (like a peach pit) that mechanically prevents the seed within from breaking through (Temple, 1977). The seeds could be made to germinate by grinding away some of the outer husk before planting the seeds. The same effect was obtained by feeding these seeds to turkeys, a bird about the size of the dodo. The dodo was a type of bird that lived on the island of Mauritius before Europeans and their domestic animals caused its extinction around 1681. Dodos ate the seeds of the tambalacoque tree, and the hard outer husk was an adaptation that permitted the seeds to survive the digestive action of the dodo's stomach. (Seed-eating birds often intentionally swallow and retain pebbles; these pebbles work like pulverizing machines in their muscular stomachs to abrade such things as tough seeds.) Later studies (e.g., Witmer and Cheke, 1991) showed that a small percentage of tambalacoque trees did reproduce successfully in nature without the dodos. Various studies have helped focus attention on the survival of the trees, and young tambalacoque trees are now growing with the aid of humans, saving the species from the threat of extinction.

18.2.3.3 Species currently in danger The list of endangered species is long and growing. Among mammals, the list includes giant pandas, gorillas, orangutans, lemurs, elephants, manatees, caribou, timber wolves, and dozens of less familiar species (Figure 18.8). There are also large numbers of endangered fishes, birds, amphibians, reptiles, insects, and plants. The International Council for Bird Preservation (an IUCN affiliate) estimates that close to 2,000 species of birds have already become extinct in the past 2,000 years, mostly driven by human agency, and that 11% of the living species (1029 of 9040) are endangered.

A few species that were once listed as endangered have, for the moment, been saved from the brink of extinction, but only when a concerted effort has been made to conserve the species and its habitat. For example, bald eagles (*Haliaeetus leucocephalus*) have been protected and in some cases released into the wild from captivity. Because their numbers are once again sufficient to maintain stable populations, they are no longer considered an endangered species.

Some species, of course, are endangered because of indiscriminate hunting, fishing, or poaching. These are mostly large and conspicuous organisms. However, a much larger threat to biodiversity lies in the destruction of natural habitats, a process that threatens thousands of species at once, including those that are inconspicuous and poorly known. Among these small and inconspicuous organisms, the number of endangered species is certainly many more than those on any official list, and many of these inconspicuous organisms will undoubtedly become extinct before they have even been discovered and named.

The whole strategy of saving individual species is being rethought. In the decades since the first legislation on endangered species, scientists have come to realize that individual species cannot be saved without saving their habitats. The focus is therefore shifting to habitat preservation.

THOUGHT QUESTIONS

1. Experts still disagree on whether the great extinctions of the past occurred suddenly or gradually. If a new article claimed new evidence for either of these hypotheses, how would you go about evaluating this evidence?
2. The presence or absence of dinosaur fossils is used to determine whether a certain bed of rock is Cretaceous or Tertiary. How would this practice influence research on the question of whether dinosaur extinction came about gradually (at different times in different places) or suddenly (simultaneously everywhere)?
3. In 1995, the U.S. Department of the Interior downgraded the legal status of bald eagles (*Haliaeetus leucocephalus*) from endangered to threatened, and in 1999 they were removed from the endangered species list altogether. Why is the U.S. government involved in this matter? Bald eagles also live in Canada and part of Northern Mexico. Does the U.S. government have any authority to declare the bald eagle endangered or threatened in those places?

(A) Titan arum or "corpse flower", *Amorphophallus titanum* (Anthophyta, Liliopsida, Arismatales).

(B) *Osmoderma lassallei* (Arthropoda, Insecta, Coleoptera).

(C) Jewel beetle, *Buprestis splendens* (Arthropoda, Insecta, Coleoptera).

(D) Togo slippery frog, *Conraua derooi* (Chordata, Amphibia, Anura).

(E) *Iguana delicatissima* (Chordata, Sauropsida, Lepidosauria).

(F) Crocodile, *Crocodilus niloticus* (Chordata, Sauropsida, Crocodilia).

(G) Egyptian vulture, *Neophron percnopteris* (Chordata, Sauropsida, Falconiformes).

(H) Koala, *Phascolarctos cinereus* (Chordata, Mammalia, Diprotodonta).

(I) Giant anteater, *Myrmecophaga tridactyla* (Chordata, Mammalia, Edentata).

(J) Coquerel's sifaka, *Propithecus coquereli* (Chordata, Mammalia, Primates).

(K) Mountain gorilla, *Gorilla beringei* (Chordata, Mammalia, Primates). ©www.gorilladoctors.org

Figure 18.8 Selected species currently threatened with extinction.

Figure 18.8 (*Continued*) Selected species currently threatened with extinction.

(L) Pangolin, *Manis javanica* (Chordata, Mammalia, Pholidota).

(M) Giant panda, *Ailuropoda melanoleuca* (Chordata, Mammalia, Carnivora).

(N) African lion, *Panthera leo* (Chordata, Mammalia, Carnivora).

(O) Greater one-horned rhinoceros, *Rhinoceros unicornis* (Chordata, Mammalia, Perissodactyla).

(P) Pampas deer, *Ozotoceras bezoarticus* (Chordata, Mammalia, Artiodactyla).

(Q) West Caucasian Tur, *Capra caucasica* (Chordata, Mammalia, Artiodactyla).

(R) Dorcas gazelle, *Gazella dorcas* (Chordata, Mammalia, Artiodactyla).

(S) African elephant, *Loxodonta africana* (Chordata, Mammalia, Proboscidea).

18.3 SOME ENTIRE HABITATS ARE THREATENED

The destruction of its habitat is among the most serious threats that any species faces. Of all the endangered species in the world, an estimated 73% are endangered for this reason. Habitat destruction threatens many species at once, often thousands at a time. Norman Myers, an ecologist, has identified 18 areas of the world as "hot spots" (Figure 18.9) where threatened habitats put many thousands

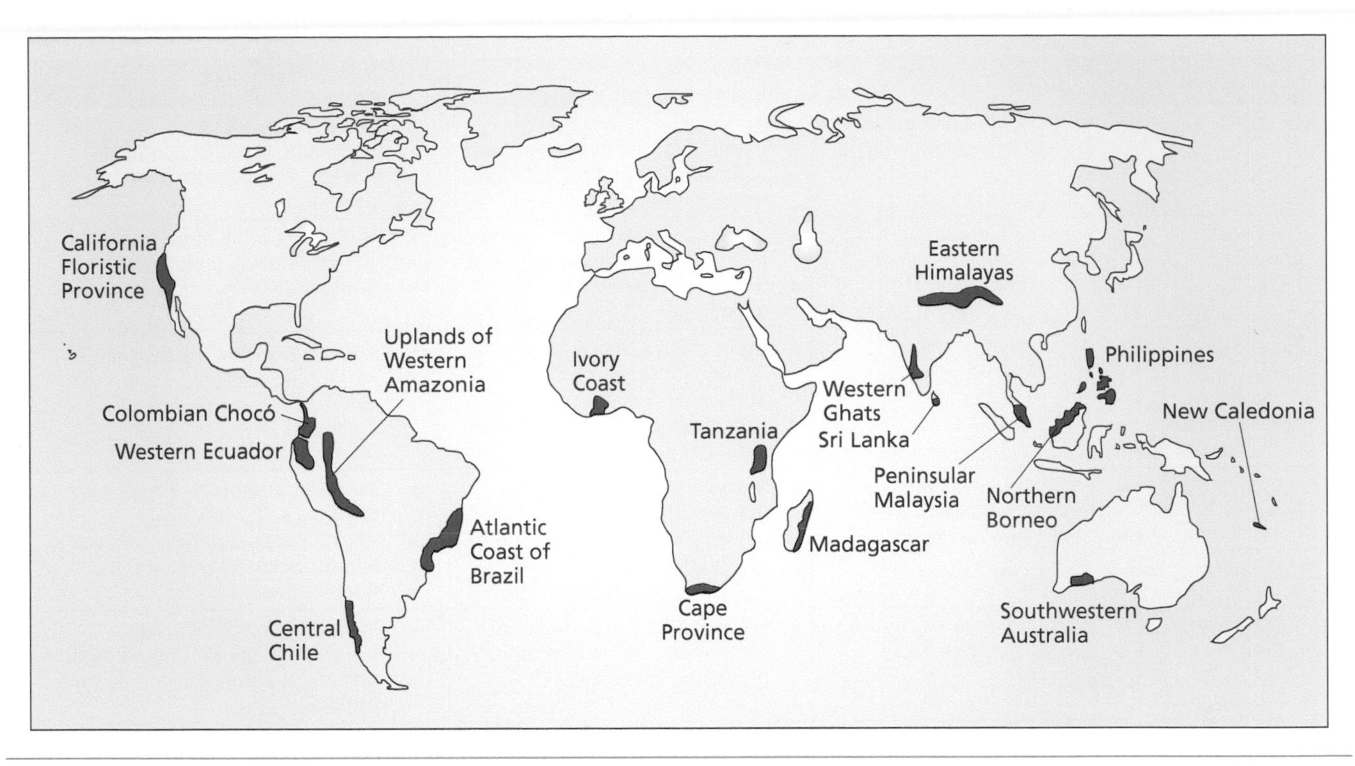

Figure 18.9 **"Hot spots": Habitats containing many species found nowhere else and threatened with extinction from human activity.**

of species at the risk of extinction simultaneously (Myers, 1991; Mittermeier et al., 1999; Kunich, 2003). One reason that so many species are simultaneously threatened is that these places contain many species that are found nowhere else.

Habitats do not exist independently, but are associated into ecosystems. Groups of ecosystems that are similar in type are called **biomes**. Each biome contains ecosystems in different parts of the world that are similar in climate and contain similar habitats and species assortments (Box 18.1). Notice that most biomes

Box 18.1 Terrestrial biomes of the world

A biome is a group of similar ecosystems around the world on different continents. Each biome has a characteristic climate and characteristic types of species, but the particular species and families will often vary from one continent to another within the same biome.

Biome	Climate	Predominant plant life	Predominant animal life
Tropical rainforest	■ Abundant rainfall year-round (200 cm or more annually) ■ Temperature above 25°C year-round	■ Many species of tall trees ■ Diverse understory and forest floor ■ Many epiphytes	■ Maximum species diversity, including many insects, mammals (including primates), birds, and other vertebrates and invertebrates ■ Many species inhabiting trees and underbrush
Tropical seasonal forest	■ Temperature remains above 20°C year-round, but seasonal rainfall varies between extremes	■ Many trees, shrubs, and other plants with adaptations to reduce water loss in the dry season	■ Large diversity of vertebrates and invertebrates, many of them arboreal
Tropical grassland (savanna)	■ Warm in most seasons ■ Rain confined to a wet season ■ Long dry season	■ Landscape dominated by grasses ■ Trees common only along watercourses or in groves	■ Diverse species, with large herbivores often forming large migratory herds (e.g., antelopes in Africa) ■ Lakes may have diverse fish populations
Tropical thornbush scrubland (sahel)	■ Warm to hot ■ Very little rain	■ Bushes often have thorns and water-conserving adaptations; trees are rare ■ Much bare ground	■ Sparse, often confined to vegetated patches, but some species wide-ranging

(Continued)

Biome	Climate	Predominant plant life	Predominant animal life
Desert	▪ Very dry, little precipitation (25 cm or less annually)	▪ No trees ▪ Plants adapted to dryness, with waxy cuticles and thick, fleshy water-storing parts	▪ Few species, including dry-adapted mammals, reptiles, scorpions ▪ Most burrow or hide in hot afternoons
Temperate grassland	▪ Warm or hot summers with low to moderate rainfall ▪ Cold, snowy winters	▪ Landscape dominated by grasses and other wind-dispersed plants ▪ Suitable for large-scale cereal agriculture	▪ Herbivores of all sizes common ▪ Predators often small in size ▪ Medium diversity.
Temperate shrubland (chaparral, Mediterranean)	▪ Warm, dry summers ▪ Mild, rainy winters	▪ Short trees and low flowers and bushes most abundant ▪ Many fire-resistant adaptations ▪ Fruit crops (wine grapes, olives, apricots) grow well	▪ Many herbivores, migratory birds ▪ Medium diversity
Temperate forest	▪ Warm summers with moderate rainfall ▪ Cold, snowy winters	▪ Many broadleaf angiosperm (hardwood) trees whose deciduous leaves contribute to rich humus	▪ Deer and other leaf-browsing herbivores ▪ Many rodents and their predators, also amphibians and reptiles ▪ Medium diversity
Taiga (boreal forest)	▪ Brief summers ▪ Long snowy bitter-cold winters	▪ Predominantly large tracts of coniferous trees (pines, spruces), harvested for lumber ▪ Low diversity	▪ Fur-bearing mammals with temperature-conserving adaptations ▪ Typically, no amphibians and few reptiles
Tundra	▪ Bitter cold much of the year ▪ Little precipitation (mostly snow)	▪ No trees ▪ Very low plants only	▪ Very few species, mostly fur-bearing mammals with cold adaptations

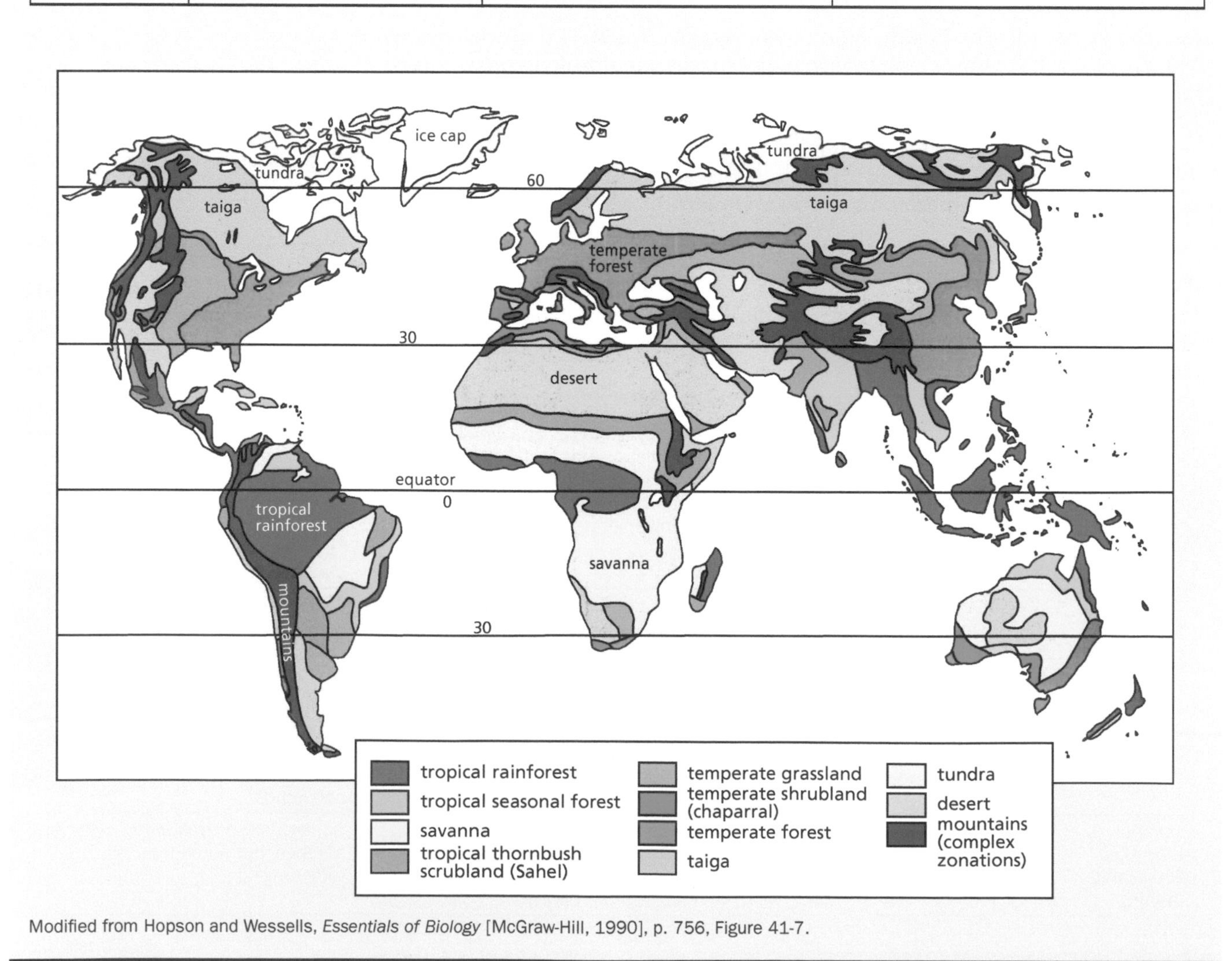

Modified from Hopson and Wessells, *Essentials of Biology* [McGraw-Hill, 1990], p. 756, Figure 41-7.

are restricted to certain bands of latitude, sometimes in both the Northern and Southern Hemispheres. Individual species differ from place to place, but a particular biome has certain proportions of ecological types, such as large herbivores and tall grasses on tropical grassland (savanna) or large stands of pine and other evergreen conifers on taiga. Temperate shrubland (also called chaparral or Mediterranean) has mild, rainy winters and supports many shrubs and small trees, such as olives. Some other examples of biomes are desert, rainforest, and tundra (a cold, treeless land with low-growing vegetation only). Marine biomes also exist, but they are less clearly defined.

What impact does habitat destruction have on human lives? How should humans value habitat? In this section we examine two specific types of habitat destruction: the destruction of tropical rainforests and the expansion of deserts.

18.3.1 The tropical rainforest biome

Tropical rainforests are a biome that encompasses habitats of low latitude (from about 15°S to 25°N), continually warm temperatures, and year-round high precipitation (Alameda and Pringle, 1988; Gay, 1993; Gentry, 1992; Newman, 2002). Average temperatures are usually about 25°C, typically fluctuating only from about 22° to 27°C with seasonal extremes no lower than 20°C. Precipitation is high throughout the year, with annual totals of 1800 mm (about 71 inches) or more. Although there is seasonal variation, no month averages less than 60 mm (2.36 inches) of rainfall. Under these conditions, humidity stays moderate to high at all times. The largest tropical rainforest, the Amazon, occupies much of Brazil and several adjoining countries; other large rainforests occupy much of tropical Asia (from Myanmar to the Philippines, including Indonesia) and tropical Africa (from Senegal to the Democratic Republic of Congo).

The most conspicuous rainforest vegetation consists of tall trees, 30–70 m (about 100–240 feet) in height, up to the height of a 20-story building. As Figure 18.10 shows, their leafy tops make a continuous canopy through which tree-living animals can roam widely without ever descending to the ground. The tallest of the trees may protrude above this canopy level. The taller trees are often buttressed at their bases with a variety of woody supports that give their base a fluted rather than cylindrical shape (Figure 18.10A,C). Much of the rainforest receives little direct sunlight below the canopy, and most of the plants here are shade tolerant. Many plants adapted to the understory have huge, dark green leaves that capture a maximum amount of light, and some are now used as indoor office and lobby decorations precisely because they don't need much sunlight.

Tropical rainforests are vital to the biosphere because almost half of all the photosynthesis that plants perform on our planet takes place in tropical rainforests. This photosynthesis is an important source of atmospheric oxygen and a safeguard against a global increase in carbon dioxide. Unfortunately, human activities are already producing carbon dioxide at a rate so fast that global photosynthesis rates cannot keep up. Scientists believe that the increasing levels of atmospheric carbon dioxide are leading to global warming (see Chapter 19). Tropical rainforests are an important global resource whose continued health benefits the entire planet, not just the countries in which the rainforests are found.

18.3.1.1 Rainforest biodiversity and ecological diversity As we have seen, the tropics contain many more species than do temperate and polar regions. Habitat destruction in the tropics therefore affects many more species than it does at higher latitudes. Tropical rainforests are particularly vulnerable because they contain so many different habitats, and they are being destroyed at an accelerating rate. Ecologist E.O. Wilson estimates that one Peruvian farmer clearing some rainforest land to grow food for his family will cut down more kinds of trees than are native to all of Europe.

Rainforests have a high diversity of both plant and animal species, with no single species forming more than a small fraction of this rich diversity. In addition to the many hundreds of species of trees and other plants, rainforests are home to several hundred kinds of birds; numerous reptiles, amphibians, fishes, and mammals; and thousands of insect species (Figure 18.11). Some biologists have

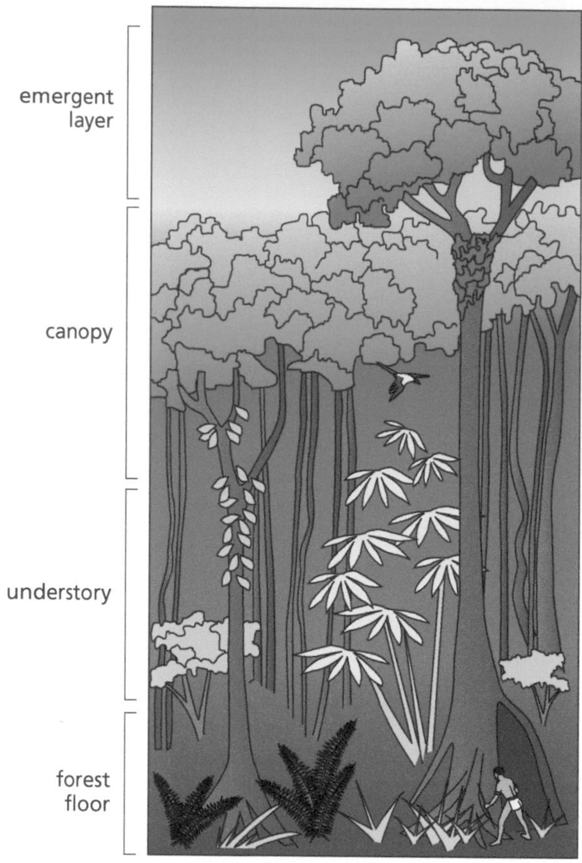

(A) Profile view of a rainforest, showing vertical stratification.

(B) The interior of a Costa Rican rainforest. Notice the marked differences in light levels from the canopy above to the understory below.

(C) Trees in tropical rainforests often have buttress supports. Those at the base of this tree have been inundated by flood waters of the Rio Negro in the Amazon rainforest.

Figure 18.10 Rainforests.

Guzmania nicaraguensis, showing bright yellow flowers surrounded by modified leaves called bracts, whose bright red color attracts the hummingbirds that pollinate this species. Bright red colors are common in bird-pollinated plants.

Spathophyllum, showing leaves damaged by the feeding activities of herbivores.

Piper, a tree whose inflorescences reach upward.

A tarantula from Costa Rica.

A walking stick insect, camouflaged to resemble the lichens that grow on many tree trunks.

A butterfly feeding on a colorful flower.

Figure 18.11 Some of the biodiversity of tropical rainforests. *(Continued)*

Bronze beetle.

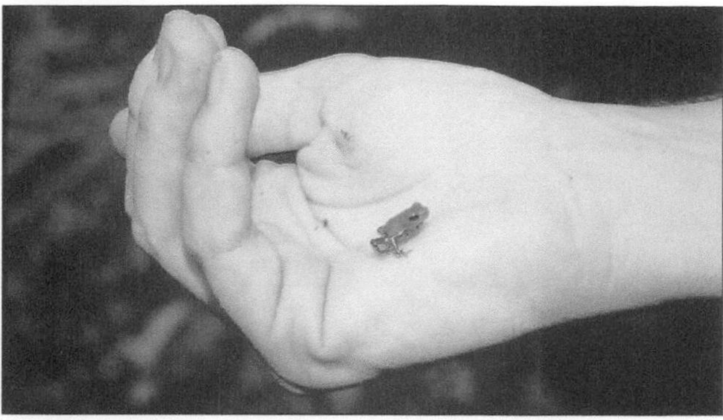

Strawberry frog, one of many poison arrow frogs (*Dendrobates*) whose skin secretes distasteful chemicals that deter predators. Many of these skin secretions are used as dart poisons or arrow poisons by Native Americans because they contain curare, a poison that can paralyze muscles by inhibiting acetylcholinesterase.

Emerald boa, one of many tropical snakes.

White-throated toucan, a large bird of the parrot family.

A tree sloth and her young, who spend most of each day hanging as you see here.

Jaguar.

Figure 18.11 (*Continued*) **Some of the biodiversity of tropical rainforests.**

estimated that more than half of the world's species live in the rainforests and nowhere else. For example, Costa Rica, a nation about the size of West Virginia, has 820 species of birds, more than the whole of the United States and Canada combined!

The variety of rainforest trees makes for a variety of habitats for other plants, such as orchids, that use these trees for support; plants that use other plants for support are called **epiphytes** (Figure 18.12). There are also many different habitats for other plant and animal species that have adapted to life on the forest floor,

Figure 18.12 Life on rainforest tree trunks.

Orchids and other epiphytes growing on a tree trunk in Madagascar.

Lianas (woody vines) growing in Costa Rica along with the aerial roots of epiphytic plants growing high above.

in the canopy, in the understory, or in the well-lit clearings created by riverbanks, landslides, or fallen trees. Numerous decomposer species, including ants, termites, fungi, and bacteria, favor the fast decomposition and recycling of nutrients. The great diversity of small-scale habitats within a rainforest contributes to the maintenance of biodiversity among species.

18.3.1.2 Complex interactions among species One result of this great diversity in tropical rainforests is the large number of ecological interactions among the many species present, including predation, competition, and **mutualism**. Mutualism in particular is a form of symbiosis from which both species benefit, and it is especially common in tropical forests. Most tropical plants have elaborate mechanisms to ensure seed dispersal by animals (dispersal by wind is much less effective amidst tall trees). For example, shrubs of the genus *Piper* (Figure 18.11) have fruits that are eaten by bats, which disperse the seeds in their feces. More elaborate are the many mutualisms between ants (phylum Arthropoda, class Insecta, order Hymenoptera) and the plants that they inhabit. Many plants provide their resident ant populations with food and with places to live. In the case of the bull's horn acacia (*Acacia cornigera*, phylum Anthophyta, class Eudicotyledonae), ants of the genus *Pseudomyrmex* live in the base of swollen thorns, and the plants provide them with sugary food from nectaries and protein-rich globules at the tips of the leaflets (Figure 18.13). In return, the ants vigorously defend the plants against other insects and swarm to bite any large herbivore that would feed on the plant more destructively. Bull's horn acacia trees whose resident ant populations have been experimentally removed are soon eaten by goats, deer, or other mammalian herbivores (Janzen, 1966, 1974).

As an example of an even more complex interaction, consider the figs of the genus *Ficus* (phylum Anthophyta, class Eudicotyledonae), an extremely successful group of tropical shrubs and trees that often grow to great heights. The edible part of a fig is an aggregate fruit called a receptacle; this fruit is unusual in that several dozen flowers are contained within it (Figure 18.14). Many of these flowers are home to the tiny fig wasps of the genus *Blastophaga* (phylum Arthropoda, class Insecta, order Hymenoptera) that pollinate the plants (Ecological Society of America, 2011).

Figs produced in the cooler months bear winter receptacles containing mostly sterile flowers but a few fertile male flowers. A female wasp lays her fertilized eggs in the winter receptacle before she dies. The wasps develop within the sterile flowers inside the fig throughout the winter months and in the spring emerge from their pupae (cocoons). The male wasps emerge first, wingless and nearly blind. They move around inside the fig looking for female wasps, which are still in their pupae. A male then chews his way into a female pupa and inseminates the female

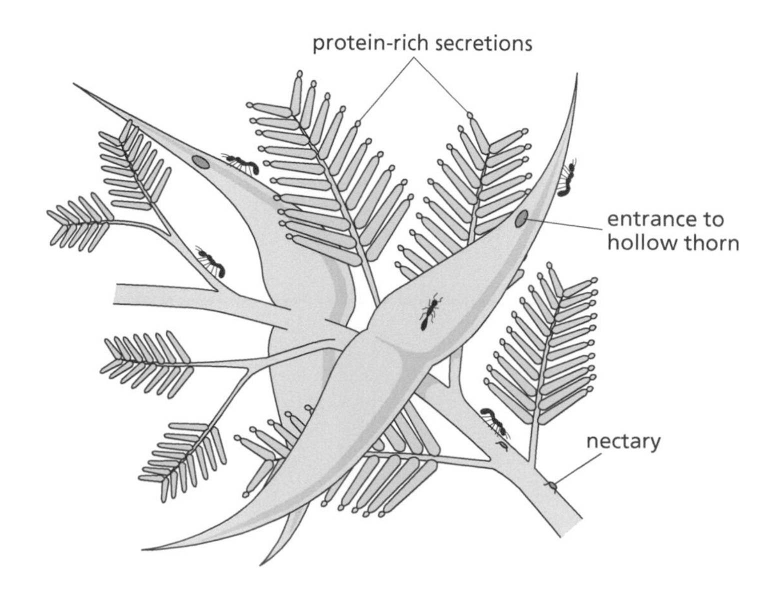

Figure 18.13 Mutualism between the bull's horn acacia (*Acacia cornigera*) and ants of the genus *Pseudomyrmex*.

Figure 18.14 **Cutaway view of a fig (*Ficus*) showing a female fig wasp (*Blastophaga*) ready to enter.** An enlarged view of the female wasp is shown; the male has a similar body but is wingless. One of the many flowers inside the fig is shown in red.

before she emerges; he then dies without ever leaving the fig. The female wasp emerges later; as she leaves the fig, she picks up pollen from the male flowers located near the exit.

The newly emerged female wasp flies around in search of fresh spring-season figs, which have a second type of receptacle containing both fertile female flowers and sterile flowers. The female wasp enters a fig and roams around inside as she lays her eggs by the hundreds and meanwhile pollinates the female flowers with pollen she has picked up from the male flowers in the winter receptacle. The tiny new wasps that emerge then repeat the process and produce a second generation of wasps.

Female wasps emerging late in the year lay their eggs in a third type of receptacle that contains only sterile flowers. The wasps emerging as this third generation lay their eggs in the winter receptacles, completing the yearly cycle. Three types of receptacles are thus home to three generations of fig wasps each year, with male flowers appearing only in the first (winter) type of receptacle and female flowers (and thus seeds) only in the second. All three types of receptacles contain sterile flowers, which alone support the development of new fig wasps. The wasps develop only within these sterile flowers, which they also use as food. The figs are pollinated only by the wasps, with each species of figs generally supporting its own species of wasps.

This story of complex interactions does not end there, for the seeds of the figs will not grow if they fall beneath the tree that bore them. The established trees have such an overwhelming competitive advantage that the offspring have little or no chance of competing successfully for moisture and nutrients if they fall and germinate near their parents. The seeds that succeed are therefore the ones that have dispersed to other locations. The seeds of some fig species germinate and grow first as epiphytes upon the branches of other trees and later grow roots reaching down into the ground; to germinate properly, the seeds of these species must find their way to above-ground perches.

The service of dispersing the fig seeds is performed by animals, with different species of animals scattering the seeds of different figs in different areas. In parts of Indonesia and Malaysia, the best dispersers of fig seeds are orangutans, *Pongo pygmaeus* (phylum Chordata, class Mammalia, order Primates, family Pongidae). These large apes practice quadrumanual clambering, a form of locomotion through the trees in which the orangutan's weight hangs from a branch, often supported by the feet as well as the hands (Figure 18.15). Quadrumanual clambering requires a lot of energy, especially for a large animal. An orangutan, if it is to avoid starving, must therefore eat enough in any one place to sustain it on its high-energy journey to the next place, which may be miles—and days—away. Wild

Figure 18.15 The orangutan, *Pongo pygmaeus*, showing quadrumanual clambering.

orangutans are nearly always hungry and are continually wandering in search of energy-rich foods, the most preferred of which are usually figs.

The wide-roaming habits of orangutans are ideal for the figs, for the orangutans consume hundreds of figs, containing tens of thousands of seeds, then wander for miles through the forest. When an orangutan defecates amidst the branches, it leaves behind hundreds of fig seeds, together with a supply of moist, nutrient-rich fertilizer that helps the seeds to sprout and establish themselves and eventually replenish the supply of fig trees in a forest.

The orangutans must cope with the fact that the fig trees flower and bear fruit at different times of the year. Because there are many species of figs, each bearing fruits in a different seasonal pattern, a resourceful orangutan can usually manage to find some trees bearing fruit in almost every month of the year. Among orangutans, there is thus a selective advantage in having a good spatial memory—a mental map of the forest covering many square miles. The orangutan with the best chance of survival is the one whose intelligence allows it to remember where to find the most fig trees, when each was last visited, how far along its figs were at the time, and when the time would be optimal to visit each particular tree again.

The interrelatedness of the lives of figs, wasps, and orangutans (and goats, birds, monkeys, humans, and other seed-dispersing species in the places where the orangutans do not live) shows how complex life may be among the species of the rainforest. Destroy a few fig trees, and many orangutans may starve. The removal of a few orangutans may decrease the ability of fig trees to disperse their seeds, which would also diminish their ability to provide homes for fig wasps and food for the insect-eating species that feed on the wasps. The destruction of a portion of the rainforest thus has consequences far beyond the portion actually cleared, for it diminishes the health of the whole ecosystem for miles around.

18.3.1.3 Slowness of ecological succession Rainforests occasionally suffer from heavy storms that topple one or more trees, creating a clearing. Human activity may create other clearings. The process of ecological succession, described earlier, ensures that the pioneer species that take advantage of such clearings are later replaced by more shade-tolerant species. It will take many years for any of these to grow back to the height of the original trees that fell. Tropical botanists have found that ecological succession in rainforests is very slow: It may take centuries for the rainforest to again reach its former canopy height and species density. Because the small-scale conditions may be different from those in previous successions, tropical rainforests, even if they regrow, may not become the same community as the forest that was there before. Small clearings in the rainforest are much more likely to grow back than large ones. Many large clearings may not grow back at all.

18.3.1.4 Deforestation Around the world, rainforests are being destroyed at an alarming rate. One expert on tropical rainforests estimates that they are disappearing at the rate of 150,000 km^2/year (410 km^2/day)—an area equivalent to the size of Manhattan Island every 3.5 hours—and that at least 40% of the world's rainforests have already been destroyed!

Tropical deforestation has many causes. These include the growth of human populations (see Chapter 9), the spread of agriculture, the desire of farmers to earn a better living, the attitude that humans are entitled to dominion over nature, and the quest for corporate profits from timber, minerals, or agricultural activities. In the twentieth century, the United Fruit Company (Chiquita Banana) cleared large rainforest areas in Central America and replaced them with banana plantations, while the Goodyear Rubber Company cleared large parts of West African rainforests to establish rubber plantations.

Among the most destructive uses of the rainforest are the exploitation of nonrenewable resources on a one-time basis. Some rainforests are cleared because of the mineral wealth that lies beneath them. In other cases, it is the trees themselves that are harvested for timber (Figure 18.16). Trees are **renewable** given a sufficient length of time, but regrowth of tropical rainforests is usually very slow. Many logging operations destroy the trees much faster than they can grow back, so that the forest cannot long sustain continued logging on such a scale. **Sustainable use** of a resource must allow the resource the opportunity to regrow and replenish itself; any use that exceeds the capacity of the resource to renew itself is nonsustainable.

The largest amount of rainforest destruction is carried out for agricultural reasons. The land cleared from the rainforests is put to agricultural use as grazing land for cattle or as farmland for crops, either for local human consumption or for commerce. In many places, new fields are cleared by cutting down and burning forest (Figure 18.16), a practice called slash-and-burn agriculture. In many cases, fields created in this way are fertile for only a decade or two, so new fields are then cleared to replace them.

It may seem odd that rainforests that appear so lush will not grow back quickly and will not support agriculture for long. This is because the soil beneath many rainforests is very poor in nutrients, for several reasons. One reason is that tropical ecosystems recycle most nutrients in the forest litter before they reach the soil. Another reason is that the continuous and heavy rainfall washes soluble and partially soluble minerals from the upper layers of the soil, a process called leaching, which leaves few mineral nutrients behind. What remains behind after extensive leaching is in many cases a dark or reddish, very hard, mineral-poor soil called laterite. Most attempts to grow crops on lateritic soils have been very disappointing because of their low nutrient content and frequent mudslides.

In rainforest regions with lateritic soils, agricultural use of the land quickly exhausts the low nutrient content of the soil. The amount of rainfall is often too high for most crops, and problems with drainage and erosion also arise in many places where the soil holds water poorly. In Madagascar, Haiti, and many other places, rainforests cleared for agricultural use have given way to widespread erosion in which thousands of tons of red, lateritic soil wash annually into the sea. Madagascar also has one of the world's fastest-growing human populations; its population doubled from 1996 to 2020, putting even more demands on precious tropical habitat. Of the 111 species of lemur found on Madagascar (and nowhere else), 105 are threatened with extinction.

Figure 18.16 Tropical deforestation.

These trees have been cut down and will soon be burned, a practice called "slash-and-burn agriculture" that depletes the soil of its nutrients, requiring the fields to be abandoned after several years and new fields to be cleared by burning more forest.

These slopes in Costa Rica were once forested, but the harvesting of trees has increased the frequency of landslides and the rate of erosion.

Timber ready for transport. These operations destroy many hectares of forest at a time; the results include a loss of topsoil and extensive erosion, often leaving the land unsuited for agriculture or human habitation.

Rainforest destruction is often permanent. Because rainforest vegetation contributes (through leaf transpiration) to the rain clouds that maintain the rainy climate, any large-scale clearing of the rainforest is bound to alter the delicately balanced water cycle. Large areas cleared from rainforests tend to suffer greatly reduced precipitation, and the land soon becomes unsuitable for crops or even for cattle grazing.

Deforestation is a problem at all latitudes, not just in the tropics. Norway, Russia, Canada, and the United States all have vast forests that consume carbon dioxide and contribute oxygen and water to the atmosphere. These forests are also being harvested for their wood (including both timber and pulpwood for making paper) and cleared for agricultural and other uses. Northern forest timber resources can be renewable if properly managed, especially if only some of the trees are cut in any one location. In contrast, removal of all the trees (clear-cutting) from large tracts of land is damaging to forest ecosystems and to the global atmosphere that all forests support. Even if the forest is replanted, it is often with a single species of tree. This is a form of monoculture (see Figure 17.16); it may maintain a supply of one tree species, but it does not restore biodiversity or habitat variety. Species displaced from their habitat may not come back because the new habitat is very different from the old habitat. Some species, such as the spotted owl mentioned earlier, are particularly sensitive to these changes and can thus be used as sentinel species that allow us to monitor the health of whole ecosystems, as explained later. Of course, species driven to extinction are lost forever.

One remarkable technique produces wood sustainably without cutting down trees. Since the fourteenth century, the Japanese have used a technique called daisugi. Under this 700-year-old practice, the trees are planted for future generations and will not be cut down, but are instead pruned as if they were giant bonsai trees. By applying this technique to cedars, the wood that can be obtained is uniform, straight, and without knots, practically perfect for construction. The daisugi pruning is an art that allows the tree to grow and germinate while using its wood, without ever cutting it down.

18.3.2 Desertification

Land that supports the richness of life of a rainforest can be transformed into a desert capable of supporting very little life. Destruction of rainforest can begin this process, which is called desertification. To understand it, we need to learn more about zones of climate and vegetation in tropical regions. In this section, we first use Africa as an example and take a close look at desertification there. We then extend our view worldwide and consider whether humans can reverse desertification.

18.3.2.1 Climatic zones of Africa The global atmospheric patterns shown in Figure 18.17 conspire to rob the regions around 30°N and 30°S latitude

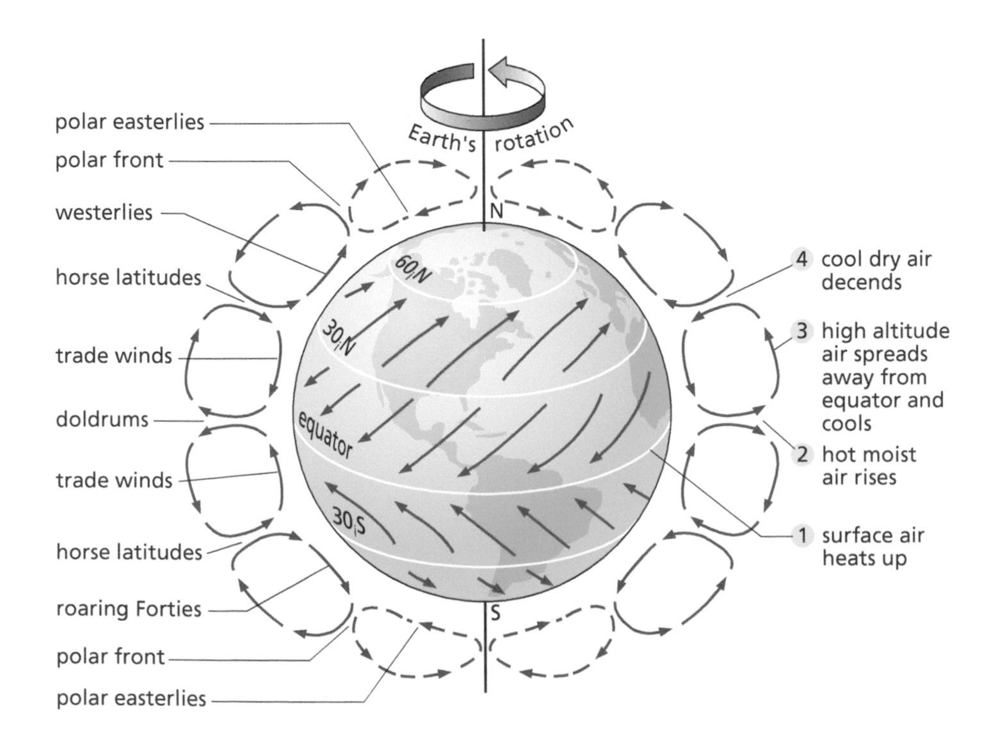

Figure 18.17 **Global patterns of prevailing winds.** Notice that winds blow away in both directions from latitudes 30°S and 30°N, carrying moisture away from these latitudes and creating desert regions.

of all their moisture on a continuing basis: Prevailing high air pressure creates winds that evaporate moisture from the land and transport it away from these regions. The northern half of Africa lies in the belt at 30°N latitude and is occupied by the world's largest desert, the Sahara.

To the south of the Sahara, much of Africa is characterized by a series of parallel zones differing in moisture and thus differing in vegetation (**Figure 18.18**): There is tropical rainforest along the coast from Sierra Leone to Gabon and continuing inland across the Democratic Republic of Congo; then, heading north, patches of forests interrupted by more open land; then a more open woodland with scattered trees and shrubs only, giving way to a tropical grassland (savanna) farther inland, then a type of dry pastureland, the Sahel; and finally the desert. Each band has more rainfall than the one to its north, and less than the band to its south. The overall pattern has local exceptions where the land is mountainous, but in general it prevails across most of Africa north of the Equator. Each band supports a distinctive kind of vegetation and a distinctive human culture. Each of these bands is also a biome, similar to other ecosystems in other parts of the world (Box 18.1).

These zones of vegetation and rainfall are not static but are slowly changing. The Sahara is very slowly advancing southward by the process of

Figure 18.18 African vegetation zones.

(A) Vegetation zones in Africa north of the Equator.

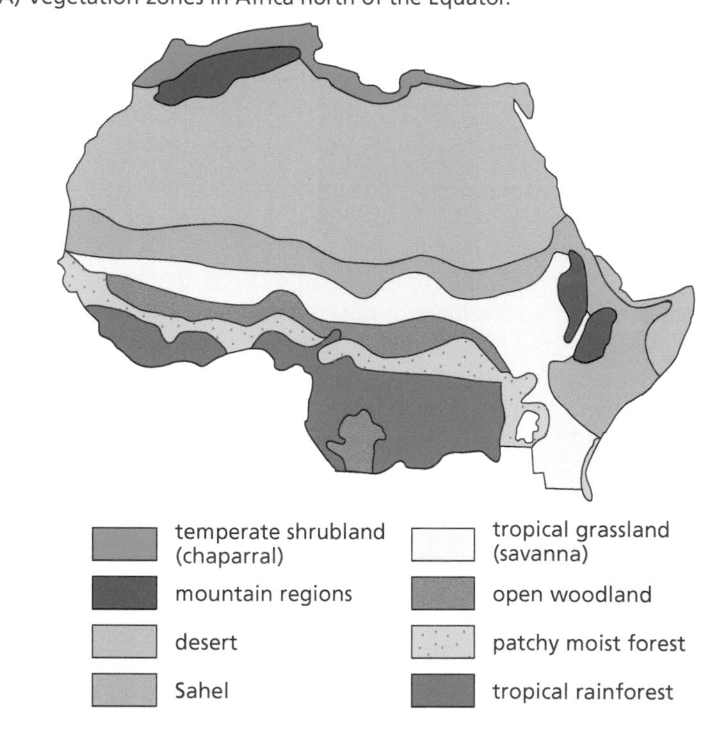

(B) Dusty sandstorm in the Sahel region, Dollo Ado, Ethiopia.

desertification, and the other vegetation zones are moving southward with it. Desertification has also taken place in other directions where the Sahara reaches westward to the shores of the Atlantic and eastward to the shores of the Red Sea and beyond into the Arabian Peninsula. Archaeological excavations confirm that these lands were all much wetter, and the vegetation was lush only a few thousand years ago, as the bones (and cave paintings) of hippopotami and crocodiles attest.

How does desertification take place? At least within the Sahel, an important factor promoting desertification is the overgrazing of pasture lands by flocks of domestic animals—goats, cattle, camels, sheep, and other species—that removes the land's vegetative cover. Without the many plant roots that held the soil and its moisture, the precious topsoil blows away. The land, which can no longer support plant life, becomes a desert.

Another important process takes place farther south, where rainforests and tropical woodlands are cleared for agricultural use, often by the slash-and-burn agriculture described earlier. After a few decades of agricultural use, the land is abandoned, and new land is cleared. There are several unfortunate consequences of this method of agriculture: The abandoned land is never totally reclaimed by forest ecosystems; the agricultural land has much less ability to retain moisture than the forests that it replaces; and the dry pasture of the Sahel replaces much of the abandoned fields.

18.3.2.2 Desertification around the world The problem of desertification is not limited to Africa, although the advance of the Sahara claims more new land each year than all the other deserts of the world combined. The Mojave Desert in Southern California and the Great Indian Desert (along the India-Pakistan border) are two other deserts that are advancing on adjacent agricultural land. The situation in India may be broadly similar to that in much of Africa. The situation in the Western United States is somewhat different because desertification is only in its early stages in most places and because most of the problems seem to be associated with the use of underground water reserves for irrigation and for domestic use in cities like Los Angeles. Farmers and ranchers throughout the Western United States use aquifers (underground water deposits) for irrigation. In many cases, these aquifers are either shrinking or becoming saltier as ocean waters (e.g., from the Gulf of California) encroach farther inland. If water use exceeds the natural capacity of aquifers to refill, desertification will result.

The immediate effects of desertification are the loss of cropland and rangeland, an effect that is felt keenly, but locally. In a few cases, there is also increasing conflict over water rights—for example, between California and Arizona over the use of the Colorado River and between Turkey and Syria over the use of the Euphrates River. The long-term effects of desertification are much more serious, however. With reduced vegetation cover, the ground retains less moisture. This means that the air above can become drier, and rain clouds are far less likely to form. The absence of rain clouds results in reduced rainfall, which in turn accelerates the process of desertification.

18.3.2.3 Prospects for reversing desertification Can desertification be arrested or turned back? Yes, but only very slowly, very expensively, and with a concerted effort over many years. Israel has had great success in "making the desert bloom," turning desert and scrubland (like the Sahel) into agricultural land. One key to this process is irrigation, using water from rivers or lakes or desalinated sea water. Irrigation is always an expensive undertaking—particularly so in a dry climate—and natural water supplies set limits to what can be sustainably farmed with the help of irrigation. Israel is a relatively prosperous nation that has the economic resources to reclaim desert lands on a scale appropriate to its small size. Many of the world's desert regions are in poor nations that do not have the financial resources to repeat Israel's successful experiment in reclaiming desert lands for agricultural use.

18.3.3 Valuing habitat

Although we have examined only two of the world's many biomes, some of the conclusions that we have drawn pertain to other types of ecosystems as well. In particular, habitat destruction threatens many ecosystems, whether they are coastal wetlands, pine forests, or the African Sahel. Why does this habitat destruction continue?

Humans often make decisions by first assigning a value, consciously or unconsciously, to such things as happiness, land, money, and even life itself. Decisions are then made by choosing the alternative that has maximum value. Under this system, bad decisions often result from attributing too much or too little value to something, and conflicts may arise if different people assign very different values to the same thing. For example, different people attribute different values to rainforest habitats and to the need to sustain the habitability of our planet.

18.3.3.1 Ways of assessing value Philosophers often distinguish between intrinsic value, the value that something has as an end in itself, and instrumental value, the value that something has as a means to some other end. Dollar bills, for example, have no intrinsic value; they are valued only because of what we can buy with them, and they would be useless in a society that did not accept them in trade. We will soon examine the instrumental value of various habitats as places where valuable resources can be obtained. Before we do so, let us also point out that many people also value other living species, and entire ecosystems, as having a high intrinsic value. The habitat that sustains living ecosystems is likewise valued intrinsically by many people. Also, in the view of many people, no species has a right to destroy another species or to deprive it of its habitat or its means for continued existence.

Value does not only include the value of something to our species alone. Another type of value may be biological value: the interdependence of species, genetic diversity, ecological diversity, and the resulting dynamic stability of the biological community.

18.3.3.2 Ethical dimensions of habitat destruction Habitat destruction can take many forms, including the clear-cutting of forests and the draining of swamps. In some cases, the destruction takes place to permit the building of housing tracts or shopping centers. In other cases, land is cleared for agricultural use. In still others, extractive industries such as mining or logging simply exploit the land on a one-time basis for its mineral wealth or its standing crop of trees. Many of the social, political, and economic forces that impinge on tropical rainforests often threaten the destruction of other habitats as well.

When we pause to consider what the many cases of habitat destruction have in common, we soon realize that the same ethical issues recur in case after case (Kellert, 1996). How important are natural communities? How important are their habitats? Is it more important to leave nature undisturbed or to feed an expanding human population? Is it more important to preserve natural habitat or to satisfy people's demands for agricultural land, timber, or housing? To what extent do the answers to the previous questions depend on the quality of the soil, the economics of the country in question, or other factors? To what extent do the answers depend on how much we value other species in addition to our own? Do other species have value apart from their relationships to humans?

All of these are basically ethical questions (see Section 1.4), or parts of a larger, all-embracing ethical question: Is it better to preserve a particular ecosystem in its "natural" state, or is it better to convert the area into agricultural or similar use? Viewed one nation at a time, the economic and political forces that push toward one alternative or the other weigh heavily against many natural ecosystems (Hurst, 1990). The pressure of an increasing human population, the need for land and food, the need for income, and the need for economic development are all obvious to the people living near the habitats that come under threat of destruction around the world. When measured against all these forces, the value of undisturbed wilderness is not always obvious or locally appreciated. Of course, things are never that simple: Many Brazilians and Malaysians want to preserve their rainforest habitats, even if this means that economic development cannot proceed quite as fast as other Brazilians and Malaysians would like.

When we consider the worldwide ecosystem of the Earth as a whole, however, the balance seems to shift in the other direction, in favor of preserving the natural environment. The advance of the Sahara or the destruction of rainforests threatens the planet with consequences far greater than the continuation of poverty and underdevelopment in any one country. The case for Brazil can easily be argued in these terms: The preservation of the Amazon rainforest is best for the planet as a whole, and the economic best interests of Brazil would be viewed as secondary if the good of the planet were given priority. Perhaps this makes sense to North American environmentalists and philosophers, but it is certain to be a very unpopular attitude in Brazil! It is the Brazilians (likewise the Indonesians and many others) who are largely in control of their own rainforest, and they are likely to resent any suggestion that they sacrifice the well-being of their nation's economy for the "greater good" of a global environment that the wealthier nations of the north have already started to destroy. A similar argument can be directed against the industrial nations of North America and Europe: A reduction of resource consumption by these nations would reduce pollution, reduce the trend toward global warming, and benefit the planet as a whole. Large tracts of land in Australia, Argentina, and the United States are devoted to cattle ranching, an activity that produces far less food per unit area than if that same land were used for growing crops. More of the world's hungry could be fed if lands now used for cattle ranching were instead used to raise wheat or corn, but ranchers (whose interests are often supported by their governments) are not likely to give up their way of life for that reason alone. They argue that much of the land now used for ranching is so used precisely because it is unsuitable for growing crops economically.

It is easy, on utilitarian principles (see Section 1.4.3), to argue that the good of the planet should take precedence over the economic well-being of any single nation or occupational class. However, by the principle of fairness, it is just as easy for Brazilians to argue that they should not bear the entire burden for a sacrifice that benefits the whole world. If the world benefits from the Brazilian rainforest, then the world should somehow pay to maintain it in its natural state. If rainforests offer such good protection against global warming, then all nations should contribute to rainforest conservation, perhaps in proportion to the amount of carbon dioxide that they generate. As of 2023, Brazil produces less than a tenth of the carbon dioxide emissions of the U.S. (The United States ranks second in emissions, with China emitting the highest amount, more than double that of the U.S. See https://worldpopulationreview.com/country-rankings/co2-emissions-by-country) Also, Brazilians can point to the logging operations that destroy forests at an alarming rate in the United States and other northern countries. Why, they ask, should one nation be told to cease cutting its forests when other nations continue to cut theirs?

18.3.3.3 Habitat destruction versus sustainable use It does not take long to realize that many forms of habitat destruction are driven by very shortsighted goals. As an example, the harvesting of slow-growing trees brings only short-term gains to only to a small number of people (those in a single industry, sometimes only a single company), but the damage that it causes both to the local economy and to the biosphere as a whole may be irreversible. The same can be said of cutting down the rainforest for the planting of those crops that grow poorly on lateritic soils.

Easter Island in the southeast Pacific Ocean shows us a particularly gruesome lesson in the consequences of habitat destruction. When humans first arrived, the island supported a rainforest that was home to many edible plant species as well as numerous species of birds and other animals. The Easter Islanders, a Polynesian people, prospered for several hundred years, building the large stone statues for which the islands are now famous. But instead of conserving the rainforest and living off its rich resources, the Easter Islanders cut much of it down, until too little was left to sustain the edible bird and plant species, which slowly disappeared along with the forest itself. Like other rainforests, the one on Easter Island had created its own rain clouds, producing the conditions for high rainfall, but when it was destroyed, less rain fell on agricultural crops. With vanishing timber supplies, the Easter Islanders could scarcely find enough wood to build the boats needed to sustain their fishing activities. As all food supplies dwindled, the Easter Islanders began to starve, and the last survivors resorted to cannibalism before they were rescued by the arrival of Europeans. This case is not unique. The Maori, a Polynesian people of New Zealand, hunted many of the native species of their islands to extinction or nearly so, and they were beginning to show signs of starvation and decline when Europeans arrived. Other islands in the Pacific were subject to similar exploitation, although nowhere else did the process go as far as it did on Easter Island.

What we need instead of destructive uses are **sustainable** uses of forests, uses that allow people to derive profit from maintaining the rainforest instead of from destroying it (Anderson, 1990; Rice, 1997; Khor, 2002). Most nations that contain rainforests are non-industrialized, and many of them are also poor. A sustainable economic use of the rainforest would be an economic incentive to maintain the forest rather than to destroy it. It is therefore in the long-term best interests of all nations and all people to help tropical nations develop such sustainable uses. This is why one provision of the Paris Agreement of 2016 (see Chapter 19) encourages wealthier nations to finance sustainable development and conservation efforts in the economically less prosperous parts of the world.

One example of sustainable use is the gathering of small amounts of high-income rainforest products such as pharmaceutical plants. For example, the anti-cancer drugs vinblastine and vincristine, derived from a rainforest plant (the Madagascar rose periwinkle), account for sales of $180 million a year. In 1991, the pharmaceutical firm of Merck and Company entered into a million-dollar agreement with Costa Rica's National Institute of Biodiversity. Scientists working for the Institute were tasked with identifying as many rainforest plants as they could (the total number is estimated to be 12,000), extracting samples from them, and sending the more promising ones to Merck for tests of their medicinal value. Merck's $1 million investment should be compared with their annual sales close to $15 billion in 1994. In 1990, Merck sold $735 million worth of just one drug, Mevacor® (lovastatin), a cholesterol-lowering drug derived from a soil fungus. If even one new drug discovered by scientists under this agreement brings in a small fraction of this amount, Merck will recoup its original investment many times over.

Certain kinds of rainforest agriculture can be sustainable, but others are not, and there is still much to be learned from experimentation and frequent reevaluation. The most promising attempts are those that employ mixed uses of rainforest habitats, allowing tall trees, shorter trees and shrubs, and smaller plants to persist side-by-side. Coffee, vanilla, cocoa, cashews, bananas, and certain spices are potential candidates for such experimental attempts, and tropical botanists can help to identify others. Many uses of rainforest plants are traditional, but they could benefit economically from improvements in harvesting, transport, and marketing. Agricultural scientists and business interests could help to develop new markets, which would bring much-needed income and provide local people with an economic incentive to maintain the forest ecosystem. Tropical plants that could easily be marketed more widely include amaranth (a nutritious and drought-resistant grain), fruits such as durians and mangosteens, and the winged bean (*Psophocarpus tetragonolobus*) of New Guinea (phylum Anthophyta, class Eudicotyledonae). This last species is a fast-growing plant that produces spinach-like leaves, young seed pods that resemble green beans, and mature seeds similar to soybeans, all without the use of fertilizers. Hundreds of other fruits are grown and eaten in the tropics but are only rarely exported.

An important goal of such efforts would be to identify which plants might profitably be grown or harvested in a given region without harm to the environment. New ways must be found to exploit the rainforests without destroying them—to develop rainforest ecosystems into sustainable resources for both local and worldwide benefit.

Sustainable use and habitat preservation make sense economically in nearly all ecosystems, not just in rainforests. The city of New York was planning to spend between $6 billion and $8 billion to build a new water filtration and treatment plant, until officials discovered that they could accomplish the same goals by spending only $1.5 billion to help preserve the natural watersheds of the Catskills and the Delaware Basin, two sources of naturally filtered water supplies. In Hawaii, the Maui Pineapple Company runs the Pu'u Kukui Watershed Preserve, in which they protect native plants that maintain both soil and groundwater that sustains nearby pineapple plantations. "Without it," says a company official, "rain would run off into the ocean."

18.3.3.4 Ecotourism If Brazil and other tropical nations are to preserve the rainforest instead of destroying it, they must have economic incentives to do so. One type of economic incentive is the small but growing market for ecologically based tourism, also called green tourism or ecotourism (Boo, 1900). Ideally, this type of tourism seeks to make as little impact on the natural environment as possible. Ecotourism is now Costa Rica's second largest source of foreign income, behind coffee and ahead of bananas. Ecotourism is also a major source of revenue in Kenya. By comparison, the rainforests of Brazil afford a largely untapped tourist resource. Of course, some land would have to be set aside for airports, roads, and hotels. Beyond this initial investment, however, ecotourism would provide economic incentives for leaving the rest of the rainforest untouched. Ecotourism not only provides a country with income; it also gives that country an economic incentive to preserve its own natural heritage for the benefit of all.

As we search for ways to stop the destruction of ecosystems, we must realize that no solution will work if the rich and poor nations continue at odds with one another. If battles continue over short-range economic interests, the planetary ecosystem will undoubtedly be the loser.

THOUGHT QUESTIONS

1. Is a biodegradable insecticide completely harmless? If it were used over a wide area of a forest, might it cause the extinction of an insect species confined to that area? What effect would such a loss have on other species? Do you think the use of such insecticides in research carries a certain amount of risk? If so, how can this risk be minimized?
2. Do you think undisturbed habitats have intrinsic value or only instrumental value? In other words, is habitat valuable as an end in itself or only because of the uses to which it might be put? Think of other things that you intrinsically value, such as close family members. Does habitat have the same kind of intrinsic value? In what ways are the values similar? In what ways are they different?
3. Are there things of intrinsic value that you personally would be willing to do without if it meant preserving more habitat for other species? Will it be possible to preserve habitat for other species without humans changing their use of habitat?
4. How much habitat destruction do you think takes place at the hands of wealthy people and how much at the hands of poor people? (The answer may differ from country to country.) What would it take to secure the cooperation of both rich and poor people in an effort to halt desertification or rainforest destruction? Could you easily appeal to both rich and poor together, or would it be easier to appeal to the two groups separately?
5. The ecologist E.O. Wilson has estimated that one Peruvian farmer clearing land to grow food cuts down more species of trees than are native to all of Europe. Do you think there may have been greater numbers of trees in Europe before human populations grew to their present density? More species of trees?
6. Are habitats being destroyed near where you live or go to college? What are they? What factors contribute to the destruction?
7. Does ecotourism sometimes do harm? How? Can the harm be minimized? Can it be eliminated entirely? Think of some examples of ecotourism and other forms of rural tourism. On the whole, do you think that the benefits of ecotourism outweigh the harm?

CONCLUDING REMARKS

Human activity is contributing to a rapid decline in biodiversity comparable to the extinction rates of the mass extinctions in the geological past. The high rate of extinction of species is made even higher by the destruction of entire habitats, such as in rainforests. The quest for corporate profits motivates some of this destruction, but so does the need for food and living space for human populations. The reduction of biodiversity makes ecosystems less stable and makes our planet less habitable for humans and for many other species. Can we value the continued existence of other species, not just humans? Can we learn that the continued existence of humans depends on our living in natural balance with other species? The survival of our planet and its ecosystems depends on the choices that we make. To many people, "long-range planning" means thinking only one year into the future, but the choices that we make today often have consequences that last for decades, centuries, or even longer. Technology may soon link humans everywhere into a single global community, but we should also remember that ecosystems link our species to the rest of life.

CHAPTER SUMMARY

- **Biodiversity** is measured by the number and variety of species, of which 70% are either insects or vascular plants.
- Each species occupies a **niche** (a way of life or a role) within a **community**. A community and its physical environment interact as an **ecosystem**. One type of community may replace another in a process called **ecological succession.**
- Speciation increases biodiversity, while **extinction** decreases it.
- **Endangered species** are those threatened with extinction, often because their populations are too small and genetically too homogeneous to adapt to change.
- **Habitats** support the interdependent lives of biological communities of species.
- Groups of ecosystems that are similar but in different geographic areas are called **biomes.**
- Destruction of a habitat threatens the survival of all the species that live in it. Overuse, rather than **sustainable** use, is endangering many habitats worldwide.
- Biodiversity is greatest in tropical rainforests and in coral reefs, largely because of the great amount of solar energy and climate stability near the Equator. The expansion of human populations and agricultural lands threatens many rainforests. Because a majority of the world's photosynthesis and oxygen production occurs in rainforests, their destruction is a worldwide threat to the atmosphere and to the entire global ecosystem or biosphere.

BIBLIOGRAPHIC REFERENCES

Bibliographic references to material in this chapter can be found online at biologytrending.routledge.com/chapter18

GLOSSARY: KEY TERMS TO KNOW

These key terms are defined at the end of each chapter as an aid for student review.

Biodiversity The number and variety of biological species, their alleles, and their communities.

Biome A group of similar ecosystems in various locations around the world.

Biosphere The ecosystem that includes the whole Earth and its atmosphere.

Community A group of species that interact in such a way that a change in the population of one species has consequences for the other species in the community.

Ecological succession An ecological process by which one community replaces another.

Ecosystem A biological community interacting with its physical environment.

Endangered species A species threatened with extinction.

Epiphyte A plant that lives upon and derives support, but not nutrition, from another plant.

Extinction Termination of a lineage without any descendants.

Habitat The place and environmental conditions in which an organism or a species lives.

Lineage A succession of species in an ancestor-to-descendant sequence.

Mutualism An interaction between species in which both species benefit from the interaction.

Niche The way of life of a species, or its role in the community.

Pseudoextinction Extinction of a taxon by its evolution into something else, thus continuing its line of descendants.

Renewable Capable of regrowing or repairing any disturbance by natural processes (within a reasonable time span).

Succession See *Ecological succession*.

Sustainable Any practice that could continue indefinitely without depleting any material whose supply is limited.

CONNECTIONS TO OTHER CHAPTERS

Chapter 1: Pollution and habitat destruction violate several ethical injunctions, including the principle to do no harm to others.
Chapter 4: Genomes are an important biodiversity resource.
Chapter 5: Species arise through evolution. Speciation increases biodiversity, while extinction reduces it.
Chapter 6: Classifications must be modified as newly discovered species are inserted.
Chapter 7: Allele frequencies change in populations.
Chapter 9: The expansion of human populations puts increased pressure on ecosystems all over the world.
Chapters 10 and 13: Many drugs, both new and traditional, and including those to treat cancer, come from plants whose habitats are threatened.
Chapter 17: Plants are energy producers in every ecosystem and thus support many other species.
Chapter 19: Pollution threatens the habitats of many species.

PRACTICE QUESTIONS

1. Which has greater biodiversity, a habitat in which there are 1500 resident species, most of which are prokaryotes, or a habitat in which there are 1500 species, most of which are plants?
2. Rainforest is being destroyed at the rate of 410 km^2/day. How many square miles per day is that? How many acres? How big is your campus, town, or city?
3. What percentage of animal species are birds? How many times more species of insects are there than species of birds?
4. What biome is your home or college located in? Where on the globe are comparable biomes located?
5. How do wind patterns contribute to deserts being located at 30°N and 30°S latitudes and rainforests being located along the Equator?
6. What percentage of the world's bird species live in Costa Rica? What percentage live in Hawaii?
7. If 500,000 genera were living in the Permian period, and if 85% of them became extinct in the last 5 million years of that period, then how many genera became extinct, on average, during each million-year interval? If there were five species per genus, on average, then how many species per million years became extinct? How does this rate compare with the estimated 100,000 species extinctions for the decade of the 1980s?

CHAPTER 19

Climate Change and Other Threats to the Biosphere

ISSUES

- Is it possible to live in the industrial world without polluting?
- How can societies limit the threats to ecosystems?
- How do economics influence pollution?
- Who should pay for cleanup and remediation?
- How can societies limit the threats to the atmosphere?
- How can people in one place help control pollution that originates elsewhere?

BIOLOGICAL CONCEPTS

- Ecosystems and biosphere
- Environmental factors (biotic, abiotic)
- Evolution (origin of life and of photosynthesis)
- Human influences on the biosphere
- Atmosphere
- Matter (water cycle, carbon cycle)
- Chemical and physical basis of biology (molecular structure, oxidation and reduction reactions)
- Energy and metabolism (autotrophs, heterotrophs)
- Valuing habitat
- Conservation biology (renewable and nonrenewable resources, sustainable and nonsustainable uses)

CHAPTER OUTLINE

DOI: 10.1201/9781003391159-19

Human populations have important impacts on ecosystems, both locally and globally. Just in order to meet basic human needs, our quest for food, drinking water, and places to live creates various disturbances in local ecosystems. Human agriculture and housing for people are two important ways in which we will always alter our local ecosystems. In addition, people's demands for industrial products and for energy create further impacts. In all these ways, the growth of human populations (Chapter 9) puts ever increasing stress on world ecosystems. Some of these impacts were described in Chapter 18; others, especially those related to pollution, are described in this chapter.

The Earth's land, water, atmosphere, and living things form a global ecosystem called the **biosphere** (see Chapter 18). As part of this ecosystem, the atmosphere supports life in the sense that all animals and many other organisms would soon die without oxygen. The atmosphere also supports life in that it maintains the Earth's surface temperature within a certain range. In another sense, however, it is life on planet Earth that supports Earth's atmosphere, because all the important atmospheric gases exist in equilibrium with the activities of living organisms: Plants and other photosynthesizing organisms produce the oxygen, while bacteria regulate the nitrogen in a cycle described in Section 17.3.2.

Human activities can have a great and varying influence on a local scale. People can cut down trees, clear land, plant fields, dig mines, or erect buildings of many kinds. Human activities can also leave behind many waste products, whether intentionally or unintentionally. On a global scale, human activities can also change the Earth's atmosphere and thus threaten the stability of many ecosystems and myriad species of life on Earth. One such change involves the destruction of atmospheric ozone (O_3), allowing more ultraviolet radiation to reach the Earth's surface, causing higher rates of skin cancer due to mutations in skin cells exposed to that radiation (see Section 10.6.4). Another change comes from the buildup of carbon dioxide, which may raise global temperatures to a point that would cause widespread extinctions, devastating crop losses, and the flooding of most coastal cities. Photosynthesis by plants, especially forest plants, may help to limit the buildup of carbon dioxide that contributes to global warming. The health of our atmosphere thus depends on the continued health of major ecosystems in the tropics and elsewhere. These are among the issues that we will examine in this chapter.

19.1 THE BIOSPHERE: LAND, WATER, ATMOSPHERE, AND LIFE

The biosphere includes the Earth and all its living species. As the largest ecosystem of all, the biosphere includes all animals, plants, and other organisms, as well as all of the Earth's land features, its oceans and fresh waters, and its atmosphere. All of these interact with one another and form an interconnected whole. Each mountain, pond, forest, and biological species helps to reshape all the rest and is in turn reshaped by them. No full understanding of any one of them can be gained without considering all the others and their past histories.

The Earth's landforms have been shaped through time by purely physical processes such as earthquakes and also by the activities of organisms. Lichens and fungi secrete chemicals that slowly dissolve certain rocks, and plant roots wedge themselves into cracks and help break up larger rocks into smaller ones. Burrowing animals help reshape the soil, while plant roots hold the soil and protect it from washing away by erosion. The activities of aquatic organisms may either loosen sediments or build up new structures such as coral reefs. The biological activities of these organisms may alter the chemistry of the water itself. Organisms also interact with the atmosphere and use its gases; slowly they may modify the composition of the atmosphere over long spans of time.

19.1.1 The development of the atmosphere and of life

The Earth's atmosphere supports all its ecosystems because most forms of life require oxygen. Species that do not need oxygen interact with those that do, and they require other atmospheric gases. The atmosphere also maintains the Earth's surface temperature within a certain range, cooler than our planet would be with a much denser atmosphere and much warmer than our planet would be with no atmosphere at all. We have only to look to our nearest neighboring planets to view some of the alternative possibilities: Venus, with a much denser atmosphere and closer to the sun, is unbelievably hot, while Mars, with a very meager atmosphere and farther from the sun, is inhospitably cold. In another sense, however, it is life on planet Earth that supports Earth's atmosphere, because all the important atmospheric gases exist in equilibrium with the activities of living organisms: Plants and other photosynthesizing organisms produce the oxygen, while multiple species of bacteria regulate the nitrogen (see Section 17.3.2).

The present atmosphere of planet Earth is about 78% nitrogen (N_2), 21% oxygen (O_2), less than 1% argon (Ar), and much smaller amounts of water vapor (H_2O), carbon dioxide (CO_2), and other gases (Figure 19.1). While the element nitrogen (N) is essential to all living things, few organisms can use nitrogen gas directly (see Section 17.3.2). Argon gas is chemically unreactive and has no known influence on the activities of organisms. On the other hand, oxygen is absolutely essential to the continued existence of animals and many other organisms alive today. Oxygen is also important for many of the chemical reactions that happen naturally at the Earth's surface, including fire and other types of combustion, as well as rust and other forms of slower oxidation. The normal form of oxygen consists of molecules of O_2, oxygen atoms bonded together in groups of two. There is also a more reactive form of oxygen called **ozone**, a form of oxygen abundant at high altitudes. Ozone molecules are O_3, oxygen atoms bonded together in groups of three.

19.1.1.1 Life has changed the atmosphere A good deal of evidence suggests that the atmosphere has not always had its present composition. Modern thoughts on the evolution of Earth's atmosphere began with Aleksandr Oparin's theory described in Section 5.5.1. According to this theory, now widely accepted, Earth's original atmosphere was a hydrogen-rich or "reducing" atmosphere containing the gases H_2 (hydrogen), CH_4 (methane), NH_3 (ammonia), and H_2O (water vapor). Our present atmosphere differs greatly from the primordial atmosphere postulated by Oparin because free oxygen (O_2) is now abundant. The oldest rocks on Earth contain several compounds that would not have persisted under oxidizing (oxygen-rich) conditions, and thus seem to have been deposited at a time when the atmosphere contained little or no oxygen.

One important finding is that chemicals formed from the breakdown of chlorophyll molecules are not present in the oldest rocks. Chlorophyll is a key molecule in photosynthesis, a reaction by which plants, algae, and blue-green bacteria (Cyanobacteria) produce O_2 (see Section 17.1.2). The breakdown products of chlorophyll first appeared in the geological record at about the time that the reducing atmosphere began to change slowly, over a period of about a billion years, to the oxidizing conditions that now prevail.

If the atmosphere now differs so drastically from the primordial atmosphere postulated by Oparin, how and when did the change occur? In particular, how did the present oxidizing conditions replace the earlier reducing conditions? Most scientists who have investigated this question have concluded that life itself is primarily responsible.

Figure 19.1 **Chemical composition of the atmosphere.**

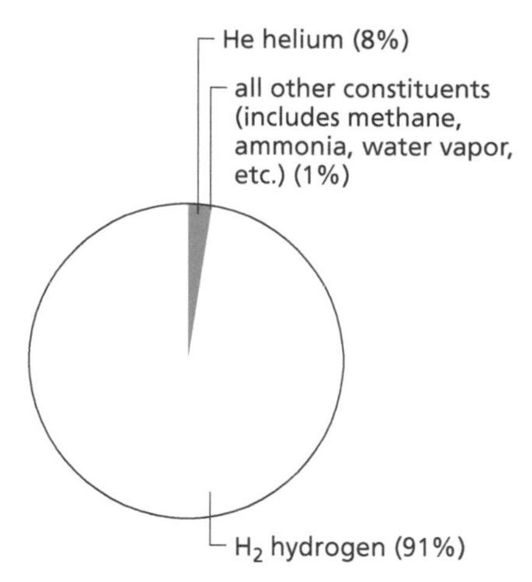

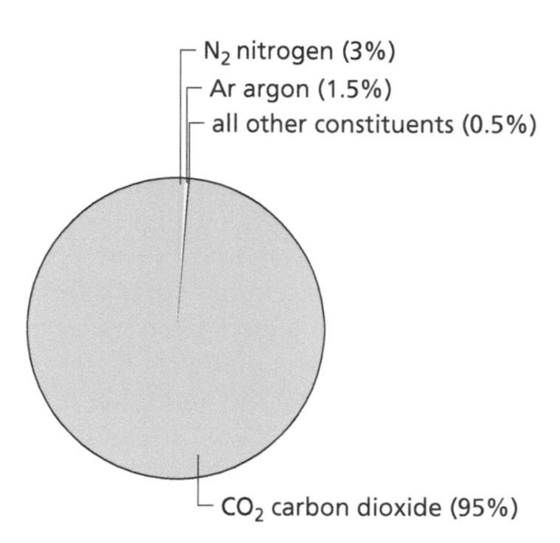

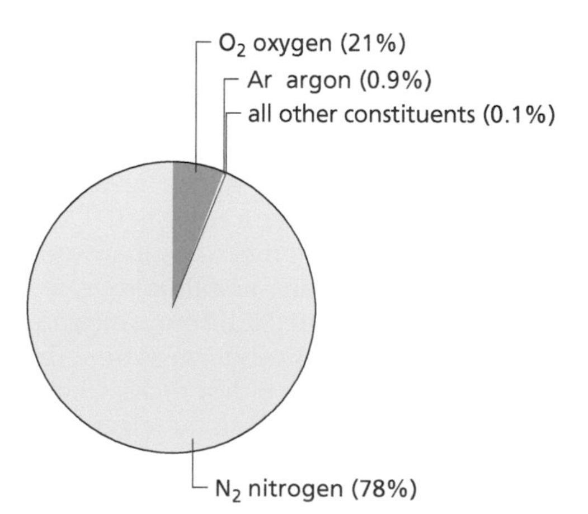

19.1.1.2 The first forms of life The first forms of life had to live under reducing conditions in which no oxygen was present, conditions that are called **anaerobic**. A variety of bacteria are capable of living under anaerobic conditions. The first bacteria or archaea were presumably enclosed by membranes and contained nucleic acids capable of passing on genetic information, but their sources of energy must have been different from those used by most organisms alive today. Most of the early bacteria were **heterotrophs**, meaning that they had to derive all their energy from the high-energy organic molecules that they found in their environs. Nowadays, such molecules are in most cases produced by other organisms, but the first organisms had to rely on the organic molecules that had formed without life (abiotically), under conditions like the ones simulated by such experiments as S.L. Miller's, described in Section 5.5.1. For many thousands or maybe millions of years, the supply of these molecules may have been adequate for heterotrophic life to expand and perhaps to flourish. (We can't tell for sure, because such organisms leave very few fossil traces of their existence.) Eventually, however, the organisms expanded to the point that the limited amount of energy-rich chemicals in the environment were just not enough. This was possibly the first global environmental crisis in the history of life on Earth.

We can imagine several possible responses to this crisis. Some organisms may have discovered a way to attack and devour other organisms, getting their nutrients from their prey. Organisms of this type continued to eat one another, but the total quantity of living organisms (the total biomass) that the planet could support remained limited. Those organisms more efficient at eating one another, or at making do with what little food they could find, did better than their less efficient competitors. Quite possibly, none of the organisms were able to adapt to this way of life, and all eventually perished. In that case, a second abiotic synthesis would have taken place all over again, and perhaps a third, until finally some group of organisms succeeded in finding a better and more permanent solution.

Even for organisms that ate other organisms, however, the possibilities would have been strictly limited. Other organisms would be encountered only so often, and only some of their constituents would be usable. Also, after a period of evolution had elapsed, more and more organisms would have evolved defenses against predators, and the amount of biomass would still have been limited. The problem may even have gotten worse, because some of the waste products of metabolism included gases like CO_2, which simply escaped into the atmosphere, taking carbon out of reach of most organisms.

19.1.1.3 Evolution of photosynthesis and its atmospheric effects

A more permanent solution to the limited supply of energy-rich chemicals occurred much later, when some organisms produced the first chlorophyll-like molecules, about 4 billion years ago. Such molecules would enable life forms to use solar energy in the form of light, which would solve the problem of using the limited number of other organisms. All forms of photosynthesis require the use of some chemical to supply hydrogen atoms and act as an electron acceptor (see Section 17.1.2). Most bacterial forms of photosynthesis use hydrogen sulfide (H_2S) for this purpose; others use iron compounds or organic molecules such as nicotinamide adenine dinucleotide (NAD). NAD was already present in organisms as derivatives of the nucleic acids. The primitive forms of photosynthesis did result in some minor changes in the atmosphere, perhaps including the use of some atmospheric CO_2 as a raw material.

The greatest change to the Earth's atmosphere resulted from the evolution of a new and more efficient kind of photosynthesis, using a new and different hydrogen donor (see Section 17.1.2). The new source of hydrogen was water, the most abundant hydrogen source on Earth. The splitting of water in photosynthesis generated a new atmospheric gas: oxygen (O_2). The first organisms to evolve this kind of photosynthesis were blue-green bacteria (Cyanobacteria). During the next 2 billion years or so, these blue-green bacteria became the dominant form of life on Earth (Schopf, 1974), reducing the abundance of atmospheric CO_2 to a small fraction of its former level and slowly generating more and more oxygen. Calculations of the photosynthetic capabilities of these blue-green bacteria show that they were indeed capable of generating all the oxygen in the Earth's atmosphere within half a billion years or so.

As the Earth's atmosphere became more and more oxygen-rich, additional changes began to occur. Capture of certain wavelengths of ultraviolet light by oxygen breaks up the O_2 molecules into a pair of highly reactive oxygen atoms (O). These will then react with the nearest other molecule to produce new compounds.

If oxygen atoms react with water, they produce hydrogen peroxide (H_2O_2), which acts as a bacterial poison. If oxygen atoms (O) react with oxygen molecules (O_2) in the presence of ultraviolet light, they produce ozone (O_3). Eventually, the production of ozone in this manner would slowly give rise to an ozone layer in the stratosphere and screen out additional ultraviolet light from reaching the Earth's surface. However, before such a layer existed, much more ultraviolet light would have penetrated the atmosphere and reached the Earth's surface, splitting more oxygen molecules and making more hydrogen peroxide. Because hydrogen peroxide is toxic to cells, early bacteria evolved enzymes that can break down this toxic chemical. Such enzymes also exist in the cells of all modern organisms that live in **aerobic** environments.

Scientists believe that the present composition of the Earth's atmosphere reflects the activities of living organisms (Figure 19.1). The earliest forms of life on Earth used up much of the methane and ammonia in the synthesis of organic compounds. Their waste products contributed to a buildup of atmospheric CO_2. The evolution of photosynthetic organisms gradually depleted the atmosphere of its carbon dioxide, replacing it with oxygen and giving our atmosphere its modern composition. Differences in density have caused the gases in the atmosphere to form several major layers that we will describe later (see Figure 19.6).

19.1.1.4 The atmosphere sustains life and is sustained by life The atmosphere of Earth and the living things on Earth have always influenced each other to such an extent that life and the Earth's atmosphere should be considered parts of an integrated whole (Holland, 1989; Schneider and Londer, 1984). The very air we breathe is a product of biological activity, and the stability of the planet's overall temperature is in part the result of atmospheric gases that are maintained in equilibrium with biological systems. This general idea has been promoted by James Lovelock and Lynn Margulis. Lovelock (1988) called it the **Gaia hypothesis**, after Gaia (or Gaea), the Greek goddess of the Earth after whom the science of geology is named (Joseph, 1990). Life on Earth has drastically changed our atmosphere from a reducing (hydrogen-rich) to an oxidizing one (oxygen-rich). Many of the chemicals that existed in reduced forms in primordial times are now more abundant in their oxidized forms: carbon as carbon dioxide rather than methane, and oxygen as molecular oxygen and ozone in addition to water. These changes, produced by living organisms, have made the planet even more habitable for the growth of other living organisms. Ozone, for example, captures ultraviolet light up in the stratosphere, greatly reducing the formation of peroxides at the surface. Several atmospheric gases, such as the CO_2 produced by many organisms, trap solar energy within the troposphere and reflect it back to Earth, producing the so-called "greenhouse effect" described later in this chapter. This greenhouse effect makes the Earth warmer than it would otherwise be and more hospitable to life as a result.

19.1.2 The water cycle

Water is essential to human life and to the lives of most other creatures on Earth. Irrigation is essential to agriculture in many places, and many thousands, if not millions, of animals and plants die each year under drought conditions.

Water evaporates from the surface of the oceans and from all bodies of fresh water. Much water is also transpired through the leaves of plants (see Section 17.2.2) or released from the respiratory activities of nearly all species of organisms. The resulting water vapor mixes into the atmosphere. At altitudes where the temperature is cold enough, much of this water vapor condenses to form ice crystals. In other places, water droplets form where the air is saturated with moisture. Ice crystals and water droplets form clouds as they accumulate. As air masses move around, some of these clouds cool down a bit and are no longer able to hold as much moisture as before. The excess moisture then returns to ground level as precipitation in the form of rain or snow. Much of this precipitation falls back into bodies of water, but some of it also falls upon the land, where it runs downhill in streams and rivers, accumulating in ponds, lakes, and oceans. Water thus recycles throughout the world's ecosystems, forming a **water cycle** (Figure 19.2).

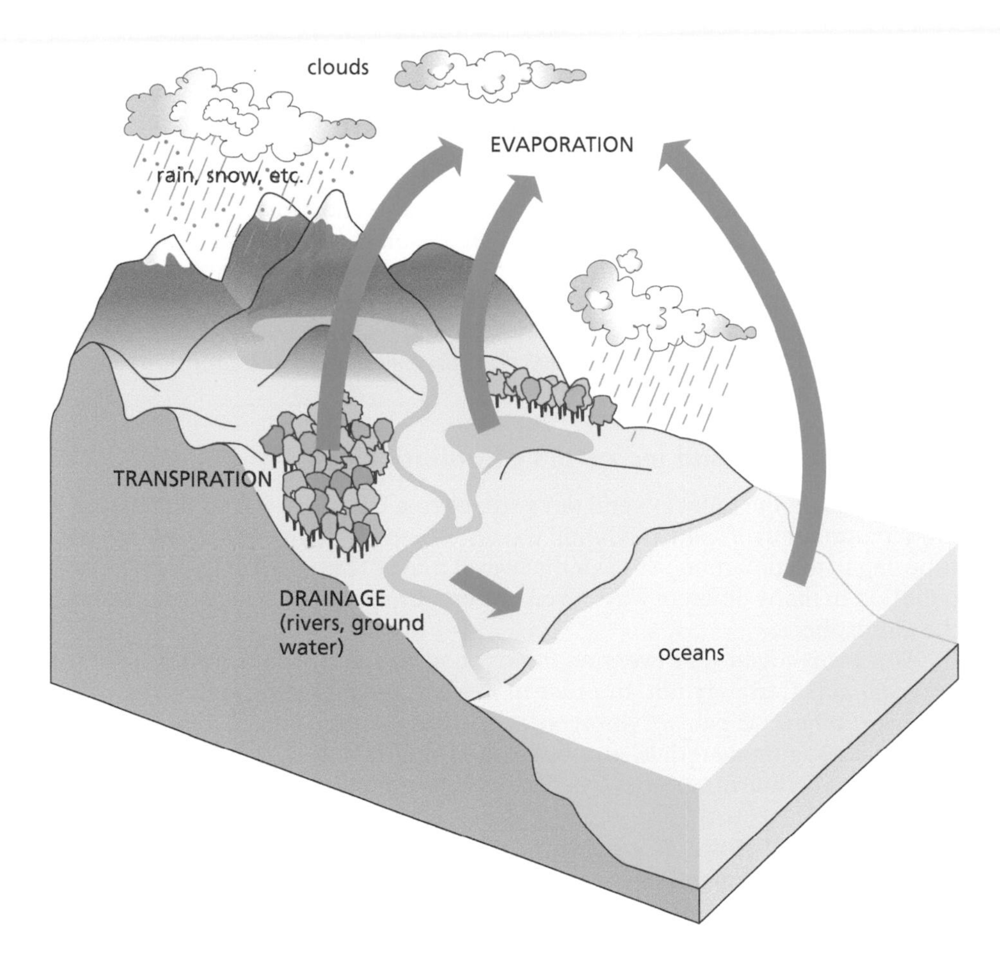

Figure 19.2 The water cycle.

THOUGHT QUESTIONS

1. In what way is the atmosphere part of the biosphere?
2. What would Earth's atmosphere be like if life had not evolved as it did? What would it be like if heterotrophic life had evolved, but not photosynthesis?
3. Which step in the early history of life do you think was the most important? How much do we know about each step in the process of life's origin? What means of investigation do you think will bring us additional knowledge?
4. How do you think the properties of life listed in Chapter 1 originated?
5. Are there limitations on the types of organisms that could evolve on Earth? How would we investigate such limitations? What limitations, if any, would apply to organisms on other planets or in other solar systems? How similar to organisms on Earth would such extraterrestrial organisms necessarily be?

19.2 POLLUTION THREATENS MUCH OF LIFE ON EARTH

Toxic dumps are places where waste disposal creates environmentally hazardous conditions. Millions of tons of waste materials are intentionally discarded every year. Other environmental problems arise from accidents. Large oil tankers run aground on occasion and spill thousands of tons of oil, and smaller oil spills occur even more frequently. Toxic dumps and oil spills are two of the many kinds of pollution that threaten our environment. Other types of pollution include groundwater contamination, fertilizer buildup in soils, and the release of harmful gases into the atmosphere from vehicles, smokestacks, and other sources.

The original meaning of the verb *pollute* was to contaminate or make dirty. Today, **pollution** may be defined as anything that is present in the wrong quantities or concentrations, in the wrong place, or at the wrong time. Although there is room for people to disagree about acceptable quantities, usually there is general agreement that pollution exists when it affects the health of humans or other organisms. Oil spilled from tankers or drilling operations can kill thousands of aquatic birds, mammals, fish, and other organisms. Lead, cadmium, and other heavy metals in drinking water, food, or house paints can cause brain damage and other neurological defects. Toxic dumps can poison people and raise cancer rates. In some of the worst cases, numerous unidentified chemicals were dumped together into the same toxic waste site, forming a "witch's brew" that underwent further and often unpredictable chemical reactions to produce additional hazards. In many countries, it is now illegal to dispose of many chemicals except by government-approved methods.

19.2.1 Sources and indicators of pollution

When we flush the toilet or send our garbage to a landfill, we are contributing to the accumulation of solid and liquid wastes. When we drive our cars, we are contributing to the pollution of the air that we all must breathe. All of us contribute to pollution in many different ways. Even our breathing releases carbon dioxide into the atmosphere.

Does this mean that every act mentioned in the previous paragraph is an immoral act? Certainly not. In order to live, we must breathe, eat, urinate, and defecate. When we eat, we throw away inedible parts (skin, bones, pits, rinds, shells), packaging materials, and unfinished remains. To go on living, we must continue to pollute in certain ways. So why is there such a fuss?

19.2.1.1 Pollution: A problem of quantities Pollution in most cases is a problem of quantities, and sometimes also of location and rates of accumulation. Clearly, you don't want your garbage to accumulate in your living room. Suppose, for the moment, that garbage disposal services were not available and you had to dispose of your household garbage yourself, as many people in rural places still do. You could perhaps bury it in the backyard, or just toss it away. If you only tossed away bones, rinds, and other biodegradable materials (things that can be broken down by bacteria, fungi, and other decomposer organisms), then you might be able to dispose of your own garbage in this way—up to a point, that is. Your backyard might have enough decomposer organisms to break down and recycle your own personal wastes, or perhaps your family's wastes. Of course, this depends in part on the size of your family and also on the size of your backyard. Clearly, your backyard would not be able to handle the garbage produced by an entire town or city.

Just about every known pollutant is harmless in some sufficiently small quantity. Pollutants become bothersome or toxic as quantities increase. Therefore, in most cases, measuring pollution means measuring its quantities.

How do we know whether a habitat is polluted, or to what extent? If pollution cleanup is attempted in a particular place, how do we measure the success of the cleanup effort? Most of the specific answers depend upon the measuring of particular chemical substances. First, a particular chemical pollutant or breakdown product must be identified as an indicator of the pollution in question. Second, the concentration of this particular chemical must be measured repeatedly at various places and times.

19.2.1.2 Sentinel species More general indicators of pollution are also needed, especially pollution from hazards that have not yet been clearly identified. A number of environmentally sensitive species have been suggested as possible general indicators of pollution. These **sentinel species** serve the same role as the canaries that coal miners often took with them into the mines. Because the canaries were extremely sensitive to methane and other dangerous gases present in coal mines, the health of the canary reassured the miners, and the sickness or death of a canary was always viewed by the miners as a danger signal. Frogs and other amphibians are sentinel species that can warn us of the deterioration of freshwater habitats, just as the spotted owl can be considered a sentinel species whose numbers are indicative of the health of old-growth Pacific forests. Dolphins have occasionally been suggested as sentinel species for marine habitats because environmental pollution can stress the immune systems of these marine mammals and raise their rate of infectious diseases. The grounding of marine mammals on beaches may also be a stress-related phenomenon that reflects marine pollution.

19.2.2 Toxic effects

The field of toxicology deals with the damage done to human or animal health by various quantities of poisons, including environmental pollutants. Toxic damage can affect any of the body's systems, but the nervous and reproductive systems are especially vulnerable in many species. Chemical tests can often detect the quantities of environmental pollutants, but biological tests (bioassays) are in many cases even more sensitive. Bioassays can even be used to monitor the effects of hazards that have not yet been chemically identified or hazards that may contain a complex mix of several chemical substances.

As an example of an environmental pollution hazard, consider the chemicals known as dioxins. Dioxins are chlorine-containing organic compounds that arise as inadvertent by-products from the use of chlorine as a bleach in the paper-making process. There are several different dioxins, of which the most toxic is 2,3,7,8-tetrachloro-dibenzo-*p*-dioxin. Dioxins are often found in freshwater ecosystems downstream from industrial sources, and in these places, they can poison and sometimes kill fish and other aquatic animals. Dioxins are chemically similar to "agent orange," a toxic substance used as a defoliant herbicide during the Vietnam War (1960–1975). Dioxins are among the many toxic substances that have estrogenic effects, meaning that they activate cell signaling in ways that are similar to those of the female hormone estrogen (see Section 9.2.1). Excessive estrogen signaling can affect the reproductive systems of both males and females of many species, including humans. Knowing of these toxic effects, many companies have been testing other types of bleaching processes for paper, and some paper is no longer bleached at all.

Other types of toxic chemicals are used intentionally, for example, as pesticides, and they then become hazardous to human health when they accumulate in unwanted places. Some long-lasting toxic pollutants accumulate in biological tissues and are subject to biomagnification (see Section 17.4.5.4), a process that increases their concentration as they pass through the food chain. Pollutants subject to biomagnification include heavy metals such as mercury and many insecticides such as DDT. Most insecticides and many heavy metals are neurological poisons. Many long-lasting pesticides, including DDT, are now banned in the United States and in many other countries. Pollution from insecticides can be greatly reduced by avoiding the long-lasting chemicals that accumulate in biological tissues and are biomagnified, switching instead to biodegradable pesticides that break down in biological systems.

19.2.3 Pollution prevention

Pollution awareness is a crucial component in the prevention of pollution. Unless we realize how we are polluting, it is unlikely that we will take any corrective action to pollute less (Harrison, 1996; Park and Labys, 1998). We should all inform ourselves about the disposal of our garbage and our industrial waste; about the emissions from our cars and from nearby (and distant) smokestacks; and about the cleanliness of our beaches, playgrounds, drinking supplies, and foods.

Because pollution is a consequence of the ways in which we live and work, there are certain things that we can each do to reduce the amount of pollution that we cause. Most of these measures also have further benefits, such as saving money or improving human health. For example, carpooling saves money, while bicycling to work saves even more money and contributes to health and fitness as well. Recycling saves money, too, and we can all recycle such items as paper, bottles, and cans. Making an aluminum soda can out of recycled materials requires only a small fraction of the electricity needed to make the same can from aluminum ore. Many industries are now responding to market forces by using recycled materials in their products and by advertising their use of biodegradable or other "Earth-friendly" materials.

19.2.3.1 Costs and benefits Because pollution is largely a problem of quantities, ethical discussions about pollution usually include discussions about quantities. The determination of both the costs and the benefits associated with pollution depends upon the measurement of quantities, whether in dollars, in lives lost, or in reduced human health. The same is true of the benefits of preventing or cleaning up pollution, which can be measured in dollars, in lives saved, or in increased health or enjoyment.

One general problem is that costs are often easier to identify and measure than benefits and are subject to much less uncertainty. The costs of curtailing or modifying a manufacturing process can easily be measured in terms of costs of new equipment, costs to operate the equipment, and either increased wages (if additional employees are needed) or reduced employment (if an activity is curtailed). These costs are fairly certain and are easily measured. The benefits to the ecosystem are less certain and are harder to measure: If a particular set of changes is implemented, will the ecosystem recover? What levels will the pollutants reach at some distance from the source? How many lives will be saved or how much disease prevented as a result? What dollar value should be placed on the prevention of disease, on the bird or fish population, or on the health of the environment?

Translating "quality" of health or enjoyment into something that can be quantified is a relatively new and important specialty within economics. One way of measuring these intangibles (certainly not the only way, but definitely one of the easiest ways) is by measuring the average dollar amount that people are willing to pay to obtain them. To take one small example, the value of a better neighborhood can be measured by the additional amount that people are willing to pay to live in it compared with some other neighborhood. The value of a clean environment can be measured by the amounts of money that people are willing to sacrifice in order to live in it, work in it, or visit it.

Attempts to measure environmental quality can sometimes be misunderstood. When we try to measure environmental quality, we need to consider the value of the entire ecosystem, not just the indicator species used as a measuring tool to monitor quality. Salmon have a certain value as a commercial food species, but they have a far greater importance as a general indicator of pollution or of the health of a river ecosystem. If salmon populations are used to measure pollution or pollution abatement in freshwater ecosystems, it is the value of the entire ecosystem that should be counted as a benefit, not just the commercial value of the salmon fishery. Likewise, bald eagles and spotted owls (neither of which have commercial value) can be used to indicate the general health of ecosystems, and it is the health of these entire ecosystems that should be counted as a benefit, not just the value that we place on the eagles or owls. The death of the canary in the mine means much more to the miners than the price of a replacement bird.

THOUGHT QUESTIONS

1. How does your town or city dispose of trash? Is there a recycling program? What gets recycled? What is not currently recycled in your community but could be?
2. Is it possible to live in the industrial world without polluting?
3. Think of at least two important sources of pollution in the region where you live or go to school. Is there a way to reduce this pollution? What practices would have to change?
4. Materials that cannot easily be recycled are generally taken to places specifically set aside as dumps. If such facilities are needed, where should they be located? Because dumps are generally unsightly, unclean (sometimes toxic), and subject to noisy traffic, people generally don't want to live near them. The motivating desire to have these facilities somewhere else is often symbolized by the phrase "not in my back yard" (NIMBY). One result is that dumps are often located wherever people have the least political clout, sometimes giving rise to charges of "ecological racism." What further social problems arise from the widespread application of the NIMBY principle? Is there any socially responsible way to locate an undesirable facility? What is the best way to reduce the need for such facilities in the first place?

19.3 HUMAN ACTIVITIES ARE AFFECTING THE BIOSHPERE

Human activities can change the biosphere in many ways. Humans clear many habitats for housing, for industrial development, and for the planting of crops (see Chapter 18). Many other human activities affect the quality and availability of water. Some human activities pollute the land, water, and atmosphere. Atmospheric pollution can impair respiration and can affect the acid content of rain and snow. Some atmospheric pollution can also result in the destruction of atmospheric ozone, a process that can raise the rates of mutation and cancer by increasing the amount of ultraviolet radiation that reaches the Earth's surface. Increased release of carbon dioxide and certain other gases can raise global temperatures. The health of the biosphere is thus at stake in what we do.

19.3.1 Aquatic pollution and its biological effects

Materials produced by human activities can enter the water directly or can be washed into the water from soil or from disposal sites. Materials dissolved and suspended in the water can be taken up by organisms and can interact with biological systems. Some materials are beneficial, but many others can cause harm. Aquatic pollution derived from human activity includes agricultural runoff, industrial wastes, human sewage, and accidental spills. In Section 17.4, we discussed how agricultural fertilizers can run off and enter aquatic ecosystems, causing algal blooms that may sometimes kill fish and other animals. Toxic chemicals that can enter aquatic ecosystems include pesticides from agricultural fields, dioxins from paper factories, polychlorinated biphenyls (PCBs) from electrical insulation, heavy metals from various industrial processes and from mining activities, and a variety of other chemicals including solvents. Certain conditions also allow pathogens to enter water and to proliferate and cause diseases like schistosomiasis, as we saw in Chapter 16.

Pollution of water supplies can limit the availability of water for irrigation or for human consumption. Water pollution may be a source of political conflict between downstream states or countries and those located more upstream. In many cases, the downstream users seek protection from upstream uses that might affect water flow or water quality. Such "water wars" are an important source of friction between California and Arizona, between Georgia and Florida, between Israel and its neighbors, and between Turkey and Iraq.

Any material can be accidentally spilled, especially during its transport, but spills of oily substances are usually of greatest concern because they are transported in very large quantities and because they spread out very rapidly in water. When oil is spilled at sea, almost none of it dissolves in the water. Crude oil (petroleum) is a mixture of many different chemical compounds, the exact mixture depending on where the oil came from and whether or not it has been refined. Most of the compounds are nonpolar and are therefore not very soluble in water, which is very polar (see Section 8.1.2). Crude oil is also less dense than water, so it floats on top, gradually spreading across the surface to form a "slick." Although oil slicks may be only a few molecules thick, they interfere with organisms that need to absorb oxygen at the water surface. Evaporation removes the smaller, lighter-weight components of the slick; the components left behind may then be denser than water and sink, eventually blocking bottom-dwelling organisms' access to oxygen. Oil pollution can clog the gills and feeding surfaces of many fishes and other aquatic organisms; it can also harm the insulating fur or feathers of marine mammals and birds, making them less able to escape, to keep warm, or to resist diseases. Evaporation, sinking, or dispersal of oil may make it disappear from view, but the oil molecules are unchanged and can still cause harm, even years later (Mitchell, 1999).

19.3.1.1 Plastic pollution As a case in point, consider the case of plastics. As an adjective, the term "plastic" means capable of being formed into any desired shape. Cellulose-based plastics such as Celluloid® were first developed in the

1860s and were used in filming early motion pictures. Bakelite®, invented in 1907, was the first all-synthetic (petroleum-based) plastic and was first used as a replacement for ivory in items such as billiard balls. The use of plastics for packaging and for inexpensive toys became widespread after about 1960, and polystyrene foam (under the brand name Styrofoam®) was introduced in 1954. All of these materials could be manufactured economically in a variety of forms and shapes, and all are durable and long-lasting under ordinary use.

As plastics became more widely used, their manufacturing costs diminished, and many new uses were found for plastic products as replacements for wood, metal, and even clothing fabrics (polyester and spandex are two examples). Cheap plastics were increasingly used for plastic packaging and for inexpensive, throw-away items like soda straws and coffee cups. Microscopic "nanoplastics" were sometimes added to clothing fabrics to make them more washable and stain-resistant. More and more plastics found their way into garbage dumps and urban litter, and the chemical durability of plastics that had once been a commercial blessing became instead an environmental curse (Freinkel, 2011). With very few exceptions, plastics cannot be recycled, and, unlike paper and food waste, plastics are not biodegradable. Most plastics are made from petroleum, which is considered a **nonrenewable resource** because there is only a finite supply that will eventually run out.

When released to the environment, plastic waste can accumulate in landfills, where its estimated lifespan is measured in thousands of years. Most plastics cannot be burned without releasing toxic by-products. Worst of all, carelessly discarded plastics find their way into waterways and eventually into the oceans, where they often accumulate on beaches (Figure 19.3) and are sometimes joined by plastics washed out of garbage dumps and landfills. Over many years, most plastics exposed to air and seawater become brittle and break into smaller and smaller pieces, called "microplastics," without chemical alteration. (A few plastics do break down chemically over time, but usually into toxic pollutants such as ethylene and methane.)

Microplastics have been detected in snow from the Alps (Bergmann et al., 2019), on remote Arctic islands, and in some human tissues. Most of the plastic in the oceans floats at or near the sea surface, but a good deal of plastic also sinks to the ocean bottom, where it is much harder to study and therefore much less understood. Since the 1980s, more and more of our ocean surface waters have become laden with tiny pieces of plastic that fish and many other sea creatures mistake for food items. Several large accumulations of these harmful plastic particles have been discovered in oceans around the world. Most of the particles are small, so the appearance is more like that of cloudy water than of floating garbage. Many birds, fish, sea turtles, and marine mammals have choked to death on plastics or become entangled in plastic fishing lines or nets; many others have starved to death with stomachs full of indigestible plastic. Plastic ingested by smaller

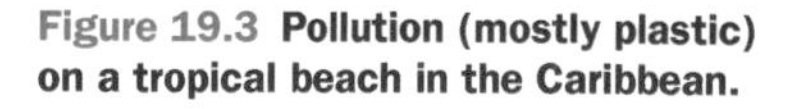
Figure 19.3 **Pollution (mostly plastic) on a tropical beach in the Caribbean.**

organisms passes up the food chain when eaten by larger organisms, sometimes finding its way into the human food supply and into human tissues.

Because of environmental problems and trash buildup, many places have banned the use of certain types of plastics. Plastic grocery bags are either outlawed or heavily taxed in China, Germany, the United Kingdom, and over 30 other countries around the world. In the United States, plastic grocery bags are similarly banned in a growing number of states including California, Hawaii, New York, Maine, Rhode Island, and North Carolina, and in several large cities as well. Cities that have banned some or all Styrofoam food and beverage containers include New York, Los Angeles, San Francisco, Seattle, Baltimore, Miami Beach, Portland (Maine), Portland (Oregon), and Washington, DC. France has banned Styrofoam containers nationwide, and Dunkin' Donuts has phased out all use of Styrofoam worldwide. Plastic straws have been banned in Seattle and San Francisco, and several large corporations (including Starbucks, Disney, Marriott, Hyatt, and American Airlines) have already eliminated their use internationally. In 2018, the European Parliament voted to ban the use of all single-use plastic items by 2021, and several island nations havc done so as well.

19.3.2 Bioremediation

Living organisms can be used to reduce some types of pollution to restore habitats. **Bioremediation** is an approach in which the decomposition activities of living organisms are put to work in cleaning up contaminated soil and water (Chakrabarty, 1996; Dixon, 1996). **Biodegradation** refers to the natural decomposition processes that go on without human intervention; bioremediation, in contrast, implies the manipulation of biodegradative processes by humans. Bioremediation has been used in the cleanup of oil and chemical spills and in wastewater treatment. The aim of bioremediation is to change the potentially harmful molecules into something harmless by using the chemical reactions carried out by decomposer organisms, some of which are very effective at this task.

19.3.2.1 Bioremediation of oil spills Bioremediation is often applied to problems of oil contamination on land, but in this section, we discuss the challenge of cleaning up aquatic oil spills. As we saw in Section 17.1, certain organisms derive their energy by breaking down complex molecules. These are the decomposers, mostly bacteria and fungi, that keep both energy and matter cycling through the biosphere.

Oil-degrading bacteria and fungi are found in all types of aquatic habitats, both freshwater and marine, and include representatives of over 70 different genera. No one species can degrade all the molecular compounds in oil. Even when bacteria capable of producing oil-degrading enzymes are present where oil has spilled, the rate and extent of biodegradation depend on many environmental factors, including temperature, the amount of oxygen, and the availability of other nutrients (Biello, 2010; Zheng et al., 2018). Because many of these factors are unpredictable, the success of biodegradation at any given site is also unpredictable.

Probably the most critical factor for biodegradation is the availability of nutrients such as nitrogen, phosphorus, and iron. These elements are necessary for microbial synthesis of proteins and nucleic acids and as enzyme cofactors, much as they are in other living organisms.

Most types of bioremediation attempt to enhance the natural processes of biodegradation. Three basic strategies are employed at a spill site: (1) enrichment of rate-limiting nutrients, (2) introduction of naturally occurring bacteria, and (3) introduction of genetically engineered bacteria.

In concept, nutrient enrichment is much like the fertilization of soil (see Section 17.4.1). It assumes that oil-degrading microorganisms adapted to the local conditions are already present. Supplying nutrients supports the faster reproduction of degradative bacteria, and the hope is that the bacterial population will grow quickly and to a great enough density to overcome the large quantity of oil. After the *Exxon Valdez* spill in 1989, the concept was extensively tested by the Environmental Protection Agency (EPA), in conjunction with Exxon and the state of Alaska. In these tests, the rate of biodegradation on a 110-mile stretch of nutrient-enriched beach was accelerated two-fold to four-fold for a period of at least 30 days. A second application of nutrients after three to five weeks accelerated the rate even more. Similar measures have been used to clean up other oil spills, such as after the Persian Gulf War of 1991 and the Deepwater Horizon oil spill of 2010 (Biello, 2010; Zheng et al., 2018).

The bacterial species indigenous to a spill area may not have the right enzymes to degrade the compounds in the type of oil in that spill. Each decomposer species is able to make specific enzymes that can degrade some specific molecules but not others. Even if the right species are present, their numbers may not be sufficient to degrade the oil at an appreciable rate. Hence, it may be advantageous to introduce a mixed bacterial population containing many species able to digest more of the oil components than any single species would. Some people further advocate using genetic engineering to produce bacteria that can make a greater range of enzymes. The first patented bacterial species was one that had been genetically engineered to degrade oil. This approach may also have potential for remediation of biohazardous wastes other than oil.

Although the concept of introducing bacteria into oil spills seems plausible, nobody has yet perfected the process. Two species of commercially available bacteria were introduced on beaches polluted by the *Exxon Valdez* oil spill of 1989, but (in contrast to the boosting of locally occurring bacteria by fertilizing) there was no significant enhancement of biodegradation compared with that on untreated beaches. It is possible that introduced bacteria might possibly increase to numbers great enough to upset the ecological balance of other organisms, but it is more likely that the introduced population would die off as soon as the oil was used up. Initial experience suggests that a greater problem than overgrowth may be getting an introduced population to survive long enough to degrade all the oil.

19.3.2.2 Bioremediation of wastewater Wastewater includes any water that has been used, for whatever purpose. All wastewater treatment depends on the biological activity of microorganisms. Sewage is water that contains fecal wastes from humans and animals. If not treated, sewage is a major route for the spread of infectious diseases (see Chapter 16). Ingestion of fecal-contaminated water spreads bacterial diseases such as cholera, shigellosis, and typhoid fever, as well as viral diseases such as hepatitis A. Swimming or wading in contaminated water puts people at risk of such parasitic diseases as schistosomiasis. Fecal-contaminated water is concentrated by filter-feeding animals, so mollusks collected from polluted waters may be highly infectious to people who eat them.

Streams and soil often contain both minerals that trap some pollutants and bacteria capable of biodegrading waste materials. For the very low human population densities of times past, natural filtering and biodegradation by streams and soil was adequate sewage treatment. Even today, in many areas where soil is suitable and population density is not too great, the wastes from homes can be collected and treated by septic tanks and leach beds (Figure 19.4). The treated water goes back into the ground, not back into the home. This type of system can only be effective when the wastewater input does not exceed the biodegradation capacity of the soil bacteria. Putting materials down the drain that kill the soil bacteria make the system nonfunctional.

A slightly larger-scale sewage system can separate fecal material from wastewater in a shallow lagoon in which the wastewater can be contained for about 30 days. This is sufficient time for solids to settle and for sunlight, air, and microorganisms to kill the bacteria and viruses from human and animal sources, including many microorganisms that cause human disease. Wastewater lagoons are actually complex ecosystems. Algae grow in a wastewater lagoon, using carbon dioxide to conduct photosynthesis and producing oxygen in the process. The oxygen produced by the algae keeps the lagoon aerated, allowing the growth of aerobic bacteria that digest organic matter and kill fecal bacteria. Simple sewage systems work well if they are not overloaded. They can be overloaded by excess rainwater, an increase in users from a growing population, or an increase in wastewater production per person.

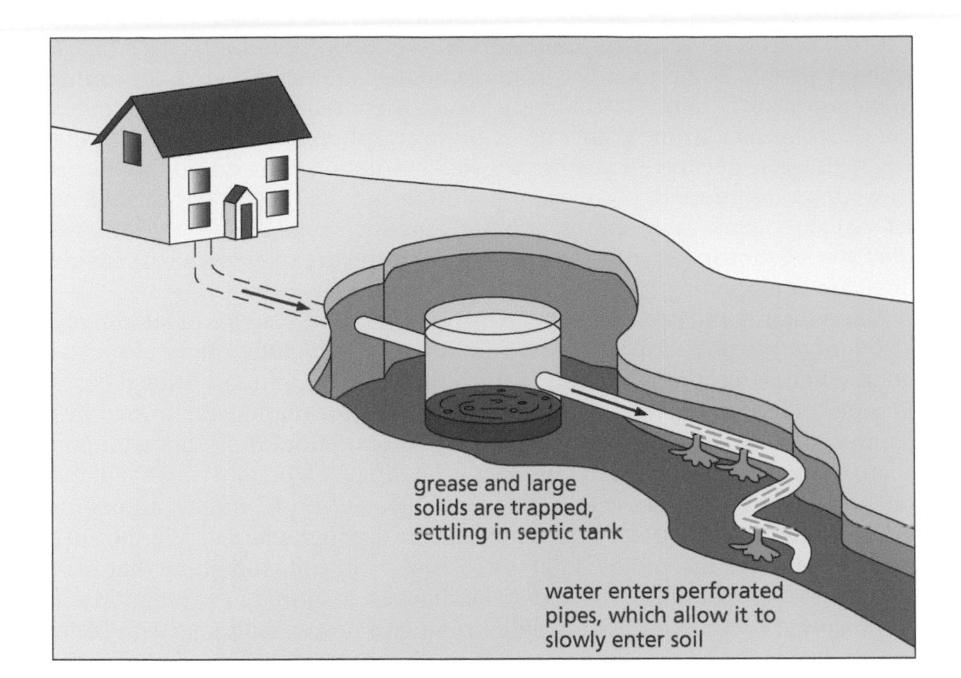

Figure 19.4 Septic tanks and leach beds.

Where simple systems prove inadequate, centralized water-treatment plants are needed. Municipal water-treatment plants use multiple treatment methods, usually in succession: settling in wastewater lagoons, bacterial biodegradation, fine filtration, and treatment with chlorine or other disinfectants. Once the water is free from bacterial and other pathogens, it is usually discharged into rivers, lakes, or oceans, but sometimes it can be diverted for agricultural use.

Some towns are experimenting with using marshlands to help treat wastewater. Because many natural wetlands have been destroyed by pollution or development, towns have created new wetlands. Wastewater goes through primary treatment and then is pumped into these marshes. Marsh plants remove nitrogen and phosphorus, break down sewage, and even filter out toxic chemicals. Preliminary treatment may be performed in greenhouses. In tanks containing such plants as water hyacinths and cattails, algae and microorganisms decompose the organic materials, which are then used as nutrients by the plants. The water then passes to other tanks containing snails and zooplankton that consume the algae and microorganisms; the zooplankton are then eaten by fish, and the water is further purified by marshlands. Such systems work well, but, again, only if the quantities of wastewater do not exceed the capacity of the treatment ecosystem. The demonstrated success of such small-scale treatment facilities emphasizes that there are many different, ecologically sound ways to treat wastewater.

19.3.3 Air pollution

The air we breathe is necessary to human life, but some human activities have begun to change the atmosphere. Natural ecosystems can generally handle the gases that we exhale. The other products of biological activities, including agriculture, are usually recycled naturally and safely as long as these products do not accumulate too much in any one place.

Air pollution affects the air that we breathe, both indoors and outdoors, and causes many unwanted changes in the atmosphere. Some of these changes are local only, but others have worldwide consequences.

Automobile exhausts and industrial plants are major sources of outdoor air pollution. Large forest fires can also pollute the atmosphere sporadically in the regions where they occur. Many outdoor air pollutants are oxides of carbon (carbon monoxide and carbon dioxide), oxides of nitrogen, and oxides of sulfur released as combustion products. Ozone is released to the air by many processes, but automobile exhausts release more ozone than any other source. Certain other gases, such as chlorine, benzene, and hydrogen sulfide, are also released by some industrial processes. When the concentrations of these pollutants reach high levels, many people begin to suffer from respiratory ailments. Smog, for example, can be particularly harmful to lungs because it contains dangerous amounts of ozone. (In contrast, ozone higher up in the stratosphere protects us from dangerously high levels of cancer-causing ultraviolet radiation.) In addition to the gases that we have mentioned, outdoor air pollution can also include industrial soot, bacterial pathogens, mold spores, and allergens such as pollen. The U.S. EPA estimates that 6.6 metric tons of particulate air pollution are released each year in the United States alone.

Air pollutants can become trapped in the ventilation systems of buildings and cause indoor air pollution. Indoor air pollution can include many of the same components as outdoor air pollution (including pollen), plus additional bacteria and other pollutants such as asbestos, benzene, or formaldehyde formed by the decomposition of materials used in building construction. Buildings with poorly designed ventilation systems can impair the health of the people working in them, a phenomenon sometimes called "sick building syndrome." Indoor air pollution may be one of the causes of a 70% increase in the incidence of allergies in the United States over the period 1980–2000. Studies in England show that people working in older buildings with open windows as a source of ventilation suffer fewer respiratory illnesses than people working in newer buildings with recirculated air. Secondhand cigarette smoke is another form of indoor air pollution. A large-scale English study showed that cancer rates were much higher among nonsmoking married women whose husbands smoked than among a matched group of women married to nonsmokers, presumably because the women in the first group were breathing the polluted indoor air containing the secondhand smoke (see Section 10.6.4). Campaigns to banish smoking from restaurants and other public places are motivated by such health considerations.

19.3.4 Acid rain

One of the most widespread and best understood pollution problems is that of acid rain (U.S. EPA, 1980; Boyle and Boyle, 1983; Yanarella and Ihara, 1985). We should really speak of "acid deposition" (NRC, 1983), because much of the problem comes from acid snow, acid fog, and dry deposition of acid dust or condensation. The acidity of a substance is measured on a standard pH scale (see Figure 8.11), where lower values indicate higher acidity (more hydrogen ions). Rainwater of pH 5 or below is considered acid rain, and values as low as 2.7 have been measured on occasion. Recall that the pH scale is a logarithmic scale, so that a lowering of pH by one unit corresponds to a 10-fold increase in the concentration of acid, and a pH of 3.0 is therefore 100 times as acidic as a pH of 5. Because most enzymes work optimally only within a very narrow pH range, and because the structure of biological molecules may change at different pH values, a small change in pH can have an enormous effect on living organisms (**Figure 19.5**).

Chemical tests of acid rain show that the source of most of the acidity is either sulfuric acid (H_2SO_4) or nitric acid (HNO_3). Acid rain across the eastern United States and the Scandinavian countries is primarily from sulfuric acid, while the acid content of acid rain in Colorado and other western United States is more than 50% nitric acid, derived primarily from the nitric oxides produced by automobile exhausts (Johnson et al., 1987).

Sulfuric acid pollution begins when materials containing sulfur are burned in air, forming sulfur dioxide (SO_2). The sulfur dioxide combines with additional oxygen to make sulfur trioxide, which then combines with water to make sulfuric acid. Sulfur is a common impurity in many coal deposits and is also present in many of the ores from which lead, zinc, nickel, and certain other minerals are commonly obtained. In many parts of the United States, Canada, England, Germany, and Poland, the mining of metals from sulfur ores generates sulfur dioxide that ends up as acid rain in such downwind locations as the northeastern United States and Sweden. The burning of high-sulfur coal in electrical power plants is an even larger source of sulfur dioxide pollution that eventually falls as acid rain. This type of acid rain is a political as well as an environmental problem

trees killed by acid rain in North Carolina

trout taken from a lake in Ontario, Canada, in 1979, at pH 5.4

trout taken from the same lake in 1982, at pH 5.1

stone sculpture dissolving in acid rain

Figure 19.5 Some of the effects of acid rain.

because the governments of New York and of Sweden are relatively powerless to control pollution that originates in Illinois, Indiana, or Germany. The United States and Canada have accused one another of being major sources of cross-border acid rain pollution.

Acid rain is a problem wherever there are large numbers of factories or automobiles. In many parts of Asia and Latin America, countries burn more coal as they industrialize and more gasoline as their populations become more prosperous. Acid rain is therefore becoming more of a problem in these countries, most acutely in China.

Acid rain erodes and slowly dissolves marble and limestone statues and buildings (Figure 19.5), including those of the famous Parthenon in Athens,

Greece. In localities where the rock formations are predominantly limestone and other carbonate rocks, acid rain is neutralized as it runs through these rocks or as it percolates through the soils derived from the weathering of these rocks. However, granitic rocks, such as those that predominate in New England, northern New York State, Scandinavia, and parts of Ontario, have little or no capacity to neutralize acid rain. As a result, acid deposition in those areas accumulates in ponds and lakes to levels that kill many fish (Figure 19.5). The Adirondacks of New York State contain hundreds of lakes and ponds that once teemed with fish, but whose fish populations have completely died out because of the acidity of the water.

19.3.5 Atmospheric ozone

The Earth's current atmosphere is organized into several major layers (**Figure 19.6**). The lowest layer, which most immediately impacts human activity, is the troposphere, containing the majority of atmospheric gases. Weather phenomena and all routine types of aircraft flight are confined to the troposphere. Above the troposphere lies the stratosphere, the zone in which ozone accumulates. Still higher are the mesosphere and thermosphere.

The two most abundant gases in the atmosphere are nitrogen and oxygen. Most oxygen consists of molecules of O_2, oxygen atoms bonded together in groups of two, but ozone (O_3) is a more reactive form in which oxygen atoms are bonded together in groups of three. If you have ever smelled a sharp, pungent, "electrical"

Figure 19.6 Structure of the Earth's atmosphere.

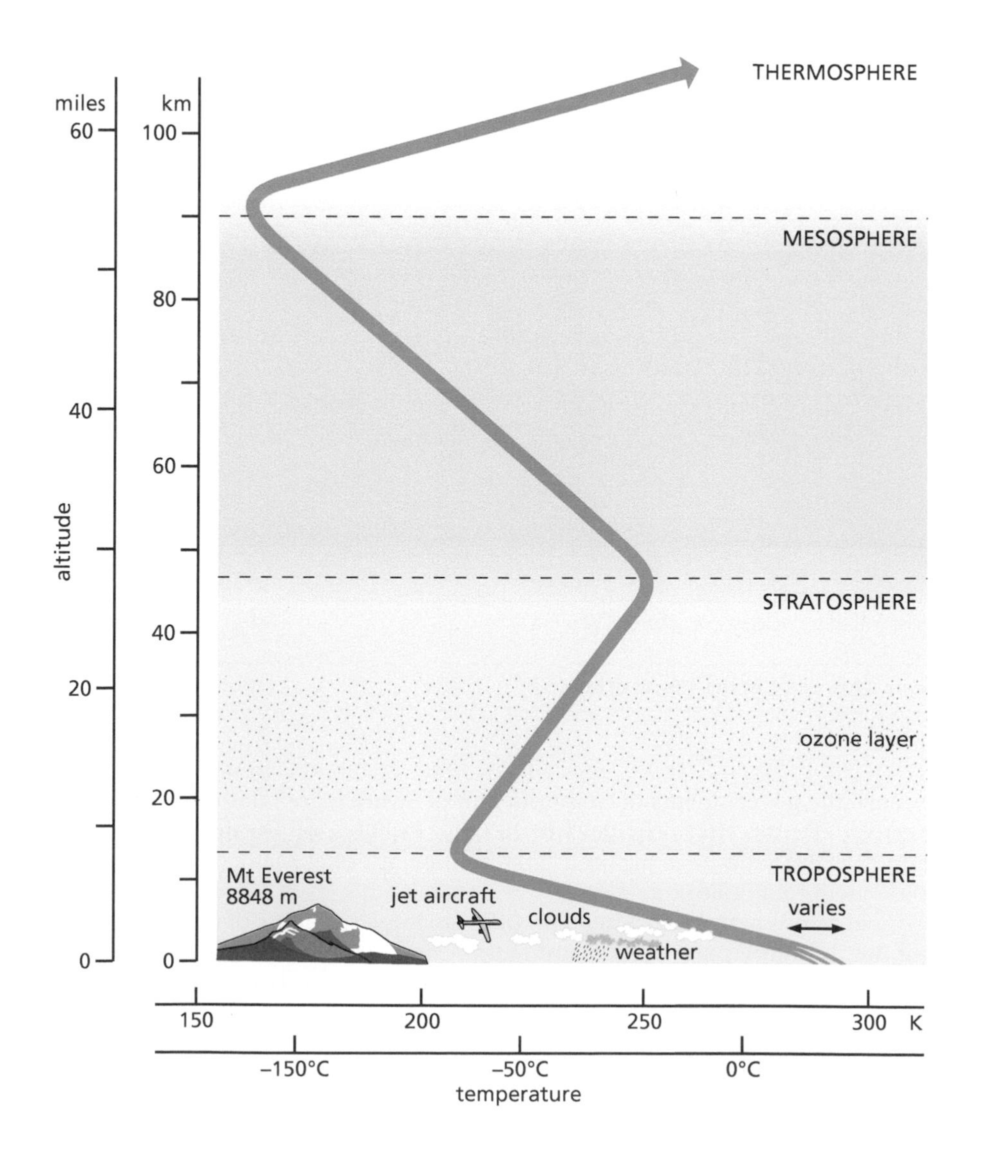

smell around a hot iron or toaster, that is the smell of the small concentration of ozone produced when electrical activity dissociates oxygen molecules:

$$O_2 \rightarrow O + O; \text{ then } O + O_2 \rightarrow O_3$$

Ozone is the most important and most dangerous constituent of urban smog, produced when exhaust from thousands of automobiles is trapped beneath a layer of warm air (a temperature inversion). When enough nitrogen oxides accumulate from this automobile exhaust, a photochemical (light-induced) reaction produces ozone as follows:

$$NO_2 + \text{light} \rightarrow NO + O$$

$$O + O_2 \rightarrow O_3$$

Ozone at ground level is hazardous to human health because it causes irritation of the eyes and respiratory passages and oxidative damage to many mucous membranes. Death may occur by respiratory failure, especially in individuals with preexisting respiratory problems. Despite the harm that ozone produces at ground level, it is beneficial—even essential—at high altitude. In the Earth's stratosphere, oxygen molecules bombarded by the sun's ultraviolet radiation slowly rearrange to form ozone. The ozone becomes concentrated into a distinctive layer within the stratosphere (Figure 19.6). The ozone layer actually contains only a small fraction of ozone, but this fraction is very distinctive because ozone is quite rare in the other layers. By absorbing ultraviolet light with wavelengths from about 40 to 320 nm, the ozone layer protects the Earth's surface and all living organisms against the harmful effects of ultraviolet light in these wavelengths (see Sections 7.4.2 and 10.6.4).

19.3.5.1 Chlorofluorocarbons Certain chemicals have the potential to do great harm to the ozone layer. This fact was first realized by some theoretical chemists who tried to determine the fate of certain industrially produced molecules. The molecules in question are called **chlorofluorocarbons** or CFCs, small molecules composed of carbon and hydrogen (hydrocarbons) to which chlorine and fluorine atoms are attached. These molecules had been used since the late 1930s as nontoxic refrigerants, fluids pumped through refrigerators and air conditioners in their cooling cycles. Early refrigerators had used either ammonia or sulfur dioxide for this purpose, but both were toxic and chemically reactive, so CFC refrigerants (commonly known as "freons") were hailed as safe substitutes when they first came into use in the 1930s. In the 1950s, other uses were developed for CFCs, especially as propellants for insecticides, hair sprays, deodorants, and other aerosol products. CFCs were preferred in these applications because they were chemically unreactive. In spray cans, the CFC propellants could be used with a wide variety of products and not react with them.

Consider, however, what becomes of the CFCs when they are released from aerosol spray cans or from discarded refrigerators and air conditioners. Remember that CFCs were used because they were chemically unreactive. When released into the atmosphere, the CFCs would not react with other atmospheric gases—they would simply dissipate into the atmosphere and mix by wind action with the other gases until they were spread across the entire planet. Moreover, they would stay in the atmosphere and not "rain out" because they were not soluble in rain water. After a period of time measured in decades, these molecules would eventually be carried aloft by upward air currents until they reached the stratosphere. Once the CFCs rose above the ozone layer (but not sooner), they would be broken down by ultraviolet light, releasing many chemically reactive breakdown products, including such free radicals as chlorine monoxide (ClO), monatomic chlorine (Cl), and monatomic oxygen (O). ("Free radicals" owe their high reactivity to unpaired electrons.) Two American scientists, F.S. Rowland and Mario Molina, theorized that these free radicals would react repeatedly with ozone, destroying one ozone molecule after another (Molina and Rowland, 1974).

The predicted destruction of ozone could not be immediately confirmed when these results were first published. However, the most alarming prediction was that it would take from 40 to 150 years for the CFCs *already released* into the lower atmosphere to migrate into the stratosphere and destroy much of the ozone layer, *even if no additional CFCs were ever released*. In other words, if these predictions were correct, the harm had already been done, though it would take decades to become apparent. The predicted destruction of the ozone layer would allow a large increase in the amount of ultraviolet light reaching the Earth's surface, causing severe itching of the eyes and a large increase in the incidence of skin cancer, particularly among light-skinned people (see Section 7.4.2 and 10.6.4). Mutation rates would also increase, and the ultraviolet light would also cause more ozone production in the atmosphere at ground level, resulting in increased respiratory problems.

When the Rowland–Molina model of ozone destruction by CFCs was first proposed, industry spokespeople were quick to point out that the model was untested and that there was little data to support it. After all, no damage to the Earth's ozone layer had yet been measured. The world's largest manufacturer of CFCs, the Dupont Corporation, in a display of corporate responsibility, pledged in 1975 that it would stop making CFCs *if* it could be shown that these chemicals did in fact threaten the Earth's ozone layer.

From the time when Rowland and Molina made their predictions until the destruction of stratospheric ozone was detected over Antarctica in 1984, various environmental activists raised cries of alarm and proposed the banning of CFCs. Industry voices countered with calls for more data and further study. Both sides examined the Molina–Rowland model, and many new scientists became increasingly interested in this area of research. As additional chemical reactions were considered, the models became more numerous and more complex. The various models predicted different *rates* of ozone depletion, but all predicted that some damage to the ozone layer would sooner or later take place. The news media began reporting "debates" over CFCs, sometimes exaggerating the extent of disagreement among scientists, but they also stirred up public interest in the ozone layer. Ordinary citizens with no interest in the details of atmospheric chemistry began to hear that some scientists thought that CFCs were damaging the ozone layer. Although industry began by dragging its feet or in some cases by actively opposing any ban, many Earth-conscious consumers began to shun spray-can products. A few small communities banned their sale. A few companies began to advertise that their products did *not* use CFCs, the first major example of what came to be known as "green marketing." In 1975, the S.C. Johnson Wax Company announced that they would no longer use CFCs in their spray cans. Once consumers started buying sprays that did not contain CFCs, other spray-can manufacturers followed suit. After a few years, very few spray cans remained that used CFC propellants, *even though no widespread ban had been implemented*. In this case, market forces had caused an industry to discontinue use of an environmentally destructive product even before there was a legislative ban, and (more remarkably) even before the scientific community had reached any general consensus on the issue! Congress finally did pass legislation against the use of CFC spray cans, but the measure was by then largely symbolic because such sprays had already all but disappeared from the market.

The case of CFCs illustrates the more general problem of making public policy decisions in the face of scientific uncertainty (see Chapter 1). In this case, there was a claim that a product was a potential long-range threat to the Earth's ozone layer, but the claim could not yet be verified. As we discussed in Section 1.4.4, controversies of this sort are now commonly divided into three parts: scientific issues, science policy issues, and policy issues. The important scientific issue in this case was the correctness of the Rowland–Molina model, for which critical data were largely unavailable because the predicted damage to the ozone layer lay many years into the future. In terms of science policy issues, the economic costs of banning CFCs were easy to measure, but the environmental costs of continued use were not. In the face of this uncertainty, most people examined the issue at the level of the science policy questions: What harm would be done if we banned

CFCs and the ban later turned out to be unnecessary? On the other hand, what harm would occur if the predictions were correct and we waited several years for more data before implementing a ban? In this case, many people decided that the costs of a CFC ban were bearable, while the possible costs to society of a large but uncertain amount of future damage to the ozone layer were great. In 1978, Congress banned the use of CFC propellants in spray cans. Most unusual and most significant was the fact that this decision was made even before convincing evidence was available of any damage to the environment.

19.3.5.2 Evidence supporting the Molina–Rowland theory The hypothesis that CFCs would damage the ozone layer was largely untested, but it did not remain that way for long. Balloons were sent aloft to measure CFC concentrations, and satellites began gathering data on ozone concentrations, but nobody noticed any changes immediately.

Late in 1984, scientists working for the British Antarctic Expedition in Halley Bay, Antarctica, noticed that their annual measurements of atmospheric constituents had been undergoing an alarming change (Farman et al., 1985; McElroy et al., 1986; Solomon, 1986): There was a definite trend toward a decline in stratospheric ozone concentration, and about 40% of the ozone over Antarctica has already been destroyed (Figure 19.7). For the next two years, scientists debated the significance of this "ozone hole" that appeared each year during the Antarctic spring in October. Other ground-based observations confirmed the existence and seasonal growth of this ozone hole, and satellite-based observations soon confirmed the results as well. In 1987, airborne recording devices confirmed the presence over Antarctica of chlorine monoxide (ClO), one of the free radicals that Rowland and Molina had predicted. Depletion of the ozone layer was already occurring, and human-made (anthropogenic) chlorine was the major culprit.

The annual appearance of the "ozone hole" is no longer disputed, nor is the role of anthropogenic chlorine compounds in destroying the ozone that produces this hole. For their theoretical model of atmospheric ozone depletion, Roland and Molina were awarded a Nobel Prize in 1995.

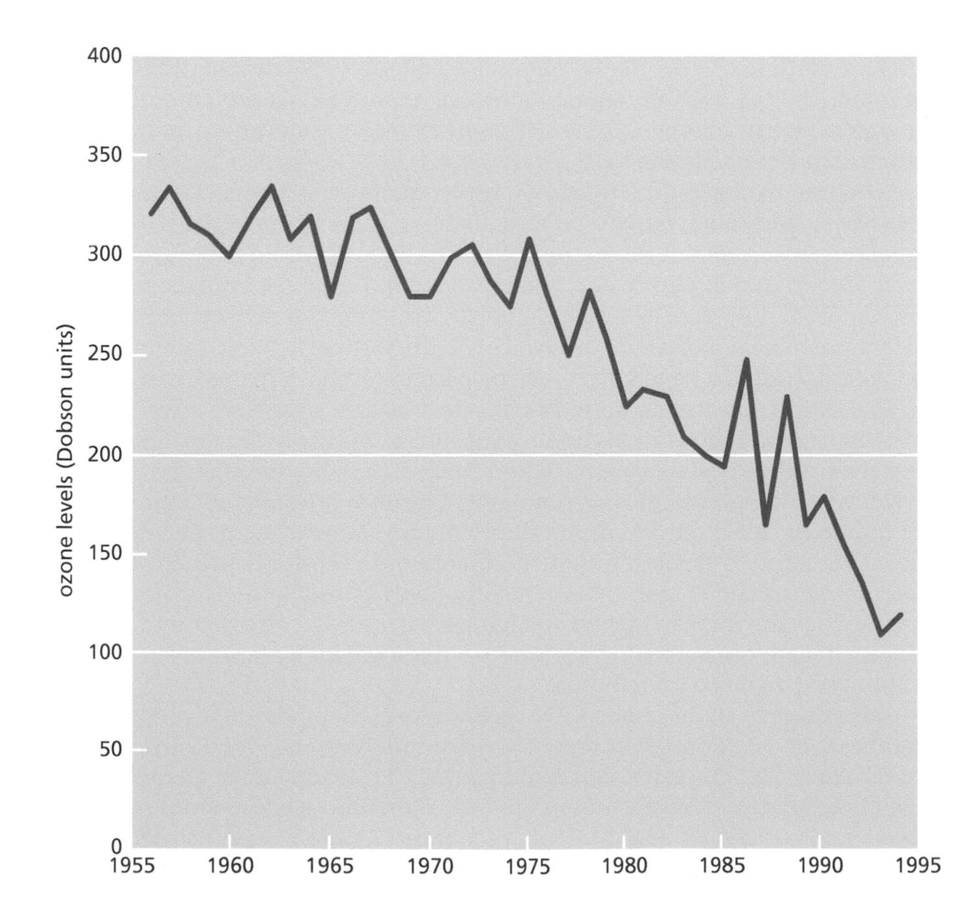

Figure 19.7 Changes in the levels of atmospheric ozone as measured at Halley Bay, Antarctica, in October of each year.

In 1987, representatives from 43 nations met in Montreal, Canada, and agreed to curtail most CFC use. The total elimination of all CFCs was stated as a long-range goal. Because of the Montreal Protocol, spray cans no longer use CFCs. In the United States and most of Europe, anyone who dismantles or recharges a refrigerator or air conditioner containing CFCs is required to capture the CFCs into waste containers and not release them into the atmosphere. However, older models are often sold to non-industrialized countries, where they are subject to no such restrictions and are probably destined someday to release their refrigerants to the atmosphere. Ozone-friendly refrigerants and propellants have been developed as alternatives to CFCs. The computer and electronics industries, which formerly used CFCs as solvents to clean their circuits, have switched to other solvents, as has the dry-cleaning industry.

Recent measurements show that the increase in atmospheric CFCs is beginning to slow down, and the Antarctic ozone hole may now be shrinking. Measurements by the National Aeronautics and Space Administration (NASA) and National Oceanic and Atmospheric Administration (NOAA) show that the Antarctic ozone hole in 2017 was the smallest that it had ever been since 1988, but still larger than it was in 1966, before the widespread use of CFC aerosols.

19.3.6 CO_2 and climate change, including global warming

In Section 17.1.2, we saw that carbon dioxide gas (CO_2) is used in photosynthesis. The carbon is used to synthesize carbohydrates and subsequently other biological molecules. When these molecules break down during respiration, CO_2 is released back to the atmosphere, completing the carbon cycle. CO_2 is also released when any carbon-containing compound is burned or when it decays and is oxidized more slowly. Most of the carbon compounds used as fuel are fossil fuels—the dead remains of plants and other forms of life fossilized many millions of years ago.

The development of life on Earth initially caused an increase in CO_2 and certain other gases because CO_2 was a product of many biological processes. The evolution of photosynthetic organisms caused a decrease in CO_2 and an increase in oxygen. The burning of fossil fuels adds greatly to atmospheric CO_2; global CO_2 levels are currently on the rise and have been blamed for causing a **global warming** trend. The U.S. Environmental Protective Agency (EPA) estimates that about 6.3 billion metric tons of CO_2 are released into the atmosphere worldwide by human activities each year.

When sunlight hits the surface of any planet, much of the energy is absorbed and warms the planet, only to be re-radiated at a later time as heat. Without a planetary atmosphere, the heat (or infrared radiation) would simply escape into space. For a planet of a given size, composition, and distance from the sun, scientists can easily calculate or measure the amount of incident solar radiation and the amount of heat that would be given off in the form of infrared radiation; these calculations (called a heat budget) can be used to predict a planet's temperature.

The Earth's atmosphere modifies this heat budget in several ways. The most important modification is that certain atmospheric gases absorb the infrared radiation given off by the planet's surface, and much of this absorbed heat is re-radiated back down to the planet. A certain amount of heat (or infrared radiation) is therefore trapped by the atmosphere and stays with the planet instead of being lost into space. The effect is similar to that of a greenhouse, and it is usually called the **greenhouse effect** (Figure 19.8). Venus, with a much denser atmosphere (mostly CO_2), has an even greater greenhouse effect and consequently a temperature much higher than would otherwise be predicted for a planet of its size, composition, and distance from the sun.

The existence of an atmospheric greenhouse effect was first proposed as a hypothesis by J.B. Fourier in the 1820s. Around 1860, the Irish physicist John Tyndall expanded this explanation and named CO_2, water vapor, and other gases as being responsible for absorbing the heat radiating upwards from the Earth's surface and radiating it back to Earth, thus warming the planet instead of letting the heat escape. The existence of such a greenhouse effect on Earth has since been

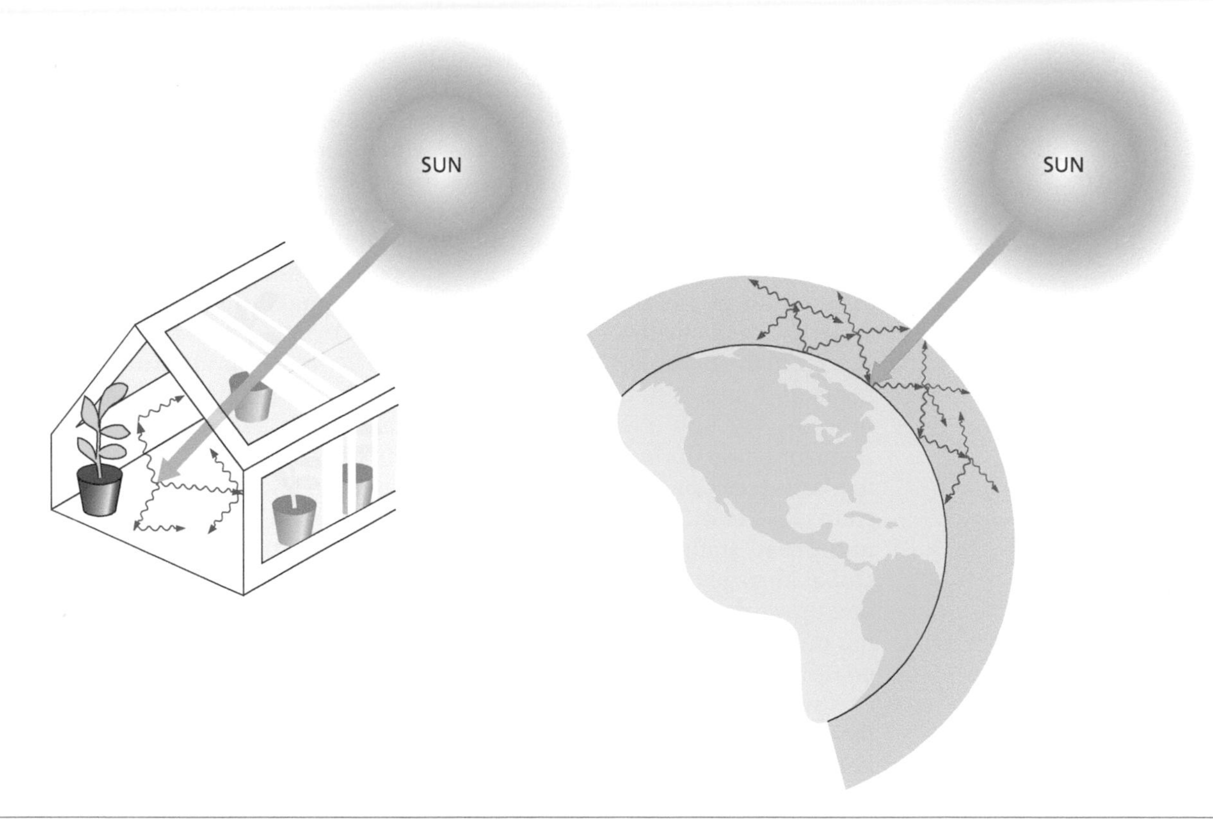

Figure 19.8 How a greenhouse captures heat. Sunlight penetrates the atmosphere or the walls of a greenhouse. When the sunlight strikes opaque objects, much of the energy is converted into infrared (heat) radiation, which becomes trapped inside the greenhouse or inside the atmosphere. (Figure is not to scale.)

substantiated by many studies, with much discussion of the consequences, for example, by Bolin et al. (1986), National Research Council (1983), Lyman (1990), Kraljic (1992), Oppenheimer and Boyle (1990), and the National Academy of Sciences (1992).

19.3.6.1 Increasing atmospheric concentrations of CO_2 In 1896, Swedish scientist Svante Arrhenius (who later won the Nobel Prize for his work on chemical equilibria) explained that, by burning wood and coal, we were "vaporizing our coal mines into the air," thus increasing the CO_2 content of the atmosphere. The burning (rapid oxidation) of any organic (carbon-containing) material, including wood, coal, or oil, causes oxygen to combine with carbon and form atmospheric CO_2. Arrhenius also concluded that this change in atmospheric CO_2 was capable of raising the Earth's temperature. The effects of such a temperature increase were not yet apparent, but they became gradually known with advances in the atmospheric sciences since the 1950s. Among other effects, scientists predicted that global warming would cause longer and more devastating droughts in semi-arid grassland areas, an expansion of the world's deserts by an accelerated rate of "desertification" (see Section 18.3.2), an increase in the number and severity of forest fires, and a partial melting of the polar ice caps. The melting of glaciers and polar ice would also cause a significant rise in sea levels, inundating many of the world's large coastal cities. Pathogens that thrive in warmer climates would become more widespread, and the epidemic diseases that they cause might become more frequent and more devastating (see Chapter 16). As climates shifted, many existing species would become extinct, and biodiversity (see Chapter 18) would suffer a decline of uncertain scope.

While nobody doubted that industry and home heating were releasing CO_2 into the atmosphere, some scientists were hopeful that the world's plants would be able to use up the extra CO_2 by stepping up photosynthesis. Other scientists argued that CO_2 would become trapped as insoluble carbonate salts (principally calcium carbonate, $CaCO_3$) in the shells of marine invertebrate animals and in underwater carbonate deposits. These hopes were dashed when oceanographer Roger Revelle and isotope chemist Hans Suess explained in 1957 that the ocean's capacity to absorb additional CO_2 was limited and that at least half of the increase in CO_2 emissions would show up as increases in atmospheric concentrations. Still, many scientists were uncertain as to how much of the global CO_2 output would be naturally recycled or absorbed and how much would accumulate and contribute to global warming. In fact, the debate continues over the role that the oceans may play in slowing down (but not stopping) the global increase of atmospheric CO_2 (Siegenthaler and Sarmiento, 1993).

From the 1950s on, C. David Keeling of the Scripps Institution of Oceanography measured atmospheric concentrations of CO_2 at various locations, at various times of day, and in various seasons. Levels of CO_2 rise everywhere during the night and drop during the day because of the daily cycle of photosynthetic activity in plants (see Chapter 17), but the daily afternoon lows and seasonal spring lows were close to 315 parts per million all over the world, and the levels recorded at mid-ocean stations (such as Mauna Loa in Hawaii) closely approximated the global averages. As Keeling continued to keep CO_2 records, he noticed a continued upward movement of the seasonal cycles: The annual lows were at 315 parts per million in 1957, but they crept upwards to 350 ppm by 1987 and continued on an upward trend that has exceeded 400 ppm since 2016 (**Figure 19.9**).

In Greenland, scientists drilling core samples through glacial ice meanwhile developed techniques for analyzing the composition of tiny air bubbles trapped beneath the surface, bubbles that contained minute atmospheric samples from past centuries. First in Greenland, and later in Antarctica, scientists were able to confirm that global CO_2 concentrations were on the rise. (Readings taken in remote areas, far from any industrial sources, are much more likely to reflect global averages.) Before the industrial revolution, CO_2 concentrations were only 280 ppm, meaning that human industry had caused more than a 40% increase in CO_2 concentrations since that time. Moreover, the record extended back for some 160,000 years, and the rise and fall of CO_2 concentrations matched the known increases and decreases in the Earth's temperature during that time, as documented by oxygen isotope ratios. (Different isotopes of oxygen have slightly different solubilities in water, and differences in the oxygen isotope ratios are preserved in the shells of marine fossils and are used to estimate temperatures going back to Jurassic times.) Although we are not sure of the causes of the earlier CO_2 fluctuations, a correlation between atmospheric CO_2 concentrations and global temperatures had finally been demonstrated: They both rise and fall together, in

Figure 19.9 Annual fluctuations and persistent long-term increases in CO_2 concentrations as measured at the Mauna Loa Observatory in Hawaii.

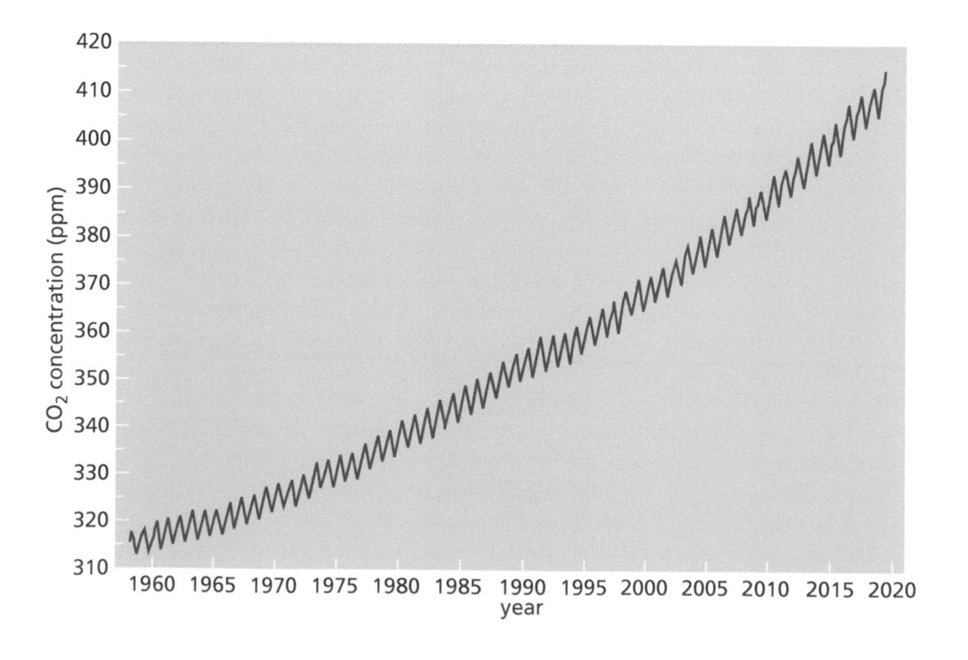

accordance with the hypothesis of a greenhouse effect. Subsequent observations continue to confirm such a link. In 1988, the United Nations established an Intergovernmental Panel on Climate Change (IPCC) to study the connections between CO_2 emissions, global warming, and climate change. In 2007, the IPCC was awarded the Nobel Peace Prize for their scientific work on investigating climate change and its effects. In 2018, the IPCC reported that, based on the best evidence available, the average global temperature has already increased 1°C above preindustrial (c. 1750) levels and, at current rates of CO_2 emissions, it will reach 1.5°C above preindustrial levels at some time between 2030 and 2052.

Meanwhile, further evidence of global warming and its effects continues to accumulate (Jones and Wigley, 1990; Houghton, 1997; Abrahamson, 1989; Schneider, 1989b; Karling, 2001; Johanson, 2002; U.S. Global Change Research Program, 2017, 2018). Direct temperature measurements (with both daily and yearly fluctuations averaged out) show a fairly steady upward trend wherever good records have been kept. Climate Central (www.climatecentral.org) is one source that continually updates such records; Climate Change Guide (www.climate-change-guide.com) is another. Early in 2018, NOAA announced that the three preceding years (2015–2017) were the hottest years ever recorded, based on worldwide averages. Later that year, California recorded new record-breaking high temperatures and also the deadliest wildfires in its history, both of which continue to break new records from year to year. A NASA press release concluded that "Since 1880, the world has warmed by 1.9 degrees Fahrenheit (1.09 degrees Celsius), with the five warmest years on record occurring in the last five years." (NASA, 2019). The summer of 2019 saw the highest temperatures ever recorded in many parts of Europe, including record-breaking highs of 45.9°C (114.6°F) in southern France. Wildfires of unprecedented scale devastated Australia in 2019–2020 and coastal Greece in 2021. Extensive flooding ravaged Germany and several nearby countries in 2021. Hurricanes have become more numerous and, in many cases, more deadly.

Making matters worse, global warming has a devastating "feedback effect" that increases the rate of further warming even faster (Woodwell and Mackenzie, 1995). The melting of polar ice exposes more rocks and soil, so more of the sun's radiant energy is absorbed by these darker surfaces instead of being reflected out to space by the ice. The melting of polar ice also releases more trapped methane, buried in compounds called clathrates, as we will explain shortly, and the methane causes even more global warming. As temperatures rise, fires become larger and more devastating, releasing even more CO_2 into the atmosphere, and destroying large quantities of the very types of trees that were helping remove CO_2 from the atmosphere and slow down the process. By all these mechanisms, global warming is a self-accelerating process.

Climate experts have repeatedly warned of the near-certain consequences of global warming, doing so with greater urgency as data have accumulated to illustrate these effects (Lynas, 2007; Ripple, 2019). Former vice president Al Gore produced two documentary films (2016, 2017) and founded an organization (www.climaterealityproject.org) to bring greater awareness to the problem (Climate Reality Project, 2016).

Global warming and carbon dioxide accumulation are the root cause of several other forms of **climate change**. While the term "global warming" only refers to the worldwide increase in overall average temperatures, "climate change" is a much broader term that includes global warming, carbon dioxide buildup, ocean acidification, melting of polar icepacks and mountain glaciers worldwide, sea level rises, increasingly devastating forest fires, and an alteration of rainfall patterns that brings more floods to some places and devastating droughts to others.

Global warming causes glaciers to melt and sea levels to rise. Alpine glaciers (those at high altitudes) have been studied since the 1800s; photographs and measurements show a consistent, worldwide pattern of increased melting (**Figure 19.10**), in some cases going back over 100 years or more. In Norway, the Engabreen Glacier retreated a full 2 km (about 1¼ miles) from 1883 to 1974. Swiss scientists estimate that the glaciers of Switzerland have diminished in area by about 25% over the past century. At lower altitudes but higher latitudes,

(A) Anderson Glacier in Olympic National Park, Washington, in 1936 and 2015. (National Park Service.)

(B) Shepard Glacier in Glacier National Park, Montana, in 1913 and 2005. (Northern Rocky Mountain Science Center.)

(C) Engebreen Glacier, Norway, in 1883 and 1974 (U.S. Geological Survey), showing a 2 km retreat.

Figure 19.10 The melting of glaciers has been documented on every continent.

continental glaciers in both Greenland and Antarctica are also melting, and large icebergs are breaking off the margins of these glaciers at an alarming rate. In 2000, the largest iceberg ever recorded (larger than the entire island country of Jamaica) broke off the coast of Antarctica, followed in 2017 by another massive iceberg larger than the state of Delaware and over half as big as Jamaica.

The melting of icebergs and glaciers is causing a slow but accelerating rise in sea levels worldwide. The rise in sea levels, which has been measured in many locations, globally averages about 18–20 cm (7–8 inches) since 1900. Rising sea levels threaten low-lying terrain all over the world. Many island nations, including Kiribati and the Maldives, will be inundated and may become uninhabitable, and a similar fate may befall much of Florida and Bangladesh, both of which have dense populations living at very low elevation. Most of the world's great cities are built along coastlines, and many of them have begun to plan how to cope with the rising sea levels that they anticipate.

Rising temperatures are also causing increased evaporation from many of the world's lakes and seas, causing several of them to shrink or even disappear. The Aral Sea, on the border between Kazakhstan and Uzbekistan, was once the fourth largest freshwater lake in the world, measuring 435 km (270 miles) from north to south and covering about 68,000 km^2 (28,000 mi^2) in 1960. Over the next 50 years, nearly all of it dried up, and several once-thriving fishing towns lie mostly abandoned, complete with rusting fishing boats marooned on dry land. A similar fate is threatening Africa's Lake Chad, which covered about 25,000 km^2 (or 10,000 mi^2) in 1960 but shrank by about 90% in the ensuing 50 years and is predicted to disappear entirely by about 2030 if present trends continue.

Rising temperatures are not uniform across the globe. Arctic regions in particular are warming faster than the rest of the planet, with increased melting of floating sea ice as one of the consequences. At the end of the Arctic winter in March 2018, 14.5 million km^2 of Arctic waters were covered with sea ice, compared to historic levels averaging 15.5 million km^2 during the 1981–2010 baseline period. Polar bears (*Ursus arctos*) rely on floating sea ice to rest, to build winter dens, and especially to give birth. As polar sea ice melts, polar bear habitat is shrinking. Another problem that affects Arctic regions is the melting of **permafrost**, soil that remains frozen year-round. **Methane clathrates**, compounds in which methane is trapped within the crystal structure of ice, occur within the permafrost layer and also in deep ocean sediments; the methane is hypothesized to be produced by various species of Archaea (see Section 6.3.1). Because methane is such a potent greenhouse gas (**Table 19.1**), the melting of permafrost is of great concern because the clathrates will release their methane content to the atmosphere and accelerate global warming even further.

Because CO_2 dissolves in water to make carbonic acid, increases in atmospheric CO_2 are slowly making ocean waters more acidic, especially in the tropics. Ocean acidification is causing the death of the colonial polyps that build coral reefs, transforming the formerly colorful and biodiverse habitats (see Figure 18.2) into dead, white, skeletal remains. This "bleaching" of coral reefs has been documented throughout the tropics and is especially evident in Australia.

Global warming is causing new and unpredictable weather patterns, bringing an increased danger of wildfires, violent coastal storms, and other catastrophic weather events. The severity (not just the count) of wildfires, droughts, violent storms, and coastal flooding, and the damage that they cause, have all increased in recent years and are all predicted to increase further as a consequence of global warming. In 2018, NOAA reported that two of the three most devastating years for U.S. wildfires in terms of acres burned occurred in 2015 and 2017. An unprecedented drought spread across Australia in 2018, and a 2018 wildfire in California became the largest in that state's history. Unusually large wildfires have also occurred in Portugal and Spain (2017) and in Greece (2018, 2021). In 2012, "Superstorm Sandy," an unusually late-season hurricane, inundated the coast of New Jersey, causing a tidal surge of 13.88 feet (4.23 m) in lower Manhattan and flooding portions of the New York City subway system. The 2017 hurricane season was the most damaging on record up to that time, causing an estimated quarter-trillion dollars in damage and devastating Texas, Cuba, Florida, Puerto Rico, and the Virgin Islands. Hurricane Maria alone knocked out electric power across

Table 19.1 Ability of gases to absorb infrared heat waves and add to greenhouse effect (compared to CO_2)

Gas	Ability
Oxygen (O_2)	Negligible
Nitrogen (N_2)	Negligible
Argon (Ar)	0.0
Carbon dioxide (CO_2)	1.0
Methane (CH_4)	20
Nitrous oxide (N_2O)	250
Chlorofluorocarbons	20,000

Puerto Rico for many months and caused an estimated 135,000 people to flee to the mainland. An article in the *New England Journal of Medicine* calculated that deaths in Puerto Rico increased by 4,645 over the previous year, due largely to such indirect consequences of the hurricane as increased disease incidence, interruption of power to medical facilities, and the inability of emergency help to reach people in need. Climatologists point out that increased ocean water temperatures in the tropics have resulted in more and larger hurricanes, while rising sea levels have made coastal storms more costly and more deadly.

Measurements continue to be made of CO_2 concentrations in the atmosphere, dissolved carbonates in the oceans, carbonate sediments, estimated biomass of the world's organisms, and the rates of exchange by which carbon travels among these and several other sources. About one-third of anthropogenic CO_2 (produced by the burning of fossil fuels and by tropical deforestation) is absorbed by the oceans, but atmospheric CO_2 continues to increase.

19.3.6.2 Remediation of excess CO_2 by plants Green plants can reduce the CO_2 concentration in the atmosphere through photosynthesis, a process that consumes CO_2 and releases oxygen (Section 17.1.2). Animals and other heterotrophs use O_2 and release CO_2 during the respiration that produces the adenosine triphosphate (ATP) they need (see Section 8.3.3). Globally, photosynthesis and respiration are balanced, although the balance shifts back and forth. In most places, photosynthesis decreases in the winter; this is why Keeling's record of CO_2 levels rises and falls annually. The largest amount of photosynthesis takes place in tropical latitudes, especially in rainforests, but algae in the open ocean also perform a significant amount of photosynthesis. The gas exchange of CO_2 for O_2 during photosynthesis is sometimes compared to the exchange of O_2 for CO_2 in our lungs, with the rainforests described metaphorically as the "lungs" of our planet.

Unfortunately, rainforests are being destroyed at the astounding rate of about three football fields a *second* (see Section 18.3.1). Much of this destruction is taking place in Brazil, a developing nation in need of agricultural land. The destruction often takes the form of burning, using fires that are so large that they can be photographed from space. Burning the rainforests not only adds further CO_2 to the atmosphere, but it also removes the very plants that could have remediated much of the increase.

Although rainforests also exist in Indonesia and many other nations, international attention has focused on Brazil because Brazil contains, in its Amazon region, approximately half of the world's rainforests. Ecologists from all over the world have urged Brazil to stop destroying its rainforests. Brazil, however, is a rapidly growing nation trying to modernize. Its population is growing at a rapid rate (see Chapter 9), and two of its cities (São Paulo and Rio de Janeiro) are among the world's largest. Mining companies and timber companies stand to make millions or even billions of dollars from the continued decimation of the rainforests. Landowners (including the Brazilian government) stand to make even more money from the development of agricultural land, and their impatience has often led to the burning of the forests, a process much more rapid than the harvesting of timber. Many Brazilians argue that they are only striving to achieve the material standard of living that the industrialized nations have enjoyed for decades. They point to America's feeble efforts to curb automobile pollution and urban smog as evidence of a lack of commitment to solving ecological problems that conflict with economic self-interest. Auto exhausts produce carbon monoxide gas and nitrogen oxides, all of which contribute to global warming. Nations like Brazil, India, and Indonesia have asked why they should stifle their development to forestall a problem created largely by the industrial nations. This argument raises again some general ethical issues of policy choice: How much suffering should one nation endure for the good of the planet? Why should any poor nation make sacrifices to its economic future, while the wealthy nations seem unwilling to make comparable sacrifices to their own economic well-being? In one example of a cooperative partnership, Papua New Guinea has agreed to a moratorium on cutting down its rainforests as a source of mahogany, and some of its former customers, including New Zealand, are subsidizing this effort.

19.3.6.3 The atmosphere as part of the global ecosystem In addition to CO_2, there are other "greenhouse gases" that contribute to global warming. The contributions of several other gases to the greenhouse effect are shown in Table 19.1. Although these other gases are much less abundant than CO_2, their ability to absorb infrared heat waves (and thus contribute to the greenhouse effect) is much greater than that of CO_2, often many hundreds or thousands of times greater (Table 19.1). Some of these gases result from the metabolic activity of living organisms, but a great amount is also anthropogenic in origin. For example, the same CFCs that destroy ozone also contribute to global warming and are thus doubly dangerous.

Human activities continue to add both CO_2 and other greenhouse gases to our atmosphere in record amounts; these other gases include methane, CFCs, and the oxides of nitrogen. Methane is added by industry and also by domestic cattle and other large animals raised for food. It is also produced in swamps, marshes, rice paddies, and other wetlands. Eastern Europe has been mentioned as a significant methane source because of natural gas leaking from its aging pipelines and storage tanks. CFCs are still used in some air conditioners and are released whenever older-model refrigerators or air conditioners are dismantled. Automobile exhausts still produce several kinds of nitrogen oxides. Pollution control laws in the United States, the world's largest user of automobiles, have started to address the question of automobile exhaust gases, but the laws are still very weak. For one thing, the laws pertain only to *new* cars, while older, more heavily polluting vehicles continue to be sold secondhand to many non-industrialized countries.

The existence of a greenhouse effect is no longer disputed among scientists, nor is it disputed any longer that we are compounding the effect by releasing CO_2 and other greenhouse gases into the atmosphere. Even Dixie Lee Ray, a steadfast critic of the environmentalist movement, writes as follows: "We need to remember that our Earth, together with its enveloping atmosphere, does indeed constitute a 'Greenhouse.' ... The Greenhouse theory holds that an increase in the concentration of any of the greenhouse gases will lead to increased warming. No one disputes this, but the question is how much will it warm and are there naturally occurring corrective phenomena?" (Ray and Guzzo, 1993, p. 17). In other words, there is general agreement that a release of CO_2 and other greenhouse gases will intensify the greenhouse effect and will increase global temperatures. One question that continues to be disputed among scientists is how severe the increase in global warming will be and how rapidly it will occur (Haley, 2002).

Meanwhile, our production of greenhouse gases continues largely unabated, and many large nations, including China and India, still generate most of their electricity by burning coal. Rainforest destruction adds to the problem by robbing the Earth of its ability to cope with the CO_2 buildup, making a natural balance ever harder to achieve. Not surprisingly, Keeling's curve of CO_2 levels snakes steadily upward to set new records each year. With the increase in CO_2 and other greenhouse gases, the global warming trend continues, and there is no immediate end in sight. The global average temperature has increased by about 1.0°C (or 1.8°F) since 1900, with most of that increase occurring since 1950. As human activities change the atmosphere, the changed atmosphere also affects human activities.

19.3.6.4 What can we do? Is there anything we can do to stop global warming and other forms of climate change? Yes, many scientists have suggested a number of things that we can do, both locally and globally, to mitigate or reverse the trend (U.S. OTA, 1991; Hawken, 2017). However, nobody really knows how much temperature change the Earth can sustain before it becomes uninhabitable.

We can all reduce CO_2 emissions by using less energy. A total reduction in energy use may be hard to achieve, but we can certainly strive for energy efficiency, and doing so would, in most cases, save money as well. In countries like the United States, more energy is used to heat homes than is used for any other single purpose, including transportation. Better insulation in homes and other buildings will cut carbon emissions and also save on heating costs. More fuel-efficient cars and trucks would also save money while cutting back on emissions of CO_2 and nitrogen oxides. Increased use of public transit would reduce emissions even further and would save money in many cases.

Here are some things that each of us can do as individuals on a local level:

- Drive less and walk or bicycle more. Doing so has many health benefits and also saves money.
- Use public transportation whenever possible.
- If you do drive, consider electric or hybrid vehicles wherever possible. Also consider ride-sharing instead of driving alone.
- Whenever possible, use things again and again instead of tossing them away after a single use; this usually saves money.
- Recycle as much as possible. Recycled products are usually much less expensive than comparable first-time materials, especially for materials like glass and aluminum.
- If you own a home, consider using solar power for heating and electricity. Not only does this save on money, it also means that you are less likely to suffer a power outage during a winter storm or a hurricane. Solar power is becoming less expensive and more cost-efficient with each passing year. If you generate more solar energy than you use, you can also sell the excess to the power company, saving you even more money.

Private businesses, colleges and universities, and local governments can do even more:

- Encourage reuse and recycling as much as possible by convenient placing of recycling bins and by supplying reusable plates, trays, and utensils in eating areas. Encourage electric vehicle and bicycle use by providing bicycle racks and electric recharging stations.
- Encourage the use of carpools with sign-up registries.
- Where appropriate, encourage public transportation by providing frequent shuttle service to the nearest public transit station.
- Insulate buildings as much as possible.
- Wherever possible, heat buildings using solar energy or wind power.

Many businesses now have a high-ranking official in charge of "sustainability," which includes implementing many of these measures. If a business or other enterprise generates more energy than they use, they no longer contribute to global CO_2 emissions and are said to be **carbon-neutral**. About one third of colleges and universities in the United States have pledged to become carbon-neutral in the near future; several colleges and universities that have already met this goal include Colby College and College of the Atlantic (both in Maine), American University (Washington, DC), and the University of Minnesota at Morris. Several large businesses, such as Google, are already carbon-neutral. Many other corporations are using solar power or other nonpolluting sources to reduce their "carbon footprint," saving energy costs in the process. Hundreds of corporations, including Walmart, have announced plans to use 100% renewable energy sources in the future, including solar power, wind power, tidal power, and hydroelectric. More and more state and local governments are doing the same, heating their buildings with renewable energy (usually solar) and replacing their vehicle fleet with electric or other nonpolluting vehicles.

Governments at all levels have a special role to play. Apart from energy uses in government buildings and government-operated vehicles, governments can also pass laws and appropriate funds to encourage both energy conservation and sustainable energy development. Among other possibilities, this may include:

- Investments in public transportation
- Incentives to encourage recycling, such as programs that charge households for disposal of nonrecyclable trash but pick up recyclables free of charge; such "Pay As You Throw" programs have proven to double or triple recycling in different localities

- Modifying building codes to encourage solar and other sustainable forms of energy and also more efficient building insulation
- Investment incentives to encourage the development of other nonpolluting energy sources including hydropower, wind power, and, in certain locations, tidal and geothermal power as well

Worldwide, over 100 cities now generate 70% or more of their electric power from nonpolluting sources such as hydropower, wind, and solar. Reykjavik, Iceland, generates all of its electricity from geothermal sources and hydropower. Basel, Switzerland, gets its electricity from hydropower and wind. Denmark generates over half of its electricity from wind power. Many large parts of South America get a substantial portion of their electricity from hydropower. Dozens of cities in the United States are planning to convert most or all of their power generation to nonpolluting sources. Burlington, Vermont, already generates 100% of its power needs from renewable resources such as hydropower, solar, and biofuels derived from agricultural and industrial wastes (https://insideclimatenews.org/news/27022018/renewable-energy-cities-clean-power-technology-cdp-report-global-warming-solutions).

In 1992, the United Nations sponsored a conference in Rio de Janeiro that resulted in a Framework Convention on Climate Change. Most industrial nations agreed to set binding targets to reduce their emissions of greenhouse gases. In 1997, the resulting agreement was signed in Kyoto, Japan, by 192 nations, including nearly all members of the United Nations. In 2016, 195 nations signed a new agreement in Paris, setting more stringent but voluntary goals and allowing each nation to set its own goal, with the intent to limit global warming to a further temperature increase of "well below 2 degrees Celsius," compared to the 1990 baseline period. (The original goal was 2°C, but several low-lying island nations vigorously argued that a more stringent limit of 1.5°C was needed to prevent their nations from disappearing beneath rising sea levels.) To date, 190 nations have formally ratified the treaty, and many states, cities, and corporations have additionally pledged to hold themselves to the Paris Accord. Many nations have already begun to implement the changes to which they agreed. The Dutch National Railways, for example, now runs entirely on electricity generated by wind power, while France and Norway plan to phase out diesel-powered and gasoline-powered automobiles over the next few decades. Morocco is planning to generate enough solar energy in its desert regions to sell the excess to Europe. China, the world's largest producer of solar photovoltaic cells, now generates over 100 gigawatts of power from solar energy, and its commitment to solar energy is steadily increasing.

THOUGHT QUESTIONS

1. How can one persuade legislators in one jurisdiction to spend money on antipollution measures if most of the damage occurs in other jurisdictions? For example, if acid rain in Norway and Sweden originates mostly in Germany or Poland, how can the people in Norway and Sweden influence legislators to change German or Polish laws or industrial practices?
2. What would be the composition of our atmosphere if plants had not evolved?
3. When oil from the tanker *Aragon* washed ashore on beaches in Spain and Portugal, nutrients were sprayed on the beaches as part of an attempt at bioremediation. The results were not as good as in the case of the *Exxon Valdez* oil spill. What factors might explain why nutrient enrichment might work for one oil spill and not for another?
4. How long has the Earth had a greenhouse effect? In what ways is the greenhouse effect helpful? In what ways is it harmful?
5. What would have a greater impact on atmospheric levels of CO_2 and other greenhouse gases: slowing down production of these gases or slowing down the destruction of trees that reduce some of these gases? How would you go about measuring the rates of the relevant processes?
6. Are there any ways that human societies can meet their needs for fuel without producing CO_2? Do modern societies need to consume fossil fuels in order to maintain their standard of living? Are there any societies with a high standard of living but low fuel consumption?
7. If half of global photosynthesis comes from sources other than tropical rainforests, why has so much attention been focused on these tropical rainforests? Are other scientific principles involved? Are nonscientific considerations involved? What kinds of information would you need to make these comparisons?
8. Hydroelectric dams often serve as barriers to the migration and dispersal of fish species. How big a problem is this? What can be done to mitigate its effects?

CONCLUDING REMARKS

The causes of pollution are many: industrial activity, agriculture, and also our own everyday activities such as driving our cars and heating our homes. As pollutants build up, more evidence continues to accumulate that we are poisoning much of the Earth's water supply and the very air we breathe, and each increase in human populations only compounds these problems. Fortunately, many forms of ecological damage can be remediated once the damage has stopped, and biologists are devising new ways to use living organisms to help in remediation.

Ozone in the stratosphere screens out the more dangerous wavelengths of ultraviolet light. Chlorofluorocarbons cause depletion of this ozone layer and will continue to do so for many decades. Conditions over Antarctica have already resulted in the appearance of an "ozone hole" over that continent every spring. Ozone is a natural product of atmospheric oxygen. Neither ozone nor oxygen existed in the atmosphere originally, but were produced by the activities of living organisms, especially blue-green bacteria and photosynthetic plants. These same plants keep carbon dioxide levels from rising too rapidly. Human activities, however, including industrial pollution and the burning of millions of acres of rainforest, have greatly added to the carbon dioxide levels and have simultaneously reduced the availability of plant life to cope with this increase. The continued buildup of carbon dioxide and the consequent global warming are altering the entire biosphere. The trend toward global warming has closely followed changes in carbon dioxide concentrations for the past 160,000 years. This global warming trend is continuing, and no immediate end for it is in sight.

The atmosphere and the living organisms of the Earth both belong to the same vast ecosystem. Changes in either one of these usually result in changes in the other because the interrelations are so numerous. For example, the world's forests, especially the rainforests, contain plants capable of reducing global carbon dioxide levels, given enough time. To give plants a chance to halt the process of global warming, we must give these natural ecosystems a chance to work. In particular, we need to limit the release of CO_2 and other greenhouse gases in order to halt the upward trend of global temperatures.

Can we learn to live in ways that pollute less? The Suquamish people of Puget Sound were once led by Chief Seattle, to whom the following words are often attributed. The words may not be authentically his, but they do express an important truth:

> *The Earth is our mother. What befalls the Earth befalls all the sons of the Earth. ... Man did not weave the web of life, he is merely a strand in it. Whatever he does to the web, he does to himself.*

CHAPTER SUMMARY

- All living species belong to a worldwide ecosystem called the **biosphere**, which also includes the land, water, and atmosphere.
- Scientists think that Earth's atmosphere was originally hydrogen-rich (**reducing**). Our present oxygen-rich (**oxidizing**) atmosphere is, in large measure, the result of photosynthesis.
- **Pollution** is a matter of quantities. Ethical decisions about pollution are also decisions about quantities because neither costs nor benefits can be measured without measuring all the quantities involved. Local sources of pollution can have widespread effects on the biosphere. Sustainable practices can reduce or prevent pollution. **Sentinel species** can be used to monitor pollution.
- Living organisms can restore polluted areas by **biodegradation** and **bioremediation**, but prevention of pollution is far less costly than restoration afterwards.

- **Ozone** in the upper atmosphere protects us from dangerous levels of ultraviolet radiation. Chlorofluorocarbons can damage this ozone layer.
- Because of the **greenhouse effect**, increases in carbon dioxide emissions have resulted in a **global warming** trend that threatens the global ecosystems on which all life depends. In order to reduce global warming and related forms of climate change, we all need to learn how to reduce CO_2 emissions in our daily lives and encourage governments and corporations to do so as well.

BIBLIOGRAPHIC REFERENCES

Bibliographic references to material in this chapter can be found online at biologytrending.routledge.com/chapter19

GLOSSARY: KEY TERMS TO KNOW

These key terms are defined at the end of each chapter as an aid for student review.

Aerobic Capable of living only in the presence of free oxygen (O_2).

Anaerobic Conditions in which free oxygen (O_2) is not present, or organisms that can live under such conditions.

Biodegradation Biological activity that breaks down materials into harmless substances.

Bioremediation The enhancement of biodegradation by adding or encouraging the growth of certain organisms.

Biosphere The land, water, air, and living things that make up the Earth's global ecosystem.

Carbon-neutral Any process or entity that releases no more carbon dioxide than it consumes.

Chlorofluorocarbons Organic chemicals derived from hydrocarbons by replacing some of the hydrogen atoms with chlorine and others with fluorine.

Clathrates See *methane clathrates.*

Climate change Any global change in Earth's temperature, rainfall, or seasonal variations.

Gaia hypothesis The hypothesis that living organisms are responsible for the conditions that make continued life on Earth possible.

Global warming An increase in the global or average temperature of a planet, such as resulting from a buildup of carbon dioxide.

Greenhouse effect The trapping of solar energy in the atmosphere of a planet (or in a greenhouse) in the form of infrared radiation or heat, much of which is reflected from the atmosphere back to the planetary surface.

Heterotroph An organism not capable of manufacturing its own energy-rich organic compounds and therefore dependent on consuming such compounds as food.

Methane clathrates Compounds in which methane (CH_4) is trapped within the crystal structure of ice.

Nonrenewable resource Anything whose quantity is finite or limited and whose continued use will eventually deplete the resource.

Ozone A reactive molecule (O_3) containing three oxygen atoms, highly reactive at Earth's surface, but more prevalent within the stratosphere.

Permafrost Permanently frozen soil; soil that remains frozen year-round.

Pollution Contamination of an environment by substances present in undesirable locations or quantities.

Renewable resource Anything that can replenish itself naturally if not overused.

Sentinel species Any species whose numbers or health can be monitored as an indicator of the overall health of an ecosystem.

Water cycle A global cycle in which atmospheric water vapor condenses into liquid and falls as rain or snow, then runs downhill into bodies of water such as lakes and oceans, from whose surface it evaporates into the atmosphere once again, supplemented also by leaf transpiration and other biological processes.

CONNECTIONS TO OTHER CHAPTERS

Chapter 1: Pollution and habitat destruction violate several ethical injunctions, including the principle to do no harm to others.

Chapter 1: Ozone depletion and global warming are examples of scientific theories that have been developed and tested.

Chapter 1: Public policy decisions must sometimes be made before all the risks and benefits are certain.

Chapter 3: Certain pollutants increase mutation rates.

Chapter 3: The Earth's ozone layer protects us against dangerously high levels of ultraviolet radiation.

Chapter 5: The Earth's organisms have co-evolved with its atmosphere.

Chapter 7: Ultraviolet light is more dangerous for light-skinned people.

Chapter 9: Population growth increases fuel consumption, CO_2 buildup, and global warming and puts increased pressure on ecosystems all over the world.

Chapter 10: Chemical pollutants and ultraviolet radiation can increase cancer rates. The ozone layer protects us from skin cancer.

Chapter 14: Chemical pollutants and ultraviolet radiation can suppress our immune systems.

Chapter 16: New infectious threats will emerge at a much higher rate if global climate change continues.

Chapter 17: Plant photosynthesis is a global source of oxygen and a process that reduces atmospheric CO_2.

Chapter 17: Bacteria and other organisms contribute to the nitrogen cycle and other cycles that maintain our present atmosphere.

Chapter 18: Photosynthesis in tropical ecosystems can help reduce CO_2 levels and limit global warming to some extent, but only if these ecosystems remain healthy. The destruction of forest plants greatly accelerates global CO_2 buildup.

PRACTICE QUESTIONS

1. How many times more H+ ions are there in acid rain at pH 5 than in water at neutral pH? How many times more at pH 5 than at pH 6?
2. Which is more acidic, pH 4 acid rain composed mostly of sulfuric acid or pH 4 acid rain composed mostly of nitric acid?
3. What are the similarities between bioremediation and biodegradation? What are the differences?
4. How does an increase in ultraviolet light increase cancer rates? For which cancers? (You might wish to review material from Chapters 7 and 10 in answering this question.)
5. In what ways could temperature affect the efficiency of biodegradation of an oil spill?

Credit Sources

For the purposes of permissions, this is considered to be a Fourth Edition of *Biology Today*.

CHAPTER 1

Figure 1.1 Credited from "Rose is Rose" © Pat Brady and Don Wimmer. Reprinted by permission of Andrews Mcmeel Syndication for UFS. All rights reserved.

Figure 1.4 (A) Courtesy of Huntington Library, San Marino, California. (B) Courtesy of Museum of the Old Brno Abbey. (C) Courtesy of MBL Archives https://hdl.handle.net/1912/16272. (D) Courtesy of Marjorie Bhavnani. (E) Courtesy of the Lawrence Berkeley National Laboratory, Berkely, CA. (F) Courtesy of Dr. J. Doudna.

Figure 1.7 Courtesy of James Qube / Pixabay.

Figure 1.8 Courtesy of Cruelty Free International.

CHAPTER 2

Figure 2.1 Portions modified from Postlewait, Hopson and Veres, *Biology! Bringing Science to Life* [McGraw-Hill, 1991], p. 137, Figure 8.3.

Figure 2.2 Portions modified from Postlewait, Hopson and Veres, *Biology! Bringing Science to Life* [McGraw-Hill, 1991], p. 137, Figure 8.3.

Figure 2.3 Redrawn after Sinnott, Dunn, and Dobzhansky, *Principles of Genetics*, 5th ed. [McGraw-Hill, 1958].

Figure 2.4 Modified from Sinnott, Dunn, and Dobshansky, *Principles of Genetics*, 5th ed. [McGraw-Hill, 1958], p. 72, Figure 6.1.

Figure 2.5A From *Molecular Biology of The Cell.* Fourth Edition by Bruce Alberts, et al., Copyright ©2002 by Bruce Alberts, Alexander Johnson, Julian Lewis, Martin Raff, Keith Roberts, and Peter Walter. ©1983, 1989, 1994 by Bruce Alberts, Dennis Bray, Julian Lewis, Martin Raff, Keith Roberts, and James D. Watson. Used by permission of W. W. Norton & Company, Inc.

Figure 2.5B Courtesy of the heirs of Josef Reischig. Published under CC BY-SA 3.0 license, https://commons.wikimedia.org/wiki/File:Mitosis_(261_13)_Pressed;_root_meristem_of_onion_(cells_in_prophase,_metaphase,_anaphase,_telophase).jpg.

Figure 2.8B Courtesy of Flatters & Co., https://commons.wikimedia.org/wiki/File:Hydra_Budding.svg.

Figure 2.12 Modified from Sinnott, Dunn, and Dobshansky, *Principles of Genetics*, 5th ed. [McGraw-Hill, 1958], p. 163, Figure 13.2.

Figure 2.16 Modified from Pelczar, Chan, and Krieg, *Microbiology Concepts and Applications* [McGraw-Hill, 1993], pp. 44–45, Figures 1.22–1.24.

Figure 2.17 Modified from Pelczar, Chan, and Krieg, *Microbiology Concepts and Applications* [McGraw-Hill, 1993], pp. 44–45, Figures 1.22–1.24.

CHAPTER 3

Figure 3.2 Modified from Postlethwait and Hopson, *The Nature of Life*, 3rd ed. [McGraw-Hill, 1992], p. 245, Figure 10.7B.

Figure 3.3 Modified from Postlethwait and Hopson, *The Nature of Life*, 3rd ed. [McGraw-Hill, 1992], p. 245, Figure 10.7B.

Figure 3.4 Used with permission of American Association for the Advancement of Science from "Visualization of Bacterial Genes in Action" by O. L. Miller, Jr.Barbara A. Hamkalo and C. A. Thomas, Jr., *Science* 24 Jul 1970; vol 169, issue 3943, pp. 392–395, DOI: 10.1126/science.169.3943.392; permission conveyed through Copyright Clearance Center, Inc.

Figure 3.5 Part A modified from https://commons.wikimedia.org/wiki/File:OSC_Microbio_10_03_tRNA.jpg [Creative Commons license].

Figure 3.8 ourtesy of Laurent J. Beuregard, Genetics Laboratory, Eastern Maine Medical Centre, Bangor, ME.

Figure 3.11 Courtesy of Laurent J. Beuregard, Genetics Laboratory, Eastern Maine Medical Centre, Bangor, ME.

Figure 3.13 The chromosomes are courtesy of Dr. Beauregard at Eastern Maine Medical Center (permission granted for all editions), and the image of the girl is from the March of Dimes National Foundation (also permission granted for all editions).

CHAPTER 5

Figure 5.3 All photos courtesy of Jonah Benningfield.

Figure 5.4 Modified from Postlethwait and Hopson, *The Nature of Life*, 3rd ed. [McGraw-Hill, 1995], p. 860, Figure 39.11.

Figure 5.5 Photographs from the estate of E.B. Ford, courtesy of J.S. Haywood.

Figure 5.6 Modified from Postlethwait and Hopson, *The Nature of Life*, 3rd ed. [McGraw-Hill, 1995], p. 371, Figure 15.7.

Figure 5.7 Modified from Moore, Lalicker and Fischer, *Invertebrate Fossils* [McGraw-Hill, 1957], pp. 336, 338, 344, 394, Figures 9-2, 9-3, 9-4, 9-47.

Figure 5.9 Modified from Moore, Lalicker and Fischer, *Invertebrate Fossils* [McGraw-Hill, 1957], pp. 345, 364, Figures 9-5, 9-23.

Figure 5.10 Modified from Colbert and Morales, *Evolution of the Vertebrates*, 4th ed. [Wiley-Liss, 1991], p. 186, Figure 14-2, as redrawn from G. Heilmann, *Origins of Birds* [D. Appleton and Co., 1927].

Figure 5.12 Modified in part from Hopson and Wessells, *Essentials of Biology* [McGraw-Hill, 1990], pp. 736, 734, Figures 40-3, 40-1.

Figure 5.13 Modified in part from Hopson and Wessells, *Essentials of Biology* [McGraw-Hill, 1990], pp. 736, 734, Figures 40-3, 40-1.

Figure 5.14 Modified from Postlethwait and Hopson, *The Nature of Life*, 3rd ed. [McGraw-Hill, 1995], p. 404, Figure 16.8.

CHAPTER 6

Figure 6.2 Courtesy of Rodney M. Donlan, PhD, and Janice Carr, USCDCP.

Figure 6.7 Portions redrawn from *Foundations of Parasitology*, 4th edition, by Schmidt/Roberts et al. [Times Mirror/Mosby College Publishing, 1988]. Courtesy of Elsevier Science & Technology Journals.

Figure 6.9 Courtesy of Robert Thomas.

Figure 6.10 (A) Permission from Ajaya Bhatkar/Shutterstock. (B) Courtesy of Barry Logan. (C) Courtesy of Robert Thomas. (D) Courtesy of Dr. I.K. Sjøtun. E, Provided with permission from Lebenskulturen.de/Shutterstock.

Figure 6.11 Courtesy of Robert Thomas.

Figure 6.12 Photos courtesy of Weltdeisgn/Pixabay, Robert Thomas, Ascyrafft Adnan "DukeAsh"/Pixabay, Barry Logan, and Roberts Botany Slide Collection, Bowdoin College, Brunswick, ME.

Figure 6.15 Photos courtesy of and with permission from Lebenskulturen.de/Shutterstock, Aernout Bouwman "Persblik"/Pixabay, and David Have/Shutterstock.

Figure 6.18 (A) Courtesy of Frederick Atwood. (B) Courtesy of ER Degginger/Science Photo Library. (C) Courtesy of CDC/Mae Melvin and CDC.

Figure 6.19 (A) Courtesy of Dr. MA Ansary/Science Photo Library. (B) Courtesy of Robert O. Schuster and Bohart Museum of Entomology, University of California, Davis. (C), Courtesy of Wilhelm Hagen. (D) Courtesy of WikiImages/Pixabay. (F) Courtesy of Ralphs_Fotos/Pixabay.

Figure 6.20 Photos courtesy of Frederick Atwood; Manuae/Wikimedia Commons, https://creativecommons.org/licenses/by-sa/3.0/deed.en, no changes were made; and Sandrine Rongère "glucosala"/Pixabay.

Figure 6.21 Courtesy of Dieter Piepenburg.

Figure 6.22 (B) Courtesy of Will Ambrose. (C) With permission from Alfredo Maiquez/Shutterstock. (D) Courtesy of Sonja Rieck "Tawnyowl"/Pixabay. (E) Courtesy of Silvia "sipa"/Pixabay. (F) Courtesy of Rene Rauschenberger "rauschenberger"/Pixabay. (G) Courtesy of Cecilie V. Quillefelt. (H) Courtesy of Hillebrand Steve, USFWS/Pixnio. (I) Courtesy of Ian Lindsay "IanZA"/Pixabay.

Figure 6.23 Photos courtesy of Onkel Ramirez/Pixabay, David Agee/AnthroPhoto, David Chivers/AnthroPhoto, Irven DeVore/AnthroPhoto, and John Harcourt/AnthroPhoto.

Figure 6.24 Courtesy Jay Kelley/AnthroPhoto, and K. Cannon-Bonventre/AnthroPhoto.

Figure 6.26 Courtesy of D. Cooper/AnthroPhoto and B. Vandermeersch/AnthroPhoto.

CHAPTER 7

Figure 7.3 Modified from Buettner-Janusch, *Origins of Man* [John-Wiley, 1966], pp. 499, 500, 501.

Figure 7.4 Portions modified from Hopson and Wessells, *Essentials of Biology* [McGraw-Hill, 1990], p. 191, Figure 11-16.

Figure 7.5 Modified from Carola, Harley, and Noback, *Human Anatomy and Physiology*, 2nd ed. [McGraw-Hill, 1992], p. 571, Figure 18.11.

Figure 7.7 Modified from Stein and Rowe, *Physical Anthropology*, 5th ed. [McGraw-Hill, 1993], p. 190, Figure 8.9.

Figure 7.8 Courtesy of Patricia Farnsworth.

Figure 7.10 Modified from Stein and Rowe, *Physical Anthropology*, 5th ed. [McGraw-Hill, 1993], pp. 126-127, Figure 6.6.

Figure 7.11 Modified from Stein and Rowe, *Physical Anthropology*, 5th ed. [McGraw-Hill, 1993], p. 177, Figure 8.2.

Figure 7.12 Courtesy of Glenbow Archives, Calgary, Canada [ND-&-714].

CHAPTER 8

Figure 8.2 Portions modified from Postlethwait, Hopson and Veres, *Biology! Bringing Science to Life* [McGraw-Hill, 1991], p. 37, Figure 2.20.

Figure 8.4 Portions modified from Postlethwait, Hopson and Veres, *Biology! Bringing Science to Life* [McGraw-Hill, 1991], p. 38, Figure 2.21.

Figure 8.5 Portions modified from Carola, Harley and Noback, *Human Anatomy and Physiology*, 2nd ed. [McGraw-Hill, 1992], p. 58, Figure 3.2, and from Postlethwait and Hopson, *The Nature of Life*, 3rd ed., [McGraw-Hill, 1995], p. 76, Figure 3.13.

Figure 8.6 From Alberts, et al., *Essentials of Cell Biology [*Garland Publishing, 1998], p. 169, Figure 5-27: photograph courtesy of Richard J. Feldman.

Figure 8.10 Portions modified from Vander, Sherman and Luciano, *Human Physiology*, 6th ed. [McGraw-Hill, 1994], p. 562, Figure 8.1, and Carola, Harley and Noback, *Human Anatomy and Physiology*, 2nd ed. [McGraw-Hill, 1992], p. 778, Figure 8.1.

Figure 8.11 Modified from Starr and Taggart, *Biology: The Unity and Diversity of Life*, 8th ed. [Wadsworth Publishing, 1998], p. 32, Figure 2.19.

Figure 8.13 From Figure 2-24, from *Essential Cell Biology*, Third Edition by Bruce Alberts, et al. Copyright © 2010, 2004 by Bruce Alberts, Dennis Bray, Karen Hopkin, Alexander Johnson, Julian Lewis, Martin Raff, Keith Roberts, and Peter Walter © 1998 by Bruce Alberts, Dennis Bray, Alexander Johnson, Julian Lewis, Martin Raff, Keith Roberts, and Peter Walter. Used by permission of W. W. Norton & Company, Inc.

Figure 8.14 Portions modified from Van Wynsberghe, Noback and Carola, *Human Anatomy and Physiology*, 3rd ed. [McGraw-Hill, 1995], p. 63, Figure 3.4.

Figure 8.15 Modified from Postlethwait, Hopson and Veres, *Biology! Bringing Science to Life* [McGraw-Hill, 1991], p. 100, Figure 5.16.

Figure 8.16 Modified from Postlethwait, Hopson and Veres, *Biology! Bringing Science to Life* [McGraw-Hill, 1991], p. 100, Figure 5.16.

Figure 8.17 Modified from Postlethwait and Hopson, *The Nature of Life*, 3rd ed. [McGraw-Hill, 1995], p. 588, Figure 25.10A.

Figure 8.18 Modified from Hopson and Wessells, *Essentials of Biology* [McGraw-Hill, 1990], p. 519, Figure 29-5.

CHAPTER 9

Figure 9.1 Photo with permission from katatonia82/Shutterstock.

Figure 9.2 Data from UN estimates published 2018 and CIA World Fact Book 2018.

Figure 9.6 Data from United Nations [2016].

Figure 9.7 Data from United Nations [2016], WHO report [2018], and other sources.

Figure 9.8 Courtesy of David Epel.

Figure 9.9 Modified from Postlethwait and Hopson, *The Nature of Life*, 3rd ed. [McGraw-Hill, 1995], p.335, Figure 14.3A.

Figure 9.10 Modified from Postlethwait and Hopson, *The Nature of Life*, 3rd ed. [McGraw-Hill, 1995], p.338, Figure 14.5A.

Figure 9.11 Modified from Postlethwait and Hopson, *The Nature of Life*, 3rd ed. [McGraw-Hill, 1995], p.338, Figure 14.5A.

Figure 9.12 Modified from Audeskirk and Audeskirk, *Biology: Life on Earth*, 4th ed. [Prentice Hall, 1996], p. 778, Figure 35-18B.

Figure 9.15 Based on data from Trussell et al., Cost effectiveness of contraceptives in the United States, *Contraception*, ISSN: 18790518, January 1, 2009, Vol. 79, Issue 1, pp. 5–14.

Figure 9.16 Courtesy of Dennis Graflin.

Figure 9.17 Photo courtesy of Frederick Atwood.

CHAPTER 10

Figure 10.2 From *Molecular Biology of the Cell*, Fifth Edition, by Bruce Alberts, et al.. Copyright ©2008, 2002 by Bruce Alberts, Alexander Johnson, Julian Lewis, Martin Raff, Keith Roberts, and Peter Walter. Used by permission of W. W. Norton & Company, Inc.

Figure 10.8 From Molecular Biology of The Cell. Fourth Edition by Bruce Alberts, et al., Copyright ©2002 by Bruce Alberts, Alexander Johnson, Julian Lewis, Martin Raff, Keith Roberts, and Peter Walter. ©1983, 1989, 1994 by Bruce Alberts, Dennis Bray, Julian Lewis, Martin Raff, Keith Roberts, and James D. Watson. Used by permission of W. W. Norton & Company, Inc.

Figure 10.9 Photo courtesy of Lan Bo Chen.

Figure 10.10 Modified from *New England Journal of Medicine* 332: 986, 1995, Figure 1.

Figure 10.11 From *Molecular Biology of The Cell*, Fifth Edition, by Bruce Alberts, et al.. Copyright ©2008, 2002 by Bruce Alberts, Alexander Johnson, Julian Lewis, Martin Raff, Keith Roberts, and Peter Walter. Used by permission of W. W. Norton & Company, Inc.

Figure 10.12 Modified from Weinberg, *The Biology of Cancer*, 2nd ed. [Taylor & Francis, 2014], Figure 12.1.

Figure 10.13 Modified from Hopson and Wessells, *Essentials of Biology* [McGraw-Hill, 1990], p. 291, Figure 17.4.

Figure 10.14 Part A modified from Hopson and Wessells, *Essentials of Biology* [McGraw Hill, 1990], p. 280, Figure 16-12b; Part B modified from Hopson and Wessells, *Essentials of Biology* [McGraw-Hill, 1990], p. 289, Figure 17-2.

Figure 10.15 Modified from Gurdon, *Gene Expression During Cell Differentiation* [Oxford University Press, 1973].

Figure 10.16 Modified from Alberts et al., *Molecular Biology of the Cell*, 4th ed. [Garland Publishing, 2002], Figure 21-30.

Figure 10.18 From *Molecular Biology of the Cell.* Fourth Edition by Bruce Alberts, et al., Copyright ©2002 by Bruce Alberts, Alexander Johnson, Julian Lewis, Martin Raff, Keith Roberts, and Peter Walter. ©1983, 1989, 1994 by Bruce Alberts, Dennis Bray, Julian Lewis, Martin Raff, Keith Roberts, and James D. Watson. Used by permission of W. W. Norton & Company, Inc.

Figure 10.19 From Cancer Research UK, https://scienceblog.cancerresearchuk.org/2018/06/20/age-the-biggest-cancer-risk-factor/, accessed June 2020.

Figure 10.20 Modified from Alberts et al., *Essential Cell Biology* [Garland Publishing, 1998], p. 201, Figure 6-25.

Figure 10.22 (A) Data from National Center for Health Statistics data as analyzed by NCI. (B) US Mortality Volumes 1930 to 1959, US Mortality Data 1960 to 2017, Data from National Center for Health Statistics, Centers for Disease Control and Prevention © 2020. (C) US Mortality Volumes 1930 to 1959, US Mortality Data 1960 to 2017, Data from National Center for Health Statistics, Centers for Disease Control and Prevention © 2020 American Cancer Society, Inc., Surveillance Research. (D) A, from National Center for Health Statistics data as analyzed by NCI; B,C from American Cancer Society, Cancer Facts & Figures, 2020; D, redrawn from several sources.

Figure 10.23 (A) Courtesy of J Morley-Smith. (B) from Van Steeg, H and Kraemer, K.H. Xeroderma pigmentosum and the role of UV-induced DNA damage in cancer. *Molecular Medicine Today* 5: 86-94, 1999. (C) Courtesy of National Cancer Institute and Miguel Rodriguez-Bigas, MD, University of Texas, MD Anderson Cancer Center.

Figure 10.24 Data from National Cancer Institute, 2019.
Figure 10.25 Modified from Weinberg, *The Biology of Cancer*, 2nd ed. [Taylor & Francis, 2014, Figure 3.22.
Figure 10.26 Modified from Weinberg, *The Biology of Cancer*, 2nd ed. [Taylor & Francis, 2014], Figure 12.12.
Figure 10.27 Data from W. J. Blot et al., *Cancer Research*, 48:3282-3287, 1998.
Figure 10.28 Modified from R. Doll and A. B. Hill, *British Medical Journal* 1:1399-1410, 1964.

CHAPTER 11

Figure 11.2 Modified from Vander, Sherman, and Luciano, *Human Physiology*, 6th ed. [McGraw Hill, 1994], p. 182, Figure 8-3.
Figure 11.3 Modified from Hopson and Wessells, *Essentials of Biology* [McGraw-Hill, 1990], p. 616, Figure 34-3[a].
Figure 11.5 Modified from Postlethwaite and Hopson, *The Nature of Life*, 3rd ed. [McGraw-Hill, 1995], p. 699, Figure 31.5.
Figure 11.9 Modified from Campbell et al., *Biology: Concepts & Connections*, 2nd ed. [Benjamin/Cummings, 1997], p. 581, Figure B.
Figure 11.12 (A-C) modified from Vander, Sherman, and Luciano, *Human Physiology*, 6th ed. [McGraw Hill, 1994], p. 260, Figure 9-36 and p. 262, Figure 9-40. (D) Courtesy of Robert Harrison, Auditory Science Laboratory, Hospital for Sick Children, Toronto, Canada.
Figure 11.14 Modified from Tullar, *The Human Species* [McGraw-Hill, 1977], p. 18, Figure 1-8.
Figure 11.15 Modified from Noback and Demarest, *The Human Nervous System*, 3rd ed. [McGraw-Hill, 1981], p. 6, Figure 1-6.
Figure 11.20 Photo courtesy of Voisin/Phanie/Science Photo Library. Other portions redrawn from multiple sources.

CHAPTER 12

Figure 12.1 Data from Gapminder.org. Courtesy of Max Roser, Our World in Data, https://ourworldindata.org/life-expectancy, licensed under CC-BY-SA, https://creativecommons.org/licenses/by/4.0/.
Figure 12.3 Photo courtesy of The Progeria Research Foundation.
Figure 12.4 Images courtesy of the National Institute on Aging/National Institutes of Health; public domain.
Figure 12.5 Image courtesy of the National Institute on Aging/National Institutes of Health; public domain.
Figure 12.6 Image courtesy of the National Institute on Aging/National Institutes of Health; public domain.
Figure 12.7 Image courtesy of the National Institute on Aging/National Institutes of Health; public domain.
Figure 12.8 O'Brien and Wong, Amyloid precursor protein processing and Alzheimer's disease. *Annual Review of Neuroscience*, 21 Jul 2011, Vol. 34, Issue 1, pages 185–204.
Figure 12.10 Image courtesy of the National Institute on Aging/National Institutes of Health; public domain.
Figure 12.12 Image courtesy of the National Institute on Aging/National Institutes of Health; public domain.
Figure 12.13 Image courtesy of the National Institute on Aging/National Institutes of Health; public domain.
Figure 12.14 Image courtesy of the National Institute on Aging/National Institutes of Health; public domain.
Figure 12.15 Image courtesy of the National Institute on Aging/National Institutes of Health; public domain.

CHAPTER 13

Figure 13.1 Portions modified from Vander, Sherman, and Luciano, *Human Physiology*, 6th ed. [McGraw-Hill, 1994], pp. 475, 477, Figures 15-1, 15-3B.

Figure 13.2 Modified from Hopson and Wessells, *Essentials of Biology* [McGraw-Hill, 1990], p. 599, Figure 33-5a and b and Postlethwait and Hopson, *The Nature of Life*, 3rd ed. [McGraw-Hill, 1995], p. 666, Figure 29.8B.

Figure 13.3 Modified from Hopson and Wessells, *Essentials of Biology* [McGraw-Hill, 1990], p. 616, Figure 34-3a.

Figure 13.4 Portions modified from https://en.wikibooks.org/wiki/Organic_Chemistry/Chirality#/media/File:Chiral_molecures_example.svg, courtesy of Calvero.

Figure 13.7 Data from U.S. Dept. Health & Human Services, 2020.

Figure 13.10 Modified from Carlson, *Patten's Foundations of Embryology*, 5th ed. [McGraw-Hill, 1988], pp. 279, 281, Figures 7-21C, 7-22.

CHAPTER 14

Figure 14.1 Courtesy of Biophoto Associates/Science Photo Library.

Figure 14.2 Modified from Van Wynsberghe, Noback and Carola, *Human Anatomy and Physiology*, 3rd ed. [McGraw-Hill, 1995], p. 711, Figure 22.1A.

Figure 14.4 Portion redrawn from Van Wynsberghe, Noback and Carola, *Human Anatomy and Physiology*, 3rd ed. [McGraw-Hill, 1995], pp. 66–67, Figure 3.5.

Figure 14.9 Modified from Hopson and Wessells, *Essentials of Biology* [McGraw-Hill, 1990], p. 548, Figure 30-14.

Figure 14.10 Modified from Hopson and Wessells, *Essentials of Biology* [McGraw-Hill, 1990], p. 626, Figure 34-14 and Hole, *Human Anatomy and Physiology*, 6th ed. [WCB/McGraw-Hill, 1993], p. 404, Figure 11.39.

Figure 14.11 Modified from Carola, Harley and Noback, *Human Anatomy and Physiology*, 2nd ed. [McGraw-Hill, 1992], p. 534, Figure 17.13.

Figure 14.12 Courtesy of David and Suzanne Felten.

CHAPTER 15

Figure 15.2 Courtesy of The Names Project Foundation. Photograph by Mark Theissen.

Figure 15.3 Modified from Van Wynsberghe, Noback and Carola, *Human Anatomy and Physiology*, 3rd ed. [McGraw-Hill, 1995], p. 752, unnumbered figure.

Figure 15.4 Modified from Postlethwait and Hopson, *The Nature of Life*, 3rd ed. [McGraw-Hill, 1995], p. 614, Figure 26.14.

Figure 15.5 Photographs courtesy of Cynthia Goldsmith, Erskine Palmer and Paul Feorino, Centers for Disease Control and Prevention [CDC].

Figure 15.6 Courtesy of Alyne Harrison, Erskine Palmer and Paul Feorino, Centers for Disease Control and Prevention [CDC].

Figure 15.8 Modified from Postlethwait and Hopson, *The Nature of Life*, 3rd ed. [McGraw-Hill, 1995], p. 232, Box 9.1, Figure 1.

Figure 15.9 Data from the Centers for Disease Control and Prevention [CDC].

Figure 15.10 Courtesy of Pam Baker.

Figure 15.11 Data from UNAIDS at https://www.unaids.org/en/resources/documents/2018/unaids-data-2018 and https://www.unaids.org/en/resources/documents/2020/unaids-data-2020.

CHAPTER 16

Figure 16.7 Courtesy of Em Unit, VLA/Science Photo Library.
Figure 16.8 Copyright © 2011 From *Protozoa and Human Disease* by M. F. Wiser/Elizabeth Owen, Figure 4.1A. Reproduced by permission of Taylor & Francis Group, LLC, a division of Informa PLC.
Figure 16.9 Courtesy of CDC.
Figure 16.10 (A) Courtesy of CDC. (B) Courtesy of Johns Hopkins University.
Figure 16.11 Courtesy of CDC.

CHAPTER 17

Figure 17.2 Modified from Hopson and Wessells, *Essentials of Biology* [McGraw-Hill, 1990], p. 139, Figure 8-6.
Figure 17.3 Modified from Hopson and Wessells, *Essentials of Biology* [McGraw-Hill, 1990], p. 138, Figure 8-4.
Figure 17.4 Modified from Hopson and Wessells, *Essentials of Biology* [McGraw-Hill, 1990], p. 143, Figure 8-9.
Figure 17.5 Modified from Starr and Taggart, *Biology: The Unity and Diversity of Life*, 8th ed. [Wadsworth Publishing, 1998], p. 115.
Figure 17.6 Modified from Hopson and Wessells, *Essentials of Biology* [McGraw-Hill, 1990], p. 485, Figure 27-1.
Figure 17.7 Modified from Hopson and Wessells, *Essentials of Biology* [McGraw-Hill, 1990], p. 486, Figure 27-2.
Figure 17.8 From Janet L. Hopson and Norman K. Wessells, *Essentials of Biology* [McGraw-Hill, 1990], p. 489, Figure 27-5.
Figure 17.9 Courtesy of Richard Gross/Biological Photography and Robert Thomas.
Figure 17.10 Modified from Chrispeels and Sadava, *Plants, Genes and Agriculture* [Jones and Bartlett, 1994], p. 214, Figure 7.16.
Figure 17.11 Courtesy of Dr. Jeremy Burgess/Science Photo Library.
Figure 17.12 Photos courtesy of Richard Gross/Biological Photography, modified from Slack, *Carnivorous Plants* [MIT Press, 1980], p. 126, and with permission from Karnwela/Shutterstock.
Figure 17.13 Courtesy of David W. Schindler.
Figure 17.15 Photo used with permission from Alen Thien/Shutterstock.
Figure 17.16 Courtesy of the Nebraska Wheat Board and of David Handley.
Figure 17.19 (A,B,C) courtesy of David Handley. (D) Richard Gross/Biological Photography.
Figure 17.20 Data from "Fifty Generations of Selection for Protein and Oil in Corn" by C. M. Woodworth, Earl R. Leng, and R. W. Jugenheimer, *Agronomy Journal* [John Wiley & Sons], Vol 44, issue 2 [Feb 1, 1952], p. 61.
Figure 17.22 Modified from Chrispeels and Sadava, *Plants, Genes and Agriculture* [Jones and Bartlett, 1994], p. 404, Figure 15.2.

CHAPTER 18

Figure 18.1 Modified from Wilson, *The Diversity of Life* [The Belknap Press of Harvard University Press], p. 134.
Figure 18.2 Courtesy of Georgette Douwma/Science Photo Library.
Figure 18.3 Modified from Moore, Lalicker and Fischer, *Invertebrate Fossils* [McGraw-Hill, 1957], portions of Figures 3-4, 4-17, 4-29, 5-9, 6-20, 6-24, 6-36, 13-7, 13-20, 18-20, 18-29, and 23-1.
Figure 18.4 Modified from Palmer and Fowler, *Fieldbook of Natural History*, 2nd ed. [McGraw-Hill, 1974], p. 112 and p. 433; Hyman, *The*

Invertebrates, vol. 5 [McGraw-Hill, 1959], p. 519, Figure 183C; and Weichert, *Anatomy of the Chordates* [McGraw-Hill, 1970], p. 27, Figure 2.18.

Figure 18.6 (A) Modified from Moore, Lalicker and Fisher, *Invertebrate Fossils* [McGraw-Hill, 1957], Figure 9-40 and 9-41. (B) Courtesy of Don Prothero. (C) Peabody Museum of Natural History, Yale University, New Haven, Connecticut, USA.

Figure 18.7 Modified from Wilson, *The Diversity of Life* [The Belknap Press of Harvard University Press], p. 252.

Figure 18.8 (A) Copyright © 2021 From *Global Biodiversity Volume 1: Selected Countries in Asia*, Chapter 5, by editor T. Pulliaiah. Reproduced by permission of Taylor & Francis Group, LLC, a division of Informa PLC. (B) Courtesy of Dr. P. Petrakis. (C) Courtesy of Dr. Nikola Rahmé. (D) Copyright © 2021 From *Global Biodiversity Volume 3: Selected Countries in Africa*, Chapter 12, by editor T. Pulliaiah. Reproduced by permission of Taylor & Francis Group, LLC, a division of Informa PLC. (E) Courtesy of Dr. Charles Knapp. (F) Copyright © 2021 From *Global Biodiversity Volume 3: Selected Countries in Africa*, Chapter 2, by editor T. Pulliaiah. Reproduced by permission of Taylor & Francis Group, LLC, a division of Informa PLC. (G) Copyright © 2021 From *Global Biodiversity Volume 1: Selected Countries in Asia*, Chapter 13, by editor T. Pulliaiah. Reproduced by permission of Taylor & Francis Group, LLC, a division of Informa PLC. (H) Photo by Teisha Brook/ Paradise Country, Gold Coast, Australia. (I) Courtesy of Ulf Drechsel/PyBio.org. (J) Courtesy of Mark Thomas/Pixabay. (K) Courtesy of © www.gorilladoctors.org. (L) From Ansar Khan Photography, Bharatpur, India. (M) Courtesy of Orythys/ Pixabay. (N) Copyright © 2021 From *Global Biodiversity Volume 3: Selected Countries in Africa*, Chapter 13, by editor T. Pulliaiah. Reproduced by permission of Taylor & Francis Group, LLC, a division of Informa PLC. (O) Copyright © 2021 From *Global Biodiversity Volume 1: Selected Countries in Asia*, Chapter 13, by editor T. Pulliaiah. Reproduced by permission of Taylor & Francis Group, LLC, a division of Informa PLC. (P) Courtesy of Ulf Drechsel/PyBio.org. (Q) Courtesy of www.bazieri.ge. (R) Copyright © 2021 From *Global Biodiversity Volume 3: Selected Countries in Africa*, Chapter 2, by editor T. Pulliaiah. Reproduced by permission of Taylor & Francis Group, LLC, a division of Informa PLC. (S) Copyright © 2021 From *Global Biodiversity Volume 3: Selected Countries in Africa*, Chapter 11, by editor T. Pulliaiah. Reproduced by permission of Taylor & Francis Group, LLC, a division of Informa PLC.

Figure 18.10 (A) Modified from Postlethwait and Hopson, *The Nature of Life*, 3rd ed. [McGraw-Hill, 1995], p. 896, Figure 41.7A. (B) Courtesy of Jane Mackarell. (C) Permission from Paralaxis/Shutterstock.

Figure 18.11 Photos courtesy of Sharon Kinsman, Frederick Atwood, and Jane Mackarell and with permission from George Arenas/ Shutterstock.

Figure 18.12 (A) Courtesy of Philippe Psaila/Science Photo Library. (B) Courtesy of Jane Mackarell.

Figure 18.15 Courtesy of Tim Laman/AnthroPhoto.

Figure 18.16 (A) Guentermanaus/Shutterstock (B,C) Courtesy of Sharon Kinsman.

Figure 18.17 Modified from Hopson and Wessells, *Essentials of Biology* [McGraw-Hill, 1990], p. 755, Figure 41-6.

Figure 18.18 Part B provided with permission from Stanley Dullea/Shutterstock.

CHAPTER 19

Figure 19.3	Courtesy of Dr. Sebastian Velez.
Figure 19.5	(A) Courtesy of Richard Gross/Biological Photography. (B) Courtesy of David Schindler from Schindler et al., Science, 228: 1395-1401 © 1985 American Association for the Advancement of Science. (C) Courtesy of Cordelia Molloy/Science Photo Library
Figure 19.9	Credit by Scripps Institution of Oceanography SIO.
Figure 19.10	(A) National Park Service/Photo 1936 by Asahel Curtis and Photo 2015 by Byron Adams. (B) U.S. Geological Survey/Photo by Alden; U.S. Geological Survey/Photo by B. Reardon. (C) U.S. Geological Survey/Photo by Charles Rabot; U.S. Geological Survey/Photo by Nils Haakensen.

Index

Note: Page numbers in *italics* refer to figures, **bold** refers to tables, and ***bold italics*** refers to box.

D

E

S

T

U

V